Fundamentals of Atmospheric Modeling

Second Edition

This well-received and comprehensive textbook on atmospheric processes and numerical methods has been thoroughly revised and updated. The new edition includes a wide range of new numerical techniques for solving problems in areas such as cloud microphysics, ocean–atmosphere exchange processes, and atmospheric radiative properties. It also contains improved descriptions of atmospheric physics, dynamics, radiation, aerosol, and cloud processes. Numerous examples and problems are included, with answers available to lecturers at http://www.cambridge.org/0521548659

Fundamentals of Atmospheric Modeling is essential reading for researchers and advanced students of atmospheric science, meteorology, and environmental science.

Mark Z. Jacobson is an associate professor of civil and environmental engineering at Stanford University. Goals of his research are to improve our understanding of physical, chemical, and dynamical processes in the atmosphere through numerical modeling and to improve the simulation of air pollution, weather, and climate. He is the author of two textbooks: *Fundamentals of Atmospheric Modeling* and *Atmospheric Pollution: History, Science, and Regulation*.

Fundamentals of Atmospheric Modeling

Second Edition

MARK Z. JACOBSON
Stanford University

CAMBRIDGE
UNIVERSITY PRESS

CAMBRIDGE UNIVERSITY PRESS
Cambridge, New York, Melbourne, Madrid, Cape Town, Singapore, São Paulo

Cambridge University Press
The Edinburgh Building, Cambridge CB2 2RU, UK

Published in the United States of America by Cambridge University Press, New York

www.cambridge.org
Information on this title: www.cambridge.org/9780521839709

First published 1998
Second edition 2005

Printed in the United Kingdom at the University Press, Cambridge

A catalog record for this book is available from the British Library

Library of Congress Cataloging in Publication data
Jacobson, Mark Z. (Mark Zachary)
Fundamentals of atmospheric modeling / Mark Z. Jacobson.
p. cm.
Includes bibliographical references and index.
ISBN 0 521 83970 X (hardback) ISBN 0 521 54865 9 (paperback)
1. Atmospheric models. 2. Atmospheric physics – Mathematical models. I. Title.
QC861.3.J33 2005
551.51′01′1 – dc22 2004057382

ISBN-13 978-0-521-83970-9
ISBN 0 521 83970 X hardback
ISBN-13 978-0521-54865-6
ISBN 0 521 54865 9 paperback

To Dionna and Daniel

Contents

Preface

Modern atmospheric science is a field that combines meteorology, physics, mathematics, chemistry, computer sciences, and to a lesser extent geology, biology, microbiology, and oceanographic sciences. Until the late 1940s scientific studies of the atmosphere were limited primarily to studies of the weather. At that time, heightened concern about air pollution caused an increase in studies of atmospheric chemistry. With the invention of the computer, modeling of weather and air pollution commenced. Since the late 1940s, the number of meteorological and air-pollution studies has increased rapidly, and many meteorological and air-pollution models have merged.

The purposes of this book are to provide (1) a physical understanding of dynamical meteorology, land- and water-surface processes, radiation, gas chemistry, aerosol microphysics and chemistry, and cloud processes, (2) a description of numerical methods and computational techniques used to simulate these processes, and (3) a catalog of steps required to construct, apply, and test a numerical model.

The first chapter of this book gives an overview of model processes and time scales. Chapter 2 describes atmospheric structure, composition, and thermodynamics. In Chapters 3–5, basic equations describing dynamical meteorology are derived. In Chapter 6, numerical methods of solving partial differential equations are discussed. A technique of solving dynamical meteorological equations is provided in Chapter 7. In Chapter 8, boundary-layer and ground processes are described. Chapter 9 introduces radiation. Chapters 10–12 focus on photochemistry and numerical methods of solving chemical equations. Chapters 13–17 describe aerosol physical and chemical processes. Chapter 18 discusses cloud thermodynamics and microphysics. Chapter 19 discusses aqueous chemistry in aerosol particles and clouds. Chapter 20 describes sedimentation and dry deposition. Chapter 21 outlines computer model development, application, and testing.

The book is designed as an upper-level undergraduate, graduate, and research text. The text assumes students have a basic physical science, mathematical, and computational background. Both Système Internationale (SI) and centimeter-gram-second (CGS) units are used. Dynamical meteorologists often use SI units, and atmospheric chemists often use CGS units. Thus, both unit systems are retained. Unit and variable conversions are given in Appendix A.

Acknowledgments

I would like to thank several colleagues who provided comments, suggestions, and/or corrections relating to the text. In particular, I am indebted to (in alphabetical order) A. April, Akio Arakawa, Mary Barth, Jaime Benitez, Merete Bilde, Steve Bryson, Bob Chatfield, Tu-Fu Chen, Johann Feichter, Frank Freedman, Ann Fridlind, A. V. Gemintern, J. Haigh, Hiroshi Hayami, Roy Harrison, James Holton, Daniel Jacob, Liming Li, Jinyou Liang, Jin-Sheng Lin, Bernd Kaercher, Gerard Ketefian, Bennert Machenhauer, Ed Measure, Gary Moore, Elmar Reiter, Doug Rotman, Roberto San Jose, Hjalti Sigurjonsson, Hanwant Singh, Jing Song, Tae-Joon Song, Amy Stuart, Azadeh Tabazadeh, Roland von Glasow, Chris Walcek, Thomas Warner, Debra Weisenstein, Don Wuebbles, and Yang Zhang.

1

Introduction

1.1 BRIEF HISTORY OF METEOROLOGICAL SCIENCES

THE history of atmospheric sciences begins with weather forecasting. Forecasting originally grew out of three needs – for farmers to produce crops, sailors to survive at sea, and populations to avoid weather-related disasters such as floods. Every society has forecast wind, rain, and other weather events. Some forecasts are embodied in platitudes and lore. Virgil stated, "Rain and wind increase after a thunderclap." The Zuni Indians had a saying, "If the first thunder is from the east, winter is over." Human experiences with the weather have led to more recent forecast rhymes, such as, "Rainbow in morning, sailors take warning. Rainbow at night, a sailor's delight."

Primitive forecasts have also been made based on animal and insect behavior or the presence of a human ailment. Bird migration was thought to predict oncoming winds. This correlation has since proved unreliable. Rheumatism, arthritis, and gout have been associated with the onset of rain, but such ailments are usually unrelated to the weather. The presence of locusts has correctly been associated with rainfall in that locusts fly downwind until they reach an area of converging winds, where rain is likely to occur.

In the 1870s, forecasting based on observations and experience became a profession. Many felt that early professional forecasting was more of an art than a science, since it was not based on scientific theory. Although the amount of data available to forecasters was large and increasing, the data were not always used. Data were gathered by observers who used instruments that measured winds, pressure, temperature, humidity, and rainfall. Many of these instruments had been developed over the previous two centuries, although ideas and crude technologies existed prior to that time.

The Greeks, around 430 BC, may have been the first to measure winds. Yet, reliable instruments to measure wind force and direction were not developed until nearly two millennia later. In 1450, the Italian mathematician Leone Battista Alberti (1404–72) developed the first known anemometer, a **swinging-plate anemometer** that consisted of a disc placed perpendicular to the wind. It was used to measure wind speed based on the angle between the disc in its original position and its displaced position. In 1667 Robert Hooke developed a similar device, the **pressure-plate anemometer**, which consisted of a sheet of metal hanging vertically. Windmills were used as early as AD 644 in Persia, but the first spinning-cup anemometer, which applies the principle of the windmill to measure wind speed, was not developed until the nineteenth century. In 1846, the Irish physicist John Thomas Romney Robinson invented a **spinning-cup anemometer** that consisted of four hemispherical cups mounted on a vertical axis. In 1892, William Henry Dines invented the **pressure-tube (Dines) anemometer**, which is a device that measures wind speed from the pressure difference arising from wind blowing in a tube versus that blowing across the tube. The pressure difference is proportional to the square of the wind speed.

In 1643, Evangelista Torricelli (1608–47) invented the **mercury barometer**, becoming the first to measure air pressure. He filled a glass tube 1.2 m long with mercury and inverted it onto a dish. He found that only a portion of the mercury flowed from the tube into the dish, and the resulting space above the mercury in the tube was devoid of air (a **vacuum**). Thus, Torricelli was also the first person to record a sustained vacuum. He suggested that the change in height of the mercury in the tube each day was caused by a change in atmospheric pressure. Air pressure balanced the pressure exerted by the column of mercury in the tube, preventing the mercury from flowing freely from the tube. The **aneroid barometer**, which represented an advance over the mercury barometer, was not developed adequately until 1843. The aneroid barometer contains no fluid. Instead, it measures pressure by gauging the expansion and contraction of a tightly sealed metal cell that contains no air.

A third important invention for meteorologists was the thermometer. Prior to 1600, Galileo Galilei (1564–1642) devised the **thermoscope**, which estimated temperature change by measuring the expansion of air. The instrument did not have a scale and was unreliable. Torricelli's mercury barometer, which contained fluid, led to the invention of the **liquid-in-glass thermometer** in Florence in the mid seventeenth century. In the early eighteenth century, useful thermometer scales were developed by Gabriel Daniel Fahrenheit of Germany (1686–1736) and Anders Celsius of Sweden (1701–1744).

A fourth important invention was the **hygrometer**, which measures humidity. In 1450, the German cardinal, philosopher, and administrator Nicolas of Cusa (Nicolas Cryfts) (1401–64) described the first hygrometer with the following:

> If someone should hang a good deal of wool, tied together on one end of a large pair of scales, and should balance it with stones at the other end in a place where the air is temperate it would be found that the weight of the wool would increase when the air became more humid, and decrease when the air tended to dryness.

(Brownawell 2004). In 1481, Leonardo da Vinci (1452–1519) drew Cryfts' hygrometer in his *Codex Atlanticus*, using a sponge instead of wool. The purpose of the hygrometer, according to da Vinci, was

> to know the qualities and thickness of the air, and when it is going to rain.

(White 2000). In 1614, Santorio Santorre developed a hygrometer that measured vapor by the contraction and elongation of cord or lyre strings. Later hygrometers were made of wood, seaweed, paper, hair, nylon, and acetate. The hair hygrometer is still used today, although another instrument, the **psychrometer**, is more accurate. A psychrometer consists of two liquid-in-glass thermometers mounted together, one with a dry bulb and the other with a bulb covered with a moistened cloth.

Following the inventions above, observations of pressure, temperature, humidity, wind force, wind direction, and rainfall became regular. By the nineteenth century, weather-station networks and meteorological tables were common. Observers gathered data and forecasters used the data to predict the weather, but neither observers nor forecasters applied significant theory in their work. Theoreticians studied physical laws of nature but did not take advantage of the abundance of data available.

One of the first weather theoreticians was Aristotle, who wrote *Meteorologica* about 340 BC. In that text, Aristotle attempted to explain the cause of winds, clouds, rain, mist, dew, frost, snow, hail, thunder, lightning, thunderstorms, halos, rainbows, and mock suns. On the subject of winds, he wrote (Lee 1951),

> These, then are the most important different winds and their positions. There are two reasons for there being more winds from the northerly than from the southerly regions. First, our inhabited region lies toward the north; second, far more rain and snow is pushed up into this region because the other lies beneath the sun and its course. These melt and are absorbed by the Earth and when subsequently heated by the sun and the Earth's own heat cause a greater and more extensive exhalation.

On the subject of thunder, he wrote,

> Let us now explain lightning and thunder, and then whirlwinds, firewinds and thunderbolts: for the cause of all of them must be assumed to be the same. As we have said, there are two kinds of exhalation, moist and dry; and their combination (air) contains both potentially. It condenses into cloud, as we have explained before, and the condensation of clouds is thicker toward their farther limit. Heat when radiated disperses into the upper region. But any of the dry exhalation that gets trapped when the air is in process of cooling is forcibly ejected as the clouds condense and in its course strikes the surrounding clouds, and the noise caused by the impact is what we call thunder.

Aristotle's monograph established a method of qualitatively explaining meteorological problems. Since Aristotle was incorrect about nearly all his meteorological conclusions, *Meteorologica* was never regarded as a significant work. Aristotle made observations, as evidenced by diagrams and descriptions in *Meteorologica*, but he did not conduct experiments. Lacking experiments, his conclusions, while rational, were not scientifically based.

Aristotle's method of rationalizing observations with little or no experiment governed meteorological theory through the seventeenth century. In 1637, René Descartes (1596–1650) wrote *Les Météores*, a series of essays attached to *Discours de la Méthode*. In some parts of this work, Descartes improved upon Aristotle's treatise by discussing experiments. In other parts, Descartes merely expanded or reformulated many of Aristotle's explanations. On the subject of northerly winds, Descartes wrote (Olscamp 1965),

> We also observe that the north winds blow primarily during the day, that they come from above to below, and that they are very violent, cold and dry. You can see the explanation of this by considering that the Earth EBFD [referring to a diagram] is covered with many clouds and mists near the poles E and F, where it is hardly heated by the sun at all; and that at B, where the sun is immediately overhead, it excites a quantity of vapors which are quite agitated by the action of its light and rise into the air very quickly, until they have risen so high that the resistance of their weight makes it easier for them to swerve, . . .

Like Aristotle, Descartes was incorrect about many explanations. Despite some of the weaknesses of his work, Descartes is credited with being one of the first in meteorological sciences to form hypotheses and then to conduct experiments.

Between the seventeenth and mid nineteenth centuries, knowledge of basic physics increased, but mathematics and physics were still not used rigorously to explain atmospheric behavior. In 1860, William Ferrel published a collection of papers that were the first to apply mathematical theory to fluid motions on a rotating Earth. This work was the impetus behind the modern-day field of **dynamical meteorology**, which uses physics and mathematics to explain atmospheric motion.

Between 1860 and the early 1900s weather forecasting and theory advanced along separate paths. In 1903, Vilhelm Bjerknes of Norway (1862–1951) promulgated the idea that weather forecasting should be based on the laws of physics. This idea was not new, but Bjerknes advanced it further than others (Nebeker 1995). Bjerknes thought that weather could be described by seven primary variables – pressure, temperature, air density, air water content, and the three components of wind velocity. He also realized that many of the equations describing the change in these variables were physical laws already discovered. Such laws included the continuity equation for air, Newton's second law of motion, the ideal-gas law, the hydrostatic equation, and the thermodynamic energy equation.

Bjerknes did not believe that meteorological equations could be solved analytically. He advocated the use of physical principles to operate on graphical observations to analyze the weather. For example, from a map of observed wind barbs, which give horizontal wind speeds and directions, he could draw a map of streamlines (lines of constant direction) and isolines (lines of constant wind speed), then use graphical differentiation and graphical algebra to determine vertical wind speeds, which would be drawn on another map. This technique was called **graphical calculus**.

Between 1913 and 1919, Lewis Fry Richardson (1881–1953) developed a different method of analyzing the analytical equations describing the weather (Richardson 1922). The method involved dividing a region of interest into rectilinear cells (grid cells), then writing a finite-difference form of the analytical meteorological equations for each grid cell and solving the equations by hand over all cells. Whereas Carle Runge and Wilhelm Kutta developed a method of finite-differencing ordinary differential equations in the 1890s, Richardson extended finite-differencing (central differencing in his case) to partial differential equations and to multiple grid cells. Richardson was not satisfied with his solution technique, though, because data available to test his method were sparse, and predictions from his method were not accurate. Nevertheless, his was the first attempt to predict the weather numerically in detail.

Until the 1940s, much of Richardson's work was ignored because of the lack of a means to carry out the large number of calculations required to implement his method. In 1946, John von Neumann (1903–57), who was associated with work to build the world's first electronic digital computer **ENIAC** (Electronic Numerical Integrator and Computer), proposed a project to make weather forecasting its main application. The project was approved, and the first computer model of the atmosphere was planned. Among the workers on von Neumann's project was Jule Charney, who became director of the project in 1948. Charney made the first numerical forecast on ENIAC with a one-dimensional model (Charney 1949, 1951). Since then, numerical models of weather prediction have become more elaborate, and computers have become faster.

1.2 BRIEF HISTORY OF AIR-POLLUTION SCIENCE

Meteorological science is an old and established field; air-pollution science has a shorter history. Natural air pollution has occurred on Earth since the planet's formation. Fires, volcanic eruptions, meteorite impacts, and high winds all cause natural air pollution. Anthropogenic air-pollution problems have existed on urban scales for centuries and have resulted from burning of wood, vegetation, coal, oil, natural gas, waste, and chemicals.

In the nineteenth and early twentieth centuries, most air pollution was due to chimney and smokestack emission of coal and chemical-factory combustion products. In 1905, Harold Antoine Des Voeux described the combination of smoke and fog he observed in cities in Great Britain as **smog**. Smog from coal and chemical combustion resulted in several air pollution episodes that killed thousands of people between 1850 and 1960. The worst of these was in December 1952, when smog resulted in over 4000 deaths in London. Pollution resulting from coal and chemical-factory combustion in the presence of fog is commonly referred to as **London-type smog**.

In the early twentieth century, the widespread use of automobiles and the increase in industrial activity increased the prevalence of another type of air pollution, called **photochemical smog**. This pollution was most noticeable and formed almost daily in Los Angeles, California. It became so serious that an Air Pollution Control District was formed in Los Angeles in 1947 to combat it. The composition of photochemical smog was not elucidated until 1951, when Arie Haagen-Smit produced ozone in a laboratory from oxides of nitrogen and reactive organic gases, in the presence of sunlight and suggested that these gases were the main constituents of Los Angeles air pollution. Photochemical smog has since been observed in most cities of the world.

Before the twentieth century, air pollution was not treated as a science but as a regulatory problem (Boubel *et al.* 1994). In Great Britain, emission from furnaces and steam engines led to the Public Health Act of 1848. Emission of hydrogen chloride from soap making led to the Alkali Act of 1863. In both cases, pollution abatement was controlled by agencies. In the nineteenth century, pollution abatement in the United States was delegated to municipalities. In most cases, regulation did not reduce pollution much, but in some cases it led to pollution control technologies, such as the electrostatic precipitator for reducing particle emission from smokestacks. In one case, the development of a pollutant-control technology, the scrubber for removing hydrochloric acid gas from chemical factory emission, provided incentive for the swift passage of a regulation, the Alkali Act of 1863. Inventions unrelated to air-pollution regulation reduced some pollution problems. For example, in the early twentieth century, the advent of the electric motor centralized sources of combustion at electric utilities, reducing local air pollution caused by the steam engine.

1.3 THE MERGING OF AIR-POLLUTION AND METEOROLOGICAL SCIENCES

In the 1950s, laboratory work was undertaken to understand better the formation of photochemical and London-type smog. Since the computer was already available, box models simulating atmospheric chemical reactions were readily implemented. In the 1960s and 1970s, air-pollution models, termed **air-quality models**, were expanded to two and three dimensions. Such models included treatment of emission, transport, gas chemistry, and gas deposition to the ground. Most models used interpolated fields of meteorological data as inputs. Today, many air quality models use meteorological fields calculated in real time as inputs.

In the 1970s, atmospheric pollution problems, aside from urban air pollution, were increasingly recognized. Such problems included regional acid deposition, global ozone reduction, Antarctic ozone depletion, and global climate change. Initially, ozone reduction and climate change problems were treated separately by dynamical meteorologists and atmospheric chemists. More recently, computer models that incorporate atmospheric chemistry and dynamical meteorology have been used to study these problems.

1.4 WEATHER, CLIMATE, AND AIR POLLUTION

A **model** is a mathematical representation of a process. An **atmospheric computer model** is a computer-coded representation of dynamical, physical, chemical, and radiative processes in the atmosphere. In atmospheric models, time-dependent processes are mathematically described by **ordinary differential equations**. Space- and time-dependent processes are described by **partial differential equations**. Ordinary and partial differential equations are replaced with finite-difference or other approximations, then computerized and solved.

Computer models also solve parameterized and empirical equations. A **parameterized equation** is an equation in which one parameter is expressed in terms of at least two other parameters. The equation of state, which relates pressure to temperature and air density, is a parameterized equation. An **empirical equation** is an equation in which one parameter is expressed as an empirical function (e.g., a polynomial fit) of at least one other parameter. Whereas parameterized equations are derived from insight, empirical equations do not always make physical sense. Instead, they reproduce observed results under a variety of conditions. In this text, computer modeling of the atmosphere is discussed. Such modeling requires solutions to ordinary differential equations, partial differential equations, parameterized equations, and empirical equations.

Since the advent of atmospheric computer modeling in 1948, models have been applied to study weather, climate, and air pollution on urban, regional, and global scales. **Weather** is the state of the atmosphere at a given time and place, and **climate** is the average of weather events over a long period. Some basic weather variables include wind speed, wind direction, pressure, temperature, relative humidity, and rainfall. Standard climate variables include mean annual temperatures and mean monthly rainfall at a given location or averaged over a region.

Air pollutants are gases, liquids, or solids suspended in the air in high enough concentration to affect human, animal, or vegetation health, or to erode structures. Standard air pollution problems include urban smog, acid deposition, Antarctic ozone depletion, global ozone reduction, and global climate change. **Urban smog** is characterized by the local concentration buildup of gases and particles emitted from automobiles, smokestacks, and other human-made sources. **Acid deposition** occurs following long-range transport of sulfur dioxide gas emitted from coal-fired power plants, conversion of the sulfur dioxide to liquid-phase sulfuric acid, and deposition of sulfuric-acid-related species to the ground by rain or another means. Acid deposition also occurs when nitric acid gas, produced chemically from automobile pollutants, dissolves into fog drops, which deposit to the ground or lungs. This form of acid deposition is **acid fog**. Acids harm soils, lakes, and forests and damage structures.

Antarctic ozone depletion and **global ozone reduction** are caused, to a large extent, by chlorine and bromine compounds that are emitted anthropogenically into the atmosphere and break down only after they have diffused to the upper atmosphere. Ozone reduction increases the intensity of ultraviolet radiation from the Sun reaching the ground. Some ultraviolet wavelengths destroy microorganisms on the surface of the Earth and cause skin cancer in humans. **Global climate change** is characterized by changes in global temperature and rainfall patterns due to increases in atmospheric carbon dioxide, methane, nitrous oxide, water vapor, and other gases that absorb infrared radiation. The addition of particles to the atmosphere may in some cases warm, and in other cases cool, climate as well.

Historically, meteorological models have been used to simulate weather, climate, and climate change. Photochemical models have been used to study urban, regional, and global air-pollution emission, chemistry, aerosol processes, and transport of

Table 1.1 Scales of atmospheric motion

Scale name	Scale dimension	Examples
Molecular scale	≪2 mm	Molecular diffusion, molecular viscosity
Microscale	2 mm–2 km	Eddies, small plumes, car exhaust, cumulus clouds
Mesoscale	2–2000 km	Gravity waves, thunderstorms, tornados, cloud clusters, local winds, urban air pollution
Synoptic scale	500–10 000 km	High- and low-pressure systems, weather fronts, tropical storms, hurricanes, Antarctic ozone hole
Planetary scale	>10 000 km	Global wind systems, Rossby (planetary) waves, stratospheric ozone reduction, global warming

pollutants. Only recently have meteorological models merged with photochemical models to tackle these problems together.

One purpose of developing a model is to understand better the physical, chemical, dynamical, and radiative properties of air pollution and meteorology. A second purpose is to improve the model so that it may be used for forecasting. A third purpose is to develop a tool that can be used for policy making. With an accurate model, policy makers can try to mitigate pollution problems.

1.5 SCALES OF MOTION

Atmospheric problems can be simulated over a variety of spatial scales. **Molecular-scale** motions occur over distances much smaller than 2 mm. Molecular diffusion is an example of a molecular-scale motion. **Microscale** motions occur over distances of 2 mm to 2 km. Eddies, or swirling motions of air, are microscale events. **Mesoscale** motions, such as thunderstorms, occur over distances of 2–2000 km. The **synoptic scale** covers motions or events on a scale of 500–10 000 km. High- and low-pressure systems and the Antarctic ozone hole occur over the synoptic scale. **Planetary-scale** events are those larger than synoptic-scale events. Global wind systems are planetary-scale motions. Some phenomena occur on more than one scale. Acid deposition is a mesoscale and synoptic-scale phenomenon. Table 1.1 summarizes atmospheric scales and motions or phenomena occurring on each scale.

1.6 ATMOSPHERIC PROCESSES

Atmospheric models simulate many processes and feedbacks among them. Figure 1.1 shows a diagram of an air pollution–weather–climate model that simulates gas, aerosol, cloud, radiative, dynamical, transport, and surface processes.

A **gas** is an individual atom or molecule suspended in the air in its own phase state. Gas molecules have diameters on the order of $2\text{–}5 \times 10^{-10}$ m.

An **aerosol** is an ensemble of solid, liquid, or mixed-phase particles suspended in air. Each particle consists of an aggregate of atoms and/or molecules bonded together. An **aerosol particle** is a single particle within an aerosol. Aerosol particles

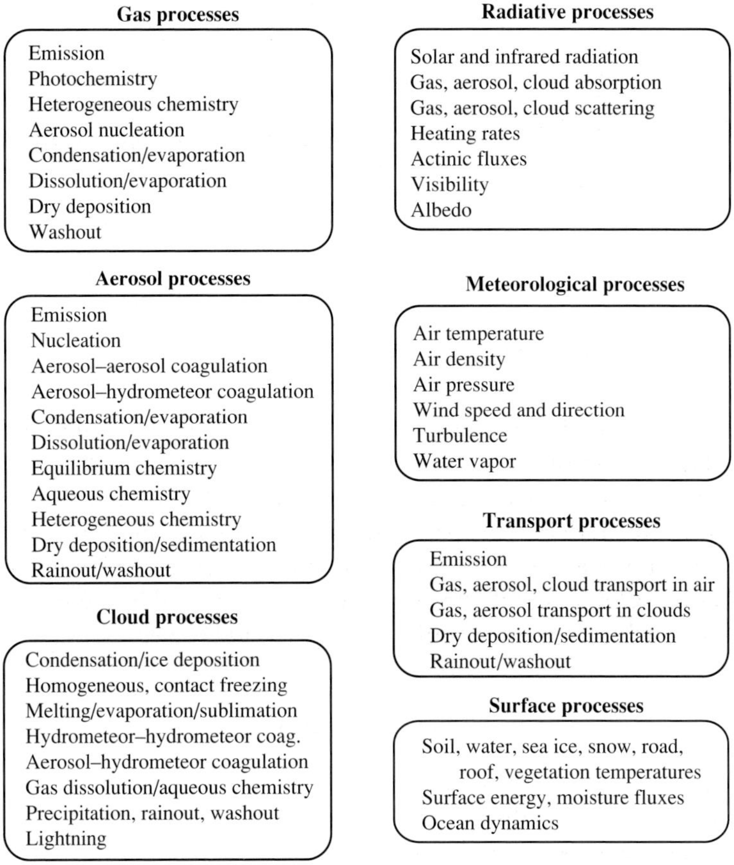

Figure 1.1 Diagram of processes simulated in an air pollution–weather–climate model.

have diameters that range in size from a few tens of gas molecules to 10 mm and can contain many components, including liquid water.

A **hydrometeor** is an ensemble of liquid, solid, or mixed-phase predominantly water-containing particles suspended in or falling through the air. A **hydrometeor particle** is a single particle within a hydrometeor. Examples of hydrometeor particles are cloud drops, ice crystals, raindrops, snowflakes, and hailstones. The main difference between an aerosol particle and a hydrometeor particle is that the latter contains much more water than the former. Hydrometeor particles generally range in diameter from 10 μm to 10 mm. In this text, the term **particle** used alone may refer to an aerosol particle or a hydrometeor particle.

Figure 1.1 lists the major processes affecting gases in the atmosphere. **Emission** is the addition of a pollutant to the atmosphere. **Photochemistry** encompasses gas kinetic chemistry and photolysis. Gas **kinetic chemistry** is the process by which reactant gases collide with each other and transform to product gases. **Photolysis** is the process by which reactant gases are broken down by sunlight to form products.

Gases are also affected by **gas-to-particle conversion**. Conversion processes include heterogeneous chemistry, nucleation, condensation/evaporation, dissolution/evaporation, and deposition/sublimation. Gases react chemically on the surfaces of particles to form gas, liquid, or solid products during **heterogeneous chemistry. Nucleation** occurs when gas molecules aggregate and change phase to a liquid or solid to form a new small aerosol particle or a cluster on an existing particle surface. **Condensation** occurs when a gas diffuses to and sticks to the surface of a particle and changes state to a liquid. **Evaporation** occurs when a liquid molecule on a particle surface changes state to a gas and diffuses away from the surface. **Dissolution** occurs when a gas molecule diffuses to and dissolves into liquid on the surface of a particle. Evaporation, in this case, is the opposite of dissolution.

Gases are physically removed from the atmosphere by dry deposition and washout. **Gas dry deposition** (different from solid deposition) is a removal process that occurs when a gas (or particle) impinges upon and sticks to a surface, such as the ground or a house. **Gas washout** is the dissolution of a gas in **precipitation** (rainfall) that falls to the ground.

Figure 1.1 lists major aerosol processes. Some, such as emission, nucleation, condensation/evaporation, dissolution/evaporation, and heterogeneous chemistry, affect gases as well. **Aerosol–aerosol coagulation** occurs when two aerosol particles collide and coalesce (stick together) to form a third, larger particle. Aerosol–hydrometeor coagulation occurs when an aerosol particle collides and coalesces with a hydrometeor particle. **Equilibrium chemistry** is reversible chemistry between or among liquids, ions, and/or solids within aerosols and hydrometeors. **Aqueous chemistry** is irreversible chemistry important in water-containing aerosols and in hydrometeors.

Aerosol particles are physically removed by dry deposition, sedimentation, rainout, and washout. **Sedimentation** is the process by which particles fall from one altitude to another or to the surface due to their weight. This differs from **aerosol dry deposition**, which occurs when particles contact a surface and stick to the surface. **Aerosol rainout** is the growth of cloud drops on aerosol particles and the eventual removal of the aerosol particle to the surface by precipitation. This differs from **aerosol washout**, which is the coagulation of aerosol particles with precipitation that subsequently falls to the ground.

Clouds are affected by several of the same processes affected by aerosol processes. In addition, clouds are affected by **ice deposition**, which is the growth of water vapor onto aerosol particles to form ice crystals. **Sublimation** is the conversion of the ice crystals back to vapor. During **freezing**, liquid water within a hydrometeor changes state to ice. **Melting** is the reverse. Hydrometeor particles may coagulate with themselves or with aerosol particles. **Lightning** occurs when ice crystals collide with then bounce off of other ice crystals, creating a charge separation that eventually leads to a flash of light.

Gases, aerosol particles, and hydrometeor particles **scatter** (redirect) and **absorb solar radiation** (emitted by the Sun) and **infrared radiation** (emitted by the Earth, atmosphere, and Sun), affecting **heating rates** (the rates at which the atmosphere heats or cools), **actinic fluxes** (a parameter used to calculate photolysis rate coefficients), and **visibility** (the distance a person can see). **Albedo** is the reflectivity of a surface. It affects radiation in the atmosphere and, in the case of sea ice, snow, and clouds, is itself affected by pollutants.

Important meteorological variables determined in a model include air temperature, air density, air pressure, wind speed and direction, turbulence, and water vapor. Winds and turbulence affect transport of gases, aerosol particles, and hydrometeor particles. Likewise, gases, aerosol particles, and hydrometeor particles feed back to temperature, pressure, and winds in several ways.

Finally, meteorology and air pollution are affected by surface temperature and energy/moisture fluxes. Surface temperature itself is affected by surface type, some of which include vegetated and bare soil, water, sea ice, roads, rooftops, and snow.

In a model, meteorological variables are simulated by solving a set of partial differential equations and parameterized equations, including the **momentum equation**, the **thermodynamic energy equation**, the **continuity equation for air**, the **equation of state**, and the **continuity equation for total water**. Heating rates and actinic fluxes are calculated with the **radiative-transfer equation.** Changes in gas, aerosol particle, and hydrometeor particle concentration are found by solving the **species continuity equation**, which describes transport with partial differential equations and chemistry/physics with ordinary differential equations.

In sum, weather, climate, and air pollution can be modeled by taking into account a fairly reasonably well-defined set of physical, chemical, and/or dynamical equations. In the rest of this book, the processes shown in Fig. 1.1 are examined.

2

Atmospheric structure, composition, and thermodynamics

THE atmosphere contains a few highly concentrated gases, such as nitrogen, oxygen, and argon, and many trace gases, among them water vapor, carbon dioxide, methane, and ozone. All such gases are constituents of air. Important characteristics of air are its pressure, density, and temperature. These parameters vary with altitude, latitude, longitude, and season and are related to each other by the equation of state. Two other fundamental equations applicable to the atmosphere are the Clausius–Clapeyron equation and the first law of thermodynamics. The Clausius–Clapeyron equation relates temperature to the quantity of water vapor over a surface at saturation. The first law of thermodynamics relates the temperature change of a gas to energy transfer and the change in work. In this chapter, atmospheric variables and gases are discussed, and basic equations describing atmospheric physics and thermodynamics are introduced.

2.1 PRESSURE, DENSITY, AND COMPOSITION

In the Earth's atmosphere, air density, pressure, and temperature change with altitude. **Air density** is the mass of air per unit volume of air, where the mass of air is summed over all gases, aerosol particles, and hydrometeor particles. Since the mass concentration of gas in the air is 500 times that of hydrometeor particles in the thickest cloud and ten million times that of aerosol particles in polluted air (Table 13.1), air density is accurately determined from the mass of gas only. Air density decreases exponentially with altitude, as shown in Fig. 2.1(a). Air density is the greatest near the surface since atmospheric mass is concentrated near the surface.

Air pressure is ideally the weight (force) of air above a horizontal plane, divided by the area of the plane. This type of air pressure is called **hydrostatic air pressure**, which is the pressure due solely to the weight of air in a column above a given altitude. The term **hydrostatic** means "fluid at rest." Air pressure is hydrostatic only if air is not accelerating vertically, which occurs either if it is at rest or if it has a constant vertical speed. The assumption that air pressure is hydrostatic is reasonable when air pressure is averaged over a large horizontal area ($>$2–3 km in diameter) and outside of a storm system, since vertical accelerations in such cases are generally small. Over small areas ($<$2–3 km in diameter) and in individual clouds, vertical accelerations can be large. When air accelerates vertically, air pressure is nonhydrostatic. **Nonhydrostatic air pressure** is discussed in Section 5.1.3.

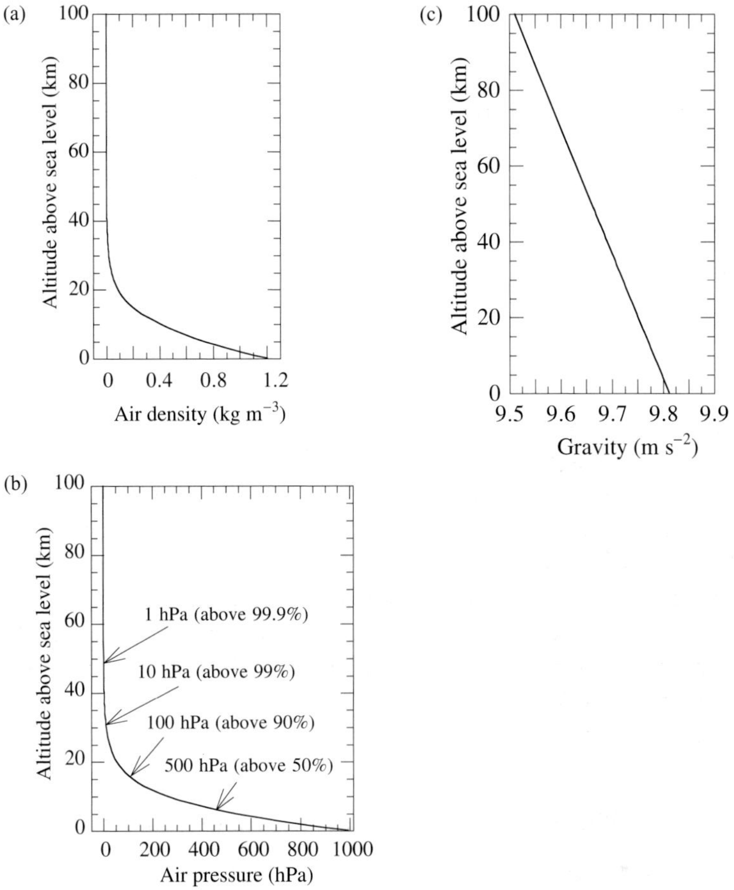

Figure 2.1 Variation of (a) air density, (b) air pressure, and (c) gravitational acceleration versus altitude in the Earth's lower atmosphere. The pressure diagram shows that 99.9 percent of the atmosphere lies below an altitude of about 48 km (1 hPa), and 50 percent lies below about 5.5 km (500 hPa).

Hydrostatic air pressure at any altitude z (m) above sea level is found by integrating the product of air density and gravity from the top of the atmosphere to altitude z with

$$p_a(z) = \int_z^\infty \rho_a(z) g(z) \mathrm{d}z \tag{2.1}$$

where $p_a(z)$ is air pressure as a function of altitude (pascal, where 1 Pa = 1 N m^{-2} = 1 kg m^{-1} s^{-2} = 0.01 hPa = 0.01 millibar or mb), $\rho_a(z)$ is air density as a function of altitude (kg m^{-3}) from Fig. 2.1(a), and $g(z)$ is gravitational acceleration (m s^{-2}). Alternative units for air pressure are given in Appendix Table A.6. Figures 2.1(b) and (c) show variations in pressure and gravitational acceleration, respectively, with altitude in the atmosphere. Tabulated values of these variables are given in Appendix Table B.1.

Table 2.1 Average composition of the lowest 100 km in the Earth's atmosphere

Gas	Volume mixing ratio (percent)	Volume mixing ratio (ppmv)
Fixed gases		
Nitrogen (N_2)	78.08	780 000
Oxygen (O_2)	20.95	209 500
Argon (Ar)	0.93	9 300
Neon (Ne)	0.0015	15
Helium (He)	0.0005	5
Krypton (Kr)	0.0001	1
Xenon (Xe)	0.000005	0.05
Variable gases		
Water vapor (H_2O)	0.00001–4.0	0.1–40 000
Carbon dioxide (CO_2)	0.0375	375
Methane (CH_4)	0.00017	1.8
Ozone (O_3)	0.000003–0.001	0.03–10

Figure 2.1(b) shows that, as altitude (z) increases, air pressure decreases exponentially. Air pressure decreases with increasing altitude because less air lies above a higher altitude than a lower altitude. Air pressure decreases exponentially with increasing altitude because density, used to derive pressure, decreases exponentially with increasing altitude (Fig. 2.1(a)).

Figure 2.1(b) also shows that 50 percent of atmospheric mass lies between sea level and 5.5 km. About 99.9 percent of mass lies below about 48 km. The Earth's radius is approximately 6370 km. Thus, almost all of Earth's atmosphere lies in a layer thinner than 1 percent of the radius of the Earth.

Standard sea-level surface pressure is 1013 hPa (or 1013 mb or 760 mm Hg). The sea-level pressure at a given location and time typically varies by +10 to −20 hPa from standard sea-level pressure. In a strong low-pressure system, such as at the center of a hurricane, the actual sea-level pressure may be more than 50 hPa lower than standard sea-level pressure. At the surface of the Earth, which averages 231.4 m above sea level globally, gravitational acceleration is 9.8060 m s^{-2}. Gravitational acceleration is discussed further in Section 4.2.3.

2.1.1 Fixed gases

Table 2.1 gives the basic composition of the bottom 100 km of the Earth's atmosphere, called the **homosphere**. In this region, the primary gases are molecular nitrogen (N_2) and molecular oxygen (O_2), which together make up over 99 percent of all air molecules. Argon (Ar), a chemically inert gas, makes up most of the remaining 1 percent. Nitrogen, oxygen, and argon are **fixed gases** in that their **volume mixing ratios** (number of molecules of each gas divided by the total number of molecules of dry air) do not change substantially in time or space. Fixed gases are well mixed in the homosphere. At any given altitude, oxygen makes up about

20.95 percent and nitrogen makes up about 78.08 percent of all non-water gas molecules by volume (23.17 and 75.55 percent, respectively, by mass).

2.1.1.1 *Molecular nitrogen*

Gas-phase molecular nitrogen (N_2) is produced biologically in soils by **denitrification**, a two-step process carried out by denitrifying bacteria in **anaerobic** (oxygen-depleted) soils. In the first step, the bacteria reduce the nitrate ion (NO_3^-) to the nitrite ion (NO_2^-). In the second, they reduce the nitrite ion to molecular nitrogen and some nitrous oxide gas (N_2O).

Molecular nitrogen is removed from the air by **nitrogen fixation**. This process is carried out by bacteria, such as *Rhizobium*, *Azotobacter*, and *Beijerinckia*, in **aerobic** (oxygen-rich) environments, and produces the ammonium ion (NH_4^+). Ammonium in soils is also produced by the bacterial decomposition of organic matter during **ammonification**. Fertilizer application also adds ammonium to soils.

The ammonium ion regenerates the nitrate ion in soils during **nitrification**, which is a two-step aerobic process. In the first step, nitrosofying (nitrite-forming) bacteria produce nitrite from ammonium. In the second step, nitrifying (nitrate-forming) bacteria produce nitrate from nitrite.

Because N_2 does not react significantly in the atmosphere and because its removal by nitrogen fixation is slower than its production by denitrification, its atmospheric concentration has built up over time.

2.1.1.2 *Molecular oxygen*

Gas-phase molecular oxygen is produced by **green-plant photosynthesis**, which occurs when carbon dioxide gas (CO_2) reacts with water (H_2O) in the presence of sunlight and green-pigmented **chlorophylls**. Chlorophylls exist in plants, trees, blue-green algae, and certain bacteria. They appear green because they absorb red and blue wavelengths of visible light and reflect green wavelengths. Products of photosynthesis reactions are carbohydrates and molecular oxygen gas. One photosynthetic reaction is

$$6CO_2 + 6H_2O + h\nu \rightarrow C_6H_{12}O_6 + 6O_2 \tag{2.2}$$

where $h\nu$ is a photon of radiation, and $C_6H_{12}O_6$ is glucose. Some bacteria that live in anaerobic environments photosynthesize carbon dioxide with hydrogen sulfide (H_2S) instead of with water to produce organic material and elemental sulfur (S). This type of photosynthesis is **anoxygenic photosynthesis** and predates the onset of green-plant photosynthesis.

2.1.1.3 *Argon and others*

Argon is a noble gas that is colorless and odorless. Like other noble gases, argon is inert and does not react chemically. It forms from the radioactive decay of potassium. Other fixed but inert gases present in trace concentrations are neon (Ne), helium (He), krypton (Kr), and xenon (Xe). The source of helium, krypton, and xenon is radioactive decay of elements in the Earth's crust, and the source of neon is volcanic outgassing. The mixing ratios of these gases are given in Table 2.1.

2.1.2 Variable gases

Gases in the atmosphere whose volume mixing ratios change significantly in time and space are **variable gases**. Thousands of variable gases are present in the atmosphere. Several dozen are very important. Four of these, water vapor, carbon dioxide, methane (CH_4), and ozone (O_3), are discussed below.

2.1.2.1 *Water vapor*

Major sources of **water vapor** to the air are evaporation from soils, lakes, streams, rivers, and oceans, sublimation from glaciers, sea ice, and snow packs, and **transpiration** from plant leaves. Water vapor is also produced during fuel combustion and many gas-phase chemical reactions. Approximately 85 percent of water in the atmosphere originates from ocean surface evaporation. Sinks of water vapor are condensation to the liquid phase, ice deposition to the solid phase, transfer to the ocean and other surfaces, and gas-phase chemical reaction. The mixing ratio of water vapor varies with location. When temperatures are low, water vapor readily condenses as a liquid or deposits as ice, so its gas-phase mixing ratio is low. Over the North and South Poles, water vapor mixing ratios are almost zero. When temperatures are high, liquid water readily evaporates and ice readily sublimates to the gas phase, so water vapor mixing ratios are high. Over equatorial waters, where temperatures are high and evaporation from ocean surfaces occurs readily, the atmosphere contains up to 4 percent or more water vapor by volume. Water vapor is not only a **greenhouse gas** (a gas that readily absorbs thermal-infrared radiation), but also a chemical reactant and carrier of latent heat.

2.1.2.2 *Carbon dioxide*

Carbon dioxide gas is produced by cellular respiration in plants and trees, biological decomposition of dead organic matter, evaporation from the oceans, volcanic outgassing, and fossil-fuel combustion. **Cellular respiration** occurs when oxygen reacts with carbohydrates in the presence of enzymes in living cells to produce CO_2, H_2O, and energy. The reverse of (2.2) is a cellular respiration reaction. **Biological decomposition** occurs when bacteria and other organisms convert dead organic matter to CO_2 and H_2O.

Like water vapor, CO_2 is a greenhouse gas. Unlike water vapor, CO_2 does not react chemically in the atmosphere. Its lifetime against chemical destruction is approximately 100–200 years. CO_2 is removed more readily by green-plant

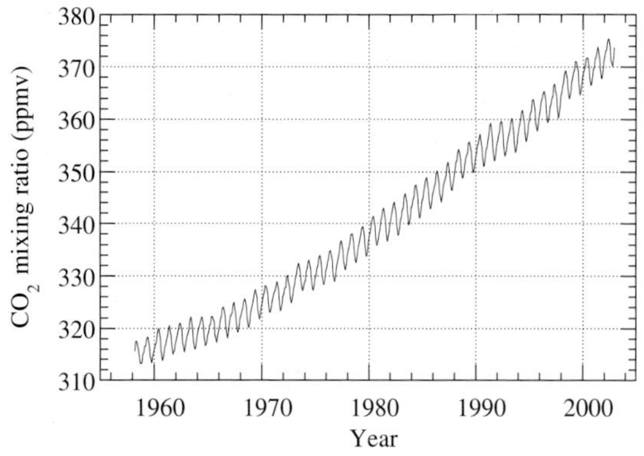

Figure 2.2 Yearly and seasonal fluctuations in CO_2 mixing ratio since 1958. Data from C. D. Keeling at the Mauna Loa Observatory, Hawaii.

photosynthesis and dissolution into ocean water and raindrops. These processes reduce its overall lifetime in the air to 30–95 years.

Average global CO_2 mixing ratios have increased from approximately 275 parts per million by volume (ppmv) in the mid 1700s to approximately 375 ppmv in 2004. Figure 2.2 shows how observed CO_2 mixing ratios have increased steadily since 1958. The yearly increases are due to increased rates of CO_2 emission from fossil-fuel combustion. The seasonal fluctuation in CO_2 mixing ratio is due to photosynthesis and biological decomposition. When annual plants grow in the spring and summer, they remove CO_2 from air by photosynthesis. When they die in the fall and winter, they return CO_2 by biological decomposition.

2.1.2.3 *Methane*

Methane, the main component of natural gas, is a variable gas. It is produced in anaerobic environments, where methane-producing bacteria (methanogens) consume organic material and excrete methane. Ripe anaerobic environments include rice paddies, landfills, wetlands, and the digestive tracts of cattle, sheep, and termites. Methane is also produced in the ground from the decomposition of fossilized carbon. This methane often leaks to the atmosphere or is harnessed and used as a source of energy. Additional methane is produced from biomass burning and as a by-product of atmospheric chemical reaction. Sinks of methane include slow chemical reaction and dry deposition. Methane's mixing ratio in the lower atmosphere is near 1.8 ppmv, which is an increase from about 0.7 ppmv in the mid 1700s. Its mixing ratio has increased steadily due to increased farming and landfill usage. Methane is important because it is a greenhouse gas that absorbs infrared radiation emitted by the Earth 21 times more efficiently, molecule for molecule, than does carbon dioxide. Mixing ratios of carbon dioxide, however, are much larger than are those of methane.

2.1.2.4 *Ozone*

Ozone is a trace gas formed by photochemical reaction and is not emitted into the atmosphere. In the stratosphere, it is produced following photolysis of molecular oxygen. Near the surface of Earth, it is produced following photolysis of nitrogen dioxide (NO_2). Photolysis of molecular oxygen does not occur near the surface, since the wavelengths of radiation required to break apart oxygen are absorbed by molecular oxygen and ozone in the upper atmosphere. In urban regions, ozone production is enhanced by organic gases. Typical ozone mixing ratios in urban air, rural surface air, and stratospheric air are 0.1, 0.04, and 10 ppmv, respectively. Thus, ozone's mixing ratios are the highest in the **stratosphere**, the location of which is shown in Fig. 2.4 (Section 2.2.2.1).

2.2 TEMPERATURE STRUCTURE

Temperature is a measure of the kinetic energy of an air molecule. At a given temperature, the speeds at which air molecules travel are distributed statistically about a Maxwellian distribution. From this distribution, one can define an **average thermal speed** ($\bar{v}_a$), a **root-mean-square speed** (v_{rms}), and a **most probable speed** (v_p) of an air molecule (m s^{-1}). These speeds are related to absolute temperature (T, K) by

$$\bar{v}_a = \sqrt{\frac{8k_B T}{\pi \bar{M}}} \qquad v_{rms} = \sqrt{\frac{3k_B T}{\bar{M}}} \qquad v_p = \sqrt{\frac{2k_B T}{\bar{M}}} \qquad (2.3)$$

respectively, where k_B is **Boltzmann's constant** (1.380658×10^{-23} kg m^2 s^{-2} K^{-1} molec.$^{-1}$; alternative units are given in Appendix A) and $\bar{M}$ is the average mass of one air molecule (4.8096×10^{-26} kg molec.$^{-1}$). Equation (2.3) can be rewritten as

$$\frac{4}{\pi}k_B T = \frac{1}{2}\bar{M}\bar{v}_a^2 \qquad \frac{3}{2}k_B T = \frac{1}{2}\bar{M}v_{rms}^2 \qquad k_B T = \frac{1}{2}\bar{M}v_p^2 \qquad (2.4)$$

The right side of each equation in (2.4) is kinetic energy. When defined in terms of $\bar{v}_a$, temperature is proportional to the kinetic energy of an air molecule traveling at its average speed. When defined in terms of v_{rms}, temperature is proportional to the average kinetic energy among all air molecules since v_{rms}^2 is determined by summing the squares of all individual speeds then dividing by the total number of speeds. When defined in terms of v_p, temperature is proportional to the kinetic energy of an air molecule traveling at its most probable speed. In this text, temperature is defined in terms of $\bar{v}_a$.

Example 2.1

What is the average thermal speed, the root-mean-square speed, and the most probable speed of an air molecule when T = 300 K? What about at 200 K? Although these speeds are faster than those of an airplane, why do air molecules hardly move relative to airplanes over the same time increment?

SOLUTION

At 300 K, $\bar{v}_a = 468.3$ m s^{-1} (1685 km hr^{-1}), $v_{rms} = 508.3$ m s^{-1}, and $v_p = 415.0$ m s^{-1}. At 200 K, $\bar{v}_a = 382.4$ m s^{-1} (1376 km hr^{-1}), $v_{rms} = 415.0$ m s^{-1}, and $v_p = 338.9$ m s^{-1}. Although air molecules move quickly, they continuously collide with and are redirected by other air molecules, so their net migration rate is slow, particularly in the lower atmosphere. Airplanes, on the other hand are not redirected when they collide with air molecules.

The bottom 100 km of Earth's atmosphere, the **homosphere**, is the lower atmosphere and is divided into four major regions in which temperature changes with altitude. These are, from bottom to top, the **troposphere**, **stratosphere**, **mesosphere**, and **thermosphere**. The troposphere is divided into the **boundary layer**, which is the region from the surface to about 500–3000-m altitude, and the **free troposphere**, which is the rest of the troposphere.

2.2.1 Boundary layer

The **boundary layer** is the portion of the troposphere influenced by the Earth's surface and that responds to surface forcing with a time scale of about an hour or less (Stull 1988). The free troposphere is influenced by the boundary layer, but on a longer time scale. Temperature varies significantly in the boundary layer during the day and between day and night. Variations are weaker in the free troposphere. Temperatures in the boundary are affected by the specific heats of soil and air and by energy transfer processes, such as conduction, radiation, mechanical turbulence, thermal turbulence, and advection. These factors are described below.

2.2.1.1 *Specific heat*

Specific heat capacity (specific heat, for short) is the energy required to increase the temperature of 1 gram of a substance 1 degree Celsius (°C). Soil has a lower specific heat than has liquid water. During the day, the addition of a fixed amount of sunlight increases soil temperature more than it increases water temperature. At night, emission of a fixed amount of infrared radiation decreases soil temperature more than it decreases water temperature. Thus, between day and night, soil temperature varies more than does water temperature.

Table 2.2 Specific heats and thermal conductivities of four media

Substance	Specific heat ($J\ kg^{-1}\ K^{-1}$)	Thermal conductivity at 298 K ($J\ m^{-1}\ s^{-1}\ K^{-1}$)
Dry air at constant pressure	1004.67	0.0256
Liquid water	4185.5	0.6
Clay	1360	0.920
Dry sand	827	0.298

Specific heat varies not only between land and water, but also among soil types. Table 2.2 shows that the specific heat of clay is greater than is that of sand, and the specific heat of liquid water is greater than is that of clay or sand. If all else is the same, sandy soil heats to a higher temperature than does clay soil, and dry soil heats to a higher temperature than does wet soil during the day. Dry, sandy soils cool to a greater extent than do wet, clayey soils at night.

2.2.1.2 *Conduction*

Specific heat is a property of a material that affects its temperature change. Temperature is also affected by processes that transfer energy within or between materials. One such process is conduction. **Conduction** is the passage of energy from one molecule to the next in a medium (the conductor). The medium, as a whole, experiences no molecular movement. Conduction occurs through soil, air, and particles. Conduction affects ground temperature by transferring energy between the soil surface and bottom molecular layers of the atmosphere, and between the soil surface and molecules of soil just below the surface. The rate of a material's conduction is determined by its **thermal conductivity**, which quantifies the rate of flow of thermal energy through a material in the presence of a temperature gradient. Empirical expressions for the **thermal conductivities of dry air and water vapor** ($J\ m^{-1}\ s^{-1}\ K^{-1}$) are

$$\kappa_d \approx 0.023807 + 7.1128 \times 10^{-5}(T - 273.15) \tag{2.5}$$

$$\kappa_v \approx 0.015606 + 8.3680 \times 10^{-5}(T - 273.15) \tag{2.6}$$

respectively, which are used in an interpolation equation to give the **thermal conductivity of moist air** (dry air plus water vapor)

$$\kappa_a \approx \kappa_d \left[1 - \left(1.17 - 1.02\frac{\kappa_v}{\kappa_d}\right)\frac{n_v}{n_v + n_d}\right] \tag{2.7}$$

(Pruppacher and Klett 1997). Here, n_v and n_d are the number of moles of water vapor and dry air, respectively, and T is temperature (K). Under atmospheric conditions, the thermal conductivity of moist air is not much different from that of dry air. Table 2.2 shows that liquid water, clay, and dry sand are more conductive than is dry air. Clay is more conductive and dry sand is less conductive than is liquid water.

The flux of energy through the atmosphere due to conduction can be estimated with the **conductive heat flux equation,**

$$H_c = -\kappa_a \frac{\Delta T}{\Delta z} \tag{2.8}$$

(W m^{-2}), where ΔT (K) is the change in temperature over an incremental height Δz (m). Adjacent to the ground, molecules of soil, water, and other surface elements transfer energy by conduction to molecules of air overlying the surface. Since the temperature gradient ($\Delta T/\Delta z$) between the surface and a thin (e.g., 1 mm) layer of air just above the surface is large, the conductive heat flux at the ground is large. Above the ground, temperature gradients are small, and the conductive heat flux is much smaller than at the ground.

Example 2.2

Find the conductive heat flux through a 1-mm thin layer of air touching the surface if $T = 298$ K, $\Delta T = -12$ K, and assuming the air is dry. Do a similar calculation for the free troposphere, where $T = 273$ K and $\Delta T/\Delta z = -6.5$ K km^{-1}.

SOLUTION

Near the surface, $\kappa_a \approx \kappa_d = 0.0256$ J m^{-1} s^{-1} K^{-1}, giving a conductive heat flux of $H_c = 307$ W m^{-2}. In the free troposphere, $\kappa_a \approx \kappa_d = 0.0238$ J m^{-1} s^{-1} K^{-1}, giving $H_c = 1.5 \times 10^{-4}$ W m^{-2}. Thus, heat conduction through the air is important only at the ground.

2.2.1.3 *Radiation*

Radiation is the transfer of energy by electromagnetic waves, which do not require a medium, such as air, for their transmission. **Solar radiation** is relatively short-wavelength radiation emitted by the Sun. **Thermal-infrared radiation** is relatively long-wavelength radiation emitted by the Earth, atmosphere, and clouds. The Earth's surface receives solar radiation during the day only, but its surface and atmosphere emit thermal-infrared radiation during day and night. Radiation is an important energy transfer process and is discussed in Chapter 9.

2.2.1.4 *Mechanical turbulence and forced convection*

Convection is a predominantly vertical motion that results in the transport and mixing of atmospheric properties. **Forced convection** is vertical motion produced by mechanical means, such as mechanical turbulence. **Mechanical turbulence** arises when winds travel over objects protruding from a surface, producing swirling motions of air, or eddies. **Turbulence** is the effect of groups of eddies of different size. Turbulence mixes energy and other variables vertically and horizontally. Strong winds produce strong eddies and turbulence. Turbulence from wind-generated eddies is mechanical turbulence.

When mechanical turbulence is the dominant process of vertical motion in the boundary layer, the boundary layer is in a state of forced convection. Mechanical turbulence is only one type of forced convection; forced convection also occurs when air rises along a topographical barrier or weather front, or when horizontal winds converge and rise.

2.2.1.5 *Thermal turbulence and free convection*

Free convection is a predominantly vertical motion produced by buoyancy. The boundary layer is in a state of free convection when thermal turbulence is the dominant process of vertical motion. **Thermal turbulence** occurs when the Sun heats the ground differentially, creating thermals. Differential heating occurs because clouds or hills block the sun in some areas but not in others or because different surfaces lie at different angles. **Thermals** are parcels of air that rise buoyantly, generally over land during the day. When the Sun heats the ground, conduction transfers energy from the ground to molecules of air adjacent to the ground. The warmed air above the ground rises buoyantly and expands. Cooler air from nearby is drawn in near the surface to replace the rising air. The cooler air heats by conduction and rises.

Free convection differs from conduction in that free convection is the mass movement of air molecules containing energy due to density difference, and conduction is the transfer of energy from molecule to molecule. Free convection occurs most readily over land when the sky is cloud-free and winds are light.

2.2.1.6 *Advection and other factors*

Advection is the horizontal propagation of the mean wind. Horizontal winds advect energy spatially just as they advect gases and particles. Advection is responsible for transferring energy on small and large scales. Other processes that affect temperature in the boundary layer are emissivity, albedo, pressure systems, and length of day.

Emissivity is the ratio of the radiation emitted by an object to the radiation emitted by a perfect emitter. Sand has an emissivity of 0.84–0.91, and clay has an emissivity of 0.9–0.98. The higher its emissivity, the faster a surface, such as soil, cools at night. **Albedo** (or reflectivity) is the ratio of reflected radiation to incident radiation. For dry sand, the albedo varies from 20 to 40 percent. For clay, it varies from 5 to 20 percent. Thus, sand reflects more solar radiation during the day and emits less infrared radiation at night than does clay, counteracting some of the effects of the low specific heat and thermal conductivity of sand.

Large-scale pressure systems also affect temperatures in the boundary layer. Within a large-scale **high-pressure system**, air descends and warms, often on top of cooler surface air, creating an **inversion**, which is an increase in temperature with

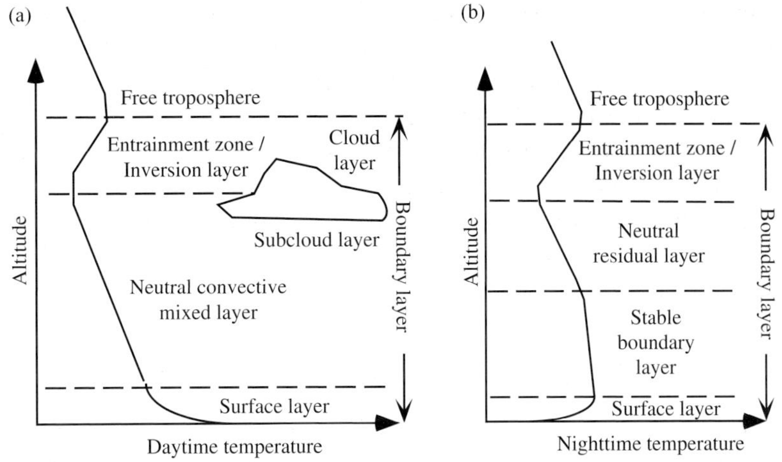

Figure 2.3 Variation of temperature with height during the (a) day and (b) night in the atmospheric boundary layer over land under a high-pressure system. Adapted from Stull (1988).

increasing height. An inversion associated with a large-scale pressure system is a **large-scale subsidence inversion.** As air warms when it sinks in a subsidence inversion, it evaporates clouds, increasing the sunlight reaching the ground, increasing ground temperature. Within a large-scale **low-pressure system**, air rises and cools, often creating clouds and decreasing sunlight reaching the surface.

Temperature is also affected by length of day. Longer days produce longer periods of surface heating by sunlight, and longer nights produce longer periods of cooling due to lack of sunlight and continuous thermal-infrared emission.

2.2.1.7 *Boundary-layer characteristics*

Figures 2.3(a) and (b) show idealized temperature variations in the boundary layer over land during the day and night, respectively, under a high-pressure system. During the day, the boundary layer consists of a surface layer, a convective mixed layer, and an entrainment zone. The **surface layer** is a region of strong wind shear that comprises the bottom 10 percent of the boundary layer. Since the boundary-layer depth ranges from 500 to 3000 m, the surface layer is about 50–300 m thick.

Over land during the day, temperature decreases rapidly with increasing altitude in the surface layer but less so in the mixed layer. In the surface layer, the strong temperature gradient is caused by rapid solar heating of the ground. The temperature gradient is usually so strong that air adjacent to the ground buoyantly rises and accelerates into the mixed layer. The atmosphere is called **unstably stratified** when it exhibits the strong decrease in temperature with increasing height required for air parcels to rise and accelerate buoyantly.

In the mixed layer, the temperature gradient is not strong enough to allow unrestrained convection but not weak enough to prevent some convection. Under such a condition, the atmosphere is **neutrally stratified**, and parcels of air can mix up or down but do not accelerate in either direction.

When a high-pressure system is present, a large-scale subsidence inversion resides above the mixed layer. Environments in which temperatures increase with increasing height (an inversion) or slightly decrease with increasing height are **stably stratified**. Thermals originating in the surface layer or mixed layer cannot easily penetrate buoyantly through an inversion. Some mixing (entrainment) between an inversion layer and mixed layer always occurs; thus, the inversion layer is also called an **entrainment zone**.

Other features of the daytime boundary layer are the cloud layer and subcloud layer. A region in which clouds appear in the boundary layer is the **cloud layer**, and the region underneath is the **subcloud layer**. A cloud forms if rising air in a thermal cools sufficiently. An inversion may prevent a cloud from rising past the mixed layer.

During the night, the surface cools radiatively, causing temperature to increase with increasing height from the ground, creating a surface inversion. Eventual cooling at the top of the surface layer cools the bottom of the mixed layer, increasing the stability of the mixed layer at its base. The portion of the daytime mixed layer that is stable at night is the **stable (nocturnal) boundary layer**. The remaining portion of the mixed layer, which stays neutrally stratified, is the **residual layer**. Because thermals do not form at night, the residual layer does not undergo significant change, except at its base. At night, the nocturnal boundary layer grows, eroding the base of the residual layer. The top of the residual layer is not affected by this growth.

Over the ocean, the boundary layer is influenced more by large-scale pressure systems than by thermal or mechanical turbulence. Since water temperature does not change significantly during the day, thermal turbulence over the ocean is not so important as over land. Since the ocean surface is relatively smooth, mechanical turbulence is also weaker than over land. Large-scale high-pressure systems still cause subsidence inversions to form over the ocean.

2.2.2 Free atmosphere

2.2.2.1 *Troposphere*

The free troposphere lies above the boundary layer. Figure 2.4 shows a standard profile of the temperature structure of the lower atmosphere, ignoring the boundary layer. The **troposphere**, which is the bottom layer of the lower atmosphere, is the region extending from the surface in which the temperature, on average, decreases with increasing altitude. The average rate of temperature decrease in the free troposphere (above the boundary layer) is about 6.5 K km^{-1}.

The temperature decreases with increasing altitude in the free troposphere for the following simplified reason: Earth's surface receives energy from the Sun daily, heating the ground. Simultaneously, the top of the troposphere continuously radiates energy upward, cooling the upper troposphere. The troposphere, itself, has relatively little capacity to absorb solar energy; thus, it relies on radiative, turbulent,

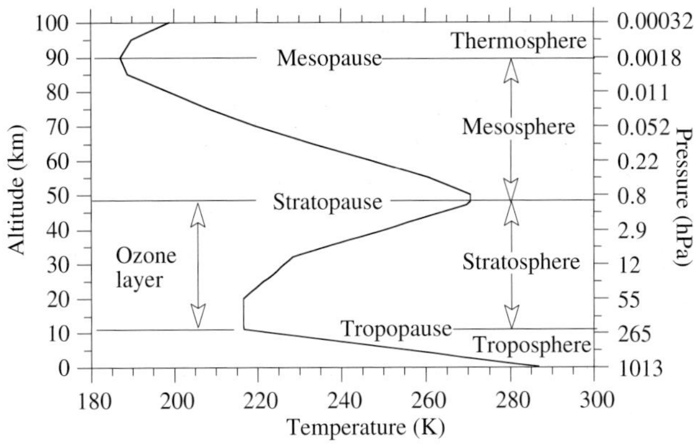

Figure 2.4 Temperature structure of the Earth's lower atmosphere, ignoring the boundary layer.

and conductive energy transfer from the ground to maintain its temperature. The most important of these transfer processes is (thermal-infrared) radiation transfer, followed by turbulence, then conduction. The last two processes transfer energy to the boundary layer only. Thermal-infrared radiation emitted by the ground is absorbed by selective gases in the air, including water vapor and carbon dioxide. These and other gases are most concentrated near the surface, thus they have the first opportunity to absorb the radiation. Once they absorb the radiation, they reemit half of it upward and the other half downward, but their emission is at a lower temperature than is the emission of the radiation they absorbed. The lower the temperature of radiant emission, the less intense the emission. Thus, not only does less thermal-infrared radiation penetrate to higher than to lower altitudes, but the radiation that does penetrate is less intense than at lower altitudes. Because less radiation is available for absorption with increasing altitude, temperatures decrease with increasing altitude in the troposphere.

Figures 2.5(a) and (b) show global latitude–altitude contour plots of zonally averaged temperatures for a generic January and July, respectively. A **zonally averaged** temperature is found from a set of temperatures, averaged over all longitudes at a given latitude and altitude. The figures indicate that near-surface tropospheric temperatures decrease from the Equator to high latitudes, which is expected, since the Earth receives the greatest quantity of incident solar radiation near the Equator.

The **tropopause** is the upper boundary of the troposphere. It is defined by the World Meteorological Organization (WMO) as the lowest altitude at which the lapse rate (rate of decrease of temperature with increasing height) decreases to 2 K km^{-1} or less, and at which the lapse rate, averaged between this altitude and any altitude within the next 2 km, does not exceed 2 K km^{-1}. Above the tropopause base, the temperature is relatively constant with increasing altitude (**isothermal**) before it increases with increasing height in the stratosphere.

Figures 2.5(a) and (b) indicate that the tropopause height decreases from 15–18 km near the Equator to 8 km near the poles. Because temperature decreases with increasing height in the troposphere, and because tropopause height is higher

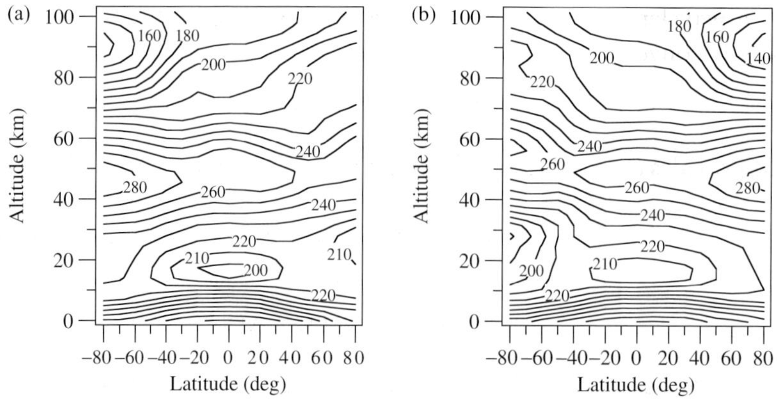

Figure 2.5 Zonally and monthly averaged temperatures for (a) January and (b) July. Data for the plots were compiled by Fleming *et al.* (1988).

over the Equator than over the poles, temperature decreases to a lower value over the Equator than over the poles. As a result, minimum tropopause temperatures occur over equatorial regions, as shown in Figs. 2.5(a) and (b). Tropopause heights are the highest over the Equator because strong vertical motions over the Equator raise the base of the ozone layer and force ozone to spread horizontally to higher latitudes. Since ozone is responsible for warming above the tropopause, pushing ozone to a greater height increases the altitude at which warming begins. Near the poles, downward motion pushes ozone downward. The resulting latitudinal gradient of the ozone-layer base decreases the tropopause height from the Equator to the poles.

Tropopause temperature is also affected by water vapor. Water vapor absorbs the Earth's thermal-infrared radiation, preventing some of it from reaching the tropopause. Water-vapor mixing ratios are much higher over the Equator than over the poles. The high mixing ratios near the surface over the Equator enhance cooling of the tropopause above.

2.2.2.2 *Stratosphere*

The stratosphere is a region of the atmosphere above the troposphere that contains 90 percent of the Earth's ozone and in which temperature increases with increasing height. The temperature profile in the stratosphere is caused by ozone interactions with radiation, discussed below.

Ozone is produced when molecular oxygen absorbs ultraviolet radiation and photodissociates by the reactions,

$$O_2 + h\nu \rightarrow \dot{O}\cdot(^1D) + \dot{O}\cdot \qquad \lambda < 175\,\text{nm} \tag{2.9}$$

$$O_2 + h\nu \rightarrow \dot{O}\cdot + \dot{O}\cdot \qquad 175 < \lambda < 245\,\text{nm} \tag{2.10}$$

where $h\nu$ is a photon of radiation, $O(^1D)$ is excited atomic oxygen, O [$=O(^3P)$] is ground state atomic oxygen, the dots identify the valence of an atom (Chapter 10), and λ is a wavelength of radiation affecting the reaction. $O(^1D)$ produced from (2.9) rapidly converts to O via

$$\cdot\dot{O}(^1D) + M \rightarrow \cdot\dot{O} + M \qquad (2.11)$$

where M provides collisional energy for the reaction, but is not created or destroyed by it. Because N_2 and O_2 are the most abundant gases in the air, M is most likely N_2 or O_2, although it can be another gas. The most important reaction creating ozone is

$$\dot{O}\cdot + O_2 + M \rightarrow O_3 + M \qquad (2.12)$$

where M, in this case, carries away energy released by the reaction.

Ozone in the stratosphere is destroyed naturally by the photolysis reactions,

$$O_3 + h\nu \rightarrow O_2 + \cdot\dot{O}(^1D) \qquad \lambda < 310\,\text{nm} \qquad (2.13)$$

$$O_3 + h\nu \rightarrow O_2 + \dot{O}\cdot \qquad \lambda > 310\,\text{nm} \qquad (2.14)$$

and the two-body reaction

$$\dot{O}\cdot + O_3 \rightarrow 2O_2 \qquad (2.15)$$

Chapman (1930) postulated that the reactions (2.10), (2.12), (2.14), and (2.15), together with

$$\dot{O}\cdot + \dot{O}\cdot + M \rightarrow O_2 + M \qquad (2.16)$$

describe the natural formation and destruction of ozone in the stratosphere. These reactions make up the **Chapman cycle**, and they simulate the process fairly well. Some Chapman reactions are more important than others. The reactions (2.10), (2.12), and (2.14) affect ozone most significantly. The non-Chapman reaction, (2.13), is also important.

The peak ozone density in the atmosphere occurs at about 25–32 km (Chapter 11, Fig. 11.3). The reason is that ozone forms most readily when sufficient oxygen and ultraviolet radiation are present to produce atomic oxygen through reactions (2.9) and (2.10). Since oxygen density, like air density, decreases exponentially with increasing altitude, and since ultraviolet radiation intensity increases with increasing altitude (because radiation attenuates as it passes through the atmosphere), an altitude exists at which the quantities of radiation and oxygen are optimal for producing an ozone maximum. This altitude is between 25 and 32 km. At higher altitudes, the oxygen density is too low to produce peak ozone, and at lower altitudes, the radiation is not intense enough to photodissociate enough oxygen to produce peak ozone.

Peak stratospheric temperatures occur at the top of the stratosphere, because this is the altitude at which ozone absorbs the shortest ultraviolet wavelengths reaching the stratosphere (about 0.175 μm). Although the concentration of ozone at the top of the stratosphere is small, each molecule there can absorb short wavelengths, increasing the average kinetic energy and temperature of all molecules. In the lower stratosphere, short wavelengths do not penetrate, and temperatures are lower than in the upper stratosphere.

2.2.2.3 *Mesosphere*

In the **mesosphere**, ozone densities are too low for ozone to absorb radiation and affect temperatures. As such, temperatures decrease with increasing altitude in a manner similar to in the upper troposphere.

2.2.2.4 *Thermosphere*

In the **thermosphere**, temperatures increase with increasing altitude because molecular oxygen and nitrogen there absorb very short wavelengths of solar radiation. Air in the thermosphere does not *feel* hot to the skin, because the thermosphere contains so few gas molecules. But each gas molecule in the thermosphere is highly energized, so the average temperature is high. Direct sunlight in the thermosphere is more intense and damaging than at the surface. Because oxygen and nitrogen absorb the shortest wavelengths of radiation in the thermosphere, such wavelengths do not penetrate to the mesopause.

2.3 EQUATION OF STATE

The **equation of state** describes the relationship among pressure, volume, and absolute temperature of a real gas. The **ideal gas law** describes this relationship for an ideal gas. An ideal gas is a gas for which the product of the pressure and volume is proportional to the absolute temperature. A real gas is ideal only to the extent that intermolecular forces are small, which occurs when pressures are low enough or temperatures are high enough for the gas to be sufficiently dilute. Under typical atmospheric temperature and pressure conditions, the ideal gas law gives an error of less than 0.2 percent for dry air and water vapor in comparison with an expanded equation of state (Pruppacher and Klett 1997). Thus, the ideal gas law can reasonably approximate the equation of state.

The **ideal gas law** is expressed as a combination of Boyle's law, Charles' law, and Avogadro's law. In 1661, Robert Boyle (1627–1691) found that doubling the pressure exerted on a gas at constant temperature reduced the volume of the gas by one-half. This relationship is embodied in **Boyle's law**,

$$p \propto \frac{1}{V} \quad \text{at constant temperature} \tag{2.17}$$

where p is the pressure exerted on the gas (hPa), and V is the volume enclosed by the gas (m^3 or cm^3). Boyle's law describes the compressibility of a gas. When high pressure is exerted on a gas, as in the lower atmosphere, the gas compresses (its volume decreases) until it exerts an equal pressure on its surroundings. When low pressure is exerted on a gas, as in the upper atmosphere, the gas expands until it exerts an equal pressure on its surroundings.

In 1787, Jacques Charles (1746–1823) found that increasing the absolute temperature of a gas at constant pressure increased the volume of the gas. This relationship is embodied in **Charles' law**,

$$V \propto T \quad \text{at constant pressure} \tag{2.18}$$

where T is the temperature of the gas (K). Charles' law states that, at constant pressure, the volume of a gas must decrease when its temperature decreases. Since gases change phase to liquids or solids before 0 K, Charles' law cannot be extrapolated to 0 K.

Amedeo Avogadro (1776–1856) hypothesized that equal volumes of different gases under the same conditions of temperature and pressure contain the same number of molecules. In other words, the volume of a gas is proportional to the number of molecules of gas present and independent of the type of gas. This relationship is embodied in **Avogadro's law**,

$$V \propto n \quad \text{at constant pressure and temperature} \tag{2.19}$$

where n is the number of gas moles. The number of molecules in a mole is constant for all gases and given by **Avogadro's number**, $A = 6.0221367 \times 10^{23}$ molec. mol^{-1}.

Combining Boyle's law, Charles' law, and Avogadro's law gives the **ideal gas law** or **simplified equation of state** as

$$p = \frac{nR^*T}{V} = \frac{nA}{V}\left(\frac{R^*}{A}\right)T = Nk_BT \tag{2.20}$$

where R^* is the **universal gas constant** (0.0831451 m^3 hPa mol^{-1} K^{-1} or 8.31451×10^4 cm^3 hPa mol^{-1} K^{-1}), $N = nA/V$ is the number concentration of gas molecules (molecules of gas per cubic meter or cubic centimeter of air), and $k_B = R^*/A$ is **Boltzmann's constant** in units of 1.380658×10^{-25} m^3 hPa K^{-1} molec.$^{-1}$ or 1.380658×10^{-19} cm^3 hPa K^{-1} molec.$^{-1}$. Appendix A contains alternative units for R^* and k_B.

Example 2.3

Calculate the number concentration of air molecules in the atmosphere at standard sea-level pressure and temperature and at a pressure of 1 hPa.

SOLUTION

At standard sea level, p = 1013 hPa and T = 288 K. Thus, from (2.20), $N = 2.55 \times 10^{19}$ molec. cm^{-3}. From Fig. 2.1(b), p = 1 hPa occurs at 48 km. At this altitude and pressure, T = 270 K, as shown in Fig. 2.4. Under such conditions, $N = 2.68 \times 10^{16}$ molec. cm^{-3}.

Equation (2.20) can be used to relate the partial pressure exerted by a gas to its number concentration. In 1803, John Dalton (1766–1844) stated that total atmospheric pressure equals the sum of the partial pressures of the individual gases in a mixture. This is **Dalton's law of partial pressure.** The **partial pressure** exerted by a gas in a mixture is the pressure the gas exerts if it alone occupies the same volume as the mixture. Mathematically, the partial pressure of gas q is

$$p_q = N_q k_B T \tag{2.21}$$

where N_q is the number concentration of the gas (molec. cm^{-3}). Total atmospheric pressure is

$$p_a = \sum_q p_q = k_B T \sum_q N_q = N_a k_B T \tag{2.22}$$

where N_a is the number concentration of the air, determined as the sum of the number concentrations of individual gases.

Total atmospheric pressure can also be written as $p_a = p_d + p_v$, where p_d is the partial pressure exerted by dry air, and p_v is the partial pressure exerted by water vapor. Similarly, the number concentration of air molecules can be written as $N_a = N_d + N_v$, where N_d is the number concentration of dry air, and N_v is the number concentration of water vapor.

Dry air consists of all gases in the atmosphere, except water vapor. Table 2.1 shows that together, N_2, O_2, Ar, and CO_2 constitute 99.996 percent of dry air by volume. The concentrations of all gases aside from these four can be ignored, without much loss in accuracy, when dry-air pressure is calculated. This assumption is convenient in that the concentrations of most trace gases vary in time and space.

Partial pressure is related to the mass density and number concentration of dry air through the **equation of state for dry air,**

$$p_d = \frac{n_d R^* T}{V} = \frac{n_d m_d}{V}\left(\frac{R^*}{m_d}\right)T = \rho_d R' T = \frac{n_d A}{V}\left(\frac{R^*}{A}\right)T = N_d k_B T \tag{2.23}$$

where p_d is dry-air partial pressure (hPa), n_d is the number of moles of dry air, m_d is the molecular weight of dry air, ρ_d is the mass density of dry air (kg m^{-3} or g cm^{-3}), and R' is the gas constant for dry air. The **molecular weight of dry air** is a volume-weighted average of the molecular weights of N_2, O_2, Ar, and CO_2. The standard value of m_d is 28.966 g mol^{-1}. The dry-air mass density, number concentration, and gas constant are, respectively,

$$\rho_d = \frac{n_d m_d}{V} \qquad N_d = \frac{n_d A}{V} \qquad R' = \frac{R^*}{m_d} \tag{2.24}$$

where R' has a value of 2.8704 m^3 hPa kg^{-1} K^{-1} or 2870.3 cm^3 hPa g^{-1} K^{-1}. Alternative units for R' are given in Appendix A.

Example 2.4

When $p_d = 1013$ hPa and $T = 288$ K, the density of dry air from (2.23) is $\rho_d = 1.23$ kg m^{-3}.

The **equation of state for water vapor** is

$$p_v = \frac{n_v R^* T}{V} = \frac{n_v m_v}{V}\left(\frac{R^*}{m_v}\right)T = \rho_v R_v T = \frac{n_v A}{V}\left(\frac{R^*}{A}\right)T = N_v k_B T \quad (2.25)$$

where p_v is the partial pressure exerted by water vapor (hPa), n_v is the number of moles of water vapor, m_v is the molecular weight of water vapor, ρ_v is the mass density of water vapor (kg m^{-3} or g cm^{-3}), and R_v is the gas constant for water vapor (4.6140 m^3 hPa kg^{-1} K^{-1} or 4614.0 cm^3 hPa g^{-1} K^{-1}). Alternative units for R_v are given in Appendix A. The water-vapor mass density, number concentration, and gas constant are, respectively,

$$\rho_v = \frac{n_v m_v}{V} \qquad N_v = \frac{n_v A}{V} \qquad R_v = \frac{R^*}{m_v} \quad (2.26)$$

Example 2.5

When $p_v = 10$ hPa and $T = 298$ K, water vapor density from (2.21) is $\rho_v = 7.27 \times 10^{-3}$ kg m^{-3}.

The equation of state for water vapor can be rewritten in terms of the dry-air gas constant as

$$p_v = \rho_v R_v T = \rho_v\left(\frac{R_v}{R'}\right)R'T = \frac{\rho_v R'T}{\varepsilon} \quad (2.27)$$

where

$$\varepsilon = \frac{R'}{R_v} = \frac{R^*}{m_d}\left(\frac{m_v}{R^*}\right) = \frac{m_v}{m_d} = 0.622 \quad (2.28)$$

The number concentration of a gas (molecules per unit volume of air) is an absolute quantity. The abundance of a gas may also be expressed in terms of a relative quantity, **volume mixing ratio**, defined as the number of gas molecules per molecule of dry air, and expressed for gas q as

$$\chi_q = \frac{N_q}{N_d} = \frac{p_q}{p_d} = \frac{n_q}{n_d} \quad (2.29)$$

where N_q, p_q, and n_q are the number concentration, partial pressure, and number of moles of gas q, respectively. Another relative quantity, mass mixing ratio, is the mass of gas per mass of dry air. The **mass mixing ratio** of gas q (kilograms of gas per kilogram of dry air) is

$$\omega_q = \frac{\rho_q}{\rho_\mathrm{d}} = \frac{m_q N_q}{m_\mathrm{d} N_\mathrm{d}} = \frac{m_q p_q}{m_\mathrm{d} p_\mathrm{d}} = \frac{m_q n_q}{m_\mathrm{d} n_\mathrm{d}} = \frac{m_q}{m_\mathrm{d}} \chi_q \tag{2.30}$$

where ρ_q is the mass density (kg m^{-3}) and m_q is the molecular weight of gas q (g mol^{-1}). Volume and mass mixing ratios may be multiplied by 10^6 and expressed in **parts per million by volume** (ppmv) or **parts per million by mass** (ppmm), respectively.

Example 2.6

Find the mass mixing ratio, number concentration, and partial pressure of ozone if its volume mixing ratio in an urban air parcel is $\chi_q = 0.10$ ppmv. Assume $T = 288$ K and $p_\mathrm{d} = 1013$ hPa.

SOLUTION

The molecular weight of ozone is $m_q = 48.0$ g mol^{-1}, and the molecular weight of dry air is $m_\mathrm{d} = 28.966$ g mol^{-1}. From (2.30), the mass mixing ratio of ozone is $\omega_q = 48.0$ g mol^{-1} $\times$ 0.10 ppmv/28.966 g mol^{-1} = 0.17 ppmm. From Example 2.3, $N_\mathrm{d} = 2.55 \times 10^{19}$ molec. cm^{-3}. Thus, from (2.29), the number concentration of ozone is $N_q = 0.10$ ppmv $\times 10^{-6} \times 2.55 \times 10^{19}$ molec. cm^{-3} = 2.55×10^{12} molec. cm^{-3}. From (2.21), the partial pressure exerted by ozone is $p_q = 0.000101$ hPa.

Combining (2.28) and (2.30) gives the **mass mixing ratio of water vapor** (kilograms of water vapor per kilogram of dry air) as

$$\omega_\mathrm{v} = \frac{\rho_\mathrm{v}}{\rho_\mathrm{d}} = \frac{m_\mathrm{v} p_\mathrm{v}}{m_\mathrm{d} p_\mathrm{d}} = \varepsilon \frac{p_\mathrm{v}}{p_\mathrm{d}} = \frac{\varepsilon p_\mathrm{v}}{p_\mathrm{a} - p_\mathrm{v}} = \varepsilon \chi_\mathrm{v} \tag{2.31}$$

Example 2.7

If the partial pressure exerted by water vapor is $p_\mathrm{v} = 10$ hPa, and the total air pressure is $p_\mathrm{a} = 1010$ hPa, the mass mixing ratio of water vapor from (2.31) is $\omega_\mathrm{v} = 0.00622$ kg kg^{-1}.

Water vapor can also be expressed in terms of relative humidity, discussed in Section 2.5, or **specific humidity**. The mass of any substance per unit mass of moist air (dry air plus water vapor) is the **moist-air mass mixing ratio** (q). Specific humidity is the moist-air mass mixing ratio of water vapor (mass of water vapor per unit mass of moist air). An expression for specific humidity (kilograms of water vapor per kilogram of moist air) is

$$q_v = \frac{\rho_v}{\rho_a} = \frac{\rho_v}{\rho_d + \rho_v} = \frac{\dfrac{p_v}{R_v T}}{\dfrac{p_d}{R'T} + \dfrac{p_v}{R_v T}} = \frac{\dfrac{R'}{R_v} p_v}{p_d + \dfrac{R'}{R_v} p_v} = \frac{\varepsilon p_v}{p_d + \varepsilon p_v} \tag{2.32}$$

where $\rho_a = \rho_d + \rho_v$ is the **mass density of moist air.** Specific humidity is related to the mass mixing ratio of water vapor by

$$q_v = \omega_v \frac{\rho_d}{\rho_a} = \frac{\omega_v}{1 + \omega_v} \tag{2.33}$$

> **Example 2.8**
>
> If $p_v = 10$ hPa and $p_a = 1010$ hPa, $p_a = p_d + p_v$ gives $p_d = 1000$ hPa. Under such conditions, (2.32) gives the specific humidity as $q_v = 0.00618$ kg kg^{-1}.

The **equation of state for moist air** is the sum of the equations of state for dry air and water vapor. Thus,

$$p_a = p_d + p_v = \rho_d R'T + \rho_v R_v T = \rho_a R'T \frac{\rho_d + \rho_v R_v / R'}{\rho_a} \tag{2.34}$$

Substituting $\varepsilon = R'/R_v$, $\rho_a = \rho_d + \rho_v$, and $\omega_v = \rho_v/\rho_d$ into (2.34) yields the equation of state for moist air as

$$p_a = \rho_a R'T \frac{\rho_d + \rho_v/\varepsilon}{\rho_d + \rho_v} = \rho_a R'T \frac{1 + \rho_v/(\rho_d \varepsilon)}{1 + \rho_v/\rho_d} = \rho_a R'T \frac{1 + \omega_v/\varepsilon}{1 + \omega_v} \tag{2.35}$$

This equation can be simplified to

$$\boxed{p_a = \rho_a R_m T = \rho_a R' T_v} \tag{2.36}$$

where

$$R_m = R' \frac{1 + \omega_v/\varepsilon}{1 + \omega_v} = R' \left(1 + \frac{1 - \varepsilon}{\varepsilon} q_v\right) = R'(1 + 0.608 q_v) \tag{2.37}$$

is the **gas constant for moist air** and

$$T_v = T \frac{R_m}{R'} = T \frac{1 + \omega_v \varepsilon}{1 + \omega_v} = T\left(1 + \frac{1 - \varepsilon}{\varepsilon} q_v\right) = T(1 + 0.608 q_v) \tag{2.38}$$

is **virtual temperature.** This quantity is the temperature of a sample of dry air at the same density and pressure as a sample of moist air. Since the gas constant for moist air is larger than that for dry air, moist air has a lower density than does dry air at the same temperature and pressure. For the dry-air density to equal the moist-air density at the same pressure, the temperature of the dry air must be higher than that of the moist air by the factor R_m/R'. The resulting temperature is the virtual temperature, which is always larger than the actual temperature. In (2.38), the liquid water content is assumed to equal zero.

Equating $R_m = R^*/m_a$ with R_m from (2.37) and noting that $R' = R^*/m_d$ give the **molecular weight of moist air** (g mol^{-1}) as

$$m_a = \frac{m_d}{1 + 0.608q_v} \tag{2.39}$$

The molecular weight of moist air is less than that of dry air.

Example 2.9

If $p_d = 1013$ hPa, $p_v = 10$ hPa, and $T = 298$ K, calculate q_v, m_a, R_m, T_v, and ρ_a.

SOLUTION

From (2.32), $q_v = 0.622 \times 10/(1013 + 0.622 \times 10) = 0.0061$ kg kg^{-1}.
From (2.39), $m_a = 28.966/(1 + 0.608 \times 0.0061) = 28.86$ g mol^{-1}.
From (2.37), $R_m = 2.8704 \times (1 + 0.608 \times 0.0061) = 2.8811$ m^3 hPa kg^{-1} K^{-1}.
From (2.38), $T_v = 298 \times (1 + 0.608 \times 0.0061) = 299.1$ K.
From (2.36), $\rho_a = p_a/(R_m T) = 1023/(2.8811 \times 298) = 1.19$ kg m^{-3}.

2.4 CHANGES OF PRESSURE WITH ALTITUDE

The variation of pressure with altitude in the atmosphere can be estimated several ways. The first is by considering the equation for hydrostatic air pressure given in (2.1). The differential form of the equation, called the **hydrostatic equation,** is

$$dp_a = -\rho_a g \, dz \tag{2.40}$$

where air pressure, air density and gravity are functions of altitude (z), but parentheses have been removed. The negative sign arises because pressure decreases with increasing altitude. The hydrostatic equation assumes that the downward force of gravity per unit volume of air ($-\rho_a g$) exactly balances an upward pressure gradient force per unit volume ($\partial p_a/\partial z$). Figure 2.1(b) shows that pressure decreases exponentially with increasing altitude, giving rise to an upward pressure gradient force (a force that moves air from high to low pressure – Section 4.2.4). If gravity did not exist, the upward pressure gradient would accelerate air to space.

Pressure at a given altitude can be estimated from the hydrostatic equation with

$$p_{a,k} \approx p_{a,k+1} - \rho_{a,k+1} g_{k+1} (z_k - z_{k+1}) \tag{2.41}$$

where $p_{a,k}$ is the pressure at any upper altitude z_k, and $p_{a,k+1}$, $\rho_{a,k+1}$, and g_{k+1} are pressure, density, and gravity, respectively, at any lower altitude, z_{k+1}. If the pressure is known at the surface, and a vertical density and gravity profile are known, (2.41) can be used to estimate the pressure at each altitude above the surface, as illustrated in Example 2.10. Appendix Table B.1 gives vertical profiles of density and gravity in the atmosphere.

Example 2.10

If pressure, density, and gravity at sea level are $p_{a,k+1} = 1013.25$ hPa, $\rho_{a,k+1} = 1.225$ kg m^{-3}, and $g_{k+1} = 9.8072$ m s^{-2}, respectively, estimate the pressure at 100-m altitude.

SOLUTION

From (2.41), $p_{a,100\,\mathrm{m}} = 1013.25\,\mathrm{hPa} - 1.225\,\frac{\mathrm{kg}}{\mathrm{m}^3}\left(9.8072\,\frac{\mathrm{m}}{\mathrm{s}^2}\right)(100 - 0\,\mathrm{m})\,\frac{\mathrm{hPa\,m\,s^2}}{100\,\mathrm{kg}} = 1001.24$ hPa, which compares well with the standard atmosphere pressure at 100 m given in Appendix Table B.1. This result also suggests that, near the surface, pressure decreases approximately 1 hPa for every 10 m increase in altitude.

Altitude versus pressure can also be determined from a pressure altimeter, which measures the pressure at an unknown altitude with an **aneroid barometer** (Section 1.1). From the barometric pressure, altitude is calculated under **standard atmospheric conditions.** Under such conditions, sea-level air pressure, sea-level air temperature, and the environmental lapse rate (average negative change in air temperature with altitude in the free troposphere) are $p_{a,s} = 1013.25$ hPa, $T_{a,s} = 288$ K, and $\Gamma_s = -\partial T/\partial z = +6.5$ K km^{-1}, respectively (List 1984).

The equation for altitude in a standard atmosphere is derived by substituting $p_a = \rho_a R_m T$ into the hydrostatic equation from (2.40). The result is

$$\frac{\partial p_a}{\partial z} = -\frac{p_a}{R_m T}g \tag{2.42}$$

Substituting $T = T_{a,s} - \Gamma_s z$, where $z = 0$ km corresponds to sea level, into (2.42), rearranging, then integrating from $p_{a,s}$ to p_a and 0 to z yields

$$\ln\left(\frac{p_a}{p_{a,s}}\right) = \frac{g}{\Gamma_s R_m}\ln\left(\frac{T_{a,s} - \Gamma_s z}{T_{a,s}}\right) \tag{2.43}$$

Rearranging again gives altitude as a function of pressure in a standard troposphere as

$$z = \frac{T_{a,s}}{\Gamma_s}\left[1 - \left(\frac{p_a}{p_{a,s}}\right)^{\frac{\Gamma_s R_m}{g}}\right] \tag{2.44}$$

Temperature variations with altitude and sea-level pressure in the real atmosphere usually differ from those in the standard atmosphere. Empirical and tabulated expressions correcting for the differences are available (List 1984). Since the corrections are not always accurate, airplanes rarely use pressure altimeters to measure altitude. Instead, they use radar altimeters, which measure altitude with radio waves.

Example 2.11

If a pressure altimeter reads $p_a = 850$ hPa and the air is dry, find the standard-atmosphere altitude.

SOLUTION

Because the air is dry, $R_m = R' = 287.04\ \text{m}^2\ \text{s}^{-2}\ \text{K}^{-1}$. Thus, $\Gamma_s R_m/g = 0.1902$, and from (2.44), $z = 1.45$ km.

A third way to estimate pressure versus altitude is with the **scale-height equation.** From the equation of state for moist air, air density is

$$\rho_a = \frac{p_a}{R'T_v} = \frac{m_d}{R^*}\frac{p_a}{T_v} = \frac{p_a}{T_v}\left(\frac{A}{R^*}\right)\frac{m_d}{A} \approx \frac{p_a}{T_v}\left(\frac{1}{k_B}\right)\bar{M} = \frac{p_a \bar{M}}{k_B T_v} \tag{2.45}$$

where $\bar{M} \approx m_d/A$ was previously defined as the **average mass of one air molecule.** Equation (2.45) can be combined with the hydrostatic equation to give

$$\frac{\mathrm{d}p_a}{p_a} = -\frac{\bar{M}g}{k_B T_v}\mathrm{d}z = -\frac{\mathrm{d}z}{H} \tag{2.46}$$

where

$$H = \frac{k_B T_v}{\bar{M}g} \tag{2.47}$$

is the **scale height** of the atmosphere at a given virtual temperature. The scale height is the height above a reference height at which pressure decreases to 1/e of its value at the reference height. Since temperature varies with altitude in the atmosphere, scale height also varies with altitude.

Example 2.12

Determine the scale height at $T_v = 298$ K and $p_a = 1013.25$ hPa. Assume dry air.

SOLUTION

From Appendix A and Appendix Table B.1, $\bar{M} = 4.8096 \times 10^{-26}$ kg, $g = 9.8072\ \text{m s}^{-2}$, and $k_B = 1.380658 \times 10^{-23}\ \text{kg m}^2\ \text{s}^{-2}\ \text{K}^{-1}\ \text{molec.}^{-1}$ at the given pressure. Thus, from (2.47), the scale height is $H = 8.72$ km.

Integrating (2.46) at constant temperature gives pressure as a function of altitude as

$$p_a = p_{a,ref} e^{-(z - z_{ref})/H} \qquad (2.48)$$

where $p_{a,ref}$ is pressure at a reference level, $z = z_{ref}$. In a model with several vertically stacked layers, each with a known average temperature, pressure at the top of each layer can be calculated from (2.48) assuming that the reference pressure at the base of the layer is the pressure calculated for the top of the next-lowest layer. At the ground, the pressure is the surface pressure, which must be known.

2.5 WATER IN THE ATMOSPHERE

Water appears in three states – gas, liquid, and solid. Sources and sinks of water vapor in the atmosphere were discussed in Section 2.1.2.1. Sources of liquid water include sea spray emission, volcanos, combustion, condensation of water vapor, and melting of ice crystals. Sinks of liquid water include evaporation, freezing, and sedimentation to the surface of aerosol particles, fog drops, drizzle, and raindrops. Ice in the atmosphere forms from freezing of liquid water and solid deposition of water vapor. Losses of ice occur by sublimation to the gas phase, melting to the liquid phase, and sedimentation to the surface.

2.5.1 Types of energy

When water changes state, it releases or absorbs energy. **Energy** is the capacity of a physical system to do work on matter. **Matter** is mass that exists as a solid, liquid, or gas. Energy takes many forms.

Kinetic energy is the energy within a body due to its motion and equals one-half the mass of the body multiplied by its speed squared. The faster a body moves, the greater its kinetic energy. To change the speed and kinetic energy of a body in motion, mechanical work must be done.

Potential energy is the energy of matter that arises due to its position, rather than its motion. Potential energy represents the amount of work that a body can do. A coiled spring, charged battery, and chemical reactant have potential energy. When an object is raised vertically, it possesses gravitational potential energy because it can potentially do work by sinking.

Internal energy is the kinetic and/or potential energy of molecules within an object, but does not include the kinetic and/or potential energy of the object as a whole.

Work is the energy added to a body by the application of a force that moves the body in the direction of the force.

Electromagnetic (radiant) energy is the energy transferred by electromagnetic waves that originate from bodies with temperatures above 0 K.

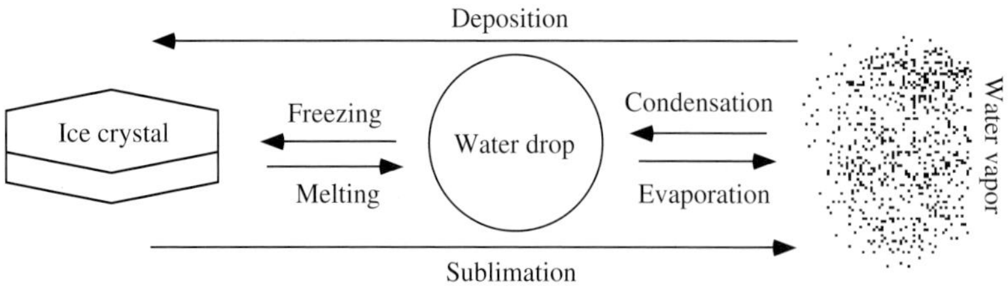

Figure 2.6 Phase changes of water. Freezing (melting) at 0 °C releases (absorbs) 333.5 J g^{-1}, deposition (sublimation) at 0 °C releases (absorbs) 2835 J g^{-1}, and condensation (evaporation) releases (absorbs) 2510 J g^{-1} at 0 °C and 2259 J g^{-1} at 100 °C.

Heat, itself, is not a form of energy. However, **heat transfer** is a term used to describe the energy transfer between two bodies that occurs, for example, when their internal energies (or temperatures) differ. **Heat capacity** (e.g., specific heat capacity) is the energy required to change the temperature of a given quantity of a substance 1 °C. Finally, **heat release (absorption)** occurs when a substance releases (absorbs) energy to (from) the surrounding environment upon a change of state. Below, heat release/absorption is discussed.

2.5.2 Latent heat

During condensation, freezing, and solid deposition of a substance, energy is released. During evaporation, melting, and sublimation, energy is absorbed. The stored energy released or the energy absorbed during such processes is called **latent heat**. Latent heat absorbed (released) during evaporation (condensation) is **latent heat of evaporation**. It varies with temperature as

$$\frac{dL_e}{dT} = c_{p,V} - c_W \tag{2.49}$$

where $c_{p,V}$ is the specific heat of water vapor at constant pressure, and c_W is the specific heat of liquid water. The latent heat absorbed (released) during sublimation (deposition) is the **latent heat of sublimation**. The latent heat absorbed (released) during melting (freezing) is the **latent heat of melting**. The latent heats of sublimation and melting vary with temperature according to

$$\frac{dL_s}{dT} = c_{p,V} - c_I \qquad \frac{dL_m}{dT} = c_W - c_I \tag{2.50}$$

respectively, where c_I is the specific heat of ice. Figure 2.6 gives the quantity of energy absorbed or released during phase changes of water.

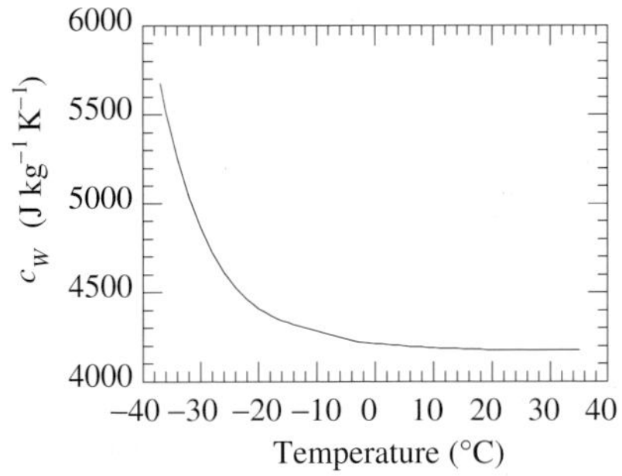

Figure 2.7 Variation of specific heat of liquid water with temperature, from (2.51).

The specific heats of liquid water, ice, and water vapor vary with temperature. Polynomial expressions for the specific heat of liquid water (J kg^{-1} K^{-1}) are

$$c_W = \begin{cases} 4187.9 - 11.319T_c - 0.097215T_c^2 + 0.018317T_c^3 + 0.0011354T_c^4 & -37 \le T_c < 0 \\ 4175.2 + 0.01297(T_c - 35)^2 + 1.5899 \times 10^{-5}(T_c - 35)^4 & 0 \le T_c < 35 \end{cases} \tag{2.51}$$

(Osborne *et al.* 1939; Angell *et al.* 1982; Pruppacher and Klett 1997), where T_c is temperature in degrees Celsius. Figure 2.7 shows c_W versus temperature from (2.51). The variation of c_W below 0 °C is large, but that above 0 °C is small. An expression for the specific heat of ice is

$$c_I = 2104.6 + 7.322T_c \qquad -40 \le T_c < 0\,^{\circ}\text{C} \tag{2.52}$$

(Giauque and Stout 1936; Flubacher *et al.* 1960; Pruppacher and Klett 1997). The specific heat of water vapor at constant pressure also varies slightly with temperature. At 298.15 K, $c_{p,V} \approx 1865.1$ (Lide 1993). At 303.15 K, it is 2 percent larger than at 243.15 K (Rogers and Yau 1989).

Because changes in c_W and $c_{p,V}$ are small for temperatures above 0 °C, these parameters may be held constant when the variation of the latent heat of evaporation with temperature is calculated. Integrating (2.49) with constant specific heat gives

$$L_e = L_{e,0} - (c_W - c_{p,V})(T - T_0) \tag{2.53}$$

where $L_{e,0}$ is the latent heat of evaporation at temperature T_0 (K). An empirical expression for the latent heat of evaporation (J kg^{-1}) is (Bolton 1980; List 1984)

$$L_e \approx 2.501 \times 10^6 - 2370T_c \tag{2.54}$$

Example 2.13

Equation (2.54) predicts that, at 0 °C, 2501 J (about 600 cal) is required to evaporate 1 g of liquid water. At 100 °C, 2264 J (about 540 cal) is required to evaporate 1 g of liquid water.

An empirical expression for the latent heat of melting (J kg^{-1}), valid for temperatures below 0 °C, is

$$L_m \approx 3.3358 \times 10^5 + T_c(2030 - 10.46T_c) \tag{2.55}$$

(List 1984). While pure water at standard pressure always melts at temperatures above 0 °C, it may or may not freeze at temperatures below 0 °C. Water that remains liquid below 0 °C is **supercooled liquid water.**

Example 2.14

Equation (2.55) predicts that, when 1 g of liquid water freezes, 333.5 J (about 80 cal) is released at 0 °C and 312.2 J (74.6 cal) is released at −10 °C.

The latent heat of sublimation (J kg^{-1}) is the sum of the latent heats of evaporation and melting. Thus,

$$L_s = L_e + L_m \approx 2.83458 \times 10^6 - T_c(340 + 10.46T_c) \tag{2.56}$$

2.5.3 Clausius–Clapeyron equation

The rates of formation and growth of liquid water drops and ice crystals depend on several factors, described in Chapters 16 and 18. One important parameter, derived here, is the **saturation vapor pressure** (SVP, also called the equilibrium vapor pressure or surface vapor pressure). This parameter is the partial pressure of a gas over a particle surface at a specific temperature when the gas is in equilibrium with its liquid or solid phase on the surface.

A simplistic way of looking as SVP is to view it as the maximum amount of vapor the air can sustain without the vapor condensing as a liquid or depositing as ice on a surface. When the air is cold, SVPs are lower than when the air is warm. At low temperatures, liquid-water molecules on a particle surface have little kinetic energy and cannot easily break free (evaporate) from the surface. Thus, in equilibrium, air just above a surface contains few vapor molecules and exerts a low SVP. When the air is warm, liquid-water molecules have higher kinetic energies, are more agitated, and break loose more readily from a surface. The resulting evaporation increases the vapor content of the air, and thus the SVP over the particle surface.

The temperature variation of the saturation vapor pressure of water over a liquid surface ($p_{v,s}$, hPa) is approximated with the **Clausius–Clapeyron equation,**

$$\frac{dp_{v,s}}{dT} = \frac{\rho_{v,s}}{T} L_e \tag{2.57}$$

where L_e is the latent heat of evaporation of water (J kg^{-1}), and $\rho_{v,s}$ is the **saturation mass density of water vapor** (kg m^{-3}). Combining $\rho_{v,s} = p_{v,s}/R_v T$ from (2.25) with (2.57) gives

$$\frac{dp_{v,s}}{dT} = \frac{L_e p_{v,s}}{R_v T^2} \tag{2.58}$$

Substituting L_e from (2.54) into (2.58) and rearranging give

$$\frac{dp_{v,s}}{p_{v,s}} = \frac{1}{R_v}\left(\frac{A_h}{T^2} - \frac{B_h}{T}\right)dT \tag{2.59}$$

where $A_h = 3.14839 \times 10^6$ J kg^{-1} and $B_h = 2370$ J kg^{-1} K^{-1}. Integrating (2.59) from $p_{v,s,0}$ to $p_{v,s}$ and T_0 to T, where $p_{v,s,0}$ is a known saturation vapor pressure at T_0, gives the **saturation vapor pressure of water over a liquid surface** as

$$p_{v,s} = p_{v,s,0} \exp\left[\frac{A_h}{R_v}\left(\frac{1}{T_0} - \frac{1}{T}\right) + \frac{B_h}{R_v} \ln\left(\frac{T_0}{T}\right)\right] \tag{2.60}$$

At $T = T_0 = 273.15$ K, $p_{v,s,0} = 6.112$ hPa. Substituting these values, $R_v = 461.91$ J kg^{-1} K^{-1}, A_h, and B_h into (2.60) gives

$$p_{v,s} = 6.112 \exp\left[6816\left(\frac{1}{273.15} - \frac{1}{T}\right) + 5.1309 \ln\left(\frac{273.15}{T}\right)\right] \tag{2.61}$$

where T is in kelvin and $p_{v,s}$ is in hPa.

Example 2.15

From (2.60), $p_{v,s} = 1.26$ hPa at $T = 253.15$ K (-20 °C) and $p_{v,s} = 31.60$ hPa at $T = 298.15$ K (25 °C).

An empirical parameterization of the saturation vapor pressure of water over a liquid surface is

$$p_{v,s} = 6.112 \exp\left(\frac{17.67 T_c}{T_c + 243.5}\right) \tag{2.62}$$

(Bolton 1980), where T_c is in degrees Celsius and $p_{v,s}$ is in hPa. The fit is valid for $-35 < T_c < 35$ °C. Equations (2.61) and (2.62) are saturation vapor pressures over flat, dilute liquid surfaces. Saturation vapor pressures are affected by surface

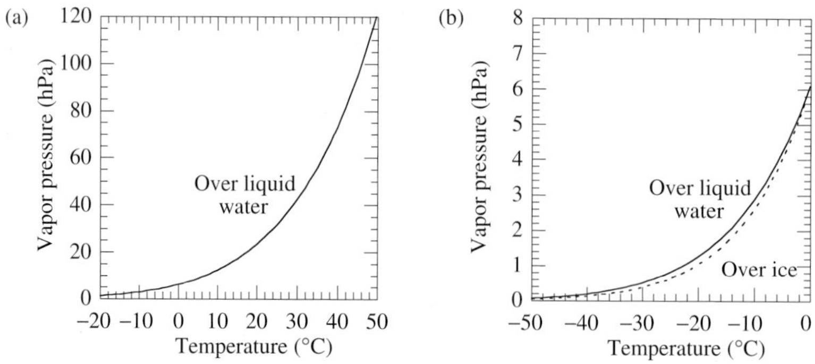

Figure 2.8 Saturation vapor pressure over (a) liquid water and (b) liquid water and ice, versus temperature.

curvature, the presence of solutes in water, and the rate of drop heating or cooling. These effects are discussed in Chapter 16.

Example 2.16

At $T = 253.15$ K (-20 °C), the saturation vapor pressure from (2.62) is $p_{v,s} = 1.26$ hPa. At $T = 298.15$ K (25 °C), $p_{v,s} = 31.67$ hPa. A comparison of these results with those from the previous examples shows that (2.61) and (2.62) give almost identical estimates for $p_{v,s}$.

Figure 2.8(a) shows saturation vapor pressure of water over a liquid surface versus temperature, obtained from (2.62). The figure indicates that saturation vapor pressure increases with increasing temperature. At 0 °C, $p_{v,s} = 6.1$ hPa, which is equivalent to 0.6 percent of the total sea-level air pressure (1013 hPa). At 30 °C, $p_{v,s} = 42.5$ hPa (4.2 percent of the total sea-level air pressure). Since the partial pressure of water vapor can rarely exceed its saturation vapor pressure, the maximum partial pressure of air varies from <1 percent to ≈4 percent of total atmospheric air pressure. Near the poles, where temperatures are low, saturation vapor pressures and partial pressures approach zero. Near the Equator, where temperatures are high, saturation vapor pressures and partial pressures can increase to 4 percent or more of the total air pressure.

The saturation vapor pressure of water over an ice surface is lower than that over a liquid surface at the same subfreezing temperature because ice molecules require more energy to sublimate than liquid molecules require to evaporate at the same temperature. The Clausius–Clapeyron equation for the saturation vapor pressure of water over ice ($p_{v,I}$, hPa) is rewritten from (2.58) as

$$\frac{dp_{v,I}}{dT} = \frac{L_s p_{v,I}}{R_v T^2} \tag{2.63}$$

Substituting the latent heat of sublimation from (2.56) into (2.63) and integrating give the **saturation vapor pressure of water over ice** as

$$p_{v,I} = 6.112\ \exp\left[4648\left(\frac{1}{273.15} - \frac{1}{T}\right) - 11.64\ln\left(\frac{273.15}{T}\right) + 0.02265(273.15 - T)\right] \qquad (2.64)$$

where $T \leq 273.15$ K is in kelvin and $p_{v,I}$ is in hPa. At $T = 273.15$ K, the saturation vapor pressure over ice equals that over liquid water ($p_{v,I,0} = 6.112$ hPa). Figure 2.8(b) shows that, at subfreezing temperatures, $p_{v,I} < p_{v,s}$. Above 273.15 K (0 °C), ice surfaces do not exist. An alternative parameterization, valid from 223.15 K to 273.15 K is

$$p_{v,I} = 6.1064\exp\left[\frac{21.88\,(T - 273.15)}{T - 7.65}\right] \qquad (2.65)$$

(Pruppacher and Klett 1997), where $T \leq 273.15$ K is in kelvin and $p_{v,I}$ is in hPa.

Example 2.17

At $T = 253.15$ K (-20 °C), (2.64) gives $p_{v,I} = 1.034$ hPa, which is less than $p_{v,s} = 1.26$ hPa at the same temperature. Thus, the saturation vapor pressure of water over liquid is greater than that over ice.

2.5.4 Condensation and deposition

Saturation vapor pressures are critical for determining the extent of liquid drop and ice crystal formation. When the air temperature is above the freezing temperature of water (273.15 K) and the partial pressure of water vapor is greater than the saturation vapor pressure of water over a liquid surface ($p_v > p_{v,s}$), water vapor generally condenses as a liquid. Liquid water evaporates when $p_v < p_{v,s}$. Figures 2.9(a) and (b) illustrate growth and evaporation.

If the air temperature falls below freezing and $p_v > p_{v,I}$, water vapor deposits as ice. When $p_v < p_{v,I}$, ice sublimates to water vapor. If liquid water and ice coexist in the same parcel of air and $p_{v,I} < p_{v,s} < p_v$, deposition onto ice particles is favored over condensation onto liquid water drops because growth is driven by the difference between partial pressure and saturation vapor pressure, and $p_v - p_{v,I} > p_v - p_{v,s}$.

When $p_{v,I} < p_v < p_{v,s}$, water evaporates from liquid surfaces and deposits onto ice surfaces, as shown in Fig. 2.10. This is the central assumption behind the **Wegener–Bergeron–Findeisen (Bergeron) process** of ice crystal growth in cold clouds. In such clouds, supercooled liquid water cloud droplets coexist with ice crystals. When the ratio of liquid cloud drops to ice crystals is less than 100 000:1, each ice crystal receives water that evaporates from less than 100 000 cloud drops, and ice crystals do not grow large or heavy enough to fall from their cloud. When the ratio is greater than 1 000 000:1, the relatively few ice crystals present grow

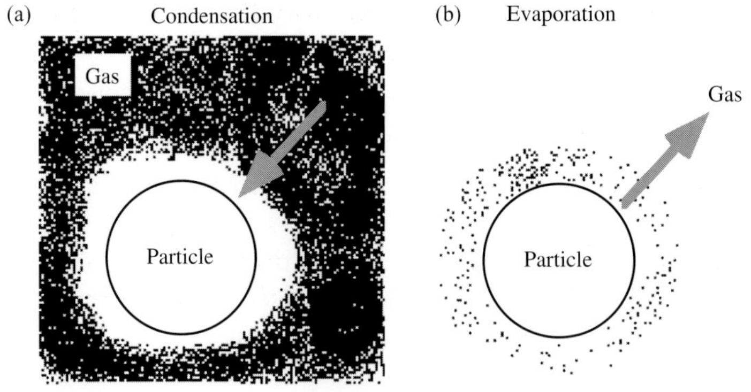

Figure 2.9 (a) Condensation occurs when the partial pressure of a gas exceeds its saturation vapor pressure over a particle surface. (b) Evaporation occurs when the saturation vapor pressure exceeds the partial pressure of the gas. The schematics are not to scale.

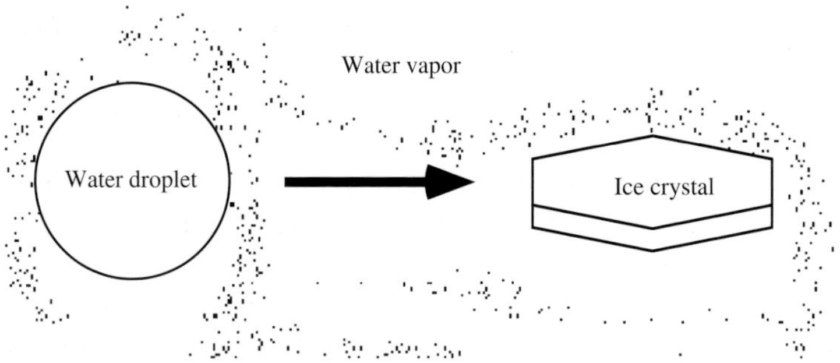

Figure 2.10 Bergeron process. When $p_{v,I} < p_v < p_{v,s}$ at a subfreezing temperature, liquid evaporates, and the resulting gas diffuses to and deposits onto ice crystals.

large and fall quickly from their cloud before much liquid water evaporates, leaving the cloud with lots of liquid drops too small to fall from the cloud. When the ratio is between 100 000:1 and 1 000 000:1, each ice crystal receives the liquid water from 100 000 to 1 000 000 droplets. These ice crystals fall from the cloud, maximizing precipitation.

Figure 2.8(b) shows that the greatest difference between the saturation vapor pressures over liquid water and ice occurs at $T_c = -15\,^\circ\mathrm{C}$. This is the temperature at which ice crystals grow the fastest. Dendrites, which are snowflake-shaped crystals, form most favorably at this temperature.

The ratio of the water vapor partial pressure to its saturation vapor pressure over a liquid surface determines whether the vapor condenses. This ratio is embodied in the **relative humidity**, defined as

$$f_r = 100\% \times \frac{\omega_v}{\omega_{v,s}} = 100\% \times \frac{p_v(p_a - p_{v,s})}{p_{v,s}(p_a - p_v)} \approx 100\% \times \frac{p_v}{p_{v,s}} \qquad (2.66)$$

by the World Meteorological Organization (WMO), where ω_v is the mass mixing ratio of water vapor in the air from (2.31), and

$$\omega_{v,s} = \frac{\varepsilon p_{v,s}}{p_a - p_{v,s}} \approx \frac{\varepsilon p_{v,s}}{p_d} \tag{2.67}$$

is the **saturation mass mixing ratio of water vapor** over a liquid surface, found by substituting $p_{v,s}$ for p_v and $\omega_{v,s}$ for ω_v in (2.31). The saturation mass mixing ratio in the equation for the relative humidity is always taken with respect to liquid water, even if $T \leq 273.15$ K (List 1984). Prior to the WMO definition, the relative humidity was defined exactly as $f_r = 100\% \times p_v / p_{v,s}$.

Example 2.18

If $T = 288$ K and $p_v = 12$ hPa, what is the relative humidity?

SOLUTION

From (2.62), $p_{v,s} = 17.04$ hPa. From (2.66), $f_r = 100$ percent $\times$ 12 hPa / 17.04 hPa = 70.4 percent.

Equation (2.66) implies that, if the relative humidity exceeds 100 percent and $T > 273.15$ K, water vapor condenses. If the relative humidity exceeds 100 percent and $T \leq 273.15$ K, water vapor may condense as a liquid or deposit as ice.

Another parameter used to predict when bulk condensation occurs is the **dew point** (T_D), which is the temperature to which air must be cooled, at constant water vapor partial pressure and air pressure, to reach saturation with respect to liquid water. Similarly, the **frost point** is the temperature to which air must be cooled, at constant water vapor partial pressure and air pressure, to reach saturation with respect to ice. If the air temperature drops below the dew point, the relative humidity increases above 100 percent and condensation occurs. If the dew point is known, the partial pressure of water can be obtained from Fig. 2.8. If the ambient temperature is known, the saturation vapor pressure can be obtained from the same figure.

Example 2.19

If $T_D = 20$ °C and $T = 30$ °C, estimate the partial pressure of water, the saturation vapor pressure of water, and the relative humidity from Fig. 2.8.

SOLUTION

From the figure, $p_v \approx 23.4$ hPa and $p_{v,s} \approx 42.5$ hPa. Thus, $f_r \approx 55$ percent.

An equation for the dew point can be derived from any equation for saturation vapor pressure by substituting p_v for $p_{v,s}$ and solving for the temperature. Applying

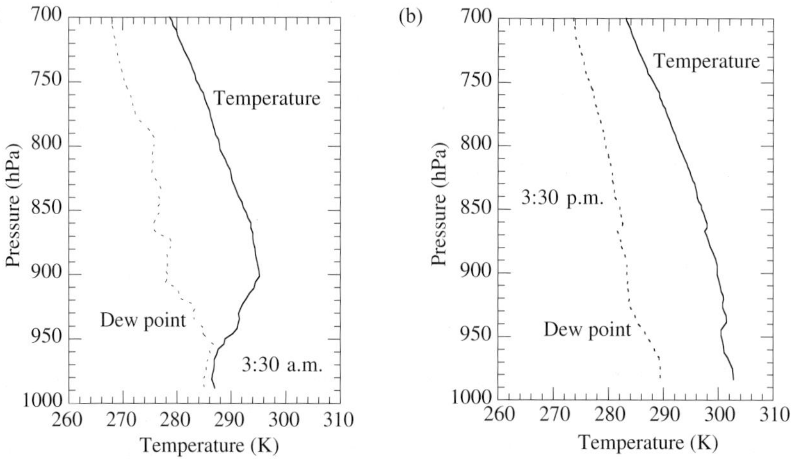

Figure 2.11 Observed vertical profiles of temperature and dew point at (a) 3:30 a.m. and (b) 3:30 p.m. on August 27, 1987 at Riverside, California. The air was nearly saturated near the surface in the morning.

this technique to (2.62) and converting the result to absolute temperature give the dew point as

$$T_D = \frac{4880.357 - 29.66 \ln p_v}{19.48 - \ln p_v} = \frac{4880.357 - 29.66 \ln(\omega_v p_d/\varepsilon)}{19.48 - \ln(\omega_v p_d/\varepsilon)} \tag{2.68}$$

where $\omega_v = \varepsilon p_v / p_d$ from (2.31), T is in kelvin, and p_v is in hPa.

Example 2.20

Calculate the dew point when $p_v = 12$ hPa.

SOLUTION

From (2.68), $T_D = 282.8$ K. Thus, if unsaturated air at 288 K cools to 282.8 K when $p_v = 12$ hPa, condensation occurs.

When the ambient temperature is close to the dew point ($T \approx T_D$), the relative humidity is high. When the ambient temperature and dew point are far apart, the relative humidity is low. Figures 2.11(a) and (b) show observed vertical profiles of temperature and dew point at Riverside, California, in the morning and afternoon, respectively, of August 27, 1987. In the morning, near the ground, the dew point and air temperature were close to each other, indicating the air was nearly saturated, the relative humidity was high, and a fog was almost present. Above 950 hPa (about 500 m above sea level) in the morning, the dew point and air temperature were far apart, indicating the relative humidity was low and the air was unsaturated. In the afternoon, when air near the surface was warm, the dew point and air temperature were also far apart, and the relative humidity was low.

2.6 FIRST LAW OF THERMODYNAMICS

As applied to the atmosphere, the first law of thermodynamics relates the change in temperature of a parcel of air to energy transfer between the parcel and its environment and work done by or on the parcel. The first law is used to derive the **thermodynamic energy equation**, which gives the time-rate-of-change of temperature in the atmosphere due to energy transfer and work. The first law is also used to derive analytical expressions for atmospheric stability.

The **first law of thermodynamics** as applied to the atmosphere is

$$\mathrm{d}Q^* = \mathrm{d}U^* + \mathrm{d}W^* \tag{2.69}$$

where $\mathrm{d}Q^*$ is called the **diabatic heating term**, which is the **energy** (J) transferred between an air parcel and its environment, $\mathrm{d}U^*$ is the change in **internal energy** (J) of the parcel, and $\mathrm{d}W^*$ is the **work** (J) done by or on the parcel. When $\mathrm{d}Q^* > 0$, energy is transferred to the parcel from the environment, and the process is **endothermic**. When $\mathrm{d}Q^* < 0$, energy is transferred to the environment from the parcel, and the process is **exothermic**. When $\mathrm{d}W^* > 0$, work is done by the parcel. When $\mathrm{d}W^* < 0$, work is done on the parcel. Equation (2.69) states that if energy is added to an air parcel, some of it is used to change the internal energy (and temperature) of the parcel, and the rest is used by the parcel to do work. The equation also states that changes in the internal energy (temperature) of a parcel result from energy transfer or work. Internal energy changes resulting from energy transfer are **diabatic** while those resulting from work are **adiabatic**. Diabatic sources or sinks of energy include conduction, radiative cooling/heating, and latent heat release/absorption. Substituting

$$\mathrm{d}Q = \frac{\mathrm{d}Q^*}{M_\mathrm{a}} \qquad \mathrm{d}U = \frac{\mathrm{d}U^*}{M_\mathrm{a}} \qquad \mathrm{d}W = \frac{\mathrm{d}W^*}{M_\mathrm{a}} \tag{2.70}$$

where $M_\mathrm{a} = M_\mathrm{d} + M_\mathrm{v}$ is the mass of a parcel of air (kg) consisting of dry air mass M_d and water vapor mass M_v, into (2.69) gives the first law in terms of energy per unit mass of air (J kg^{-1}) as

$$\mathrm{d}Q = \mathrm{d}U + \mathrm{d}W \tag{2.71}$$

Terms in this equation are discussed below.

When a gas expands, work is done by the gas. When air expands, work done by the air is $\mathrm{d}W^* = p_\mathrm{a}\,\mathrm{d}V$, and work done per unit mass of air is

$$\mathrm{d}W = \frac{\mathrm{d}W^*}{M_\mathrm{a}} = \frac{p_\mathrm{a}\,\mathrm{d}V}{M_\mathrm{a}} = p_\mathrm{a}\,\mathrm{d}\alpha_\mathrm{a} \tag{2.72}$$

In this equation, $\mathrm{d}V$ is the change in volume of the air, and

$$\alpha_\mathrm{a} = \frac{V}{M_\mathrm{a}} = \frac{1}{\rho_\mathrm{a}} \tag{2.73}$$

is the **specific volume of air**. Air expands when it rises to lower pressure. In such cases, work is done by the air, and $\mathrm{d}V > 0$. When a parcel of air sinks to higher pressure, the parcel compresses, work is done on the air in the parcel, and $\mathrm{d}V < 0$.

Energy transfer between a parcel and its environment occurs when $dQ \neq 0$. In the atmosphere, major sources (sinks) of external energy are radiative heating (cooling), condensation (evaporation), deposition (sublimation), and freezing (melting).

The change in **internal energy** of the air is its change in temperature multiplied by the energy required to change its temperature one degree Celsius (1 K) without affecting the volume or work done by or on the air. In other words,

$$dU = \left(\frac{\partial Q}{\partial T}\right)_{\alpha_a} dT = c_{v,m}\, dT \tag{2.74}$$

where $c_{v,m} = (\partial Q/\partial T)_{\alpha_a}$ is the **specific heat of moist air at constant volume.** It is the energy required to raise the temperature of 1 g of air 1 K at constant volume and varies with water-vapor mass mixing ratio. An expression for $c_{v,m}$ can be derived by noting that, at constant volume,

$$(M_d + M_v)dQ = (M_d c_{v,d} + M_v c_{v,V})dT \tag{2.75}$$

where $c_{v,d} = 717.63\ \text{J kg}^{-1}\ \text{K}^{-1}$ at 298 K is the **specific heat of dry air at constant volume** and $c_{v,V} = 1403.2\ \text{J kg}^{-1}\ \text{K}^{-1}$ is the **specific heat of water vapor at constant volume.** The specific heat of dry air at constant volume decreases by less than 0.2 percent down to 200 K. Dividing (2.75) through by $(M_d + M_v)\, dT$ gives

$$c_{v,m} = \left(\frac{\partial Q}{\partial T}\right)_{\alpha_a} = \frac{M_d c_{v,d} + M_v c_{v,V}}{M_d + M_v} = \frac{c_{v,d} + c_{v,V}\omega_v}{1 + \omega_v} = c_{v,d}(1 + 0.955 q_v) \tag{2.76}$$

where $q_v = \omega_v/(1 + \omega_v)$ from (2.33) and $c_{v,V}/c_{v,d} - 1 = 0.955$.

Substituting (2.74) and (2.72) into (2.71) gives the **first law of thermodynamics for the atmosphere** as

$$dQ = c_{v,m}\, dT + p_a\, d\alpha_a \tag{2.77}$$

Combining the equation of state, $p_a = \rho_a R_m T$, with $\alpha_a = 1/\rho_a$ gives $p_a \alpha_a = R_m T$, which can be differentiated to give

$$p_a\, d\alpha_a + \alpha_a\, dp_a = R_m\, dT \tag{2.78}$$

Combining (2.78) with (2.77) yields another form of the first law as

$$dQ = c_{p,m}\, dT - \alpha_a\, dp_a \tag{2.79}$$

where

$$\begin{aligned} c_{p,m} &= \left(\frac{dQ}{dT}\right)_{p_a} = \frac{M_d c_{p,d} + M_v c_{p,V}}{M_d + M_v} = \frac{c_{p,d} + c_{p,V}\omega_v}{1 + \omega_v} \\ &= c_{p,d}(1 + 0.856 q_v) \\ &\approx c_{p,d}(1 + 0.859\omega_v) \end{aligned} \tag{2.80}$$

is the **specific heat of moist air at constant pressure.** In this equation, $c_{p,\mathrm{V}}/c_{p,\mathrm{d}} - 1 = 0.856$, where $c_{p,\mathrm{d}} = 1004.67$ J kg^{-1} K^{-1} at 298 K is the specific heat of dry air at constant pressure, and $c_{p,\mathrm{V}} = 1865.1$ J kg^{-1} K^{-1} is the **specific heat of water vapor at constant pressure.** Like $c_{v,\mathrm{d}}$, $c_{p,\mathrm{d}}$ and $c_{p,\mathrm{V}}$ vary slightly with temperature. The specific heat of moist air at constant pressure is the energy required to increase the temperature of 1 g of air 1 K without affecting air pressure.

Differentiating (2.77) at constant pressure gives another expression for $c_{p,\mathrm{m}}$ as

$$c_{p,\mathrm{m}} = \left(\frac{\mathrm{d}Q}{\mathrm{d}T}\right)_{p_\mathrm{a}} = c_{v,\mathrm{m}} + p_\mathrm{a}\left(\frac{\mathrm{d}\alpha_\mathrm{a}}{\mathrm{d}T}\right)_{p_\mathrm{a}} = c_{v,\mathrm{m}} + p_\mathrm{a}\left(\frac{\mathrm{d}}{\mathrm{d}T}\frac{R_\mathrm{m}T}{p_\mathrm{a}}\right)_{p_\mathrm{a}} = c_{v,\mathrm{m}} + R_\mathrm{m} \tag{2.81}$$

When the air is dry, this expression simplifies to $c_{p,\mathrm{d}} = c_{v,\mathrm{d}} + R'$. Substituting $c_{v,\mathrm{m}} = c_{v,\mathrm{d}}(1 + 0.955\mathrm{q_v})$ from (2.76) and $R_\mathrm{m} = R'(1 + 0.608\mathrm{q_v})$ from (2.37) into (2.81) gives the empirical expressions in (2.80).

The first law of thermodynamics can be approximated in terms of virtual temperature instead of temperature with

$$\mathrm{d}Q = \frac{1 + 0.856\mathrm{q_v}}{1 + 0.608\mathrm{q_v}} c_{p,\mathrm{d}}\,\mathrm{d}T_\mathrm{v} - \alpha_\mathrm{a}\,\mathrm{d}p_\mathrm{a} \approx c_{p,\mathrm{d}}\,\mathrm{d}T_\mathrm{v} - \alpha_\mathrm{a}\,\mathrm{d}p_\mathrm{a} \tag{2.82}$$

which was derived by substituting $c_{p,\mathrm{m}} = c_{p,\mathrm{d}}(1 + 0.856\mathrm{q_v})$ from (2.80) and $T = T_\mathrm{v}/(1 + 0.608\mathrm{q_v})$ from (2.38) into (2.79). An advantage of (2.82) is that the specific heat is in terms of dry instead of moist air. The maximum energy error in (2.82) due to neglecting the empirical water vapor terms is 1 percent. The average error is 0.2 percent.

2.6.1 Applications of the first law of thermodynamics

Here, the first law of thermodynamics is modified for four special cases. First, for an **isobaric process** ($\mathrm{d}p_\mathrm{a} = 0$), the first law simplifies from (2.79) to

$$\mathrm{d}Q = c_{p,\mathrm{m}}\,\mathrm{d}T = \frac{c_{p,\mathrm{m}}}{c_{v,\mathrm{m}}}\,\mathrm{d}U \tag{2.83}$$

Second, for an **isothermal process** ($\mathrm{d}T = 0$), (2.79) becomes

$$\mathrm{d}Q = -\alpha_\mathrm{a}\,\mathrm{d}p_\mathrm{a} = p_\mathrm{a}\,\mathrm{d}\alpha_\mathrm{a} = \mathrm{d}W \tag{2.84}$$

Third, for an **isochoric process** ($\mathrm{d}\alpha_\mathrm{a} = 0$), the first law simplifies from (2.77) to

$$\mathrm{d}Q = c_{v,\mathrm{m}}\,\mathrm{d}T = \mathrm{d}U \tag{2.85}$$

A fourth case is for an **adiabatic process.** Under adiabatic conditions, no energy is transferred to or from a parcel of air ($\mathrm{d}Q = 0$). Instead, a parcel's temperature changes only when the parcel expands or contracts as it ascends or descends, respectively. When a parcel rises, it encounters lower pressure and expands. As it expands, the kinetic energy of the air molecules within the parcel is converted to work used to expand the air. Since temperature decreases proportionally to kinetic energy from (2.4), a rising, expanding parcel of air cools under adiabatic conditions. The expansion of an air parcel in the absence of diabatic sources or sinks of energy is called an **adiabatic expansion.**

When a parcel sinks, it compresses and warms. When no energy transfer is considered, the compression is called an **adiabatic compression.** Under adiabatic conditions, the first law can be rewritten from (2.77), (2.79), and (2.82), respectively, to

$$c_{v,\mathrm{m}}\,\mathrm{d}T = -p_\mathrm{a}\,\mathrm{d}\alpha_\mathrm{a} \tag{2.86}$$

$$c_{p,\mathrm{m}}\,\mathrm{d}T = \alpha_\mathrm{a}\,\mathrm{d}p_\mathrm{a} \tag{2.87}$$

$$c_{p,\mathrm{d}}\,\mathrm{d}T_v \approx \alpha_\mathrm{a}\,\mathrm{d}p_\mathrm{a} \tag{2.88}$$

2.6.1.1 *Dry adiabatic lapse rate*

When an air parcel rises and cools, and no condensation occurs, the parcel's rate of cooling with height is approximately 9.8 K km^{-1}, which is the **dry (unsaturated) adiabatic lapse rate.** A lapse rate is the negative change in temperature with height. A positive lapse rate indicates that temperature decreases with increasing height.

The dry adiabatic lapse rate can be derived from the hydrostatic equation and the adiabatic form of the first law of thermodynamics. Taking the negative differential of (2.88) with respect to altitude, substituting the hydrostatic equation from (2.40), and substituting $\alpha_\mathrm{a} = 1/\rho_\mathrm{a}$ give the **dry adiabatic lapse rate in terms of virtual temperature** as

$$\Gamma_\mathrm{d} = -\left(\frac{\partial T_\mathrm{v}}{\partial z}\right)_\mathrm{d} \approx -\left(\frac{\alpha_\mathrm{a}}{c_{p,\mathrm{d}}}\right)\frac{\partial p_\mathrm{a}}{\partial z} = \left(\frac{\alpha_\mathrm{a}}{c_{p,\mathrm{d}}}\right)\rho_\mathrm{a} g = \frac{g}{c_{p,\mathrm{d}}} = +9.8\ \mathrm{K\ km}^{-1} \tag{2.89}$$

where the subscript d indicates that the change is dry (unsaturated) adiabatic. Equation (2.89) states that the virtual temperature of an unsaturated air parcel cools 9.8 K for every kilometer it ascends in the atmosphere under dry adiabatic conditions. From (2.87), the **dry adiabatic lapse rate in terms of actual temperature** can be written as

$$\Gamma_\mathrm{d,m} = -\left(\frac{\partial T}{\partial z}\right)_\mathrm{d} = \frac{g}{c_{p,\mathrm{m}}} = \frac{g}{c_{p,\mathrm{d}}}\left(\frac{1+\omega_\mathrm{v}}{1+c_{p,\mathrm{V}}\omega_\mathrm{v}/c_{p,\mathrm{d}}}\right) \tag{2.90}$$

where $c_{p,\mathrm{m}}$, obtained from (2.80), varies with water vapor content. The advantage of (2.89) is that the right side of the equation does not depend on water vapor content.

2.6.1.2 *Potential temperature*

A parameter used regularly in atmospheric analysis and modeling is potential temperature. **Potential temperature** is the temperature an unsaturated air parcel attains if it is brought adiabatically from its altitude down to a pressure of 1000 hPa.

Potential temperature is derived by first substituting $\alpha_\mathrm{a} = R_\mathrm{m}T/p_\mathrm{a}$ into (2.87), giving

$$\frac{\mathrm{d}T}{T} = \left(\frac{R_\mathrm{m}}{c_{p,\mathrm{m}}}\right)\frac{\mathrm{d}p_\mathrm{a}}{p_\mathrm{a}} \tag{2.91}$$

Integrating (2.91) from T_0 to T and $p_{a,0}$ to p_a yields **Poisson's equation,**

$$T = T_0\left(\frac{p_a}{p_{a,0}}\right)^{\frac{R_m}{c_{p,m}}} = T_0\left(\frac{p_a}{p_{a,0}}\right)^{\frac{R'(1+0.608q_v)}{c_{p,d}(1+0.856q_v)}} \approx T_0\left(\frac{p_a}{p_{a,0}}\right)^{\kappa(1-0.251q_v)} \tag{2.92}$$

where

$$\kappa = \frac{R'}{c_{p,d}} = \frac{c_{p,d} - c_{v,d}}{c_{p,d}} = 0.286 \tag{2.93}$$

When $p_{a,0} = 1000$ hPa, T_0 is called the **potential temperature of moist air** ($\theta_{p,m}$), and (2.92) becomes

$$\theta_{p,m} = T\left(\frac{1000\text{ hPa}}{p_a}\right)^{\kappa(1-0.251q_v)} \tag{2.94}$$

In the absence of water vapor, $q_v = 0$ and $p_a = p_d$. In such a case, (2.94) simplifies to

$$\theta_p = T\left(\frac{1000\,\text{hPa}}{p_d}\right)^{\kappa} \tag{2.95}$$

where θ_p is the **potential temperature of dry air.** Since q_v is usually smaller than 0.03 kg kg^{-1}, neglecting q_v in (2.94) causes an error in κ of less than 0.75 percent. Thus, for simplicity, (2.95) is usually used instead of (2.94) for defining potential temperature, even when water vapor is present. Potential temperature is conserved (stays constant) if an unsaturated air parcel is displaced adiabatically.

Figure 2.12 shows vertical profiles of potential temperature in the morning and afternoon at Riverside, California on August 27, 1987.

Example 2.21

If the temperature of a dry air parcel at $p_d = 800$ hPa is $T = 270$ K, then $\theta_p = 287.8$ K from (2.95).

A parameter related to potential temperature is **potential virtual temperature** (θ_v), which is found by converting all the moisture in a parcel to dry air, then bringing the parcel to a pressure of 1000 hPa and determining its temperature. θ_v is the potential temperature of a sample of moist air as if it were dry and at the same density and pressure as the moist air. It is derived by substituting $\alpha_a = R'T_v/p_a$ into (2.88), then integrating from $T_0 = \theta_v$ to T and $p_{a,0} = 1000$ hPa to p_a. The result is

$$\theta_v = T(1 + 0.608q_v)\left(\frac{1000\text{ hPa}}{p_a}\right)^{\kappa} = T_v\left(\frac{1000\text{ hPa}}{p_a}\right)^{\kappa} \tag{2.96}$$

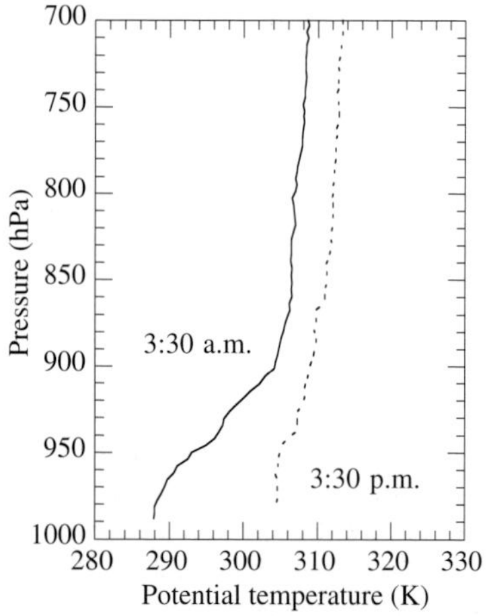

Figure 2.12 Observed vertical profiles of potential temperature at 3:30 a.m. and 3:30 p.m. on August 27, 1987 at Riverside, California. Potential temperatures correspond to actual temperatures shown in Fig. 2.11(a) and (b), respectively.

The change in **entropy** of an air parcel per unit mass (J kg^{-1} K^{-1}) due to energy transfer between the parcel and its environment is $dS = dQ/T$. During adiabatic expansion and compression, $dQ = 0$ and no change in entropy occurs. Since $dQ = 0$ along surfaces of constant potential virtual temperature, $dS = 0$ along such surfaces as well, and the surfaces are called **isentropic.** Figure 2.13 illustrates that potential virtual temperatures increase monotonically with height in the Northern-Hemisphere troposphere, causing isentropic surfaces to slant toward the North Pole.

A parameter related to potential virtual temperature is **virtual potential temperature,** which is found by bringing a moist parcel to a pressure of 1000 hPa, then converting all moisture in the parcel to dry air and determining the parcel's temperature. It is the virtual temperature of air that has been brought adiabatically to 1000 hPa. It is obtained by applying the virtual temperature correction from (2.38) to the potential temperature of moist air. The result is

$$\theta_{p,v} = \theta_{p,m}(1 + 0.608q_v) = T_v\left(\frac{1000\,\text{hPa}}{p_a}\right)^{\kappa(1-0.251q_v)} \tag{2.97}$$

The difference between potential virtual temperature and virtual potential temperature is relatively small (Brutsaert 1991).

The **Exner function,** which is used in future chapters, is defined as $c_{p,d}P$, where

$$P = \left(\frac{p_a}{1000\,\text{hPa}}\right)^{\kappa} \tag{2.98}$$

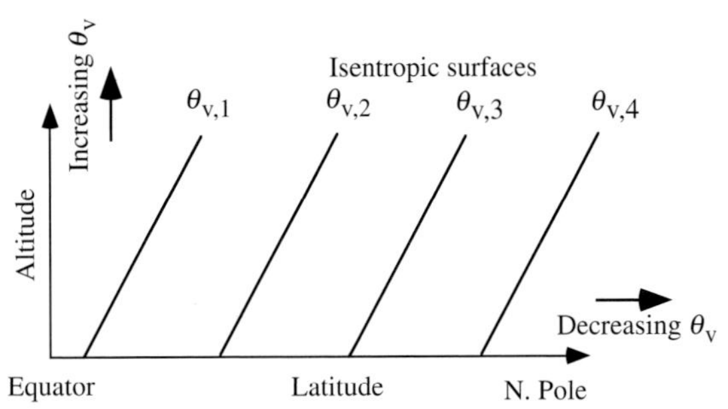

Figure 2.13 Isentropic surfaces (surfaces of constant potential virtual temperature) between the Equator and the North Pole. Sea-level temperature and, therefore, θ_v decrease from the Equator to the pole. Since the free troposphere is stably stratified with respect to unsaturated air, θ_v increases with altitude, and lines of constant θ_v slant toward the poles in the vertical.

Substituting P into (2.95) and (2.96) gives

$$T_v = \theta_v P \tag{2.99}$$

2.6.2 Stability criteria for unsaturated air

The concepts derived above from the first law of thermodynamics are useful for analyzing atmospheric stability. The atmosphere is stable (stably stratified) when a parcel of air, displaced vertically, decelerates and returns to its original position. The atmosphere is unstable when a displaced parcel accelerates in the direction in which it is displaced. The atmosphere is neutral when a parcel does not accelerate or decelerate after being displaced.

When the atmosphere is stable near the surface, pollution builds up, since air parcels cannot accelerate out of the stable layer to disperse the pollution. Stability also inhibits clouds of vertical development from forming. When the atmosphere is unstable, emitted pollutants accelerate vertically, decreasing their concentration near the surface. Clouds of vertical development can form in unstable air.

2.6.2.1 *Determining stability from the dry adiabatic lapse rate*

When the air is unsaturated with water vapor (i.e., the relative humidity is less than 100 percent), stability can be determined by comparing the environmental lapse rate with the dry adiabatic lapse rate ($\Gamma_{d,m}$ from (2.90)). The **environmental lapse rate,** $\Gamma_e = -\partial T/\partial z$, is the negative change of actual temperature with altitude. When temperature decreases with increasing altitude, $\Gamma_e > 0$. In terms of Γ_e and $\Gamma_{d,m}$, the stability criteria for unsaturated air are

$$\Gamma_e \begin{cases} > \Gamma_{d,m} & \text{unsaturated unstable} \\ = \Gamma_{d,m} & \text{unsaturated neutral} \\ < \Gamma_{d,m} & \text{unsaturated stable} \end{cases} \tag{2.100}$$

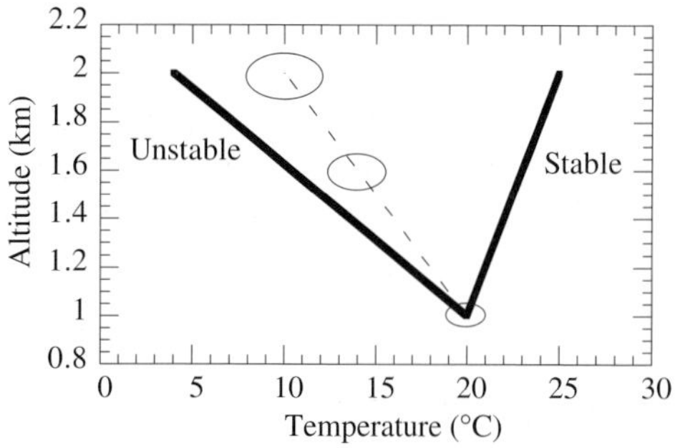

Figure 2.14 Demonstration of stability and instability in unsaturated air. When a parcel is displaced vertically, it rises and cools adiabatically (along the dashed line). If the environmental temperature profile is stable (right thick line), the rising parcel is cooler than the environment, decelerates, then sinks to its original position. If the environmental temperature profile is unstable (left thick line), the rising parcel is warmer than the air around it and continues to rise. The parcel stops rising only when it encounters air with the same temperature as itself. This occurs when the parcel reaches a layer with a new environmental temperature profile.

Figure 2.14 demonstrates how stability can be determined graphically.

Example 2.22

If the observed change in temperature with altitude is $\partial T/\partial z = -15$ K km^{-1} and the air contains no water vapor, what is the stability class of the atmosphere?

SOLUTION

The environmental lapse rate in this case is $\Gamma_e = +15$ K km^{-1}, which is greater than $\Gamma_{d,m} = +9.8$ K km^{-1}; thus, the atmosphere is unstable with respect to unsaturated air.

Equation (2.100) indicates that temperature can increase or decrease with increasing altitude in stable air but must decrease with increasing altitude in unstable air. A **temperature inversion** is an increase in temperature with increasing altitude. An inversion occurs only in stable air, but the presence of stable air does not necessarily mean that an inversion is present, as illustrated in Example 2.23.

Example 2.23

When temperature decreases slightly with increasing altitude (e.g., $\Gamma_e = +2.0\,\mathrm{K\,km^{-1}}$), the atmosphere is stable but an inversion is not present. An inversion occurs when temperature increases with altitude, as demonstrated by the line marked **stable** in Fig. 2.14.

Stability is enhanced by any process that warms air at higher altitudes relative to air at lower altitudes. At night, surface air becomes stable because the ground cools radiatively. During the day, stability is enhanced over land near the sea when cool marine air travels inland and displaces warm land-air vertically (creating warm air over cold air). Stability also increases when warm air blows over a cold surface or when air in a high-pressure system sinks, compresses, and warms on top of cool marine air below. Instability occurs when the land heats rapidly during the day or when a cold wind blows over a warm surface.

2.6.2.2 *Determining stability from potential virtual temperature*

Stability can also be determined from potential virtual temperature. Differentiating (2.96) gives

$$d\theta_v = dT_v\left(\frac{1000}{p_a}\right)^{\kappa} + T_v\kappa\left(\frac{1000}{p_a}\right)^{\kappa-1}\left(-\frac{1000}{p_a^2}\right)dp_a = \frac{\theta_v}{T_v}dT_v - \kappa\frac{\theta_v}{p_a}dp_a \tag{2.101}$$

Taking the partial derivative of (2.101) with respect to height and substituting $\partial p_a/\partial z = -\rho_a g$ and the **virtual temperature lapse rate,** $\Gamma_v = -\partial T_v/\partial z$, into the result give

$$\frac{\partial\theta_v}{\partial z} = \frac{\theta_v}{T_v}\frac{\partial T_v}{\partial z} - \kappa\frac{\theta_v}{p_a}\frac{\partial p_a}{\partial z} = -\frac{\theta_v}{T_v}\Gamma_v + \frac{R'}{c_{p,d}}\frac{\theta_v}{p_a}\rho_a g \tag{2.102}$$

Substituting $T_v = p_a/\rho_a R'$ and $\Gamma_d = g/c_{p,d}$ from (2.89) into (2.102) results in

$$\frac{\partial\theta_v}{\partial z} = -\frac{\theta_v}{T_v}\Gamma_v + \frac{\theta_v g}{T_v c_{p,d}} = \frac{\theta_v}{T_v}(\Gamma_d - \Gamma_v) \tag{2.103}$$

Equation (2.100) indicates that the air is stable when $\Gamma_d > \Gamma_v$, suggesting that $\partial\theta_v/\partial z > 0$ also indicates that the air is stable. Thus, the stability criteria in terms of potential virtual temperature are

$$\frac{\partial\theta_v}{\partial z}\begin{cases} < 0 & \text{unsaturated unstable} \\ = 0 & \text{unsaturated neutral} \\ > 0 & \text{unsaturated stable} \end{cases} \tag{2.104}$$

Figure 2.15 demonstrates how stability can be determined from a graph of potential virtual temperature versus altitude.

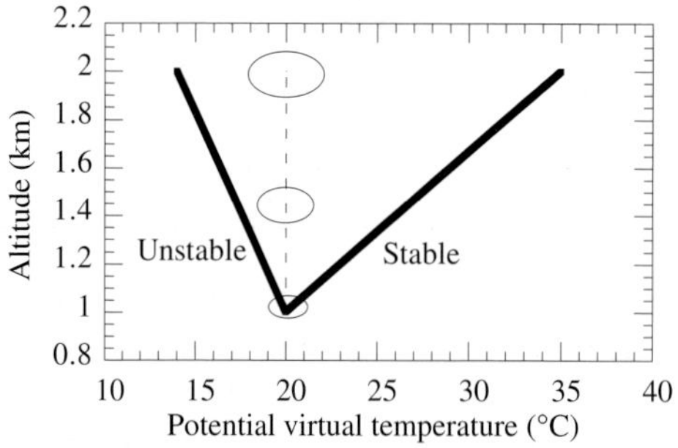

Figure 2.15 Demonstration of stability and instability from potential virtual temperature. When an unsaturated parcel is displaced vertically, it rises and cools adiabatically (along the dashed line). If the ambient potential-virtual-temperature profile slopes positively, a rising parcel is always cooler than the environment, the atmosphere is stable, and the parcel sinks back to its original position. If the ambient potential-virtual-temperature profile slopes negatively, a rising parcel is always warmer than the environment, the atmosphere is unstable, and the parcel continues rising.

The advantage of deriving stability from θ_v instead of from potential temperature of dry air is that the latter parameter does not account for water vapor whereas the former does. The advantage of deriving stability from θ_v instead of from the potential temperature of moist air or from virtual potential temperature is that, although the latter parameters include water vapor more accurately than does θ_v, their differentiation requires several additional terms of little consequence not required in the differentiation of θ_v.

Example 2.24

Given $\Gamma_v = +7$ K km^{-1}, $p_a = 925$ hPa, and $T_v = 290$ K, find $\partial\theta_v/\partial z$.

SOLUTION

From (2.96), $\theta_v = 296.5$ K. Thus, from (2.103), $\partial\theta_v/\partial z = 2.86$ K km^{-1}, and the atmosphere is stable with respect to unsaturated air.

2.6.2.3 *Determining stability from the Brunt–Väisälä frequency*

Another way to write equation (2.103) is

$$\frac{\partial \ln \theta_v}{\partial z} = \frac{1}{T_v}(\Gamma_d - \Gamma_v) \qquad (2.105)$$

Multiplying by gravitational acceleration, g, gives

$$N_{bv}^2 = g\frac{\partial \ln \theta_v}{\partial z} = \frac{g}{T_v}(\Gamma_d - \Gamma_v) \tag{2.106}$$

where N_{bv} is the **Brunt–Väisälä frequency** (or buoyancy frequency).

The Brunt–Väisälä frequency measures the static stability of the atmosphere. If θ_v increases with increasing height ($\Gamma_d > \Gamma_v$), then $N_{bv}^2 > 0$, and the atmosphere is stable. In such a case, buoyancy acts as a restoring force, causing a perturbed parcel of air to oscillate about its initial altitude with a period $\tau_{bv} = 2\pi / N_{bv}$. During the oscillation, kinetic energy is exchanged with potential energy. The oscillations arising from buoyancy restoration are **gravity waves**, discussed in Chapter 4.

If θ_v is constant with increasing altitude ($N_{bv}^2 = 0$), the atmosphere is **neutral,** and displacements occur without resistance from a restoring buoyancy force. If θ_v decreases with increasing altitude ($N_{bv}^2 < 0$), the atmosphere is **unstable,** and a parcel's displacement increases exponentially with time. In sum, the stability criteria from the Brunt–Väisälä frequency are

$$N_{bv}^2 \begin{cases} < 0 & \text{unsaturated unstable} \\ = 0 & \text{unsaturated neutral} \\ > 0 & \text{unsaturated stable} \end{cases} \tag{2.107}$$

Example 2.25

Given $\Gamma_v = +6.5 \text{ K km}^{-1}$ and $T_v = 288$ K, estimate the Brunt–Väisälä frequency and the period of oscillation of a perturbed parcel of air.

SOLUTION

Since $\Gamma_d = +9.8 \text{ K km}^{-1}$ and $g = 9.81 \text{ m s}^{-2}$, we have $N_{bv} = 0.0106 \text{ s}^{-1}$ from (2.106), and $\tau_{bv} = 593$ s. The atmosphere is statically stable with respect to unsaturated air in this case.

2.7 SUMMARY

In this chapter, the structure and composition of the atmosphere were discussed. The bottom 100 km of the atmosphere consists of four primary regions of temperature variation: the troposphere, stratosphere, mesosphere, and thermosphere. The troposphere consists of the boundary layer and the free troposphere. Three important variables describing the atmosphere, temperature, pressure, and density, are related by the equation of state. Other variables discussed include the relative humidity and potential virtual temperature. Some equations derived included the hydrostatic equation, Clausius–Clapeyron equation, and first law of thermodynamics. The equation of state and hydrostatic equation were combined to give expressions for altitude as a function of air pressure and vice versa. The Clausius–Clapeyron equation was combined with empirical relationships for latent heat to derive expressions for the saturation vapor pressure of water over liquid and ice. The first law of thermodynamics was used to derive atmospheric stability criteria. In the next chapter, the continuity and thermodynamic energy equations are discussed.

2.8 PROBLEMS

2.1 If $T = 295$ K at 1 mm above the ground and the conductive heat flux is $H_c = 250$ W m^{-2}, estimate the temperature at the ground. Assume the air is dry.

2.2 If $N_q = 1.5 \times 10^{12}$ molec. cm^{-3} for O_3 gas, $T = 285$ K, and $p_d = 980$ hPa, find the volume mixing ratio, dry-air mass mixing ratio, and partial pressure of ozone.

2.3 If $\omega_q = 1.3$ ppmm for carbon monoxide gas, $T = 285$ K, and $p_d = 980$ hPa, find the volume mixing ratio, number concentration, and partial pressure of carbon monoxide.

2.4 If $T = 268$ K and $p_d = 700$ hPa, respectively, find $p_{v,s}$ in hPa, and find the corresponding mixing ratio of water vapor in percent, ppmm, and ppmv.

2.5 Find the mass density of moist air (ρ_a) if $T = 283$ K, $f_r = 78$ percent, and $p_d = 850$ hPa.

2.6 Find the pressure exerted by moist air if $T = 288$ K, $f_r = 82$ percent, and $p_d = 925$ hPa.

2.7 Find the virtual temperature when $N_a = 2.1\times 10^{19}$ molec. cm^{-3}, $T = 295$ K, and $f_r = 92$ percent.

2.8 Find the partial pressure of water vapor if $q_v = 3$ g kg^{-1}, $T = 278$ K, and $\rho_d = 0.5$ kg m^{-3}.

2.9 If $T_v = 281$ K, $p_v = 3$ hPa, and $p_a = 972$ hPa, find the air temperature.

2.10 If the total air pressure, temperature, and relative humidity are $p_a = 945$ hPa, $T = 276$ K, and $f_r = 46$ percent, find ω_v, m_a, R_m, T_v, and ρ_a.

2.11 If dry-air pressure, temperature, and water-vapor–dry-air mass mixing ratio are $p_d = 927$ hPa, $T = 281$ K, and $\omega_v = 0.005$ kg kg^{-1}, find f_r, m_a, R_m, T_v, and ρ_a.

2.12 If the total air pressure, water-vapor volume mixing ratio, and temperature are $p_a = 966$ hPa, $\chi_v = 3000$ ppmv, and $T = 284$ K, find p_v, m_a, R_m, T_v, and ρ_a.

2.13 Find the altitude in a standard atmosphere that a pressure altimeter gives if the pressure measured by the altimeter is $p_a = 770$ hPa and the air is dry.

2.14 Estimate the scale height of the atmosphere (H) and resulting pressure at $z = 200$ m altitude if the air is dry, the pressure at $z = 100$ m is $p_d = 990$ hPa, and the average temperature between $z = 100$ m and $z = 200$ m is $T = 284$ K.

2.15 If the air is dry, $z = 10$ km, $p_d = 250$ hPa, and $T = 218$ K (base of the tropopause), estimate the scale height at $z = 10$ km and pressure at $z = 10.5$ km.

2.16 Calculate the saturation vapor pressure of water over liquid and ice if $T_c = -15$ °C. Find ω_v and p_v at this temperature if $f_r = 3$ percent and $p_d = 230$ hPa.

2.17 Calculate the dew point if $f_r = 54$ percent and $T = 263$ K.

2.18 Derive the expression for the water-vapor mass mixing ratio as a function of dew point and pressure from (2.68). If $T_D = 284$ K and $p_d = 1000$ hPa, find ω_v.

2.19 If $T_D = 279$ K, $T = 281$ K, and $p_d = 930$ hPa, calculate f_r, p_v, ω_v, and $\omega_{v,s}$.

2.20 **(a)** Estimate the diabatic energy (dQ) that needs to be added to or removed from a parcel of air to increase its virtual temperature $\mathrm{d}T_v = +2$ K when $\rho_a = 1.2$ kg m^{-3} and when the pressure change in the parcel due to adiabatic expansion is $\mathrm{d}p_a = -10$ hPa.

(b) If dQ is removed from part (a) and other conditions stay the same, what is the parcel virtual-temperature change? What type of process is this?

(c) If the parcel in part (a) does not rise or expand ($\mathrm{d}p_a = 0$), but dQ calculated from part (a) remains, what is the virtual-temperature change of the parcel? What is the name of this type of process?

(d) If the parcel in part (a) does not change virtual temperature ($\mathrm{d}T_v = 0$), but $\mathrm{d}p_a = -10$ hPa, what is the new value of dQ? What is the name of this process?

2.21 Calculate the potential virtual temperature of dry air when (a) $p_a = 900$ hPa and $T = 280$ K; (b) $p_a = 850$ hPa and $T = 278$ K. Is the air parcel between 900 and 850 hPa stable, unstable, or neutral with respect to unsaturated air?

2.22 If $\theta_p = 303$ K at $p_d = 825$ hPa, find the air temperature at this pressure.

2.23 Calculate the change in potential virtual temperature with altitude ($\partial\theta_v/\partial z$) when the ambient virtual-temperature lapse rate is $\Gamma_v = +6.2$ K km^{-1}, the air pressure is $p_a = 875$ hPa, and $T_v = 283$ K. Is this air stable, unstable, or neutral with respect to unsaturated air?

2.24 If the air is dry and the potential virtual temperature increases at the rate $\partial\theta_v/\partial z = 1$ K km^{-1}, calculate the ambient virtual-temperature lapse rate when $p_a = 925$ hPa and $T = 288$ K. Is this air stable, unstable, or neutral with respect to unsaturated air?

2.25 If the potential virtual temperature increases at the rate $\partial\theta_v/\partial z = 2$ K km^{-1} in dry air at an altitude where $p_a = 945$ hPa and $T = 287$ K, estimate T at 100 m above this altitude.

2.26 Would liquid water and/or ice particles grow when (a) $p_v = 1$ hPa, $T = -30$ K; (b) $p_v = 1.2$ hPa, $T = -20$ K; or (c) $p_v = 1$ hPa, $T = -16$ K? Use Fig. 2.8(b).

2.27 Does potential virtual temperature at sea level increase, decrease, or stay constant (on average) between the Equator and North Pole? Why? Does potential virtual temperature increase, decrease, or stay constant with altitude if $\Gamma_v = +6.5$ K km^{-1}? Why?

2.28 What might Figs. 2.3(a) and (b) look like under a low-pressure system, if all other conditions were the same?

2.9 COMPUTER PROGRAMMING PRACTICE

2.29 Write a computer script to calculate air pressure (p_a) as a function of altitude from (2.41). Assume $p_a = 1013$ hPa and $T = 288$ K at the surface, the temperature decreases at the rate of 6.5 K km^{-1}, and the air is dry. Use the program to estimate the pressure from $z = 0$ to 10 km in increments of 100 m. Calculate the density ρ_a with the equation of state at the base of each layer before each pressure calculation for the next layer. Plot the results.

2.30 Write a computer script to calculate p_a as a function of altitude from (2.44). Assume $T = 298$ K and $p_a = 1013$ hPa at the surface and the air is dry. Use the program to estimate the pressure from $z = 0$ to 10 km in increments of 100 m. Plot the results.

2.31 Write a computer script to calculate p_a as a function of altitude from (2.48). Assume $T = 298$ K and $p_a = 1013$ hPa at the surface, the air is dry, and the temperature decreases from the surface at 6.5 K km^{-1}. Use the program to estimate the pressure from $z = 0$ to 10 km in increments of 100 m, calculating the scale height for each layer. Plot the results.

2.32 Write a computer script to calculate the saturation vapor pressure of water over liquid and ice from (2.61) and (2.64), respectively. Use the program to estimate $p_{v,s}$ between -50 °C and $+50$ °C and $p_{v,I}$ between -50 °C and 0 °C, in increments of 1 °C. Plot the results.

2.33 Write a computer script to calculate $p_{v,s}$, $p_{v,I}$, and T_D versus altitude. Assume T is 298 K at $z = 0$ km and decreases 6.5 K km^{-1}. Assume also that $f_r = 90$ percent at all altitudes. Use the program to estimate parameters from $z = 0$ to 10 km in increments of 100 m. Plot the results.

3

The continuity and thermodynamic energy equations

The continuity equations for air, individual gases, and aerosol particles, and the thermodynamic energy equation are fundamental equations in atmospheric models. Continuity equations are used to simulate changes in concentration or mixing ratio of a variable over time and take account of transport, external sources, and external sinks of the variable. The thermodynamic energy equation is used to predict changes in temperature with time and takes account of transport, external sources, and external sinks of energy. In this chapter, scalars, vectors, gradient operators, local derivatives, and total derivatives are defined, and the continuity and thermodynamic energy equations are derived.

3.1 DEFINITIONS

In this section, definitions relating to wind speed and direction and differentiation are provided. The definitions will be used in subsequent sections to derive time-dependent continuity equations.

3.1.1 Wind velocity

Scalars are variables, such as temperature, air pressure, and air mass, that have magnitude but not direction. **Vectors** are variables, such as velocity, that have magnitude and direction.

Winds are described by three parameters – velocity, the scalar components of velocity, and speed. **Velocity** is a vector that quantifies the rate at which the position of a body changes over time. Total and horizontal wind velocity vectors are defined in Cartesian (rectangular) horizontal coordinates as

$$\mathbf{v} = \mathbf{i}u + \mathbf{j}v + \mathbf{k}w \qquad \mathbf{v}_\mathrm{h} = \mathbf{i}u + \mathbf{j}v \tag{3.1}$$

respectively, where **i**, **j**, and **k** are Cartesian-coordinate west–east, south–north, and vertical unit vectors, respectively, and

$$u = \frac{\mathrm{d}x}{\mathrm{d}t} \qquad v = \frac{\mathrm{d}y}{\mathrm{d}t} \qquad w = \frac{\mathrm{d}z}{\mathrm{d}t} \tag{3.2}$$

are scalar components of velocity (**scalar velocities**) (m s^{-1}). Scalar velocities have magnitude only. When applied in (3.1), positive u, v, and w correspond to winds

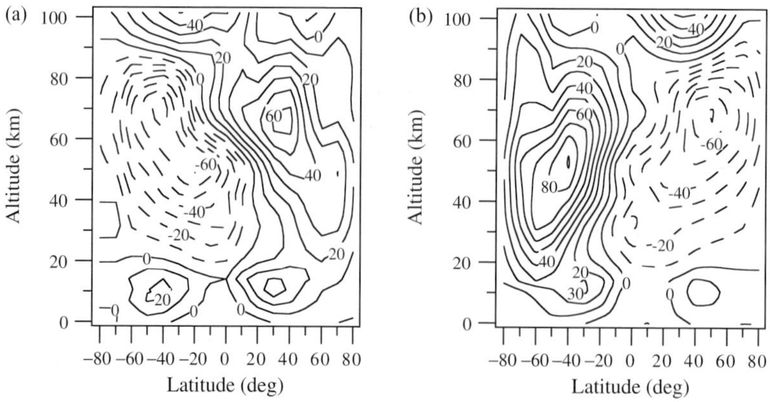

Figure 3.1 Zonally averaged west–east scalar velocities (m s^{-1}) for (a) January and (b) July. Data for the plots were compiled by Fleming *et al.* (1988).

moving from west to east, south to north, and lower to higher elevation, respectively. The vertical scalar velocity in (3.2) is written in the **altitude vertical coordinate** system. In this coordinate system (coordinate), tops and bottoms of horizontal layers are defined by surfaces of constant altitude.

The magnitude of the wind is its speed. The total and horizontal wind speeds are defined as

$$|\mathbf{v}| = \sqrt{u^2 + v^2 + w^2} \qquad |\mathbf{v}_\mathrm{h}| = \sqrt{u^2 + v^2} \tag{3.3}$$

respectively.

Wind **direction** is generally named for where a wind originates from. A westerly wind, southwesterly wind, sea breeze, and mountain breeze originate from the west, the southwest, the sea, and a mountain, respectively. A positive scalar velocity u with no south–north component is a westerly wind. A positive scalar velocity v with no west–east component is a southerly wind.

Air velocities vary in space and time. Figures 3.1(a) and (b) show global-scale latitude–altitude contour plots of zonally averaged west–east scalar velocities for January and July, respectively. The figures indicate that west–east winds in the upper troposphere almost always originate from the west. The two peaks near 10 km in each figure correspond to **subtropical jet streams.** Near the surface at the Equator and poles, winds originate from the east but are weak. Near the surface at midlatitudes (30°–60°) in both hemispheres, winds originate from the west. In the stratosphere, westerly wind speeds increase with height in the winter hemisphere (Northern Hemisphere in January; Southern Hemisphere in July), forming **polar night jets** near 60 km. Easterly wind speeds increase with increasing altitude in the summer hemisphere. Winds above the surface are driven by pressure gradients, and pressure gradients are driven by temperature gradients. Thus, strong winds aloft indicate strong temperature and pressure gradients.

3.1.2 Time and spatial rates of change

The time rate of change of a variable, such as concentration, momentum, or temperature, can be determined at a fixed location or in the frame of reference of the variable as it moves. Suppose a plume, carrying a gas with number concentration $N = N(t, x[t])$ (molec. cm^{-3}), travels with the wind from fixed point A in the west to fixed point B in the east. The time rate of change of N anywhere along the plume's trajectory is the **total derivative,** $\mathrm{d}N/\mathrm{d}t$. The total derivative can be expanded with the chain rule in Cartesian coordinates as

$$\frac{\mathrm{d}N}{\mathrm{d}t} = \frac{\partial N}{\partial t}\frac{\mathrm{d}t}{\mathrm{d}t} + \frac{\partial N}{\partial x}\frac{\mathrm{d}x}{\mathrm{d}t} = \frac{\partial N}{\partial t} + u\frac{\partial N}{\partial x} \tag{3.4}$$

where $\partial N/\partial t$ is the time rate of change of concentration at fixed point A (**local derivative**), and $u\partial N/\partial x$ is the time rate of change of concentration in the plume that results from a west–east scalar velocity transporting the plume.

The total derivative of a variable is nonzero when processes other than transport affect the variable. In the case of gases, external processes include chemistry and gas-to-particle conversion. If $\mathrm{d}N/\mathrm{d}t = 0$, the concentration of a gas does not change as it travels with the wind.

The local derivative of a variable is the difference between the total derivative and the rate of change of the variable due to transport. Thus, the local derivative is affected by external processes plus transport. If $\partial N/\partial t = 0$, the rate of production (loss) of a variable due to external processes equals the rate of loss (production) of the variable due to transport of a spatial gradient of the variable $[u\,(\partial N/\partial x)]$.

Example 3.1

Suppose the time rate of change of concentration of a gas along the path of a hot-air balloon traveling with the wind from east to west at $u = -10$ m s^{-1} is $\mathrm{d}N/\mathrm{d}t = 10^8$ molec. cm^{-3} s^{-1}. If the west–east gradient in concentration is $\partial N/\partial x = 10^{10}$ molec. cm^{-3} km^{-1} (concentration increases from west to east), determine the time rate of change of concentration at a fixed point A, which the balloon passes over.

SOLUTION

Since $u\partial N/\partial x = -10^8$ molec. cm^{-3} s^{-1}, (3.4) predicts $(\partial N/\partial t)_A \approx 2 \times 10^8$ molec. cm^{-3} s^{-1}. Thus, transport from the east ($u\partial N/\partial x$) accounts for one-half of the production rate of N at point A, and transformations along the trajectory ($\mathrm{d}N/\mathrm{d}t$) account for the other half.

A **Lagrangian** frame of reference is a frame of reference that moves relative to a fixed coordinate system. An **Eulerian** frame of reference is a frame of reference in a fixed coordinate system. The left side of (3.4) is written in terms of a Lagrangian frame of reference. The right side is written in terms of an Eulerian frame of reference. Generalizing (3.4) to three dimensions gives

$$\frac{dN}{dt} = \frac{\partial N}{\partial t} + u\frac{\partial N}{\partial x} + v\frac{\partial N}{\partial y} + w\frac{\partial N}{\partial z} \tag{3.5}$$

3.1.3 Gradient operator

A gradient operator (also called a directional derivative, nabla operator, or del operator) is a vector operator of partial derivatives. The **gradient operator in Cartesian-altitude coordinates** is

$$\nabla = \mathbf{i}\frac{\partial}{\partial x} + \mathbf{j}\frac{\partial}{\partial y} + \mathbf{k}\frac{\partial}{\partial z} \tag{3.6}$$

The **dot product** of the velocity vector with the gradient operator is a scalar operator,

$$\mathbf{v} \cdot \nabla = (\mathbf{i}u + \mathbf{j}v + \mathbf{k}w) \cdot \left(\mathbf{i}\frac{\partial}{\partial x} + \mathbf{j}\frac{\partial}{\partial y} + \mathbf{k}\frac{\partial}{\partial z}\right) = u\frac{\partial}{\partial x} + v\frac{\partial}{\partial y} + w\frac{\partial}{\partial z} \tag{3.7}$$

where $\mathbf{i} \cdot \mathbf{i} = 1$, $\mathbf{j} \cdot \mathbf{j} = 1$, and $\mathbf{k} \cdot \mathbf{k} = 1$. Cross terms are zero ($\mathbf{i} \cdot \mathbf{j} = 0$, $\mathbf{i} \cdot \mathbf{k} = 0$, and $\mathbf{j} \cdot \mathbf{k} = 0$), since the unit vectors are orthogonal. The dot product of two vectors is a scalar and symmetric (e.g., $\mathbf{a} \cdot \mathbf{v} = \mathbf{v} \cdot \mathbf{a}$). The dot product of a gradient operator with a vector is a scalar operator but not symmetric ($\nabla \cdot \mathbf{v} \neq \mathbf{v} \cdot \nabla$). Instead,

$$\nabla \cdot \mathbf{v} = \left(\mathbf{i}\frac{\partial}{\partial x} + \mathbf{j}\frac{\partial}{\partial y} + \mathbf{k}\frac{\partial}{\partial z}\right) \cdot (\mathbf{i}u + \mathbf{j}v + \mathbf{k}w) = \frac{\partial u}{\partial x} + \frac{\partial v}{\partial y} + \frac{\partial w}{\partial z} \tag{3.8}$$

which is a scalar **divergence**. When concentration is multiplied by a divergence, the result is the scalar

$$N(\nabla \cdot \mathbf{v}) = N\frac{\partial u}{\partial x} + N\frac{\partial v}{\partial y} + N\frac{\partial w}{\partial z} \tag{3.9}$$

The gradient of a scalar, such as concentration, is a vector. For example,

$$\nabla N = \left(\mathbf{i}\frac{\partial}{\partial x} + \mathbf{j}\frac{\partial}{\partial y} + \mathbf{k}\frac{\partial}{\partial z}\right) N = \mathbf{i}\frac{\partial N}{\partial x} + \mathbf{j}\frac{\partial N}{\partial y} + \mathbf{k}\frac{\partial N}{\partial z} \tag{3.10}$$

Applying the dot product of velocity with the gradient operator to N gives the scalar

$$(\mathbf{v} \cdot \nabla)N = \left(u\frac{\partial}{\partial x} + v\frac{\partial}{\partial y} + w\frac{\partial}{\partial z}\right) N = u\frac{\partial N}{\partial x} + v\frac{\partial N}{\partial y} + w\frac{\partial N}{\partial z} \tag{3.11}$$

Substituting this result into the total-derivative equation (3.5) yields

$$\frac{dN}{dt} = \frac{\partial N}{\partial t} + (\mathbf{v} \cdot \nabla)N \tag{3.12}$$

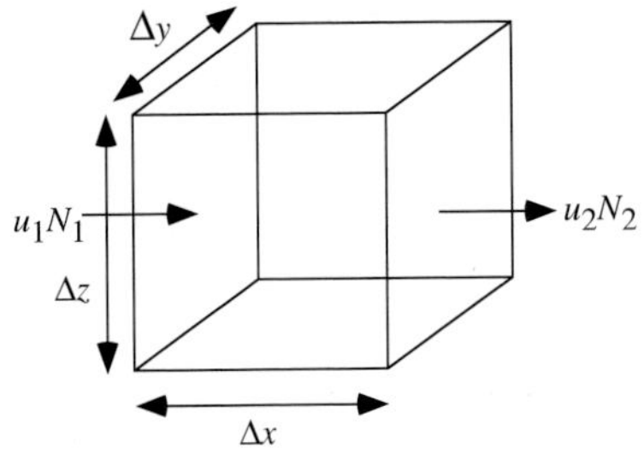

Figure 3.2 Example of mass conservation. The number of molecules entering minus the number of molecules leaving the box equals the number of molecules accumulating in the box.

Generalizing (3.12) for any variable gives the **total derivative in Cartesian-altitude coordinates** as

$$\frac{\mathrm{d}}{\mathrm{d}t} = \frac{\partial}{\partial t} + u\frac{\partial}{\partial x} + v\frac{\partial}{\partial y} + w\frac{\partial}{\partial z} = \frac{\partial}{\partial t} + \mathbf{v} \cdot \nabla \tag{3.13}$$

3.2 CONTINUITY EQUATIONS

When air circulates in an enclosed volume, and no chemical or physical processes affect it, the mass of the air, summed throughout the volume, is conserved. In an atmospheric model divided into many **grid cells (grid boxes)**, the mass of air entering one cell minus the mass leaving the cell equals the final mass minus the initial mass in the cell. The same is true for other atmospheric variables, such as gas concentrations or energy, when only transport affects these variables.

Figure 3.2 shows a grid cell with dimensions Δx, Δy, Δz (m). The west–east scalar velocities entering and leaving the cell are u_1 and u_2 (m s^{-1}), respectively. Gas concentrations at the west and east boundaries of the cell are N_1 and N_2 (molec. cm^{-3}), respectively. Mass fluxes of gas into the cell and out of the cell are $u_1 N_1$ and $u_2 N_2$ (m molec. cm^{-3} s^{-1}), respectively.

From the information given, the numbers of molecules entering, leaving, and accumulating in the box during time period Δt are $u_1 N_1 \Delta y \Delta z \Delta t$, $u_2 N_2 \Delta y \Delta z \Delta t$, and

$$\Delta N \Delta x \Delta y \Delta z = u_1 N_1 \Delta y \Delta z \Delta t - u_2 N_2 \Delta y \Delta z \Delta t \tag{3.14}$$

respectively. Dividing both sides of (3.14) by Δt and by the box volume ($\Delta x \Delta y \Delta z$) gives

$$\frac{\Delta N}{\Delta t} = -\left(\frac{u_2 N_2 - u_1 N_1}{\Delta x}\right) \tag{3.15}$$

As $\Delta x \to 0$ and $\Delta t \to 0$, this equation approaches

$$\frac{\partial N}{\partial t} = -\frac{\partial (uN)}{\partial x} \tag{3.16}$$

which is the **continuity equation** for a gas affected by velocity in one direction. This equation expands to three dimensions in Cartesian-altitude coordinates as

$$\frac{\partial N}{\partial t} = -\frac{\partial (uN)}{\partial x} - \frac{\partial (vN)}{\partial y} - \frac{\partial (wN)}{\partial z} = -\nabla \cdot (\mathbf{v}N) \tag{3.17}$$

where $\mathbf{v}N = \mathbf{i}uN + \mathbf{j}vN + \mathbf{k}wN$. A similar equation can be written for air density. Equation (3.17) states that the time rate of change of N at a fixed location equals the negative of the local spatial gradient of the flux of N. Equation (3.17) is a **flux divergence form** of the continuity equation so called because $\nabla \cdot (\mathbf{v}N)$ is a divergence of concentration.

Substituting

$$\nabla \cdot (\mathbf{v}N) = N(\nabla \cdot \mathbf{v}) + (\mathbf{v} \cdot \nabla)N \tag{3.18}$$

into (3.17) and writing a similar equation for air density give the continuity equations for gas number concentration and total air mass density as

$$\frac{\partial N}{\partial t} = -N(\nabla \cdot \mathbf{v}) - (\mathbf{v} \cdot \nabla)N \tag{3.19}$$

$$\frac{\partial \rho_a}{\partial t} = -\rho_a(\nabla \cdot \mathbf{v}) - (\mathbf{v} \cdot \nabla)\rho_a \tag{3.20}$$

respectively. Substituting

$$(\mathbf{v} \cdot \nabla)N = \frac{\mathrm{d}N}{\mathrm{d}t} - \frac{\partial N}{\partial t} \tag{3.21}$$

from (3.12) into (3.19) and (3.20) gives **velocity divergence forms** of the continuity equations as

$$\frac{\mathrm{d}N}{\mathrm{d}t} = -N(\nabla \cdot \mathbf{v}) \tag{3.22}$$

$$\frac{\mathrm{d}\rho_a}{\mathrm{d}t} = -\rho_a(\nabla \cdot \mathbf{v}) \tag{3.23}$$

where $\nabla \cdot \mathbf{v}$ is the divergence of velocity. The equations are also **advective forms** of the continuity equation in that $\mathrm{d}/\mathrm{d}t$ contains the advection term, $\mathbf{v} \cdot \nabla$. The equations state that the change of a scalar variable over time in a moving parcel equals the scalar variable multiplied by the negative local spatial gradient of velocity.

The gas number concentration N (molecules per cubic centimeter of air) is related to the moist-air mass mixing ratio, q (kilograms per kilogram of moist air), of a species with molecular weight m (g mol^{-1}) by

$$N = \frac{A\rho_a q}{m} \tag{3.24}$$

where A is Avogadro's number (molec. mol^{-1}). Substituting (3.24) into (3.19) and expanding give

$$q\left(\frac{\partial \rho_a}{\partial t} + \rho_a(\nabla \cdot \mathbf{v}) + (\mathbf{v} \cdot \nabla)\rho_a\right) + \rho_a\frac{\partial q}{\partial t} = -\rho_a(\mathbf{v} \cdot \nabla)q \tag{3.25}$$

Substituting the continuity equation for air from (3.20) into (3.25) gives the **gas continuity equation** in units of the moist-air mass mixing ratio as

$$\frac{\partial q}{\partial t} = -(\mathbf{v} \cdot \nabla)q \tag{3.26}$$

Equations (3.22) and (3.23) assume that air is **compressible**, meaning that total volume of a parcel of air changes over time. Ocean water is considered **incompressible**, meaning that the total volume of a parcel of ocean water does not change over time. Thus,

$$\frac{\partial u}{\partial x} + \frac{\partial v}{\partial y} + \frac{\partial w}{\partial z} = 0 \tag{3.27}$$

which is the **continuity equation for an incompressible fluid.** If (3.27) is not satisfied, a net divergence out of or convergence into a fluid volume occurs, causing the volume to expand or contract, respectively. Equation (3.27) states that an **incompressible fluid is nondivergent.** The equation can also be written as $\nabla \cdot \mathbf{v} = 0$. Substituting water density (ρ_w) for air density and substituting $\nabla \cdot \mathbf{v} = 0$ into (3.23) give

$$\frac{d\rho_w}{dt} = 0 \tag{3.28}$$

which states that the density of an incompressible fluid is constant along the motion of the fluid. At a fixed point in the fluid, the density may change. Substituting water density for air density and $\nabla \cdot \mathbf{v} = 0$ into (3.20) give

$$\frac{\partial \rho_w}{\partial t} = -(\mathbf{v} \cdot \nabla)\rho_w \tag{3.29}$$

which states that the change in water density at a fixed point in an incompressible fluid is the negative product of velocity and the spatial gradient of density. In sum, the density of an incompressible fluid, such as liquid water, can vary spatially, but the total volume of such a fluid is constant over time. A fluid in which density varies spatially throughout the fluid is **inhomogeneous.** Ocean water is inhomogeneous and incompressible. A fluid in which density is always constant throughout a volume ($\partial \rho_w/\partial t = 0$) is **homogeneous.** Pure liquid water is a relatively homogeneous, incompressible fluid. **Air is an inhomogeneous and compressible fluid.**

3.3 EXPANDED CONTINUITY EQUATIONS

Equation (3.17) gave the continuity equation without molecular diffusion or external source and sink terms. A more complete form of the **continuity equation for a gas** is

$$\frac{\partial N}{\partial t} = -\nabla \cdot (\mathbf{v}N) + D\nabla^2 N + \sum_{n=1}^{N_{e,t}} R_n \tag{3.30}$$

(e.g., Reynolds *et al.* 1973), where D is the **molecular diffusion coefficient** of the gas (cm^2 s^{-1}), $N_{e,t}$ is the number of external processes (e.g., chemistry, emission, etc.) affecting the gas, and R_n is the time rate of change of trace-gas concentration due to the nth external process affecting the gas (molec. cm^{-3} s^{-1}). **Molecular diffusion** is the movement of molecules due to their kinetic energy. As molecules move, they collide with other molecules and are redirected along arbitrary paths. A molecular diffusion coefficient quantifies the rate of molecular diffusion, and is defined mathematically in Section 16.2.

The squared gradient in the molecular diffusion term of (3.30) expands to

$$\begin{aligned}\nabla^2 N = (\nabla \cdot \nabla)\, N &= \left[\left(\mathbf{i}\frac{\partial}{\partial x} + \mathbf{j}\frac{\partial}{\partial y} + \mathbf{k}\frac{\partial}{\partial z}\right) \cdot \left(\mathbf{i}\frac{\partial}{\partial x} + \mathbf{j}\frac{\partial}{\partial y} + \mathbf{k}\frac{\partial}{\partial z}\right)\right] N \\ &= \frac{\partial^2 N}{\partial x^2} + \frac{\partial^2 N}{\partial y^2} + \frac{\partial^2 N}{\partial z^2}\end{aligned} \tag{3.31}$$

Substituting (3.17) and (3.31) into (3.30) gives the continuity equation for a gas as

$$\frac{\partial N}{\partial t} + \frac{\partial (uN)}{\partial x} + \frac{\partial (vN)}{\partial y} + \frac{\partial (wN)}{\partial z} = D\left(\frac{\partial^2 N}{\partial x^2} + \frac{\partial^2 N}{\partial y^2} + \frac{\partial^2 N}{\partial z^2}\right) + \sum_{n=1}^{N_{e,t}} R_n \tag{3.32}$$

3.3.1 Time and grid volume averaging

The spatial domain in a model is divided into grid cells of finite size. Time is also divided into time steps of finite size for advancing species concentrations, velocities, and other variables. Real atmospheric motions generally occur over spatial scales much smaller than the resolution of model grid cells and over temporal scales smaller than the resolution of model time steps. For example, a typical mesoscale model might have horizontal resolution 5 km × 5 km, vertical resolution 50 m, and time resolution 5 s. A global-scale model might have horizontal resolution 400 km × 400 km, vertical resolution 200 m, and time resolution 300 s. Fluctuations in atmospheric motions due to eddies occur on smaller scales in both cases. Eddies (Section 4.2.6; Section 8.4), for example, range in diameter from a couple of millimeters to hundreds of meters and on time scales of seconds to hours.

Whereas many models do not resolve eddies, some do. These are discussed in Section 8.4. To account for subgrid-scale disturbances in those models that do not resolve eddies, a process called Reynolds averaging, named after Osborne Reynolds, is used. Models that treat turbulence using Reynolds averaging are called **Reynolds-averaged models.**

During Reynolds averaging, each variable in (3.32) and in other model equations is divided into an average and perturbation component. Such a division is referred to as **Reynolds decomposition.** In the case of gases, gas number concentration is decomposed as

$$N = \bar{N} + N' \tag{3.33}$$

where N is the **actual (precise or instantaneous) concentration,** $\bar{N}$ is the **average concentration,** and N' is the instantaneous **perturbation concentration.** A precise concentration occurs at a given instant and location within a grid cell. An average concentration is obtained by integrating and averaging over a model time step and grid-cell volume. Thus,

$$\bar{N} = \frac{1}{h\Delta x\Delta y\Delta z}\int_t^{t+h}\left\{\int_x^{x+\Delta x}\left[\int_y^{y+\Delta y}\left(\int_z^{z+\Delta z} N\,\mathrm{d}z\right)\mathrm{d}y\right]\mathrm{d}x\right\}\mathrm{d}t \tag{3.34}$$

(e.g., Pielke 1984), where h is the time step, and Δx, Δy, Δz are space increments, shown in Fig. 3.2. The average concentrations are averages over one grid cell and time step and differ for each grid cell and time step. Perturbation concentrations are distributed on both sides of the average, so that the spatial and temporal average of all perturbations is zero ($\bar{N'} = 0$), which is the **Reynolds assumption.**

Scalar and vector velocities can be decomposed in a similar manner. Thus, for example,

$$u = \bar{u} + u' \qquad v = \bar{v} + v' \qquad w = \bar{w} + w' \tag{3.35}$$

where $\bar{u}$, $\bar{v}$, and $\bar{w}$ are the time- and volume-averaged scalar velocities, and u', v', and w' are perturbation scalar velocities, and

$$\mathbf{v} = \bar{\mathbf{v}} + \mathbf{v}' \tag{3.36}$$

where $\bar{\mathbf{v}} = \mathbf{i}\bar{u} + \mathbf{j}\bar{v} + \mathbf{k}\bar{w}$ is a time- and volume-averaged velocity and $\mathbf{v}' = \mathbf{i}u' + \mathbf{j}v' + \mathbf{k}w'$ is a perturbation velocity. **Advection** is the mean horizontal velocity. Thus, $\bar{u}$ and $\bar{v}$ are components of advection. Figure 3.3 shows an example of precise, mean, and perturbation scalar velocities and trace-gas concentrations.

Unsteady flow occurs when $\mathbf{v}$ varies with time, but not necessarily randomly, at a given location. **Steady flow** occurs when $\mathbf{v}$ is independent of time. **Turbulent flow** is unpredictable flow in which $\mathbf{v}$ varies randomly with time at a location. Thus, turbulent flow is unsteady, but unsteady flow is not necessarily turbulent. **Laminar flow** is nonturbulent flow in which $\mathbf{v}$ may vary, but not randomly, with time at a given location. In laminar flow, fluid particles travel along well-defined streamlines and fluid layers flow independent of each other. Laminar flow can be steady or unsteady. Nearly all flows in the atmosphere are turbulent.

Subgrid-scale effects are estimated by substituting decomposed variables into an equation, then taking a time average and grid volume average of resulting terms. Substituting (3.33) and (3.35) into the species continuity equation from (3.32) and

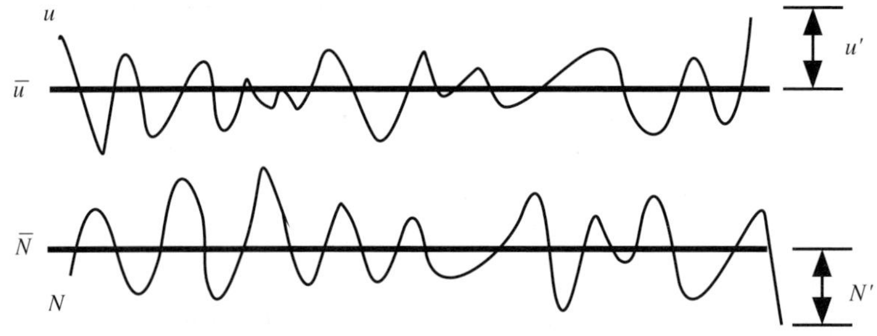

Figure 3.3 Precise, mean, and perturbation components of scalar velocity and gas concentration. The precise scalar velocity is denoted by u, the precise gas concentration is denoted by N, time- and volume-averaged values are denoted by an overbar, and perturbation components are denoted by a prime. Each point on the horizontal axis is a perturbation at a given time and location within a grid cell.

averaging terms over space and time (**Reynolds averaging**) give

$$\left[\overline{\frac{\partial(\bar{N}+N')}{\partial t}}\right]+\left[\overline{\frac{\partial(\bar{u}+u')(\bar{N}+N')}{\partial x}}\right]+\left[\overline{\frac{\partial(\bar{v}+v')(\bar{N}+N')}{\partial y}}\right]$$
$$+\left[\overline{\frac{\partial(\bar{w}+w')(\bar{N}+N')}{\partial z}}\right]=D\left\{\left[\overline{\frac{\partial^2(\bar{N}+N')}{\partial x^2}}\right]+\left[\overline{\frac{\partial^2(\bar{N}+N')}{\partial y^2}}\right]\right.$$
$$\left.+\left[\overline{\frac{\partial^2(\bar{N}+N')}{\partial z^2}}\right]\right\}+\overline{\sum_{n=1}^{N_{e,t}}R_n} \tag{3.37}$$

Since $\overline{\partial(\bar{N}+N')/\partial t}=\partial(\overline{\bar{N}+N'})/\partial t$, $\overline{\bar{N}+N'}=\bar{\bar{N}}+\bar{N'}$, $\bar{\bar{N}}=\bar{N}$, and $\bar{N'}=0$, the first term in (3.37) simplifies to

$$\left[\overline{\frac{\partial(\bar{N}+N')}{\partial t}}\right]=\frac{\partial(\bar{\bar{N}}+\bar{N'})}{\partial t}=\frac{\partial\bar{N}}{\partial t} \tag{3.38}$$

Since $\overline{u'\bar{N}}=0$, $\overline{\bar{u}N'}=0$, and $\overline{\bar{u}\bar{N}}=\bar{u}\bar{N}$, the second term simplifies to

$$\left[\overline{\frac{\partial(\bar{u}+u')(\bar{N}+N')}{\partial x}}\right]=\frac{\partial(\overline{\bar{u}\bar{N}}+\overline{\bar{u}N'}+\overline{u'\bar{N}}+\overline{u'N'})}{\partial x}=\frac{\partial(\bar{u}\bar{N}+\overline{u'N'})}{\partial x} \tag{3.39}$$

The product $\overline{u'N'}$ (m molec. cm^{-3} s^{-1}) represents the west–east transport of N' due to subgrid-scale eddies. It is a **kinematic turbulent flux** in that its units are those of concentration flux (kg [molec. cm^{-3}] m^{-2} s^{-1}) divided by air density (kg m^{-3}). When a variable or a flux is divided by the air density, it becomes a **kinematic** variable or flux. The partial derivative $\partial(\overline{u'N'})/\partial x$ (molec. cm^{-3} s^{-1}) is a **turbulent flux divergence** term.

Example 3.2

Suppose two gas concentrations ($N_1 = 8$ and $N_2 = 4$) and scalar velocities ($u_1 = 3$ and $u_2 = -1$) are measured at different locations within a grid cell at a given time. Estimate $\bar{N}$, N_1', N_2', $\bar{u}$, u_1', u_2', $\overline{u'N'}$, $\bar{u}\bar{N}$, and $\overline{uN}$. Ignore units.

SOLUTION

$$\bar{N} = (N_1 + N_2)/2 = 6 \qquad \bar{u} = (u_1 + u_2)/2 = 1$$
$$N_1' = N_1 - \bar{N} = 2 \qquad u_1' = u_1 - \bar{u} = 2$$
$$N_2' = N_2 - \bar{N} = -2 \qquad u_2' = u_2 - \bar{u} = -2$$

$$\overline{u'N'} = (u_1'N_1' + u_2'N_2')/2 = 4\bar{u}\bar{N} = 6$$
$$\overline{uN} = \bar{u}\bar{N} + \overline{u'N'} = (u_1N_1 + u_2N_2)/2 = 10$$

Substituting (3.38), (3.39), and similar terms for other directions into (3.37) gives

$$\frac{\partial \bar{N}}{\partial t} + \frac{\partial(\bar{u}\bar{N})}{\partial x} + \frac{\partial(\bar{v}\bar{N})}{\partial y} + \frac{\partial(\bar{w}\bar{N})}{\partial z} + \frac{\partial \overline{u'N'}}{\partial x} + \frac{\partial \overline{v'N'}}{\partial y} + \frac{\partial \overline{w'N'}}{\partial z} = D\left(\frac{\partial^2 \bar{N}}{\partial x^2} + \frac{\partial^2 \bar{N}}{\partial y^2} + \frac{\partial^2 \bar{N}}{\partial z^2}\right) + \sum_{n=1}^{N_{e,t}} \bar{R}_n \tag{3.40}$$

For motions larger than the molecular scale, the molecular diffusion terms in (3.40) are much smaller than are the turbulent flux divergence terms and can be removed. Thus, (3.40) simplifies to

$$\frac{\partial \bar{N}}{\partial t} + \frac{\partial(\bar{u}\bar{N})}{\partial x} + \frac{\partial(\bar{v}\bar{N})}{\partial y} + \frac{\partial(\bar{w}\bar{N})}{\partial z} + \frac{\partial \overline{u'N'}}{\partial x} + \frac{\partial \overline{v'N'}}{\partial y} + \frac{\partial \overline{w'N'}}{\partial z} = \sum_{n=1}^{N_{e,t}} \bar{R}_n \tag{3.41}$$

which compresses to the **continuity equation for a gas,**

$$\frac{\partial \bar{N}}{\partial t} + \nabla \cdot (\bar{\mathbf{v}}\bar{N}) + \nabla \cdot (\overline{\mathbf{v}'N'}) = \sum_{n=1}^{N_{e,t}} \bar{R}_n \tag{3.42}$$

An analogous equation for air density results in the **continuity equation for air,**

$$\frac{\partial \bar{\rho}_a}{\partial t} + \nabla \cdot (\bar{\mathbf{v}}\bar{\rho}_a) + \nabla \cdot (\overline{\mathbf{v}'\rho_a'}) = 0 \tag{3.43}$$

In (3.43), the external source and sink terms for air molecules are neglected because they are small in comparison with the other terms.

Equation (3.42) can be rederived in terms of moist-air mass mixing ratio. Adding source and sink terms to (3.26) yields

$$\frac{\partial \mathrm{q}}{\partial t} + \mathbf{v} \cdot \nabla \mathrm{q} = \sum_{n=1}^{N_{e,t}} R_n \tag{3.44}$$

where R_n is now in kilograms per kilogram of moist air per second. Multiplying the continuity equation for air from (3.20) by q, multiplying (3.44) by ρ_a, adding the results, and compressing give

$$\frac{\partial(\rho_a \mathrm{q})}{\partial t} + \nabla \cdot (\rho_a \mathbf{v} \mathrm{q}) = \rho_a \sum_{n=1}^{N_{e,t}} R_n \tag{3.45}$$

The moist-air mass mixing ratio, velocity, and density can be decomposed with $\mathrm{q} = \bar{\mathrm{q}} + \mathrm{q}'$, $\mathbf{v} = \bar{\mathbf{v}} + \mathbf{v}'$, and $\rho_a = \bar{\rho}_a + \rho_a'$, respectively. Density perturbations in the atmosphere are relatively small; thus, $\rho_a' \ll \bar{\rho}_a$, and $\rho_a \approx \bar{\rho}_a$. Substituting decomposed variable values into all but the R term in (3.45) gives

$$\frac{\partial[\bar{\rho}_a(\bar{\mathrm{q}} + \mathrm{q}')]}{\partial t} + \nabla \cdot \left[\bar{\rho}_a(\bar{\mathbf{v}} + \mathbf{v}')(\bar{\mathrm{q}} + \mathrm{q}')\right] = \bar{\rho}_a \sum_{n=1}^{N_{e,t}} R_n \tag{3.46}$$

Taking the time and grid volume average of this equation, eliminating zero-value terms and removing unnecessary overbars results in

$$\bar{\rho}_a \left[\frac{\partial \bar{\mathrm{q}}}{\partial t} + (\bar{\mathbf{v}} \cdot \nabla)\bar{\mathrm{q}}\right] + \bar{\mathrm{q}}\left[\frac{\partial \bar{\rho}_a}{\partial t} + \nabla \cdot (\overline{\mathbf{v}\rho_a})\right] + \nabla \cdot (\bar{\rho}_a \overline{\mathbf{v}'\mathrm{q}'}) = \bar{\rho}_a \sum_{n=1}^{N_{e,t}} \bar{R}_n \tag{3.47}$$

Equation (3.47) can be simplified by first noting that, when $\rho_a' \ll \bar{\rho}_a$, (3.43) becomes

$$\frac{\partial \bar{\rho}_a}{\partial t} + \nabla \cdot (\overline{\mathbf{v}\rho_a}) = 0 \tag{3.48}$$

Substituting this expression into (3.47) and dividing through by $\bar{\rho}_a$ give

$$\frac{\partial \bar{\mathrm{q}}}{\partial t} + (\bar{\mathbf{v}} \cdot \nabla)\bar{\mathrm{q}} + \frac{1}{\bar{\rho}_a}\nabla \cdot (\bar{\rho}_a \overline{\mathbf{v}'\mathrm{q}'}) = \sum_{n=1}^{N_{e,t}} \bar{R}_n \tag{3.49}$$

which is the **gas continuity equation** in units of the moist-air mass mixing ratio.

In (3.49), $\overline{u'\mathrm{q}'}$, $\overline{v'\mathrm{q}'}$, and $\overline{w'\mathrm{q}'}$ (m kg kg^{-1} s^{-1}) are **kinematic turbulent fluxes** of mixing ratio. Whereas models calculate spatially and temporally averaged values (e.g., $\bar{N}$, $\bar{\mathrm{q}}$, $\bar{\rho}_a$, $\bar{u}$), kinematic turbulent fluxes (e.g., $\overline{u'N'}$) are parameterized. Some parameterizations are discussed in Section 8.4. Here, a simple overview of K-theory is given.

With ***K*-theory** (gradient transport theory), kinematic turbulent fluxes are replaced with the product of a constant and the gradient of the mean value of a fluctuating variable (Calder 1949; Pasquill 1962; Monin and Yaglom 1971; Reynolds *et al.* 1973; Stull 1988). This is convenient, because models predict mean quantities.

Kinematic turbulent fluxes of gas concentration, for example, are parameterized with

$$\overline{u'N'} = -K_{h,xx}\frac{\partial \bar{N}}{\partial x} \qquad \overline{v'N'} = -K_{h,yy}\frac{\partial \bar{N}}{\partial y} \qquad \overline{w'N'} = -K_{h,zz}\frac{\partial \bar{N}}{\partial z} \tag{3.50}$$

where $K_{h,xx}$, $K_{h,yy}$, and $K_{h,zz}$ (e.g., cm^2 s^{-1}) are **eddy diffusion coefficients** in the x-, y-, and z-directions, respectively. The subscript h indicates that the **eddy diffusion coefficient for energy** (eddy thermal diffusivity) is used. The eddy diffusion coefficient for energy is used because the turbulent transport of a gas is similar to that of energy. When turbulent transport of velocity is simulated, an **eddy diffusion coefficient for momentum** (eddy viscosity) term is used. Eddy diffusion coefficients for energy and momentum differ, but not by much. Eddy diffusion coefficients represent an average diffusion coefficient for eddies of all sizes smaller than the grid cell. Eddy diffusion coefficients are also called eddy transfer, eddy exchange, turbulent transfer, and gradient transfer coefficients.

Eddy diffusion coefficients are parameterizations of subgrid scale transport of energy and momentum. In the vertical, such transport is caused by **mechanical shear** (mechanical turbulence) and/or **buoyancy** (thermal turbulence). Horizontal wind shear creates eddies that increase in size when the wind flows over rough surfaces. Buoyancy creates instability, causing shear-induced eddies to become wider and taller. Vertical motions in eddies transfer surface air upward and air aloft downward. Eddies also exchange air horizontally.

Substituting (3.50) into (3.41) gives

$$\frac{\partial \bar{N}}{\partial t} + \frac{\partial(\bar{u}\bar{N})}{\partial x} + \frac{\partial(\bar{v}\bar{N})}{\partial y} + \frac{\partial(\bar{w}\bar{N})}{\partial z} = \frac{\partial}{\partial x}\left(K_{h,xx}\frac{\partial \bar{N}}{\partial x}\right) + \frac{\partial}{\partial y}\left(K_{h,yy}\frac{\partial \bar{N}}{\partial y}\right) + \frac{\partial}{\partial z}\left(K_{h,zz}\frac{\partial \bar{N}}{\partial z}\right) + \sum_{n=1}^{N_{e,t}} \bar{R}_n \tag{3.51}$$

Compressing (3.51) and removing overbars for simplicity give the **continuity equation for an individual gas in number concentration units** and Cartesian-altitude coordinates as

$$\frac{\partial N}{\partial t} + \nabla \cdot (\mathbf{v}N) = (\nabla \cdot \mathbf{K}_h \nabla)\, N + \sum_{n=1}^{N_{e,t}} R_n \tag{3.52}$$

where

$$\mathbf{K}_h = \begin{bmatrix} K_{h,xx} & 0 & 0 \\ 0 & K_{h,yy} & 0 \\ 0 & 0 & K_{h,zz} \end{bmatrix} \tag{3.53}$$

is the **eddy diffusion tensor for energy.** The analogous **continuity equation for an individual gas in moist-air mass mixing ratio units** is

$$\frac{\partial q}{\partial t} + (\mathbf{v} \cdot \nabla) q = \frac{1}{\rho_a} (\nabla \cdot \rho_a \mathbf{K}_h \nabla) q + \sum_{n=1}^{N_{e,t}} R_n \tag{3.54}$$

The units of R_n differ in the two cases.

3.3.2 Continuity equation for air

External sources and sinks (R_n) are relatively small and can be ignored in the continuity equation for air. For most modeling applications, the turbulent flux divergence term in the equation can also be ignored because $\rho'_a \ll \bar{\rho}_a$. After removing overbars for convenience and making the above modifications, the continuity equation for air in Cartesian-altitude coordinates reduces from (3.43) to

$$\frac{\partial \rho_a}{\partial t} + \nabla \cdot (\mathbf{v} \rho_a) = 0 \tag{3.55}$$

3.3.3 Gas continuity equation

The continuity equations for trace gases and particles include several external source and sink terms. Gases enter the atmosphere from surface and elevated sources by **emission.** They are removed onto water, soil, foliage, roads, buildings, cars, and other surfaces by **dry deposition.** In many cases, gases are swept out of the atmosphere by falling raindrops during **washout.** Gases react chemically with each other and are dissociated by solar radiation during **photochemistry.** Some gases aggregate to form new particles during **homogeneous nucleation** or aggregate on existing particle surfaces during **heterogeneous nucleation.** Once a surface has nucleated, gas molecules may diffuse to and **condense** as a liquid or **deposit** as a solid on the surface. Liquid material may also **evaporate** or solid material may **sublimate** to the gas phase. A gas may also **dissolve** in liquid water on the surface of a particle. Dissolved gases may evaporate. Finally, a gas may react chemically on the surface of a particle during **heterogeneous chemistry.**

A form of the continuity equation for a gas q that accounts for the processes discussed above is

$$\begin{aligned} \frac{\partial N_q}{\partial t} + \nabla \cdot (\mathbf{v} N_q) = {} & (\nabla \cdot \mathbf{K}_h \nabla) N_q \\ & + R_{emisg} + R_{depg} + R_{washg} + R_{chemg} \\ & + R_{nucg} + R_{c/eg} + R_{dp/sg} + R_{ds/eg} + R_{hrg} \end{aligned} \tag{3.56}$$

where

R_{emisg} = rate of surface or elevated emission
R_{depg} = rate of dry deposition to the ground
R_{washg} = rate of washout to the ground or from one altitude to another
R_{chemg} = rate of photochemical production or loss
R_{nucg} = rate of gas loss due to homogeneous or heterogeneous nucleation
$R_{\mathrm{c/eg}}$ = rate of gas loss (production) due to condensation (evaporation)
$R_{\mathrm{dp/sg}}$ = rate of gas loss (production) due to depositional growth (sublimation)
$R_{\mathrm{ds/eg}}$ = rate of gas loss (production) due to dissolutional growth (evaporation)
R_{hrg} = rate of gas loss (production) due to heterogeneous reactions

All rates are expressed in units of concentration per unit time (e.g., molec. cm^{-3} s^{-1}).

3.3.4 Particle continuity equation

The continuity equation for particles is divided into two subequations. One is for particle number concentration, and the other is for particle volume component concentration. Particles contain anywhere from one to hundreds of components. The volume of each component varies over time due to physical and chemical processes. If the total volume of one particle in a size bin i is denoted by υ_i (cm^3/particle), the **volume** of component q within that particle is $\upsilon_{q,i}$. Thus, $\upsilon_{q,i}$ gives information about a component in a single particle of a given size. A variable giving information about that component summed over all particles of the same size, is more relevant. Such a parameter is **volume concentration** (cubic centimeters of the component per cubic centimeter of air), defined as

$$v_{q,i} = n_i \upsilon_{q,i} \tag{3.57}$$

where n_i is the **number concentration** of particles of size i (particles per cubic centimeter of air). If two of the three variables in (3.57) are predicted numerically, the third can be found from the equation. Typically, volume concentration and number concentration are predicted numerically. They are found from separate continuity equations, because different external sources and sinks affect the number and volume concentrations.

The continuity equation for the number concentration of particles of size i is

$$\frac{\partial n_i}{\partial t} + \nabla \cdot (\mathbf{v} n_i) = (\nabla \cdot \mathbf{K}_h \nabla) n_i + R_{\mathrm{emisn}} + R_{\mathrm{depn}} + R_{\mathrm{sedn}} + R_{\mathrm{washn}} + R_{\mathrm{nucn}} + R_{\mathrm{coagn}} \tag{3.58}$$

where

R_{emisn} = rate of surface or elevated emission
R_{depn} = rate of particle dry deposition to the surface
R_{sedn} = rate of sedimentation to the surface or between layers
R_{washn} = rate of washout to the surface or from one altitude down to another
R_{nucn} = rate of production of new particles due to homogeneous nucleation
R_{coagn} = rate of coagulation of number concentration

All rates are in units of particles cm^{-3} s^{-1}. Sources and sinks that affect particle number concentration include emission, dry deposition, sedimentation, washout, homogeneous nucleation, and coagulation. **Sedimentation** occurs when particles fall through the atmosphere due to their mass. Sedimentation by gases is negligible because gas molecules have extremely small masses. **Particle dry deposition** occurs when particles diffuse to or otherwise impact a surface by any transport process. **Particle washout** occurs when rain sweeps particles in its path to lower altitudes or the surface. **Homogeneous nucleation** is a source of new particles. **Heterogeneous nucleation** does not produce new particles but allows growth to proceed on existing particles. **Coagulation** occurs when two particles collide and stick to form a single, larger particle.

The continuity equation for the volume concentration of component q in particles of size i is

$$\frac{\partial v_{q,i}}{\partial t} + \nabla \cdot (\mathbf{v} v_{q,i}) = (\nabla \cdot \mathbf{K}_h \nabla) v_{q,i} + R_{\text{emisv}} + R_{\text{depv}} + R_{\text{sedv}} + R_{\text{washv}} + R_{\text{nucv}} + R_{\text{coagv}} + R_{\text{c/ev}} + R_{\text{dp/sv}} + R_{\text{ds/ev}} + R_{\text{eqv}} + R_{\text{aqv}} + R_{\text{hrv}} \quad (3.59)$$

where

R_{emisv} = rate of surface or elevated emission
R_{depv} = rate of dry deposition to the surface
R_{sedv} = rate of sedimentation to the surface or from one altitude to another
R_{washv} = rate of washout to the surface or from one altitude to another
R_{nucv} = rate of change due to homogeneous or heterogeneous nucleation
R_{coagv} = rate of change due to coagulation
$R_{\text{c/ev}}$ = rate of change due to condensational growth (evaporation)
$R_{\text{dp/sv}}$ = rate of change due to depositional growth (sublimation)
$R_{\text{ds/ev}}$ = rate of change due to dissolutional growth (evaporation)
R_{eqv} = rate of change due to reversible chemical equilibrium reactions
R_{aqv} = rate of change due to irreversible aqueous chemical reactions
R_{hrv} = rate of change due to heterogeneous reactions on particle surfaces

Rates in this equation have units of cubic centimeters of component q per cubic centimeter of air per second. Some processes, such as homogeneous nucleation and coagulation, affect number and volume concentrations. Others, such as heterogeneous nucleation, condensation, deposition, dissolution, heterogeneous reaction, chemical equilibrium and aqueous chemistry affect volume concentration but not number concentration.

3.3.5 Continuity equation for gas, liquid, and solid water

Water in the atmosphere appears as a gas, liquid, or solid. In a model, the total water content is estimated as

$$q_T = q_v + \sum_{i=1}^{N_B} (q_{L,i} + q_{I,i}) \tag{3.60}$$

where N_B is the number of particle size categories (bins), q_v is the specific humidity of water vapor (kilograms per kilogram of moist air), $q_{L,i}$ is the moist-air mass mixing ratio of liquid water in a size bin, and $q_{I,i}$ is the moist-air mass mixing ratio of ice in a size bin. Mass mixing ratios are determined from the continuity equations for water vapor, liquid, and ice,

$$\frac{\partial q_v}{\partial t} + (\mathbf{v} \cdot \nabla) q_v = \frac{1}{\rho_a}(\nabla \rho_a \mathbf{K}_h \nabla) q_v + R_{emisV} + R_{depV} + R_{chemV} + R_{c/eV} + R_{dp/sV} \tag{3.61}$$

$$\frac{\partial q_{L,i}}{\partial t} + (\mathbf{v} \cdot \nabla) q_{L,i} = \frac{1}{\rho_a}(\nabla \rho_a \mathbf{K}_h \nabla) q_{L,i} + R_{emisL} + R_{depL} + R_{sedL} + R_{coagL} + R_{c/eL} + R_{f/mL} \tag{3.62}$$

$$\frac{\partial q_{I,i}}{\partial t} + (\mathbf{v} \cdot \nabla) q_{I,i} = \frac{1}{\rho_a}(\nabla \rho_a \mathbf{K}_h \nabla) q_{I,i} + R_{depI} + R_{sedI} + R_{coagI} + R_{f/mI} + R_{dp/sI} \tag{3.63}$$

where

R_{emis} = rate of surface or elevated emission
R_{dep} = rate of dry deposition to the surface
R_{sed} = rate of sedimentation to the surface or from one altitude to another
R_{chem} = rate of photochemical production or loss
R_{coag} = rate of liquid or ice production or loss in a size bin due to coagulation
$R_{c/e}$ = rate of change due to condensational growth (evaporation)
$R_{dp/s}$ = rate of change due to depositional growth (sublimation)
$R_{f/m}$ = rate of change due to freezing (melting)

and the units of R are kilograms per kilogram of moist air per second.

Many meteorological models simulate liquid water and ice as **bulk parameters**. In such cases, liquid water and ice are not separated into size categories, and their number concentrations are not tracked. Instead, only the moist-air mass mixing ratios of total liquid water and ice are predicted. Since particles are not size-resolved in a bulk parameterization, some processes, such as coagulation, cannot be treated adequately. When liquid and ice content are treated as bulk parameters,

$q_T = q_v + q_L + q_I$, where the subscript i has been dropped because a bulk parameterization has no size resolution.

3.4 THERMODYNAMIC ENERGY EQUATION

Air **temperature** is affected by energy transfer and work. Energy transfer processes include conduction, mechanical turbulence, thermal turbulence, advection, and radiation, all introduced in Section 2.2. Energy is released to the air during condensation of water vapor, deposition of water vapor, freezing of liquid water, exothermic chemical reactions, and radioactive decay. Energy is removed from the air upon melting of ice, sublimation of ice, and evaporation of liquid water. Energy exchange may also occur upon the change of state of substances other than water. Because the quantities of nonwater substances changing state are relatively small, resulting energy exchanges are small. Energy, like air density and species concentrations, is conserved in a system.

An equation describing energy changes in the atmosphere can be derived by combining the first law of thermodynamics with the continuity equation for air. The first law of thermodynamics as expressed in (2.82) was $\mathrm{d}Q \approx c_{p,\mathrm{d}}\,\mathrm{d}T_\mathrm{v} - \alpha_\mathrm{a}\,\mathrm{d}p_\mathrm{a}$. Differentiating this equation with respect to time, substituting $\alpha_\mathrm{a} = 1/\rho_\mathrm{a}$, and rearranging give the **thermodynamic energy equation** as

$$\frac{\mathrm{d}T_\mathrm{v}}{\mathrm{d}t} \approx \frac{1}{c_{p,\mathrm{d}}}\frac{\mathrm{d}Q}{\mathrm{d}t} + \frac{1}{c_{p,\mathrm{d}}\rho_\mathrm{a}}\frac{\mathrm{d}p_\mathrm{a}}{\mathrm{d}t} \tag{3.64}$$

If the thermodynamic energy equation is written in terms of potential virtual temperature, the last term in (3.64) can be eliminated. Differentiating $\theta_\mathrm{v} = T_\mathrm{v}(1000/p_\mathrm{a})^\kappa$ with respect to time give

$$\frac{\mathrm{d}\theta_\mathrm{v}}{\mathrm{d}t} = \frac{\mathrm{d}T_\mathrm{v}}{\mathrm{d}t}\left(\frac{1000}{p_\mathrm{a}}\right)^\kappa + T_\mathrm{v}\kappa\left(\frac{1000}{p_\mathrm{a}}\right)^{\kappa-1}\left(-\frac{1000}{p_\mathrm{a}^2}\right)\frac{\mathrm{d}p_\mathrm{a}}{\mathrm{d}t} = \frac{\theta_\mathrm{v}}{T_\mathrm{v}}\frac{\mathrm{d}T_\mathrm{v}}{\mathrm{d}t} - \frac{\kappa\theta_\mathrm{v}}{p_\mathrm{a}}\frac{\mathrm{d}p_\mathrm{a}}{\mathrm{d}t} \tag{3.65}$$

Substituting (3.65), $\kappa = R'/c_{p,\mathrm{d}}$, and $p_\mathrm{a} = \rho_\mathrm{a} R' T_\mathrm{v}$ into (3.64), and expanding the total derivative with (3.13) give the thermodynamic energy equation in terms of potential virtual temperature as

$$\frac{\mathrm{d}\theta_\mathrm{v}}{\mathrm{d}t} = \frac{\partial\theta_\mathrm{v}}{\partial t} + (\mathbf{v}\cdot\nabla)\theta_\mathrm{v} \approx \frac{\theta_\mathrm{v}}{c_{p,\mathrm{d}}T_\mathrm{v}}\frac{\mathrm{d}Q}{\mathrm{d}t} \tag{3.66}$$

Multiplying all terms in (3.66) by $c_{p,\mathrm{d}}\rho_\mathrm{a}$, multiplying all terms in the continuity equation for air from (3.20) by $c_{p,\mathrm{d}}\theta_\mathrm{v}$, adding the two equations, and compressing give

$$\frac{\partial(c_{p,\mathrm{d}}\rho_\mathrm{a}\theta_\mathrm{v})}{\partial t} + \nabla\cdot(\mathbf{v}c_{p,\mathrm{d}}\rho_\mathrm{a}\theta_\mathrm{v}) \approx \rho_\mathrm{a}\frac{\theta_\mathrm{v}}{T_\mathrm{v}}\frac{\mathrm{d}Q}{\mathrm{d}t} \tag{3.67}$$

Substituting the **energy density** (J m^{-3}), defined as $E = c_{p,\mathrm{d}}\rho_\mathrm{a}\theta_\mathrm{v}$, into (3.67) gives the **continuity equation for energy,**

$$\frac{\partial E}{\partial t} + \nabla \cdot (\mathbf{v}E) \approx \rho_\mathrm{a}\frac{\theta_\mathrm{v}}{T_\mathrm{v}}\frac{\mathrm{d}Q}{\mathrm{d}t} \tag{3.68}$$

This equation is similar to the continuity equations for air mass density or gas number concentration. It states that the time rate of change of energy in a box equals the energy flux in minus the energy flux out plus (minus) external sources (sinks). Replacing N with E in Fig. 3.2 yields a diagram of energy fluxes into and out of a hypothetical grid cell.

In a model, subgrid eddies affect energy transport. To account for such eddies, variables in (3.67) can be decomposed as $\mathbf{v} = \bar{\mathbf{v}} + \mathbf{v}'$, $\rho_\mathrm{a} = \bar{\rho}_\mathrm{a} + \rho'_\mathrm{a}$ and $\theta_\mathrm{v} = \bar{\theta}_\mathrm{v} + \theta'_\mathrm{v}$. Since density perturbations are small ($\rho'_\mathrm{a} \ll \bar{\rho}_\mathrm{a}$), density is approximated as $\rho_\mathrm{a} \approx \bar{\rho}_\mathrm{a}$. Substituting velocity, density, and potential virtual temperature decompositions into (3.67), setting the approximation to an equal sign for simplicity, and taking the time- and grid-volume average of the result yield

$$c_{p,\mathrm{d}}\overline{\left\{\frac{\partial[\bar{\rho}_\mathrm{a}(\bar{\theta}_\mathrm{v} + \theta'_\mathrm{v})]}{\partial t}\right\}} + c_{p,\mathrm{d}}\overline{\nabla \cdot [\bar{\rho}_\mathrm{a}(\bar{\mathbf{v}}\bar{\theta}_\mathrm{v} + \bar{\mathbf{v}}\theta'_\mathrm{v} + \mathbf{v}'\bar{\theta}_\mathrm{v} + \mathbf{v}'\theta'_\mathrm{v})]} = \bar{\rho}_\mathrm{a}\overline{\frac{\theta_\mathrm{v}}{T_\mathrm{v}}\frac{\mathrm{d}Q}{\mathrm{d}t}} \tag{3.69}$$

Eliminating zero-value time and spatial derivatives and unnecessary overbars results in

$$\frac{\partial(\bar{\rho}_\mathrm{a}\bar{\theta}_\mathrm{v})}{\partial t} + \nabla \cdot (\bar{\rho}_\mathrm{a}\bar{\mathbf{v}}\bar{\theta}_\mathrm{v}) + \overline{\nabla \cdot (\bar{\rho}_\mathrm{a}\mathbf{v}'\theta'_\mathrm{v})} = \frac{\bar{\rho}_\mathrm{a}}{c_{p,\mathrm{d}}}\overline{\frac{\theta_\mathrm{v}}{T_\mathrm{v}}\frac{\mathrm{d}Q}{\mathrm{d}t}} \tag{3.70}$$

which expands to

$$\bar{\rho}_\mathrm{a}\left[\frac{\partial\bar{\theta}_\mathrm{v}}{\partial t} + (\bar{\mathbf{v}} \cdot \nabla)\bar{\theta}_\mathrm{v}\right] + \bar{\theta}_\mathrm{v}\left[\frac{\partial\bar{\rho}_\mathrm{a}}{\partial t} + \nabla \cdot (\bar{\mathbf{v}}\bar{\rho}_\mathrm{a})\right] + \nabla \cdot (\bar{\rho}_\mathrm{a}\overline{\mathbf{v}'\theta'_\mathrm{v}}) = \frac{\bar{\rho}_\mathrm{a}}{c_{p,\mathrm{d}}}\overline{\frac{\theta_\mathrm{v}}{T_\mathrm{v}}\frac{\mathrm{d}Q}{\mathrm{d}t}} \tag{3.71}$$

Substituting the continuity equation for air from (3.48) into (3.71) and dividing by $\bar{\rho}_\mathrm{a}$ give the thermodynamic energy equation as

$$\frac{\partial\bar{\theta}_\mathrm{v}}{\partial t} + (\bar{\mathbf{v}} \cdot \nabla)\bar{\theta}_\mathrm{v} + \frac{1}{\bar{\rho}_\mathrm{a}}\nabla \cdot (\bar{\rho}_\mathrm{a}\overline{\mathbf{v}'\theta'_\mathrm{v}}) = \overline{\frac{\theta_\mathrm{v}}{c_{p,\mathrm{d}}T_\mathrm{v}}\frac{\mathrm{d}Q}{\mathrm{d}t}} \tag{3.72}$$

The kinematic turbulent sensible-heat fluxes ($\overline{\mathbf{v}'\theta'_\mathrm{v}}$) can be parameterized with

$$\overline{u'\theta'_\mathrm{v}} = -K_{h,xx}\frac{\partial\bar{\theta}_\mathrm{v}}{\partial x} \qquad \overline{v'\theta'_\mathrm{v}} = -K_{h,yy}\frac{\partial\bar{\theta}_\mathrm{v}}{\partial y} \qquad \overline{w'\theta'_\mathrm{v}} = -K_{h,zz}\frac{\partial\bar{\theta}_\mathrm{v}}{\partial z} \tag{3.73}$$

where the eddy diffusion coefficients for energy are the same as those used in the continuity equation for a trace species. Substituting $\overline{\mathbf{v}'\theta'_\mathrm{v}} = -\mathbf{K}_h\nabla\bar{\theta}_\mathrm{v}$ into (3.72) and eliminating overbars for simplicity give

$$\frac{\partial\theta_\mathrm{v}}{\partial t} + (\mathbf{v} \cdot \nabla)\theta_\mathrm{v} = \frac{1}{\rho_\mathrm{a}}(\nabla \cdot \rho_\mathrm{a}\mathbf{K}_h\nabla)\theta_\mathrm{v} + \frac{\theta_\mathrm{v}}{c_{p,\mathrm{d}}T_\mathrm{v}}\frac{\mathrm{d}Q}{\mathrm{d}t} \tag{3.74}$$

The diabatic heating rate consists of the terms

$$\frac{dQ}{dt} = \sum_{n=1}^{N_{e,h}} \frac{dQ_n}{dt} = \frac{dQ_{c/e}}{dt} + \frac{dQ_{f/m}}{dt} + \frac{dQ_{dp/s}}{dt} + \frac{dQ_{solar}}{dt} + \frac{dQ_{ir}}{dt} \tag{3.75}$$

where $N_{e,h}$ is the number of diabatic energy sources and sinks. All Q's are in joules per kilogram. $dQ_{c/e}/dt$ is the rate of energy release (absorption) due to condensation (evaporation), $dQ_{f/m}/dt$ is the rate of energy release (absorption) due to freezing (melting), $dQ_{dp/s}/dt$ is the rate of energy release (absorption) due to deposition (sublimation), dQ_{solar}/dt is the rate of solar heating, and dQ_{ir}/dt is the rate of net infrared heating (cooling). Substituting (3.75) into (3.74) gives the **thermodynamic energy equation** as

$$\frac{\partial \theta_v}{\partial t} + (\mathbf{v} \cdot \nabla)\theta_v = \frac{1}{\rho_a}(\nabla \cdot \rho_a \mathbf{K}_h \nabla)\theta_v + \frac{\theta_v}{c_{p,d} T_v} \sum_{n=1}^{N_{e,h}} \frac{dQ_n}{dt} \tag{3.76}$$

3.5 SUMMARY

In this chapter, local and total derivatives were defined, and the continuity and thermodynamic energy equations were derived. Continuity equations included those for air, trace gases, aerosol number concentration, and aerosol volume concentration. These equations treat subgrid eddy motions with kinematic turbulent flux terms, which are generally parameterized. A common type of parameterization is a *K*-theory parameterization. Equations described in this chapter are necessary for simulating the transport and transformations of total air, gases, aerosol particles, and energy. An equation used for predicting wind speed and direction, the momentum equation, is discussed next.

3.6 PROBLEMS

3.1 Expand the total derivative of the u-scalar velocity (i.e., substitute u for N in (3.5)) when the air flow is (a) steady, (b) unsteady.

3.2 Explain why (3.17) differs from (3.26).

3.3 What is the purpose of Reynolds averaging?

3.4 If $u = -5$ m s^{-1} and $v = +5$ m s^{-1}, write out the horizontal velocity vector, determine the horizontal wind speed, and name the wind.

3.5 Assume that a grid cell has dimension $\Delta x = 5$ km, $\Delta y = 4$ km, and $\Delta z = 0.1$ km and that the west, east, south, north, and lower scalar velocities are $u_1 = +3$, $u_2 = +4$, $v_3 = -3$, $v_4 = +2$, and $w_5 = +0.2$ m s^{-1}. If the atmosphere is incompressible, what is w at the top of the cell?

3.6 **(a)** A grid cell has dimensions $\Delta x = 5$ km, $\Delta y = 4$ km, and $\Delta z = 0.1$ km. Assume the gas concentration and scalar velocity at the west boundary of the cell are $N_1 = 1\times10^{11}$ molec. cm^{-3} and $u_1 = +7$ m s^{-1}, respectively, and those at the east boundary of the cell are $N_2 = 5 \times 10^{11}$ molec. cm^{-3} and $u_2 = +8$ m s^{-1}, respectively. (i) Assuming external sources and sinks and eddy diffusion are absent, estimate N at the cell center after 60 s if the initial N is an average of the two boundary N-values and boundary parameters remain constant. (ii) Calculate the time after the start at which N at the cell center becomes zero.

(b) Assume that the gas concentration and scalar velocity at the south boundary of the grid cell in part (a) are $N_3 = 1 \times 10^{11}$ molec. cm^{-3} and $v_3 = -2$ m s^{-1}, respectively, and those at the north boundary are $N_4 = 7 \times 10^{11}$ molec. cm^{-3} and $v_4 = +1$ m s^{-1}, respectively. Calculate (i) N at the cell center after 60 s and (ii) the time after the start at which N at the center becomes zero. Assume fluxes operate in four directions, and the initial N at the center of the cell is the average of all four grid-boundary N-values.

(c) Convert the gas number concentrations from part (a) (N_1 and N_2) to moist-air mass mixing ratios, assuming that the gas is ozone, $T_v = 298$ K, and $p_a = 1013$ hPa.

(d) Re-solve parts (a) (i) and (a) (ii) with (3.26) using moist-air mass mixing ratios instead of number concentration units. Assume that the west–east velocity for this question is an average of the grid-cell boundary velocities. Convert the mass mixing ratio from the 60-s case back to number concentration units. How does the result compare with that found in part (a) (i)? If it differs, why does it differ?

3.7 A grid cell has dimensions $\Delta x = 5$ km, $\Delta y = 4$ km, and $\Delta z = 0.1$ km. Assume that the potential virtual temperature, pressure, and scalar velocity at the west boundary of the grid cell are $\theta_{v,1} = 302$ K, $p_{a,1} = 1004$ hPa, and $u_1 = +7$ m s^{-1}, respectively, and those at the east boundary of the grid cell are $\theta_{v,2} = 299$ K, $p_{a,2} = 1008$ hPa, and $u_2 = +8$ m s^{-1}, respectively.

(a) Calculate the virtual temperature and air density at the west and east boundaries of the grid cell.

(b) Calculate the energy density E at each boundary.

(c) Assuming diabatic energy sources and sinks and eddy diffusion are absent, calculate the potential virtual temperature at the center of the grid cell after 10 s.

3.7 COMPUTER PROGRAMMING PRACTICE

3.8 Assume that grid-cell size, boundary conditions, and N are initially the same as in Problem 3.6(a). Write a computer script to calculate the final N at the grid-cell center after a time step h. After each time step, set the east-boundary gas concentration (N_2) equal to the final concentration at the center of the cell. Set $h = 3$ s, and run the program for a simulation period of one hour. Plot the grid center concentration versus time.

3.9 Assume that grid-cell size, boundary conditions, and initial θ_v are the same as in Problem 3.7. Write a computer script to calculate the final θ_v at the cell center after a time step h. After each step, set $\theta_{v,2}$ equal to θ_v at the cell center. Set $h = 3$ s, and run the program for six hours. Plot θ_v versus time at the cell center.

4

The momentum equation in Cartesian and spherical coordinates

THE **momentum equation** (equation of motion) describes the movement of air. In a model, it is used to predict wind velocity (speed and direction). In this chapter, the momentum equation and terms within it are derived. These terms include local acceleration, the Earth's centripetal acceleration (apparent centrifugal force), the Coriolis acceleration (apparent Coriolis force), the gravitational force, the pressure-gradient force, the viscous force, and turbulent-flux divergence. The equation is derived for Cartesian and spherical horizontal coordinate systems. Cartesian coordinates are often used over microscale and mesoscale domains, where the Earth's curvature may be neglected. Spherical coordinates are used over global-, synoptic-, and many mesoscale, domains, where curvature cannot be neglected. Equations for the geostrophic wind, gradient wind, and surface wind are derived from the momentum equation. Atmospheric waves are also discussed. Important waves include acoustic, Lamb, gravity, inertia Lamb, inertia gravity, and Rossby waves.

4.1 HORIZONTAL COORDINATE SYSTEMS

Atmospheric modeling equations can be derived for a variety of horizontal coordinate systems. In this section, some of these systems are briefly discussed. Equations for the conversion of variables from Cartesian to spherical horizontal coordinates are then given.

4.1.1 Cartesian, spherical, and other coordinate systems

Many atmospheric models use Cartesian or spherical horizontal coordinates. **Cartesian** (rectangular) coordinates are used on the microscale and mesoscale to simulate flow, for example, in street canyons, downwind of smokestacks, in cities, and in clouds. Over short distances (<500 km), the Earth's curvature is relatively small, and Earth's surface is often divided into rectangles for modeling. Over long distances, curvature prevents the accurate division of the Earth's surface into a contiguous set of rectangles. Nevertheless, over such distances, it is possible to envelop the Earth with many rectangular meshes, each with a different origin and finite overall length and width, and where each mesh partly overlaps other meshes. This is the idea behind the **universal transverse Mercator** (UTM) coordinate system, which is a type of Cartesian coordinate system. In the UTM system, separate, overlapping meshes of rectangular grid cells are superimposed over the globe. UTM coordinate

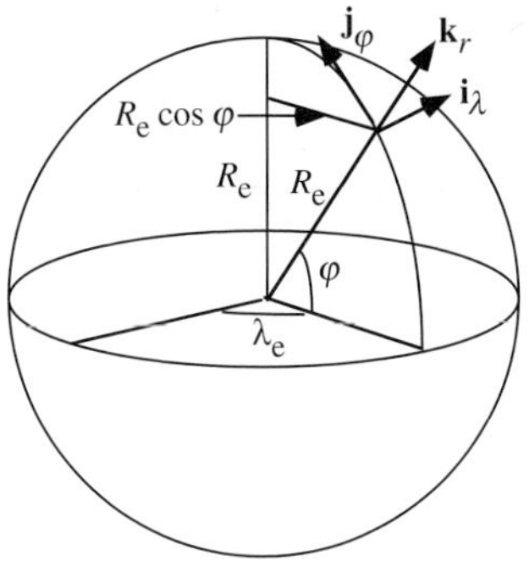

Figure 4.1 Spherical coordinate symbols. R_e is the Earth's radius, φ is latitude, λ_e is longitude, and $\mathbf{i}_\lambda$, $\mathbf{j}_\varphi$, and $\mathbf{k}_r$ are west–east, south–north, and vertical unit vectors, respectively.

locations are mapped back to spherical coordinate locations with UTM-to-spherical conversion equations (U.S. Department of the Army 1958).

For model simulations on small or large scales, the use of **spherical coordinates** (Fig. 4.1) is more natural than the use of Cartesian coordinates. The spherical coordinate system divides the Earth into **longitudes** (meridians), which are south–north lines extending from the South Pole to the North Pole, and **latitudes** (parallels), which are west–east lines parallel to each other extending around the globe. The **Prime Meridian**, which runs through Greenwich, United Kingdom, is defined to have longitude 0°. Meridians extend westward to −180° (180W) longitude and eastward to +180° (180E) longitude. The **Equator** is defined to have latitude 0°. Parallels extend from −90° (90S) latitude to +90° (90N) latitude. On the spherical-coordinate grid, the west–east distance between meridians is the greatest at the Equator and converges to zero at both poles. In fact, all meridians converge to a single point at the poles. Thus, the poles are **singularities**. The presence of a singularity at the poles presents a boundary-condition problem when the spherical coordinate system is used for global atmospheric or ocean (in the case of the North Pole since no ocean exists over the South Pole) simulations. This problem is addressed in Chapter 7. The main advantage of the spherical coordinate system is that it takes into account the Earth's curvature since the Earth is close to spherical.

Some other horizontal map projections are Mercator, stereographic, and Lambert conformal. A **Mercator projection** is one in which each rhumb line on a sphere is represented as a straight line. A **rhumb line** is a curve on the surface of a sphere that cuts all meridians at the same angle. Since the angle can be any angle, a rhumb line can spiral to the poles. A **stereographic projection** is one in which points on a sphere correspond exactly to points on an extended plane and in which the North Pole on the sphere corresponds to infinity on the plane. A **Lambert conformal**

projection is one in which meridians are represented as straight lines converging toward the nearer pole and parallels are represented as arc segments of concentric circles. Although these three projections are used in cartography, they are used less frequently in atmospheric modeling. Snyder (1987) presents equations for converting among these and other coordinate systems.

Spherical and Cartesian coordinate grids are regular grids. In a **regular grid**, grid cells are aligned in a lattice or fixed geometric pattern. Such grid cells do not need to be rectangular. For example, spherical-coordinate grid cells are nonrectangular but are distributed in a fixed pattern. Another type of grid is an irregular grid. In an **irregular grid**, grid cells are not aligned in a lattice or fixed pattern and may have irregular shape and size. Irregular grids are useful for modeling applications in which real boundaries do not match up well with regular-grid boundaries. For example, in ocean modeling, coastlines are uneven boundaries that do not match up well with Cartesian or spherical coordinate boundaries. Irregular grids are also useful for modeling the North and South Poles in a global model to avoid the singularity problem that arises with the spherical coordinate system. Finally, irregular grids are useful for treating some regions in a model at high resolution and others at lower resolution to save computer time.

Regular grids can also be applied to some regions at high resolution and others at low resolution through the use of grid stretching and nesting. **Grid stretching** is the gradual decrease then increase in west–east and/or south–north grid spacing on a spherical-coordinate grid to enable higher resolution in some locations. **Nesting** is the placement of a fine-resolution grid within a coarse-resolution grid that provides boundary conditions to the fine-resolution grid. Nesting is discussed in Chapter 21.

Flows over irregular grids are generally solved with **finite-element methods** (Section 6.5) or **finite-volume methods** (Section 6.6). **Finite-difference methods** (Section 6.4) are challenging (but not impossible) to implement over irregular grids. Flows over regular grids are generally solved with finite-difference methods although the finite-element and finite-volume methods are often used as well. Celia and Gray (1992) describe finite-element formulations on an irregular grid. Durran (1999) describes finite-element and finite-volume formulations on an irregular grid. In this text, the formulation of the equations of atmospheric dynamics is limited to regular grids.

4.1.2 Conversion from Cartesian to spherical coordinates

Figure 4.1 shows the primary components of the spherical coordinate system on a spherical Earth. Although the Earth is an oblate spheroid, slightly bulging at the Equator, the difference between the equatorial and polar radii is small enough (21 km) that the Earth can be considered to be a sphere for modeling purposes. The spherical-coordinate unit vectors for the Earth are $\mathbf{i}_\lambda$, $\mathbf{j}_\varphi$ and $\mathbf{k}_r$, which are west–east, south–north, and vertical unit vectors, respectively. Because the Earth's surface is curved, spherical-coordinate unit vectors have a different orientation at each horizontal location on a sphere. On a Cartesian grid, **i**, **j**, and **k** are oriented in the same direction everywhere on the grid.

In spherical coordinates, west–east and south–north distances are measured in terms of changes in **longitude** (λ_e) and **latitude** (φ), respectively. The vertical coordinate for now is the altitude (z) coordinate. It will be converted in Chapter 5 to the pressure (p), sigma–pressure (σ–p), and sigma–altitude (σ–z) coordinates.

Conversions between increments of distance in Cartesian coordinates and increments of longitude or latitude in spherical coordinates, along the surface of Earth, are obtained from the equation for arc length around a circle. In the west–east and south–north directions, these conversions are

$$\mathrm{d}x = (R_e \cos\varphi)\mathrm{d}\lambda_e \qquad \mathrm{d}y = R_e\,\mathrm{d}\varphi \tag{4.1}$$

respectively, where $R_e \approx 6371$ km is the radius of the Earth, $\mathrm{d}\lambda_e$ is a west–east longitude increment (radians), $\mathrm{d}\varphi$ is a south–north latitude increment (radians), and $R_e \cos\varphi$ is the distance from the Earth's axis of rotation to the surface of the Earth at latitude φ, as shown in Fig. 4.1. Since $\cos\varphi$ is maximum at the Equator (where $\varphi = 0$) and minimum at the poles, $\mathrm{d}x$ decreases from Equator to pole when $\mathrm{d}\lambda_e$ is constant.

Example 4.1

If a grid cell has dimensions $\mathrm{d}\lambda_e = 5^\circ$ and $\mathrm{d}\varphi = 5^\circ$, centered at $\varphi = 30^\circ$ N latitude, find $\mathrm{d}x$ and $\mathrm{d}y$ at the grid cell latitudinal center.

SOLUTION

First, $\mathrm{d}\lambda_e = \mathrm{d}\varphi = 5^\circ \times \pi/180^\circ = 0.0873$ radians. Substituting these values into (4.1) gives $\mathrm{d}x = (6371)(0.866)(0.0873) = 482$ km and $\mathrm{d}y = (6371)(0.0873) = 556$ km.

The **local velocity vector** and **local horizontal velocity vector** in spherical coordinates are

$$\mathbf{v} = \mathbf{i}_\lambda u + \mathbf{j}_\varphi v + \mathbf{k}_r w \qquad \mathbf{v}_h = \mathbf{i}_\lambda u + \mathbf{j}_\varphi v \tag{4.2}$$

respectively. When spherical coordinates are used, **horizontal scalar velocities** can be redefined by substituting (4.1) into (3.2), giving

$$u = \frac{\mathrm{d}x}{\mathrm{d}t} = R_e \cos\varphi \frac{\mathrm{d}\lambda_e}{\mathrm{d}t} \qquad v = \frac{\mathrm{d}y}{\mathrm{d}t} = R_e \frac{\mathrm{d}\varphi}{\mathrm{d}t} \qquad w = \frac{\mathrm{d}z}{\mathrm{d}t} \tag{4.3}$$

The third term in (4.3) is the altitude-coordinate vertical scalar velocity, which is the same on a spherical horizontal grid as on a Cartesian horizontal grid.

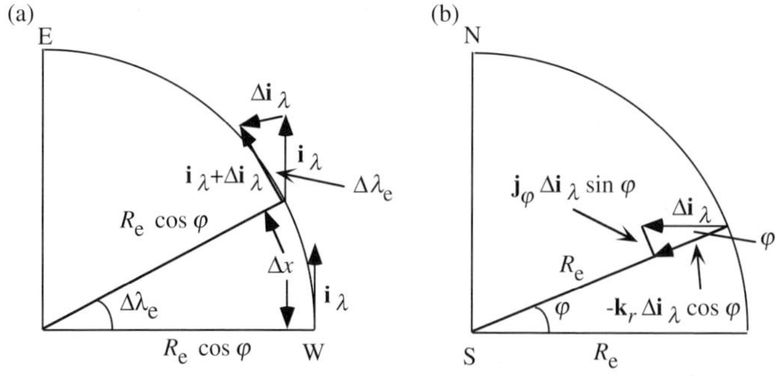

Figure 4.2 (a) Polar and (b) equatorial views of the Earth, showing unit vectors used to determine $\partial \mathbf{i}_\lambda / \partial \lambda_e$. Adapted from Holton (1992).

The **gradient operator in spherical-altitude coordinates** is

$$\nabla = \mathbf{i}_\lambda \frac{1}{R_e \cos\varphi} \frac{\partial}{\partial \lambda_e} + \mathbf{j}_\varphi \frac{1}{R_e} \frac{\partial}{\partial \varphi} + \mathbf{k}_r \frac{\partial}{\partial z} \tag{4.4}$$

which is found by substituting (4.1) into (3.6) and replacing Cartesian- with spherical-coordinate unit vectors. In spherical coordinates, the dot product of the gradient operator with a scalar can be written from (4.4) with no more than three terms. However, the dot product of the gradient operator with a vector requires more than three terms because unit vectors change orientation at different locations on a sphere. For example, expanding $\nabla \cdot \mathbf{v}$ in spherical coordinates gives

$$\begin{aligned}
\nabla \cdot \mathbf{v} &= \left(\mathbf{i}_\lambda \frac{1}{R_e \cos\varphi} \frac{\partial}{\partial \lambda_e} + \mathbf{j}_\varphi \frac{1}{R_e} \frac{\partial}{\partial \varphi} + \mathbf{k}_r \frac{\partial}{\partial z} \right) \cdot (\mathbf{i}_\lambda u + \mathbf{j}_\varphi v + \mathbf{k}_r w) \\
&= \left(\frac{1}{R_e \cos\varphi} \frac{\partial u}{\partial \lambda_e} + \mathbf{i}_\lambda u \frac{1}{R_e \cos\varphi} \frac{\partial \mathbf{i}_\lambda}{\partial \lambda_e} + \mathbf{i}_\lambda v \frac{1}{R_e \cos\varphi} \frac{\partial \mathbf{j}_\varphi}{\partial \lambda_e} + \mathbf{i}_\lambda w \frac{1}{R_e \cos\varphi} \frac{\partial \mathbf{k}_r}{\partial \lambda_e} \right) \\
&\quad + \left(\frac{1}{R_e} \frac{\partial v}{\partial \varphi} + \mathbf{j}_\varphi u \frac{1}{R_e} \frac{\partial \mathbf{i}_\lambda}{\partial \varphi} + \mathbf{j}_\varphi v \frac{1}{R_e} \frac{\partial \mathbf{j}_\varphi}{\partial \varphi} + \mathbf{j}_\varphi w \frac{1}{R_e} \frac{\partial \mathbf{k}_r}{\partial \varphi} \right) \\
&\quad + \left(\frac{\partial w}{\partial z} + \mathbf{k}_r u \frac{\partial \mathbf{i}_\lambda}{\partial z} + \mathbf{k}_r v \frac{\partial \mathbf{j}_\varphi}{\partial z} + \mathbf{k}_r w \frac{\partial \mathbf{k}_r}{\partial z} \right)
\end{aligned} \tag{4.5}$$

where some terms were eliminated because $\mathbf{i}_\lambda \cdot \mathbf{j}_\varphi = 0$, $\mathbf{i}_\lambda \cdot \mathbf{k}_r = 0$, and $\mathbf{j}_\varphi \cdot \mathbf{k}_r = 0$. Partial derivatives of the unit vectors in (4.5) can be derived graphically. From Fig. 4.2(a) and the equation for arc length around a circle, we have

$$|\Delta \mathbf{i}_\lambda| = |\mathbf{i}_\lambda| \, \Delta\lambda_e = \Delta\lambda_e \tag{4.6}$$

where $|\Delta \mathbf{i}_\lambda|$ is the magnitude of the change in the west–east unit vector per unit change in longitude, $\Delta\lambda_e$. Figure 4.2(b) indicates that, when $\Delta\lambda_e$ is small,

$$\Delta \mathbf{i}_\lambda = \mathbf{j}_\varphi \, |\Delta \mathbf{i}_\lambda| \sin\varphi - \mathbf{k}_r \, |\Delta \mathbf{i}_\lambda| \cos\varphi \tag{4.7}$$

Substituting (4.6) into (4.7), dividing by $\Delta\lambda_e$, and letting $\Delta\mathbf{i}_\lambda \to 0$ and $\Delta\lambda_e \to 0$ give

$$\frac{\partial \mathbf{i}_\lambda}{\partial \lambda_e} \approx \frac{\mathbf{j}_\varphi \Delta\lambda_e \sin\varphi - \mathbf{k}_r \Delta\lambda_e \cos\varphi}{\Delta\lambda_e} \approx \mathbf{j}_\varphi \sin\varphi - \mathbf{k}_r \cos\varphi \tag{4.8}$$

Similar derivations for other derivatives yield

$$\begin{array}{lll} \dfrac{\partial \mathbf{i}_\lambda}{\partial \lambda_e} = \mathbf{j}_\varphi \sin\varphi - \mathbf{k}_r \cos\varphi & \dfrac{\partial \mathbf{i}_\lambda}{\partial \varphi} = 0 & \dfrac{\partial \mathbf{i}_\lambda}{\partial z} = 0 \\ \dfrac{\partial \mathbf{j}_\varphi}{\partial \lambda_e} = -\mathbf{i}_\lambda \sin\varphi & \dfrac{\partial \mathbf{j}_\varphi}{\partial \varphi} = -\mathbf{k}_r & \dfrac{\partial \mathbf{j}_\varphi}{\partial z} = 0 \\ \dfrac{\partial \mathbf{k}_r}{\partial \lambda_e} = \mathbf{i}_\lambda \cos\varphi & \dfrac{\partial \mathbf{k}_r}{\partial \varphi} = \mathbf{j}_\varphi & \dfrac{\partial \mathbf{k}_r}{\partial z} = 0 \end{array} \tag{4.9}$$

Substituting these expressions into (4.5) gives

$$\nabla \cdot \mathbf{v} = \frac{1}{R_e \cos\varphi} \frac{\partial u}{\partial \lambda_e} + \frac{1}{R_e \cos\varphi} \frac{\partial}{\partial \varphi}(v \cos\varphi) + \frac{1}{R_e^2} \frac{\partial}{\partial z}\left(w R_e^2\right) \tag{4.10}$$

which simplifies to

$$\boxed{\nabla \cdot \mathbf{v} = \frac{1}{R_e \cos\varphi} \frac{\partial u}{\partial \lambda_e} + \frac{1}{R_e \cos\varphi} \frac{\partial}{\partial \varphi}(v \cos\varphi) + \frac{\partial w}{\partial z}} \tag{4.11}$$

when R_e is held constant. Since the incremental distance z above the Earth's surface is much smaller (<60 km) for most modeling applications than the radius of Earth (6371 km), the assumption of a constant R_e for use in (4.11) gives only a small error.

4.2 NEWTON'S SECOND LAW OF MOTION

The momentum equation is derived from **Newton's second law of motion,** $F = Ma$, where F is force (N), M is mass (kg), and a is acceleration (m s^{-2}). Newton's second law states that the acceleration of a body due to a force is proportional to the force, inversely proportional to the mass of the body, and in the direction of the force. When applied to the atmosphere, the second law can be written in vector form as

$$\mathbf{a}_i = \frac{1}{M_a} \sum \mathbf{F} \tag{4.12}$$

where $\mathbf{a}_i$ is the **total or inertial acceleration,** which is the rate of change of velocity of a parcel of air in motion relative to a coordinate system fixed in space (outside the Earth–atmosphere system), M_a is the mass of the air parcel, and $\sum \mathbf{F}$ is the sum of the force vectors acting on the parcel. A reference frame at rest or that moves in a straight line at a constant velocity is an **inertial reference frame.** A reference frame that either accelerates or rotates is a **noninertial reference frame.** A spaceship accelerating or a car at rest, at constant velocity, or accelerating on a rotating. Earth

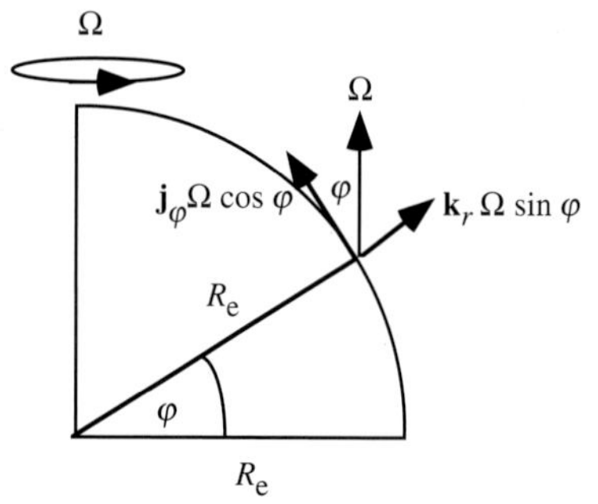

Figure 4.3 Components of the Earth's angular velocity vector.

is in a noninertial reference frame. An observer at a fixed point in space is in an inertial reference frame with respect to any body on a rotating Earth, even if the body is at rest on the surface of Earth.

Inertial acceleration is derived by considering that, to an observer fixed in space, the **absolute velocity** (m s^{-1}) of a body in motion near the surface of the Earth is

$$\mathbf{v}_A = \mathbf{v} + \boldsymbol{\Omega} \times \mathbf{R}_e \tag{4.13}$$

where **v** is the local velocity, defined in (4.2), of the body relative to the Earth's surface, $\boldsymbol{\Omega}$ is the angular velocity vector for Earth, $\mathbf{R}_e$ is the radius vector for the Earth, and $\boldsymbol{\Omega} \times \mathbf{R}_e$ is the rate of change in position of the body due to the Earth's rotation. The Earth's **angular velocity vector** (rad s^{-1}) and **radius vector** (m) are defined as

$$\boldsymbol{\Omega} = \mathbf{j}_\varphi \Omega \cos\varphi + \mathbf{k}_r \Omega \sin\varphi \qquad \mathbf{R}_e = \mathbf{k}_r R_e \tag{4.14}$$

where $\Omega = 2\pi$ rad/86 164 s $= 7.292 \times 10^{-5}$ rad s^{-1} is the magnitude of the angular velocity, and 86 164 is the number of seconds that the Earth takes to make one revolution around its axis (23 h 56 m 4 s). The angular velocity vector acts perpendicular to the equatorial plane of the Earth, as shown in Fig. 4.3. It does not have a west–east component.

Inertial acceleration is defined mathematically as

$$\mathbf{a}_i = \frac{d\mathbf{v}_A}{dt} + \boldsymbol{\Omega} \times \mathbf{v}_A \tag{4.15}$$

Substituting (4.13) into (4.15) and noting that $\boldsymbol{\Omega}$ is independent of time give

$$\mathbf{a}_i = \frac{d\mathbf{v}}{dt} + \boldsymbol{\Omega} \times \frac{d\mathbf{R}_e}{dt} + \boldsymbol{\Omega} \times \mathbf{v} + \boldsymbol{\Omega} \times (\boldsymbol{\Omega} \times \mathbf{R}_e) \tag{4.16}$$

The total derivative of $\mathbf{R}_e$ is

$$\frac{d\mathbf{R}_e}{dt} = R_e \frac{d\mathbf{k}_r}{dt} = \mathbf{i}_\lambda u + \mathbf{j}_\varphi v \approx \mathbf{v} \tag{4.17}$$

where $d\mathbf{k}_r/dt$ is derived shortly in (4.28). Substituting (4.17) into (4.16) yields

$$\mathbf{a}_i = \frac{d\mathbf{v}}{dt} + 2\boldsymbol{\Omega} \times \mathbf{v} + \boldsymbol{\Omega} \times (\boldsymbol{\Omega} \times \mathbf{R}_e) = \mathbf{a}_l + \mathbf{a}_c + \mathbf{a}_r \tag{4.18}$$

where

$$\mathbf{a}_l = \frac{d\mathbf{v}}{dt} \qquad \mathbf{a}_c = 2\boldsymbol{\Omega} \times \mathbf{v} \qquad \mathbf{a}_r = \boldsymbol{\Omega} \times (\boldsymbol{\Omega} \times \mathbf{R}_e) \tag{4.19}$$

are the local, Coriolis, and Earth's centripetal accelerations, respectively. **Local acceleration** is the rate of change of velocity of a parcel of air in motion relative to a coordinate system fixed on Earth, **Coriolis acceleration** is the rate of change of velocity of a parcel due to the rotation of a spherical Earth underneath the parcel, and the **Earth's centripetal acceleration** is the inward-directed rate of change of velocity of a parcel due to its motion around the Earth's axis.

When Reynolds decomposition is applied to the precise local acceleration term in (4.19), the term becomes $\mathbf{a}_l = \bar{\mathbf{a}}_l + \mathbf{a}'_l$, where $\bar{\mathbf{a}}_l$ is a mean local acceleration and $\mathbf{a}'_l$ is a perturbation component, called a **turbulent-flux divergence of momentum.** This term accounts for perturbations to the mean flow of wind, such as those caused by **mechanical shear** (mechanical turbulence), **thermal buoyancy** (thermal turbulence), and **atmospheric waves.** For now, the precise local acceleration term is retained. It will be decomposed later in this section.

Whereas the centripetal and Coriolis effects are viewed as accelerations from an inertial frame of reference, they are viewed as apparent forces from a noninertial frame of reference. An **apparent** (or **inertial**) **force** is a fictitious force that appears to exist when an observation is made in a noninertial frame of reference. For example, when a car rounds a curve, a passenger within, who is in a noninertial frame of reference, appears to be pulled outward by a local **apparent centrifugal force,** which is equal and opposite to local centripetal acceleration multiplied by mass. On the other hand, an observer in an inertial frame of reference sees the passenger and car accelerating inward as the car rounds the curve. Similarly, as the Earth rotates, an observer in a noninertial frame of reference, such as on the Earth's surface, views the Earth and atmosphere being pushed away from the Earth's axis of rotation by an apparent centrifugal force. On the other hand, an observer in an inertial frame of reference, such as in space, views the Earth and atmosphere accelerating inward.

The Coriolis effect can also be viewed from different reference frames. In a noninertial frame of reference, moving bodies appear to feel the **Coriolis force** pushing them to the right in the Northern Hemisphere and to the left in the Southern Hemisphere. In an inertial frame of reference, such as in space, rotation of the

Earth underneath a moving body makes the body appear to accelerate toward the right in the Northern Hemisphere or left in the Southern Hemisphere. In sum, the centripetal and Coriolis effects can be treated as either accelerations or apparent forces, depending on the frame of reference considered.

The terms on the right side of (4.12) are real forces. Real forces that affect local acceleration of a parcel of air include the force of gravity (**true gravitational force**), the force arising from spatial pressure gradients (**pressure-gradient force**), and the force arising from air molecules exchanging momentum with each other (**viscous force**). Substituting inertial acceleration terms from (4.18) into (4.12) and expanding the right side give

$$\mathbf{a}_l + \mathbf{a}_c + \mathbf{a}_r = \frac{1}{M_a}(\mathbf{F}_g^* + \mathbf{F}_p + \mathbf{F}_v) \tag{4.20}$$

where $\mathbf{F}_g^*$ represents true gravitational force, $\mathbf{F}_p$ represents the pressure gradient force, and $\mathbf{F}_v$ represents the viscous force. Atmospheric models usually require expressions for local acceleration; thus, the momentum equation is written most conveniently in a reference frame fixed on the surface of the Earth rather than fixed outside the Earth–atmosphere system. In such a case, only local acceleration is treated as an acceleration. The Coriolis acceleration is treated as a Coriolis force per unit mass ($\mathbf{a}_c = \mathbf{F}_c/M_a$), and the Earth's centripetal acceleration is treated as an apparent centrifugal (negative centripetal) force per unit mass ($\mathbf{a}_r = -\mathbf{F}_r/M_a$). Combining these terms with (4.20) gives the **momentum equation from a reference frame fixed on Earth's surface** as

$$\mathbf{a}_l = \frac{1}{M_a}(\mathbf{F}_r - \mathbf{F}_c + \mathbf{F}_g^* + \mathbf{F}_p + \mathbf{F}_v) \tag{4.21}$$

In the following subsections, terms in (4.21) are discussed.

4.2.1 Local acceleration

The **local acceleration**, or the total derivative of velocity, expands to

$$\mathbf{a}_l = \frac{d\mathbf{v}}{dt} = \frac{\partial \mathbf{v}}{\partial t} + (\mathbf{v} \cdot \nabla)\mathbf{v} \tag{4.22}$$

This equation states that the local acceleration along the motion of a parcel equals the local acceleration at a fixed point plus changes in local acceleration due to fluxes of velocity gradients.

In Cartesian-altitude coordinates, the left side of (4.22) expands to

$$\frac{d\mathbf{v}}{dt} = \frac{d(\mathbf{i}u + \mathbf{j}v + \mathbf{k}w)}{dt} = \mathbf{i}\frac{du}{dt} + \mathbf{j}\frac{dv}{dt} + \mathbf{k}\frac{dw}{dt} \tag{4.23}$$

and the right side expands to

$$\frac{\partial \mathbf{v}}{\partial t} + (\mathbf{v} \bullet \nabla)\,\mathbf{v} = \left(\frac{\partial}{\partial t} + u\frac{\partial}{\partial x} + v\frac{\partial}{\partial y} + w\frac{\partial}{\partial z}\right)(\mathbf{i}u + \mathbf{j}v + \mathbf{k}w) \tag{4.24}$$

$$= \mathbf{i}\left(\frac{\partial u}{\partial t} + u\frac{\partial u}{\partial x} + v\frac{\partial u}{\partial y} + w\frac{\partial u}{\partial z}\right) + \mathbf{j}\left(\frac{\partial v}{\partial t} + u\frac{\partial v}{\partial x} + v\frac{\partial v}{\partial y} + w\frac{\partial v}{\partial z}\right)$$

$$+ \mathbf{k}\left(\frac{\partial w}{\partial t} + u\frac{\partial w}{\partial x} + v\frac{\partial w}{\partial y} + w\frac{\partial w}{\partial z}\right)$$

In spherical-altitude coordinates, the left side of (4.22) expands, with the chain rule, to

$$\frac{d\mathbf{v}}{dt} = \frac{d(\mathbf{i}_\lambda u + \mathbf{j}_\varphi v + \mathbf{k}_r w)}{dt} = \left(\mathbf{i}_\lambda\frac{du}{dt} + u\frac{d\mathbf{i}_\lambda}{dt}\right) + \left(\mathbf{j}_\varphi\frac{dv}{dt} + v\frac{d\mathbf{j}_\varphi}{dt}\right) + \left(\mathbf{k}_r\frac{dw}{dt} + w\frac{d\mathbf{k}_r}{dt}\right) \tag{4.25}$$

Time derivatives of the unit vectors are needed to complete this equation. Substituting (4.1) into the total derivative in Cartesian-altitude coordinates from (3.13) gives the **total derivative in spherical-altitude coordinates** as

$$\boxed{\frac{d}{dt} = \frac{\partial}{\partial t} + u\frac{1}{R_e\cos\varphi}\frac{\partial}{\partial\lambda_e} + v\frac{1}{R_e}\frac{\partial}{\partial\varphi} + w\frac{\partial}{\partial z}} \tag{4.26}$$

Applying (4.26) to $\mathbf{i}_\lambda$ yields

$$\frac{d\mathbf{i}_\lambda}{dt} = \frac{\partial\mathbf{i}_\lambda}{\partial t} + u\frac{1}{R_e\cos\varphi}\frac{\partial\mathbf{i}_\lambda}{\partial\lambda_e} + v\frac{1}{R_e}\frac{\partial\mathbf{i}_\lambda}{\partial\varphi} + w\frac{\partial\mathbf{i}_\lambda}{\partial z} \tag{4.27}$$

Since $\mathbf{i}_\lambda$ does not change in time at a given location, $\partial\mathbf{i}_\lambda/\partial t = 0$. Substituting $\partial\mathbf{i}_\lambda/\partial t = 0$ and terms from (4.9) into (4.27) and into like expressions for $d\mathbf{j}_\varphi/dt$ and $d\mathbf{k}_r/dt$ gives

$$\frac{d\mathbf{i}_\lambda}{dt} = \mathbf{j}_\varphi\frac{u\tan\varphi}{R_e} - \mathbf{k}_r\frac{u}{R_e} \tag{4.28}$$

$$\frac{d\mathbf{j}_\varphi}{dt} = -\mathbf{i}_\lambda\frac{u\tan\varphi}{R_e} - \mathbf{k}_r\frac{v}{R_e}$$

$$\frac{d\mathbf{k}_r}{dt} = \mathbf{i}_\lambda\frac{u}{R_e} + \mathbf{j}_\varphi\frac{v}{R_e}$$

Finally, substituting (4.28) into (4.25) results in

$$\frac{d\mathbf{v}}{dt} = \mathbf{i}_\lambda\left(\frac{du}{dt} - \frac{uv\tan\varphi}{R_e} + \frac{uw}{R_e}\right) + \mathbf{j}_\varphi\left(\frac{dv}{dt} + \frac{u^2\tan\varphi}{R_e} + \frac{vw}{R_e}\right) + \mathbf{k}_r\left(\frac{dw}{dt} - \frac{u^2}{R_e} - \frac{v^2}{R_e}\right) \tag{4.29}$$

Example 4.2

If $u = 20$ m s^{-1}, $v = 10$ m s^{-1}, and $w = 0.01$ m s^{-1}, and if du/dt scales as $u/(\Delta x/u)$, estimate the value of each term on the right side of (4.29) at $\varphi =$ 45° N latitude assuming $\Delta x = 500$ km, $\Delta y = 500$ km, and $\Delta z = 10$ km for large-scale motions.

SOLUTION

From the values given,

$$\frac{du}{dt} \approx 8 \times 10^{-4}\,\mathrm{m\,s^{-2}} \qquad \frac{uv\tan\varphi}{R_e} \approx 3.1 \times 10^{-5}\,\mathrm{m\,s^{-2}} \qquad \frac{uw}{R_e} \approx 3.1 \times 10^{-8}\,\mathrm{m\,s^{-2}}$$

$$\frac{dv}{dt} \approx 2 \times 10^{-4}\,\mathrm{m\,s^{-2}} \qquad \frac{u^2\tan\varphi}{R_e} \approx 6.3 \times 10^{-5}\,\mathrm{m\,s^{-2}} \qquad \frac{vw}{R_e} \approx 1.6 \times 10^{-8}\,\mathrm{m\,s^{-2}}$$

$$\frac{dw}{dt} \approx 1 \times 10^{-8}\,\mathrm{m\,s^{-2}} \qquad \frac{u^2}{R_e} \approx 6.3 \times 10^{-5}\,\mathrm{m\,s^{-2}} \qquad \frac{v^2}{R_e} \approx 1.6 \times 10^{-5}\,\mathrm{m\,s^{-2}}$$

uw/R_e and vw/R_e are small for large- and small-scale motions. dw/dt is also small for large-scale motions.

Example 4.2 shows that uw/R_e and vw/R_e are small for large-scale motions and can be removed from (4.29). If these terms are removed, u^2/R_e and v^2/R_e must also be removed from the vertical term to avoid a false addition of energy to the system. Fortunately, these latter terms are small in comparison with the gravitational and pressure-gradient forces per unit mass. Implementing these simplifications in (4.29) gives the **local acceleration in spherical-altitude coordinates** as

$$\frac{d\mathbf{v}}{dt} = \mathbf{i}_\lambda\left(\frac{du}{dt} - \frac{uv\tan\varphi}{R_e}\right) + \mathbf{j}_\varphi\left(\frac{dv}{dt} + \frac{u^2\tan\varphi}{R_e}\right) + \mathbf{k}_r\frac{dw}{dt} \tag{4.30}$$

Expanding the total derivative in (4.30) gives the right side of (4.22) in spherical-altitude coordinates as

$$\begin{aligned}\frac{\partial\mathbf{v}}{\partial t} + (\mathbf{v}\cdot\nabla)\mathbf{v} = {} & \mathbf{i}_\lambda\left(\frac{\partial u}{\partial t} + \frac{u}{R_e\cos\varphi}\frac{\partial u}{\partial\lambda_e} + \frac{v}{R_e}\frac{\partial u}{\partial\varphi} + w\frac{\partial u}{\partial z} - \frac{uv\tan\varphi}{R_e}\right)\\ & + \mathbf{j}_\varphi\left(\frac{\partial v}{\partial t} + \frac{u}{R_e\cos\varphi}\frac{\partial v}{\partial\lambda_e} + \frac{v}{R_e}\frac{\partial v}{\partial\varphi} + w\frac{\partial v}{\partial z} + \frac{u^2\tan\varphi}{R_e}\right)\\ & + \mathbf{k}_r\left(\frac{\partial w}{\partial t} + \frac{u}{R_e\cos\varphi}\frac{\partial w}{\partial\lambda_e} + \frac{v}{R_e}\frac{\partial w}{\partial\varphi} + w\frac{\partial w}{\partial z}\right)\end{aligned} \tag{4.31}$$

In the **horizontal**, local accelerations have magnitude on the order of 10^{-4} m s^{-2}. These accelerations are less important than Coriolis accelerations or than the pressure-gradient force per unit mass, but greater than accelerations due to the viscous force, except adjacent to the ground. In the **vertical**, local accelerations over large horizontal distances are on the order of 10^{-7} m s^{-2} and can be neglected,

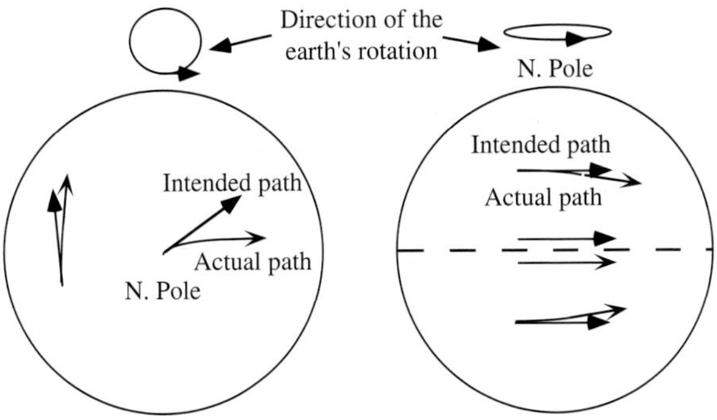

Figure 4.4 Example of Coriolis deflections. The Coriolis force deflects moving bodies to the right in the Northern Hemisphere and to the left in the Southern Hemisphere. The deflection is zero at the Equator. Deflections in the figure are exaggerated.

since gravity and pressure-gradient accelerations are a factor of 10^8 larger. Over small horizontal distances (<3 km), local accelerations in the vertical arc important and cannot be ignored.

4.2.2 Coriolis force

The second term in the momentum equation is the Coriolis force. In a noninertial frame of reference, the Coriolis force appears to push moving bodies to the right in the Northern Hemisphere and to the left in the Southern Hemisphere. In the Northern Hemisphere, it acts 90° to the right of the direction of motion, and in the Southern Hemisphere, it acts 90° to the left of the direction of motion. The Coriolis force is only apparent: no force really acts. Instead, the rotation of a spherical Earth below a moving body makes the body accelerate to the right in the Northern Hemisphere or left in the Southern Hemisphere when viewed from an inertial frame of reference, such as from space. The acceleration is zero at the Equator, maximum near the poles, and zero for bodies at rest. Moving bodies include winds, ocean currents, airplanes, and baseballs. Figure 4.4 gives an example of Coriolis deflections.

In terms of an apparent force per unit mass, the Coriolis term in (4.19) expands in spherical-altitude coordinates to

$$\frac{\mathbf{F}_\mathrm{c}}{M_\mathrm{a}} = 2\boldsymbol{\Omega} \times \mathbf{v} = 2\Omega \begin{vmatrix} \mathbf{i}_\lambda & \mathbf{j}_\varphi & \mathbf{k}_r \\ 0 & \cos\varphi & \sin\varphi \\ u & v & w \end{vmatrix}$$
$$= \mathbf{i}_\lambda 2\Omega\,(w\cos\varphi - v\sin\varphi) + \mathbf{j}_\varphi 2\Omega u\sin\varphi - \mathbf{k}_r 2\Omega u\cos\varphi \tag{4.32}$$

If only a **zonal** (west–east) **wind** is considered, (4.32) simplifies to

$$\frac{\mathbf{F}_\mathrm{c}}{M_\mathrm{a}} = 2\boldsymbol{\Omega} \times \mathbf{v} = \mathbf{j}_\varphi 2\Omega u\sin\varphi - \mathbf{k}_r 2\Omega u\cos\varphi \tag{4.33}$$

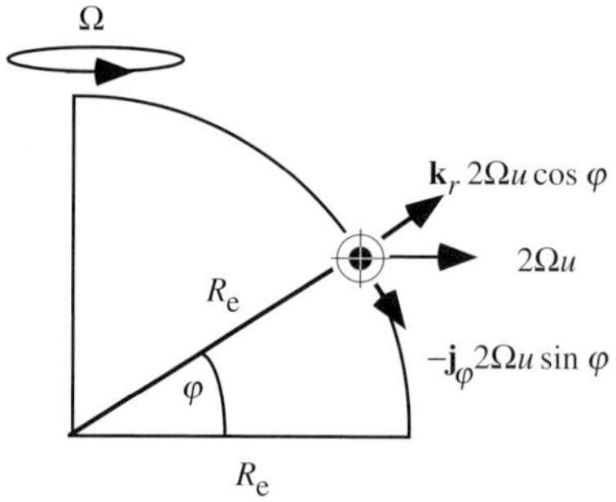

Figure 4.5 Coriolis acceleration components that result when the Coriolis force acts on a west-to-east wind traveling around the Earth, parallel to the Equator, and into the page (denoted by arrow tail). See Example 4.3 for a discussion.

Example 4.3 uses (4.33) to show that moving bodies are deflected to the right in the Northern Hemisphere.

Example 4.3

The fact that the Coriolis effect appears to deflect moving bodies to the right in the Northern Hemisphere can be demonstrated graphically. Consider only local acceleration and the Coriolis force per unit mass in (4.21). In such a case, $\mathbf{a}_l = -\mathbf{F}_c/M_a$. Substituting (4.33) and $\mathbf{a}_l = \mathrm{d}\mathbf{v}/\mathrm{d}t$ from (4.19) into this expression when a west–east wind is present gives

$$\frac{\mathrm{d}\mathbf{v}}{\mathrm{d}t} = -\mathbf{j}_\varphi 2\Omega u \sin\varphi + \mathbf{k}_r 2\Omega u \cos\varphi$$

Figure 4.5 shows the terms on the right side of this equation and the magnitude and direction of the resulting acceleration. The figure shows that the Coriolis effect acts perpendicular to a wind blowing from the west ($+u$), forcing the wind toward the south and vertically. At $\varphi = 0^\circ$ N, only the vertical component of the Coriolis effect remains, and the wind is not turned horizontally. At $\varphi = 90^\circ$ N, only the horizontal component remains, and the wind is not turned vertically.

Because vertical scalar velocities are much smaller than horizontal scalar velocities, $\mathbf{i}_\lambda 2\Omega w \cos\varphi$ may be removed from (4.32). Because the vertical component of the Coriolis force is smaller than other vertical components in the momentum equation (e.g., gravity and pressure-gradient terms), $\mathbf{k}_r 2\Omega u \cos\varphi$ may also be removed. With these changes, the **Coriolis force vector** per unit mass in spherical-altitude coordinates simplifies to

$$\frac{\mathbf{F}_c}{M_a} = 2\boldsymbol{\Omega} \times \mathbf{v} \approx -\mathbf{i}_\lambda 2\Omega v \sin\varphi + \mathbf{j}_\varphi 2\Omega u \sin\varphi \tag{4.34}$$

Defining the **Coriolis parameter** as

$$f = 2\Omega \sin\varphi \tag{4.35}$$

gives another form of the Coriolis term as

$$\frac{\mathbf{F}_c}{M_a} \approx -\mathbf{i}_\lambda f v + \mathbf{j}_\varphi f u = f \begin{vmatrix} \mathbf{i}_\lambda & \mathbf{j}_\varphi & \mathbf{k}_r \\ 0 & 0 & 1 \\ u & v & 0 \end{vmatrix} = f\mathbf{k}_r \times \mathbf{v}_h \tag{4.36}$$

Equation (4.36) can be approximated in Cartesian coordinates by substituting **i**, **j**, and **k** for $\mathbf{i}_\lambda$, $\mathbf{j}_\varphi$, and $\mathbf{k}_r$. The magnitude of (4.36) is $|\mathbf{F}_c|/M_a = f\,|\mathbf{v}_h| = f\sqrt{u^2 + v^2}$, where $|\mathbf{v}_h|$ is the horizontal wind speed (m s^{-1}).

Example 4.4

A mean wind speed of $|\mathbf{v}_h| = 10$ m s^{-1} at the North Pole results in a Coriolis acceleration magnitude of about $|\mathbf{F}_c|/M_a = 0.001454$ m s^{-2}.

4.2.3 Gravitational force

Gravity is a real force that acts on a parcel of air. The gravity that we experience is really a combination of true gravitational force and the Earth's apparent centrifugal force. **True gravitational force** acts toward the center of the Earth. The Earth's **apparent centrifugal force,** which acts away from the axis of rotation of the Earth, slightly displaces the direction and magnitude of the true gravitational force. The sum of the true gravitational and apparent centrifugal force vectors gives an **effective gravitational force vector,** which acts normal to the surface of the Earth but not toward its center.

The Earth's apparent centrifugal force (or centripetal acceleration) arises because the Earth rotates. To an observer fixed in space, objects moving with the surface of a rotating Earth exhibit an inward centripetal acceleration. The object, itself, feels as if it is being pushed outward, during rotation, by an apparent centrifugal force. The force is the greatest at the Equator, where the component of the Earth's angular velocity normal to the Earth's axis of rotation is the greatest, and zero at the poles, where the normal component of the Earth's angular velocity is zero. Over time, the apparent centrifugal force has caused the Earth to bulge at the Equator and compress at the poles. The equatorial radius of Earth is now about 21 km longer than the polar radius, making the Earth an **oblate spheroid.**

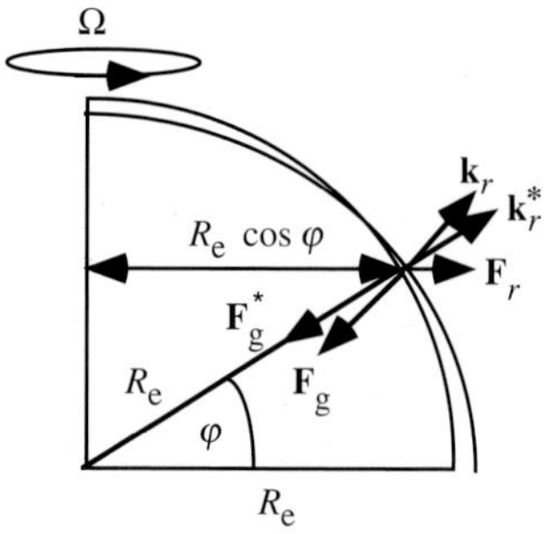

Figure 4.6 Gravitational force components for the Earth. True gravitational force acts toward the center of the Earth, and apparent centrifugal force acts away from its axis of rotation. The effective gravitational force, which is the sum of the true gravitational and apparent centrifugal forces, acts normal to the true surface but not toward the center of the Earth. The apparent centrifugal force (negative centripetal acceleration) has caused the Earth to bulge at the Equator, as shown in the diagram, making the Earth an oblate spheroid. Vectors corresponding to a true sphere are marked with asterisks to distinguish them from those corresponding to the oblate spheroid.

The **true gravitational force vector** per unit mass, which acts toward the center of Earth, is

$$\frac{\mathbf{F}_g^*}{M_a} = -\mathbf{k}_r^* g^* \tag{4.37}$$

where g^* is the **true gravitational acceleration**. The vectors $\mathbf{i}_\lambda^*$, $\mathbf{j}_\varphi^*$, and $\mathbf{k}_r^*$ are unit vectors on a true sphere. The vectors $\mathbf{i}_\lambda$, $\mathbf{j}_\varphi$, and $\mathbf{k}_r$ are unit vectors on the Earth, which is an oblate spheroid. Figure 4.6 shows the orientation of true gravitational force and vertical unit vectors for Earth and for a perfect sphere.

True gravitational acceleration is derived from **Newton's law of universal gravitation**. This law gives the gravitational force vector between two bodies as

$$\mathbf{F}_{12,g} = -\mathbf{r}_{21} \frac{G M_1 M_2}{r_{21}^3} \tag{4.38}$$

where G is the **universal gravitational constant** (6.6720×10^{-11} m^3 kg^{-1} s^{-2}), M_1 and M_2 are the masses of the two bodies, respectively, $\mathbf{F}_{12,\mathrm{g}}$ is the vector force exerted on M_2 by M_1, $\mathbf{r}_{21}$ is the distance vector pointing from body 2 to body 1, r_{21} is the distance between the centers of the two bodies, and the negative sign indicates that the force acts in a direction opposite to that of the vector $\mathbf{r}_{21}$. The magnitude of the gravitational force vector is $F_\mathrm{g} = GM_1M_2/r_{21}^2$.

In the case of the Earth, $\mathbf{F}_{12,\mathrm{g}} = \mathbf{F}_\mathrm{g}^*$, $\mathbf{r}_{21} = \mathbf{R}_\mathrm{e} = \mathbf{k}_r^* R_\mathrm{e}$, $M_1 = M_\mathrm{a}$, $M_2 = M_\mathrm{e}$, and $r_{21} = R_\mathrm{e}$, where M_a is the mass of a parcel of air, and M_e is the mass of the Earth. Substituting these values into (4.38) gives the **true gravitational force vector per unit mass** as

$$\frac{\mathbf{F}_\mathrm{g}^*}{M_\mathrm{a}} = -\mathbf{k}_r^* \frac{GM_\mathrm{e}}{R_\mathrm{e}^2} \tag{4.39}$$

Equating (4.39) with (4.37) and taking the magnitude give

$$\frac{|\mathbf{F}_\mathrm{g}^*|}{M_\mathrm{a}} = g^* = \frac{GM_\mathrm{e}}{R_\mathrm{e}^2} \tag{4.40}$$

Example 4.5

The mass of the Earth is about $M_\mathrm{e} = 5.98 \times 10^{24}$ kg, and the mean radius is $R_\mathrm{e} = 6.37 \times 10^6$ m. Thus, from (4.40), the true gravitational acceleration has a magnitude of about $g^* = 9.833$ m s^{-2}.

For a true sphere, the apparent centrifugal force per unit mass expands to

$$\frac{\mathbf{F}_\mathrm{r}}{M_\mathrm{a}} = -\mathbf{a}_\mathrm{r} = -\boldsymbol{\Omega} \times (\boldsymbol{\Omega} \times \mathbf{R}_\mathrm{e}) = -\Omega \begin{vmatrix} \mathbf{i}_\lambda^* & \mathbf{j}_\varphi^* & \mathbf{k}_r^* \\ 0 & \cos\varphi & \sin\varphi \\ R_\mathrm{e}\cos\varphi & 0 & 0 \end{vmatrix}$$
$$= -\mathbf{j}_\varphi^* R_\mathrm{e}\Omega^2 \cos\varphi \sin\varphi + \mathbf{k}_r^* R_\mathrm{e}\Omega^2 \cos^2\varphi \tag{4.41}$$

where

$$\boldsymbol{\Omega} = \mathbf{j}_\varphi^*\Omega\cos\varphi + \mathbf{k}_r^*\Omega\sin\varphi \qquad \mathbf{R}_\mathrm{e} = \mathbf{k}_r^* R_\mathrm{e} \tag{4.42}$$

are the angular velocity vector and radius vector, respectively, of the Earth as if it were a true sphere, $R_\mathrm{e}\cos\varphi$ is the perpendicular distance between the axis of rotation of Earth and the surface of Earth at latitude φ, as shown in Fig. 4.6, and

$$\boldsymbol{\Omega} \times \mathbf{R}_\mathrm{e} = \Omega \begin{vmatrix} \mathbf{i}^* & \mathbf{j}^* & \mathbf{k}^* \\ 0 & \cos\varphi & \sin\varphi \\ 0 & 0 & R_\mathrm{e} \end{vmatrix} = \mathbf{i}^* R_\mathrm{e}\Omega\cos\varphi \tag{4.43}$$

The magnitude of (4.41) is $|\mathbf{F}_r|/M_a = R_e\Omega^2\cos\varphi$. Adding (4.41) to (4.37) gives the effective gravitational force vector per unit mass on the Earth as

$$\frac{\mathbf{F}_g}{M_a} = \frac{\mathbf{F}_g^*}{M_a} + \frac{\mathbf{F}_r}{M_a} = -\mathbf{j}_\varphi^* R_e\Omega^2\cos\varphi\sin\varphi + \mathbf{k}_r^*(R_e\Omega^2\cos^2\varphi - g^*) = -\mathbf{k}_r g \quad (4.44)$$

where $\mathbf{k}_r$ is the unit vector normal to the oblate spheroid surface of the Earth, and

$$g = [(R_e\Omega^2\cos\varphi\sin\varphi)^2 + (g^* - R_e\Omega^2\cos^2\varphi)^2]^{1/2} \quad (4.45)$$

is the magnitude of the gravitational force per unit mass, or **effective gravitational acceleration** (effective gravity). The effective gravity at sea level varies from $g = 9.799$ m s^{-2} at the Equator to $g = 9.833$ m s^{-2} at the poles. These values are much larger than accelerations due to the Coriolis effect. The effective gravity at the poles equals the true gravitational acceleration there, since the apparent centrifugal acceleration does not act at the poles. Centripetal acceleration affects true gravitational acceleration by about 0.34 percent at the Equator. The difference between the equatorial and polar radii of Earth is about 21 km, or 0.33 percent of an average Earth's radius. Thus, apparent centrifugal force appears to account for the bulging of the Earth at its Equator.

The **globally averaged effective gravity** at the Earth's topographical surface, which averages 231.4 m above sea level, is approximately $g_0 = 9.8060$ m s^{-2}. Since the variation of the effective gravity g with latitude is small, g is often approximated with g_0 in models of the Earth's lower atmosphere.

Figure 2.1(c) and Appendix Table B.1 give the globally averaged effective gravity versus altitude. The figure and data were derived by replacing R_e with $R_e + z$ in (4.40), combining the result with (4.45), and averaging the result globally.

Example 4.6

Both g^* and g vary with altitude in the Earth's atmosphere. Equation (4.40) predicts that, 100 km above the equator, $g^* \approx 9.531$ m s^{-2}, or 3.1 percent lower than its surface value. Equation (4.45) predicts that, at 100 km, $g \approx 9.497$ m s^{-2}, also 3.1 percent lower than its surface value. Thus, the variation of gravity with altitude is more significant than is the variation of gravity due to centripetal acceleration (0.34 percent).

Effective gravity is used to calculate **geopotential**, the work done against gravity to raise a unit mass of air from sea level to a given altitude. Geopotential is a scalar that is a measure of the gravitational potential energy of air per unit mass. The **magnitude of geopotential** (m^2 s^{-2}) is

$$\Phi(z) = \int_0^z g(z)\,\mathrm{d}z \quad (4.46)$$

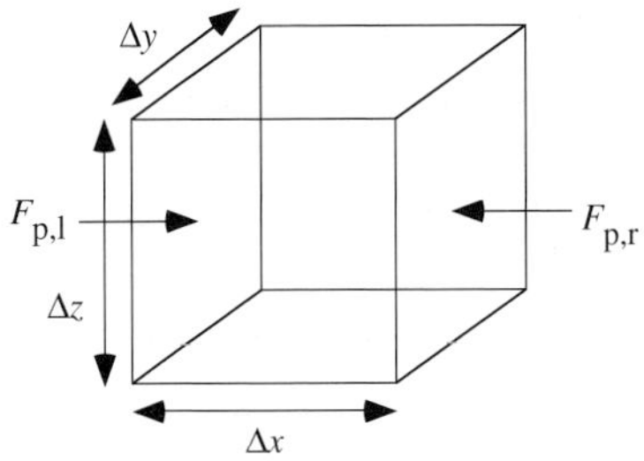

Figure 4.7 Example of pressure-gradient forces acting on both sides of a parcel of air. $F_{p,r}$ is the force acting on the right side, and $F_{p,l}$ is the force acting on the left side.

where $z = 0$ corresponds to sea level and $g(z)$ is gravitational acceleration as a function of height above the Earth's surface (see Appendix Table B.1).

Geopotential height is defined as

$$Z = \frac{\Phi(z)}{g_0} \tag{4.47}$$

Near the Earth's surface, geopotential height approximately equals altitude ($Z \approx z$) since $g(z) \approx g_0$. Geopotential height differs from actual altitude by about 1.55 percent at 100 km. At 25 km, the difference is 0.39 percent. The assumptions $Z \approx z$ and $g(z) \approx g_0 = g$ are often made in models of the lowest 100 km of the atmosphere. Under these assumptions, the magnitude and gradient of geopotential height are

$$\Phi(z) \approx gz \quad \nabla\Phi(z) = \nabla\Phi = \mathbf{k}_r \frac{\partial \Phi(z)}{\partial z} \approx \mathbf{k}_r g \tag{4.48}$$

respectively. Substituting (4.48) into (4.44) gives the **effective gravitational force per unit mass** in spherical coordinates as

$$\frac{\mathbf{F}_g}{M_a} = -\mathbf{k}_r g = -\nabla\Phi \tag{4.49}$$

This equation can be written in Cartesian-altitude coordinates by substituting **k** for $\mathbf{k}_r$.

4.2.4 Pressure-gradient force

The **pressure-gradient force** is a real force that causes air to move from regions of high pressure to regions of low pressure. The force results from pressure differences. Suppose a cubic parcel of air has volume $\Delta x \Delta y \Delta z$, as shown in Fig. 4.7. Suppose

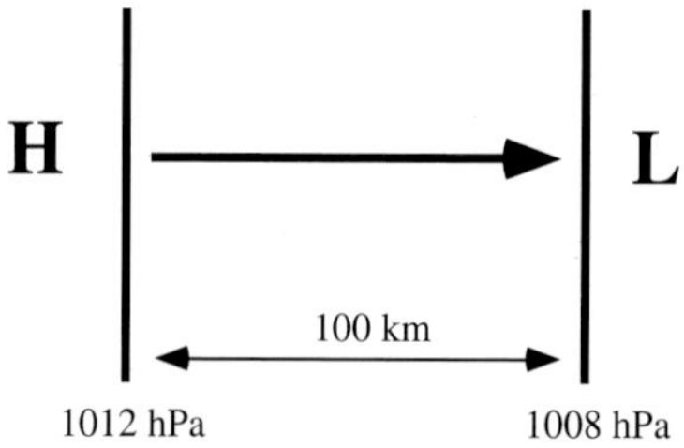

Figure 4.8 Example of a pressure gradient. The difference in pressure over a 100-km distance is 4 hPa. The letters H and L indicate high and low pressure, respectively. The thick arrow indicates the direction of the pressure-gradient force.

also that air pressures on the right and left sides of the parcel impart the forces

$$F_{\mathrm{p,r}} = -\left(p_{\mathrm{c}} + \frac{\partial p}{\partial x}\frac{\Delta x}{2}\right)\Delta y \Delta z \qquad F_{\mathrm{p,l}} = \left(p_{\mathrm{c}} - \frac{\partial p}{\partial x}\frac{\Delta x}{2}\right)\Delta y \Delta z \tag{4.50}$$

respectively, where p_{c} is the pressure at the center of the parcel. Dividing the sum of these forces by the mass of the parcel, $M_{\mathrm{a}} = \rho_{\mathrm{a}}\Delta x \Delta y \Delta z$, and allowing Δx, Δy, and Δz to approach zero give the pressure-gradient force per unit mass in the x-direction as

$$\frac{F_{\mathrm{p},x}}{M_{\mathrm{a}}} = -\frac{1}{\rho_{\mathrm{a}}}\frac{\partial p_{\mathrm{a}}}{\partial x} \tag{4.51}$$

Example 4.7

Figure 4.8 shows two **isobars**, or lines of constant pressure, 100 km apart. The pressure difference between the isobars is 4 hPa. Assuming $\rho_{\mathrm{a}} = 1.2$ kg m^{-3}, the magnitude of the horizontal pressure-gradient force per unit mass is approximately

$$\frac{1}{\rho_{\mathrm{a}}}\frac{\partial p_{\mathrm{a}}}{\partial x} \approx \frac{1}{1.2\,\mathrm{kg\,m^{-3}}}\left(\frac{1012 - 1008\,\mathrm{hPa}}{10^5\,\mathrm{m}}\right)\frac{100\,\mathrm{kg\,m^{-1}\,s^{-2}}}{\mathrm{hPa}} = 0.0033\,\mathrm{m\,s^{-2}}$$

which is much smaller than the force per unit mass due to gravity, but on the same scale as the Coriolis force per unit mass.

The **pressure-gradient force per unit mass** can be generalized for three directions in Cartesian-altitude coordinates with

$$\frac{\mathbf{F}_{\mathrm{p}}}{M_{\mathrm{a}}} = -\frac{1}{\rho_{\mathrm{a}}}\nabla p_{\mathrm{a}} = -\frac{1}{\rho_{\mathrm{a}}}\left(\mathbf{i}\frac{\partial p_{\mathrm{a}}}{\partial x} + \mathbf{j}\frac{\partial p_{\mathrm{a}}}{\partial y} + \mathbf{k}\frac{\partial p_{\mathrm{a}}}{\partial z}\right) \tag{4.52}$$

and in spherical-altitude coordinates with

$$\frac{\mathbf{F}_{\mathrm{P}}}{M_{\mathrm{a}}} = -\frac{1}{\rho_{\mathrm{a}}}\nabla p_{\mathrm{a}} = -\frac{1}{\rho_{\mathrm{a}}}\left(\mathbf{i}_{\lambda}\frac{1}{R_{\mathrm{e}}\cos\varphi}\frac{\partial p_{\mathrm{a}}}{\partial \lambda_{\mathrm{e}}} + \mathbf{j}_{\varphi}\frac{1}{R_{\mathrm{e}}}\frac{\partial p_{\mathrm{a}}}{\partial \varphi} + \mathbf{k}_{r}\frac{\partial p_{\mathrm{a}}}{\partial z}\right) \tag{4.53}$$

Example 4.8

In the vertical, the pressure-gradient force per unit mass is much larger than is that in the horizontal. Pressures at sea level and 100-m altitude are $p_{\mathrm{a}} \approx 1013$ and 1000 hPa, respectively. The average air density at 50 m is $\rho_{\mathrm{a}} \approx 1.2\ \mathrm{kg\,m^{-3}}$. In this case, the pressure-gradient force per unit mass in the vertical is

$$\frac{1}{\rho_{\mathrm{a}}}\frac{\partial p_{\mathrm{a}}}{\partial z} \approx \frac{1}{1.2\,\mathrm{kg\,m^{-3}}}\left(\frac{1013 - 1000\,\mathrm{hPa}}{100\,\mathrm{m}}\right)\frac{100\,\mathrm{kg\,m^{-1}s^{-2}}}{\mathrm{hPa}} = 10.8\,\mathrm{m\,s^{-2}}$$

which is over 3000 times greater than the horizontal pressure-gradient force per unit mass calculated in Example 4.7.

4.2.5 Viscous force

Molecular viscosity is a property of a fluid that increases its resistance to motion. This resistance arises for a different reason in liquids than in gases. In liquids, viscosity is an internal friction that arises when molecules collide with each other and briefly bond, for example by hydrogen bonding, van der Waals forces, or attractive electric charge. Kinetic energy of the molecules is converted to energy required to break the bonds, slowing the flow of the liquid. At high temperatures, bonds between liquid molecules are easier to break, thus viscosity is lower and liquids move more freely than at low temperatures.

In gases, viscosity is the transfer of momentum between colliding molecules. As gas molecules collide, they generally do not bond, so there is little net loss of energy. When a fast molecule collides with a slow molecule, the faster molecule is slowed down, the slower molecule increases in speed, and both molecules are redirected. As a result, viscosity reduces the spread of a plume of gas. As temperatures increase, the viscosity of a gas increases because high temperatures increase the kinetic energy of each gas molecule, thereby increasing the probability that a molecule will collide with and exchange momentum with another gas molecule. Due to the increase in resistance to flow due to enhanced collision and redirection experienced by a gas at high temperature, a flame burning in air, for example, will not rise so high at a high temperature as it will at a low temperature. In sum, the effect of heating on viscosity is opposite for gases than for liquids.

When gas molecules collide with a stationary surface, they impart momentum to the surface, causing a net loss of energy among gas molecules. The dissipation of kinetic energy contained in gas molecules at the ground due to viscosity causes the wind speed at the ground to be zero and to increase logarithmically above the ground. In the absence of molecular viscosity, the wind speed immediately above the ground would be large.

The molecular viscosity of air can be quantified with

$$\eta_a = \frac{5}{16 A d_a^2}\sqrt{\frac{m_a R^* T}{\pi}} \approx 1.8325 \times 10^{-5}\left(\frac{416.16}{T+120}\right)\left(\frac{T}{296.16}\right)^{1.5} \quad (4.54)$$

(kg m^{-1} s^{-1}), called the **dynamic viscosity of air**. The first expression is based on gas kinetic theory and can be extended to any gas, and the second expression is based on experiment and referred to as Sutherland's equation (List 1984). In the equations, m_a is the molecular weight of air (28.966 g mol^{-1}), R^* is the universal gas constant (8314.51 g m^2 s^{-2} mol^{-1} K^{-1}), T is absolute temperature (K), A is Avogadro's number (molec. mol^{-1}), and d_a is the **average diameter of an air molecule** (m). Equating the two expressions at 296.16 K gives $d_a \approx 3.673 \times 10^{-10}$ m. A related parameter is the **kinematic viscosity of air**

$$\nu_a = \frac{\eta_a}{\rho_a} \quad (4.55)$$

(m^2 s^{-1}), which is a molecular diffusion coefficient for air, analogous to the molecular diffusion coefficient for a trace gas.

Viscous interactions among air molecules sliding over each other give rise to a viscous force, which is an **internal force** caused by molecular interactions within a parcel. The change of wind speed with height (i.e., $\partial u/\partial z$) is **wind shear**. As layers of air slide over one another at different speeds due to wind shear, each layer exerts a **viscous stress (shearing stress)**, or force per unit area, on the other. The stress acts parallel to the direction of motion and over a plane normal to the direction of shear. If wind shear in the z-direction exerts a force in the x-direction per unit area of the x–y plane, the resulting shearing stress is

$$\tau_{zx} = \eta_a \frac{\partial u}{\partial z} \quad (4.56)$$

(N m^{-2} or kg m^{-1} s^{-2}), where η_a is the dynamic viscosity of air from (4.54) and is also the ratio of shearing stress to shear. Shearing stress results when momentum is transported down a gradient of velocity, just as gas molecules are transported by molecular diffusion down a gradient of gas concentration. A cubic parcel of air experiences a shearing stress on its top and bottom, as shown in Fig. 4.9.

The net **viscous force** on a parcel of air equals the shearing stress on the top minus the shearing stress on the bottom, multiplied by the area over which the stress acts. The force acts parallel to the direction of motion. If τ_{zx} is the shearing stress in the middle of the parcel, and if $\partial \tau_{zx}/\partial z$ is the vertical gradient of shearing stress, the shearing stresses at the top and bottom, respectively, of the parcel are

$$\tau_{zx,\text{top}} = \tau_{zx} + \frac{\partial \tau_{zx}}{\partial z}\frac{\Delta z}{2} \qquad \tau_{zx,\text{bot}} = \tau_{zx} - \frac{\partial \tau_{zx}}{\partial z}\frac{\Delta z}{2} \quad (4.57)$$

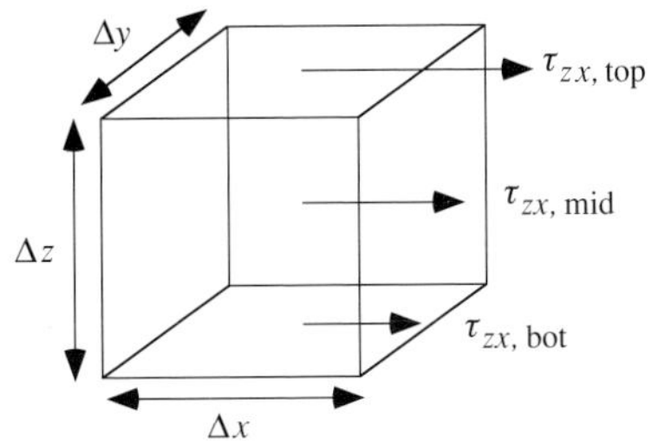

Figure 4.9 Example of shearing stress in the *x*-direction on a volume of air.

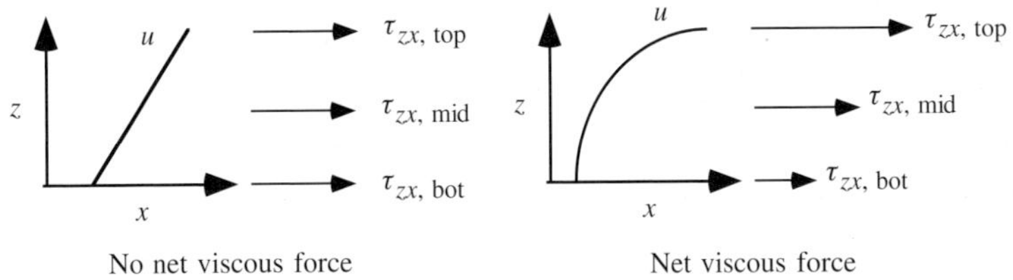

Figure 4.10 A linear vertical wind shear results in a constant shearing stress at all heights and no net viscous force. A nonlinear wind shear results in a change of shearing stress with height and a net viscous force.

Subtracting the shearing stress at the bottom from that at the top of the parcel, multiplying by area, and dividing by air parcel mass, $M_a = \rho_a \Delta x \Delta y \Delta z$, give the net viscous force per unit mass as

$$\frac{F_{v,zx}}{M_a} = (\tau_{zx,\mathrm{top}} - \tau_{zx,\mathrm{bot}})\frac{\Delta x \Delta y}{\rho_a \Delta x \Delta y \Delta z} = \frac{1}{\rho_a}\frac{\partial \tau_{zx}}{\partial z} \tag{4.58}$$

Substituting shearing stress from (4.57) into (4.58), and assuming η_a is invariant with altitude, give

$$\frac{F_{v,zx}}{M_a} = \frac{1}{\rho_a}\frac{\partial}{\partial z}\left(\eta_a \frac{\partial u}{\partial z}\right) \approx \frac{\eta_a}{\rho_a}\frac{\partial^2 u}{\partial z^2} \tag{4.59}$$

This equation suggests that, if wind speed does not change with height or changes linearly with height and η_a is constant, the viscous force per unit mass in the *x*-direction due to shear in the *z*-direction is zero. In such cases, shearing stress is constant with height, and no net viscous force occurs. When wind speed changes nonlinearly with height, the shearing stress at the top of a parcel differs from that at the bottom, and the net viscous force is nonzero. Figure 4.10 illustrates these two cases.

Expanding (4.59) gives the **viscous-force vector** per unit mass as

$$\frac{\mathbf{F}_v}{M_a} = \frac{\eta_a}{\rho_a}\nabla^2 \mathbf{v} = \nu_a \nabla^2 \mathbf{v} \tag{4.60}$$

where ν_a was given in (4.55). The gradient squared term expands in Cartesian-altitude coordinates to

$$\nabla^2 \mathbf{v} = (\nabla \cdot \nabla)\,\mathbf{v} = \mathbf{i}\left(\frac{\partial^2 u}{\partial x^2} + \frac{\partial^2 u}{\partial y^2} + \frac{\partial^2 u}{\partial z^2}\right) + \mathbf{j}\left(\frac{\partial^2 v}{\partial x^2} + \frac{\partial^2 v}{\partial y^2} + \frac{\partial^2 v}{\partial z^2}\right) + \mathbf{k}\left(\frac{\partial^2 w}{\partial x^2} + \frac{\partial^2 w}{\partial y^2} + \frac{\partial^2 w}{\partial z^2}\right) \quad (4.61)$$

Substituting (4.1) and spherical coordinate unit vectors converts (4.61) to spherical-altitude coordinates.

Viscous forces in the atmosphere are small, except adjacent to a surface, where collision of molecules with the surface results in a loss of kinetic energy to the surface. The loss of momentum at the surface causes winds to be zero at the surface but to increase logarithmically immediately above the surface. The relatively large, nonlinear wind-speed variation over a short distance increases the magnitude of the viscous force.

Example 4.9

Viscous forces away from the ground are small. Suppose $u_1 = 10$ m s^{-1} at altitude $z_1 = 1000$ m, $u_2 = 14$ m s^{-1} at $z_2 = 1250$ m, and $u_3 = 20$ m s^{-1} at $z_3 = 1500$ m. If the average temperature and air density at z_2 are $T = 280$ K and $\rho_a = 1.085$ kg m^{-3}, respectively, the dynamic viscosity of air is $\eta_a = 1.753 \times 10^{-5}$ kg m^{-1} s^{-1}. The resulting viscous force per unit mass in the x-direction due to wind shear in the z-direction is approximately

$$\frac{F_{\mathrm{V},zx}}{M_a} \approx \frac{\eta_a}{\rho_a}\frac{1}{(z_3 - z_1)/2}\left(\frac{u_3 - u_2}{z_3 - z_2} - \frac{u_2 - u_1}{z_2 - z_1}\right) = 5.17 \times 10^{-10}\,\mathrm{m\,s^{-2}}$$

which is much smaller than the pressure-gradient force per unit mass from Example 4.7.

Example 4.10

Viscous forces near the ground are often significant. Suppose $u_1 = 0$ m s^{-1} at altitude $z_1 = 0$ m, $u_2 = 0.4$ m s^{-1} at $z_2 = 0.05$ m, and $u_3 = 1$ m s^{-1} at $z_3 = 0.1$ m. If the average temperature and air density at z_2 are $T = 288$ K and $\rho_a = 1.225$ kg m^{-3}, the dynamic viscosity of air is $\eta_a = 1.792 \times 10^{-5}$ kg m^{-1} s^{-1}. The resulting viscous force per unit mass in the x-direction due to wind shear in the z-direction is

$$\frac{F_{\mathrm{V},zx}}{M_a} \approx \frac{\eta_a}{\rho_a}\frac{1}{(z_3 - z_1)/2}\left(\frac{u_3 - u_2}{z_3 - z_2} - \frac{u_2 - u_1}{z_2 - z_1}\right) = 1.17 \times 10^{-3}\,\mathrm{m\,s^{-2}}$$

which is comparable in magnitude with the horizontal pressure-gradient force per unit mass calculated in Example 4.7.

4.2.6 Turbulent-flux divergence

At the ground, wind speeds are zero. In the **surface layer,** which is a 50–300-m-thick region of the atmosphere adjacent to the surface, wind speeds increase logarithmically with increasing height, creating wind shear. Wind shear produces a shearing stress that enhances collisions on a molecular scale. On larger scales, wind shear produces rotating air motions, or eddies. Wind shear can arise when, for example, moving air encounters an obstacle, such as a rock, tree, structure, or mountain. The obstacle slows down the wind beyond it, but the wind above the obstacle is still fast, giving rise to a vertical gradient in wind speed that produces a rotating eddy. Eddies created downwind of obstacles are called **turbulent wakes.** Eddies can range in size from a few millimeters in diameter to the size of the boundary layer (several hundred meters in diameter).

Eddies are created not only by wind shear but also by buoyancy. Heating of surface air causes the air to rise in a thermal, forcing air aloft to sink to replace the rising air, creating a circulation system. **Turbulence** consists of many eddies of different size acting together. Wind shear arising from obstacles is called **mechanical shear,** and the resulting turbulence is called **mechanical turbulence. Thermal turbulence** is turbulence due to eddies of different size arising from buoyancy. Mechanical turbulence is most important in the surface layer, and thermal turbulence is most important in the mixed layer. Thermal turbulence magnifies the effect of mechanical turbulence by enabling eddies to extend to greater heights, increasing their ability to exchange air between the surface and the mixed layer (Stull 1988).

Because eddies are circulations of air, they transfer momentum, energy, gases, and particles vertically and horizontally. For example, eddies transfer fast winds aloft toward the surface, creating gusts near the surface, and slow winds near the surface, aloft. Both effects reduce wind shear. The transfer of momentum by eddies is analogous to the smaller-scale transfer of momentum by molecular viscosity. As such, turbulence due to eddies is often referred to as **eddy viscosity.**

If a numerical model has grid-cell resolution on the order of a few millimeters, it resolves the flow of eddies of all sizes without the need to treat such eddies as subgrid phenomena. Such models, called **direct numerical simulation** (DNS) models (Section 8.4) cannot ignore the viscous force because, at that resolution, the viscous force is important. All other models must parameterize the subgrid eddies to some degree (Section 8.4).

In models that parameterize subgrid-scale eddies, the parameterization can be obtained as follows. Multiplying the precise acceleration of air from (4.22) by ρ_a, multiplying the continuity equation for air from (3.20) by $\mathbf{v}$, and adding the results yield

$$\rho_a \mathbf{a}_l = \rho_a \left[\frac{\partial \mathbf{v}}{\partial t} + (\mathbf{v} \cdot \nabla)\mathbf{v} \right] + \mathbf{v} \left[\frac{\partial \rho_a}{\partial t} + \nabla \cdot (\mathbf{v}\rho_a) \right] \tag{4.62}$$

The variables in this equation are decomposed as $\mathbf{v} = \bar{\mathbf{v}} + \mathbf{v}'$ and $\rho_a = \bar{\rho}_a + \rho_a'$. Because density perturbations are generally small ($\rho_a' \ll \bar{\rho}_a$), the density simplifies to $\rho_a \approx \bar{\rho}_a$. Substituting the decomposed values into (4.62) gives

$$\bar{\rho}_a \mathbf{a}_l = \bar{\rho}_a \left\{ \frac{\partial(\bar{\mathbf{v}}+\mathbf{v}')}{\partial t} + [(\bar{\mathbf{v}}+\mathbf{v}')\cdot\nabla](\bar{\mathbf{v}}+\mathbf{v}') \right\} + (\bar{\mathbf{v}}+\mathbf{v}') \left\{ \frac{\partial \bar{\rho}_a}{\partial t} + \nabla\cdot[(\bar{\mathbf{v}}+\mathbf{v}')\bar{\rho}_a] \right\} \tag{4.63}$$

Taking the time and grid-volume average of this equation, eliminating zero-value terms, and removing unnecessary overbars give

$$\bar{\rho}_a \mathbf{a}_l = \bar{\rho}_a \left[\frac{\partial \bar{\mathbf{v}}}{\partial t} + (\bar{\mathbf{v}}\cdot\nabla)\bar{\mathbf{v}} \right] + \bar{\mathbf{v}} \left[\frac{\partial \bar{\rho}_a}{\partial t} + \nabla\cdot(\bar{\mathbf{v}}\bar{\rho}_a) \right] + \overline{\bar{\rho}_a (\mathbf{v}'\cdot\nabla)\mathbf{v}' + \mathbf{v}'\nabla\cdot(\mathbf{v}'\bar{\rho}_a)} \tag{4.64}$$

Substituting the time- and grid-volume-averaged continuity equation for air from (3.48) into (4.64) and dividing through by $\bar{\rho}_a$ yield $\mathbf{a}_l = \bar{\mathbf{a}}_l + \mathbf{a}'_l$, where

$$\bar{\mathbf{a}}_l = \frac{\partial \bar{\mathbf{v}}}{\partial t} + (\bar{\mathbf{v}}\cdot\nabla)\bar{\mathbf{v}} \qquad \mathbf{a}'_l = \frac{\mathbf{F}_t}{M_a} = \frac{1}{\bar{\rho}_a}\left[\overline{\bar{\rho}_a (\mathbf{v}'\cdot\nabla)\mathbf{v}' + \mathbf{v}'\nabla\cdot(\mathbf{v}'\bar{\rho}_a)} \right] \tag{4.65}$$

The second term is treated as a force per unit mass in the momentum equation, where $\mathbf{F}_t$ is a turbulent-flux divergence vector multiplied by air mass. Since $\mathbf{F}_t$ originates on the left side of (4.12), it must be subtracted from the right side when treated as a force. $\mathbf{F}_t$ accounts for mechanical shear, buoyancy, and other eddy effects, such as waves. Expanding the **turbulent-flux divergence** term in (4.65) in Cartesian-altitude coordinates gives

$$\begin{aligned} \frac{\mathbf{F}_t}{M_a} = & \mathbf{i}\frac{1}{\rho_a}\left[\frac{\partial(\rho_a \overline{u'u'})}{\partial x} + \frac{\partial(\rho_a \overline{v'u'})}{\partial y} + \frac{\partial(\rho_a \overline{w'u'})}{\partial z} \right] \\ & + \mathbf{j}\frac{1}{\rho_a}\left[\frac{\partial(\rho_a \overline{u'v'})}{\partial x} + \frac{\partial(\rho_a \overline{v'v'})}{\partial y} + \frac{\partial(\rho_a \overline{w'v'})}{\partial z} \right] \\ & + \mathbf{k}\frac{1}{\rho_a}\left[\frac{\partial(\rho_a \overline{u'w'})}{\partial x} + \frac{\partial(\rho_a \overline{v'w'})}{\partial y} + \frac{\partial(\rho_a \overline{w'w'})}{\partial z} \right] \end{aligned} \tag{4.66}$$

Averages, such as $\overline{u'u'}$ and $\overline{w'v'}$ (m m s^{-1} s^{-1}), are **kinematic turbulent fluxes of momentum**, since they have units of momentum flux (kg m s^{-1} m^{-2} s^{-1}) divided by air density (kg m^{-3}). Each such average must be parameterized. A parameterization introduced previously and discussed in Chapter 8 is ***K*-theory**. With K-theory, vertical kinematic turbulent fluxes of west–east and south–north momentum are approximated with

$$\overline{w'u'} = -K_{m,zx}\frac{\partial \bar{u}}{\partial z} \qquad \overline{w'v'} = -K_{m,zy}\frac{\partial \bar{v}}{\partial z} \tag{4.67}$$

respectively, where the K_m's are **eddy diffusion coefficients for momentum** (m^2 s^{-1} or cm^2 s^{-1}). In all, nine eddy diffusion coefficients for momentum are needed. Only three were required for energy. Since $\overline{u'v'} = \overline{v'u'}$, $\overline{u'w'} = \overline{w'u'}$, and $\overline{v'w'} = \overline{w'v'}$, the number of coefficients for momentum can be reduced to six. Substituting (4.67) and other like terms into (4.66) and dropping overbars for simplicity give the turbulent-flux divergence as

$$\begin{aligned}\frac{\mathbf{F}_t}{M_a} = &-\mathbf{i}\frac{1}{\rho_a}\left[\frac{\partial}{\partial x}\left(\rho_a K_{m,xx}\frac{\partial u}{\partial x}\right)+\frac{\partial}{\partial y}\left(\rho_a K_{m,yx}\frac{\partial u}{\partial y}\right)+\frac{\partial}{\partial z}\left(\rho_a K_{m,zx}\frac{\partial u}{\partial z}\right)\right]\\ &-\mathbf{j}\frac{1}{\rho_a}\left[\frac{\partial}{\partial x}\left(\rho_a K_{m,xy}\frac{\partial v}{\partial x}\right)+\frac{\partial}{\partial y}\left(\rho_a K_{m,yy}\frac{\partial v}{\partial y}\right)+\frac{\partial}{\partial z}\left(\rho_a K_{m,zy}\frac{\partial v}{\partial z}\right)\right]\\ &-\mathbf{k}\frac{1}{\rho_a}\left[\frac{\partial}{\partial x}\left(\rho_a K_{m,xz}\frac{\partial w}{\partial x}\right)+\frac{\partial}{\partial y}\left(\rho_a K_{m,yz}\frac{\partial w}{\partial y}\right)+\frac{\partial}{\partial z}\left(\rho_a K_{m,zz}\frac{\partial w}{\partial z}\right)\right]\end{aligned} \tag{4.68}$$

This equation can be transformed to spherical-altitude coordinates by substituting (4.1) and spherical-coordinate unit vectors into it. Each term in (4.68) represents an acceleration in one direction due to transport of momentum normal to that direction. For example, the *zx* term is an acceleration in the *x*-direction due to a gradient in wind shear and transport of momentum in the *z*-direction. Kinematic turbulent fluxes are analogous to shearing stresses in that both result when momentum is transported down a gradient of velocity. In the case of viscosity, momentum is transported by molecular diffusion. In the case of turbulence, momentum is transported by eddy diffusion. As with the viscous forces per unit mass in (4.60), the turbulent-flux divergence terms in (4.68) equal zero when the $\rho_a K_m$ terms are constant in space and the wind shear is either zero or changes linearly with distance. Usually, the $\rho_a K_m$ terms vary in space.

The eddy diffusion coefficients in (4.68) can be written in tensor form as

$$\mathbf{K}_m = \begin{bmatrix} K_{m,xx} & 0 & 0 \\ 0 & K_{m,yx} & 0 \\ 0 & 0 & K_{m,zx} \end{bmatrix} \text{for } u, \quad \begin{bmatrix} K_{m,xy} & 0 & 0 \\ 0 & K_{m,yy} & 0 \\ 0 & 0 & K_{m,zy} \end{bmatrix} \text{for } v,$$

$$\begin{bmatrix} K_{m,xz} & 0 & 0 \\ 0 & K_{m,yz} & 0 \\ 0 & 0 & K_{m,zz} \end{bmatrix} \text{for } w \tag{4.69}$$

where a different tensor is used depending on whether the *u*, *v*, or *w* momentum equation is being solved. In vector and tensor notation, (4.68) simplifies to

$$\frac{\mathbf{F}_t}{M_a} = -\frac{1}{\rho_a}(\nabla \cdot \rho_a \mathbf{K}_m \nabla)\mathbf{v} \tag{4.70}$$

where the choice of tensor K_m depends on whether the scalar in vector **v** is *u*, *v*, or *w*.

Example 4.11

What is the west–east acceleration of wind at height 300 m due to the downward transfer of westerly momentum from 400 m under the following conditions: $u_1 = 10$ m s^{-1} at altitude $z_1 = 300$ m, $u_2 = 12$ m s^{-1} at $z_2 = 350$ m, and $u_3 = 15$ m s^{-1} at $z_3 = 400$ m? Assume also that a typical value of K_m in the vertical direction in the middle of the boundary layer is 50 m^2 s^{-1} and that the diffusion coefficient and air density remain constant between 300 and 400 m.

SOLUTION

$$\frac{F_{t,zx}}{M_a} = \frac{1}{\rho_a}\frac{\partial}{\partial z}\left(\rho_a K_{m,zx}\frac{\partial u}{\partial z}\right) \approx \frac{K_{m,zx}}{(z_3 - z_1)/2}\left(\frac{u_3 - u_2}{z_3 - z_2} - \frac{u_2 - u_1}{z_2 - z_1}\right) = 0.02\,\mathrm{m\,s^{-2}}$$

Example 4.12

What is the deceleration of westerly wind in the south due to transfer of momentum to the north under the following conditions: $u_1 = 10$ m s^{-1} at location $y_1 = 0$ m, $u_2 = 9$ m s^{-1} at $y_2 = 500$ m, and $u_3 = 7$ m s^{-1} at $y_3 = 1000$ m? Assume also that a typical value of K_m in the horizontal is about $K_m = 100$ m^2 s^{-1} and that the diffusion coefficient and air density remain constant between y_1 and y_3.

SOLUTION

$$\frac{F_{t,yx}}{M_a} = \frac{1}{\rho_a}\frac{\partial}{\partial y}\left(\rho_a K_{m,yx}\frac{\partial u}{\partial y}\right) \approx \frac{K_{m,yx}}{(y_3 - y_1)/2}\left(\frac{u_3 - u_2}{y_3 - y_2} - \frac{u_2 - u_1}{y_2 - y_1}\right)$$
$$= -0.0004\,\mathrm{m\,s^{-2}}$$

4.2.7 Complete momentum equation

Table 4.1 summarizes the accelerations and forces per unit mass derived above and gives approximate magnitudes of the terms. Noting that $\mathbf{a}_l = \bar{\mathbf{a}}_l + \mathbf{F}_t/M_a$, removing overbars for simplicity, and substituting terms from Table 4.1 into (4.21) give the **vector form of the momentum equation** as

$$\frac{d\mathbf{v}}{dt} = -f\mathbf{k} \times \mathbf{v} - \nabla\Phi - \frac{1}{\rho_a}\nabla p_a + \frac{\eta_a}{\rho_a}\nabla^2\mathbf{v} + \frac{1}{\rho_a}(\nabla \cdot \rho_a \mathbf{K}_m \nabla)\,\mathbf{v} \qquad (4.71)$$

Table 4.1 Terms in the momentum equation and their horizontal and vertical magnitudes

Term	Acceleration or force/mass expression	Horizontal acceleration (m s^{-2})	Vertical acceleration (m s^{-2})
Local acceleration	$\bar{a}_l = \frac{d\bar{v}}{dt} = \frac{\partial \bar{v}}{\partial t} + (\bar{v} \bullet \nabla)\bar{v}$	10^{-4}	[a] 10^{-7}–1
Coriolis force per unit mass	$\frac{F_c}{M_a} = f\mathbf{k} \times \mathbf{v}$	10^{-3}	0
Effective gravitational force per unit mass	$\frac{F_g}{M_a} = \frac{F_g^*}{M_a} + \frac{F_r}{M_a} = -\nabla\Phi$	0	10
Pressure-gradient force per unit mass	$\frac{F_p}{M_a} = -\frac{1}{\rho_a}\nabla p_a$	10^{-3}	10
Viscous force per unit mass	$\frac{F_v}{M_a} = \frac{\eta_a}{\rho_a}\nabla^2 \mathbf{v}$	[b] 10^{-12}–10^{-3}	[b] 10^{-15}–10^{-5}
Turbulent-flux divergence of momentum	$\frac{F_t}{M_a} = -\frac{1}{\rho_a}(\nabla \bullet \rho_a \mathbf{K}_m \nabla)\mathbf{v}$	[c] 0–0.005	[c] 0–1

[a] Low value for large-scale motions, high value for small-scale motions (<3 km).
[b] Low value for free atmosphere, high value for air adjacent to the surface.
[c] Low value for no wind shear, high value for large wind shear.

Table 4.1 shows that some terms in these equations are unimportant, depending on the scale of motion. Three dimensionless parameters – the Ekman number, Rossby number, and Froude number – are used for scale analysis to estimate the importance of different processes. These parameters are

$$\text{Ek} = \frac{\nu_a u/x^2}{uf} \qquad \text{Ro} = \frac{u^2/x}{uf} \qquad \text{Fr}^2 = \frac{w^2/z}{g} \tag{4.72}$$

respectively. The **Ekman number** gives the ratio of the viscous force to the Coriolis force. Above the ground, viscous terms are unimportant relative to Coriolis terms. Thus, the Ekman number is small, allowing the viscous term to be removed from the momentum equation without much loss in accuracy. The **Rossby number** gives the ratio of the local acceleration to the Coriolis force per unit mass. Local accelerations are more important than viscous accelerations, but less important than Coriolis accelerations. Thus, the Rossby number is much larger than the Ekman number, but usually less than unity. The **Froude number** gives the ratio of local acceleration to gravitational acceleration in the vertical. Over large horizontal scales (>3 km), vertical accelerations are small in comparison with gravitational accelerations. In such cases, the vertical acceleration term is often removed from the momentum equation, resulting in the hydrostatic assumption.

Example 4.13

Over large horizontal scales $\nu_a \approx 10^{-6}$ m^2 s^{-1}, $u \approx 10$ m s^{-1}, $x \approx 10^6$ m, $f \approx 10^{-4}$ s^{-1}, $w \approx 0.01$ m s^{-1}, and $z \approx 10^4$ m. The Ekman number under these conditions is Ek $= 10^{-14}$, indicating that viscous forces are small. The Rossby number has a value of about Ro $= 0.1$, indicating that local accelerations are an order of magnitude smaller than Coriolis accelerations. The Froude number is Fr $= 3 \times 10^{-5}$. Thus, local accelerations in the vertical are unimportant over large horizontal scales.

Since air viscosity is negligible for most atmospheric scales, it can be ignored in the momentum equation. Removing viscosity from (4.71) and expanding the equation in **Cartesian-altitude coordinates** give

$$\frac{du}{dt} = \frac{\partial u}{\partial t} + u\frac{\partial u}{\partial x} + v\frac{\partial u}{\partial y} + w\frac{\partial u}{\partial z} = fv - \frac{1}{\rho_a}\frac{\partial p_a}{\partial x} + \frac{1}{\rho_a}\left[\frac{\partial}{\partial x}\left(\rho_a K_{m,xx}\frac{\partial u}{\partial x}\right) + \frac{\partial}{\partial y}\left(\rho_a K_{m,yx}\frac{\partial u}{\partial y}\right) + \frac{\partial}{\partial z}\left(\rho_a K_{m,zx}\frac{\partial u}{\partial z}\right)\right] \tag{4.73}$$

$$\frac{dv}{dt} = \frac{\partial v}{\partial t} + u\frac{\partial v}{\partial x} + v\frac{\partial v}{\partial y} + w\frac{\partial v}{\partial z} = -fu - \frac{1}{\rho_a}\frac{\partial p_a}{\partial y} + \frac{1}{\rho_a}\left[\frac{\partial}{\partial x}\left(\rho_a K_{m,xy}\frac{\partial v}{\partial x}\right) + \frac{\partial}{\partial y}\left(\rho_a K_{m,yy}\frac{\partial v}{\partial y}\right) + \frac{\partial}{\partial z}\left(\rho_a K_{m,zy}\frac{\partial v}{\partial z}\right)\right] \tag{4.74}$$

$$\frac{dw}{dt} = \frac{\partial w}{\partial t} + u\frac{\partial w}{\partial x} + v\frac{\partial w}{\partial y} + w\frac{\partial w}{\partial z} = -g - \frac{1}{\rho_a}\frac{\partial p_a}{\partial z} + \frac{1}{\rho_a}\left[\frac{\partial}{\partial x}\left(\rho_a K_{m,xz}\frac{\partial w}{\partial x}\right) + \frac{\partial}{\partial y}\left(\rho_a K_{m,yz}\frac{\partial w}{\partial y}\right) + \frac{\partial}{\partial z}\left(\rho_a K_{m,zz}\frac{\partial w}{\partial z}\right)\right] \tag{4.75}$$

The momentum equation in spherical-altitude coordinates requires additional terms. When a reference frame fixed on the surface of the Earth (noninertial reference frame) is used, the spherical-altitude coordinate conversion terms from (4.30), $uv \tan\varphi / R_e$ and $u^2 \tan\varphi / R_e$, should be treated as apparent forces and are moved to the right side of the momentum equation. Implementing this change and substituting (4.1) into (4.73)–(4.75) give approximate forms of the directional momentum equations in **spherical-altitude coordinates** as

$$\frac{\partial u}{\partial t}+\frac{u}{R_e\cos\varphi}\frac{\partial u}{\partial\lambda_e}+\frac{v}{R_e}\frac{\partial u}{\partial\varphi}+w\frac{\partial u}{\partial z}
=\frac{uv\tan\varphi}{R_e}+fv-\frac{1}{\rho_a R_e\cos\varphi}\frac{\partial p_a}{\partial\lambda_e}+\frac{1}{\rho_a}\left[\frac{1}{R_e^2\cos\varphi}\frac{\partial}{\partial\lambda_e}\left(\frac{\rho_a K_{m,xx}}{\cos\varphi}\frac{\partial u}{\partial\lambda_e}\right)+\frac{1}{R_e^2}\frac{\partial}{\partial\varphi}\left(\rho_a K_{m,yx}\frac{\partial u}{\partial\varphi}\right)+\frac{\partial}{\partial z}\left(\rho_a K_{m,zx}\frac{\partial u}{\partial z}\right)\right] \quad (4.76)$$

$$\frac{\partial v}{\partial t}+\frac{u}{R_e\cos\varphi}\frac{\partial v}{\partial\lambda_e}+\frac{v}{R_e}\frac{\partial v}{\partial\varphi}+w\frac{\partial v}{\partial z}
=-\frac{u^2\tan\varphi}{R_e}-fu-\frac{1}{\rho_a R_e}\frac{\partial p_a}{\partial\varphi}+\frac{1}{\rho_a}\left[\frac{1}{R_e^2\cos\varphi}\frac{\partial}{\partial\lambda_e}\left(\frac{\rho_a K_{m,xy}}{\cos\varphi}\frac{\partial v}{\partial\lambda_e}\right)+\frac{1}{R_e^2}\frac{\partial}{\partial\varphi}\left(\rho_a K_{m,yy}\frac{\partial v}{\partial\varphi}\right)+\frac{\partial}{\partial z}\left(\rho_a K_{m,zy}\frac{\partial v}{\partial z}\right)\right] \quad (4.77)$$

$$\frac{\partial w}{\partial t}+\frac{u}{R_e\cos\varphi}\frac{\partial w}{\partial\lambda_e}+\frac{v}{R_e}\frac{\partial w}{\partial\varphi}+w\frac{\partial w}{\partial z}
=-g-\frac{1}{\rho_a}\frac{\partial p_a}{\partial z}+\frac{1}{\rho_a}\left[\frac{1}{R_e^2\cos\varphi}\frac{\partial}{\partial\lambda_e}\left(\frac{\rho_a K_{m,xz}}{\cos\varphi}\frac{\partial w}{\partial\lambda_e}\right)+\frac{1}{R_e^2}\frac{\partial}{\partial\varphi}\left(\rho_a K_{m,yz}\frac{\partial w}{\partial\varphi}\right)+\frac{\partial}{\partial z}\left(\rho_a K_{m,zz}\frac{\partial w}{\partial z}\right)\right] \quad (4.78)$$

4.3 APPLICATIONS OF THE MOMENTUM EQUATION

The continuity equation for air, the species continuity equation, the thermodynamic energy equation, the three momentum equations (one for each direction), and the equation of state are referred to here as the **equations of atmospheric dynamics.** Removing the species continuity equation from this list and substituting the hydrostatic equation in place of the full vertical momentum equation give the **primitive equations,** which are a basic form of the Eulerian equations of fluid motion. Many atmospheric motions, including the geostrophic wind, surface winds, the gradient wind, surface winds around high- and low- pressure centers, and atmospheric

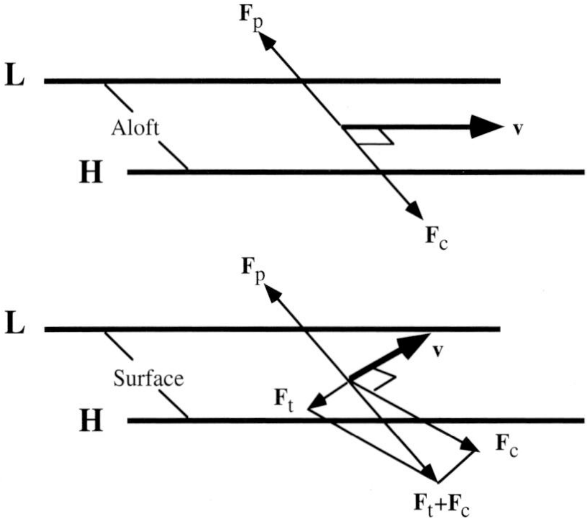

Figure 4.11 Force and wind vectors aloft and at the surface in the Northern Hemisphere. The parallel lines are isobars.

waves, can be understood by looking at simplified forms of the equations of atmospheric dynamics.

4.3.1 Geostrophic wind

The momentum equation can be simplified to isolate air motions, such as the geostrophic wind. The **geostrophic** (Earth-turning) **wind** arises when the Coriolis force exactly balances the pressure-gradient force. Such a balance occurs following a geostrophic adjustment process, described in Fig. 4.19 (Section 4.3.5.3). Many motions in the free troposphere are close to being geostrophic because, in the free troposphere, horizontal local accelerations and turbulent accelerations are usually much smaller than Coriolis and pressure-gradient accelerations. Removing local acceleration and turbulence terms from (4.73) and (4.74), and solving give the **geostrophic scalar velocities** in Cartesian-altitude coordinates as

$$v_g = \frac{1}{f\rho_a}\frac{\partial p_a}{\partial x} \qquad u_g = -\frac{1}{f\rho_a}\frac{\partial p_a}{\partial y} \tag{4.79}$$

In geostrophic equilibrium, the pressure-gradient force is equal in magnitude to and opposite in direction to the Coriolis force. The resulting geostrophic wind flows 90° to the left of the Coriolis force in the Northern Hemisphere, as shown in the top portion of Fig. 4.11. In the Southern Hemisphere, the geostrophic wind flows 90° to the right of the Coriolis force.

Example 4.14

If $\varphi = 30°$ N, $\rho_a = 0.76$ kg m^{-3}, and the pressure gradient is 4 hPa per 150 km in the south–north direction, estimate the west–east geostrophic wind speed.

SOLUTION

From (4.35), $f = 7.292 \times 10^{-5}$ rad s^{-1}. From (4.79), $u_g = 48.1$ m s^{-1}.

In vector form, the geostrophic velocity in Cartesian-altitude coordinates is

$$\mathbf{v}_g = \mathbf{i}u_g + \mathbf{j}v_g = \frac{1}{f\rho_a}\left(-\mathbf{i}\frac{\partial p_a}{\partial y} + \mathbf{j}\frac{\partial p_a}{\partial x}\right) = \frac{1}{f\rho_a}\begin{vmatrix} \mathbf{i} & \mathbf{j} & \mathbf{k} \\ 0 & 0 & 1 \\ \dfrac{\partial p_a}{\partial x} & \dfrac{\partial p_a}{\partial y} & 0 \end{vmatrix}$$

$$= \frac{1}{f\rho_a}\mathbf{k} \times \nabla_z p_a \tag{4.80}$$

where

$$\nabla_z = \left(\mathbf{i}\frac{\partial}{\partial x}\right)_z + \left(\mathbf{j}\frac{\partial}{\partial y}\right)_z = \mathbf{i}\frac{\partial}{\partial x} + \mathbf{j}\frac{\partial}{\partial y} \tag{4.81}$$

is the **horizontal gradient operator in Cartesian-altitude coordinates.** The subscript z indicates that the partial derivative is taken along a surface of constant altitude; thus, $\mathbf{k}\partial/\partial z = 0$. Equation (4.80) indicates that the geostrophic wind flows parallel to lines of constant pressure (isobars).

4.3.2 Surface-layer winds

In steady state, winds in the surface layer are affected primarily by the pressure-gradient force, Coriolis force, mechanical turbulence, and thermal turbulence. Objects protruding from the surface, such as blades of grass, bushes, trees, structures, hills, and mountains, slow winds near the surface. Aloft, fewer obstacles exist, and wind speeds are generally higher than at the surface. In terms of the momentum equation, forces that slow winds near the surface appear in the turbulent-flux divergence vector $\mathbf{F}_t$.

Near the surface and in steady state, the sum of the Coriolis force and turbulent flux vectors balance the pressure-gradient force vector. Aloft, the Coriolis and pressure-gradient force vectors balance each other. Figure 4.11 shows forces and resulting winds in the Northern Hemisphere in both cases. On average, friction near the surface shifts winds about 30° counterclockwise in comparison with winds aloft in the Northern Hemisphere, and 30° clockwise in comparison with winds aloft in the Southern Hemisphere. The variation in wind direction may be 45° or more, depending on the roughness of the surface. In both hemispheres, surface winds are tilted toward low pressure.

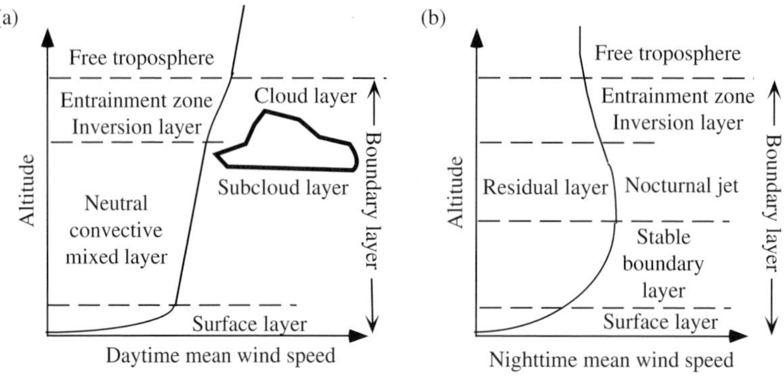

Figure 4.12 Variation of wind speed with height during the (a) day and (b) night in the atmospheric boundary layer. Adapted from Stull (1988).

In the presence of near-surface turbulence, the steady-state horizontal momentum equations simplify in Cartesian-altitude coordinates from (4.73)–(4.74) to

$$-fv = -\frac{1}{\rho_a}\frac{\partial p_a}{\partial x} + \frac{1}{\rho_a}\frac{\partial}{\partial z}\left(\rho_a K_{m,zx}\frac{\partial u}{\partial z}\right)$$

$$fu = -\frac{1}{\rho_a}\frac{\partial p_a}{\partial y} + \frac{1}{\rho_a}\frac{\partial}{\partial z}\left(\rho_a K_{m,zy}\frac{\partial v}{\partial z}\right) \tag{4.82}$$

Figure 4.12 shows idealized variations of wind speed with increasing height in the boundary layer during day and night. During the day, wind speeds increase logarithmically with height in the surface layer. In the mixed layer, temperature and wind speeds are relatively uniform with height. Above the entrainment zone, wind speeds increase to their geostrophic values.

At night, wind speeds near the surface increase logarithmically with increasing height but are lower than during the day. The reduced mixing in the stable boundary layer increases wind speed toward the top of the layer, creating a **nocturnal** or **low-level jet** that is faster than the geostrophic wind. In the residual layer, wind speeds decrease with increasing height and approach the geostrophic wind speed.

Figures 4.13(a) and (b) show measurements of u- and v-scalar velocities at Riverside, California, during a summer night and day, respectively. Riverside lies between the coast and a mountain range. At night, the u- and v-winds were small. In the lower boundary layer, during the day, the u-winds were strong due to a sea breeze. Aloft, flow toward the ocean was strong due to a thermal low inland, which increased pressure and easterly winds aloft inland. The v-component of wind was relatively small.

4.3.3 The gradient wind

When air rotates around a center of low pressure, as in a midlatitude cyclone, tornado, or hurricane, or when it rotates around a center of high pressure, as in an

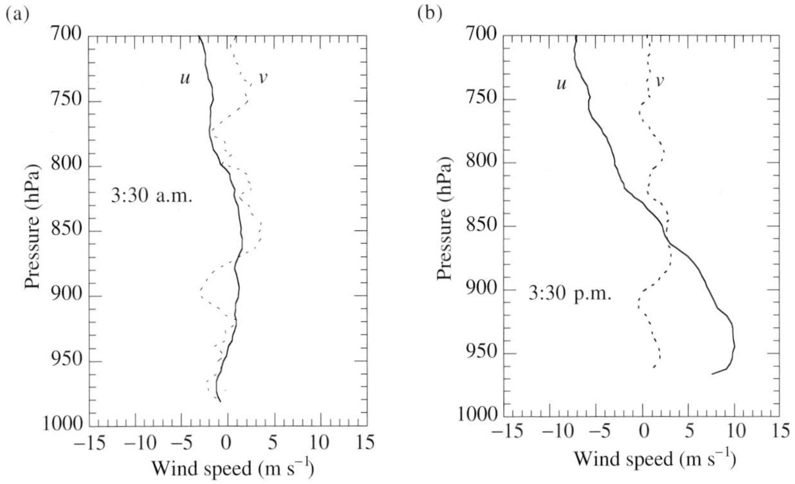

Figure 4.13 Measured variation of u and v scalar velocities with height at (a) 3:30 a.m. and (b) 3:30 p.m. on August 27, 1987, at Riverside, California.

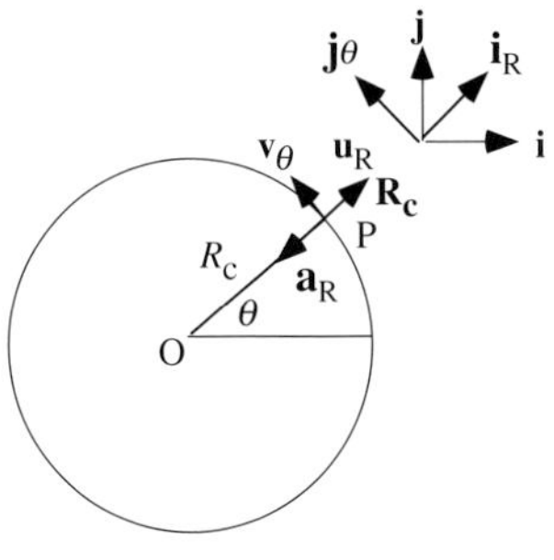

Figure 4.14 Cylindrical coordinate components (**i** and **j** are in Cartesian coordinates).

anticyclone, the horizontal momentum equations are more appropriately written in cylindrical coordinates.

In Cartesian coordinates, distance variables are x, y, and z, and unit vectors are **i**, **j**, and **k**. In cylindrical coordinates, distance variables are R_c, θ, and z, where R_c is the radius of curvature, or distance from point O to P in Fig. 4.14, and θ is the angle (radians) measured counterclockwise from the positive x-axis. Distances x and y in Cartesian coordinates are related to R_c and θ in cylindrical coordinates by

$$x = R_c \cos\theta \qquad y = R_c \sin\theta \tag{4.83}$$

Thus,

$$R_c^2 = x^2 + y^2 \qquad \theta = \tan^{-1}\left(\frac{y}{x}\right) \tag{4.84}$$

The unit vectors in cylindrical coordinates are

$$\mathbf{i}_\mathrm{R} = \mathbf{i}\cos\theta + \mathbf{j}\sin\theta \qquad \mathbf{j}_\theta = -\mathbf{i}\sin\theta + \mathbf{j}\cos\theta \qquad \mathbf{k} = \mathbf{k} \tag{4.85}$$

which are directed normal and tangential, respectively, to the circle in Fig. 4.14. The radial vector, directed normal to the circle, is

$$\mathbf{R}_\mathrm{c} = \mathbf{i}_\mathrm{R} R_\mathrm{c} \tag{4.86}$$

The radial velocity, tangential velocity, and angular velocity vectors in cylindrical coordinates are

$$\mathbf{u}_\mathrm{R} = \mathbf{i}_\mathrm{R} u_\mathrm{R} \qquad \mathbf{v}_\theta = \mathbf{j}_\theta v_\theta \qquad \mathbf{w}_z = \mathbf{k}\omega_z \tag{4.87}$$

respectively, where

$$u_\mathrm{R} = \frac{\mathrm{d}R_\mathrm{c}}{\mathrm{d}t} \qquad v_\theta = R_\mathrm{c}\frac{\mathrm{d}\theta}{\mathrm{d}t} \qquad \omega_z = \frac{\mathrm{d}\theta}{\mathrm{d}t} = \frac{v_\theta}{R_\mathrm{c}} \tag{4.88}$$

are the radial, tangential, and angular scalar velocities, respectively.

The tangential and angular velocity vectors in cylindrical coordinates are related to each other by

$$\mathbf{v}_\theta = \mathbf{w}_z \times \mathbf{R}_\mathrm{c} = \begin{vmatrix} \mathbf{i}_\mathrm{R} & \mathbf{j}_\theta & \mathbf{k} \\ 0 & 0 & v_\theta/R_\mathrm{c} \\ R_\mathrm{c} & 0 & 0 \end{vmatrix} = \mathbf{j}_\theta v_\theta \tag{4.89}$$

The local apparent centrifugal force per unit mass, which acts away from the center of curvature, is equal and opposite to the local centripetal acceleration vector,

$$\mathbf{a}_\mathrm{R} = -\frac{\mathbf{F}_\mathrm{R}}{M_\mathrm{a}} = \mathbf{w}_z \times \mathbf{v}_\theta = \begin{vmatrix} \mathbf{i}_\mathrm{R} & \mathbf{j}_\theta & \mathbf{k} \\ 0 & 0 & v_\theta/R_\mathrm{c} \\ 0 & v_\theta & 0 \end{vmatrix} = -\mathbf{i}_\mathrm{R}\frac{v_\theta^2}{R_\mathrm{c}} = -\mathbf{i}_\mathrm{R} a_\mathrm{R} \tag{4.90}$$

where $a_\mathrm{R} = v_\theta^2/R_\mathrm{c}$ is the scalar centripetal acceleration.

In cylindrical coordinates and in the absence of eddy diffusion, the horizontal momentum equations transform from (4.73) and (4.74) to

$$\frac{\mathrm{d}u_\mathrm{R}}{\mathrm{d}t} = f v_\theta - \frac{1}{\rho_\mathrm{a}}\frac{\partial p_\mathrm{a}}{\partial R_\mathrm{c}} + \frac{v_\theta^2}{R_\mathrm{c}} \qquad \frac{\mathrm{d}v_\theta}{\mathrm{d}t} = -f u_\mathrm{R} - \frac{u_\mathrm{R} v_\theta}{R_\mathrm{c}} \tag{4.91}$$

respectively, where v_θ^2/R_c and $-u_\mathrm{R}v_\theta/R_\mathrm{c}$ are centripetal accelerations, treated as apparent centrifugal forces per unit mass, that arise from the transformation from Cartesian to cylindrical coordinates. These terms are implicitly included in the Cartesian-coordinate momentum equations.

When air flows around a center of low or high pressure aloft, as in Fig. 4.15(a) or (b), respectively, the primary forces acting on the air are the Coriolis, pressure gradient, and apparent centrifugal forces. If local acceleration is removed from the first equation in (4.91), the resulting wind is the **gradient wind**. Assuming $\mathrm{d}u_\mathrm{R}/\mathrm{d}t = 0$ in the first equation in (4.91) and solving for the gradient-wind scalar

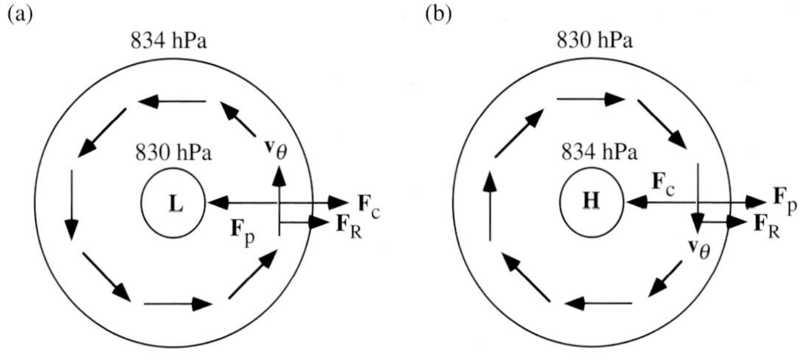

Figure 4.15 Gradient winds around a center of (a) low and (b) high pressure in the Northern Hemisphere, and the forces affecting them.

velocity give

$$v_\theta = -\frac{R_c f}{2} \pm \frac{R_c}{2}\sqrt{f^2 + 4\frac{1}{R_c \rho_a}\frac{\partial p_a}{\partial R_c}} \tag{4.92}$$

In this solution, the positive square root is correct and results in v_θ that is either positive or negative. The negative root is incorrect and results in an unphysical solution.

Example 4.15

Suppose the pressure gradient a distance $R_c = 70$ km from the center of a hurricane is $\partial p_a/\partial R_c = 45$ hPa per 100 km. Assume $\varphi = 15°$ N, $p_a = 850$ hPa, and $\rho_a = 1.06$ kg m^{-3}. These values give $v_\theta = 52$ m s^{-1} and $v_g = 1123$ m s^{-1}. Since $v_\theta < v_g$, the apparent centrifugal force slows down the geostrophic wind.

If the same conditions are used for the high-pressure center case (except the pressure gradient is reversed), the quadratic has an imaginary root, which is an unphysical solution. Reducing $\partial p_a/\partial R_c$ to -0.1 hPa per 100 km in this example gives $v_\theta = -1.7$ m s^{-1}. Thus, the magnitude of pressure gradients and resulting wind speeds around high-pressure centers are smaller than those around comparative low-pressure centers.

4.3.4 Surface winds around highs and lows

The deceleration of winds due to drag must be included in the momentum equation when flow around a low- or high-pressure center near the surface occurs. Figures 4.16(a) and (b) show forces acting on the wind in such cases. In cylindrical coordinates, the horizontal momentum equations describing flow in the presence of the

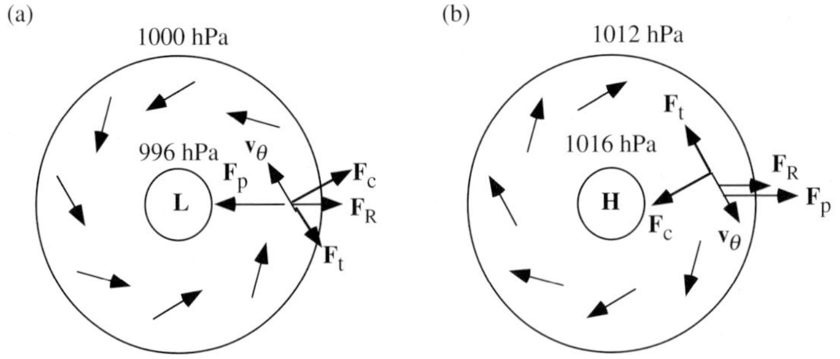

Figure 4.16 Surface winds around centers of (a) low and (b) high pressure in the Northern Hemisphere and the forces affecting them.

Coriolis force, the pressure gradient force, and friction are

$$\frac{du_R}{dt} = fv_\theta - \frac{1}{\rho_a}\frac{\partial p_a}{\partial R_c} + \frac{v_\theta^2}{R_c} + \frac{1}{\rho_a}\frac{\partial(\rho_a\overline{w'u_R'})}{\partial z}$$

$$\frac{dv_\theta}{dt} = -fu_R - \frac{u_R v_\theta}{R_c} + \frac{1}{\rho_a}\frac{\partial(\rho_a\overline{w'v_\theta'})}{\partial z} \tag{4.93}$$

where u_R' and v_θ' are the perturbation components of u_R and v_θ, respectively.

4.3.5 Atmospheric waves

Atmospheric waves are oscillations in pressure and/or velocity that propagate in space and time. Because waves perturb air parcels, which contain gas molecules, aerosol particles, and energy, waves result in spatial and temporal changes in momentum, concentration, and temperature. Several wave types occur in the atmosphere, including acoustic (sound), gravity (buoyancy), and Rossby (planetary) waves. Waves are important to understand from a modeling point of view, because they are often the fastest motion in a model domain and must be filtered from or treated with special numerical methods in the equations of atmospheric dynamics. Such filtering is discussed in Section 5.1. In this section, analytical wave equation solutions are derived from the equations of atmospheric dynamics.

An atmospheric wave consists of a **group** of individual waves of different characteristics superimposed upon each other. A set of individual waves can be considered as a group under the **superposition principle**, which states that a group of two or more waves can traverse the same space simultaneously, and the displacement of a medium due to the group is the sum of the displacements of each individual wave in the group. As a result of superposition, the shape of the sum of all waves in a group, called the **envelope**, differs from the shape of each individual wave in the group.

Individual waves are transverse, longitudinal, or both. A **transverse wave** is a wave that has a sinusoidal wave shape perpendicular to the direction of propagation of the wave. If one end of a spring is fixed to a floor and the other end is lifted vertically and then oscillated horizontally, a sinusoidal transverse wave propagates down the spring to the floor. Pure gravity waves, light waves, and violin-string waves are transverse waves. A **longitudinal wave** is one in which the disturbance in wave shape flows along (parallel to) the direction of propagation of the wave. If one end of a spring is fixed to the floor and the other end is lifted vertically and then oscillated vertically, a longitudinal wave propagates to the floor. Pure acoustic waves are longitudinal. Water waves at the top of the ocean surface are a combination of transverse and longitudinal waves.

The main characteristics of an individual wave are its wavelength, wavenumber, frequency of oscillation, phase speed, and amplitude. A **wavelength** is the distance (m) between two crests or troughs in a wave. On a Cartesian grid, wavelengths in the x-, y-, and z-directions are denoted by $\lambda_{\alpha,x}$, $\lambda_{\alpha,y}$, and $\lambda_{\alpha,z}$, respectively. The **wavenumber** of a wave is defined here as 2π divided by the wavelength, or the number of wavelengths in a circle of unit radius. In three dimensions, the **wavenumber vector** is

$$\tilde{\mathbf{K}} = \mathbf{i}\tilde{k} + \mathbf{j}\tilde{l} + \mathbf{k}\tilde{m} \qquad (4.94)$$

where individual wavenumbers in the x-, y-, and z-directions are related to wavelength by

$$\tilde{k} = \frac{2\pi}{\lambda_{\alpha,x}} \qquad \tilde{l} = \frac{2\pi}{\lambda_{\alpha,y}} \qquad \tilde{m} = \frac{2\pi}{\lambda_{\alpha,z}} \qquad (4.95)$$

(m^{-1}) respectively. The magnitude of the wavenumber vector is

$$|\tilde{\mathbf{K}}| = \sqrt{\tilde{k}^2 + \tilde{l}^2 + \tilde{m}^2} \qquad (4.96)$$

The **frequency of oscillation** (ν_α, s^{-1}) (the angular frequency in this case) of a wave is the number of wavelengths that pass through a given point in a circle of unit radius per unit time. The **phase speed** (c_α, $\mathrm{m\ s}^{-1}$) (or wave speed) of a wave is the speed at which a mathematical surface of constant phase travels. In other words, it is the speed at which all components of the individual wave travel along the direction of propagation. The frequency of oscillation is related to the phase speed and wavenumber by the **dispersion relationship,**

$$\nu_\alpha = c_\alpha\sqrt{\tilde{k}^2 + \tilde{l}^2 + \tilde{m}^2} = c_\alpha|\tilde{\mathbf{K}}| \qquad (4.97)$$

The **amplitude** (A_w, m) of a wave is the magnitude of its maximum displacement. In the case of transverse waves, the **displacement** (D, m) is the height of the wave shape normal to the direction of propagation and varies between $\pm A_\mathrm{w}$. A common wave equation for the propagation of a transverse wave is

$$D(x,t) = A_\mathrm{w}\sin\left(\tilde{k}x - \nu_\alpha t\right) \qquad (4.98)$$

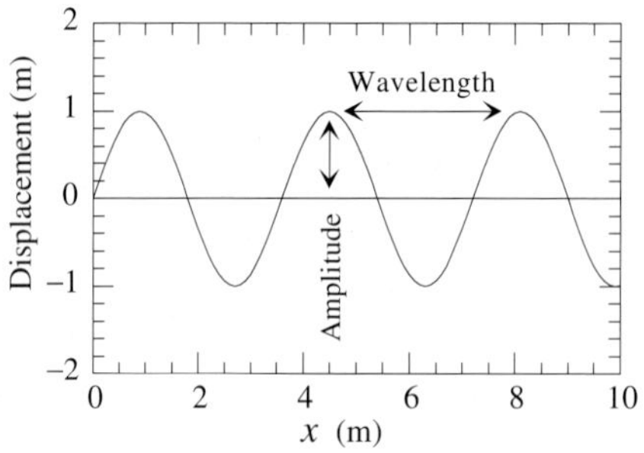

Figure 4.17 Characteristics of a wave generated from wave equation (4.98) with $A_w = 1$ m, $k = 1.745$ m^{-1}, and $\lambda_{\alpha,x} = 3.601$ m at $t = 0$.

where the displacement is a function of distance x along the direction of propagation and time t. The argument in (4.98) is in radians. Figure 4.17 shows characteristics of a wave generated from (4.98).

When individual waves propagate as a group, the **group velocity** is the velocity of the envelope of the group. The three-dimensional group velocity vector is

$$\mathbf{c}_g = \mathbf{i}c_{g,x} + \mathbf{j}c_{g,y} + \mathbf{k}c_{g,z} \tag{4.99}$$

where

$$c_{g,x} = \frac{\partial \nu_\alpha}{\partial \tilde{k}} \qquad c_{g,y} = \frac{\partial \nu_\alpha}{\partial \tilde{l}} \qquad c_{g,z} = \frac{\partial \nu_\alpha}{\partial \tilde{m}} \tag{4.100}$$

are **group scalar velocities.** Substituting (4.97) into (4.100) gives

$$c_{g,x} = c_\alpha \frac{\tilde{k}}{|\tilde{\mathbf{K}}|} + |\tilde{\mathbf{K}}| \frac{\partial c_\alpha}{\partial \tilde{k}} \qquad c_{g,y} = c_\alpha \frac{\tilde{l}}{|\tilde{\mathbf{K}}|} + |\tilde{\mathbf{K}}| \frac{\partial c_\alpha}{\partial \tilde{l}}$$
$$c_{g,z} = c_\alpha \frac{\tilde{m}}{|\tilde{\mathbf{K}}|} + |\tilde{\mathbf{K}}| \frac{\partial c_\alpha}{\partial \tilde{m}} \tag{4.101}$$

The magnitude of the group velocity is the **group speed**,

$$|\mathbf{c}_g| = \sqrt{c_{g,x}^2 + c_{g,y}^2 + c_{g,z}^2} \tag{4.102}$$

If waves traveling as a group propagate with a group speed equal to the phase speed of each wave within the group, the shape of the group does not deform over time, and the medium in which the group propagates is called a **nondispersive** medium. For example, free space is a nondispersive medium for electromagnetic waves (Chapter 9), and air is a nondispersive medium for sound waves (Section 4.3.5.1). If air were a dispersive medium for sound waves, then high-frequency and low-frequency sounds from a piano would reach a person's ear at different times, and the resulting sound would not be harmonious.

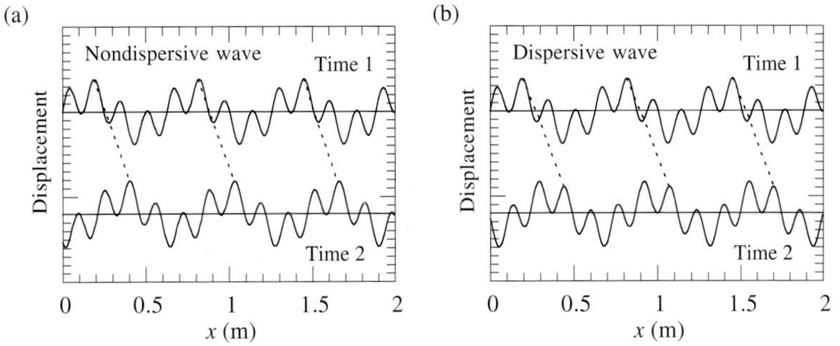

Figure 4.18 Wave pulses in (a) nondispersive and (b) dispersive media. In the nondispersive case, $\nu_\alpha = c_s\tilde{k}$ (sound waves), where c_s is the speed of sound (346 m s^{-1} here). In the dispersive case, $\nu_\alpha = (g\tilde{k})^{1/2}$ (deep-water waves). In both cases, two waves of wavenumber $\tilde{k}_1 = 10$ m^{-1} and $\tilde{k}_2 = 40$ m^{-1}, respectively, and amplitude $A_{w,1} = A_{w,2} = 1$, are summed to form a group with displacement $D(x,t) = A_{w,1}\sin(\tilde{k}_1 x - \nu_{\alpha,1}t) + A_{w,2}\sin(\tilde{k}_2 x - \nu_{\alpha,2}t)$. Time 1 in the figures occurs when $t = 0$ s in both cases, and time 2 occurs when $t = 0.5$ s in the nondispersive case and 4 s in the dispersive case. In the nondispersive case, the group speed equals the phase speed of each harmonic wave in the group and the wave group does not change shape over time as it travels over distance x. In the dispersive case, the group speed differs from the phase speed of each harmonic wave, and the group changes shape over time.

For a medium to be nondispersive, the phase speed of each wave in a group must be independent of wavenumber (or wavelength). In other words,

$$\frac{\partial c_\alpha}{\partial \tilde{k}} = \frac{\partial c_\alpha}{\partial \tilde{l}} = \frac{\partial c_\alpha}{\partial \tilde{m}} = 0 \tag{4.103}$$

Substituting these terms into (4.101) gives the group scalar velocities in a nondispersive medium as

$$c_{g,x} = c_\alpha \frac{\tilde{k}}{|\tilde{\mathbf{K}}|} \qquad c_{g,y} = c_\alpha \frac{\tilde{l}}{|\tilde{\mathbf{K}}|} \qquad c_{g,z} = c_\alpha \frac{\tilde{m}}{|\tilde{\mathbf{K}}|} \tag{4.104}$$

Substituting (4.104) into (4.102) gives $|\mathbf{c}_g| = c_\alpha$. Thus, in a nondispersive medium, the group speed equals the phase speed of individual waves. In the case of sound waves originating from a stationary source, the phase speed (c_α) is the speed of sound (c_s).

If waves traveling as a group propagate with a group speed that differs from the phase speed of any individual wave in the group, the shape of the group deforms over time, and the medium in which the group propagates is called a **dispersive** medium. Dispersive-medium waves arise when the phase speed is a function of wavenumber. Water waves are examples of dispersive waves (Example 4.16). Figure 4.18 shows examples of wave pulses in nondispersive and dispersive media.

Example 4.16

Determine the phase speed and group speed of deep-water waves, which have a dispersion relation, $\nu_\alpha = \sqrt{g\tilde{k}}$.

SOLUTION

Substituting the dispersion relationship into (4.97) gives the phase speed of deep-water waves as

$$c_\alpha = \frac{\nu_\alpha}{\tilde{k}} = \sqrt{\frac{g}{\tilde{k}}}$$

Substituting the dispersion relationship into (4.100) and the result into (4.102) gives the group speed as

$$\left|\mathbf{c}_\mathrm{g}\right| = c_{\mathrm{g},x} = \frac{\partial \nu_\alpha}{\partial \tilde{k}} = \frac{1}{2}\sqrt{\frac{g}{\tilde{k}}} = \frac{1}{2}c_\alpha$$

Thus, individual deep-water waves in a group move twice as fast as the envelope of the group.

In the following subsections, atmospheric wave types and their characteristic dispersion relationships are identified and briefly described.

4.3.5.1 *Pure acoustic waves*

Pure **acoustic (sound) waves** occur when a vibration causes alternating adiabatic compression and expansion of a compressible fluid, such as air. During compression and expansion, air pressure oscillates, causing acceleration to oscillate along the direction of propagation of the wave. Pure acoustic waves are longitudinal because wave disturbances travel parallel to the direction of wave propagation. Acoustic waves transport not only sound, but also parcels of air containing energy, gases, and particles.

The dispersion relationship for pure acoustic waves can be derived in three dimensions by solving the u-, v-, and w-momentum equations (ignoring the Coriolis, gravitational, viscous, and eddy diffusion terms), the continuity equation for air, and the thermodynamic energy equation (ignoring diabatic energy sources and sinks). If the direction of wave propagation is assumed to be along the x-axis, the dispersion relationship can be simplified by setting v- and w-terms to zero. In this case, the u-momentum, continuity, and thermodynamic energy equations simplify from (4.73), (3.23), and (3.66) to

$$\frac{\mathrm{d}u}{\mathrm{d}t} = -\frac{1}{\rho_\mathrm{a}}\frac{\partial p_\mathrm{a}}{\partial x} \tag{4.105}$$

$$\frac{\mathrm{d}\rho_\mathrm{a}}{\mathrm{d}t} = -\rho_\mathrm{a}\frac{\partial u}{\partial x} \tag{4.106}$$

$$\frac{1}{\theta_\mathrm{v}}\frac{\mathrm{d}\theta_\mathrm{v}}{\mathrm{d}t} = \frac{\mathrm{d}\ln\theta_\mathrm{v}}{\mathrm{d}t} = 0 \tag{4.107}$$

respectively. Substituting $\theta_v = T_v(1000/p_a)^\kappa$ and $p_a = \rho_a R' T_v$ into (4.107) gives

$$\frac{d\rho_a}{dt} = \frac{\rho_a}{\gamma}\frac{d\ln p_a}{dt} \tag{4.108}$$

where $\gamma = 1/(1-\kappa) = c_{p,d}/c_{p,d}c_{v,d} \approx 1.4$. Substituting (4.108) into (4.106) gives

$$\frac{d\ln p_a}{dt} = \frac{1}{p_a}\frac{dp_a}{dt} = -\gamma\frac{\partial u}{\partial x} \tag{4.109}$$

Taking the time derivative of (4.109), then substituting in (4.105), $p_a = \rho_a R' T_v$, $p_a = \bar{p}_a + p'_a$, $\rho_a = \bar{\rho}_a + \rho'_a$, and $u = \bar{u} + u'$, and eliminating products of perturbation variables yield the **acoustic wave equation,**

$$\frac{d^2 p'_a}{dt^2} = \left(\frac{\partial}{\partial t} + \bar{u}\frac{\partial}{\partial x}\right)^2 p'_a = c_s^2 \frac{\partial^2 p'_a}{\partial x^2} \tag{4.110}$$

where

$$c_s = \pm\sqrt{\gamma R' T_v} \tag{4.111}$$

is the **adiabatic speed of sound.**

Example 4.16

When $T_v = 298$ K, $c_s = 346$ m s^{-1}. When $T_v = 225$ K, $c_s = 301$ m s^{-1}.

A sinusoidal solution to (4.110) is

$$p'_a = p'_{a,0}\sin(\tilde{k}x - \nu_\alpha t) \tag{4.112}$$

where x is the distance along the direction of phase propagation, t is time, and $p'_{a,0}$ is the amplitude of the oscillation. Substituting this solution into (4.110) gives the **dispersion relationship** for pure acoustic waves,

$$\nu_\alpha = (\bar{u} \pm c_s)\tilde{k} \tag{4.113}$$

Substituting (4.113) into (4.97) and solving give the phase speed of sound waves. Taking the partial derivative of (4.113) with respect to $\tilde{k}$ gives the group scalar velocity in the x-direction. The result in both cases is identical:

$$|\mathbf{c}_g| = c_{g,x} = c_\alpha = \bar{u} \pm c_s \tag{4.114}$$

Since their group speed equals their phase speed, **acoustic waves are nondispersive-medium waves.** The group speed equals the group scalar velocity in the x-direction only because sound waves do not propagate in the y- and z-directions in this case.

4.3.5.2 *Gravity, acoustic-gravity, and Lamb waves*

When the atmosphere is stably stratified and a parcel of air is displaced vertically, buoyancy restores the parcel to its equilibrium position in an oscillatory manner. The frequency of oscillation, N_{bv}, is the **Brunt–Väisälä frequency**, defined in Chapter 2. The period of oscillation is $\tau_{bv} = 2\pi / N_{bv}$. The wave resulting from this oscillation is a **gravity (buoyancy) wave.** Air-parcel displacements that give rise to gravity waves may result from forced convection, air flow over mountains, wind shear in frontal regions, wind shear associated with jet-stream flow, or perturbations associated with the geostrophic wind. Because gravity waves displace parcels of air, they perturb temperatures, gas and particle concentrations, vertical scalar velocities, and other parameters associated with the parcel. Gravity-wave motion may increase vertical transfer of ozone between the lower stratosphere and upper troposphere (e.g., Lindzen 1981; van Zandt and Fritts 1989). Gravity waves may also cause large periodic fluctuations in ozone at specific locations in the troposphere (Langford *et al.* 1996).

The dispersion relationship for gravity waves can be isolated from a broader relationship for acoustic-gravity waves. In three dimensions, the **acoustic-gravity-wave dispersion relationship** is found by solving the three momentum equations (retaining gravity in the vertical), the continuity equation for air, and the thermodynamic energy equation. If only the x–z plane is considered, a simplified acoustic-gravity wave relationship is obtained by solving

$$\frac{\mathrm{d}u}{\mathrm{d}t} = -\frac{1}{\rho_a}\frac{\partial p_a}{\partial x} \qquad \frac{\mathrm{d}w}{\mathrm{d}t} = -\frac{1}{\rho_a}\frac{\partial p_a}{\partial z} - g$$
$$\frac{\mathrm{d}\rho_a}{\mathrm{d}t} = -\rho_a\left(\frac{\partial u}{\partial x} + \frac{\partial w}{\partial z}\right) \qquad \frac{\mathrm{d}\rho_a}{\mathrm{d}t} = \frac{\rho_a}{\gamma}\frac{\mathrm{d}\ln p_a}{\mathrm{d}t} \tag{4.115}$$

written from (4.73), (4.75), (3.23), and (4.108), respectively. When $p_a = \bar{p}_a + p'_a$, $\rho_a = \bar{\rho}_a + \rho'_a$, $u = \bar{u} + u'$, and $w = \bar{w} + w'$, (4.115) can be solved analytically to obtain the **acoustic-gravity wave dispersion relationship,**

$$\frac{N_{bv}^2}{(\nu_\alpha - \bar{u}\tilde{k})^2}\tilde{k}^2 + \frac{(\nu_\alpha - \bar{u}\tilde{k})^2}{c_s^2} = \tilde{m}^2 + \tilde{k}^2 + \frac{\nu_c^2}{c_s^2} \tag{4.116}$$

where

$$\nu_c = \frac{c_s}{2H} \tag{4.117}$$

is the **acoustic cutoff frequency** (s^{-1}). In this equation, H is the scale height of the atmosphere, obtained from (2.47) under the assumption that the atmosphere is isothermal and stably stratified. The acoustic cutoff frequency is a frequency above which acoustic waves may propagate vertically if other conditions are right. Below this frequency, acoustic waves cannot propagate vertically.

Example 4.17

For an isothermal layer at $T_v = 298$ K, find the acoustic cutoff and Brunt–Väisälä frequencies.

SOLUTION

From Appendix A, $\bar{M} = 4.8096 \times 10^{-26}$ kg, $g = 9.81$ m s^{-2}, and $k_B = 1.3807 \times 10^{-23}$ kg m^2 s^{-2} K^{-1} molec.$^{-1}$. From (2.47), the scale height is $H = 8.72$ km. Thus, from (4.111) $c_s = 346$ m s^{-1} and from (4.117), $\nu_c = 0.0198$ s^{-1}. From (2.106), the Brunt–Väisälä frequency is $N_{bv} = \sqrt{g(\Gamma_d - \Gamma_v)/T_v}$. Substituting $\Gamma_v = 0$ (since the atmosphere is isothermal) and $\Gamma_d = +9.8$ K km^{-1} into this equation gives $N_{bv} = 0.0180$ s^{-1}. Thus, the acoustic cutoff frequency slightly exceeds the Brunt–Väisälä frequency ($\nu_c > N_{bv}$) in this case.

Dispersion relationships for gravity waves, acoustic waves modified by the effects of stable stratification, and mixed waves can be isolated from the acoustic-gravity-wave dispersion relationship by considering certain frequency and wavenumber regimes. For example, (4.116) simplifies for several cases to

$$\nu_\alpha = \begin{cases} \bar{u}\tilde{k} + \dfrac{N_{bv}\tilde{k}}{\left(\tilde{k}^2 + \tilde{m}^2 + \nu_c^2/c_s^2\right)^{1/2}} & \nu_\alpha^2 \ll \nu_c^2 & \left\{\text{low-frequency gravity waves}\right. \\ \bar{u}\tilde{k} + \left(c_s^2\tilde{k}^2 + c_s^2\tilde{m}^2 + \nu_c^2\right)^{1/2} & \nu_\alpha^2 \gg N_{bv}^2 & \left\{\text{high-frequency acoustic waves}\right. \\ \bar{u}\tilde{k} + \dfrac{N_{bv}\tilde{k}}{(\tilde{k}^2 + \tilde{m}^2)^{1/2}} & \tilde{k} \to \infty & \left\{\text{mountain lee waves}\right. \\ (\bar{u} + c_s)\tilde{k} & \left\{\begin{array}{l} \nu_\alpha^2 \ll \nu_c^2,\ \tilde{m}^2 = 0,\ \tilde{k} \to 0 \\ \nu_\alpha^2 \gg N_{bv}^2,\ \tilde{m}^2 = 0,\ \tilde{k} \to \infty \end{array}\right. & \left\{\text{Lamb waves}\right. \end{cases} \tag{4.118}$$

When $\nu_\alpha^2 \ll \nu_c^2$, (4.116) simplifies to the dispersion relationship for **low-frequency gravity waves**. The same result is found by assuming the atmosphere is incompressible ($c_s \to \infty$). Figure 4.19 shows that vertically and horizontally propagating, low-frequency gravity waves occur when $\tilde{m}^2 > 0$, $\nu_\alpha - \bar{u}\tilde{k} < c_s\tilde{k}$ as $\tilde{k} \to 0$, and $\nu_\alpha - \bar{u}\tilde{k} < N_{bv}$ as $\tilde{k} \to \infty$. Such waves are **internal gravity waves**. An **internal wave** is a wave that exhibits vertical motion trapped beneath a surface. The

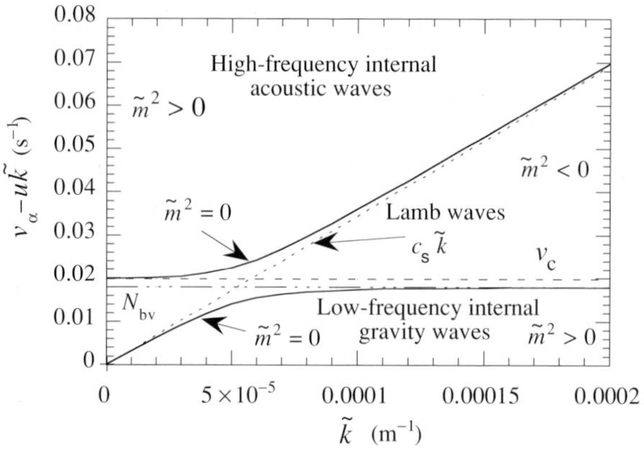

Figure 4.19 Vertical wavenumber squared, $\tilde{m}^2$, as a function of $\tilde{k}$ and ν_α when $H = 8.72$ km, $c_s = 346$ m s^{-1}, and $N_{bv} = 0.0180$ s^{-1} (see Example 4.17). The upper region where $\tilde{m}^2 > 0$ represents conditions that give rise to vertically propagating high-frequency acoustic waves. The lower region where $\tilde{m}^2 > 0$ represents conditions that give rise to vertically propagating low-frequency gravity waves. The region in which $\tilde{m}^2 < 0$ represents conditions that give rise to horizontally propagating acoustic-gravity and Lamb waves. Lamb waves connect the point $\tilde{k} = 0, \tilde{m}^2 = 0$ to the point $\tilde{k} = \infty, \tilde{m}^2 = 0$. The acoustic cutoff frequency slightly exceeds the Brunt–Väisälä frequency.

Brunt–Väisälä frequency demarks the upper frequency for internal gravity waves. Equating (4.97) with the first term on the right side of (4.118) gives the phase speed of internal gravity waves as

$$c_\alpha = \bar{u} + \frac{N_{bv}}{\sqrt{\tilde{k}^2 + \tilde{m}^2 + \nu_c^2/c_s^2}} \tag{4.119}$$

Applying (4.100) to this phase speed gives group scalar velocities of internal gravity waves as

$$c_{g,x} = \bar{u} + \frac{N_{bv}\left(\tilde{m}^2 + \nu_c^2/c_s^2\right)}{\left(\tilde{k}^2 + \tilde{m}^2 + \nu_c^2/c_s^2\right)^{3/2}} \qquad c_{g,z} = -\frac{N_{bv}\tilde{k}\tilde{m}}{\left(\tilde{k}^2 + \tilde{m}^2 + \nu_c^2/c_s^2\right)^{3/2}} \tag{4.120}$$

Thus, **internal gravity waves are dispersive-medium waves**, since their phase speed differs from the magnitude of their group velocity vector.

In the region of Fig. 4.19 where $\tilde{m}^2 < 0$, waves are **external acoustic-gravity waves** and propagate only horizontally. An **external wave** is a wave that has its maximum amplitude at the external boundary (free surface) of a fluid. Such waves have no influence in the vertical, since their amplitudes decrease exponentially with distance from the surface.

When $\nu_\alpha^2 \gg N_{bv}^2$, the frequency from (4.118) simplifies to that of acoustic waves, modified by the effects of stable stratification. Figure 4.19 shows that vertically and horizontally propagating acoustic waves occur when $\tilde{m}^2 > 0$, $\nu_\alpha - \bar{u}\tilde{k} > \nu_c$ as $\tilde{k} \to 0$, and $\nu_\alpha - \bar{u}\tilde{k} > c_s\tilde{k}$ as $\tilde{k} \to \infty$. Such waves are **high-frequency internal acoustic waves.** The dispersion relationship for internal acoustic waves differs from that for pure acoustic waves only in that the latter ignores the effect of buoyancy in a stably-stratified atmosphere.

The dotted line in Fig. 4.19 connecting the points between $\tilde{m}^2 = 0$, $\tilde{k} \to 0$ for internal gravity waves and $\tilde{m}^2 = 0$, $\tilde{k} \to \infty$ for internal acoustic waves modified by stratification describes the dispersion relationship for external **Lamb waves** (Lamb 1910). The frequency of a Lamb wave approaches $\nu_\alpha \approx (\bar{u} + c_s)\tilde{k}$, which is the same frequency as that of a pure acoustic wave, as shown in (4.113). Lamb waves are nondispersive and propagate horizontally at the speed of sound. Whereas their energy decreases exponentially with increasing altitude, their amplitude may decrease or increase exponentially with altitude, depending on the value of $\tilde{m}$.

Mountain lee waves are gravity waves formed as a wind is perturbed vertically as it flows over a mountain in a stable atmosphere. Such waves have a wavelength on the order of 10 km, which results in a wavenumber of $\tilde{k} \approx 6.3 \times 10^{-4}$ m^{-1}. This is large in comparison with $\nu_c/c_s \approx 5.7 \times 10^{-5}$ m^{-1}, obtained from Example 4.17. Thus, in the case of mountain lee waves, $\tilde{k} \gg \nu_c/c_s$, and the acoustic-gravity-wave dispersion relationship simplifies from that of low-frequency gravity waves to that of **mountain lee waves**, as shown in (4.118).

4.3.5.3 *Inertia gravity waves and inertia Lamb waves*

After pressure and buoyancy oscillations, a third type of oscillation is an inertial oscillation. When a parcel of air moving from west to east with the geostrophic wind is perturbed in the south–north direction, the Coriolis force propels the parcel toward its original latitude in an **inertially stable** atmosphere and away from its original latitude in an **inertially unstable** atmosphere. In the former case, the parcel subsequently oscillates about its initial latitude in an **inertial oscillation.**

Whether the atmosphere is inertially stable or unstable can be estimated by solving the horizontal momentum equations,

$$\frac{du}{dt} = fv = f\frac{dy}{dt} \tag{4.121}$$

$$\frac{dv}{dt} = f(u_g - u) \tag{4.122}$$

where only acceleration, the Coriolis force, and the south–north pressure gradient (embodied in u_g) are considered. If a parcel of air moving with the geostrophic wind is displaced in the south–north direction a distance Δy, (4.121) can be integrated between y_0 and $y_0 + \Delta y$ to give

$$u(y_0 + \Delta y) - u_g(y_0) \approx f\Delta y \tag{4.123}$$

The geostrophic wind at point y_0 can be found by taking a first-order Taylor series expansion of the geostrophic wind at $y_0 + \Delta y$, giving

$$u_g(y_0 + \Delta y) \approx u_g(y_0) + \frac{\partial u_g}{\partial y}\Delta y \tag{4.124}$$

Substituting (4.124) into (4.123) yields

$$u_g(y_0 + \Delta y) - u(y_0 + \Delta y) \approx -\left(f - \frac{\partial u_g}{\partial y}\right)\Delta y \tag{4.125}$$

which substitutes into (4.122) to give

$$\boxed{\frac{dv}{dt} = -f\left(f - \frac{\partial u_g}{\partial y}\right)\Delta y} \tag{4.126}$$

This equation states that, if $f - \partial u_g/\partial y > 0$ and a parcel of air is displaced northward of its original latitude in the Northern Hemisphere ($\Delta y > 0$), then $dv/dt < 0$, and the parcel decelerates, turns around, and begins to accelerate southward. As the parcel passes south of its original latitude ($\Delta y < 0$), then $dv/dt > 0$, and the parcel decelerates, turns around, and accelerates northward again, and so on. Thus, when $f - \partial u_g/\partial y > 0$ a south–north perturbation to an air parcel causes it to oscillate inertially about its original latitude. Under such circumstances, the atmosphere is inertially stable.

If $f - \partial u_g/\partial y < 0$ and an air parcel is displaced northward ($\Delta y > 0$), then $dv/dt > 0$, and the parcel continues to accelerate northward. Similarly, if it is displaced southward, $dv/dt < 0$, and it continues to accelerate southward. Under these conditions, the atmosphere is inertially unstable.

Finally, when $f - \partial u_g/\partial y = 0$, a perturbation to a parcel has no effect on south–north acceleration ($dv/dt = 0$). The atmosphere is inertially neutral in this scenario.

In sum, **inertial stability criteria for the Northern Hemisphere**, where $f > 0$, can be defined with

$$f - \frac{\partial u_g}{\partial y} \begin{cases} < 0 & \text{inertially unstable} \\ = 0 & \text{inertially neutral} \\ > 0 & \text{inertially stable} \end{cases} \tag{4.127}$$

which are analogous to the buoyancy-related stability criteria discussed in Chapter 2.

The atmosphere is almost always inertially stable, except near the Equator, where the Coriolis parameter is small. When inertial oscillations occur, their frequency f is on the order of 10^{-4} s^{-1}. This is much lower than the acoustic cutoff frequency (ν_c) and Brunt–Väisälä frequency (N_{bv}), which have values on the order of 10^{-2} s^{-1} (Example 4.17). Thus, inertial oscillations are important only for perturbations involving low-frequency (long-wavelength) waves.

Since vertically propagating acoustic waves occur only at high frequency ($>\nu_c$, as shown in Fig. 4.19), they are not affected much by inertial oscillations. When the frequency of a Lamb wave is below ν_c, it may be affected by inertial oscillations. Under such a condition, the dispersion relationship for **inertia Lamb waves** is rewritten from (4.118) to

$$\nu_\alpha^2 = f^2 + c_s^2 \tilde{k}^2 \tag{4.128}$$

Similarly, the dispersion relationship for **inertia gravity waves** is rewritten from that of low-frequency gravity waves in (4.118) to

$$\nu_\alpha^2 = f^2 + \frac{N_{bv}^2 \tilde{k}^2}{\tilde{k}^2 + \tilde{m}^2 + \nu_c^2 / c_s^2} \tag{4.129}$$

Inertia gravity waves are driven by inertial and buoyant oscillations and occur when the atmosphere is inertially and buoyantly stable. Inertia Lamb waves are driven by inertial and pressure oscillations.

Inertia Lamb waves and inertia gravity waves participate in the restoration of the geostrophic wind following a perturbation to it. When the geostrophic wind in midlatitudes, which flows from west to east, is perturbed to the north or south, horizontally propagating inertia Lamb waves act to restore vertically integrated geostrophic velocity and surface pressure fields while vertically and horizontally propagating inertia gravity waves act to restore vertical shear and temperature fields to geostrophic balance (Arakawa 1997). The restoration to geostrophic balance is called **geostrophic adjustment.**

During geostrophic adjustment, the primary factor that determines whether the pressure (mass) field adjusts to the velocity field or vice-versa is the size of the length scale of horizontal motion (L) relative to the **Rossby radius of deformation.** For an atmospheric wave, the length scale is its horizontal wavelength. The Rossby radius of deformation is

$$\lambda_R = \frac{\sqrt{g h_e}}{f} \tag{4.130}$$

where

$$h_e = \begin{cases} \dfrac{c_s^2}{g} & \text{inertia Lamb waves} \\[2ex] \dfrac{N_{bv}^2 / g}{\tilde{m}^2 + \nu_c^2 / c_s^2} & \text{inertia gravity waves} \end{cases} \tag{4.131}$$

is the **equivalent depth** of the atmosphere. A typical value of λ_R for inertia Lamb waves is 3000 km, and typical values for inertia gravity waves are 100–1000 km.

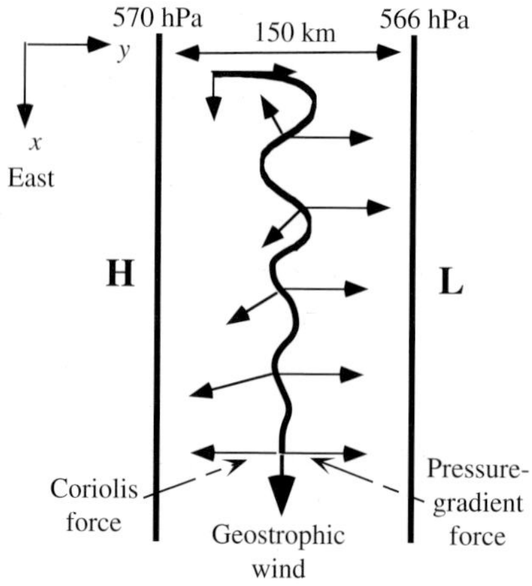

Figure 4.20 Geostrophic adjustment of the velocity field to the pressure field at one altitude when $L > \lambda_R$ and when the domain is assumed to be large enough to allow the energy of the oscillations to disperse and decay. In reality, the vertical mean velocity field, not the velocity field at a given altitude, adjusts to the surface pressure. A perturbation to the geostrophic wind causes waves to restore the system to geostrophic balance along an oscillatory path.

Example 4.18

From the information in Example 4.17, calculate the equivalent depth and the Rossby radius of deformation for inertia gravity waves assuming $\lambda_{\alpha,z} = 5$ km and $\varphi = 30°$ N latitude.

SOLUTION

From Example 4.17, $c_s = 346$ m s^{-1}, $\nu_c = 0.0198$ s^{-1}, and $N_{bv} = 0.0180$ s^{-1}, since $\tilde{m} = 2\pi/\lambda_{\alpha,z} = 0.0013$ m^{-1}. From (4.131), $h_e = 19.5$ m. At $\varphi = 30°$ N, we have $f = 7.292\times 10^{-5}$ s^{-1} and $\lambda_R = 190$ km.

When inertia Lamb waves are present and $L > \lambda_R$, the perturbed vertical mean velocity field adjusts to the pressure field, as roughly demonstrated in two dimensions in Fig. 4.20. When $L < \lambda_R$, the pressure field adjusts to the velocity field. When $L = \lambda_R$, a mutual adjustment between the pressure and velocity fields occurs.

When inertia gravity waves are present and $L > \lambda_R$, the wind shear field adjusts to the temperature field. Otherwise the temperature field adjusts to the wind shear field or mutual adjustment occurs (Arakawa 1997).

4.3.5.4 *Rossby waves*

Rossby (planetary) waves (Rossby *et al.* 1939) arise from a change in the Coriolis parameter $f = 2\Omega \sin\varphi$ with latitude. These waves are important for synoptic and planetary-scale meteorology. If an air parcel moving from west to east is perturbed to the south or north, the change in f with latitude provides a restoring force that gives rise to freely propagating Rossby waves. The presence of topography can initiate forced topographic Rossby waves. Rossby waves differ from pure inertial oscillations in that pure inertial oscillations do not require a variation of f with latitude.

In an incompressible, frictionless, adiabatic atmosphere of constant depth, absolute vorticity of air is conserved. Rossby waves can be described in terms of the conservation of absolute vorticity. **Vorticity** is the measure of spin around an axis. The faster a body spins, the greater its vorticity. Air spins relative to the surface of the Earth, and the surface of the Earth spins relative to its own axis of rotation. The spin of air relative to the surface of the Earth is **relative vorticity**, and the spin of the Earth relative to its own axis is the **Earth's vorticity**. Positive vorticity corresponds to counterclockwise spin when the spin is viewed from above the North Pole. Since the Earth rotates counterclockwise, the Earth's vorticity is positive. For a body at the North Pole, the Earth's vorticity is maximum, since the body spins around its own vertical axis. For a body standing at the Equator, the Earth's vorticity is zero, since the body does not spin around its own vertical axis. The vertical component of the Earth's vorticity is the Coriolis parameter f. At the Equator, North Pole, and South Pole, $f = 0$, 2Ω, and -2Ω, respectively.

The **relative vorticity** of air is the vorticity of air relative to the Earth. Mathematically, the relative vorticity is the curl of velocity. The relative vorticity vector is

$$\zeta_r = \nabla \times \mathbf{v} = \begin{vmatrix} \mathbf{i} & \mathbf{j} & \mathbf{k} \\ \partial/\partial x & \partial/\partial y & \partial/\partial z \\ u & v & w \end{vmatrix} = \left(\frac{\partial w}{\partial y} - \frac{\partial v}{\partial z}\right)\mathbf{i} - \left(\frac{\partial w}{\partial x} - \frac{\partial u}{\partial z}\right)\mathbf{j} + \left(\frac{\partial v}{\partial x} - \frac{\partial u}{\partial y}\right)\mathbf{k} \tag{4.132}$$

The **absolute vorticity** is the sum of the Earth's and the relative vorticity. The vertical component of absolute vorticity is $\zeta_{a,z} = f + \zeta_{r,z}$, where $\zeta_{r,z}$ is the vertical component of relative vorticity. In a frictionless, incompressible, and adiabatic atmosphere, the vertical component of absolute vorticity divided by the depth of the atmosphere is constant. In other words,

$$P_v = \frac{f + \zeta_{r,z}}{\Delta z_t} = \frac{f + \dfrac{\partial v}{\partial x} - \dfrac{\partial u}{\partial y}}{\Delta z_t} = \text{constant} \tag{4.133}$$

where P_v is a simplified form of **potential vorticity** and Δz_t is the depth of the atmosphere from the surface to the tropopause. If the depth of the atmosphere is constant and v is nonzero, a freely propagating Rossby wave can develop as follows: As air moves north, f increases. To conserve potential vorticity, $\zeta_{r,z}$ decreases. For

$\zeta_{r,z}$ to decrease, u and v must change in a way that forces the particle south, and so on. The sinusoidal motion resulting from the continuous changes of u, v, and f is a freely propagating Rossby wave.

The dispersion relationship for freely propagating Rossby waves in a frictionless, incompressible, and adiabatic atmosphere is found by solving the u and v momentum equations and the continuity equation for air. For this derivation, atmospheric depth is allowed to vary. The constant-depth result is easily extracted from the solution.

If only inertial, Coriolis, and pressure-gradient terms are considered, the horizontal momentum equations simplify from (4.73) and (4.74) to

$$\frac{\mathrm{d}u}{\mathrm{d}t} = fv - \frac{1}{\rho_a}\left(\frac{\partial p_a}{\partial x}\right)_z \tag{4.134}$$

$$\frac{\mathrm{d}v}{\mathrm{d}t} = -fu - \frac{1}{\rho_a}\left(\frac{\partial p_a}{\partial y}\right)_z \tag{4.135}$$

where the subscript z indicates the partial derivative is taken along a surface of constant altitude. At least three sets of terms can be substituted into these equations.

The first is the **midlatitude beta-plane approximation,**

$$f = f_0 + \beta(y - y_0) \tag{4.136}$$

which is an approximation for the variation of the Coriolis parameter with latitude. In this equation, y_0 is the south–north distance from the Equator at which $f = f_0$, y is the south–north distance of interest from the Equator, $\beta\,(y - y_0) \ll f_0$, and

$$\boxed{\beta = \frac{\partial f}{\partial y} = 2\Omega\frac{\partial\varphi}{\partial y}\cos\varphi \approx \frac{2\Omega}{R_e}\cos\varphi} \tag{4.137}$$

where $\partial y/\partial_\varphi = R_e$.

The second is a pair of equations relating pressure gradients along surfaces of constant altitude to geopotential gradients along surfaces of constant pressure. The equations, derived in Section 5.3, are

$$\left(\frac{\partial p_a}{\partial x}\right)_z = \rho_a\left(\frac{\partial \Phi}{\partial x}\right)_p \tag{4.138}$$

$$\left(\frac{\partial p_a}{\partial y}\right)_z = \rho_a\left(\frac{\partial \Phi}{\partial y}\right)_p \tag{4.139}$$

where p indicates that a partial derivative is taken along a surface of constant pressure.

The third is a set of equations dividing scalars (u, v, Φ) into **geostrophic** (u_g, v_g, Φ_g) and **ageostrophic** (u_a, v_a, Φ_a) components. Substituting (4.138), (4.136), $u = u_g + u_a$, and $\Phi = \Phi_g + \Phi_a$ into (4.134) gives

$$\frac{\mathrm{d}(u_g + u_a)}{\mathrm{d}t} = [f_0 + \beta(y - y_0)]\,(v_g + v_a) - \left[\frac{\partial(\Phi_g + \Phi_a)}{\partial x}\right]_p \tag{4.140}$$

Substituting (4.139), (4.136), $v = v_g + v_a$, and $\Phi = \Phi_g + \Phi_a$ into (4.135) gives

$$\frac{\mathrm{d}(v_g + v_a)}{\mathrm{d}t} = -[f_0 + \beta(y - y_0)](u_g + u_a) - \left[\frac{\partial(\Phi_g + \Phi_a)}{\partial y}\right]_p \tag{4.141}$$

These equations can be simplified by combining (4.138) and (4.139) with (4.79) at latitude y_0 to give **geostrophic scalar velocities** along surfaces of constant pressure as

$$v_g = \frac{1}{f_0}\left(\frac{\partial \Phi_g}{\partial x}\right)_p \qquad u_g = -\frac{1}{f_0}\left(\frac{\partial \Phi_g}{\partial y}\right)_p \tag{4.142}$$

Substituting these terms into (4.140) and (4.141) and assuming $\mathrm{d}u_a/\mathrm{d}t$, $\mathrm{d}v_a/\mathrm{d}t$, $\beta(y - y_0)u_a$, and $\beta(y - y_0)v_a$ are small give the **quasigeostrophic horizontal momentum equations,**

$$\frac{\mathrm{d}u_g}{\mathrm{d}t} = f_0 v_a + \beta(y - y_0)v_g - \left(\frac{\partial \Phi_a}{\partial x}\right)_p \tag{4.143}$$

$$\frac{\mathrm{d}v_g}{\mathrm{d}t} = -f_0 u_a - \beta(y - y_0)u_g - \left(\frac{\partial \Phi_a}{\partial y}\right)_p \tag{4.144}$$

Subtracting $\partial/\partial y$ of (4.143) from $\partial/\partial x$ of (4.144) yields

$$\frac{\mathrm{d}}{\mathrm{d}t}\left(\frac{\partial v_g}{\partial x} - \frac{\partial u_g}{\partial y}\right) = -f_0\left(\frac{\partial u_a}{\partial x} + \frac{\partial v_a}{\partial y}\right) - \beta v_g \tag{4.145}$$

where all remaining terms cancel. For example, $\partial u_g/\partial x + \partial v_g/\partial y = 0$, which indicates that the geostrophic wind is nondivergent.

The vertical scalar velocity in the altitude coordinate is

$$w = \frac{\mathrm{d}z}{\mathrm{d}t} = \frac{1}{g}\frac{\mathrm{d}\Phi}{\mathrm{d}t} \tag{4.146}$$

Substituting this expression, $u = u_g + u_a$, $v = v_g + v_a$, and $\partial u_g/\partial x + \partial v_g/\partial y = 0$ into the incompressible continuity equation from (3.27), and assuming $\mathrm{d}\Phi_a/\mathrm{d}t$ is small give

$$\frac{1}{g}\frac{\partial}{\partial z}\left(\frac{\mathrm{d}\Phi_g}{\mathrm{d}t}\right) = -\left(\frac{\partial u_a}{\partial x} + \frac{\partial v_a}{\partial y}\right) \tag{4.147}$$

Integrating (4.147) from the surface ($z = 0$) to the mean tropopause height, $z = \Delta z_t$, gives

$$\frac{\mathrm{d}\Phi_g}{\mathrm{d}t} = -g\Delta z_t\left(\frac{\partial u_a}{\partial x} + \frac{\partial v_a}{\partial y}\right) \tag{4.148}$$

When $\mathrm{d}\Phi_g/\mathrm{d}t \neq 0$, the atmosphere has a free surface and average depth Δz_t. When $\mathrm{d}\Phi_g/\mathrm{d}t = 0$, the atmosphere has a rigid lid with constant depth Δz_t. Substituting

(4.148) into (4.145) gives

$$\frac{\mathrm{d}}{\mathrm{d}t}\left(\zeta_{\mathrm{g}} - \frac{f_0}{g\Delta z_{\mathrm{t}}}\Phi_{\mathrm{g}}\right) = -\beta v_{\mathrm{g}} \tag{4.149}$$

where

$$\zeta_{\mathrm{g}} = \frac{\partial v_{\mathrm{g}}}{\partial x} - \frac{\partial u_{\mathrm{g}}}{\partial y} = \frac{1}{f_0}\left(\frac{\partial^2 \Phi_{\mathrm{g}}}{\partial x^2} + \frac{\partial^2 \Phi_{\mathrm{g}}}{\partial y^2}\right)_p = \frac{\nabla_p^2 \Phi_{\mathrm{g}}}{f_0} \tag{4.150}$$

is the **geostrophic potential vorticity.**

Substituting (4.150) and v_{g} from (4.142) into (4.149) and expanding the total derivative give the **quasigeostrophic potential vorticity equation,**

$$\left(\frac{\partial}{\partial t} + \bar{u}\frac{\partial}{\partial x}\right)\left(\nabla_p^2 \Phi_{\mathrm{g}} - \frac{f_0^2}{g\Delta z_{\mathrm{t}}}\Phi_{\mathrm{g}}\right) + \beta\left(\frac{\partial \Phi_{\mathrm{g}}}{\partial x}\right)_p = 0 \tag{4.151}$$

where u and v were replaced by $u = \bar{u} + u'$ and $v \approx v'$, respectively, and the equation was Reynolds-averaged. The assumption $v \approx v'$ is reasonable for midlatitudes, where u dominates over v. Equation (4.151) has a sinusoidal wave solution,

$$\Phi_{\mathrm{g}} = \Phi_{\mathrm{g},0} \sin(\tilde{k}x + \tilde{l}y - \nu_\alpha t) \tag{4.152}$$

where $\Phi_{\mathrm{g},0}$ is the amplitude of the wave. Substituting (4.152) into (4.151) and solving for ν_α give the **dispersion relationship for freely propagating Rossby waves** in a frictionless, incompressible, and adiabatic atmosphere as

$$\nu_\alpha = \left(\bar{u} - \frac{\beta}{\tilde{k}^2 + \tilde{l}^2 + \lambda_{\mathrm{R}}^{-2}}\right)\tilde{k} \tag{4.153}$$

where $\lambda_{\mathrm{R}} = \sqrt{g\Delta z_{\mathrm{t}}}/f_0$ is another form of the **Rossby radius of deformation.** For an atmosphere of constant depth, $\mathrm{d}\Phi_{\mathrm{g}}/\mathrm{d}t = 0$ in (4.149), and (4.153) simplifies to

$$\nu_\alpha = \left(\bar{u} - \frac{\beta}{\tilde{k}^2 + \tilde{l}^2}\right)\tilde{k} \tag{4.154}$$

Both dispersion relationships suggest that Rossby-wave phase speeds, which are less than zero in the equations, propagate from east to west (Example 4.19). However, since the observed mean zonal wind $\bar{u}$ is from west to east with a magnitude greater than the Rossby-wave phase speed, Rossby waves plus the mean wind travel from west to east.

Example 4.19

From (4.154), estimate the Rossby-wave phase speed when $\lambda_{\alpha,z} \approx \lambda_{\alpha,y} = 6000$ km at $\varphi = 45°$ N.

SOLUTION

From (4.95), $\tilde{k} \approx \tilde{l} = 2\pi/\lambda_{\alpha,z} = 1.05 \times 10^{-6}$ m^{-1}. From (4.137), $\beta = 1.62 \times 10^{-11}$ m^{-1} s^{-1}. From (4.154), $c_\alpha = -\beta/\left(\tilde{k}^2 + \tilde{l}^2\right) = -7.3$ m s^{-1}. Thus, Rossby-wave phase speed propagates from east to west.

The wave solution for Rossby waves accounts for horizontally propagating waves in an incompressible atmosphere. Rossby waves propagate vertically as well as horizontally. To account for vertical propagation and air compressibility, the dispersion relationship for Rossby waves can be found by using the divergent continuity equation for air (in three dimensions) and the thermodynamic energy equation from (4.115), instead of the nondivergent continuity equation from (3.27). In the new case, the dispersion relationship is the same as (4.153), except that the Rossby radius of deformation $\lambda_R = \sqrt{g\hbar_e}/f_0$, where the equivalent depth is the same as that for inertia gravity waves, given in (4.131). The resulting dispersion relationship for **vertically and horizontally propagating Rossby waves in a compressible atmosphere** is

$$\nu_\alpha = \bar{u}\tilde{k} - \frac{\beta\tilde{k}}{\tilde{k}^2 + \tilde{l}^2 + \dfrac{f_0^2}{N_{bv}^2}\left(\tilde{m}^2 + \dfrac{\nu_c^2}{c_s^2}\right)} \tag{4.155}$$

For **Rossby waves that propagate only horizontally in a compressible atmosphere,** the dispersion relationship is given by (4.153) with $\lambda_R = \sqrt{g\hbar_e}/f_0$, where $\hbar_e = c_s^2/g$.

Rossby waves can be initiated, not only by a perturbation to the geostrophic velocity in the south–north direction, but also by airflow over large topographical barriers, such as the Rocky or Himalayan mountain ranges. As seen from (4.133), a change in the depth of the atmosphere forces f, u, and v to change, initiating **forced topographic Rossby waves.**

4.4 SUMMARY

In this chapter, elements of the momentum equation, including the local acceleration, the Coriolis acceleration, the Earth's centripetal acceleration, the gravitational force, the pressure-gradient force, the viscous force, and the turbulent-flux divergence, were derived. For a coordinate system fixed on the surface of the Earth, the Coriolis and the Earth's centripetal acceleration are treated as apparent forces.

The momentum equation was simplified to derive expressions for the geostrophic wind and surface winds. It was also transformed to cylindrical coordinates to derive an expression for the gradient wind. Atmospheric waves were also discussed. Important waves include acoustic, Lamb, gravity, inertia Lamb, inertia gravity, and Rossby waves. In the next chapter, the equations of atmospheric dynamics are converted to pressure, sigma-pressure, and sigma-altitude vertical coordinates.

4.5 PROBLEMS

4.1 Assume a spherical coordinate grid cell, centered at $\varphi = 60°$ S, has dimension $d\lambda_e = 2.5°$ and $d\varphi = 2°$. Calculate dx at the north and south boundaries of the cell, and calculate dy.

4.2 If $u = 10$ m s^{-1}, $v = 10$ m s^{-1}, $w = 0.01$ m s^{-1}, and du/dt scales as $u/(\Delta x/u)$, estimate each term on the right side of (4.29) at $\varphi = 30°$ N. Let $\Delta x = 5$ km, $\Delta y = 5$ km, and $\Delta z = 1$ km.

4.3 In Cartesian coordinates, show that $(\mathbf{v} \cdot \nabla)\mathbf{v} = \nabla(\mathbf{v}^2/2) + (\nabla \times \mathbf{v}) \times \mathbf{v}$.

4.4 Assume a grid cell has dimension $\Delta x = 5$ km, $\Delta y = 4$ km, and $\Delta z = 0.1$ km, and assume the west, east, south, north, lower, and upper boundary scalar velocities are $u_1 = +2$, $u_2 = +3$, $v_3 = +1$, $v_4 = -3$, $w_5 = +0.03$, and $w_6 = +0.04$ m s^{-1}, respectively. Estimate the magnitude of the divergence term of the local acceleration $(\mathbf{v} \cdot \nabla\mathbf{v})$.

4.5 If $u = 30$ m s^{-1}, $v = 10$ m s^{-1}, and $\varphi = 45°$ N, find the Coriolis-force magnitude per unit mass.

4.6 Assume a grid cell has dimension $\Delta x = 500$ km and $\Delta y = 400$ km, centered at $\varphi = 30°$ N latitude. Assume potential virtual temperature and air pressure at the west, east, south, and north boundaries are $\theta_{v,1} = 302$ K and $p_{a,1} = 520$ hPa, $\theta_{v,2} = 304$ K and $p_{a,2} = 530$ hPa, $\theta_{v,3} = 302$ K and $p_{a,3} = 500$ hPa, and $\theta_{v,4} = 304$ K and $p_{a,4} = 540$ hPa, respectively. Estimate the geostrophic scalar velocities and geostrophic wind speed.

4.7 In a region to the east of the center of an intense hurricane at $\varphi = 20°$ N latitude, a west–east pressure gradient of 50 hPa per 125 km and a gradient wind scalar velocity of $v = 70$ m s^{-1} are observed. What is the distance between the center of the hurricane and the observation? Assume $T_v = 280$ K and $p_a = 930$ hPa.

4.8 **(a)** Assume $u_1 = 1$ m s^{-1} at $x_1 = 0$ m, $u_2 = 2$ m s^{-1} at $x_2 = 2500$ m, $u_3 = 3$ m s^{-1} at $x_3 = 5000$ m, and $K_{m,xx} = 2.5 \times 10^3$ m^2 s^{-1}. If density and $K_{m,xx}$ are constant, estimate the change in u, due to eddy diffusion alone, at the cell center after one hour.

(b) Assume $w_1 = 0.02$ m s^{-1} at location $z_1 = 0$ m, $w_2 = 0.02$ m s^{-1} at $z_2 = 50$ m, $w_3 = 0.04$ m s^{-1} at $z_3 = 1000$ m, and $K_{m,zz} = 50$ m^2 s^{-1}. If density and $K_{m,zz}$ are constant, estimate the change in w at the cell center after 100 s, due to eddy diffusion alone.

4.9 Calculate the viscous acceleration force per unit mass when the air is dry, $T = 298$ K, $p_a = 995$ hPa, and $u = 1.0$ m s^{-1} at 10 cm above the ground and 0.7 m s^{-1} at 5 cm above the ground.

4.10 Derive (4.155).

4.6 COMPUTER PROGRAMMING PRACTICE

4.11 **(a)** Assume $\Delta x = 5$ km for a grid cell, $v = 5$ m s^{-1} at the center of the cell, $\varphi = 35°$ N, $T = 285$ K, $\partial u/\partial y = 0$, $\partial u/\partial z = 0$, $p_{a,1} = 1010$ hPa and $p_{a,2} = 1006$ hPa at the west and east boundaries of the cell, and $u_1 = +3$ m s^{-1} and $u_2 = +4$ m s^{-1} at the west and east boundaries of the cell, respectively. Write a computer script to find u over time from (4.73), assuming that viscous and diffusive terms are zero. Assume $\partial u/\partial x$ in the $u\partial u/\partial x$ term is constant throughout the simulation, but u is not. Set the time step to 1 s and run the program for one hour. Plot the results.

(b) Change $p_{a,2}$ to (i) 1009 hPa and (ii) 1011 hPa. Plot the results in both cases. For $p_{a,2} = 1009$ hPa, change u_1 to $+3.5$ m s^{-1} and plot the results. Comment on the sensitivity of the solution to changes in pressure and velocity.

4.12 Write a computer program to replicate the results in Fig. 4.18 for both nondispersive and dispersive waves and plot the results.

4.13 Write a computer program to replicate the results in Fig. 4.19 and plot the results.

5

Vertical-coordinate conversions

THE equations of atmospheric dynamics can be solved either in fully compressible form or after they are simplified with the hydrostatic, anelastic, or Boussinesq approximation. In many cases, the equations are adapted to a vertical coordinate other than the altitude coordinate. Such coordinates include the pressure, sigma-pressure, sigma-altitude, and isentropic coordinates. The first three are discussed in this chapter. A disadvantage of the altitude and pressure coordinates is that both allow model layers to intersect ground topography. The sigma-pressure and sigma-altitude coordinates are terrain-following and do not allow model layers to intersect ground topography. A disadvantage of the pressure and sigma-pressure coordinates is that both assume the atmosphere is in hydrostatic balance. Such an assumption is reasonable for global and most mesoscale models but causes inaccuracy for fine horizontal grid resolution, where the atmosphere is often nonhydrostatic. For nonhydrostatic simulations, the altitude or sigma-altitude coordinate can be used.

5.1 HYDROSTATIC AND NONHYDROSTATIC MODELS

Explicit numerical schemes require a time step smaller than the minimum grid spacing divided by the speed of the fastest motion in the grid domain. When acoustic waves are present, they are the fastest motion in the domain, and the time step required to resolve these waves is generally limited by the vertical grid spacing (which is usually smaller than the horizontal grid spacing) divided by the speed of sound. As shown in (4.118), vertically propagating acoustic waves arise as an analytical solution to the equations of atmospheric dynamics when the equations include the horizontal and vertical momentum equations, the continuity equation for air, and the thermodynamic energy equation.

To eliminate the need for a short time step, vertically propagating acoustic waves must be removed (filtered out) as a solution to the basic equations. This can be accomplished by removing the total derivative ($\mathrm{d}w/\mathrm{d}t$) from the vertical momentum equation or removing the local derivative ($\partial\rho_a/\partial t$) from the continuity equation for air. Removing $\mathrm{d}w/\mathrm{d}t$ from the vertical momentum equation gives the **hydrostatic approximation.** Removing $\partial\rho_a/\partial t$ from the continuity equation for air gives the **anelastic approximation** (e.g., Ogura and Phillips 1962; Pielke 1984; Bannon 1996). A subset of the anelastic approximation, used for shallow flows only, is the **Boussinesq approximation.**

A model that uses the hydrostatic equation instead of the full vertical momentum equation is a **hydrostatic model**. The hydrostatic approximation filters out vertically propagating acoustic waves in a hydrostatic model. A model that retains the full vertical momentum equation is a **nonhydrostatic model**. A nonhydrostatic model may be **anelastic** (it solves the anelastic continuity equation for air) or **elastic** (solves the full continuity equation for air and is fully compressible). The anelastic and Boussinesq approximations filter out vertically propagating acoustic waves in an anelastic nonhydrostatic model. Vertically propagating acoustic waves are retained in an elastic nonhydrostatic model.

In a hydrostatic model and in an anelastic or Boussinesq nonhydrostatic model, horizontally propagating acoustic waves still exist but vertically propagating acoustic waves do not. Thus, when either model is solved explicitly, its time step is limited by the smaller of the minimum horizontal grid spacing divided by the speed of sound and the minimum vertical grid spacing divided by the maximum vertical scalar velocity. In most cases, the smaller value is the minimum horizontal grid spacing divided by the speed of sound.

Example 5.1

When $T_v = 298$ K, the maximum vertical scalar velocity is 10 cm s^{-1}, the minimum vertical grid spacing is 10 m, and the minimum horizontal grid spacing is (a) 500 km, (b) 5 km, and (c) 5 m, find the maximum time step in a (i) hydrostatic model and (ii) elastic nonhydrostatic model when both models are solved explicitly.

SOLUTION

From (4.111), the adiabatic speed of sound at 298 K is $c_s = 346$ m s^{-1}.

(i) For the hydrostatic model, the time step is the smaller of 100 s (the vertical dimension divided by the vertical scalar velocity) and (a) 1450 s, (b) 14.5 s, or (c) 0.014 s (the horizontal grid spacing divided by the speed of sound), giving the time step in each case as (a) 100 s, (b) 14.5 s, or (c) 0.014 s.

(ii) For the nonhydrostatic model, the time step is the smaller of 0.028 s (the vertical dimension divided by the speed of sound) and (a) 1450 s, (b) 14.5 s, or (c) 0.014 s (the horizontal dimension divided by the speed of sound), giving the time step in each case as (a) 0.028 s, (b) 0.028 s, and (c) 0.014 s.

Thus, the time step required in an elastic nonhydrostatic model solved explicitly is much smaller than is that in a hydrostatic model unless the horizontal grid spacing is equal to or less than the vertical grid spacing in both models.

In an elastic nonhydrostatic model solved explicitly, the time step is limited by the minimum horizontal or vertical grid spacing divided by the speed of sound. In most mesoscale models, the minimum horizontal grid spacing is 10–50 times greater than the vertical grid spacing, so the time step in an explicit elastic nonhydrostatic model is proportionately smaller than in a hydrostatic or elastic model. Below, the equations required in a hydrostatic, an elastic nonhydrostatic, an anelastic nonhydrostatic, and a Boussinesq nonhydrostatic model are discussed briefly.

5.1.1 Hydrostatic model

A hydrostatic model is a model in which $dw/dt = 0$ in the vertical momentum equation. Such a model requires the solution of the following equations. The potential virtual temperature is found from the thermodynamic energy equation given in (3.76). The water vapor specific humidity is found from the species continuity equation given in (3.61). Horizontal scalar velocities are determined from the momentum equations given in (4.73)/(4.74) and (4.76)/(4.77). Air pressure is extracted diagnostically from the hydrostatic equation given in (2.40). Air temperature is calculated diagnostically from potential virtual temperature with (2.96). Density is found as a function of pressure and temperature from the equation of state given in (2.36). Once density is known, w is extracted diagnostically from the continuity equation for air, given in (3.55). An advantage of the hydrostatic approximation is that it removes vertically propagating acoustic waves as a possible solution to the equations of atmospheric dynamics. A disadvantage is that it does not resolve vertical accelerations well. Such accelerations become more important as grid spacing becomes finer.

5.1.2 Elastic nonhydrostatic model

An elastic nonhydrostatic model is a model that solves the fully compressible equations of atmospheric dynamics without the anelastic or hydrostatic assumption. The equations solved are the same as for the hydrostatic model, except that the vertical scalar velocity is found from the vertical momentum equation, (4.75) or (4.78), air density is found from the continuity equation for air, and air pressure is found from the equation of state.

When an explicit (as opposed to implicit or semiimplicit – Chapter 6) time-differencing scheme is used to solve elastic equations, a short time step is required to resolve acoustic-wave perturbations. A method to reduce computational time in such cases is to solve terms producing acoustic waves with a shorter time step than other terms (e.g., Klemp and Wilhelmson 1978; Chen 1991; Dudhia 1993; Wicker and Skamarock 1998, 2002). Alternatively, an implicit or semiimplicit scheme can be used to integrate terms responsible for acoustic waves implicitly (e.g., Tapp and White 1976; Tanguay *et al.* 1990; Golding 1992; Skamarock and Klemp 1992).

5.1.3 Anelastic nonhydrostatic model

An anelastic nonhydrostatic model is a model in which $\partial \rho_a / \partial t = 0$ in the continuity equation for air. The potential virtual temperature, specific humidity, horizontal scalar velocities, temperature, and density are found from the same equations as in a hydrostatic model, but pressure and the vertical scalar velocity are derived from new equations.

The pressure is determined from an equation for **nonhydrostatic pressure**, derived by assuming that the pressure at a given location can be decomposed as

$$p_a = \hat{p}_a + p''_a \tag{5.1}$$

In this equation, $\hat{p}_a$ is an average pressure, integrated horizontally over a large-scale, hydrostatic environment, and p''_a ($\ll \hat{p}_a$) is the difference between the actual pressure and the large-scale pressure. In a hydrostatic model, $p''_a = 0$. Other variables, including density (ρ_a), specific volume ($\alpha_a = 1/\rho_a$), and potential virtual temperature (θ_v), are decomposed in a similar manner. In the large-scale environment, the atmosphere is assumed to be in hydrostatic balance. Thus, the following equality holds:

$$\frac{1}{\hat{\rho}_a}\frac{\partial \hat{p}_a}{\partial z} = \hat{\alpha}_a \frac{\partial \hat{p}_a}{\partial z} = -g \tag{5.2}$$

At any given location, the air is not in hydrostatic balance, and the full vertical momentum equation must be considered. Decomposing gravitational and pressure-gradient terms in the vertical momentum equation of (4.75) and substituting (5.2) into the result give

$$g + \frac{1}{\rho_a}\frac{\partial p_a}{\partial z} = g + (\hat{\alpha}_a + \alpha''_a)\frac{\partial}{\partial z}(\hat{p}_a + p''_a) \approx \hat{\alpha}_a \frac{\partial p''_a}{\partial z} - \frac{\alpha''_a}{\hat{\alpha}_a} g \tag{5.3}$$

where the product of perturbation terms has been removed because it is small. Substituting (5.3) into (4.75) and ignoring eddy diffusion for now give

$$\frac{\partial w}{\partial t} = -u\frac{\partial w}{\partial x} - v\frac{\partial w}{\partial y} - w\frac{\partial w}{\partial z} - \hat{\alpha}_a \frac{\partial p''_a}{\partial z} + \frac{\alpha''_a}{\hat{\alpha}_a} g \tag{5.4}$$

Taking the divergence of the sum of (5.4) and the horizontal momentum equations, (4.73) and (4.74), results in

$$\frac{\partial}{\partial t}\nabla \cdot (\mathbf{v}\hat{\rho}_a) = -\nabla \cdot [\hat{\rho}_a (\mathbf{v} \cdot \nabla)\mathbf{v}] - \nabla \cdot (\hat{\rho}_a f \mathbf{k} \times \mathbf{v})$$
$$-\nabla_z^2 \hat{p}_a - \nabla^2 p''_a + g\frac{\partial}{\partial z}\left(\frac{\alpha''_a}{\hat{\alpha}_a^2}\right) \tag{5.5}$$

where ∇ and ∇_z are total and horizontal gradient operators from (3.6) and (4.81), respectively. Equation (5.5) can be simplified by differentiating $p_a\alpha_a = R'T_v$ to obtain $d\alpha_a/\alpha_a = dT_v/T_v - dp_a/p_a$, substituting $T_v = \theta_v(p_a/1000\text{ hPa})^\kappa$ from (2.96) into the result, and replacing differential and nondifferential terms with perturbation and large-scale terms, respectively. These steps yield

$$\frac{\alpha''_a}{\hat{\alpha}_a} \approx \frac{\theta''_v}{\hat{\theta}_v} - \frac{c_{v,d}}{c_{p,d}}\frac{p''_a}{\hat{p}_a} \tag{5.6}$$

Another simplification to (5.5) is obtained by removing the local derivative in the continuity equation for air and assuming that density in the equation is large-scale density. The result is the **anelastic approximation to the continuity equation,**

$$\nabla \cdot (\mathbf{v}\hat{\rho}_a) = 0 \qquad (5.7)$$

Substituting (5.6) and (5.7) into (5.5), adding the eddy diffusion term, and solving for perturbation pressure give the **diagnostic equation for nonhydrostatic pressure** as

$$\nabla^2 p_a'' - g\frac{c_{v,d}}{c_{p,d}}\frac{\partial}{\partial z}\left(\hat{\rho}_a\frac{p_a''}{\hat{p}_a}\right) = -\nabla \cdot [\hat{\rho}_a(\mathbf{v}\cdot\nabla)\mathbf{v}] - \nabla \cdot [\hat{\rho}_a f\mathbf{k}\times\mathbf{v}] - \nabla_z^2\hat{p}_a + g\frac{\partial}{\partial z}\left(\hat{\rho}_a\frac{\theta_v''}{\hat{\theta}_v}\right) + \nabla \cdot (\nabla \cdot \hat{\rho}_a\mathbf{K}_m\nabla)\mathbf{v} \qquad (5.8)$$

(e.g., Pielke 1984). Since this equation is diagnostic, it does not have a time derivative. The pressure is calculated from other variables in a manner that ensures that the continuity equation given in (5.7) is satisfied.

When pressure is calculated from (5.8), vertical scalar velocities can be extracted diagnostically from (5.7), since u and v are known from the horizontal momentum equations and $\hat{\rho}_a$ is available from (5.2). In sum, in an anelastic nonhydrostatic model, the vertical momentum equation is not solved directly. Instead, a diagnostic equation for nonhydrostatic pressure, whose solution is computationally intensive, is solved.

5.1.4 Boussinesq approximation

The **Boussinesq approximation** is a special case of the anelastic approximation, valid for cases where the vertical scale of circulation is much less than the scale height of the atmosphere (shallow circulation). With the Boussinesq approximation, density (ρ_a) is replaced with a constant boundary-layer-averaged value ($\hat{\rho}_a$) in all the equations of atmospheric dynamics, except in the buoyancy term of the vertical momentum equation. This differs from the anelastic approximation, in which density is horizontally averaged but varies for each layer in the vertical. Because density is constant everywhere with the Boussinesq approximation, the continuity equation for air is nondivergent and given by $\nabla \cdot \mathbf{v} = 0$ rather than by (5.7). With the Boussinesq approximation, the vertical momentum equation is the same as in (5.4), but with the eddy diffusion term added back in, and the diagnostic equation for nonhydrostatic pressure is the same as in (5.8), except that the second term on the left side of the equation is eliminated. The resulting equation is a **Poisson partial differential equation.** Lilly (1996) discusses advantages and disadvantages of the anelastic and Boussinesq approximations.

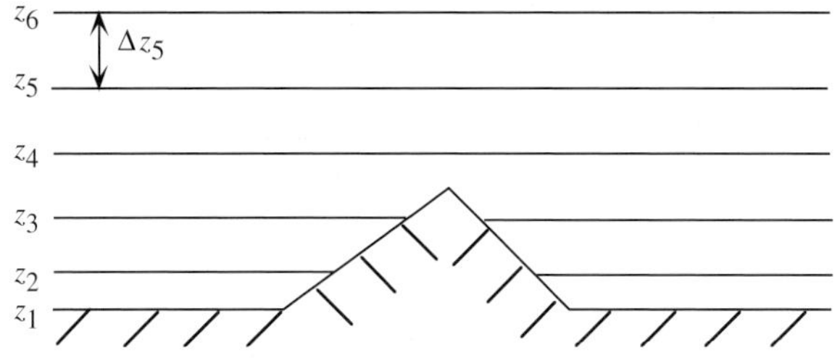

Figure 5.1 Heights of altitude coordinate surfaces.

In the following sections, the conversion of the equations of atmospheric dynamics from the altitude coordinate to the pressure, sigma-pressure, and sigma-altitude coordinates is discussed. Whereas the altitude and sigma-altitude coordinates are useful for solving hydrostatic and nonhydrostatic model equations, the pressure and sigma-pressure coordinates are used for solving hydrostatic model equations only. An additional vertical coordinate, the isentropic coordinate, in which layer tops and bottoms are defined as surfaces of constant potential temperature, is discussed in Kasahara (1974).

5.2 ALTITUDE COORDINATE

Equations describing basic atmospheric processes were written in Chapters 3 and 4 in Cartesian and spherical horizontal coordinates and the altitude vertical coordinate. In the altitude (z) coordinate, layer tops and bottoms are defined as surfaces of constant altitude, and pressure varies in the x- and y-directions along these surfaces. Altitude is an independent variable, and pressure is a dependent variable. This coordinate causes boundary-condition problems when surface elevations are not uniform. In such cases, planes of constant altitude intercept topography, as shown in Fig. 5.1. Special boundary conditions are needed in this situation.

The altitude coordinate can be used to simulate hydrostatic or nonhydrostatic model equations. In both cases, some variables are solved for prognostically whereas others are solved for diagnostically. A **prognostic equation** is one in which a time derivative is solved. The thermodynamic energy equation, species continuity equation, and horizontal momentum equations are prognostic equations for the potential virtual temperature, species mixing ratio, and horizontal scalar velocities, respectively. A **diagnostic equation** is one in which a time derivative is not solved. The equation of state is a diagnostic equation for any parameter within it for hydrostatic and nonhydrostatic models. The hydrostatic equation is a diagnostic equation in a hydrostatic model.

5.3 PRESSURE COORDINATE

In the **pressure (p or isobaric) coordinate**, layer tops and bottoms are defined as surfaces of constant pressure. Since altitude is a function of pressure in the

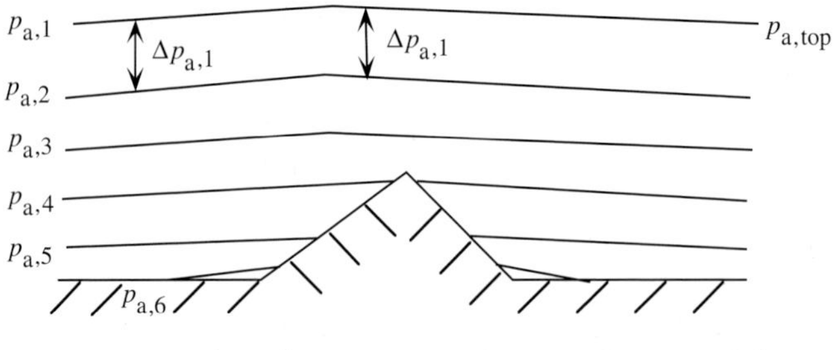

Figure 5.2 Heights of pressure coordinate surfaces. Each line is a surface of constant pressure. Pressure thicknesses of each layer are constant throughout the layer.

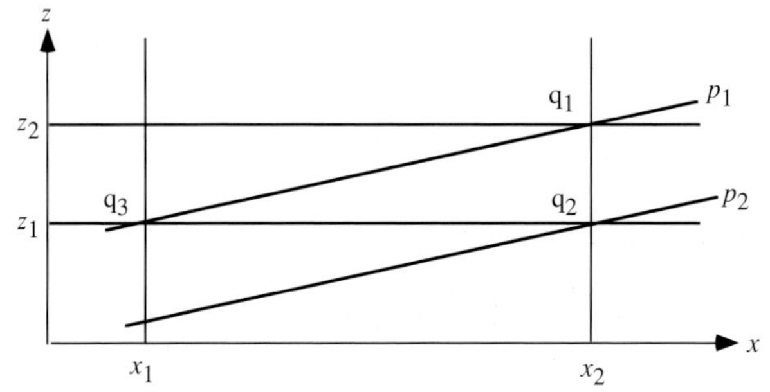

Figure 5.3 Intersection of pressure (p) with altitude (z) surfaces. The moist-air mass mixing ratio (q) varies along each surface of constant pressure.

x- and y-directions, pressure is the independent variable and altitude is a dependent variable. The pressure coordinate does not erase problems associated with surface topography since surfaces of constant pressure intercept topography in the pressure coordinate (Fig. 5.2). Equations in the pressure coordinate are derived from the hydrostatic assumption. This assumption removes vertically propagating acoustic waves as a solution to the equations of atmospheric dynamics but reduces the accuracy of the solution when horizontal grid cells are less than 3 km in width. An advantage of the pressure coordinate is that the continuity equation for air in the coordinate requires neither a density nor a time-derivative term. In this section, hydrostatic model equations for the pressure coordinate are derived.

5.3.1 Gradient conversion from altitude to pressure coordinate

The gradient conversion from the altitude to the pressure coordinate can be obtained diagrammatically. Figure 5.3 shows two lines of constant pressure (**isobars**) intersecting two lines of constant altitude in the x–z plane, where x is a Cartesian coordinate distance. Suppose the moist-air mass mixing ratio q varies along each surface of constant pressure, as shown in the figure. In this case, q varies with distance as

$$\frac{q_2 - q_3}{x_2 - x_1} = \frac{q_1 - q_3}{x_2 - x_1} + \left(\frac{p_2 - p_1}{x_2 - x_1}\right)\left(\frac{q_1 - q_2}{p_1 - p_2}\right) \tag{5.9}$$

which is an exact equivalence.

As $x_2 - x_1 \rightarrow 0$ and $p_1 - p_2 \rightarrow 0$, the differences in (5.9) can be approximated by differentials. From Fig. 5.3, the approximations for each term are

$$\left(\frac{\partial q}{\partial x}\right)_z = \frac{q_2 - q_3}{x_2 - x_1} \qquad \left(\frac{\partial q}{\partial x}\right)_p = \frac{q_1 - q_3}{x_2 - x_1}$$

$$\left(\frac{\partial p_a}{\partial x}\right)_z = \frac{p_2 - p_1}{x_2 - x_1} \qquad \left(\frac{\partial q}{\partial p_a}\right)_x = \frac{q_1 - q_2}{p_1 - p_2} \tag{5.10}$$

Substituting these terms into (5.9) gives

$$\left(\frac{\partial q}{\partial x}\right)_z = \left(\frac{\partial q}{\partial x}\right)_p + \left(\frac{\partial p_a}{\partial x}\right)_z\left(\frac{\partial q}{\partial p_a}\right)_x \tag{5.11}$$

which generalizes for any variable as

$$\left(\frac{\partial}{\partial x}\right)_z = \left(\frac{\partial}{\partial x}\right)_p + \left(\frac{\partial p_a}{\partial x}\right)_z\left(\frac{\partial}{\partial p_a}\right)_x \tag{5.12}$$

A similar equation can be written for the y-direction. Combining the x- and y-equations in vector form yields the **horizontal gradient conversion from Cartesian-altitude to Cartesian-pressure coordinates** as

$$\nabla_z = \nabla_p + \nabla_z(p_a)\frac{\partial}{\partial p_a} \tag{5.13}$$

where ∇_z is the horizontal gradient operator in Cartesian-altitude coordinates given in (4.81), and

$$\nabla_p = \mathrm{i}\left(\frac{\partial}{\partial x}\right)_p + \mathrm{j}\left(\frac{\partial}{\partial y}\right)_p \tag{5.14}$$

is the **horizontal gradient operator in Cartesian-pressure coordinates.** The gradient operator in Cartesian-pressure coordinates is used to take partial derivatives of a variable along a surface of constant pressure. The derivative with respect to pressure in (5.13) is taken in a fixed column in the x–y plane.

In (5.12), any dimension can be substituted for x. Substituting time t for x gives the **time-derivative conversion between the altitude and the pressure coordinate** as

$$\left(\frac{\partial}{\partial t}\right)_z = \left(\frac{\partial}{\partial t}\right)_p + \left(\frac{\partial p_a}{\partial t}\right)_z\left(\frac{\partial}{\partial p_a}\right)_t \tag{5.15}$$

Equation (5.13) may be used to obtain the relationship between the variation of geopotential along a surface of constant pressure and the variation of pressure along a surface of constant altitude. From (4.48), the geopotential was defined as $\Phi = gz$. The geopotential is constant along surfaces of constant altitude ($\nabla_z \Phi = 0$). Substituting $\Phi = gz$, $\partial p_a / \partial z = -\rho_a g$, and $\partial \Phi / \partial z = g$ into (5.13) and rearranging give

$$\nabla_z (p_a) = -\frac{\partial p_a}{\partial \Phi} \nabla_p \Phi = -\frac{\partial p_a}{g \partial z} \nabla_p \Phi = \rho_a \nabla_p \Phi \tag{5.16}$$

which relates changes in geopotential along surfaces of constant pressure to changes in pressure along surfaces of constant altitude. This equation expands to

$$\left(\frac{\partial p_a}{\partial x}\right)_z = \rho_a \left(\frac{\partial \Phi}{\partial x}\right)_p \qquad \left(\frac{\partial p_a}{\partial y}\right)_z = \rho_a \left(\frac{\partial \Phi}{\partial y}\right)_p \tag{5.17}$$

5.3.2 Continuity equation for air in the pressure coordinate

Equation (3.20) gave the continuity equation for air in Cartesian-altitude coordinates. Dividing (3.20) into horizontal and vertical components gives

$$\left(\frac{\partial \rho_a}{\partial t}\right)_z = -\rho_a \left(\nabla_z \cdot \mathbf{v}_h + \frac{\partial w}{\partial z}\right) - (\mathbf{v}_h \cdot \nabla_z)\rho_a - w \frac{\partial \rho_a}{\partial z} \tag{5.18}$$

where the subscript z indicates that the value is taken along a surface of constant altitude, and $\mathbf{v}_h = \mathbf{i}u + \mathbf{j}v$. Applying (5.13) to $\mathbf{v}_h$ gives

$$\nabla_z \cdot \mathbf{v}_h = \nabla_p \cdot \mathbf{v}_h + \nabla_z (p_a) \cdot \frac{\partial \mathbf{v}_h}{\partial p_a} \tag{5.19}$$

Substituting (5.19) and $\mathrm{d}z = -\mathrm{d}p_a / \rho_a g$ into (5.18) gives

$$\left(\frac{\partial \rho_a}{\partial t}\right)_z = -\rho_a \left(\nabla_p \cdot \mathbf{v}_h + \nabla_z (p_a) \cdot \frac{\partial \mathbf{v}_h}{\partial p_a}\right) - (\mathbf{v}_h \cdot \nabla_z)\, \rho_a + \rho_a g \frac{\partial (w \rho_a)}{\partial p_a} \tag{5.20}$$

The **vertical scalar velocity in the pressure coordinate** (hPa s^{-1}) is defined as

$$\boxed{\mathrm{w}_p = \frac{\mathrm{d}p_a}{\mathrm{d}t} = \left(\frac{\partial p_a}{\partial t}\right)_z + (\mathbf{v} \cdot \nabla) p_a = \left(\frac{\partial p_a}{\partial t}\right)_z + (\mathbf{v}_h \cdot \nabla_z) p_a + w \frac{\partial p_a}{\partial z}} \tag{5.21}$$

Substituting $\partial p_a / \partial z = -\rho_a g$ into (5.21) yields

$$\mathrm{w}_p = -\left(\rho_a g \frac{\partial z}{\partial t}\right)_z + (\mathbf{v}_h \cdot \nabla_z) p_a - w \rho_a g \tag{5.22}$$

If horizontal and temporal variations in pressure are ignored, the vertical scalar velocity in the pressure coordinate simplifies to $\mathrm{w}_p = -w \rho_a g$. Thus, a positive vertical scalar velocity in the pressure coordinate corresponds to downward motion, or a negative vertical scalar velocity in the altitude coordinate. Horizontal and temporal variations in pressure cannot be excluded when the continuity equation for air is derived. Taking the partial derivative of w_p from (5.22) with respect to altitude gives

$$\frac{\partial w_p}{\partial z} = -g\left(\frac{\partial \rho_a}{\partial t}\right)_z + \nabla_z(p_a) \cdot \frac{\partial \mathbf{v}_h}{\partial z} + (\mathbf{v}_h \cdot \nabla_z)\frac{\partial p_a}{\partial z} - g\frac{\partial(w\rho_a)}{\partial z} \tag{5.23}$$

Substituting the hydrostatic equation throughout yields

$$\rho_a \frac{\partial w_p}{\partial p_a} = \left(\frac{\partial \rho_a}{\partial t}\right)_z + \rho_a \nabla_z(p_a) \cdot \frac{\partial \mathbf{v}_h}{\partial p_a} + (\mathbf{v}_h \cdot \nabla_z)\rho_a - \rho_a g\frac{\partial(w\rho_a)}{\partial p_a} \tag{5.24}$$

Adding (5.20) to (5.24) and compressing give the **continuity equation for air in Cartesian-pressure coordinates** as

$$\nabla_p \cdot \mathbf{v}_h + \frac{\partial w_p}{\partial p_a} = 0 \tag{5.25}$$

which expands to

$$\left(\frac{\partial u}{\partial x} + \frac{\partial v}{\partial y}\right)_p + \frac{\partial w_p}{\partial p_a} = 0 \tag{5.26}$$

Equation (5.25) shows that, in the pressure coordinate, the continuity equation for air does not depend on air density, nor does it require a time-derivative term.

Example 5.1

Assume a grid cell in the pressure coordinate has dimension $\Delta x = 5$ km, $\Delta y = 5$ km, and $\Delta p_a = -10$ hPa, and that west, east, south, north, and lower boundary scalar velocities are $u_1 = -3$, $u_2 = -1$ m s^{-1}, $v_3 = +2$, $v_4 = -2$ m s^{-1}, and $w_{p,5} = +0.02$ hPa s^{-1}, respectively. Use the continuity equation in the pressure coordinate to estimate the pressure-coordinate vertical scalar velocity at the cell's top.

SOLUTION

Applying the velocities and incremental distances to (5.26) gives

$$\frac{(-1+3)\,\text{m s}^{-1}}{5000\,\text{m}} + \frac{(-2-2)\,\text{m s}^{-1}}{5000\,\text{m}} + \frac{(w_{p,6} - 0.02)\,\text{hPa s}^{-1}}{-10\,\text{hPa}} = 0$$

which has solution $w_{p,6} = +0.016$ hPa s^{-1}.

5.3.3 Total derivative in the pressure coordinate

The total derivative in the pressure coordinate may be used to derive the species continuity equation, the thermodynamic energy equation, and the momentum equations in the pressure coordinate. The total derivative in Cartesian-altitude coordinates was

$$\frac{\mathrm{d}}{\mathrm{d}t} = \left(\frac{\partial}{\partial t}\right)_z + (\mathbf{v}_\mathrm{h} \cdot \nabla_z) + w\frac{\partial}{\partial z} \tag{5.27}$$

Substituting time and horizontal gradient conversions from (5.15) and (5.13), respectively, into (5.27) gives

$$\frac{\mathrm{d}}{\mathrm{d}t} = \left(\frac{\partial}{\partial t}\right)_p + \left(\frac{\partial p_\mathrm{a}}{\partial t}\right)_z \frac{\partial}{\partial p_\mathrm{a}} + (\mathbf{v}_\mathrm{h} \cdot \nabla_p) + [(\mathbf{v}_\mathrm{h} \cdot \nabla_z)\, p_\mathrm{a}]\frac{\partial}{\partial p_\mathrm{a}} + w\frac{\partial}{\partial z} \tag{5.28}$$

From (5.21), the vertical scalar velocity in the altitude coordinate can be written as a function of the vertical scalar velocity in the pressure coordinate as

$$w = \frac{\left(\frac{\partial p_\mathrm{a}}{\partial t}\right)_z + (\mathbf{v}_\mathrm{h} \cdot \nabla_z)p_\mathrm{a} - \mathrm{w}_p}{\rho_\mathrm{a} g} \tag{5.29}$$

Substituting this equation and the hydrostatic equation into (5.28) and simplifying give the **total derivative in Cartesian-pressure coordinates** as

$$\frac{\mathrm{d}}{\mathrm{d}t} = \left(\frac{\partial}{\partial t}\right)_p + (\mathbf{v}_\mathrm{h} \cdot \nabla_p) + \mathrm{w}_p\frac{\partial}{\partial p_\mathrm{a}} \tag{5.30}$$

5.3.4 Species continuity equation in the pressure coordinate

Applying the total derivative in Cartesian-pressure coordinates to (3.54) gives the **species continuity equation in Cartesian-pressure coordinates** as

$$\frac{\mathrm{dq}}{\mathrm{d}t} = \left(\frac{\partial \mathrm{q}}{\partial t}\right)_p + (\mathbf{v}_\mathrm{h} \cdot \nabla_p)\mathrm{q} + \mathrm{w}_p\frac{\partial \mathrm{q}}{\partial p_\mathrm{a}} = \frac{(\nabla \cdot \rho_\mathrm{a}\mathbf{K}_h\nabla)\,\mathrm{q}}{\rho_\mathrm{a}} + \sum_{n=1}^{N_{\mathrm{e,t}}} R_n \tag{5.31}$$

If the concentration of a species is given as molecules per cubic centimeter of air (N), particles per cubic centimeter of air (n_i), or cubic centimeters of particle component per cubic centimeter of air ($v_{q,i}$), it can be converted to the moist-air mass mixing ratio for use in (5.31) with

$$\mathrm{q} = \frac{Nm}{\rho_\mathrm{a} A} \qquad \mathrm{q} = \frac{n_i \upsilon_i \rho_\mathrm{p}}{\rho_\mathrm{a}} \qquad \mathrm{q} = \frac{v_{q,i}\rho_\mathrm{p}}{\rho_\mathrm{a}} \tag{5.32}$$

respectively, where m is molecular weight (g mol^{-1}), A is Avogadro's number (molec. mol^{-1}), υ_i is the volume of one particle in size bin i (cm^3), and ρ_p is particle mass density (g cm^{-3}).

Since R_n does not include spatial derivatives, it does not need to be transformed to the pressure coordinate. The eddy diffusion term in (5.31) includes spatial derivatives. Its transformation is performed by breaking the term into directional components, applying the gradient conversion from (5.13), and applying $\mathrm{d}z = \mathrm{d}p_\mathrm{a}/\rho_\mathrm{a} g$.

The result is

$$(\nabla\bullet\rho_a\mathbf{K}_h\nabla)\,q=\left(\frac{\partial}{\partial x}\right)_z\left[\rho_a K_{h,xx}\left(\frac{\partial q}{\partial x}\right)_z\right]+\left(\frac{\partial}{\partial y}\right)_z\left[\rho_a K_{h,yy}\left(\frac{\partial q}{\partial y}\right)_z\right]+\frac{\partial}{\partial z}\left(\rho_a K_{h,zz}\frac{\partial q}{\partial z}\right)$$
$$=\left[\left(\frac{\partial}{\partial x}\right)_p+\left(\frac{\partial p_a}{\partial x}\right)_z\frac{\partial}{\partial p_a}\right]\left\{\rho_a K_{h,xx}\left[\left(\frac{\partial q}{\partial x}\right)_p+\left(\frac{\partial p_a}{\partial x}\right)_z\frac{\partial q}{\partial p_a}\right]\right\}$$
$$+\left[\left(\frac{\partial}{\partial y}\right)_p+\left(\frac{\partial p_a}{\partial y}\right)_z\frac{\partial}{\partial p_a}\right]\left\{\rho_a K_{h,yy}\left[\left(\frac{\partial q}{\partial y}\right)_p+\left(\frac{\partial p_a}{\partial y}\right)_z\frac{\partial q}{\partial p_a}\right]\right\}$$
$$+\rho_a g^2\frac{\partial}{\partial p_a}\left(\rho_a^2 K_{h,zz}\frac{\partial q}{\partial p_a}\right)\qquad(5.33)$$

5.3.5 Thermodynamic energy equation in the pressure coordinate

Applying the total derivative in Cartesian-pressure coordinates to (3.76) gives the **thermodynamic energy equation in Cartesian-pressure coordinates** as

$$\left(\frac{\partial\theta_v}{\partial t}\right)_p+(\mathbf{v}_h\bullet\nabla_p)\theta_v+w_p\frac{\partial\theta_v}{\partial p_a}=\frac{(\nabla\bullet\rho_a\mathbf{K}_h\nabla)\theta_v}{\rho_a}+\frac{\theta_v}{c_{p,d}T_v}\sum_{n=1}^{N_{e,h}}\frac{dQ_n}{dt}\qquad(5.34)$$

The eddy diffusion terms here are the same as those used in the species continuity equation, except that θ_v is used here instead of q.

5.3.6 Horizontal momentum equations in the pressure coordinate

Applying the total derivative in Cartesian-pressure coordinates to velocity, neglecting viscous terms, and substituting $\nabla_z(p_a)=\rho_a\nabla_p\Phi$ from (5.16) into (4.71) give the **horizontal momentum equation in Cartesian-pressure coordinates** as

$$\left(\frac{\partial\mathbf{v}_h}{\partial t}\right)_p+(\mathbf{v}_h\bullet\nabla_p)\mathbf{v}_h+w_p\frac{\partial\mathbf{v}_h}{\partial p_a}=-f\mathbf{k}\times\mathbf{v}_h-\nabla_p\Phi+\frac{(\nabla\bullet\rho_a\mathbf{K}_m\nabla)\,\mathbf{v}_h}{\rho_a}\qquad(5.35)$$

The eddy diffusion term expands to $(\nabla\bullet\rho_a\mathbf{K}_m\nabla)\mathbf{v}_h=\mathbf{i}\,(\nabla\bullet\rho_a\mathbf{K}_m\nabla)\,u+\mathbf{j}\,(\nabla\bullet\rho_a\mathbf{K}_m\nabla)\,v$. Applying the gradient conversion from (5.13) to the u-term gives

$$(\nabla\bullet\rho_a\mathbf{K}_m\nabla)\,u=\left[\left(\frac{\partial}{\partial x}\right)_p+\left(\frac{\partial p_a}{\partial x}\right)_z\frac{\partial}{\partial p_a}\right]\left\{\rho_a K_{m,xx}\left[\left(\frac{\partial u}{\partial x}\right)_p+\left(\frac{\partial p_a}{\partial x}\right)_z\frac{\partial u}{\partial p_a}\right]\right\}$$
$$+\left[\left(\frac{\partial}{\partial y}\right)_p+\left(\frac{\partial p_a}{\partial y}\right)_z\frac{\partial}{\partial p_a}\right]\left\{\rho_a K_{m,yx}\left[\left(\frac{\partial u}{\partial y}\right)_p+\left(\frac{\partial p_a}{\partial y}\right)_z\frac{\partial u}{\partial p_a}\right]\right\}$$
$$+\rho_a g^2\frac{\partial}{\partial p_a}\left(\rho_a^2 K_{m,zx}\frac{\partial u}{\partial p_a}\right)\qquad(5.36)$$

The equation for the v-direction is similar. The three terms on the right side of (5.36) represent accelerations in the x-direction due to west–east, south–north, and vertical turbulent transport, respectively, of west–east momentum.

5.3.7 Vertical momentum equation in the pressure coordinate

The vertical momentum equation in the pressure coordinate is the hydrostatic equation, since the conversion from the altitude to the pressure coordinate requires the hydrostatic assumption. Over large horizontal scales (>3 km), the hydrostatic assumption is reasonable. Over smaller scales and in the presence of large-scale convective clouds, vertical accelerations are important, and the hydrostatic assumption is not so good. In such cases, the altitude or sigma-altitude coordinate is generally used, and a vertical momentum equation that includes a local acceleration term is solved.

When the hydrostatic assumption is used, the vertical momentum equation in the altitude coordinate is $\partial p_a/\partial z = -\rho_a g$. Substituting $g = \partial \Phi/\partial z$ from (4.48), $p_a = \rho_a R' T_v$ from (2.36), and $T_v = \theta_v P$ from (2.99) into the hydrostatic equation gives the **hydrostatic equation in the pressure coordinate** as

$$\frac{\partial \Phi}{\partial p_a} = -\frac{R' T_v}{p_a} = -\frac{R' \theta_v P}{p_a} = -\frac{R' \theta_v}{p_a}\left(\frac{p_a}{1000\text{ hPa}}\right)^{\kappa} \tag{5.37}$$

Substituting $\kappa = R'/c_{p,d}$ into (5.37) and rearranging give another form of the equation as

$$d\Phi = -c_{p,d}\theta_v\, d\left[\left(\frac{p_a}{1000\text{ hPa}}\right)^{\kappa}\right] = -c_{p,d}\theta_v\, dP \tag{5.38}$$

A finite-difference discretization of this equation is given in Chapter 7.

5.3.8 Geostrophic wind in the pressure coordinate

Substituting (5.17) into (4.79) gives **geostrophic scalar velocities in Cartesian-pressure coordinates** as

$$v_g = \frac{1}{f}\left(\frac{\partial \Phi}{\partial x}\right)_p \qquad u_g = -\frac{1}{f}\left(\frac{\partial \Phi}{\partial y}\right)_p \tag{5.39}$$

The vector form of (5.39) is

$$\mathbf{v}_g = \mathbf{i}u_g + \mathbf{j}v_g = -\mathbf{i}\frac{1}{f}\left(\frac{\partial \Phi}{\partial y}\right)_p + \mathbf{j}\frac{1}{f}\left(\frac{\partial \Phi}{\partial x}\right)_p = \frac{1}{f}(\mathbf{k} \times \nabla_p \Phi) \tag{5.40}$$

which indicates that the geostrophic wind flows parallel to lines of constant geopotential. Lines of constant geopotential (**contour lines**) on a constant-pressure surface are analogous to lines of constant pressure (isobars) on a constant-altitude surface.

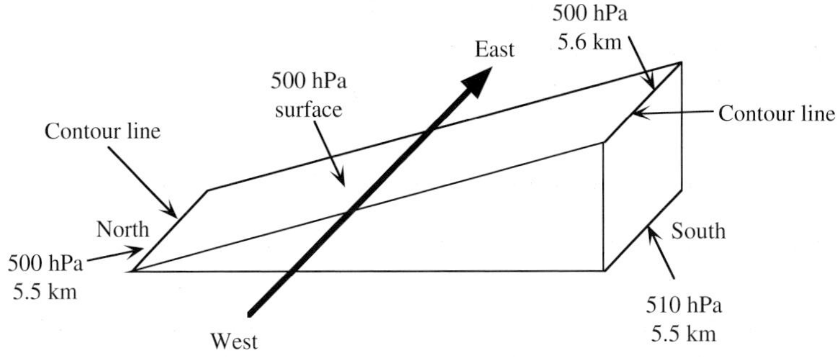

Figure 5.4 The geostrophic wind (arrow) flows parallel to contour lines on a constant-pressure surface. In this figure, contour lines at 5.5 and 5.6 km are shown on a 500-hPa surface. The resulting geostrophic wind originates from the west.

Since the geostrophic wind flows parallel to isobars, it also flows parallel to contour lines. Figure 5.4 shows two west–east contour lines at 5.5 and 5.6 km on a surface of constant pressure. The geostrophic wind in this case flows from the west.

5.4 SIGMA-PRESSURE COORDINATE

A vertical coordinate in which surfaces of the coordinate variable do not intersect ground topography is the **sigma-pressure (σ-p) coordinate** (Phillips 1957). In this coordinate, layer tops and bottoms are defined as surfaces of constant σ, where σ equals the difference between the layer bottom and model top pressures divided by the difference between the model bottom and model top pressures. Since the hydrostatic assumption is used to derive equations in the sigma-pressure coordinate, vertically propagating acoustic waves are filtered out as a solution to these equations. Nonhydrostatic flows are generally not simulated in the sigma-pressure coordinate. In this section, hydrostatic model equations for the sigma-pressure coordinate are derived.

5.4.1 Definitions

In the sigma-pressure coordinate system, layer tops and bottoms are defined as surfaces of constant σ, given by

$$\sigma = \frac{p_a - p_{a,top}}{p_{a,surf} - p_{a,top}} = \frac{p_a - p_{a,top}}{\pi_a} \tag{5.41}$$

where p_a is the air pressure at the altitude of interest, $p_{a,surf}$ is the model surface pressure, $p_{a,top}$ is the model top pressure, and $\pi_a = p_{a,surf} - p_{a,top}$ is the pressure difference between the model surface and top (**π-value**). From (5.41), the pressure at a σ-level is

$$p_a = p_{a,top} + \sigma\pi_a \tag{5.42}$$

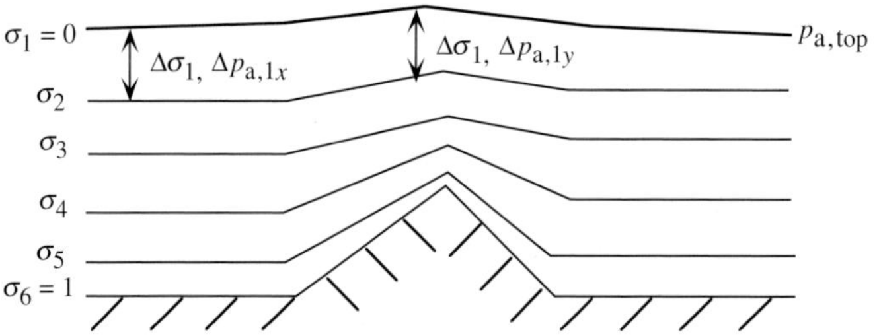

Figure 5.5 Heights of sigma-pressure coordinate surfaces. Each layer has the same σ-thickness but a different pressure thickness. Subscripts x and y denote two horizontal locations. Surface pressures vary horizontally.

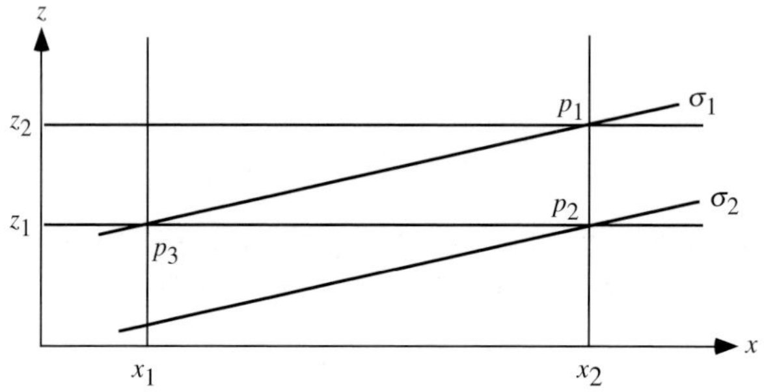

Figure 5.6 Intersection of σ-p with z surfaces and values of pressure at each intersection.

The parameter σ is a fraction ≤ 1. At the model top, $\sigma = 0$. At the model surface, $\sigma = 1$. Layer boundaries correspond to σ-values that are constant in time. The model top pressure, $p_{a,top}$, is constant in space and time along a boundary corresponding to $\sigma = 0$. The model surface pressure, $p_{a,surf}$, varies in space and time along the surface corresponding to $\sigma = 1$. Since $p_{a,surf}$ varies, p_a and π_a must also vary in space and time. Figure 5.5 shows heights of constant-σ surfaces in the sigma-pressure coordinate.

5.4.2 Gradient conversion from the altitude to the sigma-pressure coordinate

The conversion between pressure in the altitude coordinate and pressure in the sigma-pressure coordinate can be derived from Fig. 5.6. The figure shows the intersection of σ with z-surfaces and the values of pressure at each intersection point on an x–z plane. From the figure, the change in pressure per unit distance is

$$\frac{p_2 - p_3}{x_2 - x_1} = \frac{p_1 - p_3}{x_2 - x_1} + \left(\frac{\sigma_2 - \sigma_1}{x_2 - x_1}\right)\left(\frac{p_1 - p_2}{\sigma_1 - \sigma_2}\right) \tag{5.43}$$

which is an exact equivalence. Substituting

$$\left(\frac{\partial p_a}{\partial x}\right)_z = \frac{p_2 - p_3}{x_2 - x_1} \qquad \left(\frac{\partial p_a}{\partial x}\right)_\sigma = \frac{p_1 - p_3}{x_2 - x_1}$$
$$\left(\frac{\partial \sigma}{\partial x}\right)_z = \frac{\sigma_2 - \sigma_1}{x_2 - x_1} \qquad \left(\frac{\partial p_a}{\partial \sigma}\right)_x = \frac{p_1 - p_2}{\sigma_1 - \sigma_2} \tag{5.44}$$

into (5.43) gives

$$\left(\frac{\partial p_a}{\partial x}\right)_z = \left(\frac{\partial p_a}{\partial x}\right)_\sigma + \left(\frac{\partial \sigma}{\partial x}\right)_z \left(\frac{\partial p_a}{\partial \sigma}\right)_x \tag{5.45}$$

A similar expression exists for the y-direction. The vector sum of the two equations is

$$\nabla_z(p_a) = \nabla_\sigma(p_a) + (\nabla_z\sigma)\frac{\partial p_a}{\partial \sigma} \tag{5.46}$$

where

$$\nabla_\sigma = \left(\mathbf{i}\frac{\partial}{\partial x}\right)_\sigma + \left(\mathbf{j}\frac{\partial}{\partial y}\right)_\sigma \tag{5.47}$$

is the **horizontal gradient operator in Cartesian-sigma-pressure coordinates.** Equation (5.46) can be generalized for any variable with

$$\nabla_z = \nabla_\sigma + \nabla_z(\sigma)\frac{\partial}{\partial \sigma} \tag{5.48}$$

which is the **gradient conversion from Cartesian-altitude to Cartesian-sigma-pressure coordinates.**

The gradient of σ defined in (5.41), along a surface of constant altitude is

$$\nabla_z\sigma = (p_a - p_{a,top})\nabla_z\left(\frac{1}{\pi_a}\right) + \frac{\nabla_z(p_a)}{\pi_a} = -\frac{\sigma}{\pi_a}\nabla_z(\pi_a) + \frac{\nabla_z(p_a)}{\pi_a} \tag{5.49}$$

Substituting (5.49) into (5.48) yields another form of the gradient conversion,

$$\nabla_z = \nabla_\sigma - \left[\frac{\sigma}{\pi_a}\nabla_z(\pi_a) - \frac{\nabla_z(p_a)}{\pi_a}\right]\frac{\partial}{\partial \sigma} \tag{5.50}$$

5.4.3 Gradient conversion from the pressure to the sigma-pressure coordinate

The gradient conversion from the pressure to the sigma-pressure coordinate is derived from Fig. 5.7. The figure shows intersections of pressure, altitude, and σ surfaces in an x–z plane. From the figure, the change in moist-air mass mixing ratio over distance is

$$\frac{q_1 - q_3}{x_2 - x_1} = \frac{q_2 - q_3}{x_2 - x_1} + \left(\frac{\sigma_1 - \sigma_2}{x_2 - x_1}\right)\left(\frac{q_1 - q_2}{\sigma_1 - \sigma_2}\right) \tag{5.51}$$

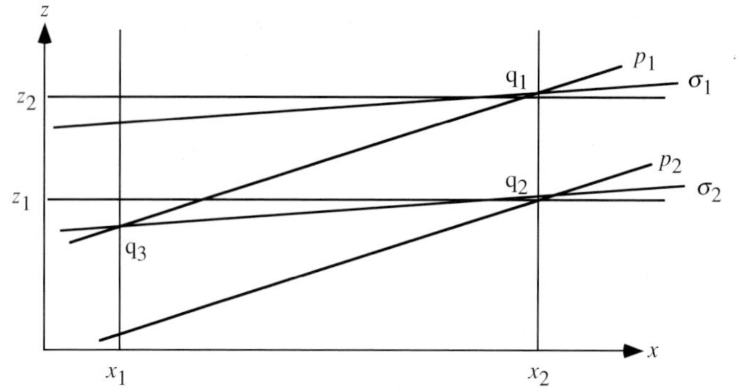

Figure 5.7 Intersection of p, z, and σ surfaces in an x–z plane. Moist-air mass mixing ratios (q) on constant-pressure surfaces are also shown.

which is an exact equivalence. Making substitutions similar to those in (5.44) gives

$$\left(\frac{\partial q}{\partial x}\right)_p = \left(\frac{\partial q}{\partial x}\right)_\sigma + \left(\frac{\partial \sigma}{\partial x}\right)_p \left(\frac{\partial q}{\partial \sigma}\right)_x \tag{5.52}$$

A similar equation is written for the y-direction. Writing the x- and y-equations in gradient operator form and generalizing for any variable give the **gradient conversion from Cartesian-pressure to Cartesian-sigma-pressure coordinates** as

$$\nabla_p = \nabla_\sigma + \nabla_p(\sigma)\frac{\partial}{\partial \sigma} \tag{5.53}$$

Taking the gradient of σ along a surface of constant pressure and noting that $\nabla_p(p_a) = 0$ and $\nabla_p(p_{top}) = 0$ yield

$$\nabla_p(\sigma) = (p_a - p_{top})\nabla_p\left(\frac{1}{\pi_a}\right) + \frac{\nabla_p(p_a - p_{top})}{\pi_a} = -\frac{\sigma}{\pi_a}\nabla_p(\pi_a) \tag{5.54}$$

Since horizontal gradients of π_a are independent of the vertical coordinate system, $\nabla_p(\pi_a) = \nabla_\sigma(\pi_a) = \nabla_z(\pi_a)$. Substituting this expression and (5.54) into (5.53) gives the **gradient conversion from Cartesian-pressure to Cartesian-sigma-pressure coordinates** as

$$\nabla_p = \nabla_\sigma - \frac{\sigma}{\pi_a}\nabla_\sigma(\pi_a)\frac{\partial}{\partial \sigma} \tag{5.55}$$

5.4.4 Continuity equation for air in the sigma-pressure coordinate

5.4.4.1 *Derivation*

The continuity equation for air in the sigma-pressure coordinate is derived from the continuity equation for air in the pressure coordinate, the hydrostatic equation, and the total derivative in the sigma-pressure coordinate. The partial derivative of pressure with respect to σ is $\partial p_a/\partial\sigma = \pi_a$. Substituting this expression and the gradient conversion from (5.55) into the continuity equation for air in the pressure coordinate of (5.25) gives

$$\nabla_\sigma \cdot \mathbf{v}_h - \frac{\sigma}{\pi_a}\nabla_\sigma(\pi_a) \cdot \frac{\partial \mathbf{v}_h}{\partial \sigma} + \frac{1}{\pi_a}\frac{\partial \mathrm{w}_p}{\partial \sigma} = 0 \tag{5.56}$$

An expression for $\partial \mathrm{w}_p/\partial\sigma$ is now needed.

The **vertical scalar velocity in the sigma-pressure coordinate** (σ s^{-1}) is

$$\dot{\sigma} = \frac{d\sigma}{dt} \tag{5.57}$$

The relationship between $\dot{\sigma}$ and w_p is found by substituting $p_a = p_{a,top} + \pi_a\sigma$ and (5.57) into (5.21) to yield

$$\mathrm{w}_p = \frac{dp_a}{dt} = \sigma\frac{d\pi_a}{dt} + \frac{d\sigma}{dt}\pi_a = \sigma\frac{d\pi_a}{dt} + \dot{\sigma}\pi_a \tag{5.58}$$

If π_a is assumed to be constant with time, and horizontal gradients of π_a are ignored, (5.22) simplifies to $\mathrm{w}_p = -w\rho_a g$. Substituting this term into (5.58) with constant π_a yields $\dot{\sigma} = -w\rho_a g/\pi_a$. Thus, positive vertical velocities in the sigma-pressure coordinate result in downward motions. Because π_a varies in time and space, (5.58) is not simplified in the present derivation.

The **total derivative in Cartesian-sigma-pressure coordinates** is found by replacing p_a with σ and w_p with $\dot{\sigma}$ in (5.30). The result is

$$\frac{d}{dt} = \left(\frac{\partial}{\partial t}\right)_\sigma + \mathbf{v}_h \cdot \nabla_\sigma + \dot{\sigma}\frac{\partial}{\partial \sigma} \tag{5.59}$$

Substituting the total derivative of π_a into (5.58) gives

$$\mathrm{w}_p = \sigma\left[\left(\frac{\partial \pi_a}{\partial t}\right)_\sigma + (\mathbf{v}_h \cdot \nabla_\sigma)\pi_a\right] + \dot{\sigma}\pi_a \tag{5.60}$$

where $\partial\pi_a/\partial\sigma = 0$. The partial derivative of this equation with respect to σ is

$$\frac{\partial \mathrm{w}_p}{\partial \sigma} = \left(\frac{\partial \pi_a}{\partial t}\right)_\sigma + (\mathbf{v}_h \cdot \nabla_\sigma)\pi_a + \sigma\nabla_\sigma(\pi_a) \cdot \frac{\partial \mathbf{v}_h}{\partial \sigma} + \pi_a\frac{\partial \dot{\sigma}}{\partial \sigma} \tag{5.61}$$

Substituting (5.61) into (5.56), canceling terms, and compressing the result yield the **continuity equation for air in Cartesian-sigma-pressure coordinates** as

$$\left(\frac{\partial \pi_a}{\partial t}\right)_\sigma + \nabla_\sigma \cdot (\mathbf{v}_h\pi_a) + \pi_a\frac{\partial \dot{\sigma}}{\partial \sigma} = 0 \tag{5.62}$$

Substituting $\mathbf{v}_h\pi_a$ for $\mathbf{v}$ in the horizontal component of (4.11), substituting the result into (5.62), and multiplying through by $R_e^2\cos\varphi$ give the **continuity equation for**

air in spherical-sigma-pressure coordinates as

$$R_e^2 \cos\varphi \left(\frac{\partial \pi_a}{\partial t}\right)_\sigma + \left[\frac{\partial}{\partial \lambda_e}(u\pi_a R_e) + \frac{\partial}{\partial \varphi}(v\pi_a R_e \cos\varphi)\right]_\sigma + \pi_a R_e^2 \cos\varphi \frac{\partial \dot{\sigma}}{\partial \sigma} = 0 \tag{5.63}$$

5.4.4.2 *Pressure and vertical scalar velocity from the continuity equation for air*

The time rate of change of air pressure and vertical scalar velocity can be calculated from the continuity equation for air. The integral of (5.62) over all σ-layers is

$$\int_0^1 \left(\frac{\partial \pi_a}{\partial t}\right)_\sigma \mathrm{d}\sigma = -\nabla_\sigma \bullet \int_0^1 (\mathbf{v}_h \pi_a)\mathrm{d}\sigma - \pi_a \int_0^0 \mathrm{d}\dot{\sigma} \tag{5.64}$$

The integral limits are $\sigma = 0$ and $\dot{\sigma} = 0$ at the model top and $\sigma = 1$ and $\dot{\sigma} = 0$ at the model bottom. **At the ground, vertical scalar velocities are always zero,** and at the model top, velocities in the σ-p coordinate are assumed to be zero to close a boundary condition. In reality, $\dot{\sigma} \neq 0$ at the model top. Integrating (5.64) gives a **prognostic equation for column pressure** in Cartesian-sigma-pressure coordinates as

$$\left(\frac{\partial \pi_a}{\partial t}\right)_\sigma = -\nabla_\sigma \bullet \int_0^1 (\mathbf{v}_h \pi_a)\mathrm{d}\sigma \tag{5.65}$$

The analogous equation in spherical-sigma-pressure coordinates is

$$R_e^2 \cos\varphi \left(\frac{\partial \pi_a}{\partial t}\right)_\sigma = -\int_0^1 \left[\frac{\partial}{\partial \lambda_e}(u\pi_a R_e) + \frac{\partial}{\partial \varphi}(v\pi_a R_e \cos\varphi)\right]_\sigma \mathrm{d}\sigma \tag{5.66}$$

A numerical solution to this equation is shown in Chapter 7.

Once (5.66) has been written, an equation for the vertical scalar velocity at any altitude can be derived. Integrating (5.62) with respect to σ from the model top to any σ-level, and rearranging give

$$\pi_a \int_0^{\dot{\sigma}} \mathrm{d}\dot{\sigma} = -\nabla_\sigma \bullet \int_0^\sigma (\mathbf{v}_h \pi_a)\mathrm{d}\sigma - \int_0^\sigma \left(\frac{\partial \pi_a}{\partial t}\right)_\sigma \mathrm{d}\sigma \tag{5.67}$$

Integrating (5.67) yields

$$\dot{\sigma}\pi_a = -\nabla_\sigma \bullet \int_0^\sigma (\mathbf{v}_h \pi_a)\mathrm{d}\sigma - \sigma\left(\frac{\partial \pi_a}{\partial t}\right)_\sigma \tag{5.68}$$

which is the vertical scalar velocity at any σ-level. This equation is solved after $(\partial \pi_a/\partial t)_\sigma$ has been obtained from (5.65). Applying the spherical-coordinate transformation from (4.11) to (5.68) and multiplying through by $R_e^2 \cos\varphi$ give a

diagnostic equation for vertical scalar velocity at any σ-level as

$$\dot{\sigma}\pi_a R_e^2 \cos\varphi = -\int_0^\sigma \left[\frac{\partial}{\partial \lambda_e}(u\pi_a R_e) + \frac{\partial}{\partial \varphi}(v\pi_a R_e \cos\varphi)\right]_\sigma \mathrm{d}\sigma - \sigma R_e^2 \cos\varphi \left(\frac{\partial \pi_a}{\partial t}\right)_\sigma \tag{5.69}$$

5.4.5 Species continuity equation in the sigma-pressure coordinate

Applying the total derivative from (5.59) to (3.54) gives the **species continuity equation in Cartesian-sigma-pressure coordinates** as

$$\frac{\mathrm{d}q}{\mathrm{d}t} = \left(\frac{\partial q}{\partial t}\right)_\sigma + (\mathbf{v}_h \cdot \nabla_\sigma)q + \dot{\sigma}\frac{\partial q}{\partial \sigma} = \frac{(\nabla \cdot \rho_a \mathbf{K}_h \nabla)q}{\rho_a} + \sum_{n=1}^{N_{e,t}} R_n \tag{5.70}$$

The eddy diffusion term expands in Cartesian-sigma-pressure coordinates to

$$\begin{aligned}(\nabla \cdot \rho_a \mathbf{K}_h \nabla)q = &\left[\left(\frac{\partial}{\partial x}\right)_\sigma + \left(\frac{\partial \sigma}{\partial x}\right)_z \frac{\partial}{\partial \sigma}\right]\left\{\rho_a K_{h,xx}\left[\left(\frac{\partial q}{\partial x}\right)_\sigma + \left(\frac{\partial \sigma}{\partial x}\right)_z \frac{\partial q}{\partial \sigma}\right]\right\} \\ &+ \left[\left(\frac{\partial}{\partial y}\right)_\sigma + \left(\frac{\partial \sigma}{\partial y}\right)_z \frac{\partial}{\partial \sigma}\right]\left\{\rho_a K_{h,yy}\left[\left(\frac{\partial q}{\partial y}\right)_\sigma + \left(\frac{\partial \sigma}{\partial y}\right)_z \frac{\partial q}{\partial \sigma}\right]\right\} \\ &+ \frac{\rho_a g^2}{\pi_a^2}\frac{\partial}{\partial \sigma}\left(\rho_a^2 K_{h,zz}\frac{\partial q}{\partial \sigma}\right)\end{aligned} \tag{5.71}$$

The vertical term was found by substituting $\partial p_a/\partial\sigma = \pi_a$ into the vertical term of (5.33).

Multiplying (5.70) by π_a, multiplying the continuity equation for air from (5.62) by q, and summing the two equations give the flux-form species continuity equation as

$$\left[\frac{\partial(\pi_a q)}{\partial t}\right]_\sigma + \nabla_\sigma \cdot (\mathbf{v}_h \pi_a q) + \pi_a \frac{\partial(\dot{\sigma}q)}{\partial \sigma} = \pi_a \left[\frac{(\nabla \cdot \rho_a \mathbf{K}_h \nabla)q}{\rho_a} + \sum_{n=1}^{N_{e,t}} R_n\right] \tag{5.72}$$

Applying the spherical coordinate transformation from (4.11) to (5.72) and multiplying through by $R_e^2 \cos\varphi$ give the **flux form of the species continuity equation in spherical-sigma-pressure coordinates** as

$$\begin{aligned}&R_e^2\cos\varphi\left[\frac{\partial}{\partial t}(\pi_a q)\right]_\sigma + \left[\frac{\partial}{\partial \lambda_e}(u\pi_a q R_e) + \frac{\partial}{\partial \varphi}(v\pi_a q R_e \cos\varphi)\right]_\sigma \\ &\quad + \pi_a R_e^2 \cos\varphi \frac{\partial}{\partial \sigma}(\dot{\sigma}q) = \pi_a R_e^2 \cos\varphi \left[\frac{(\nabla \cdot \rho_a \mathbf{K}_h \nabla)q}{\rho_a} + \sum_{n=1}^{N_{e,t}} R_n\right]\end{aligned} \tag{5.73}$$

5.4.6 Thermodynamic equation in the sigma-pressure coordinate

Applying the total derivative from (5.59) to (4.76) gives the **thermodynamic energy equation in Cartesian-sigma-pressure coordinates** as

$$\left(\frac{\partial \theta_v}{\partial t}\right)_\sigma + (\mathbf{v}_h \cdot \nabla_\sigma)\theta_v + \dot{\sigma}\frac{\partial \theta_v}{\partial \sigma} = \frac{(\nabla \cdot \rho_a \mathbf{K}_h \nabla)\theta_v}{\rho_a} + \frac{\theta_v}{c_{p,d} T_v}\sum_{n=1}^{N_{e,h}} \frac{dQ_n}{dt} \tag{5.74}$$

Multiplying (5.74) by π_a, multiplying (5.62) by θ_v, adding the two equations, and compressing terms give the flux form of the thermodynamic energy equation as

$$\left[\frac{\partial(\pi_a \theta_v)}{\partial t}\right]_\sigma + \nabla_\sigma \cdot (\mathbf{v}_h \pi_a \theta_v) + \pi_a \frac{\partial(\dot{\sigma}\theta_v)}{\partial \sigma} = \pi_a\left[\frac{(\nabla \cdot \rho_a \mathbf{K}_h \nabla)\theta_v}{\rho_a} + \frac{\theta_v}{c_{p,d} T_v}\sum_{n=1}^{N_{e,h}} \frac{dQ_n}{dt}\right] \tag{5.75}$$

The eddy diffusion term in this equation is the same as that in (5.71) except θ_v is used in the present case instead of q. Applying the spherical coordinate transformation from (4.11) to (5.75) and multiplying through by $R_e^2 \cos\varphi$ give the **flux form of the thermodynamic energy equation in spherical-sigma-pressure coordinates** as

$$\begin{aligned} &R_e^2 \cos\varphi\left[\frac{\partial}{\partial t}(\pi_a \theta_v)\right]_\sigma + \left[\frac{\partial}{\partial \lambda_e}(u\pi_a \theta_v R_e) + \frac{\partial}{\partial \varphi}(v\pi_a \theta_v R_e \cos\varphi)\right] \\ &+ \pi_a R_e^2 \cos\varphi \frac{\partial}{\partial \sigma}(\dot{\sigma}\theta_v) = \pi_a R_e^2 \cos\varphi\left[\frac{(\nabla \cdot \rho_a \mathbf{K}_h \nabla)\theta_v}{\rho_a} + \frac{\theta_v}{c_{p,d} T_v}\sum_{n=1}^{N_{e,h}} \frac{dQ_n}{dt}\right] \end{aligned} \tag{5.76}$$

5.4.7 Horizontal momentum equations in the sigma-pressure coordinate

Applying the total derivative from (5.59) to the horizontal momentum equation in Cartesian-altitude coordinates from (4.71) gives

$$\left(\frac{\partial \mathbf{v}_h}{\partial t}\right)_\sigma + (\mathbf{v}_h \cdot \nabla_\sigma)\mathbf{v}_h + \dot{\sigma}\frac{\partial \mathbf{v}_h}{\partial \sigma} + f\mathbf{k} \times \mathbf{v}_h = -\frac{1}{\rho_a}\nabla_z(p_a) + \frac{(\nabla \cdot \rho_a \mathbf{K}_m \nabla)\mathbf{v}_h}{\rho_a} \tag{5.77}$$

The pressure-gradient term in (5.77) can be converted to the sigma-pressure coordinate by combining (5.16) with (5.55). The result is

$$\frac{1}{\rho_a}\nabla_z(p_a) = \nabla_p \Phi = \nabla_\sigma \Phi - \frac{\sigma}{\pi_a}\nabla_\sigma(\pi_a)\frac{\partial \Phi}{\partial \sigma} \tag{5.78}$$

Substituting (5.78) into (5.77) gives the **horizontal momentum equation in Cartesian-sigma-pressure coordinates** as

$$\left(\frac{\partial \mathbf{v}_h}{\partial t}\right)_\sigma + (\mathbf{v}_h \cdot \nabla_\sigma)\mathbf{v}_h + \dot{\sigma}\frac{\partial \mathbf{v}_h}{\partial \sigma} + f\mathbf{k} \times \mathbf{v}_h$$
$$= -\nabla_\sigma \Phi + \frac{\sigma}{\pi_a}\nabla_\sigma(\pi_a)\frac{\partial \Phi}{\partial \sigma} + \frac{(\nabla \cdot \rho_a \mathbf{K}_m \nabla)\mathbf{v}_h}{\rho_a} \quad (5.79)$$

5.4.8 Coupling horizontal and vertical momentum equation

The hydrostatic equation is used in lieu of a prognostic vertical momentum equation in the sigma-pressure coordinate because the conversion of altitude to sigma-pressure coordinate requires the hydrostatic assumption. Substituting $p_a = \rho_a R' T_v$ and $\partial p_a / \partial \sigma = \pi_a$ into (5.37) gives the **hydrostatic equation in the sigma-pressure coordinate** as

$$\frac{\partial \Phi}{\partial \sigma} = -\frac{\pi_a R' T_v}{p_a} = -\frac{\pi_a}{\rho_a} = -\alpha_a \pi_a \quad (5.80)$$

Substituting (5.80) into (5.79) gives the horizontal momentum equation as

$$\left(\frac{\partial \mathbf{v}_h}{\partial t}\right)_\sigma + (\mathbf{v}_h \cdot \nabla_\sigma)\mathbf{v}_h + \dot{\sigma}\frac{\partial \mathbf{v}_h}{\partial \sigma} = -f\mathbf{k} \times \mathbf{v}_h - \nabla_\sigma \Phi - \sigma \alpha_a \nabla_\sigma(\pi_a) + \frac{(\nabla \cdot \rho_a \mathbf{K}_m \nabla)\mathbf{v}_h}{\rho_a} \quad (5.81)$$

The term α_a should be modified to make the finite differencing of (5.81) in Chapter 7 consistent with that of geopotential. Combining $\alpha_a = R' T_v / p_a$ with $R' = \kappa c_{p,d}$, $T_v = \theta_v P$, $\partial p_a / \partial \sigma = \pi_a$, and $\partial P / \partial p_a = \kappa P / p_a$ gives

$$\alpha_a = \frac{R' T_v}{p_a} = \frac{\kappa c_{p,d} \theta_v P}{p_a} = c_{p,d}\theta_v \frac{\partial P}{\partial p_a} = \frac{c_{p,d}\theta_v}{\pi_a}\frac{\partial P}{\partial \sigma} \quad (5.82)$$

Substituting (5.82) into (5.81), multiplying the result by π_a, multiplying (5.62) by $\mathbf{v}_h$, and summing the two equations give the flux form of the horizontal momentum equation in Cartesian-sigma-pressure coordinates as

$$\left[\frac{\partial(\mathbf{v}_h \pi_a)}{\partial t}\right]_\sigma + \mathbf{v}_h \nabla_\sigma \cdot (\mathbf{v}_h \pi_a) + \pi_a(\mathbf{v}_h \cdot \nabla_\sigma)\mathbf{v}_h + \pi_a \frac{\partial}{\partial \sigma}(\dot{\sigma}\mathbf{v}_h)$$
$$= -\pi_a f\mathbf{k} \times \mathbf{v}_h - \pi_a \nabla_\sigma \Phi - \sigma c_{p,d}\theta_v \frac{\partial P}{\partial \sigma}\nabla_\sigma(\pi_a) + \pi_a \frac{(\nabla \cdot \rho_a \mathbf{K}_m \nabla)\mathbf{v}_h}{\rho_a} \quad (5.83)$$

The advection terms in this equation expand to

$$\mathbf{v}_h \nabla_\sigma \cdot (\mathbf{v}_h \pi_a) = \mathbf{i}u\left[\frac{\partial(u\pi_a)}{\partial x} + \frac{\partial(v\pi_a)}{\partial y}\right] + \mathbf{j}v\left[\frac{\partial(u\pi_a)}{\partial x} + \frac{\partial(v\pi_a)}{\partial y}\right] \quad (5.84)$$

$$\pi_a(\mathbf{v}_h \cdot \nabla_\sigma)\mathbf{v}_h = \mathbf{i}\pi_a\left(u\frac{\partial u}{\partial x} + v\frac{\partial u}{\partial y}\right) + \mathbf{j}\pi_a\left(u\frac{\partial v}{\partial x} + v\frac{\partial v}{\partial y}\right) \quad (5.85)$$

Substituting these terms into (5.83), applying the transformation from (4.31), applying (4.1), and multiplying through by $R_e^2 \cos\varphi$ give the **flux forms of the horizontal momentum equations in spherical-sigma-pressure coordinates** as

$$R_e^2 \cos\varphi \left[\frac{\partial}{\partial t}(\pi_a u) \right]_\sigma + \left[\frac{\partial}{\partial \lambda_e}(\pi_a u^2 R_e) + \frac{\partial}{\partial \varphi}(\pi_a uv R_e \cos\varphi) \right]_\sigma + \pi_a R_e^2 \cos\varphi \frac{\partial}{\partial \sigma}(\dot{\sigma} u)$$

$$= \pi_a uv R_e \sin\varphi + \pi_a f v R_e^2 \cos\varphi - R_e \left(\pi_a \frac{\partial \Phi}{\partial \lambda_e} + \sigma c_{p,d} \theta_v \frac{\partial P}{\partial \sigma} \frac{\partial \pi_a}{\partial \lambda_e} \right)_\sigma$$

$$+ R_e^2 \cos\varphi \frac{\pi_a}{\rho_a} (\nabla \cdot \rho_a \mathbf{K}_m \nabla) u \qquad (5.86)$$

$$R_e^2 \cos\varphi \left[\frac{\partial}{\partial t}(\pi_a v) \right]_\sigma + \left[\frac{\partial}{\partial \lambda_e}(\pi_a uv R_e) + \frac{\partial}{\partial \varphi}(v^2 \pi_a R_e \cos\varphi) \right]_\sigma + \pi_a R_e^2 \cos\varphi \frac{\partial}{\partial \sigma}(\dot{\sigma} v)$$

$$= -\pi_a u^2 R_e \sin\varphi - \pi_a f u R_e^2 \cos\varphi - R_e \cos\varphi \left(\pi_a \frac{\partial \Phi}{\partial \varphi} + \sigma c_{p,d} \theta_v \frac{\partial P}{\partial \sigma} \frac{\partial \pi_a}{\partial \varphi} \right)_\sigma$$

$$+ R_e^2 \cos\varphi \frac{\pi_a}{\rho_a} (\nabla \cdot \rho_a \mathbf{K}_m \nabla) v \qquad (5.87)$$

The west–east eddy diffusion term expands in Cartesian-sigma-pressure coordinates to

$$(\nabla \cdot \rho_a \mathbf{K}_m \nabla) u = \left[\left(\frac{\partial}{\partial x} \right)_\sigma + \left(\frac{\partial \sigma}{\partial x} \right)_z \frac{\partial}{\partial \sigma} \right] \left\{ \rho_a K_{m,xx} \left[\left(\frac{\partial u}{\partial x} \right)_\sigma + \left(\frac{\partial \sigma}{\partial x} \right)_z \frac{\partial u}{\partial \sigma} \right] \right\}$$

$$+ \left[\left(\frac{\partial}{\partial y} \right)_\sigma + \left(\frac{\partial \sigma}{\partial y} \right)_z \frac{\partial}{\partial \sigma} \right] \left\{ \rho_a K_{m,yx} \left[\left(\frac{\partial u}{\partial y} \right)_\sigma + \left(\frac{\partial \sigma}{\partial y} \right)_z \frac{\partial u}{\partial \sigma} \right] \right\}$$

$$+ \frac{\rho_a g^2}{\pi_a^2} \frac{\partial}{\partial \sigma} \left(\rho_a^2 K_{m,zx} \frac{\partial u}{\partial \sigma} \right) \qquad (5.88)$$

The expression for $(\nabla \cdot \rho_a \mathbf{K}_m \nabla) v$ is similar to (5.88), except v, $K_{m,xy}$, $K_{m,yy}$, and $K_{m,zy}$ are used in the new term instead of u, $K_{m,xx}$, $K_{m,yx}$, and $K_{m,zx}$, respectively.

5.5 SIGMA-ALTITUDE COORDINATE

The **sigma-altitude** (*s-z*) **coordinate** is defined such that layer tops and bottoms are surfaces of constant *s*, where *s* equals the difference between the model top altitude and the bottom altitude of a layer divided by the difference between the model top and the model bottom altitude (e.g., Kasahara 1974). This coordinate is used to simulate nonhydrostatic or hydrostatic flows. In the sigma-altitude coordinate, layer thicknesses do not change, whereas in the sigma-pressure coordinate, layer thicknesses change continuously. An advantage of the sigma-altitude and sigma-pressure coordinates in comparison with the altitude and pressure coordinates is that the former do not permit model layers to intercept ground topography, whereas the latter do. In this section, elastic nonhydrostatic model equations for the sigma-altitude coordinate are derived. Conversion to anelastic and hydrostatic equations is also discussed.

5.5.1 Definitions

Layer tops and bottoms in the sigma-altitude coordinate are defined as surfaces of constant s, where s is defined as

$$s = \frac{z_{\mathrm{top}} - z}{z_{\mathrm{top}} - z_{\mathrm{surf}}} = \frac{z_{\mathrm{top}} - z}{Z_{\mathrm{t}}} \tag{5.89}$$

In this equation, z is the altitude at a layer bottom boundary, z_{surf} is the model bottom (surface) altitude, z_{top} is the model top altitude, $Z_{\mathrm{t}} = z_{\mathrm{top}} - z_{\mathrm{surf}}$ is the altitude difference between the model top and bottom, and s is a fraction ≤ 1. At the model top, $s = 0$. At the model bottom, $s = 1$. From (5.89), the altitude at a given s-level is

$$z = z_{\mathrm{top}} - Z_{\mathrm{t}} s \tag{5.90}$$

Figure 5.5, which shows heights of constant sigma-pressure surfaces, can also be used to show heights of constant sigma-altitude surfaces by replacing σ's with s's. In the s-z coordinate, layer tops and bottoms correspond to s-values that are constant in time. Altitudes, including z, z_{surf}, and z_{top}, vary in space, but not in time, along surfaces of constant s. Pressure (p_a) varies in space and time along surfaces of constant s.

5.5.2 Gradient conversion from altitude to sigma-altitude coordinate

The horizontal **gradient conversion of a variable from Cartesian-altitude to Cartesian-sigma-altitude coordinates** is obtained in a manner similar to that for the conversion from Cartesian-altitude to Cartesian-pressure coordinates. The result is

$$\nabla_z = \nabla_s + \nabla_z(s)\frac{\partial}{\partial s} \tag{5.91}$$

Taking the horizontal gradient of (5.89) and noting that $\nabla_z(z) = \nabla_z(z_{\mathrm{top}}) = 0$ give

$$\nabla_z(s) = -\frac{z_{\mathrm{top}} - z}{Z_{\mathrm{t}}^2}\nabla_z(Z_{\mathrm{t}}) = -\frac{s}{Z_{\mathrm{t}}}\nabla_z(Z_{\mathrm{t}}) \tag{5.92}$$

Substituting this result into (5.91) yields another form of the gradient conversion as

$$\nabla_z = \nabla_s - \frac{s}{Z_{\mathrm{t}}}\nabla_z(Z_{\mathrm{t}})\frac{\partial}{\partial s} \tag{5.93}$$

The time-derivative conversion from the altitude to sigma-altitude coordinate is

$$\left(\frac{\partial}{\partial t}\right)_z = \left(\frac{\partial}{\partial t}\right)_s \tag{5.94}$$

which was derived by substituting s for p in (5.15) and noting that $(\partial s/\partial t)_z = 0$.

The **vertical scalar velocity in the sigma-altitude coordinate** ($s\ \mathrm{s}^{-1}$) is

$$\dot{s} = \frac{\mathrm{d}s}{\mathrm{d}t} = (\mathbf{v}_\mathrm{h} \cdot \nabla_z)\, s + w\frac{\partial s}{\partial z} = (\mathbf{v}_\mathrm{h} \cdot \nabla_z)\, s - \frac{w}{Z_\mathrm{t}} \tag{5.95}$$

which has a zero local time derivative, since $(\partial s/\partial t)_z = 0$. Equation (5.95) was derived under the assumption that $\partial s/\partial z = -1/Z_\mathrm{t}$, obtained by differentiating (5.89). A positive vertical scalar velocity in the sigma-altitude coordinate corresponds to downward motion, just as in the sigma-pressure and pressure coordinates. Substituting $\dot{s}$ for w_p and s for p in (5.30) gives the **total derivative in Cartesian-sigma-altitude coordinates** as

$$\frac{\mathrm{d}}{\mathrm{d}t} = \left(\frac{\partial}{\partial t}\right)_s + (\mathbf{v}_\mathrm{h} \cdot \nabla_s) + \dot{s}\frac{\partial}{\partial s} \tag{5.96}$$

5.5.3 Continuity equation for air in the sigma-altitude coordinate

The continuity equation for air in Cartesian-sigma-altitude coordinates is derived from the continuity equation for air in Cartesian-altitude coordinates. Splitting (3.20) into horizontal and vertical components and applying (5.91) to $\mathbf{v}_\mathrm{h}$ and ρ_a give

$$\left(\frac{\partial \rho_\mathrm{a}}{\partial t}\right)_s = -\rho_\mathrm{a}\left[\nabla_s \cdot \mathbf{v}_\mathrm{h} + \nabla_z(s)\,\frac{\partial \mathbf{v}_\mathrm{h}}{\partial s} + \frac{\partial w}{\partial z}\right] - \mathbf{v}_\mathrm{h} \cdot \left[\nabla_s(\rho_\mathrm{a}) + \nabla_z(s)\,\frac{\partial \rho_\mathrm{a}}{\partial s}\right] - w\frac{\partial \rho_\mathrm{a}}{\partial z} \tag{5.97}$$

Rewriting (5.95) as $w = Z_\mathrm{t}\,[\mathbf{v}_\mathrm{h} \cdot \nabla_z\,(s) - \dot{s}]$ and differentiating give

$$\frac{\partial w}{\partial z} = Z_\mathrm{t}\left[\nabla_z(s)\,\frac{\partial \mathbf{v}_\mathrm{h}}{\partial z} + (\mathbf{v}_\mathrm{h} \cdot \nabla_z)\frac{\partial s}{\partial z} - \frac{\partial \dot{s}}{\partial z}\right] \tag{5.98}$$

Substituting $\mathrm{d}z = -Z_\mathrm{t}\,\mathrm{d}s$ into (5.98) yields

$$\frac{\partial w}{\partial z} = \frac{\partial \dot{s}}{\partial s} - \nabla_z(s)\,\frac{\partial \mathbf{v}_\mathrm{h}}{\partial s} + \frac{1}{Z_\mathrm{t}}(\mathbf{v}_\mathrm{h} \cdot \nabla_z)\, Z_\mathrm{t} \tag{5.99}$$

Substituting $w = Z_\mathrm{t}[\mathbf{v}_\mathrm{h} \cdot \nabla_z(s) - \dot{s}]$, (5.99), and $\partial s/\partial z = -1/Z_\mathrm{t}$ into (5.97) gives

$$\left(\frac{\partial \rho_\mathrm{a}}{\partial t}\right)_s = -\rho_\mathrm{a}\left[\nabla_s \cdot \mathbf{v}_\mathrm{h} + \frac{\partial \dot{s}}{\partial s} + \frac{1}{Z_\mathrm{t}}(\mathbf{v}_\mathrm{h} \cdot \nabla_z)Z_\mathrm{t}\right] - (\mathbf{v}_\mathrm{h} \cdot \nabla_s)\rho_\mathrm{a} - \dot{s}\frac{\partial \rho_\mathrm{a}}{\partial s} \tag{5.100}$$

Substituting $\nabla_z(Z_t) = \nabla_s(Z_t)$ into (5.100) gives the **elastic nonhydrostatic continuity equation for air in Cartesian-sigma-altitude coordinates** as

$$\left(\frac{\partial \rho_a}{\partial t}\right)_s = -\frac{1}{Z_t}\nabla_s \cdot (\mathbf{v}_h \rho_a Z_t) - \frac{\partial}{\partial s}(\dot{s}\rho_a) = -\frac{1}{Z_t}\left[\frac{\partial(u\rho_a Z_t)}{\partial x} + \frac{\partial(v\rho_a Z_t)}{\partial y}\right]_s - \frac{\partial(\dot{s}\rho_a)}{\partial s} \tag{5.101}$$

When (5.101) is used to predict air density, vertically propagating acoustic waves are one solution to the equations of atmospheric dynamics. To eliminate these waves and enhance numerical stability, the **anelastic approximation** is made to (5.101) by setting $\partial\rho_a/\partial t = 0$ (Section 5.1).

If hydrostatic balance is assumed, vertically propagating acoustic waves are also eliminated as a possible solution to the basic equations. Substituting $\partial s/\partial z = -1/Z_t$ into (2.40) gives the **hydrostatic equation in the sigma-altitude coordinate** as

$$\rho'_a = -\frac{1}{g}\frac{\partial p'_a}{\partial z} = \frac{1}{Z_t g}\frac{\partial p'_a}{\partial s} \tag{5.102}$$

where primes have been added to indicate that density and pressure are in hydrostatic balance. Substituting (5.102) into (5.101) gives the **hydrostatic continuity equation in Cartesian-sigma-altitude coordinates** as

$$\frac{\partial}{\partial t}\left(\frac{\partial p'_a}{\partial s}\right) = -\nabla_s \cdot \left(\mathbf{v}_h \frac{\partial p'_a}{\partial s}\right) - \frac{\partial}{\partial s}\left(\dot{s}\frac{\partial p'_a}{\partial s}\right) \tag{5.103}$$

5.5.4 Species continuity equation in the sigma-altitude coordinate

The **species continuity equation in Cartesian-sigma-altitude coordinates** is obtained by applying the total derivative from (5.96) to (3.54). The result is

$$\left(\frac{d\mathrm{q}}{dt}\right)_s = \left(\frac{\partial \mathrm{q}}{\partial t}\right)_s + (\mathbf{v}_h \cdot \nabla_s)\mathrm{q} + \dot{s}\frac{\partial \mathrm{q}}{\partial s} = \frac{(\nabla \cdot \rho_a \mathbf{K}_h \nabla)\mathrm{q}}{\rho_a} + \sum_{n=1}^{N_{e,t}} R_n \tag{5.104}$$

The eddy diffusion term in (5.104) expands in Cartesian-sigma-altitude coordinates to

$$\begin{aligned}(\nabla \cdot \rho_a \mathbf{K}_h \nabla)\mathrm{q} = & \left[\left(\frac{\partial}{\partial x}\right)_s + \left(\frac{\partial s}{\partial x}\right)_z \frac{\partial}{\partial s}\right]\left\{\rho_a K_{h,xx}\left[\left(\frac{\partial \mathrm{q}}{\partial x}\right)_s + \left(\frac{\partial s}{\partial x}\right)_z \frac{\partial \mathrm{q}}{\partial s}\right]\right\} \\ & + \left[\left(\frac{\partial}{\partial y}\right)_s + \left(\frac{\partial s}{\partial y}\right)_z \frac{\partial}{\partial s}\right]\left\{\rho_a K_{h,yy}\left[\left(\frac{\partial \mathrm{q}}{\partial y}\right)_s + \left(\frac{\partial s}{\partial y}\right)_z \frac{\partial \mathrm{q}}{\partial s}\right]\right\} \\ & + \frac{1}{Z_t^2}\frac{\partial}{\partial s}\left(\rho_a K_{h,zz}\frac{\partial \mathrm{q}}{\partial s}\right)\end{aligned} \tag{5.105}$$

where $\partial s/\partial z = -1/Z_t$ was used to obtain the vertical term.

5.5.5 Thermodynamic equation in the sigma-altitude coordinate

The **thermodynamic energy equation in Cartesian-sigma-altitude coordinates** is obtained by applying (5.96) to (3.76). The result is

$$\left(\frac{\partial \theta_v}{\partial t}\right)_s + (\mathbf{v}_h \cdot \nabla_s)\theta_v + \dot{s}\frac{\partial \theta_v}{\partial s} = \frac{(\nabla \cdot \rho_a \mathbf{K}_h \nabla)\theta_v}{\rho_a} + \frac{\theta_v}{c_{p,d} T_v} \sum_{n=1}^{N_{e,h}} \frac{dQ_n}{dt} \tag{5.106}$$

Eddy diffusion is treated as in (5.105), except θ_v is used in the new term instead of q.

5.5.6 Horizontal momentum equations in the sigma-altitude coordinate

Applying (5.96) to (4.71) gives the **horizontal momentum equation in Cartesian-sigma-altitude coordinates** as

$$\left(\frac{\partial \mathbf{v}_h}{\partial t}\right)_s + (\mathbf{v}_h \cdot \nabla_s)\mathbf{v}_h + \dot{s}\frac{\partial \mathbf{v}_h}{\partial s} + f\mathbf{k} \times \mathbf{v}_h = -\frac{1}{\rho_a}\nabla_z(p_a) + \frac{(\nabla_z \cdot \rho_a \mathbf{K}_m \nabla_z)\mathbf{v}_h}{\rho_a} \tag{5.107}$$

Applying the coordinate conversion from (5.93) to pressure gives

$$\nabla_z(p_a) = \nabla_s(p_a) - \frac{s}{Z_t}\nabla_z(Z_t)\frac{\partial p_a}{\partial s} \tag{5.108}$$

Substituting (5.108) into (5.107) gives the horizontal momentum equation as

$$\left(\frac{\partial \mathbf{v}_h}{\partial t}\right)_s + (\mathbf{v}_h \cdot \nabla_s)\mathbf{v}_h + \dot{s}\frac{\partial \mathbf{v}_h}{\partial s} = -f\mathbf{k} \times \mathbf{v}_h - \frac{1}{\rho_a}\left[\nabla_s(p_a) - \frac{s}{Z_t}\nabla_z(Z_t)\frac{\partial p_a}{\partial s} - (\nabla \cdot \rho_a \mathbf{K}_m \nabla)\mathbf{v}_h\right] \tag{5.109}$$

Equation (5.109) expands in the x- and y-directions to

$$\left(\frac{\partial u}{\partial t} + u\frac{\partial u}{\partial x} + v\frac{\partial u}{\partial y}\right)_s + \dot{s}\frac{\partial u}{\partial s} = fv - \frac{1}{\rho_a}\left[\left(\frac{\partial p_a}{\partial x}\right)_s - \frac{s}{Z_t}\left(\frac{\partial Z_t}{\partial x}\right)_z \frac{\partial p_a}{\partial s} - (\nabla \cdot \rho_a \mathbf{K}_m \nabla)u\right] \tag{5.110}$$

$$\left(\frac{\partial v}{\partial t} + u\frac{\partial v}{\partial x} + v\frac{\partial v}{\partial y}\right)_s + \dot{s}\frac{\partial v}{\partial s} = -fu - \frac{1}{\rho_a}\left[\left(\frac{\partial p_a}{\partial y}\right)_s - \frac{s}{Z_t}\left(\frac{\partial Z_t}{\partial y}\right)_z \frac{\partial p_a}{\partial s} - (\nabla \cdot \rho_a \mathbf{K}_m \nabla)v\right] \tag{5.111}$$

respectively. The eddy diffusion term in the x-direction is

$$(\nabla \bullet \rho_a \mathbf{K}_m \nabla)u = \left[\left(\frac{\partial}{\partial x}\right)_s + \left(\frac{\partial s}{\partial x}\right)_z \frac{\partial}{\partial s}\right]\left\{\rho_a K_{m,xx}\left[\left(\frac{\partial u}{\partial x}\right)_s + \left(\frac{\partial s}{\partial x}\right)_z \frac{\partial u}{\partial s}\right]\right\}$$
$$+ \left[\left(\frac{\partial}{\partial y}\right)_s + \left(\frac{\partial s}{\partial y}\right)_z \frac{\partial}{\partial s}\right]\left\{\rho_a K_{m,yx}\left[\left(\frac{\partial u}{\partial y}\right)_s + \left(\frac{\partial \sigma}{\partial y}\right)_z \frac{\partial u}{\partial s}\right]\right\}$$
$$+ \frac{1}{Z_t^2}\frac{\partial}{\partial s}\left(\rho_a K_{m,zx}\frac{\partial u}{\partial s}\right) \tag{5.112}$$

The expression for $(\nabla \bullet \rho_a \mathbf{K}_m \nabla)v$ is similar.

5.5.7 Vertical momentum equation in the sigma-altitude coordinate

Substituting $\partial s/\partial z = -1/Z_t$ into the vertical momentum equation in Cartesian-altitude coordinates from (4.75) and expanding the total derivative of w with (5.96) give

$$\left(\frac{\partial w}{\partial t} + u\frac{\partial w}{\partial x} + v\frac{\partial w}{\partial y}\right)_s + \dot{s}\frac{\partial w}{\partial s} = -g + \frac{1}{Z_t \rho_a}\frac{\partial p_a}{\partial s} + \frac{(\nabla \bullet \rho_a \mathbf{K}_m \nabla)\, w}{\rho_a} \tag{5.113}$$

Substituting $w = Z_t\left[\mathbf{v}_h \bullet \nabla_z(s) - \dot{s}\right]$ from (5.95) into (5.113) gives the **vertical momentum equation in Cartesian-sigma-altitude coordinates** as

$$\left[\left(\frac{\partial}{\partial t}\right)_s + u\left(\frac{\partial}{\partial x}\right)_s + v\left(\frac{\partial}{\partial y}\right)_s + \dot{s}\frac{\partial}{\partial s}\right]\left[Z_t u\left(\frac{\partial s}{\partial x}\right)_z + Z_t v\left(\frac{\partial s}{\partial y}\right)_z - Z_t \dot{s}\right]$$
$$= -g + \frac{1}{Z_t \rho_a}\frac{\partial p_a}{\partial s} + \frac{1}{\rho_a}(\nabla \bullet \rho_a \mathbf{K}_m \nabla)\left[Z_t u\left(\frac{\partial s}{\partial x}\right)_z + Z_t v\left(\frac{\partial s}{\partial y}\right)_z - Z_t \dot{s}\right] \tag{5.114}$$

The diffusion term $(\nabla \bullet \rho_a \mathbf{K}_m \nabla)(Z_t \dot{s})$ is similar to $(\nabla \bullet \rho_a \mathbf{K}_m \nabla)u$ from (5.112), except that $Z_t \dot{s}$, $K_{m,xz}$, $K_{m,yz}$, and $K_{m,zz}$ are used in the new term instead of u, $K_{m,xx}$, $K_{m,yx}$, and $K_{m,zx}$.

5.5.8 Basic equations in spherical-sigma-altitude coordinates

The elastic nonhydrostatic sigma-altitude equations derived above can be converted from Cartesian to spherical horizontal coordinates with (4.1), (4.11), and (4.31). The resulting equations are given as follows:

Elastic continuity equation for air from (5.101)

$$R_e^2 \cos\varphi\left(\frac{\partial \rho_a}{\partial t}\right)_s = -\frac{1}{Z_t}\left[\frac{\partial}{\partial \lambda_e}(u\rho_a Z_t R_e) + \frac{\partial}{\partial \varphi}(v\rho_a Z_t R_e \cos\varphi)\right]_s - R_e^2 \cos\varphi \frac{\partial}{\partial s}(\dot{s}\rho_a) \tag{5.115}$$

Species continuity equation from (5.104)

$$\left(\frac{\partial q}{\partial t} + \frac{u}{R_e \cos\varphi}\frac{\partial q}{\partial \lambda_e} + \frac{v}{R_e}\frac{\partial q}{\partial \varphi}\right)_s + \dot{s}\frac{\partial q}{\partial s} = \frac{(\nabla \bullet \rho_a \mathbf{K}_h \nabla)q}{\rho_a} + \sum_{n=1}^{N_{e,t}} R_n \tag{5.116}$$

Thermodynamic energy equation from (5.106)

$$\left(\frac{\partial \theta_{\mathrm{v}}}{\partial t}+\frac{u}{R_{\mathrm{e}}\cos\varphi}\frac{\partial \theta_{\mathrm{v}}}{\partial \lambda_{\mathrm{e}}}+\frac{v}{R_{\mathrm{e}}}\frac{\partial \theta_{\mathrm{v}}}{\partial \varphi}\right)_s+\dot{s}\frac{\partial \theta_{\mathrm{v}}}{\partial s}=\frac{(\nabla\cdot\rho_{\mathrm{a}}\mathbf{K}_h\nabla)\,\theta_{\mathrm{v}}}{\rho_{\mathrm{a}}}+\frac{\theta_{\mathrm{v}}}{c_{p,\mathrm{d}}T_{\mathrm{v}}}\sum_{n=1}^{N_{\mathrm{e,h}}}\frac{\mathrm{d}Q_n}{\mathrm{d}t} \tag{5.117}$$

West–east momentum equations from (5.110)

$$\begin{aligned}&\left(\frac{\partial u}{\partial t}+\frac{u}{R_{\mathrm{e}}\cos\varphi}\frac{\partial u}{\partial \lambda_{\mathrm{e}}}+\frac{v}{R_{\mathrm{e}}}\frac{\partial u}{\partial \varphi}\right)_s+\dot{s}\frac{\partial u}{\partial s}\\&\quad=\frac{uv\tan\varphi}{R_{\mathrm{e}}}+fv-\frac{1}{\rho_{\mathrm{a}}R_{\mathrm{e}}\cos\varphi}\left[\left(\frac{\partial p_{\mathrm{a}}}{\partial \lambda_{\mathrm{e}}}\right)_s-\frac{s}{Z_{\mathrm{t}}}\left(\frac{\partial Z_{\mathrm{t}}}{\partial \lambda_{\mathrm{e}}}\right)_z\frac{\partial p_{\mathrm{a}}}{\partial s}\right]\\&\quad\quad+\frac{(\nabla\cdot\rho_{\mathrm{a}}\mathbf{K}_m\nabla)\,u}{\rho_{\mathrm{a}}}\end{aligned} \tag{5.118}$$

South–north momentum equation from (5.111)

$$\begin{aligned}&\left(\frac{\partial v}{\partial t}+\frac{u}{R_{\mathrm{e}}\cos\varphi}\frac{\partial v}{\partial \lambda_{\mathrm{e}}}+\frac{v}{R_{\mathrm{e}}}\frac{\partial v}{\partial \varphi}\right)_s+\dot{s}\frac{\partial v}{\partial s}\\&\quad=-\frac{u^2\tan\varphi}{R_{\mathrm{e}}}-fu-\frac{1}{R_{\mathrm{e}}\rho_{\mathrm{a}}}\left[\left(\frac{\partial p_{\mathrm{a}}}{\partial \varphi}\right)_s-\frac{s}{Z_{\mathrm{t}}}\left(\frac{\partial Z_{\mathrm{t}}}{\partial \varphi}\right)_z\frac{\partial p_{\mathrm{a}}}{\partial s}\right]+\frac{(\nabla\cdot\rho_{\mathrm{a}}\mathbf{K}_m\nabla)\,v}{\rho_{\mathrm{a}}}\end{aligned} \tag{5.119}$$

Vertical momentum equation from (5.114)

$$\begin{aligned}&\left[\left(\frac{\partial}{\partial t}\right)_s+\frac{u}{R_{\mathrm{e}}\cos\varphi}\left(\frac{\partial}{\partial \lambda_{\mathrm{e}}}\right)_s+\frac{v}{R_{\mathrm{e}}}\left(\frac{\partial}{\partial \varphi}\right)_s+\dot{s}\frac{\partial}{\partial s}\right]\\&\quad\times\left[\frac{Z_{\mathrm{t}}u}{R_{\mathrm{e}}\cos\varphi}\left(\frac{\partial s}{\partial \lambda_{\mathrm{e}}}\right)_z+\frac{Z_{\mathrm{t}}v}{R_{\mathrm{e}}}\left(\frac{\partial s}{\partial \varphi}\right)_z-Z_{\mathrm{t}}\dot{s}\right]=-g+\frac{1}{Z_{\mathrm{t}}\rho_{\mathrm{a}}}\frac{\partial p_{\mathrm{a}}}{\partial s}\\&\quad+\frac{1}{\rho_{\mathrm{a}}}(\nabla\cdot\rho_{\mathrm{a}}\mathbf{K}_m\nabla)\left[\frac{Z_{\mathrm{t}}u}{R_{\mathrm{e}}\cos\varphi}\left(\frac{\partial s}{\partial \lambda_{\mathrm{e}}}\right)_z+\frac{Z_{\mathrm{t}}v}{R_{\mathrm{e}}}\left(\frac{\partial s}{\partial \varphi}\right)_z-Z_{\mathrm{t}}\dot{s}\right]\end{aligned} \tag{5.120}$$

As described in more detail in Section 5.1.2, these equations can be solved (a) explicitly with a small time step in every term, (b) explicitly with a small time step in terms producing acoustic waves and a large time step in all other terms, or (c) implicitly or semiimplicitly with a large time step in terms producing acoustic waves and explicitly with a large time step in all other terms.

Alternatively, the elastic equations above can be converted to anelastic form by deriving a diagnostic equation for nonhydrostatic pressure (Section 5.1.3). In such a case, the vertical momentum equation from (5.114) is not solved, since the vertical scalar velocity is determined diagnostically from the anelastic continuity equation, discussed following (5.101). In that case, the potential virtual temperature is still found from the thermodynamic energy equation, the horizontal scalar velocities are still found from the horizontal momentum equations, and the specific humidity is still found from the species continuity equation.

5.6 SUMMARY

The equations of atmospheric dynamics include the continuity equation for air, the species continuity equation, the thermodynamic energy equation, the horizontal momentum equation, and the vertical momentum equation. In this chapter, the vertical coordinate in the equations was transformed from the altitude to the pressure, sigma-pressure, and sigma-altitude coordinates. With the altitude (and pressure) coordinate, model layers may intersect surface topography. With the sigma-altitude and sigma-pressure coordinates, model layers are terrain-following and cannot intersect topography. When the pressure or sigma-pressure coordinate is used, the atmosphere is assumed to be in hydrostatic balance. This assumption is reasonable when coarse horizontal scales are simulated. For finer scales, the altitude or sigma-altitude coordinate should be used so that nonhydrostatic motions can be simulated.

5.7 PROBLEMS

5.1 **(a)** Assume $\Delta x = 5$ km, $\Delta y = 5$ km, and $\Delta p_a = -10$ hPa for a grid cell in the pressure coordinate and that the west, east, south, north, and lower boundary scalar velocities are $u_1 = -2$ m s^{-1}, $u_2 = +1$ m s^{-1}, $v_3 = +1$ m s^{-1}, $v_4 = -2$ m s^{-1}, and $w_5 = +0.03$ m s^{-1}, respectively. Convert w_5 to the pressure coordinate with $w_p = -w\rho_a g$, assuming $T = 284$ K and $p_a = 980$ hPa. Use the pressure-coordinate continuity equation for air to estimate w_p at the top of the cell.

(b) For the same cell as in part (a), assume the west, east, south, north, lower, and upper boundary mass mixing ratios of a gas are 0.004, 0.005, 0.003, 0.004, 0.0045, and 0.0055 kg kg^{-1}, respectively. Estimate the change in q at the center of the cell after 500 s ignoring eddy diffusion and external sources/sinks.

5.2 Assume a horizontal grid cell has dimension $\Delta x = 500$ km and $\Delta y = 400$ km, centered at $\varphi = 30°$ N. Assume that the cell is on a surface of constant pressure at $p_a = 500$ hPa and that the altitudes of the west, east, south, and north boundaries are 5.5, 5.4, 5.6, and 5.3 km, respectively. Calculate geostrophic scalar velocities (u_g and v_g) and the geostrophic wind speed.

5.3 Assume that a cell in the σ-p coordinate has dimension $\Delta x = 4$ km, $\Delta y = 5$ km, and $\Delta\sigma = 0.05$. The west, east, south, and north boundary u- and π_a-values are $u_1 = -2$ m s^{-1} and $\pi_{a,1} = 748$ hPa, $u_2 = +1$ m s^{-1} and $\pi_{a,2} = 752$ hPa, $v_3 = -1$ m s^{-1} and $\pi_{a,3} = 749$ hPa, and $v_4 = -2$ m s^{-1} and $\pi_{a,4} = 753$ hPa, respectively. Assume $p_{a,top} = 250$ hPa, $\sigma = 0.9$ at the cell bottom, the grid-cell center π_a-value is an average of the four boundary values, $T_v = 298$ K, the lower boundary vertical scalar velocity in the altitude coordinate is $w_5 = +0.02$ m s^{-1}, and the air is dry.

(a) Convert vertical scalar velocity from the altitude to the sigma-pressure coordinate with $\dot{\sigma} = -w\rho_a g/\pi_a$. Use the continuity equation for air in the sigma-pressure coordinate to estimate the sigma-pressure coordinate vertical scalar velocity at the top of the cell, assuming $\partial\pi_a/\partial t = 0$.

(b) Assume that the west, east, south, north, lower, and upper boundary values of θ_v are 299, 297, 304, 301, 300, and 302 K, respectively, and that no eddy diffusion or external sources/sinks exist. Estimate the value of θ_v at the center of the grid cell after 200 s.

5.4 Assume that a horizontal grid cell has dimension $\Delta x = 5$ km and $\Delta y = 4$ km and that θ_v and p_a on the west, east, south, and north boundaries of the cell are $\theta_{v,1} = 298$ K and $p_{a,1} = 1010$ hPa, $\theta_{v,2} = 304$ K and $p_{a,2} = 1004$ hPa, $\theta_{v,3} = 302$ K and $p_{a,3} = 1000$ hPa, and $\theta_{v,4} = 301$ K and $p_{a,4} = 1006$ hPa, respectively. What is the change in the u- and v-component scalar velocities after 10 min due to the pressure gradient force alone? Assume the air is dry.

5.5 Derive the continuity equation for air in Cartesian-sigma-pressure coordinates from the continuity equation for air in Cartesian-altitude coordinates.

5.6 Derive the horizontal momentum equation in Cartesian-pressure coordinates from that in Cartesian-sigma-pressure coordinates. Ignore conversion of the eddy diffusion term.

5.8 COMPUTER PROGRAMMING PRACTICE

5.7 Write a computer script to set up a model grid over the globe. Assume the grid stretches from φ centered at $-88°$ S to $+88°$ N, where $d\varphi = 4°$ and from λ_e centered at $-177.5°$ W to $+177.5°$ E, where $d\lambda_e = 5°$. Calculate dx and dy at the southern and western boundaries, respectively, of each grid cell, and print the values to a table.

6

Numerical solutions to partial differential equations

ATMOSPHERIC models simulate physical processes described by ordinary and partial differential equations. For example, gas and aqueous chemistry and gas-to-particle conversion processes are described by ordinary differential equations, and transport processes are described by partial differential equations. In this chapter, ordinary and partial differential equations are defined, and numerical methods of solving partial differential equations are discussed. Methods of solving partial differential equations include finite-difference, series expansion, and finite-volume methods. A special case of the finite-difference method is the semi-Lagrangian method. Two series expansion methods are finite-element and pseudospectral methods. Below, several solution methods are applied to the advection–diffusion equation, which is a unidirectional form of the species continuity equation. In addition, time-stepping schemes and their stability characteristics are discussed. Such schemes include the Forward Euler, Implicit, Crank–Nicolson, Leapfrog, Matsuno, Heun, Adams–Bashforth, and Runge–Kutta schemes. Finally, necessary characteristics of schemes that solve the advection–diffusion equation in three-dimensional models are discussed.

6.1 ORDINARY AND PARTIAL DIFFERENTIAL EQUATIONS

An **ordinary differential equation** (ODE) is an equation with one independent variable, such as time, and a **partial differential equation** (PDE) is an equation with more than one independent variable, such as time and space. ODEs and PDEs are classified by their order and degree. The **order** is the highest derivative rank of the equation, and the **degree** is the highest polynomial exponent of the highest derivative. A **homogeneous** differential equation is an equation that does not contain a term involving the independent variable. A **linear** differential equation is one in which the dependent variable and its derivatives do not appear in second-degree or higher terms and in which the dependent variable is not multiplied by other derivatives of itself.

Table 6.1 shows ordinary and partial differential equations of varying orders and degrees. Table 6.1 equations (a), (b), (d), (e), and (f) are homogeneous, and the remaining equations are inhomogeneous. Table 6.1 (b) and (e) are linear, and the rest are nonlinear. Chemical equations are first-order, first-degree, homogeneous ODEs (e.g., Table 6.1 (a) and (b)). These equations are either linear or nonlinear. The species continuity equation and the thermodynamic energy equation are

Table 6.1 Examples of the orders and degrees of ordinary and partial differential equations

Order, degree	Ordinary differential equations	Partial differential equations
First-order, first-degree	(a) $\frac{dN}{dt} = 16 - 4N^2$	(e) $\frac{\partial N}{\partial t} + \frac{\partial (uN)}{\partial x} = 0$
First-order, first-degree	(b) $\frac{dN}{dt} = 3AB - 4NC$	(f) $\frac{\partial u}{\partial t} + u\frac{\partial u}{\partial x} + v\frac{\partial u}{\partial y} = 0$
Second-order, first-degree	(c) $\frac{d^2N}{dt^2} + \frac{dN}{dt} + 5t = 0$	(g) $\frac{\partial^2 N}{\partial t^2} + \frac{\partial^2 N}{\partial x^2} = 3t^2 + x$
Second-order, second-degree	(d) $\left(\frac{d^2N}{dt^2}\right)^2 + \frac{dN}{dt} + 4 = 0$	(h) $\left(\frac{\partial^2 N}{\partial t^2}\right)^2 + \frac{\partial N}{\partial x} = t - x$

The variable t is time, x is west–east distance, y is south–north distance, N, A, B, and C are concentrations, u is west–east scalar velocity, and v is south–north scalar velocity.

first-order, first-degree, homogeneous, linear PDEs (e.g., Table 6.1 (e)). The momentum equation is a first-order, first-degree, homogeneous, nonlinear PDE (e.g., Table 6.1 (f)).

Boundary conditions for ODEs and PDEs must be specified. When conditions are known at one end of a domain but not the other, an **initial value problem** arises. If concentrations (N) are known at time $t = 0$, if time is the independent variable, and if concentration is the dependent variable, then the solution to a set of ODEs is an initial value problem. When conditions are known at both ends of a domain, the solution to a set of ODEs is a **boundary value problem**. If time and west–east direction (x) are independent variables, if concentration is the dependent variable, and if the concentrations are known everywhere at $t = 0$ and at both ends of the spatial domain at all times, the solution to a set of PDEs is an initial value problem with respect to time and a boundary value problem with respect to space.

6.2 OPERATOR SPLITTING

Major processes in an atmospheric model are often solved separately from each other. Suppose a model treats dynamics, transport, and gas chemistry. Each of these processes may be solved sequentially during a common time interval with a unique numerical scheme that takes a unique number of time steps. A **time step** is an increment in time for a given process. A **time interval** is the period during which several time steps of a process are solved without interference by another process. Suppose the time step for dynamics is 6 s, that for transport is 300 s, that for chemistry is variable, and the time interval common to all processes is 300 s. During the time interval, 50 dynamics time steps are taken, followed by 1 transport time step, followed by a variable number of chemistry time steps. After the dynamics time interval, the resulting wind speeds are used as inputs into the transport calculation. During the transport time interval, which equals the transport time step, gases are

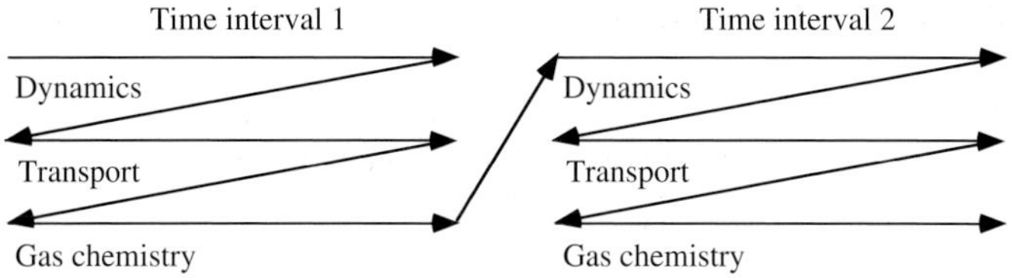

Figure 6.1 Example of operator-splitting scheme. During the first time interval, dynamics, transport, and gas chemistry are solved sequentially. Values determined from the end of a time interval after one process are used to initialize values at the beginning of the same time interval for another process. Values from the end of the last process in one time interval are used at the beginning of the first process in the next time interval.

moved around the grid. Final concentrations from the transport time interval are used as initial values for the first chemistry time step of the gas chemistry time interval. Final values from the chemistry time interval are used as initial values for the first dynamics time step in the next dynamics time interval. Figure 6.1 illustrates this example.

The isolation of individual processes during a time interval is called **time splitting** or **operator splitting.** Operator splitting is used because computers today cannot solve all model ODEs and PDEs simultaneously in three dimensions. Yanenko (1971) discusses the theoretical basis behind operator splitting with respect to certain mathematical equations.

6.3 ADVECTION–DIFFUSION EQUATIONS

First-order, first-degree, homogeneous, linear or nonlinear partial differential equations solved in atmospheric models include the species continuity equation, the thermodynamic energy equation, and the directional momentum equations. In Chapter 7, a method of solving these equations together is given. In this chapter, methods of solving **advection–diffusion equations**, which are operator-split forms of the species continuity equation, are discussed. Advection–diffusion equations are derived by considering that the four-dimensional (t, x, y, z) species continuity equation can be divided into three two-dimensional partial differential equations $([t, x]$, $[t, y]$, and $[t, z])$ and a single one-dimensional (t) ordinary differential equation. The sequential solution to the four operator-split equations approximates the solution to the original four-dimensional equation. This method of operator splitting a mathematical equation is called the **locally one-dimensional** (LOD) **procedure** or the **method of fractional steps** (e.g., Yanenko 1971; Mitchell 1969) and has been used widely in atmospheric models (e.g., Reynolds *et al.* 1973; Carmichael *et al.* 1986; Toon *et al.* 1988).

From the four-dimensional species continuity equation given in (3.52), the west–east, south–north, and vertical **unidirectional advection–diffusion equations** can be written in number concentration units as

$$\frac{\partial N}{\partial t} + \frac{\partial (uN)}{\partial x} - \frac{\partial}{\partial x}\left(K_{h,xx}\frac{\partial N}{\partial x}\right) = 0 \tag{6.1}$$

$$\frac{\partial N}{\partial t} + \frac{\partial (vN)}{\partial y} - \frac{\partial}{\partial y}\left(K_{h,yy}\frac{\partial N}{\partial y}\right) = 0 \tag{6.2}$$

$$\frac{\partial N}{\partial t} + \frac{\partial (wN)}{\partial z} - \frac{\partial}{\partial z}\left(K_{h,zz}\frac{\partial N}{\partial z}\right) = 0 \tag{6.3}$$

respectively. The solution order of these equations may be reversed each time interval to improve accuracy (e.g., Yanenko 1971). Thus, if the equations are solved in the order (6.1), (6.2), (6.3) during one time interval, they may be solved in the order (6.3), (6.2), (6.1) during the next time interval. Fractional-step schemes associated with order reversal are called **alternating-directions schemes**.

The remaining terms in the species continuity equation are external source/sink terms. These terms may be operator-split from the advection–diffusion equations as a single ordinary differential equation

$$\frac{\partial N}{\partial t} = \sum_{n=1}^{N_{e,t}} R_n \tag{6.4}$$

or split into several ODEs. Equation (6.4) can be solved before or after (6.1)–(6.3) are solved.

In moist-air mass-mixing-ratio units, the operator-split west–east advection–diffusion equation can be written from (3.54) as

$$\frac{\partial q}{\partial t} + u\frac{\partial q}{\partial x} - \frac{1}{\rho_a}\frac{\partial}{\partial x}\left(\rho_a K_{h,xx}\frac{\partial q}{\partial x}\right) = 0 \tag{6.5}$$

Analogous equations can be written for the south–north and vertical directions and for external sources/sinks. In the following subsections, finite-difference, series expansion, and finite-volume methods of approximating derivatives and of solving advection–diffusion equations are discussed.

6.4 FINITE-DIFFERENCE APPROXIMATIONS

Approximate solutions to partial differential equations, such as advection–diffusion equations, can be found with finite-difference, series expansion, or finite-volume methods. The purpose of using an approximation is to reduce the solution space for each continuous differential function from an infinite to a finite number of spatial or temporal nodes in order to speed up computation of the differential equation.

A **finite-difference** approximation involves the replacement of each continuous differential operator (d) with a discrete difference analog (Δ). This analog is an approximation written in terms of a finite number of values of the variable being operated on at each temporal or spatial node. If the west–east scalar velocity is a continuous function in space at a given time, its values can be mapped from the

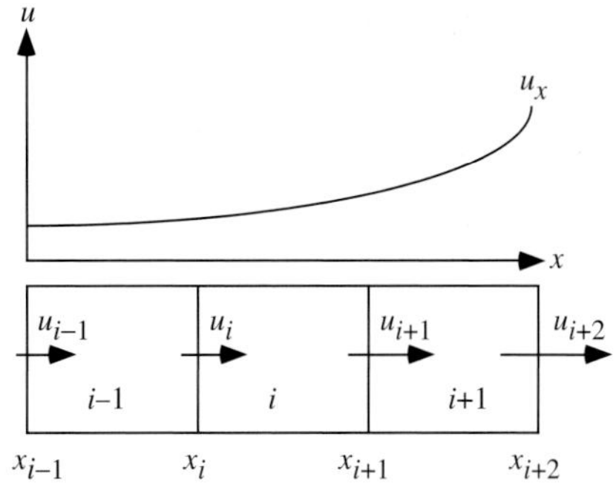

Figure 6.2 Discretization of a continuous west–east scalar velocity u_x. The west–east grid is broken into discrete cells, and u-values are mapped from the continuous function to the edge of each cell. The arrows in the cells represent magnitudes of the wind speed. Distances along the x-axis are also mapped to the cells.

function to a discretized west–east grid, as shown in Fig. 6.2. The grid consists of several **grid cells** (also called **grid boxes, grid points**, or **nodes**) placed any distance apart.

Table 6.1 (e) and (f) show partial differential equations commonly simulated in atmospheric models. Such equations are written with respect to time and space. The solution to Table 6.1 (e) requires finite-difference analogs for $\partial N/\partial t$ and $\partial(uN)/\partial x$. The solution to Table 6.1 (f) requires finite-difference analogs for $\partial u/\partial t$, $\partial u/\partial x$, and $\partial u/\partial y$.

6.4.1 Consistency, convergence, and stability

A numerical solution can replicate an exact solution to a partial differential equation if several criteria are met. First, a finite-difference analog in space or time must **converge** to its differential expression when terms in the analog approach zero. For example, if $\Delta N/\Delta x$ is a finite-difference analog of $\partial N/\partial x$, the convergence condition

$$\frac{\partial N}{\partial x} = \lim_{\Delta x \to 0} \left\| \frac{\Delta N}{\Delta x} \right\| \tag{6.6}$$

must be met for the approximation to be accurate.

Second, a finite-difference analog must be consistent. The finite-difference analog $\Delta N/\Delta x$ in (6.6) is obtained from a Taylor series expansion. In the expansion, high-order terms are neglected to reduce the computational burden of the approximation. The difference between the full Taylor series expansion and the truncated approximation is the **truncation error**. A finite-difference approximation of a derivative is **consistent** if the truncation error of the approximation approaches zero as Δx (or Δt) approaches zero. Consistency occurs when

$$\lim_{\Delta x \to 0} \left\| \mathrm{TE}\left(\frac{\Delta N}{\Delta x}\right) \right\| = 0 \tag{6.7}$$

where TE is the truncation error of the approximation $\Delta N/\Delta x$.

Third, if a finite-difference approximation is consistent, the rate at which its truncation error approaches zero depends on the order of approximation. The **order of approximation** is the order of the lowest-order term in the Taylor series expansion neglected in the approximation. The higher the order of approximation, the faster the truncation error converges toward zero upon an increase in spatial (or temporal) resolution. Thus, with the same Δx, a high-order approximation is more accurate than a low-order approximation. For the same truncation error, a low-order approximation requires a smaller Δx than does a high-order approximation. In sum, a high-order approximation with a large Δx can have the same truncation error as a low-order approximation with a small Δx. Because a high-order approximation includes more terms, it requires more computations than does a low-order approximation with the same Δx. Obtaining high order with respect to one variable, such as space, is useful only if the order of the other variable, such as time, is also high. Otherwise, low accuracy in the time derivative swamps the high accuracy in the space derivative. An optimal finite-difference solution has similar order in space and time.

Fourth, while individual finite-difference analogs must converge toward exact differentials, the overall numerical solution to a PDE must converge to an exact solution when spatial and temporal differences decrease toward zero. If $N_{e,x,t}$ is an exact solution, and $N_{f,x,t}$ is a finite-difference approximation of a PDE, **overall convergence** occurs when

$$\lim_{\Delta x, \Delta t \to 0} \| N_{e,x,t} - N_{f,x,t} \| = 0 \tag{6.8}$$

If a numerical solution is **nonconvergent,** it is not useful.

Fifth, for a numerical method to be successful, it must be stable. **Stability** occurs if the absolute-value difference between the numerical and exact solutions does not grow over time. Thus,

$$\lim_{t \to \infty} \| N_{e,x,t} - N_{f,x,t} \| \le C \tag{6.9}$$

where C is a constant. Stability often depends on the time-step size used. If a numerical solution is stable for any time step smaller than a specified value, the solution is **conditionally stable.** If a solution is stable, regardless of the time step, it is **unconditionally stable.** If a solution is unstable, regardless of the time step, it is **unconditionally unstable.**

A scheme that is unconditionally unstable cannot be convergent overall, but individual finite-difference analogs in an unstable scheme may converge and may be consistent. In other words, consistency and convergence of individual analogs do not guarantee stability. On the other hand, stability is guaranteed if a scheme is convergent overall and its finite-difference analogs are convergent and consistent.

Other problems arising from finite-difference and other solutions to partial differential equations are **numerical diffusion** (artificial spreading of peak values across several grid cells) and **numerical dispersion** (waves appearing ahead of and behind peak values). These problems can usually be mitigated by increasing the resolution of the spatial grid (e.g., decreasing Δx), decreasing the time step, or increasing the order of approximation of the finite-difference analog.

6.4.2 Low-order approximations of derivatives

A finite-difference approximation of a differential, such as $\partial u/\partial x$, involves the replacement of individual differential expressions, such as du or dx, with finite-difference analogs, such as Δu or Δx, respectively. Suppose $\partial u/\partial x$ is discretized over a west–east grid, as shown in Fig. 6.2, where all grid cells are rectangular. Each cell is denoted by an index number i, and the distance from the western edge of the entire grid to the western edge of cell i is x_i.

On the grid layout just defined, the differential scalar velocity du at point x_i can be approximated as $\Delta u_i = u_{i+1} - u_{i-1}$, $\Delta u_i = u_{i+1} - u_i$, or $\Delta u_i = u_i - u_{i-1}$, which are the **central-**, **forward-**, and **backward-difference** approximations, respectively. The corresponding discretizations of dx are $\Delta x_i = x_{i+1} - x_{i-1}$, $\Delta x_i = x_{i+1} - x_i$, and $\Delta x_i = x_i - x_{i-1}$, respectively. In the central-difference case, the slope of the tangent at point x_i in Fig. 6.2 is approximately

$$\frac{\partial u}{\partial x} \approx \frac{\Delta u_i}{\Delta x_i} = \frac{u_{i+1} - u_{i-1}}{x_{i+1} - x_{i-1}} \tag{6.10}$$

Similar equations can be written for the forward- and backward-difference cases.

The approximations just discussed can be derived from a Taylor series expansion. If gas concentration is a continuous function of west–east distance, as shown in Fig. 6.3, the values of N at points $x + \Delta x$ and $x - \Delta x$, respectively, are determined from **Taylor's theorem** as

$$N_{x+\Delta x} = N_x + \Delta x \frac{\partial N_x}{\partial x} + \frac{1}{2}\Delta x^2 \frac{\partial^2 N_x}{\partial x^2} + \frac{1}{6}\Delta x^3 \frac{\partial^3 N_x}{\partial x^3} + \frac{1}{24}\Delta x^4 \frac{\partial^4 N_x}{\partial x^4} + \cdots \tag{6.11}$$

$$N_{x-\Delta x} = N_x - \Delta x \frac{\partial N_x}{\partial x} + \frac{1}{2}\Delta x^2 \frac{\partial^2 N_x}{\partial x^2} - \frac{1}{6}\Delta x^3 \frac{\partial^3 N_x}{\partial x^3} + \frac{1}{24}\Delta x^4 \frac{\partial^4 N_x}{\partial x^4} - \cdots \tag{6.12}$$

If grid spacing is uniform (Δx is constant), the sum of (6.11) and (6.12) is

$$N_{x+\Delta x} + N_{x-\Delta x} = 2N_x + \Delta x^2 \frac{\partial^2 N_x}{\partial x^2} + \frac{1}{12}\Delta x^4 \frac{\partial^4 N_x}{\partial x^4} + \cdots \tag{6.13}$$

Rearranging (6.13) gives

$$\frac{\partial^2 N_x}{\partial x^2} = \frac{N_{x+\Delta x} - 2N_x + N_{x-\Delta x}}{\Delta x^2} + O(\Delta x^2) \tag{6.14}$$

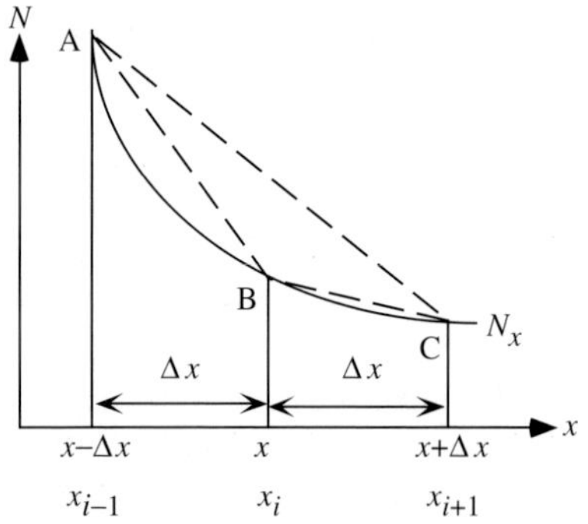

Figure 6.3 Derivative approximations at a point on a continuous function. The derivative at point B is approximated with chords AC, BC, or AB, which give the slope of the tangent at point B for the central-, forward-, and backward-difference approximations, respectively.

where

$$O(\Delta x^2) = -\frac{1}{12}\Delta x^2 \frac{\partial^4 N_x}{\partial x^4} - \cdots \tag{6.15}$$

includes all terms of order Δx^2 and higher. If $O(\Delta x^2)$ is small, (6.14) simplifies to

$$\frac{\partial^2 N_x}{\partial x^2} \approx \frac{N_{x+\Delta x} - 2N_x + N_{x-\Delta x}}{\Delta x^2} \tag{6.16}$$

where $O(\Delta x^2)$ is now the **truncation error.** Equation (6.16) is a **second-order central-difference approximation of** $\partial^2 N_x/\partial x^2$. The equation is second-order because the lowest-order exponent in the truncation error is two. It is a central-difference approximation because it relies on equally weighted values of N on each side of node x.

Subtracting (6.12) from (6.11) gives

$$N_{x+\Delta x} - N_{x-\Delta x} = 2\Delta x\frac{\partial N_x}{\partial x} + \frac{1}{3}\Delta x^3 \frac{\partial^3 N_x}{\partial x^3} + \cdots \tag{6.17}$$

Rearranging this equation results in

$$\frac{\partial N_x}{\partial x} = \frac{N_{x+\Delta x} - N_{x-\Delta x}}{2\Delta x} + O(\Delta x^2) \tag{6.18}$$

where

$$O(\Delta x^2) = -\frac{1}{6}\Delta x^2 \frac{\partial^3 N_x}{\partial x^3} - \cdots \tag{6.19}$$

includes all terms of order Δx^2 and higher. If $O(\Delta x^2)$ is small, (6.18) simplifies to

$$\frac{\partial N_x}{\partial x} \approx \frac{N_{x+\Delta x} - N_{x-\Delta x}}{2\Delta x} = \frac{N_{i+1} - N_{i-1}}{2\Delta x} \tag{6.20}$$

where $i + 1$ and $i - 1$ are surrogates for $x + \Delta x$ and $x - \Delta x$, respectively. This equation is a **second-order central-difference approximation of the first derivative of** N_x. Equation (6.20) gives the slope of the tangent (represented by chord AC) of N_x at point B in Fig. 6.3.

Another approximation of the first derivative of N_x is obtained from the first two terms of (6.11). Rearranging these terms gives

$$\frac{\partial N_x}{\partial x} \approx \frac{N_{x+\Delta x} - N_x}{\Delta x} = \frac{N_{i+1} - N_i}{\Delta x} \tag{6.21}$$

The truncated portion of the approximation includes terms first-order and higher [$O(\Delta x)$]; thus, (6.21) is a **first-order forward-difference approximation of the first derivative** of N_x. The slope of this derivative is represented by chord BC in Fig. 6.3.

Rearranging the first two terms of (6.12) gives

$$\frac{\partial N_x}{\partial x} \approx \frac{N_x - N_{x-\Delta x}}{\Delta x} = \frac{N_i - N_{i-1}}{\Delta x} \tag{6.22}$$

which is the **first-order backward-difference approximation of the first derivative** of N_x, represented by chord AB in Fig. 6.3.

If time, not space, is the independent variable, the second-order central-, first-order forward-, and first-order backward-difference approximations of $\partial N_t/\partial t$ are

$$\frac{\partial N_t}{\partial t} \approx \frac{N_{t+h} - N_{t-h}}{2h} \qquad \frac{\partial N_t}{\partial t} \approx \frac{N_{t+h} - N_t}{h} \qquad \frac{\partial N_t}{\partial t} \approx \frac{N_t - N_{t-h}}{h} \tag{6.23}$$

respectively, where $h = \Delta t$ is the **time-step size**, t is the current time, $t + h$ is one time step forward, and $t - h$ is one time step backward. These equations are derived in the same manner as (6.20), (6.21), and (6.22), respectively.

6.4.3 Arbitrary-order approximations of derivatives

Finite-difference approximations of arbitrary order can be obtained systematically (e.g., Celia and Gray 1992). The approximation of $\partial^m N/\partial x^m$, which is the mth derivative of N, can be obtained by expanding the derivative across q discrete nodes in the x-direction. If the independent variable is time, the derivative can be expanded along q time steps. The minimum number of nodes allowed in the

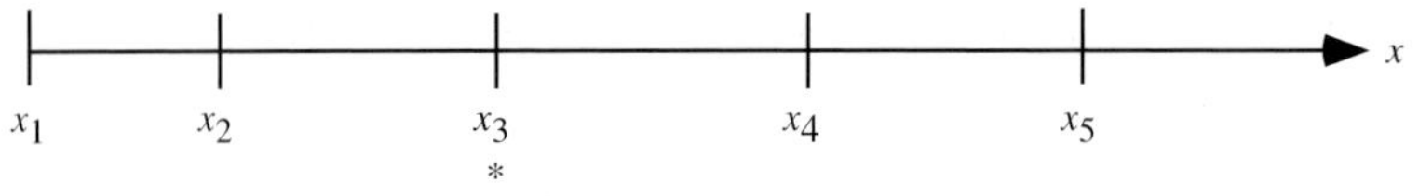

Figure 6.4 Grid spacing of an arbitrary-spaced grid where $q = 5$. The derivative is taken at node point x_3, marked*.

expansion is $m + 1$. In general, the maximum **order of approximation** of a finite-difference solution is $q - m$, although it may be smaller or larger for some individual cases. For instance, when m is even and the grid spacing is constant, the order of approximation can by increased to $q - m + 1$.

Figure 6.4 shows the arbitrary grid spacing for the derivation to come. The location at which the derivative is taken does not need to correspond to a node point, although in the figure the derivative is assumed to be taken at node point x_3. The distance between two node points is $\Delta x_i = x_{i+1} - x_i$, where i varies from 1 to $q - 1$.

The finite-difference solution to the mth derivative across q nodes is approximately

$$\frac{\partial^m N}{\partial x^m} \approx \sum_{i=1}^{q} \gamma_i N_i = \gamma_1 N_1 + \gamma_2 N_2 + \cdots + \gamma_q N_q \tag{6.24}$$

where the γ_i's are constants to be determined. A Taylor series expansion of N at node i across the point at which the derivative is taken (*) is

$$N_i = N_* + (x_i - x_*)\frac{\partial N_*}{\partial x} + \frac{1}{2}(x_i - x_*)^2\frac{\partial^2 N_*}{\partial x^2} + \frac{1}{6}(x_i - x_*)^3\frac{\partial^3 N_*}{\partial x^3} + \cdots \tag{6.25}$$

Combining (6.24) with (6.25) and gathering terms gives

$$\frac{\partial^m N}{\partial x^m} \approx \sum_{i=1}^{q} \gamma_i N_i = \sum_{i=1}^{q} \gamma_i N_* + \sum_{i=1}^{q} \gamma_i (x_i - x_*)\frac{\partial N_*}{\partial x} + \sum_{i=1}^{q} \gamma_i \frac{1}{2}(x_i - x_*)^2\frac{\partial^2 N_*}{\partial x^2} + \cdots \tag{6.26}$$

This equation can be rewritten as

$$\sum_{i=1}^{q} \gamma_i N_i = B_0 N_* + B_1 \frac{\partial N_*}{\partial x} + B_2 \frac{\partial^2 N_*}{\partial x^2} + \cdots \tag{6.27}$$

where

$$B_n = \sum_{i=1}^{q} \gamma_i \frac{1}{n!}(x_i - x_*)^n \qquad \text{for } n = 0 \cdots q - 1 \tag{6.28}$$

Equation (6.28) represents a matrix of q equations and unknowns. Multiplying (6.28) by $n!$ gives

$$\begin{bmatrix} 1 & 1 & 1 & \cdots & 1 \\ (x_1 - x_*) & (x_2 - x_*) & (x_3 - x_*) & \cdots & (x_q - x_*) \\ (x_1 - x_*)^2 & (x_2 - x_*)^2 & (x_3 - x_*)^2 & \cdots & (x_q - x_*)^2 \\ \vdots & \vdots & \vdots & & \vdots \\ (x_1 - x_*)^{q-1} & (x_2 - x_*)^{q-1} & (x_3 - x_*)^{q-1} & \cdots & (x_q - x_*)^{q-1} \end{bmatrix} \begin{bmatrix} \gamma_1 \\ \gamma_2 \\ \gamma_3 \\ \vdots \\ \gamma_q \end{bmatrix} = \begin{bmatrix} 0!B_0 \\ 1!B_1 \\ 2!B_2 \\ \vdots \\ (q-1)!B_{q-1} \end{bmatrix} \tag{6.29}$$

The highest-order derivative is found when $B_n = 1$ for $n = m$ and $B_n = 0$ for all other n.

The **first-order backward-difference approximation** of $\partial N/\partial x(m = 1)$ is found from (6.29) by discretizing $\partial N/\partial x$ across two equally spaced grid cells ($q = 2$), setting $B_1 = 1$, and setting $B_n = 0$ for all other n. The resulting matrix is

$$\begin{bmatrix} 1 & 1 \\ -\Delta x & 0 \end{bmatrix} \begin{bmatrix} \gamma_{i-1} \\ \gamma_i \end{bmatrix} = \begin{bmatrix} 0 \\ 1 \end{bmatrix} \tag{6.30}$$

where the subscript $i - 1$ indicates one node to the left of i. The matrix has solution $\gamma_{i-1} = -1/\Delta x$ and $\gamma_i = 1/\Delta x$. Substituting these coefficients into

$$\frac{\partial N}{\partial x} \approx \gamma_1 N_1 + \gamma_2 N_2 = \gamma_{i-1} N_{i-1} + \gamma_i N_i \tag{6.31}$$

from (6.24) gives the approximation shown in (6.22) and Table 6.2 (a).

Second-order central- and backward-difference approximations of $\partial N/\partial x$ ($m = 1$) are found by discretizing $\partial N/\partial x$ across three nodes ($q = 3$). The resulting matrices are

$$\begin{bmatrix} 1 & 1 & 1 \\ -\Delta x & 0 & \Delta x \\ (-\Delta x)^2 & 0 & (\Delta x)^2 \end{bmatrix} \begin{bmatrix} \gamma_{i-1} \\ \gamma_i \\ \gamma_{i+1} \end{bmatrix} = \begin{bmatrix} 0 \\ 1 \\ 0 \end{bmatrix} \begin{bmatrix} 1 & 1 & 1 \\ -2\Delta x & -\Delta x & 0 \\ (-2\Delta x)^2 & (-\Delta x)^2 & 0 \end{bmatrix} \begin{bmatrix} \gamma_{i-2} \\ \gamma_{i-1} \\ \gamma_i \end{bmatrix} = \begin{bmatrix} 0 \\ 1 \\ 0 \end{bmatrix} \tag{6.32}$$

respectively. Substituting solutions to these matrices into (6.24) gives the approximation shown in Table 6.2 (c) and (d), respectively. The **second-order forward-difference approximation** of $\partial N/\partial x$ is found by discretizing around the first column in (6.32). The result is shown Table 6.2 (e). Forward- and backward-difference discretizations are negatively symmetric to each other.

Third-order backward- and forward-difference approximations of $\partial N/\partial x$ are found in a similar manner. The results are shown in Table 6.2 (f) and (g), respectively, where the discretizations are around four cells.

Table 6.2 Finite-difference approximations of $\partial N/\partial x$ and $\partial^2 N/\partial x^2$

Order	m	q	Approximation
(a) First-order backward	1	2	$\frac{\partial N}{\partial x} \approx \frac{N_i - N_{i-1}}{\Delta x}$
(b) First-order forward	1	2	$\frac{\partial N}{\partial x} \approx \frac{N_{i+1} - N_i}{\Delta x}$
(c) Second-order central	1	3	$\frac{\partial N}{\partial x} \approx \frac{N_{i+1} - N_{i-1}}{2\Delta x}$
(d) Second-order backward	1	3	$\frac{\partial N}{\partial x} \approx \frac{N_{i-2} - 4N_{i-1} + 3N_i}{2\Delta x}$
(e) Second-order forward	1	3	$\frac{\partial N}{\partial x} \approx \frac{-3N_i + 4N_{i+1} - N_{i+2}}{2\Delta x}$
(f) Third-order backward	1	4	$\frac{\partial N}{\partial x} \approx \frac{N_{i-2} - 6N_{i-1} + 3N_i + 2N_{i+1}}{6\Delta x}$
(g) Third-order forward	1	4	$\frac{\partial N}{\partial x} \approx \frac{-2N_{i-1} - 3N_i + 6N_{i+1} - N_{i+2}}{6\Delta x}$
(h) Fourth-order central	1	5	$\frac{\partial N}{\partial x} \approx \frac{N_{i-2} - 8N_{i-1} + 8N_{i+1} - N_{i+2}}{12\Delta x}$
(i) Fourth-order backward (I)	1	5	$\frac{\partial N}{\partial x} \approx \frac{-N_{i-3} + 6N_{i-2} - 18N_{i-1} + 10N_i + 3N_{i+1}}{12\Delta x}$
(j) Fourth-order forward (I)	1	5	$\frac{\partial N}{\partial x} \approx \frac{-3N_{i-1} - 10N_i + 18N_{i+1} - 6N_{i+2} + N_{i+3}}{12\Delta x}$
(k) Fourth-order backward (II)	1	5	$\frac{\partial N}{\partial x} \approx \frac{-3N_{i-4} + 16N_{i-3} - 36N_{i-2} + 48N_{i-1} - 25N_i}{12\Delta x}$
(l) Fourth-order forward (II)	1	5	$\frac{\partial N}{\partial x} \approx \frac{25N_i - 48N_{i+1} + 36N_{i+2} - 16N_{i+3} + 3N_{i+4}}{12\Delta x}$
(m) Second-order central	2	3	$\frac{\partial^2 N}{\partial x^2} \approx \frac{N_{i+1} - 2N_i + N_{i-1}}{\Delta x^2}$
(n) Fourth-order central	2	5	$\frac{\partial^2 N}{\partial x^2} \approx \frac{-N_{i-2} + 16N_{i-1} - 30N_i + 16N_{i+1} - N_{i+2}}{12\Delta x^2}$

A **fourth-order central-difference approximation** of $\partial N/\partial x$ is found from

$$\begin{bmatrix} 1 & 1 & 1 & 1 & 1 \\ -2\Delta x & -\Delta x & 0 & \Delta x & 2\Delta x \\ (-2\Delta x)^2 & (-\Delta x)^2 & 0 & (\Delta x)^2 & (2\Delta x)^2 \\ (-2\Delta x)^3 & (-\Delta x)^3 & 0 & (\Delta x)^3 & (2\Delta x)^3 \\ (-2\Delta x)^4 & (-\Delta x)^4 & 0 & (\Delta x)^4 & (2\Delta x)^4 \end{bmatrix} \begin{bmatrix} \gamma_{i-2} \\ \gamma_{i-1} \\ \gamma_i \\ \gamma_{i+1} \\ \gamma_{i+2} \end{bmatrix} = \begin{bmatrix} 0 \\ 1 \\ 0 \\ 0 \\ 0 \end{bmatrix} \quad (6.33)$$

The approximation resulting from this matrix is shown in Table 6.2 (h). One **fourth-order backward-difference approximation** of $\partial N/\partial x$ is obtained by solving (6.33) after discretizing around the fourth instead of the third column in the equation. The result is shown in Table 6.2 (i). The corresponding **fourth-order forward-difference approximation** is shown in Table 6.2 (j). Another fourth-order backward-difference approximation of $\partial N/\partial x$ is obtained by solving (6.33) after discretizing around the fifth column in the equation. The result appears in Table 6.2 (k). The corresponding fourth-order forward-difference approximation appears in

Table 6.2 (l). A **fourth-order central-difference approximation** of $\partial^2 N/\partial x^2$ is obtained by solving (6.33), but setting $B_2 = 1$ and $B_n = 0$ for all other n. The solution is shown in Table 6.2 (n).

6.4.4 Time-stepping schemes for the advection–diffusion equation

Finite-difference approximations can be applied to the temporal and spatial derivatives of the advection–diffusion equations given in (6.1)–(6.3). In the following subsections, approximations to temporal derivatives are discussed with respect to the west–east form of the advection–diffusion equation.

6.4.4.1 *Courant–Friedrichs–Lewy stability criterion*

Some time-stepping schemes are explicit, whereas others are implicit or semiimplicit. In an **explicit** time-stepping scheme, final terms (time t) are calculated explicitly from known values (e.g., from values at times $t - h$, $t - 2h$, etc.). In an **implicit** time-stepping scheme, final terms (time t) are evaluated from other terms at time t, which are initially unknown but which are solved simultaneously with the desired terms. In a **semiimplicit** scheme, final terms are evaluated from some terms that are known (times $t - h$, $t - 2h$, etc.) and other terms that are unknown (time t).

Whereas explicit and semiimplicit time-stepping schemes are generally conditionally stable (stable for any time step below a specified value), some implicit schemes may be unconditionally stable (stable regardless of time step). Such schemes, by nature, do not require iteration. Other implicit schemes require iteration, and these are conditionally stable.

The time step limitation for an explicit solution to the advection equation can be approximated with the **Courant–Friedrichs–Lewy (CFL) stability criterion** (Courant *et al.* 1928). The **advection equation** in the west–east direction, for example, is found by removing the diffusion term from (6.1). When the resulting equation is solved explicitly, stability is generally maintained when the CFL criterion,

$$h < \Delta x_{\text{min}}/|u_{\text{max}}| \tag{6.34}$$

is met, where $|u_{\text{max}}|$ is the maximum west–east wind speed, and Δx_{min} is the minimum west–east grid-cell length in the domain. If the maximum wind speed is $|u_{\text{max}}| = 20$ m s^{-1}, for example, and the minimum grid-cell length is $\Delta x_{\text{min}} = 5$ km, the CFL criterion predicts that the maximum time step for maintaining stability is $h = 250$ s. To maintain stability, a parcel of air is not allowed to travel across a grid cell during a single time step.

For equations more general than the advection–diffusion equation, $|u_{\text{max}}|$ should be replaced by $|c_{\text{max}}|$, where $|c_{\text{max}}|$ is the maximum speed of propagation in the domain. In the case of the primitive equations, the maximum speed of propagation is the speed of horizontally propagating acoustic waves (the speed of sound) (Section 5.1).

If advection is ignored, (6.1) simplifies to the west–east **diffusion equation.** A stability criterion for this equation, analogous to (6.34), is $h < \Delta x_{\text{min}}^2/K_{\text{max}}$, where

K_{max} is the largest eddy diffusion coefficient in the domain. For a typical vertical eddy diffusion coefficient of 50 m^2 s^{-1}, for example, this stability criterion suggests that the time step for eddy diffusion through an altitude of 100 m needs to be less than 200 s. For a typical horizontal eddy diffusion coefficient of 2500 m^2 s^{-1}, the time step for eddy diffusion across a 5-km cell must be less than 10 000 s. Explicit solutions to the diffusion equation are more likely to become unstable in the vertical than in the horizontal in an atmospheric model.

6.4.4.2 *Forward Euler scheme*

A basic time-discretization scheme for the advection–diffusion equation is the forward Euler scheme. If u varies and $K = K_{h,xx}$ is constant along x, and if u and K are constant during a time step, the time, advection, and diffusion derivatives in (6.1) can be discretized with (6.23), (6.20), and (6.16), respectively, to yield

$$\frac{N_{i,t} - N_{i,t-h}}{h} + \frac{(uN)_{i+1,t-h} - (uN)_{i-1,t-h}}{2\Delta x} - K\frac{N_{i+1,t-h} - 2N_{i,t-h} + N_{i-1,t-h}}{\Delta x^2} = 0 \quad (6.35)$$

The temporal and spatial derivatives in this equation are first- and second-order approximations, respectively. The equation is called the **forward-in-time, centered-in-space** (FTCS) approximation because the time derivative uses information from one previous time step, and the advection terms are central-difference expressions. Since all terms in (6.35), except the final concentration, are evaluated at time $t - h$, the FTCS approximation is explicit. When a time derivative is first-order and spatial derivatives are determined explicitly, as in the example above, the time scheme is a **forward Euler** one. For all values of u, this equation is unconditionally unstable for $K = 0$ and for large values of K and conditionally stable for small values of K, except when $K = 0$ (Mesinger and Arakawa 1976).

Because values on the right side of (6.35) are known, the equation can be solved immediately for $i = 1, \ldots, I$ (where I is the number of west–east nodes). If the grid contains lateral boundaries, the solution depends on the terms $(uN)_{0,t-h}$, $(uN)_{I+1,t-h}$, $(KN)_{0,t-h}$, and $(KN)_{I+1,t-h}$, which lie beyond the boundaries. Outside boundary values may be set equal to values just inside the boundary (e.g., at nodes $i = 1$ and $i = I$) from the previous time step. When the grid has **periodic boundary conditions** (e.g., a grid that has no lateral boundaries because it wraps around on itself), node $i = 0$ is also node $i = I$, and node $i = 1$ is also node $i = I + 1$. In such a case, outside boundary values are not needed.

6.4.4.3 *Implicit scheme*

Equation (6.35) can be solved implicitly by evaluating all terms on the right side at time t. In implicit form, (6.35) becomes

$$\frac{N_{i,t} - N_{i,t-h}}{h} + \frac{(uN)_{i+1,t} - (uN)_{i-1,t}}{2\Delta x} - K\frac{N_{i+1,t} - 2N_{i,t} + N_{i-1,t}}{\Delta x^2} = 0 \quad (6.36)$$

The solution to (6.36) along $i = 1, \ldots, I$ is obtained by rearranging the equation as

$$A_i N_{i-1,t} + B_i N_{i,t} + D_i N_{i+1,t} = N_{i,t-h} \tag{6.37}$$

where

$$A_i = -h\left(\frac{u}{2\Delta x} + \frac{K}{\Delta x^2}\right)_{i-1} \qquad B_i = 1 + h\left(\frac{2K}{\Delta x^2}\right)_i \qquad D_i = h\left(\frac{u}{2\Delta x} - \frac{K}{\Delta x^2}\right)_{i+1} \tag{6.38}$$

For a **limited-area domain** (a domain with boundaries at both ends), the matrix arising from (6.37) is

$$\begin{bmatrix} B_1 & D_1 & 0 & 0 & \cdots & 0 & 0 & 0 \\ A_2 & B_2 & D_2 & 0 & \cdots & 0 & 0 & 0 \\ 0 & A_3 & B_3 & D_3 & \cdots & 0 & 0 & 0 \\ 0 & 0 & A_4 & B_4 & \cdots & 0 & 0 & 0 \\ \vdots & \vdots & \vdots & \vdots & & \vdots & \vdots & \vdots \\ 0 & 0 & 0 & 0 & \cdots & B_{I-2} & D_{I-2} & 0 \\ 0 & 0 & 0 & 0 & \cdots & A_{I-1} & B_{I-1} & D_{I-1} \\ 0 & 0 & 0 & 0 & \cdots & 0 & A_I & B_I \end{bmatrix} \begin{bmatrix} N_{1,t} \\ N_{2,t} \\ N_{3,t} \\ N_{4,t} \\ \vdots \\ N_{I-2,t} \\ N_{I-1,t} \\ N_{I,t} \end{bmatrix} = \begin{bmatrix} N_{1,t-h} \\ N_{2,t-h} \\ N_{3,t-h} \\ N_{4,t-h} \\ \vdots \\ N_{I-2,t-h} \\ N_{I-1,t-h} \\ N_{I,t-h} \end{bmatrix} - \begin{bmatrix} A_1 N_{0,t} \\ 0 \\ 0 \\ 0 \\ \vdots \\ 0 \\ 0 \\ D_I N_{I+1,t} \end{bmatrix} \tag{6.39}$$

where $A_1 N_{0,t}$ and $D_I N_{I+1,t}$ are outside boundary values. Outside values are assumed to be known in advance although they carry the subscript t. Equation (6.39) is a **tridiagonal matrix**, which is solved by matrix decomposition and backsubstitution in the order,

Decomposition

$$\gamma_1 = -\frac{D_1}{B_1} \qquad \gamma_i = -\frac{D_i}{B_i + A_i\gamma_{i-1}} \qquad \text{for } i = 2, \ldots, I$$

$$\alpha_1 = \frac{R_1}{B_1} \qquad \alpha_i = \frac{R_i - A_i\alpha_{i-1}}{B_i + A_i\gamma_{i-1}} \qquad \text{for } i = 2, \ldots, I \tag{6.40}$$

Backsubstitution

$$N_{I,t} = \alpha_I \qquad N_{i,t} = \alpha_i + \gamma_i N_{i+1,t} \qquad \text{for } i = I-1, \ldots, 1, -1 \tag{6.41}$$

where R_i represents the right side of (6.39). The solution to (6.39) is mass conserving and unconditionally stable for all values of u and K, but it is numerically diffusive. To obtain the solution, (6.36) was converted from a partial differential equation (dependent on time and space) to an ordinary differential equation (dependent on time only). The linear ODE was then solved by matrix decomposition and backsubstitution.

For a domain with periodic boundary conditions (Section 6.4.4.2), the matrix for (6.39) becomes

$$\begin{bmatrix} B_1 & D_1 & 0 & 0 & \cdots & 0 & 0 & A_1 \\ A_2 & B_2 & D_2 & 0 & \cdots & 0 & 0 & 0 \\ 0 & A_3 & B_3 & D_3 & \cdots & 0 & 0 & 0 \\ 0 & 0 & A_4 & B_4 & \cdots & 0 & 0 & 0 \\ \vdots & \vdots & \vdots & \vdots & & \vdots & \vdots & \vdots \\ 0 & 0 & 0 & 0 & \cdots & B_{I-2} & D_{I-2} & 0 \\ 0 & 0 & 0 & 0 & \cdots & A_{I-1} & B_{I-1} & D_{I-1} \\ D_I & 0 & 0 & 0 & \cdots & 0 & A_I & B_I \end{bmatrix} \begin{bmatrix} N_{1,t} \\ N_{2,t} \\ N_{3,t} \\ N_{4,t} \\ \vdots \\ N_{I-2,t} \\ N_{I-1,t} \\ N_{I,t} \end{bmatrix} = \begin{bmatrix} N_{1,t-h} \\ N_{2,t-h} \\ N_{3,t-h} \\ N_{4,t-h} \\ \vdots \\ N_{I-2,t-h} \\ N_{I-1,t-h} \\ N_{I,t-h} \end{bmatrix} \tag{6.42}$$

where values at node I are adjacent to those at node 1. The solution is obtained by solving (6.40), followed by

$$\chi_1 = -\frac{A_1}{B_1} \qquad \chi_i = -\frac{A_i\chi_{i-1}}{B_i + A_i\gamma_{i-1}} \qquad \text{for } i = 2, \ldots, I$$

$$\psi_I = 1 \qquad \psi_i = \gamma_i\psi_{i+1} + \chi_i \qquad \text{for } i = I-1, \ldots, 1, -1$$

$$\beta_I = 0 \qquad \beta_i = \gamma_i\beta_{i+1} + \alpha_i \qquad \text{for } i = I-1, \ldots, 1, -1$$

$$N_{I,t} = \frac{\alpha_I - \dfrac{\beta_1 D_I}{B_I + A_I\gamma_{I-1}}}{1 + \dfrac{D_I\psi_1 + A_I\chi_I}{B_I + A_I\gamma_{I-1}}}$$

$$N_{i,t} = \alpha_i + \gamma_i N_{i+1,t} + \chi_i N_{I,t} \qquad \text{for } i = I-1, \ldots, 1, -1 \tag{6.43}$$

This solution is mass-conserving but does not require outside boundary information.

6.4.4.4 *Crank–Nicolson scheme*

The implicit approximation just described was first order in time and second order in space. The order of approximation in time can be improved to second order with the **Crank–Nicolson** (trapezoidal) **scheme** (Crank and Nicolson 1947). This scheme is semiimplicit since some terms on the right side are evaluated at time t and others are evaluated at time $t - h$. With the Crank–Nicolson scheme, spatial derivatives are weighted 50 percent between the initial and final times. Rewriting (6.35) gives

$$\begin{aligned} &\frac{N_{i,t} - N_{i,t-h}}{h} + \left[\mu_c \frac{(uN)_{i+1,t} - (uN)_{i-1,t}}{2\Delta x} + (1-\mu_c)\frac{(uN)_{i+1,t-h} - (uN)_{i-1,t-h}}{2\Delta x}\right] \\ &\quad -K\left[\mu_c \frac{N_{i+1,t} - 2N_{i,t} + N_{i-1,t}}{\Delta x^2} + (1-\mu_c)\frac{N_{i+1,t-h} - 2N_{i,t-h} + N_{i-1,t-h}}{\Delta x^2}\right] = 0 \end{aligned} \tag{6.44}$$

where μ_c is the **Crank–Nicolson parameter.** When a finite-difference equation is written in terms of μ_c, the equation is in **Crank–Nicolson form.** When $\mu_c = 0.5$,

(6.44) reduces to the Crank–Nicolson scheme, which is second order in time and unconditionally stable for all values of u and K. When $\mu_c = 0$, (6.44) reduces to (6.35), the forward Euler scheme, and when $\mu_c = 1$, (6.44) reduces to (6.36), the implicit scheme. Equation (6.44) can be rewritten as

$$A_i N_{i-1,t} + B_i N_{i,t} + D_i N_{i+1,t} = E_i N_{i-1,t-h} + F_i N_{i,t-h} + G_i N_{i+1,t-h} \tag{6.45}$$

where

$$\begin{aligned} A_i &= -\mu_c h\left(\frac{u}{2\Delta x} + \frac{K}{\Delta x^2}\right)_{i-1} & B_i &= 1 + \mu_c h\left(\frac{2K}{\Delta x^2}\right)_i \\ D_i &= \mu_c h\left(\frac{u}{2\Delta x} - \frac{K}{\Delta x^2}\right)_{i+1} & E_i &= (1-\mu_c)\,h\left(\frac{u}{2\Delta x} + \frac{K}{\Delta x^2}\right)_{i-1} \\ F_i &= 1 - (1-\mu_c)\,h\left(\frac{2K}{\Delta x^2}\right)_i & G_i &= -(1-\mu_c)\,h\left(\frac{u}{2\Delta x} - \frac{K}{\Delta x^2}\right)_{i+1} \end{aligned} \tag{6.46}$$

When a grid has lateral boundaries, the matrix arising from (6.45) is the same as (6.39), except that the right side of (6.39) is replaced with

$$= \begin{bmatrix} F_1 & G_1 & 0 & 0 & \cdots & 0 & 0 & 0 \\ E_2 & F_2 & G_2 & 0 & \cdots & 0 & 0 & 0 \\ 0 & E_3 & F_3 & G_3 & \cdots & 0 & 0 & 0 \\ 0 & 0 & E_4 & F_4 & \cdots & 0 & 0 & 0 \\ \vdots & \vdots & \vdots & \vdots & & \vdots & \vdots & \vdots \\ 0 & 0 & 0 & 0 & \cdots & F_{I-2} & G_{I-2} & 0 \\ 0 & 0 & 0 & 0 & \cdots & E_{I-1} & F_{I-1} & G_{I-1} \\ 0 & 0 & 0 & 0 & \cdots & 0 & E_I & F_I \end{bmatrix} \begin{bmatrix} N_{1,t-h} \\ N_{2,t-h} \\ N_{3,t-h} \\ N_{4,t-h} \\ \vdots \\ N_{I-2,t-h} \\ N_{I-1,t-h} \\ N_{I,t-h} \end{bmatrix} + \begin{bmatrix} E_1 N_{0,t-h} - A_1 N_{0,t} \\ 0 \\ 0 \\ 0 \\ \vdots \\ 0 \\ 0 \\ G_I N_{I+1,t-h} - D_I N_{I+1,t} \end{bmatrix} \tag{6.47}$$

Equations (6.39) and (6.47) are solved with (6.40) and (6.41). For a domain with periodic boundary conditions, (6.39) and (6.47) are solved after the rightmost column in (6.47) is removed, after A_1 and D_I are placed in the top right and bottom left corners, respectively, of (6.39), and after E_1 and G_I are placed in the top right and bottom left corners, respectively, of (6.47).

6.4.4.5 *Leapfrog scheme*

Another scheme that increases the order of approximation in time to second order is the **leapfrog scheme**. This scheme uses information from two previous time steps to predict information for a third. More specifically, spatial derivatives from time $t - h$ are used to evaluate differences between times t and $t - 2h$. The leapfrog solution to the west–east advection–diffusion equation is

$$\frac{N_{i,t} - N_{i,t-2h}}{2h} + \frac{(uN)_{i+1,t-h} - (uN)_{i-1,t-h}}{2\Delta x} - K\frac{N_{i+1,t-h} - 2N_{i,t-h} + N_{i-1,t-h}}{\Delta x^2} = 0 \tag{6.48}$$

where values from time $t - h$ and $t - 2h$ are determined from previous time steps.

The discretization above is second-order in time and space. When used alone, the leapfrog scheme is unconditionally unstable for all nonzero values of K. When $K = 0$, the leapfrog scheme is conditionally stable for linear equations. For nonlinear equations, the scheme destabilizes over time. To suppress such instability, computations from another scheme must be inserted every few leapfrog steps (Mesinger and Arakawa 1976). A scheme used to stabilize the leapfrog scheme is the Matsuno scheme.

6.4.4.6 *Matsuno scheme*

The **Matsuno scheme** (Matsuno 1966) is an explicit time-stepping scheme commonly used to stabilize and initialize the leapfrog scheme. With the Matsuno scheme, time derivatives are estimated with a forward-difference approximation. The estimated values (subscript "est") are substituted into the spatial derivatives to estimate final values. The estimation and correction steps are

$$\frac{N_{i,\mathrm{est}} - N_{i,t-h}}{h} + \frac{(uN)_{i+1,t-h} - (uN)_{i-1,t-h}}{2\Delta x} - K\frac{N_{i+1,t-h} - 2N_{i,t-h} + N_{i-1,t-h}}{\Delta x^2} = 0 \tag{6.49}$$

$$\frac{N_{i,t} - N_{i,t-h}}{h} + \frac{(uN)_{i+1,\mathrm{est}} - (uN)_{i-1,\mathrm{est}}}{2\Delta x} - K\frac{N_{i+1,\mathrm{est}} - 2N_{i,\mathrm{est}} + N_{i-1,\mathrm{est}}}{\Delta x^2} = 0 \tag{6.50}$$

respectively. Although the Matsuno scheme requires twice as many computations as does either the forward Euler or leapfrog scheme per time step, the Matsuno scheme is still a first-order approximation in time. It is conditionally stable for all values of u when K is zero or small but absolutely unstable for large values of K (Mesinger and Arakawa 1976). When combined with the leapfrog scheme to solve the equations of atmospheric dynamics, Matsuno steps are usually taken every 5–15 leapfrog steps.

6.4.4.7 *Heun scheme*

With the **Heun scheme**, time derivatives are estimated with a forward-difference approximation that uses initial values in the spatial derivative. Final time derivatives are determined with an average spatial derivative. The average is taken as one-half the spatial derivative determined from initial values plus one-half the spatial derivative determined from estimated values. With respect to the advection–diffusion equation, the Heun scheme involves solving (6.49) followed by

$$\frac{N_{i,t} - N_{i,t-h}}{h} + \frac{1}{2}\frac{(uN)_{i+1,\mathrm{est}} - (uN)_{i-1,\mathrm{est}}}{2\Delta x} - \frac{K}{2}\frac{N_{i+1,\mathrm{est}} - 2N_{i,\mathrm{est}} + N_{i-1,\mathrm{est}}}{\Delta x^2}$$
$$+ \frac{1}{2}\frac{(uN)_{i+1,t-h} - (uN)_{i-1,t-h}}{2\Delta x} - \frac{K}{2}\frac{N_{i+1,t-h} - 2N_{i,t-h} + N_{i-1,t-h}}{\Delta x^2} = 0 \tag{6.51}$$

The Heun scheme is a second-order approximation in time. For all values of u, this scheme is unconditionally unstable when $K = 0$ and when K is large, and conditionally stable when K is small and nonzero (Mesinger and Arakawa 1976).

6.4.4.8 *Adams–Bashforth scheme*

Another time-differencing scheme is a simplified version of the **Adams–Bashforth scheme**. Like the leapfrog scheme, this scheme is explicit, uses three time levels, and is a second-order approximation in time. The scheme discretizes the advection–diffusion equation as

$$\frac{N_{i,t}-N_{i,t-h}}{h}+\frac{3}{2}\frac{(uN)_{i+1,t-h}-(uN)_{i-1,t-h}}{2\Delta x}-\frac{3}{2}K\frac{N_{i+1,t-h}-2N_{i,t-h}+N_{i-1,t-h}}{\Delta x^2}$$
$$-\frac{1}{2}\frac{(uN)_{i+1,t-2h}-(uN)_{i-1,t-2h}}{2\Delta x}+\frac{1}{2}K\frac{N_{i+1,t-2h}-2N_{i,t-2h}+N_{i-1,t-2h}}{\Delta x^2}=0 \tag{6.52}$$

where the $t-h$ time level is favored over the $t-2h$ time level. For all values of u, the Adams–Bashforth scheme is unconditionally unstable when $K=0$ and when K is large, and conditionally stable when K is small and nonzero (Mesinger and Arakawa 1976). This scheme is useful for short integration periods when a small time step is taken.

6.4.4.9 *Fourth-order Runge–Kutta scheme*

The last time-differencing scheme discussed is the fourth-order Runge–Kutta scheme (e.g., Press *et al.* 1992). This scheme is explicit and requires information from one time step backward only, but makes three guesses before forecasting final values for the time step. When the Runge–Kutta scheme is applied to the advection–diffusion equation, the concentration at time t is calculated with

$$N_{i,t}=N_{i,t-h}+\frac{k_1}{6}+\frac{k_2}{3}+\frac{k_3}{3}+\frac{k_4}{6} \tag{6.53}$$

where

$$\begin{aligned}
k_1&=h\left[-\frac{(uN)_{i+1,t-h}-(uN)_{i-1,t-h}}{2\Delta x}+K\frac{N_{i+1,t-h}-2N_{i,t-h}+N_{i-1,t-h}}{\Delta x^2}\right]\\
k_2&=h\left[-\frac{(u_{t-h}N_{\text{est}1})_{i+1}-(u_{t-h}N_{\text{est}1})_{i-1}}{2\Delta x}+K\frac{N_{i+1,\text{est}1}-2N_{i,\text{est}1}+N_{i-1,\text{est}1}}{\Delta x^2}\right]\\
k_3&=h\left[-\frac{(u_{t-h}N_{\text{est}2})_{i+1}-(u_{t-h}N_{\text{est}2})_{i-1}}{2\Delta x}+K\frac{N_{i+1,\text{est}2}-2N_{i,\text{est}2}+N_{i-1,\text{est}2}}{\Delta x^2}\right]\\
k_4&=h\left[-\frac{(u_{t-h}N_{\text{est}3})_{i+1}-(u_{t-h}N_{\text{est}3})_{i-1}}{2\Delta x}+K\frac{N_{i+1,\text{est}3}-2N_{i,\text{est}3}+N_{i-1,\text{est}3}}{\Delta x^2}\right]
\end{aligned} \tag{6.54}$$

and

$$N_{i,\text{est}1}=N_{i,t-h}+\frac{k_1}{2}\qquad N_{i,\text{est}2}=N_{i,t-h}+\frac{k_2}{2}\qquad N_{i,\text{est}3}=N_{i,t-h}+k_3 \tag{6.55}$$

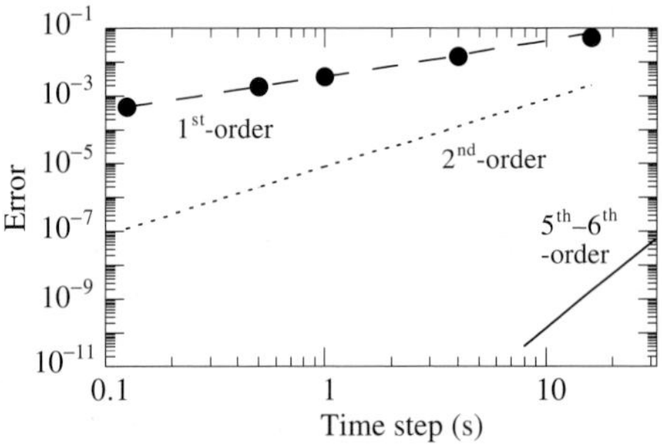

Figure 6.5 Comparison of convergence among four schemes applied to the one-dimensional advection–diffusion equation for a tracer concentration. —— Runge–Kutta, ----- Adams–Bashforth, – – – Matsuno, • forward Euler. The wind speed and diffusion coefficient were constant at $5\,\mathrm{m\,s^{-1}}$ and $1000\,\mathrm{m^2\,s^{-1}}$, respectively. Boundary conditions were periodic. The actual orders of approximation determined by the curves are shown. From Ketefian and Jacobson (2005a).

When $K = 0$, the scheme is stable when $h \le 2\sqrt{2}\Delta x/|u_{\max}|$, where $|u_{\max}|$ is the maximum west–east wind speed. When $|u_{\max}| = 0$, the scheme is stable when $h \le C\Delta x^2/K_{\max}$, where $K_{\max}$ is the maximum diffusion coefficient on the grid and C is a constant between 0.25 and 0.3. When $K > 0$ and $|u_{\max}| > 0$, the stability of the scheme is a complex function of $|u_{\max}|$, $K_{\max}$, and grid spacing (Ketefian 2005; Ketefian and Jacobson 2005a).

Figure 6.5 compares the convergence of the fourth-order Runge–Kutta, Adams–Bashforth, Matsuno, and forward Euler schemes. The Matsuno and forward Euler schemes are first-order approximations in time, as illustrated in the figure, which shows that when the time step is reduced by half, errors of the schemes decrease by half. The Adams–Bashforth scheme is a second-order approximation in time. Thus, a factor of two reduction in the time step reduces its error by a factor of four. A factor of two reduction in the fourth-order Runge–Kutta time step ideally reduces its error by a factor of 16. The figure, though, shows that a reduction in the time step by a factor of two reduced the error by a factor of 36–41, making this scheme a fifth–sixth order scheme for this application. The reason is that the order of approximation is the order of the lowest-order term in a Taylor series expansion neglected in the approximation. In the present case, the lowest-order term neglected was negligible, thus the order of approximation was governed by the second-lowest-order term neglected.

Figure 6.5 shows that not only does a high-order scheme decrease the error with decreasing time step to a greater extent than does a low-order scheme, but a high-order scheme also results in a lower error at any given time step than does a low-order scheme.

6.4.5 Fourth-order in space solution to the advection–diffusion equation

The time-difference schemes discussed above can be applied with higher-order spatial finite-difference approximations. For example, substituting fourth-order expansions of the advection and diffusion terms from Table 6.2 (h) and (n), respectively, into (6.1) gives a fully implicit, unconditionally stable form of the advection–diffusion equation as

$$\frac{N_{i,t} - N_{i,t-h}}{h} + \frac{(uN)_{i-2,t} - 8(uN)_{i-1,t} + 8(uN)_{i+1,t} - (uN)_{i+2,t}}{12\Delta x}$$

$$- K\frac{-N_{i-2,t} + 16N_{i-1,t} - 30N_{i,t} + 16N_{i+1,t} - N_{i+2,t}}{12\Delta x^2} = 0 \qquad (6.56)$$

This equation can be written in banded-matrix form with five terms on both sides and solved with a banded matrix method (e.g., Press *et al.* 1992). Such methods may combine matrix decomposition and backsubstitution with a sparse-matrix technique to reduce the number of computations.

6.4.5.1 *Variable grid spacing and eddy diffusion coefficients*

The previous solutions to the advection–diffusion equation were obtained by assuming constant grid spacing and eddy diffusion coefficients. Here a solution that assumes variable grid spacing and diffusion coefficients is considered.

The west–east advection term in (6.1) can be discretized with (6.24) as

$$\frac{\partial(uN)}{\partial x} = \gamma_{a,i-1}(uN)_{i-1} + \gamma_{a,i}(uN)_i + \gamma_{a,i+1}(uN)_{i+1} \qquad (6.57)$$

When grid spacing is variable, the second-order central-difference approximation coefficients for this equation are obtained by solving the matrix equation from (6.29) with $B_1 = 1$ and all other $B_n = 0$,

$$\begin{bmatrix} 1 & 1 & 1 \\ -(x_i - x_{i-1}) & 0 & (x_{i+1} - x_i) \\ (x_i - x_{i-1})^2 & 0 & (x_{i+1} - x_i)^2 \end{bmatrix} \begin{bmatrix} \gamma_{a,i-1} \\ \gamma_{a,i} \\ \gamma_{a,i+1} \end{bmatrix} = \begin{bmatrix} 0 \\ 1 \\ 0 \end{bmatrix} \qquad (6.58)$$

to yield

$$\gamma_{a,i-1} = \frac{-(x_{i+1} - x_i)}{(x_i - x_{i-1})(x_{i+1} - x_{i-1})}$$

$$\gamma_{a,i} = \frac{(x_{i+1} - x_i) - (x_i - x_{i-1})}{(x_{i+1} - x_i)(x_i - x_{i-1})} \qquad (6.59)$$

$$\gamma_{a,i+1} = \frac{x_i - x_{i-1}}{(x_{i+1} - x_i)(x_{i+1} - x_{i-1})}$$

The west–east diffusion term in (6.1) expands to

$$\frac{\partial}{\partial x}\left(K\frac{\partial N}{\partial x}\right) = \frac{\partial K}{\partial x}\frac{\partial N}{\partial x} + K\frac{\partial^2 N}{\partial x^2} \tag{6.60}$$

The second-order central-difference approximations to the terms on the right side of this equation are

$$\begin{aligned} \frac{\partial K}{\partial x} &\approx \gamma_{\mathrm{a},i-1}K_{i-1} + \gamma_{\mathrm{a},i}K_i + \gamma_{\mathrm{a},i+1}K_{i+1} \\ \frac{\partial N}{\partial x} &\approx \gamma_{\mathrm{a},i-1}N_{i-1} + \gamma_{\mathrm{a},i}N_i + \gamma_{\mathrm{a},i+1}N_{i+1} \\ K\frac{\partial^2 N}{\partial x^2} &\approx K_i(\gamma_{\mathrm{d},i-1}N_{i-1} + \gamma_{\mathrm{d},i}N_i + \gamma_{\mathrm{d},i+1}N_{i+1}) \end{aligned} \tag{6.61}$$

where the γ_a terms are the same as those in (6.59). The γ_d terms are found by solving the matrix equation for a second-order central-difference approximation from (6.29) with $B_2 = 1$ and all other $B_n = 0$,

$$\begin{bmatrix} 1 & 1 & 1 \\ -(x_i - x_{i-1}) & 0 & (x_{i+1} - x_i) \\ (x_i - x_{i-1})^2 & 0 & (x_{i+1} - x_i)^2 \end{bmatrix} \begin{bmatrix} \gamma_{\mathrm{d},i-1} \\ \gamma_{\mathrm{d},i} \\ \gamma_{\mathrm{d},i+1} \end{bmatrix} = \begin{bmatrix} 0 \\ 0 \\ 2 \end{bmatrix} \tag{6.62}$$

to yield

$$\begin{aligned} \gamma_{\mathrm{d},i-1} &= \frac{2}{(x_i - x_{i-1})(x_{i+1} - x_{i-1})} \\ \gamma_{\mathrm{d},i} &= \frac{-2}{(x_{i+1} - x_i)(x_i - x_{i-1})} \\ \gamma_{\mathrm{d},i+1} &= \frac{2}{(x_{i+1} - x_i)(x_{i+1} - x_{i-1})} \end{aligned} \tag{6.63}$$

Substituting (6.61) into (6.60) gives

$$\frac{\partial}{\partial x}\left(K\frac{\partial N}{\partial x}\right) \approx \beta_{K,i-1}N_{i-1} + \beta_{K,i}N_i + \beta_{K,i+1}N_{i+1} \tag{6.64}$$

where

$$\begin{aligned} \beta_{K,i-1} &= (\gamma_{\mathrm{a},i-1}K_{i-1} + \gamma_{\mathrm{a},i}K_i + \gamma_{\mathrm{a},i+1}K_{i+1})\gamma_{\mathrm{a},i-1} + K_i\gamma_{\mathrm{d},i-1} \\ \beta_{K,i} &= (\gamma_{\mathrm{a},i-1}K_{i-1} + \gamma_{\mathrm{a},i}K_i + \gamma_{\mathrm{a},i+1}K_{i+1})\gamma_{\mathrm{a},i} + K_i\gamma_{\mathrm{d},i} \\ \beta_{K,i+1} &= (\gamma_{\mathrm{a},i-1}K_{i-1} + \gamma_{\mathrm{a},i}K_i + \gamma_{\mathrm{a},i+1}K_{i+1})\gamma_{\mathrm{a},i+1} + K_i\gamma_{\mathrm{d},i+1} \end{aligned} \tag{6.65}$$

Applying (6.57) and (6.65) to (6.1) gives a second order in space solution to the advection–diffusion equation that allows for variable grid spacing, wind speeds, and eddy diffusion coefficients. In Crank–Nicolson form, the advection–diffusion equation is

$$\frac{N_{i,t} - N_{i,t-h}}{h} = -\mu_c\{[(\gamma_a u - \beta_K)N]_{i-1} + [(\gamma_a u - \beta_K)N]_i + [(\gamma_a u - \beta_K)N]_{i+1}\}_t$$
$$- (1 - \mu_c)\{[(\gamma_a u - \beta_K)N]_{i-1} + [(\gamma_a u - \beta_K)N]_i + [(\gamma_a u - \beta_K)N]_{i+1}\}_{t-h} \quad (6.66)$$

This equation can be written in tridiagonal matrix form and solved.

6.4.6 Finite-differencing in two directions

Combining the west–east and south–north terms in the advection–diffusion equation gives the horizontal form of the equation as

$$\frac{\partial N}{\partial t} + \frac{\partial (uN)}{\partial x} + \frac{\partial (vN)}{\partial y} - \frac{\partial}{\partial x}\left(K_{h,xx}\frac{\partial N}{\partial x}\right) - \frac{\partial}{\partial y}\left(K_{h,yy}\frac{\partial N}{\partial y}\right) = 0 \quad (6.67)$$

When constant grid spacing and eddy diffusion coefficients are assumed, the implicit expansion of (6.67) to second order in space is

$$\frac{N_{i,j,t} - N_{i,j,t-h}}{h} + \left[\frac{(uN)_{i+1,j} - (uN)_{i-1,j}}{2\Delta x} + \frac{(vN)_{i,j+1} - (vN)_{i,j-1}}{2\Delta y}\right]_t$$
$$- \left(K_{h,xx}\frac{N_{i-1,j} - 2N_{i,j} + N_{i+1,j}}{\Delta x^2} + K_{h,yy}\frac{N_{i,j-1} - 2N_{i,j} + N_{i,j+1}}{\Delta y^2}\right)_t = 0 \quad (6.68)$$

where j is the grid index in the north–south direction. This equation is linear for all i and j and can be solved implicitly in matrix form, just as with the one-dimensional case. Although the matrix is not banded, it may be solved by decomposition and backsubstitution. Equation (6.66) can also be extended to three dimensions, put in Crank–Nicolson form, and solved with variable grid spacing and eddy diffusion coefficients.

The primary disadvantage of solving (6.68) implicitly is that the matrix order quickly becomes large. For a 100×100 horizontal grid, the required matrix is $10\,000 \times 10\,000$. The advantage of solving the equation implicitly is that the solution is unconditionally stable for all values of u, v, $K_{h,xx}$, and $K_{h,yy}$. If (6.68) is solved explicitly, it is unconditionally unstable for all values of u and v when $K = 0$ or K is large and conditionally stable for all values of u and v for other values of K.

6.4.7 The semi-Lagrangian method

A special case of the finite-difference method is the **semi-Lagrangian method** (e.g., Pepper *et al.* 1979; Robert 1982; Staniforth and Cote 1991; Makar and Karpik 1996; Yabe *et al.* 2001; Bermejo and Conde 2002; Nair *et al.* 2002). With this method, the value of a variable at a specific location and time is obtained by first tracing back where the air parcel containing the variable came from during the last time step. Suppose it is desired to find the concentration of a gas at time t at point B, located at the center of a model grid cell. If the wind speed near point B is 5 m s^{-1} and the model time step is $h = 300$ s, the concentration at point B and time t can be estimated as the concentration at time $t - h$ a distance $\Delta x =$ 5 m s^{-1} $\times$ 300 s = 1500 m to the west of point B, defined now as point A. During the time interval, the wind advects the gas from point A to B. Since point A is not necessarily located at the center of a model grid cell, the gas concentration at point A must be found by interpolating concentrations from adjacent grid-cell centers.

In sum, with the semi-Lagrangian method, the concentration of a gas at time t and location x is estimated as

$$N_{x,t} = N_{x-uh,t-h} \tag{6.69}$$

where N at location $x - uh$ and time $t - h$ is interpolated from values of N at nearby node points. Mixing ratios, potential temperatures, and u- and v-scalar velocities can be estimated in a similar manner. Equation (6.69) can be written for one, two, or three directions.

Several methods exist to interpolate a variable to point A from nearby grid-cell center values. One method is to estimate the slope of the variable between two grid cells surrounding point A and to interpolate linearly between the two cells. A more complex method is to fit a polynomial through three or more adjacent cells surrounding point A and to calculate the value at point A from the polynomial. When the polynomial is a cubic, the method is the **cubic spline method** (Price and MacPherson 1973; Purnell 1976).

An advantage of semi-Lagrangian schemes is that they can be run with a long time step without concern for stability. Advection of tracers, such as water vapor and potential temperature, is sometimes more accurate with a semi-Lagrangian scheme than with a pure finite-difference scheme. A disadvantage of semi-Lagrangian schemes is that, unless proper steps are taken, the mass of a trace species is not conserved during advection. Transported mass can be conserved during a time step if the fitted spatial concentration curve, used for interpolation, is normalized so that the integral of mass under the curve equals the total mass in the system at the beginning of the time step.

6.5 SERIES EXPANSION METHODS

With a finite-difference method, each differential in a PDE is replaced with a difference analog written in terms of a finite number of values along a temporal or spatial direction. With a **series expansion method**, a dependent variable (e.g., u, v, w, N) in a PDE is replaced with a finite series that approximates its value. If the

PDE arising from the west–east advection equation at node i is

$$\frac{\partial N_i}{\partial t} + \frac{\partial (uN)_i}{\partial x} = 0 \tag{6.70}$$

a series-expansion approximation of the number concentration at node i is

$$N_i \approx \mathbf{N}_i(x) = \sum_j N_j \mathrm{e}_j(x) \tag{6.71}$$

where $\mathbf{N}_i(x)$ is called a **trial function.** Assume for now that u is constant. The set of j nodes over which the trial function is approximated is the **trial space.** The trial function is the sum, over each node in the trial space, of the true concentration, N_j, multiplied by a **basis function** $\mathrm{e}_j(x)$. The difference between (6.70) when $\mathbf{N}_i(x)$ is used and that when the summation over $N_j\mathrm{e}_j(x)$ is used is the **residual,** $R_i(x)$. A residual is the difference between an approximate and an exact function.

A series expansion method that uses a local basis function is a **finite-element method.** A series expansion method that uses a global basis function orthogonal to the residual is a **spectral method.** A common finite-element method is the **Galerkin finite-element method.** With this method, the local basis function is also orthogonal to the residual. The basis functions of other finite-element methods may or may not be orthogonal to the residual. Below, the Galerkin finite-element method is discussed, and a type of spectral method is briefly described.

6.5.1 Finite-element method

With the Galerkin finite-element method, basis functions are treated like weight functions in that they weight a residual at each of several nodes along a spatial grid (**test space**). The sum, over the test space, of the residual multiplied by the weight is zero (e.g., Pepper *et al.* 1979; Celia and Gray 1992). Thus,

$$\int_x R_i(x)\mathrm{e}_i(x)\mathrm{d}x = 0 \tag{6.72}$$

The purpose of this constraint is to minimize $R_i(x)$ by forcing its weighted average over the domain to zero. With the Galerkin method, the weight function $\mathrm{e}_i(x)$ and test space in (6.72) are the same as the basis function $\mathrm{e}_j(x)$ and trial space, respectively, in (6.71). For other methods, $\mathrm{e}_i(x)$ and $\mathrm{e}_j(x)$ may differ, and the test space may differ from the trial space. Methods in which the test and trial spaces differ are Petrov–Galerkin methods.

The residual of (6.70) is

$$R_i(x) = \left[\frac{\partial \mathbf{N}_i(x)}{\partial t} + u\frac{\partial \mathbf{N}_i(x)}{\partial x}\right] - \left[\frac{\partial N_i}{\partial t} + u\frac{\partial N_i}{\partial x}\right]$$

$$= \left[\frac{\partial \mathbf{N}_i(x)}{\partial t} + u\frac{\partial \mathbf{N}_i(x)}{\partial x}\right] - 0 \tag{6.73}$$

which is the difference between the approximate and exact form of the equation. Substituting (6.73) and (6.71) into (6.72) and assuming constant u gives

$$\int_x \left[\frac{\partial \mathbf{N}_i(x)}{\partial t} + u \frac{\partial \mathbf{N}_i(x)}{\partial x} \right] \mathrm{e}_i(x)\mathrm{d}x$$

$$= \int_x \left[\frac{\partial}{\partial t} \left(\sum_j N_j \mathrm{e}_j(x) \right) + u \frac{\partial}{\partial x} \left(\sum_j N_j \mathrm{e}_j(x) \right) \right] \mathrm{e}_i(x)\mathrm{d}x$$

$$= \sum_j \left(\frac{\partial N_j}{\partial t} \int_x \mathrm{e}_j(x)\mathrm{e}_i(x)\mathrm{d}x \right) + u \sum_j \left(N_j \int_x \frac{\mathrm{d}\mathrm{e}_j(x)}{\mathrm{d}x} \mathrm{e}_i(x)\mathrm{d}x \right) = 0 \quad (6.74)$$

Expanding the basis function in (6.74) over three nodes centered at i ($j = i - 1, \ldots, i + 1$), and taking a first-order forward-difference approximation in time of the result gives

$$\frac{N_{i-1,t} - N_{i-1,t-h}}{h} \int_{x_{i-1}}^{x_i} \mathrm{e}_{i-1}(x)\mathrm{e}_i(x)\mathrm{d}x + \frac{N_{i,t} - N_{i,t-h}}{h} \int_{x_{i-1}}^{x_{i+1}} \mathrm{e}_i(x)\mathrm{e}_i(x)\mathrm{d}x$$

$$+ \frac{N_{i+1,t} - N_{i+1,t-h}}{h} \int_{x_i}^{x_{i+1}} \mathrm{e}_{i+1}(x)\mathrm{e}_i(x)\mathrm{d}x + u \left(N_{i-1,t} \int_{x_{i-1}}^{x_i} \frac{\mathrm{d}\mathrm{e}_{i-1}(x)}{\mathrm{d}x} \mathrm{e}_i(x)\mathrm{d}x \right.$$

$$\left. + N_{i,t} \int_{x_{i-1}}^{x_{i+1}} \frac{\mathrm{d}\mathrm{e}_i(x)}{\mathrm{d}x\mathrm{e}_i(x)\mathrm{d}x + N_{i+1,t} \int_{x_i}^{x_{i+1}} \frac{\mathrm{d}\mathrm{e}_{i+1}(x)}{\mathrm{d}x} \mathrm{e}_i(x)\mathrm{d}x \right) = 0 \quad (6.75)$$

One set of basis functions is the **chapeau (hat) function,**

$$\mathrm{e}_i(x) = \begin{cases} \dfrac{x - x_{i-1}}{x_i - x_{i-1}} & x_{i-1} \le x \le x_i \\ \dfrac{x_{i+1} - x}{x_{i+1} - x_i} & x_i < x \le x_{i+1} \\ 0 & \text{all other cases} \end{cases} \quad (6.76)$$

These weightings favor the center cell (i) and decrease to zero at $i - 1$ and $i + 1$, giving them the appearance of a peaked hat. The functions are the same for $\mathrm{e}_j(x)$. When chapeau functions are used, the integral in the first term of (6.75) simplifies to

$$\int_{x_{i-1}}^{x_i} \mathrm{e}_{i-1}(x)\mathrm{e}_i(x)\mathrm{d}x = \int_{x_{i-1}}^{x_i} \left(\frac{x_i - x}{x_i - x_{i-1}} \right) \left(\frac{x - x_{i-1}}{x_i - x_{i-1}} \right) \mathrm{d}x = \frac{x - x_{i-1}}{6} \quad (6.77)$$

After other terms have been integrated, (6.75) becomes

$$\frac{(N_{i-1,t} - N_{i-1,t-h})\Delta x_i + (N_{i,t} - N_{i,t-h})2(\Delta x_{i+1} + \Delta x_i) + (N_{i+1,t} - N_{i+1,t-h})\Delta x_{i+1}}{6h}$$

$$+ u \frac{N_{i+1,t} - N_{i-1,t}}{2} = 0 \quad (6.78)$$

where $\Delta x_i = x_i - x_{i-1}$ and $\Delta x_{i+1} = x_{i+1} - x_i$.

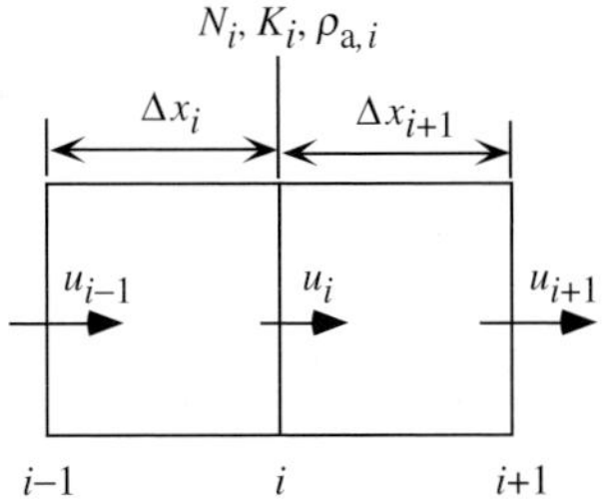

Figure 6.6 Locations of variables along a west–east grid for a Galerkin finite-element scheme.

Equation (6.75) assumes that u is constant. When u varies, its trial function is $U_i(x) = \sum_j u_j e_j(x)$. A similar equation applies to the eddy diffusion coefficient. Toon *et al.* (1988) and Pepper *et al.* (1979) show a Galerkin method with chapeau functions as finite-elements and a Crank–Nicolson time-stepping scheme. The method uses a fourth-order approximation in space and a second-order approximation in time and solves the unidirectional advection–diffusion equation given in (6.1). The method allows variable scalar velocities, grid spacing, and eddy diffusion coefficients. The solution is found by setting up $i = 2, \ldots, I - 1$ equations of the form,

$$\frac{(N_{i-1,t} - N_{i-1,t-h})\Delta x_i + (N_{i,t} - N_{i,t-h})2(\Delta x_{i+1} + \Delta x_i) + (N_{i+1,t} - N_{i+1,t-h})\Delta x_{i+1}}{h} + \mu_c(\gamma_i N_{i+1} + \beta_i N_i - \alpha_i N_{i-1})_t + (1 - \mu_c)(\gamma_i N_{i+1} + \beta_i N_i - \alpha_i N_{i-1})_{t-h} = 0 \tag{6.79}$$

where $\mu_c = 1$, 0.5, or 0 implies an implicit, Crank–Nicolson, or forward Euler solution, respectively, and

$$\alpha_i = (u_i + 2u_{i-1}) + \frac{3(\rho_{a,i} K_i + \rho_{a,i-1} K_{i-1})}{\rho_{a,i-1}\Delta x_i}$$
$$\beta_i = (-u_{i-1} + u_{i+1}) + \frac{3[(\rho_{a,i-1} K_{i-1} + \rho_{a,i} K_i)\Delta x_{i+1} + (\rho_{a,i} K_i + \rho_{a,i+1} K_{i+1})\Delta x_i]}{\rho_{a,i}\Delta x_i \Delta x_{i+1}}$$
$$\gamma_i = (u_i + 2u_{i+1}) - \frac{3(\rho_{a,i} K_i + \rho_{a,i+1} K_{i+1})}{\rho_{a,i+1}\Delta x_{i+1}} \tag{6.80}$$

Velocities, eddy diffusion coefficients, and densities in (6.80) are located at grid-cell boundaries, as shown in Fig. 6.6. Equation (6.79) can be written in tridiagonal form as

$$A_i N_{i-1,t} + B_i N_{i,t} + D_i N_{i+1,t} = E_i N_{i-1,t-h} + F_i N_{i,t-h} + G_i N_{i+1,t-h} \tag{6.81}$$

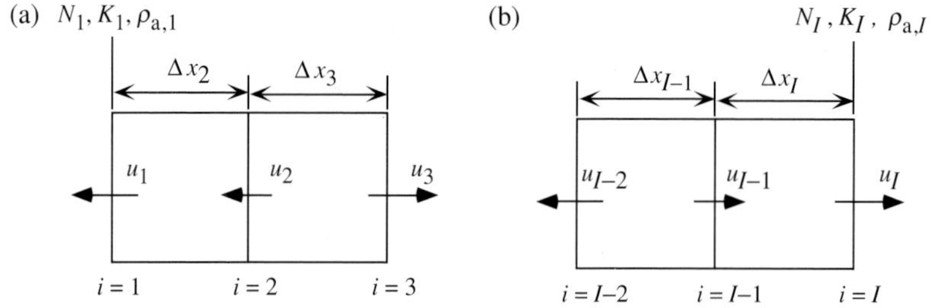

Figure 6.7 Location of grid variables at (a) low and (b) high boundaries.

where

$$\begin{aligned} A_i &= \Delta x_i - h\mu_c\alpha_i & E_i &= \Delta x_i + (1-\mu_c)\alpha_i \\ B_i &= 2(\Delta x_i + \Delta x_{i+1}) + h\mu_c\beta_i & F_i &= 2(\Delta x_i + \Delta x_{i+1}) - h(1-\mu_c)\beta_i \\ D_i &= \Delta x_{i+1} + h\mu_c\gamma_i & G_i &= \Delta x_{i+1} - h(1-\mu_c)\gamma_i \end{aligned} \tag{6.82}$$

Like (6.45), Equation (6.81) may be solved with a tridiagonal matrix technique.

Figure 6.7 shows locations of variables for boundaries on a limited-area grid. The coefficients for **outflow from the high boundary** are

$$\begin{aligned} A_I &= \Delta x_I - h\mu_c\alpha_I & E_I &= \Delta x_I + h(1-\mu_c)\alpha_I \\ B_I &= 2\Delta x_I + h\mu_c\beta_I & F_I &= 2\Delta x_I - h(1-\mu_c)\beta_I \\ D_I &= 0 & G_I &= 0 \end{aligned} \tag{6.83}$$

where

$$\begin{aligned} \alpha_I &= u_I + 2u_{I-1} + \frac{3(\rho_{a,I}K_I + \rho_{a,I-1}K_{I-1})}{\rho_{a,I-1}\Delta x_I} \\ \beta_I &= 4u_I - u_{I-1} + \frac{9\rho_{a,I}K_I + 3\rho_{a,I-1}K_{I-1}}{\rho_{a,I}\Delta x_I} \end{aligned} \tag{6.84}$$

The coefficients for **outflow from the low boundary** are

$$\begin{aligned} A_1 &= 0 & E_1 &= 0 \\ B_1 &= 2\Delta x_1 + h\mu_c\beta_1 & F_1 &= 2\Delta x_2 - h(1-\mu_c)\beta_1 \\ D_1 &= \Delta x_2 + h\mu_c\gamma_1 & G_1 &= \Delta x_2 - h(1-\mu_c)\gamma_1 \end{aligned} \tag{6.85}$$

where

$$\beta_1 = -4u_1 + u_2 + \frac{9\rho_{a,1}K_1 + 3\rho_{a,2}K_2}{\rho_{a,1}\Delta x_2} \qquad \gamma_1 = u_1 + 2u_2 - \frac{3(\rho_{a,1}K_1 + \rho_{a,2}K_2)}{\rho_{a,2}\Delta x_2} \tag{6.86}$$

Inflow boundary equations are obtained by extending (6.81) one node beyond the boundary and estimating concentrations and wind speeds in this virtual node. Virtual-node values may be set to values just inside the boundary or extrapolated from two or three nodes inside the boundary. Another option is to set virtual-node

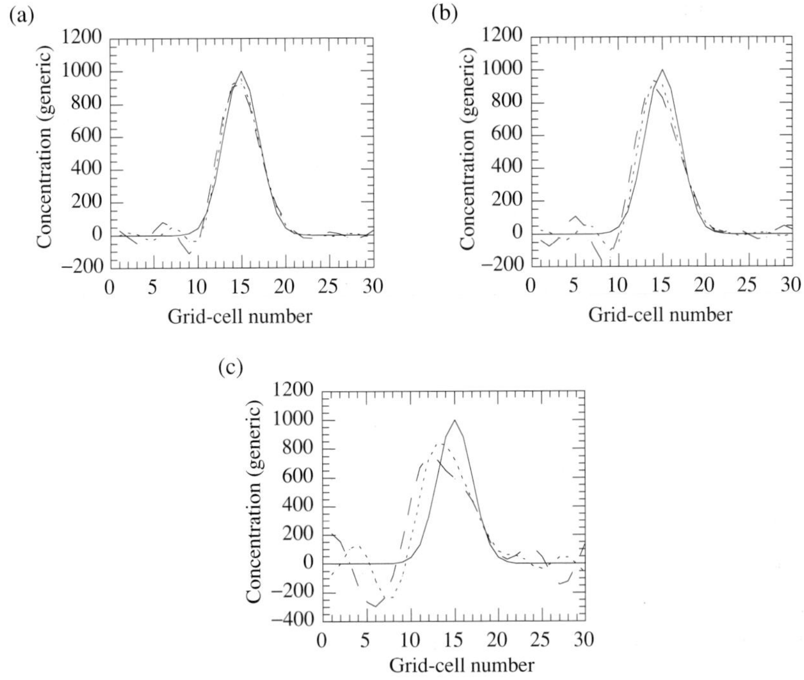

Figure 6.8 Preservation of a Gaussian peak during finite-element transport over a grid with periodic boundary conditions when (a) $uh/\Delta x = 0.02$, (b) $uh/\Delta x = 0.25$, (c) $uh/\Delta x = 0.6$. Solid lines are initial values, short-dashed lines are values after four revolutions, and long-dashed lines are values after eight revolutions.

values equal to an average of values just inside the boundary from the current and the previous time step.

Equation (6.81) can also be solved with periodic boundary conditions. Figures 6.8 (a)–(c) show results from the finite-element method described above when a Gaussian plume was advected over a grid with such boundary conditions. Grid spacing and wind speeds were uniform for the test. Each figure shows a result when a different value of $uh/\Delta x$ was used. The figures indicate that $uh/\Delta x$ should be 0.25 or less to minimize **numerical diffusion**, which is the artificial spreading of the peak, and **numerical dispersion**, which is oscillations upwind and downwind of the peak. If grid spacing is 5 km and the average wind speed is 10 m s^{-1}, the time step in this case should be no larger than 0.25×5000 m/10 m s^{-1} = 125 s.

6.5.2 Pseudospectral method

Like the finite-element method, the **pseudospectral method** involves the replacement of the spatial differential operator by a finite series of basis functions. In the case of the finite-element method, the basis functions are local functions. In the case of the pseudospectral method, the basis functions are a finite series of orthogonal functions. The difference between a spectral and a pseudospectral method is that, with the former, time and space derivatives are approximated with a finite series.

With the latter, spatial derivatives are approximated with a finite series, but time derivatives are approximated with an explicit Taylor series expansion or another method.

If wind speed in the x-direction is constant, the west–east advection equation for gas number concentration is

$$\frac{\partial N}{\partial t} + u\frac{\partial N}{\partial x} = 0 \tag{6.87}$$

The pseudospectral solution to this equation can be found by representing $N(x,t)$ over the interval $0 \leq x \leq L$ by the **Fourier series**

$$N(x, t) = \sum_{k=0}^{\infty} a_k(t)\,\mathrm{e}^{\mathrm{i}k2\pi x/L} \tag{6.88}$$

(Orszag 1971; Wengle and Seinfeld 1978; Hack 1992), where k is the wavenumber and $a_k(t)$ are **complex Fourier coefficients**. At $t = 0$, N is a known function of x. Values of $a_k(0)$ are found by integrating both sides of (6.88) from $0 \leq x \leq L$. The result is

$$a_k(0) = \frac{1}{L}\int_0^L N(x, 0)\mathrm{e}^{-\mathrm{i}k2\pi x/L}\mathrm{d}x \tag{6.89}$$

For practical application, the infinite series in (6.88) is truncated to a finite number of wavenumbers, K, giving

$$N(x, t) = \sum_{k=0}^{K} a_k(t)\mathrm{e}^{\mathrm{i}k2\pi x/L} \tag{6.90}$$

The larger the value of K, the more accurate the estimate of N.

A pseudospectral solution to (6.87) can be found by taking a second-order, central-difference approximation of (6.90) with respect to time and the partial derivative of (6.90) with respect to space. The resulting expressions are

$$\frac{\partial N}{\partial t} \approx \frac{1}{2h}\left(\sum_{k=0}^{K} a_{k,t}\mathrm{e}^{\mathrm{i}k2\pi x/L} - \sum_{k=0}^{K} a_{k,t-2h}\mathrm{e}^{\mathrm{i}k2\pi x/L}\right) \tag{6.91}$$

$$\frac{\partial N}{\partial x} = \sum_{k=0}^{K} \frac{\mathrm{i}k2\pi a_{k,t-h}}{L}\mathrm{e}^{\mathrm{i}k2\pi x/L} \tag{6.92}$$

respectively. Substituting these into (6.87) yields

$$\frac{1}{2h}\sum_{k=0}^{K}(a_{k,t} - a_{k,t-2h})\mathrm{e}^{\mathrm{i}k2\pi x/L} = -u\sum_{k=0}^{K}\frac{\mathrm{i}k2\pi a_{k,t-h}}{L}\mathrm{e}^{\mathrm{i}k2\pi x/L} \tag{6.93}$$

which can be separated into K equations of the form

$$\frac{a_{k,t} - a_{k,t-2h}}{2h} = -\frac{u\mathrm{i}k2\pi a_{k,t-h}}{L} \tag{6.94}$$

Equation (6.94) is explicit and can be solved immediately, since values of a_k at time $t - h$ and $t - 2h$ are known from previous time steps. Fourier coefficients for the first time step ($t = h$) are found by taking a forward- instead of central-difference approximation in (6.91). Once Fourier coefficients have been determined from (6.94), they are substituted back into (6.90) to give an estimate of N at time t for any value of x.

An advantage of a pseudospectral scheme over a finite-difference approximation is that only K equations need to be solved in the pseudospectral scheme. In a finite-difference scheme, I finite-difference equations need to be solved per time step, where I is the number of grid cells in one direction. Usually, $I > K$. Whereas the pseudospectral solution to the linear advection problem is easy to implement, it is not readily applied to nonlinear problems, such as when u varies in space or when u is a prognostic variable. In such cases, a separate basis function for u is required. The multiplication of two finite series, such as one for u and one for N, results in additional terms, slowing the pseudospectral numerical solution. One way to avoid the multiplication of spectral-basis-function products is with the **spectral transform method** (Eliasen *et al.* 1970; Orszag 1970).

For global modeling, the basis functions used are the **spherical harmonics**. These functions are a combination of sine and cosine functions along the zonal (west–east) direction and Legendre functions along the meridianal (south–north) direction on a sphere. They are computationally fast in comparison with some other basis functions. Spectral and pseudospectral methods are discussed in more detail in Orszag (1970), Washington and Parkinson (1986), Holton (1992), Hack (1992), and Krishnamurti *et al.* (1998). Pseudospectral techniques are commonly used to discretize horizontal advection terms in global models.

6.6 FINITE-VOLUME METHODS

Finite-volume methods are methods of solving partial differential equations that divide space into discrete volumes. Partial differential equations are integrated over each volume. Variables are then approximated in each volume by averaging across the volume. Variables are exchanged between neighboring volumes at volume interfaces in a flux-conserving manner. Finite-volume methods are generally applied over irregular grids (Section 4.1). A detailed discussion of the finite-volume method is given in Durran (1999).

6.7 ADVECTION SCHEMES USED IN AIR-QUALITY MODELS

Advection schemes are useful in three-dimensional air quality models only if they are not allowed to generate unphysical extreme concentrations (they are **bounded**), do not generate artificial oscillations (**nonoscillatory**), and preserve gradients of mixing ratio (**monotonic**). When a scheme is bounded, the mixing ratio in the current grid cell can never increase above or decrease below the mixing ratio in any adjacent grid cell after an advection time step. When such a condition is satisfied over a grid domain as a whole, no mixing ratio can fall below the lowest or rise above the highest initial mixing ratio anywhere in the domain. Schemes that are bounded are generally nonoscillatory and monotonic as well.

A problem with nonbounded schemes is that they can produce negative mixing ratios (or concentrations), which are not physical, or mixing ratios above the maximum possible, which may lead to inaccurate predictions of pollution levels. Figure 6.8, for example, shows a scheme in which concentrations drop below zero. This scheme, in its original form, is unbounded and oscillatory. The scheme can be made bounded by limiting the minimum and maximum mixing ratios upwind and downwind of each grid cell, then adjusting fluxes in the horizontal to conserve mass. Although such adjustments may increase numerical diffusion for some schemes, they remove oscillations, spurious peaks, and negative mixing ratios.

In previous sections, several schemes used for solving unidirectional advection–diffusion equations were given. Below, additional schemes are described. Some of the schemes, as originally implemented, are bounded, nonoscillatory, and monotonic, whereas others are not.

The forward Euler Taylor–Galerkin method (Donea 1984) is a Galerkin method that uses the forward Euler rather than the Crank–Nicolson time discretization. A Galerkin method that uses a chapeau basis function but a modified weighting function is the Petrov–Galerkin method (Hughes and Brooks 1979). The accurate space derivative (ASD) scheme of Gazdag (1973) is a pseudospectral scheme in which time and spatial derivatives are approximated with truncated Taylor series expansions and truncated Fourier series expansions, respectively. The scheme of Smolarkiewicz (1983) is an iterative, positive-definite, backward-difference approximation scheme that reduces numerical diffusion by correcting the velocity with an "antidiffusion velocity," derived from the truncated terms in a Taylor-series expansion of the advection equation.

Collela and Woodward (1984), Prather (1986), Tremback *et al.* (1987), Bott (1989), Carpenter *et al.* (1990), and Easter (1993) derived solutions to the advection–diffusion equation in which mixing ratios are expanded in space with parabolic functions or higher-order polynomials. Each polynomial is derived from mixing ratios in several adjacent grid cells. In the case of Prather (1986), first and second derivatives of the parabolic function are stored for use during subsequent advection time steps. These schemes generally modify fluxes at grid-cell interfaces to ensure mass conservation. Related schemes include those by Yamartino (1993), Rasch (1994), and Thuburn (1996, 1997).

Walcek and Aleksic (1998) and Walcek (2000) developed a bounded, nonoscillatory, and monotonic scheme in which spatial gradients in mixing ratio are fitted with linear functions, rather than polynomials, optimized to preserve local peaks and reduce numerical diffusion. Fluxes across grid-cell interfaces are limited to ensure mass conservation. Figure 6.9 shows an example of the conservation and low numerical diffusion resulting from this scheme when a tracer with mixing ratio represented by several shapes is advected on a two-dimensional grid. The smooth background field is a product of the bounded and nonoscillatory nature of the scheme. The relative preservation of the shapes is an indicator of the highly monotonic nature of the scheme.

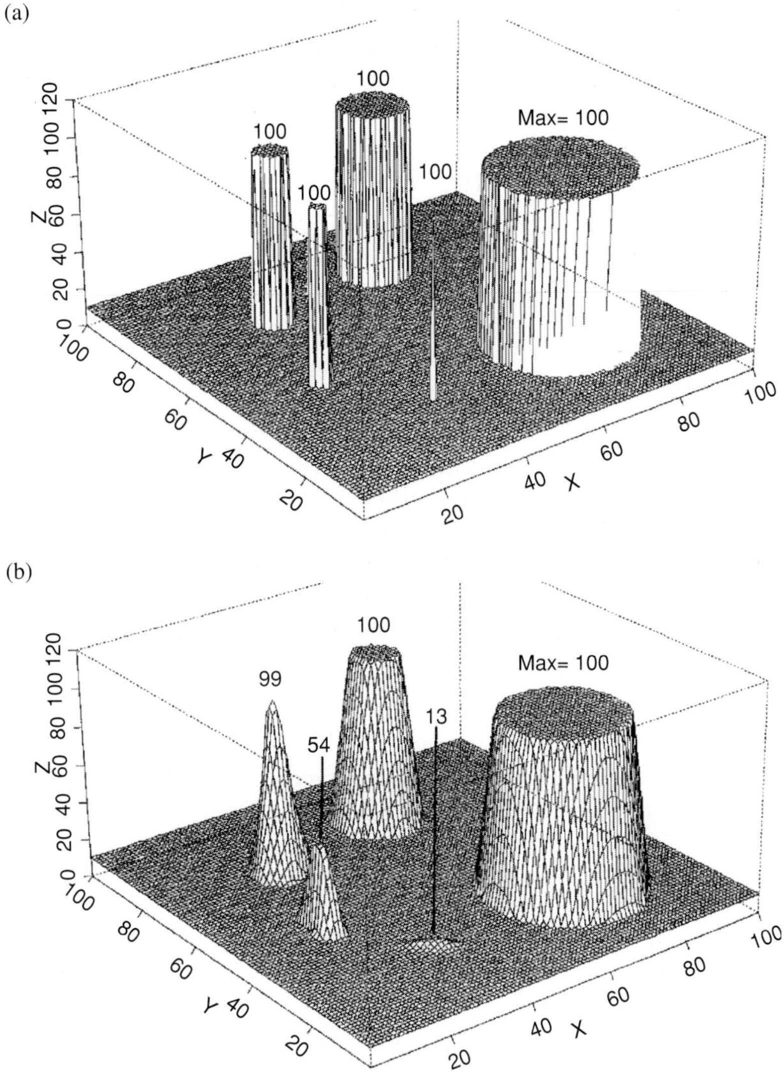

Figure 6.9 Comparison of (a) initial and expected exact final shapes and (b) actual final shapes after six rotations with 628 time steps per rotation of advection around a two-dimensional 100×100 horizontal domain center using periodic boundary conditions. The maximum Courant number was near 0.5. The numbers represent peak mixing ratios for each shape. The scheme and diagram originate from Walcek (2000).

In the trajectory grid scheme (Chock *et al.* 1996), advection is solved with a fully Lagrangian method and eddy diffusion is solved with an Eulerian diffusion scheme. In the scheme of Nguyen and Dabdub (2001), spatial derivatives of a tracer are approximated with quintic splines and time derivatives are approximated with a Taylor series expansion.

Many of the schemes described above have been compared with each other when used alone (e.g., Chock 1991; Walcek 2000) or coupled with chemistry (e.g., Chock and Winkler 1994; Dabdub and Seinfeld 1994).

6.8 SUMMARY

In this chapter, methods of solving partial differential equations, and specifically the advection–diffusion equation, were discussed. The methods include finite-difference, series expansion, and finite-volume methods. A finite-difference approximation involves the replacement of a continuous differential operator with a discrete difference analog along a predetermined number of spatial or temporal nodes. The order of approximation increases with the number of nodes along which the differential is discretized. Forward-, backward-, and central-difference approximations of the first and second derivatives of a variable were derived for a variety of orders under the assumption of constant grid spacing. Finite-difference approximations of the advection–diffusion equation were also discussed for cases of nonuniform grid spacing and variable eddy diffusion coefficients. The semi-Lagrangian method, which is in the family of finite-difference methods, was also discussed, and series expansion methods were described. These methods, which include finite-element and pseudospectral methods, involve the replacement of a dependent variable with a finite series that approximates the variable. In addition, several time-stepping schemes and their stability characteristics were described. Finally, characteristics necessary for schemes solving the advection–diffusion equation in three-dimensional air pollution models were discussed.

6.9 PROBLEMS

6.1 Identify five characteristics of a good numerical approximation, and explain which of these characteristics you think the forward Euler scheme (Section 6.4.4.2) with no diffusion has. (Hint: the characteristics of the scheme may be obtained by estimation or by writing a one-dimensional code with periodic boundary conditions and testing the effect of changes in time step and grid spacing).

6.2 If an implicit approximation such as (6.36) is unconditionally stable, why is it less accurate, for the same grid spacing and time step, than (6.44) when $\mu_c = 0.5$?

6.3 What advantage does the finite-element scheme of fourth-order in space and second-order in time shown in this chapter have over a finite-difference scheme of the same orders?

6.4 Find a sixth-order central-difference approximation of $\partial N/\partial x$. Assume constant grid spacing.

6.5 Find a fifth-order backward-difference approximation of $\partial N/\partial x$. Assume constant grid spacing.

6.6 Suppose the horizontal wind in a large domain shifted from a 5 m s^{-1} southwesterly wind to a 10 m s^{-1} westerly wind to an 8 m s^{-1} northwesterly wind back to a 10 m s^{-1} westerly wind wind 3, 2, and 1 hour ago, respectively. Draw the trajectory of a Lagrangian pollution parcel carried by the wind, and calculate the straight-line distance the parcel traveled between 4 hours ago and the present time. Ignore curvature of the Earth.

6.10 COMPUTER PROGRAMMING PRACTICE

6.7 Set up a west–east grid with periodic boundary conditions that has 100 uniformly spaced grid cells. Assume the last grid cell to the east meets the first grid cell to the west. Set $\Delta x = 5$ km in each cell and $u = 5$ m s^{-1} at each cell west–east boundary. Initialize the gas concentration on the grid with the Gaussian distribution $N_i = N_0 e^{-(i-50)^2/8}$, where $N_0 = 10^{12}$ molec. cm^{-3} is a peak concentration and i corresponds to a cell boundary. Solve (6.36), assuming the eddy diffusion coefficient is zero, with the matrix given in (6.42). Use a time step of 5 s, and solve until 10 revolutions around the grid have been completed. Plot concentration versus grid cell at the end of each revolution. Discuss the numerical diffusion of the peak over time.

6.8 Do the same as in Problem 6.7, but instead solving with the fourth-order Runge–Kutta technique. Use a time step of 0.1 s to obtain an "exact" solution. Plot the difference between the "exact" and modeled concentration at one point along the grid at a specific time during the simulation versus time step when different time steps are used to demonstrate the fourth-order nature of the solution.

7

Finite-differencing the equations of atmospheric dynamics

Many numerical models have been developed to predict the weather on regional and global scales. The first was that of Richardson (1922), which was solved by hand (Section 1.1). The first computer model was the one-dimensional model of Charney (1949), developed as part of the project to apply the world's first computer (Section 1.1). Three-dimensional computer models today usually discretize time and vertical spatial derivatives with finite-difference approximations. Horizontal advection terms are discretized with finite-difference, spectral, finite-element, or semi-Lagrangian approximations. In this chapter, one numerical solution to the equations of atmospheric dynamics, a three-dimensional finite-difference solution in spherical-sigma-pressure coordinates, is discussed. The equations solved include the continuity equation for air, species continuity equation, thermodynamic energy equation, horizontal momentum equations, and hydrostatic equation. The solution scheme originates from Arakawa and Lamb (1977) and Arakawa and Suarez (1983). Although more recent and advanced versions of the scheme have been developed, and although many other numerical solutions exist, the version presented was chosen for illustration purposes. A modeling project at the end of the chapter allows students to develop a basic regional- or global-scale dynamics model from the equations given.

7.1 VERTICAL MODEL GRID

A complete weather-prediction model requires all the components shown in Fig. 1.1. In this chapter, only the meteorological component is addressed. A numerical technique is described to predict wind speed and direction, air pressure, energy, and moisture. Techniques for solving equations for other processes are given in other chapters of this text. Chapter 21 describes how different components of a model are integrated and tested.

The first step in the development of a dynamics module is to define the vertical coordinate system and location of variables within this system. For the present application, the sigma-pressure vertical coordinate will be used. Figure 7.1 shows the location of model variables in the vertical with this coordinate system. Vertical scalar velocities are located at the top and bottom of each layer. Horizontal scalar velocities and other variables (e.g., gas and particle mixing ratios and potential virtual temperature) are located at the vertical midpoint of the layer. This type

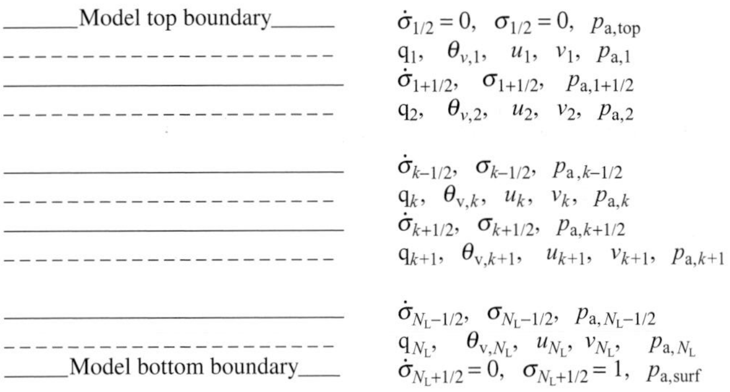

Figure 7.1 Location of variables in the vertical direction of a sigma-pressure coordinate Lorenz grid. Integer subscripts correspond to layer centers, and fraction subscripts correspond to layer boundaries.

of grid is called the **Lorenz grid**, after Lorenz (1960). A similar grid that stores potential virtual temperature at layer top and bottom boundaries instead of centers is called the **Charney–Phillips grid**, after Charney and Phillips (1953). Both grids are described in more detail in Arakawa and Konor (1995). In Fig. 7.1, layer center indices increase from 1, at the top of the model, to N_L, at the bottom, and layer boundary indices increase from 1/2, at the top, to $N_L + 1/2$ at the bottom.

Sigma-pressure coordinate vertical scalar velocities at the bottom of the model are set to zero, since the model bottom is assumed to be a material surface, namely, the ground surface. Setting the model bottom vertical scalar velocities to zero also helps to filter out **Lamb waves**, which are horizontally propagating acoustic waves that give rise to pressure, velocity, and density perturbations (Chapter 4). Since the amplitude of Lamb-wave oscillations is usually greatest at the lowest model level, setting the vertical scalar velocity to zero at that level decreases perturbations due to the oscillations.

At the model top, vertical scalar velocities are also set to zero for lack of a better assumption. Zero velocities at the top are not realistic, unless the model top is the top of the atmosphere. An alternative to setting top vertical scalar velocities to zero is to estimate them, which is also an error-prone process. In either case, errors can be reduced by raising the model top to a higher altitude. This may require adding more model layers, increasing computational requirements.

In the sigma-pressure coordinate, the σ**-thickness** of a layer is the difference in σ-values between the top and bottom of the layer. Thus,

$$\Delta\sigma_k = \sigma_{k+1/2} - \sigma_{k-1/2} \qquad \text{for } k = 1, \ldots, N_L \tag{7.1}$$

where the σ-value at the top and bottom of each layer is predetermined and constant throughout the model domain and during the simulation. The σ-value at the top of the model usually equals zero ($\sigma_{1/2} = \sigma_{top} = 0$), and that at the ground surface usually equals unity ($\sigma_{N_L+1/2} = \sigma_{surf} = 1$) although these limits are arbitrary. The setting of σ-values at the boundaries of other layers is described below.

When a model extends from the ground to the top of the stratosphere, the choice of σ-values is important for maintaining numerical stability. In the stratosphere, small pressure changes correspond to large altitude changes. Near the surface, small pressure changes correspond to small altitude changes. Thus, the σ- (and pressure-) thickness of a layer near the model top should not be too large and that near the ground should not be too small.

One way to select σ-values is first to set up a test column and assume that the altitude thickness of all layers in the test column is the same. To set up the test column, it is necessary to define the **model top pressure**, $p_{a,top}$ (hPa), which is uniformly constant throughout the model domain and over time, and a mean surface altitude $z_{surf,test}$ (m), which is just an average topographical altitude above sea level over the model domain. A typical top pressure for a regional model is 250 hPa. That for a global model is 0.5 hPa. As a default, the mean surface altitude can be set to zero.

From the model top pressure, the model top altitude (m) in the test column can be estimated with

$$z_{top,test} = z_{below} + \frac{p_{a,below} - p_{a,top}}{\rho_{a,below} g_{below}} \tag{7.2}$$

derived from (2.41), where $p_{a,below}$ (Pa), $\rho_{a,below}$ (kg m^{-3}) and g_{below} (m s^{-2}) are air pressure, air density, and gravity corresponding to the air pressure in Appendix Table B.1 just greater than (at a lower altitude than) $p_{a,top}$ (Pa). The altitude at the bottom of each test column layer can then be calculated from

$$z_{k+1/2,test} = z_{surf,test} + (z_{top,test} - z_{surf,test})\left(1 - \frac{k}{N_L}\right) \quad \text{for } k = 0, \ldots, N_L \tag{7.3}$$

where k is the layer index number. The pressure $p_{a,k+1/2,test}$ at each altitude $z_{k+1/2,test}$ is then determined from (2.41) using interpolated density and gravity values from Appendix Table B.1. Finally, the σ-value at the bottom boundary of each model layer is calculated with

$$\sigma_{k+1/2} = \frac{p_{a,k+1/2,test} - p_{a,top}}{p_{a,N_L+1/2,test} - p_{a,top}} \quad \text{for } k = 1, \ldots, N_L \tag{7.4}$$

At the model top (k = 0), $p_{a,1/2,test} = p_{a,top}$. The σ-value at the boundary of each real model layer is then set equal to that at each test layer.

Once σ-values have been determined, they are used to calculate the pressure at the bottom boundary of each model layer. This can be done only after surface pressure $p_{a,surf}$ (Pa) has been defined for each column in the model. For realistic simulations, surface pressures should be initialized with interpolated measurements. For idealized simulations in the absence of topography, surface pressure can be initialized everywhere with standard sea-level surface pressure (1013.25 hPa). In idealized simulations with topography, topographical surface air pressure can be estimated by applying (2.41) between sea level ($z = 0$) and the topographical altitude using air density and gravity estimates from Appendix Table B.1. From the model surface and top pressures, the pressure thickness of each model column is defined as

$$\pi_a = p_{a,surf} - p_{a,top} \tag{7.5}$$

The pressure at the bottom boundary of each model layer is then

$$p_{a,k+1/2} = p_{a,top} + \sigma_{k+1/2}\pi_a \tag{7.6}$$

Although $p_{a,top}$ is the same in each model column, $p_{a,surf}$ differs for each column; thus, π_a and $p_{a,k+1/2}$ also differ for each column and grid cell, respectively.

The vertical midpoint of a layer can be defined as the center of height, the center of mass, or the pressure at which the mass-weighted mean of a variable is located. If it is defined as the center of mass, the air pressure at the vertical center is

$$p_{a,k} = p_{a,k-1/2} + 0.5(p_{a,k+1/2} - p_{a,k-1/2}) \tag{7.7}$$

Alternatively, the vertical midpoint can be defined such that the finite-difference expression for the hydrostatic equation, given in Section 7.6, becomes exact for an isentropic atmosphere (e.g., where potential virtual temperature increases monotonically with height, as shown in Fig. 2.13). In such a case, the midpoint is the pressure at which the mass-weighted mean of P, the potential temperature factor defined in (2.98), is located (Arakawa and Suarez 1983). The mass-weighed mean of P in layer k is

$$\begin{aligned} P_k &= \frac{1}{p_{a,k+1/2} - p_{a,k-1/2}} \int_{p_{a,k-1/2}}^{p_{a,k+1/2}} P \, \mathrm{d}p_a \\ &= \frac{1}{1+\kappa}\left(\frac{P_{k+1/2}p_{a,k+1/2} - P_{k-1/2}p_{a,k-1/2}}{p_{a,k+1/2} - p_{a,k-1/2}}\right) \end{aligned} \tag{7.8}$$

where values of p_a and P at layer boundaries are $p_{a,k+1/2}$ and

$$P_{k+1/2} = \left(\frac{p_{a,k+1/2}}{1000\,\text{hPa}}\right)^{\kappa} \tag{7.9}$$

respectively. Once P_k has been found from (7.8), the pressure corresponding to P_k is

$$p_{a,k} = (1000\,\text{hPa})\,P_k^{1/\kappa} \tag{7.10}$$

In the grid described in Fig. 7.1, potential virtual temperature is evaluated at the vertical midpoint of a layer. However, values are needed at layer tops and bottoms as well to calculate vertical energy fluxes. If (7.10) is used to locate the vertical pressure-midpoint of a layer, a consistent formulation for θ_v at a layer boundary is

$$\theta_{v,k+1/2} = \frac{(P_{k+1/2} - P_k)\theta_{v,k} + (P_{k+1} - P_{k+1/2})\theta_{v,k+1}}{P_{k+1} - P_k} \tag{7.11}$$

Example 7.1

Given the following pressure and potential virtual temperature profiles, calculate the pressure at the midpoints, with respect to mass and to the mass integral of P, of layers k and $k+1$ and potential virtual temperature at the layer boundary, $k+1/2$. Assume the air is dry.

$p_{a,k-1/2} = 700$ hPa
$\theta_{v,k} = 308$ K
$p_{a,k+1/2} = 750$ hPa
$\theta_{v,k+1} = 303$ K
$p_{a,k+3/2} = 800$ hPa

SOLUTION

Substituting values above into (7.7) gives $p_{a,k} = 725$ hPa and $p_{a,k+1} = 775$ hPa. From (7.9), $P_{k-1/2} = 0.9030214$, $P_{k+1/2} = 0.9210167$, and $P_{k+3/2} = 0.9381747$. Substituting these into (7.8) gives $P_k = 0.9120929$ and $P_{k+1} = 0.9296616$. From (7.10), $p_{a,k} = 724.897$ hPa and $p_{a,k+1} = 774.904$ hPa. In sum, pressure at the midpoint with respect to the mass integral of P is slightly lower than pressure at the mass center of a layer. From (7.11), $\theta_{v,k+1/2} = 305.54$ K.

7.2 THE CONTINUITY EQUATION FOR AIR

In the sigma-pressure coordinate, the continuity equation for air is used prognostically to calculate changes in total column pressure and diagnostically to calculate vertical scalar velocities at the top and bottom of each grid cell. From (5.66), the prognostic equation for the change in column pressure was

$$R_e^2 \cos\varphi \left(\frac{\partial \pi_a}{\partial t}\right)_\sigma = -\int_0^1 \left[\frac{\partial}{\partial \lambda_e}(u\pi_a R_e) + \frac{\partial}{\partial \varphi}(v\pi_a R_e \cos\varphi)\right]_\sigma d\sigma \tag{7.12}$$

Replacing d, dt, dλ_e, dφ, and dσ with Δ, h, $\Delta\lambda_e$, $\Delta\varphi$, and $\Delta\sigma$, respectively, and multiplying through by $\Delta\lambda_e\Delta\varphi$ give a first-order in time and second-order in space

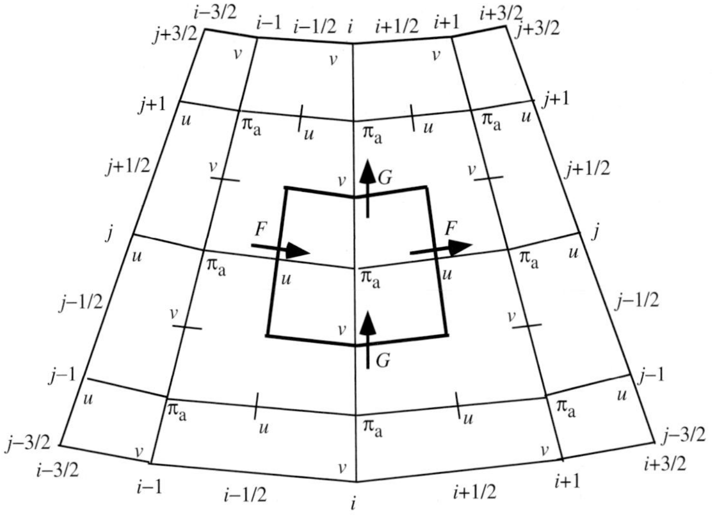

Figure 7.2 The Arakawa C-grid. The grid is centered around π_a for finite-differencing of the continuity equation. The figure shows exact locations of u, v, and π_a relative to lateral boundaries.

approximation for the column pressure as

$$\left(R_e^2 \cos\varphi\, \Delta\lambda_e \Delta\varphi\right)_{i,j} \left(\frac{\pi_{a,t} - \pi_{a,t-h}}{h}\right)_{i,j}$$

$$= -\sum_{k=1}^{N_L} \left[\frac{(u\pi_a R_e \Delta\varphi \Delta\lambda_e \Delta\sigma)_{i+1/2,j} - (u\pi_a R_e \Delta\varphi \Delta\lambda_e \Delta\sigma)_{i-1/2,j}}{\Delta\lambda_e}\right]_{k,t-h}$$

$$- \sum_{k=1}^{N_L} \left[\frac{(v\pi_a R_e \cos\varphi\, \Delta\varphi \Delta\lambda_e \Delta\sigma)_{i,j+1/2} - (v\pi_a R_e \cos\varphi\, \Delta\varphi \Delta\lambda_e \Delta\sigma)_{i,j-1/2}}{\Delta\varphi}\right]_{k,t-h} \tag{7.13}$$

The horizontal finite-difference grid chosen for discretizing this equation is the **Arakawa C-grid** (Arakawa and Lamb 1977), shown in Fig. 7.2. In this grid, column pressures (π_a) are evaluated at π_a-points, which are bounded to the west and east by u-scalar velocity points and to the south and north by v-scalar velocity points. The left sides of the second and third numerators in (7.13) are fluxes entering the solid box shown in Fig. 7.2, and the right sides are fluxes leaving the box.

The grid shown in Fig. 7.2 is a **limited-area grid**, which is a grid with lateral boundaries; u-points lie on western and eastern boundaries, and v-points lie on southern and northern boundaries. Boundary conditions for a limited-area grid are discussed in this chapter. If the limited-area grid is **nested** within a global grid, boundary conditions for the limited-area grid are obtained from the global grid, as described in Section 21.1.11. A **global grid** converges to a singularity at the poles in the south–north direction. As such, the poles must be treated carefully; otherwise mass converges at the poles and computational instabilities arise.

One method of treating the poles on a global grid is to set v-scalar velocities to zero at the southernmost and northernmost boundaries. As a result, winds advect from west to east or east to west around the poles ($u \neq 0$), avoiding the singularity. In the west–east direction on a global grid, boundary conditions are **periodic** (Section 6.4.4.2), so values on the west boundary are the same as values on the east boundary at u-, v-, and π_a-points.

Equation (7.13) can be solved to give a **prognostic equation for the column pressure,**

$$\pi_{a,i,j,t} = \pi_{a,i,j,t-h} - \frac{h}{\left(R_e^2 \cos\varphi\, \Delta\lambda_e \Delta\varphi\right)_{i,j}} \times \sum_{k=1}^{N_L} \left[(F_{i+1/2,j} - F_{i-1/2,j} + G_{i,j+1/2} - G_{i,j-1/2})_{k,t-h} \Delta\sigma_k \right] \quad (7.14)$$

where F and G are fluxes defined at u- and v-points, respectively, as shown in Fig. 7.2. At interior points in the domain, F and G are

$$F_{i+1/2,j,k,t-h} = \left[\frac{\pi_{a,i,j} + \pi_{a,i+1,j}}{2} (u R_e \Delta\varphi)_{i+1/2,j,k} \right]_{t-h} \quad (7.15)$$

$$G_{i,j+1/2,k,t-h} = \left[\frac{\pi_{a,i,j} + \pi_{a,i,j+1}}{2} (v R_e \cos\varphi\, \Delta\lambda_e)_{i,j+1/2,k} \right]_{t-h} \quad (7.16)$$

where the division by two indicates that the π_a values at i, j points have been averaged to obtain approximate π_a values at $i + 1/2$, j and i, $j + 1/2$ points. At eastern and northern lateral boundaries, F and G are

$$F_{I+1/2,j,k,t-h} = [\pi_{a,I,j} (u R_e \Delta\varphi)_{I+1/2,j,k}]_{t-h} \quad (7.17)$$

$$G_{i,J+1/2,k,t-h} = [\pi_{a,i,J} (v R_e \cos\varphi\, \Delta\lambda_e)_{i,J+1/2,k}]_{t-h} \quad (7.18)$$

respectively, where $I + 1/2$ and $J + 1/2$ are the easternmost and northernmost u- and v-points, respectively (i.e., points $i + 3/2$ and $j + 3/2$, respectively, in Fig. 7.2). Similar equations can be written for the western and southern boundaries. Once new column pressures have been calculated from (7.14) at the end of a time step, they are used in (7.6) to obtain new pressures at the top and bottom boundaries of each grid cell. Layer vertical midpoint pressures are then obtained from (7.8)–(7.10).

After new column pressures have been determined, vertical scalar velocities at the top and bottom of each grid cell are calculated diagnostically from (5.69), given as

$$\dot{\sigma} \pi_a R_e^2 \cos\varphi = -\int_0^{\sigma} \left[\frac{\partial}{\partial \lambda_e} (u \pi_a R_e) + \frac{\partial}{\partial \varphi} (v \pi_a R_e \cos\varphi) \right]_{\sigma} d\sigma - \sigma R_e^2 \cos\varphi \left(\frac{\partial \pi_a}{\partial t} \right)_{\sigma} \quad (7.19)$$

Replacing differential and integral operators in this equation with finite-difference analogs and multiplying through by $\Delta\lambda_e \Delta\varphi$ yield the **vertical scalar velocity** at the bottom of a layer as

$$\left(\dot{\sigma}\pi_a R_e^2 \cos\varphi\, \Delta\lambda_e \Delta\varphi\right)_{i,j,k+1/2,t}$$

$$= -\sum_{l=1}^{k}\left[\frac{(u\pi_a R_e \Delta\lambda_e \Delta\varphi\Delta\sigma)_{i-1/2,j} - (u\pi_a R_e \Delta\lambda_e \Delta\varphi\Delta\sigma)_{i+1/2,j}}{\Delta\lambda_e}\right]_{l,t-h}$$

$$-\sum_{l=1}^{k}\left[\frac{(v\pi_a R_e \cos\varphi\, \Delta\lambda_e \Delta\varphi\Delta\sigma)_{i,j-1/2} - (v\pi_a R_e \cos\varphi\, \Delta\lambda_e \Delta\varphi\Delta\sigma)_{i,j+1/2}}{\Delta\varphi}\right]_{l,t-h}$$

$$-\sigma_{k+1/2}\left(R_e^2 \cos\varphi\, \Delta\lambda_e \Delta\varphi\right)_{i,j}\left(\frac{\pi_{a,t} - \pi_{a,t-h}}{h}\right)_{i,j} \tag{7.20}$$

Substituting fluxes from (7.15)–(7.18) into (7.20) gives the vertical scalar velocity as

$$\dot{\sigma}_{i,j,k+1/2,t} = -\frac{1}{\left(\pi_a R_e^2 \cos\varphi\, \Delta\lambda_e \Delta\varphi\right)_{i,j,t}} \times \sum_{l=1}^{k}\left[(F_{i+1/2,j} - F_{i-1/2,j} + G_{i,j+1/2} - G_{i,j-1/2})_{l,t-h}\Delta\sigma_l\right] - \sigma_{k+1/2}\left(\frac{\pi_{a,t} - \pi_{a,t-h}}{h\pi_{a,t}}\right)_{i,j} \tag{7.21}$$

Equation (7.21) is solved in the order, $k = 1, \ldots, N_L - 1$. At $k = 0$, $\dot{\sigma}_{1/2} = \dot{\sigma}_{top} = 0$ by definition, and at $k = N_L$, the equation predicts a vertical scalar velocity at the lowest model boundary of zero ($\dot{\sigma}_{N_L+1/2} = 0$). This characteristic is demonstrated mathematically by combining (7.14) and (7.21) with $k = N_L$ in the latter equation.

7.3 THE SPECIES CONTINUITY EQUATION

Transport of each gas and aerosol-particle component in a model is simulated with the species continuity equation. This equation can be solved as three one-dimensional equations, a two-dimensional and a one-dimensional equation, or one three-dimensional equation. In Chapter 6, techniques for solving the one- and two-dimensional equations were given. Here, an explicit solution to the three-dimensional equation is shown.

The flux form of the **species continuity equation** in spherical-sigma-pressure coordinates was given in (5.73) as

$$R_e^2 \cos\varphi\left[\frac{\partial}{\partial t}(\pi_a q)\right]_\sigma + \left[\frac{\partial}{\partial\lambda_e}(u\pi_a q R_e) + \frac{\partial}{\partial\varphi}(v\pi_a q R_e \cos\varphi)\right]_\sigma + \pi_a R_e^2 \cos\varphi \frac{\partial}{\partial\sigma}(\dot{\sigma} q)$$

$$= \pi_a R_e^2 \cos\varphi\left[\frac{(\nabla \cdot \rho_a \mathbf{K}_h \nabla)q}{\rho_a} + \sum_{n=1}^{N_{e,t}} R_n\right] \tag{7.22}$$

where q is the **moist-air mass mixing ratio** of the species (or **specific humidity,** in the case of water vapor). Replacing differential operators with finite-difference analogs and multiplying through by $\Delta\lambda_e \Delta\varphi$ give

$$
\begin{aligned}
&\left(R_e^2 \cos\varphi\, \Delta\lambda_e \Delta\varphi\right)_{i,j} \left(\frac{\pi_{a,t} q_t - \pi_{a,t-h} q_{t-h}}{h}\right)_{i,j,k} \\
&\quad + \frac{(u\pi_a q R_e \Delta\lambda_e \Delta\varphi)_{i+1/2,j,k,t-h} - (u\pi_a q R_e \Delta\lambda_e \Delta\varphi)_{i-1/2,j,k,t-h}}{\Delta\lambda_e} \\
&\quad + \frac{(v\pi_a q R_e \cos\varphi\, \Delta\lambda_e \Delta\varphi)_{i,j+1/2,k,t-h} - (v\pi_a q R_e \cos\varphi\, \Delta\lambda_e \Delta\varphi)_{i,j-1/2,k,t-h}}{\Delta\varphi} \\
&\quad + \left[\pi_{a,t} R_e^2 \cos\varphi\, \Delta\lambda_e \Delta\varphi \frac{(\dot{\sigma}_t q_{t-h})_{k+1/2} - (\dot{\sigma}_t q_{t-h})_{k-1/2}}{\Delta\sigma_k}\right]_{i,j} \\
&= \left\{\pi_a R_e^2 \cos\varphi\, \Delta\lambda_e \Delta\varphi \left[\frac{(\nabla_z \cdot \rho_a \mathbf{K}_h \nabla_z)\, q}{\rho_a} + \sum_{n=1}^{N_{e,t}} R_n\right]\right\}_{i,j,k,t-h}
\end{aligned}
\tag{7.23}
$$

where the eddy diffusion term is not differenced here for simplicity. In this equation, $\pi_a \dot{\sigma}$ at time t, calculated from (7.21), is required. Substituting (7.15)–(7.18) into (7.23) and interpolating q's to u- and v-points where necessary give a **prognostic form of the species continuity equation** as

$$
\begin{aligned}
q_{i,j,k,t} = {} & \frac{(\pi_a q)_{i,j,k,t-h}}{\pi_{a,i,j,t}} + \frac{h}{\left(\pi_{a,t} R_e^2 \cos\varphi\, \Delta\lambda_e \Delta\varphi\right)_{i,j}} \\
& \times \left\{ \begin{pmatrix} F_{i-1/2,j} \dfrac{q_{i-1,j} + q_{i,j}}{2} - F_{i+1/2,j} \dfrac{q_{i,j} + q_{i+1,j}}{2} \\ + G_{i,j-1/2} \dfrac{q_{i,j-1} + q_{i,j}}{2} - G_{i,j+1/2} \dfrac{q_{i,j} + q_{i,j+1}}{2} \end{pmatrix}_{k,t-h} \right. \\
& + \left[\pi_{a,t} R_e^2 \cos\varphi\, \Delta\lambda_e \Delta\varphi \frac{(\dot{\sigma}_t q_{t-h})_{k-1/2} - (\dot{\sigma}_t q_{t-h})_{k+1/2}}{\Delta\sigma_k}\right]_{i,j} \\
& \left. + \left[\pi_a R_e^2 \cos\varphi\, \Delta\lambda_e \Delta\varphi \left(\frac{(\nabla_z \cdot \rho_a \mathbf{K}_h \nabla_z)\, q}{\rho_a} + \sum_{n=1}^{N_{e,t}} R_n\right)\right]_{i,j,k,t-h} \right\}
\end{aligned}
\tag{7.24}
$$

where the fluxes are the same as those shown in Fig. 7.2. In this equation, q values outside lateral boundaries (e.g., at $(0, j)$, $(I + 1, j)$ $(i, 0)$, and $(i, J + 1)$ points) on a nonnested, limited-area grid can be specified, set to nearest q values inside the boundary from the current time step, or set to nearest q values inside the boundary from the previous time step. Arakawa (1984) and Lu (1994) discuss a more detailed treatment of lateral boundary conditions in which fluxes of scalars at the boundaries are derived from a mass-conservation relationship.

Mixing ratios for (7.24) at the top and bottom of a layer can be interpolated from layer midpoint values with

$$q_{i,j,k-1/2} = \frac{\ln q_{i,j,k-1} - \ln q_{i,j,k}}{(1/q_{i,j,k}) - (1/q_{i,j,k-1})} \qquad q_{i,j,k+1/2} = \frac{\ln q_{i,j,k} - \ln q_{i,j,k+1}}{(1/q_{i,j,k+1}) - (1/q_{i,j,k})} \tag{7.25}$$

respectively. Mixing ratios at the surface and model top are not needed, since $\dot{\sigma} = 0$ at those locations.

7.4 THE THERMODYNAMIC ENERGY EQUATION

The finite-difference form of the thermodynamic energy equation is similar to that of the species continuity equation. From (5.76), the flux form of the **thermodynamic energy equation in spherical-sigma-pressure coordinates** was

$$R_e^2 \cos\varphi \left[\frac{\partial}{\partial t}(\pi_a \theta_v)\right]_\sigma + \left[\frac{\partial}{\partial \lambda_e}(u\pi_a\theta_v R_e) + \frac{\partial}{\partial \varphi}(v\pi_a\theta_v R_e \cos\varphi)\right] + \pi_a R_e^2 \cos\varphi \frac{\partial}{\partial \sigma}(\dot{\sigma}\theta_v)$$
$$= \pi_a R_e^2 \cos\varphi \left[\frac{(\nabla \cdot \rho_a \mathbf{K}_h \nabla)\theta_v}{\rho_a} + \frac{\theta_v}{c_{p,d} T_v} \sum_{n=1}^{N_{e,h}} \frac{dQ_n}{dt}\right] \tag{7.26}$$

Replacing differential operators with finite-difference analogs, multiplying through by $\Delta\lambda_e\Delta\varphi$, taking the finite-difference of each term in (7.26), and substituting fluxes from (7.15)–(7.18) give potential virtual temperature as

$$\theta_{v,i,j,k,t} = \frac{(\pi_a\theta_v)_{i,j,k,t-h}}{\pi_{a,i,j,t}} + \frac{h}{\left(\pi_{a,t} R_e^2 \cos\varphi\ \Delta\lambda_e\Delta\varphi\right)_{i,j}}$$
$$\times \left\{ \left(\begin{array}{l} F_{i-1/2,j}\dfrac{\theta_{v,i-1,j} + \theta_{v,i,j}}{2} - F_{i+1/2,j}\dfrac{\theta_{v,i,j} + \theta_{v,i+1,j}}{2} \\ + G_{i,j-1/2}\dfrac{\theta_{v,i,j-1} + \theta_{v,i,j}}{2} - G_{i,j+1/2}\dfrac{\theta_{v,i,j} + \theta_{v,i,j+1}}{2} \end{array} \right)_{k,t-h} \right.$$
$$+ \left[\pi_{a,t} R_e^2 \cos\varphi\ \Delta\lambda_e\Delta\varphi \frac{(\dot{\sigma}_t\theta_{v,t-h})_{k-1/2} - (\dot{\sigma}_t\theta_{v,t-h})_{k+1/2}}{\Delta\sigma_k}\right]_{i,j}$$
$$\left. + \left(\pi_a R_e^2 \cos\varphi\ \Delta\lambda_e\Delta\varphi \left[\frac{(\nabla_z \cdot \rho_a \mathbf{K}_h \nabla_z)\theta_v}{\rho_a} + \frac{\theta_v}{c_{p,d} T_v} \sum_{n=1}^{N_{e,h}} \frac{dQ_n}{dt}\right]\right)_{i,j,k,t-h} \right\} \tag{7.27}$$

where the eddy diffusion term is not differenced for simplicity. θ_v values outside lateral boundaries (e.g., at $(0, j)$, $(I + 1, j)$ $(i, 0)$, and $(i, J + 1)$ points) can be specified, set to nearest θ_v values inside the boundary from the current time step, or set to nearest θ_v values inside the boundary from the previous time step. The parameters $\theta_{v,i,j,k-1/2}$ and $\theta_{v,i,j,k+1/2}$ are potential virtual temperatures at the top and bottom of a layer, respectively, found from (7.11). Values of θ_v at the surface and top are not needed, since $\dot{\sigma} = 0$ at those locations.

7.5 THE HORIZONTAL MOMENTUM EQUATIONS

The horizontal momentum equations are finite-differenced with a scheme that is more accurate than the differencing schemes described in the previous sections. Following Arakawa and Lamb (1977), a second-order scheme is discussed that conserves the domain-integrated kinetic energy from inertial processes and the vertical components of absolute vorticity and enstrophy during advection by the nondivergent part of horizontal velocity. **Kinetic energy** (J = kg m^2 s^{-2}) is

$$\mathrm{KE} = \frac{1}{2}\rho_a V(u^2 + v^2) \tag{7.28}$$

where V is a control volume (e.g., the volume of a grid cell). **Absolute vorticity** (Section 4.3.5.4) is the sum of **relative vorticity** (ζ_r) and the Earth's vorticity (f). The vertical component of absolute vorticity (s^{-1}) is

$$\zeta_{a,z} = \zeta_{r,z} + f = \frac{\partial v}{\partial x} - \frac{\partial u}{\partial y} + f \tag{7.29}$$

Enstrophy (s^{-2}) is one-half the square of absolute vorticity,

$$\mathrm{ENST} = \frac{1}{2}\zeta_{a,z}^2 \tag{7.30}$$

Conserving kinetic energy, absolute vorticity, and enstrophy over a model domain is important for reducing nonlinear computational instabilities. The kinetic energy constraint reduces error by preventing computational cascades of kinetic energy into small-scale motions (Arakawa and Lamb 1977). Conserving these properties requires strict adherence to the horizontal interpolations presented below and permits solutions to remain stable for long integration periods.

Figure 7.3 shows the conservation properties of the scheme of Arakawa and Lamb (1977) described in this chapter, but written for **two-dimensional nondivergent** ($\nabla_h \cdot \mathbf{v} = 0$), **irrotational** ($f = 0$) flow and in terms of the vorticity and streamfunction equations in place of the horizontal momentum equations and the continuity equation for air. The **vorticity equation** is $\partial\zeta_{r,z}/\partial t + \mathbf{v}_h \cdot \nabla\zeta_{r,z} = 0$, and the **streamfunction equation** is $\zeta_{r,z} = \nabla_h^2\psi$, where ψ is **streamfunction** (m^2 s^{-1}), which is related to u and v by $u = \partial\psi/\partial y$ and $v = \partial\psi/\partial x$, respectively. Density and temperature were assumed constant. The figure shows that the scheme conserved vorticity, enstrophy, and kinetic energy in this case to about 10 decimal places.

For three-dimensional divergent, rotational flow in spherical-sigma-pressure coordinates, the **flux form of the momentum equation in the u-direction** from

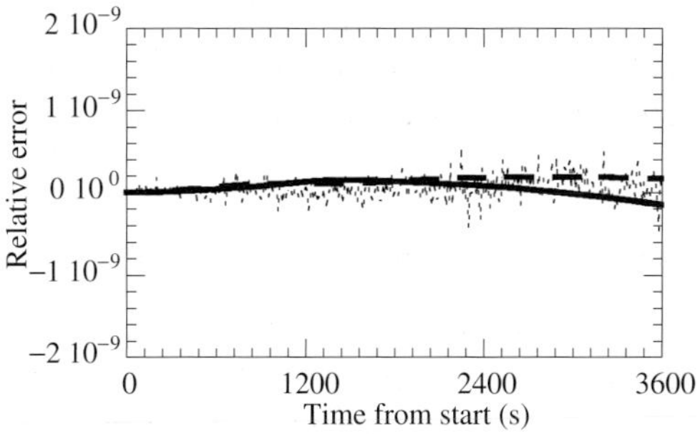

Figure 7.3 Time-dependent change in relative vorticity (ζ) ---- (since $f = 0$), enstrophy (ENST) —, and kinetic energy (KE) —, summed over the model domain, divided by the initial summed values of these respective variables, for two-dimensional nondivergent, irrotational flow in the Cartesian coordinate, assuming constant density and temperature. The equations solved were the vorticity and streamfunction equations (replacing the horizontal momentum equations and the continuity equation for air) (Arakawa and Lamb 1977). The fourth-order Runge–Kutta scheme was used for time stepping. The grid domain consisted of 16 × 16 grid cells, each 1.25 km × 1.25 km on edge. The time step was 10 s. The boundary conditions were periodic. From Ketefian and Jacobson (2005a).

(5.86) was

$$R_e^2 \cos\varphi \left[\frac{\partial}{\partial t}(\pi_a u)\right]_\sigma + \left[\frac{\partial}{\partial \lambda_e}(\pi_a u^2 R_e) + \frac{\partial}{\partial \varphi}(\pi_a u v R_e \cos\varphi)\right]_\sigma + \pi_a R_e^2 \cos\varphi \frac{\partial}{\partial \sigma}(\dot{\sigma} u)$$
$$= \pi_a u v R_e \sin\varphi + \pi_a f v R_e^2 \cos\varphi$$
$$- R_e \left(\pi_a \frac{\partial \Phi}{\partial \lambda_e} + \sigma c_{p,d} \theta_v \frac{\partial P}{\partial \sigma} \frac{\partial \pi_a}{\partial \lambda_e}\right)_\sigma + R_e^2 \cos\varphi \frac{\pi_a}{\rho_a} (\nabla \cdot \rho_a \mathbf{K}_m \nabla)\, u \qquad (7.31)$$

The finite-difference form of this equation is obtained by assuming fluxes enter or leave a grid cell through each of eight points, as shown in Fig. 7.4. This differs from Fig. 7.2, where fluxes entered or left through each of four points. The locations of variables and index values in Fig. 7.4 are the same as those in Fig. 7.2.

Substituting differential operators for difference operators in (7.28), multiplying through by $\Delta\lambda_e \Delta\varphi$, and interpolating variables in a manner consistent with kinetic energy and enstrophy conservation give the momentum equation in the u-direction as

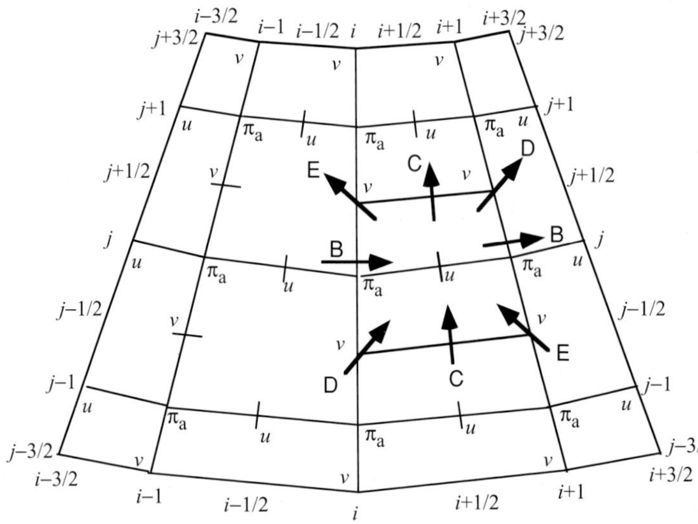

Figure 7.4 Location of fluxes for finite-differencing the momentum equation in the u-direction. The grid is the same as that in Fig. 7.2. The figure shows exact locations of u, v, and π_a relative to lateral boundaries.

Time difference term

$$u_{i+1/2,j,k,t} = \frac{(\pi_{a,t-h}\Delta A)_{i+1/2,j}}{(\pi_{a,t}\Delta A)_{i+1/2,j}} u_{i+1/2,j,k,t-h} + \frac{h}{(\pi_{a,t}\Delta A)_{i+1/2,j}} \times \Bigg\{ \tag{7.32}$$

Horizontal advection terms

$$\left(\begin{aligned} & B_{i,j}\frac{u_{i-1/2,j}+u_{i+1/2,j}}{2} - B_{i+1,j}\frac{u_{i+1/2,j}+u_{i+3/2,j}}{2} \\ & + C_{i+1/2,j-1/2}\frac{u_{i+1/2,j-1}+u_{i+1/2,j}}{2} - C_{i+1/2,j+1/2}\frac{u_{i+1/2,j}+u_{i+1/2,j+1}}{2} \\ & + D_{i,j-1/2}\frac{u_{i-1/2,j-1}+u_{i+1/2,j}}{2} - D_{i+1,j+1/2}\frac{u_{i+1/2,j}+u_{i+3/2,j+1}}{2} \\ & + E_{i+1,j-1/2}\frac{u_{i+3/2,j-1}+u_{i+1/2,j}}{2} - E_{i,j+1/2}\frac{u_{i+1/2,j}+u_{i-1/2,j+1}}{2} \end{aligned} \right)_{k,t-h} \tag{7.33}$$

Vertical transport of horizontal momentum

$$+\frac{1}{\Delta\sigma_k}(\pi_{a,t}\Delta A\dot{\sigma}_{k-1/2,t}u_{k-1/2,t-h} - \pi_{a,t}\Delta A\dot{\sigma}_{k+1/2,t}u_{k+1/2,t-h})_{i+1/2,j} \tag{7.34}$$

Coriolis and spherical grid conversion terms

$$
+\frac{R_e(\Delta\lambda_e\Delta\varphi)_{i+1/2,j}}{2} \times \left[\begin{array}{l} \pi_{a,i,j}\dfrac{v_{i,j-1/2}+v_{i,j+1/2}}{2}\left(f_j R_e\cos\varphi_j+\dfrac{u_{i-1/2,j}+u_{i+1/2,j}}{2}\sin\varphi_j\right) \\ +\pi_{a,i+1,j}\dfrac{v_{i+1,j-1/2}+v_{i+1,j+1/2}}{2}\left(f_j R_e\cos\varphi_j+\dfrac{u_{i+1/2,j}+u_{i+3/2,j}}{2}\sin\varphi_j\right) \end{array} \right]_{k,t-h} \tag{7.35}
$$

Pressure gradient terms

$$
-R_e\Delta\varphi_{i+1/2,j} \times \left[\begin{array}{l} (\Phi_{i+1,j,k}-\Phi_{i,j,k})\dfrac{\pi_{a,i,j}+\pi_{a,i+1,j}}{2}+(\pi_{a,i+1,j}-\pi_{a,i,j}) \\ \times\dfrac{c_{p,d}}{2}\left(\begin{array}{l} \left[\theta_{v,k}\dfrac{\sigma_{k+1/2}(P_{k+1/2}-P_k)+\sigma_{k-1/2}(P_k-P_{k-1/2})}{\Delta\sigma_k}\right]_{i,j} \\ +\left[\theta_{v,k}\dfrac{\sigma_{k+1/2}(P_{k+1/2}-P_k)+\sigma_{k-1/2}(P_k-P_{k-1/2})}{\Delta\sigma_k}\right]_{i+1,j} \end{array} \right) \end{array} \right]_{t-h} \tag{7.36}
$$

Eddy diffusion term

$$
+\left[(\pi_{a,t-h}\Delta A)_{i+1/2,j}\left[\frac{(\nabla_z\cdot\rho_a\mathbf{K}_m\nabla_z)u}{\rho_a}\right]_{i+1/2,j,k,t-h}\right]\Bigg\} \tag{7.37}
$$

All $\dot{\sigma}$ and some π_a values are evaluated at time t. All other terms are evaluated at time $t-h$. The interpolation of the second pressure gradient term in (7.36) is differenced in a manner consistent with the differencing of geopotential, discussed shortly.

The equations above require additional interpolations for certain terms. First, column pressure multiplied by grid-cell area at a u-point is interpolated with

$$
(\pi_a\Delta A)_{i+1/2,j}=\frac{1}{8}\left\{ \begin{array}{l} (\pi_a\Delta A)_{i,j+1}+(\pi_a\Delta A)_{i+1,j+1} \\ +2[(\pi_a\Delta A)_{i,j}+(\pi_a\Delta A)_{i+1,j}] \\ +(\pi_a\Delta A)_{i,j-1}+(\pi_a\Delta A)_{i+1,j-1} \end{array} \right\} \tag{7.38}
$$

where

$$
\Delta A = R_e^2\cos\varphi\,\Delta\lambda_e\Delta\varphi \tag{7.39}
$$

is the **horizontal area of a grid cell.** Equation (7.38) relies on column pressures (at times t or $t-h$) and grid areas at π_a points. An interpolation similar to (7.38), required for (7.34), is

$$(\pi_{a,t}\Delta A\dot{\sigma}_{k-1/2,t})_{i+1/2,j} = \frac{1}{8}\begin{bmatrix} (\pi_{a,t}\Delta A\dot{\sigma}_{k-1/2,t})_{i,j+1} + (\pi_{a,t}\Delta A\dot{\sigma}_{k-1/2,t})_{i+1,j+1} \\ +2\left\{(\pi_{a,t}\Delta A\dot{\sigma}_{k-1/2,t})_{i,j} + (\pi_{a,t}\Delta A\dot{\sigma}_{k-1/2,t})_{i+1,j}\right\} \\ +(\pi_{a,t}\Delta A\dot{\sigma}_{k-1/2,t})_{i,j-1} + (\pi_{a,t}\Delta A\dot{\sigma}_{k-1/2,t})_{i+1,j-1} \end{bmatrix} \tag{7.40}$$

All values at time t on the right side of this equation are known from (7.14) and (7.21) at the time the equation is evaluated.

A third set of interpolations, required for the fluxes in (7.33) shown in Fig. 7.4, is

$$\begin{aligned} \mathrm{B}_{i,j} = \frac{1}{12}[&F_{i-1/2,j-1} + F_{i+1/2,j-1} + 2(F_{i-1/2,j} + F_{i+1/2,j}) \\ &+ F_{i-1/2,j+1} + F_{i+1/2,j+1}] \end{aligned} \tag{7.41}$$

$$\begin{aligned} \mathrm{C}_{i+1/2,j-1/2} = \frac{1}{12}[&G_{i,j-3/2} + G_{i+1,j-3/2} + 2(G_{i,j-1/2} + G_{i+1,j-1/2}) \\ &+ G_{i,j+1/2} + G_{i+1,j+1/2}] \end{aligned} \tag{7.42}$$

$$\begin{aligned} \mathrm{D}_{i,j+1/2} = \frac{1}{24}(&G_{i,j-1/2} + 2G_{i,j+1/2} + G_{i,j+3/2} + F_{i-1/2,j} + F_{i-1/2,j+1} \\ &+ F_{i+1/2,j} + F_{i+1/2,j+1}) \end{aligned} \tag{7.43}$$

$$\begin{aligned} \mathrm{E}_{i,j+1/2} = \frac{1}{24}(&G_{i,j-1/2} + 2G_{i,j+1/2} + G_{i,j+3/2} - F_{i-1/2,j} - F_{i-1/2,j+1} \\ &- F_{i+1/2,j} - F_{i+1/2,j+1}) \end{aligned} \tag{7.44}$$

where F and G are the fluxes defined in (7.15) and (7.16), respectively. Finally, u-values at the bottom boundary of each layer, required for (7.34), may be interpolated vertically with

$$u_{i+1/2,j,k+1/2,t-h} = \frac{\Delta\sigma_{k+1}u_{i+1/2,j,k,t-h} + \Delta\sigma_k u_{i+1/2,j,k+1,t-h}}{\Delta\sigma_k + \Delta\sigma_{k+1}} \tag{7.45}$$

For nonnested, limited-area grids, special horizontal boundary conditions are necessary for the momentum equation. A virtual row is added beyond the horizontal boundary for each variable. For example, if $i+1$ in Fig. 7.4 is the eastern π_a-point boundary for values of π_a, Φ, and $\dot{\sigma}$, an additional row just outside the boundary at π_a-point $i+2$ is needed. A virtual row of u points at $i+5/2$ outside the boundary is also needed. In the pressure gradient term of the momentum equation, Φ_{t-h} and $\pi_{a,t-h}$ values in the virtual row outside the boundary are set to Φ_{t-2h} and $\pi_{a,t-2h}$ values just inside the boundary (e.g., Lu 1994). In such a case, (7.36) is replaced with

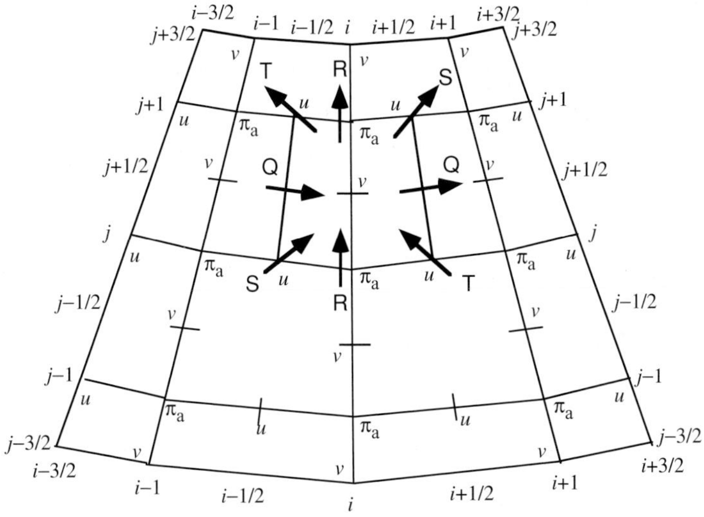

Figure 7.5 Location of fluxes for the finite-differencing of the momentum equation in the v-direction. The grid is the same as that in Fig. 7.2. The figure shows exact locations of u, v, and π_a relative to lateral boundaries.

Pressure gradient terms

$$-R_e\Delta\varphi_{I+1/2,j}\left[\begin{array}{l}(\Phi_{I,j,k,t-2h}-\Phi_{I,j,k,t-h})\pi_{a,I,j,t-h}+(\pi_{a,I,j,t-2h}-\pi_{a,I,j,t-h})\\ \times c_{p,d}\left[\theta_{v,k}\dfrac{\sigma_{k+1/2}(P_{k+1/2}-P_k)+\sigma_{k-1/2}(P_k-P_{k-1/2})}{\Delta\sigma_k}\right]\end{array}\right]_{I,j,t-h} \tag{7.46}$$

This boundary condition dampens the effect of gravity waves, which rapidly create instabilities when the boundary condition is absent.

Virtual boundary values of $\dot{\sigma}$ and u can be set to nearest inside boundary values from the current time step or previous time step. Arakawa (1984) and Lu (1994) contain a more rigorous treatment of these variables at the boundaries. In that treatment, momentum fluxes at the boundaries are derived so as to conserve mass at the boundaries.

The momentum equation in the south–north direction is differenced in a manner similar to that in the west–east direction. Figure 7.5 shows the locations of fluxes for finite differencing this equation. The finite-difference form of the ***v*-momentum equation,** given in (5.87), is

Time difference term

$$v_{i,j+1/2,k,t}=\frac{(\pi_{a,t-h}\Delta A)_{i,j+1/2}}{(\pi_{a,t}\Delta A)_{i,j+1/2}}v_{i,j+1/2,k,t-h}+\frac{h}{(\pi_{a,t}\Delta A)_{i,j+1/2}}\times\Bigg\{ \tag{7.47}$$

Horizontal advection terms

$$\left(\begin{array}{l} Q_{i-1/2,j+1/2}\dfrac{v_{i-1,j+1/2}+v_{i,j+1/2}}{2}-Q_{i+1/2,j+1/2}\dfrac{v_{i,j+1/2}+v_{i+1,j+1/2}}{2} \\ +R_{i,j}\dfrac{v_{i,j-1/2}+v_{i,j+1/2}}{2}-R_{i,j+1}\dfrac{v_{i,j+1/2}+v_{i,j+3/2}}{2} \\ +S_{i-1/2,j}\dfrac{v_{i-1,j-1/2}+v_{i,j+1/2}}{2}-S_{i+1/2,j+1}\dfrac{v_{i,j+1/2}+v_{i+1,j+3/2}}{2} \\ +T_{i+1/2,j}\dfrac{v_{i+1,j-1/2}+v_{i,j+1/2}}{2}-T_{i-1/2,j+1}\dfrac{v_{i,j+1/2}+v_{i-1,j+3/2}}{2} \end{array}\right)_{k,t-h} \quad (7.48)$$

Vertical transport of horizontal momentum

$$+\frac{1}{\Delta\sigma_k}\left(\pi_{a,t}\Delta A\dot{\sigma}_{k-1/2,t}v_{k-1/2,t-h}-\pi_{a,t}\Delta A\dot{\sigma}_{k+1/2,t}v_{k+1/2,t-h}\right)_{i,j+1/2} \quad (7.49)$$

Coriolis and spherical grid conversion terms

$$-\frac{R_e(\Delta\lambda_e\Delta\varphi)_{i,j+1/2}}{2}\times\left[\pi_{a,i,j}\frac{u_{i-1/2,j}+u_{i+1/2,j}}{2}\left(f_j R_e\cos\varphi_j + \frac{u_{i-1/2,j}+u_{i+1/2,j}}{2}\sin\varphi_j\right)+\pi_{a,i,j+1}\frac{u_{i-1/2,j+1}+u_{i+1/2,j+1}}{2}\times\left(f_{j+1}R_e\cos\varphi_{j+1}+\frac{u_{i-1/2,j+1}+u_{i+1/2,j+1}}{2}\sin\varphi_{j+1}\right)\right]_{k,t-h} \quad (7.50)$$

Pressure gradient terms

$$-R_e(\cos\varphi\,\Delta\lambda_e)_{i,j+1/2}\times\left[\begin{array}{l}(\Phi_{i,j+1,k}-\Phi_{i,j,k})\dfrac{\pi_{a,i,j}+\pi_{a,i,j+1}}{2}+(\pi_{a,i,j+1}-\pi_{a,i,j}) \\ \times\dfrac{c_{p,d}}{2}\left(\begin{array}{l}\left[\theta_{v,k}\dfrac{\sigma_{k+1/2}(P_{k+1/2}-P_k)+\sigma_{k-1/2}(P_k-P_{k-1/2})}{\Delta\sigma_k}\right]_{i,j} \\ +\left[\theta_{v,k}\dfrac{\sigma_{k+1/2}(P_{k+1/2}-P_k)+\sigma_{k-1/2}(P_k-P_{k-1/2})}{\Delta\sigma_k}\right]_{i,j+1}\end{array}\right)\end{array}\right]_{t-h} \quad (7.51)$$

Eddy diffusion term

$$\left.+(\pi_{a,t-h}\Delta A)_{i,j+1/2}\left[\frac{(\nabla_z\cdot\rho_a\mathbf{K}_m\nabla_z)\,u}{\rho_a}\right]_{i,j+1/2,k,t-h}\right\} \quad (7.52)$$

In the equations above, column pressure multiplied by the area of a grid cell at a v-point is interpolated with

$$(\pi_a\Delta A)_{i,j+1/2}=\frac{1}{8}\left[\begin{array}{l}(\pi_a\Delta A)_{i+1,j}+(\pi_a\Delta A)_{i+1,j+1} \\ +2[(\pi_a\Delta A)_{i,j}+(\pi_a\Delta A)_{i,j+1}] \\ +(\pi_a\Delta A)_{i-1,j}+(\pi_a\Delta A)_{i-1,j+1}\end{array}\right] \quad (7.53)$$

In addition,

$$(\pi_{a,t}\Delta A\dot{\sigma}_{k-1/2,t})_{i,j+1/2} = \frac{1}{8}\begin{bmatrix}(\pi_{a,t}\Delta A\dot{\sigma}_{k-1/2,t})_{i+1,j} + (\pi_{a,t}\Delta A\dot{\sigma}_{k-1/2,t})_{i+1,j+1} \\ +2[(\pi_{a,t}\Delta A\dot{\sigma}_{k-1/2,t})_{i,j} + (\pi_{a,t}\Delta A\dot{\sigma}_{k-1/2,t})_{i,j+1}] \\ +(\pi_{a,t}\Delta A\dot{\sigma}_{k-1/2,t})_{i-1,j} + (\pi_{a,t}\Delta A\dot{\sigma}_{k-1/2,t})_{i-1,j+1}\end{bmatrix} \tag{7.54}$$

Fluxes in (7.48) are found with

$$\mathrm{Q}_{i-1/2,j+1/2} = \frac{1}{12}[F_{i-3/2,j} + F_{i-3/2,j+1} + 2(F_{i-1/2,j} + F_{i-1/2,j+1}) + F_{i+1/2,j} + F_{i+1/2,j+1}] \tag{7.55}$$

$$\mathrm{R}_{i,j} = \frac{1}{12}[G_{i-1,j-1/2} + G_{i-1,j+1/2} + 2(G_{i,j-1/2} + G_{i,j+1/2}) + G_{i+1,j-1/2} + G_{i+1,j+1/2}] \tag{7.56}$$

$$\mathrm{S}_{i+1/2,j} = \frac{1}{24}(G_{i,j-1/2} + G_{i,j+1/2} + G_{i+1,j-1/2} + G_{i+1,j+1/2} + F_{i-1/2,j} + 2F_{i+1/2,j} + F_{i+3/2,j}) \tag{7.57}$$

$$\mathrm{T}_{i+1/2,j} = \frac{1}{24}(G_{i,j-1/2} + G_{i,j+1/2} + G_{i+1,j-1/2} + G_{i+1,j+1/2} - F_{i-1/2,j} - 2F_{i+1/2,j} - F_{i+3/2,j}) \tag{7.58}$$

Finally, the south–north scalar velocity at the bottom of a layer can be interpolated with

$$v_{i,j+1/2,k+1/2,t-h} = \frac{\Delta\sigma_{k+1}v_{i,j+1/2,k,t-h} + \Delta\sigma_k v_{i,j+1/2,k+1,t-h}}{\Delta\sigma_k + \Delta\sigma_{k+1}} \tag{7.59}$$

Boundary conditions for the v-equation are similar to those for the u-equation. For instance, pressure-gradient boundary conditions in the v-equation are

Pressure gradient terms

$$-R_e(\cos\varphi\Delta\lambda_e)_{i,J+1/2} \times \begin{bmatrix}(\Phi_{i,J,k,t-2h} - \Phi_{i,J,k,t-h})\pi_{a,i,J,t-h} + (\pi_{a,i,J,t-2h} - \pi_{a,i,J,t-h}) \\ \times c_{p,d}\left[\theta_{v,k}\dfrac{\sigma_{k+1/2}(P_{k+1/2} - P_k) + \sigma_{k-1/2}(P_k - P_{k-1/2})}{\Delta\sigma_k}\right]\end{bmatrix}_{i,J,t-h} \tag{7.60}$$

7.6 THE HYDROSTATIC EQUATION

When a grid-cell dimension is greater than 3 km, the model atmosphere is generally in hydrostatic equilibrium, and the geopotential used in the horizontal momentum equations can be calculated diagnostically with the hydrostatic equation. From (5.38), the hydrostatic equation was $d\Phi = -c_{p,d}\theta_v\, dP$. A finite-difference form of this equation giving the geopotential at the vertical midpoint of the bottom layer at time $t - h$ is

$$\Phi_{i,j,\mathrm{N_L},t-h} = \Phi_{i,j,\mathrm{N_L}+1/2} - c_{p,\mathrm{d}}[\theta_{\mathrm{v},\mathrm{N_L}}(P_{\mathrm{N_L}} - P_{\mathrm{N_L}+1/2})]_{i,j,t-h} \tag{7.61}$$

(Arakawa and Suarez 1983) where $P_{\mathrm{N_L}+1/2}$ and $P_{\mathrm{N_L}}$ are found from (7.9) and (7.8), respectively. The geopotential at the surface, $\Phi_{i,j,\mathrm{N_L}+1/2}$, is the topographical surface altitude multiplied by gravity. The geopotential at the bottom and vertical midpoint of each layer above the surface are found from

$$\Phi_{i,j,k+1/2,t-h} = \Phi_{i,j,k+1,t-h} - c_{p,\mathrm{d}}[\theta_{\mathrm{v},k+1}(P_{k+1/2} - P_{k+1})]_{i,j,t-h} \tag{7.62}$$

$$\Phi_{i,j,k,t-h} = \Phi_{i,j,k+1/2,t-h} - c_{p,\mathrm{d}}[\theta_{\mathrm{v},k}(P_k - P_{k+1/2})]_{i,j,t-h} \tag{7.63}$$

respectively. The geopotential is solved for from the model surface to top. The boundary conditions for the geopotential were discussed in Section 7.5.

7.7 ORDER OF CALCULATIONS

The equations in the preceding sections should be solved during a time step in the following order: column pressure from (7.14), vertical scalar velocity from (7.21), specific humidity and moist-air mass mixing ratios of gases and aerosol-particle components from (7.24), potential virtual temperature from (7.27), geopotential from (7.61)–(7.63), west–east scalar velocities from (7.32)–(7.37), and south–north scalar velocities from (7.47)–(7.52). Quantities in the equations are used explicitly (from time $t - h$), except for all $\dot{\sigma}$-values and specified π_{a}-values. Since most equations rely on $\dot{\sigma}$ and some π_{a}-values at time t, these parameters must be determined first and second, respectively.

7.8 TIME-STEPPING SCHEMES

In this chapter, prognostic equations were written as explicit, first-order approximations in time (in forward Euler form). Since the forward Euler is unconditionally unstable for any wind speed in the absence of eddy diffusion, as discussed in Chapter 6, the forward Euler causes solutions to break down quickly. To increase stability and the length of simulation, a better time-stepping scheme is needed. Stability and accuracy can be improved by using the Crank–Nicolson time-stepping scheme (Section 6.4.4.4), the Matsuno scheme alone (Section 6.4.4.6), the Matsuno scheme combined with the leapfrog scheme (Section 6.4.4.5), the Heun scheme (Section 6.4.4.7), the Adams–Bashforth scheme (Section 6.4.4.8), or the fourth-order Runge–Kutta scheme (Section 6.4.4.9). Figure 6.5 shows that the fourth-order Runge–Kutta is generally the most accurate among these schemes when applied to the one-dimensional advection–diffusion equation. It is also the most accurate and the most conservative when applied to the equations of atmospheric dynamics (e.g., Ketefian and Jacobson 2005a,b). The tradeoff is that it requires more computer time per time step than the other schemes because it requires four evaluations of all equations solved each time step rather than one or two evaluations, in the case of the other schemes listed.

Due to computer time constraints, most atmospheric models use a simpler time-stepping scheme. The **Matsuno scheme** is first-order in time but conditionally stable. This scheme, alone, can be used to solve the equations of atmospheric dynamics. During a Matsuno time step, variable values at time t are first estimated from values at time $t - h$. If the time derivative of the moist-air mass mixing ratio of a species is

$$\frac{\partial \mathrm{q}}{\partial t} = f(\mathrm{q}) \tag{7.64}$$

then q is estimated explicitly with

$$\mathrm{q}_{\mathrm{est}} = \mathrm{q}_{t-h} + hf(\mathrm{q}_{t-h}) \tag{7.65}$$

The estimate is then used to calculate the spatial derivative in a second equation that also relies on information from one time step backward. The final value of q from the second equation is

$$\mathrm{q}_{t} = \mathrm{q}_{t-h} + hf(\mathrm{q}_{\mathrm{est}}) \tag{7.66}$$

Thus, with the Matsuno scheme, the right sides of all equations are evaluated twice during each time step.

The time-derivative approximation can be improved to second order by using the **leapfrog scheme**. Since the leapfrog scheme destabilizes over time when used alone to solve nonlinear equations, it cannot be used alone. Thus, the Matsuno scheme or another scheme must be inserted every 5–15 leapfrog steps to maintain stability. Combining the Matsuno scheme, which is a first-order approximation, with the leapfrog scheme, which is a second-order approximation, gives a resulting approximation between first and second order.

With the leapfrog scheme, time derivatives leap over a time step in the spatial derivative terms. The mixing ratio at the end of a leapfrog time step is

$$\mathrm{q}_{t+h} = \mathrm{q}_{t-h} + 2hf(\mathrm{q}_{t}) \tag{7.67}$$

which requires information from two time steps backward. The Matsuno scheme is used not only to stabilize the leapfrog scheme but also to initialize a new simulation or time interval. Figure 7.6 shows a sequence of dynamics calculations when Matsuno and leapfrog schemes are used together. In Chapter 6, a **time step** was defined as an increment in time for a given algorithm, and a **time interval** was defined as the period during which several time steps of a process are completed without interference by another process. In Fig. 7.6, M1, L2, L3, . . . , L6 are time

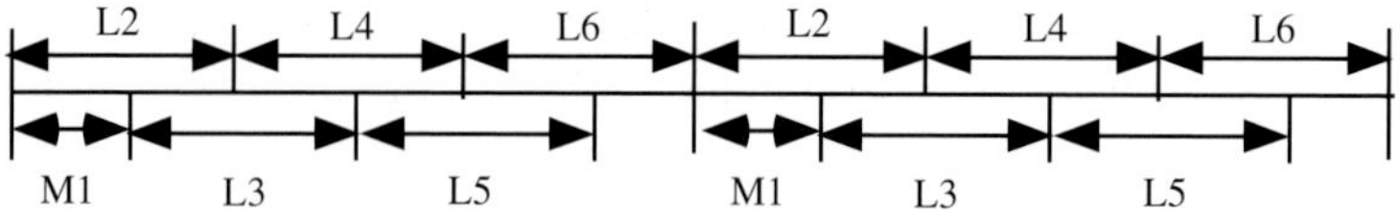

Figure 7.6 Example of a dynamical time-stepping scheme that uses a Matsuno (M) and a leapfrog (L) scheme. Two interruptions with a Matsuno step are shown – one to start the simulation and one after several leapfrog steps. The number next to each letter indicates the order in which the step is taken during a sequence.

steps, and each set of M1, . . . , L6 is a time interval. A typical dynamic time step in a global model is 60–300 s. That in a mesoscale model is 3–30 s.

7.9 SUMMARY

In this chapter, finite-difference solutions to the equations of atmospheric dynamics, including the continuity equation for air, the species continuity equation, the thermodynamic energy equation, the horizontal momentum equations, and the hydrostatic equation, were given. The thermodynamic energy equation is used to calculate the potential virtual temperature. The horizontal momentum equations are used to obtain horizontal scalar velocities. When the atmosphere is in hydrostatic equilibrium, the continuity equation for air is used to solve for vertical scalar velocity and column pressure, and the hydrostatic equation is used to solve for geopotential. A common time-stepping scheme in dynamical models is the combination of a Matsuno and leapfrog scheme.

7.10 PROBLEMS

7.1 For the same time step and grid spacing, should the model accuracy increase or decrease when Matsuno but no leapfrog steps are taken versus when both Matsuno and leapfrog steps are taken? Why? How is the computer time affected in each case?

7.2 **(a)** Assume a grid cell, centered at $\varphi = 45°$ N, has dimensions $\Delta\lambda_e = 5°$, $\Delta\varphi = 4°$, and $\Delta\sigma = 0.05$. Calculate q at the center of the cell after one 5-min time step if $u_1 = +3$ m s^{-1}, $\pi_{a,1} = 745$ hPa, $q_1 = 0.0062$ kg kg^{-1}; $u_2 = +1$ m s^{-1}, $\pi_{a,2} = 752$ hPa, $q_2 = 0.0047$ kg kg^{-1}; $v_3 = +\ 2$ m s^{-1}, $\pi_{a,3} = 754$ hPa, $q_3 = 0.0064$ kg kg^{-1}; $v_4 = +\ 1$ m s^{-1}, $\pi_{a,4} = 746$ hPa, $q_4 = 0.0054$ kg kg^{-1}, $\dot\sigma_5 = -4.58 \times 10^{-6}\sigma$ s^{-1}, $q_5 = 0.005$ kg kg^{-1}; and $\dot\sigma_6 = -6.11 \times 10^{-6}\sigma$ s^{-1}, $q_6 = 0.006$ kg kg^{-1}, where subscripts 1, . . . , 6 denote the west, east, south, north, lower, and upper boundaries, respectively. Average the four π_a-values and six q-values to obtain center values. Neglect eddy diffusion and external sources and sinks. Assume the air is dry and the π_a-values stay constant for the time step.

(b) For the conditions of part (a), calculate θ_v at the center of the cell after 30 minutes if the west, east, south, north, lower, and upper grid-cell boundary values are $\theta_v = 297$, 298, 300, 299, 298.3, and 298.7 K, respectively. Average the six boundary values to obtain an initial center value of θ_v.

7.11 COMPUTER PROGRAMMING PRACTICE

7.3 Set up a 20-layer (21-boundary) sigma-pressure-coordinate vertical model grid between $p_{a,top} = 250$ hPa and $p_{a,surf} = 1000$ hPa at sea level. Assume each layer has equal sigma thickness, and assume $T = 288$ K at the surface. Assume also that temperature decreases with increasing altitude at 6.5 K km^{-1}. Calculate pressure at the center of each layer with (7.10). Plot pressure versus altitude.

7.12 MODELING PROJECT

The purpose of this project is to develop a basic regional- or global-scale model that solves the equations of atmospheric dynamics, except for the water-vapor continuity equation. Diabatic energy sources and sinks and eddy diffusion are ignored. Winds are driven primarily by pressure gradients.

(a) Choose a southwest-corner latitude and longitude anywhere on Earth away from the poles.

(b) Set a model top pressure everywhere of 250 hPa (near the tropopause).

(c) Set NLAT, NLONG, and NVERT, which denote the numbers of latitudinal, longitudinal, and vertical grid centers in the model. Note that 1, . . . , NLAT, 1, . . . , NLONG, and 1, . . . , NVERT increase from south to north, west to east, and top to bottom, respectively. NLAT + 1, NLONG + 1, and NVERT + 1 equal the numbers of boundaries in the respective directions. Allow parameters to be variable, but set them initially to NLAT = 40, NLONG = 50, and NVERT = 15.

(d) Use spherical horizontal coordinates and the sigma-pressure vertical coordinate. Select $\Delta\lambda_e = 0.05^\circ$ and $\Delta\varphi = 0.05^\circ$. Select values of $\Delta\sigma$ for each layer using the method discussed in this chapter. Ensure that the bottom layer is at least 150 m thick (1 hPa $\approx$ 10 m).

(e) Initialize surface pressure by creating a small hill of pressure in the center of the grid superimposed on background pressure. First, select a background surface pressure ($p_{a,base}$) of 1000 hPa and a peak incremental surface pressure ($\Delta p_{a,peak}$) of 3 hPa. Calculate the surface pressure at the horizontal center of each grid cell i, j with the Gaussian distribution,

$$p_{a,surf,i,j} = p_{a,base} + \Delta p_{a,peak} \times \exp\left[-\frac{\left[R_e \cos\left(\frac{\varphi_{i,j}+\varphi_c}{2}\right)(\lambda_{i,j}-\lambda_c)\right]^2}{2} - \frac{[R_e(\varphi_{i,j}-\varphi_c)]^2}{2}\right]$$

where λ_c and φ_c are the longitude and latitude (radians), respectively, at the center of your grid. Calculate the initial column pressure at the horizontal center of each grid cell with $\pi_{i,j} - p_{a,surf,i,j} - p_{top}$.

(f) Use (7.6) to calculate pressure at the horizontal midpoint/vertical boundary of each cell. Use (7.10) to calculate pressure at the horizontal midpoint/vertical midpoint of each cell from horizontal midpoint/vertical boundary values.

(g) Initialize temperatures at the horizontal midpoint/vertical midpoint of each cell in one corner column of the model. Estimate temperatures from Appendix Table B.1 given the pressures from part (f).

(h) Set the temperature at the horizontal midpoint/vertical midpoint of each cell in all other columns equal to that in the initialized column. Calculate potential virtual temperature at the horizontal midpoint/vertical midpoint of each cell. Assume the air is dry. Interpolate potential virtual temperature to the horizontal midpoint/vertical boundary of each cell with (7.11). No values are needed for the model top or bottom boundaries ($k = 1/2$ or $N_L + 1/2$).

(i) Use the equation of state to calculate the density at the horizontal midpoint/vertical midpoint of each cell.

(j) Use (7.14) to solve for the column pressure at the horizontal midpoint of each column in the model.

(k) Use (7.21) to solve for vertical velocities at the horizontal midpoint/vertical boundary of each cell.

(l) Use (7.27) to solve for the potential virtual temperature at the horizontal midpoint/vertical midpoint of each cell. Ignore diabatic sources and sinks and ignore eddy diffusion.

(m) Solve for the geopotential in each cell with (7.61)–(7.63). Assume the surface geopotential is zero.

(n) Use (7.32)–(7.37) and (7.47)–(7.52) to solve for u and v at locations shown in Figs. 7.4 and 7.5, respectively. Ignore eddy diffusion. Assume u, v, and $\dot{\sigma}$ are initially zero everywhere.

(o) Use a Matsuno or Matsuno-plus-leapfrog time-stepping scheme instead of the forward Euler scheme to advance all prognostic equations for this problem. Recalculate all diagnostic variables during each Matsuno and/or leapfrog step.

(p) Debug, and run simulations. Change conditions by changing the location and/or magnitude of peak surface pressure.

(q) Test the effect of adding topography by changing the geopotential at the surface.

(r) Plot horizontal and/or vertical fields of temperature, pressure, and velocities.

(s) Instead of using horizontal limited-area boundary conditions as described in the text, it may be more convenient, for this project, to use periodic boundary conditions. To accomplish this, set up the limited-area model grid so that it is symmetric in the south–north direction about the Equator, allow flow out of the north boundary edge to enter the south boundary edge and vice versa, and allow flow out of the east boundary edge to enter the west boundary edge and vice versa. For this example case, the Coriolis parameter, f, may be set to zero, otherwise it will have opposite signs on the north and south boundaries, which connect to each other. Allowing f to vary, however, does ensure enstrophy and energy conservation. Although not necessarily realistic, this setup will simplify the coding, and allow for useful sensitivities of the numerical properties of the code.

8

Boundary-layer and surface processes

THE **boundary layer** is the region of the atmosphere between the Earth's surface and 500–3000 m height that is influenced substantially by energy and moisture from the surface. The bottom ten percent of the boundary layer is the **surface layer**. Some parameters that affect the surface layer and boundary layer are ground temperature, soil moisture, and turbulent fluxes. In this chapter, expressions for turbulent fluxes of momentum, energy, and moisture are given in terms of bulk aerodynamic formulae, Monin–Obukhov similarity theory, K-theory, and higher-order parameterizations. From these equations, eddy diffusion coefficients for momentum and energy are derived. Analytical equations for the vertical profile of wind speed, potential virtual temperature, and moisture in the surface layer are also provided. Prognostic equations for surface temperature and soil moisture are then given. Types of surfaces treated include soil, water, roads, rooftops, sea ice, vegetation over soil, and snow over all surfaces. Boundary-layer and surface processes affect weather and air pollution from urban to global scales.

8.1 TURBULENT FLUXES OF MOMENTUM, ENERGY, AND MOISTURE

Turbulence is due to wind shear (mechanical turbulence) and buoyancy (thermal turbulence). Turbulence mixes gradients of momentum, energy, moisture, gases, and particles vertically and horizontally. In a model, the degree of vertical mixing due to turbulence can be quantified with a turbulent flux term. In the case of vertical mixing of horizontal momentum, the term is a function of the **kinematic vertical turbulent momentum flux,** $\overline{w'u'}$ and $\overline{w'v'}$ (m^2 s^{-1}) (Section 4.2.6). In the case of vertical mixing of energy, it is a function of the **kinematic vertical turbulent sensible-heat flux,** $\overline{w'\theta'_v}$ (m K s^{-1}) (Section 3.4). In the case of vertical mixing of moisture, it is a function of the **kinematic vertical moisture flux,** $\overline{w'q'_v}$ (m kg s^{-1} kg^{-1}). Analogous fluxes can be developed for other gases and particles.

Kinematic turbulent fluxes of momentum are negatively proportional to **Reynolds stresses.** A **stress** is a force per unit area that causes a body to deform. A Reynolds stress, which arises when a fluid undergoes turbulent motion, causes a parcel of air to deform. Suppose an air parcel fluctuates randomly in time due

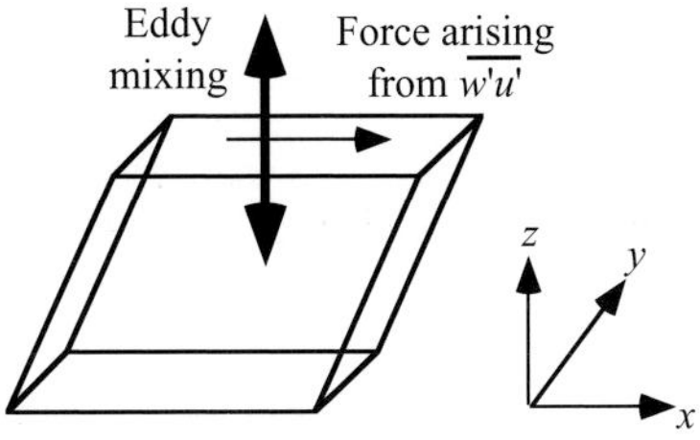

Figure 8.1 Deformation of a cubic air parcel caused by a kinematic vertical momentum flux $\overline{w'u'}$, as described in the text. Adapted from Stull (1988).

to mechanical shear and buoyancy. Precise scalar velocities, w and u, in the parcel have mean components $\bar{w}$ and $\bar{u}$ and eddy components w' and u', respectively. A w'-velocity has the effect of mixing u'-velocities in the z-direction. The vertical mixing of the u'-velocity exerts a force in the x-direction over an area normal to the z-direction. The effect of the force per unit area, or stress, is to induce a drag on $\bar{u}$ and to cause the air parcel to deform, as shown in Fig. 8.1.

In the example above, the scalar component of Reynolds stress in the x-direction along a plane normal to the z-direction is

$$\tau_{zx} = -\rho_a \overline{w'u'} \tag{8.1}$$

(kg m^{-1} s^{-2} or N m^{-2}), where ρ_a is air density (kg m^{-3}). This stress results from the vertical transport of a west–east gradient of momentum. Since w' mixes u' in the same way that u' mixes w', we have $\overline{u'w'} = \overline{w'u'}$, $\tau_{zx} = \tau_{xz}$, and $\tau_{xz} = -\rho_a \overline{u'w'}$. The Reynolds stress in the y-direction along a plane normal to the z-direction is $\tau_{zy} = -\rho_a \overline{w'v'}$. Both τ_{zx} and τ_{zy} have the effect of transporting gradients of horizontal momentum vertically. Combining the two gives the magnitude of the **vertical turbulent flux of horizontal momentum** as

$$|\tau_z| = \rho_a \left[(\overline{w'u'})^2 + (\overline{w'v'})^2 \right]^{1/2} \tag{8.2}$$

(kg m^{-1} s^{-2}). From (8.1), the kinematic vertical turbulent fluxes of west–east and south–north momentum are

$$\overline{w'u'} = -\frac{\tau_{zx}}{\rho_a} \qquad \overline{w'v'} = -\frac{\tau_{zy}}{\rho_a} \tag{8.3}$$

Fluxes such as these are needed in the turbulent-flux divergence term of the momentum equation given in (4.66). The fluxes can be measured, but in a model, they need to be parameterized, as discussed in Section 8.4.

Example 8.1

Reynolds stresses are usually much larger than viscous stresses. From (4.56), the shearing stress in the x-direction over the x–y plane due to wind shear in the z-direction was $\tau_{zx} = \eta_a \partial u/\partial z$. If $T = 288$ K and $u = 5$ m s^{-1} at $z = 10$ m, then $\eta_a = 1.792 \times 10^{-5}$ kg m^{-1} s^{-1} and $\partial u/\partial z = 0.5$ s^{-1}. The resulting shearing stress due to viscosity is $\tau_{zx} = 8.96 \times 10^{-6}$ kg m^{-1} s^{-2}.

A typical value of $\overline{w'u'}$ is 0.4 m^2 s^{-2}. For a surface air density of ρ_a = 1.25 kg m^{-3}, (8.1) predicts a Reynolds stress due to turbulence of τ_{zx} = 0.5 kg m^{-1} s^{-2}.

Gradients of energy, like momentum, are transferred vertically by turbulence. The **vertical turbulent sensible-heat flux** is defined as

$$H_f = \rho_a c_{p,d} \overline{w'\theta'_v} \tag{8.4}$$

(W m^{-2}) where $c_{p,d}$ is the specific heat of dry air at constant pressure (J kg^{-1} K^{-1}), θ'_v is the turbulent fluctuation of potential virtual temperature (K), and $\overline{w'\theta'_v}$ (m K s^{-1}) is the **kinematic vertical turbulent sensible-heat flux.** Rewriting (8.4) gives

$$\overline{w'\theta'_v} = \frac{H_f}{\rho_a c_{p,d}} \tag{8.5}$$

which is needed to calculate the effect of turbulence on mixing of energy gradients in the thermodynamic energy equation of (3.72). The kinematic flux can be calculated from observed quantities of θ'_v and w' or parameterized, as discussed in Section 8.4.

Gradients of water vapor are also transferred vertically by turbulence. The **vertical turbulent flux of water vapor** is

$$E_f = \rho_a \overline{w'q'_v} \tag{8.6}$$

(kg m^{-2} s^{-1}) where q'_v is the turbulent fluctuation of water-vapor specific humidity (kg kg^{-1}), and $\overline{w'q'_v}$ (m kg s^{-1} kg^{-1}) is the **kinematic vertical turbulent moisture flux.** Rewriting (8.6) gives

$$\overline{w'q'_v} = \frac{E_f}{\rho_a} \tag{8.7}$$

which is needed to calculate the effect of turbulence on mixing of moisture gradients in the water vapor continuity equation of (3.49).

8.2 FRICTION WIND SPEED

When horizontal winds flow over roughness elements protruding from a surface, drag slows the wind near the surface relative to the wind aloft, creating vertical wind shear. Wind shear produces eddies that exchange momentum, energy, gases, and aerosol particles vertically. The greater the height that roughness elements protrude from a surface and the greater the horizontal wind speed, the greater the resulting wind shear and vertical flux of horizontal momentum. A scaling parameter that provides a measure of the vertical flux of horizontal momentum in the surface layer (denoted by the subscript s) is the **friction wind speed,**

$$u_* = \left[(\overline{w'u'})_s^2 + (\overline{w'v'})_s^2\right]^{1/4} = \left(\frac{|\tau_z|}{\rho_a}\right)_s^{1/2} \qquad (8.8)$$

The greater the friction wind speed, the greater **mechanical turbulence**, and the faster that momentum, energy, and pollutants from aloft mix down to the surface and vice versa. Typical roughness elements at the surface include rocks, trees, buildings, grass, and sand. The friction wind speed can be parameterized or found from field experiments in which u', v', and w' are measured.

Example 8.2

If two measurements of west–east and vertical scalar velocity yield $u_1 = 10$ m s^{-1}, $w_1 = 0.1$ m s^{-1}, and $u_2 = 6$ m s^{-1}, $w_2 = 0.2$ m s^{-1}, respectively, estimate u_*. Assume the south–north scalar velocity is zero.

SOLUTION

$$\bar{u} = (u_1 + u_2)/2 = 8 \text{ m s}^{-1} \qquad \bar{w} = (w_1 + w_2)/2 = 0.15 \text{ m s}^{-1}$$
$$u_1' = u_1 - \bar{u} = 2 \text{ m s}^{-1} \qquad w_1' = w_1 - \bar{w} = -0.05 \text{ m s}^{-1}$$
$$u_2' = u_2 - \bar{u} = -2 \text{ m s}^{-1} \qquad w_2' = w_2 - \bar{w} = 0.05 \text{ m s}^{-1}$$

Therefore,

$$\overline{w'u'} = (w_1'u_1' + w_2'u_2')/2 = -0.1\text{m}^{-2}\text{s}^{-2} \qquad u_* = (\overline{w'u'})_s^{1/2} = 0.32 \text{ m s}^{-1}$$

8.3 SURFACE ROUGHNESS LENGTHS

Three variables used frequently in boundary-layer parameterizations are the surface roughness lengths for momentum, energy, and moisture. The **surface roughness length for momentum** ($z_{0,m}$), or **aerodynamic roughness**, is the height above a surface at which the logarithmic profile of wind speed versus altitude extrapolates to zero wind speed. It gives a measure of vertical turbulence that occurs when a horizontal wind flows over a rough surface. The greater $z_{0,m}$, the greater the magnitude of turbulence that arises when wind passes over a roughness element. For a perfectly smooth surface, the roughness length is zero, and mechanical turbulence is minimized. For other surfaces, it is sometimes approximated as 1/30th the height of the average roughness element protruding from the surface. For surfaces with sparsely placed roughness elements, $z_{0,m}$ is the height above the base of the roughness elements (Brutsaert 1991). For densely placed roughness elements of average

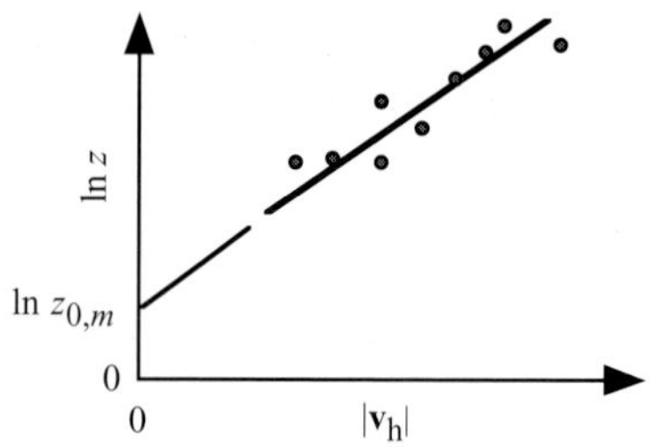

Figure 8.2 The surface roughness length for momentum is calculated by taking wind speed measurements at several heights at a given location when the wind is strong. The data are plotted as the natural log of height versus wind speed. Wind speeds from higher altitudes on the plot are then extrapolated until a zero wind speed intersects the altitude axis. The altitude at which this occurs is the surface roughness length for momentum. Points in the diagram are measurements.

height h_c, $z_{0,m}$ is the height above a displacement height (d_c). Displacement heights usually lie between 0 and h_c (Section 8.4.2.8).

Figure 8.2 describes a method of estimating $z_{0,m}$ from observed wind speed profiles. Parameterizations of $z_{0,m}$ from field data have also been developed. For smooth surfaces, such as over a smooth ocean with low wind speeds, one parameterization is

$$z_{0,m} \approx 0.11\frac{\nu_a}{u_*} = 0.11\frac{\eta_a}{\rho_a u_*} \tag{8.9}$$

(e.g., Hinze 1975; Garratt 1992), where $z_{0,m}$ is in m, $\nu_a = \eta_a/\rho_a$ is the kinematic viscosity of air (m^2 s^{-1}), η_a is the dynamic viscosity of air (kg m^{-1} s^{-2}) from (4.54), and ρ_a is air density (kg m^{-3}). Over a rough ocean with high wind speeds,

$$z_{0,m} \approx \alpha_c \frac{u_*^2}{g} \tag{8.10}$$

which is the **Charnock relation** (Charnock 1955), where $\alpha_c \approx 0.016$ is the **Charnock constant** (e.g., Garratt 1992). Over urban areas containing structures,

$$z_{0,m} \approx 0.5\frac{h_0 S_0}{A_0} \tag{8.11}$$

(Lettau 1969; Petersen 1997), where 0.5 is the average drag coefficient over the structures, h_0 (m) is the average structure height, A_0 (m^2) is the area over which $z_{0,m}$ is estimated, and S_0 (m^2) is the silhouette area, measured in a vertical plane normal to the mean wind, of all structures in surface area A_0.

A **canopy** is the vegetation cover provided by a group of plants or trees. An expression for the surface roughness length (m) over a vegetation canopy is

$$z_{0,m} = h_c(1 - 0.91e^{-0.0075L_T}) \tag{8.12}$$

(Sellers *et al.* 1996), where h_c is the **canopy height** (height in meters above the ground of the top of a vegetation canopy) and L_T is the **one-sided leaf area index** (square meters of leaf surfaces per square meter of underlying ground). The one-sided leaf area index measures a canopy area density and is calculated by integrating the **foliage area density**, which is the area of plant surface per unit volume of air, from the ground to the canopy height. The upper limit of L_T may be $7f_v$, where f_v is the fraction of land surface covered by vegetation. Deardorff (1978) argued that, at higher values of L_T, insufficient light is available to support further growth in a canopy. Table 8.1 gives $z_{0,m}$ for several surfaces, including those with vegetation. Canopy heights and upper limits to leaf area indices ($L_{T,\max}$) are given where applicable.

Example 8.3

If $T = 288$ K, $p_d = 1013$ hPa, $u_* = 0.165$ m s^{-1}, and the air is dry, estimate $z_{0,m}$ over a smooth ocean. Find $z_{0,m}$ over a rough sea when $u_* = 0.5$ m s^{-1}.

SOLUTION

From (4.54), $\eta_a = 0.0000179$ kg m^{-1} s^{-1}; from (2.23), $\rho_a = \rho_d = 1.225$ kg m^{-3}; and from (4.55), $\nu_a = 0.0000146$ m^2 s^{-1}. Thus, from (8.9), $z_{0,m} \approx 1 \times 10^{-5}$ m over a smooth sea. From (8.10), $z_{0,m} \approx 0.0004$ m over a rough sea.

The surface roughness length for momentum characterizes the ability of surface elements to absorb momentum. Surface roughness lengths for energy and water vapor characterize the ability of elements protruding from the surface to absorb energy and moisture, relative to their ability to absorb momentum (Kaimal and Finnigan 1994). Both lengths are integration constants, used to derive vertical profiles of potential virtual temperature and specific humidity, respectively, in the surface layer. Brutsaert (1991) gives several parameterizations of $z_{0,h}$ and $z_{0,v}$ based on measurements over different terrain. More general expressions for the **energy and moisture roughness lengths** are

$$z_{0,h} = \frac{D_h}{ku_*} \qquad z_{0,v} = \frac{D_v}{ku_*} \tag{8.13}$$

respectively (Garratt and Hicks 1973), where D_h is the **molecular thermal diffusion coefficient** (molecular thermal diffusivity), D_v is the **molecular diffusion coefficient of water vapor**, and k is the **von Kármán constant.** The von Kármán constant has a value between 0.35 and 0.43 (e.g., Hogstrom 1988). A value of 0.40 is assumed

Table 8.1 Aerodynamic roughnesses ($z_{0,m}$), structure or canopy heights (h_c), displacement heights (d_c), and maximum one-sided leaf area indices ($L_{T,max}$) for several surfaces

Surface type	$z_{0,m}$(m)	h_c(m)	d_c(m)	$L_{T,max}$ ($m^2\ m^{-2}$)	**Reference**
Smooth sea	0.00001				(8.9)
Rough sea	0.000015–0.0015				(8.10)
Ice	0.00001				Oke (1978)
Snow	0.00005–0.0001				Oke (1978)
Level desert	0.0003				Sehmel (1980)
Short grass	0.003–0.01	0.02–0.1	<0.075		Oke (1978)
Long grass	0.04–0.1	0.25–1.0	0.19–0.75		Oke (1978)
Savannah	0.4	8	4.8		Garratt (1992)
Agricultural crops	0.04–0.2	0.4–2	0.27–1.3		Oke (1978)
Orchard	0.5–1.0	5–10	3.3–6.7		Oke (1978)
Coniferous forest	0.28–3.9	10.4–27.5	6.3–25.3		Jarvis *et al.* (1976)
Tropical forest	2.2	35	29.8		Shuttleworth (1989)
Broadleaf evergreen forest	4.8	35	26.3	7	Sellers *et al.* (1996)
Broadleaf deciduous trees	2.7	20	15	7	Sellers *et al.* (1996)
Broad- and needleleaf trees	2.8	20	15	7.5	Sellers *et al.* (1996)
Needleleaf evergreen trees	2.4	17	12.8	8	Sellers *et al.* (1996)
Needleleaf deciduous trees	2.4	17	12.8	8	Sellers *et al.* (1996)
Short vegetation/C4 grassland	0.12	1	0.75	5	Sellers *et al.* (1996)
Broadleaf shrubs w/bare soil	0.06	0.5	0.38	5	Sellers *et al.* (1996)
Dwarf trees and shrubs	0.07	0.6	0.45	5	Sellers *et al.* (1996)
Agriculture/C3 grassland	0.12	1	0.75	5	Sellers *et al.* (1996)
2500-m^2 lot with a building 8 m high and 160-m^2 silhouette	0.26	8			(8.11)
25 000-m^2 lot with a building 80 m high and 3200-m^2 silhouette	5.1	80			(8.11)

Displacement heights in most cases were calculated as $0.75 \times h_c$.

here. Expressions for D_h and D_v (m^2 s^{-1}) are

$$D_h = \frac{\kappa_a}{\rho_a c_{p,m}} \qquad D_v = 2.11 \times 10^{-5}\left(\frac{T}{273.15\,K}\right)^{1.94}\left(\frac{1013.25\,hPa}{p_a}\right) \tag{8.14}$$

respectively (Pruppacher and Klett 1997), where κ_a is the thermal conductivity of moist air (J m^{-1} s^{-1} K^{-1}), ρ_a is the density of air (kg m^{-3}), $c_{p,m}$ is the specific heat of moist air at constant pressure (J kg^{-1} K^{-1}), T is temperature (K), and p_a is air pressure (hPa). For very smooth surfaces, $z_{0,m}$ is slightly less than $z_{0,h}$ and $z_{0,v}$. For rough surfaces, $z_{0,m}$ is up to 10^5 times larger than $z_{0,h}$ and $z_{0,v}$ (Brutsaert 1991). A typical ratio is about 100 (Garratt and Hicks 1973). For rough surfaces, surface elements are more efficient at absorbing momentum than they are at absorbing energy or water vapor.

Example 8.4

Assuming $u_* = 0.4$ m s^{-1}, $p_d = 1013$ hPa, and the air is dry, calculate the surface roughness lengths for energy and water vapor when $T = 298$ K.

SOLUTION

At $T = 298$ K, $\kappa_a \approx \kappa_d = 0.0256$ J m^{-1} s^{-1} K^{-1} from (2.5) and $\rho_a \approx \rho_d = 1.18$ kg m^{-3} from (2.23). Thus, $D_h = 2.16 \times 10^{-5}$ m^2 s^{-1} and $D_v = 2.50 \times 10^{-5}$ m^2 s^{-1} from (8.14). Substituting these values into (8.13) gives $z_{0,h} = 0.000135$ m and $z_{0,v} = 0.000156$ m, respectively.

8.4 PARAMETERIZATIONS OF KINEMATIC TURBULENT FLUXES

Most atmospheric turbulence is produced by mechanical shear and buoyancy in the form of large eddies. Large eddies are the size of the boundary layer, which is on the order of hundreds of meters to three thousand meters in diameter. This size range of eddies is called the **energy-containing subrange**. Eddies carry **turbulent kinetic energy** (TKE), which is the mean kinetic energy per unit mass associated with eddies in turbulent flow. Large eddies shrink over time due to **dissipation**, which is the conversion of turbulence into heat by molecular viscosity. As a large eddy shrinks to a small eddy, TKE stored in the eddy decreases, ultimately to zero, in a process called **inertial cascade**. Thus, TKE is not conserved over time. The rate of TKE transfer to small eddies is generally proportional to the dissipation rate. The dissipation rate is greatest for the smallest eddies, which are on the order of millimeters in diameter. Medium-sized eddies, which are created by dissipation of large eddies but equally destroyed by dissipation to small eddies, are in the **inertial subrange** of eddy size.

Atmospheric models with horizontal grid size greater than a few hundred meters on edge do not resolve the large eddies in turbulence. These models need to parameterize kinematic turbulent fluxes of momentum, sensible heat, and moisture, defined in (8.3), (8.5), and (8.7), respectively, used in the momentum, thermodynamic energy, and water vapor continuity equations. Such models are referred to as **Reynolds-averaged models** (Section 3.3.1).

In the other extreme are models that resolve eddies of all sizes. The smallest eddies in the atmosphere must be larger than the **Kolmogorov scale,**

$$\eta_k = \left(\frac{\nu_a^3}{\varepsilon_d}\right)^{1/4} \tag{8.15}$$

(m) which is the length scale of turbulent motion below which the effects of molecular viscosity are nonnegligible. In this equation, ν_a is the kinematic viscosity of air from (4.55) (m^2 s^{-1}) and ε_d is the **dissipation rate of turbulent kinetic energy** (m^2 s^{-3}), which is the rate of conversion of turbulence into heat by molecular viscosity. A typical cloud-free dissipation rate is 0.0005 m^2 s^{-3}. Example 8.5 shows that this dissipation rate gives a Kolmogorov scale of about 1.6 mm. Thus, the smallest eddies in the atmosphere are a few millimeters in diameter. Models that have dimension on the order of the Kolmogorov scale resolve all eddies and are called **direct numerical simulation** (DNS) models. Since these models resolve turbulence, they do not parameterize it. DNS models generally solve the unsteady, incompressible equations of atmosphere dynamics (e.g., Kim *et al.* 1987; Joslin *et al.* 1993; Coleman 1999).

Example 8.5

Calculate a typical Kolmogorov scale in the cloud-free atmosphere when $T = 288$ K and $p_d = 1013$ hPa.

SOLUTION

From Example 8.3, $\nu_a = 0.0000146$ m^2 s^{-1} at the temperature and pressure given. A typical free-atmosphere turbulent dissipation rate is $\varepsilon_d = 0.0005$ m^2 s^{-3}. Thus, from (8.15), the Kolmogorov scale is about 1.6 mm.

Models that have resolution between a few hundred meters and a few meters (the scale of the inertial subrange) resolve large eddies but not small ones. These models are referred to as **large-eddy simulation** (LES) models (e.g., Deardorff 1972; Moeng 1984). They resolve turbulence on the large scales but need to parameterize it on the small scales.

Reynolds-averaged models must parameterize kinematic turbulent fluxes. Ideally, these equations are solved prognostically. However, prognostic equations for kinematic turbulent fluxes contain more unknowns than equations available to solve for them. When a new equation is written for an unknown, the equation creates more unknowns forcing at least some terms in the equation to be parameterized as a function of known or derived variables. The inability to close equations for turbulent flux terms without parameterizing some terms is called the problem of closure. The **order of closure** is the highest order of a prognostic equation retained in a parameterization. If, for example, a kinematic turbulent flux equation is prognostic to first order, and all higher-order terms are parameterized, the equation is a first-order closure equation. Stull (1988) contains more details about the closure problem.

In the surface layer, kinematic turbulent fluxes are often parameterized with bulk aerodynamic formulae or Monin–Obukhov similarity theory. Both methods are **zero-order closure** techniques in that the resulting equations are fully parameterized and have no prognostic parts. Gradient transport theory (K-theory) is generally a **first-order closure** technique. K-theory assumes that the kinematic turbulent flux of a quantity is negatively proportional to a constant diffusion coefficient multiplied by the gradient of the mean value of the quantity. When bulk aerodynamic formulae or scaling parameters from similarity theory are used to calculate eddy diffusion coefficients in a K-theory expression for kinematic turbulent flux, the result is also a first-order closure equation. When the diffusion coefficient in K-theory is solved as a function of several other variables, the resulting K-theory expressions can be made second order or higher. In the next subsections, closure techniques of several orders are described.

8.4.1 Bulk aerodynamic formulae

Bulk aerodynamic formulae are equations for surface-layer kinematic vertical turbulent fluxes that assume a constant drag coefficient. A constant drag coefficient is most applicable when the boundary layer is well mixed, the surface layer is thin, and wind speeds, potential virtual temperature, and moisture do not change significantly with height above the surface layer. Below, parameterizations of kinematic vertical turbulent fluxes and eddy diffusion coefficients from bulk aerodynamic theory are given for momentum, energy, and moisture.

8.4.1.1 *Momentum fluxes*

In the absence of wind shear above the thin surface layer, the vertical turbulent transfer of horizontal momentum is affected primarily by skin drag, form drag, and wave drag. **Skin drag** is the near-surface drag that results from molecular diffusion of momentum across the surface–air interface. **Form drag** is the near-surface drag resulting from turbulence and vertical momentum transfer that occurs when winds hit large obstacles, such as rocks or trees. **Wave drag** is the near-surface drag that results from vertical transfer of momentum by gravity waves, which propagate vertically and horizontally, as discussed in Chapter 4. These three types of drag are embodied in the dimensionless coefficient of drag C_D, which is used in expressions for **bulk aerodynamic kinematic turbulent momentum fluxes** in the surface layer,

$$(\overline{w'u'})_s = -C_D|\bar{\mathbf{v}}_h(z_r)|[\bar{u}(z_r) - \bar{u}(z_{0,m})] \tag{8.16}$$

$$(\overline{w'v'})_s = -C_D|\bar{\mathbf{v}}_h(z_r)|[\bar{v}(z_r) - \bar{v}(z_{0,m})] \tag{8.17}$$

In these equations, $\bar{u}(z_{0,m}) = \bar{v}(z_{0,m}) = 0$ are the mean scalar velocities at height $z_{0,m}$, z_r is a **reference height** above the surface (usually 10 m), and $|\bar{\mathbf{v}}_h(z_r)|$ is the mean horizontal wind speed at the reference height. The coefficient of drag, which is evaluated at the reference height, ranges from 0.001 for smooth surfaces to 0.02 for rough surfaces. Over the ocean a typical value is 0.0015. In tropical storms and hurricanes, C_D varies with wind speed over the ocean, since increased wind speeds increase ocean surface roughness (e.g., Krishnamurti *et al.* 1998).

From (4.67), the kinematic vertical turbulent fluxes of momentum in the surface layer from *K*-theory were defined as

$$(\overline{w'u'})_s = -K_{m,zx}\frac{\partial \bar{u}}{\partial z} \qquad (\overline{w'v'})_s = -K_{m,zy}\frac{\partial \bar{v}}{\partial z} \tag{8.18}$$

The finite-difference form of the scalar-velocity gradient in each case is

$$\frac{\partial \bar{u}}{\partial z} = \frac{\bar{u}(z_r) - \bar{u}(z_{0,m})}{z_r - z_{0,m}} \qquad \frac{\partial \bar{v}}{\partial z} = \frac{\bar{v}(z_r) - \bar{v}(z_{0,m})}{z_r - z_{0,m}} \tag{8.19}$$

respectively. Substituting (8.19), (8.16), and (8.17) into (8.18) and solving give the *K*-theory **eddy diffusion coefficient for momentum** ($m^2\ s^{-1}$) in the surface layer in terms of bulk aerodynamic formulae as

$$K_{m,zx} = K_{m,zy} \approx C_D\,|\bar{\mathbf{v}}_h(z_r)|\,(z_r - z_{0,m}) \tag{8.20}$$

Example 8.6

If $C_D = 0.02$ (rough surface), $u = 10\ m\ s^{-1}$, and $v = 0\ m\ s^{-1}$ at $z = 10$ m, estimate $K_{m,zx}$.

SOLUTION

Assuming $z_{0,m}$ is negligible relative to z, we have $K_{m,zx} = 2\ m^2\ s^{-1}$ from (8.20).

8.4.1.2 *Energy fluxes*

From bulk aerodynamic formulae the kinematic vertical turbulent **sensible-heat flux** in the surface layer is

$$(\overline{w'\theta'_v})_s = -C_H\,|\bar{\mathbf{v}}_h(z_r)|\,[\bar{\theta}_v(z_r) - \bar{\theta}_v(z_{0,h})] \tag{8.21}$$

where C_H is the **bulk heat-transfer coefficient** or **Stanton number** (dimensionless), and $\bar{\theta}_v(z_{0,h})$ is the mean potential virtual temperature at height $z_{0,h}$. In a model, $\bar{\theta}_v$ at the ground is often used instead of $\bar{\theta}_v(z_{0,h})$ for convenience. The ground value of $\bar{\theta}_v$ can be estimated with a soil model, as discussed in Section 8.6. C_H depends on the molecular thermal diffusivity across the surface–air interface (skin drag). Its value ranges from 0.001 for smooth surfaces to 0.02 for rough surfaces. C_H differs from C_D because momentum is affected by skin, form, and wave drag while energy is affected primarily by skin drag.

Substituting (8.21) into (8.4) gives the vertical turbulent sensible heat flux near the surface as

$$H_f \approx \rho_a c_{p,d} C_H \,|\bar{\mathbf{v}}_h(z_r)|\, [\bar{\theta}_v(z_{0,h}) - \bar{\theta}_v(z_r)] \tag{8.22}$$

(W m^{-2}). This equation states that energy is transferred upward when $\bar{\theta}_v$ at $z_{0,h}$ is greater than that at z_r.

From (3.73), the kinematic vertical turbulent sensible-heat flux in the surface layer from *K*-theory was defined as

$$(\overline{w'\theta'_v})_s = -K_{h,zz}\frac{\partial \bar{\theta}_v}{\partial z} \tag{8.23}$$

A finite-difference form of the gradient of potential virtual temperature is

$$\frac{\partial \bar{\theta}_v}{\partial z} = \frac{\bar{\theta}_v(z_r) - \bar{\theta}_v(z_{0,h})}{z_r - z_{0,h}} \tag{8.24}$$

Substituting (8.21) and (8.24) into (8.23) gives the *K*-theory **vertical eddy diffusion coefficient for energy** (m^2 s^{-1}) in the surface layer in terms of bulk aerodynamic formulae as

$$K_{h,zz} \approx C_H|\bar{\mathbf{v}}_h(z_r)|(z_r - z_{0,h}) \tag{8.25}$$

8.4.1.3 *Moisture fluxes*

A bulk aerodynamic equation for the kinematic vertical turbulent water-vapor flux is

$$(\overline{w'q'_v})_s = -C_E|\bar{\mathbf{v}}_h(z_r)|\, [\bar{q}_v(z_r) - \bar{q}_v(z_{0,v})] \tag{8.26}$$

where C_E (dimensionless) is the **bulk transfer coefficient for water vapor** and $\bar{q}_v(z_{0,v})$ is the mean specific humidity at height $z_{0,v}$. In a model, the specific humidity in the top molecular soil or water surface layers is often used instead of $\bar{q}_v(z_{0,v})$. C_E is usually set equal to C_H. Kinematic vertical turbulent fluxes of gases other than water vapor and aerosol particles are written in a manner similar to (8.26).

Substituting (8.26) into (8.6) gives the vertical turbulent moisture flux near the surface as

$$E_f \approx \rho_a C_E|\bar{\mathbf{v}}_h(z_r)|\, [\bar{q}_v(z_{0,v}) - \bar{q}_v(z_r)] \tag{8.27}$$

(kg m^{-2} s^{-1}). This equation states that moisture is transferred upward when $\bar{q}_v$ at $z_{0,v}$ is greater than that at z_r. The vertical eddy diffusion coefficient for moisture is assumed to be the same as that for energy since C_E is set to C_H.

8.4.2 Monin–Obukhov similarity theory

In the presence of strong wind shear above the surface, bulk aerodynamic formulae are not useful. A better method of parameterizing kinematic fluxes near the surface is with Monin–Obukhov similarity theory. **Similarity theory** is a method by which variables are first combined into dimensionless groups. Experiments are then conducted to obtain values for each variable in the dimensionless group. The dimensionless group, as a whole, is then fitted, as a function of some parameter, with an empirical equation. The experiment is repeated. Usually, equations obtained from later experiments are similar to those from the first experiment. Hence, this method of obtaining an empirical equation for the dimensionless group is called similarity theory, and the relationship between the empirical equation and the dimensionless group is a **similarity relationship**. When similarity theory is applied to the surface layer, it is usually called **Monin–Obukhov similarity theory** or **surface-layer similarity theory** (Monin and Obukhov 1954; Stull 1988).

In this subsection, kinematic vertical turbulent fluxes and eddy diffusion coefficients are derived from similarity theory. The derivation requires the discussion of two similarity relationships and some parameters, discussed first.

8.4.2.1 *Dimensionless wind shear*

One similarity relationship is that for the **dimensionless wind shear**,

$$\frac{\phi_m}{k} = \frac{z}{u_*}\frac{\partial|\bar{\mathbf{v}}_\mathrm{h}|}{\partial z} \tag{8.28}$$

The right side of this equation, as a whole, is dimensionless. Individual factors, such as the wind shear ($\partial|\bar{\mathbf{v}}_\mathrm{h}|/\partial z$) and u_*, are found from field experiments. Wind shear is measured directly, and u_* is found from (8.8), the terms of which are measured. The parameter ϕ_m/k is determined as a function of z/L by substituting measurements of $\partial|\bar{\mathbf{v}}_\mathrm{h}|/\partial z$ and u_* into (8.28) for different values of z/L and fitting curves to the resulting data. L is the **Monin–Obukhov length** (m), discussed shortly, and z/L is a dimensionless group. The von Kármán constant k is found by substituting measurements of $\partial|\bar{\mathbf{v}}_\mathrm{h}|/\partial z$ and u_* into (8.28) under neutral conditions, when $\phi_m = 1$, then solving for k. Businger *et al.* (1971) derived ϕ_m from field data when $k = 0.35$ as

$$\phi_m = \begin{cases} 1 + \beta_m \dfrac{z}{L} & \dfrac{z}{L} > 0 \quad \text{stable} \\ \left(1 - \gamma_m \dfrac{z}{L}\right)^{-1/4} & \dfrac{z}{L} < 0 \quad \text{unstable} \\ 1 & \dfrac{z}{L} = 0 \quad \text{neutral} \end{cases} \tag{8.29}$$

where $\beta_m = 4.7$ and $\gamma_m = 15.0$. When $k = 0.4$, the values $\beta_m = 6.0$ and $\gamma_m = 19.3$ should be used instead to obtain the same values of ϕ_m as when $k = 0.35$ (Hogstrom 1988). Equation (8.29) was derived for the range, $|z/L| < 2$, but has been used successfully beyond the range under unstable conditions (San Jose *et al.* 1985). Other similarity relationships for ϕ_m include those by Dyer (1974) and Dyer and Bradley (1982), among others.

Integrating both sides of (8.28) between $z_{0,m}$ and z_r, solving for u_*, and noting that $|\bar{\mathbf{v}}_h(z_{0,m})| = 0$ give

$$u_* = \frac{k|\bar{\mathbf{v}}_h(z_r)|}{\displaystyle\int_{z_{0,m}}^{z_r} \phi_m \frac{dz}{z}} \tag{8.30}$$

where

$$\int_{z_{0,m}}^{z_r} \phi_m \frac{dz}{z}$$

$$= \begin{cases} \ln\dfrac{z_r}{z_{0,m}} + \dfrac{\beta_m}{L}(z_r - z_{0,m}) & \dfrac{z}{L} > 0 \quad \text{stable} \\ \ln\dfrac{\left(1-\gamma_m\dfrac{z_r}{L}\right)^{1/4} - 1}{\left(1-\gamma_m\dfrac{z_r}{L}\right)^{1/4} + 1} - \ln\dfrac{\left(1-\gamma_m\dfrac{z_{0,m}}{L}\right)^{1/4} - 1}{\left(1-\gamma_m\dfrac{z_{0,m}}{L}\right)^{1/4} + 1} & \\ +2\tan^{-1}\left(1-\gamma_m\dfrac{z_r}{L}\right)^{1/4} - 2\tan^{-1}\left(1-\gamma_m\dfrac{z_{0,m}}{L}\right)^{1/4} & \dfrac{z}{L} < 0 \quad \text{unstable} \\ \ln\dfrac{z_r}{z_{0,m}} & \dfrac{z}{L} = 0 \quad \text{neutral} \end{cases} \tag{8.31}$$

This integral exceeds zero and increases with increasing z/L. Thus, for the same wind speed, u_* from (8.30) is larger in unstable air ($z/L < 0$) than in stable air ($z/L > 0$). Increasing values of $z_{0,m}$ also increase u_* by decreasing the integral in (8.31).

8.4.2.2 *Monin–Obukhov length*

The **Monin–Obukhov length** (L) is a length scale (m) proportional to the height above the surface at which buoyant production of turbulence first dominates mechanical (shear) production of turbulence. Mathematically,

$$L = -\frac{u_*^3 \bar{\theta}_v}{kg\left(\overline{w'\theta'_v}\right)_s} = \frac{u_*^2 \bar{\theta}_v}{kg\theta_*} \tag{8.32}$$

where θ_* is a **potential temperature scale** (K), discussed shortly, and the second expression is derived by substituting the similarity-theory approximation

$$\left(\overline{w'\theta'_v}\right)_s \approx -u_*\theta_* \tag{8.33}$$

into the first expression.

The parameter θ_* is proportional to $\bar{\theta}_v(z_r) - \bar{\theta}_v(z_{0,h})$, the vertical difference in potential virtual temperature. The greater $\bar{\theta}_v$ at $z_{0,h}$ in comparison with its value at z_r, the more negative the change in $\bar{\theta}_v$ with increasing height, and the greater the instability of the surface layer. In such cases, L is negative but has a small magnitude, since it is inversely proportional to θ_*. When L is negative with a small magnitude, z/L is negative with a large magnitude. Such values of z/L correspond to large instability due to buoyancy. Positive values of z/L correspond to increasing $\bar{\theta}_v$ with altitude and stable stratification.

8.4.2.3 *Dimensionless potential temperature gradient*

An expression for the potential temperature scale, θ_*, can be obtained from a similarity relationship for the **dimensionless potential temperature gradient**,

$$\frac{\phi_h}{k} \approx \frac{z}{\theta_*}\frac{\partial \bar{\theta}_v}{\partial z} \tag{8.34}$$

where $\partial\bar{\theta}_v/\partial z$ is the change in mean potential virtual temperature with height. Businger *et al.* (1971) performed experiments to find ϕ_h for different stability regimes when $\bar{\theta}_p$ was used instead of $\bar{\theta}_v$ and $k = 0.35$. The resulting parameterization was

$$\phi_h = \begin{cases} \mathrm{Pr_t} + \beta_h \dfrac{z}{L} & \dfrac{z}{L} > 0 \quad \text{stable} \\ \mathrm{Pr_t}\left(1 - \gamma_h \dfrac{z}{L}\right)^{-1/2} & \dfrac{z}{L} < 0 \quad \text{unstable} \\ \mathrm{Pr_t} & \dfrac{z}{L} = 0 \quad \text{neutral} \end{cases} \tag{8.35}$$

where $\beta_h = 4.7$, $\gamma_h = 9.0$, and

$$\mathrm{Pr_t} = \frac{K_{m,zx}}{K_{h,zz}} \tag{8.36}$$

is the **turbulent Prandtl number**, which approximates the ratio of the eddy diffusion coefficient for momentum to that for energy. For $k = 0.35$, Businger *et al.* estimated $\mathrm{Pr_t} \approx 0.74$. Hogstrom (1988) noted that, when $k = 0.4$, Businger *et al.*'s constants should be modified to $\beta_h = 7.8$, $\gamma_h = 11.6$, and $\mathrm{Pr_t} \approx 0.95$ to obtain the same relationship as when $k = 0.35$. Integrating both sides of (8.34) between $z_{0,h}$ and z_r and solving for θ_* give

$$\theta_* = \frac{k[\bar{\theta}_v(z_r) - \bar{\theta}_v(z_{0,h})]}{\displaystyle\int_{z_{0,h}}^{z_r} \phi_h \frac{dz}{z}} \tag{8.37}$$

where

$$\int_{z_{0,h}}^{z_r} \phi_h \frac{dz}{z}$$

$$= \begin{cases} \mathrm{Pr_t} \ln \dfrac{z_r}{z_{0,h}} + \dfrac{\beta_h}{L}(z_r - z_{0,h}) & \dfrac{z}{L} > 0 \quad \text{stable} \\ \mathrm{Pr_t}\left[\ln \dfrac{\left(1-\gamma_h \dfrac{z_r}{L}\right)^{1/2} - 1}{\left(1-\gamma_h \dfrac{z_r}{L}\right)^{1/2} + 1} - \ln \dfrac{\left(1-\gamma_h \dfrac{z_{0,h}}{L}\right)^{1/2} - 1}{\left(1-\gamma_h \dfrac{z_{0,h}}{L}\right)^{1/2} + 1}\right] & \dfrac{z}{L} < 0 \quad \text{unstable} \\ \mathrm{Pr_t} \ln \dfrac{z_r}{z_{0,h}} & \dfrac{z}{L} = 0 \quad \text{neutral} \end{cases} \tag{8.38}$$

In sum, u_* can be determined from (8.30) and (8.31) if L is known. If L is not known, L, u_*, and θ_* must be determined simultaneously by solving (8.30), (8.32), and (8.37).

8.4.2.4 *Noniterative parameterization for momentum and potential temperature scales*

A noniterative method of determining u_*, θ_*, and L is with the parameterization of Louis (1979). The first step is to calculate the **bulk Richardson number,**

$$\mathrm{Ri_b} = \frac{g[\bar{\theta}_v(z_r) - \bar{\theta}_v(z_{0,h})](z_r - z_{0,m})^2}{\bar{\theta}_v(z_{0,h})[\bar{u}(z_r)^2 + \bar{v}(z_r)^2](z_r - z_{0,h})} \tag{8.39}$$

which quantifies the ratio of buoyancy to mechanical shear. The second step is to calculate u_* and θ_* as

$$u_* \approx \frac{k|\bar{\mathbf{v}}_h(z_r)|}{\ln(z_r/z_{0,m})}\sqrt{G_m} \qquad \theta_* \approx \frac{k^2|\bar{\mathbf{v}}_h(z_r)|\left[\bar{\theta}_v(z_r) - \bar{\theta}_v(z_{0,h})\right]}{u_* \mathrm{Pr_t} \ln^2(z_r/z_{0,m})} G_h \tag{8.40}$$

where

$$G_m = 1 - \frac{9.4\mathrm{Ri_b}}{1 + \dfrac{70k^2(|\mathrm{Ri_b}|\, z_r/z_{0,m})^{0.5}}{\ln^2(z_r/z_{0,m})}} \qquad \mathrm{Ri_b} \le 0$$

$$G_h = 1 - \frac{9.4\mathrm{Ri_b}}{1 + \dfrac{50k^2(|\mathrm{Ri_b}|\, z_r/z_{0,m})^{0.5}}{\ln^2(z_r/z_{0,m})}} \qquad \mathrm{Ri_b} \le 0$$

$$G_m, G_h = \frac{1}{(1 + 4.7\mathrm{Ri_b})^2} \qquad \mathrm{Ri_b} > 0 \tag{8.41}$$

The surface roughness length for momentum is used for momentum and energy terms in (8.41) to maintain consistency. From values of u_* and θ_*, L can be determined diagnostically with (8.32).

Example 8.7

Given

$$z_{0,m} = 0.01\,\text{m} \qquad z_{0,h} = 0.0001\,\text{m} \qquad z_\text{r} = 10\,\text{m}$$
$$\bar{u}(z_\text{r}) = 10\,\text{m s}^{-1} \qquad \bar{v}(z_\text{r}) = 5\,\text{m s}^{-1} \qquad k = 0.4$$
$$\bar{\theta}_\text{v}(z_\text{r}) = 285\,\text{K} \qquad \bar{\theta}_\text{v}(z_{0,h}) = 288\,\text{K} \qquad \text{Pr}_\text{t} = 0.95$$

calculate u_*, θ_*, and L.

SOLUTION

From (8.39)–(8.41),

$$|\bar{\mathbf{v}}_\text{h}(z_\text{r})| = 11.18\,\text{m s}^{-1} \qquad \text{Ri}_\text{b} = -8.15 \times 10^{-3} \qquad G_m = 1.046$$
$$G_h = 1.052 \qquad u_* = 0.662\,\text{m s}^{-1} \qquad \theta_* = -0.188\,\text{K}$$
$$L = -1.69\,\text{m}$$

8.4.2.5 *Gradient Richardson number*

The bulk Richardson number is used for practical application in meteorological modeling. It is derived from the **gradient Richardson number,**

$$\text{Ri}_\text{g} = \frac{\dfrac{g}{\bar{\theta}_\text{v}}\dfrac{\partial \bar{\theta}_\text{v}}{\partial z}}{\left(\dfrac{\partial \bar{u}}{\partial z}\right)^2 + \left(\dfrac{\partial \bar{v}}{\partial z}\right)^2} \tag{8.42}$$

The gradient and bulk Richardson numbers give the ratio of turbulence due to buoyancy relative to that due to shear. When $\text{Ri}_\text{g}, \text{Ri}_\text{b} < 0$, the potential virtual temperature decreases with increasing altitude, and the atmosphere is buoyantly unstable and turbulent. When $\text{Ri}_\text{g}, \text{Ri}_\text{b}$ are small and negative, wind shear is large in comparison with buoyancy, and turbulence due to mechanical shear dominates turbulence due to buoyancy. When $\text{Ri}_\text{g}, \text{Ri}_\text{b}$ are large and negative, turbulence due to free convection dominates turbulence due to forced convection. When $\text{Ri}_\text{g}, \text{Ri}_\text{b} > 0$, the potential virtual temperature gradient exceeds zero, the atmosphere is buoyantly stable, and turbulence due to free convection does not occur. When $\text{Ri}_\text{g}, \text{Ri}_\text{b}$ are small and positive, wind shear is large in comparison with buoyant stability, and turbulence due to forced convection occurs. When $\text{Ri}_\text{g}, \text{Ri}_\text{b}$ are large and positive, wind shear is low in comparison with buoyant stability, and turbulence due to forced or free convection does not occur. Instead, air flow is laminar. Table 8.2 summarizes the flow regimes obtained from different Richardson numbers.

Table 8.2 Characteristics of vertical flow of air for different values of Ri_b and Ri_g

Ri_b or Ri_g	Type of flow	Level of turbulence due to buoyancy	Level of turbulence due to shear
Large, negative	Turbulent	Large	Small
Small, negative	Turbulent	Small	Large
Small, positive	Turbulent	None (weakly stable)	Large
Large, positive	Laminar	None (strongly stable)	Small

The level of turbulence due to buoyancy is relative to that due to shear, and vice versa.

When Ri_b and Ri_g are large and positive and decrease to less than a **critical Richardson number** (Ri_c) of 0.25, laminar flow becomes turbulent and wind shear increases. When Ri_b and Ri_g are small and positive and increase to above the **termination Richardson number** (Ri_T) of 1.0, turbulent flow becomes laminar.

8.4.2.6 *Momentum fluxes from similarity theory*

Kinematic vertical turbulent fluxes of momentum and eddy diffusion coefficients for momentum can be derived from similarity theory. Substituting (8.16) and (8.17) into (8.8) gives the relationship between friction wind speed and the coefficient of drag as

$$u_* = |\bar{\mathbf{v}}_h(z_r)|\sqrt{C_D} \tag{8.43}$$

Example 8.8

If $C_D = 0.001$ and $|\bar{\mathbf{v}}_h(z_r)| = 10$ m s^{-1}, then $u_* = 0.32$ m s^{-1} from (8.43). If $C_D = 0.02$ and $|\bar{\mathbf{v}}_h(z_r)| = 10$ m s^{-1}, then $u_* = 1.41$ m s^{-1}. These are typical values of u_* over smooth and rough surfaces, respectively, when the surface layer is well mixed.

Substituting the expression for C_D from (8.43) into (8.16) and (8.17) gives the **kinematic vertical turbulent fluxes of momentum from similarity theory** as

$$(\overline{w'u'})_s = -\frac{u_*^2}{|\bar{\mathbf{v}}_h(z_r)|}\bar{u}(z_r) \tag{8.44}$$

$$(\overline{w'v'})_s = -\frac{u_*^2}{|\bar{\mathbf{v}}_h(z_r)|}\bar{v}(z_r) \tag{8.45}$$

Substituting (8.44), (8.45), and (8.19) into (8.18) and solving give the **eddy diffusion coefficients for momentum in the surface layer from similarity theory** as

$$K_{m,zx} = K_{m,zy} \approx \frac{u_*^2}{|\bar{\mathbf{v}}_h(z_r)|}(z_r - z_{0,m}) \tag{8.46}$$

Further combining

$$\frac{\partial|\bar{\mathbf{v}}_\mathrm{h}|}{\partial z} \approx \frac{|\bar{\mathbf{v}}_\mathrm{h}(z_\mathrm{r})|}{z_\mathrm{r} - z_{0,m}} \tag{8.47}$$

and the dimensionless wind shear equation from (8.28) with (8.46) gives the diffusion coefficient under stable, unstable, or neutral conditions as

$$K_{m,zx} = K_{m,zy} \approx \frac{kzu_*}{\phi_m} \tag{8.48}$$

where $kz = \lambda_\mathrm{e}$ is the **mixing length of an eddy** near the surface. The eddy mixing length is the average distance an eddy travels before it exchanges momentum with surrounding eddies. Near the surface, mixing is limited by the ground. Under neutral conditions, $\phi_m = 1$, and (8.48) simplifies to $K_{m,zx} = kzu_* = K_{m,zy}$.

8.4.2.7 *Energy and moisture fluxes from similarity theory*

The kinematic vertical turbulent fluxes of sensible heat and moisture from similarity theory are

$$(\overline{w'\theta'_\mathrm{v}})_\mathrm{s} = -u_*\theta_* \tag{8.49}$$

$$(\overline{w'\mathrm{q}'_\mathrm{v}})_\mathrm{s} = -u_*\mathrm{q}_* \tag{8.50}$$

where θ_* is the potential temperature scale (K) and q_* is the water-vapor scale (kg kg^{-1}). Like θ_*, q_* is found from a similarity relationship,

$$\frac{\phi_q}{k} = \frac{z}{\mathrm{q}_*}\frac{\partial\bar{\mathrm{q}}_\mathrm{v}}{\partial z} \tag{8.51}$$

where $\phi_q \approx \phi_h$ is a **dimensionless specific humidity gradient.** Substituting $\phi_h \approx \phi_q$ into (8.51), integrating between $z_{0,v}$ and z_r, and solving for q_* give

$$\mathrm{q}_* = \frac{k\left[\bar{\mathrm{q}}_\mathrm{v}(z_\mathrm{r}) - \bar{\mathrm{q}}_\mathrm{v}(z_{0,v})\right]}{\displaystyle\int_{z_{0,v}}^{z_\mathrm{r}} \phi_h \frac{\mathrm{d}z}{z}} \tag{8.52}$$

This integral of ϕ_h is found from (8.38), but with $z_{0,v}$ substituted for $z_{0,h}$.

Substituting (8.49) and (8.50) into (8.4) and (8.6), respectively, gives the surface-layer **vertical turbulent sensible heat and moisture fluxes** as

$$H_\mathrm{f} \approx -\rho_\mathrm{a} c_{p,\mathrm{d}} u_*\theta_* \qquad E_\mathrm{f} \approx -\rho_\mathrm{a} u_*\mathrm{q}_* \tag{8.53}$$

(W m^{-2} and kg m^{-2} s^{-1}, respectively). Substituting (8.49) and (8.34) into (8.23) gives the **vertical eddy diffusion coefficient for energy** from similarity theory as

$$K_{h,zz} = \frac{u_* \theta_*}{(\partial \bar{\theta}_v / \partial z)} \approx \frac{kzu_*}{\phi_h} \tag{8.54}$$

(m^2 s^{-1}). The eddy diffusion coefficients for water vapor and other species are set equal to those for energy.

Combining (8.48) and (8.38) with $\mathrm{Pr_t} = K_{m,zx}/K_{h,zz}$ from (8.36) gives the similarity-theory expression,

$$\mathrm{Pr_t} = \frac{\phi_h}{\phi_m} \tag{8.55}$$

Another parameterization for $\mathrm{Pr_t}$ was developed by Tjernstrom (1993), who estimated

$$\mathrm{Pr_t} \approx (1 + 4.47\mathrm{Ri_g})^{1/2} \tag{8.56}$$

from turbulence observations in the boundary layer. In this equation, $\mathrm{Ri_g}$ varies from 0.01 to 10.

Example 8.9

Given the conditions in Example 8.7, find $K_{h,zz}$, $K_{m,zx}$, $K_{m,zy}$, and $K_{m,zx}/K_{h,zz}$.

SOLUTION

Substituting u_*, θ_*, $\partial \bar{u}/\partial z$, $\partial \bar{v}/\partial z$, $|\bar{\mathbf{v}}_h(z_r)|$, and $\partial \bar{\theta}_v/\partial z$ into (8.46) and (8.54) gives

$$K_{h,zz} = 0.41\ \mathrm{m^2\ s^{-1}} \qquad K_{m,zx} = 0.39\ \mathrm{m^2\ s^{-1}}$$
$$K_{m,zy} = 0.39\ \mathrm{m^2\ s^{-1}} \qquad K_{m,zx}/K_{h,zz} = 0.95$$

Thus, the Louis equations predict $\mathrm{Pr_t} = K_m/K_h$ consistently with the value of $\mathrm{Pr_t}$ used in (8.40).

8.4.2.8 *Vertical profiles of wind speed, potential virtual temperature, and moisture*

Similarity theory can be used to derive profiles of vertical wind speed, potential virtual temperature, and specific humidity in the surface layer. From (8.28) and (8.34), the vertical gradients of wind speed and potential virtual temperature in the surface layer are

$$\frac{\partial |\bar{\mathbf{v}}_{\mathrm{h}}(z)|}{\partial z} = \frac{u_*}{kz}\phi_m = \frac{u_*}{kz}[1-(1-\phi_m)] \tag{8.57}$$

$$\frac{\partial \bar{\theta}_{\mathrm{v}}}{\partial z} = \frac{\theta_*}{kz}\phi_h = \frac{\theta_*}{kz}[1-(1-\phi_h)] \tag{8.58}$$

respectively. Integrating the first equation between $z_{0,m}$ and z and the second equation between $z_{0,h}$ and z gives wind speed and potential virtual temperature versus altitude as

$$|\bar{\mathbf{v}}_{\mathrm{h}}(z)| = \frac{u_*}{k}\left[\ln\left(\frac{z}{z_{0,m}}\right) - \psi_m\right] \tag{8.59}$$

$$\bar{\theta}_{\mathrm{v}}(z) = \bar{\theta}_{\mathrm{v}}(z_{0,h}) + \mathrm{Pr}_{\mathrm{t}}\frac{\theta_*}{k}\left[\ln\left(\frac{z}{z_{0,h}}\right) - \psi_h\right] \tag{8.60}$$

respectively, where

$$\psi_m = \int_{z_{0,m}}^{z}(1-\phi_m)\frac{\mathrm{d}z}{z} \qquad \psi_h = \int_{z_{0,h}}^{z}(1-\phi_h)\frac{\mathrm{d}z}{z} \tag{8.61}$$

are **influence functions for momentum and energy**. Integrating (8.61) with values of ϕ_m and ϕ_h from (8.29) and (8.35), respectively, gives

$$\psi_m = \begin{cases} -\dfrac{\beta_m}{L}(z - z_{0,m}) & \dfrac{z}{L} > 0 \quad \text{stable} \\ \ln\dfrac{[1+\phi_m(z)^{-2}][1+\phi_m(z)^{-1}]^2}{[1+\phi_m(z_{0,m})^{-2}][1+\phi_m(z_{0,m})^{-1}]^2} & \\ -2\tan^{-1}[\phi_m(z)]^{-1} + 2\tan^{-1}[\phi_m(z_{0,m})]^{-1} & \dfrac{z}{L} < 0 \quad \text{unstable} \\ 0 & \dfrac{z}{L} = 0 \quad \text{neutral} \end{cases} \tag{8.62}$$

$$\psi_h = \begin{cases} -\dfrac{1}{\mathrm{Pr}_{\mathrm{t}}}\dfrac{\beta_h}{L}(z - z_{0,h}) & \dfrac{z}{L} > 0 \quad \text{stable} \\ 2\ln\dfrac{1+\phi_h(z)^{-1}}{1+\phi_h(z_{0,h})^{-1}} & \dfrac{z}{L} < 0 \quad \text{unstable} \\ 0 & \dfrac{z}{L} = 0 \quad \text{neutral} \end{cases} \tag{8.63}$$

The influence function for momentum accounts for the difference between a logarithmic wind speed profile and an actual profile under stable and unstable conditions. The influence function for energy is analogous to that for momentum.

Under neutral conditions, $\phi_m = 1$, and (8.59) reduces to a standard **logarithmic wind profile** for a neutrally stratified surface layer,

$$|\bar{\mathbf{v}}_{\mathrm{h}}(z)| = \frac{u_*}{k}\ln\frac{z}{z_{0,m}} \tag{8.64}$$

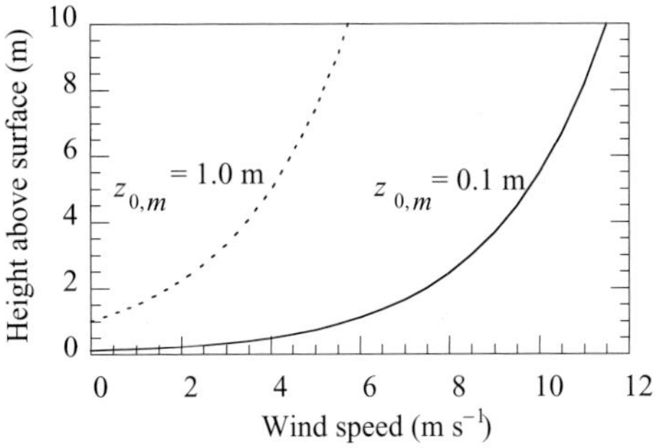

Figure 8.3 Logarithmic wind profiles in the surface layer from (8.64) when $u_* = 1$ m s^{-1}.

This equation states that the wind speed at $z_{0,m}$ is zero but increases logarithmically with altitude. Figure 8.3 shows two examples of logarithmic wind profiles.

The **vertical profile of specific humidity in the surface layer** is derived in the same way as that for potential virtual temperature. The result is

$$\bar{q}_v(z) = \bar{q}_v(z_{0,v}) + \mathrm{Pr_t}\frac{q_*}{k}\left[\ln\left(\frac{z}{z_{0,v}}\right) - \psi_h\right] \qquad (8.65)$$

where the energy influence function from (8.63) is used, but with $z_{0,v}$ instead of $z_{0,h}$. In (8.60) and (8.65), values of the potential virtual temperature and specific humidity at the ground surface are often substituted for values at the surface roughness length.

In a canopy, such as in a field of crops, an orchard, or a forest, the canopy top affects wind speed, potential virtual temperature, and water vapor more than does the ground. In the presence of a canopy, $z_{0,m}$, $z_{0,h}$, and $z_{0,v}$ are displaced a vertical distance d_c above the ground. This height is the **displacement height,** which usually lies within 70 to 80 percent of the **canopy height** h_c (Deardorff 1978; Pielke 1984; Kaimal and Finnigan 1994). When a displacement height exists, $z_{0,m}$, $z_{0,h}$, and $z_{0,v}$ are defined as heights above the displacement height, and the mean wind speed extrapolates to zero at the height $d_c + z_{0,m}$. Wind speed, potential virtual temperature, and specific humidity profiles in a canopy are redefined as

$$|\bar{\mathbf{v}}_h(z)| = \frac{u_*}{k}\left[\ln\left(\frac{z-d_c}{z_{0,m}}\right) - \psi_m\left(\frac{z-d_c}{L}\right)\right] \qquad (8.66)$$

$$\bar{\theta}_v(z) = \bar{\theta}_v(d_c + z_{0,h}) + \mathrm{Pr_t}\frac{\theta_*}{k}\left[\ln\left(\frac{z-d_c}{z_{0,h}}\right) - \psi_h\left(\frac{z-d_c}{L}\right)\right] \qquad (8.67)$$

$$\bar{q}_v(z) = \bar{q}_v(d_c + z_{0,v}) + \mathrm{Pr_t}\frac{q_*}{k}\left[\ln\left(\frac{z-d_c}{z_{0,v}}\right) - \psi_h\left(\frac{z-d_c}{L}\right)\right] \qquad (8.68)$$

respectively. Figure 8.4 shows the relationship among d_c, h_c, and $z_{0,m}$ and discusses how to calculate d_c. Table 8.1 gives values of h_c and d_c for some surfaces.

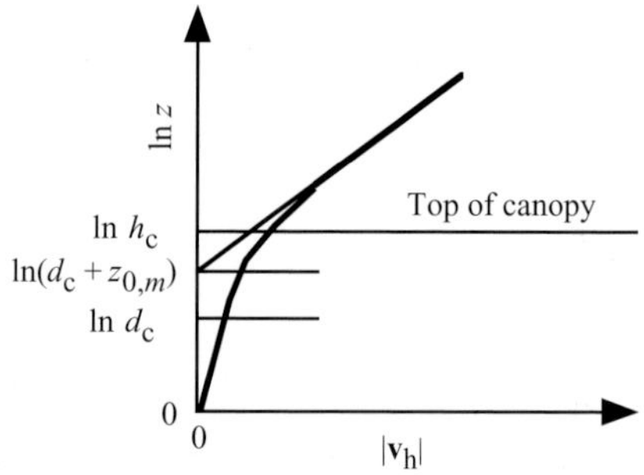

Figure 8.4 Relationship among d_c, h_c, and $z_{0,m}$. The displacement height is found by plotting wind speed over a canopy versus $\ln z$. The plotted wind speed is extrapolated to zero to obtain $d_c + z_{0,m}$. Different values of d_c are substituted into $d_c + z_{0,m}$ to estimate $z_{0,m}$. Both d_c and $z_{0,m}$ are substituted into (8.66) until the predicted curve of wind speed with height matches the logarithmic curve shown in the diagram. This method works best when winds aloft are strong.

Another common way to estimate the vertical profile of wind speed in the boundary layer is with a power-law profile,

$$|\bar{\mathbf{v}}_h(z)| = |\bar{\mathbf{v}}_h(z_r)| \left(\frac{z}{z_r} \right)^{\alpha} \tag{8.69}$$

where α is typically set to a constant, such as 1/7 (e.g., Elliott *et al.* 1986; Arya 1988). Both the power law and logarithmic profile can replicate observed wind profiles in idealized cases. However, real wind profiles are often far from ideal. For some purposes, it is thus useful to fit observed wind profiles with analytical expressions (e.g., Archer and Jacobson 2003).

8.5 EDDY DIFFUSION ABOVE THE SURFACE LAYER

Similarity theory expressions for eddy diffusion coefficients are used for the surface layer. Above the surface layer, turbulent transport of momentum and energy can be parameterized with a hybrid local/nonlocal closure scheme or with a scheme that solves the turbulent kinetic energy (TKE) equation to high order. Below, both hybrid and TKE schemes are discussed.

8.5.1 Hybrid scheme

In a hybrid model, vertical turbulent transport is treated differently for stable and weakly unstable conditions versus for strongly unstable conditions. For stable and weakly unstable conditions, turbulent transport is simulated with K-theory using an **eddy diffusion coefficient for momentum** that depends on mechanical shear and buoyancy. One such diffusion coefficient is

$$K_{m,zx} \approx K_{m,zy} \approx \lambda_e^2 \sqrt{\left(\frac{\Delta \bar{u}}{\Delta z}\right)^2 + \left(\frac{\Delta \bar{v}}{\Delta z}\right)^2} \frac{\mathrm{Ri_c} - \mathrm{Ri_b}}{\mathrm{Ri_c}} \tag{8.70}$$

(m^2 s^{-1}) (e.g., Blackadar 1976; Stull 1988), where z is in meters and

$$\lambda_e = \frac{kz}{1 + kz/\lambda_m} \tag{8.71}$$

is an expression for the mixing length that simplifies to kz near the surface (small z) and to the **free-atmospheric mixing length,** λ_m = 70–200 m, above the surface layer (large z). Near the surface, turbulence and mixing length are limited by the ground. In the free atmosphere, turbulence is limited by a maximum mixing length. Equation (8.70) is applicable when $\mathrm{Ri_b} < \mathrm{Ri_c} \approx 0.25$. When $\mathrm{Ri_b} \geq \mathrm{Ri_c}$, the atmosphere is strongly stable, (8.70) predicts no turbulence (a negative value of $K_{m,zx}$), and the diffusion coefficient is set to a minimum positive value.

When $\mathrm{Ri_c} > \mathrm{Ri_b} > 0$, turbulence is due to mechanical shear, and the atmosphere is weakly stable. When $\mathrm{Ri_b}$ is small and negative, turbulence is due to shear and weak buoyancy, and the atmosphere is weakly unstable. Under both conditions, (8.70) can be used.

In a jet, wind speeds increase and then decrease with altitude, minimizing wind shear in the jet's peak. In such a case, and in the absence of buoyancy, (8.70) predicts a small $K_{m,zx}$. Some observations of turbulence near a jet's maximum speed indicate that turbulence peaks near the maximum (Lenschow *et al.* 1988; Tjernstrom 1993), while other observations indicate that turbulence is lowest near the jet maximum (e.g., Mahrt *et al.* 1979; Lenschow *et al.* 1988; Tjernstrom 1993). In cases when the turbulence peaks near the jet maximum, (8.70) may not properly predict $K_{m,zx}$ (Shir and Bornstein 1976).

The eddy diffusion coefficient for energy is related to that for momentum by $K_{m,zx} \approx K_{m,zy} \approx \mathrm{Pr_t} K_{h,zz}$. Eddy diffusion coefficients for trace gases or particles are usually set equal to those for energy.

When $\mathrm{Ri_b}$ is large and negative, the atmosphere is strongly unstable, and free convection occurs. Since (8.70) captures effects of small eddies but not large eddies that arise during free convection (Stull 1988), it is not used to simulate free convection. Instead, vertical turbulent transport under strong, unstable conditions is often simulated with a **free convective plume scheme** (Blackadar 1978; Zhang and Anthes 1982; Lu 1994). A free convective plume scheme assumes buoyant plumes from the surface rise and mix with air in every level of the boundary layer, exchanging momentum, energy, moisture, and gases. The result of mixing is to distribute these parameters evenly throughout the boundary layer in a short time. A plume scheme differs from a K-theory parameterization in that the former mixes all layers simultaneously whereas the latter mixes material between adjacent layers.

A turbulence scheme that mixes material among all layers simultaneously is called a **nonlocal closure scheme**. One that mixes material between adjacent grid layers is called a **local closure scheme**.

8.5.2 TKE schemes

Several local-closure turbulence parameterizations have been developed to predict diffusion coefficients as a function of turbulent kinetic energy (TKE) for use in K-theory equation models. In such cases, TKE itself is modeled with a time-dependent equation. Two major types of TKE models have been developed, E-λ_e and E-ε_d models, where E is TKE, λ_e is the mixing length, and ε_d is the dissipation rate of turbulent kinetic energy. The main difference between the two types is that E-λ_e models solve prognostic equations for E and λ_e, whereas E-ε_d models solve prognostic equations for E and ε_d. Both types of models are local closure models, generally of order 2, 2.5, 3, or 4 (e.g., Mellor and Yamada 1974, 1982; Andre *et al.* 1978; Enger 1986; Briere 1987; Andrén 1990; Apsley and Castro 1997; Abdella and McFarlane 1997; Liu and Leung 2001; Cheng *et al.* 2002; Freedman and Jacobson 2002, 2003).

8.5.2.1 E-λ_e *schemes*

In this subsection, the order 2.5 E-λ_e model of Mellor and Yamada (1982) is discussed. In this scheme, TKE is modeled prognostically with

$$\frac{\partial E}{\partial t} - \frac{\partial}{\partial z}\left(s_q \lambda_e \sqrt{2E}\frac{\partial E}{\partial z}\right) = P_s + P_b - \varepsilon_d \tag{8.72}$$

where $E = (u'^2 + v'^2 + w'^2)/2$ is TKE per unit mass (m^2 s^{-2}), s_q is a constant set to 0.2, λ_e is the eddy mixing length (m), $s_q\lambda_e\sqrt{2E}$ is a diffusion coefficient for turbulent transport of the gradient of TKE (m^2 s^{-1}), P_s and P_b are the production rates of TKE due to shear and buoyancy, respectively (m^2 s^{-3}), and ε_d is the dissipation rate of TKE to heat due to molecular viscosity (m^2 s^{-2} s^{-1} = m^2 s^{-3}).

The **eddy mixing length** is determined from the prognostic equation for $E\lambda_e$,

$$\frac{\partial(2E\lambda_e)}{\partial t} - \frac{\partial}{\partial z}\left(s_l \lambda_e \sqrt{2E}\frac{\partial(2E\lambda_e)}{\partial z}\right) = \lambda_e e_1(P_s + P_b) - \lambda_e\varepsilon_d\left[1 + e_2\left(\frac{\lambda_e}{kz}\right)^2\right] \tag{8.73}$$

where s_l is a constant set to 0.2, $s_l\lambda_e\sqrt{2E}$ is a diffusion coefficient for turbulent transport of the gradient of $E\lambda_e$ (m^2 s^{-1}), $e_1 = 1.8$ and $e_2 = 1.33$ are constants, $k = 0.4$ is the von Kármán constant, and z is the height above the surface (m). The mixing length is extracted from this equation with $\lambda_e = E\lambda_e/E$, where E in the denominator is determined from (8.72). As a simplification, the mixing length can also be determined diagnostically from (8.71).

In (8.72) and (8.73), the turbulent **production rates of shear and buoyancy** are

$$P_s = K_m\left[\left(\frac{\partial \bar{u}}{\partial z}\right)^2 + \left(\frac{\partial \bar{v}}{\partial z}\right)^2\right] \tag{8.74}$$

$$P_b = -\frac{g}{\bar{\theta}_v} K_h \frac{\partial \bar{\theta}_v}{\partial z} \tag{8.75}$$

respectively, where K_m and K_h are diffusion coefficients for momentum and energy ($m^2\ s^{-1}$) from the previous time step of calculation. The **dissipation rate of TKE** is

$$\varepsilon_d = \frac{(2E)^{3/2}}{B_1 \lambda_e} \tag{8.76}$$

where B_1 is a constant set to 16.6. Once E and λ_e have been calculated, new diffusion coefficients are determined with

$$K_m = S_M \lambda_e \sqrt{2E} \tag{8.77}$$

$$K_h = S_H \lambda_e \sqrt{2E} \tag{8.78}$$

where S_M and S_H are calculated from the following equations:

$$S_M = 2E\left(\frac{0.74P_3 + 0.6992P_1}{P_1P_4 - P_3P_2}\right) \tag{8.79}$$

$$S_H = 2E\left(\frac{0.74P_4 + 0.6992P_2}{P_1P_4 - P_3P_2}\right) \tag{8.80}$$

$$P_1 = 2E - 30.5916G_H \tag{8.81}$$

$$P_2 = 4.0848G_M \tag{8.82}$$

$$P_3 = 16.284G_H \tag{8.83}$$

$$P_4 = 6.1272G_H - 5.0784G_M - 2E \tag{8.84}$$

$$G_M = \frac{\lambda_e^2}{2E}\left[\left(\frac{\partial \bar{u}}{\partial z}\right)^2 + \left(\frac{\partial \bar{v}}{\partial z}\right)^2\right] \tag{8.85}$$

$$G_H = -\frac{\lambda_e^2}{2E}\frac{g}{\bar{\theta}_v}\frac{\partial \bar{\theta}_v}{\partial z} \tag{8.86}$$

The last two terms are related to the gradient Richardson number by

$$\mathrm{Ri_g} = -\frac{G_H}{G_M} \tag{8.87}$$

Diffusion coefficients are used not only to recalculate production rates of shear and buoyancy for the next time step but also to determine kinematic vertical turbulent fluxes of momentum and sensible heat.

8.5.2.2 *E*-ε_d *schemes*

The main difference between an E-λ_e and an E-ε_d scheme is that, in the latter, a prognostic equation is solved for the dissipation rate of TKE instead of for the mixing length. Prognostic equations for TKE are solved in both cases. One form of the **prognostic equation for the dissipation rate** is

$$\frac{\partial \varepsilon_d}{\partial t} - \frac{\partial}{\partial z}\left(\frac{K_m}{\sigma_\varepsilon}\frac{\partial \varepsilon_d}{\partial z}\right) = c_{\varepsilon 1}\frac{\varepsilon_d}{E}(P_s + P_b) - c_{\varepsilon 2}\frac{\varepsilon_d^2}{E} \tag{8.88}$$

where

$$K_m = c_\mu \frac{E^2}{\varepsilon_d} \tag{8.89}$$

is an expression for the eddy diffusion coefficient for momentum, and σ_ε, $c_{\varepsilon 1}$, $c_{\varepsilon 2}$, and c_μ are constants determined either from experiment or from enforcing consistency with Monin–Obukhov similarity theory (e.g., Freedman and Jacobson 2002, 2003). The mixing length for use in the TKE equation (8.72) is calculated diagnostically from the most recent value of the TKE and the dissipation rate with

$$\lambda_e = c_\mu^{3/4}\frac{E^{3/2}}{\varepsilon_d} \tag{8.90}$$

8.6 GROUND SURFACE TEMPERATURE AND SOIL MOISTURE

Ground temperature and soil moisture (liquid-water content) affect the fluxes of energy and moisture, respectively, from the ground to the boundary layer. The fluxes, in turn, affect mixing height, wind speed, and pollutant concentration. For example, low soil moisture increases ground temperature, increasing thermal turbulence, increasing mixing height, increasing wind speed, and decreasing primary pollutant concentration (Jacobson 1999a). The increase in near-surface wind speed is due to the increased turbulent transport of momentum from aloft, where the wind is fast, to the surface, where it is generally slower. In this section, factors that affect surface temperature and moisture content are discussed, and equations describing soil energy and liquid-water transport are described.

8.6.1 Factors affecting soil temperature

Soil temperature is affected by several parameters, discussed briefly below.

8.6.1.1 *Specific heat*

Dry soil contains solid soil and air (voids). Since the specific heat of dry air is lower than that of solid soil (Table 2.2), the average specific heat of soil plus air is less than that of soil alone. When liquid water is added to soil, it replaces air. Since the specific heat of liquid water is much larger than that of air, the specific heat of a soil–water–air mixture is greater than that of a soil–air mixture of the same volume. Thus, a wet, sandy soil heats up less during the day and cools down less during the night than does a dry, sandy soil when only specific heat is considered.

8.6.1.2 *Soil moisture*

Soil moisture also affects the rate of evaporation. The lower the liquid-water content of soil, the lower the rate of evaporation of water to the air (or greater the rate of condensation of water vapor to the ground), the lesser the latent heat flux to the air, and the lesser the cooling (greater the warming) of the ground due to absorption (release) of latent heat by liquid water when it evaporates (condenses).

8.6.1.3 *Conduction*

Conduction between surface soil and molecules below the surface affects soil temperature. During the night, the soil surface cools radiatively, creating a temperature gradient in the top soil layers, forcing energy stored below the surface to conduct upward to replenish the lost energy. The greater the thermal conductivity of the soil, the faster the energy transfer occurs. Table 2.2 shows that clay is more conductive than sand. At night, the replenishment of energy to a clay surface from the subsoil is faster than for a sandy surface. During the day, conduction of absorbed radiation from the surface to the subsoil is faster for clay than for sand.

8.6.1.4 *Additional factors*

Additional factors affect soil moisture and ground temperatures. Solar radiation heats the ground during the day. Infrared emission cools the ground during the day and the night. Vegetation cover reduces the solar radiation reaching the ground. Transpiration removes liquid water from deep soil covered by vegetation. The water is drawn from soil through roots and xylem (water-conducting tissue within plants) to the leaves. Water vapor escapes through plant leaf stomata (pores). Finally, a major source of liquid water in soil is precipitation.

Below, equations for calculating bare-soil temperature and moisture are given. Subsequently, equations for calculating temperature and moisture of vegetated soil and other surfaces are described.

8.6.2 Ground temperature and moisture over bare soil

Soil models predict soil surface and subsurface temperature and liquid water content by dividing the soil near the surface into multiple layers. In this subsection, equations for temperature and moisture changes among multiple soil layers are described for when the soil surface is bare.

At the soil surface, conductive, radiative, sensible, and evaporative energy fluxes affect soil temperature. Below the surface, conduction is the most important factor affecting temperature. Temperature changes due to conduction in a homogeneous soil below the surface are estimated with the **heat conduction equation,**

$$\frac{\partial T_s}{\partial t} = \frac{1}{\rho_g c_G} \frac{\partial}{\partial z}\left(\kappa_s \frac{\partial T_s}{\partial z}\right) \qquad (8.91)$$

where T_s is the soil temperature, κ_s is the thermal conductivity of the soil–water–air mixture (J m^{-1} s^{-1} K^{-1}), ρ_g is the density of the mixture (kg m^{-3}), c_G is the specific heat of the mixture (J kg^{-1} K^{-1}), and $\kappa_s \partial T_s/\partial z$ is the conductive heat flux through the soil–water–air mixture (J m^{-2} s^{-1}). The **thermal conductivity** (J m^{-1} s^{-1} K^{-1}) of a soil–water–air mixture may be approximated with

$$\kappa_s = \max\left(418 e^{-\log_{10}|\psi_p| - 2.7}, 0.172\right) \qquad (8.92)$$

(Al Nakshabandi and Konhke 1965; McCumber and Pielke 1981), where ψ_p is the moisture potential (cm) or soil water tension and "max" indicates the larger of the two values. The **moisture potential** is the potential energy required to extract water from capillary and adhesive forces in the soil. Clapp and Hornberger (1978) parameterized the moisture potential as

$$\psi_p = \psi_{p,s}\left(\frac{w_{g,s}}{w_g}\right)^b \qquad (8.93)$$

where $\psi_{p,s}$ is the moisture potential when the soil is saturated with liquid water (cm), w_g is the **volumetric water content of the soil** (**soil moisture**) in cubic meters of liquid water per cubic meter of soil–water–air mixture, $w_{g,s}$ is the maximum volumetric water content that a given soil type can hold (m^3 m^{-3}), and b is a coefficient required to fit (8.93) to data. Values of $\psi_{p,s}$, $w_{g,s}$, and b are given in Table 8.3 for different soil types. An alternative to the Clapp and Hornberger moisture potential equation is one developed by van Genuchten (1980) that has been widely used in soil-physics studies (Cuenca *et al.* 1996).

The product of mass density and specific heat of a soil–water–air mixture is

$$\rho_g c_G = (1 - w_{g,s})\rho_s c_S + w_g \rho_w c_W \qquad (8.94)$$

where ρ_s is the density of solid soil (kg m^{-3}), c_S is the specific heat of solid soil (J kg^{-1} K^{-1}), ρ_w is the density of liquid water (1000 kg m^{-3}), c_W is the specific heat of liquid water (J kg^{-1} K^{-1}), $1 - w_{g,s}$ is the volumetric content of solid soil, w_g is the volumetric content of water in the soil, and $w_{g,s} - w_g$ is the volumetric air content in the soil–water–air mixture. In the equation, the product of mass density and specific heat of air is neglected, since the mass density of air is much smaller than that of soil or water. Values of $\rho_s c_S$ are given in Table 8.3 for different soil types.

Table 8.3 Soil parameters for 11 soil types

Soil type	b	$w_{g,s}$ ($m^3\ m^{-3}$)	w_{fc} ($m^3\ m^{-3}$)	w_{wilt} ($m^3\ m^{-3}$)	$\psi_{p,s}$ (cm)	$K_{g,s}$ ($m\ s^{-1}$)	$\rho_s c_S$ ($J\ m^{-3}\ K^{-1}$)
Sand	4.05	0.395	0.135	0.068	−12.1	1.76(−4)	1.47(6)
Loamy sand	4.38	0.410	0.150	0.075	−9.0	1.56(−4)	1.41(6)
Sandy loam	4.90	0.435	0.195	0.114	−21.8	3.41(−5)	1.34(6)
Silt loam	5.30	0.485	0.255	0.179	−78.6	7.20(−6)	1.27(6)
Loam	5.39	0.451	0.240	0.155	−47.8	7.00(−6)	1.21(6)
Sandy clay loam	7.12	0.420	0.255	0.175	−29.9	6.30(−6)	1.18(6)
Silty clay loam	7.75	0.477	0.322	0.218	−35.6	1.70(−6)	1.32(6)
Clay loam	8.52	0.476	0.325	0.250	−63.0	2.50(−6)	1.23(6)
Sandy clay	10.40	0.426	0.310	0.219	−15.3	2.20(−6)	1.18(6)
Silty clay	10.40	0.492	0.370	0.283	−49.0	1.00(−6)	1.15(6)
Clay	11.40	0.482	0.367	0.286	−40.5	1.30(−6)	1.09(6)

Adapted from Clapp and Hornberger (1978), Pielke (1984), Noilhan and Planton (1989), and Mahfouf and Noilan (1996). 1.00(6) means 1.00×10^6.

The time rate of change of volumetric water content of soil below the surface can be approximated with the **water transport equation,**

$$\frac{\partial w_g}{\partial t} = \frac{\partial}{\partial z}\left[K_g\left(\frac{\partial \psi_p}{\partial z} + 1\right)\right] = \frac{\partial}{\partial z}\left(D_g\frac{\partial w_g}{\partial z} + K_g\right) \tag{8.95}$$

(McCumber and Pielke 1981) where K_g is the coefficient of permeability of liquid water through soil (hydraulic conductivity), D_g is the diffusion coefficient for water in soil, and $K_g\partial(\psi_p + z)/\partial z$ is the kinematic flux of liquid water through the soil (m [$m^3\ m^{-3}$] s^{-1}). Liquid water feeds back to soil temperature through the thermal conductivity and specific heat terms in (8.91).

The **hydraulic conductivity** ($m\ s^{-1}$), which accounts for gravity drainage through a viscous soil, is affected by water viscosity and the shapes and sizes of voids between soil particles. Clapp and Hornberger (1978) parameterized it as

$$K_g = K_{g,s}\left(\frac{w_g}{w_{g,s}}\right)^{2b+3} \tag{8.96}$$

where $K_{g,s}$ is the hydraulic conductivity at saturation ($m\ s^{-1}$), and values of b and $w_{g,s}$ are shown in Table 8.3. An alternative parameterization is given by van Genuchten (1980). The **diffusion coefficient for water in soil** ($m^2\ s^{-1}$) is

$$D_g = K_g\frac{\partial \psi_p}{\partial w_g} = -\frac{bK_{g,s}\psi_{p,s}}{w_g}\left(\frac{w_g}{w_{g,s}}\right)^{b+3} = -\frac{bK_{g,s}\psi_{p,s}}{w_{g,s}}\left(\frac{w_g}{w_{g,s}}\right)^{b+2} \tag{8.97}$$

where the second expression was obtained by substituting (8.93) and (8.96) into the first.

At the soil surface, (8.91) and (8.95) can be modified to

$$\frac{\partial T_s}{\partial t} = \frac{1}{\rho_g c_G}\frac{\partial}{\partial z}\left(\kappa_s \frac{\partial T_s}{\partial z} + F_{n,g} - H_f - L_e E_f\right) \tag{8.98}$$

$$\frac{\partial w_g}{\partial t} = \frac{\partial}{\partial z}\left(D_g \frac{\partial w_g}{\partial z} + K_g + \frac{E_f - P_g}{\rho_w}\right) \tag{8.99}$$

respectively, where $\kappa_s \partial T_s/\partial z$ (W m^{-2}) is the conductive heat flux between the soil surface and the layer of soil just below the surface, $F_{n,g}$ (W m^{-2}) is the net downward minus upward solar plus infrared radiative flux at the surface (positive $F_{n,g}$ is down), H_f (W m^{-2}) is the vertical turbulent sensible-heat flux at the soil surface (positive is up), E_f (kg m^{-2} s^{-1}) is the evaporation rate at the surface (positive is up), L_e (J kg^{-1}) is the latent heat of evaporation, and P_g (kg m^{-2} s^{-1}) is the net flux of liquid water reaching the soil surface (precipitation minus runoff) (positive is down). The product $L_e E_f$ is the net **latent heat flux** between the soil surface and the atmosphere. $F_{n,g}$ is determined from a radiative-transfer calculation (Chapter 9).

H_f and E_f can be estimated from similarity theory with (8.53) or from bulk aerodynamic equations with (8.22) and (8.27), respectively. H_f depends on the potential virtual temperature of air at height $z_{0,h}$, and E_f depends on the specific humidity of air at height $z_{0,v}$. Two methods of calculating the specific humidity at height $z_{0,v}$ are

$$\bar{q}_v(z_{0,v}) = \alpha_g q_{v,s}(T_g) \tag{8.100}$$

$$\bar{q}_v(z_{0,v}) = \beta_g q_{v,s}(T_g) + (1 - \beta_g)\bar{q}_v(z_r) \tag{8.101}$$

where $q_{v,s}(T_g)$ is the saturation specific humidity at the ground temperature, and α_g and β_g are wetness functions. The function α_g is like a relative humidity adjacent to the water in soil pores. Philip (1957) estimated $\alpha_g \approx \exp[\psi_p g/R_v T_g]$, but Wetzel and Chang (1987), Avissar and Mahrer (1988), Kondo *et al.* (1990), Lee and Pielke (1992), and Mihailovic *et al.* (1995) pointed out that this equation is incorrect. These papers and Mahfouf and Noilhan (1991) give alternative formulations for α_g. An expression for β_g is

$$\beta_g = \begin{cases} \frac{1}{4}\left[1 - \cos\left(\frac{w_g}{w_{fc}}\pi\right)\right]^2 & w_g < w_{fc} \text{ and } \bar{q}_v(z_r) \le \bar{q}_v(z_{0,v}) \\ 1 & w_g \ge w_{fc} \text{ or } \bar{q}_v(z_r) > \bar{q}_v(z_{0,v}) \end{cases} \tag{8.102}$$

(Lee and Pielke 1992), who obtained the fit from data of Kondo *et al.* (1990), where w_g (m^3 m^{-3}) is the liquid water content (LWC) of the top soil layer and w_{fc} (m^3 m^{-3}) is the **soil LWC at field capacity** (Table 8.3), which is the LWC after all macropores of soil have emptied due to gravitational drainage and liquid water remains only in soil micropores. When condensation occurs, and for water, ice, and snow surfaces, $\beta_g = 1$.

Together, (8.91), (8.95), (8.98), and (8.99) can be solved numerically among several soil layers (e.g., McCumber and Pielke, 1981). The solution is generally obtained explicitly for below-surface layers assuming (8.99) applies to the middle of the top soil layer (replacing (8.95) in that layer). At the soil surface, ground temperature is then diagnosed from the **surface energy balance equation,**

$$\kappa_{s,1}\frac{\partial T_s}{\partial z} + F_{n,g} - H_f - L_e E_f = 0 \tag{8.103}$$

which is obtained by assuming (8.98) is in steady state. In this equation, $\kappa_{s,1}$ (W m^{-1} K^{-1}) is the thermal conductivity of a soil–water–air mixture in the top model-layer of soil. The soil-top temperature from the equation is used as an upper boundary term during the next time step in the calculation of subsurface temperatures from (8.91).

An alternative to a multilayer soil model is a two-compartment soil model (e.g., Noilhan and Planton 1989; Mahfouf and Noilhan 1996). This model is a generalization of the **force-restore method,** which involves forcing the temperature and liquid-water content in the top centimeter of soil over a short time and restoring these variables with deep-soil values over a longer time (Bhumralkar 1975; Blackadar 1976; Deardorff 1977).

8.6.3 Temperature and moisture in vegetated soil

When soil is covered with vegetation, energy and moisture from the soil must pass through the vegetation before reaching the free atmosphere. Conversely, energy and moisture from the atmosphere must pass through the vegetation before reaching the ground. In this section, energy and moisture transfer through and storage within vegetation are discussed. The numerical treatment is described in more detail in Jacobson (2001a).

Vegetation, as defined here, includes trees, shrubs, grass, and plants. A **canopy** is the vegetation cover provided by a group of plants or trees. **Foliage** is a cluster of leaves within a canopy.

Ground temperature in a model grid cell that contains bare soil and soil covered by vegetation can be calculated with the **surface energy balance equation** from (8.103), modified with the following fluxes:

$$\bar{F}_{n,g} = f_s\bar{F}_s + \bar{F}_{i}\!\downarrow - \varepsilon_s\sigma_B\bar{T}_g^4 \tag{8.104}$$

$$\bar{H}_f = -f_s\frac{\bar{\rho}_a c_{p,d}}{R_a}\left[\bar{\theta}_p(z_r) - \frac{\bar{T}_g}{\bar{P}_g}\right] - f_v\frac{\bar{\rho}_a c_{p,d}}{R_f}\left[\frac{T_{af}}{\bar{P}_f} - \frac{\bar{T}_g}{\bar{P}_g}\right] \tag{8.105}$$

$$L_e\bar{E}_f = -f_s L_e\frac{\bar{\rho}_a}{R_a}\beta_g[\bar{q}_v(z_r) - q_{v,s}(\bar{T}_g)] - f_v L_e\frac{\bar{\rho}_a}{R_f}\beta_g[q_{af} - q_{v,s}(\bar{T}_g)] \tag{8.106}$$

where

$\bar{F}_{n,g}$	Grid-cell averaged net downward minus upward solar plus infrared radiative flux absorbed by bare soil (positive is down) (W m^{-2})
$\bar{H}_f$	Grid-cell averaged vertical turbulent sensible-heat flux. It is calculated as the flux between bare soil and open air plus that between vegetated soil and canopy air (positive is up) (W m^{-2})
$\bar{E}_f$	Grid-cell averaged water vapor flux between bare soil and open air plus that between vegetated soil and canopy air (positive is up) (kg m^{-2} s^{-1})
L_e	Latent heat of evaporation (J kg^{-1})
f_s, f_v	Fractions of bare and vegetated soil, respectively, in a grid cell ($f_s + f_v = 1$)
$\bar{F}_s$	Grid-cell averaged net downward minus upward solar irradiance at the top of the canopy plus that over bare soil (positive is downward) (W m^{-2})
$f_s\bar{F}_s$	Net downward minus upward solar irradiance over bare soil in a grid cell (positive is downward) (W m^{-2})
$\bar{F}_i\!\downarrow$	Grid-cell averaged downward (not net downward minus upward) thermal-IR irradiance absorbed by the ground (positive is downward) (W m^{-2})
$\bar{T}_g$	Grid-cell averaged soil surface temperature (the surface value of T_s) (K)
σ_B	Stefan–Boltzmann constant (5.67051 × 10^{-8} W m^{-2} K^{-4})
ε_s	Thermal-IR emissivity of soil (–)
$\sigma_B\varepsilon_s\bar{T}_g^4$	Grid-cell averaged upward thermal-IR irradiance emitted by the ground (W m^{-2})
$\bar{\rho}_a$	Grid-cell averaged air density immediately above the surface (kg m^{-3})
$c_{p,d}$	Specific heat of dry air at constant pressure (J kg^{-1} K^{-1})
R_a	Aerodynamic resistance (s m^{-1}) between bare soil outside of a canopy and a reference height z_r
R_f	Aerodynamic resistance between soil and overlying vegetation (s m^{-1})
$\bar{\theta}_p(z_r)$	Grid-cell averaged potential temperature at the reference height (K)
$\bar{P}_g$	$[\bar{p}_{a,g}/1000\ \text{hPa}]^\kappa$ converts temperature at the ground to potential temperature
$\bar{p}_{a,g}$	Air pressure at the ground (hPa)
κ	$R'/c_{p,d} = 0.286$
$\bar{P}_f$	$[\bar{p}_{a,f}/1000\ \text{hPa}]^\kappa$ converts temperature in the foliage to potential temperature
T_{af}	Temperature of air in foliage (K)
β_g	Wetness function from (8.92)
$\bar{q}_v(z_r)$	Grid-cell averaged water vapor specific humidity at the reference height (kg kg^{-1})
$q_{v,s}(T_g)$	Saturation specific humidity at the ground temperature (kg kg^{-1})
q_{af}	Specific humidity of water vapor in foliage air (kg kg^{-1})
z_r	Reference height (10 m)

An expression for the aerodynamic resistance (s m^{-1}) over bare soil is given in (20.12). An expression for the **aerodynamic resistance in the foliage** (s m^{-1}) is

$$R_f = \frac{1}{c_f u_{af}} \tag{8.107}$$

where $u_{af} \approx 0.83u_*$ (m s^{-1}) is the wind speed in the foliage and

$$c_f = 0.01 - \frac{0.003}{u_{af}} \tag{8.108}$$

(u_{af} in m s^{-1}) is the dimensionless heat-transfer coefficient (e.g., Deardorff 1978; Pielke 1984). Expressions for **temperature and specific humidity of air in the foliage** are

$$T_{af} = 0.3\bar{T}_a(z_r) + 0.6T_f + 0.1\bar{T}_g \tag{8.109}$$

$$q_{af} = 0.3\bar{q}_v(z_r) + 0.6q_f + 0.1\bar{q}_g \tag{8.110}$$

(Deardorff 1978), where

$\bar{T}_a(z_r)$ Grid-cell averaged temperature at the reference height (K)
$\bar{q}_v(z_r)$ Grid-cell averaged water vapor specific humidity at reference height (kg kg^{-1})
T_f Temperature in leaf stomata (K)
q_f Specific humidity in leaf stomata (kg kg^{-1})
$\bar{T}_g$ Grid-cell averaged ground temperature (K)
$\bar{q}_g$ Grid-cell averaged specific humidity in soil macropores (kg kg^{-1})

The **water vapor specific humidity in leaf stomata** is

$$q_f = \min[\gamma q_{v,s}(T_f) + (1-\gamma)q_{af}, q_{v,s}(T_f)], \tag{8.111}$$

where

$$\gamma = \begin{cases} \dfrac{R_f}{R_f + R_{st}}\left[1 - \left(\dfrac{W_c}{W_{c,max}}\right)^{2/3}\right] & q_{af} < q_{v,s}(T_f) \\ 0 & q_{af} \geq q_{v,s}(T_f) \end{cases} \tag{8.112}$$

(Deardorff 1978). In this equation, W_c (m^3 m^{-2} or m) is the depth of liquid water on an individual leaf surface (m) multiplied by the one-sided leaf area index, L_T (square meters of leaf surfaces per square meter of underlying ground), $W_{c,max}$ (m) is the maximum possible value of W_c (approximated as $0.0002L_T$ (m) from Dickinson 1984), and

$$R_{st} = \frac{R_{min}}{F_c}\left[1 + \left(\frac{200}{\bar{F}_s + 0.1}\right)^2\right]\frac{400}{(T_f - 273.15)(313.15 - T_f)} \tag{8.113}$$

is the **leaf stomata resistance** (s m^{-1}), which is infinite when $T_f \leq 273.15$ K and $T_f \geq 313.15$ K, and large at night, when $\bar{F}_s = 0$ (Baldocchi *et al.* 1987). In this equation, R_{min} is the minimum bulk canopy stomata resistance (Appendix Table B.11), and

$$F_c = \max\left[\min\left(\frac{w_{g,avg} - w_{wilt}}{w_{cr} - w_{wilt}}, 1.0\right), 1.0 \times 10^{-12}\right] \tag{8.114}$$

is a factor accounting for reduction in transpiration due to drying up of the soil toward the **wilting point,** w_{wilt} (m^3 m^{-3}) (e.g., Noilhan and Planton 1989), which is the liquid water content of soil when a plant is permanently wilted. Wilting points of different soils are given in Table 8.3. In this equation, $w_{g,avg}$ (m^3 m^{-3}) is the average liquid water content in the root layers of soil, and w_{cr} (m^3 m^{-3}) is a critical liquid-water content taken as $0.75w_{g,s}$ (Thompson *et al.* 1981), where $w_{g,s}$ (m^3 m^{-3}) is the soil moisture content at saturation.

The **foliage temperature** T_f is found by solving iteratively a foliage energy balance equation that considers a net solar flux at the top of the foliage, a net thermal-IR flux at the top of the foliage and at ground level, and sensible and latent heat fluxes at the top of the foliage and at ground level (e.g., Deardorff 1978; McCumber 1980; Pielke 1984; Jacobson 2001a). One such equation is

$$f_v\left[\bar{F}_s + \varepsilon_v \bar{F}_i\!\downarrow + \frac{\varepsilon_v\varepsilon_s}{\varepsilon_v + \varepsilon_s - \varepsilon_v\varepsilon_s}\sigma_B \bar{T}_g^4 - \frac{\varepsilon_v + 2\varepsilon_s - \varepsilon_v\varepsilon_s}{\varepsilon_v + \varepsilon_s - \varepsilon_v\varepsilon_s}\varepsilon_v\sigma_B T_f^4\right] = H_v + L_e E_d + L_e E_t \tag{8.115}$$

where ε_v is the **thermal-IR emissivity of vegetation,** $\bar{F}_i\!\downarrow$ (W m^{-2}) is the downward thermal-IR irradiance at the top of the canopy, H_v (W m^{-2}) is the sensible heat flux between air in the foliage and leaves, E_d (kg m^{-2} s^{-1}) is the turbulent moisture flux due to direct evaporation from/condensation to leaves in the foliage, and E_t (kg m^{-2} s^{-1}) is the turbulent moisture flux due to transpiration from leaf stomata.

The sensible heat flux, direct evaporation/condensation, and transpiration terms in (8.114) are

$$H_v = -1.1 L_T \frac{\bar{\rho}_a c_{p,d}}{R_f \bar{P}_f}\left[T_{af} - T_f\right] \tag{8.116}$$

$$E_d = -L_T \frac{\bar{\rho}_a \beta_d}{R_f}\left[q_{af} - q_{v,s}(T_f)\right] \tag{8.117}$$

$$E_t = -L_T \frac{\bar{\rho}_a (1 - \beta_d)}{R_f + R_{st}}\left[q_{af} - q_{v,s}(T_f)\right] \tag{8.118}$$

respectively, where the factor 1.1 in (8.116) accounts for the effects of stalks, stems, twigs, and limbs that exchange energy but do not transpire, L_T is limited to no greater than $7f_v$, and

$$\beta_d = \begin{cases} \left(\dfrac{W_c}{W_{c,max}}\right)^{2/3} & q_{af} < q_{v,s}(T_f) \\ 1 & q_{af} \geq q_{v,s}(T_f) \end{cases} \tag{8.119}$$

accounts for the decrease in the evaporation rate when the leaves contain little water (Monteith and Szeicz 1962; Deardorff 1978). During condensation, direct moisture flux to leaf surfaces via (8.117) occurs at the potential rate ($\beta_d = 1$). When dew condenses on leaves, $1 - \beta_d = 0$ and no transpiration occurs by (8.118). When direct evaporation occurs, the transpiration rate is limited by the fraction of leaf surfaces not covered by liquid water. The saturation specific humidities in (8.117) and (8.118) are obtained at the temperature of the foliage.

The iterative solution to T_f is found by substituting (8.116)–(8.118) into (8.114), then linearizing T_f^4 and $q_{v,s}(T_f)$ with

$$T_{f,t,n+1}^4 = T_{f,t,n}^4 + 4T_{f,t,n}^3(T_{f,t,n+1} - T_{f,t,n}) \tag{8.120}$$

$$q_{v,s}(T_{f,t,n+1}) = q_{v,s}(T_{f,t,n}) + \frac{dq_{v,s}(T_{f,t,n})}{dT}(T_{f,t,n+1} - T_{f,t,n}) \tag{8.121}$$

respectively (Deardorff 1978), where the subscript t indicates a value at the end of the time step and the subscript n indicates iteration number. Applying these steps gives the foliage temperature as

$$T_{f,t,n+1} = \frac{\left\{\begin{array}{l} f_v\left[\begin{array}{l}\bar{F}_s + \varepsilon_v \bar{F}_{i\downarrow} + \dfrac{\varepsilon_v\varepsilon_s}{\varepsilon_v + \varepsilon_s - \varepsilon_v\varepsilon_s}\sigma_B \bar{T}_{g,t-h}^4 \\ +3\dfrac{\varepsilon_v + 2\varepsilon_s - \varepsilon_v\varepsilon_s}{\varepsilon_v + \varepsilon_s - \varepsilon_v\varepsilon_s}\varepsilon_v\sigma_B T_{f,t,n}^4\end{array}\right] \\ +1.1L_T\dfrac{\bar{\rho}_a c_{p,d}}{R_f \bar{P}_f}T_{af} + L_e L_T \bar{\rho}_a\left\{\dfrac{\beta_d}{R_f} + \dfrac{\bar{\rho}_a\beta_t}{R_f + R_{st}}\right\} \\ \times\left[q_{af} - \left\{q_{v,s}(T_{f,t,n}) - \dfrac{dq_{v,s}(T_{f,t,n})}{dT}T_{f,t,n}\right\}\right]\end{array}\right\}}{\left\{\begin{array}{l} 4\dfrac{\varepsilon_v + 2\varepsilon_s - \varepsilon_v\varepsilon_s}{\varepsilon_v + \varepsilon_s - \varepsilon_v\varepsilon_s}\varepsilon_v\sigma_B T_{f,t,n}^3 + 1.1L_T\dfrac{\bar{\rho}_a c_{p,d}}{R_f\bar{P}_f} \\ + L_e L_T\bar{\rho}_a\left\{\dfrac{\beta_d}{R_f} + \dfrac{\bar{\rho}_a(1-\beta_d)}{R_f + R_{st}}\right\}\dfrac{dq_{v,s}(T_{f,t,n})}{dT}\end{array}\right\}} \tag{8.122}$$

(Jacobson 2001a), where the subscript $t - h$ indicates a value at the beginning of a time step. This equation needs to be iterated only four times to achieve significant convergence.

A prognostic equation for **water on leaf surfaces** is

$$\frac{\partial W_c}{\partial t} = \bar{P}_r - \frac{E_d}{\rho_w} - R \tag{8.123}$$

where $\bar{P}_r$ (m s^{-1}) is the grid-cell averaged precipitation rate, ρ_w (kg m^{-3}) is the density of liquid water, and R (m s^{-1}) is the runoff rate from leaf surfaces. Combining (8.117), (8.119), and (8.123) and finite-differencing the result over time step h (s) give

$$W_{c,t} = W_{c,t-h} + \begin{cases} h\left(\bar{P}_r + \dfrac{L_T}{R_f}\dfrac{\bar{\rho}_a}{\rho_w}\left(\dfrac{W_{c,t-h}}{W_{c,\max}}\right)^{\frac{2}{3}}[q_{af} - q_{v,s}(T_f)]\right) & q_{af} < q_{v,s}(T_f) \\ h\left(\bar{P}_r + \dfrac{L_T}{R_f}\dfrac{\bar{\rho}_a}{\rho_w}[q_{af} - q_{v,s}(T_f)]\right) & q_{af} \ge q_{v,s}(T_f) \end{cases} \tag{8.124}$$

The solution is limited by $W_c = \min(\max[W_c, 0], W_{c,\max})$, which accounts for the fact that any liquid water over depth $W_{c,\max}$ is runoff (thus R does not appear in the equation). Each solution from (8.124) is substituted into (8.119), which is itself substituted into (8.117) to give the direct evaporation rate and into (8.118) to give the transpiration rate.

The **grid-cell averaged soil-top temperature**, accounting for bare and vegetated soil, is now calculated by substituting (8.104)–(8.106) into the energy balance equation (8.103), and linearizing. The result is

$$\bar{T}_{g,t,n} = \bar{T}_{g,t,n-1} + \frac{\left\{\begin{array}{l} f_s \dfrac{\bar{\rho}_a c_{p,d}}{R_a}\left[\bar{\theta}_p(z_r) - \dfrac{\bar{T}_{g,t,n-1}}{\bar{P}_g}\right] \\ f_v \dfrac{\bar{\rho}_a c_{p,d}}{R_f}\left[\dfrac{T_{af,t}}{\bar{P}_f} - \dfrac{\bar{T}_{g,t,n-1}}{\bar{P}_g}\right] \\ f_s \dfrac{\bar{\rho}_a L_e}{R_a}\beta_g[\bar{q}_v(z_r) - q_{v,s}(\bar{T}_{g,t,n-1})] \\ f_v \dfrac{\bar{\rho}_a L_e}{R_f}\beta_g[q_{af} - q_{v,s}(\bar{T}_{g,t,n-1})] \\ f_s \bar{F}_s + \bar{F}_{i\downarrow} - \sigma_B \varepsilon_s \bar{T}^4_{g,t,n-1} \\ \dfrac{\kappa_{s,1}}{D_1}(T_{1,t} - \bar{T}_{g,t,n-1}) \end{array}\right\}}{f_s \dfrac{\bar{\rho}_a c_{p,d}}{R_a \bar{P}_g} + f_v \dfrac{\bar{\rho}_a c_{p,d}}{R_f \bar{P}_g} + 4\varepsilon_s \sigma_B \bar{T}^3_{g,t,n-1} + \dfrac{\kappa_{s,1}}{D_1}} \tag{8.125}$$

where D_1 (m) is the positive distance between the surface and the middle of the first soil layer, and $T_{1,t}$ (K) is the temperature at the middle of the top soil layer, from (8.91). Equation (8.125) is iterated four times without updating the saturation specific humidity in the soil during each iteration.

Finally, the grid-cell averaged fluxes of sensible heat (J m^{-2} s^{-1}) and moisture (kg m^{-2} s^{-1}) to the boundary layer, accounting for bare and vegetated surfaces, are

$$\bar{H}_f = H_f - f_v \frac{\bar{\rho}_a c_{p,d}}{R_a}\left[\bar{\theta}_p(z_r) - \frac{T_{af}}{\bar{P}_f}\right] \tag{8.126}$$

$$\bar{E}_f = E_f - f_v \frac{\bar{\rho}_a}{R_a}\beta_g[\bar{q}_v(z_r) - q_{af}] \tag{8.127}$$

respectively (positive is up in both cases), where H_f and E_f are the values over bare soil from (8.105) and (8.106), respectively.

Figure 8.5 shows time series predictions of foliage air temperature (T_{af}), foliage temperature (T_f), soil-top temperature ($\bar{T}_g$), and above-canopy air temperature ($\bar{T}_a$) over sandy loam in a grid cell of an atmospheric model containing a near-surface air temperature monitoring site at Lodi, California. The Lodi site is rural and located in the San Joaquin Valley, California. The Lodi cell contains about 58.4 percent sandy loam and 41.6 percent clay loam. Figure 8.5(a) indicates that during the day, foliage temperatures exceeded foliage air temperatures, which exceeded ground temperatures, which exceeded air temperatures. Figure 8.5(b) shows the modeled differences in temperatures between sandy loam and clay loam in the cell. Differences in soil-top temperature between the two soil types were up to 10 °C

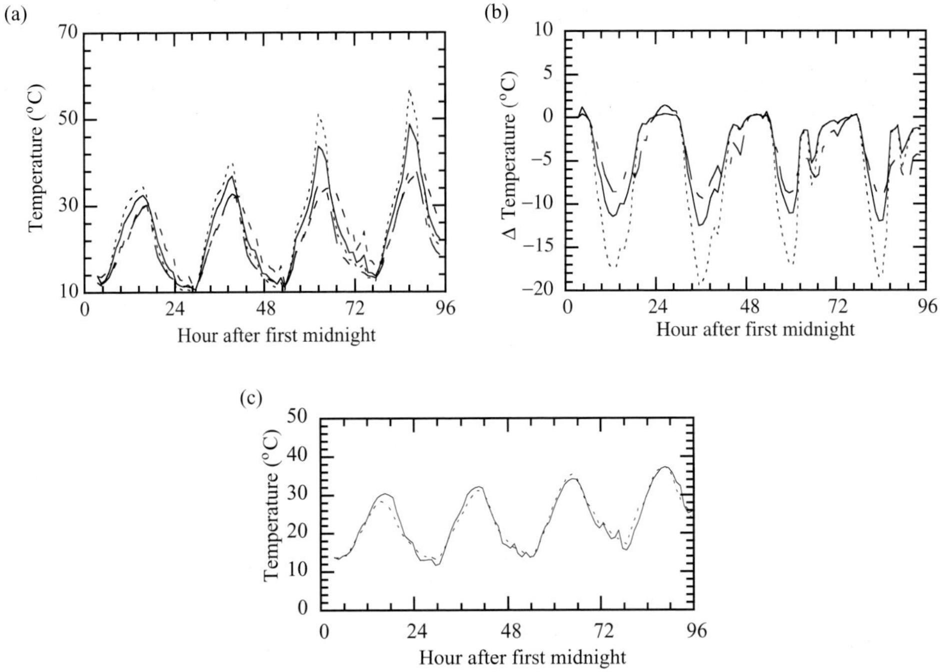

Figure 8.5 (a) Time series plot of predicted foliage air temperature (T_{af}) —, foliage temperature (T_f) ·······, soil-top temperature ($\bar{T}_g$) ——--, and air temperature above the canopy (T_a) – – – over sandy loam soil in a model grid cell containing the Lodi (LOD) monitoring site. (b) Difference between temperatures over sandy loam and those over clay loam in the same model grid cell. (c) Predicted, —, versus measured, ----, air temperatures at 5 m at the Lodi monitoring site. The predicted values were obtained from bilinear interpolation among four grid cells near or containing the Lodi site. From Jacobson (2001a).

during the day. Figure 8.5(c) compares modeled with measured near-surface air temperatures at Lodi when temperatures were averaged proportionately over both soil types in the grid cell.

8.6.4 Road and rooftop surfaces

In this section, temperatures of road and rooftop surfaces are discussed. Construction materials, such as asphalt, concrete, wood, brick, and composites cover a large fraction of urban surfaces. Oke *et al.* (1999) found that 25 percent of surfaces affecting the energy balance in a densely built-up central Mexico City site consisted of impervious ground material (e.g., roads, sidewalks), 32 percent consisted of rooftops, 42 percent consisted of walls, and 1 percent consisted of vegetation. A Vancouver site consisted of 32 percent impervious ground material, 37 percent rooftops, 27 percent walls, and 4 percent vegetation. Vegetation fractions were 94 percent for a rural site, 70 percent for an urban residential site, and 16 percent for an urban commercial site.

8.6.4.1 *Road surfaces*

Road surfaces often consist of asphalt or concrete. An average asphalt road is about 6 cm thick and overlies soil. The **temperature at the top of an asphalt surface** can be solved iteratively from the equilibrium energy balance equation with

$$\bar{T}_{g,t,n} = \bar{T}_{g,t,n-1} + \frac{\left\{ \begin{array}{l} \dfrac{\bar{\rho}_a c_{p,d}}{R_a}\left[\bar{\theta}_p(z_r) - \dfrac{\bar{T}_{g,t,n-1}}{\bar{P}_g}\right] \\ \dfrac{\bar{\rho}_a L_e}{R_a}\beta_d[\bar{q}_v(z_r) - q_{v,s}(\bar{T}_{g,t,n-1})] \\ \bar{F}_s + \bar{F}_{i}\!\downarrow - \sigma_B \varepsilon_{as} \bar{T}^4_{g,t,n-1} \\ \dfrac{\kappa_{as}}{D_1}(T_{1,c,t} - \bar{T}_{g,t,n-1}) \end{array} \right\}}{\dfrac{\bar{\rho}_a c_{p,d}}{R_a \bar{P}_g} + 4\varepsilon_{as}\sigma_B \bar{T}^3_{g,t,n-1} + \dfrac{\kappa_{as}}{D_1}} \qquad (8.128)$$

(Jacobson 2001a) where $\varepsilon_{as} = 0.95$ is the thermal-IR emissivity of asphalt (Oke 1978), D_1 (m) is the distance between the ground surface and the middle of the first asphalt layer, and β_d is from (8.119), which requires the calculation of W_c from (8.124). Here, W_c accounts for condensation/evaporation and precipitation/runoff over an asphalt surface. When precipitation or condensation (dew formation) occurs, all liquid water above $W_{c,max}$ is treated as runoff, and the liquid water content below $W_{c,max}$ is allowed to evaporate.

Below the asphalt top, temperatures and moisture can be solved with the heat-conduction and water transport equations, (8.91) and (8.95), respectively. In that case, model layers, down to 6 cm, may be treated as asphalt, and the remaining layers as soil containing liquid water. The asphalt is treated as impermeable to water, thus, the liquid water content of the asphalt is set to zero, and the water transport equation is solved only under the asphalt base. The heat conduction equation may be solved assuming a mean thermal conductivity of various asphalts of $\kappa_{as} = 1.7$ W m^{-1} K^{-1} (Anandakumar 1999).

8.6.4.2 *Rooftops*

Many types of roofing exist, and rooftop types generally differ between residential and commercial buildings. In the United States, four out of five residential rooftops contain asphalt shingles. Asphalt-shingle roof systems consist of asphalt shingles overlying saturated felt, overlying a water-impermeable membrane, overlying a roof deck, overlying insulation or air. Asphalt shingles are either organic or fiberglass based. Organic-based shingles contain a base of cellulose fibers, which is saturated with an asphalt coating and surfaced with weather-resistant mineral granules. Fiberglass-based shingles contain a base of glass fibers surfaced with an asphalt coating and weather-resistant mineral granules. The saturated felt is

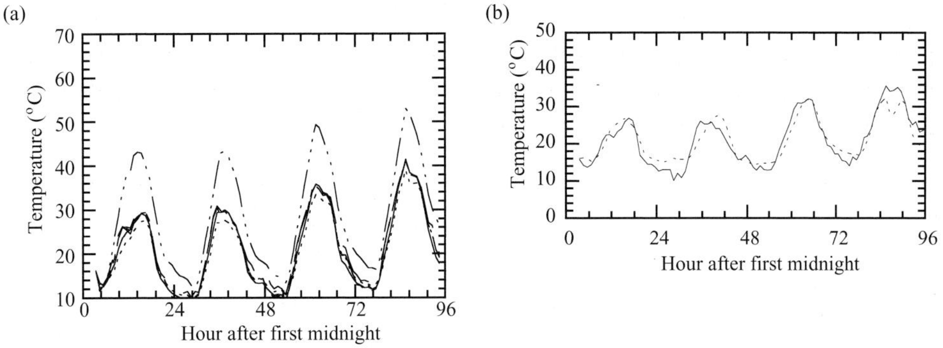

Figure 8.6 (a) Predicted ground temperatures over four soil types, road surfaces, and rooftops at Fremont: — silt loam, ······ clay, --- loam, — ······ road, – – – – silt clay loam, **—** roof. (b) Comparison of modeled, —, with measured, ----, near-surface air temperature at Fremont when all surface types were considered in the model (from Jacobson 2001a).

an asphalt-impregnated, organic-based felt between the roofing material and the waterproof membrane (Asphalt Roofing Manufacturers Association 1999). Other residential rooftops consist of wood shake, concrete tiles, or slate over a membrane over a deck and insulation over air.

Many commercial rooftops also consist of asphalt shingles, but others consist of two to four layers of bitumen strengthened with a fabric, such as polyester or fiberglass, or with a felt. Gravel or granules are usually embedded on the top layer of strengthened bitumen. The bitumen layers usually overlie insulation, a vapor retarder, and a deck. Bitumen is a black or dark-colored cement-like, solid, semi-solid, or viscous substance composed of high-molecular-weight hydrocarbons and is found in asphalts, tars, pitches, and asphaltines.

Roof surface temperatures can be solved with (8.128) and subsurface material temperatures can be solved with the heat conduction equation from (8.91), but with a different configuration of materials than in the case of road surfaces. One possible configuration is to assume that the rooftop consists of 1.5 cm of asphalt shingles over 1 cm of saturated felt (asphalt-felt composite) over 1.25 cm of plywood base over air. No liquid water is permitted to transmit through or below the roof, and the air below the roof contains no liquid water, so the water transport equation is not necessary to solve in this case.

8.6.4.3 *Application*

Figure 8.6(a) shows modeled soil, road, and roof surface temperatures at Fremont, California. Roads were the hottest surfaces during the day, since asphalt contains no water, and energy conduction to soil below asphalt is slow. Peak temperatures over asphalt varied from 43 to 52 °C. Anandakumar (1999) shows a peak August temperature over an asphalt road at Vienna, Austria, of 44 °C. Vienna is 10° latitude further north than Fremont. Rooftop temperatures in Fig. 8.6 were cooler than road temperatures because air under a rooftop convects energy conducted to it from the roof base away faster than soil conducts energy conducted to it by the asphalt base away. Figure 8.6(b) compares modeled with measured near-surface air temperatures at Fremont when all surface types were considered.

8.6.5 Snow on soil, vegetation, and asphalt

The depth of snow over soil, vegetation, roads, and rooftops is affected by snowfall, deposition/sublimation, and melting. A **prognostic equation for snow depth** (m) accounting for these factors is

$$
\begin{aligned}
D_{s,t} &= D_{s,t-h} + h\bar{P}_s \\
&\quad + h\frac{\bar{\rho}_a}{\rho_{sn}}\left\{ f_s\frac{[\bar{q}_v(z_r) - q_{v,s}(\min(T_{g,t}, T_{s,m}))]}{R_a} + f_v\frac{[q_{af} - q_{v,s}(\min(T_{g,t}, T_{s,m}))]}{R_f}\right\} \\
&\quad + h\frac{\left\{\begin{array}{l} -f_s\dfrac{\bar{\rho}_a c_{p,d}}{R_a}\left[\bar{\theta}_p(z_r) - \dfrac{T_{s,m}}{\bar{P}_g}\right] - f_v\dfrac{\bar{\rho}_a c_{p,d}}{R_f}\left[\dfrac{T_{af,t}}{\bar{P}_f} - \dfrac{T_{s,m}}{\bar{P}_g}\right] \\ -f_s\dfrac{\bar{\rho}_a L_s}{R_a}[\bar{q}_v(z_r) - q_{v,s}(T_{s,m})] - f_v\dfrac{\bar{\rho}_a L_s}{R_f}[q_{af} - q_{v,s}(T_{s,m})] \\ -f_s\bar{F}_s - \bar{F}_i\!\downarrow + \sigma_B\varepsilon_{sn}T_{s,m}^4 - \dfrac{\kappa_{sn}}{D_1}(T_{1,t} - T_{s,m}) \end{array}\right\}}{\rho_{sn}L_m}
\end{aligned}
\tag{8.129}
$$

where $\bar{P}_s$ is the snowfall rate (m s^{-1}), L_m is the latent heat of melting (J kg^{-1}), L_s is the latent heat of sublimation (J kg^{-1}), $q_{v,s}(T_{s,m})$ is the saturation specific humidity (kg kg^{-1}) over ice at the melting point of snow, $T_{s,m} = 273.15$ K, ρ_{sn} is the mass density of snow (kg m^{-3}), $\varepsilon_{sn} = 0.99$ is the emissivity of snow, and $\kappa_{sn} = 0.08$ W m^{-1} K^{-1} (Oke 1978) is the thermal conductivity of snow. The first term in the equation accounts for snowfall, the second accounts for deposition/sublimation, and the third accounts for melting. The melting term is applied only if the term is less than zero (energy is added to the system) and the snow temperature exceeds the melting point of snow.

The temperature at the top of a snow surface is calculated with (8.125) or (8.128), except (a) $\beta_g = 1$ and $\beta_d = 1$, (b) $q_{v,s}$ is defined as the saturation specific humidity over ice, not liquid water, (c) L_s replaces L_e, (d) ε_{sn} is used instead of ε_s, ε_v, or ε_{as}, (e) κ_{sn} is used instead of κ_s or κ_{as}, and (f) transpiration through leaf surfaces is set to zero ($E_t = 0$).

8.6.6 Water, sea ice, and snow over sea ice

Over oceans, inland seas, and lakes, water-top temperature is affected by sensible, latent, and radiative heat fluxes as well as energy advection (due to water circulation). In the **mixed layer of the ocean**, which is its top layer of depth D_l (typically 50–100 m), in which energy and salinity are well mixed, the **water temperature** can be calculated noniteratively in time with

$$\bar{T}_{g,t} = \bar{T}_{g,t-h} + h \frac{\left\{ \begin{array}{l} \dfrac{\bar{\rho}_a c_{p,d}}{R_a}\left[\bar{\theta}_p(z_r) - \dfrac{\bar{T}_{g,t-h}}{\bar{P}_g}\right] \\ \dfrac{\bar{\rho}_a L_e}{R_a}[\bar{q}_v(z_r) - q_{v,s}(\bar{T}_{g,t-h})] \\ \bar{F}_s + \bar{F}_i\!\downarrow - \sigma_B \varepsilon_w \bar{T}^4_{g,t-h} \end{array} \right\}}{\rho_{sw} c_{p,sw} D_l} \tag{8.130}$$

where $q_{v,s}(T_{g,t-h})$ (kg kg^{-1}) is the saturation specific humidity over liquid water at the temperature of the water, $\varepsilon_w = 0.97$ is the thermal-IR emissivity of liquid water (Oke 1978), ρ_{sw} is the density of seawater (1028 kg m^{-3} at 0 °C in the presence of 35 parts per 1000 of salinity; Lide 2003), and $c_{p,sw}$ is the temperature-dependent specific heat of liquid water at constant pressure (3986.5 J kg^{-1} K^{-1} at 0 °C in the presence of 35 parts per 1000 of salinity).

Equation (8.130) accounts for all energy fluxes to and from the ocean mixed layer except for horizontal advective fluxes and diffusive fluxes to the deep ocean. Such fluxes need to be calculated with an **ocean circulation model**, which solves equations similar to the equations of atmospheric dynamics, but for ocean water. Circulation models can be two- or three-dimensional. The two-dimensional models resolve the mixed layer only. The three-dimensional models account for transfer between the mixed layer and deep ocean as well.

When the water temperature falls below the freezing point of seawater in the presence of 35 parts per 1000 of salinity, $T_{i,f} = 271.23$ K (Lide 2003), sea ice begins to form on top of ocean water. The **sea ice surface temperature** can be determined with the four-iteration surface-energy balance calculation,

$$T_{g,t,n} = T_{g,t,n-1} + \frac{\left\{ \begin{array}{l} \dfrac{\bar{\rho}_a c_{p,d}}{R_a}\left[\bar{\theta}_p(z_r) - \dfrac{\bar{T}_{g,t,n-1}}{\bar{P}_g}\right] \\ \dfrac{\bar{\rho}_a L_s}{R_a}[\bar{q}_v(z_r) - q_{v,s}(\bar{T}_{g,t,n-1})] \\ \bar{F}_s + \bar{F}_i\!\downarrow - \sigma_B \varepsilon_i \bar{T}^4_{g,t,n-1} \\ \dfrac{\kappa_i}{D_{i,t-h}}(T_{i,f} - \bar{T}_{g,t,n-1}) \end{array} \right\}}{\dfrac{\bar{\rho}_a c_{p,d}}{R_a \bar{P}_g} + 4\varepsilon_i \bar{T}^3_{g,t,n-1} + \dfrac{\kappa_i}{D_{i,t-h}}} \tag{8.131}$$

where $q_{v,s}(T_{g,t,n-1})$ is the saturation specific humidity over ice, $\varepsilon_i = 0.97$ is the thermal-IR emissivity of ice, and $\kappa_i = 2.20$ W m^{-1} K^{-1} is the thermal conductivity of ice (Lide 2003). The **change in sea ice thickness (m) at the sea ice/air interface** due to sublimation/deposition and melting is

$$D_{i,t} = D_{i,t-h} + h\frac{\bar{\rho}_a}{\rho_i R_a}[\bar{q}_v(z_r) - q_{v,s}(\min(\bar{T}_{g,t}, T_{i,m}))] + h\frac{\left\{\begin{array}{l} -\frac{\bar{\rho}_a c_{p,d}}{R_a}\left[\bar{\theta}_p(z_r) - \frac{T_{i,m}}{\bar{P}_g}\right] - \frac{\bar{\rho}_a L_s}{R_a}[\bar{q}_v(z_r) - q_{v,s}(T_{i,m})] \\ -\bar{F}_s - \bar{F}_i\!\downarrow + \sigma_B \varepsilon_i T_{i,m}^4 - \frac{\kappa_i}{D_{i,t-h}}(T_{i,f} - T_{i,m}) \end{array}\right\}}{\rho_i L_m} \quad (8.132)$$

where ρ_i (kg m^{-3}) is the density of ice (916.7 kg m^{-3}) at its melting point, $T_{i,m} = 273.05$ K. The melting (second) term applies only if the term is less than zero (energy is added to the system) and if the temperature from (8.131) exceeds the melting point.

The **change in sea ice thickness due to freezing/melting** at the water/sea ice interface is

$$D_{i,t} = D_{i,t-h} + h\frac{\bar{F}_b - \frac{\kappa_i}{D_{i,t-h}}(T_{i,f} - \bar{T}_{g,t})}{\rho_i L_m} \quad (8.133)$$

where $\bar{F}_b$ (W m^{-2}) is the net flux of energy from below the interface, estimated as 2 W m^{-2} for Arctic waters and 25 W m^{-2} for Antarctic waters (Parkinson and Washington 1979), and the second term accounts for conduction between the interface and the top of the ice. Maximum ice thicknesses are typically 4 m in the Arctic and 1.5 m in the Antarctic.

When sea ice exists and snowfall occurs, the snow covers the ice. The **temperature at the top of the snow overlying the sea ice layer** can be modeled with

$$\bar{T}_{g,t,n} = \bar{T}_{g,t,n-1} + \frac{\left\{\begin{array}{l} \frac{\bar{\rho}_a c_{p,d}}{R_a}\left[\bar{\theta}_p(z_r) - \frac{\bar{T}_{g,t,n-1}}{\bar{P}_g}\right] \\ \frac{\bar{\rho}_a L_s}{R_a}[\bar{q}_v(z_r) - q_{v,s}(\bar{T}_{g,t,n-1})] \\ \bar{F}_s + \bar{F}_i\!\downarrow - \sigma_B \varepsilon_{sn} \bar{T}_{g,t,n-1}^4 \\ \frac{\kappa_{sn}\kappa_i}{\kappa_{sn} D_{i,t-h} + \kappa_i D_{s,t-h}}(T_{i,f} - \bar{T}_{g,t,n-1}) \end{array}\right\}}{\frac{\bar{\rho}_a c_{p,d}}{R_a \bar{P}_g} + 4\varepsilon_{sn}\bar{T}_{g,t,n-1}^3 + \frac{\kappa_{sn}\kappa_i}{\kappa_{sn} D_{i,t-h} + \kappa_i D_{s,t-h}}} \quad (8.134)$$

where the thermal conductivity term accounts for conduction through ice and snow layers. An estimate of the temperature at the snow–ice interface is

$$T_I = \frac{\kappa_{sn} D_{i,t-h}\bar{T}_{g,t} + \kappa_i D_{s,t-h} T_{i,f}}{\kappa_{sn} D_{i,t-h} + \kappa_i D_{s,t-h}} \quad (8.135)$$

(Parkinson and Washington 1979).

8.7 SUMMARY

In this chapter, boundary-layer and surface processes were discussed. Expressions for kinematic turbulent fluxes of momentum, energy, and moisture were derived. For the surface layer, such fluxes were described in terms of bulk aerodynamic formulae and Monin–Obukhov similarity theory. Some similarity-theory parameters are friction wind speed, the potential temperature scale, and the Monin–Obukhov length. Fluxes of momentum, energy, and moisture above the surface layer were parameterized with a hybrid model and with a high-order turbulence closure technique. Equations describing soil temperature and moisture content were then discussed. Ground temperatures depend on the specific heat, thermal conductivity, and moisture content of soil and on fluxes of radiative, sensible, and latent heat between the soil and air. Prognostic equations for temperature and moisture over soil, vegetation-covered soil, rooftops, road surfaces, water, sea ice, and snow over all surfaces were described.

8.8 PROBLEMS

8.1 Calculate the surface roughness length for momentum over a rough ocean assuming $u_* = 0.3$ m s^{-1} and over vegetated land assuming a forest canopy with top $h_c = 20$ m and $L_T = 7$ m^2 m^{-2}. Calculate the wind speed 100 m above the ocean surface and above h_c assuming a displacement height of $d_c = 15$ m for the canopy. Assume a neutral boundary layer and logarithmic wind profile in both cases. In which case is the wind speed higher. Why?

8.2 If conditions at the surface are $z_r = 10$ m, $z_{0,m} = 0.05$ m, $z_{0,h} = 0.0005$ m, $\bar{u}(z_r) = 8$ m s^{-1}, $\bar{v}(z_r) = 2$ m s^{-1}, $\bar{T}(z_r) = 285$ K, $\bar{T}(z_{0,h}) = 286$ K, $p_a(z_r) = 1004$ hPa, and $p_a(z_{0,h}) = 1005$ hPa, calculate u_*, θ_*, L, $\mathrm{Ri_b}$, and $K_{m,zx}$ at the reference height. Assume dry air.

8.3 Assume $z_{0,m} = 0.01$ m, $u_* = 1$ m s^{-1}, $\bar{u}(z_r) = 10$ m s^{-1}, and $\bar{v}(z_r) = 5$ m s^{-1} at $z_r = 10$ m. Calculate the eddy diffusion coefficient for momentum, $K_{m,zx}$.

8.4 Assuming $u_* = 0.1$ m s^{-1}, $p_a = 998$ hPa, and the air is dry, calculate the surface roughness lengths for energy and water vapor when $\bar{T} = 288$ K.

8.5 Compare vertical turbulent sensible-heat fluxes over the ocean from bulk aerodynamic formulae and Monin–Obukhov similarity theory. Assume the same conditions as in Problem 8.2, except assume $z_{0,m} = 0.000\,01$ m. Assume $\bar{\theta}_v$ at the roughness length for energy equals that at the roughness length for momentum. Discuss differences in results.

8.6 Using conditions from Problems 8.2 and 8.5, calculate the eddy diffusion coefficient for energy with bulk aerodynamic formulae and similarity theory. Discuss differences in results.

8.7 Assume all conditions are the same as in Problem 8.2, except that the relative humidity at the reference height is now 85 percent, and the specific humidity at the roughness length for moisture equals the saturation specific humidity at the roughness length for energy. Estimate the vertical turbulent flux of moisture from similarity theory.

8.8 Calculate the thermal conductivity of a soil–water–air mixture for a sandy-loam soil when the volumetric water content is 0.2 and 0.05 $m^3\ m^{-3}$. In which case is the mixture more conductive? Why?

8.9 Simplify (8.130) to calculate the temperature of the mixed layer of the ocean in the absence of sensible and latent heat fluxes and assuming the only source or sink of thermal-infrared radiation is the sink due to upward emission ($\sigma_B \varepsilon_W T^4_{g,t-h}$). Calculate the temperature after two hours if the initial temperature is 290 K, the downward solar irradiance is 800 W m^{-2}, and the thickness of the mixed layer is 70 m. Should the temperature change increase or decrease if the mixing depth increases?

8.9 COMPUTER PROGRAMMING PRACTICE

8.10 If $z_r = 10$ m, $\bar{u}(z_r) = 3$ m s^{-1}, $\bar{v}(z_r) = 16$ m s^{-1}, $\bar{T}(z_r) = 293$ K, $\bar{T}(z_{0,h}) = 292.5$ K, $p_a(z_r) = 1002$ hPa, and $p_a(z_{0,h}) = 1003$ hPa, write a script to calculate u_*, θ_*, L, $\mathrm{Ri_b}$, and $K_{m,zz}$ as a function of $z_{0,m}$ when the air is dry. Calculate values for 0.000 01 m $< z_{0,m} <$ 5 m. Assume $z_{0,h} \approx z_{0,m}/100$. Plot the results for each variable.

8.11 Write a computer script to calculate the change of wind speed and potential virtual temperature with height in the surface layer from similarity theory. Assume the same conditions as in Problem 8.2. Plot the results from the ground surface up to 10 m altitude. Change $\bar{T}(z_r)$ to 286 K, and replot the results. Discuss the differences between the two cases.

8.12 Write a computer program to calculate the temperature over a single asphalt layer 6 cm thick from (8.128) assuming air temperature at the reference height (10 m) is a constant 295 K, the initial ground temperature is 285 K, the constant soil temperature below the ground is 290 K, the surface air pressure is 1013 hPa, the pressure at the reference height is 1012 hPa, the incident solar radiation varies with a sine function between 06:00 and 18:00, with a peak value of 1000 W m^{-2} at 12:00 and zero values before 06:00 and after 18:00, and the downward thermal-IR irradiance is zero. In addition, assume $R_a = 0.74 \ln(z_r/z_{0,h})/ku_*$, where $z_{0,h} = 0.000\,135$ m and $u_* = 0.4$ m s^{-1}. Finally, assume $\bar{q}_v(z_r) = 0$. Use the program to calculate the ground temperature from 03:00 to 21:00 on a single day. Plot and discuss the result. (Hint: take a time step of 1 s. Iterate the equation four times for each time step. At the end of each time step, update temperatures and solar radiation on the right side of the equation).

9

Radiative energy transfer

RADIATION through the atmosphere affects temperature, pollutant concentration, visibility, and color. Temperature is affected by heating and cooling from ultraviolet, visible, and infrared radiation interactions with the ground, gases, aerosol particles, and hydrometeor particles. Pollution is affected by interactions of ultraviolet and some visible radiation with gases. Visibility and colors are affected by interactions of visible radiation with gases, aerosol particles, and hydrometeor particles. In this chapter, radiation laws, optical properties of gases and particles, light processes, and a solution to the radiative transfer equation are discussed. The radiative transfer equation is used to determine rates of heating, cooling, and molecular photolysis. Important radiative laws discussed here include Planck's law, Wien's law, and the Stefan–Boltzmann law. Equations quantifying the level of attenuation and redirection of radiation by gases, particles, and cloud drops are also described.

9.1 ENERGY TRANSFER PROCESSES

Radiation is the emission or propagation of energy in the form of a photon or electromagnetic wave. A **photon** is a particle or quantum of electromagnetic energy that has no mass, no electric charge, and an indefinite lifetime. An **electromagnetic wave** is a disturbance traveling through a medium, such as air or space, that transfers energy from one object to another without permanently displacing the medium itself. Electromagnetic waves may be considered as dual transverse waves in that they consist of an electric wave and a magnetic wave in phase with and at right angles to each other and to the direction of propagation. Radiation is emitted by all bodies in the Universe that have a temperature above absolute zero (0 K). Once emitted, radiation passes through space or air to another body. Upon reaching the second body, the radiation can be reflected, scattered, absorbed, refracted, dispersed, or transmitted. Each of these processes is discussed in this chapter.

When a body, such as air, emits more radiation than it absorbs, its temperature decreases. When a body absorbs more radiation than it emits, its temperature increases. Other processes that affect air temperatures include advection, forced convection, turbulence, and latent-heat exchange. The thermodynamic energy equation in (3.76) takes these processes into account.

Figure 9.1 shows the relative importance of several processes on the atmospheric energy budget. Solar radiation provides energy for atmospheric, cloud, and surface

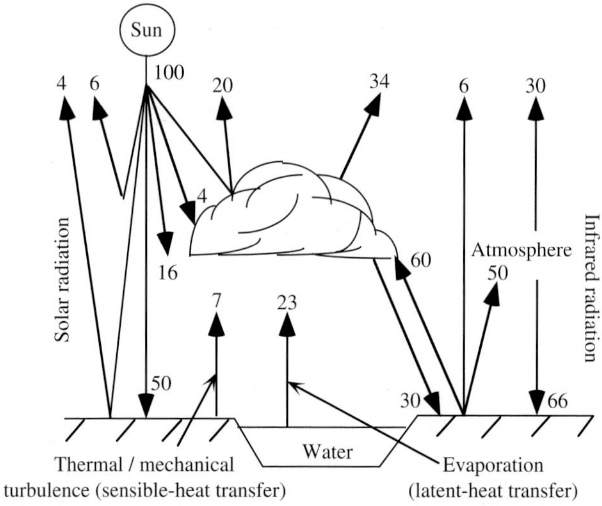

Figure 9.1 Energy balance for Earth–atmosphere system. Values are dimensionless relative quantities of energy. The sum of sources minus sinks for clouds, the atmosphere, or the Earth equals zero. For example, clouds are in radiative balance, since they absorb and emit 64 units of radiation.

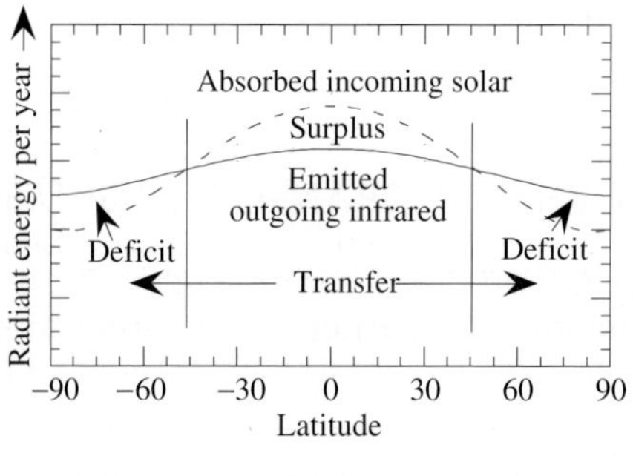

Figure 9.2 Schematic showing that a globally and yearly averaged energy surplus at the Equator and deficit at the poles is compensated for by energy transfer from the Equator toward the poles.

heating. The Earth's surface emits infrared radiation, much of which is absorbed by greenhouse gases and clouds and some of which escapes to space. The atmosphere and clouds emit infrared radiation in all directions. Turbulence transfers energy from the surface to the troposphere. Evaporation from the surface releases water vapor, which stores latent heat. The water vapor travels to the free troposphere, where it may condense to form clouds, releasing the stored latent heat.

The **transport of latent heat** via water vapor is an important process by which solar energy absorbed near the Equator is transferred poleward. Figure 9.2 shows that, at the Equator, the surface absorbs more solar radiation than it emits infrared radiation, causing an energy surplus. At the poles, the reverse is true, and an energy deficit occurs. In the absence of energy transfer between the Equator and poles, temperatures near the poles would continuously decrease, and those near the Equator would continuously increase. On Earth, energy is continuously supplied from the Equator to the poles by three processes. One is poleward transport of energy by winds, the second is poleward transport of energy by ocean currents, and the third is the poleward transport of water vapor and its stored latent heat by winds. When the water vapor condenses, mostly at midlatitudes, it releases its latent heat to the air around it.

9.2 ELECTROMAGNETIC SPECTRUM

Whether radiation is considered an electromagnetic wave or photon, it travels at the speed of light. If radiation propagates as a wave, its **wavelength** is

$$\lambda = \frac{c}{\nu} = \frac{1}{\tilde{\nu}} \tag{9.1}$$

where ν is the wave's **frequency of oscillation**, or number of wavelengths that pass through a point per unit time (s^{-1}), c is the speed of light (e.g., m s^{-1}) and $\tilde{\nu} = 1/\lambda$ is the **wavenumber** (e.g., m^{-1}), defined here as the number of wavelengths per unit distance. This definition of wavenumber, applied here to electromagnetic waves, differs from that given in (4.89), which was applied to atmospheric waves. Units for wavelength are micrometers (1 μm = 10^{-6} m), nanometers (1 nm = 10^{-9} m), centimeters (cm), or meters (m). In a vacuum, the speed of light is $c = 2.9979 \times 10^8$ m s^{-1}.

In 1900, Max Planck theorized that the total radiative energy (J) emitted or absorbed by a substance was

$$E_t = nh\nu = n\frac{hc}{\lambda} \tag{9.2}$$

where n is an integer, called a **quantum number**, and $h = 6.6256 \times 10^{-34}$ J s is **Planck's constant**. Equation (9.2) states that substances do not emit or absorb radiative energy continuously, but in packets or quanta. Although Planck believed that light was emitted or absorbed discontinuously, he held that it traveled through space as an electromagnetic wave. In 1905, Albert Einstein postulated that light energy traveled through space in concentrated bundles, and the energy of each bundle was

$$E_p = h\nu = \frac{hc}{\lambda} \tag{9.3}$$

In 1926, Gilbert Lewis termed these bundles **photons**. Equation (9.3) represents the energy emitted or absorbed per photon (J $photon^{-1}$) and depends on the wavelength (or frequency) of radiation. The greater the frequency, the shorter the wavelength, and the greater the energy of a quantum. The electromagnetic and photon theories of light are interrelated by (9.2).

Example 9.1

Find the energy emitted per photon, the frequency, and the wavenumber of a $\lambda =$ 0.5-μm and $\lambda =$ 10-μm wavelength of energy.

SOLUTION

When $\lambda = 0.5$ μm, $E = 3.97 \times 10^{-19}$ J photon^{-1}, $\nu = 5.996\times 10^{14}$ s^{-1}, and $\tilde{\nu} =$ 2 μm^{-1}. When $\lambda = 10$ μm, $E = 1.98 \times 10^{-20}$ J photon^{-1}, $\nu = 2.998 \times 10^{13}$ s^{-1}, and $\tilde{\nu} = 0.1$ μm^{-1}. Shorter wavelengths generate more energy per photon than do longer wavelengths.

From (9.2), Planck derived an equation relating the intensity of radiant emission from a perfectly emitting substance to the absolute temperature of the substance and the wavelength of emission. A perfectly emitting substance is one that, in thermodynamic equilibrium, emits all radiation that it absorbs. **Absorption** occurs when electromagnetic energy enters a substance and is converted to internal energy. If a body's emission is less than its absorption, the body is out of equilibrium, and its temperature rises.

A **blackbody** is a substance that absorbs all radiation that is incident upon it. No incident radiation is reflected by a blackbody. No bodies are true blackbodies, although the Earth and the Sun are close, as are black carbon, platinum black, and black gold (e.g., Siegel and Howell 1992). The term blackbody was coined because good absorbers of visible radiation generally appear black. However, good absorbers of infrared radiation are not necessarily black. For example, one such absorber is white oil-based paint.

Blackbodies not only absorb, but also emit the maximum possible intensity of radiant energy at a given wavelength and temperature. Planck determined this intensity as

$$B_{\lambda,T} = \frac{2hc^2}{\lambda^5\left[\exp\left(\frac{hc}{\lambda k_B T}\right) - 1\right]} \tag{9.4}$$

now called **Planck's law**, where $B_{\lambda,T}$ is **radiant intensity** or **radiance** (W m^{-2} μm^{-1} sr^{-1}), and k_B is Boltzmann's constant (1.38×10^{-23} J K^{-1} = W s K^{-1}). Radiance is the energy emitted per unit area per unit time per unit wavelength per incremental solid angle (units of steradians). Since the radiance in (9.4) is defined for individual wavelengths, it is called a **spectral radiance**. Incremental solid angle is defined shortly.

Table 9.1 Emissivities of different surface types for a typical infrared wavelength

Surface type	Emissivity (fraction)	Surface type	Emissivity (fraction)
Liquid water	1.0[a]	Soil	0.9–0.98[b]
Fresh snow	0.99[b]	Grass	0.9–0.95[b]
Old snow	0.82[b]	Desert	0.84–0.91[b]
Liquid water clouds	0.25–1.0[c]	Forest	0.95–0.97[b]
Cirrus clouds	0.1–0.9[c]	Concrete	0.71–0.9[b]
Ice	0.96[d]	Urban	0.85–0.87[a]

[a] Seaman *et al.* (1989), [b] Oke (1978), [c] Liou (2002), [d] Sellers (1965).

Equation (9.4) applies to any blackbody with a temperature above absolute zero (0 K). Real substances are generally not perfect emitters. Instead, they usually emit a fraction of the radiance that a blackbody emits. The radiance (W m^{-2} μm^{-1} sr^{-1}) actually emitted by any substance is approximately

$$e_\lambda = \varepsilon_\lambda B_{\lambda,T} \tag{9.5}$$

where ε_λ (dimensionless) is the **emissivity** of the substance. Emissivity is the fraction (≤ 1) of $B_{\lambda,T}$ actually emitted. Emissivities depend on wavelength. Table 9.1 gives emissivities for some surface types in the infrared part of the electromagnetic spectrum.

Absorptivity (a_λ) is the fraction (≤ 1) of incident radiation that a substance actually absorbs. **Kirchoff's law** (named for Gustav Kirchoff, 1824–87) states that, in thermodynamic equilibrium, the emissivity and absorptivity of a substance equal each other ($a_\lambda = \varepsilon_\lambda$). Thus, the efficiency at which a substance absorbs radiation equals that at which it emits radiation, and a perfect emitter of radiation ($\varepsilon_\lambda = 1$) is also a perfect absorber of radiation ($a_\lambda = 1$) and a blackbody. Suppose a blackbody object is placed in a vacuum enclosed by blackbody walls. Over time, the temperatures of the object and walls equalize. If Kirchoff's law did not hold at this point, a net heat transfer would occur between the object and the walls even though the object and wall temperatures are the same. Such a heat transfer violates the second law of thermodynamics. Thus, the absorptivity and emissivity of a substance must equal each other in equilibrium.

An **incremental solid angle**, $d\Omega_a$, used in the definition of radiance, is an incremental surface area on a unit sphere, which is a sphere with radius normalized to unity. The equation for incremental solid angle is

$$d\Omega_a = \frac{dA_s}{r_s^2} \tag{9.6}$$

where dA_s is an incremental surface area, and r_s is the radius of a true sphere. Incremental solid angle has units of steradians (sr), which is analogous to units of radians for a circle.

The incremental surface area in (9.6) can be found from Fig. 9.3. In the figure, a line is drawn from the center of the sphere to the center of an incremental area

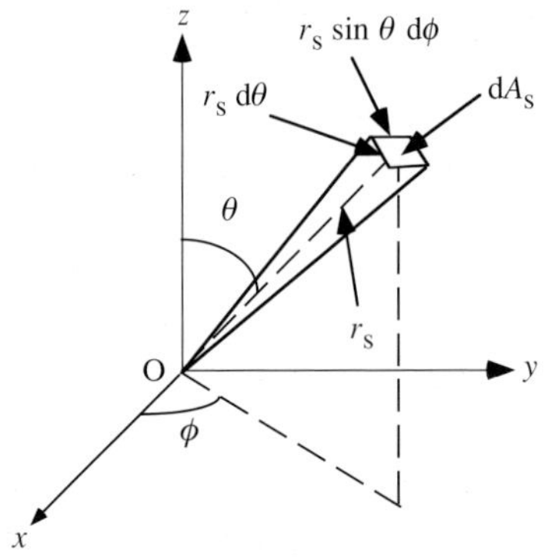

Figure 9.3 Radiance, emitted from point (O) on a horizontal plane, passes through an incremental area dA_s at a distance r_s from the point of emission. The angle between the z-axis and the angle of emission is the zenith angle (θ), and the horizontal angle between a reference axis (x-axis) and the line of emission is the azimuth angle (ϕ). The size of the incremental surface area is exaggerated.

dA_s, which is a distance r_s from the sphere's center. The line is directed at a **zenith angle** θ from the surface normal (where the surface is on the x–y plane). The line is also located at an **azimuth angle** ϕ, directed counterclockwise from the positive x-axis to a horizontal line dropped from the line. From the geometry shown, the incremental surface area is

$$dA_s = (r_s\, d\theta)(r_s \sin\theta\, d\phi) = r_s^2 \sin\theta\, d\theta\, d\phi \tag{9.7}$$

where $d\theta$ and $d\phi$ are incremental zenith and azimuth angles, respectively. Substituting (9.7) into (9.6) gives the **incremental solid angle** as

$$d\Omega_a = \sin\theta\, d\theta\, d\phi \tag{9.8}$$

Integrating $d\Omega_a$ over all possible solid angles around the center of a sphere gives the solid angle around a sphere as

$$\Omega_a = \int_{\Omega_a} d\Omega_a = \int_0^{2\pi}\int_0^{\pi} \sin\theta\, d\theta\, d\phi = 4\pi \text{ steradians} \tag{9.9}$$

The solid angle around the center of the base of a hemisphere is 2π steradians.

Planck's law gives the spectral radiance, $B_{\lambda,T}$ (W m^{-2} μm^{-1} sr^{-1}) emitted by a blackbody. In terms of Fig. 9.3, imagine that the x–y plane is the surface of an object. The radiance emitted from point O can travel in any direction above the x–y plane. If it travels through area dA_s, it passes through an incremental solid angle, $d\Omega_a$, which is through a cone originating at point O. $B_{\lambda,T}$ is the spectral radiance emitted by a blackbody at the body's surface. Radiance changes with distance through space. I_λ is defined here as the **spectral radiance** at a given point in space, regardless of the source of the radiance. At the surface of a blackbody, $I_\lambda = B_{\lambda,T}$.

In atmospheric models, two quantities related to spectral radiance are commonly calculated. These are spectral actinic flux and spectral irradiance (e.g., Madronich 1987). **Spectral actinic flux** (also called spherical intensity or actinic irradiance) is the integral of spectral radiance over all solid angles of a sphere. It is used for calculating photolysis coefficients of gases in the atmosphere. Since gas molecules can absorb radiation, regardless of the direction the radiation originates from, radiance is integrated over a sphere to obtain actinic flux.

Spectral actinic flux is determined by integrating incremental actinic flux over all solid angles around the center of a sphere. The **incremental actinic flux** is

$$dE_\lambda = I_\lambda \, d\Omega_a \qquad (9.10)$$

where E_λ is in units of W m^{-2} μm^{-1} but is typically converted to units of photons cm^{-2} s^{-1} μm^{-1} for photolysis calculations (Section 9.8.4). The integral of incremental actinic flux over a sphere is

$$E_\lambda = \int_{\Omega_a} dE_\lambda = \int_{\Omega_a} I_\lambda \, d\Omega_a = \int_0^{2\pi}\int_0^{\pi} I_\lambda \sin\theta \, d\theta \, d\phi \qquad (9.11)$$

When radiation propagates with equal intensity in all directions, I_λ is independent of direction. In such cases, the radiance is called **isotropic**, and (9.11) simplifies to

$$E_\lambda = I_\lambda \int_0^{2\pi}\int_0^{\pi} \sin\theta \, d\theta \, d\phi = 4\pi \, I_\lambda \qquad (9.12)$$

Equation (9.12) states that the isotropic spectral actinic flux equals 4π multiplied by the spectral radiance, where the units of 4π are steradians.

Spectral irradiance (also called the net flux or energy flux) is the vertical component of spectral radiance, originating from all directions above a flat plane, that passes across the plane, per unit surface area, time, and wavelength. In terms of Fig. 9.3, it is the vertical component of radiance, integrated over the hemisphere above the x–y plane. For a zenith angle of $\theta = 0°$, the irradiance impinging on the x–y plane is maximum. For a zenith angle of $\theta = 90°$, the irradiance impinging on the x–y plane is zero.

Spectral irradiance is calculated by integrating the component of radiance normal to the x–y plane over all solid angles of the hemisphere above the x–y plane. If I_λ is the spectral radiance passing through point O in Fig. 9.3, $I_\lambda \cos\theta$ is the component of radiance normal to the x–y plane. Multiplying this quantity by incremental solid angle gives the **incremental spectral irradiance** normal to the x–y plane as

$$dF_\lambda = I_\lambda \cos\theta \, d\Omega_a \qquad (9.13)$$

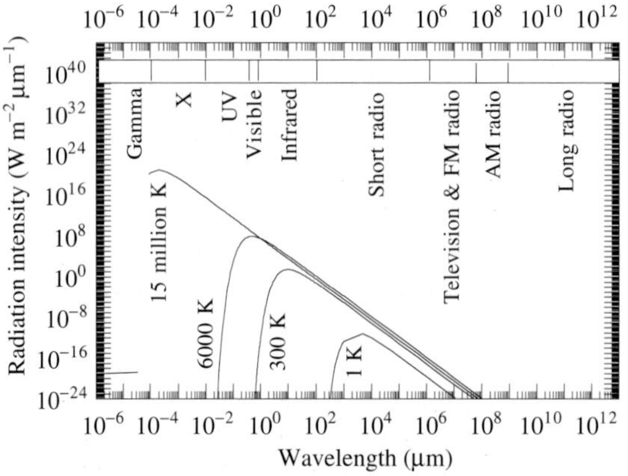

Figure 9.4 Spectral irradiance as a function of wavelength and temperature resulting from emission by a blackbody. Data from (9.16) and (9.4).

(W m^{-2} µm^{-1}). Integrating (9.13) over the hemisphere above the *x–y* plane in Fig. 9.3 gives

$$F_\lambda = \int_{\Omega_a} \mathrm{d}F_\lambda = \int_{\Omega_a} I_\lambda \cos\theta \, \mathrm{d}\Omega_a = \int_0^{2\pi}\int_0^{\pi/2} I_\lambda \cos\theta \sin\theta \, \mathrm{d}\theta \, \mathrm{d}\phi \qquad (9.14)$$

For isotropic emission, (9.14) simplifies to

$$F_\lambda = I_\lambda \int_0^{2\pi}\int_0^{\pi/2} \cos\theta \sin\theta \, \mathrm{d}\theta \, \mathrm{d}\phi = \pi I_\lambda \qquad (9.15)$$

Thus, the isotropic spectral irradiance is π multiplied by the isotropic radiance, where units of π are steradians. At the surface of a blackbody, the **spectral irradiant emission** is

$$F_\lambda = \pi I_\lambda = \pi B_{\lambda,T} \qquad (9.16)$$

Figure 9.4 shows irradiant emission from a blackbody versus wavelength for different temperatures, obtained from (9.16) and (9.4). The figure shows that hotter bodies emit much more energy than do cooler bodies.

The sources of irradiant energy in the Earth's atmosphere are the Sun, the surface of the Earth, and the atmosphere itself. The temperature at the center of the Sun is about 15 million K, and the Sun emits primarily gamma rays and X-rays, as shown in Fig. 9.4. Most of these waves are redirected or absorbed during their random walk toward the outside of the Sun. Solar radiation that reaches the Earth originates mainly from the visible surface of the Sun, the **photosphere**, which has an effective temperature near 6000 K (closer to 5800 K). This temperature results in the **solar spectrum** of radiation shown in Fig. 9.5. The solar spectrum includes

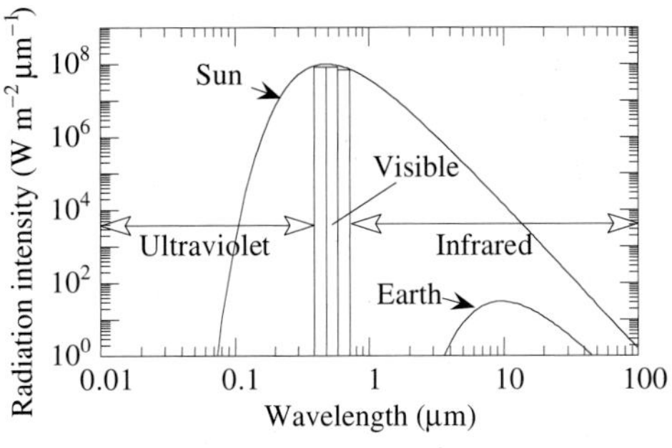

Figure 9.5 Irradiant emission versus wavelength for the Sun and the Earth when both are considered perfect emitters. Data from (9.16) and (9.4).

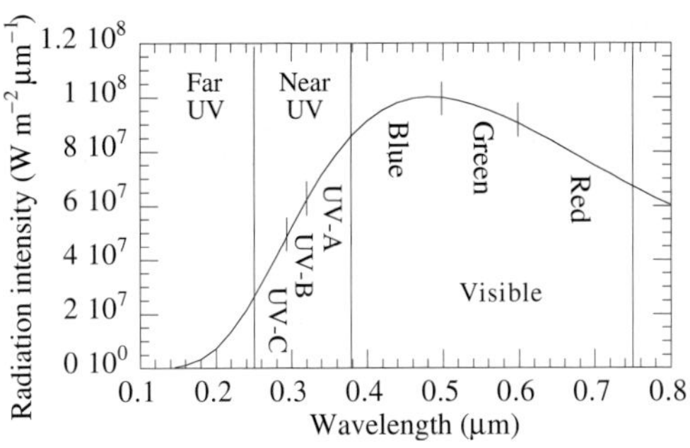

Figure 9.6 Ultraviolet and visible portions of the solar spectrum. Data from (9.16) and (9.4).

ultraviolet, visible, and infrared wavelength regions. The **ultraviolet region** consists of wavelengths < 0.38 μm, the **infrared region** consists of wavelengths > 0.75 μm, and the **visible region** consists of wavelengths 0.38–0.75 μm. The infrared spectrum is divided into the **near- or solar-infrared** (0.75–4 μm) and the **far- or thermal-infrared** (>4 μm) spectra. Most infrared radiation from the Sun that reaches the Earth is solar-infrared radiation.

Figure 9.5 shows the irradiant emission spectrum of the Earth. The effective temperature at the surface of Earth is approximately 288 K; thus, almost all radiation emission from the Earth is thermal-infrared radiation. Even in the stratosphere, where temperatures can drop to below 200 K, and in hot deserts, where temperatures can exceed 315 K, emission is thermal-infrared. Thus, with respect to studies of the Earth's atmosphere, incoming wavelengths of interest are ultraviolet, visible and solar-infrared, and outgoing wavelengths of interest are thermal-infrared.

Figure 9.6 shows the ultraviolet (UV) and visible portions of the solar spectrum. The UV spectrum is divided into the **far-UV** (0.01 to 0.25 μm) and **near-UV** (0.25 to 0.38 μm) spectra, as shown in Fig. 9.6. The near-UV spectrum is further

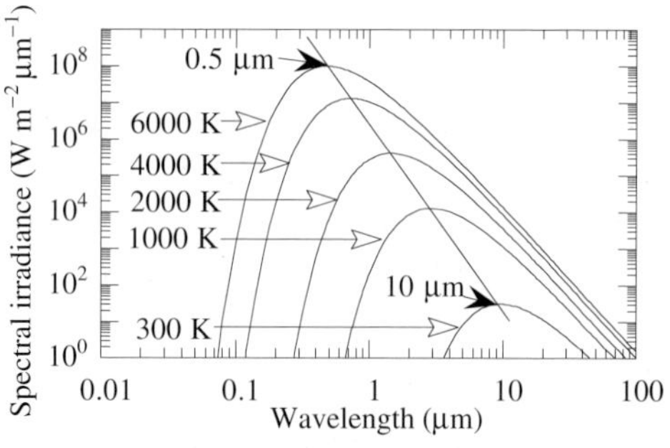

Figure 9.7 Radiation spectrum as a function of temperature and wavelength. Data from (9.16) and (9.4). The line through the peaks was obtained from Wien's law.

divided into **UV-A** (0.32 to 0.38 μm), **UV-B** (0.29 to 0.32 μm), and **UV-C** (0.25 to 0.29 μm) wavelengths. All solar radiation below 0.28 μm is absorbed by the Earth's atmosphere above the troposphere. Nitrogen gas (N_2) absorbs wavelengths less than 0.1 μm in the thermosphere, and oxygen absorbs wavelengths less than 0.245 μm in the thermosphere, mesosphere, and stratosphere. Ozone in the stratosphere and troposphere absorbs much radiation from 0.17 to 0.35 μm. A fraction of UV-A, UV-B, and UV-C radiation longer than 0.28 μm passes through the stratosphere and troposphere to the ground.

UV-A radiation (tanning radiation) is relatively harmless to most humans. UV-B radiation can cause sunburn. UV-C radiation can cause severe damage to many forms of life when received in high dosages. Since ozone in the stratosphere absorbs most UV-B and UV-C radiation, ozone reduction in the stratosphere would allow more UV-B and UV-C radiation to reach the surface, damaging exposed life forms.

The visible spectrum consists of wavelengths corresponding to the colors of the rainbow. For convenience, visible light is divided into blue (0.38–0.5 μm), green (0.5–0.6 μm), and red (0.6–0.75 μm) light.

The peak radiation wavelength emitted at a given temperature can be found from **Wien's displacement law** (Wien's law). This law is derived by differentiating (9.4) with respect to wavelength at a constant temperature and setting the derivative equal to zero. The result is

$$\lambda_p(\mu\text{m}) \approx \frac{2897}{T(\text{K})} \tag{9.17}$$

where λ_p is the peak wavelength of emission from a blackbody. The law states that the hotter a body, the shorter the peak wavelength of radiation emitted. Figure 9.7 shows how peak wavelengths of radiation at different temperatures can be connected by a line on a log–log scale, where the line is described by (9.17).

Example 9.2

In the Sun's photosphere, the peak wavelength of emission is near $\lambda_p = 2897/5800 = 0.5\ \mu m$, which is in the visible part of the solar spectrum. At the Earth's surface, the peak wavelength of emission is $\lambda_p = 2897/288 = 10.1\ \mu m$, which is in the infrared part of the spectrum. Both peak wavelengths are shown in Fig. 9.7.

Integrating the spectral irradiance in (9.16) over all wavelengths gives the **Stefan–Boltzmann law**,

$$F_b = \pi \int_0^\infty B_{\lambda,T}\, d\lambda = \sigma_B T^4 \tag{9.18}$$

where F_b is the total irradiance emitted by a blackbody at a given temperature and $\sigma_B = 2k_B^4\pi^4/15h^3c^2 = 5.67 \times 10^{-8}$ W m^{-2} K^{-4} is the **Stefan–Boltzmann constant.** F_b is the area under each curve shown in Figs. 9.4–9.7.

Example 9.3

For an effective temperature of T = 5800 K, the irradiance emitted by the photosphere is $F_b = 64 \times 10^6$ W m^{-2}. For an effective temperature of $T = 288$ K, the irradiance emitted by the Earth's surface is $F_b = 390$ W m^{-2}.

9.3 LIGHT PROCESSES

Absorption is one process that affects electromagnetic radiation in the Earth's atmosphere. Other processes include reflection, refraction, dispersion, diffraction, particle scattering, and gas scattering. These processes are discussed here.

9.3.1 Reflection

Reflection occurs when a wave or photon of radiation is absorbed by an object and reemitted at an angle (called angle of reflection) equal to the angle of incidence. No energy is lost during absorption and reemission. Figure 9.8 shows an example of reflection.

The reflectivity of a surface is called its **albedo**, which is the fraction of sunlight incident on a surface that is reflected. Albedos are wavelength-dependent. Table 9.2 gives albedos, averaged over the visible spectrum, for several surface types. The table shows that the albedo of the Earth and atmosphere together (**planetary albedo**) is about 30 percent. Two-thirds of Earth's surface is covered with water, which has an albedo of 5–20 percent (typical value of 8 percent), depending

Table 9.2 Albedos averaged over the visible spectrum for several surface types

Surface type	Albedo (fraction)	Surface type	Albedo (fraction)
Earth and atmosphere	0.3[a]	Soil	0.05–0.2[b]
Liquid water	0.05–0.2[b]	Grass	0.16–0.26[c]
Fresh snow	0.75–0.95[d]	Desert	0.20–0.40[b]
Old snow	0.4–0.7[d]	Forest	0.10–0.25[b]
Thick clouds	0.3–0.9[b]	Asphalt	0.05–0.2[c]
Thin clouds	0.2–0.7[b]	Concrete	0.1–0.35[c]
Sea ice	0.25–0.4[b]	Urban	0.1–0.27[c]

[a] Liou (2002), [b] Hartmann (1994), [c] Oke (1978), [d] Sellers (1965).

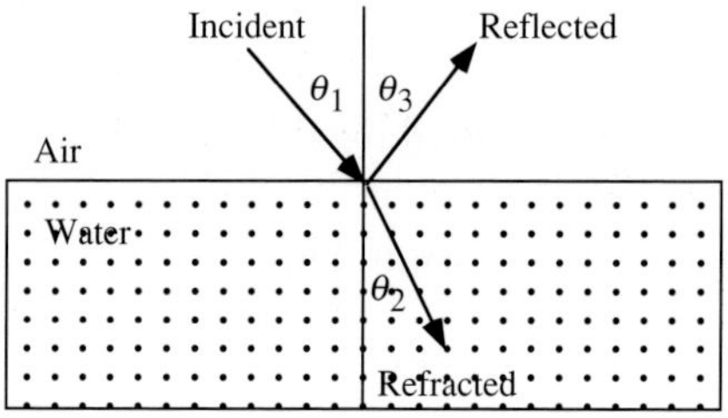

Figure 9.8 Examples of reflection and refraction. During reflection, $\theta_1 = \theta_3$. During refraction, the angles of incidence and refraction are related by Snell's law.

largely on the angle of the Sun. Soils and forests also have low albedos. Much of the Earth–atmosphere reflectivity is due to clouds and ice, which have high albedos.

UV-B albedos are much smaller than are visible albedos over grassland and other nonsnow surfaces, but UV-B albedos are slightly larger than are visible albedos over snow (Blumthaler and Ambach 1988). Harvey *et al.* (1977) found the UV-A–UV-B albedo to be 3 percent for water and 0.8 percent for grassland.

9.3.2 Refraction

Refraction occurs when a wave or photon travels through a medium of one density then bends as it enters a medium of another density. During the transition, the speed of the wave changes, changing the angle of the incident wave relative to a surface normal, as shown in Fig. 9.8. If a wave travels from a medium of one density to a medium of a higher density, it bends (refracts) toward the surface normal. The angle of refraction is related to the angle of incidence by **Snell's law**,

$$\frac{n_2}{n_1} = \frac{\sin\theta_1}{\sin\theta_2} \tag{9.19}$$

Table 9.3 Real indices of refraction of air and liquid water versus wavelength

Wavelength (μm)	n_{air}	n_{water}	Wavelength (μm)	n_{air}	n_{water}
0.2	1.000324	1.396	1.0	1.000274	1.327
0.3	1.000292	1.349	4.0	1.000273	1.351
0.4	1.000283	1.339	7.0	1.000273	1.317
0.5	1.000279	1.335	10.0	1.000273	1.218
0.6	1.000277	1.332	20.0	1.000273	1.480
0.7	1.000276	1.331			

Data for air were obtained from (9.21), and data for liquid water were obtained from Hale and Querry (1973).

In this equation, n is the real index of refraction (dimensionless), θ is the angle of incidence or refraction, and subscripts 1 and 2 refer to incident and refracted light, respectively. The **real index of refraction** is the ratio of the speed of light in a vacuum (c) to that in a different medium (c_1). Thus,

$$n_1 = c/c_1 \tag{9.20}$$

Since light cannot travel faster than its speed in a vacuum, the real index of refraction of a medium other than a vacuum must exceed unity. The index of refraction is wavelength dependent. The real index of refraction of air as a function of wavelength is approximately

$$n_{a,\lambda} - 1 = 10^{-8}\left(8342.13 + \frac{2\,406\,030}{130 - \lambda^{-2}} + \frac{15\,997}{38.9 - \lambda^{-2}}\right) \tag{9.21}$$

(Edlen 1966) where λ is in micrometers. Real indices of refraction of air and liquid water are given in Table 9.3 for several wavelengths.

Example 9.4

Find the angle of refraction in water if light of wavelength $\lambda = 0.5$ μm enters water from air at an incident angle θ_1 of 45°. Also find the speed of light in air and water at this wavelength.

SOLUTION

At $\lambda = 0.5$ μm, $n_{air} = 1.000\,279$ and $n_{water} = 1.335$ from Table 9.3. Thus, from (9.19), the angle of refraction in water is $\theta_2 = 32°$. From (9.20), the speeds of light in air and water at $\lambda = 0.5$ μm are 2.9971×10^8 m s^{-1} and 2.2456×10^8 m s^{-1}, respectively.

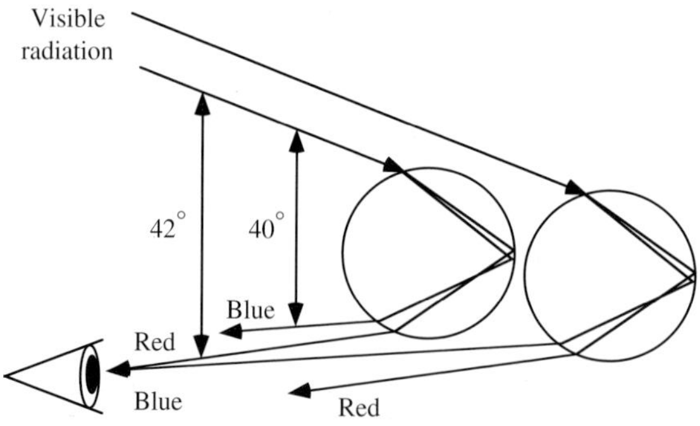

Figure 9.9 Geometry of a primary rainbow.

9.3.3 Dispersion

Refraction affects atmospheric optics in several ways. **Rainbows** appear when incident light is refracted, dispersed, reflected, then refracted again, as shown in Fig. 9.9. As a beam of visible light enters a raindrop, all wavelengths bend toward the surface normal due to refraction. Blue light bends the most and red light, the least. The separation of white light into individual colors by selective refraction is called **dispersion** (or **dispersive refraction**). When individual wavelengths hit the back of a drop, they reflect internally. When reflected waves reach the front edge of the drop, they leave the drop and refract away from the surface normal. Only one wavelength from each raindrop impinges upon a viewer's eye. Thus, a rainbow appears when individual waves from many raindrops are seen simultaneously. Red light appears on the top and blue (or violet) light appears on the bottom of a **primary rainbow** (Fig. 9.9). The overall angle between the incident beam and the beam reaching a viewer's eye following interaction with rain is 40° for blue light and 42° for red light (Fig. 9.9). A **secondary rainbow** can occur if a second reflection occurs inside each drop. The geometry of a rainbow is such that the Sun must be at a viewer's back for the viewer to see a rainbow.

9.3.4 Diffraction

Diffraction is a process by which the direction that a wave propagates changes when the wave encounters an obstruction. Diffraction bends waves as they pass by the edge of an obstruction, such as an aerosol particle, cloud drop, or raindrop.

Diffraction can be explained in terms of **Huygens' principle**, which states that each point of an advancing wavefront may be considered the source of a new series of secondary waves. A **wavefront** is a surface of constant phase in a wave's motion. If a stone is dropped in a tank of water, waves move out horizontally in all directions, and wavefronts are seen as concentric circles around the stone. If a point source emits waves in three dimensions, wavefronts are concentric spherical

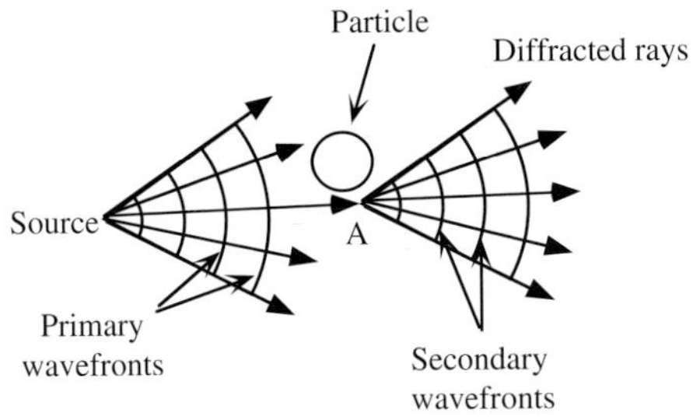

Figure 9.10 Diffraction around a particle. Any point along a wavefront may be taken as the source of a new series of secondary waves. Rays emitted from point A appear to cause waves from the original source to bend around the particle.

surfaces. When a wavefront encounters an obstacle, such as in Fig. 9.10, waves appear to bend (diffract) around the obstacle because a series of secondary concentric waves is emitted at the surface of the obstacle (and at other points along the primary wavefronts). New waves do not appear in the backward direction, because the intensity of the secondary wavelet depends on angle and is zero in the backward direction, as Kirchoff demonstrated.

A **corona** is a set of prismatic colored rings around the Sun or Moon that arises when sunlight or moonlight diffract around spherical liquid water drops in thin clouds slightly blocking the Sun or Moon. Sometimes, the corona around the Moon appears white with alternating bands of light and dark. These alternating bands are caused by constructive and destructive interference, which arise due to the bending of some waves more than others during diffraction. **Constructive interference** results when the crests or troughs of two waves meet, producing a band of bright light. **Destructive interference** results when a crest meets a trough, producing a band of darkness.

9.3.5 Particle scattering

Particle scattering is the combination of the effects of reflection, refraction, and diffraction. When a wave approaches a spherical particle, such as a cloud drop, it can reflect off the particle, diffract around the edge of the particle, or refract into the particle. Once in the particle, the wave can be absorbed, transmit through the particle and refract out, or reflect internally one or more times and then refract out. Figure 9.11 illustrates these processes, except for absorption, which is not a scattering process. The processes that affect particle scattering the most are diffraction and double refraction, identified by rays C and B, respectively. Thus, particles scatter light primarily in the forward direction. They also scatter some light to the side and in the backward direction. **Backscattered** light results primarily

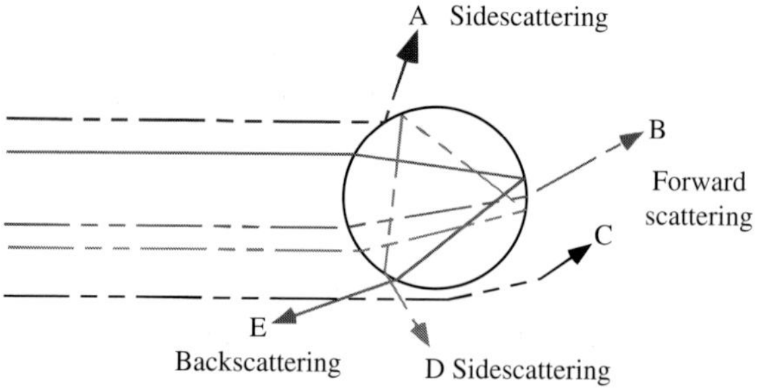

Figure 9.11 Radiative scattering by a sphere. Ray A is reflected, B is refracted twice, C is diffracted, D is refracted, internally reflected twice, then refracted, and E is refracted, reflected once, then refracted. Rays A, B, C, and D scatter in the forward or sideward direction. E scatters in the backward direction.

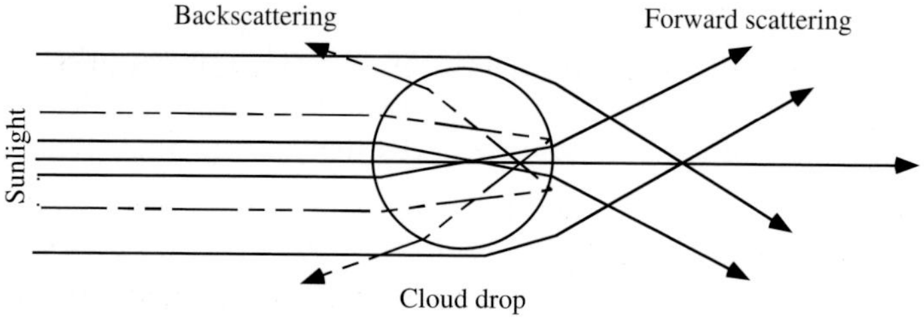

Figure 9.12 Forward scattering and backscattering by a cloud drop. Forward scattering is due primarily to diffraction and double refraction. Backscattering is due primarily to a single internal reflection.

from a single internal reflection (ray E). Figure 9.12 shows the primary processes by which a cloud drop might scatter radiation in the forward and backward directions.

Whether one or more internal reflections occurs within a particle depends on the angle at which a wave within the particle strikes a surface. Suppose light in Fig. 9.8 travels from water to air instead of from air to water. In the figure, **total internal reflection** within the water occurs only if $\theta_1 \geq 90^\circ$. From (9.19), $\theta_1 = 90^\circ$ when

$$\theta_{2,\mathrm{c}} = \sin^{-1}\left(\frac{n_1}{n_2}\sin 90^\circ\right) \tag{9.22}$$

which is the **critical angle**. When $\theta_2 < \theta_{2,\mathrm{c}}$, some light reflects internally, but most refracts out of the water to the air. When $\theta_2 \geq \theta_{2,\mathrm{c}}$, total internal reflection occurs within the water. Total internal reflection can occur only when light originates from the medium of lower index of refraction.

Example 9.5

Find the critical angle of light at $\lambda = 0.5$ μm in a liquid water drop suspended in air.

SOLUTION

From Table 9.3, the indices of refraction of water and air are 1.335 and 1.000 279, respectively. Substituting these values into (9.22) gives $\theta_{2,c} = 48.53$. Thus, total internal reflection within the drop occurs when $\theta_2 \geq 48.53°$.

9.3.6 Gas scattering

Gas scattering is the redirection of radiation by a gas molecule without a net transfer of energy to the molecule. When a gas molecule scatters, incident radiation is redirected symmetrically in the forward and backward direction and somewhat off to the side. This differs from cloud drops and aerosol particles, which scatter mostly in the forward direction. Visible radiation does not diffract readily around gas molecules, because visible wavelengths are much larger than is the diameter of a gas molecule.

Another difference between gas scattering and particle scattering is that gases selectively scatter the shortest (blue) wavelengths of light whereas particles scatter all wavelengths relatively equally. Selective scattering by gas molecules explains why the Sun appears white at noon, yellow in the afternoon, and red at sunset and why the sky is blue.

White sunlight that enters the Earth's atmosphere travels a shorter distance before reaching a viewer's eye at noon than at any other time during the day, as illustrated in Fig. 9.13. During white light's travel through the atmosphere, blue wavelengths are preferentially scattered out of the direct beam by gas molecules, but not enough blue light is scattered for a person looking at the Sun to notice that the incident beam has changed its color from white. Thus, a person looking at the Sun at noon often sees a **white Sun**. The blue light that scatters out of the direct beam is scattered by gas molecules multiple times, and some of it eventually enters the viewer's eye when the viewer looks away from the Sun. As such, a viewer looking away from the Sun sees a **blue sky**.

In the afternoon, light takes a longer path through the air than it does at noon; thus, more blue and some green light is scattered out of the direct solar beam in the afternoon than at noon. Although a single gas molecule is less likely to scatter a green than a blue wavelength, the number of gas molecules along a viewer's line of sight is so large in the afternoon that the probability of green light scattering is sizable. Nearly all red and some green are still transmitted to the viewer's eye in the afternoon, causing the **Sun to appear yellow**. In clean air, the Sun can remain yellow until just before it reaches the horizon.

When the Sun reaches the horizon at sunset, sunlight traverses its longest distance through the atmosphere, and all blue and green and some red wavelengths are

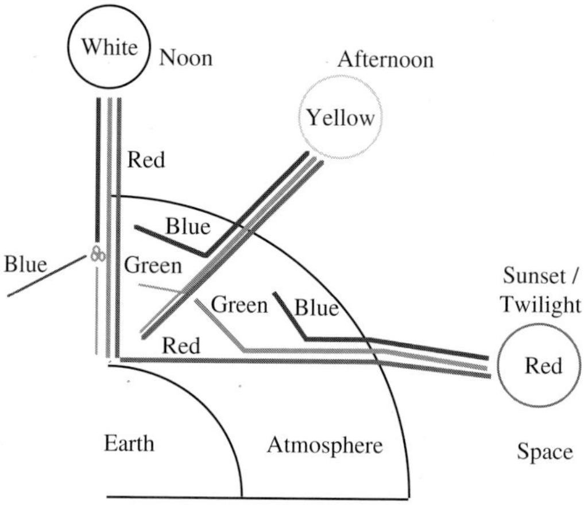

Figure 9.13 Colors of the Sun. At noon, the Sun appears white because red, green, and some blue light transmit to a viewer's eye. In the afternoon, sunlight traverses a longer path in the atmosphere, removing more blue. At sunset, most green is removed from the line of sight, leaving a red Sun. After sunset, the sky appears red due to refraction between space and the Earth's atmosphere.

scattered out of the Sun's direct beam. Only some direct red light transmits, and a viewer sees a **red Sun**. Sunlight can be seen after sunset because sunlight refracts as it enters Earth's atmosphere, as illustrated in Fig. 9.13.

9.4 ABSORPTION AND SCATTERING BY GASES AND PARTICLES

Whereas scattering redirects radiation, absorption removes radiation from an incident beam. In both cases, radiation in the beam is attenuated, reducing the quantity of radiation transmitted. Aerosol particles, cloud drops, and gases scatter and absorb as a function of wavelength. In polluted, cloud-free air, the main process reducing visibility is aerosol particle scattering. Absorption of visible light is important only when soot (black carbon plus organic matter) is present. Gas absorption is important only when nitrogen dioxide concentrations are high. Gas scattering always occurs but is important relative to other processes only in clean air. In this section, gas absorption, gas scattering, particle absorption, and particle scattering are discussed in some detail.

9.4.1 Gas absorption

Gases selectively absorb radiation as a function of wavelength. In this subsection, gas absorption in the solar and infrared spectra are discussed, the gas absorption extinction coefficient is defined, and the effect of gas absorption on visibility is analyzed.

Table 9.4 Wavelengths of absorption in the visible and UV spectra by several gases

Gas name	Chemical formula	Absorption wavelengths (μm)
Visible/near-UV/far-UV absorbers		
Ozone	O_3	<0.35, 0.45–0.75
Nitrate radical	NO_3	<0.67
Nitrogen dioxide	NO_2	<0.71
Near-UV/far-UV absorbers		
Nitrous acid	HONO	<0.4
Dinitrogen pentoxide	N_2O_5	<0.38
Formaldehyde	HCHO	<0.36
Hydrogen peroxide	H_2O_2	<0.35
Acetaldehyde	CH_3CHO	<0.345
Peroxynitric acid	HO_2NO_2	<0.33
Nitric acid	HNO_3	<0.33
Peroxyacetyl nitrate	$CH_3CO_3NO_2$	<0.3
Far-UV absorbers		
Molecular oxygen	O_2	<0.245
Nitrous oxide	N_2O	<0.24
CFC-11	$CFCl_3$	<0.23
CFC-12	CF_2Cl_2	<0.23
Methyl chloride	CH_3Cl	<0.22
Carbon dioxide	CO_2	<0.21
Water vapor	H_2O	<0.21
Molecular nitrogen	N_2	<0.1

9.4.1.1 *Gas absorption in the solar spectrum*

Gas absorption in the solar spectrum is important for three reasons: (1) absorption heats the air and prevents some solar radiation, including harmful UV radiation, from reaching the ground, (2) absorption often breaks gases into smaller molecules or atoms during **photolysis**, driving atmospheric photochemistry, and (3) absorption affects visibility.

When a gas absorbs solar radiation, it converts electromagnetic energy to internal energy, increasing the thermal speed of the gas, thereby raising its temperature. Table 9.4 shows several gases that absorb UV and visible radiation. The most noticeable effects of gas absorption on air temperature are the stratospheric inversion, which is caused entirely by absorption of UV light by **ozone** (O_3), and the thermospheric inversion, caused by absorption of UV light by **molecular nitrogen** (N_2) and **molecular oxygen** (O_2) (Fig. 2.4).

In 1880, M. J. Chappuis found that ozone absorbs visible radiation of 0.45–0.75 μm (now the **Chappuis bands**). In 1916, English physicists Alfred Fowler (1868–1940) and Robert John Strutt (1875–1947, the son of Lord Rayleigh) showed that ozone also weakly absorbs at 0.31–0.35 μm (now the **Huggins bands**). When ozone absorbs in the Huggins and Chappuis bands, it often photolyzes to $O_2 + O(^3P)$. In 1881, John Hartley suggested that ozone was present in the upper atmosphere and hypothesized that the reason that the Earth's surface received little radiation shorter than 0.31 μm was because ozone absorbed these wavelengths.

The absorption bands of ozone below 0.31 μm are now the **Hartley bands.** When ozone absorbs in the Hartley bands, it often photolyzes to $O_2 + O(^1D)$. Ozone absorption prevents nearly all UV wavelengths 0.245–0.28 μm and most wavelengths 0.28–0.32 μm from reaching the troposphere, protecting the surface of the Earth.

Molecular oxygen absorbs wavelengths < 0.245 μm and molecular nitrogen absorbs wavelengths < 0.1 μm, together preventing nearly all solar wavelengths < 0.245 μm from reaching the troposphere. Oxygen absorption at wavelengths < 0.175 μm (an absorption wavelength region called the **Schumann–Runge system**) often results in its photolysis to $O(^1D) + O(^3P)$. Absorption from 0.175 to 0.245 μm (the **Herzberg continuum**) often results in its photolysis to $2O(^3P)$.

Although the gases in Table 9.4, aside from O_3, O_2, and N_2, absorb UV radiation, their mixing ratios are too low to reduce it significantly. For instance, stratospheric mixing ratios of **water vapor** (0 to 6 ppmv) are much lower than are those of O_2, which absorbs many of the same wavelengths as does water vapor. Thus, water vapor has little effect on UV attenuation in the stratosphere. Similarly, stratospheric mixing ratios of **carbon dioxide** (370 ppmv) are much lower than are those of O_2 or N_2, both of which absorb the same wavelengths as does carbon dioxide. Although water vapor mixing ratios are much higher in the troposphere than in the stratosphere, wavelengths that water absorbs do not reach the troposphere. Wavelengths reaching the troposphere are longer than 0.28 μm.

Of the gases listed in Table 9.4, only ozone, **nitrogen dioxide** (NO_2), and the **nitrate radical** (NO_3) absorb in the visible spectrum. The rest absorb in the ultraviolet spectrum. Absorption by ozone is weak in the visible spectrum, and concentrations of the nitrate radical are relatively low, except at night, when sunlight is absent. Thus, in polluted air, nitrogen dioxide is the only gas that absorbs enough radiation to affect visibility. Nitrogen dioxide appears yellow, brown, or red because it absorbs blue light preferentially and green light to a lesser extent.

9.4.1.2 *Gas absorption in the infrared spectrum: the greenhouse effect*

Gas absorption in the solar-infrared and thermal-infrared affects the Earth's energy balance. Such absorption does not affect photolysis significantly or visibility at all.

Figure 9.14 shows the fraction of radiation incident at the top of Earth's atmosphere that is transmitted to the surface, versus wavelength, in the presence of individual gases and the combination of multiple gases. The gases listed in the figure (H_2O, CO_2, CH_4, CO, O_3, O_2, N_2O, CH_3Cl, $CFCl_3$, CF_2Cl_2, and CCl_4) are **greenhouse gases**, which are gases that are relatively transparent to incoming visible radiation but opaque to selective wavelengths of outgoing infrared radiation.

The **natural greenhouse effect** is the warming of the Earth's lower atmosphere due to natural greenhouse gases. Greenhouse gases cause a net warming of the Earth's atmosphere like a glass house causes a net warming of its interior. Because most incoming solar radiation can penetrate a glass house but a portion of outgoing thermal-infrared radiation cannot, air inside a glass house warms during the day

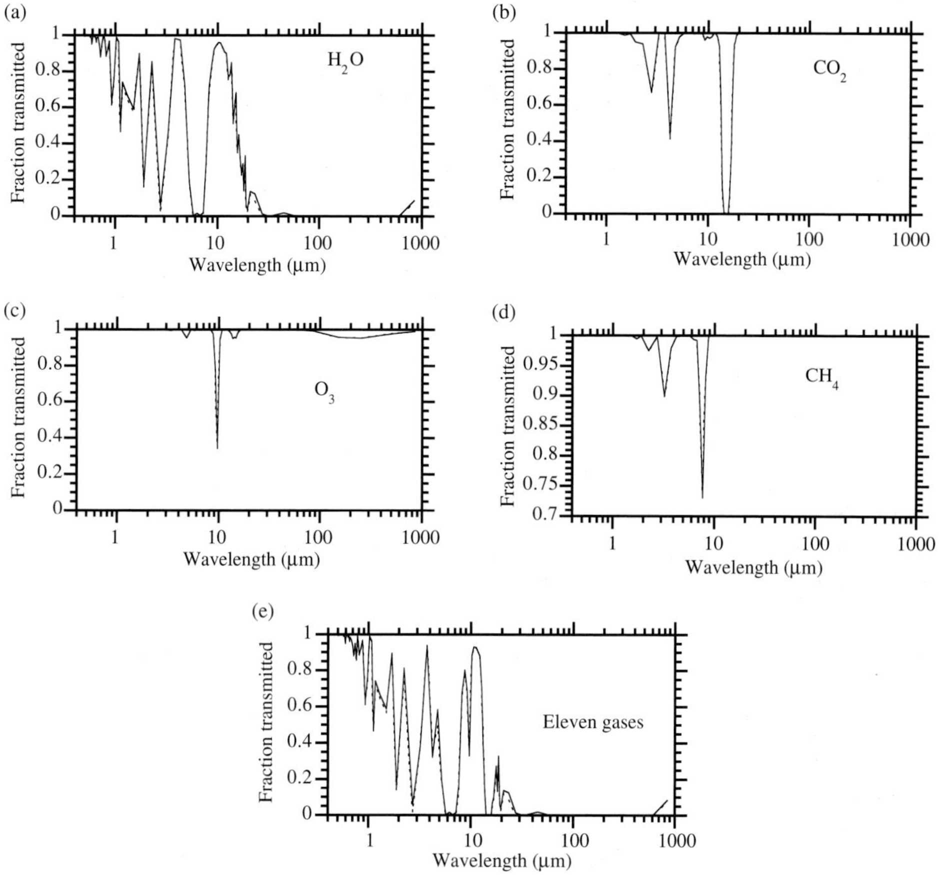

Figure 9.14 Comparison of the wavelength-dependent fraction of transmitted incident radiation, calculated from line-by-line data (—) (Rothman *et al.* 2003) with that calculated from a model (-----) (Jacobson 2005a) when absorption by (a) H_2O, (b) CO_2, (c) O_3, (d) CH_4, and (e) 11 absorbing gases together, is considered. Mass pathlengths (g cm^{-2}) of the gases, defined in (9.26), were as follows: H_2O: 4.7275; CO_2: 0.58095; O_3: 0.00064558; CH_4: 0.00092151; O_2: 239.106; N_2O: 0.000461839; CO: 8.5111×10^{-5}; CH_3Cl: 9.100641×10^{-7}; CCl_4: 4.69957×10^{-7}; $CFCl_3$: 8.53059×10^{-7}; CF_2Cl_2: 1.39333×10^{-6}. These pathlengths represent pathlengths through the entire atmosphere, but the calculations were performed at a constant temperature (270 K) and pressure (322.15 hPa).

so long as mass (such as plant mass) is present to absorb solar and reemit thermal-infrared radiation. The surface of the Earth, like plants, absorbs solar and reemits thermal-infrared radiation. Greenhouse gases, like glass, are transparent to most solar radiation but absorb a portion of thermal-infrared.

The natural greenhouse effect is responsible for about 33 K of the Earth's average near-surface air temperature of 288 K. Without the natural greenhouse effect, Earth's average near-surface temperature would be about 255 K, which is too cold to sustain most life. Thus, the presence of natural greenhouse gases is beneficial. Human emission of greenhouse gases has increased the concentrations of

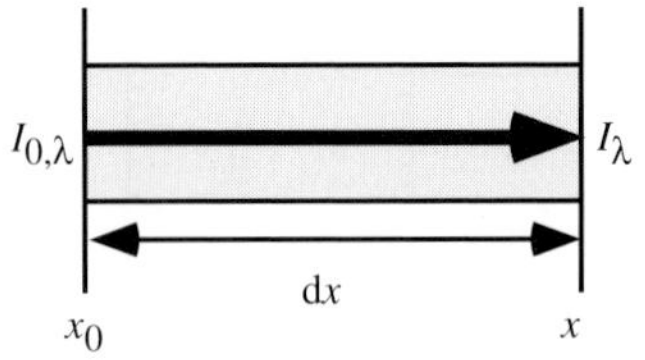

Figure 9.15 Attenuation of incident radiance I_0 due to absorption in a column of gas.

such gases, causing global warming. **Global warming** is the increase in the Earth's temperature above that from the natural greenhouse effect due to the addition of anthropogenically emitted greenhouse gases and particulate black carbon to the air.

The most important greenhouse gas is water vapor, which accounts for approximately 90 percent of the 33-K temperature increase due to natural greenhouse warming. Figure 9.14(a) shows the infrared wavelengths at which water vapor absorbs (and transmits). Carbon dioxide is the second most important and abundant natural greenhouse gas. Its transmission spectrum is shown in Fig. 9.14(b). Both water vapor and carbon dioxide are transparent to most visible radiation. Most infrared absorption by water vapor and carbon dioxide occurs outside the wavelength region, 8–12 μm, called the **atmospheric window**. The atmospheric window can be seen in Fig. 9.14(e), which shows the transmission through all important infrared absorbers in the atmosphere. In the atmospheric window, gases are relatively transparent to thermal-infrared radiation, allowing the radiation to radiate to higher altitudes. Ozone (Fig. 9.14(c)), several chlorofluorocarbons, and methyl chloride absorb radiation within the atmospheric window. Nitrous oxide and methane (Fig. 9.4(d)) absorb at the edges of the window. Increases in concentration of these gases enhance global warming by strengthening thermal-infrared absorption at wavelengths that usually radiate to higher altitudes.

9.4.1.3 *Gas absorption extinction coefficient*

The **gas absorption extinction coefficient,** $\sigma_{a,g,\lambda}$ (cm^{-1}, m^{-1}, or km^{-1}) quantifies the extent of absorption by a gas. An **extinction coefficient** describes the loss of electromagnetic radiation due to a specific process, per unit distance, and may be determined as the product of an effective cross section and a number concentration. In the case of gas absorption, the extinction coefficient through a uniformly mixed gas q is $\sigma_{a,g,q,\lambda,T} = N_q b_{a,g,q,\lambda,T}$, where N_q is the number concentration of the gas (molec. cm^{-3}) and $b_{a,g,q,\lambda,T}$ is the absorption cross section of the gas (cm^2) at wavelength λ and temperature T. The **absorption cross section** of a gas is an effective cross section that results in radiance reduction by absorption.

Suppose incident radiance $I_{0,\lambda}$ travels a distance dx through the uniformly mixed absorbing gas shown in Fig. 9.15. The reduction in radiance with distance through the gas is

$$\frac{dI_\lambda}{dx} = -N_q b_{a,g,q,\lambda,T} I_\lambda = -\sigma_{a,g,q,\lambda,T} I_\lambda \tag{9.23}$$

Integrating (9.23) from $I_{0,\lambda}$ to I_λ and x_0 to x gives

$$I_\lambda = I_{0,\lambda} e^{-N_q b_{a,g,q,\lambda,T}(x-x_0)} = I_{0,\lambda} e^{-\sigma_{a,g,q,\lambda,T}(x-x_0)} \tag{9.24}$$

Equation (9.24) states that incident radiation decreases from $I_{0,\lambda}$ to I_λ over distance dx by gas absorption. The absorption extinction coefficient quantifies the rate of absorption per unit distance.

Absorption cross-section data can be extracted mathematically from (9.24) if the attenuation of radiance from $I_{0,\lambda}$ to I_λ through a column of gas of **molecular pathlength** $N_q(x - x_0)$ (molec. cm^{-2}) is measured experimentally. DeMore *et al.* (1997) and Atkinson *et al.* (1997) provide absorption cross-section data as a function of wavelength and temperature calculated in this manner.

For radiative transfer calculations, the extinction coefficient due to gas absorption, summed over all absorbing gases, is required. This quantity is

$$\sigma_{a,g,\lambda} = \sum_{q=1}^{N_{ag}} N_q b_{a,g,q,\lambda,T} = \sum_{q=1}^{N_{ag}} \sigma_{a,g,q,\lambda,T} \tag{9.25}$$

where N_{ag} is the number of absorbing gases, and the subscript T was dropped in $\sigma_{a,g,\lambda}$. In the infrared spectrum, the gas absorption extinction coefficient may be written in terms of a **mass absorption coefficient** $k_{a,g,q,\lambda}$(cm^2 g^{-1}),

$$\sigma_{a,g,\lambda} = \sum_{q=1}^{N_{ag}} \rho_q k_{a,g,q,\lambda} = \sum_{q=1}^{N_{ag}} \frac{N_q m_q}{A} k_{a,g,q,\lambda} = \sum_{q=1}^{N_{ag}} \frac{u_q}{(x - x_0)} k_{a,g,q,\lambda} \tag{9.26}$$

where $\rho_q = N_q m_q / A$ is the mass density (g cm^{-3}), N_q is the number concentration (molec. cm^{-3}), m_q is the molecular weight (g mol^{-1}), $u_q = \rho_q (x - x_0)$ is the **mass pathlength** (g cm^{-2}) of gas q in air, and A is Avogadro's number (molec. mol^{-1}). Mass absorption coefficients are cross sections, averaged over a wavelength increment, per unit mass of absorbing gas.

Mass absorption coefficients can be obtained from high-resolution spectral data, such as from the HITRAN database (Rothman *et al.* 2003). This particular database provides absorption parameters for over one million narrow absorption line distributions in the solar- and thermal-infrared spectra. **Line distributions** are functions describing the intensity of gas absorption over a narrow region in wavelength space. The width and peak intensity of absorption of each line distribution of a gas varies with temperature and pressure. Line distributions may overlap one another. Line-distribution data are commonly referred to as **line-by-line** data.

The shape of a line distribution is often modeled with the **Lorentz profile** (Lorentz 1906; Rothman *et al.* 2003), examples of which are shown in Fig. 9.16. With this

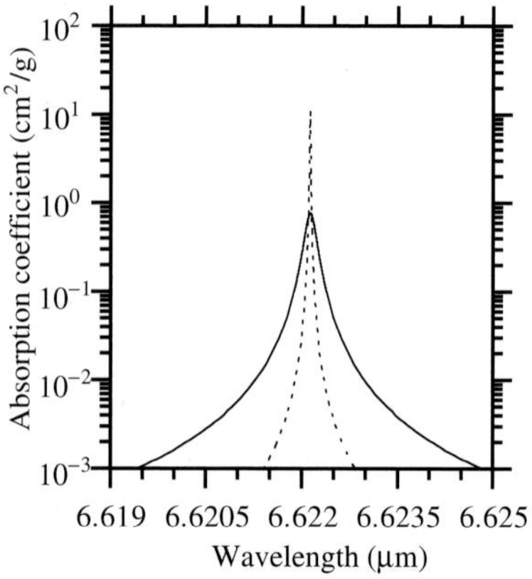

Figure 9.16 Absorption coefficient of water vapor versus wavelength over one line distribution with a Lorentzian shape profile at two pressures, — 322.15 hPa, ----22.57 hPa and at a temperature of 270 K. The curve was calculated with (9.27) with data from Rothman *et al.* (2003).

profile, the absorption coefficient ($cm^2\,g^{-1}$) of gas q at wavelength λ is calculated with

$$k_{a,g,q,\lambda} = \frac{A}{m_q}\frac{1}{\pi}\left\{\frac{S_q(T)\gamma_q(p_a, T)}{\gamma_q(p_a, T)^2 + [\tilde{\nu}_\lambda - (\tilde{\nu}_q - \delta_q(p_{ref})p_a)]^2}\right\} \tag{9.27}$$

where A is Avogadro's number (molec. mol^{-1}), m_q is the molecular weight of gas q ($g\,mol^{-1}$), $S_q(T)$ is the temperature-dependent **intensity** of the line distribution ($cm^{-1}/(molec.^{-1}cm^{-2})$), $\gamma_q(p_a, T)$ is the **pressure-broadened halfwidth** of the line distribution (cm^{-1}), $\tilde{\nu}_\lambda$ is the **wavenumber** (cm^{-1}) at wavelength λ, $\tilde{\nu}_q$ is the **central wavenumber** (cm^{-1}) of the line distribution before an air-broadened pressure shift, $\delta_q(p_{ref})$ is the **air-broadened pressure shift** ($cm^{-1}\,atm^{-1}$) of wavenumber $\tilde{\nu}_q$ at pressure $p_{ref} = 1$ atm, p_a is current atmospheric pressure (atm), and $\tilde{\nu}_q - \delta_q(p_{ref})p_a$ is the central wavenumber of the distribution after adjustment for pressure (cm^{-1}). The pressure-broadened halfwidth is

$$\gamma_q(p_a, T) = \left(\frac{T_{ref}}{T}\right)^{n_q}[\gamma_{air,q}(p_{ref}, T_{ref})(p_a - p_q) + \gamma_{self,q}(p_{ref}, T_{ref})p_q] \tag{9.28}$$

where $T_{ref} = 296$ K, T is current temperature (K), n_q is a coefficient of temperature dependence, $\gamma_{air,q}(p_{ref}, T_{ref})$ is the **air-broadened halfwidth** ($cm^{-1}\,atm^{-1}$) at T_{ref} and p_{ref}, p_q is the partial pressure of gas q (atm), and $\gamma_{self,q}$ is the **self-broadened halfwidth** ($cm^{-1}\,atm^{-1}$) of gas q at T_{ref} and p_{ref}. The parameter $S_q(T)$ is calculated as a function of temperature from $S_q(T_{ref})$, which, along with n_q, $\gamma_{air,q}(p_{ref}, T_{ref})$, and

$\gamma_{\text{self},q}(p_{\text{ref}}, T_{\text{ref}})$, are available from the HITRAN database (Rothman *et al.* 2003). Figure 9.16 shows absorption coefficient versus wavelength, calculated from (9.27), from one line distribution at two pressures. At high pressure, the line distribution is broader than at low pressure.

Greenhouse gases, such as H_2O and CO_2, absorb in several tens of thousands of line distributions, such as those shown in Fig. 9.14, across the solar- and thermal-infrared spectra. Because atmospheric models discretize the wavelength spectrum into only tens to hundreds of wavelength intervals, a method is needed to gather line-distribution information into each model interval for practical use.

Merely averaging line-by-line absorption coefficients from several line distributions over each model wavelength interval, though, overestimates absorption. This can be demonstrated as follows. The monochromatic (single-wavelength) **transmission** of radiation through an absorbing gas is the fraction of incident radiance that penetrates through the gas, and is defined as

$$T_{\text{a,g},q,\lambda}(u_q) = \frac{I_\lambda}{I_0} = \mathrm{e}^{-u_q k_{\text{a,g},q,\lambda}} = \mathrm{e}^{-\sigma_{\text{a,g},q,\lambda}(x-x_0)} \tag{9.29}$$

where $u_q = \rho_q(x - x_0)$ is the mass pathlength of the gas (g cm^{-2}), and $\sigma_{\text{a,g},q,\lambda} = \rho_q k_{\text{a,g},q,\lambda}$. Suppose a gas has two nonoverlapping absorption line distributions, equally spaced within a model wavenumber interval, $\tilde{\nu}$ to $\tilde{\nu} + \Delta\tilde{\nu}$, but of different strength, k_x and k_y (where the subscripts a, g, and q were omitted). The exact transmission through pathlength u of the gas within the interval is

$$T_{\tilde{\nu},\tilde{\nu}+\Delta\tilde{\nu}}(u) = 0.5(\mathrm{e}^{-k_x u} + \mathrm{e}^{-k_y u}) \tag{9.30}$$

If the absorption coefficients are first averaged, the resulting transmission is

$$T_{\tilde{\nu},\tilde{\nu}+\Delta\tilde{\nu}}(u) = \mathrm{e}^{-0.5(k_x+k_y)u} \tag{9.31}$$

which is always smaller (more absorbing) than the exact solution, (9.30), except when $k_x = k_y$. Thus, averaging multiple absorption coefficients overestimates absorption.

A method of addressing the overabsorption problem is the ***k*-distribution method** (e.g., Ambartzumiam 1936; Kondratyev 1969; Yamamoto *et al.* 1970; Arking and Grossman 1972; Lacis and Hansen 1974; Liou 2002). With this method, absorption line distributions in each model wavenumber interval are grouped into typical **probability intervals** according to absorption coefficient strength. For example, the strongest absorption coefficients are grouped together, the next strongest are grouped together, and so on. The absorption coefficients within each probability interval are then averaged, and the **transmission in the wavenumber interval** as a whole is calculated by integrating the transmission through each probability interval with the rightmost expression in

$$T_{\tilde{\nu},\tilde{\nu}+\Delta\tilde{\nu}}(u) = \frac{1}{\Delta\tilde{\nu}}\int_{\tilde{\nu}}^{\tilde{\nu}+\Delta\tilde{\nu}} \mathrm{e}^{-k_{\tilde{\nu}} u}\,\mathrm{d}\tilde{\nu} \approx \int_0^\infty \mathrm{e}^{-k' u} f_{\tilde{\nu}}(k')\mathrm{d}k' \tag{9.32}$$

where u is the pathlength of the gas through the layer of air considered, $k_{\tilde{\nu}}$ is the monochromatic (single-wavenumber) absorption coefficient, and k' is the absorption coefficient within a probability interval of **normalized probability** $f_{\tilde{\nu}}(k')$ (or **differential probability** $f_{\tilde{\nu}}(k')\mathrm{d}k'$) of occurring. The integral of the normalized probability, over all possible absorption coefficient strengths, is unity:

$$\int_0^\infty f_{\tilde{\nu}}(k')\mathrm{d}k' = 1 \tag{9.33}$$

The middle expression in (9.32) is the exact solution, but requires integration over tens of thousands of individual spectral subintervals. The rightmost integral is generally approximated as a summation over 4–16 probability intervals. With the k-distribution method, probability intervals of a gas differ for each layer of the atmosphere.

An extension of the original method is the **correlated *k*-distribution method** (e.g., Lacis *et al.* 1979; Goody *et al.* 1989; West *et al.* 1990; Lacis and Oinas 1991; Fu and Liou 1992; Stam *et al.* 2000; Liou 2002). With this method, the wavenumber order of absorption coefficients is first rearranged by strength into a **cumulative wavenumber frequency distribution,**

$$g_{\tilde{\nu}}(k) = \int_0^k f_{\tilde{\nu}}(k')\mathrm{d}k' \tag{9.34}$$

where $g_{\tilde{\nu}}(k)$ varies between 0 and 1 and is a monotonically increasing function of the absorption coefficient k. The cumulative distribution is then divided up into probability intervals. The probability intervals are then assumed to be the same (correlated) at each altitude for a given gas. Finally, radiative transfer is calculated through the atmosphere for each probability interval, and the resulting transmission is weighted among all probability intervals to obtain a net transmission.

When multiple gases are considered, the complexity deepens because the relative strength of absorption of one gas at a given wavelength depends on the quantity of radiation removed by all other gases at the same wavelength. Methods of treating absorption among multiple gases are given in West *et al.* (1990), Lacis and Oinas (1991), and Jacobson (2005a). With the last method, a major absorber is selected for each model wavelength interval. Absorption coefficients calculated in each model subinterval for the major absorber are then grouped by strength in each probability interval. Finally, model subintervals containing absorption coefficients for other gases are grouped into the same probability intervals as those of the major gas according to wavenumber. In other words, if a subinterval of the major gas of a given wavenumber is placed in a given probability interval, the subintervals of all other gases at that same wavenumber are placed in the same probability interval. This technique, called the **multiple-absorber correlated-*k*-distribution method,** ensures exact correlation of absorption wavelengths (wavenumbers) within a probability interval for all gases through all layers of the atmosphere when multiple gases are considered. The technique is physical since all gases in reality are correlated in wavelength space. Figure 9.14 compares modeled with line-by-line

Table 9.5 Extinction coefficients ($\sigma_{a,g}$) and meteorological ranges ($x_{a,g}$) due to NO_2 absorption at selected wavelength intervals ($\lambda \pm 0.005$ μm) and concentrations

		0.01 ppmv NO_2		**0.25 ppmv NO_2**		
λ (μm)	b (10^{-19} cm^2)	$\sigma_{a,g}$ (10^8 cm^{-1})	$x_{a,g}$ (km)	$\sigma_{a,g}$ (10^8 cm^{-1})	$x_{a,g}$ (km)	$x_{s,g}$ (km)
0.42	5.39	13.2	296	330	11.8	112
0.45	4.65	11.4	343	285	13.7	148
0.50	2.48	6.10	641	153	25.6	227
0.55	0.999	2.46	1590	61.5	63.6	334
0.60	0.292	0.72	5430	18.0	217	481
0.65	0.121	0.30	13000	7.5	520	664

Absorption-cross-section data (b) for NO_2 are from Schneider *et al.* (1987). Also shown is the meteorological range due to Rayleigh scattering only ($x_{s,g}$). $T = 298$ K and $p_a = 1$ atm.

absorption coefficients for individual gases and multiple gases together from this technique.

Once an absorption coefficient of an individual gas is calculated, it is applied in (9.26) to determine the extinction coefficient of the gas at the current mass density of the gas.

9.4.1.4 *Gas absorption effects on visibility*

In the solar spectrum, gas absorption, primarily by nitrogen dioxide, affects visibility. The furthest distance a typical eye can see at a given wavelength may be estimated with the **Koschmieder equation,** $x_\lambda = 3.912/\sigma_{ext,\lambda}$, derived in Section 9.5. In the equation, x_λ is the **meteorological range** and $\sigma_{ext,\lambda}$ is the total extinction coefficient, which includes extinction due to gas absorption, gas scattering, particle absorption, and particle scattering.

Table 9.5 gives extinction coefficients and meteorological ranges due to NO_2 absorption, alone. It shows that NO_2 absorbs more strongly at shorter (blue) than at longer (green or red) wavelengths. At low concentrations (0.01 ppmv), the effect of NO_2 absorption on visibility is less than that of gas scattering at all wavelengths. At typical polluted-air concentrations (0.1–0.25 ppmv), NO_2 reduces visibility significantly for wavelengths < 0.50 μm and moderately for wavelengths 0.5–0.6 μm.

Most effects of NO_2 on visibility are limited to times when its concentration peaks. Results from a project studying Denver's **brown cloud**, for example, showed that NO_2 accounted for about 6.6 percent of total extinction averaged over all sampling periods, and 37 percent of extinction during periods of maximum NO_2 concentration. Scattering and absorption by particles caused most remaining extinction (Groblicki *et al.* 1981).

Figure 9.17 shows extinction coefficients due to NO_2 and ozone absorption at different mixing ratios. The figure indicates that NO_2 affects extinction (and therefore radiative transfer and visibility) only at high mixing ratios and at wavelengths below about 0.5 μm. In polluted air, such as in Los Angeles, NO_2 mixing ratios typically range from 0.01 to 0.1 ppmv and peak near 0.15 ppmv during the

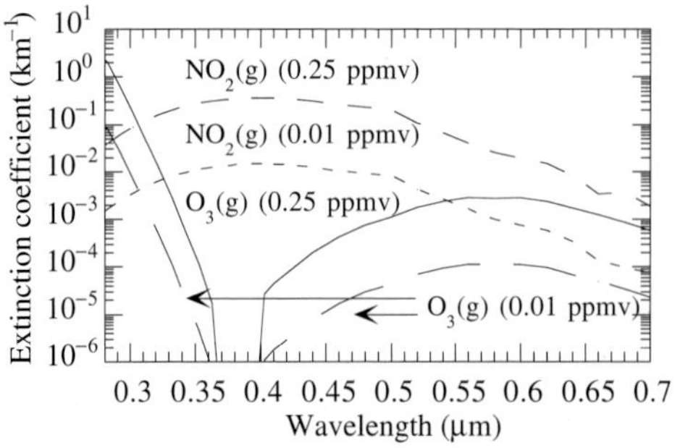

Figure 9.17 Extinction coefficients due to NO_2 and O_3 absorption when $T = 298$ K and $p_a = 1013$ hPa. Cross-section data for NO_2 are from DeMore *et al.* (1997) and interpolated from Schneider *et al.* (1987). Data for ozone are from Atkinson *et al.* (1997).

morning. A typical value is 0.05 ppmv, which results in an extinction coefficient of about 0.07 km^{-1} at 0.4 μm and 0.01 km^{-1} at 0.55 μm. The visibility (meteorological range) is about 390 km with this latter value. Ozone has a larger effect on extinction than does nitrogen dioxide at wavelengths below about 0.32 μm. Ozone mixing ratios in polluted air usually peak between 0.05 and 0.25 ppmv. Nevertheless, the cumulative effect of ozone, nitrogen dioxide, and other gases on extinction is small in comparison with the effects of scattering and absorption by particles.

9.4.2 Gas scattering

The only gas-scattering process in the atmosphere is **Rayleigh scattering**, which is the scattering of radiation by gas molecules (primarily N_2 and O_2). Rayleigh scattering gives the sky its blue color. It causes dark objects a few kilometers away to appear behind a blue haze of scattered light, and bright objects more than 30 km away to appear reddened (Waggoner *et al.* 1981).

A Rayleigh scatterer has molecular radius r much smaller than the wavelength λ of interest ($2\pi r/\lambda \ll 1$). The radius of a gas molecule is sufficiently small in relation to visible-light wavelengths to meet this criterion. Since oxygen and nitrogen are the most abundant gases in the atmosphere, they are the most abundant Rayleigh scatterers.

The extinction coefficient (cm^{-1}) due to Rayleigh scattering is

$$\sigma_{s,g,\lambda} = N_a b_{s,g,\lambda} \tag{9.35}$$

where N_a is the number concentration of air molecules (molec. cm^{-3}) at a given altitude, and $b_{s,g,\lambda}$ is the **scattering cross section** of a typical air molecule. Since the number concentration of air molecules decreases exponentially with increasing altitude, (9.35) suggests that extinction due to Rayleigh scattering also decreases exponentially with increasing altitude.

The scattering cross section can be estimated with

$$b_{s,g,\lambda} = \frac{8\pi^3(n_{a,\lambda}^2 - 1)^2}{3\lambda^4 N_{a,0}^2} f(\delta_*) \approx \frac{32\pi^3(n_{a,\lambda} - 1)^2}{3\lambda^4 N_{a,0}^2} f(\delta_*) \tag{9.36}$$

where $n_{a,\lambda}$ is the real index of refraction of air, obtained from (9.21), $N_{a,0} = 2.55 \times 10^{19}$ cm^{-3} is the number concentration of air molecules at standard temperature and pressure (288 K and 1 atm), and $f(\delta_*)$ is the **anisotropic correction term.** This term accounts for deviations of Rayleigh scattering from perfect isotropic scattering and can be estimated with

$$f(\delta_*) = \frac{6 + 3\delta_*}{6 - 7\delta_*} \approx 1.05 \tag{9.37}$$

where $\delta_* \approx 0.0279$ (Young 1980). $f(\delta_*)$ is a function of wavelength and varies from 1.08 at $\lambda = 0.2$ μm to 1.047 at $\lambda = 1.0$ μm (Liou 2002). The second expression in (9.36) was found by assuming $n_{a,\lambda}^2 - 1 \approx 2\,(n_{a,\lambda} - 1)$, which is possible because $n_{a,\lambda} \approx 1$.

Equation (9.36) suggests that the Rayleigh scattering cross section varies inversely with the fourth power of wavelength. In other words, gases scatter short wavelengths much more effectively than they do long wavelengths. Since the meteorological range is inversely proportional to the extinction coefficient, the meteorological range should then increase with increasing wavelength when only gas scattering is considered. Table 9.5 confirms this supposition. It shows that, at short visible wavelengths (e.g., 0.42 μm), the meteorological range due to Rayleigh scattering is much smaller than that at long wavelengths (e.g., 0.65 μm).

Example 9.6

For $\lambda = 0.5$ μm, $p_a = 1$ atm (sea level), and $T = 288$ K, $\sigma_{s,g,\lambda} = 1.72 \times 10^{-7}$ cm^{-1} from (9.35). This extinction coefficient results in a meteorological range of $x = 3.912/\sigma_{s,g,\lambda} = 227$ km.

For $\lambda = 0.55$ μm, the extinction coefficient decreases to $\sigma_{s,g,\lambda} = 1.17 \times 10^{-7}$ cm^{-1} and the meteorological range increases to $x = 334$ km. Waggoner *et al.* (1981) reported a total extinction at Bryce Canyon, Utah, corresponding to a meteorological range of within a few percent of 400 km for a wavelength of 0.55 μm.

9.4.3 Particle absorption and scattering

Particle scattering is the most important solar radiation attenuation process in polluted air, followed, in order, by particulate absorption, gas absorption, and gas scattering. All particle components scatter solar and infrared radiation and absorb infrared radiation, but few absorb solar radiation. For several species, the ability to absorb solar radiation increases from the visible to ultraviolet spectra. Below, important particle absorbers of radiation are discussed. Subsequently, methods of calculating particle scattering and absorption in an atmospheric model are described.

9.4.3.1 *Important particle absorbers*

The strongest particle absorber of solar radiation is **black carbon**, the main component of soot. Other absorbers of solar radiation include **hematite** (Fe_2O_3) and **aluminum oxide** (alumina, Al_2O_3). Hematite is found in soil-dust particles. Some forms of aluminum oxide are found in soil-dust particles and others are found in combustion particles.

Particle absorbers in the UV spectrum include black carbon, hematite, aluminum oxide, and certain organic compounds. The organics absorb UV radiation but less visible radiation. The strongest **near-UVabsorbing organics** include certain nitrated aromatics, polycyclic aromatic hydrocarbons (PAHs), benzaldehydes, benzoic acids, aromatic polycarboxylic acids, and phenols (Jacobson 1999c). The strong absorptivity of nitrated aromatics occurs because substitution of a nitrate group onto a benzene ring shifts the peak absorption 0.057 μm toward a longer wavelength. The strong absorptivity of benzaldehydes and benzoic acids results because the addition of an aldehyde or an acid group to a benzene ring shifts the peak absorption 0.046 or 0.0255 μm, respectively, toward a longer wavelength (Dean 1992). Inorganic particle components with a near-UV absorption peak include the nitrate ion (0.302 μm), ammonium nitrate (0.308 μm), and sodium nitrate (0.297 μm) (Cleaver *et al.* 1963; Sommer 1989). Although near-UV radiation (0.25–0.38 μm) makes up only 5 percent of total solar radiation, near-UV wavelengths are responsible for most gas photolysis.

Most particle components are weak absorbers of visible and UV radiation. **Silicon dioxide** (SiO_2), which is the white, colorless, crystalline compound found in quartz, sand, and other minerals, is an example. **Sodium chloride** (NaCl), **ammonium sulfate** ($(NH_4)_2SO_4$), and **sulfuric acid** (H_2SO_4) are also weak absorbers of visible and UV radiation. Soil-dust particles, which contain SiO_2, Al_2O_3, Fe_2O_3, $CaCO_3$, $MgCO_3$, and other compounds, absorb both UV and visible radiation, with stronger absorption in the UV (e.g., Gillette *et al.* 1993; Sokolik *et al.* 1993). During heavy dust storms, particularly over desert regions, concentrations of soil dust can decrease visibility to a few meters or tens of meters.

9.4.3.2 *Particle refractive indices*

Figure 9.18 shows a possible path of radiant energy through a particle. If scattering is ignored, the attenuation, due to absorption, of radiant energy as it propagates through the particle is approximately

$$\frac{\mathrm{d}I}{\mathrm{d}x} = -\frac{4\pi\kappa}{\lambda}I \tag{9.38}$$

Table 9.6 Real and imaginary indices of refraction for some substances at $\lambda = 0.51$ and $10.0\ \mu m$

	0.5 μm		**10 μm**	
	Real (n_λ)	**Imaginary** (κ_λ)	**Real** (n_λ)	**Imaginary** (κ_λ)
$H_2O(aq)$[a]	1.34	1.0×10^{-9}	1.22	0.05
Soot(s)[b]	1.82	0.74	2.40	1.0
Organic C(s)[b]	1.45	0.001	1.77	0.12
$H_2SO_4(aq)$[b]	1.43	1.0×10^{-8}	1.89	0.46
$(NH_4)_2SO_4(s)$[b]	1.52	0.0005	2.15	0.02
NaCl(s)[b]	1.45	0.00015	1.53	0.033

[a] Hale and Querry (1973), [b] Krekov (1993).

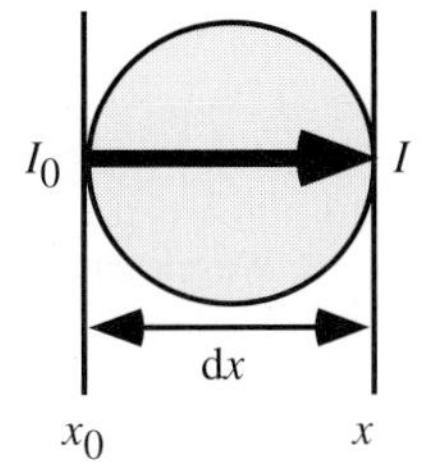

Figure 9.18 Attenuation of incident radiance I_0 due to absorption in a particle.

where κ is the **imaginary index of refraction,** which quantifies the extent to which a substance absorbs radiation. The term $4\pi\kappa/\lambda$ is an absorption extinction coefficient for a single particle. Although κ is a function of wavelength, the wavelength subscript in (9.38) was dropped. Integrating (9.38) from I_0 to I and x_0 to x gives

$$I = I_0 e^{-4\pi\kappa(x-x_0)/\lambda} \tag{9.39}$$

The optical properties of a particle component can generally be derived from a combination of its real and imaginary refractive indices. These parameters give information about the particle's ability to refract and absorb, respectively. Both parameters are embodied in the **complex refractive index,**

$$m_\lambda = n_\lambda - i\kappa_\lambda \tag{9.40}$$

Table 9.6 gives the real (n_λ) and imaginary (κ_λ) indices of refraction for some substances at $\lambda = 0.50$ and $10\ \mu m$. Soot has the largest imaginary index of refraction in the visible spectrum among the substances shown.

Table 9.7 Light transmission through soot and liquid water particles at $\lambda = 0.50$ μm

	Transmission (I/I_0)	
Particle diameter (μm)	**Soot ($\kappa = 0.74$)**	**Water ($\kappa = 10^{-9}$)**
0.1	0.16	0.999999997
1.0	8.0×10^{-9}	0.99999997
10.0	0	0.9999997

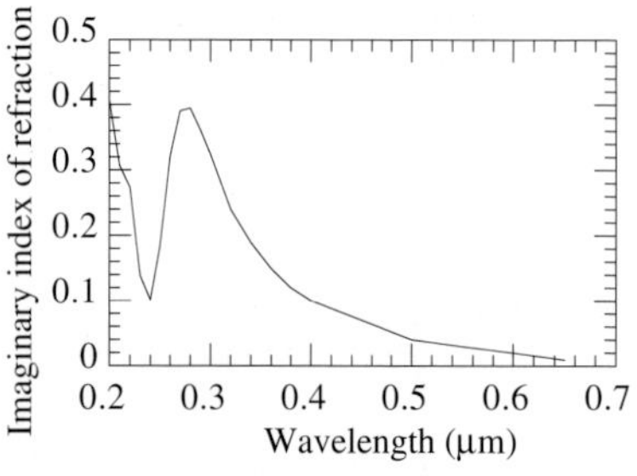

Figure 9.19 Imaginary index of refraction of liquid nitrobenzene versus wavelength. Data from Foster (1992).

Figure 9.19 shows the imaginary index of refraction of liquid nitrobenzene as a function of wavelength. Although liquid nitrobenzene has a low concentration in atmospheric particles, the shape of its index-of-refraction curve is important because it suggests that some organic particle components with peak absorption wavelengths below 0.3 μm absorb moderately between 0.3 and 0.4 μm as well, thereby affecting photolysis. Most organic atmospheric particles have peak absorption wavelengths below 0.3 μm.

Table 9.7 shows the extent of transmission of light through soot and liquid water particles at $\lambda = 0.50$ μm. These calculations were obtained from (9.39). The **transmission** (I/I_0) is the fraction of incident radiation passing through a substance. The table shows that a 0.1-μm-diameter soot particle transmits 16 percent of incident 0.50-μm radiation, and a 0.1-μm water drop transmits nearly 100 percent of the radiation if only absorption is considered. A 10-μm liquid water drop absorbs a negligible amount of incident 0.50-μm radiation. Thus, soot is a strong absorber of visible radiation, and liquid water is not.

9.4.3.3 *Absorption and scattering efficiencies*

In a model, absorption and scattering extinction coefficients must be considered for both aerosol particles and hydrometeor particles. Aerosol absorption, aerosol scattering, cloud absorption, and cloud scattering extinction coefficients are calculated with

$$\sigma_{a,a,\lambda} = \sum_{i=1}^{N_B} n_i b_{a,a,i,\lambda} \qquad \sigma_{s,a,\lambda} = \sum_{i=1}^{N_B} n_i b_{s,a,i,\lambda}$$
$$\sigma_{a,h,\lambda} = \sum_{i=1}^{N_B} n_i b_{a,h,i,\lambda} \qquad \sigma_{s,h,\lambda} = \sum_{i=1}^{N_B} n_i b_{s,h,i,\lambda} \tag{9.41}$$

respectively, where n_i is the number concentration of aerosol particles or hydrometeor particles of a given size (particles cm^{-3}), N_B is the number of size bins in either case, $b_{a,a,i,\lambda}$ and $b_{a,h,i,\lambda}$ are the effective **absorption cross sections** of a single aerosol particle and hydrometeor particle, respectively, and $b_{s,a,i,\lambda}$ and $b_{s,h,i,\lambda}$ are the effective **scattering cross sections** of a single aerosol particle and hydrometeor particle, respectively.

For spherical aerosol particles, the effective absorption and scattering cross sections are

$$b_{a,a,i,\lambda} = \pi r_i^2 Q_{a,i,\lambda} \qquad b_{s,a,i,\lambda} = \pi r_i^2 Q_{s,i,\lambda} \tag{9.42}$$

respectively, where πr_i^2 is the actual aerosol cross section (cm^2), $Q_{a,i,\lambda}$ (dimensionless) is the single-particle absorption efficiency, and $Q_{s,i,\lambda}$ is the single-particle scattering efficiency. The absorption and scattering efficiencies are functions of the complex index of refraction m_λ, which depends on wavelength and particle composition, and of the **size parameter**

$$\alpha_{i,\lambda} = \frac{2\pi r_i}{\lambda} \tag{9.43}$$

which depends on particle size and wavelength. Absorption cross sections for hydrometeor particles are calculated as in (9.42), but with single-particle absorption and scattering efficiencies for hydrometeor particles rather than for aerosol particles.

The **single-particle scattering efficiency** is the ratio of the effective scattering cross section of a particle to its actual cross section. The scattering efficiency can exceed unity, since a portion of the radiation diffracting around a particle can be intercepted and scattered by the particle. Scattering efficiencies above unity account for this additional scattering. The greater the absorption by a particle, the lesser the scattering efficiency, since absorption hinders scattering. The scattering efficiency is highest when the wavelength of light is close to the particle radius.

The **single-particle absorption efficiency** is the ratio of the effective absorption cross section of a particle to its actual cross section. The absorption efficiency can exceed unity, since a portion of the radiation diffracting around a particle can be intercepted and absorbed by the particle. Absorption efficiencies above unity account for this additional absorption. The larger the imaginary refractive index of a particle, the greater its absorption efficiency. The real refractive index affects the angle that radiation bends upon entering a particle, affecting the distance it travels and its cumulative absorption within the particle.

Single-particle absorption and scattering efficiencies vary with particle size and radiation wavelength. When a particle's diameter is much smaller than the wavelength of light ($d_i < 0.03\lambda$, or $\alpha_{i,\lambda} < 0.1$), the particle is in the **Rayleigh regime** and is called a **Tyndall absorber** or **scatterer**. Such particles have an absorption efficiency of

$$Q_{a,i,\lambda} = -4\frac{2\pi r_i}{\lambda}\mathrm{Im}\left(\left|\frac{m_\lambda^2 - 1}{m_\lambda^2 + 2}\right|^2\right) \approx \frac{2\pi r_i}{\lambda}\left[\frac{24 n_\lambda \kappa_\lambda}{\left(n_\lambda^2 + \kappa_\lambda^2\right)^2 + 4\left(n_\lambda^2 - \kappa_\lambda^2 + 1\right)}\right] \quad (9.44)$$

As $\kappa_\lambda \to 0$, (9.44) approaches

$$Q_{a,i,\lambda} = \frac{2\pi r_i}{\lambda}\left[\frac{24 n_\lambda \kappa_\lambda}{\left(n_\lambda^2 + 2\right)^2}\right] \quad (9.45)$$

which is a linear function of κ_λ. When $r_i \ll \lambda$, the absorption efficiency is small. Thus, small particles are relatively inefficient absorbers of radiation.

The **Tyndall scattering efficiency** is

$$Q_{s,i,\lambda} = \frac{8}{3}\left(\frac{2\pi r_i}{\lambda}\right)^4 \left|\frac{m_\lambda^2 - 1}{m_\lambda^2 + 2}\right|^2 \quad (9.46)$$

Equations (9.46) and (9.44) show that the scattering and absorption efficiencies of small particles are proportional to $(r_i/\lambda)^4$ and r_i/λ, respectively. Thus, at a sufficiently short wavelength, the absorption efficiency of an absorbing particle (large κ) is much greater than is the scattering efficiency of the same particle. Figures 9.20 and 9.21 show the single-particle absorption and scattering efficiencies, including Tyndall efficiencies, of soot, a strong absorber, and of liquid water, a weak absorber.

Example 9.7

At $\lambda = 0.5$ μm and $r_i = 0.01$ μm, find the single-particle scattering and absorption efficiencies of liquid water.

SOLUTION

From Table 9.6, $n_\lambda = 1.34$ and $\kappa_\lambda = 1.0 \times 10^{-9}$ at $\lambda = 0.5$ μm. From (9.46) and (9.45), respectively, $Q_{s,i,\lambda} = 2.92 \times 10^{-5}$ and $Q_{a,i,\lambda} = 2.8 \times 10^{-10}$. Since liquid water is nonabsorbing, its single-particle absorption efficiency is expected to be small in comparison with its single-particle scattering efficiency.

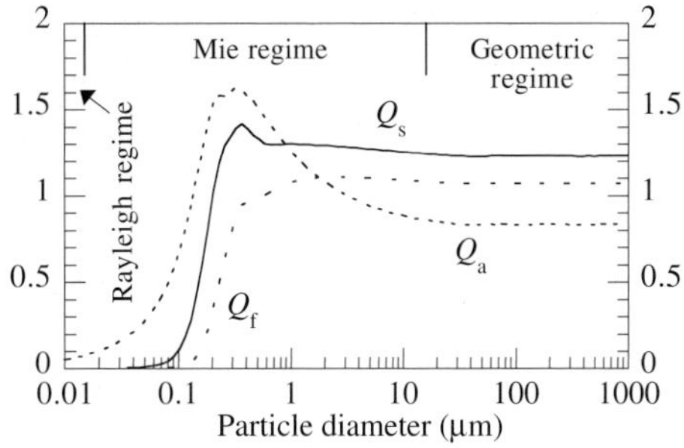

Figure 9.20 Single-particle absorption (Q_a), total scattering (Q_s), and forward scattering (Q_f) efficiencies of soot particles (which contain mostly black carbon) of different sizes at $\lambda = 0.50$ μm ($n_\lambda = 1.94$, $\kappa_\lambda = 0.66$). The efficiency for soot at any other wavelength λ_1 in the visible or UV spectrum and diameter d_1 is the efficiency in the figure located at diameter $d = 0.5\ \mu\text{m} \times d_1/\lambda_1$.

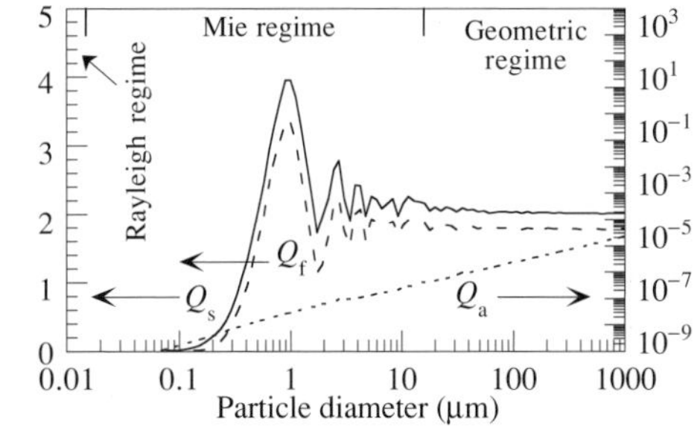

Figure 9.21 Single-particle absorption (Q_a), total scattering (Q_s), and forward scattering (Q_f) efficiencies of liquid water drops of different sizes at $\lambda = 0.50$ μm ($n_\lambda = 1.335$, $\kappa_\lambda = 1.0 \times 10^{-9}$). The efficiency for water at any other wavelength λ_1 in the visible or UV spectrum and diameter d_1 is the efficiency in the figure located at diameter $d = 0.5\ \mu\text{m} \times d_1/\lambda_1$.

When a particle's diameter is near the wavelength of light ($0.03\lambda < d_i < 32\lambda$ or $0.1 < \alpha_{i,\lambda} < 100$), the particle is in the **Mie regime.** In this regime, absorption and scattering efficiencies are approximated with Mie's solution to Maxwell's equations. The **scattering efficiency of a particle in the Mie regime** is determined by

$$Q_{s,i,\lambda} = \frac{2}{\alpha_{i,\lambda}} \sum_{k=1}^{\infty} (2k+1)(|a_k|^2 + |b_k|^2) \tag{9.47}$$

where a_k and b_k are complex functions (e.g., van de Hulst 1957; Kerker 1969; Toon and Ackerman 1981; Bohren and Huffman 1983; Liou 2002). The single-particle absorption efficiency of a particle in the Mie regime is estimated from $Q_{a,i,\lambda} = Q_{e,i,\lambda} - Q_{s,i,\lambda}$, where

$$Q_{e,i,\lambda} = \frac{2}{\alpha_{i,\lambda}} \sum_{k=1}^{\infty} (2k+1)\mathrm{Re}(a_k + b_k) \tag{9.48}$$

is the single-particle **total extinction efficiency** in the Mie regime. Figures 9.20 and 9.21 show $Q_{a,i,\lambda}$ and $Q_{s,i,\lambda}$ for soot and liquid water, as a function of particle size at $\lambda = 0.50$ μm. The figures also show the **forward scattering efficiency** $Q_{f,i,\lambda}$, which is the efficiency with which a particle scatters light in the forward direction. The forward scattering efficiency is always less than the total scattering efficiency. The proximity of $Q_{f,i}$ to $Q_{s,i}$ in both figures indicates that aerosol particles scatter strongly in the forward direction. The difference between $Q_{s,i}$ and $Q_{f,i}$ is the scattering efficiency in the backward direction.

Figure 9.20 shows that visible-light absorption efficiencies of soot particles peak when the particles are 0.2 to 0.4 μm in diameter. Such particles are in the **accumulation mode** (0.1–2 μm in diameter, Chapter 13) with respect to particle size and in the Mie regime with respect to the ratio of particle size to the wavelength of light. Figure 9.21 shows that water particles 0.3 to 2 μm in diameter scatter visible light more efficiently than do smaller or larger particles. These particles are also in the accumulation mode with respect to particle size and in the Mie regime with respect to the ratio of particle size to the wavelength of light.

Because the accumulation mode contains a relatively high particle number concentration, and because particles in this mode have high scattering and absorption (with respect to soot) efficiencies, the accumulation mode almost always causes more light reduction than do smaller or larger modes (Waggoner *et al.* 1981). In many urban regions, 20 to 50 percent of the accumulation mode mass is sulfate. Thus, sulfate is correlated with particle scattering more closely than is any other particulate species, aside from liquid water.

When a particle's diameter is much larger than the wavelength of light ($d_i > 32\lambda$ or $\alpha_{i,\lambda} > 100$), the particle is in the **geometric regime**. Such particles reflect, refract, and diffract light significantly. As $\alpha_{i,\lambda}$ increases from the Mie regime to the geometric regime, the scattering efficiency approaches a constant in a sinusoidal fashion, as shown in Fig. 9.21. As $\alpha_{i,\lambda} \to \infty$, the scattering efficiency approaches another constant, given by Chylek (1977) as

$$\lim_{\alpha_{i,\lambda} \to \infty} Q_{s,i,\lambda} = 1 + \left| \frac{m_\lambda - 1}{m_\lambda + 2} \right| \tag{9.49}$$

Although cloud drops, which have diameters >5 μm, have lower scattering efficiencies and a lower number concentration than do accumulation-mode aerosol particles, clouds can obscure visibility to a greater extent than can aerosol particles because cloud drops have much larger cross-sectional areas than do aerosol particles.

Example 9.8

At $\lambda = 0.5$ μm, calculate the scattering efficiency of liquid water as $\alpha_{i,\lambda} \rightarrow \infty$.

SOLUTION

From Table 9.6, $n_\lambda = 1.335$ at $\lambda = 0.5$ μm for liquid water. From (9.49), $Q_{s,i,\lambda} = 1.1$. Hence, the scattering efficiency at $\lambda = 0.5$ μm, shown in Fig. 9.21, decreases from 2 to 1.1 as the diameter approaches infinity. The absorption efficiency also approaches 1.1 at infinite diameter.

Deirmendjian (1969) showed that, as $\alpha_{i,\lambda} \rightarrow \infty$, a particle's absorption efficiency converges to its scattering efficiency regardless of how weak the imaginary index of refraction is. Since all liquids and solids scatter in the geometric regime, they must all absorb at large enough size, as well. In other words, **large particles absorb radiation, regardless of their imaginary index of refraction.** Figure 9.21, for example, shows how the absorption efficiency of liquid water increases with increasing diameter even when $\kappa_\lambda = 1.0 \times 10^{-9}$. The figure also shows that the absorption efficiency becomes important only when liquid drops reach raindrop-size (>1000 μm in diameter). The **absorptivity of raindrops causes the bottoms of precipitating clouds to appear gray or black.**

9.4.3.4 *Optics of particle mixtures*

In the previous discussion, particles were assumed to be **externally mixed,** or in isolation from one another. In reality, particles are often **internally mixed,** meaning they contain multiple components. Several methods have been developed to treat the effective refractive index of a mixture. These treatments are called **effective medium approximations.** A more detailed discussion and comparison of approximations is given in Bohren and Huffman (1983) and Chylek *et al.* (1988).

The simplest way to account for the internal mixing of particle components in a model is to average the complex refractive index of each component within a particle by volume. This method is called the **volume average refractive index mixing rule** and is quantified with

$$m_\lambda = \sum_{q=1}^{N_V} \frac{v_q}{v} m_{\lambda,q} = \sum_{q=1}^{N_V} \frac{v_q}{v} (n_\lambda - i\kappa_\lambda) \tag{9.50}$$

where N_V is the number of volume components in a particle, v is the volume of a single particle (cm^3 particle^{-1}), and v_q is the volume of component q within the particle (cm^3 particle^{-1}). The ratio v_q/v is the **volume fraction** of component q within the particle.

A second method is to average the dielectric constant of each component by volume (e.g., Chylek *et al.* 1988). A **dielectric** is a nonconducting material that can sustain a steady electric field and serve as an insulator. The **dielectric constant** (also called the permittivity) is a property of a material that determines how much electrostatic energy can be stored per unit volume of the material when unit voltage is applied. It is the ratio of the capacitance of a capacitor with the material to the same capacitor when the dielectric is a vacuum. The **complex dielectric constant** is related to the complex refractive index of a material by

$$\varepsilon_\lambda = m_\lambda^2 = (n_\lambda - \mathrm{i}\kappa_\lambda)^2 = n_\lambda^2 - \kappa_\lambda^2 - \mathrm{i}2n_\lambda\kappa_\lambda = \varepsilon_{\mathrm{r},\lambda} - \mathrm{i}\varepsilon_{\mathrm{i},\lambda} \tag{9.51}$$

where

$$\varepsilon_{\mathrm{r},\lambda} = n_\lambda^2 - \kappa_\lambda^2 \qquad \varepsilon_{\mathrm{i},\lambda} = 2n_\lambda\kappa_\lambda \tag{9.52}$$

are the **real and imaginary components, respectively, of the complex dielectric constant.** The real and imaginary refractive indices are related to these components by

$$n_\lambda = \sqrt{\frac{\sqrt{\varepsilon_{\mathrm{r},\lambda}^2 + \varepsilon_{\mathrm{i},\lambda}^2} + \varepsilon_{\mathrm{r},\lambda}}{2}} \qquad \kappa_\lambda = \sqrt{\frac{\sqrt{\varepsilon_{\mathrm{r},\lambda}^2 + \varepsilon_{\mathrm{i},\lambda}^2} - \varepsilon_{\mathrm{r},\lambda}}{2}} \tag{9.53}$$

respectively. The **volume average dielectric constant mixing rule** is

$$m_\lambda^2 = \sum_{q=1}^{N_V} \frac{v_q}{v} m_{\lambda,q}^2 = \sum_{q=1}^{N_V} \frac{v_q}{v} \varepsilon_{\lambda,q} = \sum_{q=1}^{N_V} \frac{v_q}{v} (\varepsilon_{\mathrm{r},\lambda} - \mathrm{i}\varepsilon_{\mathrm{i},\lambda}) \tag{9.54}$$

A third method is the **Maxwell Garnett mixing rule** (Maxwell Garnett 1904). This rule gives the effective complex refractive index of absorbing inclusions (subscript A) embedded within a matrix of a single homogeneous substance (subscript M) as

$$m_\lambda^2 = \varepsilon_\lambda = \varepsilon_{\lambda,\mathrm{M}} \left[1 + \frac{3\sum_{q=1}^{N_A} \frac{v_{\mathrm{A},q}}{v} \left(\frac{\varepsilon_{\lambda,\mathrm{A},q} - \varepsilon_{\lambda,\mathrm{M}}}{\varepsilon_{\lambda,\mathrm{A},q} + 2\varepsilon_{\lambda,\mathrm{M}}} \right)}{1 - \sum_{q=1}^{N_A} \frac{v_{\mathrm{A},q}}{v} \left(\frac{\varepsilon_{\lambda,\mathrm{A},q} - \varepsilon_{\lambda,\mathrm{M}}}{\varepsilon_{\lambda,\mathrm{A},q} + 2\varepsilon_{\lambda,\mathrm{M}}} \right)} \right] \tag{9.55}$$

where N_A is the number of absorbing substances embedded in the matrix, $\varepsilon_{\lambda,\mathrm{A},q}$ is the complex dielectric constant of each absorbing substance, and $\varepsilon_{\lambda,\mathrm{M}}$ is the complex dielectric constant of the mixture. Equation (9.55) is solved by first substituting $\varepsilon = n^2 - \kappa^2 - \mathrm{i}2n\kappa$ from (9.51) into each complex dielectric constant on the right side of the equation, rewriting the equation in the form $\varepsilon_\lambda = (a - \mathrm{i}b)/(c - \mathrm{i}d)$, then multiplying through by $(c + \mathrm{i}d)/(c + \mathrm{i}d)$ to obtain $\varepsilon_\lambda = \varepsilon_{\mathrm{r},\lambda} - \mathrm{i}\varepsilon_{\mathrm{i},\lambda}$, where $\varepsilon_{\mathrm{r},\lambda} = (ac + bd)/(c^2 - d^2)$ and $\varepsilon_{\mathrm{i},\lambda} = (bc - ad)/(c^2 - d^2)$. Finally, the average real and imaginary refractive indices are obtained from $\varepsilon_{\mathrm{r},\lambda}$ and $\varepsilon_{\mathrm{i},\lambda}$ from (9.53).

A related mixing rule is the inverted Maxwell Garnett mixing rule, which is the same as (9.55) but with the subscripts M and A reversed in all terms.

A fourth method is the **Bruggeman mixing rule** (Bruggeman 1935), which is derived from the same base equation as the Maxwell Garnett mixing rule, but using a different approximation, and also assumes one or more absorbing inclusions embedded within a matrix. With this rule, the effective complex dielectric constant, ε_λ, is solved iteratively from

$$\sum_{q=1}^{N_A} \frac{\upsilon_{A,q}}{\upsilon}\left(\frac{\varepsilon_{\lambda,A,q}-\varepsilon_\lambda}{\varepsilon_{\lambda,A,q}+2\varepsilon_\lambda}\right)+\left(1-\sum_{q=1}^{N_A}\frac{\upsilon_{A,q}}{\upsilon}\right)\left(\frac{\varepsilon_{\lambda,M}-\varepsilon_\lambda}{\varepsilon_{\lambda,M}+2\varepsilon_\lambda}\right)=0 \qquad (9.56)$$

The complex refractive index is then obtained from ε_λ with (9.51).

All the effective medium theories discussed result in an average complex refractive index for an entire particle. Optical properties of the particle are then calculated from this average refractive index. However, when an absorbing solid, such as soot, exists in an otherwise liquid particle, the soot is a distinct component and not distributed equally throughout the particle. The optical properties of the particle in this case depend on radiative interactions between the liquid and solid, which the effective medium approximations do not capture.

A more physical, but still somewhat limited, way of treating absorption in a multicomponent particle is to treat the absorbing component as a core material surrounded by a shell with a volume-averaged refractive index. Theoretical calculations with this method suggest that, when an absorbing particle component, such as soot, becomes coated by a relatively nonabsorbing material, such as sulfuric acid or organic carbon, the **absorption efficiency of the soot increases** (e.g., Toon and Ackerman 1981; Jacobson 1997a,b, 2000, 2001b; Lesins *et al.* 2002). Figure 9.22 shows the theoretical enhancement of absorption due to the addition of a coating onto particles containing black carbon. The theory has been supported experimentally by Schnaiter *et al.* (2003), who found that the addition of just a few monolayers of a condensed organic coating to a soot particle increased its absorption coefficient by 35 percent. Subsequent experiments showed that condensation of more layers increased absorption by 200 percent, a finding consistent with Fig. 9.22. The enhancement of absorption due to the coating of soot in the atmosphere may be sufficiently great to make soot the second most important component of global warming, after CO_2 and ahead of CH_4, in terms of its direct radiative effect (Jacobson 2000, 2001b).

The reasons for the enhanced absorption of a soot particle when it is coated versus when it is not are as follows: (1) In the case of particles larger than the wavelength of light, geometric optics implies that n_λ^2 more light is incident on a small sphere when it is at the center of a much larger transparent sphere with real refractive index n_λ than when it is in air (e.g., Bohren 1986; Twohy *et al.* 1989). This is called the **optical focusing effect**, because more light is focused into the core of a sphere due to refraction when a shell surrounds the core compared with when it does not. (2) In the case of particles near to or smaller than the wavelength of light, enhanced diffraction at the edge of a particle increases the exposure of the core to waves in comparison with exposure of the core to waves in the absence of a shell.

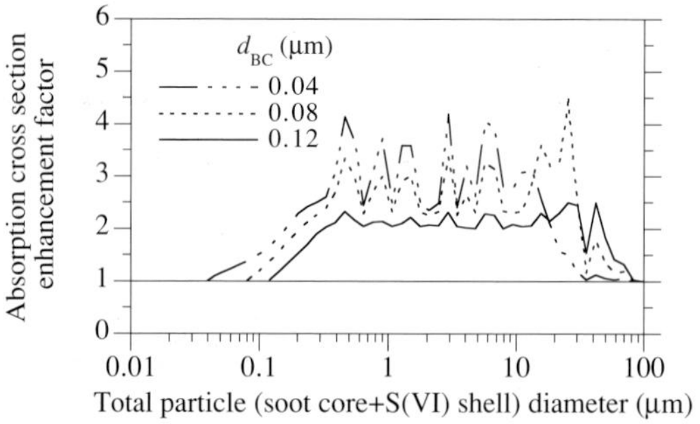

Figure 9.22 Absorption cross section enhancement factors at $\lambda = 0.5\,\mu m$, due to coating a black carbon (BC) core with a shell comprised of sulfuric acid–water (S(VI)). The enhancement factor is the ratio of the absorption cross section of a total (coated) particle with a BC core (either 40, 80, or 120 nm in diameter) and an S(VI) shell to the larger of the absorption cross section of BC alone (at either 40, 80, or 120 nm in diameter) or of pure S(VI) at the same size as the total core + shell particle (which affects the factor only at large sizes). The leftmost values in the figure are the enhancement factors (1.0) corresponding to the pure core. The rightmost values (1.0) are those corresponding to a large pure S(VI) drop that absorbs since all particles with a nonzero imaginary refractive index become absorbing at some size (Section 9.4.3.3).

Although treating soot as a concentric core is not always realistic because soot is an amorphous, randomly shaped agglomerate of many individual spherules, such treatment becomes more realistic as more material condenses. The reason is that soot aggregates collapse under the weight of a heavy coating, becoming more spherical in the process (Schnaiter *et al.* 2003; Wentzel *et al.* 2003). Other techniques of treating soot inclusions within particles, aside from the concentric core treatment, include treating a composite of several whole soot particles distributed in an internal mixture (applicable to cloud drops) (Chylek *et al.* 1988), embedding soot inclusions at random locations in a particle (Chylek *et al.* 1995; Fuller 1995), treating soot as a subsurface inclusion or grain on a particle (Fuller *et al.* 1999), and treating soot as a coated chain aggregate (Fuller *et al.* 1999).

9.4.3.5 *Modeled aerosol extinction*

Figures 9.23 (a) and (b) show modeled vertical profiles of extinction coefficients over a polluted airshed. Both figures indicate that extinction coefficients increased from the free troposphere (altitudes above 900 hPa) to the boundary layer. The first figure indicates that, at $\lambda = 0.32$ μm (ultraviolet light), extinction due to aerosol scattering in the boundary layer was less than that due to aerosol absorption. At

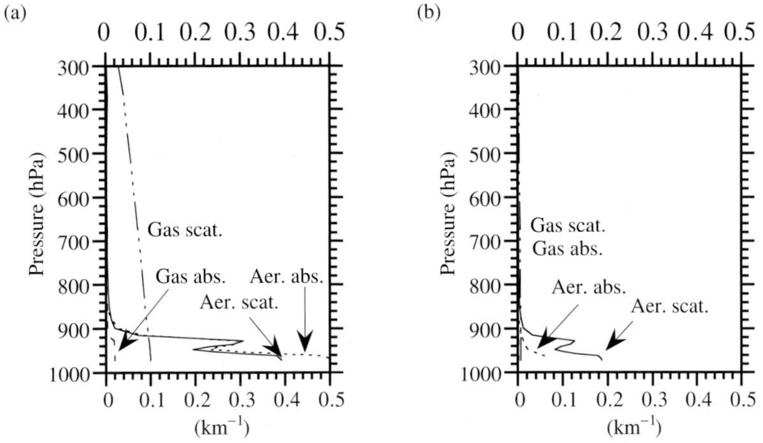

Figure 9.23 Modeled profiles from Claremont, California (11.30 on August 27, 1997) of extinction coefficients (km^{-1}) due to gas scattering, aerosol scattering, gas absorption, and aerosol absorption at (a) $\lambda = 0.32$ and (b) $\lambda = 0.61\,\mu$m. The peaks at 920 hPa indicate elevated aerosol layers. From Jacobson (1998b).

$\lambda = 0.61\ \mu$m (visible light), the reverse was true because the aerosol absorptivity decreased from the ultraviolet to visible spectra. Above the boundary layer, gas scattering dominated aerosol scattering at 0.32 and 0.61 μm.

9.5 VISIBILITY

A result of aerosol buildup is visibility degradation. Although unnatural visibility degradation, itself, causes no adverse health effects, it usually indicates the presence of pollutants, which are often harmful.

Visibility is a measure of how far we can see through the air. Even in the cleanest air, our ability to see along the Earth's horizon is limited to a few hundred kilometers by background gases and aerosol particles. If we look up through the sky at night, however, we can discern light from stars that are millions of kilometers away. The difference between looking horizontally and vertically is that more gas molecules and aerosol particles lie in front of us in the horizontal than in the vertical.

Several terms describe maximum visibility. Two subjective terms are visual range and prevailing visibility. **Visual range** is the actual distance at which a person can discern an ideal dark object against the horizon sky. **Prevailing visibility** is the greatest visual range a person can see along 50 percent or more of the horizon circle (360°), but not necessarily in continuous sectors around the circle. It is determined by a person who identifies landmarks known distances away in a full 360° circle around an observation point. The greatest visual range observed over 180° or more of the circle (not necessarily in continuous sectors) is the prevailing visibility. Thus, half the area around an observation point may have visibility worse than the prevailing visibility, which is important because most prevailing visibility observations are made at airports. If the visual range in a sector is significantly different from

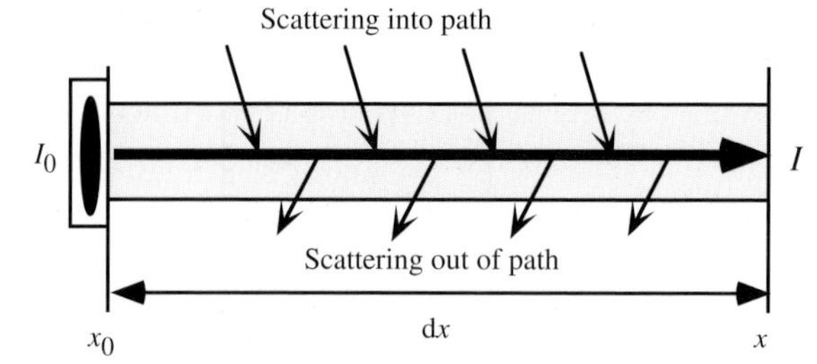

Figure 9.24 Change of radiation intensity along a beam. A radiation beam originating from a dark object has intensity $I = 0$ at point x_0. Over a distance dx, the beam's intensity increases due to scattering of background light into the beam. This added intensity is diminished somewhat by absorption along the beam and scattering out of the beam. At point x, the net intensity of the beam has increased close to that of the background intensity.

the prevailing visibility, the observer at an airport usually denotes this information in the observation record.

A less subjective and, now, a regulatory definition of visibility is the meteorological range. **Meteorological range** can be explained in terms of the following example. Suppose a perfectly absorbing dark object lies against a white background at a point x_0, as shown in Fig. 9.24. Because the object is perfectly absorbing, it reflects and emits no visible radiation; thus, its visible radiation intensity (I) at point x_0 is zero, and it appears black. As a viewer backs away from the object, background white light of intensity I_B scatters into the field of view, increasing the intensity of light in the viewer's line of sight. Although some of the added background light is scattered out of or absorbed along the field of view by gases and aerosol particles, at some distance away from the object, so much background light has entered the path between the viewer and the object that the viewer can barely discern the black object against the background light.

The meteorological range is a function of the **contrast ratio**, defined as

$$C_{\text{ratio}} = \frac{I_B - I}{I_B} \tag{9.57}$$

The contrast ratio gives the difference between the background intensity and the intensity in the viewer's line of sight, all relative to the background intensity. If the contrast ratio is unity, then an object is perfectly visible. If it is zero, then the object cannot be differentiated from background light.

The meteorological range is the distance from an object at which the contrast ratio equals the liminal contrast ratio of 0.02 (2 percent). The **liminal** or **threshold contrast ratio** is the lowest visually perceptible brightness contrast a person can see. It varies from individual to individual. Koschmieder (1924) selected a value of 0.02. Middleton (1952) tested 1000 people and found a threshold contrast range of between 0.01 and 0.20, with the mode of the sample between 0.02 and 0.03. Campbell and Maffel (1974) found a liminal contrast of 0.003 in laboratory studies of monocular vision. Nevertheless, 0.02 has become an accepted liminal contrast value for meteorological range calculations. In sum, the meteorological range is the distance from an ideal dark object at which the object has a 0.02 liminal contrast ratio against a white background.

The meteorological range can be derived from the equation for the change in object intensity along the path described in Fig. 9.24. This equation is

$$\frac{\mathrm{d}I}{\mathrm{d}x} = \sigma_t(I_B - I) \tag{9.58}$$

where all wavelength subscripts have been removed, σ_t is the total extinction coefficient, $\sigma_t I_B$ accounts for the scattering of background light radiation into the path, and $-\sigma_t I$ accounts for the attenuation of radiation along the path due to scattering out of the path and absorption along the path. A **total extinction coefficient** is the sum of extinction coefficients due to scattering and absorption by gases and aerosol particles. Thus,

$$\sigma_t = \sigma_{a,g} + \sigma_{s,g} + \sigma_{a,a} + \sigma_{s,a} \tag{9.59}$$

Integrating Equation (7.9) from $I = 0$ at point $x_0 = 0$ to I at point x with constant σ_t yields the equation for the contrast ratio,

$$C_{ratio} = \frac{I_B - I}{I_B} = e^{-\sigma_t x} \tag{9.60}$$

When $C_{ratio} = 0.02$ at a wavelength of 0.55 μm, the resulting distance x is the meteorological range (also called the **Koschmieder equation**).

$$x = \frac{3.912}{\sigma_t} \tag{9.61}$$

In polluted tropospheric air, the only important gas-phase visible-light attenuation processes are Rayleigh scattering and absorption by nitrogen dioxide (Waggoner *et al.* 1981). Several studies have found that scattering by particles, particularly those containing sulfate, organic carbon, and nitrate, may cause 60–95 percent of visibility reduction and absorption by soot may cause 5–40 percent of visibility reduction in polluted air (Cass 1979; Tang *et al.* 1981; Waggoner *et al.* 1981).

Table 9.8 shows meteorological ranges derived from extinction coefficient measurements for a polluted and less-polluted day in Los Angeles. Particle scattering dominated light extinction on both days. On the less-polluted day, gas absorption, particle absorption, and gas scattering all had similar small effects. On the polluted day, the most important visibility reducing processes were particle scattering, particle absorption, gas absorption, and gas scattering, in that order.

Table 9.8 Meteorological ranges (km) resulting from gas scattering, gas absorption, particle scattering, particle absorption, and all processes at a wavelength of 0.55 μm on a polluted and less-polluted day in Los Angeles

Day	Gas scattering	Gas absorption	Particle scattering	Particle absorption	All
Polluted (8/25/83)	366	130	9.6	49.7	7.42
Less-polluted (4/7/83)	352	326	151	421	67.1

Meteorological ranges derived from extinction coefficients of Larson *et al.* (1984).

9.6 OPTICAL DEPTH

A purpose of simulating atmospheric radiation is to estimate the spectral irradiance (F_λ) that reaches a given layer of the atmosphere. The first step in such a calculation is to determine the **total spectral extinction coefficient** at each wavelength with

$$\sigma_\lambda = \sigma_{\mathrm{s,g},\lambda} + \sigma_{\mathrm{a,g},\lambda} + \sigma_{\mathrm{s,a},\lambda} + \sigma_{\mathrm{a,a},\lambda} + \sigma_{\mathrm{s,h},\lambda} + \sigma_{\mathrm{a,h},\lambda} \tag{9.62}$$

where the individual coefficients are for scattering by gases, absorption by gases, scattering by aerosol particles, absorption by aerosol particles, scattering by hydrometeor particles, and absorption by hydrometeor particles, respectively. The next step is to define the vertical component of the incremental distance along a beam of interest through which radiation travels. When the beam of interest is the solar beam, this component is

$$\mathrm{d}z = \cos\theta_\mathrm{s}\,\mathrm{d}S_\mathrm{b} = \mu_\mathrm{s}\,\mathrm{d}S_\mathrm{b} \tag{9.63}$$

where $\mathrm{d}S_\mathrm{b}$ is the **incremental distance along the beam,** θ_s is the **solar zenith angle,** and

$$\mu_\mathrm{s} = \cos\theta_\mathrm{s} \tag{9.64}$$

The solar zenith angle is the angle between the surface normal (a line directed from the center of the Earth that extends vertically above the surface) and the direction of the Sun. Figure 9.25 shows the relationship among $\mathrm{d}z$, $\mathrm{d}S_\mathrm{b}$, and θ_s.

Integrating the spectral extinction coefficient over an incremental distance gives an **optical depth** (dimensionless). For radiative transfer calculations, the distance of interest is in the vertical. In such cases, optical depth quantifies scattering and absorbing that occurs between the top of the atmosphere and a given altitude. The optical depth increases from zero at the top of the atmosphere to a maximum at the ground. The **incremental optical depth** is

$$\mathrm{d}\tau_\lambda = -\sigma_\lambda\,\mathrm{d}z = -\sigma_\lambda\mu_\mathrm{s}\,\mathrm{d}S_\mathrm{b} \tag{9.65}$$

which increases in the opposite direction from incremental altitude. Integrating (9.65) from the top of the atmosphere ($z = S_\mathrm{b} = \infty$) to any altitude z, which

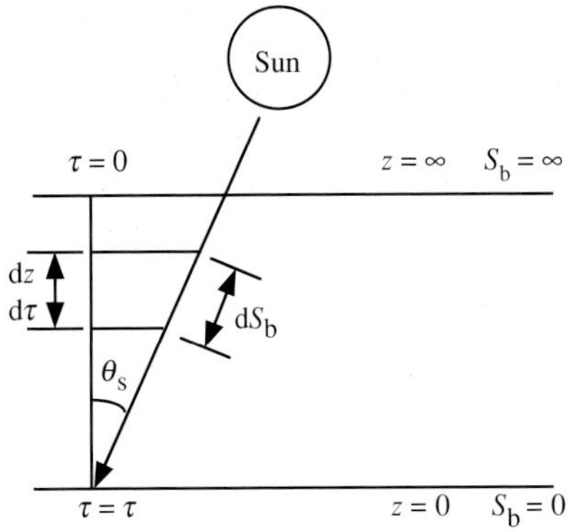

Figure 9.25 Relationship among incremental optical depth ($d\tau$), incremental altitude (dz), solar zenith angle (θ_s), and incremental distance along the beam (dS_b).

corresponds to a location S_b along the beam of interest, gives the optical depth as

$$\tau_\lambda = \int_\infty^z \sigma_\lambda \, dz = \int_\infty^{S_b} \sigma_\lambda \mu_s \, dS_b \tag{9.66}$$

9.7 SOLAR ZENITH ANGLE

The **solar zenith angle** can be determined from

$$\cos\theta_s = \sin\varphi \sin\delta + \cos\varphi \cos\delta \cos H_a \tag{9.67}$$

where φ is the latitude, δ is the solar declination angle, and H_a is the local hour angle of the Sun, as illustrated in Figs. 9.26 (a) and (b). The **declination angle** is the angle between the Equator and the north or south latitude of the **subsolar point,** which is the point at which the Sun is directly overhead. The **local hour angle** is the angle, measured westward, between the longitude (meridian) of the subsolar point and the longitude of the location of interest.

Equation (9.67) is obtained from Fig. 9.26(a) by applying the law of cosines to triangle APB. Since arcangle AOP $= 90^\circ - \varphi$, arcangle BOP $= 90^\circ - \delta$, and the distances OP, OA, and OB are known, the arc lengths AP and BP can be determined from the law of sines. With these distances and the fact that angle APB $= H_a$, the arc length AB can be obtained from the law of cosines. Since the distances AB, OA, and OB are known, the solar zenith angle (AOB) can be determined from the law of sines (e.g., Hartman 1994).

The **solar declination angle** is found from

$$\delta = \sin^{-1}(\sin\varepsilon_{ob} \sin\lambda_{ec}) \tag{9.68}$$

where ε_{ob} is the obliquity of the ecliptic and λ_{ec} is the ecliptic longitude of the Sun. The **ecliptic** is the mean plane of the Earth's orbit around the Sun. It is fixed in

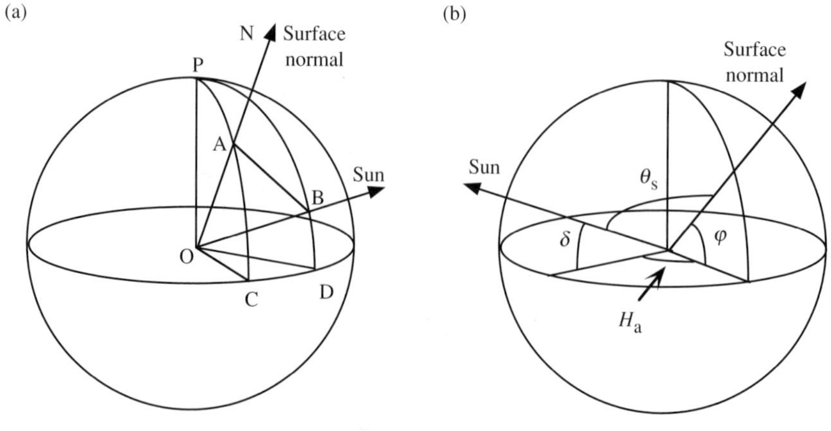

Figure 9.26 (a) Geometry for zenith angle calculations on a sphere. The ray OAN is the surface normal above the point of interest. The ray OB points to the Sun, which is incident directly over point B. Point B is the subsolar point. Angle AOB is the solar zenith angle (θ_s). Angle BOD is the solar declination (δ), angle AOC is the latitude (φ) of the surface normal, angle CAB is the solar azimuth angle (ϕ_s), and angles COD = CPD = APB are hour angles (H_a). (b) Geometry for a different solar zenith angle.

space relative to the Earth, and the Earth rotates through it. It cuts through the Tropic of Capricorn on one side of the Earth, the Equator in the middle, and the Tropic of Cancer on the other side of the Earth. The **obliquity of the ecliptic** is the angle between the plane of the Earth's Equator and the plane of the ecliptic, approximated as

$$\varepsilon_{ob} = 23^\circ.439 - 0^\circ.0000004 N_{JD} \tag{9.69}$$

(NAO 1993), where

$$N_{JD} = 364.5 + (Y - 2001) \times 365 + D_L + D_J \tag{9.70}$$

$$D_L = \begin{cases} \mathrm{INT}(Y - 2001)/4 & Y \geq 2001 \\ \mathrm{INT}(Y - 2000)/4 - 1 & Y < 2001 \end{cases}$$

is the number of days from the beginning of Julian year 2000. In (9.70), Y is the current year, D_L is the number of leap days since or before the year 2000, D_J is the **Julian day of the year,** which varies from 1 on January 1 to 365 (for nonleap years) or 366 (for leap years) on December 31. Leap years occur every year evenly divisible by 4.

The **ecliptic longitude of the Sun** is approximately

$$\lambda_{ec} = L_M + 1^\circ.915 \sin g_M + 0^\circ.020 \sin 2g_M \tag{9.71}$$

where

$$L_M = 280^\circ.460 + 0^\circ.9856474 N_{JD} \qquad g_M = 357^\circ.528 + 0^\circ.9856003 N_{JD} \tag{9.72}$$

are the mean longitude of the Sun and the mean anomaly of the Sun, respectively. The **mean anomaly of the Sun** is the angular distance, as seen by the Sun, of the

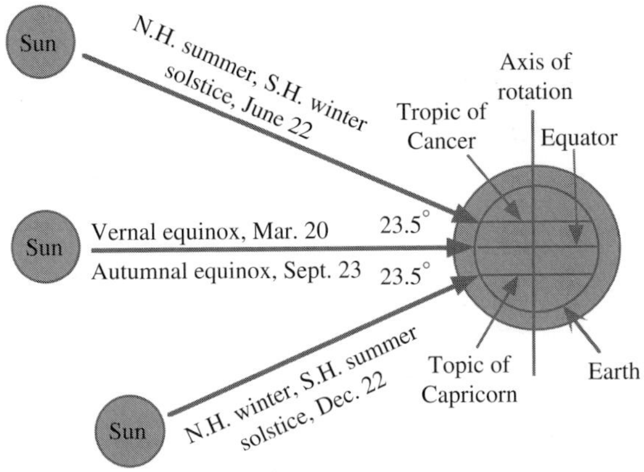

Figure 9.27 Solar declination angles during solstices and equinoxes. Of the four times shown, the Earth–Sun distance is greatest at the summer solstice.

Earth from its **perihelion,** which is the point in the Earth's orbit at which the Earth is closest to the Sun assuming the Earth's orbit is perfectly circular and the Earth is moving at a constant speed. The mean anomaly at the perihelion is 0°.

The **local hour angle** (in radians) is

$$H_a = \frac{2\pi t_s}{86\,400} \tag{9.73}$$

where t_s is the number of seconds past local noon, and 86 400 is the number of seconds in a day. At noon, when the Sun is highest, the local hour angle is zero, and (9.67) simplifies to $\cos\theta_s = \sin\varphi\sin\delta + \cos\varphi\cos\delta$. When the Sun is over the Equator, the declination angle and latitude are zero, and (9.67) simplifies to $\cos\theta_s = \cos H_a$. Figure 9.27 shows that the Sun reaches its maximum declination ($\pm 23.5°$) at the summer and winter solstices and its minimum declination (0°) at the vernal and autumnal equinoxes.

Example 9.9

Calculate the solar zenith angle at 1:00 p.m. PST, on February 27, 1994 at a latitude of $\varphi = 35°$ N.

SOLUTION

February 27 corresponds to Julian day $D_J = 58$. From (9.70), $N_{JD} = -2134.5$. From (9.72), $L_M = -1823.40°$ and $g_M = -1746.23°$. Thus, from (9.71), (9.69), and (9.68), $\lambda_{ec} = -1821.87°$, $\varepsilon_{ob} = 23.4399°$, and $\delta = -8.52°$, respectively. Equation (9.73) gives $H_a = 15.0°$. Thus, from (9.67), $\cos\theta_s = \sin 35° \sin(-8.52°) + \cos 35° \cos(-8.52°) \cos 15.0° \rightarrow \theta_s = 45.8°$.

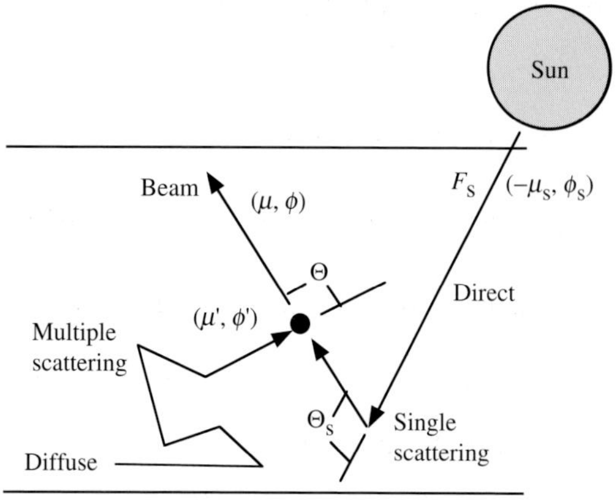

Figure 9.28 Single scattering of direct solar radiation and multiple scattering of diffuse radiation adds to the intensity along a beam of orientation μ, ϕ. The parameter $\mu = \cos\theta$ is always positive, but when a ray is directed upward, $+\mu$ is used, and when a ray is directed downward, $-\mu$ is used. Adapted from Liou (2002).

9.8 THE RADIATIVE TRANSFER EQUATION

The **radiative transfer equation** gives the change in radiance and/or irradiance along a beam of electromagnetic energy at a point in the atmosphere. Radiances are used to calculate actinic fluxes, which are used in photolysis coefficient equations. Irradiances are used to calculate heating rates, which are used in temperature calculations. The processes affecting radiation along a beam are scattering of radiation out of the beam, absorption of radiation along the beam, multiple scattering of indirect, diffuse radiation into the beam, single scattering of direct, solar radiation into the beam, and emission of infrared radiation into the beam. **Single scattering** occurs when a photon of radiation is redirected into a beam after it collides with a particle or gas molecule, as shown in Fig. 9.28. **Multiple scattering** occurs when a photon enters a beam after colliding sequentially with several particles or gas molecules, each of which redirects the photon. Solar radiation that has not yet been scattered is **direct** radiation. Radiation, either solar or infrared, that has been scattered is **diffuse** radiation.

The change in spectral radiance over the distance dS_b along a beam is

$$dI_\lambda = -dI_{so,\lambda} - dI_{ao,\lambda} + dI_{si,\lambda} + dI_{Si,\lambda} + dI_{ei,\lambda} \tag{9.74}$$

where

$$dI_{so,\lambda} = I_\lambda \sigma_{s,\lambda}\, dS_b \tag{9.75}$$

represents **scattering of radiation out of the beam,**

$$dI_{ao,\lambda} = I_\lambda \sigma_{a,\lambda}\, dS_b \tag{9.76}$$

represents **absorption of radiation along the beam,**

$$dI_{si,\lambda} = \left[\sum_k \left(\frac{\sigma_{s,k,\lambda}}{4\pi}\int_0^{2\pi}\int_{-1}^{1} I_{\lambda,\mu',\phi'} P_{s,k,\lambda,\mu,\mu',\phi,\phi'}\, d\mu'\, d\phi'\right)\right] dS_b \tag{9.77}$$

represents **multiple scattering of diffuse radiation into the beam,**

$$dI_{Si,\lambda} = \left[\sum_k \left(\frac{\sigma_{s,k,\lambda}}{4\pi} P_{s,k,\lambda\mu,-\mu_s,\phi,\phi_s}\right)\right] F_{s,\lambda} e^{-\tau_\lambda/\mu_s}\, dS_b \tag{9.78}$$

represents **single scattering of direct solar radiation into the beam,** and

$$dI_{ei,\lambda} = \sigma_{a,\lambda} B_{\lambda,T}\, dS_b \tag{9.79}$$

represents **emission of infrared radiation into the beam.** In (9.77) and (9.78), the summations are over all scattering processes ($k = \mathrm{g}$ for gases, a for aerosol particles, and h for hydrometeor particles), and the P_s factors are scattering phase functions, to be defined shortly. In (9.75), (9.76), and (9.79), the factors,

$$\sigma_{s,\lambda} = \sigma_{s,g,\lambda} + \sigma_{s,a,\lambda} + \sigma_{s,h,\lambda} \qquad \sigma_{a,\lambda} = \sigma_{a,g,\lambda} + \sigma_{a,a,\lambda} + \sigma_{a,h,\lambda} \tag{9.80}$$

are extinction coefficients due to total scattering and absorption, respectively. The extinction coefficient due to total scattering plus absorption is

$$\sigma_\lambda = \sigma_{s,\lambda} + \sigma_{a,\lambda} \tag{9.81}$$

In the solar spectrum, single scattering of solar radiation is more important than is radiative emission. In the infrared spectrum, the reverse is true. Thus, (9.78) is used for solar wavelengths and (9.79) is used for infrared wavelengths.

9.8.1 Phase function and asymmetry parameter

In (9.77), $P_{s,k,\lambda,\mu,\mu',\phi,\phi'}$ is the **scattering phase function,** which gives the angular distribution of scattered energy as a function of direction. It relates how diffuse radiation, which has direction μ', ϕ', is redirected by gases or particles toward the beam of interest, which has direction μ, ϕ, as shown in Fig. 9.28. In the figure, $\mu' = \cos\theta'$ and $\mu = \cos\theta$, where θ' and θ are the zenith angles of the diffuse radiation and the beam of interest, respectively. Similarly, ϕ' and ϕ are the azimuth angles of the diffuse radiation and the beam of interest, respectively. The integral in (9.77) is over all possible angles of incoming multiple-scattered radiation. Scattering phase functions vary with wavelength and differ for gases, aerosol particles, and cloud drops.

In (9.78), $P_{s,k,\lambda,\mu,-\mu_s,\phi,\phi_s}$ is the scattering phase function for direct radiation. The function relates how direct solar radiation, with direction $-\mu_s(= -\cos\theta_s)$, ϕ_s, is redirected by gases or particles to μ, ϕ, as shown in Fig. 9.28. This phase function is not integrated over all solid angles, since single-scattered radiation originates from one angle.

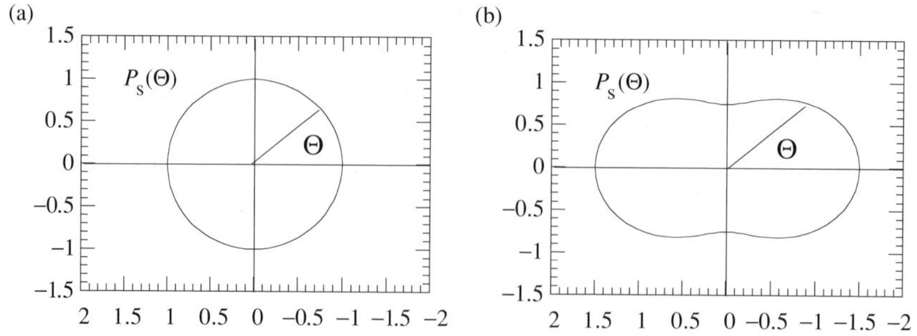

Figure 9.29 Scattering phase function distribution in polar coordinates for (a) isotropic and (b) Rayleigh scattering. The diagrams were generated from (9.84) and (9.85), respectively.

The scattering phase function is defined such that

$$\frac{1}{4\pi}\int_{4\pi} P_{s,k,\lambda}(\Theta)\,d\Omega_a = 1 \tag{9.82}$$

In this equation, Θ is the angle between directions μ', ϕ' and μ, ϕ, as shown in Fig. 9.28. Thus, $P_{s,k,\lambda,\mu,\mu',\phi,\phi'} = P_{s,k,\lambda}(\Theta)$. Here, Θ_s is the angle between the solar beam $(-\mu_s, \phi_s)$ and the beam of interest (μ, ϕ). Thus, $P_{s,k,\lambda,\mu,-\mu_s,\phi,\phi_s} = P_{s,k,\lambda}(\Theta_s)$.

Substituting $d\Omega_a = \sin\Theta\, d\Theta\, d\phi$ from (9.8) into (9.82) gives

$$\frac{1}{4\pi}\int_0^{2\pi}\int_0^{\pi} P_{s,k,\lambda}(\Theta)\sin\Theta\, d\Theta\, d\phi = 1 \tag{9.83}$$

The integral limits are defined so that the integration is over a full sphere. For **isotropic scattering**, the phase function is

$$P_{s,k,\lambda}(\Theta) = 1 \tag{9.84}$$

which satisfies (9.82). For **Rayleigh (gas) scattering**, the phase function is

$$P_{s,k,\lambda}(\Theta) = \frac{3}{4}(1 + \cos^2\Theta) \tag{9.85}$$

which also satisfies (9.82). Figures 9.29 (a) and (b) show scattering phase functions for isotropic and Rayleigh scattering, respectively, versus scattering angle. The phase function for isotropic scattering projects equally in all directions, and that for Rayleigh scattering is symmetric, but projects mostly in the forward and backward directions.

An approximation to the phase function for Mie scattering by aerosol particles and hydrometeor particles is the **Henyey–Greenstein function** (Henyey and Greenstein 1941),

$$P_{s,k,\lambda}(\Theta) = \frac{1 - g_{a,k,\lambda}^2}{\left(1 + g_{a,k,\lambda}^2 - 2g_{a,k,\lambda}\cos\Theta\right)^{3/2}} \tag{9.86}$$

where $g_{a,k,\lambda}$ is the asymmetry parameter, defined shortly. This equation is valid primarily for scattering that is not strongly peaked in the forward direction (Liou 2002), and requires advance knowledge of the asymmetry parameter, which is a function of the phase function itself.

The **asymmetry parameter**, or first moment of the phase function, is a parameter derived from the phase function that gives the relative direction of scattering by particles or gases. Its analytical form is

$$g_{a,k,\lambda} = \frac{1}{4\pi}\int_{4\pi} P_{s,k,\lambda}(\Theta)\cos\Theta\, d\Omega_a \tag{9.87}$$

The asymmetry parameter approaches +1 for scattering strongly peaked in the forward direction and −1 for scattering strongly peaked in the backward direction. If the asymmetry parameter is zero, scattering is equal in the forward and backward directions. In sum,

$$g_{a,k,\lambda}\begin{cases} > 0 & \text{forward (Mie) scattering} \\ = 0 & \text{isotropic or Rayleigh scattering} \\ < 0 & \text{backward scattering} \end{cases} \tag{9.88}$$

Expanding (9.83) with $d\Omega_a = \sin\Theta\, d\Theta\, d\phi$ yields

$$g_{a,k,\lambda} = \frac{1}{4\pi}\int_0^{2\pi}\int_0^{\pi} P_{s,k,\lambda}(\Theta)\cos\Theta\sin\Theta\, d\Theta\, d\phi \tag{9.89}$$

For **isotropic scattering**, where $P_{s,k,\lambda}(\Theta) = 1$, the asymmetry parameter simplifies to

$$g_{a,k,\lambda} = \frac{1}{4\pi}\int_0^{2\pi}\int_0^{\pi}\cos\Theta\sin\Theta\, d\Theta\, d\phi = -\frac{1}{2}\int_1^{-1}\mu\, d\mu = 0 \tag{9.90}$$

where $\mu = \cos\Theta$. The zero asymmetry parameter for isotropic scattering is expected, since isotropic scattering distributes radiation equally in all directions.

Substituting the phase function for **Rayleigh scattering** into (9.85) gives

$$\begin{aligned} g_{a,k,\lambda} &= \frac{1}{4\pi}\int_0^{2\pi}\int_0^{\pi}\frac{3}{4}(1+\cos^2\Theta)\cos\Theta\sin\Theta\, d\Theta\, d\phi \\ &= -\frac{3}{8}\int_0^{2\pi}\int_1^{-1}(\mu+\mu^3)d\mu\, d\phi = 0 \end{aligned} \tag{9.91}$$

The Rayleigh scattering intensity is evenly distributed between forward and backward directions, but Rayleigh scattering is not isotropic, since the radiation is not scattered equally in all directions.

The asymmetry parameter for aerosol particles and hydrometeor particles can be found with

$$g_{a,k,\lambda} = Q_{f,i,\lambda}/Q_{s,i,\lambda} \tag{9.92}$$

where $Q_{f,i,\lambda}$ is the single-particle forward scattering efficiency and $Q_{s,i,\lambda}$ is the single-particle total scattering efficiency, both of which can be obtained from a Mie-scattering algorithm. Curves for both parameters were shown in Figs. 9.20 and 9.21 for two substances. These figures indicate that the asymmetry parameters for soot and liquid water aerosol particles are typically between 0.6 and 0.85. Measured asymmetry parameters in clouds range from 0.7 to 0.85 (e.g., Gerber *et al.* 2000).

Another parameter derived from the phase function is the **backscatter ratio,**

$$b_{a,k,\lambda} = \frac{\int_0^{2\pi} \int_{\pi/2}^{\pi} P_{s,k,\lambda}(\Theta) \sin\Theta \, d\Theta \, d\phi}{\int_0^{2\pi} \int_0^{\pi} P_{s,k,\lambda}(\Theta) \sin\Theta \, d\Theta \, d\phi} \tag{9.93}$$

This parameter gives the fraction of total scattered radiation that is scattered in the backward hemisphere of the particle. Typical values of the backscatter ratio are 0.08–0.2 (e.g., Marshall *et al.* 1995), suggesting that 80–92 percent of energy scattered by particles is scattered in the forward direction. Particle shape has a significant impact on the backscatter ratio through its effect on the phase function (e.g., Pollack and Cuzzi 1980; Fridlind and Jacobson 2003).

9.8.2 Incident solar radiation

In (9.78), $F_{s,\lambda}$ is the incident solar radiation at the top of the atmosphere in a wavelength interval centered at λ. The radiation intensity at the top of the atmosphere depends on the solar luminosity and the Earth–Sun distance. The **solar luminosity** (L_p) is the total energy per unit time emitted by the Sun's photosphere and is approximately 3.9×10^{26} W. The irradiance, or luminosity emitted per unit area (W m^{-2}) at the photosphere, is approximately

$$F_p = \frac{L_p}{4\pi R_p^2} = \sigma_B T_p^4 \tag{9.94}$$

where $R_p = 6.96 \times 10^8$ m is the **radius of the Sun,** from its center to the photosphere, T_p is the temperature of the Sun at the photosphere, and (9.94) is a summation over all wavelength intervals.

Example 9.10

For a photosphere temperature of T_p = 5796 K, $F_p \approx 6.4 \times 10^7$ W m^{-2} from (9.94).

Irradiance decreases proportionally to the inverse square distance from its source. The yearly mean solar irradiance reaching the top of Earth's atmosphere,

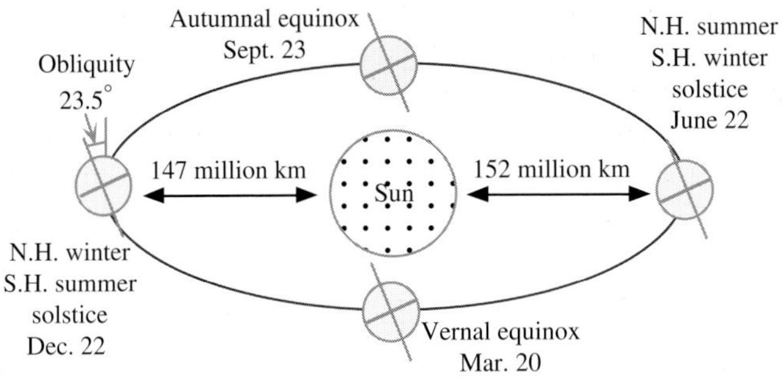

Figure 9.30 Relationship between the Sun and Earth during the solstices and equinoxes.

or **solar constant,** is related to F_p by

$$\bar{F}_s = \left(\frac{R_p}{\bar{R}_{es}}\right)^2 F_p = \left(\frac{R_p}{\bar{R}_{es}}\right)^2 \sigma_B T_p^4 \tag{9.95}$$

where $\bar{R}_{es} = 1.5 \times 10^{11}$ m is the mean distance from the center of the Sun to the Earth. Equation (9.95), with the parameters specified above, gives $\bar{F}_s \approx 1379$ W m^{-2}, which is close to measurements of $\bar{F}_s = 1365$ W m^{-2}. The value of $\bar{F}_s$ varies by ± 1 W m^{-2} over each 11-year sunspot cycle.

The actual total irradiance at the top of the atmosphere at a given time is

$$F_s = \left(\frac{R_{es}}{\bar{R}_{es}}\right)^2 \bar{F}_s \tag{9.96}$$

where R_{es} is the actual distance between the Sun and the Earth on a given day, and

$$\left(\frac{R_{es}}{\bar{R}_{es}}\right)^2 \approx 1.00011 + 0.034221 \cos\theta_J + 0.00128 \sin\theta_J + 0.000719 \cos 2\theta_J + 0.000077 \sin 2\theta_J \tag{9.97}$$

(Spencer 1971). In (9.97), $\theta_J = 2\pi D_J / D_Y$, where D_Y is the number of days in a year (365 for nonleap years and 366 for leap years), and D_J is the Julian day of the year. Figure 9.30 shows that the Earth is further from the Sun in the Northern Hemisphere summer than in the winter by about 3.4 percent.

Example 9.11

On December 22, the total irradiance at the top of Earth's atmosphere is $F_s = 1365 \times 1.034 = 1411$ W m^{-2}, and on June 22, the irradiance is $F_s = 1365 \times 0.967 = 1321$ W m^{-2}. Thus, the irradiance varies by 90 W m^{-2} (6.6 percent) between December and June.

In a model, the actual total solar irradiance reaching the top of Earth's atmosphere (F_s) (W m^{-2}) is related to the actual solar irradiance in each wavelength interval ($F_{s,\lambda}$) (W m^{-2} μm^{-1}), the mean solar irradiance in each wavelength interval ($\bar{F}_{s,\lambda}$) (W m^{-2} μm^{-1}), and the solar constant ($\bar{F}_s$) (W m^{-2}) by

$$F_s = \sum_\lambda (F_{s,\lambda}\Delta\lambda) = \left(\frac{R_{es}}{\bar{R}_{es}}\right)^2 \sum_\lambda (\bar{F}_{s,\lambda}\Delta\lambda) = \left(\frac{R_{es}}{\bar{R}_{es}}\right)^2 \bar{F}_s \tag{9.98}$$

where $\Delta\lambda$ is a wavelength interval (μm). Values for $\bar{F}_{s,\lambda}$ are given in Appendix Table B.2.

The **solar constant** $\bar{F}_s$ can be used to determine Earth's equilibrium temperature in the absence of the greenhouse effect. The total energy (W) absorbed by the Earth–atmosphere system is about

$$E_{in} = \bar{F}_s(1 - A_{e,0})(\pi R_e^2) \tag{9.99}$$

where R_e is the Earth's radius, πR_e^2 is the Earth's cross-sectional area (effective area that solar irradiance impinges upon), and $A_{e,0}$ is the globally and wavelength-averaged **Earth–atmosphere albedo** ($\approx$0.30). The outgoing energy from the Earth is

$$E_{out} = \varepsilon_{e,0}\sigma_B T_e^4(4\pi R_e^2) \tag{9.100}$$

where T_e is the temperature of the Earth's surface, and $\varepsilon_{e,0}$ is Earth's globally and wavelength-averaged **surface emissivity**. The actual average emissivity is 0.96–0.98, but if the Earth is considered a blackbody, $\varepsilon_{e,0} = 1$. Equating incoming and outgoing energy gives

$$T_e = \left[\frac{\bar{F}_s(1 - A_{e,0})}{4\varepsilon_{e,0}\sigma_B}\right]^{1/4} \tag{9.101}$$

which is the **equilibrium temperature of the Earth** in the absence of a greenhouse effect.

Example 9.12

For $F_s = 1365$ W m^{-2}, $A_{e,0} = 0.3$, and $\varepsilon_{e,0} = 1$, (9.101) predicts $T_e = 254.8$ K. The actual average surface temperature on the Earth is about 288 K, and the difference is due primarily to absorption by greenhouse gases.

9.8.3 Solutions to the radiative transfer equation

Equations (9.74)–(9.79) can be combined to give the radiative transfer equation as

$$\frac{dI_{\lambda,\mu,\phi}}{dS_b} = -I_{\lambda,\mu,\phi}(\sigma_{s,\lambda} + \sigma_{a,\lambda}) + \sum_k \left(\frac{\sigma_{s,k,\lambda}}{4\pi}\int_0^{2\pi}\int_{-1}^{1} I_{\lambda,\mu',\phi'} P_{s,k,\lambda,\mu,\mu',\phi,\phi'}\, d\mu'\, d\phi'\right) + F_{s,\lambda}e^{-\tau_\lambda/\mu_s}\sum_k \left(\frac{\sigma_{s,k,\lambda}}{4\pi} P_{s,k,\lambda,\mu,-\mu_s,\phi,\phi_s}\right) + \sigma_{a,\lambda}B_{\lambda,T} \tag{9.102}$$

The fraction of total extinction due to scattering is the **single-scattering albedo,**

$$\omega_{s,\lambda} = \frac{\sigma_{s,\lambda}}{\sigma_\lambda} = \frac{\sigma_{s,g,\lambda} + \sigma_{s,a,\lambda} + \sigma_{s,h,\lambda}}{\sigma_{s,g,\lambda} + \sigma_{a,g,\lambda} + \sigma_{s,a,\lambda} + \sigma_{a,a,\lambda} + \sigma_{s,h,\lambda} + \sigma_{a,h,\lambda}} \tag{9.103}$$

Substituting (9.103), $\sigma_\lambda = \sigma_{s,\lambda} + \sigma_{a,\lambda}$ from (9.81) and $d\tau_\lambda = -\sigma_\lambda \mu_s \, dS_b$ from (9.65) into (9.102) gives the radiative transfer equation as

$$\mu \frac{dI_{\lambda,\mu,\phi}}{d\tau_\lambda} = I_{\lambda,\mu,\phi} - J^{\text{diffuse}}_{\lambda,\mu,\phi} - J^{\text{direct}}_{\lambda,\mu,\phi} - J^{\text{emis}}_{\lambda,\mu,\phi} \tag{9.104}$$

where

$$J^{\text{diffuse}}_{\lambda,\mu,\phi} = \frac{1}{4\pi} \sum_k \left(\frac{\sigma_{s,k,\lambda}}{\sigma_\lambda} \int_0^{2\pi} \int_{-1}^{1} I_{\lambda,\mu',\phi'} P_{s,k,\lambda,\mu,\mu',\phi,\phi'} \, d\mu' \, d\phi' \right) \tag{9.105}$$

$$J^{\text{direct}}_{\lambda,\mu,\phi} = \frac{1}{4\pi} F_{s,\lambda} e^{-\tau_\lambda/\mu_s} \sum_k \left(\frac{\sigma_{s,k,\lambda}}{\sigma_\lambda} P_{s,k,\lambda,\mu,-\mu_s,\phi,\phi_s} \right) \tag{9.106}$$

$$J^{\text{emis}}_{\lambda,\mu,\phi} = (1 - \omega_{s,\lambda}) B_{\lambda,T} \tag{9.107}$$

9.8.3.1 *Analytical solutions*

The radiative transfer equation must be solved numerically, except for idealized cases. For example, an analytical solution can be derived when absorption is considered but scattering and emission are neglected. In this case, (9.104) simplifies for the upward and downward directions to

$$\mu \frac{dI_{\lambda,\mu,\phi}}{d\tau_{a,\lambda}} = I_{\lambda,\mu,\phi} \qquad -\mu \frac{dI_{\lambda,-\mu,\phi}}{d\tau_{a,\lambda}} = I_{\lambda,-\mu,\phi} \tag{9.108}$$

respectively, where $+\mu$ is used for upward radiation, $-\mu$ is used for downward radiation, and $\tau_{a,\lambda}$ is the optical depth due to absorption only. Integrating the **upward** equation from a lower optical depth $\tau_{a,\lambda,b}$ to the optical depth of interest, $\tau_{a,\lambda}$, gives

$$I_{\lambda,\mu,\phi}(\tau_{a,\lambda}) = I_{\lambda,\mu,\phi}(\tau_{a,\lambda,b}) e^{(\tau_{a,\lambda} - \tau_{a,\lambda,b})/\mu} \tag{9.109}$$

Integrating the **downward** equation from an upper optical depth $\tau_{a,\lambda,t}$ to the optical depth of interest, $\tau_{a,\lambda}$, gives

$$I_{\lambda,-\mu,\phi}(\tau_{a,\lambda}) = I_{\lambda,-\mu,\phi}(\tau_{a,\lambda,t}) e^{-(\tau_{a,\lambda} - \tau_{a,\lambda,t})/\mu} \tag{9.110}$$

Equations (9.109) and (9.110) describe **Beer's law,** which states that the absorption of radiation increases exponentially with the optical depth of the absorbing species.

A second analytical solution can be derived when absorption and infrared emission are considered, but scattering is neglected. In this case, (9.104) simplifies for the upward and downward directions as

$$\mu\frac{dI_{\lambda,\mu,\phi}}{d\tau_{a,\lambda}} = I_{\lambda,\mu,\phi} - B_{\lambda,T} \qquad -\mu\frac{dI_{\lambda,-\mu,\phi}}{d\tau_{a,\lambda}} = I_{\lambda,-\mu,\phi} - B_{\lambda,T} \tag{9.111}$$

respectively, which are **Schwartzchild's equations.** Integrating (9.111) gives upward and downward radiances as

$$I_{\lambda,\mu,\phi}(\tau_{a,\lambda}) = I_{\lambda,\mu,\phi}(\tau_{a,\lambda,b})e^{(\tau_{a,\lambda}-\tau_{a,\lambda,b})/\mu} - \frac{1}{\mu}\int_{\tau_{a,\lambda,b}}^{\tau_{a,\lambda}} \left[B_{\lambda,T(\tau'_{a,\lambda})}e^{(\tau_{a,\lambda}-\tau'_{a,\lambda})/\mu}\right]d\tau'_{a,\lambda} \tag{9.112}$$

$$I_{\lambda,-\mu,\phi}(\tau_{a,\lambda}) = I_{\lambda,-\mu,\phi}(\tau_{a,\lambda,t})e^{-(\tau_{a,\lambda}-\tau_{a,\lambda,t})/\mu} + \frac{1}{\mu}\int_{\tau_{a,\lambda,t}}^{\tau_{a,\lambda}} \left[B_{\lambda,T(\tau'_{a,\lambda})}e^{-(\tau_{a,\lambda}-\tau'_{a,\lambda})/\mu}\right]d\tau'_{a,\lambda} \tag{9.113}$$

respectively. In each equation, the Planck function, which varies with temperature, must be integrated between the two optical depths.

9.8.3.2 *Numerical solutions*

When particle scattering is included in the radiative transfer equation, analytical solutions become difficult to obtain and numerical solutions are needed. One numerical solution is found with the **two-stream method.** With this method, radiance is divided into an upward (↑) and a downward (↓) component, each of which is approximated with a forward and a backward scattering term in the diffuse phase function integral of (9.105). One approximation to the integral is

$$\frac{1}{4\pi}\int_0^{2\pi}\int_{-1}^{1} I_{\lambda,\mu',\phi'}P_{s,k,\lambda,\mu,\mu',\phi,\phi'}\,d\mu'\,d\phi' \approx \begin{cases} \dfrac{(1+g_{a,k,\lambda})}{2}I\uparrow + \dfrac{(1-g_{a,k,\lambda})}{2}I\downarrow & \text{upward} \\ \dfrac{(1+g_{a,k,\lambda})}{2}I\downarrow + \dfrac{(1-g_{a,k,\lambda})}{2}I\uparrow & \text{downward} \end{cases} \tag{9.114}$$

where $I\downarrow$ is the **downward radiance,** $I\uparrow$ is the **upward radiance,** and $(1+g_{a,k,\lambda})/2$ and $(1-g_{a,k,\lambda})/2$ are integrated fractions of the forward- and backward-scattered energy, respectively. Wavelength subscripts have been omitted on $I\uparrow$ and $I\downarrow$. Equation (9.114) is the **two-point quadrature** approximation to the phase-function integral (Liou 1974, 2002; Meador and Weaver 1980). Substituting (9.114) into

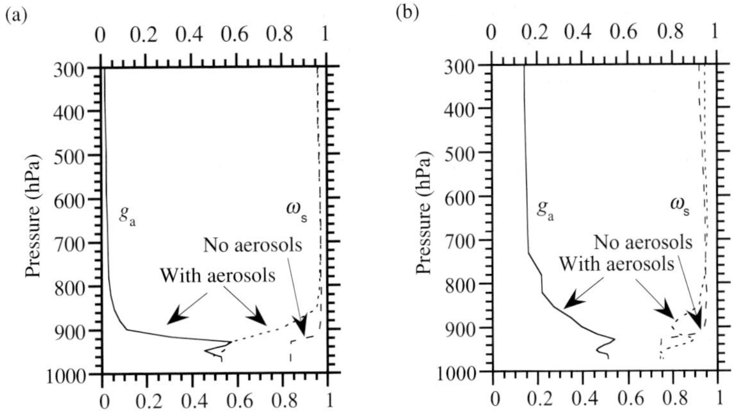

Figure 9.31 Modeled vertical profiles from Claremont, California (11.30 on August 27, 1997) of the effective asymmetry parameter and the single-scattering albedo at (a) 0.32 and (b) 0.61 μm when aerosol particles were and were not included in the simulations. The overall asymmetry parameter was zero when aerosol particles were absent. The peaks at 920 hPa indicate elevated aerosol layers. From Jacobson (1998b).

(9.105) gives

$$\frac{1}{4\pi}\sum_k\left(\frac{\sigma_{s,k,\lambda}}{\sigma_\lambda}\int_0^{2\pi}\int_{-1}^{1} I_{\lambda,\mu',\phi'}P_{s,k,\lambda,\mu,\mu',\phi,\phi'}\,d\mu'\,d\phi'\right)$$
$$\approx\begin{cases}\omega_{s,\lambda}(1-b_\lambda)I\uparrow+\omega_{s,\lambda}b_\lambda I\downarrow\\ \omega_{s,\lambda}(1-b_\lambda)I\downarrow+\omega_{s,\lambda}b_\lambda I\uparrow\end{cases} \tag{9.115}$$

where $\omega_{s,\lambda}$ is the single-scattering albedo from (9.103),

$$1-b_\lambda=\frac{1+g_{a,\lambda}}{2} \qquad b_\lambda=\frac{1-g_{a,\lambda}}{2} \tag{9.116}$$

are effective integrated fractions of forward- and backward-scattered energy, and

$$g_{a,\lambda}=\frac{\sigma_{s,a,\lambda}g_{a,a,\lambda}+\sigma_{s,c,\lambda}g_{a,h,\lambda}}{\sigma_{s,g,\lambda}+\sigma_{s,a,\lambda}+\sigma_{s,h,\lambda}} \tag{9.117}$$

is an **effective asymmetry parameter**, which is the weighted sum of asymmetry parameters for gases ($g_{a,g,\lambda}$), aerosol particles ($g_{a,a,\lambda}$), and hydrometeor particles ($g_{a,h,\lambda}$). Asymmetry parameters for gases are zero.

Figures 9.31 (a) and (b) show vertical profiles of effective asymmetry parameters and single-scattering albedos over an urban airshed from two model simulations. In one simulation, aerosol particles were assumed to be present, and in the other, they were ignored. Clouds were not present in either case. The figures indicate that effective asymmetry parameters increased from the free troposphere to the boundary layer when aerosol particles were assumed to be present. Above the boundary layer, the effective asymmetry parameters were small because Rayleigh scattering, which has a zero asymmetry parameter, dominated scattering extinction (see

Fig. 9.23). Within the boundary layer, effective asymmetry parameters increased because aerosol-scattering extinction dominated Rayleigh-scattering extinction.

The single-scattering albedo curve in Fig. 9.31(a) indicates that, above the boundary layer, ultraviolet extinction was due almost entirely to gas scattering when aerosol particles were ignored, and to gas and particle scattering when aerosol particles were present. Within the boundary layer, aerosol absorption played a larger role than it did above the boundary layer.

The **phase function for the single scattering of solar radiation,** used in (9.106), can be estimated with the Eddington approximation of the solar phase function (Eddington 1916)

$$P_{s,k,\lambda}(\Theta_s) \approx 1 \pm 3g_{a,\lambda}\mu_1\mu_s \tag{9.118}$$

where μ_1 is the diffusivity factor, set to $\mu_1 = 1/\sqrt{3}$ when the quadrature approximation for diffuse radiation is used (Liou 1974).

With the parameters above, the **spectral radiance** from (9.104) can be written for solar wavelengths (where the emission term is neglected) in terms of an upward and a downward component as

$$\mu_1 \frac{dI\uparrow}{d\tau} = I\uparrow - \omega_s(1-b)I\uparrow - \omega_s bI\downarrow - \frac{\omega_s}{4\pi}(1 - 3g_a\mu_1\mu_s)F_s e^{-\tau/\mu_s} \tag{9.119}$$

$$-\mu_1 \frac{dI\downarrow}{d\tau} = I\downarrow - \omega_s(1-b)I\downarrow - \omega_s bI\uparrow - \frac{\omega_s}{4\pi}(1 + 3g_a\mu_1\mu_s)F_s e^{-\tau/\mu_s} \tag{9.120}$$

respectively (Liou 2002), where all wavelength subscripts have been omitted. The term $-\mu_1$ is used for direct solar radiation on the right side of (9.119) because solar radiation is downward relative to $I\uparrow$. The term $+\mu_1$ is used on the right side of (9.120) for direct solar radiation since solar radiation is in the same direction as $I\downarrow$.

Equations (9.119) and (9.120) can be written in terms of **spectral irradiance** with the conversions $F\uparrow = 2\pi\mu_1 I\uparrow$ and $F\downarrow = 2\pi\mu_1 I\downarrow$, and generalized for the two-point quadrature or another two-stream approximation. The resulting equations are

$$\frac{dF\uparrow}{d\tau} = \gamma_1 F\uparrow - \gamma_2 F\downarrow - \gamma_3\omega_s F_s e^{-\tau/\mu_s}$$
$$\frac{dF\downarrow}{d\tau} = -\gamma_1 F\downarrow + \gamma_2 F\uparrow + (1-\gamma_3)\omega_s F_s e^{-\tau/\mu_s} \tag{9.121}$$

(Meador and Weaver 1980; Toon *et al.* 1989b; Liou 2002), where the γ's, defined in Table 9.9, are coefficients derived from the two-point quadrature and Eddington approximations of the diffuse phase-function integral. The Eddington approximation of the diffuse phase function integral is discussed in Irvine (1968, 1975),

Table 9.9 Coefficients for the two-stream method for two approximations of the diffuse phase function

Approximation	γ_1	γ_2	γ_3
Quadrature	$\frac{1-\omega_s(1+g_a)/2}{\mu_1}$	$\frac{\omega_s(1-g_a)}{2\mu_1}$	$\frac{1-3g_a\mu_1\mu_s}{2}$
Eddington	$\frac{7-\omega_s(4+3g_a)}{4}$	$-\frac{1-\omega_s(4-3g_a)}{4}$	$\frac{2-3g_a\mu_s}{4}$

Sources: Liou (2002), Meador and Weaver (1980), Toon *et al.* (1989a).

Kawata and Irvine (1970), and Meador and Weaver (1980). Replacing the solar term in (9.121) with an emission term gives an equation for infrared irradiance as

$$\frac{dF\uparrow}{d\tau} = \gamma_1 F\uparrow - \gamma_2 F\downarrow - 2\pi(1-\omega_s)B_T$$
$$\frac{dF\downarrow}{d\tau} = -\gamma_1 F\downarrow + \gamma_2 F\uparrow + 2\pi(1-\omega_s)B_T \qquad (9.122)$$

Equations 9.119–9.122 may be solved in a model after the equations are discretized, a vertical grid is defined, and boundary conditions are defined. Figure 7.1 shows a model atmosphere in which layer bottoms and tops are defined by indices, $k \pm 0.5$. For that grid, upward and downward irradiances at the bottom of a layer k are $F\uparrow_{k+1/2}$ and $F\downarrow_{k+1/2}$, respectively, where $k = 0, \ldots, N_L$. The derivatives in (9.119)–(9.122) can be discretized over this grid with a finite-difference expansion of any order. A discretization requires boundary conditions at the ground and top of the atmosphere. At the ground, boundary conditions for irradiance at a given wavelength are

$$F\uparrow_{N_L+1/2} = A_e F\downarrow_{N_L+1/2} + \begin{cases} A_e\mu_s F_s e^{-\tau_{N_L+1/2}/\mu_s} & \text{solar} \\ \varepsilon\pi B_T & \text{infrared} \end{cases} \qquad (9.123)$$

where A_e is the albedo and ε is the emissivity. Equation (9.123) states that the upward irradiance at the surface equals the reflected downward diffuse irradiance plus the reflected direct solar or emitted infrared irradiance. At the top of the atmosphere, the boundary condition is $F\downarrow_{1/2} = \mu_s F_s$ for solar radiation and $F\downarrow_{1/2} = F\uparrow_{1/2}$ for infrared radiation. The boundary conditions for radiance are similar. Terms from the discretization of (9.119)–(9.122) can be placed in a matrix. If a second-order central difference discretization is used, the matrix is tridiagonal and can be solved noniteratively to obtain the upward and downward fluxes at the boundary of each layer (e.g., Toon *et al.* 1989a).

When the atmosphere absorbs significantly, the two-stream and Eddington approximations underestimate forward scattering because the expansion of the phase function is too simple to obtain a strong peak in the scattering efficiency. As a partial remedy, the effective asymmetry parameter, single-scattering albedo, and optical depth can be adjusted with the **delta functions**

$$g'_a = \frac{g_a}{1 + g_a} \qquad \omega'_s = \frac{\left(1 - g_a^2\right)\omega_s}{1 - \omega_s g_a^2} \qquad \tau' = \left(1 - \omega_s g_a^2\right)\tau \tag{9.124}$$

respectively (Hansen 1969; Potter 1970; Joseph *et al.* 1976; Wiscombe 1977; Liou 2002), which replace terms in (9.104)–(9.107) and in Table 9.9.

More advanced methods of solving the radiative transfer equation are four-stream and higher-stream techniques (e.g., Liou 1974, 2002; Cuzzi *et al.* 1982; Fu and Liou 1993; Fu *et al.* 1997). A **four-stream approximation** is one in which radiance is divided into two upward and two downward components instead of one upward and one downward component, as was done for the two-stream approximation. In addition, the phase-function integral is expanded into four terms instead of the two terms in (9.114), and each of the four radiances is solved using one of the four phase-function terms. The final upward and downward radiances with the four-stream approximation are weighted averages of the two upward and two downward radiances, respectively.

9.8.4 Heating rates and photolysis coefficients

Spectral irradiances are used to estimate changes in air temperature due to radiative heating and cooling. If only radiative effects are considered, the local time rate of change of temperature of an atmospheric layer is found from the net flux divergence equation,

$$\left(\frac{\partial T}{\partial t}\right)_r = \frac{1}{c_{p,m}}\left(\frac{dQ_{solar}}{dt} + \frac{dQ_{ir}}{dt}\right) = \frac{1}{c_{p,m}\rho_a}\frac{\partial F_n}{\partial z} \tag{9.125}$$

where $F_n = \int_0^\infty (F{\downarrow}_\lambda - F{\uparrow}_\lambda)d\lambda$ is the net downward minus upward radiative flux (W m^{-2}), summed over all wavelengths, and Q_{solar} and Q_{ir} are solar and infrared **radiative heating rates** (J kg^{-1}), respectively, from the thermodynamic energy equation of (3.75). At the ground, $F_n = F_{n,g}$ from (8.104). Positive values of F_n correspond to net downward radiation. The partial derivative of F_n with respect to altitude can be discretized over layer k with

$$\frac{\partial F_{n,k}}{\partial z} \approx \frac{\sum_\lambda \left[\left(F{\downarrow}_{\lambda,k-1/2} - F{\uparrow}_{\lambda,k-1/2}\right) - \left(F{\downarrow}_{\lambda,k+1/2} - F{\uparrow}_{\lambda,k+1/2}\right)\right]}{z_{k-1/2} - z_{k+1/2}} \tag{9.126}$$

where k identifies the center of a layer, $k - 1/2$ identifies the top of a layer, and $k + 1/2$ identifies the bottom of a layer. The resulting change in temperature in the layer due to diabatic radiative heating is

$$\Delta T_k \approx \frac{1}{c_{p,m}\rho_a}\frac{\partial F_n}{\partial z}h \tag{9.127}$$

where h is the time step.

Photolysis rate coefficients are found by solving (9.119) and (9.120) for the solar spectral radiance. The spectral radiance is multiplied by 4π steradians to obtain the spectral actinic flux, as defined in (9.12). The photolysis rate coefficient (s^{-1}) of a species q producing product set p at the bottom of layer k (denoted by subscript $k + 1/2$) is

$$J_{q,p,k+1/2} = \int_0^\infty 4\pi I_{p,\lambda,k+1/2} b_{a,g,q,\lambda,T} Y_{q,p,\lambda,T} \, d\lambda \qquad (9.128)$$

where $b_{a,g,q,\lambda,T}$, previously defined, is the temperature- and wavelength-dependent absorption cross section of gas q (cm^2 molec.$^{-1}$), $Y_{q,p,\lambda,T}$ is the temperature- and wavelength-dependent quantum yield of q producing product set p (molec. photon^{-1}), and

$$I_{p,\lambda,k+1/2} = (I\downarrow_{\lambda,k+1/2} - I\uparrow_{\lambda,k+1/2}) \left(10^{-10} \frac{m^3}{cm^2\ \mu m}\right) \frac{\lambda}{hc} \qquad (9.129)$$

(photons $cm^{-2}\ s^{-1}\ \mu m^{-1}\ sr^{-1}$) is the spectral radiance at the bottom of layer k. The units of $I\uparrow\downarrow$, λ, c, and h (Planck's constant) for use in this equation are W $m^{-2}\ \mu m^{-1}\ sr^{-1}$, μm, m s^{-1}, and J s, respectively. The **quantum yield** is the fractional number (≤ 1) of molecules of a specific product formed per photon of radiation absorbed at a given wavelength. A photolysis reaction may produce different sets of products. Each set of products has its own set of wavelength-dependent quantum yields, as discussed in Chapter 10.

Example 9.14

Convert $I_\lambda = 12$ W $m^{-2}\ \mu m^{-1}\ sr^{-1}$ in the wavelength region, $0.495\ \mu m < \lambda < 0.505\ \mu m$, to photons $cm^{-2}\ s^{-1}\ sr^{-1}$.

SOLUTION

The solution to this problem is $I_{p,\lambda} = \Delta\lambda I_\lambda 10^{-10} \lambda/hc$, where the wavelength interval is $\Delta\lambda = 0.01\ \mu m$, and the mean wavelength is $\lambda = 0.5\ \mu m$. Solving gives $I_{p,\lambda} = 3.02 \times 10^{13}$ photons $cm^{-2}\ s^{-1}\ sr^{-1}$.

Gases, aerosol particles, and hydrometeor particles affect climate, weather, and air quality through their effects on heating and photolysis. For example, some studies have found that cloud drops, which are highly scattering, increase the backscattered fraction of incident solar radiation, increasing photolysis within and above a cloud and decreasing it below the cloud (e.g., Madronich 1987; van Weele and Duynkerke 1993; de Arellano *et al.* 1994). Other studies have found that aerosol particle absorption decreases photolysis below the region of absorption, decreasing ozone (e.g., Jacobson 1997b, 1998b; Castro *et al.* 2001). Conversely, aerosol particle scattering increases photolysis, potentially increasing ozone (e.g., Dickerson *et al.* 1997; Jacobson 1998b).

9.9 SUMMARY

In this chapter, radiation laws, radiation processes, optical properties, and the radiative transfer equation were described. The laws include Planck's law, Wien's law, and the Stefan–Boltzmann law. The radiation processes include reflection, refraction, scattering, absorption, diffraction, and transmission. Gases, aerosol particles, and hydrometeor particles attenuate radiation by absorbing and scattering. All gases scatter ultraviolet and short visible wavelengths. Some gases selectively absorb ultraviolet, visible, and infrared wavelengths. All aerosol particle components scatter visible light, but only a few absorb such radiation. Hydrometeor particles almost exclusively scatter visible light, except at large size. For example, large raindrops absorb visible light. The radiative transfer equation determines the change in radiance and irradiance along a beam due to scattering out of the beam, absorption along the beam, multiple scattering of diffuse radiation into the beam, single scattering of direct solar radiation into the beam, and emission of infrared radiation into the beam. Changes in radiance and irradiance are used to calculate photolysis and heating rates, respectively.

9.10 PROBLEMS

9.1 Calculate the radiance and irradiance from the Planck function at T = 273 K and (a) $\lambda = 0.4$ μm, (b) $\lambda = 1.0$ μm, and (c) $\lambda = 15$ μm.

9.2 Calculate the equilibrium surface temperature of the Earth if its emissivity and albedo were 70 and 25 percent, respectively. Does this represent an increase or decrease with respect to the equilibrium temperature (when the emissivity and albedo are 100 and 30 percent, respectively)?

9.3 Calculate the transmission of light through a 0.5-μm-diameter particle made of ammonium sulfate at a wavelength of $\lambda = 0.5$ μm (use the data from Table 9.6).

9.4 Calculate the extinction coefficient, meteorological range, and optical depth in a 1-km region of the atmosphere resulting from nitrogen dioxide absorption at a wavelength of $\lambda = 0.55$ μm when the volume mixing ratio of NO_2 is $\chi = 0.05$ ppmv, $T = 288$ K, and $p_d = 980$ hPa.

9.5 Calculate the real and imaginary refractive indices of a two-component mixture using (a) the volume average mixing rule and (b) the volume average dielectric constant mixing rule when the volume fraction of each component is 0.5 and the index of refraction of the matrix component is $m_1 = 1.34 - i1.0 \times 10^{-9}$ and that of the absorbing inclusion is $m_2 = 1.82 - i0.74$. Also, calculate the real and complex parts of the dielectric constant for each of the two components in the mixture.

9.6 Find the meteorological ranges at $\lambda = 0.53$ μm under the following conditions:

(a) In a rain shower with 1-mm-diameter water drops and a mass loading of 1 g m^{-3}.
(b) In a fog with 10-μm-diameter water drops and a mass loading of 1 g m^{-3}.
(c) In a haze with 0.5-μm-diameter particles, 40 percent ammonium sulfate by volume, 60 percent liquid water by volume, and a total mass loading of 50 μg m^{-3}.
(d) Behind a diesel exhaust, with 50 μg m^{-3} of 0.5-μm-diameter soot particles.

In all cases, assume particles are spherical, neglect Rayleigh scattering, and neglect gas absorption. In the case of soot and raindrops, assume particle scattering and absorption occur. In the other cases, assume only particle scattering occurs. In the case of haze, calculate the scattering efficiency by weighting the volume fractions of each component by the corresponding scattering efficiency. Use Figs. 9.20 and 9.21 to determine the efficiencies. Assume the scattering efficiency of ammonium sulfate is the same as that of water. Assume the mass densities of soot, ammonium sulfate, and liquid water are 1.25, 1.77, and 1.0 g cm^{-3}, respectively.

9.11 COMPUTER PROGRAMMING PRACTICE

9.7 Write a script to calculate radiance and irradiance from the Planck function versus temperature and wavelength. Use the program to calculate values between $\lambda = 0.01$ μm and $\lambda = 100$ mm for $T = 6000$ K and $T = 300$ K. Plot the results.

9.8 Divide the atmosphere from $z = 0$ to 10 km into 100 vertical layers. Assume $T = 288$ K and $p_d = 1013$ hPa at the surface, and assume the temperature decreases 6.5 K km^{-1}. Use (2.41) to estimate the air pressure in each layer. Calculate the extinction coefficient and optical depth in each layer due to Rayleigh scattering at $\lambda = 0.4$ μm. Estimate the cumulative optical depth at the surface.

10

Gas-phase species, chemical reactions, and reaction rates

THE atmosphere contains numerous gases that undergo chemical reaction. Because many chemical pathways are initiated by sunlight, atmospheric reactions are collectively called **photochemical reactions**. Lightning, changes in temperature, and molecular collisions, though, also initiate reactions. Photochemistry is responsible for the transformation of gases in all regions of the atmosphere. It converts nitrogen oxide and reactive organic gases, emitted during fuel combustion, to ozone, peroxyacetyl nitrate (PAN), and other products. It produces and destroys ozone in the free troposphere and stratosphere. Photochemistry also converts dimethyl sulfide (DMS) to sulfuric acid over the oceans. In this chapter, chemical species, structures, reactions, rate coefficients, rates, and lifetimes are discussed. In Chapter 11, photochemical reactions important in different regions of the atmosphere are described. In Chapter 12, numerical methods of solving chemical ordinary differential equations arising from such reactions are given.

10.1 ATMOSPHERIC GASES AND THEIR MOLECULAR STRUCTURES

Gases consist of neutral or charged single atoms or molecules. Elements that make up most gases are hydrogen (H), carbon (C), nitrogen (N), oxygen (O), fluorine (F), sulfur (S), chlorine (Cl), and bromine (Br). Particles contain sodium (Na), magnesium (Mg), aluminum (Al), silicon (Si), potassium (K), calcium (Ca), and/or the elements found in gases.

Gas molecules consist of atoms with covalent bonds between them. A **covalent bond** is a bond consisting of one or more pairs of electrons shared between two atoms. Gilbert Lewis (1875–1946) suggested that atoms could be held together by shared electron pairs. Such sharing occurs in the outer shell (**valence shell**) of an atom. Lewis proposed the use of **electron-dot symbols (Lewis symbols)** to describe the configuration of electrons in the valence shell of an atom. Lewis symbols are drawn by arranging electrons around an element as dots. A single dot represents an unpaired electron and two dots adjacent to each other represent an electron pair. Table 10.1 shows Lewis symbols for several elements in the periodic table.

Hydrogen (H), the first element of the periodic table, has one electron in its valence shell, as shown by its Lewis symbol in Table 10.1. The electron in H is unpaired. The number of unpaired electrons in the valence shell of an element is the **valence** of the element. If a hydrogen atom covalently bonds with another

Table 10.1 Elements in periods 1–4 and groups I–VIII of the periodic table and their Lewis symbols.

Group / Period	I	II	III	IV	V	VI	VII	VIII
1	1 1 H· 1.008							2 0 He: 4.003
2	3 1 Li· 6.941	4 2 ·Be· 9.012	5 3 ·B· 10.81	6 4 ·C· 12.01	7 3 ·N· 14.01	8 2 :O· 16.00	9 1 :F: 19.00	10 0 :Ne: 20.18
3	11 1 Na· 22.99	12 2 ·Mg· 24.30	13 3 ·Al· 26.98	14 4 ·Si· 28.09	15 3 ·P· 30.97	16 2 :S· 32.07	17 1 :Cl· 35.45	18 0 :Ar: 39.95
4	19 1 K· 39.10	20 2 ·Ca· 40.08	31 3 ·Ga· 69.72	32 4 ·Ge· 72.61	33 3 ·As· 74.92	34 2 :Se· 78.96	35 1 :Br· 79.90	36 0 :Kr: 83.80

Note: The top left number in each entry is the atomic number. The top right number is the principal valence of the element. The dots are the Lewis symbols of the element. The bottom number is the atomic weight of the element (g mol^{-1}). The group number is also the number of valence-shell electrons of each element, except for helium, which has two valence-shell electrons.

hydrogen atom, each atom contributes one electron to the valence shell of the other, giving each atom two electrons (an electron pair) in its valence shell. The resulting molecule in this case is H:H (H_2, molecular hydrogen), where the two dots indicate an electron pair. The maximum number of electrons in the valence shell of a hydrogen atom is two, so a hydrogen atom can bond with a maximum of one other hydrogen atom (or another atom). The maximum number of hydrogen atoms that can bond with one atom of an element before the valence shell of the element is full is called the **principal valence** of the element. The principal valence of hydrogen is one (Table 10.1).

Helium (He), the second element of the periodic table, has two electrons in its outer shell (Table 10.1), and they are paired together. Since helium's electrons are paired, helium has a valence of zero, and it cannot covalently bond with any hydrogen atoms, so helium is chemically inert (unreactive), and its principal valence is zero.

Whereas hydrogen and helium, which are in period (row) 1 of the periodic table, can have a maximum of two electrons in their outer shell, elements in periods 2, 3, and 4 can hold up to eight electrons in their outer shell. In these three periods, elements in groups I, II, III, and IV (Table 10.1) have 1, 2, 3, or 4 unpaired electrons, respectively, in their outer shell. Thus, the elements in groups I, II, III, and IV can bond with 1, 2, 3, or 4 hydrogen atoms, respectively, resulting in 2, 4, 6, or 8 electrons, respectively, in their outer shells.

Elements in groups V, VI, VII, and VIII (Table 10.1) have 1 paired and 3 unpaired electrons, 2 paired and 2 unpaired electrons, 3 paired and 1 unpaired electrons, and 4 paired and 0 unpaired electrons, respectively, in their outer shells. Thus, each element can bond with 3, 2, 1, or 0 hydrogen atoms, respectively, resulting in eight outer-shell electrons in all cases.

A valence shell with eight electrons is called an **octet**. This is the most stable form of a valence shell. Atoms that have their valence shells filled with eight electrons (Ne, Ar, Kr) are so stable, they are chemically inert. Atoms with unpaired electrons in their valence shells try to share electrons with other atoms (through chemical reaction) to become more stable, so they are chemically reactive.

When atoms form covalent bonds during production of a molecule, the configuration of the molecule can be illustrated with a **Lewis structure**, which consists of a combination of Lewis symbols. Appendix Table B.3 shows the Lewis structures of many gases of atmospheric importance. The table is divided into inorganic and organic gases. **Inorganic gases** are those that contain any element, including hydrogen (H) or carbon (C), but not both H and C. **Organic gases** are those that contain both H and C, but may also contain other elements. Organic gases that contain only H and C are **hydrocarbons**. Hydrocarbons include alkanes, cycloalkanes, alkenes, alkynes, aromatics, and terpenes. When methane, which is fairly unreactive, is excluded from the list of hydrocarbons, the remaining hydrocarbons are **nonmethane hydrocarbons** (NMHCs). Oxygenated functional groups, such as aldehydes, ketones, alcohols, acids, and nitrates, are added to hydrocarbons to produce **oxygenated hydrocarbons**. Nonmethane hydrocarbons and oxygenated hydrocarbons, together, are **reactive organic gases** (ROGs) or **volatile organic carbon** (VOC). Nonmethane hydrocarbons and **carbonyls** (aldehydes plus ketones), together, are **nonmethane organic carbon** (NMOC). **Total organic gas** (TOG) is the sum of ROGs and methane. Below, the Lewis structures and characteristics of a few inorganic and organic gases are discussed.

10.1.1 Molecular hydrogen (H_2)

Molecular hydrogen is emitted from volcanos, produced biologically, and produced chemically in the atmosphere. It has a typical mixing ratio of about 0.6 ppmv in the troposphere and consists of two hydrogen atoms with a single covalent bond between them. Two Lewis structures for molecular hydrogen are

$$\mathrm{H:H} \qquad \mathrm{H{-}H}$$

The two dots and the line between atoms indicate a covalently bonding electron pair.

10.1.2 Molecular oxygen (O_2)

Molecular oxygen is produced by green-plant photosynthesis and has a mixing ratio of 20.95 percent by volume throughout the homosphere. Its Lewis structures are

$$:\ddot{\mathrm{O}}::\ddot{\mathrm{O}}: \qquad \mathrm{O{=}O}$$

where the double bond indicates two shared electron pairs. In the second structure, all unshared pairs of electrons are ignored for convenience.

10.1.3 Molecular nitrogen (N_2)

Sources of molecular nitrogen were discussed in Chapter 2. Its mixing ratio is typically 78.08 percent by volume throughout the homosphere. Molecular nitrogen has the Lewis structures,

$$:\mathrm{N}:::\mathrm{N}: \qquad \mathrm{N}\equiv\mathrm{N}$$

where the triple bond indicates three shared electron pairs.

10.1.4 Hydroxyl radical (OH)

Molecular hydrogen and molecular oxygen have no unpaired electrons. Molecules that have one unpaired electron are **free radicals**. One free radical is the hydroxyl radical, which has the Lewis structures,

$$:\dot{\underset{\cdot\cdot}{\mathrm{O}}}:\mathrm{H} \qquad \dot{\mathrm{O}}-\mathrm{H}$$

The single dot in the second structure indicates that the unpaired electron is associated with the oxygen atom. The hydroxyl radical is produced chemically in the atmosphere. It breaks down many other gases and is referred to as a **scavenger**. OH is discussed in more detail in Chapter 11.

10.1.5 Nitric oxide (NO)

Nitric oxide is also a free radical. It is a colorless gas emitted from soils, plants, and combustion processes and produced by lightning and chemical reaction. Combustion sources include aircraft, automobiles, oil refineries, and biomass burning. The primary sink of NO is chemical reaction. A typical mixing ratio of NO in the background troposphere near sea level is 5 pptv. In the upper troposphere NO mixing ratios increase to 20–60 pptv. In urban regions, NO mixing ratios reach 0.1 ppmv in the early morning but decrease significantly by midmorning. The Lewis structures of NO can be drawn so that either N or O has seven electrons in its valence shell. Thus,

$$:\dot{\mathrm{N}}::\ddot{\mathrm{O}}: \longleftrightarrow {}^{-}:\ddot{\mathrm{N}}::\dot{\mathrm{O}}:{}^{+} \qquad \dot{\mathrm{N}}=\mathrm{O} \longleftrightarrow {}^{-}\mathrm{N}=\dot{\mathrm{O}}^{+}$$

where each structure is a resonance structure. **A resonance structure** is one of two or more Lewis structures used to represent a molecule that cannot be represented correctly by one Lewis structure. Multiple Lewis structures arise because electrons are not local to one atom. The charge distribution shown on the second Lewis structure of each pair indicates that the molecule may be polar (charged oppositely on either end). Since total charges balance, there is no net charge on the molecule as a whole.

10.1.6 Nitrogen dioxide (NO_2)

Nitrogen dioxide is a brown gas because it absorbs the shortest wavelengths of the Sun's visible radiation. It absorbs almost all blue light and some green light, allowing the remaining green light and all red light to scatter and transmit. The combination of red and some green light is brown. The major source of atmospheric NO_2 is photochemical oxidation of NO. NO_2 is also produced by other reaction pathways and is emitted during combustion. NO_2 is more prevalent during mid-morning than during midday or afternoon, since sunlight breaks NO_2 down past midmorning. Mixing ratios of NO_2 just above sea level in the free troposphere range from 20 to 50 pptv. In the upper troposphere, mixing ratios increase to 30–70 pptv. In urban regions, they range from 0.1 to 0.25 ppmv. The resonance structures of NO_2 are

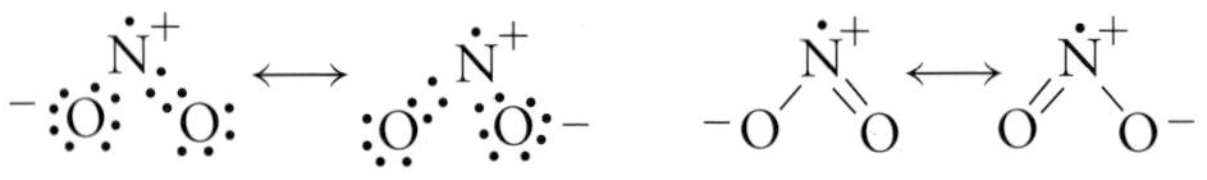

During nitrogen dioxide formation, a net negative charge is transferred to the oxygen atoms from the nitrogen atom, resulting in the charge distribution shown.

10.1.7 Ozone (O_3)

Ozone is a colorless gas that exhibits an odor, even in small concentrations. In urban areas, ozone affects human health in the short term by causing headache (>0.15 ppmv), chest pain (>0.25 ppmv), and sore throat and cough (>0.30 ppmv). Ozone decreases lung function for those who exercise steadily for over an hour while exposed to concentrations above 0.30 ppmv. Above 1 ppmv (high above ambient concentrations), ozone can temporarily narrow passages deep in the lung, increasing airway resistance and inhibiting breathing. Small decreases in lung function affect those with asthma, chronic bronchitis, and emphysema. Ozone may also accelerate the aging of lung tissue. Above 0.1 ppmv, ozone affects animals by increasing their susceptibility to bacterial infection. It also interferes with the growth of plants and trees and deteriorates organic materials, such as rubber, textile dyes and fibers, and some paints and coatings (USEPA 1978).

In the free troposphere, ozone mixing ratios range from 20 to 40 ppbv near sea level and from 30 to 70 ppbv at higher altitudes. In urban areas, ozone mixing ratios range from 0.01 ppmv at night to 0.35 ppmv during smoggy afternoons, with typical values of 0.15 ppmv during moderately polluted afternoons. Ozone is chemically reactive but not a free radical. The resonance structures of ozone are

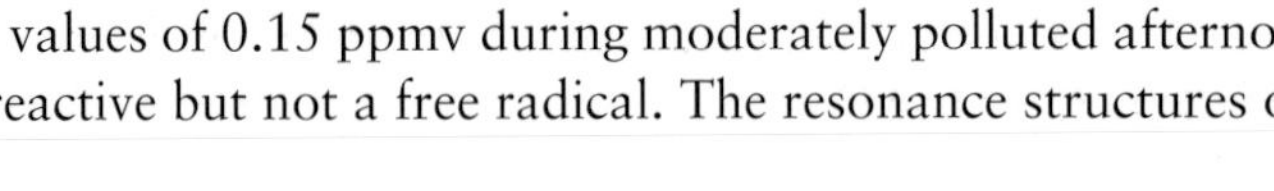

10.1.8 Carbon monoxide (CO)

Carbon monoxide is a tasteless, colorless, and odorless gas that is toxic to humans and animals exposed to it for one hour at mixing ratios above about 700 ppmv. Exposure to 300 ppmv for one hour causes headaches. In urban regions away from freeways, CO mixing ratios are typically 2–10 ppmv. On freeways and in traffic tunnels, mixing ratios can rise to more than 100 ppmv. In the free troposphere, CO mixing ratios vary from 50 to 150 ppbv. A major source of CO is incomplete combustion by automobiles, trucks, and airplanes. Wood burning and grass burning are also important sources. Natural sources of CO are plants and biological activity in the oceans. Although CO is the most abundantly emitted pollutant gas in urban air, it does not play a major role in photochemical smog formation. Photochemical smog is characterized by the buildup of ozone and related products. CO does not produce much ozone in urban air. The major sink of CO is chemical conversion to carbon dioxide (CO_2). CO can be represented by

$$^{-}\!:\!C\!:::\!O\!:^{+} \qquad ^{-}C\equiv O^{+}$$

10.1.9 Carbon dioxide (CO_2)

Carbon dioxide is an odorless and inert gas. Its major sources and sinks were described in Chapter 2. It is well mixed throughout the troposphere and stratosphere, with a current mixing ratio of about 375 ppmv. The Lewis structures of CO_2 are

$$:\ddot{\underset{\cdot\cdot}{O}}::C::\underset{\cdot\cdot}{O}: \qquad O{=}C{=}O$$

These structures indicate that CO_2 is a linear molecule but not a free radical. CO_2 is very stable: its lifetime against chemical destruction is over 100 years.

10.1.10 Sulfur dioxide (SO_2)

Sulfur dioxide is a colorless gas that exhibits an odor and taste at high concentrations. It is emitted from coal-fired power plants, automobile tailpipes, and volcanos. It is also produced chemically in the atmosphere from biologically emitted precursors, such as dimethylsulfide (DMS) and hydrogen sulfide (H_2S). In the background troposphere, SO_2 mixing ratios range from 20 pptv to 1 ppbv. In moderately polluted air, they range from 10 to 30 ppbv. SO_2 is removed from the atmosphere by gas-phase reaction, dissolution into clouds and rain, and deposition to the ground. Common Lewis structures of SO_2 are

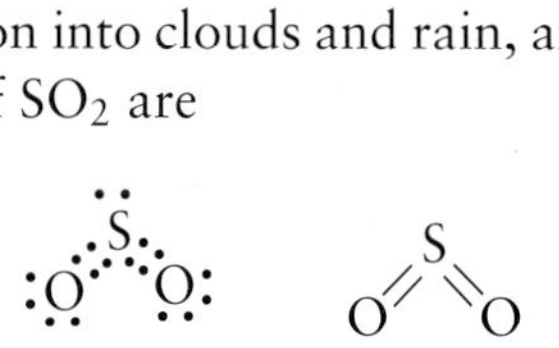

10.1.11 Methane (CH_4)

Sources and sinks of methane were discussed in Chapter 2. It is the most unreactive hydrocarbon in the atmosphere. Because methane has a long lifetime against chemical loss (about 10 years), it is well diluted throughout the free troposphere, with a mixing ratio of about 1.8 ppmv. The Lewis structures of methane are

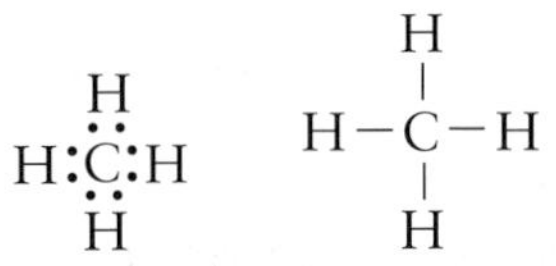

10.1.12 Peroxyacetyl nitrate (PAN) ($CH_3C(O)OONO_2$)

PAN is an eye irritant, initially discovered in the laboratory as a product of smog-forming chemical reactions. Its peak mixing ratio of about 10–20 ppbv in polluted air occurs at the same time during the afternoon as does ozone's peak mixing ratio. PAN mixing ratio in the free troposphere ranges from 2 to 100 pptv. PAN can be represented by

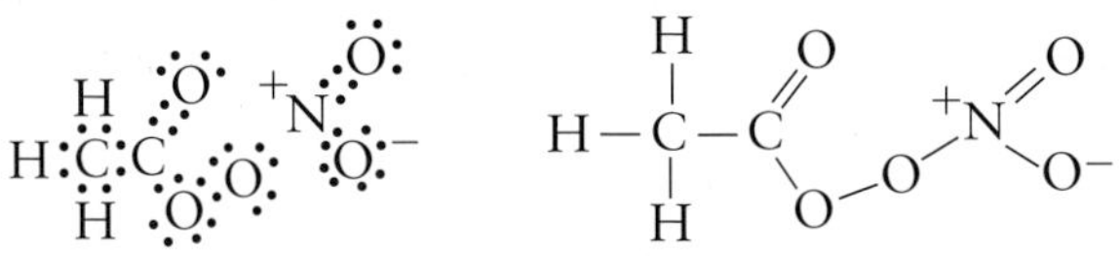

10.2 CHEMICAL REACTIONS AND PHOTOPROCESSES

A single chemical reaction with no intermediate products is an **elementary reaction.** Elementary homogeneous gas-phase chemical reactions in the atmosphere are conveniently divided into **photolysis reactions** (also called photoprocesses, photodissociation reactions, or photolytic reactions) and **chemical kinetic reactions.** Photolysis reactions are **unimolecular** (one-body) reactions initiated when a photon of radiation strikes a molecule and breaks it into two or more products. Elementary chemical kinetic reactions are **bimolecular** (two-body) or **termolecular** (three-body). Reactants and products of photolysis and kinetic reactions are neutral or charged atoms or molecules.

Photolysis reactions are unimolecular. An example of a photolysis reaction is

$$\dot{N}O_2 + h\nu \longrightarrow \dot{N}O + \dot{O}\cdot \qquad \lambda < 420\,\text{nm} \qquad (10.1)$$

where $h\nu$ is a single photon of radiation, O [= $O(^3P)$] is ground-state atomic oxygen, and paired electrons and bonds between atoms are not shown (only unpaired electrons are shown).

Elementary bimolecular reactions include thermal decomposition, isomerization, and standard collision reactions. Thermal decomposition and isomerization reactions occur when a reactant molecule collides with an air molecule. The kinetic energy of the collision elevates the reactant to a high enough vibrational energy state that it can decompose or isomerize. **Thermal decomposition** occurs when the excited reactant dissociates into two or more products. **Isomerization** occurs when the excited reactant changes chemical structure but not composition or molecular weight.

An example of a **bimolecular thermal decomposition reaction** is

$$N_2O_5 + M \longrightarrow \dot{N}O_2 + N\dot{O}_3 + M \tag{10.2}$$

where N_2O_5 is dinitrogen pentoxide, NO_3 is the nitrate radical, and M is the molecule that provides the collisional energy. In the atmosphere, M can be any molecule. Because oxygen and nitrogen, together, make up more than 99 percent of the gas molecules in the atmosphere, the molecule that M represents is usually oxygen or nitrogen. For rate calculation purposes, the concentration of M is usually set equal to the concentration of total air (Equation 10.20).

Since M in (10.2) does not change concentration, the reaction can be written in nonelementary form as

$$N_2O_5 \xrightarrow{M} \dot{N}O_2 + N\dot{O}_3 \tag{10.3}$$

Thermal decomposition reactions are temperature dependent. At high temperatures, they proceed faster than at low temperatures. Isomerization reactions are similar to (10.2) and (10.3), except that an isomerization reaction has one product, which is another form of the reactant.

Elementary **bimolecular collision reactions** are the most common types of kinetic reaction and may occur between any two chemically active reactants that collide. A prototypical collision reaction is

$$CH_4 + \dot{O}H \longrightarrow \dot{C}H_3 + H_2O \tag{10.4}$$

where CH_3 is the methyl radical, and H_2O is water vapor. In some cases, bimolecular reactions result in **collision complexes** that ultimately break into products. Such reactions have the form $A + B \rightleftharpoons AB^* \rightarrow D + F$, where AB^* is a molecule that has weak bonds and stays intact slightly longer than the characteristic time of the molecule's vibrations and rotations. Other reactions that may form collision complexes are pressure-dependent termolecular reactions.

Termolecular reactions often consist of pairs of elementary bimolecular reactions. Consider the termolecular **combination reaction**

$$\dot{N}O_2 + N\dot{O}_3 + M \longrightarrow N_2O_5 + M \tag{10.5}$$

The sequence of elementary bimolecular reactions resulting in (10.5) is $A + B \rightleftharpoons AB^*$ followed by $AB^* + M \rightarrow AB + M$, where M is a **third body** whose purpose is to carry away energy released during the second reaction. In the absence of M, the energy release causes AB^* to dissociate back to A and B. The purpose of M in (10.5) differs from that in (10.2), which was to provide collisional energy for the reaction. In both cases, the concentration of M is taken as that of total air for rate calculation purposes. Reactions (10.2) and (10.5) are pressure-dependent reactions because the concentration of M is proportional to air pressure. Since M in (10.5) does not change concentration, (10.5) can also be written as

$$\dot{N}O_2 + \dot{N}O_3 \xrightarrow{M} N_2O_5 \tag{10.6}$$

Elementary **termolecular collision reactions** are rare, since the probability that three trace gases collide and change form is not large. One possible reaction of this type is

$$\dot{N}O + \dot{N}O + O_2 \longrightarrow \dot{N}O_2 + \dot{N}O_2 \tag{10.7}$$

10.3 REACTION RATES

A **reaction rate** is the time rate of change of concentration of any reactant in a reaction. The rate of an elementary photolysis, collision, isomerization, thermal decomposition, or combination reaction equals a rate coefficient multiplied by the concentration of each reactant. A **rate coefficient** relates concentrations to a reaction rate. Elementary unimolecular reactions have **first-order rate coefficients**, meaning that such coefficients are multiplied by one reactant concentration. Elementary bimolecular and termolecular reactions have **second-** and **third-order** rate coefficients, respectively. If reactant concentrations are expressed in units of molecules of gas per cubic centimeter of air, the rate of reaction is in units of molec. cm^{-3} s^{-1}, regardless of whether the reaction has a first-, second-, or third-order rate coefficient. Rate expressions for reactions with first-, second-, and third-order rate coefficients are

$$\text{Rate} = k_F[A] \qquad \text{Rate} = k_S[A][B] \qquad \text{Rate} = k_T[A][B][C] \tag{10.8}$$

respectively, where brackets denote number concentration, and k_F, k_S, and k_T are first-, second-, and third-order rate coefficients, in units of s^{-1}, cm^3 molec.$^{-1}$ s^{-1}, and cm^6 molec.$^{-2}$ s^{-1}, respectively. For a photolysis reaction, the rate expression is

$$\text{Rate} = J[A] \tag{10.9}$$

where J is a first-order **photolysis rate coefficient** of species A (s^{-1}).

If the concentration of one reactant, such as [M], is invariant during a reaction, the concentration can be premultiplied by the rate coefficient. In such cases, second-order rate coefficients become **pseudo-first-order coefficients** and third-order coefficients become **pseudo-second-order coefficients**.

The time rate of change of concentration of a reactant equals the negative of its rate of reaction. For a photolysis reaction of the form $A + h\nu \rightarrow D + G$, the loss rate of A is

$$\frac{d[A]}{dt} = -\text{Rate} = -J[A] \tag{10.10}$$

For a bimolecular thermal decomposition reaction of the form, $A \xrightarrow{M} D + E$, the loss rate of A is

$$\frac{d[A]}{dt} = -\text{Rate} = -k_F[A] \tag{10.11}$$

where $k_F = k_S[M]$ is a pseudo-first-order rate coefficient. For a bimolecular collision reaction of the form $A + B \rightarrow D + F$ or a termolecular combination reaction of the form $A + B \xrightarrow{M} E$, the loss rates of A and B are

$$\frac{d[A]}{dt} = \frac{d[B]}{dt} = -\text{Rate} = -k_S[A][B] \tag{10.12}$$

where k_S is a second-order rate coefficient for the first reaction, and $k_S = k_T[M]$ is a pseudo-second-order rate coefficient for the second reaction. For a bimolecular collision reaction of the form $A + A \rightarrow E + F$, the loss rate of A is

$$\frac{d[A]}{dt} = -2\text{Rate} = -2k_S[A]^2 \tag{10.13}$$

where k_S is a second-order rate coefficient. For a termolecular reaction, such as $A + B + C \rightarrow E + F$, the loss rate of A, B, and C is

$$\frac{d[A]}{dt} = \frac{d[B]}{dt} = \frac{d[C]}{dt} = -\text{Rate} = -k_T[A][B][C] \tag{10.14}$$

For a termolecular reaction such as $2A + B \rightarrow E + F$,

$$\frac{d[A]}{dt} = 2\frac{d[B]}{dt} = -2\text{Rate} = -2k_T[A]^2[B] \tag{10.15}$$

In general, reactions of the form $aA + bB \rightarrow eE + fF$ have the rate expression

$$\text{Rate} = k_r[A]^a[B]^b \tag{10.16}$$

where k_r denotes a rate coefficient of order a + b. The rates of loss of species A and B are

$$\frac{d[A]}{dt} = -a\text{Rate} = -ak_r[A]^a[B]^b \qquad \frac{d[B]}{dt} = -b\text{Rate} = -bk_r[A]^a[B]^b \tag{10.17}$$

respectively, and the rates of production of E and F are

$$\frac{d[E]}{dt} = e\text{Rate} = ek_r[A]^a[B]^b \qquad \frac{d[F]}{dt} = f\text{Rate} = fk_r[A]^a[B]^b \tag{10.18}$$

respectively. When M is treated as a reactant and product, such as in the reaction, $A + B + M \rightarrow E + M$, the rate of change of M is zero. Thus,

$$\frac{d[M]}{dt} = k_T[A][B][M] - k_T[A][B][M] = 0 \tag{10.19}$$

The rate of production of E and the rates of loss of A and B are nonzero in the same reaction.

The concentration (molecules per cubic centimeter of air) of M is that of total air,

$$[M] = N_a = \frac{p_a}{k_B T} \tag{10.20}$$

where p_a is total air pressure (hPa), k_B is Boltzmann's constant $(1.380658 \times 10^{-19}$ cm^3 hPa K^{-1} molec.$^{-1})$, and T is absolute temperature (K). Total air concentration and total air pressure are the sums of the concentrations and partial pressures, respectively, of dry air and water vapor. Thus, $N_a = N_d + N_v$ and $p_a = p_d + p_v$.

The concentration (molec. cm^{-3}) of molecular nitrogen and molecular oxygen in the air, often needed to calculate reaction rates, are

$$[N_2] = \chi_{N_2} N_d \qquad [O_2] = \chi_{O_2} N_d \tag{10.21}$$

where $\chi_{N_2} = 0.7808$ and $\chi_{O_2} = 0.2095$ are the volume mixing ratios of molecular nitrogen and oxygen, respectively, from Table 2.1.

Example 10.1

Find the number concentrations of M (as total air), N_2, O_2, when $T = 278$ K and $p_a = 920$ hPa and the air is dry.

SOLUTION

From (10.20), the concentration of total air is $[M] = 2.40 \times 10^{19}$ molec. cm^{-3}. From (10.21), $[N_2] = 1.87 \times 10^{19}$ molec. cm^{-3} and $[O_2] = 5.02 \times 10^{18}$ molec. cm^{-3}.

10.4 REACTION RATE COEFFICIENTS

10.4.1 Determining rate coefficients

Rate coefficients of kinetic reactions, which vary with temperature and pressure, are determined experimentally. A common method of determining coefficients for elementary bimolecular reactions at a given temperature and pressure is with **kinetic analysis.** With this method, a small amount of one substance is exposed to a large amount of another substance, and the rate of decay of the less abundant substance is measured. Consider the bimolecular reaction $A + B \rightarrow D + F$. If a small quantity of A is exposed to a large quantity of B, the maximum loss of B is [A]. Since $[A] \ll [B]$, [B] can be held constant, and the rate of loss of A is approximately

$$\frac{d[A]_t}{dt} = -k_F[A]_t = -k_S[A]_t[B]_0 \tag{10.22}$$

where the subscript t indicates that the concentration changes with time, the subscript 0 indicates that the concentration is fixed to its initial value, and $k_F = k_S[B]_0$

is a pseudo-first-order rate coefficient. Integrating (10.22) from time $t = 0$ to h and from $[\mathrm{A}]_0$ to $[\mathrm{A}]_h$ and solving gives

$$k_{\mathrm{S}} = -\frac{1}{[\mathrm{B}]_0\, h} \ln \frac{[\mathrm{A}]_{t=h}}{[\mathrm{A}]_0} \tag{10.23}$$

If the concentration of A at any time $t = h$ is measured, the rate coefficient of the reaction can be calculated from (10.23). For elementary uni- and termolecular reactions, similar calculations can be performed. For example, for unimolecular reactions of the form $\mathrm{A} \rightarrow \mathrm{D} + \mathrm{E}$ and termolecular reactions of the form $\mathrm{A} + \mathrm{B} + \mathrm{C} \rightarrow \mathrm{E} + \mathrm{F}$, rate coefficients can be calculated with

$$k_{\mathrm{F}} = -\frac{1}{h} \ln \frac{[\mathrm{A}]_{t=h}}{[\mathrm{A}]_0} \qquad k_{\mathrm{T}} = -\frac{1}{[\mathrm{B}]_0\,[\mathrm{C}]_0 h} \ln \frac{[\mathrm{A}]_{t=h}}{[\mathrm{A}]_0} \tag{10.24}$$

respectively. In the latter case, $[\mathrm{A}]_0 \ll [\mathrm{B}]_0$ and $[\mathrm{A}]_0 \ll [\mathrm{C}]_0$. More advanced methods of calculating reaction rate coefficients include fast flow systems, flash photolysis, static reaction systems, pulse radiolysis, the cavity ringdown method, and static techniques (e.g., Finlayson-Pitts and Pitts 2000).

10.4.2 Temperature dependence of reactions

First-, second-, and third-order reaction rate coefficients vary with temperature. In many cases, the temperature dependence is estimated with the equation proposed by Svante Arrhenius in 1889. Arrhenius is also known as the person who first proposed the theory of global warming, in 1896. The **Arrhenius equation** is found by integrating

$$\frac{\mathrm{d}(\ln k_{\mathrm{r}})}{\mathrm{d}T} = \frac{E_{\mathrm{r}}}{R^* T^2} \tag{10.25}$$

where k_{r} is the rate coefficient, E_{r} is the activation energy of the reaction (J mol^{-1}), T is temperature (K), and R^* is the universal gas constant.

The **activation energy** is the smallest amount of energy required for reacting species to form an activated complex or transition state before forming products. Activation energies are determined by first integrating (10.25) as

$$\ln k_{\mathrm{r}} = \ln A_{\mathrm{r}} - \frac{E_{\mathrm{r}}}{R^* T} \tag{10.26}$$

where A_{r} is a constant of integration, called the **collisional prefactor** (frequency factor). The collisional prefactor is proportional to the frequency of those collisions with proper orientation for producing a reaction. It equals a collision frequency multiplied by a steric (efficiency) factor. The collision frequency depends on the relative size, charge, kinetic energies, and molecular weights of the reactant molecules. The **steric factor** gives the fraction of collisions that result in an effective

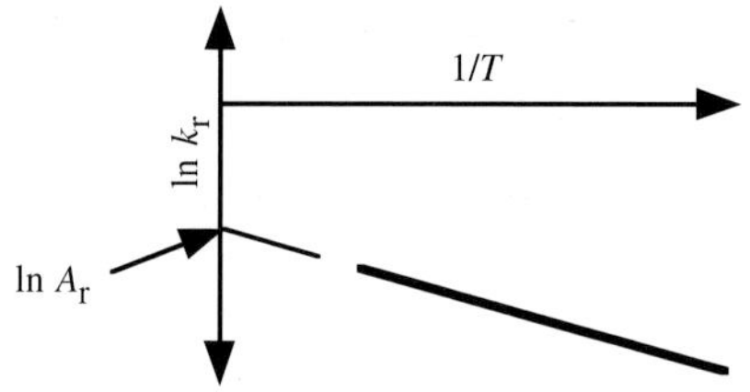

Figure 10.1 Plot of ln k_r versus $1/T$ (thick line). The slope of the line is E_r/R^*. The line can be extrapolated to $1/T \rightarrow 0$ ($T \rightarrow \infty$) to obtain ln A_r.

reaction. For first- and second-order reactions, collisional prefactors and rate coefficients have units of s^{-1} and cm^3 molec.$^{-1}$ s^{-1}, respectively.

The activation energy and collisional prefactor in (10.26) are found experimentally. If the rate coefficient (k_r) is measured at different temperatures, a graph of ln k_r versus $1/T$ can be plotted. E_r and A_r are then extracted from the plot, as shown in Fig. 10.1.

Solving (10.26) for k_r gives

$$k_r = A_r \exp\left(-\frac{E_r}{R^*T}\right) = A_r \exp\left(\frac{C_r}{T}\right) \tag{10.27}$$

where $\exp(-E_r/R^*T)$ is the fraction of reactant molecules having the critical energy, E_r, required for the reaction to occur, and $C_r = -E_r/R^*$.

The collisional prefactor is usually a weak function of temperature, but when the activation energy is near zero, it is a strong function of temperature. In such cases, a temperature factor B_r is added to (10.27), so that

$$k_r = A_r \left(\frac{300}{T}\right)^{B_r} \exp\left(\frac{C_r}{T}\right) \tag{10.28}$$

B_r is found by fitting the expression in (10.28) to data. Many combination reactions have the form of (10.28). Rate coefficients for such reactions are given in Appendix Table B.4.

Example 10.2

At $T = 298$ K, the rate coefficients (cm^3 molec.$^{-1}$ s^{-1} and cm^6 molec.$^{-2}$ s^{-1}) for a bi- and a termolecular reaction are, respectively,

$$\dot{N}O + O_3 \longrightarrow \dot{N}O_2 + O_2 \qquad k_1 = 1.80 \times 10^{-12} \exp(-1370/T) = 1.81 \times 10^{-14}$$

$$\dot{O}\cdot + O_2 + M \longrightarrow O_3 + M \qquad k_2 = 6.00 \times 10^{-34}(300/T)^{2.3} = 6.09 \times 10^{-34}$$

10.4.3 Pressure dependence of reactions

Thermal decomposition and combination reactions, both of which include an M, are **pressure dependent** because [M] varies with pressure. Such reactions have the forms

$$\mathrm{A} \xrightarrow{\mathrm{M}} \mathrm{D} + \mathrm{E} \qquad \mathrm{A} + \mathrm{B} \xrightarrow{\mathrm{M}} \mathrm{E} \tag{10.29}$$

respectively. Overall rate coefficients for these reactions may be interpolated between a low- and a high-pressure limit rate coefficient. In some reactions, such as $O + O_2 + M \longrightarrow O_3 + M$, the rate coefficient is not interpolated.

The interpolation formula for the overall rate coefficient of a pressure-dependent reaction is

$$k_r = \frac{k_{\infty,T}k_{0,T}[\mathrm{M}]}{k_{\infty,T} + k_{0,T}[\mathrm{M}]} F_c^{\left[1+\left(\log_{10}\frac{k_{0,T}[\mathrm{M}]}{k_{\infty,T}}\right)^2\right]^{-1}} \tag{10.30}$$

(Troe 1979) where F_c is called the **broadening factor** of the falloff curve and is determined theoretically, $k_{0,T}$ is the low-pressure limit rate coefficient, and $k_{\infty,T}$ is the high-pressure limit rate coefficient. For the reactions in (10.29), k_r has units of s^{-1} and cm^3 molec.$^{-1}$ s^{-1}, respectively. Values of $k_{0,T}$, $k_{\infty,T}$, and F_c are given in Table B.4 of Appendix B for applicable pressure-dependent reactions.

The **low-pressure limit rate coefficient** $k_{0,T}$, multiplied by [M], is a laboratory-determined coefficient of the overall reaction as [M] approaches zero. Thus,

$$k_{0,T}[\mathrm{M}] = \lim_{[\mathrm{M}]\to 0} k_r \tag{10.31}$$

where [M] has units of molec. cm^{-3}, and $k_{0,T}$ has units of cm^3 molec.$^{-1}$ s^{-1} or cm^6 molec.$^{-2}$ s^{-1}, respectively, for the reactions in (10.29).

The **high-pressure limit rate coefficient** $k_{\infty,T}$ is the rate of the overall reaction at infinite M concentration. Thus,

$$k_{\infty,T} = \lim_{[\mathrm{M}]\to \infty} k_r \tag{10.32}$$

Since the high-pressure limit rate coefficients are independent of [M], the reactions corresponding to (10.29) at high pressure are

$$\mathrm{A} \to \mathrm{D} + \mathrm{E} \qquad \mathrm{A} + \mathrm{B} \to \mathrm{E} \tag{10.33}$$

respectively. The high-pressure limit rate coefficients have units of s^{-1} and cm^3 molecule^{-1} s^{-1} for these two reactions, respectively.

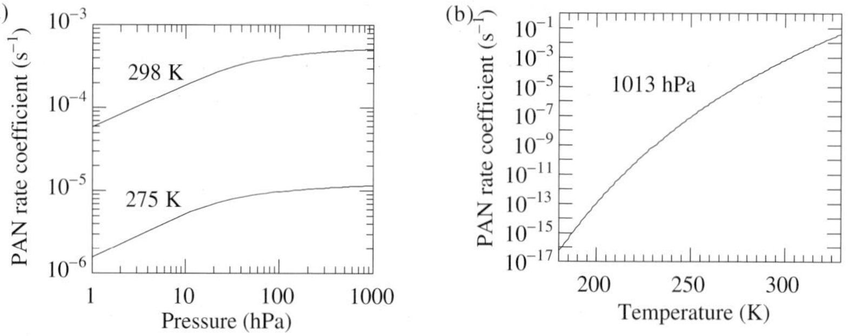

Figure 10.2 Rate coefficient of PAN decomposition (a) as a function of pressure for two temperatures and (b) as a function of temperature for one pressure.

Example 10.3

When $p_a = 140$ hPa, $T = 216$ K (stratosphere), and the air is dry, find the rate coefficient of

$$\dot{O}H + \dot{N}O_2 \xrightarrow{M} HNO_3 \qquad (10.34)$$

SOLUTION

From (10.20), $[M] = 4.69 \times 10^{18}$ molec. cm^{-3}. From Appendix Table B.4, $k_{0,T} = 2.60 \times 10^{-30}(300/T)^{2.9}$ cm^6 molec.$^{-2}$ s^{-1}, $k_{\infty,T} = 6.70 \times 10^{-11}(300/T)^{0.6}$ cm^3 molec.$^{-1}$ s^{-1}, and $F_c = 0.43$. Thus, at 216 K, $k_{0,T}[M] = 3.16 \times 10^{-11}$, $k_{\infty,T} = 8.16 \times 10^{-11}$, and $k_r = 1.11 \times 10^{-11}$ cm^3 molec.$^{-1}$ s^{-1}.

Figures 10.2 (a) and (b) show the pressure and temperature dependences of the rate of PAN decomposition by the reaction

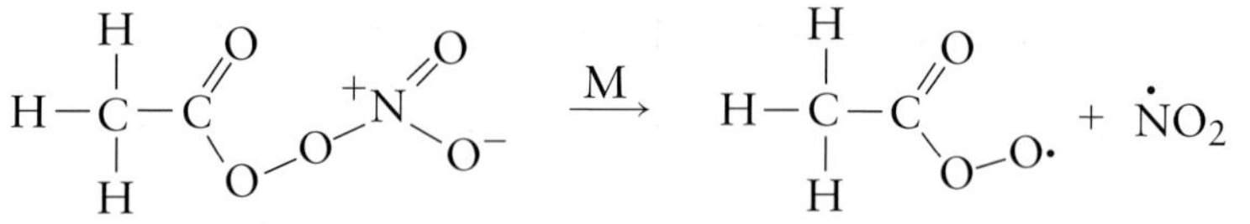

(10.34)

Peroxyacetyl nitrate — Peroxyacetyl radical

The graphs indicate that the pseudo-first-order rate coefficient for PAN decomposition varies by an order of magnitude when pressure ranges from 1 to 1000 hPa. The rate coefficient varies by 15 orders of magnitude when temperatures range from 180 to 330 K. Temperature affects PAN thermal decomposition much more than does pressure.

10.4.4 Photolysis reactions

Photolysis reactions initiate many atmospheric chemical pathways. From (9.128), the **photolysis coefficient** (s^{-1}) of gas q producing product set p was

$$J_{q,p} = \int_0^\infty 4\pi I_{p,\lambda} b_{a,g,q,\lambda,T} Y_{q,p,\lambda,T}\, d\lambda \tag{10.35}$$

where $4\pi I_{p,\lambda}$ is the actinic flux (photons cm^{-2} μm^{-1} s^{-1}) in the wavelength interval $d\lambda$ (μm), $b_{a,g,q,\lambda,T}$ is the average absorption cross section of the gas in the interval (cm^2 molec.$^{-1}$), and $Y_{q,p,\lambda,T}$ is the average quantum yield of the photoprocess in the interval (molec. photon^{-1}). Experimental absorption cross section and quantum yield data vary with wavelength and temperature. References for such data are given in Appendix Table B.4.

Photolysis of a molecule may produce one or more sets of products. Photolysis of the nitrate radical, for example, produces two possible sets of products,

$$\dot{N}O_3 + h\nu \longrightarrow \begin{cases} \dot{N}O_2 + \dot{O}\cdot & 410\,\text{nm} < \lambda < 670\,\text{nm} \\ \dot{N}O + O_2 & 590\,\text{nm} < \lambda < 630\,\text{nm} \end{cases} \tag{10.36}$$

The probability of each set of products is embodied in the quantum yield, defined in Chapter 9. Whereas the absorption cross section of a gas is the same for each set of products, the quantum yield of a gas differs for each set. Figures 10.3 (a) and (b) show modeled photolysis-coefficient profiles for several photoprocesses under specified conditions.

10.5 SETS OF REACTIONS

Atmospheric chemical problems require the determination of gas concentrations when many reactions occur at the same time. A difficulty arises because a species is usually produced and/or destroyed by several reactions. Consider the following four reactions and corresponding rate expressions:

$$\dot{N}O + O_3 \longrightarrow \dot{N}O_2 + O_2 \qquad \text{Rate}_1 = k_1[NO][O_3] \tag{10.37}$$

$$\dot{O}\cdot + O_2 + M \longrightarrow O_3 + M \qquad \text{Rate}_2 = k_2[O][O_2][M] \tag{10.38}$$

$$\dot{N}O_2 + h\nu \longrightarrow \dot{N}O + \dot{O}\cdot \qquad \text{Rate}_3 = J[NO_2] \tag{10.39}$$

$$\dot{N}O_2 + O \longrightarrow \dot{N}O + O_2 \qquad \text{Rate}_4 = k_3[NO_2][O] \tag{10.40}$$

Time derivatives of NO, NO_2, O, and O_3 concentrations from the reactions are

$$\begin{aligned}\frac{d[NO]}{dt} &= P_c - L_c = \text{Rate}_3 + \text{Rate}_4 - \text{Rate}_1 \\ &= J[NO_2] + k_3[NO_2][O] - k_1[NO][O_3]\end{aligned} \tag{10.41}$$

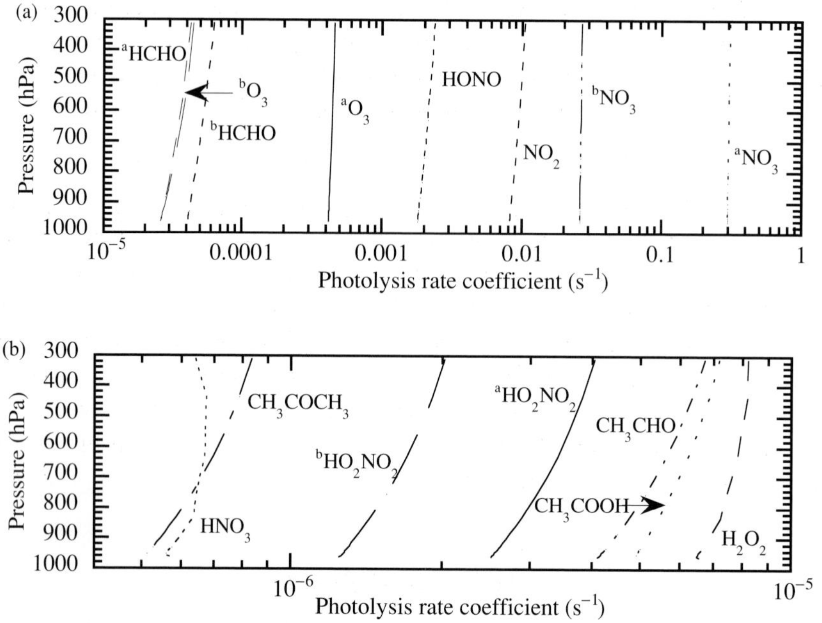

Figure 10.3 Modeled photolysis coefficient profiles at 11:30 a.m. on August 27, 1987, at Temecula, California (33.49° N, 117.22° W). Cross-section and quantum-yield data are referenced in Appendix Table B.4. The photoprocesses are as follows:

${}^{a}O_3 + h\nu \rightarrow O_2 + O$	${}^{a}HCHO + h\nu \rightarrow 2H + CO$
${}^{b}O_3 + h\nu \rightarrow O_2 + O(^1D)$	${}^{b}HCHO + h\nu \rightarrow H_2 + CO$
${}^{a}NO_3 + h\nu \rightarrow NO_2 + O$	${}^{a}HO_2NO_2 + h\nu \rightarrow HO_2 + NO_2$
${}^{b}NO_3 + h\nu \rightarrow NO + O_2$	${}^{b}HO_2NO_2 + h\nu \rightarrow OH + NO_3$
$NO_2 + h\nu \rightarrow NO + O$	$HONO + h\nu \rightarrow OH + NO$
$CH_3COCH_3 + h\nu \rightarrow CH_3 + COCH_3$	$H_2O_2 + h\nu \rightarrow 2OH$
$HNO_3 + h\nu \rightarrow OH + NO_2$	$CH_3OOH + h\nu \rightarrow CH_3O + OH$

The rate coefficient for HNO_3 decreases with increasing height near 400 hPa because the absorption cross section of HNO_3 decreases with decreasing temperature, and this factor becomes important at 400 hPa. The superscripts "a" and "b" identify different sets of products for the same photodissociating molecule.

$$\frac{d[NO_2]}{dt} = P_c - L_c = \text{Rate}_1 - \text{Rate}_3 - \text{Rate}_4$$
$$= k_1[NO][O_3] - J[NO_2] - k_3[NO_2][O] \qquad (10.42)$$

$$\frac{d[O]}{dt} = P_c - L_c = \text{Rate}_3 - \text{Rate}_2 - \text{Rate}_4$$
$$= J[NO_2] - k_2[O][O_2][M] - k_3[NO_2][O] \qquad (10.43)$$

$$\frac{d[O_3]}{dt} = P_c - L_c = \text{Rate}_2 - \text{Rate}_1 = k_2[O][O_2][M] - k_1[NO][O_3] \qquad (10.44)$$

respectively. These equations are first-order, first-degree, homogeneous ordinary differential equations (ODEs), as defined in section 6.1. In (10.38) and (10.40), O_2 is not affected significantly by the reactions, and no expression for the rate of change of O_2 is needed. Also, M is neither created nor destroyed, so no expression for the rate of change of M is needed either. The O_2 and M concentrations are still included in the rate expression of (10.38). In (10.41)–(10.44), P_c and L_c are the **total rates of chemical production and loss,** respectively. In the case of NO,

$$P_c = J[NO_2] + k_3[NO_2][O] \qquad L_c = k_1[NO][O_3] \tag{10.45}$$

Thousands of reactions occur in the atmosphere simultaneously. Because computer resources are limited, an atmospheric model's chemical mechanism needs to be limited to the most important reactions. Whether a reaction is important can be determined from an analysis of its rate. The rate of a reaction depends on the rate coefficient and reactant concentrations. Rate coefficients vary with temperature, pressure, and/or solar radiation. Reactant concentrations vary with time of day, season, and location. A low rate coefficient and low reactant concentrations render a reaction unimportant. If the rate coefficient and reactant concentrations are expected to be consistently small for a reaction during a simulation, the reaction can be ignored.

10.6 STIFF SYSTEMS

Gas and aqueous chemical reaction sets are **stiff.** Stiff systems of reactions are more difficult to solve numerically than nonstiff systems. A **stiff system** of reactions is one in which the **lifetimes** (or time scales) of species taking part in the reactions differ significantly from one another. Species lifetimes are expressed as e-folding or half-lifetimes. An **e-folding lifetime** is the time required for a species concentration to decrease to 1/e its original value. A **half-lifetime** is the time required for a species concentration to decrease to 1/2 its original value.

The **overall chemical lifetime** of a species is determined by calculating the lifetime of the species against loss from individual reactions and applying

$$\tau_A = \frac{1}{\dfrac{1}{\tau_{A1}} + \dfrac{1}{\tau_{A2}} + \cdots \dfrac{1}{\tau_{An}}} \tag{10.46}$$

where τ_A is the overall chemical lifetime of species A, and $\tau_{A1}, \ldots, \tau_{An}$ are the lifetimes of A due to loss from reactions $1, \ldots, n$, respectively. Equation (10.46) applies to e-folding or half-lifetime calculations.

The e-folding lifetime of species A due to a unimolecular reaction of the form A → products is calculated from $d[A]/dt = -k_F[A]$. Integrating this equation from concentration $[A]_0$ at $t = 0$ to $[A]$ at $t = h$ gives $[A] = [A]_0 e^{-k_F h}$. The e-folding lifetime is the time at which

$$\frac{[A]}{[A]_0} = \frac{1}{e} = e^{-k_F h} \tag{10.47}$$

This occurs when

$$\boxed{\tau_{A1} = h = \frac{1}{k_F}} \tag{10.48}$$

In the case of a bi- or termolecular reaction, the e-folding lifetime is calculated by assuming that each concentration, except for that of the species of interest, equals the initial concentration of the species. Thus, the integration is reduced to that of a unimolecular reaction. For instance, the rate of loss of A from the **bimolecular reaction** A + B → products can be described by the linearized ODE, $d[A]/dt = -k_S[A][B]_0$, where k_S is a second-order rate coefficient of the reaction, and $[B]_0$ is the initial concentration of species B. This equation gives A an e-folding lifetime of

$$\tau_{A2} = \frac{1}{k_S[B]_0} \tag{10.49}$$

Similarly, the loss rate of A from the **termolecular reaction** A + B + C → products can be described by a linearized ODE that results in an e-folding lifetime for A of

$$\tau_{A3} = \frac{1}{k_T[B]_0[C]_0} \tag{10.50}$$

where k_T is the rate coefficient of the reaction, and $[C]_0$ is the initial concentration of C.

The **half-lifetime** of a unimolecular reaction is determined in a manner similar to its e-folding lifetime. The half-lifetime is the time at which

$$\frac{[A]}{[A]_0} = \frac{1}{2} = e^{-k_F h} \tag{10.51}$$

This occurs when

$$\tau_{(1/2)A1} = h = \frac{0.693}{k_F} \tag{10.52}$$

Analogous equations for bi- and termolecular reactions are, respectively,

$$\tau_{(1/2)A2} = \frac{0.693}{k_S\,[B]_0} \qquad \tau_{(1/2)A3} = \frac{0.693}{k_T[B]_0[C]_0} \tag{10.53}$$

Example 10.4

The e-folding lifetimes of atmospheric gases vary significantly. The main loss of methane is to the reaction $CH_4 + OH \longrightarrow CH_3 + H_2O$, where $k_r = 6.2 \times 10^{-15}$ cm^3 molec.$^{-1}$ s^{-1} at 298 K. When $[OH] = 5.0 \times 10^5$ molec. cm^{-3}, $\tau_{CH_4} = 1/(k_r\,[OH]) = 10.2$ years.

For $O(^1D) + N_2 \longrightarrow O + N_2$, $k_r = 2.6 \times 10^{-11}$ cm^3 molec.$^{-1}$ s^{-1} at 298 K. When $[N_2] = 1.9 \times 10^{19}$ molec. cm^{-3}, the e-folding lifetime of $O(^1D)$ is $\tau_{O(^1D)} = 1/(k_r\,[N_2]) = 2 \times 10^{-9}$ s.

Example 10.4 shows that the ratio of lifetimes between CH_4 and $O(^1D)$ is about 17 orders of magnitude. A system of equations that includes species with a wide range of lifetimes is said to be stiff. Stiffness depends on reaction rates and species concentrations. At low concentrations, some species lose their stiffness. For almost all atmospheric cases, the range in species lifetimes is large enough for chemical equations to be stiff.

10.7 SUMMARY

In this chapter, chemical species, structures, reactions, reaction rate coefficients, and reaction rates were discussed. Elementary reactions are unimolecular, bimolecular, or termolecular and give rise to first-order, second-order, and third-order rate coefficients, respectively. Photolysis reactions are unimolecular and give rise to first-order rate coefficients. Bimolecular reactions include thermal decomposition, isomerization, and basic collision reactions. Termolecular reactions include combination and collision reactions. A combination reaction consists of a pair of elementary bimolecular reactions. In the atmosphere, reactions occur simultaneously, and the lifetimes of species against chemical loss vary by orders of magnitude. Thus, ordinary differential equations describing atmospheric chemistry are stiff.

10.8 PROBLEMS

10.1 When $T = 265$ K and $p_d = 223$ hPa, calculate the second-order rate coefficient for

$$\dot{H} + O_2 \xrightarrow{M} H\dot{O}_2$$

10.2 When $T = 298$ K and $p_d = 1013$ hPa, calculate the first-order rate coefficient for

$$N_2O_5 \xrightarrow{M} \dot{N}O_2 + N\dot{O}_3$$

Repeat for $T = 288$ K. Discuss temperature effects on the reaction rate coefficient.

10.3 Estimate the e-folding lifetimes of CO, NO, O_3, SO_2, HNO_3, ISOP (isoprene), and HO_2 against loss by OH if [OH] $= 1.0 \times 10^6$ molec. cm^{-3}, $T = 288$ K, and $p_d = 1010$ hPa. The rate coefficients are listed in Appendix Table B.4. Order the species from shortest to longest lifetimes. Which species will most likely reach the stratosphere if only OH reaction is considered?

10.4 Write rate expressions for the reactions

$$\dot{C}l + O_3 \longrightarrow Cl\dot{O} + O_2 \qquad \text{and} \qquad Cl\dot{O} + \dot{O}\cdot \longrightarrow \dot{C}l + O_2$$

Write the time derivative of each species in the reactions, assuming the reactions are solved together. What is the expression for the steady-state concentration of ClO?

10.5 Given the following observed rate coefficients as a function of temperature, find the activation energy and collisional prefactor of the associated bimolecular reaction. Show your work. In the table, 1.303 (−14) means 1.303×10^{-14}.

T (K)	278	288	298	308	318
k_r (cm^3 molec.$^{-1}$ s^{-1})	1.303 (−14)	1.547 (−14)	1.814 (−14)	2.106 (−14)	2.422 (−14)

10.9 COMPUTER PROGRAMMING PRACTICE

10.6 Write a computer script to calculate the first-order rate coefficient for

$$N_2O_5 \xrightarrow{M} \dot{N}O_2 + \dot{N}O_3$$

as a function of temperature and pressure. Draw graphs of the rate coefficient versus temperature when $p_d = 1013$ hPa and $p_d = 800$ hPa, respectively, and versus pressure when $T = 298$ K and $T = 275$ K, respectively. Discuss the results.

10.7 Write a computer script to read in reactions and rate-coefficient data from a computer file. Use the script to calculate the rate coefficients for the first 15 reactions in Appendix Table B.4 when $T = 288$ K and $p_d = 980$ hPa. Calculate reaction rates when $[O_3] = 2.45 \times 10^{12}$, $[O] = 1.0 \times 10^3$, $[O(^1D)] = 1.0 \times 10^{-3}$, $[H] = 1.0 \times 10^0$, $[OH] = 1.0 \times 10^6$, $[H_2] = 1.5 \times 10^{13}$, $[HO_2] = 1.0 \times 10^8$, $[N_2O] = 7.6 \times 10^{12}$, and $[H_2O] = 2.0 \times 10^{17}$ molec. cm^{-3}.

11

Urban, free-tropospheric, and stratospheric chemistry

DIFFERENT regions of the atmosphere are affected by different sets of chemical reactions. The free troposphere is affected primarily by reactions among inorganic, light organic, and some heavy organic gases. Urban regions are affected by inorganic, light organic, and heavy organic gases. The stratosphere is affected primarily by inorganic, light organic, and chlorinated/brominated gases. The importance of a reaction also varies between day and night and among seasons. In this chapter, chemical reaction pathways are described for the free troposphere, urban regions, and the stratosphere. Special attention is given to the marine sulfur cycle, ozone production in urban air, and ozone destruction cycles in the global and polar stratosphere. Heterogeneous reactions of gases on particle surfaces are also described.

11.1 FREE-TROPOSPHERIC PHOTOCHEMISTRY

Photochemistry in the troposphere outside of urban regions is governed primarily by reactions among inorganic and low-molecular-weight organic gases. High-molecular-weight organic gases emitted anthropogenically, such as toluene and xylene, break down chemically over hours to a few days, so they rarely penetrate far from urban regions. Air over the tropics and other vegetated regions, such as the southeastern United States, is affected by reaction products of isoprene, a hemiterpene, and other terpenes emitted from biogenic sources. Reaction pathways for these gases are discussed in Section 11.2. In the following subsections, inorganic and low-molecular-weight organic reaction pathways for the free troposphere are described.

11.1.1 Photostationary-state relationship

In many regions of the troposphere, the ozone (O_3) mixing ratio is controlled by a set of three tightly coupled chemical reactions involving itself, nitric oxide (NO), and nitrogen dioxide (NO_2). These reactions are

$$\dot{N}O + O_3 \longrightarrow \dot{N}O_2 + O_2 \qquad (11.1)$$

$$\dot{N}O_2 + h\nu \longrightarrow \dot{N}O + \dot{O}\cdot \qquad \lambda < 420\ \text{nm} \qquad (11.2)$$

$$\dot{O}\cdot + O_2 + M \longrightarrow O_3 + M \qquad (11.3)$$

In the free troposphere, the mixing ratios of O_3 (20–60 ppbv) are much higher than are those of NO (1–60 pptv) or NO_2 (5–70 pptv) (e.g., Singh *et al.* 1996), so (11.1) does not deplete ozone during day or night. At night in urban regions, NO mixing ratios may exceed those of O_3, and (11.1) can deplete local ozone.

Assuming (11.1) and (11.2) have rate coefficients k_1 and J, respectively, the time rate of change of the nitrogen dioxide concentration from these reactions is

$$\frac{\mathrm{d}[NO_2]}{\mathrm{d}t} = k_1[NO][O_3] - J[NO_2] \qquad (11.4)$$

If the time rate of change of NO_2 is small compared with those of the other terms in (11.4), NO_2 is nearly in steady state and (11.4) simplifies to the **photostationary-state relationship,**

$$[O_3] = \frac{J[NO_2]}{k_1[NO]} \qquad (11.5)$$

Equation (11.5) does not state that ozone in the free troposphere is affected by only [NO] and $[NO_2]$. Instead, it provides a relationship among $[O_3]$, $[NO_2]$, and [NO]. If two of the three concentrations are known, the third can be found from the equation.

Example 11.1

Find the photostationary-state mixing ratio of O_3 at midday when p_a = 1013 hPa, T = 298 K, $J \approx 0.01\ \mathrm{s}^{-1}$, χ_{NO} = 5 pptv, χ_{NO_2} = 10 pptv (typical free-tropospheric mixing ratios), and the air is dry.

SOLUTION

At T = 298 K, $k_1 \approx 1.8 \times 10^{-14}\ \mathrm{cm^3\ molec.^{-1}\ s^{-1}}$ from Appendix Table B.4. Since the conversion from mixing ratio to number concentration is the same for each gas, $[NO_2]/[NO] = \chi_{NO_2}/\chi_{NO}$. From (11.5), $[O_3] = 1.1 \times 10^{12}$ molec. $\mathrm{cm^{-3}}$. From (10.20), $N_d = N_a = 2.46 \times 10^{19}$ molec. $\mathrm{cm^{-3}}$. Dividing $[O_3]$ by N_d gives χ_{O_3} = 44.7 ppbv, which is a typical free tropospheric ozone mixing ratio.

Two important reactions aside from (11.1)–(11.3) that affect ozone are

$$O_3 + h\nu \longrightarrow O_2 + \cdot\dot{O}(^1D) \qquad \lambda < 310\ \mathrm{nm} \qquad (11.6)$$

$$O_3 + h\nu \longrightarrow O_2 + \dot{O}\cdot \qquad \lambda > 310\ \mathrm{nm} \qquad (11.7)$$

where $O(^1D)$ is **excited atomic oxygen**. In the free troposphere, the e-folding lifetimes of ozone against destruction by these reactions are about 0.7 and 14 h, respectively. The lifetime of ozone against destruction by NO in (11.1) was 126 h under the conditions of Example 11.1. Thus, in the daytime free troposphere, photolysis destroys ozone faster than does reaction with NO. When the ozone concentration changes due to photolysis, the $[NO_2]/[NO]$ ratio in (11.5) changes to adapt to the new ozone concentration.

The photostationary-state relationship in (11.5) is useful for free-tropospheric analysis. In urban air, though, the relationship often breaks down because reactions of NO with organic gas radicals provide an additional important source of NO_2 not included in (11.4). When organic-radical concentrations are large, as they are during the morning in urban air, the photostationary-state relationship does not hold. In the afternoon, though, organic gas concentrations in urban air decrease, and the relationship holds better.

When excited atomic oxygen forms, as in (11.6), it rapidly produces O by

$$\dot{\mathrm{O}}(^1D) \xrightarrow{\mathrm{M}} \dot{\mathrm{O}}\cdot \qquad (11.8)$$

and O rapidly produces O_3 by (11.3). The reactions (11.3) and (11.6)–(11.8) cycle oxygen atoms quickly among $O(^1D)$, O, and O_3. Losses of O from the cycle, such as from conversion of NO to NO_2 by (11.1), are slower than are transfers of O within the cycle.

11.1.2 Hydroxyl radical

The hydroxyl radical (OH) is an important chemical in the atmosphere because it decomposes (scavenges) many gases. Its globally averaged tropospheric concentration is about 8×10^5 molec. cm^{-3} (Singh 1995). Its daytime concentration at any given location in the clean free troposphere ranges from 2×10^5 to 3×10^6 molec. cm^{-3}. When clean air is exposed to an urban plume, OH concentrations increase to 6×10^6 molec. cm^{-3} or more (Comes *et al.* 1997). In urban air, OH concentrations range from 10^6 to 10^7 molec. cm^{-3}.

The primary free-tropospheric source of OH is

$$\cdot\dot{\mathrm{O}}(^1D) + \mathrm{H_2O} \longrightarrow 2\dot{\mathrm{O}}\mathrm{H} \qquad (11.9)$$

Sources of water vapor were discussed in Section 2.1.2.1. In the upper troposphere, H_2O is scarce, limiting the ability of (11.9) to produce OH. Under such conditions, other reactions, which are normally minor, may be important sources of OH. These reactions are discussed in Section 11.1.10.

Minor sources of OH in the free troposphere are photolysis of gases, some of which are produced by OH itself. For example, **nitrous acid** (HONO), produced during the day by the reaction of NO with OH (Table 11.1), rapidly photolyzes soon after by

$$\mathrm{HONO} + h\nu \longrightarrow \dot{\mathrm{O}}\mathrm{H} + \dot{\mathrm{N}}\mathrm{O} \qquad \lambda < 400\ \mathrm{nm} \tag{11.10}$$

HONO concentrations are high only during the early morning because the photolysis rate increases as sunlight becomes more intense during the day. During the night, HONO is not produced from gas-phase reactions, since OH, required for its production, is absent at night. HONO is produced from gas-particle reactions and emitted from vehicles during day and night.

Nitric acid (HNO_3), produced during the day by the reaction of NO_2 with OH (Table 11.1), photolyzes slowly to reproduce OH by

$$\mathrm{HNO_3} + h\nu \longrightarrow \dot{\mathrm{O}}\mathrm{H} + \dot{\mathrm{N}}\mathrm{O_2} \qquad \lambda < 335\ \mathrm{nm} \tag{11.11}$$

The e-folding lifetime of nitric acid against destruction by photolysis is 15–80 days, depending on season and latitude. Since this lifetime is fairly long, HNO_3 is a temporary reservoir of OH. Nitric acid is soluble, and much of it dissolves in cloud or aerosol water and reacts on aerosol surfaces. HNO_3 mixing ratios are 5–200 pptv in the free troposphere.

Hydrogen peroxide (H_2O_2), produced by HO_2 self-reaction (Section 11.1.3), photolyzes to OH by

$$\mathrm{H_2O_2} + h\nu \longrightarrow 2\dot{\mathrm{O}}\mathrm{H} \qquad \lambda < 355\ \mathrm{nm} \tag{11.12}$$

H_2O_2's e-folding lifetime against photolysis is 1–2 days. H_2O_2 is soluble and is often removed from the atmosphere by chemical reaction in clouds and dissolution in precipitation. It is also lost by reaction with OH with an e-folding lifetime of 3–14 days in the free troposphere (Table 11.1).

Peroxynitric acid (HO_2NO_2), produced by the reaction of NO_2 with HO_2 (Section 11.1.3), photolyzes to produce OH by

$$\mathrm{HO_2NO_2} + h\nu \longrightarrow \begin{cases} \mathrm{H}\dot{\mathrm{O}}_2 + \dot{\mathrm{N}}\mathrm{O_2} & \lambda < 330\ \mathrm{nm} \\ \dot{\mathrm{O}}\mathrm{H} + \mathrm{N}\dot{\mathrm{O}}_3 & \lambda < 330\ \mathrm{nm} \end{cases} \tag{11.13}$$

The e-folding lifetimes of both reactions are 2–5 days. HO_2NO_2 is slightly soluble and is removed by clouds and precipitation. It is also lost by reaction with OH (Table 11.1) and thermal decomposition (Section 11.1.3).

After its formation, OH reacts with and decomposes many inorganic and organic gases. The overall e-folding lifetime of OH against chemical destruction is about 0.1–1 s. The e-folding lifetimes of gases reacting with OH vary, as shown in Table 11.1. The e-folding lifetimes of additional organic gases against destruction by OH are given in Table 11.5 (Section 11.2).

Table 11.1 The e-folding lifetimes of several gases against destruction by OH in clean free-tropospheric air when N_{OH} is low and high

Reaction	e-folding lifetime $N_{OH} = 5 \times 10^5$ molec. cm^{-3}	$N_{OH} = 2 \times 10^6$ molec. cm^{-3}	Equation
$\dot{O}H + O_3 \longrightarrow H\dot{O}_2 + O_2$	346 d	86.5 d	(11.14)
$\dot{O}H + H_2 \longrightarrow H_2O + \dot{H}$	9.5 y	2.4 y	(11.15)
$\dot{O}H + H\dot{O}_2 \longrightarrow H_2O + O_2$	5.1 h	1.3 h	(11.16)
$\dot{O}H + H_2O_2 \longrightarrow H\dot{O}_2 + H_2O$	13.6 d	3.4 d	(11.17)
$\dot{O}H + \dot{N}O \xrightarrow{M} HONO$	2.4 d	14 h	(11.18)
$\dot{O}H + \dot{N}O_2 \xrightarrow{M} HNO_3$	1.9 d	11.4 h	(11.19)
$\dot{O}H + HO_2NO_2 \longrightarrow H_2O + \dot{N}O_2 + O_2$	4.6 d	1.2 d	(11.20)
$\dot{O}H + \dot{S}O_2 \xrightarrow{M} HS\dot{O}_3$	26 d	6.5 d	(11.21)
$\dot{O}H + CO \longrightarrow \dot{H} + CO_2$	111 d	28 d	(11.22)
$\dot{O}H + CH_4 \longrightarrow H_2O + \dot{C}H_3$	10.2 y	2.6 y	(11.23)
$\dot{O}H + C_2H_6 \longrightarrow H_2O + \dot{C}_2H_5$	93 d	23 d	(11.24)
$\dot{O}H + C_3H_8 \longrightarrow H_2O + \dot{C}_3H_7$	21 d	5.3 d	(11.25)
$\dot{O}H + CH_3OOH \longrightarrow H_2O + CH_3\dot{O}_2$	6.4 d	1.6 d	(11.26)

$T = 298$ K and $p_a = 1013$ hPa. HO_2 = hydroperoxy radical, H_2O_2 = hydrogen peroxide, HONO = nitrous acid, HNO_3 = nitric acid, NO_3 = nitrate radical, HO_2NO_2 = peroxynitric acid, CH_3 = methyl radical, HSO_3 = bisulfite radical, C_2H_6 = ethane, C_2H_5 = ethoxy radical, C_3H_8 = propane, C_3H_7 = propoxy radical, CH_3OOH = methyl hydroperoxide, and CH_3O_2 methylperoxy radical.

11.1.3 Hydroperoxy radical

Like OH, the hydroperoxy radical (HO_2) is a scavenger. Whereas OH is present during daytime only, HO_2 is present during day and night. Concentrations of HO_2 usually exceed those of OH. Chemical sources of HO_2 include OH + O_3 (Table 11.1), OH + H_2O_2 (Table 11.1), photolysis of HO_2NO_2 (11.13),

$$\dot{H} + O_2 \xrightarrow{M} H\dot{O}_2 \qquad (11.27)$$

and the thermal decomposition of peroxynitric acid,

$$HO_2NO_2 \xrightarrow{M} H\dot{O}_2 + \dot{N}O_2 \qquad (11.28)$$

This reaction occurs within seconds at high temperature near the surface, but in the cold upper troposphere, it is slow. The H in (11.27) is produced primarily from formaldehyde photolysis (Section 11.1.8.1) and the OH + CO reaction (Table 11.1).

The rate of HO_2 loss depends on the NO mixing ratio. In the presence of high NO (>10 pptv), HO_2 reacts mostly with NO and NO_2. When NO is lower

(3–10 pptv), HO_2 reacts mostly with ozone. When NO is very low (<3 pptv), HO_2 reacts mostly with itself (Finlayson-Pitts and Pitts 2000). The reactions corresponding to these conditions are

$$\left.\begin{array}{l} H\dot{O}_2 + \dot{N}O \longrightarrow \dot{O}H + \dot{N}O_2 \\ H\dot{O}_2 + \dot{N}O_2 \xrightarrow{M} HO_2NO_2 \end{array}\right\} \quad > 10 \text{ pptv NO} \qquad (11.29)$$

$$H\dot{O}_2 + O_3 \longrightarrow \dot{O}H + 2O_2 \qquad 3\text{–}10 \text{ pptv NO} \qquad (11.30)$$

$$H\dot{O}_2 + H\dot{O}_2 \longrightarrow H_2O_2 + O_2 \qquad < 3\text{pptv NO} \qquad (11.31)$$

11.1.4 Nighttime nitrogen chemistry

During the day in the free troposphere, NO and NO_2 are involved in the photostationary state cycle. Some losses from the cycle include the reactions OH + NO and OH + NO_2 (Table 11.1). During the night, these loss processes shut down since OH is absent at night. In addition, the photostationary relationship breaks down because the photolysis reaction (11.2) shuts off, eliminating the source of O for ozone production in (11.3) and the source of NO for ozone destruction in (11.1). Because NO_2 photolysis shuts down, NO_2 becomes available at night to produce the **nitrate radical** (NO_3), **dinitrogen pentoxide** (N_2O_5), and **aqueous nitric acid** ($HNO_3(aq)$) by the sequence

$$\dot{N}O_2 + O_3 \longrightarrow N\dot{O}_3 + O_2 \qquad (11.32)$$

$$\dot{N}O_2 + N\dot{O}_3 \underset{}{\overset{M}{\rightleftharpoons}} N_2O_5 \qquad (11.33)$$

$$N_2O_5 + H_2O(aq) \longrightarrow 2HNO_3(aq) \qquad (11.34)$$

Reaction (11.32) occurs during the day as well, but, during the day, it is less important than NO + O_3, and the NO_3 produced from the reaction is destroyed almost immediately by sunlight.

Reaction (11.33) is a reversible reaction. The forward reaction is a three-body, pressure-dependent reaction. The reverse reaction is a temperature-dependent thermal decomposition reaction. At high temperature, such as during the day and in the lower atmosphere, the reverse reaction occurs within seconds. At low temperature, such as at night and at high altitudes, it occurs within hours to days or even months.

Reaction (11.34) is a **heterogeneous reaction**, in that it involves a gas reacting with a chemical on an aerosol particle or hydrometeor particle surface. In this case, the chemical on the surface is liquid water. The reaction can occur on ice surfaces as well, as described in Section 11.3.6. In the absence of liquid water or ice, the corresponding **homogeneous reaction** (gas-phase only), $N_2O_5 + H_2O$, can occur, but the reaction is very slow.

After sunrise, NO_3 photolyzes almost immediately (with an e-folding lifetime of seconds) by

$$\dot{N}O_3 + h\nu \longrightarrow \begin{cases} \dot{N}O_2 + \dot{O}\cdot & 410\text{ nm} < \lambda < 670\text{ nm} \\ \dot{N}O + O_2 & 590\text{ nm} < \lambda < 630\text{ nm} \end{cases} \tag{11.35}$$

so NO_3 is not important during the day. Since N_2O_5 is not produced during the day and thermally decomposes within seconds by the reverse of (11.33) after sunrise, it is also unimportant during the day. A slower daytime loss of N_2O_5 is photolysis (with an e-folding lifetime of hours),

$$N_2O_5 + h\nu \longrightarrow \dot{N}O_2 + \dot{N}O_3 \qquad \lambda < 385\text{ nm} \tag{11.36}$$

Another loss is mechanism heterogeneous reaction by (11.34).

11.1.5 Carbon monoxide production of ozone

The mixing ratio of ozone in the free troposphere, controlled primarily by the photostationary-state relationship, is enhanced slightly by carbon monoxide (CO), methane (CH_4), and nonmethane organic gases. CO, with a typical tropospheric mixing ratio of 100 ppbv, produces ozone by the sequence

$$CO + \dot{O}H \longrightarrow CO_2 + \dot{H} \tag{11.37}$$

$$\dot{H} + O_2 \xrightarrow{M} H\dot{O}_2 \tag{11.38}$$

$$\dot{N}O + H\dot{O}_2 \longrightarrow \dot{N}O_2 + \dot{O}H \tag{11.39}$$

$$\dot{N}O_2 + h\nu \longrightarrow \dot{N}O + \dot{O}\cdot \qquad \lambda < 420\text{ nm} \tag{11.40}$$

$$\dot{O}\cdot + O_2 + M \longrightarrow O_3 + M \tag{11.41}$$

Because the lifetime of CO against breakdown by (11.37) in the free troposphere is 28–110 days, the rate of ozone production by this sequence is slow. The mechanism affects the photostationary-state relationship only slightly through Reaction (11.39).

11.1.6 Methane production of ozone

Methane, with a mixing ratio of 1.8 ppmv, is the most abundant organic gas in the Earth's atmosphere. Table 11.1 indicates that its free-tropospheric e-folding lifetime is about 10 years. This long lifetime has enabled it to mix uniformly up to the tropopause. From this height upward, its mixing ratio gradually decreases. Methane's only important loss is the reaction,

$$CH_4 + \dot{O}H \longrightarrow \dot{C}H_3 + H_2O \tag{11.42}$$

This reaction, which produces the **methyl radical** (CH_3), sets in motion the sequence of reactions,

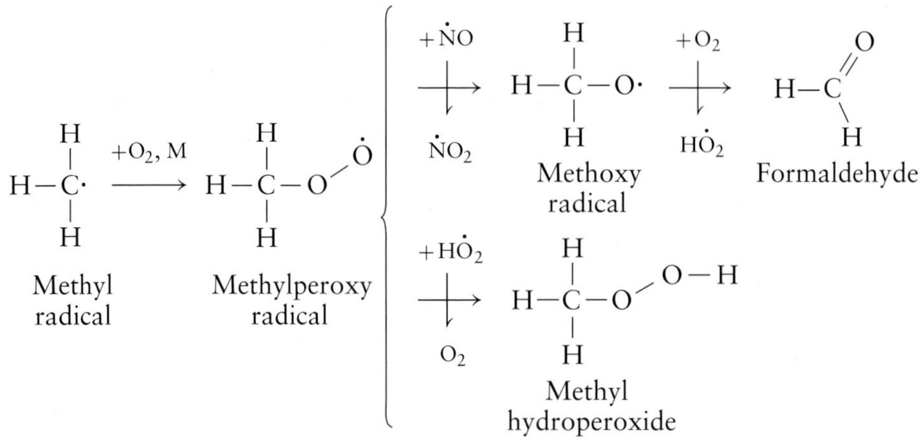

(11.43)

which leads to ozone production, but the incremental quantity of ozone produced is small compared with the photostationary quantity of ozone.

The first pathway of the **methylperoxy radical** (CH_3O_2) reaction produces NO_2 and **formaldehyde** (HCHO). Both produce ozone. NO_2 produces ozone by (11.2)–(11.3). Formaldehyde produces ozone as described in Section 11.1.8. The e-folding lifetime of the **methoxy radical** (CH_3O) intermediary against destruction by O_2 is 10^{-4} s; thus, its conversion to formaldehyde is almost instantaneous.

The second pathway of CH_3O_2 reaction produces **methyl hydroperoxide** (CH_3OOH), which stores OH and HO_2 radicals. CH_3OOH releases OH and HCHO during photolysis and releases CH_3O_2 during reaction with OH. The e-folding lifetime of CH_3OOH against photolysis is 1.5–2.5 days, and that against OH reaction is 1.6–6.4 days at 298 K (Table 11.1). The methyl hydroperoxide decomposition reactions are

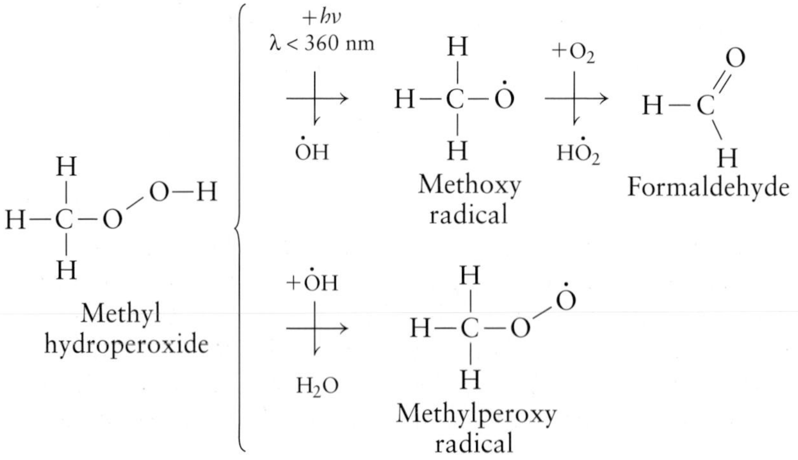

(11.44)

11.1.7 Ethane and propane production of ozone

The primary nonmethane hydrocarbons in the free troposphere are ethane (C_2H_6), propane (C_3H_8), ethene (C_2H_4), and propene (C_3H_6). Free-tropospheric mixing ratios of these gases are 0–2.5 ppbv for ethane, 0–1.0 ppbv for propane, 0–1.0 ppbv for ethene, and 0–1.0 ppbv for propene (Singh *et al.* 1988; Bonsang *et al.* 1991). The primary oxidant of ethane, propane, and other alkanes is OH. Photolysis, reaction with O_3, reaction with HO_2, and reaction with NO_3 do not affect alkane concentrations significantly. In this subsection, ethane and propane oxidation and production of ozone are discussed. Ethene and propene oxidation pathways are described in Section 11.2.2.

The hydroxyl radical attacks **ethane** by

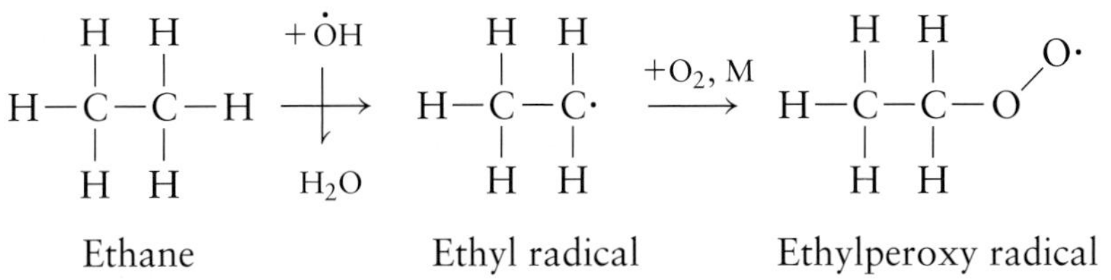

(11.45)

The e-folding lifetime for the OH reaction is about 90 days in the free troposphere. The **ethylperoxy radical** ($C_2H_5O_2$) produced by the reaction takes one of two courses,

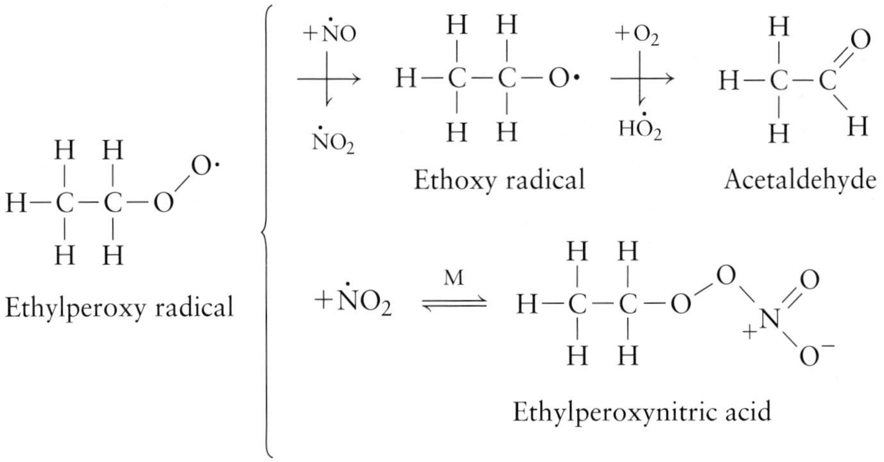

(11.46)

The first pathway leads to NO_2, HO_2, and **acetaldehyde** (CH_3CHO). NO_2 produces ozone through (11.2)–(11.3), HO_2 produces ozone through (11.39)–(11.41), and acetaldehyde produces ozone as described in Section 11.1.8. The **ethoxy radical** (C_2H_5O) produced in the first pathway also reacts with NO to produce **ethyl nitrite** (C_2H_5ONO) and with NO_2 to produce **ethyl nitrate** ($C_2H_5ONO_2$), which are storage reservoirs for nitrogen. Ethyl nitrate can also be produced directly by reaction of the ethylperoxy radical with NO (not shown), but the e-folding lifetime of $C_2H_5O_2$ against this reaction is relatively long (48–192 days).

The second pathway leads to **ethylperoxynitric acid** ($C_2H_5O_2NO_2$), which is a temporary storage reservoir for nitrogen since the reaction is reversible. The e-folding lifetime of $C_2H_5O_2$ against loss by this reaction is 1.7–6.7 days.

The hydroxyl radical attacks **propane** by

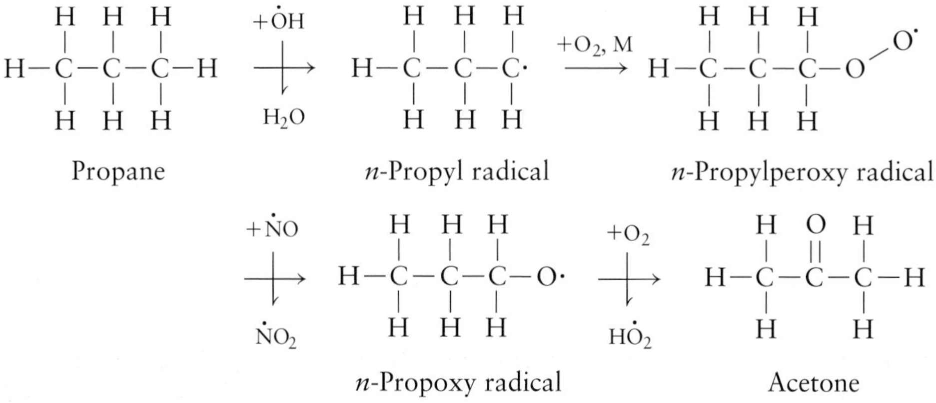

(11.47)

The e-folding lifetime of propane against loss by OH is about 20 days in the free troposphere, more than four times shorter than the lifetime of ethane. The NO_2 from propane oxidation forms ozone by (11.2)–(11.3). The **acetone** (CH_3COCH_3) formed from (11.47) is a long-lived species whose fate is discussed in Section 11.1.9. The **propylperoxy radical** ($C_3H_7O_2$) competitively reacts with NO to form **propyl nitrate** ($C_3H_7ONO_2$), but the reaction is slow. The **propoxy radical** (C_3H_7O) reacts with NO to form **propyl nitrite** (C_3H_7ONO) and with NO_2 to form propyl nitrate. The free-tropospheric e-folding lifetimes of C_3H_7O against loss by these two reactions are both 4–16 h.

11.1.8 Formaldehyde and acetaldehyde production of ozone

Formaldehyde is a carcinogen in high concentrations, an eye irritant, and an important ozone precursor. It is a colorless gas with a strong odor at mixing ratios >0.05 ppmv. Typical mixing ratios in urban air are <0.1 ppmv. It is the most abundant aldehyde in the air and moderately soluble in water. It is produced chemically from (11.43) and decomposes by photolysis, reaction with OH, reaction with HO_2, and reaction with NO_3. Other sources include incomplete combustion and emission from plywood, resins, adhesives, carpeting, particleboard, and fiberboard. Acetaldehyde is a precursor to ozone and peroxyacetyl nitrate (PAN). It is produced by ethoxy-radical oxidation and destroyed by photolysis, and reaction with OH, HO_2, and NO_3. Below, formaldehyde and acetaldehyde decomposition is discussed.

11.1.8.1 *Aldehyde photolysis*

Formaldehyde and acetaldehyde photolyze during the day by

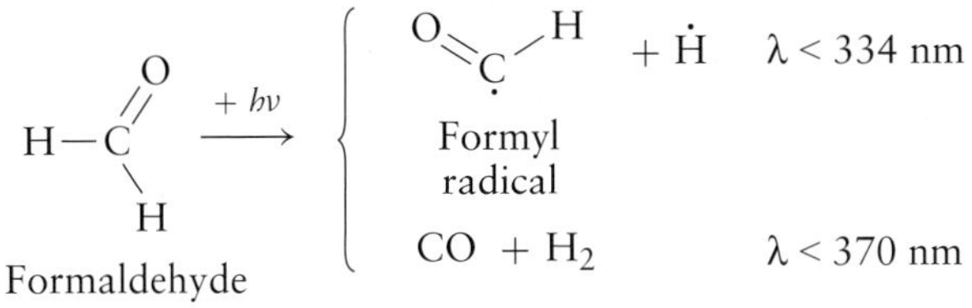

$$
H-C\begin{array}{c}O\\\\H\end{array} \xrightarrow{+\,h\nu} \left\{\begin{array}{ll} O{=}C{-}H \;+\; \dot{H} & \lambda < 334\ \mathrm{nm} \\ \text{Formyl radical} & \\ CO \;+\; H_2 & \lambda < 370\ \mathrm{nm} \end{array}\right.
\tag{11.48}
$$

Formaldehyde

$$
\underset{\text{Acetaldehyde}}{H-C-C\begin{array}{c}O\end{array}} \xrightarrow{+\,h\nu} \underset{\text{Methyl radical}}{H-\dot{C}-H} \;+\; \underset{\text{Formyl radical}}{\dot{C}\begin{array}{c}O\\H\end{array}} \qquad \lambda < 325\ \mathrm{nm}
\tag{11.49}
$$

respectively, producing H, CH_3, and the **formyl radical** (HCO). H reacts with O_2 to form HO_2 by (11.27), which produces ozone by (11.39)–(11.41). CH_3 is oxidized in (11.43) eventually producing ozone. HCO produces CO and HO_2 by

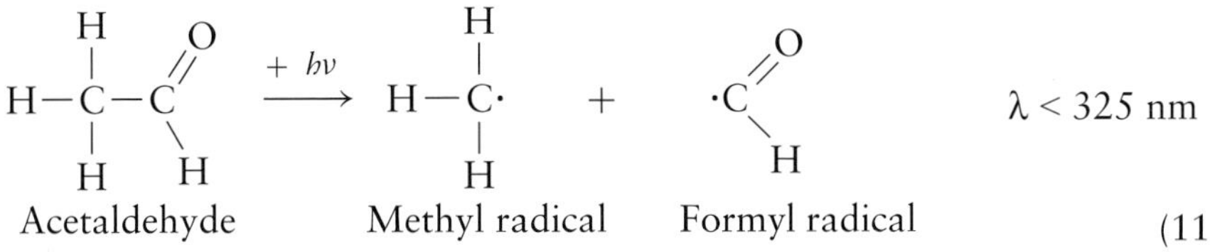

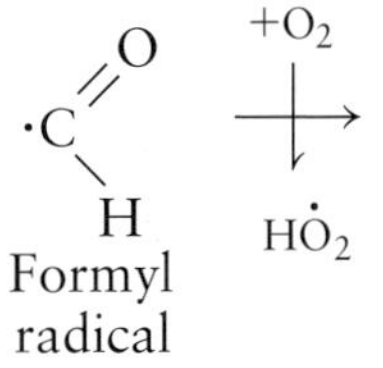

$$
\dot{C}\begin{array}{c}O\\H\end{array} \xrightarrow[\;HO_2\;]{+O_2} CO
\tag{11.50}
$$

Formyl radical

CO and HO_2 form ozone by CO oxidation (Section 11.1.5).

11.1.8.2 *Aldehyde reaction with the hydroxyl radical*

Formaldehyde and acetaldehyde react with the hydroxyl radical to produce ozone and peroxyacetyl nitrate, respectively. The formaldehyde process is

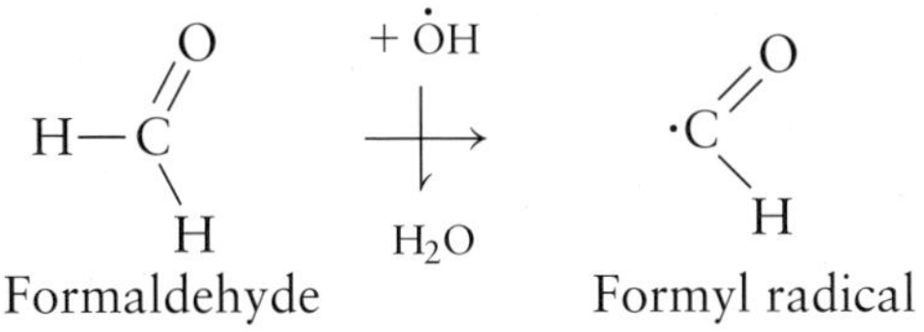

$$
\underset{\text{Formaldehyde}}{H-C\begin{array}{c}O\\H\end{array}} \xrightarrow[\;H_2O\;]{+\,\dot{O}H} \underset{\text{Formyl radical}}{\dot{C}\begin{array}{c}O\\H\end{array}}
\tag{11.51}
$$

HCO produces CO in (11.50), which produces ozone as described in Section 11.1.5.

The acetaldehyde process is

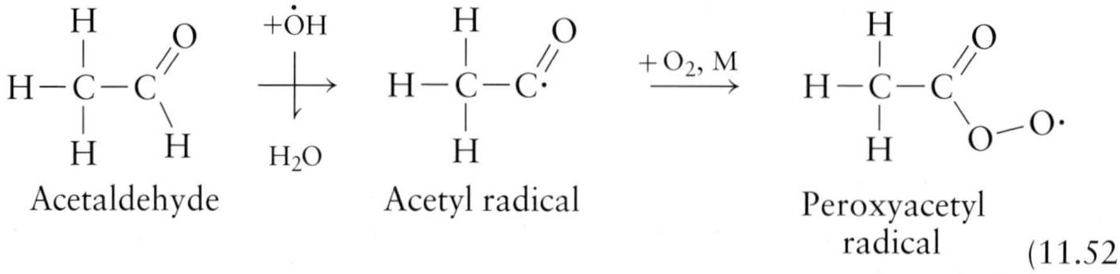

$$\text{Acetaldehyde} \quad \xrightarrow{+\dot{O}H,\ -H_2O} \quad \text{Acetyl radical} \quad \xrightarrow{+O_2,\ M} \quad \text{Peroxyacetyl radical} \tag{11.52}$$

$$\text{Peroxyacetyl radical} \quad \begin{cases} \xrightarrow{+\dot{N}O,\ -\dot{N}O_2} \text{Acetyloxy radical} \\[2ex] \xrightleftharpoons{+\dot{N}O_2,\ M} \text{Peroxyacetyl nitrate} \end{cases} \tag{11.53}$$

The NO_2 formed from this process produces ozone. The process also produces **peroxyacetyl nitrate** (PAN). PAN does not cause severe health effects, even in polluted air, but it is an eye irritant and damages plants by discoloring their leaves. Mixing ratios of PAN in clean air are typically 2–100 pptv. Those in rural air downwind of urban sites are up to 1 ppbv. Polluted air mixing ratios increase to 35 ppbv, with typical values of 10 to 20 ppbv. PAN mixing ratios peak during the afternoon, the same time that ozone mixing ratios peak. PAN is not an important constituent of air at night or in regions of heavy cloudiness. PAN was discovered during laboratory experiments of photochemical smog formation (Stephens *et al.* 1956). Its only source is chemical reaction in the presence of sunlight. At 300 K and at surface pressure, PAN's e-folding lifetime against thermal decomposition is about 25 minutes. At 280 K, its lifetime increases to 13 hours.

11.1.8.3 *Aldehyde reaction with nitrate*

At night, acetaldehyde reacts with NO_3 by the sequence

$$\text{Acetaldehyde} \quad \xrightarrow{+\dot{N}O_3,\ -HNO_3} \quad \text{Acetyl radical} \quad \xrightarrow{+O_2,\ M} \quad \text{Peroxyacetyl radical} \tag{11.54}$$

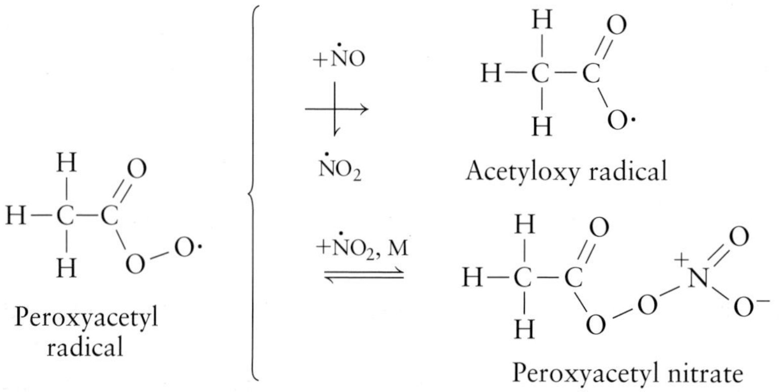

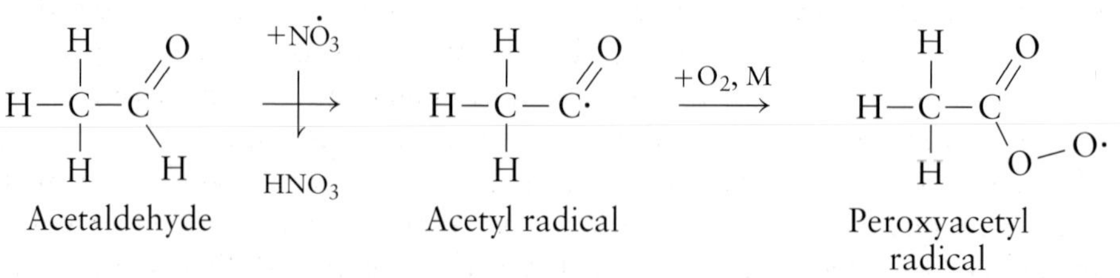

Since the peroxyacetyl radical forms PAN, as shown in (11.53), Reaction (11.54) is a nighttime source of PAN.

11.1.9 Acetone reactions

Acetone is a long-lived ketone produced from the OH oxidation of propane (through 11.47), *i*-butane, or *i*-pentane (e.g., Chatfield *et al.* 1987). The mixing ratio of acetone in the free troposphere is 200–700 pptv (Singh *et al.* 1995). Acetone decomposes by reaction with OH and photolysis. The OH reaction produces CH_3COCH_2 and H_2O. The e-folding lifetime of acetone against this reaction is 27 days when $T = 298$ K and $[OH] = 5 \times 10^5$ molec. cm^{-3}. The photolysis reaction is

$$
\begin{array}{ccc}
\underset{\substack{| \\ H}}{\overset{\substack{H \quad O \quad H}}{H-C-C-C-H}} + h\nu & \longrightarrow & \underset{\substack{| \\ H}}{\overset{\substack{H \quad O}}{H-C-C\cdot}} \;+\; \underset{\substack{| \\ H}}{\overset{H}{H-C\cdot}} \\
\text{Acetone} & & \text{Acetyl radical} \quad \text{Methyl radical}
\end{array}
$$

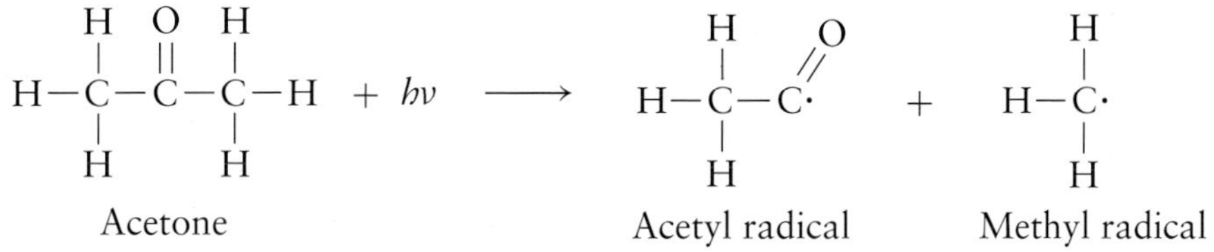

(11.55)

For the location and time given in Fig. 10.3(b), the e-folding lifetime of acetone against photolysis is 14–23 days, with the shorter lifetime corresponding to upper tropospheric conditions and the longer lifetime corresponding to lower tropospheric conditions. The acetyl radical from (11.55) forms PAN through (11.52)–(11.53). The methyl radical is oxidized in (11.43), producing HO_2, HCHO, and CH_3OOH, among other products. HCHO produces HO_2 through (11.48) and (11.50). CH_3OOH decomposes, forming HO_2 and OH in the process, through (11.44). Thus, acetone photolysis yields primarily PAN, OH, HO_2, and intermediate products.

OH production from the $O(^1D)$–H_2O reaction in the upper troposphere is less significant than in the lower troposphere because the concentration of H_2O is low in the upper troposphere. In the upper troposphere, acetone photolysis and subsequent reaction may be an important source of OH (Singh *et al.* 1995).

11.1.10 Sulfur photochemistry

In the free troposphere, several naturally emitted gases contain sulfur (S). These gases are important because they oxidize to sulfur dioxide, which is oxidized further to sulfuric acid. Sulfuric acid condenses onto or forms new aerosol particles. Aerosol particles affect radiation and serve as sites on which new cloud drops form, and clouds affect radiation. Radiation affects temperatures, which affect pressures, and winds. Thus, natural gases containing sulfur feed back to weather and climate.

Sulfur is emitted naturally in several forms. Anaerobic bacteria in marshes emit dimethyl sulfide (DMS) (CH_3SCH_3), dimethyl disulfide (DMDS) (CH_3SSCH_3), methanethiol (CH_3SH), and hydrogen sulfide (H_2S). Phytoplankton in the oceans emit DMS, DMDS, and other products. Volcanos emit carbonyl sulfide (OCS), carbon disulfide (CS_2), H_2S, and sulfur dioxide (SO_2).

Dimethyl sulfide is produced by bacteria in some soils and plants, and it is the most abundant sulfur-containing compound emitted from the oceans. DMS is produced from DMSP (dimethyl sulfonium propionate), which is emitted by many phytoplankton (Bates *et al.* 1994). When phytoplankton feed, DMSP is exuded and cleaved by enzymes to produce DMS and other products. Before DMS evaporates from the ocean surface, much of it chemically reacts or is consumed by microorganisms. Over the oceans, DMS mixing ratios vary between <10 pptv and 1 ppbv. The latter value occurs over eutrophic waters (Berresheim *et al.* 1995). Average near-surface DMS mixing ratios over oceans and land are about 100 and 20 pptv, respectively.

DMS is lost chemically in the atmosphere through hydroxyl radical abstraction and addition. **Abstraction** is the process by which a radical, such as the hydroxyl radical, removes an atom from a compound. **Addition** is the process by which a radical bonds to a compound. Above 285 K, hydroxyl radical abstraction is the dominant DMS–OH reaction pathway. Below 285 K, addition is more important (Hynes *et al.* 1986). Yin *et al.* (1990) and Tyndall and Ravishankara (1991) present detailed oxidation pathways of DMS.

Hydroxyl radical abstraction of DMS results in the formation of the **methanethiolate radical** (CH_3S) and formaldehyde by the sequence

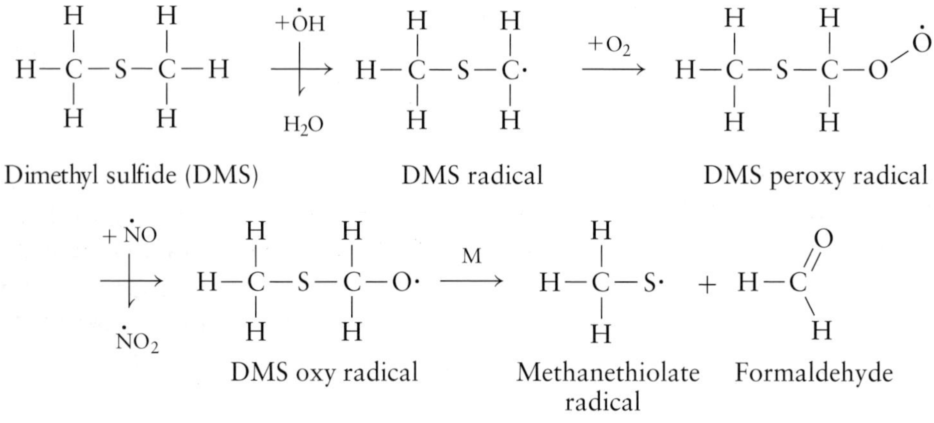

(11.56)

The methanethiolate radical reacts with O_2 by

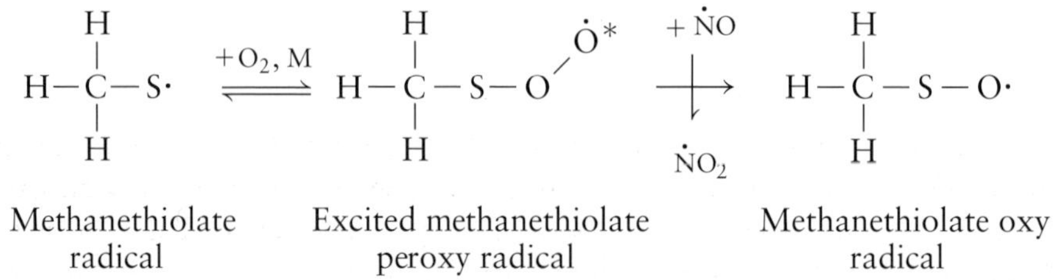

(11.57)

where the intermediate product is short-lived. The **methanethiolate oxy radical** (CH_3SO) decays to **sulfur monoxide** (SO) or forms the **methanethiolate peroxy radical** (CH_3SO_2) by

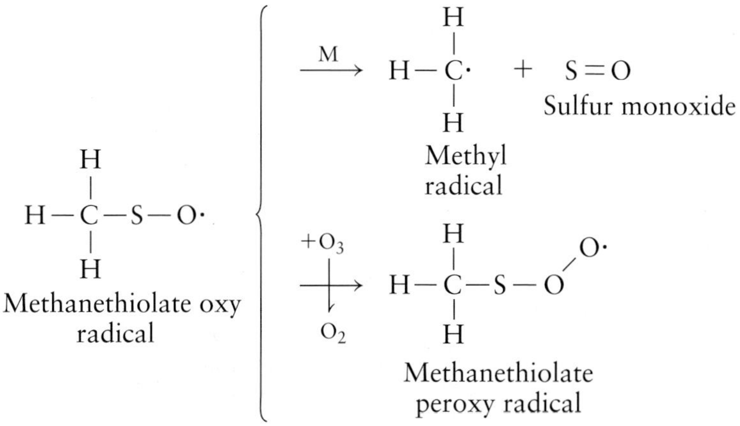

(11.58)

Sulfur monoxide forms sulfur dioxide by

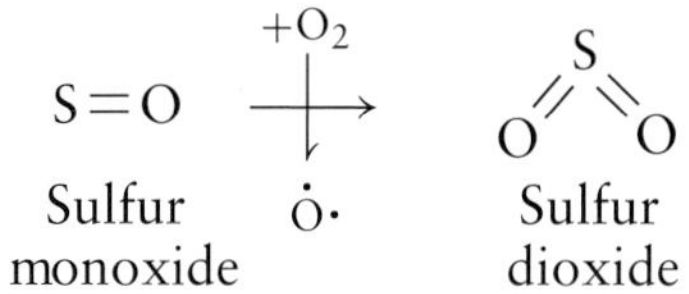

(11.59)

The e-folding lifetime of sulfur monoxide against destruction by O_2 is about 0.0005 s at 298 K and 1 atm pressure. CH_3SO_2 from (11.58) forms sulfur dioxide by

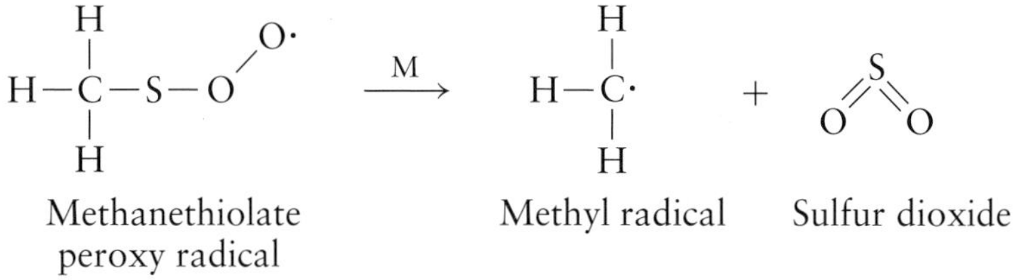

(11.60)

CH_3SO_2 also reacts with O_3, NO_2, and HO_2 to produce CH_3SO_3, which either breaks down to $CH_3 + SO_3$ or abstracts a hydrogen atom from an organic gas to form $CH_3S(O)_2OH$ (**methanesulfonic acid**, MSA). In sum, OH abstraction of DMS results in the formation of sulfur dioxide, methanesulfonic acid, and other products.

The **DMS-addition pathway** initiates when OH bonds to the sulfur atom in DMS. The reaction sequence is

$$
\begin{array}{c}
\text{H}\quad\text{H} \\
|\qquad| \\
\text{H}-\text{C}-\text{S}-\text{C}-\text{H} \\
|\qquad| \\
\text{H}\quad\text{H} \\
\text{Dimethyl sulfide (DMS)}
\end{array}
\xrightarrow{+\dot{\text{O}}\text{H}}
\begin{array}{c}
\text{H}\;\;\text{OH}\;\text{H} \\
|\quad|\quad| \\
\text{H}-\text{C}-\text{S}-\text{C}-\text{H} \\
|\quad\;\cdot\;\quad| \\
\text{H}\qquad\text{H} \\
\text{DMS-OH adduct}
\end{array}
$$

$$
\left\{
\begin{array}{l}
\xrightarrow{\;\text{M}\;}
\begin{array}{c}
\text{H}\qquad\qquad\text{H} \\
|\qquad\qquad| \\
\text{H}-\text{C}-\text{S}-\text{O}\;/^{\text{H}} + \text{H}-\text{C}\cdot \\
|\qquad\qquad| \\
\text{H}\qquad\qquad\text{H} \\
\text{Methanesulfenic}\quad\text{Methyl} \\
\text{acid}\qquad\quad\text{radical}
\end{array} \\[2em]
\xrightarrow[2\text{H}\dot{\text{O}}_2]{+\dot{\text{O}}\text{H},\,2\text{O}_2}
\begin{array}{c}
\text{H}\;\;\text{O}\;\;\text{H} \\
|\quad||\quad| \\
\text{H}-\text{C}-\text{S}-\text{C}-\text{H} \\
|\quad||\quad| \\
\text{H}\;\;\text{O}\;\;\text{H} \\
\text{Dimethyl sulfone (DMSO}_2\text{)}
\end{array}
\end{array}
\right.
$$

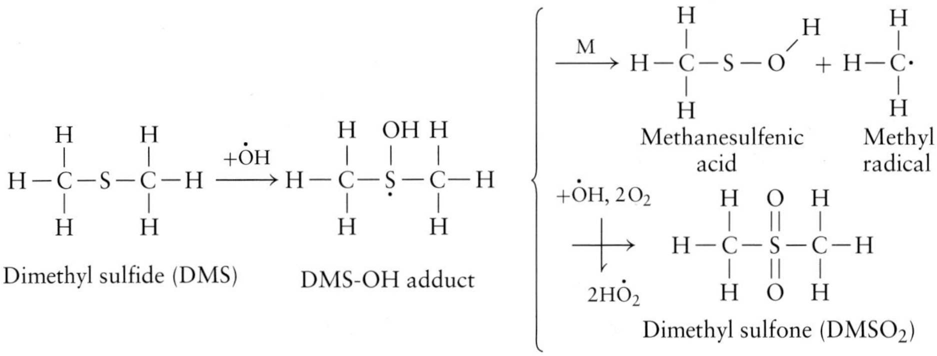

$$(11.61)$$

The **dimethyl sulfone** from this sequence does not react further. The **methanesulfenic acid** (CH_3SOH) produces CH_3SO by

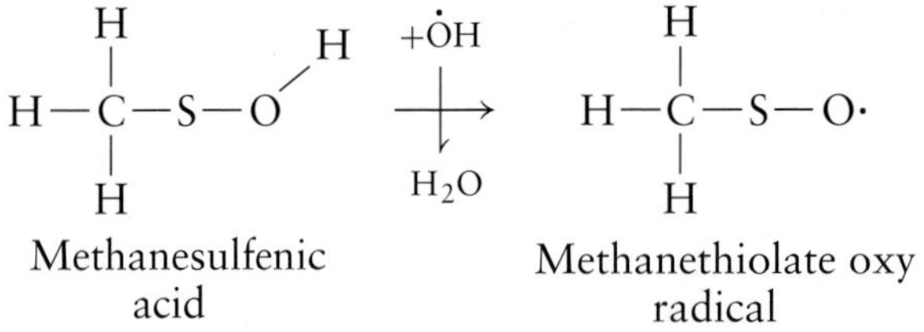

$$
\begin{array}{c}
\text{H} \\
| \\
\text{H}-\text{C}-\text{S}-\text{O}\;/^{\text{H}} \\
| \\
\text{H} \\
\text{Methanesulfenic} \\
\text{acid}
\end{array}
\x!\xrightarrow[\text{H}_2\text{O}]{+\dot{\text{O}}\text{H}}
\begin{array}{c}
\text{H} \\
| \\
\text{H}-\text{C}-\text{S}-\text{O}\cdot \\
| \\
\text{H} \\
\text{Methanethiolate oxy} \\
\text{radical}
\end{array}
\qquad(11.62)
$$

CH_3SO reacts via (11.58)–(11.60) to produce sulfur dioxide.

Dimethyl disulfide (DMDS) oxidation proceeds almost exclusively by OH addition. The sequence is

$$
\begin{array}{c}
\text{H}\qquad\text{H} \\
|\qquad| \\
\text{H}-\text{C}-\text{S}-\text{S}-\text{C}-\text{H} \\
|\qquad| \\
\text{H}\qquad\text{H} \\
\text{Dimethyl disulfide (DMDS)}
\end{array}
\xrightarrow{+\dot{\text{O}}\text{H}}
\begin{array}{c}
\text{H} \\
| \\
\text{H}-\text{C}-\text{S}-\text{O}\;/^{\text{H}} \\
| \\
\text{H} \\
\text{Methanesulfenic} \\
\text{acid}
\end{array}
\; + \;
\begin{array}{c}
\text{H} \\
| \\
\text{H}-\text{C}-\text{S}\cdot \\
| \\
\text{H} \\
\text{Methanethiolate} \\
\text{radical}
\end{array}
$$

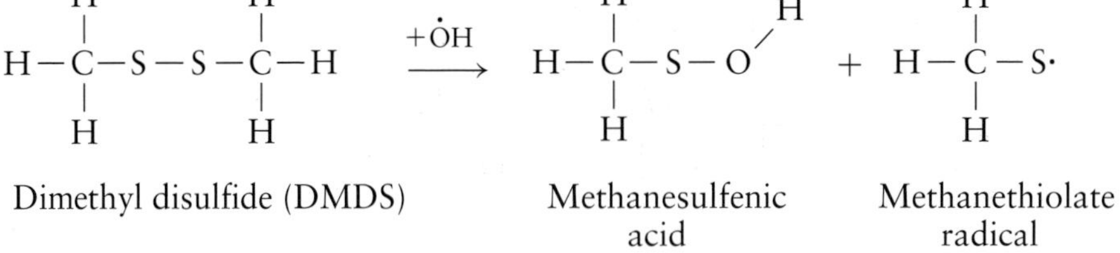

$$(11.63)$$

CH_3SOH and CH_3S react in (11.62) and (11.57), respectively, to form SO_2, MSA, and other products. DMDS also photolyzes by

$$
\begin{array}{c}
\text{H}\qquad\text{H} \\
|\qquad| \\
\text{H}-\text{C}-\text{S}-\text{S}-\text{C}-\text{H} \\
|\qquad| \\
\text{H}\qquad\text{H} \\
\text{Dimethyl disulfide (DMDS)}
\end{array}
\; + \; h\nu \longrightarrow
2\;
\begin{array}{c}
\text{H} \\
| \\
\text{H}-\text{C}-\text{S}\cdot \\
| \\
\text{H} \\
\text{Methanethiolate} \\
\text{radical}
\end{array}
$$

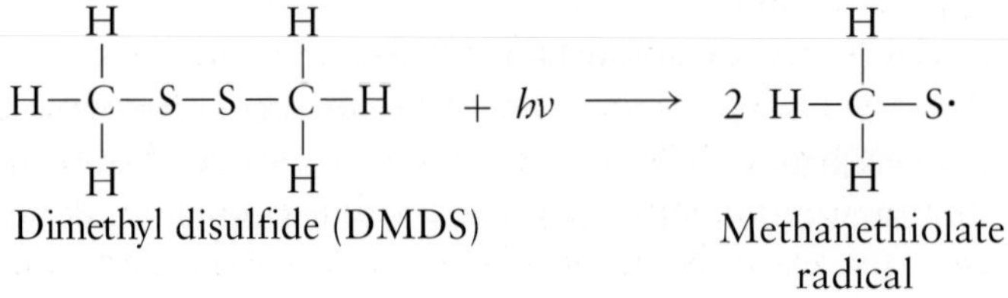

Hydrogen sulfide (H_2S), which has the odor of rotten eggs, is emitted from anaerobic soils, plants, paper manufacturing sources, and volcanos. It is also produced in the deep ocean but does not evaporate from surface ocean water before it is oxidized. Over the remote ocean, H_2S mixing ratios range from 5 to 15 pptv. Over shallow coastal waters, where more evaporation occurs, its mixing ratios reach 100–300 pptv. Over land, its mixing ratios are 5–150 pptv. Downwind of industrial sources, they increase to 1–100 ppbv (Berresheim *et al.* 1995). In the air, hydrogen sulfide reacts with OH by

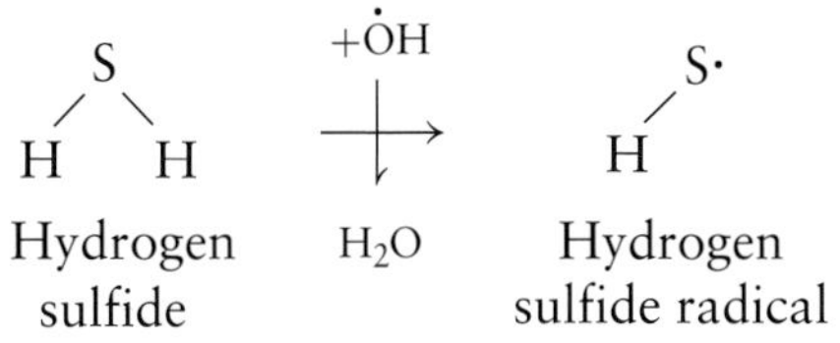

$$\begin{array}{ccc} \underset{\substack{\text{Hydrogen} \\ \text{sulfide}}}{\overset{\displaystyle S}{\underset{H \quad H}{\diagup\diagdown}}} & \overset{+\dot{O}H}{\underset{H_2O}{+}} & \underset{\substack{\text{Hydrogen} \\ \text{sulfide radical}}}{\overset{\displaystyle S\cdot}{\underset{H}{\diagup}}} \end{array} \qquad (11.65)$$

to form the **hydrogen sulfide radical** (HS), which subsequently reacts with O_3, NO_2, and O_2. The e-folding lifetimes of HS against loss by these three gases are 0.27 s, 14 s, and 0.49 s, respectively when $T = 298$ K, $p_d = 1013$ hPa, $[O_3] = 40$ ppbv, and $[NO_2] = 50$ pptv. The HS–O_2 reaction is

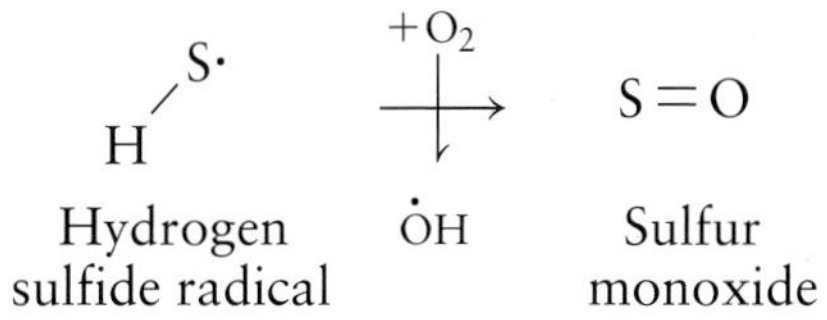

$$\begin{array}{ccc} \underset{\substack{\text{Hydrogen} \\ \text{sulfide radical}}}{\overset{\displaystyle S\cdot}{\underset{H}{\diagup}}} & \overset{+O_2}{\underset{\dot{O}H}{+}} & \underset{\substack{\text{Sulfur} \\ \text{monoxide}}}{S=O} \end{array} \qquad (11.66)$$

Sulfur monoxide produces SO_2 by (11.59). The HS–O_3 and HS–NO_2 reactions produce HSO, which reacts again with O_3 and NO_2 to produce HSO_2, which reacts further with O_2 to form SO_2 and HO_2.

Methanethiol (methyl sulfide, CH_3SH), emitted by bacteria, reacts with OH to produce the methanethiolate radical by

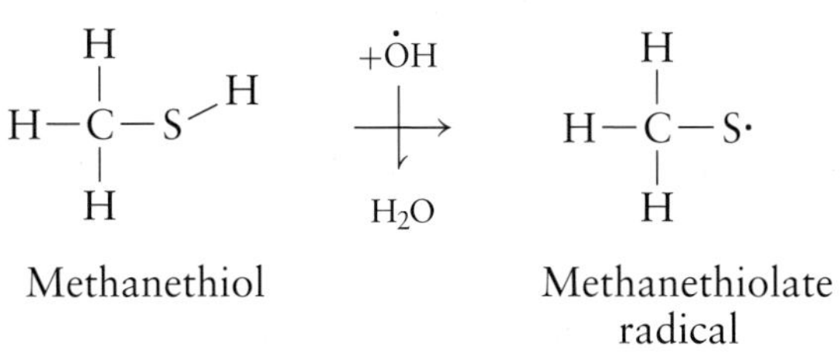

$$\begin{array}{ccc} \underset{\text{Methanethiol}}{\overset{\displaystyle H}{\underset{\displaystyle H}{H-\overset{|}{\underset{|}{C}}-S{\diagup}^{\displaystyle H}}}} & \overset{+\dot{O}H}{\underset{H_2O}{+}} & \underset{\substack{\text{Methanethiolate} \\ \text{radical}}}{\overset{\displaystyle H}{\underset{\displaystyle H}{H-\overset{|}{\underset{|}{C}}-S\cdot}}} \end{array} \qquad (11.67)$$

CH_3S initiates SO_2 production in (11.57).

Carbonyl sulfide (OCS) and **carbon disulfide** (CS_2) are emitted from volcanos. Both react with OH, but the OCS-OH reaction is slow, and photolysis of OCS occurs only in the stratosphere. OCS mixing ratios have increased over time and, at 500 pptv, are the highest among background sulfur-containing compounds. Mixing ratios of OCS are relatively uniform between the surface and tropopause, at which point photolysis reduces them. The OH and photolysis reactions of OCS are, respectively,

$$\mathrm{O{=}C{=}S} + \dot{\mathrm{O}}\mathrm{H} \longrightarrow \mathrm{H{-}S\cdot} + \mathrm{CO_2} \tag{11.68}$$

Carbonyl sulfide → Hydrogen sulfide radical

$$\mathrm{O{=}C{=}S} + h\nu \longrightarrow \mathrm{CO} + \dot{\mathrm{S}}\cdot \qquad \lambda < 260\ \mathrm{nm} \tag{11.69}$$

Carbonyl sulfide → Carbon monoxide + Atomic sulfur

Atomic sulfur from (11.69) forms SO by

$$\dot{\mathrm{S}}\cdot \xrightarrow{+\mathrm{O_2},\ -\dot{\mathrm{O}}\cdot} \mathrm{S{=}O} \tag{11.70}$$

Atomic sulfur → Sulfur monoxide

SO forms SO_2 by (11.59).

Carbon disulfide is broken down by reaction with OH and photolysis. Because the CS_2–OH and CS_2 photolysis reactions are faster than are the OCS–OH and OCS photolysis reactions, respectively, CS_2 mixing ratios of 2–200 pptv are lower than are OCS mixing ratios in the background atmosphere. The CS_2–OH and CS_2 photolysis reactions are

$$\mathrm{S{=}C{=}S} + \dot{\mathrm{O}}\mathrm{H} \longrightarrow \mathrm{H{-}S\cdot} + \mathrm{O{=}C{=}S} \tag{11.71}$$

Carbon disulfide → Hydrogen sulfide radical + Carbonyl sulfide

$$\mathrm{S{=}C{=}S} + h\nu \longrightarrow {}^{-}\mathrm{C{\equiv}S}^{+} + \dot{\mathrm{S}}\cdot \qquad \lambda < 340\ \mathrm{nm} \tag{11.72}$$

Carbon disulfide → Carbon monosulfide + Atomic sulfur

respectively. **Carbon monosulfide** (CS) produces OCS by

$$^{-}\mathrm{C{\equiv}S}^{+} + \mathrm{O_2} \longrightarrow \mathrm{O{=}C{=}S} + \dot{\mathrm{O}}\cdot \tag{11.73}$$

Carbon monosulfide → Carbonyl sulfide

Once sulfur dioxide has been produced chemically in or emitted into the atmosphere, it finds its way into aerosol particles and clouds by one of two mechanisms. The first is dissolution of SO_2 into water-containing aerosol particles and cloud drops followed by its aqueous oxidation to sulfuric acid ($H_2SO_4(aq)$). This mechanism is discussed in Chapter 19. The second mechanism is gas-phase oxidation of

SO_2 to sulfuric acid gas (H_2SO_4), which readily condenses onto aerosol particles to form aqueous sulfuric acid. The gas phase conversion of sulfur dioxide to sulfuric acid requires three steps. The first is the conversion of SO_2 to **bisulfite** (HSO_3). Bisulfite quickly reacts with oxygen to form **sulfur trioxide** (SO_3). Sulfur trioxide then reacts with water to form **sulfuric acid** gas. The reaction sequence is

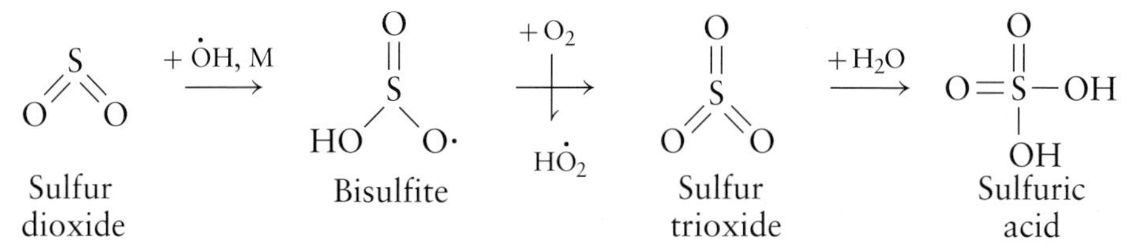

(11.74)

With both mechanisms, the aqueous sulfuric acid formed dissociates into ions, as described in Chapter 17.

In sum, free-tropospheric chemistry is governed by inorganic and light organic reactions. Important inorganic constituents in the free troposphere are NO, NO_2, O_3, OH, HO_2, CO, and SO_2. Organic species of interest include CH_4, C_2H_6, C_3H_8, C_2H_4, C_3H_6, HCHO, CH_3CHO, and CH_3COCH_3. In the free troposphere, the relationship among NO, NO_2, and O_3 can be quantified reasonably with the photostationary-state relationship. Some gases in the free troposphere originate from natural sources while others originate from combustion and other emission sources. Urban photochemistry is discussed next.

11.2 URBAN PHOTOCHEMISTRY

Two general categories of urban air pollution have been observed. The first is London-type smog and the second is called photochemical smog.

London-type smog results from the burning of coal and other raw materials in the presence of a fog or a strong temperature inversion. The pollution consists of a high concentration of directly emitted particles containing metals, ash, and acids or, when fog is present, fog drops laden with these constituents. Several deadly London-type smog events occurred in London in the nineteenth and twentieth centuries, including one in 1952 in which 4000 deaths above average were recorded. Deadly events in Glasgow and Edinburgh, Scotland (1909, 1000 deaths), Meuse Valley, Belgium (December, 1930, 63 deaths), and Donora, Pennsylvania (October, 1948, 20 deaths) have been attributed to London-type smog.

Photochemical smog results from the emission of organic gases and oxides of nitrogen in the presence of sunlight. The pollution consists of a soup of gases and aerosol particles. Some of the particles are directly emitted, whereas others are produced by gas-to-particle conversion. Today, many cities exhibit photochemical smog, including, among others, Mexico City, Santiago, Los Angeles, Tehran,

Calcutta, Beijing, Tokyo, Johannesburg, and Athens. All these cities experience some degree of London-type together with photochemical smog.

11.2.1 Basic characteristics of photochemical smog

Photochemical smog differs from background air in two ways. First, mixing ratios of nitrogen oxides and organic gases are higher in polluted air than in background air, causing ozone levels to be higher in urban air than in the background. Second, photochemical smog contains higher concentrations of high molecular weight organic gases, particularly aromatic gases, than does background air. Because such gases break down quickly in urban air, most are unable to survive transport to the background troposphere.

Photochemical smog involves reactions among **nitrogen oxides** ($NO_x = NO + NO_2$) and **reactive organic gases** (ROGs, total organic gases minus methane) in the presence of sunlight. The most recognized gas-phase by-product of smog reactions is ozone because ozone has harmful health effects and is an indicator of the presence of other pollutants.

On a typical morning, NO and ROGs are emitted by automobiles, power plants, and other combustion sources. Emitted pollutants are **primary pollutants.** ROGs are oxidized to organic peroxy radicals, denoted by RO_2, which react with NO to form NO_2. Pre-existing ozone also converts NO to NO_2. Sunlight then breaks down NO_2 to NO and O. Finally, O reacts with molecular oxygen to form ozone. The basic reaction sequence is thus

$$\dot{N}O + R\dot{O}_2 \longrightarrow \dot{N}O_2 + R\dot{O} \qquad (11.75)$$

$$\dot{N}O + O_3 \longrightarrow \dot{N}O_2 + O_2 \qquad (11.76)$$

$$\dot{N}O_2 + h\nu \longrightarrow \dot{N}O + \dot{O}\cdot \qquad \lambda < 420 \text{ nm} \qquad (11.77)$$

$$\dot{O}\cdot + O_2 + M \longrightarrow O_3 + M \qquad (11.78)$$

Pollutants, such as ozone, that form chemically or physically in the air are **secondary pollutants.** NO_2, which is emitted and forms chemically, is both a primary and secondary pollutant.

Because RO_2 competes with O_3 to convert NO to NO_2 in urban air, and because the photostationary state relationship is based on the assumption that only O_3 converts NO to NO_2, the **photostationary relationship is usually not valid in urban air.** In the afternoon, the relationship holds better than it does in the morning because RO_2 mixing ratios are lower in the afternoon than in the morning.

Figure 11.1 shows ozone mixing ratios resulting from different initial mixtures of NO_x and ROGs. This type of plot is an **ozone isopleth.** The isopleth shows that, at low NO_x, ozone is relatively insensitive to ROGs levels. At high NO_x, an increase in ROGs increases ozone. Also, at low ROGs, increases in NO_x above 0.05 ppmv decrease ozone. At high ROGs, increases in NO_x always increase ozone.

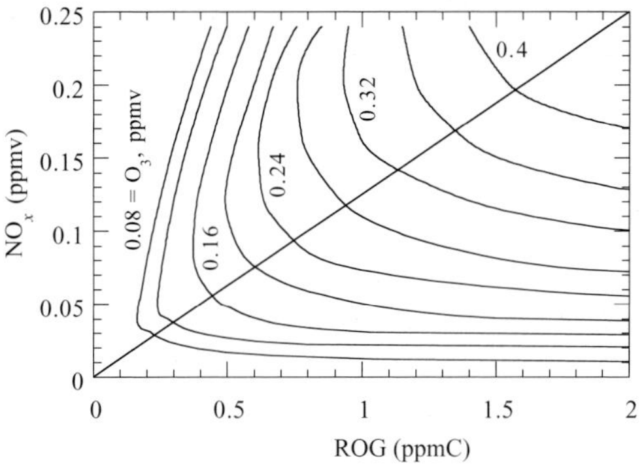

Figure 11.1 Peak ozone mixing ratios resulting from different initial mixing ratios of NO_x and ROGs. The ROG:NO_x ratio along the line through zero is 8:1. Adapted from Finlayson-Pitts and Pitts (2000).

An isopleth is useful for regulatory control of ozone. If ROG mixing ratios are high (e.g., 2 ppmC) and NO_x mixing ratios are moderate (e.g., 0.06 ppmv), the plot indicates that the most effective way to reduce ozone is to reduce NO_x. Reducing ROGs under these conditions has little effect on ozone. If ROG mixing ratios are low (e.g., 0.7 ppmC), and NO_x mixing ratios are high (e.g., 0.2 ppmv), the most effective way to reduce ozone is to reduce ROGs. Reducing NO_x under these conditions increases ozone. In many polluted urban areas, the ROG:NO_x ratio is lower than 8:1, indicating that limiting ROG emission should be the most effective method of controlling ozone. Because ozone mixing ratios depend not only on chemistry but also on meteorology, washout, dry deposition, and gas-to-particle conversion, such a conclusion is not always clearcut.

11.2.2 Meteorological factors affecting smog

Meteorological factors also affect the development of air pollution; Los Angeles is a textbook example. The Los Angeles basin is bordered on its southwestern side by the Pacific Ocean and on all other sides by mountain ranges. During the day, a **sea breeze** blows inland. The sea breeze is at its strongest in the afternoon, when the temperature difference between land and ocean is the greatest. At night, a reverse **land breeze** (from land to sea) occurs, but it is often weak. Figure 11.2(a) shows the variation of sea- and land-breeze wind speeds at Hawthorne, which is near the coast in the Los Angeles basin, during a three day period in 1987.

The sea breeze is instrumental in advecting primary pollutants, emitted mainly on the west side of the Los Angeles basin, toward the east side, where they arrive as secondary pollutants. During transport, primary pollutants, such as NO, are converted to secondary pollutants, such as O_3. Whereas NO mixing ratios peak on the west side of Los Angeles, as shown in Fig. 11.2(b), O_3 mixing ratios peak

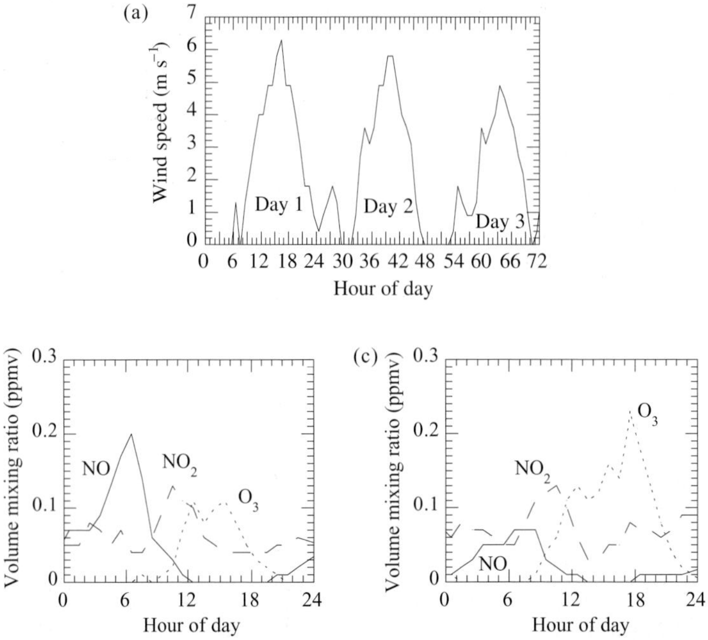

Figure 11.2 (a) Wind speeds at Hawthorne from August 26 to 28, 1987. The other panels show the evolution of the NO, NO_2, and O_3 mixing ratios at (b) central Los Angeles and (c) San Bernardino on August 28. Central Los Angeles is closer to the coast than is San Bernardino. As the sea breeze picks up during the day, primary pollutants, such as NO, are transported from the western side of the Los Angeles basin (e.g., central Los Angeles) toward the eastern side (e.g., San Bernardino). As the pollution travels, organic peroxy radicals convert NO to NO_2, which forms ozone, a secondary pollutant.

on the east side, as shown in Fig. 11.2(c). The west side of the basin is a **source region** and the east side is a **receptor region** of photochemical smog.

Other factors that exacerbate air pollution in the Los Angeles basin are its location relative to the Pacific high-pressure system and its exposure to sunlight. The Pacific high suppresses vertical air movement, inhibiting clouds other than stratus from forming. The subsidence inversion caused by the high prevents pollution from rising easily over the mountains surrounding the basin. Also, because the basin is further south (about 34° N latitude) than most United States cities, Los Angeles receives more daily radiation than do most cities, enhancing its rate of photochemical smog formation relative to other cities.

11.2.3 Emission of photochemical smog precursors

Gases emitted in urban air include nitrogen oxides, reactive organic gases, carbon monoxide, and sulfur oxides ($SO_x = SO_2 + SO_3$). Of these, NO_x and ROGs are

Table 11.2 Gas-phase emission for August 27, 1987 in a 400 × 150 km region of the Los Angeles basin

Substance	Emission (tons day^{-1})	Percentage of total
Carbon monoxide (CO)	9796	69.3
Nitric oxide (NO)	754	
Nitrogen dioxide (NO_2)	129	
Nitrous acid (HONO)	6.5	
Total NO_x + HONO	889.5	6.3
Sulfur dioxide (SO_2)	109	
Sulfur trioxide (SO_3)	4.5	
Total SO_x	113.5	0.8
Alkanes	1399	
Alkenes	313	
Aldehydes	108	
Ketones	29	
Alcohols	33	
Aromatics	500	
Hemiterpenes	47	
Total ROGs	2429	17.2
Methane (CH_4)	904	6.4
Total Emission	14 132	100

Source: Allen and Wagner (1992).

Table 11.3 Organic gases emitted in the greatest quantity in the Los Angeles basin

1. Methane	10. Propylene	19. Acetone	28. Methylcyclohexane
2. Toluene	11. Chloroethylene	20. *n*-Pentadecane	29. Nonane
3. Pentane	12. Acetylene	21. Cyclohexane	30. Methylalcohol
4. Butane	13. Hexane	22. Methylethylketone	31. 1-Hexane
5. Ethane	14. Propane	23. Acetaldehyde	32. Methylcyclopentane
6. Ethene	15. Benzene	24. Trimethylbenzene	33. Methylpentane
7. Octane	16. Methylchloroform	25. Ethylbenzene	34. Dimethylhexane
8. Xylene	17. Pentene	26. Methylvinylketone	35. Cyclopentene
9. Heptane	18. *n*-Butylacetate	27. Naphtha	

Source: Pilinis and Seinfeld (1988).

the main precursors of photochemical smog. Table 11.2 shows emission rates of several pollutants in the Los Angeles basin on a summer day in 1987. CO was the most abundantly emitted gas. About 85 percent of NO_x was emitted as NO, and almost all SO_x was emitted as SO_2. Of the ROGs, toluene, pentane, butane, ethane, ethene, octane, and xylene were emitted in the greatest abundance (Table 11.3). The abundance of a gas does not necessarily translate into proportional smog production. A combination of abundance and reactivity is essential for an ROG to be an important smog producer.

Table 11.4 Percentage emission of several gases by source category in Los Angeles in 1987

Source category	CO	NO_x	SO_x	ROG
Stationary	2	24	38	50
Mobile	98	76	62	50
Total	100	100	100	100

Source: Chang *et al.* (1991).

Table 11.4 shows the percentage emission of several gases by source category in Los Angeles. Emissions originate from point, area, and mobile sources. A **point source** is an individual pollutant source, such as a smokestack, fixed in space. A **mobile source** is a moving individual pollutant source, such as the exhaust of a motor vehicle or an airplane. An **area source** is an area, such as a city block, an agricultural field, or an industrial facility, over which many fixed sources aside from smokestacks exist. Together, point and area sources are **stationary sources.** Table 11.4 shows that CO, the most abundantly emitted gas in the basin, originated almost entirely (98 percent) from mobile sources. Oxides of nitrogen were emitted mostly (76 percent) by mobile sources. The thermal combustion reaction in automobiles that produces nitric oxide at a high temperature is

$$N{\equiv}N + O = O + \text{heat} \longrightarrow 2\dot{N} = O \tag{11.79}$$

Table 11.4 also shows that stationary and mobile sources each accounted for 50 percent of ROGs emitted in the basin. Mobile sources accounted for 62 percent of SO_x emission. The mass of SO_x emission was one-eighth that of NO_x emission. Sulfur emission in Los Angeles is low relative to that in many other cities worldwide because Los Angeles has relatively few coal-fired power plants, which are heavy emitters of SO_x.

11.2.4 Breakdown of ROGs

Once organic gases are emitted, they are broken down chemically into free radicals. **Six major processes break down hydrocarbons and other ROGs** – photolysis and reaction with OH, HO_2, O, NO_3, and O_3. Reaction of organics with OH and O occurs only during the day, because OH and O require photolysis for their production and are short-lived. NO_3 is present only at night because it photolyzes quickly during the day. O_3 and HO_2 may be present during both day and night.

OH is produced in urban air by some of the same reactions that produce it in the free troposphere. An early morning source of OH in urban air is photolysis of HONO. Since HONO may be emitted by automobiles, it is more abundant in urban air than in the free troposphere. Midmorning sources of OH in urban air are aldehyde photolysis and oxidation. The major afternoon source of OH is ozone photolysis. In sum, the three major reaction mechanisms that produce the hydroxyl radical in urban air are

Early morning source

$$\mathrm{HONO} + h\nu \longrightarrow \dot{\mathrm{O}}\mathrm{H} + \dot{\mathrm{N}}\mathrm{O} \qquad \lambda < 400\ \mathrm{nm} \tag{11.80}$$

Midmorning source

$$\mathrm{HCHO} + h\nu \longrightarrow \dot{\mathrm{H}} + \mathrm{H}\dot{\mathrm{C}}\mathrm{O} \qquad \lambda < 334\ \mathrm{nm} \tag{11.81}$$

$$\dot{\mathrm{H}} + \mathrm{O_2} \xrightarrow{\mathrm{M}} \mathrm{H}\dot{\mathrm{O}}_2 \tag{11.82}$$

$$\mathrm{H}\dot{\mathrm{C}}\mathrm{O} + \mathrm{O_2} \longrightarrow \mathrm{H}\dot{\mathrm{O}}_2 + \mathrm{CO} \tag{11.83}$$

$$\dot{\mathrm{N}}\mathrm{O} + \mathrm{H}\dot{\mathrm{O}}_2 \longrightarrow \dot{\mathrm{N}}\mathrm{O}_2 + \dot{\mathrm{O}}\mathrm{H} \tag{11.84}$$

Afternoon source

$$\mathrm{O_3} + h\nu \longrightarrow \mathrm{O_2} + \cdot\dot{\mathrm{O}}(^1D) \qquad \lambda < 310\,\mathrm{nm} \tag{11.85}$$

$$\cdot\dot{\mathrm{O}}(^1D) + \mathrm{H_2O} \longrightarrow 2\dot{\mathrm{O}}\mathrm{H} \tag{11.86}$$

ROGs emitted in urban air include alkanes, alkenes, alkynes, aldehydes, ketones, alcohols, aromatics, and hemiterpenes. Table 11.5 shows lifetimes of these ROGs against breakdown by six processes. The table shows that photolysis breaks down aldehydes and ketones, OH breaks down all eight groups during the day, HO_2 breaks down aldehydes during the day and night, O breaks down alkenes and terpenes during the day, NO_3 breaks down alkanes, alkenes, aldehydes, aromatics, and terpenes during the night, and O_3 breaks down alkenes and terpenes during the day and night.

The breakdown of ROGs produces radicals that lead to ozone formation. Table 11.6 shows the most important ROGs in Los Angeles during the summer of 1987 in terms of a combination of abundance and reactive ability to form ozone. The table shows that *m*- and *p*-xylene, both aromatic hydrocarbons, were the most important gases in terms of generating ozone. Although alkanes are emitted in greater abundance than are other organics, they are less reactive in producing ozone than are aromatics, alkenes, or aldehydes.

In the following subsections, photochemical smog processes involving the chemical breakdown of organic gases to produce ozone are discussed.

11.2.5 Ozone production from alkanes

Table 11.6 shows that *i*-pentane and butane are the most effective alkanes with respect to the combination of concentration and reactivity in producing ozone in

Table 11.5 Estimated lifetimes of reactive organic gases representing alkanes, alkenes, alkynes, aldehydes, ketones, aromatics, and terpenes against photolysis and oxidation in urban and free-tropospheric air

		Lifetime in polluted urban air at sea level				
ROG species	Photolysis	[OH] 5×10^6 molec. cm^{-3}	[HO_2] 2×10^9 molec. cm^{-3}	[O] 8×10^4 molec. cm^{-3}	[NO_3] 1×10^{10} molec. cm^{-3}	[O_3] 5×10^{12} molec. cm^{-3}
n-Butane	–	22 h	1000 y	18 y	29 d	650 y
trans-2-Butene	–	52 m	4 y	6.3 d	4 m	17 m
Acetylene	–	3.0 d	–	2.5 y	–	200 d
Toluene	–	9.0 h	–	6 y	33 d	200 d
Isoprene	–	34 m	–	4 d	5 m	4.6 h
Formaldehyde	7 h	6.0 h	1.8 h	2.5 y	2.0 d	3200 y
Acetone	23 d	9.6 d	–	–	–	–
		Lifetime in free-tropospheric air at sea level				
ROG Species	**Photolysis**	[OH] 5×10^5 molec. cm^{-3}	[HO_2] 3×10^8 molec. cm^{-3}	[O] 3×10^3 molec. cm^{-3}	[NO_3] 5×10^8 molec. cm^{-3}	[O_3] 1×10^{12} molec. cm^{-3}
n-Butane	–	9.2 d	6700 y	480 y	1.6 y	3250 y
trans-2-Butene	–	8.7 h	27 y	168 d	1.3 h	1.4 h
Acetylene	–	30 d	–	67 y	–	2.7 y
Toluene	–	3.8 d	–	160 y	1.8 y	2.7 y
Isoprene	–	5.7 h	–	106 d	1.7 h	23 d
Formaldehyde	7 h	2.5 d	11.7 h	67 y	40 d	16,000 y
Acetone	23 d	96 d	–	–	–	–

Estimated lifetimes for some species in urban air were recalculated from Finlayson-Pitts and Pitts (2000). Lifetimes of other species were obtained from rate-coefficient data. Photolysis rate coefficients were obtained from Figs. 10.3(a) and (b). Gas concentrations are typical, but not necessarily average values for each region. Units: m, minutes; h, hours; d, days; y, years; —, no data or insignificant loss.

Table 11.6 Ranking of the most abundant species in terms of reactivity during the summer Southern California Air Quality Study in 1987

1. *m*- and *p*-Xylene	8. *o*-Xylene	15. *m*-Ethyltoluene	22. *p*-Ethyltoluene
2. Ethene	9. Butane	16. Pentanal	23. C_4 Olefin
3. Acetaldehyde	10. Methylcyclopentane	17. Propane	24. 3-Methylpentane
4. Toluene	11. 2-Methylpentane	18. Propanal	25. *o*-Ethyltoluene
5. Formaldehyde	12. Pentane	19. *i*-Butane	
6. *i*-Pentane	13. 1,2,4-Trimethylbenzene	20. C_6 Carbonyl	
7. Propene	14. Benzene	21. Ethylbenzene	

Source: Lurmann *et al.* (1992). The ranking was determined by multiplying the weight fraction of each organic present in the atmosphere by a species-specific reactivity scaling factor developed by Carter (1991).

Los Angeles air. As in the free troposphere, the main pathway of alkane decomposition in urban air is OH attack. Photolysis and reaction with O_3, HO_2, and NO_3 have little effect on alkane concentrations. Of all alkanes, methane is the least reactive and the least important with respect to urban air pollution. Methane is more important with respect to free-tropospheric and stratospheric chemistry. The oxidation pathway of methane was given in Section 11.1.6, and those of ethane and propane were given in Section 11.1.7.

11.2.6 Ozone production from alkenes

Table 11.6 shows that alkenes, such as ethene and propene, are important ozone precursors in photochemical smog. Table 11.5 indicates that alkenes react most rapidly with OH, O_3, and NO_3. In the following subsections, these reaction pathways are discussed.

11.2.6.1 *Alkene reaction with the hydroxyl radical*

When ethene reacts with the hydroxyl radical, the radical substitutes into ethene's double bond to produce an **ethanyl radical** in an OH **addition** process. The ethanyl radical then reacts to produce $NO_2(g)$. The sequence is

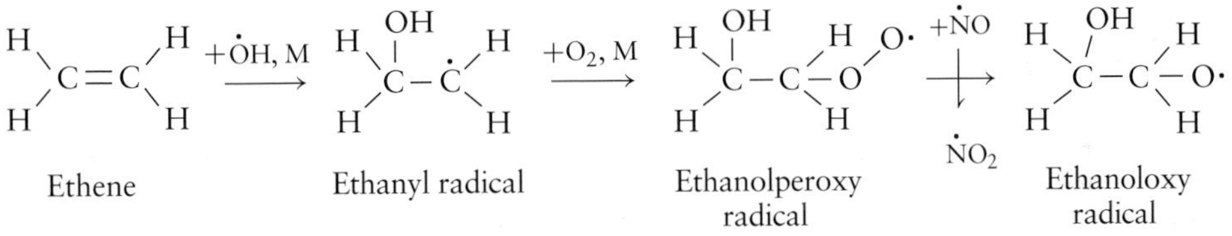

(11.87)

NO_2 produces ozone by (11.2)–(11.3). The **ethanoloxy radical** ($HOCH_2CH_2O$), a by-product of ethene oxidation, produces formaldehyde and **glycol aldehyde** ($HOCH_2CHO$) by

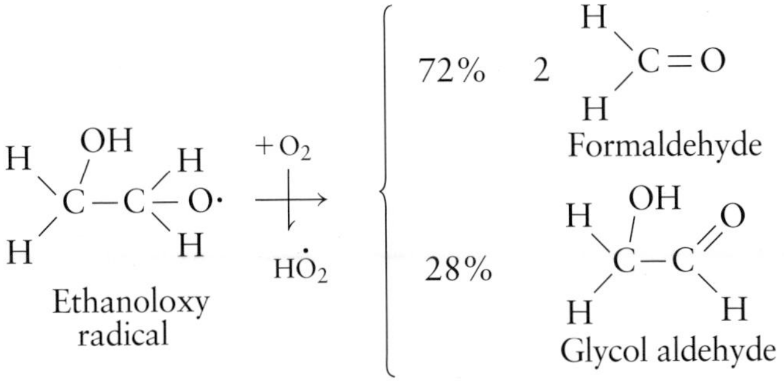

(11.88)

Formaldehyde decomposition produces ozone, as discussed in Section 11.1.8.1. Like other aldehydes, glycol aldehyde is decomposed by photolysis and reaction with OH.

11.2.6.2 *Alkene reaction with ozone*

When ethene or propene reacts with ozone, the ozone substitutes into ethene's double bond to form an unstable **ethene** or **propene molozonide**. The molozonide quickly decomposes to products that are also unstable. The reaction of ethene with ozone is

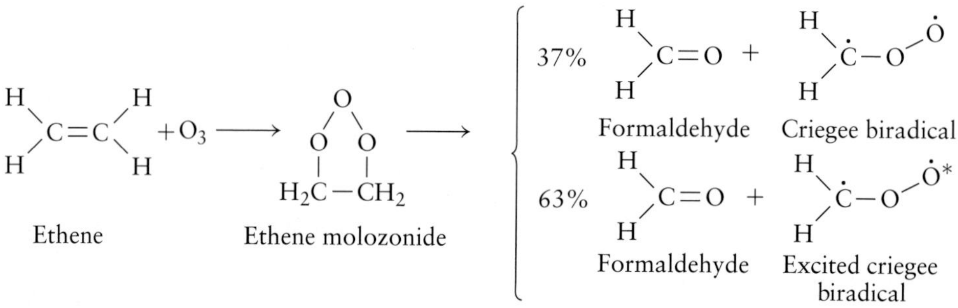

(11.89)

Formaldehyde produces ozone as described in Section 11.1.8. The **criegee biradical** forms NO_2 by

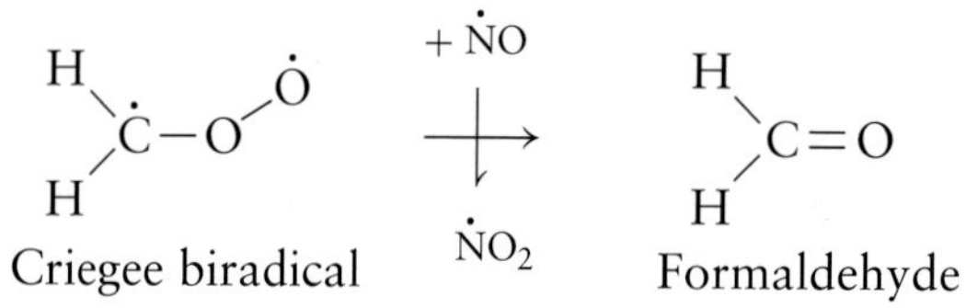

(11.90)

The **excited criegee biradical** isomerizes, and its product, excited formic acid, thermally decomposes by

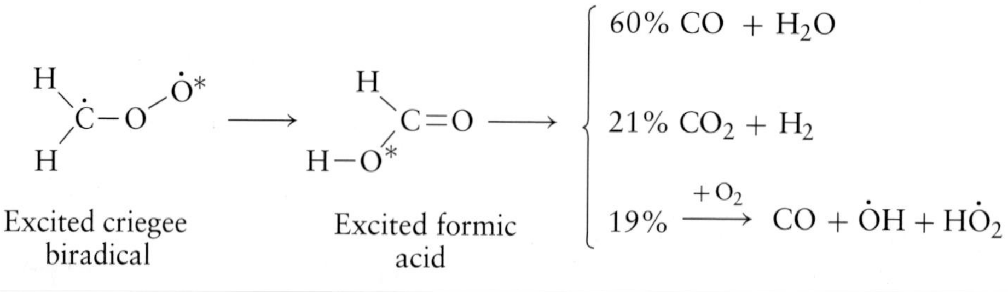

(11.91)

where the fractions are valid at room temperature (Atkinson *et al.* 1997). In sum, ozone attack on ethene produces HCHO, HO_2, CO, and NO_2. These gases not only reform the original ozone lost, but also produce new ozone.

Ozone oxidizes **propene** by the sequence

$$
\text{Propene} \quad \underset{H_3C}{\overset{H}{>}}C{=}CH_2 + O_3 \longrightarrow \text{Propene molozonide} \longrightarrow
\begin{cases}
7.5\% & \text{Formaldehyde} + \text{Methyl criegee biradical} \\
42.5\% & \text{Formaldehyde} + \text{Excited methyl criegee biradical} \\
18.5\% & \text{Acetaldehyde} + \text{Criegee biradical} \\
31.5\% & \text{Acetaldehyde} + \text{Excited criegee biradical}
\end{cases}
$$

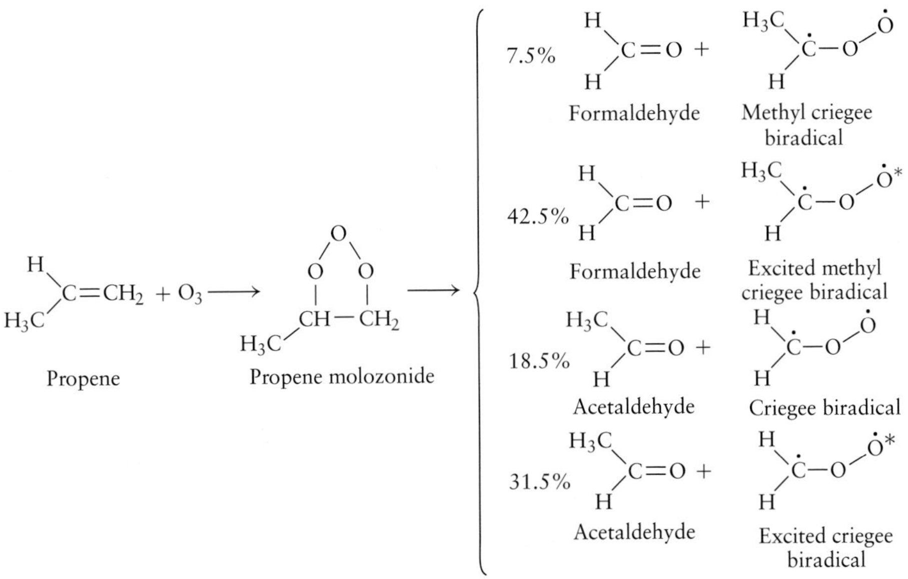

(11.92)

(Atkinson *et al.* 1997). The **methyl criegee biradical** reacts with NO by

$$
\text{Methyl criegee biradical} \;\; \overset{+\,\dot{N}O}{\underset{\dot{N}O_2}{\longrightarrow}} \;\; \text{Acetaldehyde}
$$

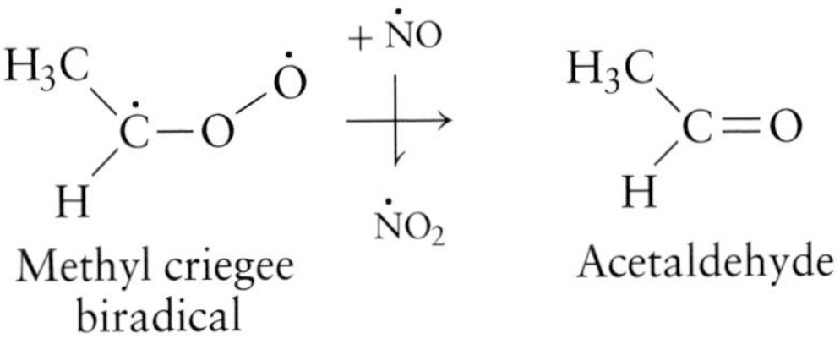

(11.93)

The **excited methyl criegee biradical** isomerizes to excited acetic acid, which thermally decomposes via

$$
\text{Excited methyl criegee biradical} \longrightarrow \text{Excited acetic acid} \longrightarrow
\begin{cases}
16\% & CH_4 + CO_2 \\
64\% & \dot{C}H_3 + CO + \dot{O}H \\
20\% & CH_3\dot{O} + H\dot{O}_2 + CO
\end{cases}
$$

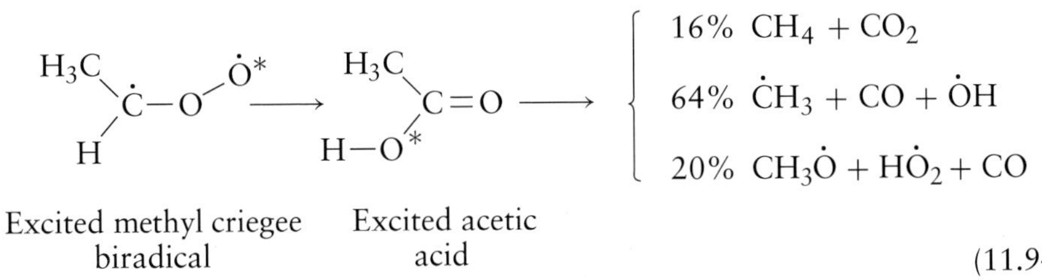

(11.94)

where the fractions are valid at room temperature (Atkinson *et al.* 1997). Thus, propene oxidation by ozone produces OH, HO_2 NO_2, HCHO, and CH_3CHO, all of which react further to reform ozone.

11.2.6.3 *Alkene reaction with nitrate*

Table 11.5 shows that the reaction of NO_3 with an alkene is rapid. Such breakdown, which occurs only at night, leads to a morning buildup of organic peroxy radicals. The reaction sequences with respect to ethene and propene oxidation by NO_3 are

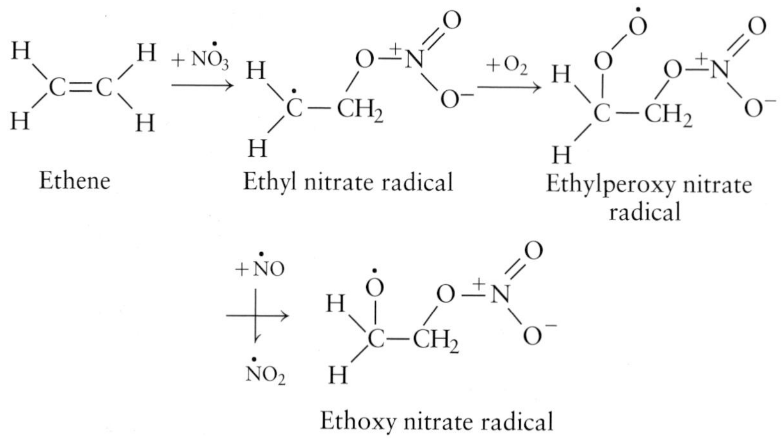

(11.95)

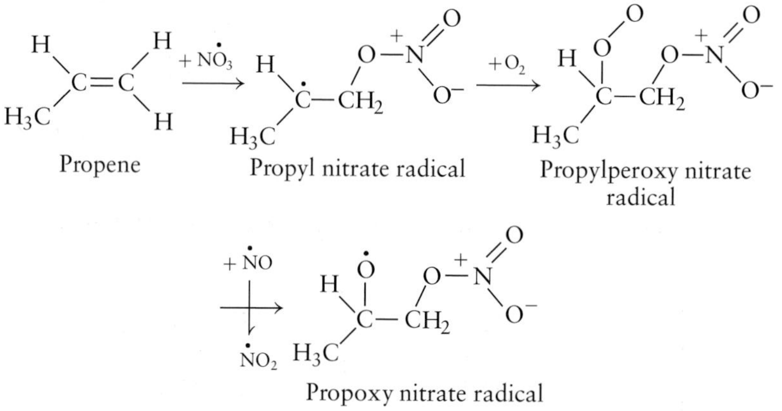

(11.96)

respectively. NO_2 produced from these reactions during the night leads to ozone formation in the morning.

11.2.7 Ozone production from aromatics

Toluene ($C_6H_5CH_3$) originates from gasoline combustion, biomass burning, petroleum refining, detergent production, paint, and building materials. After methane, it is the second most abundantly emitted organic gas in Los Angeles air and the fourth most important gas in terms of abundance and chemical reactivity (Table 11.6). Mixing ratios of toluene in polluted air range from 1 to 30 ppbv.

Table 11.5 shows that toluene is decomposed almost exclusively by OH, which breaks down toluene by abstraction and addition. The respective pathways are

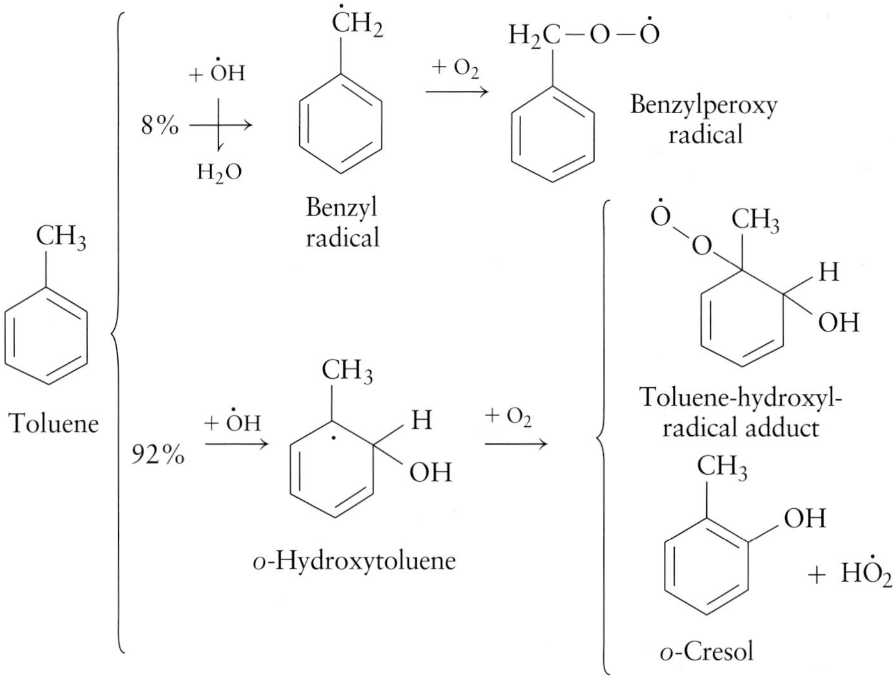

(11.97)

The benzylperoxy radical from the abstraction pathway reacts with NO by

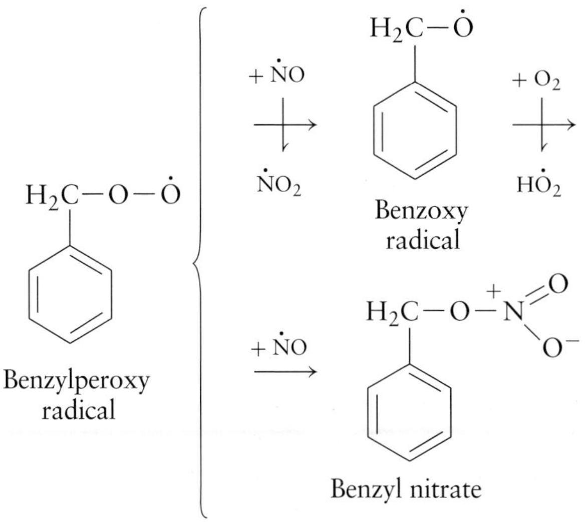

(11.98)

Benzaldehyde, like formaldehyde and acetaldehyde, produces ozone. The toluene-hydroxyl-radical adduct from the toluene addition pathway forms NO_2 by

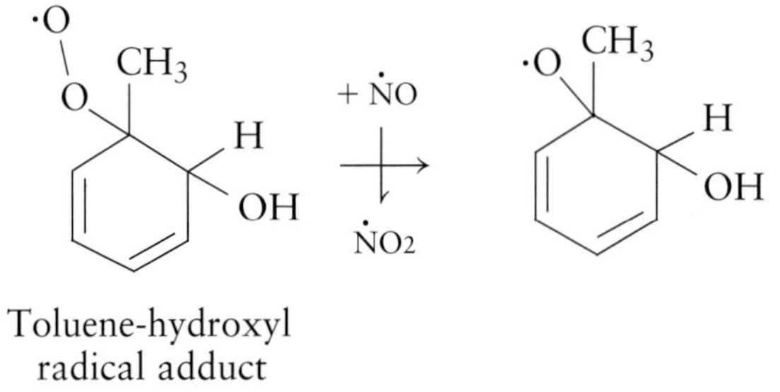

(11.99)

Cresol, from toluene addition, reacts with OH by

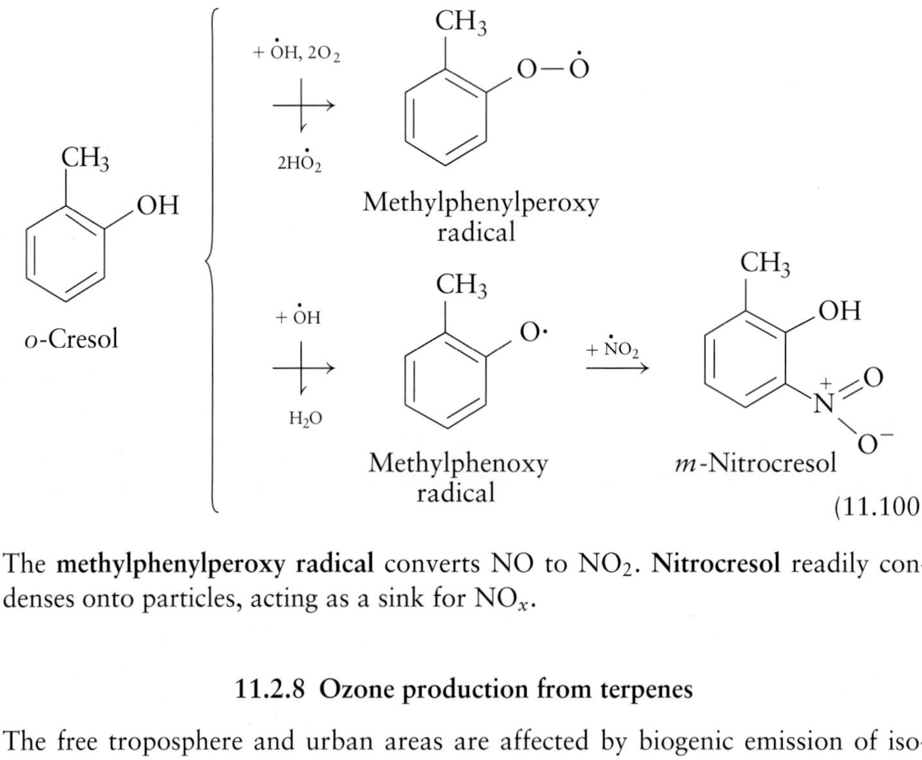

(11.100)

The **methylphenylperoxy radical** converts NO to NO_2. **Nitrocresol** readily condenses onto particles, acting as a sink for NO_x.

11.2.8 Ozone production from terpenes

The free troposphere and urban areas are affected by biogenic emission of isoprene and other terpenes. **Biogenic emissions** are emissions produced from biological sources, such as plants, trees, algae, bacteria, and animals. Strictly speaking, **terpenes** are hydrocarbons that have the formula $C_{10}H_{16}$. Loosely speaking, they are a class of compounds that include hemiterpenes (C_5H_8), such as **isoprene**; monoterpenes ($C_{10}H_{16}$), such as **α-pinene, β-pinene**, and ***d*-limonene**; sesquiterpenes ($C_{15}H_{24}$); and diterpenes ($C_{20}H_{32}$). Isoprene is emitted by sycamore, oak, aspen, spruce, willow, balsam, and poplar trees; α-pinene is emitted by pines, firs, cypress, spruce, and hemlock trees; β-pinene is emitted by loblolly pine, spruce, redwood, and California black sage trees; and *d*-limonene is emitted by loblolly pine, eucalyptus, and California black sage trees and by lemon fruit.

Table 11.5 shows that OH, O_3, and NO_3 decompose isoprene. Although isoprene does not have a long chemical lifetime, some of its by-products last longer.

11.2.8.1 *Terpene reaction with the hydroxyl radical*

The reaction pathways of isoprene with OH produce at least six peroxy radicals. The pathways are

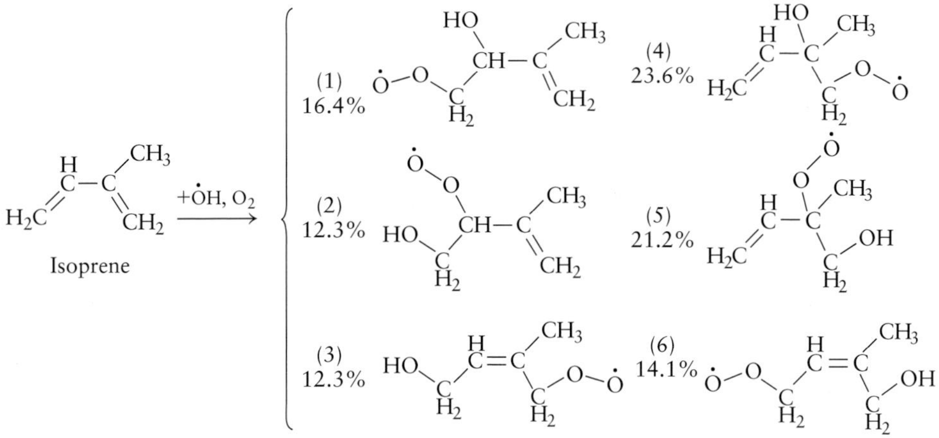

Isoprene peroxy radicals

(11.101)

(Paulson and Seinfeld 1992). The e-folding lifetime of isoprene against reaction with OH is about 30 minutes when [OH] = 5.0×10^6 molec. cm^{-3}. All six radicals convert NO to NO_2. The second and fifth radicals also create **methacrolein** and **methylvinylketone** by

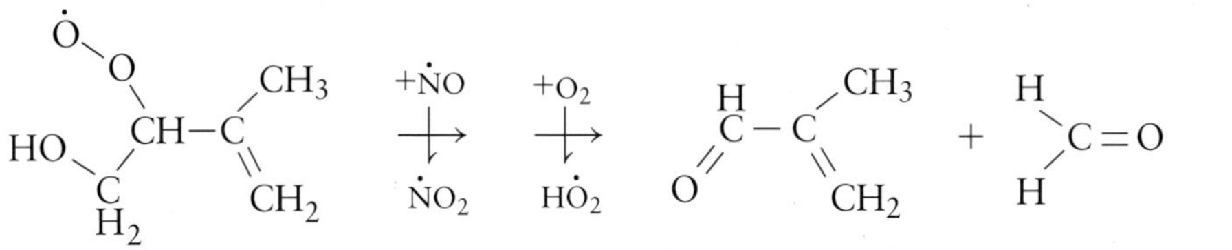

Isoprene peroxy radical Methacrolein Formaldehyde

(11.102)

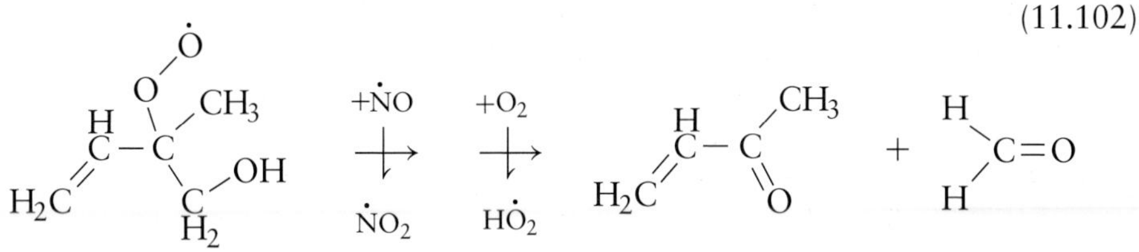

Isoprene peroxy radical Methylvinylketone Formaldehyde

(11.103)

respectively. The NO_2 from these reactions produces ozone. Methacrolein and methylvinylketone react with OH and O_3 to form additional products that convert NO to NO_2, resulting in more ozone, as shown in Appendix Table B.4.

11.2.8.2 *Terpene reaction with ozone*

When O_3 reacts with isoprene, it attacks isoprene at either end of either double bond:

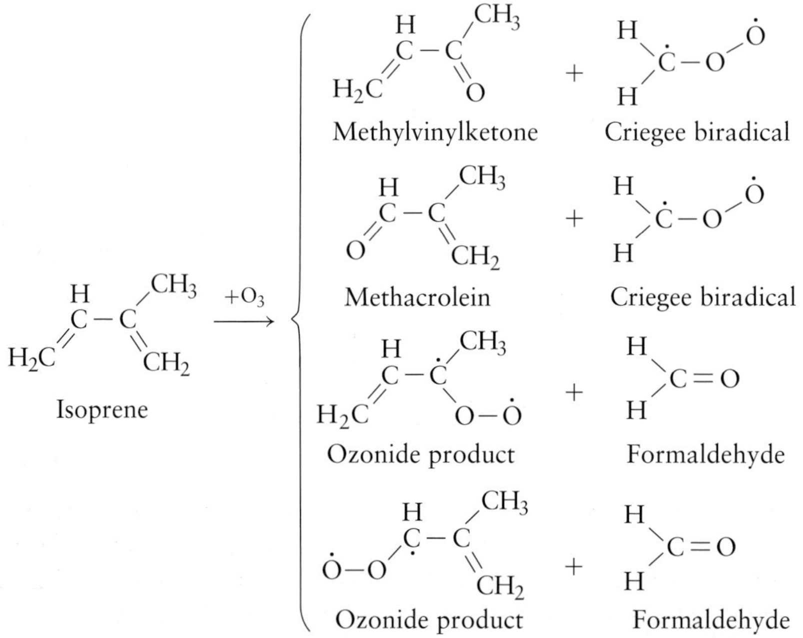

(11.104)

The criegee biradical reacts further by (11.90). Formaldehyde reacts as shown in (11.48) and (11.51). Both reactions produce by-products that form ozone.

11.2.9 Ozone production from alcohols

Two alternative motor-vehicle fuels are methanol and ethanol. Methanol oxidation produces formaldehyde and ozone, and ethanol oxidation produces acetaldehyde, a precursor to PAN. The reaction of **methanol** with OH is

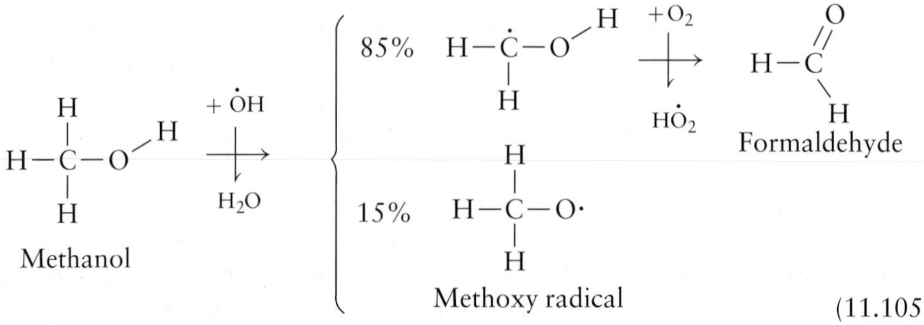

(11.105)

Methanol has an e-folding lifetime against reaction with OH of 71 days when $[OH] = 5.0 \times 10^6$ molec. cm^{-3}; thus, the reaction is not rapid. The organic product of the first reaction is formaldehyde, and that of the second reaction is the methoxy radical, which produces formaldehyde by (11.43). Formaldehyde is an ozone precursor.

Ethanol oxidation by OH produces the branched reactions

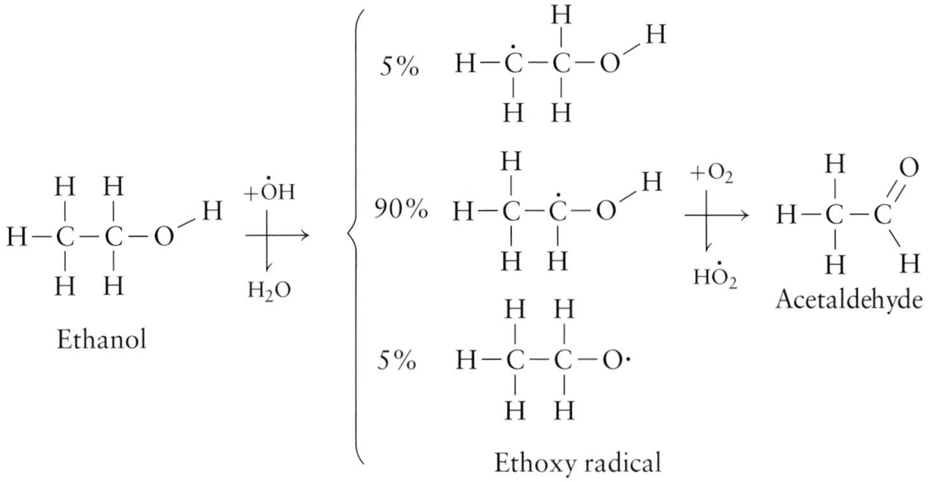

(11.106)

Ethanol lost from the most-probable (middle) reaction has an e-folding lifetime of about 19 hours when $[OH] = 5.0 \times 10^6$ molec. cm^{-3}. Acetaldehyde, formed from the middle reaction, produces PAN and ozone. Cities in Brazil have experienced high PAN mixing ratios since the introduction of their alcohol-fuel program.

11.2.10 Condensed mechanisms for organic chemistry

The number of chemical reactions involving organic gases in urban air is large. Explicit chemical mechanisms with thousands of organic reactions have been developed (e.g., Madronich and Calvert 1989; Jenkin *et al.* 2003; Saunders *et al.* 2003). Although such mechanisms can now be solved in a three-dimensional atmospheric model for a period of a few days (e.g., Liang and Jacobson 2000), the computational demand for long-term and most practical simulations requires that the number of species and reactions be reduced.

Three methods of reducing the number of organic reactions in a model are the carbon-bond lumping method (e.g., Whitten *et al.* 1980; Gery *et al.* 1989), the surrogate-species method (e.g., Atkinson *et al.* 1982; Lurmann *et al.* 1987; Griffin *et al.* 2002), and the lumped-species method (e.g., Stockwell 1986; Carter 1990, 2000).

With the **carbon-bond lumping method,** individual organic gases are segregated into one or more bond groups that have similar chemical reactivity. For example, a butane molecule, which has four carbons connected by single bonds, is divided into four single carbon atoms, each represented by the **paraffin** (PAR) bond group.

Table 11.7 Carbon-bond representations of several organic gases

Chemical name **Carbon bond group** **Chemical structure**			
Ethane 0.4 PAR + 1.6 UNR 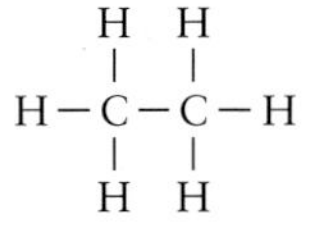	*n*-Butane 4 PAR 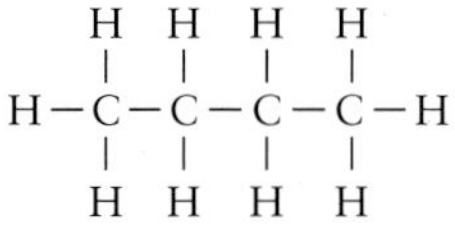	2,2,4-Trimethylpentane 8 PAR 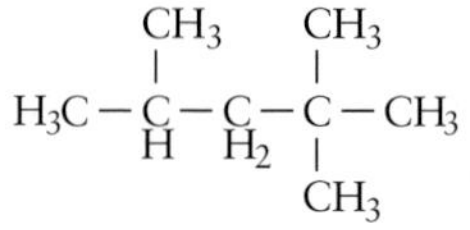	Cyclopentane 5 PAR 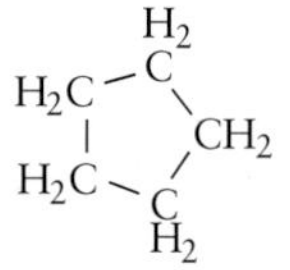
Ethene 1 ETH 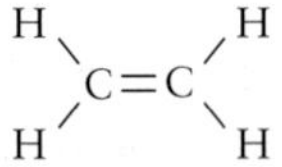	*Trans* 2-butene 2 ALD2 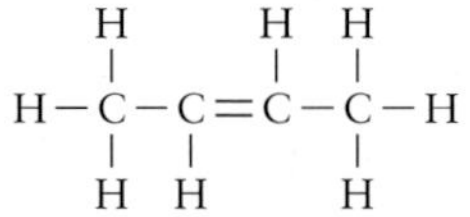	Propene 1 PAR + 1 OLE 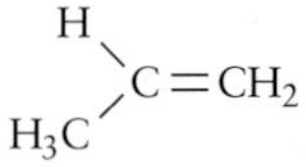	Ethyne 1 PAR + 1 UNR $H-C\equiv C-H$
Formaldehyde 1 FORM 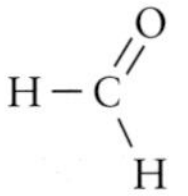	Acetaldehyde 1 ALD2 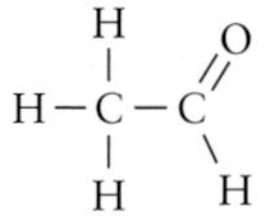	Propionaldehyde 1 PAR + 1 ALD2 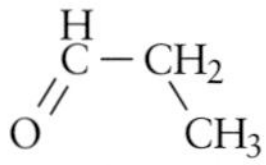	Benzaldehyde 1 ALD2 + 5 UNR 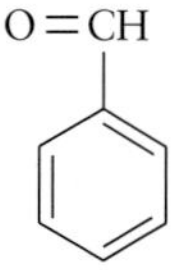
Toluene 1 TOL 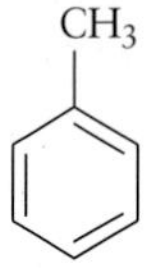	Ethylbenzene 1 PAR + 1 TOL 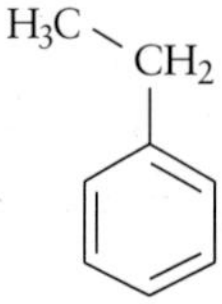	*m*-Xylene 1 XYL 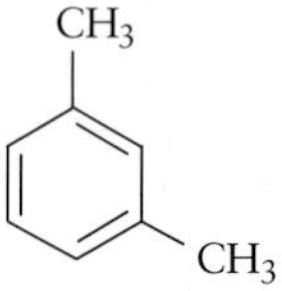	1,2,3-Trimethylbenzene 1 PAR + 1 XYL 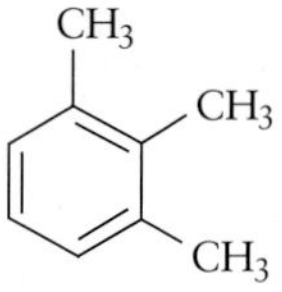
Benzene 1 PAR + 5 UNR 	Methylethylketone 3 PAR + 1 KET 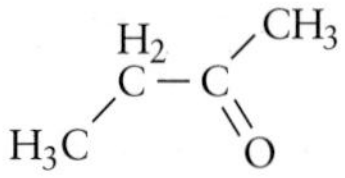	Cyclopentene 1 PAR + 2 ALD2 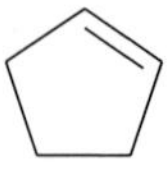	Cyclohexene 2 PAR + 2 ALD2 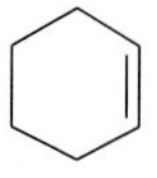

Cyclopentane, which has five single-bonded carbons, is broken into five PAR bond groups. All PAR bonds are assumed to have the same chemical reactivity, regardless of whether they originated from butane or cyclopentane. A terminal carbon atom pair with a double bond between the two atoms is represented by an **olefin** (OLE). Nonterminal carbon-atom pairs with a double bond attached to one of the carbons and terminal two-carbon carbonyl groups [C–C(=O)H] are lumped as ALD2. Single-carbon ketone groups (C=O) are lumped as KET, seven-carbon aromatics are simulated as **toluene** (TOL), eight-carbon aromatics are simulated as ***m*-xylene** (XYL), and terpenes are grouped as **isoprene** (ISOP). **Methane** (CH_4), **ethene** (ETH), **formaldehyde** (FORM), **methanol** (MEOH), **ethanol** (ETOH), **acetone** (AONE), and several other species are not lumped. Nonreactive carbon atoms in certain organic gases are labeled **unreactive** (UNR). In some cases, carbon atoms with a double or triple bond that have similar reactivity to carbon atoms with a single bond are labeled as PAR. Table 11.7 gives the carbon-bond representation of a few organic gases. Several of the reactions in Appendix Table B.4 include reactions among carbon-bond species.

With the **surrogate-species method**, all species of similar reactivity are grouped together. Propane and pentane are assumed to have the same reactivity as *n*-butane, and all three species are grouped as one surrogate species. With the **lumped-species method**, species of similar reactivity are lumped together, just as with the surrogate species method. The difference is that with the surrogate-species method the reaction rate coefficient for each surrogate species is set equal to that of a particular gas. The reaction rate coefficient of a lumped species is determined before a model simulation by taking a mole-fraction-weighted average of the reaction rates of each species in the lumped group.

11.2.11 Summary of urban photochemistry

Photochemical smog production is governed by emission of oxides of nitrogen and reactive organic gases. Emitted gases, called primary pollutants, react in the presence of sunlight to produce secondary pollutants, such as ozone and peroxyacetyl nitrate. The radicals that break down emitted reactive organic gases are OH, HO_2, O_3, NO_3, and O. Photolysis also breaks down certain organics. Because reactive organic gas radicals compete with O_3 to produce NO_2 from NO, the photostationary-state relationship does not usually hold in urban air. Because gas-phase organic chemistry involves reactions among thousands of species, condensed reaction mechanisms have been developed to simplify the simulation of organic chemistry in numerical models.

11.3 STRATOSPHERIC PHOTOCHEMISTRY

Whereas ozone molecules in urban air are harmful to humans, animals, plants, trees, and structures, the same ozone molecules in the stratosphere are beneficial in that they shield the Earth from harmful ultraviolet radiation. Figure 11.3 shows a typical variation of ozone mixing ratio (χ_{O_3}), ozone number concentration ($N_{O_3} = \chi_{O_3} N_d$), and dry air number concentration (N_d) with altitude. The ozone number concentration (molecules of ozone per cubic centimeter of air) in the stratosphere generally peaks at 25 to 32 km altitude. The ozone mixing ratio peaks at a higher altitude than does the ozone number concentration. The peak ozone number concentration in the stratosphere is close to that in polluted urban air. The peak ozone mixing ratio in the stratosphere (near 10 ppmv) is much higher than is that in polluted urban air (0.2–0.35 ppmv) or free-tropospheric air (0.02–0.04 ppmv).

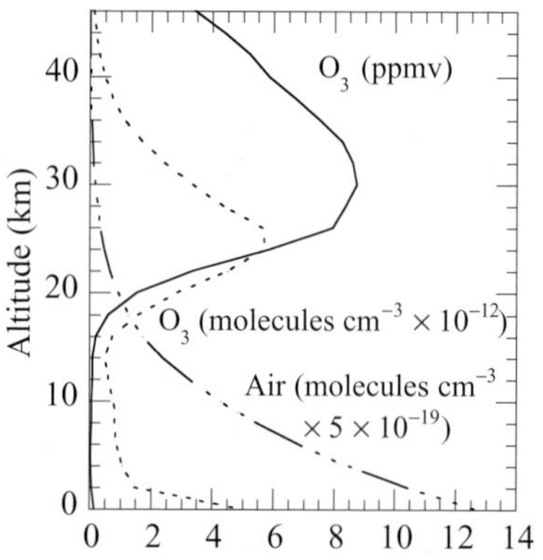

Figure 11.3 Example vertical variation in ozone mixing ratio, ozone number concentration, and air number concentration with altitude. The ozone mixing ratio at the surface is 0.20 ppmv, the level of a Stage 1 smog alert in the United States.

Approximately 90 percent of all ozone molecules in the atmosphere reside in the stratosphere because the stratosphere is approximately 32–40 km thick. Pollutants in urban regions generally mix to depths of only 0.3–1.5 km.

Another measure of ozone is its column abundance, which is the sum of all ozone molecules above a square centimeter of surface between the ground and the top of the atmosphere. When this number is divided by 2.7×10^{16}, the result is the column abundance in **Dobson units** (DUs). Thus, 1 DU is equivalent to 2.7×10^{16} molecules of ozone per square centimeter of surface (or 0.001 atm cm). In 2000, the globally-averaged column abundance of ozone from 90° S to 90° N was 293.4 DU. This column abundance contains the same number of molecules as a column of air 2.93-mm high at 1 atm of pressure and 273 K (near-surface conditions), as illustrated in Fig. 11.4.

11.3.1 Ozone formation from oxygen

The altitude of the stratospheric ozone concentration peak in Fig. 11.3 occurs where the concentration of oxygen and the quantity of ultraviolet radiation are sufficiently high to maximize ozone formation (Section 2.2.2.2). Above the ozone peak, the concentration of oxygen is too low to produce peak ozone. Below the peak, the quantity of radiation is too low to produce peak ozone. In this subsection, the oxygen cycle involved in producing the stratospheric ozone peak is revisited.

A major difference between ozone formation in the stratosphere and troposphere is that ozone formation in the stratosphere is driven by the photolysis of molecular oxygen whereas that in the troposphere is driven by photolysis of nitrogen dioxide. Nitrogen dioxide photolysis is unimportant in the stratosphere because NO_2 concentrations there are too low to produce much ozone. Oxygen photolysis

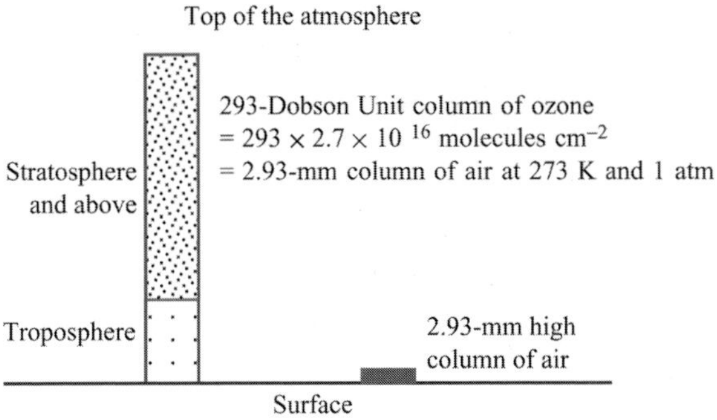

Figure 11.4 Example of globally-averaged column abundance of ozone. The number of ozone molecules per unit area of surface in a 293-DU column of ozone is equivalent to the number of air molecules in a 2.93-mm high column near the surface. (The figure is not to scale.)

is unimportant in the troposphere because ultraviolet radiation intensity required to break down O_2 there is too weak.

Ozone formation in the stratosphere is initiated by the oxygen photolysis reactions,

$$O_2 + h\nu \longrightarrow \cdot\dot{\mathrm{O}}(^1D) + \dot{\mathrm{O}}\cdot \qquad \lambda < 175\,\mathrm{nm} \qquad (11.107)$$

$$O_2 + h\nu \longrightarrow \dot{\mathrm{O}}\cdot + \dot{\mathrm{O}}\cdot \qquad 175 < \lambda < 245\,\mathrm{nm} \qquad (11.108)$$

The first reaction is important only at the top of the stratosphere, since wavelengths shorter than 0.175 μm do not penetrate deeper. The second reaction occurs down to the lower stratosphere, but not to the troposphere since wavelengths below 245 nm do not reach the troposphere. Since the oxygen concentration decreases and radiation intensity increases with increasing height, an altitude exists where the production of atomic oxygen is maximized. This is the altitude where ozone peaks in Fig. 11.3.

As in the troposphere, ozone in the stratosphere forms from

$$\dot{\mathrm{O}}\cdot + O_2 + \mathrm{M} \longrightarrow O_3 + \mathrm{M} \qquad (11.109)$$

Most of the atomic oxygen for this reaction comes directly from molecular oxygen photolysis. Some of it comes from quenching of excited atomic oxygen to the ground state by

$$\cdot\dot{\mathrm{O}}(^1D) \xrightarrow{\mathrm{M}} \dot{\mathrm{O}}\cdot \qquad (11.110)$$

In the upper stratosphere, most excited atomic oxygen comes from (11.107). In the rest of the stratosphere (and in the troposphere), most of it comes from the first ozone photolysis reaction,

$$O_3 + h\nu \longrightarrow O_2 + \dot{O}\cdot(^1D) \qquad \lambda < 310 \text{ nm} \qquad (11.111)$$

$$O_3 + h\nu \longrightarrow O_2 + \dot{O}\cdot \qquad \lambda > 310 \text{ nm} \qquad (11.112)$$

These photolysis reactions help to limit ozone formation primarily in the upper stratosphere, where radiation intensity is high, but also in the lower stratosphere.

11.3.2 Effect of nitrogen on the natural ozone layer

Oxides of nitrogen naturally destroy ozone in the upper stratosphere, helping to shave the vertical ozone profile. In the troposphere, the major sources of NO are surface emission and lightning. The major source of NO in the stratosphere is transport from the troposphere and the breakdown of **nitrous oxide** (N_2O) (laughing gas) by

$$N_2O + \cdot\dot{O}(^1D) \longrightarrow \begin{cases} 64\% & 2\dot{N}O \\ 36\% & N_2 + O_2 \end{cases} \qquad (11.113)$$

The $O(^1D)$ for this reaction comes mostly from ozone and oxygen photolysis. N_2O is a colorless gas emitted by bacteria in fertilizers, sewage, and the oceans and during biomass burning and automobile combustion. It affects not only stratospheric ozone, but also climate because it is a greenhouse gas. Its mixing ratio up to 15–20 km is constant (0.31 ppmv) before decreasing. About 10 percent of nitrous oxide loss in the stratosphere is due to (11.113) and the rest is due to photolysis by

$$N_2O + h\nu \longrightarrow N_2 + \cdot\dot{O}(^1D) \qquad \lambda < 240 \text{ nm} \qquad (11.114)$$

which gives N_2O an e-folding lifetime of about 1.3 yr at 25 km.

NO naturally reduces ozone in the upper stratosphere by

$$\dot{N}O + O_3 \longrightarrow \dot{N}O_2 + O_2 \qquad (11.115)$$

$$\dot{N}O_2 + \dot{O}\cdot \longrightarrow \dot{N}O + O_2 \qquad (11.116)$$

$$\dot{O}\cdot + O_3 \longrightarrow 2O_2 \quad \text{(net process)} \qquad (11.117)$$

This sequence destroys one molecule of ozone, but no molecules of NO or NO_2. It is called the **NO_x catalytic ozone destruction cycle** because ozone is lost but NO_x ($NO + NO_2$) is recycled. The number of times the cycle is executed before NO_x is removed from the cycle by reaction with another gas is the **chain length**. In the upper stratosphere, the chain length of this cycle is about 10^5 (Lary 1997). Thus, 10^5 molecules of O_3 are destroyed before one NO_x molecule is removed from the cycle. In the lower stratosphere, the chain length decreases to near 10.

In the absence of chlorine, the major reactions removing NO_x from the cycle are

$$\dot{N}O_2 + \dot{O}H \xrightarrow{M} HNO_3 \qquad (11.118)$$

$$HO_2 + NO_2 \longrightarrow HO_2NO_2 \qquad (11.119)$$

The products of these reactions, nitric acid and peroxynitric acid, photolyze back to the reactants that formed them by (11.11) and (11.13), respectively, but the rates are slow. Peroxynitric acid also decomposes thermally by (11.28), but thermal decomposition is slow in the stratosphere because temperatures are low there.

The natural NO_x catalytic cycle erodes the ozone layer above ozone's peak altitude shown in Fig. 11.3. Although the NO_x catalytic cycle is largely natural, an unnatural source of stratospheric NO_x and, therefore, ozone destruction, is stratospheric aircraft emission of NO and NO_2. In the 1970s and 1980s, scientists were concerned that the introduction of a fleet of supersonic transport (SST) jets into the stratosphere would enhance NO_x sufficiently to damage the ozone layer. However, the plan to introduce a fleet of stratospheric jets never materialized.

11.3.3 Effect of hydrogen on the natural ozone layer

Hydrogen-containing compounds, particularly OH and HO_2, are responsible for shaving the ozone profile in the lower stratosphere. The hydroxyl radical is produced in the stratosphere by several reactions,

$$\cdot\dot{O}(^1D) + \begin{cases} H_2O \longrightarrow 2\dot{O}H \\ CH_4 \longrightarrow \dot{C}H_3 + \dot{O}H \\ H_2 \longrightarrow \dot{H} + \dot{O}H \end{cases} \qquad (11.120)$$

OH participates in the **HO_x catalytic ozone destruction cycle**, where HO_x = OH + HO_2. The most effective HO_x cycle, which has a chain length in the lower stratosphere of 1 to 40 (Lary 1997), is

$$\dot{O}H + O_3 \longrightarrow H\dot{O}_2 + O_2 \qquad (11.121)$$

$$H\dot{O}_2 + O_3 \longrightarrow \dot{O}H + 2O_2 \qquad (11.122)$$

$$2O_3 \longrightarrow 3O_2 \quad \text{net process} \qquad (11.123)$$

HO_x can be removed temporarily from catalytic cycles by (11.118), (11.119), and

$$H\dot{O}_2 + \dot{O}H \longrightarrow H_2O + O_2 \qquad (11.124)$$

This mechanism is particularly efficient because it removes two HO_x molecules at a time.

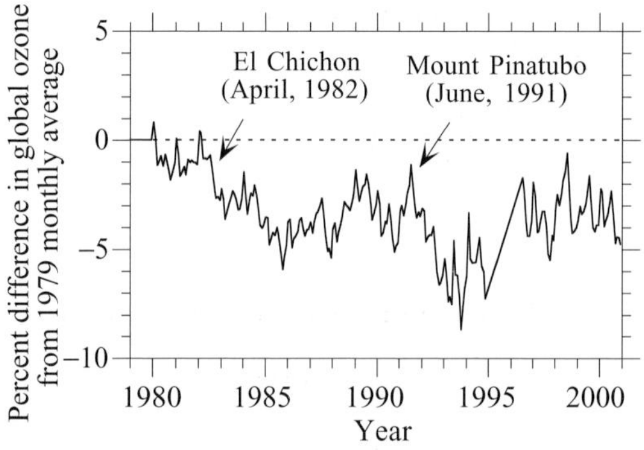

Figure 11.5 Percentage change in the monthly averaged global (90° S to 90° N) ozone column abundance between a given month since 1979 and the same month in 1979.

11.3.4 Effect of carbon on the natural ozone layer

Carbon monoxide and methane produce ozone in the stratosphere by Reactions (11.37)–(11.44). The contributions of CO and CH_4 to ozone production in the stratosphere, though, are small. A by-product of methane oxidation in the stratosphere is water vapor, produced by

$$CH_4 + \dot{O}H \longrightarrow \dot{C}H_3 + H_2O \qquad (11.125)$$

Because the mixing ratio of water vapor in the stratosphere is low and transport of water vapor from the troposphere to stratosphere is slow, this reaction is a relatively important source of water in the stratosphere.

11.3.5 Chlorine and bromine photochemistry

Between 1979 and 2000, the global stratospheric ozone column abundance decreased by approximately 3.5 percent (from 304.0 to 293.4 DU), as shown in Fig. 11.5. Unusual decreases in global ozone occurred following the El Chichón (Mexico) volcanic eruption in April 1982, and the Mount Pinatubo (Philippines) eruption in June 1991. These eruptions injected particles into the stratosphere. On the surfaces of these particles, chemical reactions involving chlorine took place that contributed to ozone loss. Over time, however, the concentration of these particles decreased, and the global ozone layer partially recovered. Because volcanic particles were responsible for only temporary ozone losses, the net loss of ozone over the globe from 1979 to 2000 was still about 3.5 percent.

Between 1950 and 1980, no measurements from three ground-based stations in the Antarctic showed ozone levels less than 220 DU, a threshold for defining Antarctic ozone depletion. Every Southern Hemisphere spring (September–November) since 1980, measurements of stratospheric ozone have shown a depletion. Farman *et al.* (1985) first reported depletions of more than 30 percent

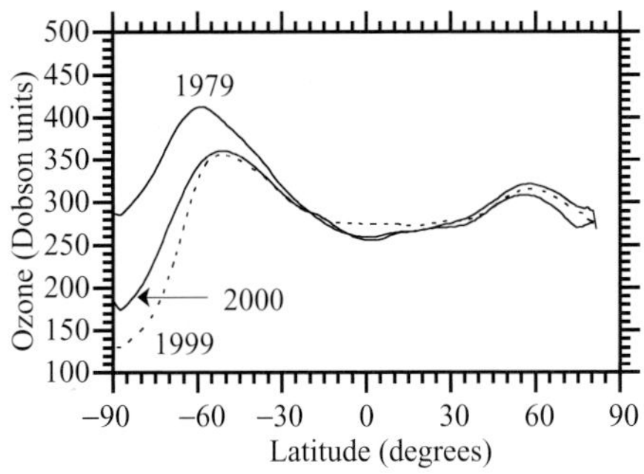

Figure 11.6 Variation with latitude of October monthly and zonally averaged column abundances of ozone in 1979, 1999, and 2000. Data were obtained from the satellite-based Total Ozone Mapping Spectrometer (TOMS) and made available by NASA Goddard Space Flight Center, Greenbelt, Maryland. No data were available from December 1994 to July 1996.

relative to pre-1980 measurements. Since then, measurements over the South Pole have indicated depletions of up to 70 percent of the column ozone for a period of a week in early October. The largest average depletion for the month of September from 60 to 90° S since 1979 was 32.8 percent and occurred in 2000. The largest depletion for the month of October from 60 to 90° S since 1979 was 38.3 percent and occurred in 1998 (Fig. 11.6). Most ozone depletion has occurred between altitudes of 12 and 20 km. The large reduction of stratospheric ozone over the Antarctic in the Southern Hemisphere spring each year is the **Antarctic ozone hole.** The areal extent of the ozone hole is now greater than the size of North America.

11.3.5.1 *Chlorofluorocarbons and related compounds*

Molina and Rowland (1974) recognized that anthropogenic chlorine compounds could destroy stratospheric ozone. Since then, scientists have strengthened the links among global ozone reduction, Antarctic ozone depletion, and the presence of chlorine and other halogenated compounds in the stratosphere.

The compounds that play the most important role in reducing stratospheric ozone are **chlorofluorocarbons (CFCs)**. Important CFCs are identified in Table 11.8. CFCs are gases formed synthetically by replacing all hydrogen atoms in methane (CH_4) or ethane (C_2H_6) with chlorine and/or fluorine atoms. For example, **CFC-12** (CF_2Cl_2) is formed by replacing the four hydrogen atoms in methane with two chlorine and two fluorine atoms.

CFC-12 was invented in 1928 by **Thomas Midgley** and his assistants on the same day that a representative of General Motors' Frigidaire division asked Midgley to find a nontoxic, nonflammable substitute for an existing refrigerant, ammonia, a

Table 11.8 Mixing ratios and lifetimes of selected chlorocarbons, bromocarbons, and fluorocarbons

Chemical formula	Trade name	Chemical name	Tropospheric mixing ratio (pptv)	Estimated overall atmospheric lifetime (yrs)
Chlorocarbons and chlorine compounds				
Chlorofluorocarbons (CFCs)				
$CFCl_3$	CFC-11	Trichlorofluoromethane	270	45
CF_2Cl_2	CFC-12	Dichlorodifluoromethane	550	100
$CFCl_2CF_2Cl$	CFC-113	1-Fluorodichloro-2-difluorochloroethane	70	85
CF_2ClCF_2Cl	CFC-114		15	220
CF_2ClCF_3	CFC-115		5	550
Hydrochlorofluorocarbons (HCFCs)				
CF_2ClH	HCFC-22	Chlorodifluoromethane	130	11.8
CH_3CFCl_2	HCFC-141b		6	9.2
CH_3CF_2Cl	HCFC-142b	2-Difluorochloroethane	8	18.5
Other chlorocarbons				
CCl_4		Carbon tetrachloride	100	35
CH_3CCl_3		Methyl chloroform	90	4.8
CH_3Cl		Methyl chloride	610	1.3
Other chlorinated compounds				
HCl		Hydrochloric acid	1–1000	<1
Bromocarbons				
Halons				
CF_3Br	H-1301	Trifluorobromomethane	2	65
CF_2ClBr	H-1211	Difluorochlorobromomethane	2	11
CF_2BrCF_2Br	H-2402	1-Difluorobromo-2-difluorobromoethane	1.5	22–30
Other bromocarbons				
CH_3Br		Methyl bromide	12	0.7
Fluorocarbons and fluorine compounds				
Hydrofluorocarbons (HFCs)				
CH_2FCF_3	HFC-134a	1-Fluoro-2-trifluoroethane	4	13.6
Perfluorocarbons (PFCs)				
C_2F_6		Perfluoroethane	4	10 000
Other fluorinated compounds				
SF_6		Sulfur hexafluoride	3.7	3200

Sources: Shen *et al.* (1995); Singh (1995); WMO (1998); Mauna Loa Data Center (2001).

flammable and toxic gas. CFC-12 and subsequent CFCs were inexpensive, nontoxic, nonflammable, nonexplosive, insoluble, and chemically unreactive under tropospheric conditions; thus, they became popular.

In 1931, **CFC-12** was produced by the DuPont chemical manufacturer under the trade name **Freon**. Its first use was in small ice cream cabinets. In 1934, it was used in refrigerators and whole-room coolers. Soon after, it was used in household and automotive air conditioning systems. In 1932, **CFC-11** was first produced for use in large air conditioning units. CFCs became airborne only when coolants leaked or were drained.

In 1943, Goodhue and Sullivan of the United States Department of Agriculture developed a method to use CFC-11 and -12 as a propellant in **spray cans**. CFCs flowed out of a spray can's nozzle, carrying with them a mist containing other ingredients. Spray cans were used to propel hair sprays, paints, deodorants, disinfectants, polishes, and insecticides.

CFC-11 and 12 have also been used as blowing agents in foam production. Foam is used in insulation, disposable cups and cartons, and fire extinguishers. CFCs are released to the air during foam-production, itself. CFCs in the air spaces of foam are usually confined and not an important source of atmospheric CFCs.

Chlorofluorocarbons are a subset of **chlorocarbons**, which are compounds containing carbon and chlorine. **Hydrochlorofluorocarbons (HCFCs)** are another subset of chlorocarbons. HCFCs are similar to CFCs, except that HCFCs have at least one hydrogen atom. The hydrogen atom allows HCFCs to be broken down in the troposphere by reaction with OH, a chemical that does not readily break down CFCs. Because HCFCs break down more readily than do CFCs, a smaller percentage of emitted HCFCs than CFCs reaches the stratosphere. Nevertheless, because HCFCs contain chlorine and some HCFCs reach the stratosphere, HCFCs are still a danger to stratospheric ozone. HCFC-22, first produced in 1943, is the most abundant HCFC in the air today. HCFC-22 has been used as a refrigerant, spray-can propellant, and blowing agent in foam production.

Other chlorocarbons include **carbon tetrachloride** (CCl_4), **methyl chloroform** (CH_3CCl_3), and **methyl chloride** (CH3Cl). Carbon t etrachloride is used as an in termediate in the production of CFC s and HCFCs, and as a solvent and grain fumigant. Methyl chloroform is used as a degreasing agent, a dry-cleaning solvent, and an industrial solvent. Methyl chloride is produced synthetically only in small quantities and used in the production of silicones and tetramethyl lead intermediates (Singh 1995). Most methyl chloride in the air is produced biogenically in the oceans.

Another chlorine-containing gas in the troposphere is **hydrochloric acid** (HCl). HCl has larger natural than anthropogenic sources. Natural sources include evaporation of chloride from sea-spray and volcanic emission.

Although chlorine-containing compounds are more abundant than are bromine-containing compounds, the latter compounds are more efficient, molecule for molecule, at destroying ozone. The primary source of stratospheric bromine is **methyl bromide** (CH_3Br), which is produced biogenically in the oceans and emitted as a soil fumigant. Other sources of bromine are a group of synthetically produced compounds termed **Halons**, which are used in fire extinguishers and as fumigants. The most common Halons are **H-1301** (CF_3Br), **H-1211** (CF_2ClBr) and **H-2402** (CF_2BrCF_2Br). Methyl bromide and Halons are **bromocarbons** because they contain both bromine and carbon.

Compounds that contain hydrogen, fluorine, and carbon but not chlorine or bromine are **hydrofluorocarbons** (HFCs). HFCs were produced in abundance only recently as a replacement for CFCs and HCFCs. Because the fluorine in HFCs has

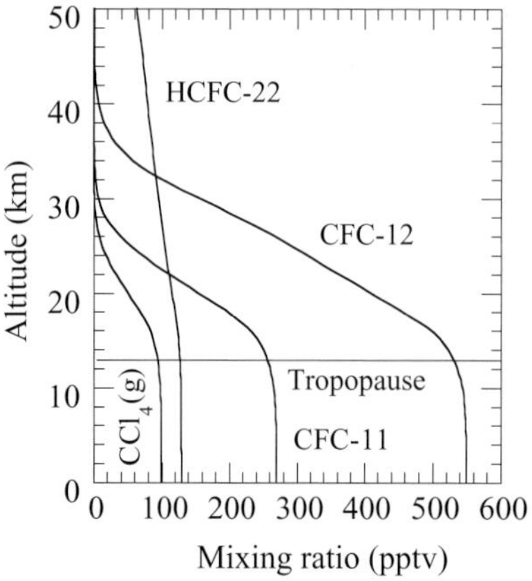

Figure 11.7 Variation of CFC-11, CFC-12, HCFC-22, and CCl_4(g) with altitude at 30° N latitude. Smoothed and scaled from Jackman *et al.* (1996) to present-day near-surface mixing ratios.

little effect on ozone, production of HFCs may increase in the future. Unfortunately, because they absorb thermal-infrared radiation, HFCs will enhance global warming if their use increases. The most abundantly emitted HFC to date has been **HFC-134a** (CH_2FCF_3(g)). Related to HFCs are **perfluorocarbons** (PFCs), such as perfluoroethane (C_2F_6), and **sulfur hexafluoride** (SF_6).

11.3.5.2 *Stratospheric breakdown of chlorinated compounds*

Once emitted, CFCs take about one year to mix up to the tropopause. Because they are chemically unreactive, and ultraviolet wavelengths that reach the troposphere are too weak to break them down, CFCs are not removed chemically from the troposphere. Instead, they become well mixed there. Today, the tropospheric mixing ratios of CFC-11 and CFC-12, the two most abundant CFCs, are about 270 and 550 pptv, respectively (Table 11.8 and Fig. 11.7). CFCs break down only when they reach the stratosphere.

Because the stratosphere is one large temperature inversion, vertical transport of ozone to the stratosphere is slow. About 10 Mt of chlorine in the form of CFCs reside in the troposphere, and the transfer rate of CFC-chlorine from the troposphere to the middle stratosphere is about 0.1 Mt per year. In this simplified scenario, the average time required for a CFC molecule to migrate from the troposphere to the middle stratosphere is about 100 years.

CFCs break down only when they reach the stratosphere, where they are exposed to far-UV radiation (wavelengths of 0.01 to 0.25 μm). This exposure occurs at an altitude of 12 to 20 km and higher. At such altitudes, far-UV wavelengths photolyze

CFC-11 and CFC-12 by

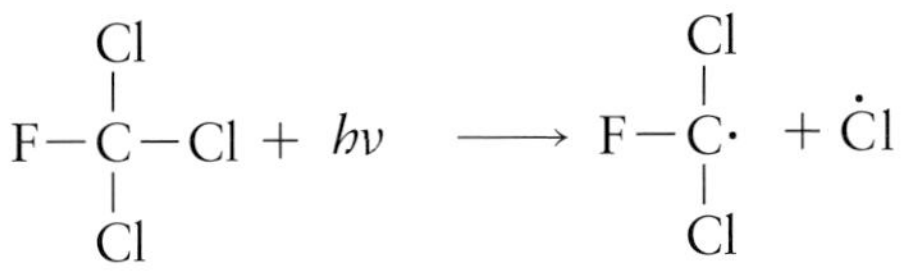

$$\text{F}-\overset{\displaystyle \text{Cl}}{\underset{\displaystyle \text{Cl}}{\text{C}}}-\text{Cl} + h\nu \longrightarrow \text{F}-\overset{\displaystyle \text{Cl}}{\underset{\displaystyle \text{Cl}}{\text{C}}}\cdot + \dot{\text{Cl}} \qquad \lambda < 250 \text{ nm} \qquad (11.126)$$

$$\text{F}-\overset{\displaystyle \text{Cl}}{\underset{\displaystyle \text{F}}{\text{C}}}-\text{Cl} + h\nu \longrightarrow \text{F}-\overset{\displaystyle \text{Cl}}{\underset{\displaystyle \text{F}}{\text{C}}}\cdot + \dot{\text{Cl}} \qquad \lambda < 226 \text{ nm} \qquad (11.127)$$

decreasing CFC mixing ratios in the stratosphere, as seen in Fig. 11.7. At 25 km, the e-folding lifetimes of CFC-11 and CFC-12 against photolysis under maximum-sunlight conditions are on the order of 23 and 251 days, respectively. Average lifetimes are on the order of two to three times these values. Thus, the time required to break down CFCs in the middle stratosphere (on the order of a year) is much less than the time required for CFCs to get there from the upper troposphere (on the order of decades). Table 11.8 indicates that the overall lifetimes of CFC-11 and CFC-12 between release at the surface and destruction in the middle stratosphere are about 45 and 100 years, respectively. The lifetime of CFC-12 is longer than that of CFC-11, partly because the former compound must climb to a higher altitude in the stratosphere before breaking apart than must the latter.

HCFC-22 has a lower emission rate and reacts faster with OH in the troposphere than do CFC-11 or CFC-12, so HCFC-22's mixing ratio and overall lifetime are lower than are those of CFC-11 or CFC-12. However, HCFC-22 photolyzes slower than do CFC-11 or CFC-12, so once HCFC-22 reaches the middle stratosphere, its mixing ratio does not decrease so rapidly with height as do mixing ratios of CFC-11 or CFC-12, as seen in Fig. 11.7.

In sum, the limiting factor in CFC decomposition in the stratosphere is not transport from the surface to the tropopause or photochemical breakdown in the stratosphere, but transport from the tropopause to the middle stratosphere. Because of their long overall lifetimes, some CFCs emitted in the 1930s through 1950s are still present in the stratosphere. Those emitted today are likely to remain in the air until the second half of the twenty-first century.

11.3.5.3 *Anthropogenic versus natural sources of chlorine*

Of the chlorine compounds reaching the stratosphere, >80 percent is anthropogenic, about 15 percent is natural methyl chloride (emitted biogenically from the oceans), and about 3 percent is natural hydrochloric acid (emitted volcanically and evaporated from sea spray) (WMO 1995).

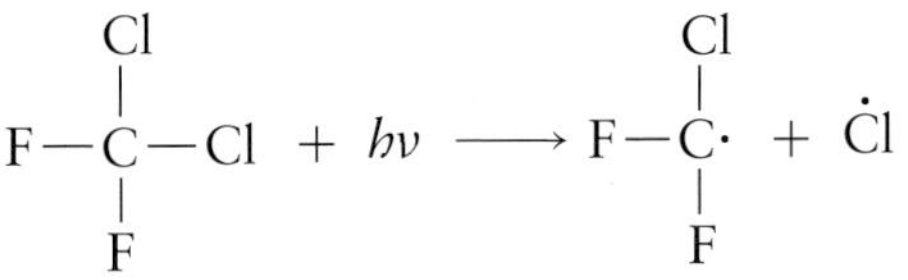

Although the oceans emit a tremendous amount of methyl chloride, the e-folding lifetime against its loss by

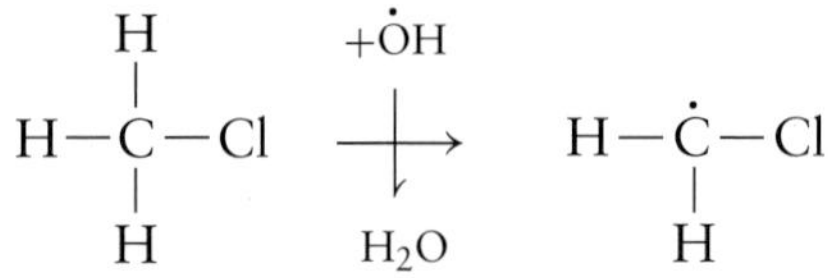

$$\begin{array}{c} \text{H} \\ | \\ \text{H} - \overset{|}{\underset{|}{\text{C}}} - \text{Cl} \\ | \\ \text{H} \end{array} \quad \xrightarrow[\text{H}_2\text{O}]{+\dot{\text{O}}\text{H}} \quad \begin{array}{c} \dot{} \\ \text{H} - \overset{|}{\underset{|}{\text{C}}} - \text{Cl} \\ | \\ \text{H} \end{array} \qquad (11.128)$$

is only 1.5 years, so a relatively small portion of it reaches the stratosphere. In the stratosphere, methyl chloride photolyzes by

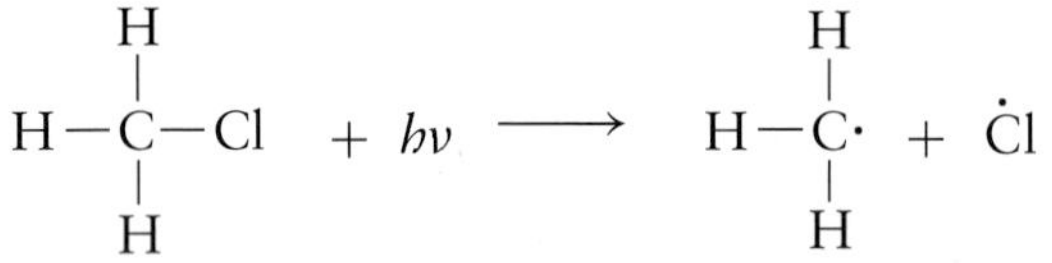

$$\begin{array}{c} \text{H} \\ | \\ \text{H} - \overset{|}{\underset{|}{\text{C}}} - \text{Cl} \\ | \\ \text{H} \end{array} + h\nu \longrightarrow \begin{array}{c} \text{H} \\ | \\ \text{H} - \overset{|}{\underset{|}{\text{C}}} \cdot \\ | \\ \text{H} \end{array} + \dot{\text{C}}\text{l} \qquad \lambda < 220 \text{ nm} \qquad (11.129)$$

with an e-folding lifetime at 25 km of 2.4 yr.

Volcanos emit lots of HCl, but most of it is either removed by clouds and rain or lost chemically before reaching the stratosphere (Lazrus *et al.* 1979; Pinto *et al.* 1989; Tabazadeh and Turco 1993a). HCl is extremely soluble in water, so any contact with clouds or precipitation removes it from the gas phase immediately. HCl also has a relatively short (15–30 day) e-folding lifetime against reaction with OH, so HCl that is not rained out is generally destroyed chemically. Because of the short lifetime of HCl (<0.1 yr), most HCl in the stratosphere is not transported there but produced chemically there by CFC by-products.

In sum, the main natural source of chlorine into the stratosphere is methyl chloride, but the quantity of its emission into the stratosphere is much lower than that of anthropogenic chlorine compounds.

11.3.5.4 *Catalytic destruction of ozone by chlorine*

Once released from CFC and non-CFC parent compounds in the stratosphere by photolysis, chlorine atoms may react in a **chlorine catalytic ozone destruction cycle,**

$$\dot{\text{C}}\text{l} + \text{O}_3 \longrightarrow \text{Cl}\dot{\text{O}} + \text{O}_2 \qquad (11.130)$$
$$\text{Cl}\dot{\text{O}} + \dot{\text{O}}\cdot \longrightarrow \dot{\text{C}}\text{l} + \text{O}_2 \qquad (11.131)$$

$$\dot{\text{O}}\cdot + \text{O}_3 \longrightarrow 2\text{O}_2 \quad \text{(net process)} \qquad (11.132)$$

At midlatitudes, the chain length of this cycle increases from about 10 in the lower stratosphere to about 1000 in the middle and upper stratosphere (Lary 1997). The Cl + ClO, where ClO is **chlorine monoxide**, that takes part in catalytic ozone destruction is called **active chlorine.**

The primary removal mechanisms of active chlorine from the catalytic cycle are chemical reactions that produce **chlorine reservoirs,** which are gases that store active chlorine and prevent it from taking part in catalytic cycles. The major chlorine reservoirs in the stratosphere are **hydrochloric acid** (HCl) and **chlorine nitrate** (ClONO$_2$).

The HCl reservoir forms by

$$\dot{Cl} + \begin{cases} CH_4 \longrightarrow HCl + \dot{C}H_3 \\ H\dot{O}_2 \longrightarrow HCl + O_2 \\ H_2 \longrightarrow HCl + \dot{H} \\ H_2O_2 \longrightarrow HCl + H\dot{O}_2 \end{cases} \tag{11.133}$$

and the ClONO$_2$ reservoir forms by

$$Cl-\dot{O} + \dot{N}O_2 \xrightarrow{\text{M}} \tag{11.134}$$

Chlorine	Chlorine
monoxide	nitrate

At any time, about 1 percent of the chlorine in the stratosphere is in the form of active chlorine. Most of the rest is in the form of a chlorine reservoir.

The HCl chlorine reservoir leaks slowly. HCl reproduces active chlorine by photolysis, reaction with OH, and reaction with O. These reactions are

$$HCl + \begin{cases} h\nu \longrightarrow \dot{H} + \dot{Cl} & \lambda < 220 \text{ nm} \\ \dot{O}H \longrightarrow \dot{Cl} + H_2O \\ \dot{O}\cdot \longrightarrow \dot{Cl} + \dot{O}H \end{cases} \tag{11.135}$$

The e-folding lifetime of HCl against photolysis is about 1.5 yr at 25 km. HCl also diffuses back to the troposphere, where it can be absorbed by clouds.

The ClONO$_2$ reservoir leaks primarily by the photolysis reactions

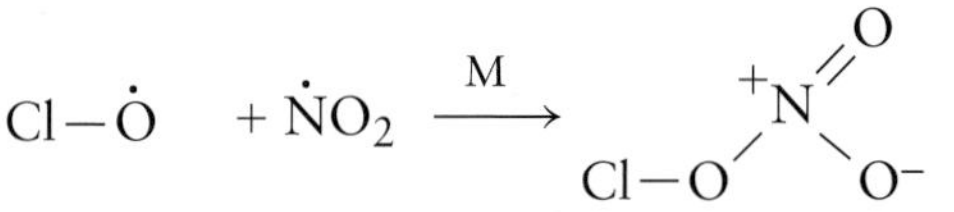

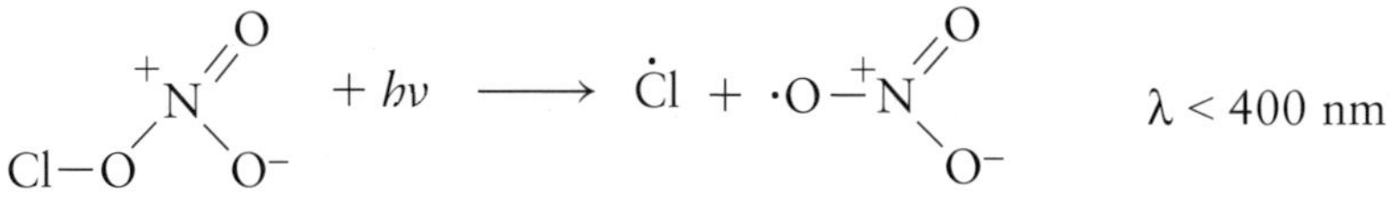

$$\cdots + h\nu \longrightarrow \dot{Cl} + \cdot O - \overset{+}{N} \qquad \lambda < 400 \text{ nm}$$

Chlorine	Nitrate radical
nitrate	

$$\tag{11.136}$$

which occurs with an e-folding lifetime of about 4.5 h at 25 km.

11.3.5.5 *Catalytic destruction of ozone by bromine*

Like chlorine, bromine affects stratospheric ozone. The primary sources of stratospheric bromine are **methyl bromide** (CH$_3$Br) and Halons. The tropospheric mixing ratios of the most common Halons, CF$_2$ClBr (H-1211) and CF$_3$Br (H-1301), are both about 2 pptv, less than 1 percent of the mixing ratios of CFC-11 and -12.

Nevertheless, the efficiency of ozone destruction by the bromine catalytic cycle is greater than is that by the chlorine catalytic cycle.

Methyl bromide and Halons photolyze in the stratosphere to produce atomic bromine. Photolysis of methyl bromide occurs above 20 km by

$$
\begin{array}{ccc}
\text{H} & & \text{H} \\
| & & | \\
\text{H}-\text{C}-\text{Br} + h\nu & \longrightarrow & \text{H}-\overset{\cdot}{\text{C}}\cdot + \overset{\cdot}{\text{Br}} \qquad \lambda < 260 \text{ nm} \\
| & & | \\
\text{H} & & \text{H}
\end{array}
\qquad (11.137)
$$

The e-folding lifetime of CH_3Br against loss by this reaction is about 10 days at 25 km.

Once released, atomic bromine takes part in the **bromine catalytic ozone destruction cycle**,

$$
\begin{array}{ll}
\overset{\cdot}{\text{Br}} + O_3 \longrightarrow \text{Br}\overset{\cdot}{\text{O}} + O_2 & (11.138) \\
\text{Br}\overset{\cdot}{\text{O}} + \overset{\cdot}{\text{O}}\cdot \longrightarrow \overset{\cdot}{\text{Br}} + O_2 & (11.139) \\
\hline
\overset{\cdot}{\text{O}}\cdot + O_3 \longrightarrow 2O_2 \quad \text{(net process)} & (11.140)
\end{array}
$$

where **bromine monoxide** (BrO) is the partner of Br in this cycle. The chain length of this cycle increases from about 100 at 20 km to about 10^4 at 40–50 km (Lary 1997). The chain length of the bromine catalytic cycle is longer than that of the chlorine catalytic cycle because Br is chemically removed more slowly from the bromine cycle than Cl is removed from the chlorine cycle. In addition, the bromine reservoirs break down more quickly than do the chlorine reservoirs.

Atomic bromine is removed from its catalytic cycle primarily by

$$
\overset{\cdot}{\text{Br}} + \begin{cases} \text{H}\overset{\cdot}{\text{O}_2} \rightarrow \text{HBr} + O_2 \\ H_2O_2 \rightarrow \text{HBr} + \text{H}\overset{\cdot}{\text{O}_2} \end{cases} \qquad (11.141)
$$

where HBr is **hydrobromic acid**. When BrO is removed, it forms **bromine nitrate** (BrONO$_2$) by

$$
\text{Br}-\overset{\cdot}{\text{O}} + \overset{\cdot}{\text{N}}O_2 \xrightarrow{\text{M}} \quad \begin{array}{c} {}^{+}\text{N}\!\!\nearrow^{\text{O}} \\ \text{Br}-\text{O}\diagup \quad \diagdown\text{O}^{-} \end{array} \qquad (11.142)
$$

Bromine   Bromine
monoxide   nitrate

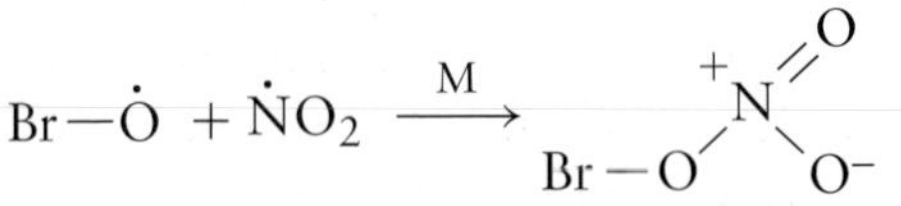

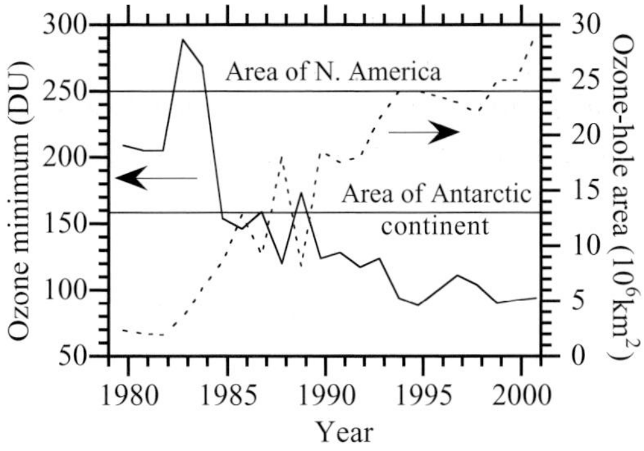

Figure 11.8 Minimum ozone column abundances and areal extent of the ozone hole over the Antarctic region from 1979 to 2000. Data from NASA Goddard Space Flight Center. For comparison, the area of the Antarctic is about 13×10^6 km^2 and the area of North America is about 24×10^6 km^2.

The HBr and $BrONO_2$ reservoirs leak through the respective reactions

$$HBr + \dot{O}H \rightarrow \dot{B}r + H_2O \qquad (11.143)$$

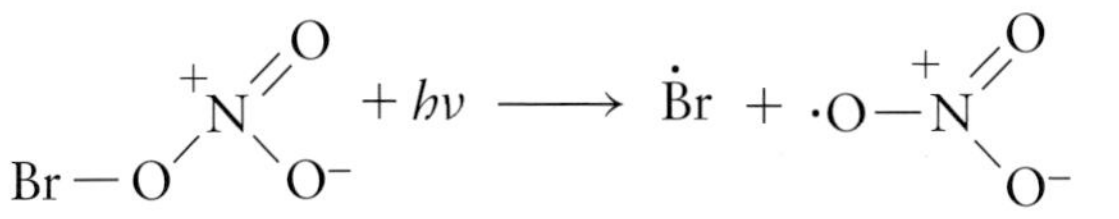

$\lambda < 390$ nm (11.144)

Bromine nitrate   Nitrate radical

The e-folding lifetime of $BrONO_2$ against the photolysis reaction is about 10 min at 25 km.

11.3.6 Antarctic ozone hole

Every September through November (Southern Hemisphere spring) since 1980, the minimum ozone column abundance over the Antarctic has decreased below its yearly average. Figure 11.8 shows that, in 2000, the lowest measured column abundance over the Antarctic was about 94 DU (occurring on September 29, 2000), which was 68 percent less than 293.4 DU, the globally and year-2000 averaged ozone column abundance. Between 1981 and 2000, the area over which ozone depletes (the **Antarctic ozone hole**), increased. The ozone hole area is defined as the area of the globe over which the ozone column abundance decreases below 220 DUs. The ozone hole in 2000 covered nearly 30×10^6 km^2, an area larger than the size of North America. Antarctic ozone depletion occurs between 12 and 24 km in altitude.

During the Northern Hemisphere late winter and spring (March–May), an **Arctic ozone dent** (reduction in ozone to 240 to 260 DU), smaller in magnitude than the Antarctic ozone hole, appears. The hole and dent are caused by a set of interlinked factors. One factor linking global ozone reductions to polar ozone depletion is the presence of chlorine and bromine in the stratosphere.

11.3.6.1 *Polar stratospheric cloud formation*

The ozone hole over the Antarctic appears in part because the Antarctic winter (June–September) is very cold. Temperatures are low because much of the polar region is exposed to 24 hours of darkness each day during the winter, and a wind system, the **polar vortex**, circles the Antarctic. The vortex is a polar-front jet-stream wind system that flows around the Antarctic continent, trapping cold air within the polar region and preventing an influx of warm air from outside this region.

Because temperatures are low in the Antarctic stratosphere, optically thin clouds, called **polar stratospheric clouds (PSCs)** form. These clouds have few particles per unit volume of air in comparison with tropospheric clouds. Two major types of clouds form. When temperatures drop to below about 195 K, nitric acid and water vapor grow on small sulfuric acid–water aerosol particles (Toon *et al.* 1986). Initially, it was thought that nitric acid and water molecules deposited to the ice phase in the ratio 1:3. Such ice crystals have the composition $HNO_3 \cdot 3H_2O$ and are called **nitric acid trihydrate** (NAT) crystals. More recently, it was found that these particles contain a variety of phases. Some contain **nitric acid dihydrate** (NAD) (Worsnop *et al.* 1993), and others contain supercooled liquid water (liquid water present at temperatures below the freezing point of water), sulfuric acid, and nitric acid (Tabazadeh *et al.* 1994). Together, nitrate containing cloud particles that form at temperatures below about 195 K in the winter polar stratosphere are called **Type I polar stratospheric clouds.**

When temperatures drop below the frost point of water, which is about 187 K under typical polar stratospheric conditions, a second type of cloud forms. These clouds contain pure water ice and are **Type II polar stratospheric clouds**. Usually, about 90 percent of PSCs are Type I and 10 percent are Type II (Turco *et al.* 1989). Typical diameters and number concentrations of a Type I PSC are 1 μm and ≤ 1 particle cm^{-3}, respectively, although diameters vary from 0.01 to 3 μm. Typical diameters and number concentrations of a Type II PSC are 20 μm and ≤ 0.1 particle cm^{-3}, respectively, although diameters vary from 1 to 100 μm.

11.3.6.2 *PSC surface reactions*

Once PSC particles form, chemical reactions take place on their surfaces. Reactions involving a gas reacting on a particle surface are **heterogeneous reactions**. In the present case, the surfaces are those of frozen particles, and the reactions occur after a gas has diffused to and adsorbed to the particle surface. **Adsorption** is the process by which a gas collides with and bonds to a liquid or solid surface. Adsorption differs from **absorption**, which is a process by which a gas penetrates into the inner structure of a liquid or solid. One type of adsorption bonding is **ion–dipole** bonding, which occurs when the surface is charged and the gas is polar. A second type is **dipole–dipole** bonding, which occurs when the surface and gas are polar. Once a gas molecule adsorbs to a surface, it can **desorb** (break away) from the surface. If it stays adsorbed to the surface, it can **diffuse** to another site on the surface. During the diffusion process, the adsorbed gas may collide and chemically react with another adsorbed molecule. A gas molecule suspended above a particle surface may also collide and react with the adsorbed molecule. In both cases, an adsorbed product is formed. This product can either desorb from the surface, diffuse on the surface, or participate in additional chemical reactions.

The primary heterogeneous reactions that occur on Types I and II PSC surfaces are

$$\mathrm{ClONO_2(g) + H_2O(a) \longrightarrow HOCl(g) + HNO_3(a)} \tag{11.145}$$

$$\boxed{\mathrm{ClONO_2(g) + HCl(a) \longrightarrow Cl_2(g) + HNO_3(a)}} \tag{11.146}$$

$$\mathrm{N_2O_5(g) + H_2O(a) \longrightarrow 2HNO_3(a)} \tag{11.147}$$

$$\mathrm{N_2O_5(g) + HCl(a) \longrightarrow ClNO_2(g) + HNO_3(a)} \tag{11.148}$$

$$\mathrm{HOCl(g) + HCl(a) \longrightarrow Cl_2(g) + H_2O(a)} \tag{11.149}$$

In these reactions, (g) denotes a gas and (a) denotes an adsorbed species. Additional reactions exist for bromine. Laboratory studies show that HCl readily coats the surfaces of Types I and II PSCs. When $ClONO_2$, N_2O_5, or HOCl impinges upon the surface of a Type I or II PSC, it can react with H_2O or HCl already on the surface. The products of these reactions are adsorbed species, some of which stay adsorbed, whereas others desorb to the vapor phase.

In sum, heterogeneous reactions convert relatively inactive forms of chlorine in the stratosphere, such as HCl and $ClONO_2$, to photochemically active forms, such as Cl_2, HOCl, and $ClNO_2$. This conversion process is **chlorine activation.** The most important heterogeneous reaction is (11.146) (Solomon *et al.* 1986; McElroy *et al.* 1986), which generates gas-phase molecular chlorine, Cl_2. Reaction (11.147) does not activate chlorine. Its only effect is to remove nitric acid from the gas phase. When nitric acid adsorbs to a Type II PSC, which is larger than a Type I PSC, the nitric acid can sediment out along with the PSC to lower regions of the stratosphere. This removal process is **stratospheric denitrification.** Denitrification is important because it removes nitrogen that might otherwise reform Type I PSCs or tie up active chlorine as $ClONO_2$.

11.3.6.3 *Reaction probabilities*

Rate coefficients for (11.145)–(11.149) have the form

$$k_{s,q} = \frac{1}{4}\bar{v}_q \gamma_q a \tag{11.150}$$

Table 11.9 Estimated reaction probabilities for the gases in the reactions (11.145)–(11.149) on Types I and II PSC surfaces

	Reaction probability	
Reaction	**Type I PSCs**	**Type II PSCs**
$ClONO_2(g) + H_2O(a)$	0.001	0.3
$ClONO_2(g) + HCl(a)$	0.1	0.3
$N_2O_5(g) + H_2O(a)$	0.0003	0.01
$N_2O_5(g) + HCl(a)$	0.003	0.03
$HOCl(g) + HCl(a)$	0.1	0.3

Source: DeMore *et al.* (1997) and references therein.

where $\bar{v}_q$ is the **thermal speed** of the impinging gas (cm s^{-1}), γ_q is the **reaction probability** (also called the **uptake coefficient**) of the gas (dimensionless), and a is the **surface-area concentration** (square centimeters of surface per cubic centimeter of air) of all particles on which reactions occur. Equation (11.150) implicitly includes the concentration of the adsorbed reactant but not of the gas reactant; thus, it is a pseudo-first-order rate coefficient (s^{-1}). When multiplied by the gas-phase reactant concentration, (11.150) gives a rate (molec. cm^{-3} s^{-1}).

The average thermal speed of a gas from (2.3) is

$$\bar{v}_q = \sqrt{\frac{8k_B T}{\pi \bar{M}_q}} \tag{11.151}$$

where $\bar{M}_q = m_q/A$ is the mass (g) of one gas molecule. Example 2.1 gives the average thermal speed of air molecules from this equation.

A **reaction probability** is the laboratory-measured fractional loss of a species from the gas phase due to reaction with a particle surface, and it takes into account diffusion of the gas to the surface as well as reaction with the surface. Estimated reaction probabilities for the gases in (11.145)–(11.149) on Types I and II PSCs are given in Table 11.9.

Reaction probabilities can be derived by considering adsorption, desorption, and reaction on a particle surface (e.g., Tabazadeh and Turco 1993b; Adamson 1990). The rate of change of species concentration in the first molecular layer over a surface is

$$\frac{dn_{s,q}}{dt} = k_{a,q} p_q (n_m - n_{s,q}) - k_{d,q} n_{s,q} - k_{s,q} n_{s,q} \tag{11.152}$$

where $n_{s,q}$ is the surface-area concentration of adsorbed gas q (molec. cm^{-2}), $k_{a,q}$ is the rate at which gas molecules adsorb to the surface (molec. hPa^{-1} s^{-1}), p_q is the partial pressure of the gas over the surface (hPa), $n_m \approx 10^{15}$ sites cm^{-2} is approximately the number concentration of adsorption sites on the surface, $k_{d,q}$

is the rate at which adsorbed molecules desorb from the surface (s^{-1}), and $k_{s,q}$ is the first-order rate coefficient for the loss of adsorbed species q due to chemical reaction with another adsorbed species or with a gas (s^{-1}). Equation (11.152) states that the change in concentration of molecules adsorbed to the surface of a particle equals an adsorption flux minus a desorption flux minus a reaction flux.

The **rate of molecular adsorption** is

$$k_{a,q} = \frac{\alpha_q \sigma_0}{2\pi \bar{M}_q k_B T} \exp\left(-\frac{E_{a,q} + \eta\theta_{f,q}}{R^* T}\right) \tag{11.153}$$

where α_q is the **mass accommodation coefficient** of the adsorbing gas, $\sigma_0 \approx 10^{-15}$ cm^2 is the molecular surface area of one site, $\bar{M}_q$ is the mass (g) of one adsorbed molecule, $E_{a,q}$ is the activation energy of adsorption (J mol^{-1}), η is a factor that allows $E_{a,q}$ to change with surface coverage, and $\theta_{f,q}$ is fractional surface coverage. The mass accommodation coefficient (sticking coefficient) is the fractional number of collisions of gas q with a particle that results in the gas sticking to the particle's surface.

The **rate of molecular desorption** is

$$k_{d,q} = \nu_0 \exp\left(-\frac{\Delta G_{d,q} + E_{d,q}}{R^* T}\right) \tag{11.154}$$

where $\nu_0 \approx 10^{13}$ s^{-1} is the frequency of atomic vibrations on the surface, $\Delta G_{d,q}$ is the change in Gibbs free energy of desorption (J mol^{-1}), and $E_{d,q}$ is the activation energy of desorption (J mol^{-1}). The **free energy of desorption** is related to the free energy of adsorption and entropy of adsorption by

$$-\Delta G_{d,q} = \Delta G_{a,q} = \Delta H_{a,q}(\theta_{f,q}) - T\Delta S_{a,q}(\theta_{f,q}) \tag{11.155}$$

where $\Delta H_{a,q}(\theta_{f,q})$ is the **enthalpy of adsorption** (J mol^{-1}), and $\Delta S_{a,q}(\theta_{f,q})$ is the **entropy of adsorption** (J mol^{-1} K^{-1}), both of which depend on the fractional surface coverage. If the adsorbed species diffuses on the surface, the entropy of adsorption is

$$\Delta S_{a,q}(\theta_{f,q}) = -R^* \ln\theta_{f,q} + R^* \ln(\bar{M}_q T \sigma_0) - R^* \ln\left(T^{5/2} \bar{M}_q^{3/2}\right) + 96.65 \tag{11.156}$$

Surface coverage is the fraction of available surface sites filled with adsorbed molecules. The steady-state surface coverage of an adsorbed substance is obtained by setting (11.152) to zero. The result is

$$\boxed{\theta_{f,q} = \frac{n_{s,q}}{n_m} = \frac{k_{a,q} p_q}{k_{d,q} + k_{s,q} + k_{a,q} p_q}} \tag{11.157}$$

The adsorption minus desorption flux in (11.152) can be written in terms of a net diffusive flux from the gas phase and a reaction probability:

$$\frac{1}{4}\bar{v}_q \gamma_q N_q = k_{a,q} p_q (n_m - n_{s,q}) - k_{d,q} n_{s,q} \tag{11.158}$$

where $N_q = p_q/k_B T$ is the number concentration of gas molecules suspended over the particle surface (molec. cm^{-3}). Substituting (11.158) into (11.152) gives

$$\frac{dn_{s,q}}{dt} = \frac{1}{4}\bar{v}_q \gamma_q N_q - k_{s,q} n_{s,q} \tag{11.159}$$

Setting (11.159) equal to zero and solving for the steady-state reaction probability gives

$$\gamma_q = \frac{4k_{s,q} n_{s,q}}{\bar{v}_q N_q} = \frac{4k_{s,q}\theta_{f,q} n_m}{\bar{v}_q N_q} \tag{11.160}$$

Parameters required for $\theta_{f,q}$, such as $E_{a,q}$, $E_{d,q}$, η, and $k_{s,q}$, are difficult to obtain. Thus, laboratory-derived reaction probabilities are generally used and substituted into (11.150) to approximate $k_{s,q}$.

11.3.6.4 *Springtime polar chemistry*

Chlorine activation occurs during the winter over the polar stratosphere. When the Sun appears over the horizon in the early spring, Cl-containing gases created by PSC reactions photolyze by

$$Cl_2 + h\nu \rightarrow 2\dot{Cl} \qquad \lambda < 450 \text{ nm} \tag{11.161}$$

$$HOCl + h\nu \longrightarrow \dot{Cl} + \dot{O}H \qquad \lambda < 375 \text{ nm} \tag{11.162}$$

$$ClNO_2 + h\nu \longrightarrow \dot{Cl} + \dot{N}O_2 \qquad \lambda < 372 \text{ nm} \tag{11.163}$$

Once Cl has been released, it attacks ozone. The catalytic cycles that destroy ozone in the springtime polar stratosphere differ from that shown in (11.130)–(11.132), which reduces ozone on a global scale. One polar stratosphere catalytic ozone destruction cycle is

$$2 \times (\dot{Cl} + O_3 \longrightarrow Cl\dot{O} + O_2) \tag{11.164}$$

$$Cl\dot{O} + Cl\dot{O} \xrightarrow{M} Cl_2O_2 \tag{11.165}$$

$$Cl_2O_2 + h\nu \longrightarrow ClO\dot{O} + \dot{Cl} \qquad \lambda < 360 \text{ nm} \tag{11.166}$$

$$ClO\dot{O} \xrightarrow{M} \dot{Cl} + O_2 \tag{11.167}$$

$$2O_3 \longrightarrow 3O_2 \tag{11.168}$$

where Cl_2O_2 is **dichlorine dioxide** and ClOO is a **chlorine peroxy radical.** This mechanism, called the **dimer mechanism** (Molina and Molina 1986), is important

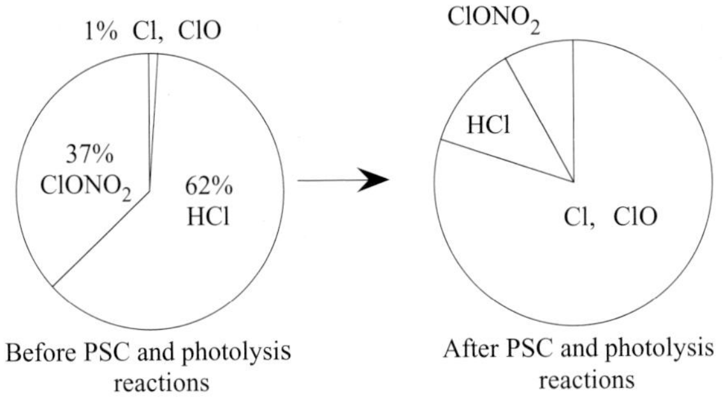

Figure 11.9 Pie chart showing conversion of chlorine reservoirs to active chlorine. During chlorine activation on PSCs, HCl and $ClONO_2$ are converted to potentially active forms of chlorine that are broken down by sunlight in springtime to form Cl. Cl forms ClO, both of which react catalytically to destroy ozone.

in the springtime polar stratosphere because, at that location and time, the ClO required for (11.165) is concentrated enough for the reaction to proceed rapidly. A second cycle is

$$\dot{Cl} + O_3 \longrightarrow Cl\dot{O} + O_2 \qquad (11.169)$$

$$\dot{Br} + O_3 \longrightarrow Br\dot{O} + O_2 \qquad (11.170)$$

$$Br\dot{O} + Cl\dot{O} \longrightarrow \dot{Br} + \dot{Cl} + O_2 \qquad (11.171)$$

$$2O_3 \longrightarrow 3O_2 \qquad (11.172)$$

(McElroy *et al.* 1986), which is important in the polar lower stratosphere. In sum, chlorine activation and springtime photochemical reactions convert chlorine from reservoir forms, such as HCl and $ClONO_2$, to active forms, such as Cl and ClO, as shown in Fig. 11.9. The active forms of chlorine destroy ozone in catalytic cycles.

Every November, the Antarctic warms up sufficiently for the polar vortex to break down and PSCs to melt, evaporate, and sublimate. Ozone from outside the polar region advects into the region. Ozone also regenerates chemically, and chlorine reservoirs of $ClONO_2$ and HCl reestablish themselves. Thus, the Antarctic ozone hole is an annual, regional phenomenon that is controlled primarily by the temperature of the polar stratosphere and the presence of chlorine and bromine. The radial extent of the hole has grown and minimum ozone values have decreased steadily over the past decades. The ozone dent over the Arctic is not nearly so large or regular as is that over the Antarctic because the vortex around the Arctic is much weaker and temperatures over the Arctic do not drop so low as they do over the Antarctic. Thus, PSC formation and subsequent chemical reaction are less widespread over the Arctic than over the Antarctic.

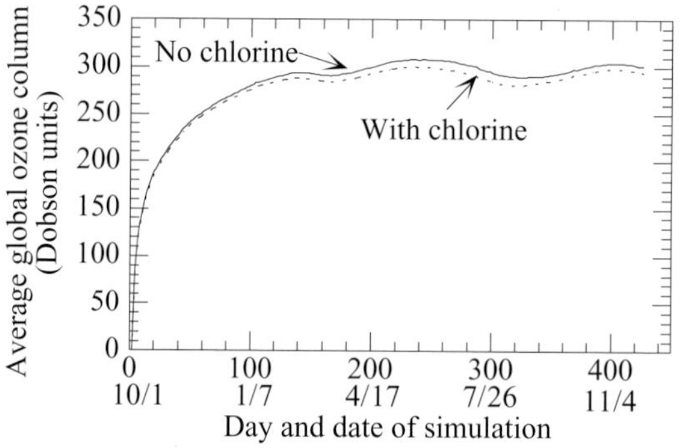

Figure 11.10 Change in ozone column abundance, averaged over the globe, during two global model simulations in which chlorine was present and absent. In both cases, ozone was initially removed from the model atmosphere on October 1, 1988. Bromine was not included in either simulation. In the chlorine case, the column abundance regenerated to a maximum of 300.4 DU, which compares with an observed maximum in 1989 of about 300 DU. In the no-chlorine case, the ozone layer regenerated to a maximum of 308.5 DU, which compares with an observed average abundance between 1964 and 1980 of 306.4 DU (Bojkov and Fioletov 1995).

11.3.7 Recovery of stratospheric ozone

In September 1987, an international agreement, the **Montreal Protocol**, was signed limiting the production of CFCs and Halons. The Montreal Protocol has since been modified several times to accelerate the phase-out of CFCs. The major effect of the Montreal Protocol and later amendments has been nearly to eliminate the emission of CFCs. The reduction in CFC emission will, in the long term, allow the stratospheric ozone layer to recover, but because of the long lifetimes of CFCs the recovery may take decades.

However, once CFCs and the active chlorine they produce are removed from the stratosphere, the ozone layer should recover rapidly. Suppose that all ozone in the stratosphere were destroyed, all ozone-destroying compounds were removed, but all oxygen remained. How long would the ozone layer take to regenerate? An estimate can be obtained from Fig. 11.10, which shows results from two global-model simulations of the atmosphere in which all ozone was initially removed, but oxygen was not. In the first simulation, ozone regeneration was simulated in the absence of chlorine and bromine. In the second, ozone regeneration was simulated in the presence of 1989 concentrations of chlorine, but in the absence of bromine. In both simulations, the globally averaged column abundance of ozone regenerated to relatively steady values in less than a year. Regeneration during the simulation in which chlorine was initially present was about 2 to 3 percent less than that during

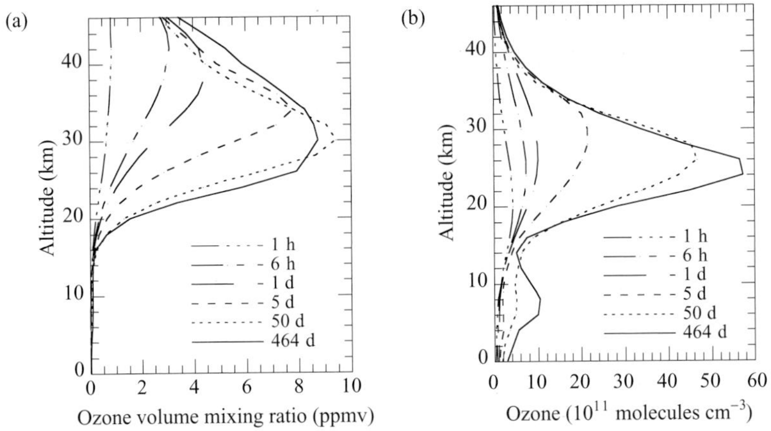

Figure 11.11 Time evolution of the vertical profile of ozone (a) volume mixing ratio and (b) number concentration at 34° N latitude, starting with zero ozone on June 21, 1988, as predicted by a one-dimensional photochemical–radiative model. Mixing ratios regenerated starting from the top of the stratosphere. Number concentrations regenerated starting from the bottom.

the no-chlorine case, consistent with the estimated global reduction in ozone of 2 to 3 percent between the 1970s and 1989 due to chlorine-containing compounds.

Figures 11.11 (a) and (b) show the time-evolution of the vertical mixing ratio and number concentration, respectively, of ozone from June 21, 1988 through September 26, 1989 (464 days later) in a one-dimensional column calculation in which all ozone was removed initially from the atmosphere. The figure shows that, within a few hours after the start, ozone mixing ratio regenerated starting from the top of the stratosphere, and number concentrations regenerated starting from the bottom. Number concentration is an absolute quantity, whereas mixing ratio is a relative quantity. In sum, not only does the total column abundance of ozone regenerate quickly after removal of chlorine and bromine, but so does the vertical profile of ozone.

11.4 SUMMARY

In this chapter, chemistry of the free troposphere, urban areas, the global stratosphere, and the polar stratosphere were discussed. Important chemical equations and reaction pathways were given. The focus of many atmospheric pollution studies is on ozone formation and destruction. In urban regions, the problem is ozone formation. In the stratosphere, the problem is ozone destruction. The chemical pathways that form and destroy ozone in each region are different. In the free troposphere, ozone is governed by NO reaction with O_3 and by NO_2 photolysis. In urban regions, ozone is governed by NO reaction with ROGs and O_3 and NO_2 photolysis. Ozone production in the stratosphere is governed by O_2 photolysis. Appendix Table B.4 summarizes most of the reactions described in this chapter. In the next chapter, methods of solving chemical ordinary differential equations are described.

Table 11.10 Mixing-ratio data from the Los Angeles basin for Problem 11.1

Station	χ_{NO} **(ppmv)**	χ_{NO_2} **(ppmv)**	$\frac{\chi_{NO_2}}{\chi_{NO}}$	$\chi_{O_3,m}$ **(ppmv) modeled**	$\chi_{O_3,p}$ **(ppmv) photostationary**	$\frac{\chi_{O_3,m}-\chi_{O_3,p}}{\chi_{O_3,p}} \times 100\%$
8:30 a.m.						
Anaheim	0.1757	0.0760		0.0058		
Costa Mesa	0.0607	0.0446		0.0119		
Hawthorne	0.0654	0.0525		0.0104		
La Habra	0.0764	0.0501		0.0092		
Lynwood	0.1209	0.0625		0.0073		
Pasadena	0.1179	0.0740		0.0089		
Reseda	0.0704	0.0645		0.0125		
Simi Valley	0.0232	0.0355		0.0204		
1:30 p.m.						
Anaheim	0.0088	0.0374		0.0859		
Costa Mesa	0.0047	0.0131		0.0617		
Hawthorne	0.0169	0.0394		0.0471		
La Habra	0.0108	0.0461		0.0879		
Lynwood	0.0096	0.0372		0.0778		
Pasadena	0.0321	0.0693		0.0491		
Reseda	0.0091	0.0495		0.1216		
Simi Valley	0.0031	0.0148		0.0990		

11.5 PROBLEMS

11.1 **(a)** Fill in Table 11.10, assuming that the photolysis rates of NO_2 at 8:30 a.m. and 1:30 p.m. were $J = 0.008\ s^{-1}$ and $J = 0.01\ s^{-1}$, respectively, and $T = 291$ K at 8:30 a.m. and $T = 298$ K at 1:30 p.m. Assume $p_d = 1005$ hPa in both cases.

(b) Modeled ozone mixing ratios in Table 11.10 were calculated with a chemical model that included all the urban reactions in Appendix Table B.4, including organic reactions. Explain the difference between modeled and photostationary-state mixing ratios, if any, between the morning and the afternoon. Why do morning and afternoon values differ less at Costa Mesa than at other locations?

11.2 **(a)** Calculate the thermal speed of chlorine nitrate when $T = 192$ K. At this temperature, what type of PSC might be present?

(b) Calculate the pseudo-first-order rate coefficient of chlorine nitrate on the surfaces of these cloud particles, assuming the surface is coated with HCl. (Hint: Use a typical number concentration and diameter, given in the text, of the chosen PSC type.)

(c) If $\chi_{ClONO_2} = 1.0$ ppbv initially, and only surface chemistry is considered, how much chlorine nitrate will remain after 15 days due to the reaction from Problem 11.2(b). What is the resulting mixing ratio of Cl_2 if none existed initially?

11.3 Write the rate expressions for the reactions (11.164)–(11.168), and write the first-derivative expressions for each species (except for O_2 and M) in the reactions (ignore the factor of two in the first reaction). Assume M is already included in the rate coefficients. What is the steady-state expression for the Cl concentration from the rate expression for Cl? If $T = 197$ K and $p_d = 25$ hPa, calculate the rate coefficient of each of the kinetic reactions.

11.4 Write out all the reactions in Appendix Table B.4 that destroy molecular hydrogen (H_2). Calculate the e-folding lifetime against loss in all cases assuming $T = 290$ K, [OH] $= 5.5 \times 10^5$ molec. cm^{-3}, [O(^1D)] $= 1.0 \times 10^{-3}$ molec. cm^{-3}, and [Cl] $= 1.0 \times 10^3$ molec. cm^{-3}. Based on the result, which of these is the most important reaction destroying molecular hydrogen in the atmosphere?

11.6 COMPUTER PROGRAMMING PRACTICE

11.5 **(a)** Input all the inorganic kinetic reactions and either the (i) chlorine, (ii) bromine, (iii) sulfur, or (iv) isoprene kinetic reactions from Appendix Table B.4 into the computer file developed for Problem 10.7. If the isoprene mechanism is chosen, many nonaromatic organic reactions should also be included to account for the chemistry of isoprene by-products. Calculate rate coefficients for each reaction. For the chlorine and bromine cases, assume $T = 192$ K and $p_d = 25$ hPa. For the isoprene and sulfur cases, assume $T = 298.15$ K and $p_d = 1013$ hPa.

(b) Input the necessary photolysis reactions from Appendix Table B.4 to complete the mechanism chosen in part (a). If a stratospheric mechanism was chosen, input the stratospheric (25 km) peak photolysis rate coefficients from the table into the file. Otherwise, input the surface (0 km) peak rate coefficients. Write a script to scale the peak rate coefficients to any time of day. Assume the coefficients are zero before 6:00 a.m. and after 6:00 p.m. but vary as a sine function from 6:00 a.m. to 6:00 p.m., peaking at noon. Calculate rate coefficients for 9:00 a.m. and 5:00 p.m.

12

Methods of solving chemical ordinary differential equations

SEVERAL methods have been developed to solve chemical ordinary differential equations (ODEs). Those discussed in this chapter include analytical, Taylor series, forward Euler, backward Euler, simple exponential, quasi-steady-state, multistep implicit–explicit, backward differentiation (Gear), and family methods. Additional methods include hybrid predictor–corrector methods (e.g., Young and Boris 1977), parameterization methods (e.g., Jacob *et al.* 1989b), Runge–Kutta–Rosenbrock schemes (Hairer and Wanner 1991; Sandu *et al.* 1997), iterative backward Euler methods (e.g., Curtiss and Hirschfelder 1952; Shimazaki and Laird 1970; Rosenbaum 1976; Hertel *et al.* 1993; Huang and Chang 2001), hybrid Newton–Raphson iterative schemes (Gong and Cho 1993), Gauss–Seidel methods (Verwer 1994), extrapolation techniques (Dabdub and Seinfeld 1995), and projection methods (Sandu 2001) among others. The choice of a method for solving atmospheric chemical problems depends on several factors, including stability, accuracy, mass conservation, positivity, and speed. These factors are discussed with respect to different solvers. Techniques of optimizing Gear's method over a three-dimensional grid are also described.

12.1 CHARACTERISTICS OF CHEMICAL ODES

Gas-phase chemical reactions are described by first-order, first-degree, homogeneous ordinary differential equations. Sets of gas-phase reactions are stiff in that the chemical e-folding lifetimes of individual gases vary by many orders of magnitude. Because sets of chemical ODEs are stiff, some classical numerical methods are not useful for solving them. For example, the fourth-order Runge–Kutta method (Section 6.4.4.9) and the Richardson extrapolation/Bulirsch–Stoer method (Stoer and Bulirsch 1980) are explicit methods that result in inefficient (slow) solutions to stiff ODE problems. With an **explicit method**, final concentrations at time t are obtained by evaluating derivatives at the beginning of the current and previous time steps (e.g., at times $t - h$, $t - 2h$, . . .), where a **time step** (s) h is the difference between the current time (t) and the time of the previous time step ($t - h$). When an explicit technique is used to solve a stiff set of equations, the time step (h) is limited by the e-folding lifetime of the shortest-lived chemical. This lifetime may be 10^{-6} s or less. Time steps longer than the lifetime of the shortest-lived chemical may destabilize the solution scheme. When a time step is always small, integration of many reactions over days to months and over a large three-dimensional grid requires a significant amount of computational power and is often impractical.

Efficient solvers of stiff ODEs are **semiimplicit** in that their solutions at current time t depend on derivatives evaluated at the current time, the beginning of the current time step, and/or the beginning of previous time steps (times t, $t - h$, $t - 2h, \ldots$). Semiimplicit solvers can take time steps much longer than the e-folding lifetime of the shortest-lived species and remain stable. Some accurate semiimplicit schemes used for solving stiff ODEs are Gear's method (Gear 1971), Runge–Kutta–Rosenbrock schemes (Kaps and Rentrop 1979; Hairer and Wanner 1991; Press *et al.* 1992), and semiimplicit Bulirsch–Stoer schemes (Bader and Deuflhard 1983; Press *et al.* 1992), among others.

12.1.1 Initial value problems

Problems requiring the use of chemical ordinary differential equations are **initial value problems** whereby the initial concentration of each species is known at time $t = 0$, and a solution is desired at a final time, $t_f > 0$. Solutions are found by integrating a set of chemical ODEs one time step (h) at a time between $t = 0$ and $t = t_f$.

In the case of gas chemistry, concentrations are denoted by N (molec. cm^{-3}). At any given time t, the concentration of a species i is $N_{i,t}$, and the vector of concentrations of a set of K species is

$$\hat{N}_t = [N_{1,t}, N_{2,t}, \ldots, N_{i,t}, \ldots, N_{K,t}] \tag{12.1}$$

Concentrations of an individual species and a set of species one time step backward are $N_{i,t-h}$ and $\hat{N}_{t-h}$, respectively. At the beginning of the first time step, concentrations of all species are set to initial concentrations. Thus,

$$N_{i,t-h} = N_{i,0} \qquad \text{for } i = 1, \ldots, K \tag{12.2}$$

where $N_{i,0}$ is the initial number concentration of gas i.

During a time step, an ODE scheme solves for all $N_{i,t}$. Once values at time t are found, t is replaced with $t - h$ and a new time step is solved for. With some techniques, solutions depend on concentrations from several time steps backward (times $t - h$, $t - 2h, \ldots$). If the technique relies on values from two time steps backward, the first time step is solved for with a technique that depends on only one time step backward, and subsequent time steps are solved for with the scheme that depends on two time steps backward.

12.1.2 Properties of ODE solvers

For a chemical ODE solution scheme to be useful, it must be stable, accurate, mass-conserving, positive definite, and computationally fast. In Section 6.4.1, a numerical scheme was defined to be **stable** if the absolute-value difference between the numerical and the exact solution did not grow over time. This definition, and the definitions of unconditional stability, conditional stability, and unconditional instability, given in the same section, apply to ODE and PDE solution techniques alike.

Solutions must be **accurate** as well as stable. A method of testing a solver's accuracy is to compare the time-dependent solution from it with an exact solution. A **normalized gross error** (NGE) is then calculated as

$$\mathrm{NGE} = \frac{1}{N_{\mathrm{tim}}} \sum_{j=1}^{N_{\mathrm{tim}}} \left(\frac{1}{K_{\mathrm{s},t_j}} \sum_{i=1}^{K_{\mathrm{s},t_j}} \frac{\left| N_{i,t_j} - E_{i,t_j} \right|}{E_{i,t_j}} \right) \times 100\% \qquad (12.3)$$

where N_{tim} is the number of time steps, N_{i,t_j} and E_{i,t_j} are predicted and exact concentrations, respectively, of species i at time t_j, and K_{s,t_j} is the number of exact concentrations above a minimum cutoff concentration at time t_j. Typical cutoff concentrations for atmospheric gas chemistry range from 10^{-3} to 10^{3} molec. cm^{-3}. In (12.3), concentrations may be compared at evenly spaced time intervals, such as every one-half hour, instead of every time step. Exact solutions are obtained by solving the equations with an integrator of known high accuracy or with the given scheme using an extremely small time step. Good solvers of chemical ODEs produce NGEs less than 1 percent.

A solver of chemical ODEs should be **mass-conserving**. A scheme is mass-conserving if the mass of each element (e.g., N, O, H, or C) summed over all species at the beginning of a simulation equals the mass of the element summed over all species at the end of the simulation, provided no external sources or sinks exist. Mass conservation cannot occur if individual reactions are not written in a mass-conserving manner. The reaction $NO + O_3 \rightarrow NO_2 + O_2$ is written in a mass-conserving manner, but the reaction $NO + O_3 \rightarrow O_2$ is not. If all reactions are mass-conserving, the ODE solution scheme can be mass-conserving. Explicit ODE solution schemes are by nature mass-conserving, since the addition of mass to one species is accompanied by a loss of the same mass from another species. However, explicit schemes may produce a concentration less than zero, in which case the concentration is often set to zero or a small number, causing the scheme to gain mass. In addition, since an inaccurate explicit scheme can still be mass-conserving, mass conservation does not imply accuracy. Implicit or semiimplicit schemes have the potential to be non-mass-conserving. **It is more important for a scheme to be accurate than exactly mass-conserving.**

Concentrations predicted by chemical ODE solvers must exceed or equal zero. This should not be a surprise, since in the atmosphere, gas concentrations exceed or equal zero. If a scheme always predicts nonnegative concentrations, it is **positive definite**. An accurate chemical ODE solver is usually positive definite, since correct solutions are always nonnegative. If a solver is unstable and/or inaccurate, concentrations may fall below zero. If at least one concentration falls below zero after a time step, the time step must either be re-solved with a shorter time step or with a new solution method, or the concentration must be set to zero or above, in which case mass is gained. A good solver of chemical ODEs can predict the time step required to keep solutions stable and positive-definite.

When used in atmospheric models, chemical ODE solvers should be **fast** as well as accurate. All chemical ODE solvers are exactly accurate at a small enough time step. A good solver can take a long step and maintain accuracy. For three-dimensional modeling, accuracy with long steps is generally not enough. Accurate solutions to chemical ODEs must be found over large model grids. In the following sections, methods of solving chemical ODEs are discussed in light of their stability, accuracy, mass conservation, positive definiteness, and/or speed.

12.2 ANALYTICAL SOLUTIONS TO ODES

The most accurate solution to a set of chemical ODEs is an **analytical solution.** Analytical solutions to a single equation or a small set of equations are readily found. Suppose nitrogen dioxide is lost by the photolysis reaction,

$$\dot{N}O_2 + h\nu \rightarrow \dot{N}O + \dot{O}\cdot \quad (J) \tag{12.4}$$

which has photolysis coefficient $J\,(s^{-1})$. The ODE describing this reaction is

$$\frac{d[NO_2]}{dt} = -J\,[NO_2] \tag{12.5}$$

Integrating (12.5) gives its analytical solution at time t as

$$[NO_2]_t = [NO_2]_{t-h}\,e^{-Jh} \tag{12.6}$$

This equation states that the nitrogen dioxide concentration decays exponentially if only photolysis is considered.

Example 12.1

For an initial concentration of $[NO_2]_{t-h} = 10^{10}$ molec. cm^{-3} and a photolysis coefficient of 0.02 s^{-1}, (12.6) reduces to $[NO_2]_t = 10^{10}e^{-0.02h}$, which is an exact solution for the conditions given.

Whereas the solution to (12.5) was found easily, analytical solutions to a set of more than a few equations are usually impractical to obtain. Thus, sets of chemical ordinary differential equations are not solved analytically in atmospheric models.

12.3 TAYLOR SERIES SOLUTION TO ODES

Because analytical solutions are difficult to derive for chemical ODEs, numerical solutions are needed. A useful method of solving sets of ODEs would appear to be an explicit Taylor series expansion of species concentrations. In an **explicit Taylor series expansion**, the concentration of species i at time t is approximated as

$$N_{i,t} = N_{i,t-h} + h\frac{dN_{i,t-h}}{dt} + \frac{h^2}{2}\frac{d^2N_{i,t-h}}{dt^2} + \frac{h^3}{6}\frac{d^3N_{i,t-h}}{dt^3} + \cdots \tag{12.7}$$

One difficulty with using the Taylor series expansion is that expressions for higher derivatives require combinations of many lower-derivative terms, increasing the computational burden. Suppose three species – NO, NO_2, and O_3 – are considered. The explicit Taylor series expansion of the NO concentration is

$$[\mathrm{NO}]_t = [\mathrm{NO}]_{t-h} + h\frac{\mathrm{d}[\mathrm{NO}]_{t-h}}{\mathrm{d}t} + \frac{h^2}{2}\frac{\mathrm{d}^2[\mathrm{NO}]_{t-h}}{\mathrm{d}t^2} + \frac{h^3}{6}\frac{\mathrm{d}^3[\mathrm{NO}]_{t-h}}{\mathrm{d}t^3} + \cdots \quad (12.8)$$

Similar equations can be written for O_3 and NO_2. If one reaction,

$$\dot{\mathrm{N}}\mathrm{O} + \mathrm{O}_3 \rightarrow \dot{\mathrm{N}}\mathrm{O}_2 + \mathrm{O}_2 \quad (12.9)$$

is considered, the first, second, and third time derivatives of NO concentration arising from this reaction are, respectively,

$$\frac{\mathrm{d}[\mathrm{NO}]}{\mathrm{d}t} = -k_b[\mathrm{NO}][\mathrm{O}_3] \quad (12.10)$$

$$\frac{\mathrm{d}^2[\mathrm{NO}]}{\mathrm{d}t^2} = -k_b\frac{\mathrm{d}[\mathrm{NO}]}{\mathrm{d}t}[\mathrm{O}_3] - k_b[\mathrm{NO}]\frac{\mathrm{d}[\mathrm{O}_3]}{\mathrm{d}t} \quad (12.11)$$

$$\frac{\mathrm{d}^3[\mathrm{NO}]}{\mathrm{d}t^3} = -k_b\frac{\mathrm{d}^2[\mathrm{NO}]}{\mathrm{d}t^2}[\mathrm{O}_3] - 2k_b\frac{\mathrm{d}[\mathrm{NO}]}{\mathrm{d}t}\frac{\mathrm{d}[\mathrm{O}_3]}{\mathrm{d}t} - k_b[\mathrm{NO}]\frac{\mathrm{d}^2[\mathrm{O}_3]}{\mathrm{d}t^2} \quad (12.12)$$

The first, second, and third derivatives of O_3 and NO_2 are related to those of NO by

$$\frac{\mathrm{d}[\mathrm{O}_3]}{\mathrm{d}t} = -\frac{\mathrm{d}[\mathrm{NO}_2]}{\mathrm{d}t} = \frac{\mathrm{d}[\mathrm{NO}]}{\mathrm{d}t} \qquad \frac{\mathrm{d}^2[\mathrm{O}_3]}{\mathrm{d}t^2} = -\frac{\mathrm{d}^2[\mathrm{NO}_2]}{\mathrm{d}t^2} = \frac{\mathrm{d}^2[\mathrm{NO}]}{\mathrm{d}t^2} \qquad \frac{\mathrm{d}^3[\mathrm{O}_3]}{\mathrm{d}t^3} = -\frac{\mathrm{d}^3[\mathrm{NO}_2]}{\mathrm{d}t^3} = \frac{\mathrm{d}^3[\mathrm{NO}]}{\mathrm{d}t^3} \quad (12.13)$$

respectively. Thus, to solve for changes in concentration due to one reaction only, many terms are necessary. With the addition of tens or hundreds of reactions, the number of terms increases.

A second problem with the explicit Taylor series method is that stiffness of the system of equations prevents a stable and accurate solution unless the time step used is small or the order of approximation is large. For these requirements to be met, an explicit Taylor series expansion requires many short time steps, many high-order terms, or both. In either case, the number of computations required to maintain stability in a model with 10^3–10^6 grid cells and dozens of reactions is prohibitively large.

12.4 FORWARD EULER SOLUTION TO ODES

If concentrations are approximated with the first two terms of an explicit Taylor series expansion, the approximation is first order. The **order of approximation** was defined in Section 6.4.1 as the order of the lowest-order term in a Taylor series expansion of a derivative that is neglected when the approximation is made.

Dividing (12.7) by h and eliminating all terms of order h and higher gives a first-order approximation of an explicit Taylor series expansion, called the **forward Euler**. The forward Euler solution for one species among a set of chemical ODEs is

$$N_{i,t} = N_{i,t-h} + h\frac{\mathrm{d}N_{i,t-h}}{\mathrm{d}t} \tag{12.14}$$

where the first derivative is a function of N_i and the concentrations of all other species at time $t - h$. In other words, $\mathrm{d}N_{i,t-h}/\mathrm{d}t = f(N_{t-h})$.

Suppose the concentrations of NO, NO_2, and O_3 are affected by the reactions,

$$\dot{\mathrm{N}}\mathrm{O} + \mathrm{O_3} \rightarrow \dot{\mathrm{N}}\mathrm{O_2} + \mathrm{O_2} \qquad (k_1) \tag{12.15}$$

$$\dot{\mathrm{N}}\mathrm{O_2} + \mathrm{O_3} \rightarrow \dot{\mathrm{N}}\mathrm{O_3} + \mathrm{O_2} \qquad (k_2) \tag{12.16}$$

$$\dot{\mathrm{N}}\mathrm{O_2} + h\nu \rightarrow \dot{\mathrm{N}}\mathrm{O} + \dot{\mathrm{O}}\cdot \qquad (J) \tag{12.17}$$

The explicit time derivative of NO_2 among these reactions is

$$\frac{\mathrm{d}[\mathrm{NO_2}]_{t-h}}{\mathrm{d}t} = k_1[\mathrm{NO}]_{t-h}[\mathrm{O_3}]_{t-h} - k_2[\mathrm{NO_2}]_{t-h}[\mathrm{O_3}]_{t-h} - J[\mathrm{NO_2}]_{t-h} \tag{12.18}$$

Substituting (12.18) into (12.14) gives

$$[\mathrm{NO_2}]_t = [\mathrm{NO_2}]_{t-h} + h(k_1[\mathrm{NO}]_{t-h}[\mathrm{O_3}]_{t-h} - k_2[\mathrm{NO_2}]_{t-h}[\mathrm{O_3}]_{t-h} - J[\mathrm{NO_2}]_{t-h}) \tag{12.19}$$

which is the forward Euler NO_2 concentration after one time step. This solution can be generalized by defining total production and loss rates, respectively, of NO_2 as

$$P_{\mathrm{c,NO_2},t-h} = k_1[\mathrm{NO}]_{t-h}[\mathrm{O_3}]_{t-h} \tag{12.20}$$

$$L_{\mathrm{c,NO_2},t-h} = k_2[\mathrm{NO_2}]_{t-h}[\mathrm{O_3}]_{t-h} + J[\mathrm{NO_2}]_{t-h} \tag{12.21}$$

Substituting these terms into (12.19) gives another form of the forward Euler solution,

$$[\mathrm{NO_2}]_t = [\mathrm{NO_2}]_{t-h} + h(P_{\mathrm{c,NO_2},t-h} - L_{\mathrm{c,NO_2},t-h}) \tag{12.22}$$

This notation can be generalized for any species i as

$$\boxed{N_{i,t} = N_{i,t-h} + h(P_{\mathrm{c},i,t-h} - L_{\mathrm{c},i,t-h})} \tag{12.23}$$

An advantage of the forward Euler solution is that it requires minimal time for computing production and loss terms. In addition, like the explicit Taylor series expansion, it is exactly mass-conserving. For example, suppose reaction (12.15) is solved alone. The forward Euler solutions for nitric oxide, ozone, nitrogen dioxide, and oxygen are

$$[\mathrm{NO}]_t = [\mathrm{NO}]_{t-h} - hk_1[\mathrm{NO}]_{t-h}[\mathrm{O}_3]_{t-h} \tag{12.24}$$

$$[\mathrm{O}_3]_t = [\mathrm{O}_3]_{t-h} - hk_1[\mathrm{NO}]_{t-h}[\mathrm{O}_3]_{t-h} \tag{12.25}$$

$$[\mathrm{NO}_2]_t = [\mathrm{NO}_2]_{t-h} + hk_1[\mathrm{NO}]_{t-h}[\mathrm{O}_3]_{t-h} \tag{12.26}$$

$$[\mathrm{O}_2]_t = [\mathrm{O}_2]_{t-h} + hk_1[\mathrm{NO}]_{t-h}[\mathrm{O}_3]_{t-h} \tag{12.27}$$

respectively. For this set of four equations to conserve mass, the following relationships must hold for nitrogen and oxygen, respectively:

$$([\mathrm{NO}_2] + [\mathrm{NO}])_t = ([\mathrm{NO}_2] + [\mathrm{NO}])_{t-h} \tag{12.28}$$

$$(2[\mathrm{O}_2] + 3[\mathrm{O}_3] + 2[\mathrm{NO}_2] + [\mathrm{NO}])_t = (2[\mathrm{O}_2] + 3[\mathrm{O}_3] + 2[\mathrm{NO}_2] + [\mathrm{NO}])_{t-h} \tag{12.29}$$

Substituting (12.24)–(12.27) into these relationships gives exact equalities, indicating that the forward Euler conserves mass exactly.

The main disadvantage of the forward Euler is that it requires a small time step to prevent concentrations of short-lived species from falling below zero. As such, it requires many more time steps to complete a simulation than a semiimplicit scheme. Because of the severity of this limitation, the forward Euler approximation is not useful on its own to solve sets of stiff chemical equations.

12.5 BACKWARD EULER SOLUTION TO ODES

An approximation that avoids the time-step stability constraint of the forward Euler solution is the **linearized backward Euler** solution. A linearized solution is one in which concentrations of all species, except the species being solved for, are set to concentrations from previous time steps (e.g., $t-h$, $t-2h$, etc.) in derivative terms. The concentration of the species of interest is set to time t and assumed to be unknown when used in derivatives. In the backward Euler, only one time step backward is considered. The linearized form of the backward Euler equation for a species is

$$N_{i,t} = N_{i,t-h} + h\frac{\mathrm{d}N_{i,t,t-h}}{\mathrm{d}t} \tag{12.30}$$

where the first derivative is evaluated at time $t-h$ for all species except for the species of interest, which is evaluated at time t. Thus, $\mathrm{d}N_{i,t,t-h}/\mathrm{d}t = f(N_{i,t}, N_{t-h})$.

The backward Euler solution can be illustrated with reactions (12.15)–(12.17). The **linearized first derivative** of NO_2 from these reactions is

$$\frac{\mathrm{d}[\mathrm{NO}_2]_{t,t-h}}{\mathrm{d}t} = k_1[\mathrm{NO}]_{t-h}[\mathrm{O}_3]_{t-h} - k_2[\mathrm{NO}_2]_t[\mathrm{O}_3]_{t-h} - J[\mathrm{NO}_2]_t \tag{12.31}$$

where NO_2 is evaluated at time t. Substituting (12.31) into (12.30) gives

$$[\mathrm{NO}_2]_t = [\mathrm{NO}_2]_{t-h} + h(k_1[\mathrm{NO}]_{t-h}[\mathrm{O}_3]_{t-h} - k_2[\mathrm{NO}_2]_t[\mathrm{O}_3]_{t-h} - J[\mathrm{NO}_2]_t) \tag{12.32}$$

The production and loss terms in this equation are

$$P_{c,NO_2,t-h} = k_1[NO]_{t-h}[O_3]_{t-h} \tag{12.33}$$

$$L_{c,NO_2,t,t-h} = k_2[NO_2]_t[O_3]_{t-h} + J[NO_2]_t \tag{12.34}$$

respectively. All loss rates contain $[NO_2]_t$ so the loss term can be divided by this concentration to give an **implicit loss coefficient,**

$$\Lambda_{c,NO_2,t-h} = \frac{L_{c,NO_2,t,t-h}}{[NO_2]_t} = k_2[O_3]_{t-h} + J \tag{12.35}$$

Substituting (12.33) and (12.35) back into (12.32) yields

$$[NO_2]_t = [NO_2]_{t-h} + h(P_{c,NO_2,t-h} - \Lambda_{c,NO_2,t-h}[NO_2]_t) \tag{12.36}$$

Since both sides of (12.36) contain $[NO_2]_t$, $[NO_2]_t$ can be gathered on the left side of the equation to produce

$$[NO_2]_t = \frac{[NO_2]_{t-h} + hP_{c,NO_2,t-h}}{1 + h\Lambda_{c,NO_2,t-h}} \tag{12.37}$$

which is the **linearized backward Euler solution.** This solution can be generalized for any species i with

$$N_{i,t} = \frac{N_{i,t-h} + hP_{c,i,t-h}}{1 + h\Lambda_{c,i,t-h}} \tag{12.38}$$

The advantage of the linearized backward Euler solution is that the final concentration cannot fall below zero, regardless of the time step. As such, the scheme is unconditionally stable. The disadvantage is that the method is not mass-conserving. For example, suppose reaction (12.15) is solved alone. The linearized backward Euler solutions for nitric oxide, ozone, nitrogen dioxide, and oxygen are

$$[NO]_t = [NO]_{t-h} - hk_1[NO]_t[O_3]_{t-h} \tag{12.39}$$

$$[O_3]_t = [O_3]_{t-h} - hk_1[NO]_{t-h}[O_3]_t \tag{12.40}$$

$$[NO_2]_t = [NO_2]_{t-h} + hk_1[NO]_{t-h}[O_3]_{t-h} \tag{12.41}$$

$$[O_2]_t = [O_2]_{t-h} + hk_1[NO]_{t-h}[O_3]_{t-h} \tag{12.42}$$

Substituting these equations into the nitrogen and oxygen mass-conservation relationships (12.28) and (12.29), respectively, gives inequalities, indicating that the linearized backward Euler does not conserve mass. The reason is that reaction rates are linearized differently when different species are solved, and for mass to be conserved, reaction rates must be identical when different species are solved. Because mass is not conserved, errors accumulate over time with the linearized backward Euler.

12.6 SIMPLE EXPONENTIAL AND QUASI-STEADY-STATE SOLUTIONS TO ODES

A method of solving ODEs, similar to the backward Euler, is the **simple exponential approximation**. This solution is obtained by integrating a linearized first derivative. The advantage of the simple exponential approximation is that the final concentration cannot fall below zero; thus, the scheme is unconditionally stable. The disadvantage is that, like the backward Euler approximation, the simple exponential approximation is not mass-conserving.

The simple exponential solution is illustrated with the linearized first derivative from the backward Euler example. Writing (12.36) in differential form gives

$$\frac{\mathrm{d}[\mathrm{NO_2}]_t}{\mathrm{d}t} = P_{\mathrm{c,NO_2},t-h} - \Lambda_{\mathrm{c,NO_2},t-h}[\mathrm{NO_2}]_t \tag{12.43}$$

The integral of this equation is the **simple exponential solution**,

$$[\mathrm{NO_2}]_t = [\mathrm{NO_2}]_{t-h}\,\mathrm{e}^{-h\Lambda_{\mathrm{c,NO_2},t-h}} + \frac{P_{\mathrm{c,NO_2},t-h}}{\Lambda_{\mathrm{c,NO_2},t-h}}\left(1-\mathrm{e}^{-h\Lambda_{\mathrm{c,NO_2},t-h}}\right) \tag{12.44}$$

which is written for any species i as

$$N_{i,t} = N_{i,t-h}\mathrm{e}^{-h\Lambda_{\mathrm{c},i,t-h}} + \frac{P_{\mathrm{c},i,t-h}}{\Lambda_{\mathrm{c},i,t-h}}\left(1-\mathrm{e}^{-h\Lambda_{\mathrm{c},i,t-h}}\right) \tag{12.45}$$

When the implicit loss coefficient is zero (no loss), (12.45) simplifies to $N_{i,t} = N_{i,t-h} + hP_{\mathrm{c},i,t-h}$. When the production term is zero, the solution simplifies to $N_{i,t} = N_{i,t-h}\mathrm{e}^{-\Lambda_{\mathrm{c},i,t-h}}$. When the implicit loss coefficient is large (short-lived species), the solution simplifies to $N_{i,t} = P_{\mathrm{c},i,t-h}/\Lambda_{\mathrm{c},i,t-h}$, which is the steady-state solution to the original ODE.

The simple exponential solution is similar to the linearized backward Euler solution, except that the former contains an analytical component and is slightly more accurate than the latter. Because the simple exponential approximation requires the computation of an exponent, it requires more computer time, per time step, than does the backward Euler.

A method related to the simple exponential is the **quasi-steady-state approximation** (QSSA) method (e.g., Hesstvedt *et al.* 1978). With this method, the parameter $h\Lambda_{\mathrm{c},i,t-h}$ is used to determine how a species' concentration is solved during a time step. When $h\Lambda_{\mathrm{c},i,t-h} < 0.01$, the species' lifetime is long, and its final concentration is calculated with the forward Euler equation. When $0.01 \le h\Lambda_{\mathrm{c},i,t-h} \le 10$, the species' lifetime is moderate, and its concentration is calculated with the simple exponential method. When $h\Lambda_{\mathrm{c},i,t-h} > 10$, the species' lifetime is short, and its concentration is calculated with the steady state equation. In sum,

$$N_{i,t} = \begin{cases} N_{i,t-h} + h(P_{\mathrm{c},i,t-h} - L_{\mathrm{c},i,t-h}) & h\Lambda_{\mathrm{c},i,t-h} < 0.01 \\ N_{i,t-h}\mathrm{e}^{-h\Lambda_{\mathrm{c},i,t-h}} + \dfrac{P_{\mathrm{c},i,t-h}}{\Lambda_{\mathrm{c},i,t-h}}\left(1-\mathrm{e}^{-h\Lambda_{\mathrm{c},i,t-h}}\right) & 0.01 \le h\Lambda_{\mathrm{c},i,t-h} \le 10 \\ P_{\mathrm{c},i,t-h}/\Lambda_{\mathrm{c},i,t-h} & h\Lambda_{\mathrm{c},i,t-h} > 10 \end{cases} \tag{12.46}$$

Final concentrations at the end of a time step are found after several iterations of (12.46). The iterative QSSA scheme can be accurate for many cases, but it is mass-conserving only for long-lived species.

In sum, the forward Euler scheme is mass-conserving but positive-definite for small time steps only; the backward Euler and simple exponential schemes are always positive-definite, but not mass-conserving; and the iterative QSSA scheme is often stable but generally not mass-conserving.

12.7 MULTISTEP IMPLICIT–EXPLICIT (MIE) SOLUTION TO ODES

A positive-definite, mass-conserving, unconditionally stable iterative technique that takes advantage of the forward and backward Euler methods is the **multistep implicit–explicit** (MIE) method (Jacobson 1994; Jacobson and Turco 1994). With this method, concentrations are estimated with an iterated backward Euler calculation, and the estimates are applied to reaction rates used in a forward Euler calculation of final concentrations. Upon iteration **the forward Euler converges to the backward Euler.** Since backward Euler solutions are always positive, forward Euler solutions must converge to positive values as well. A technique related to the MIE method is merely iterating the backward Euler equation until convergence occurs for all species. Such a method requires many more iterations than does the MIE method.

The steps for determining final concentrations with the MIE method are described below. To illustrate, four species, NO, NO_2, O_3, and O, and two reactions,

$$\dot{N}O + O_3 \rightarrow \dot{N}O_2 + O_2 \qquad (k_1) \tag{12.47}$$

$$O_3 + h\nu \rightarrow O_2 + \dot{O}\cdot \qquad (J) \tag{12.48}$$

are considered. The change in O_2 concentration by these reactions is ignored for the illustration.

The first step in the MIE solution is to initialize backward Euler concentrations (molec. cm^{-3}) and maximum backward Euler concentrations for each active species $i = 1, \ldots, K$ with concentrations from the beginning of the simulation or, in the case of a new time step, with concentrations from the end of the last time step. Thus,

$$N_{i,\mathrm{B},1} = N_{i,t-h} \tag{12.49}$$

$$N_{i,\mathrm{MAX},1} = N_{i,t-h} \tag{12.50}$$

where the subscript B indicates a backward Euler concentration, and the subscript 1 indicates the first iteration of a new time step. In subsequent equations, the iteration number is denoted by m. Here, $m = 1$. A maximum backward Euler concentration is required to prevent backward Euler concentrations from blowing up to large values upon iteration and is updated each iteration, as discussed shortly.

The second step is to estimate reaction rates by multiplying rate coefficients by backward Euler concentrations. Examples of two-body and photolysis reaction rates are

$$R_{c,n,B,m} = k_n N_{i,B,m} N_{j,B,m} \qquad R_{c,n,B,m} = J_n N_{i,B,m} \tag{12.51}$$

respectively, where R_c is the rate of reaction (molec. cm^{-3} s^{-1}), n is the reaction-rate number, m is the iteration number (= 1 for the first iteration), k_n is the kinetic rate coefficient of the nth reaction, and J_n is the photolysis rate coefficient of the nth reaction. In the two-reaction example, the backward Euler rates are

$$R_{c,1,B,m} = k_1[\mathrm{NO}]_{B,m}[\mathrm{O_3}]_{B,m} \qquad R_{c,2,B,m} = J[\mathrm{O_3}]_{B,m} \tag{12.52}$$

respectively.

The third step is to estimate production rates, loss rates, and implicit loss coefficients for each species from the reaction rates just calculated. The **backward Euler production rate** (molec. cm^{-3} s^{-1}) of species i is

$$P_{c,i,B,m} = \sum_{l=1}^{N_{\mathrm{prod},i}} R_{c,n_P(l,i),B,m} \tag{12.53}$$

where $N_{\mathrm{prod},i}$ is the number of reactions in which species i is produced, and $R_{c,n_P(l,i),B,m}$ is the lth backward Euler production rate of species i. The array $n_P(l,i)$ gives the reaction number corresponding to the lth production term of species i. In the two-reaction example, the summed production rates of O and NO_2 are

$$P_{c,\mathrm{O},B,m} = J[\mathrm{O_3}]_{B,m} \qquad P_{c,\mathrm{NO_2},B,m} = k_1[\mathrm{NO}]_{B,m}[\mathrm{O_3}]_{B,m} \tag{12.54}$$

respectively, where the two active species produced have only one production term each.

The **backward Euler loss rate** (molec. cm^{-3} s^{-1}) of a species is

$$L_{c,i,B,m} = \sum_{l=1}^{N_{\mathrm{loss},i}} R_{c,n_L(l,i),B,m} \tag{12.55}$$

where $N_{\mathrm{loss},i}$ is the number of reactions in which species i is lost, and $R_{c,n_L(l,i),B,m}$ is the lth backward Euler loss rate of species i. The array $n_L(l,i)$ gives the reaction number corresponding to the lth loss term of species i. In the two-reaction example, the summed loss rates of NO and O_3 are

$$L_{c,\mathrm{NO},B,m} = k_1[\mathrm{NO}]_{B,m}[\mathrm{O_3}]_{B,m} \qquad L_{c,\mathrm{O_3},B,m} = k_1[\mathrm{NO}]_{B,m}[\mathrm{O_3}]_{B,m} + J[\mathrm{O_3}]_{B,m} \tag{12.56}$$

The **backward Euler implicit loss coefficient** (s^{-1}) of a species is

$$\Lambda_{c,i,B,m} = \frac{L_{c,i,B,m}}{N_{i,B,m}} \tag{12.57}$$

In the two-reaction example, the implicit loss coefficients of NO and O_3 are

$$\Lambda_{c,\mathrm{NO},B,m} = k_1[\mathrm{O_3}]_{B,m} \qquad \Lambda_{c,\mathrm{O_3},B,m} = k_1[\mathrm{NO}]_{B,m} + J \tag{12.58}$$

Since a reactant concentration can equal zero, computing implicit loss coefficients with (12.57) can result in a division by zero. If the coefficients are, instead, computed directly from (12.58), division by zero is avoided.

The fourth step is to calculate **backward Euler** concentrations for all species at iteration $m+1$ with

$$N_{i,\mathrm{B},m+1} = \frac{N_{i,t-h} + hP_{\mathrm{c},i,\mathrm{B},m}}{1 + h\Lambda_{\mathrm{c},i,\mathrm{B},m}} \tag{12.59}$$

Such estimates are used to calculate production and loss terms during the next iteration.

The fifth step is to calculate **forward Euler** concentrations for all species at iteration $m+1$ with

$$N_{i,\mathrm{F},m+1} = N_{i,t-h} + h(P_{\mathrm{c},i,\mathrm{B},m} - L_{\mathrm{c},i,\mathrm{B},m}) \tag{12.60}$$

The sixth step is to check convergence. **Convergence** is determined by first checking, at the end of each iteration during a time step, whether all forward Euler concentrations from (12.60) exceed or equal zero. If they do, a counter, n_P, initialized to zero before the first iteration of the time step, is incremented by one and another iteration is solved. If a single forward Euler concentration falls below zero during a given iteration, the counter n_P is reset to zero, and a new iteration is solved to try to update n_P again. Thus, whether the counter is updated or reset to zero is determined by the following criteria:

$$\begin{aligned} \hat{N}_{\mathrm{F},m+1} &\geq 0 \rightarrow n_P = n_P + 1 \\ N_{i,\mathrm{F},m+1} &< 0 \rightarrow n_P = 0 \end{aligned} \tag{12.61}$$

where $\hat{N}_{\mathrm{F},m+1}$ is the entire set of forward Euler concentrations and $N_{i,\mathrm{F},m+1}$ is an individual concentration. If all forward Euler concentrations ≥ 0 for N_P iterations in a row (e.g., if $n_P = N_P$), convergence is said to have occurred. N_P is a constant that depends on the number of reactions and their stiffness. Typical values are 5 for large sets of equations and 30–50 for small sets of equations.

For faster solutions, the criteria can be modified so that, when $h\Lambda_{\mathrm{c},i,\mathrm{B},m} \geq L_\mathrm{T}$ for a species during an iteration, the forward Euler concentration of the species does not need to exceed zero for n_P to avoid being reset to zero. In such cases, the species is short-lived, and the backward Euler solution is more accurate than the forward Euler solution. L_T is a constant between 10^2 and 10^6. Values of 10^2 speed solutions but increase errors. Values of 10^6 slow solutions but decrease errors. In sum, the following criteria identify conditions under which the counter is updated or reset to zero under the modified convergence method:

$$\begin{aligned} \hat{N}_{\mathrm{F},m+1} \geq 0 \text{ for all species with } h\Lambda_{\mathrm{c},i,\mathrm{B},m} < L_\mathrm{T} &\rightarrow n_P = n_P + 1 \\ N_{i,\mathrm{F},m+1} < 0 \text{ and } h\Lambda_{\mathrm{c},i,\mathrm{B},m} < L_\mathrm{T} &\rightarrow n_P = 0 \end{aligned} \tag{12.62}$$

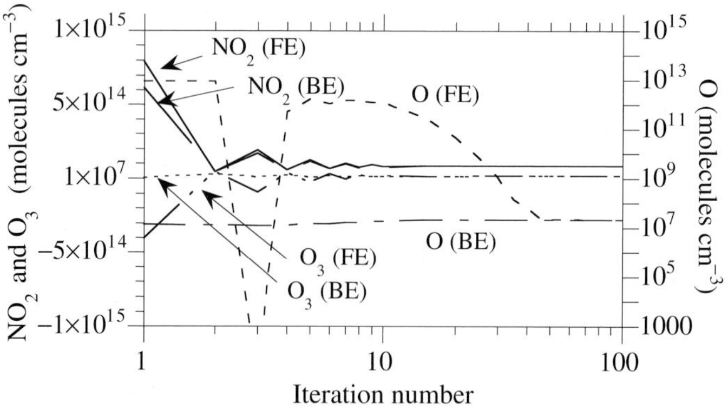

Figure 12.1 Concentrations of three species iterated until convergence with forward Euler (FE) and backward Euler (BE) calculations during a MIE simulation where $h = 10$ s.

With the modified method, convergence is again obtained when $n_P = N_P$. When the modified criterion is met, the final concentrations are

$$N_{i,t} = \begin{cases} N_{i,\mathrm{B},m+1} & \text{(backward Euler)} \quad h\Lambda_{\mathrm{c},i,\mathrm{B},m} \geq L_{\mathrm{T}} \\ N_{i,\mathrm{F},m+1} & \text{(forward Euler)} \quad h\Lambda_{\mathrm{c},i,\mathrm{B},m} < L_{\mathrm{T}} \end{cases} \tag{12.63}$$

If the modified convergence criterion is not met, iterations continue.

Before iterations continue, a seventh step, to update maximum backward Euler concentrations and to limit current backward Euler concentrations, is required. Maximum backward Euler concentrations are limited to the larger of the current backward Euler concentration and the initial concentration from the time step. Current backward Euler concentrations are then limited to the smaller of their current value and the maximum concentration from the last iteration. In other words,

$$N_{i,\mathrm{MAX},m+1} = \max(N_{i,\mathrm{B},m+1}, N_{i,t-h}) \tag{12.64}$$

$$N_{i,\mathrm{B},m+1} = \min(N_{i,\mathrm{B},m+1}, N_{i,\mathrm{MAX},m}) \tag{12.65}$$

for each species. Note that the value of $N_{i,\mathrm{MAX}}$ used in (12.65) is from the previous iteration, while that calculated in (12.64) is from the current iteration. After these updates, the iteration returns to (12.51) until convergence is obtained.

Here, it is shown that **iterated backward Euler solutions converge to iterated forward Euler solutions and positive numbers.** At the end of any iteration, the backward Euler concentration is

$$N_{i,\mathrm{B},m+1} = N_{i,t-h} + h(P_{\mathrm{c},i,\mathrm{B},m} - \Lambda_{\mathrm{c},i,\mathrm{B},m} N_{i,\mathrm{B},m+1}) \tag{12.66}$$

which is simply (12.59) rearranged. Equation (12.66) must be positive, since (12.59) cannot be negative. At the end of the same iteration, the forward Euler yields

$$N_{i,\mathrm{F},m+1} = N_{i,t-h} + h(P_{\mathrm{c},i,\mathrm{B},m} - \Lambda_{\mathrm{c},i,\mathrm{B},m} N_{i,\mathrm{B},m}) \tag{12.67}$$

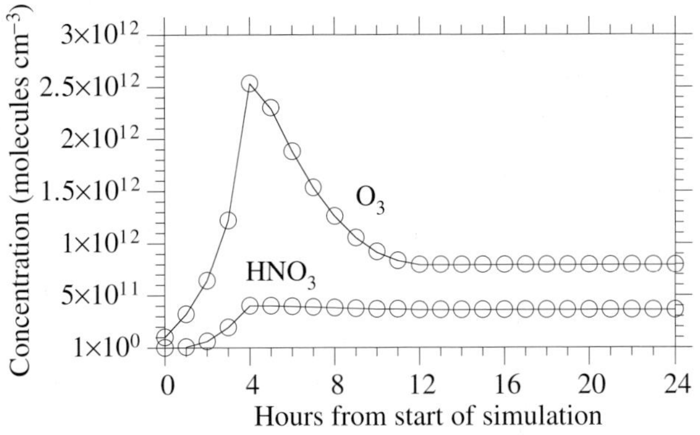

Figure 12.2 Comparison of MIE solution (circles) with an exact solution (lines) for two out of 92 species during a 24-h simulation period. The time step taken with MIE was 10 s. Photolysis rates changed every one-half hour during the first 12 h and were zero during the second 12 h. MIE and exact solutions are shown each one-half hour.

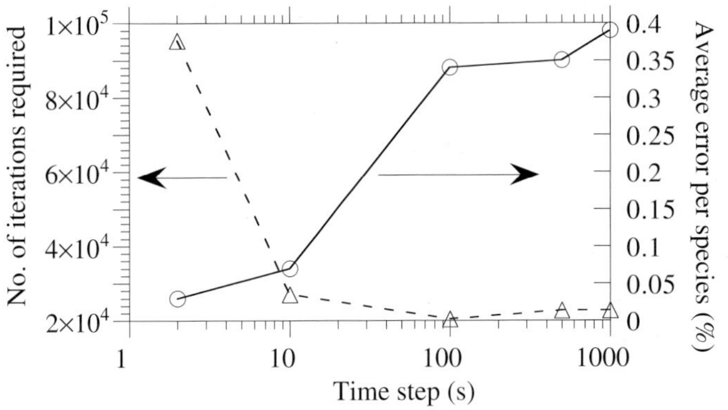

Figure 12.3 Effects of changing h with the MIE method when a set of 92 ODEs were solved. Time steps of h = 2, 10, 100, 500, and 1000 s are compared. The simulation time interval was 10 000 s. Small time steps resulted in a low normalized gross error after 10 000 s but required more iterations to complete the interval.

where $\Lambda_{c,i,B,m} N_{i,B,m} = L_{c,i,B,m}$. For the forward and backward Euler solutions to converge to each other,

$$N_{i,B,m+1} = N_{i,B,m} \tag{12.68}$$

must be satisfied. This always occurs upon iteration of the backward Euler solution. Thus, when the backward Euler solution converges, the forward Euler solution must converge to the same value, and both must converge to positive numbers.

Figure 12.1 shows convergence of forward Euler to backward Euler concentrations versus iteration number for four species. Figure 12.2 compares MIE simulation results with an exact solution for two species when a set of 92 ODEs is

Table 12.1 Values of α and β for use in (12.69)

s	$\alpha_{s,1}$	$\alpha_{s,2}$	$\alpha_{s,3}$	$\alpha_{s,4}$	$\alpha_{s,5}$	$\alpha_{s,6}$	β_s
1	1						1
2	1	1 / 3					2 / 3
3	1	6 / 11	1 / 11				6 / 11
4	1	35 / 50	10 / 50	1 / 50			24 / 50
5	1	225 / 274	85 / 274	15 / 274	1 / 274		120 / 274
6	1	1624 / 1764	735 / 1764	175 / 1764	21 / 1764	1 / 1764	720 / 1764

solved simultaneously. Figure 12.3 shows how a change in the time step affects the number of iterations and accuracy of the MIE solution in the 92-ODE case.

12.8 GEAR'S SOLUTION TO ODES

A more advanced method of solving stiff chemical ODEs is **Gear's method** (Gear 1971). Gear's method is accurate and elegant. A drawback of the original method was its need to solve equations containing large matrices of partial derivatives. This drawback prevented the use of the original code in three-dimensional models. The application of **sparse-matrix** techniques improved the speed of Gear's code tremendously (e.g., Sherman and Hindmarsh 1980; Hindmarsh 1983; Jacobson 1994, 1995, 1998a; Jacobson and Turco 1994), but the use of sparse-matrix techniques alone still did not permit the use of Gear's code in three-dimensional models. Only when sparse-matrix techniques were combined with computer optimization techniques could Gear's code be used to study problems in three dimensions (Jacobson 1994, 1995, 1998a; Jacobson and Turco 1994).

Gear's method solves the **backward differentiation formula** (BDF). The BDF is obtained by first discretizing the time derivative of an ODE with

$$\frac{\mathrm{d}N_{i,t}}{\mathrm{d}t} \approx \frac{N_{i,t} - \alpha_{s,1}N_{i,t-h} - \alpha_{s,2}N_{i,t-2h} \cdots - \alpha_{s,s}N_{i,t-sh}}{h\beta_s} = \frac{N_{i,t} - \sum_{j=1}^{s}\alpha_{s,j}N_{i,t-jh}}{h\beta_s} \tag{12.69}$$

where s is the order of approximation of the method, and α and β are scalar multipliers, given in Table 12.1, that depend on the order of the method. Rearranging (12.69) for an individual species and set of species gives

$$0 = -N_{i,t} + \sum_{j=1}^{s}\alpha_{s,j}N_{i,t-jh} + h\beta_s\frac{\mathrm{d}N_{i,t}}{\mathrm{d}t} \qquad 0 = -\hat{N}_t + \sum_{j=1}^{s}\alpha_{s,j}\hat{N}_{t-jh} + h\beta_s\frac{\mathrm{d}\hat{N}_t}{\mathrm{d}t} \tag{12.70}$$

respectively. The solution to (12.70) is found by writing

$$\mathbf{P}_t(\hat{N}_{t,m+1} - \hat{N}_{t,m}) = -\hat{N}_{t,m} + \sum_{j=1}^{s}\alpha_{s,j}\hat{N}_{t-jh} + h\beta_s\frac{\mathrm{d}\hat{N}_{t,m}}{\mathrm{d}t} \tag{12.71}$$

where $\mathbf{P}_t$ is a **predictor matrix** at time t, and $d\hat{N}_{t,m}/dt$ is the first derivative of $\hat{N}$ at iteration m. Equation (12.71) can be rewritten as

$$\mathbf{P}_t \Delta\hat{N}_{t,m} = \hat{B}_{t,m} \tag{12.72}$$

where $\Delta\hat{N}_{t,m} = \hat{N}_{t,m+1} - \hat{N}_{t,m}$ and

$$\hat{B}_{t,m} = -\hat{N}_{t,m} + \sum_{j=1}^{s} \alpha_{s,j}\hat{N}_{t-jh} + h\beta_s \frac{d\hat{N}_{t,m}}{dt} \tag{12.73}$$

$\Delta\hat{N}_{t,m}$ is solved from (12.72) by matrix decomposition and backsubstitution during each iteration. Once it is solved, concentrations for the next iteration are calculated with

$$\hat{N}_{t,m+1} = \hat{N}_{t,m} + \Delta\hat{N}_{t,m} \tag{12.74}$$

These values are substituted into (12.73) after each iteration, but only infrequently into the predictor matrix. Upon iteration of (12.72)–(12.74), $\Delta\hat{N}_{t,m}$ approaches zero. Convergence is checked with a local error test after each iteration and with a global error test after the completion of all iterations of a time step. In the **local error test**, a normalized root-mean-square error

$$\text{NRMS}_{t,m} = \sqrt{\frac{1}{K}\sum_{i=1}^{K}\left(\frac{\Delta N_{i,t,m}}{R_{\text{tol}} N_{i,t,1} + A_{\text{tol},t}}\right)^2} \tag{12.75}$$

is calculated, where $\Delta N_{i,t,m}$ (molec. cm^{-3}) is the change in concentration of species i at time step t during iteration m, K is the number of species, R_{tol} is a constant **relative error tolerance**, $A_{\text{tol},t}$ is an **absolute error tolerance** for time step t (molec. cm^{-3}), and $N_{i,t,1}$ is a concentration at the start of a time step. The relative error tolerance controls errors relative to $N_{i,t,1}$, and the absolute error tolerance controls errors relative to fixed concentrations. For pure relative-error-tolerance control, $A_{\text{tol},t} = 0$. For pure absolute-error-tolerance control, $R_{\text{tol}} = 0$. Typically, $R_{\text{tol}} = 10^{-3}$. For gas chemistry, $A_{\text{tol},t} \approx 10^3$–$10^7$ molec. cm^{-3}. Proper selection or prediction of $A_{\text{tol},t}$ reduces the computer time in large models. Jacobson (1998a) gives a method of predicting $A_{\text{tol},t}$.

If $\text{NRMS}_{t,m}$ is less than a specified parameter that varies with the order of approximation, the local error test succeeds. If the local error test fails but $\text{NRMS}_{t,m}$ is decreasing relative to $\text{NRMS}_{t,m-1}$, iterations continue until the local error test succeeds. If $\text{NRMS}_{t,m}$ increases upon iteration, convergence is not occurring, and the matrix of partial derivatives is updated, and iterations continue. If the local error test fails again, the time step is reduced until the test succeeds. Once the local error test has succeeded, a **global error test** is performed to check whether a cumulative normalized root-mean-square error, determined as

$$\mathrm{NRMS}_t = \sqrt{\frac{1}{K}\sum_{i=1}^{K}\left(\frac{\sum_m \Delta N_{i,t,m}}{R_{\mathrm{tol}} N_{i,t,1} + A_{\mathrm{tol},t}}\right)^2} \tag{12.76}$$

exceeds another parameter value that depends on the order of approximation. In this equation, $\sum_m \Delta N_{i,t,m}$ is the net change in species concentration during a time step. If the global error check fails, a new time step is predicted at the same or one order lower approximation. If the error check continues to fail, the time step is reduced. If the global test succeeds, the time step was successful, and $\hat{N}_{t,m+1}$-values from the last iteration are set to final concentrations. Every few successful time steps, the time step and order of approximation are recalculated with a time-step estimation scheme.

The predictor matrix on the left side of (12.72) expands to

$$\mathbf{P}_t \approx I - h\beta_s J_t \tag{12.77}$$

where I is the identity matrix, J_t is a Jacobian matrix of partial derivatives, and β_s, which depends on the order of the method, was defined earlier. The **Jacobian matrix** is

$$J_t = \left[\frac{\partial^2 \hat{N}_{i,t,m}}{\partial N_{k,t,m}\partial t}\right]_{i,k=1}^{K,K} \tag{12.78}$$

where K is the order of the matrix (number of species and ODEs). Matrices of partial derivatives for chemical ODEs are sparse, and sparse-matrix techniques are useful for treating them. Below, a matrix from a small equation set is shown, and a technique of reducing matrix computations is discussed.

The reactions used for the example matrix are

$$\dot{\mathrm{N}}\mathrm{O} + \mathrm{O_3} \rightarrow \dot{\mathrm{N}}\mathrm{O_2} + \mathrm{O_2} \qquad (k_1) \tag{12.79}$$

$$\dot{\mathrm{O}}\cdot + \mathrm{O_2} + \mathrm{M} \rightarrow \mathrm{O_3} + \mathrm{M} \qquad (k_2) \tag{12.80}$$

$$\dot{\mathrm{N}}\mathrm{O_2} + h\nu \rightarrow \dot{\mathrm{N}}\mathrm{O} + \dot{\mathrm{O}}\cdot \qquad (J) \tag{12.81}$$

The corresponding first derivatives of NO, NO_2, O, and O_3 are

$$\frac{\mathrm{d[NO]}}{\mathrm{d}t} = J[\mathrm{NO_2}] - k_1[\mathrm{NO}][\mathrm{O_3}] \qquad \frac{\mathrm{d[NO_2]}}{\mathrm{d}t} = k_1[\mathrm{NO}][\mathrm{O_3}] - J[\mathrm{NO_2}]$$

$$\frac{\mathrm{d[O]}}{\mathrm{d}t} = J[\mathrm{NO_2}] - k_2[\mathrm{O}][\mathrm{O_2}][\mathrm{M}] \qquad \frac{\mathrm{d[O_3]}}{\mathrm{d}t} = k_2[\mathrm{O}][\mathrm{O_2}][\mathrm{M}] - k_1[\mathrm{NO}][\mathrm{O_3}] \tag{12.82}$$

The corresponding partial derivatives are as follows:

Partial derivatives of NO

$$\frac{\partial^2[\mathrm{NO}]}{\partial[\mathrm{NO}]\partial t}=-k_1[\mathrm{O_3}]\qquad \frac{\partial^2[\mathrm{NO}]}{\partial[\mathrm{NO_2}]\partial t}=J\qquad \frac{\partial^2[\mathrm{NO}]}{\partial[\mathrm{O_3}]\partial t}=-k_1[\mathrm{NO}]\quad(12.83)$$

Partial derivatives of NO_2

$$\frac{\partial^2[\mathrm{NO_2}]}{\partial[\mathrm{NO}]\partial t}=k_1[\mathrm{O_3}]\qquad \frac{\partial^2[\mathrm{NO_2}]}{\partial[\mathrm{NO_2}]\partial t}=-J\qquad \frac{\partial^2[\mathrm{NO_2}]}{\partial[\mathrm{O_3}]\partial t}=k_1[\mathrm{NO}]\quad(12.84)$$

Partial derivatives of O

$$\frac{\partial^2[\mathrm{O}]}{\partial[\mathrm{NO_2}]\partial t}=J\qquad \frac{\partial^2[\mathrm{O}]}{\partial[\mathrm{O}]\partial t}=-k_2[\mathrm{O_2}][\mathrm{M}]\quad(12.85)$$

Partial derivatives of O_3

$$\frac{\partial^2[\mathrm{O_3}]}{\partial[\mathrm{NO}]\partial t}=-k_1[\mathrm{O_3}]\qquad \frac{\partial^2[\mathrm{O_3}]}{\partial[\mathrm{O}]\partial t}=k_2[\mathrm{O_2}][\mathrm{M}]\qquad \frac{\partial^2[\mathrm{O_3}]}{\partial[\mathrm{O_3}]\partial t}=-k_1[\mathrm{NO}]\quad(12.86)$$

All other partial derivatives are zero.

For the set of four active species, the predictor matrix ($\mathbf{P}_t$) has the form

$$\begin{array}{c} \\ \mathrm{NO} \\ \mathrm{NO_2} \\ \mathrm{O} \\ \mathrm{O_3} \end{array}
\begin{array}{c}
\begin{array}{cccc} \mathrm{NO} & \mathrm{NO_2} & \mathrm{O} & \mathrm{O_3} \end{array} \\
\begin{bmatrix}
1-h\beta_s\frac{\partial^2[\mathrm{NO}]}{\partial[\mathrm{NO}]\partial t} & -h\beta_s\frac{\partial^2[\mathrm{NO}]}{\partial[\mathrm{NO_2}]\partial t} & -h\beta_s\frac{\partial^2[\mathrm{NO}]}{\partial[\mathrm{O}]\partial t} & -h\beta_s\frac{\partial^2[\mathrm{NO}]}{\partial[\mathrm{O_3}]\partial t} \\
-h\beta_s\frac{\partial^2[\mathrm{NO_2}]}{\partial[\mathrm{NO}]\partial t} & 1-h\beta_s\frac{\partial^2[\mathrm{NO_2}]}{\partial[\mathrm{NO_2}]\partial t} & -h\beta_s\frac{\partial^2[\mathrm{NO_2}]}{\partial[\mathrm{O}]\partial t} & -h\beta_s\frac{\partial^2[\mathrm{NO_2}]}{\partial[\mathrm{O_3}]\partial t} \\
-h\beta_s\frac{\partial^2[\mathrm{O}]}{\partial[\mathrm{NO}]\partial t} & -h\beta_s\frac{\partial^2[\mathrm{O}]}{\partial[\mathrm{NO_2}]\partial t} & 1-h\beta_s\frac{\partial^2[\mathrm{O}]}{\partial[\mathrm{O}]\partial t} & -h\beta_s\frac{\partial^2[\mathrm{O}]}{\partial[\mathrm{O_3}]\partial t} \\
-h\beta_s\frac{\partial^2[\mathrm{O_3}]}{\partial[\mathrm{NO}]\partial t} & -h\beta_s\frac{\partial^2[\mathrm{O_3}]}{\partial[\mathrm{NO_2}]\partial t} & -h\beta_s\frac{\partial^2[\mathrm{O_3}]}{\partial[\mathrm{O}]\partial t} & 1-h\beta_s\frac{\partial^2[\mathrm{O_3}]}{\partial[\mathrm{O_3}]\partial t}
\end{bmatrix}
\end{array}\quad(12.87)$$

Substituting the partial derivatives from (12.83)–(12.86) into (12.87) gives

$$\begin{array}{c} \\ \mathrm{NO} \\ \mathrm{NO_2} \\ \mathrm{O} \\ \mathrm{O_3} \end{array}
\begin{array}{c}
\begin{array}{cccc} \mathrm{NO} & \mathrm{NO_2} & \mathrm{O} & \mathrm{O_3} \end{array} \\
\begin{bmatrix}
1-h\beta_s(-k_1[\mathrm{O_3}]) & -h\beta_s(J) & 0 & -h\beta_s(-k_1[\mathrm{NO}]) \\
-h\beta_s(k_1[\mathrm{O_3}]) & 1-h\beta_s(-J) & 0 & -h\beta_s(k_1[\mathrm{NO}]) \\
0 & -h\beta_s(J) & 1-h\beta_s(-k_2[\mathrm{O_2}][M]) & 0 \\
-h\beta_s(-k_1[\mathrm{O_3}]) & 0 & -h\beta_s(k_2[\mathrm{O_2}][M]) & 1-h\beta_s(-k_1[\mathrm{NO}])
\end{bmatrix}
\end{array}\quad(12.88)$$

In (12.88), 5 out of the 16 matrix positions contain zeros. Thus, the initial matrix fill-in is 11/16. For larger chemical reaction sets, the order of the matrix is larger, but the initial percentage fill-in is much smaller. A typical set of 90 species and 200 chemical reactions has an initial fill-in of about 8 percent. The method of implementing a sparse-matrix technique, described shortly, can be applied to a matrix with any percentage of initial fill-in.

Substituting (12.88) into (12.72) gives

$$\begin{bmatrix} 1 - h\beta_s(-k_1[O_3]) & -h\beta_s(J) & 0 & -h\beta_s(-k_1[NO]) \\ -h\beta_s(k_1[O_3]) & 1 - h\beta_s(-J) & 0 & -h\beta_s(k_1[NO]) \\ 0 & -h\beta_s(J) & 1 - h\beta_s(-k_2[O_2][M]) & 0 \\ -h\beta_s(-k_1[O_3]) & 0 & -h\beta_s(k_2[O_2][M]) & 1 - h\beta_s(-k_1[NO]) \end{bmatrix}$$
$$\times \begin{bmatrix} \Delta[NO]_{t,m} \\ \Delta[NO_2]_{t,m} \\ \Delta[O]_{t,m} \\ \Delta[O_3]_{t,m} \end{bmatrix} = \begin{bmatrix} B_{NO,t,m} \\ B_{NO_2,t,m} \\ B_{O,t,m} \\ B_{O_3,t,m} \end{bmatrix} \tag{12.89}$$

which is solved either by Gaussian elimination or by matrix decomposition and backsubstitution (e.g., Press *et al.* 1992). Because Gear's method reuses the same decomposed matrix for several backsubstitutions over the right side of (12.89), **decomposition plus backsubstitution is favored over Gaussian elimination.**

During the decomposition of matrix equation (12.89), many multiplications by zero occur. One sparse-matrix technique is to **remove multiplications by zero** when such multiplications are known in advance. That is not enough. To reduce computer time significantly, the **matrix should be reordered** before multiplications by zero are eliminated. An efficient way of reordering is to place species with the most partial-derivative terms at the bottom and those with the fewest at the top of the matrix (Jacobson 1994; Jacobson and Turco 1994). The use of this reordering scheme requires that partial pivoting be avoided during decomposition. Such a restriction is reasonable in that, if the decomposition should fail, the step size in the numerical solver can be reduced. Reordering (12.89) gives

$$\begin{bmatrix} 1 - h\beta_s(-k_2[O_2][M]) & -h\beta_s(J) & 0 & 0 \\ 0 & 1 - h\beta_s(-J) & -h\beta_s(k_1[O_3]) & -h\beta_s(k_1[NO]) \\ 0 & -h\beta_s(J) & 1 - h\beta_s(-k_1[O_3]) & -h\beta_s(-k_1[NO]) \\ -h\beta_s(k_2[O_2][M]) & 0 & -h\beta_s(-k_1[O_3]) & 1 - h\beta_s(-k_1[NO]) \end{bmatrix}$$
$$\times \begin{bmatrix} \Delta[O]_{t,m} \\ \Delta[NO_2]_{t,m} \\ \Delta[NO]_{t,m} \\ \Delta[O_3]_{t,m} \end{bmatrix} = \begin{bmatrix} B_{O,t,m} \\ B_{NO_2,t,m} \\ B_{NO,t,m} \\ B_{O_3,t,m} \end{bmatrix} \tag{12.90}$$

Reordering maximizes the number of multiplications by zero and minimizes matrix fill-in during decomposition and backsubstitution. Table 12.2 shows that sparse-matrix techniques can reduce the number of multiplications for a reaction set of nearly 4000 reactions by about 99.995 percent, resulting in a computer speedup of 15 000.

Although sparse-matrix techniques speed Gear's solution method, **computer optimization techniques** are also often needed to increase the speed of the chemistry solution in large models with 50–500 chemical ODEs, 10^3–10^6 grid cells, and integration times of several years. One optimization technique, useful on vector, scalar, and parallel computers, is to divide the grid domain into blocks of 100–500

Table 12.2 Reduction in array space and matrix operations due to reordering and removing multiplications by zero for a chemistry set with 1427 gases and 3911 reactions under daytime and nighttime conditions

		After sparse reductions			
Quantity	**Initial**	**Day**	**Percent reduction**	**Night**	**Percent reduction**
Order of matrix	1 427	1 427	0	1 427	0
Initial fill-in	2 036 329	14 276	99.30	12 820	99.37
Final fill-in	2 036 329	17 130	99.16	14 947	99.27
Decomp. 1	967 595 901	47 596	99.995	36 974	99.996
Decomp. 2	1 017 451	9 294	99.09	7 393	99.27
Backsub. 1	1 017 451	9 294	99.09	7 393	99.27
Backsub. 2	1 017 451	6 409	99.37	6 127	99.40

The initial and final fill-in are the initial and final numbers of matrix positions filled. The last four rows show the number of operations in each of four loops of matrix decomposition and backsubstitution. The 'Initial' column shows values before sparse-matrix reductions. The solver was SMVGEAR II.
Source: Liang and Jacobson (2000).

grid cells each. For example, 100 000 cells can be divided into 200 blocks of 500 cells each. Chemical ODEs can be solved in blocks of 500 cells at a time instead of serially, one at a time. When a grid is divided into blocks, every inner loop of the computer program should be the grid-cell loop, with length 500.

The following example (in Fortran) demonstrates how array references are minimized when the inner loop is the grid-cell loop. In this example, NBIMOLEC is the number (e.g., 150) of bimolecular reaction rates that are calculated, JSP1 is the species number of the first reactant, JSP2 is the species number of the second reactant, KTLOOP is the number (e.g., 500) of grid cells in a grid block, RRATE is the rate coefficient, CONC is a gas concentration, and TRATE is the reaction rate. In nested loop A, where the inner loop is the grid-cell loop, only 150 references are made to arrays JPROD1 and JPROD2.

Nested loop A

```
      DO 105 NK         = 1, NBIMOLEC
         JSP1           = JPROD1(NK)
         JSP2           = JPROD2(NK)
         DO 100 K       = 1, KTLOOP
            TRATE(K,NK) = RRATE(K,NK) * CONC(K,JSP1) * CONC(K,JSP2)
100      CONTINUE
105   CONTINUE
```

Nested loop B

```
      DO 105 K          = 1, KTLOOP
         DO 100 NK      = 1, NBIMOLEC
            JSP1        = JPROD1(NK)
            JSP2        = JPROD2(NK)
            TRATE(K,NK) = RRATE(K,NK) * CONC(K,JSP1) * CONC(K,JSP2)
100      CONTINUE
105   CONTINUE
```

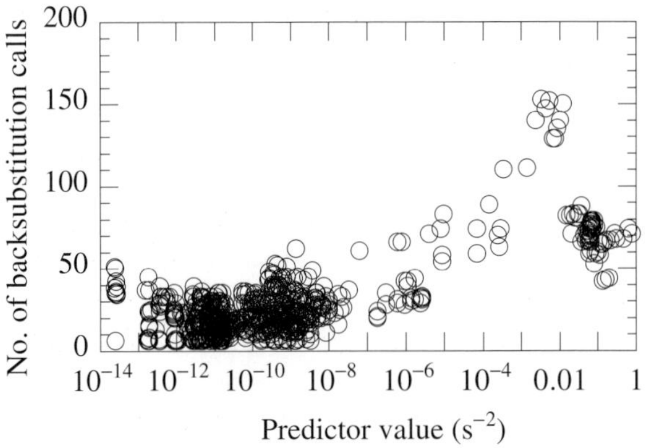

Figure 12.4 Resulting number of backsubstitution calls for different stiffness-predictor values from (12.91). The more backsubstitution calls, the greater the stiffness of the set of equations. The calculations were performed with the computer code SMVGEAR II (Jacobson 1995).

In nested loop B, where the grid-cell loop is on the outside, the number of references to JPROD1 and JPROD2 is 150 × 500. Thus, one pass through nested loop A is much faster than one pass through nested loop B, regardless of the machine. On vector machines, nested loop A also vectorizes better than does nested loop B.

When a grid domain is divided into blocks of grid cells, as described above, another issue must be addressed. Chemistry in each grid cell of the block is iterated in the same manner as in each other grid cell in the block. As such, iterations must continue in all grid cells in the block until the grid cell with stiffest chemistry converges. This results in many unnecessary iterations in grid cells where convergence occurs faster than in other grid cells. Excess iterations can be minimized by reordering grid cells among all grid blocks each time interval, according to stiffness. For example, if 20 000 cells are solved 500 at a time, the 500 stiffest cells can be placed together in a block, the next 500 stiffest can be placed together, and so on. This way, cells that require the most iterations are grouped together.

To reorder cells according to stiffness each time interval, a **stiffness predictor** is needed. One such predictor is

$$S_p = \frac{1}{K}\sum_{i=1}^{K}\left(\frac{\mathrm{d}N_{i,t}/\mathrm{d}t}{N_{i,t} + A_{\mathrm{tol},t}}\right)^2 \tag{12.91}$$

(Jacobson 1995), where S_p is proportional to stiffness, K is the number of ODEs (and species), $A_{\mathrm{tol},t}$ is the absolute error tolerance at time t, $N_{i,t}$ is the species concentration, and $\mathrm{d}N_{i,t}/\mathrm{d}t$ is the first derivative of $N_{i,t}$. The larger the value of S_p, the stiffer the equations in the grid cell. Figure 12.4 shows how (12.91) predicted the stiffness in an example simulation.

In sum, Gear's method is accurate for solving atmospheric chemistry problems. When the method is combined with sparse-matrix and computer optimization techniques, the equations can be solved over large, three-dimensional model domains for long integration periods.

12.9 FAMILY SOLUTION TO ODES

The last numerical method of solving chemical ODEs discussed is the **family method** (e.g., Crutzen 1971; Turco and Whitten 1974; Austin 1991; Elliott *et al.* 1993; Jacobson 1994). This method is less accurate than Gear's method for the same time step and must be tuned for each application. It is fast and useful for several applications.

The theory behind the family method is that some groups of gases, or **families**, exist in which atoms transfer quickly among species in the family but are lost only slowly from the family. Example families are the **odd oxygen, odd hydrogen, odd nitrogen**, and **odd chlorine** families. Some of the major species in these families are

Odd oxygen:	$[O_T] = [O] + [O(^1D)] + [O_3] + [NO_2]$	(12.92)
Odd hydrogen:	$[HO_T] = [OH] + [HO_2] + [H_2O_2]$	(12.93)
Odd nitrogen:	$[NO_T] = [NO] + [NO_2] + [NO_3]$	(12.94)
Odd chlorine	$[Cl_T] = [Cl] + [ClO] + [ClO_2]$	(12.95)

For example, oxygen atoms in the odd-oxygen family cycle among the species atomic oxygen, excited atomic oxygen, and ozone by the reactions,

$$\begin{aligned} O_3 + h\nu &\rightarrow O_2 + \cdot\dot{O}(^1D) \\ O_2 + h\nu &\rightarrow \dot{O}\cdot + \dot{O}\cdot \\ \cdot\dot{O}(^1D) &\xrightarrow{M} \dot{O}\cdot \\ \dot{O}\cdot + O_2 + M &\rightarrow O_3 + M \end{aligned} \tag{12.96}$$

Cycling of oxygen atoms among these reactions is fast, whereas loss of oxygen atoms out of this group of reactions is slower.

The family solution to ODEs for a time step requires four steps. First, rates of production and loss of individual species are calculated from initial concentrations. These rates are then summed across a family. The family concentration is then advanced with a forward Euler approximation applied to the summed production and loss terms. Finally, species concentrations are repartitioned in the family for the next time step.

Suppose a system contains four species (A, B, C, and D) and two families,

$$[Fam_1] = [A] + [B] + [C] \qquad [Fam_2] = [D] \tag{12.97}$$

If the reactions

$$A \rightarrow B \quad (k_a) \quad B \rightarrow C \quad (k_b) \quad C \rightarrow D \quad (k_c) \tag{12.98}$$

occur in the system, the first step in the family solution method is to calculate rates of production and loss of the individual species, A, B, C, and D:

$$\frac{d[A]}{dt} = -k_a[A] \qquad \frac{d[B]}{dt} = k_a[A] - k_b[B] \qquad \frac{d[C]}{dt} = k_b[B] - k_c[C] \qquad \frac{d[D]}{dt} = k_c[C] \tag{12.99}$$

The second step is to sum rates of production and loss across each family. Taking the time derivative of (12.97) and substituting (12.99) into the result gives

$$\frac{d[Fam_1]}{dt} = \frac{d[A]}{dt} + \frac{d[B]}{dt} + \frac{d[C]}{dt} = -k_c[C] \qquad \frac{d[Fam_2]}{dt} = \frac{d[D]}{dt} = k_c[C] \tag{12.100}$$

The third step is to advance the family concentration over a time step with a forward Euler. For example,

$$[Fam]_t = [Fam]_{t-h} + h\frac{d[Fam]_{t-h}}{dt} \tag{12.101}$$

Substituting (12.100) into (12.101) gives

$$[Fam_1]_t = [Fam_1]_{t-h} - hk_c[C]_{t-h} \qquad [Fam_2]_t = [Fam_2]_{t-h} + hk_c[C]_{t-h} \tag{12.102}$$

Fourth, species concentrations must be extracted from family concentrations. One method of extracting species concentrations is to calculate and apply partitioning ratios. Such ratios are calculated by summing the concentrations of the species in the family as

$$[Fam_1]_t = [A]_t + [B]_t + [C]_t = [A]_t\left(1 + \frac{[B]_t}{[A]_t} + \frac{[C]_t}{[A]_t}\right) \tag{12.103}$$

Individual species concentrations are then found from the family concentrations with

$$[A]_t = \frac{[Fam_1]_t}{1 + \dfrac{[B]_t}{[A]_t} + \dfrac{[C]_t}{[A]_t}} \qquad [B]_t = [A]_t\frac{[B]_t}{[A]_t} \qquad [C]_t = [A]_t\frac{[C]_t}{[A]_t} \tag{12.104}$$

In these equations, the **partitioning ratios** $[B]_t/[A]_t$ and $[C]_t/[A]_t$ are currently unknown. In the case of species *D*, its final concentration is that of the family it resides in, since no other species is in the family. Thus, $[D]_t = [Fam_2]_t = hk_c[C]_{t-h}$.

The simplest but least accurate way to estimate partitioning ratios is to assume that each species in the family is in steady state. With this method, the individual rates of reaction of species B and C are set to zero:

$$\frac{d[B]}{dt} = k_a[A] - k_b[B] = 0 \qquad \frac{d[C]}{dt} = k_b[B] - k_c[C] = 0 \tag{12.105}$$

The partitioning ratios of B and C are then estimated as

$$\frac{[B]_t}{[A]_t} \approx \frac{[B]}{[A]} = \frac{k_a}{k_b} \qquad \frac{[C]_t}{[A]_t} \approx \frac{[C]}{[A]} = \frac{[C]}{[B]}\frac{[B]}{[A]} = \frac{k_b}{k_c}\frac{k_a}{k_b} = \frac{k_a}{k_c} \tag{12.106}$$

respectively. Substituting (12.106) into (12.104) gives the final species concentrations of A, B, and C, respectively, as

$$[\mathrm{A}]_t = \frac{[\mathrm{Fam}_1]_t}{1+\dfrac{k_a}{k_b}+\dfrac{k_a}{k_c}} \qquad [\mathrm{B}]_t = \frac{[\mathrm{Fam}_1]_t\dfrac{k_a}{k_b}}{1+\dfrac{k_a}{k_b}+\dfrac{k_a}{k_c}} \qquad [\mathrm{C}]_t = \frac{[\mathrm{Fam}_1]_t\dfrac{k_a}{k_c}}{1+\dfrac{k_a}{k_b}+\dfrac{k_a}{k_c}} \tag{12.107}$$

A second way to estimate partitioning ratios is to linearize the ODEs of the individual species as

$$\frac{d[\mathrm{A}]_t}{dt} = -k_a[\mathrm{A}]_t \qquad \frac{d[\mathrm{B}]_t}{dt} \approx k_a[\mathrm{A}]_{t-h} - k_b[\mathrm{B}]_t \qquad \frac{d[\mathrm{C}]_t}{dt} \approx k_b[\mathrm{B}]_{t-h} - k_c[\mathrm{C}]_t \tag{12.108}$$

and then to integrate each equation. The results are

$$[\mathrm{A}]_t = [\mathrm{A}]_{t-h}e^{-k_a h} \qquad [\mathrm{B}]_t \approx [\mathrm{B}]_{t-h}e^{-k_b h} + \frac{k_a[\mathrm{A}]_{t-h}}{k_b}\left(1-e^{-k_b h}\right)$$
$$[\mathrm{C}]_t \approx [\mathrm{C}]_{t-h}e^{-k_c h} + \frac{k_b[\mathrm{B}]_{t-h}}{k_c}\left(1-e^{-k_c h}\right) \tag{12.109}$$

These estimates are substituted into the right sides of (12.104).

A third way to estimate partitioning ratios is to finite-difference the ODEs as

$$\frac{[\mathrm{A}]_t - [\mathrm{A}]_{t-h}}{h} = -k_a[\mathrm{A}]_t \qquad \frac{[\mathrm{B}]_t - [\mathrm{B}]_{t-h}}{h} = k_a[\mathrm{A}]_t - k_b[\mathrm{B}]_t$$
$$\frac{[\mathrm{C}]_t - [\mathrm{C}]_{t-h}}{h} = k_b[\mathrm{B}]_t - k_c[\mathrm{C}]_t \tag{12.110}$$

and then to rearrange the equations into a matrix equation as

$$\begin{bmatrix} 1+hk_a & 0 & 0 \\ hk_a & 1+hk_b & 0 \\ 0 & hk_b & 1+hk_c \end{bmatrix} \begin{bmatrix} [\mathrm{A}]_t \\ [\mathrm{B}]_t \\ [\mathrm{C}]_t \end{bmatrix} = \begin{bmatrix} [\mathrm{A}]_{t-h} \\ [\mathrm{B}]_{t-h} \\ [\mathrm{C}]_{t-h} \end{bmatrix} \tag{12.111}$$

Solving the matrix equation for $[\mathrm{A}]_t$, $[\mathrm{B}]_t$, and $[\mathrm{C}]_t$ then substituting the results into the right sides of (12.104) gives final partitioned concentrations.

In the example above, the reactions were unimolecular. Most chemical reactions are bimolecular. When bimolecular reactions are used, the first derivatives need to be linearized for use in (12.111). If the bimolecular reaction $\mathrm{A}+\mathrm{C} \rightarrow \mathrm{B}+\mathrm{D}$ (k_{ac}) is added to (12.98), the linearized finite-difference forms for A, B, and C are

$$\frac{[\mathrm{A}]_t - [\mathrm{A}]_{t-h}}{h} = -k_a[\mathrm{A}]_t - k_{ac}[\mathrm{A}]_t[\mathrm{C}]_{t-h} \tag{12.112}$$

$$\frac{[\mathrm{B}]_t - [\mathrm{B}]_{t-h}}{h} = k_a[\mathrm{A}]_t - k_b[\mathrm{B}]_t + 0.5k_{ac}([\mathrm{A}]_t[\mathrm{C}]_{t-h} + [\mathrm{A}]_{t-h}[\mathrm{C}]_t) \tag{12.113}$$

$$\frac{[\mathrm{C}]_t - [\mathrm{C}]_{t-h}}{h} = k_b[\mathrm{B}]_t - k_c[\mathrm{C}]_t - k_{ac}[\mathrm{A}]_{t-h}[\mathrm{C}]_t \tag{12.114}$$

respectively. The resulting matrix is (Jacobson 1994)

$$\begin{bmatrix} 1+hk_a+hk_{ac}[\mathrm{C}]_{t-h} & 0 & 0 \\ -hk_a-0.5(hk_{ac}[\mathrm{C}]_{t-h}) & 1+hk_b & -0.5(hk_{ac}[\mathrm{A}]_{t-h}) \\ 0 & -hk_b & 1+hk_c+hk_{ac}[\mathrm{A}]_{t-h} \end{bmatrix} \begin{bmatrix} [\mathrm{A}]_t \\ [\mathrm{B}]_t \\ [\mathrm{C}]_t \end{bmatrix} = \begin{bmatrix} [\mathrm{A}]_{t-h} \\ [\mathrm{B}]_{t-h} \\ [\mathrm{C}]_{t-h} \end{bmatrix} \tag{12.115}$$

Solving this matrix equation gives concentrations that can be used to estimate partitioning ratios.

The advantages of the family method are that it is fast, since it can use a long time step, and it may be accurate for moderate- to low-stiffness systems. The disadvantages are that the families need to be designed carefully and validated for each set of chemistry, and the accuracy of the method decreases with increasing stiffness. Also, although families are generally long-lived, the forward Euler solution for the family often results in negative concentrations. In such cases, the time step must be reduced and the family concentration must be recalculated. This reduces the speed advantage of the family method.

12.10 SUMMARY

In this chapter, methods of solving chemical first-order ordinary differential equations were described. These included analytical, explicit Taylor series, forward Euler, backward Euler, simple exponential, quasi-steady-state, multistep implicit–explicit, Gear, and family methods. A good solution scheme is stable, accurate, mass-conserving, positive definite, and fast. One code that fits these qualities is Gear's code combined with sparse-matrix and vectorization techniques. Stiff ODE codes can be used to solve aqueous as well as gas-phase chemical equations. Such equations are discussed in Chapter 19.

12.11 PROBLEMS

12.1 Derive an analytical solution for the loss of molecular oxygen via the reaction $O_2 + h\nu \rightarrow O + O$. If oxygen is destroyed by this reaction but not re-created, how long will it take for its concentration to decrease to 10 percent of its initial value? Use a constant stratospheric photolysis rate coefficient from Appendix Table B.4.

12.2 Given reactions $A + h\nu \rightarrow B + C$ (J_1) and $B + h\nu \rightarrow C + C$ (J_2), where J_1 and J_2 are photolysis coefficients and where [A], [B], and [C] are concentrations of A, B, and C:

(a) Write out the chemical rate equations for A, B, and C (e.g., d[A]/dt . . .)

(b) Find the time-dependent analytical solutions for each species from the initial values $[\mathrm{A}]_{t-h}$, $[\mathrm{B}]_{t-h}$, and $[\mathrm{C}]_{t-h}$. Assume J_1 and J_2 are constant in time.

(c) As time approaches infinity ($h \rightarrow \infty$), what is [C]? What is $[\mathrm{C}]_{h\rightarrow\infty}$ when $J_1 = 0$?

Table 12.3 Reactions and rate coefficients for Problem 12.3

Reaction	Rate-coefficient expression
$O \xrightarrow{O_2,M} O_3$	$k_1 = 1.4 \times 10^3 e^{1175/T}$
$NO + O_3 \rightarrow NO_2 + O_2$	$k_2 = 1.8 \times 10^{-12} e^{-1370/T}$
$NO_2 + h\nu \rightarrow NO + O$	$J = 1.7 \times 10^{-2}$

In the first reaction, only O is included in the rate-coefficient expression, since the rate coefficient has already been multiplied by O_2 and M. T is absolute temperature (K).

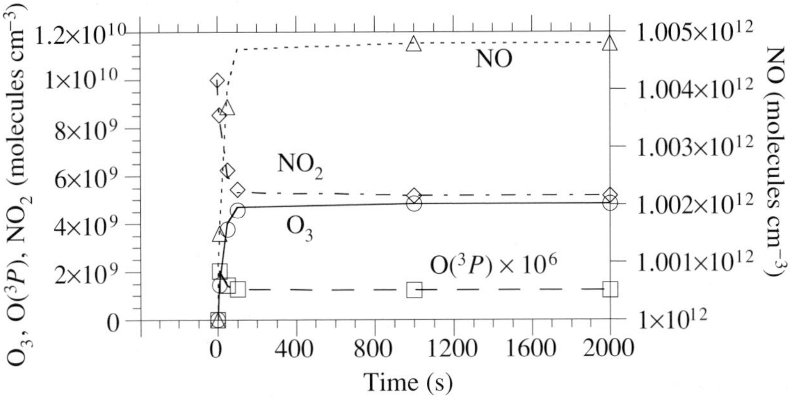

Figure 12.5 Exact and modeled solution to Problem 12.3. Exact concentrations are represented by continuous lines. Solutions from the MIE method are represented by symbols.

12.12 COMPUTER PROGRAMMING PRACTICE

12.3 Write a program to solve chemical ODEs with the MIE algorithm. Test the program with the three reactions shown in Table 12.3, assuming that $T =$ 298 K. Assume that the initial concentrations of O_3, $O(^3P)$, NO, and NO_2 are 0, 0, 10^{12}, and 10^{10} molec. cm^{-3}, respectively. Do not solve for O_2. Use a time step of $h = 10$ s. The results should exactly match those shown in Fig. 12.5.

12.4 Write a program to solve chemical ODEs with the simple exponential method. Test the program with the three reactions in Table 12.3, assuming $T = 298$ K. Assume that initial concentrations of O_3, $O(^3P)$, NO, and NO_2 are 0, 0, 10^{12}, and 10^{10} molec. cm^{-3}, respectively. Find a time step size, if any, that gives the results shown in Fig. 12.5. Discuss what happens to the solution at increasingly larger time steps.

12.5 Repeat Problem 12.4 with a forward Euler approximation. Determine the time step size at which the solution becomes unstable (concentrations

Table 12.4 Typical volume-mixing ratios of several gases at two altitudes

Gas	Volume mixing ratio (fraction) 25 km	0 km	Gas	Volume mixing ratio (fraction) 25 km	0 km
N_2	0.7808	0.7808	CH_3COOH	—	3.0 (−9)
O_2	0.2095	0.2095	CH_3COCH_3	—	7.5 (−10)
O_3	8.0 (−6)	4.0 (−8)	C_5H_8 (ISOP)	—	2.0 (−10)
H_2	6.0 (−7)	6.0 (−7)	CH_3Cl	3.6 (−10)	6.0 (−10)
H_2O	3.1 (−6)	0.01	CH_3CCl_3	2.0 (−11)	1.4 (−10)
H_2O_2	2.1 (−10)	3.0 (−9)	CCl_4	1.3 (−11)	1.1 (−10)
NO	1.1 (−9)	5.0 (−12)	$CFCl_3$	5.0 (−11)	2.7 (−10)
NO_2	1.9 (−9)	4.0 (−11)	CF_2Cl_2	2.2 (−10)	4.8 (−10)
N_2O	1.5 (−7)	3.0 (−7)	CF_2ClH	7.0 (−11)	1.2 (−10)
HNO_3	6.0 (−9)	5.0 (−10)	$CFCl_2CF_2Cl$	3.0 (−11)	7.0 (−11)
HO_2NO_2	3.4 (−10)	1.0 (−11)	HCl	1.3 (−9)	9.0 (−11)
CO	4.0 (−8)	1.1 (−7)	$ClONO_2$	8.0 (−10)	—
CO_2	3.75 (−4)	3.6 (−4)	HOCl	3.3 (−11)	—
CH_4	1.0 (−6)	1.8 (−6)	CH_3Br	1.0 (−11)	1.2 (−11)
C_2H_6	—	6.0 (−10)	$BrONO_2$	2.0 (−12)	—
C_3H_8	—	4.8 (−11)	HBr	3.0 (−12)	—
C_2H_4	—	2.1 (−11)	SO_2	1.0 (−11)	5.0 (−11)
C_3H_6	—	6.0 (−12)	CH_3SCH_3	—	1.0 (−10)
HCHO	—	2.0 (−10)	H_2S	—	5.0 (−11)
CH_3CHO	—	1.6 (−10)	OCS	1.0 (−10)	5.0 (−10)
HCOOH	—	1.8 (−9)	CS_2	2.0 (−11)	1.0 (−10)
CH_3OOH	—	1.2 (−9)			

6.0 (−6) means 6.0×10^{-6}. Multiply the volume mixing ratio of a gas (molecules of gas per molecule of dry air) by the number concentration of dry air (molec. cm^{-3}) to obtain the number concentration of the gas. The 0-km mixing ratios correspond to clean tropospheric conditions.

oscillate between positive and negative values or explode to large positive or negative numbers).

12.6 Repeat Problem 12.4 with a backward Euler approximation. Discuss differences between the results from Problems 12.5 and 12.6.

12.13 MODELING PROJECT

Using the program developed for Problem 12.3, read in the equations and rate coefficients from the computer file developed for Problems 11.4 and 10.7. Run a four-day simulation, starting at 5:00 a.m. on the first day. Calculate initial number concentrations for necessary species from volume mixing ratios shown in Table 12.4. If a stratospheric mechanism was chosen for Problem 11.4, assume $T = 192$ K and $p_d = 25$ hPa, and use the mixing ratios at 25 km in Table 12.4. If a near-surface mechanism was chosen, assume $T = 298.15$ K and $p_d = 1013$ hPa, and use the mixing ratios at 0 km, which correspond to clean lower-tropospheric conditions. Assume all initial mixing ratios not given in the table are zero. Once a baseline simulation has been run, vary temperature, pressure, and initial mixing ratios to test the sensitivity of model predictions to these parameters. Display time-series plots of species mixing ratios from the baseline and sensitivity test cases on the same graphs. Discuss results and their significance.

13

Particle components, size distributions, and size structures

PARTICLES in the atmosphere vary in size and composition. In this chapter, their importance and treatment in numerical models are discussed. The variation of particle concentration with size can be simulated with a continuous or discrete size distribution. Three continuous distributions discussed here are the lognormal, Marshall–Palmer, and modified gamma distributions. These distributions can be discretized over a model size grid, which consists of size bin increments in radius space. The variation of a size distribution with time is simulated with a size structure. Five structures discussed here are the full-stationary, full-moving, quasistationary, hybrid, and moving-center structures. In later chapters, the treatment of time-dependent processes that affect particle size distributions and composition are examined. These include emission, nucleation, coagulation, growth, evaporation, chemical equilibrium, interaction with clouds, aqueous chemistry, rainout, washout, sedimentation, and dry deposition.

13.1 INTRODUCTION TO PARTICLES

An **aerosol** is an ensemble of solid, liquid, or mixed-phase particles suspended in air. An **aerosol particle** is a single such particle. A **hydrometeor** is an ensemble of liquid, solid, or mixed-phase particles containing primarily water, suspended in or falling through the air. A **hydrometeor particle** is a single such particle. Examples of hydrometeor particles are fog drops, cloud drops, drizzle, raindrops, ice crystals, graupel, snowflakes, and hailstones. The main difference between an aerosol particle and a hydrometeor particle is that the latter contains much more water than does the former. Whereas almost all mass within a cloud drop and raindrop is liquid water, aerosol particles contain large fractions of other material as well. The sizes and number concentrations of aerosol particles and hydrometeor particles also differ. Table 13.1 compares characteristics of gas molecules, aerosol particles, and hydrometeor particles.

Aerosol particles are important because they affect health, air quality, cloud formation, meteorology, and climate. Submicrometer aerosol particles (those smaller than 1 μm in diameter) affect human health by penetrating to the deepest part of human lungs. Aerosol particles 0.2–1 μm in diameter that contain sulfate, nitrate, and organic carbon scatter light efficiently. Aerosol particles smaller than 1 μm that contain black carbon absorb light efficiently. Aerosol absorption and scattering

Table 13.1 Characteristics of gases, aerosol particles, and hydrometeor particles

	Typical diameter (μm)	Number concentration (cm^{-3})	Mass concentration ($\mu g\ m^{-3}$)
Gas molecules	0.0005	2.45×10^{19}	1.2×10^{9}
Aerosol particles			
Small	<0.2	10^{3}–10^{6}	<1
Medium	0.2–1.0	1–10^{4}	<250
Large	1.0–100	<1–10	<500
Hydrometeor particles			
Fog drops	10–20	1–500	10^{4}– 5×10^{5}
Average cloud drops	10–200	<10–1000	<10^{5}–5×10^{6}
Large cloud drops	200	<1–10	<10^{5}–5×10^{6}
Drizzle	400	0.1	10^{5}–5×10^{6}
Small raindrops	1000	0.01	10^{5}–5×10^{6}
Medium raindrops	2000	0.001	10^{5}–5×10^{6}
Large raindrops	8000	<0.001	10^{5}–5×10^{6}

Data are for typical lower-tropospheric conditions.

affect (1) radiative energy fluxes, which affect temperatures, and (2) photolysis, which affects the composition of the atmosphere. Aerosol particles also serve as sites on which cloud drops form. In fact, without aerosol particles, clouds would rarely form in the atmosphere. Finally, aerosol particles serve as sites on which chemical reactions take place and as sites for trace gases to condense upon or dissolve within.

13.2 AEROSOL, FOG, AND CLOUD COMPOSITION

Aerosol particles enter the atmosphere in one of only two ways: emission and homogeneous nucleation (Chapter 14). New particles emitted into the atmosphere are **primary particles.** New particles formed by homogeneous nucleation of gases are **secondary particles.** Homogeneous nucleation is a **gas-to-particle conversion** process because it results in the change of state of a gas to a liquid or solid. Other gas-to-particle conversion processes include heterogeneous nucleation (Chapter 14), condensation (Chapter 16), dissolution (Chapter 17), and heterogeneous reaction (Chapter 17). These processes, though, do not result in the formation of new particles. They result in the addition of mass and new chemicals to existing particles.

When new particles are emitted or formed by homogeneous nucleation, they are externally mixed. **Externally mixed** particles are those that contain only the components they are emitted with or homogeneously nucleated with. As particles age, **coagulation** (the collision and coalescence of particles) and gas-to-particle conversion internally mix particles. **Internally mixed** particles are particles that contain chemicals in addition to those the particles originally contained from emission or homogeneous nucleation.

Table 13.2 Liquids, ions, and solids found in atmospheric aerosol particles

Liquids			
Sulfuric acid	$H_2SO_4(aq)$	Formic acid	$HCOOH(aq)$
Nitric acid	$HNO_3(aq)$	Acetic acid	$CH_3COOH(aq)$
Hydrochloric acid	$HCl(aq)$	Ammonia	$NH_3(aq)$
Carbonic acid	$H_2CO_3(aq)$	Hydrogen peroxide	$H_2O_2(aq)$
Sulfurous acid	$H_2SO_3(aq)$	Organic matter	C, H, O, N(aq)
Ions			
Bisulfate ion	HSO_4^-	Sulfite ion	SO_3^{2-}
Sulfate ion	SO_4^{2-}	Hydrogen ion	H^+
Nitrate ion	NO_3^-	Ammonium ion	NH_4^+
Chloride ion	Cl^-	Sodium ion	Na^+
Bicarbonate ion	HCO_3^-	Calcium ion	Ca^{2+}
Carbonate ion	CO_3^{2-}	Magnesium ion	Mg^{2+}
Bisulfite ion	HSO_3^-	Potassium ion	K^+
Solids			
Ammonium bisulfate	$NH_4HSO_4(s)$	Potassium sulfate	$K_2SO_4(s)$
Ammonium sulfate	$(NH_4)_2SO_4(s)$	Potassium nitrate	$KNO_3(s)$
Triammonium bisulfate	$(NH_4)_3H(SO_4)_2(s)$	Potassium chloride	$KCl(s)$
Ammonium nitrate	$NH_4NO_3(s)$	Potassium bicarbonate	$KHCO_3(s)$
Ammonium chloride	$NH_4Cl(s)$	Potassium carbonate	$K_2CO_3(s)$
Sodium bisulfate	$NaHSO_4(s)$	Black carbon	BC(s)
Sodium sulfate	$Na_2SO_4(s)$	Organic matter	C, H, O, N(s)
Sodium nitrate	$NaNO_3(s)$	Silicon dioxide	$SiO_2(s)$
Sodium chloride	$NaCl(s)$	Iron (III) oxide	$Fe_2O_3(s)$
Calcium sulfate	$CaSO_4(s)$	Aluminum oxide	$Al_2O_3(s)$
Calcium nitrate	$Ca(NO_3)_2(s)$	Lead suboxide	$Pb_2O(s)$
Calcium chloride	$CaCl_2(s)$	Tire particles	
Calcium carbonate	$CaCO_3(s)$	Pollen	
Magnesium sulfate	$MgSO_4(s)$	Spores	
Magnesium nitrate	$Mg(NO_3)_2(s)$	Bacteria	
Magnesium chloride	$MgCl_2(s)$	Viruses	
Magnesium carbonate	$MgCO_3(s)$	Plant debris	
Potassium bisulfate	$KHSO_4(s)$	Meteoric debris	

Aerosol particles can contain chemicals in the liquid and/or solid phases and molecules that are inorganic and/or organic. When the relative humidity is high, the most abundant chemical in aerosol particles is typically liquid water. A solvent is a liquid in which gases dissolve. Because of its quantity within aerosol particles and the ability of many gases to dissolve in it, liquid water is generally considered the main solvent in aerosol particles. A substance that dissolves in a solvent is a **solute**. Solute and solvent, together, make up a **solution**. Dissolved substances can dissociate or remain undissociated in solution. A **dissociated species** has a positive (cationic) or negative (anionic) electric charge. An **undissociated species** has a neutral charge. Suspended materials, such as solids, may be mixed in a solution but are not part of the solution.

Table 13.2 lists several undissociated liquids, ions, and solids found in atmospheric aerosol particles. Appendix Table B.5 gives a more complete list that also includes components found in fog and clouds and identifies gas-phase precursors. The sources of aqueous sulfuric acid are homogeneous nucleation (Chapter 14), condensation (Chapter 16), aqueous reaction (Chapter 19), and some direct emission (Chapter 14). The sources of aqueous nitric acid are primarily dissolution of gas-phase nitric acid (Chapter 17) and some emission. The sources of aqueous hydrochloric acid are emission and dissolution of gas-phase hydrochloric acid. Carbonic acid originates mostly from dissolution of carbon dioxide gas into water within aerosol particles. Sulfurous acid originates primarily from dissolution of sulfur dioxide gas. Formic acid, acetic acid, ammonia, and hydrogen peroxide similarly originate from dissolution of their gas-phase counterparts. Liquid organic matter originates from emission, condensation, and dissolution.

Ions within aerosol particles originate from dissociation of liquids or solids. Many solids (e.g., ammonium sulfate, potassium nitrate) form from chemical reactions among ions in solution. Some solids (e.g., calcium carbonate, silicon dioxide, some iron (III) oxide) are emitted naturally in wind-blown soil dust. Other solids (e.g., pollen, spores, bacteria, viruses, plant debris) are biological matter lifted by the wind. Still others (e.g., black carbon, lead suboxide, some iron (III) oxide) are emitted anthropogenically.

13.3 DISCRETE SIZE DISTRIBUTIONS

The first step in designing algorithms to simulate particle size and composition in an atmospheric model is to choose an initial particle size distribution. A particle **size distribution** gives the number concentration, surface-area concentration, volume concentration, or mass concentration of particles as a function of radius or diameter. Particle radii are usually expressed in micrometers (microns) for aerosol particles and cloud drops, and millimeters for raindrops.

A size distribution may be continuous or discrete (e.g., Friedlander 1977). A **continuous distribution** is one in which the variation of particle concentration with radius is represented by a continuous function. A **discrete distribution** is one in which concentrations are distributed over increments in radius space, called **size bins** (or sections).

Some atmospheric models assume the aerosol size distribution is continuous (e.g., Hulburt and Katz 1964; McGraw and Saunders 1984; Pratsinis 1988; Binkowski and Shankar 1995; Terry *et al.* 2001). Others assume the aerosol size distribution is discrete (e.g., Turco *et al.* 1979; Gelbard and Seinfeld 1980; Toon *et al.* 1988; Jacobson 1997a,b; Meng *et al.* 1998; Zhang *et al.* 2004). Zhang *et al.* (1999) compare some algorithms developed under these different assumptions. The main advantage of a continuous distribution is that it reduces computer time because relatively few parameters are stored and calculated relative to a discrete distribution. The accuracy of aerosol dynamics with a continuous distribution can also be high for many idealized cases because analytical solutions are available in such cases. However, when aerosol particles contain multiple components and several physical and chemical processes are treated simultaneously, the relatively sparse information stored with continuous-distribution techniques increases the error of such techniques relative to discrete-distribution techniques. For complex aerosol interactions where computer time is not a major constraint, the use of a discrete distribution with high resolution is ideal for treating aerosol dynamics.

A size bin in a discrete distribution has a lower and an upper edge radius (or diameter). All particles with radii between the lower and upper edges of a size bin are assumed to reside in the bin and to have the same diameter as each other. In reality, each particle in the atmosphere has a distinct diameter. The lower atmosphere contains on the order of 10^3–10^6 particles per cubic centimeter of air, and particle diameters vary by over six orders of magnitude, from 0.001 μm (nucleated aerosol particles) to 5000 μm (large dust particles). Each particle can contain from 1 to 1000 components.

Example 13.1

An idealized particle size distribution might consist of 10 000 particles of radius between 0.005 and 0.5 μm, 100 particles of radius between 0.5 and 5.0 μm, and 10 particles of radius between 5.0 and 50 μm. The particles can be naturally distributed into a model that uses three size bins.

The number of particle size bins and components in a model is limited for two reasons. First, a typical three-dimensional model contains 10^3–10^6 grid cells, and the inclusion of too many particle sizes and/or components results in computer-time requirements that exceed those available. Second, the minimum storage requirement for particle concentrations over a model grid domain is one array with a dimension equal to the number of grid cells in the domain multiplied by the number of size bins and the number of components. This dimension can quickly surpass the central memory capabilities of many computers.

Example 13.2

With 10^5 grid cells, 100 size bins, and 100 components per size bin, the central memory requirement, just to store one array for concentration, is one gigaword (10^9 words). This translates to 8 gigabytes when 1 word = 8 bytes.

Limiting the number of sizes (or size bins) in a model is often the easiest method of reducing computer-time and memory limits. Reducing the number of size bins without losing resolution is difficult because particle sizes vary by many orders of magnitude. To cover all particle sizes with a fixed number of bins, the bins should be spread geometrically over the size range of interest. A discrete **geometric distribution**, called the **volume-ratio size distribution**, is discussed below. With

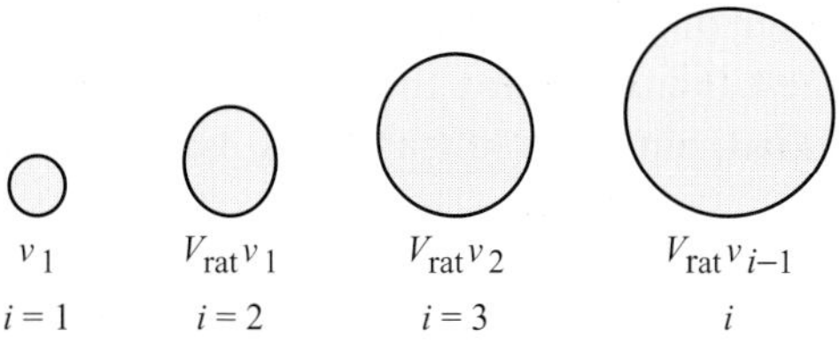

Figure 13.1 Variation in particle size with the volume-ratio size distribution. Below each particle is an expression for its single-particle volume.

this distribution, the volume of particles in a size bin equals the volume of particles in the next smallest size bin multiplied by a constant volume ratio, V_{rat}. Thus,

$$v_i = V_{rat} v_{i-1} \qquad \text{for} \quad i = 1, \ldots, N_B \tag{13.1}$$

where v_i is the volume (cm^3) of a single particle in a size bin i, v_{i-1} is the volume of a single particle in the next smaller size bin, and N_B is the number of size bins (e.g., Toon *et al.* 1988). The volume in a size bin can also be expressed as

$$v_i = v_1 V_{rat}^{i-1} \qquad \text{for} \quad i = 1, \ldots, N_B \tag{13.2}$$

where v_1 is the volume of a single particle in the smallest size bin. Figure 13.1 illustrates how particle sizes are distributed in the volume-ratio distribution.

For many model applications, it is convenient to consider particles as spherical. In such cases, the average volume of a single particle in any bin i is related to its average diameter d_i by $v_i = \pi d_i^3/6$. In reality, particles have a variety of shapes. Soot particles are porous and amorphous, whereas ice particles are crystalline. A highly resolved model can consider several particle types, each with a different size distribution, composition distribution, and shape distribution. However, particles are often (but not always) assumed to be spherical, since that may be closer to an average particle shape in the atmosphere than any other shape. Spherical geometry is also easier to treat numerically than other geometries.

In (13.1) and (13.2), V_{rat} must be specified in advance. A disadvantage of the equations is that if V_{rat} is specified incorrectly for a given N_B, the resulting volume of particles in the largest size bin, v_{N_B} may be smaller or larger than desired. To avoid this problem, V_{rat} can be precalculated from a specified v_{N_B} with

$$V_{rat} = \left(\frac{v_{N_B}}{v_1}\right)^{1/(N_B-1)} = \left(\frac{d_{N_B}}{d_1}\right)^{3/(N_B-1)} \tag{13.3}$$

where d_1 is the diameter of particles in the smallest bin, and d_{N_B} is the diameter of particles in the largest bin, assuming the particles are spherical. If, instead, v_{N_B}, d_{N_B}, and V_{rat} are specified, but N_B is not, then N_B can be found with

$$N_B = 1 + \frac{\ln\left[(d_{N_B}/d_1)^3\right]}{\ln V_{rat}} \tag{13.4}$$

Example 13.3

Equation (13.3) predicts that, for 30 size bins to cover a preset diameter range from 0.01 to 1000 μm, we have $V_{rat} = 3.29$.

Example 13.4

If $d_1 = 0.01$ μm, $d_{N_B} = 1000$ μm, and $V_{rat} = 4$, (13.4) implies that 26 size bins are needed. If $V_{rat} = 2$, then 51 size bins are needed to cover the same diameter range.

The volume width of a size bin in the volume-ratio distribution is derived by first assuming that the average single-particle volume in a bin is

$$v_i = \frac{1}{2}(v_{i,\text{hi}} + v_{i,\text{lo}}) \tag{13.5}$$

where $v_{i,\text{hi}}$ and $v_{i,\text{lo}}$ are the volumes of the largest and smallest particles in the bin, respectively. The volume of the largest particles in the bin is limited by

$$v_{i,\text{hi}} = V_{rat} v_{i,\text{lo}} \tag{13.6}$$

Substituting (13.6) into (13.5) gives

$$v_{i,\text{lo}} = \frac{2v_i}{1 + V_{rat}} \tag{13.7}$$

The **volume width** of a size bin is

$$\Delta v_i = v_{i,\text{hi}} - v_{i,\text{lo}} = \frac{2v_{i+1}}{1 + V_{rat}} - \frac{2v_i}{1 + V_{rat}} = \frac{2v_i(V_{rat} - 1)}{1 + V_{rat}} \tag{13.8}$$

and the corresponding **diameter width** is

$$\Delta d_i = d_{i,\text{hi}} - d_{i,\text{lo}} = \left(\frac{6}{\pi}\right)^{1/3} \left(v_{i,\text{hi}}^{1/3} - v_{i,\text{lo}}^{1/3}\right) = d_i 2^{1/3} \frac{V_{rat}^{1/3} - 1}{(1 + V_{rat})^{1/3}} \tag{13.9}$$

The average volume of one particle in a size bin is one parameter used in a discrete size distribution. Four other parameters are the number, volume, area, and mass concentrations. The number concentration n_i (particles cm^{-3}) is the number of particles in a size bin per unit volume of air, the volume concentration v_i (cm^3 cm^{-3}) is the volume of particles in a bin per unit volume of air, the surface-area concentration a_i (cm^2 cm^{-3}) is the surface area of particles in a bin per unit volume of air, and the mass concentration m_i (μg m^{-3}) is the mass of particles in a bin per unit volume of air. **Number concentration** is related to volume concentration and single-particle volume in a bin by

$$n_i = \frac{v_i}{\upsilon_i} \tag{13.10}$$

The total number concentration of particles, summed over all size bins in a distribution, is

$$N_D = \sum_{i=1}^{N_B} n_i \tag{13.11}$$

Individual particles usually consist of a mixture of a few to hundreds of components. For modeling purposes, the relationship between **total volume concentration** and **component volume concentrations** $v_{q,i}$ (cm^3 cm^{-3}) is

$$v_i = \sum_{q=1}^{N_V} v_{q,i} \tag{13.12}$$

where N_V is the number of components within each particle.

Assuming that particles are spherical, the **surface-area concentration** (cm^2 cm^{-3}) of particles in a size bin is

$$a_i = n_i 4\pi r_i^2 = n_i \pi d_i^2 \tag{13.13}$$

where $r_i = d_i/2$ is particle radius. The surface-area concentration is used for estimating particle surface area available for gas condensation or heterogeneous reaction.

The **mass concentration** of particles in a size bin is

$$m_i = \sum_{q=1}^{N_V} m_{q,i} = c_m \sum_{q=1}^{N_V} \rho_q v_{q,i} = c_m \rho_{p,i} \sum_{q=1}^{N_V} v_{q,i} = c_m \rho_{p,i} v_i \tag{13.14}$$

where $m_{q,i} = c_m \rho_q v_{q,i}$ is the mass concentration (micrograms per cubic meter of air) of component q in particles of size i, c_m (= 10^{12}) converts g cm^{-3} to μg m^{-3}, ρ_q is the mass density of q (grams per cubic centimeter of component), and $\rho_{p,i}$ is the **volume-averaged mass density** of a particle in bin i (grams per cubic centimeter of particle). The volume-averaged density changes continuously and is found by equating the third and fourth terms in (13.14). The result is

$$\rho_{p,i} = \sum_{q=1}^{N_V} (\rho_q v_{q,i}) \Big/ \sum_{q=1}^{N_V} v_{q,i} \tag{13.15}$$

Example 13.5

If the mass concentrations of water and sulfate are $m_{q,i}$ = 3.0 and 2.0 μg m^{-3}, respectively, in particles with an average diameter of d_i = 0.5 μm, calculate the volume concentration of each component and the mass, volume, area, and number concentrations of whole particles in the size bin. Assume the densities of water and sulfate are ρ_q = 1.0 and 1.83 g cm^{-3}, respectively.

SOLUTION

From $m_{q,i} = c_m v_{q,i} \rho_q$, the volume concentrations of water and sulfate are $v_{q,i} = 3 \times 10^{-12}$ and 1.09×10^{-12} cm^3 cm^{-3}, respectively. From (13.14), m_i = 5.0 μg m^{-3}. From (13.12), $v_i = 4.09 \times 10^{-12}$ cm^3 cm^{-3}. From $v_i = \pi d_i^3/6$, $v_i = 6.54 \times 10^{-14}$ cm^3. From (13.10), n_i = 62.5 particles cm^{-3}, and from (13.13), $a_i = 4.9 \times 10^{-7}$ cm^2 cm^{-3}.

13.4 CONTINUOUS SIZE DISTRIBUTIONS

Once a discrete model size distribution has been laid out, the initial particle number, volume, and mass concentrations must be distributed among model size bins. This can be accomplished by fitting measurements to a continuous size distribution, then discretizing the continuous distribution over the model bins. Three continuous distributions available for this procedure are the lognormal, Marshall–Palmer, and modified gamma distributions.

13.4.1 The lognormal distribution

A unimodal **lognormal distribution** is a bell-curve distribution on a log scale, as shown in Fig. 13.2(a). Figure 13.2(b) shows the same curve on a linear scale. When two, three, or four lognormal modes are superimposed on each other, the resulting lognormal distribution is bi-, tri-, or quadrimodal, respectively.

A lognormal distribution has a characteristic geometric-mean diameter (geometric-mean mass, volume, surface area, or number diameter), a geometric standard deviation, and a total mass, volume, surface area, or number concentration associated with it. Fifty percent of the area under a lognormal distribution lies below the **geometric-mean diameter**. It is analogous to the mean diameter of a normal distribution. The **geometric standard deviation** is defined so that about

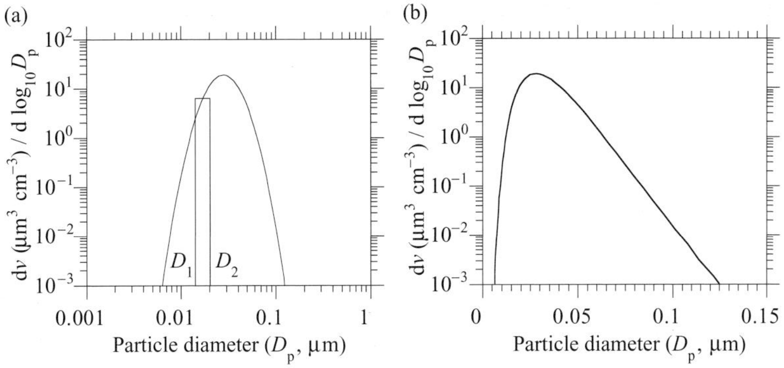

Figure 13.2 (a) A lognormal particle volume distribution. The volume concentration (dv) of material between any two diameters (D_1 and D_2) is estimated by multiplying the average value from the curve between the two diameters by $\log_{10} D_2 - \log_{10} D_1$. (b) The lognormal curve shown in (a), drawn on a linear scale.

68 percent of the area under the lognormal distribution lies within one geometric standard deviation of the geometric-mean diameter. About 95 percent of the area falls within two standard deviations of the geometric-mean diameter. The geometric standard deviation of a lognormal distribution is analogous to the standard deviation of a normal distribution.

13.4.1.1 *Obtaining parameters for a lognormal distribution*

Parameters for lognormal curves may be obtained from measurements. A low-pressure impactor is an instrument that collects particles of different size. An impactor with seven stages, for example, collects particles in seven size regimes, such as 0.05–0.075, 0.075–0.12, 0.12–0.26, 0.26–0.5, 0.5–1.0, 1–2, and 2–4 μm in diameter (e.g., Hering *et al.* 1979).

Suppose particles collected on a seven-stage impactor are weighed and their mass is plotted. If the plot is somewhat lognormal (e.g., if it looks like Fig. 13.2(a)), the **geometric-mean mass diameter** of the size distribution can be found from

$$\ln \bar{D}_M = \frac{1}{\mathrm{M}_L}\sum_{j=1}^{I_s}(m_j \ln d_j) \qquad \mathrm{M}_L = \sum_{j=1}^{I_s} m_j \tag{13.16}$$

where I_s is the number of stages (7 for a seven-stage impactor), M_L is the total mass concentration of particles (μg m^{-3}), summed over all impactor stages, and m_j is the mass concentration of particles with average diameter d_j in impactor stage j. The **geometric-mean volume diameter** of the distribution is then

$$\ln \bar{D}_V = \frac{1}{\mathrm{V}_L}\sum_{j=1}^{I_s}(v_j \ln d_j) \qquad \mathrm{V}_L = \sum_{j=1}^{I_s} v_j \qquad v_j = \frac{m_j}{c_{\mathrm{m}}\rho_j} \tag{13.17}$$

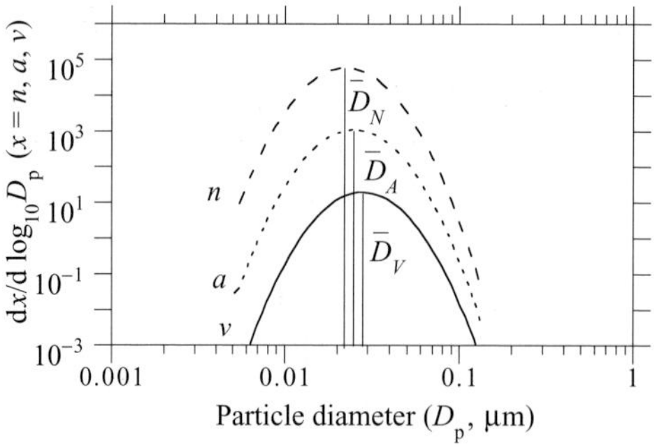

Figure 13.3 Number (n, particles cm^{-3}), area (a, cm^2 cm^{-3}), and volume (v, cm^3 cm^{-3}) concentrations and corresponding geometric-mean diameters for a lognormal distribution.

where V_L is the total volume concentration (cm^3 cm^{-3}) of particles, summed over all stages, v_j is the volume concentration of particles in impactor stage j, and ρ_j is the average density of particles in stage j. If the average density of particles is constant across all stages, $\bar{D}_V = \bar{D}_M$.

The geometric-mean surface-area diameter ($\bar{D}_A$) is always smaller than $\bar{D}_V$, and the geometric-mean number diameter ($\bar{D}_N$) is always smaller than $\bar{D}_A$ (Fig. 13.3). These latter parameters are given by

$$\ln \bar{D}_A = \frac{1}{\mathrm{A}_L}\sum_{j=1}^{I_s}(a_j \ln d_j) \qquad \mathrm{A}_L = \sum_{j=1}^{I_s} a_j \qquad a_j = \frac{3m_j}{c_\mathrm{m}\rho_j r_j} \tag{13.18}$$

$$\ln \bar{D}_N = \frac{1}{\mathrm{N}_L}\sum_{j=1}^{I_s}(n_j \ln d_j) \qquad \mathrm{N}_L = \sum_{j=1}^{I_s} n_j \qquad n_j = \frac{m_j}{c_\mathrm{m}\rho_j \upsilon_j} \tag{13.19}$$

where A_L is the total area concentration (cm^2 cm^{-3}) of particles, summed over all stages, a_j is the surface area concentration of particles in stage j, N_L is the total number concentration (particles cm^{-3}) of particles, summed over all stages, and n_j is the number concentration of particles in stage j.

From the geometric-mean diameters, the **geometric standard deviation** of the distribution is calculated with

$$\ln \sigma_\mathrm{g} = \sqrt{\frac{1}{\mathrm{M}_L}\sum_{j=1}^{I_s}\left(m_j \ln^2 \frac{d_j}{\bar{D}_M}\right)} = \sqrt{\frac{1}{\mathrm{V}_L}\sum_{j=1}^{I_s}\left(v_j \ln^2 \frac{d_j}{\bar{D}_V}\right)}$$

$$= \sqrt{\frac{1}{\mathrm{A}_L}\sum_{j=1}^{I_s}\left(a_j \ln^2 \frac{d_j}{\bar{D}_A}\right)} = \sqrt{\frac{1}{\mathrm{N}_L}\sum_{j=1}^{I_s}\left(n_j \ln^2 \frac{d_j}{\bar{D}_N}\right)} \tag{13.20}$$

Thus, σ_g is the same for a mass, volume, surface-area, or number distribution.

13.4.1.2 *Creating a model lognormal distribution*

Once parameters have been found for a lognormal mode, the mode can be discretized over a model size grid. The mass concentration of particles discretized from the continuous lognormal distribution to model size bin i is

$$m_i = \frac{\mathrm{M}_L \Delta d_i}{d_i \sqrt{2\pi} \ln \sigma_g} \exp\left[-\frac{\ln^2(d_i/\bar{D}_M)}{2 \ln^2 \sigma_g}\right] \tag{13.21}$$

where Δd_i was defined in (13.9) for the volume-ratio size distribution. Once mass concentration is known, the volume concentration, surface-area concentration, and number concentration for a size bin can be calculated from (13.14), (13.13), and (13.10), respectively. Alternatively, if V_L, A_L, or N_L is known for a lognormal mode, the volume, surface-area, or number concentration in a size bin can be calculated from

$$v_i = \frac{\mathrm{V}_L \Delta d_i}{d_i \sqrt{2\pi} \ln \sigma_g} \exp\left[-\frac{\ln^2(d_i/\bar{D}_V)}{2 \ln^2 \sigma_g}\right] \tag{13.22}$$

$$a_i = \frac{\mathrm{A}_L \Delta d_i}{d_i \sqrt{2\pi} \ln \sigma_g} \exp\left[-\frac{\ln^2(d_i/\bar{D}_A)}{2 \ln^2 \sigma_g}\right] \tag{13.23}$$

$$n_i = \frac{\mathrm{N}_L \Delta d_i}{d_i \sqrt{2\pi} \ln \sigma_g} \exp\left[-\frac{\ln^2(d_i/\bar{D}_N)}{2 \ln^2 \sigma_g}\right] \tag{13.24}$$

respectively. Figure 13.3 shows a plot of number, surface area, and volume concentrations for a lognormal distribution calculated from (13.22)–(13.24).

The total volume and number concentrations in a lognormal mode are related by

$$\mathrm{V}_L = \int_0^\infty v_d \, \mathrm{d}d = \frac{\pi}{6} \int_0^\infty n_d d^3 \, \mathrm{d}d = \frac{\pi}{6} \bar{D}_N^3 \exp\left(\frac{9}{2} \ln^2 \sigma_g\right) \mathrm{N}_L \tag{13.25}$$

Thus, if N_L is known, V_L and v_i can be found from (13.25) and (13.22), respectively.

13.4.1.3 *Multiple lognormal modes and particle components*

Particle distributions in the atmosphere are often described by up to four lognormal modes. Such modes may include a nucleation mode, two subaccumulation modes, and a coarse-particle mode. The **nucleation mode** (geometric-mean diameter <0.1 μm) contains homogeneously nucleated or small emitted particles. Sulfuric acid and water homogeneously nucleate together under the right conditions. Hot organic and inorganic vapors from combustion nucleate quickly to form small particles. Automobiles also emit soot particles (black carbon plus organic matter) in the nucleation mode. Small nucleated or emitted particles increase in size by coagulation and growth.

Growth and coagulation move particles into the **accumulation mode**, where diameters are 0.1–2 μm. Some of these particles are lost by rainout and washout, but they are too light to sediment out significantly. Particles in the nucleation and accumulation modes, together, are **fine particles**. The accumulation mode often consists of two sub-modes with geometric-mean mass diameters near 0.2 μm and 0.5–0.7 μm, respectively (Hering and Friedlander 1982; John *et al.* 1989). Hering and Friedlander observed that the mass-median diameter of particles containing sulfate in an urban area was about 0.20 μm on dry days (f_r = 17–68 percent) and 0.54 μm on moist days (f_r = 26–100 percent).

The **coarse-particle mode** consists of particles larger than 2-μm diameter. These particles originate from wind-blown dust, sea spray, volcanos, plants, and other sources and are generally heavy enough to sediment out (Whitby 1978). The emission sources and deposition sinks of fine particles differ from those of coarse particles. Fine particles usually do not grow by condensation much larger than 1 μm, indicating that coarse-mode particles originate primarily from emission.

The mass of individual components within particles can be described with lognormal modes different from those of the whole particle. Mylonas *et al.* (1991) found that the distribution of organonitrates in ambient aerosol particles is typically bimodal. Venkataraman and Friedlander (1994) found that ambient black carbon had two distinct modes, one between 0.05 and 0.12 μm and the other between 0.5 and 1 μm in diameter. Sulfate in coarse-mode particles has been observed in Los Angeles to have a larger geometric-mean mass diameter than nitrate (Noll *et al.* 1990). Sodium and chloride appear mostly in large sea-spray particles. Kritz and Rancher (1980) found the geometric-mean volume diameter particles containing NaCl over the ocean to be about 6.9 μm.

To initialize a model size distribution with several particle components, each with multiple lognormal modes, the following steps may be taken. First, lognormal parameters, such as the total mass concentration $\mathrm{M}_{L,q,k}$, geometric-mean mass diameter $\bar{D}_{M,q,k}$, and geometric standard deviation $\sigma_{g,q,k}$, are obtained for each mode k and component q with (13.16) and (13.20). Second, the mass concentration of each component ($m_{q,i,k}$) is distributed into each size bin i from each lognormal mode with (13.21). When a highly resolved model size distribution is used (e.g., as V_{rat} approaches unity), (13.21) conserves mass. Thus,

$$\sum_{i=1}^{N_B} m_{q,i,k} \approx \mathrm{M}_{L,q,k} \tag{13.26}$$

Third, the mass concentration of component q in bin i is summed over all lognormal modes. When four lognormal modes are used, the summation is

$$m_{q,i} = m_{q,i,1} + m_{q,i,2} + m_{q,i,3} + m_{q,i,4} \tag{13.27}$$

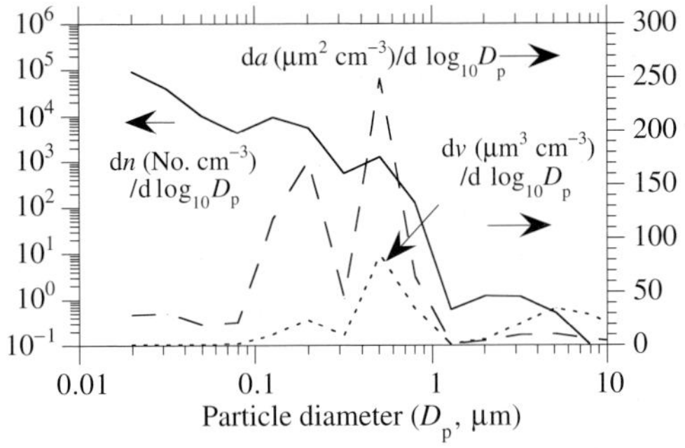

Figure 13.4 Number-, area-, and volume-concentration size distribution of particles at Claremont, California, during the morning of August 27, 1987. Sixteen model size bins and four lognormal modes were used to simulate the distribution (Jacobson 1997a).

Fourth, the volume concentration of component q in size bin i is calculated from the summed mass concentration as

$$v_{q,i} = \frac{m_{q,i}}{c_{\rm m}\rho_q} \tag{13.28}$$

The number concentration in bin i is then determined from

$$n_i = \frac{1}{\upsilon_i}\sum_{i=1}^{N_{\rm V}} v_{q,i} \tag{13.29}$$

which was obtained by substituting (13.12) into (13.10).

Figure 13.4 shows a quadrimodal lognormal distribution, fitted from data at Claremont, California, for the morning of August 27, 1987. The particles contained organic carbon, black carbon, ammonium, nitrate, sodium, chloride, liquid water, and crustal material (e.g., Mg, Al, Si, K, Ca, and Fe). All four modes are most noticeable in the number concentration distribution. The nucleation mode is less noticeable in the surface-area concentration distribution and invisible in the volume concentration distribution.

Table 13.3 shows trimodal lognormal parameters for a size distribution that describes typical continental particles. The table shows that the nucleation mode has the highest particle number concentration, the accumulation mode has the highest surface-area concentration, and the coarse mode has the highest volume concentration of particles.

13.4.2 Marshall–Palmer distribution

A second type of continuous size distribution, used to parameterize raindrops, is the **Marshall–Palmer distribution** (Marshall and Palmer 1948). In discretized form,

Table 13.3 Lognormal parameters for typical continental aerosol particles

Parameter	Nucleation mode	Accumulation mode	Coarse-particle mode
σ_g	1.7	2.03	2.15
N_L (particles cm^{-3})	7.7×10^4	1.3×10^4	4.2
$\bar{D}_N$ (μm)	0.013	0.069	0.97
A_L (μm^2 cm^{-3})	74	535	41
$\bar{D}_A$ (μm)	0.023	0.19	3.1
V_L (μm^3 cm^{-3})	0.33	22	29
$\bar{D}_V$ (μm)	0.031	0.31	5.7

Source: Whitby (1978).

the distribution gives the drop number concentration (particles cm^{-3}) between diameters d_i (μm) and $d_i + \Delta d_i$ as

$$n_i = \Delta d_i n_0 e^{-\lambda_r d_i} \tag{13.30}$$

where $\Delta d_i n_0$ is the value of n_i at $d_i = 0$, and λ_r is an empirical parameter that depends on the rainfall rate. Marshall and Palmer found $n_0 = 8 \times 10^{-6}$ particles cm^{-3} μm^{-1} and $\lambda_r = 4.1 \times 10^{-3} R^{-0.21}$ μm^{-1}, where R is the rainfall rate in millimeters per hour. Typical rainfall rates are 1–25 mm h^{-1}. If (13.30) is applied to a volume-ratio size distribution, Δd_i is found from (13.9). The total number concentration (particles cm^{-3}) and liquid water content (g m^{-3}) in a Marshall–Palmer distribution are $n_T = n_0/\lambda_r$ and $w_L = 10^{-6}\rho_w \pi n_0/\lambda_r^4$, respectively, where ρ_w is the liquid-water density (g cm^{-3}) and 10^{-6} converts units of the variables defined. (Pruppacher and Klett 1997).

Example 13.6

Estimate the number concentration of raindrops, in the diameter range $d_i =$ 1.0 mm to $d_i + \Delta d_i = 2.0$ mm of a Marshall–Palmer distribution, when the rainfall rate is $R = 5$ mm h^{-1}. What are the total number concentration and liquid water content in the distribution?

SOLUTION

From (13.30), $n_i = 4.3 \times 10^{-4}$ particles cm^{-3} in the given size range. The total number concentration and liquid water content are thus $n_T = 0.0027$ particles cm^{-3} and $w_L = 0.34$ g m^{-3}, respectively.

Table 13.4 Modified gamma parameters, liquid water content, and total drop number concentration for several cloud types

Cloud type	A_g	α_g	γ_g	$r_{c,g}$ (μm)	Liquid-water content (g m^{-3})	Number conc. (particles cm^{-3})
Stratocumulus base	0.2823	5.0	1.19	5.33	0.141	100
Stratocumulus top	0.19779	2.0	2.46	10.19	0.796	100
Stratus base	0.97923	5.0	1.05	4.70	0.114	100
Stratus top	0.38180	3.0	1.3	6.75	0.379	100
Nimbostratus base	0.080606	5.0	1.24	6.41	0.235	100
Nimbostratus top	1.0969	1.0	2.41	9.67	1.034	100
Cumulus congestus	0.5481	4.0	1.0	6.0	0.297	100
Light rain	4.97×10^{-8}	2.0	0.5	70.0	0.117	0.001

When (13.31) is summed over all sizes, the result equals the value in the last column.
Source: Welch *et al.* (1980).

13.4.3 Modified gamma distribution

A third type of continuous size distribution is the **modified gamma distribution** (Deirmendjian 1969), which is used to approximate cloud drop and raindrop number concentration as a function of size. In discretized form, the distribution gives the drop number concentration (particles cm^{-3}) in size bin i as

$$n_i = \Delta r_i A_g r_i^{\alpha_g} \exp\left[-\frac{\alpha_g}{\gamma_g}\left(\frac{r_i}{r_{c,g}}\right)^{\gamma_g}\right] \tag{13.31}$$

where Δr_i is the radius width of the bin (μm), centered at a mean radius of r_i (μm), $r_{c,g}$ is a critical radius (μm), around which the entire size distribution is centered, and A_g, α_g, and γ_g are parameterized coefficients obtained from measurement. Table 13.4 shows parameters for several cloud types.

Example 13.7

Use the modified gamma size distribution equation to approximate the number concentration of drops between 14 and 16 μm in radius at the base of a stratus cloud.

SOLUTION

The mean radius of this size bin is $r_i \approx 15$ μm, and the radius width of the bin is $\Delta r_i = 2$ μm. From Table 13.4 and (13.31), the resulting number concentration of stratus base drops in the size bin is $n_i = 0.1506$ particles cm^{-3}.

13.5 EVOLUTION OF SIZE DISTRIBUTIONS OVER TIME

A **size structure** is defined here as a size distribution that evolves over time. Five size structures include the full-stationary, full-moving, quasistationary, hybrid, and moving-center structures. These are discussed below. In all structures, each size bin contains n_i particles per cubic centimeter of air, each particle in a size bin has the same average single-particle volume (υ_i) and composition as each other particle in the bin, and $v_i = n_i \upsilon_i$.

13.5.1 Full-stationary structure

In the **full-stationary size structure** (e.g., Turco *et al.* 1979; Gelbard and Seinfeld 1980), the average volume of particles in a size bin (υ_i) stays constant, but n_i and v_i change throughout a model simulation. During growth, υ_i does not increase; instead, n_i and v_i decrease in one size bin and increase in a larger bin. Thus, particles in a bin grow numerically, not by increasing in volume in their own bin, but by moving across diameter space to a larger size bin.

An advantage of the full-stationary structure is that it covers a wide size distribution in diameter space with relatively few size bins. During nucleation, new particles are placed into an existing size bin, usually the smallest. During emission, new particles are placed in or partitioned between existing size bins. During coagulation, particles from one size bin collide with and stick to particles from another bin, forming a larger particle, a fraction of which is partitioned by number between two bins. During transport, particles in one size bin and grid cell advect and replace particles with the same υ_i in adjacent grid cells. The full-stationary structure is convenient for nucleation, emission, coagulation, and transport because υ_i is always constant.

During growth with the full-stationary size structure, particles are "advected" in diameter space across size-bin boundaries just as they are advected in real space by the winds (Chapter 6). During growth (advection) in diameter space, some particles remain in the same size bin and others move to the next bin, where they are averaged with particles remaining in the new bin (some of the particles in the new bin also grow to the next larger bin). Thus, when growth occurs, information about the original composition of the growing particle is lost due to averaging, producing errors in the size distribution upon evaporation. For example, when liquid water condenses onto aerosol particles to form a fog, the original aerosol material from many small size bins is grown into a few large bins, and the composition of the material in each large bin is now an average of material from several small bins. Upon evaporation of the fog, aerosol material does not "remember" where it originated from, and evaporation results in a distribution of core aerosol material different from that prior to growth. Figure 13.5 illustrates this problem.

A second disadvantage of the full-stationary structure is that particle growth leads to numerical diffusion in diameter space just as Eulerian advection leads to numerical diffusion in real space. When particles in size bin A grow, the number of particles in a larger bin (e.g., bin B) increases. The new particles in bin B are

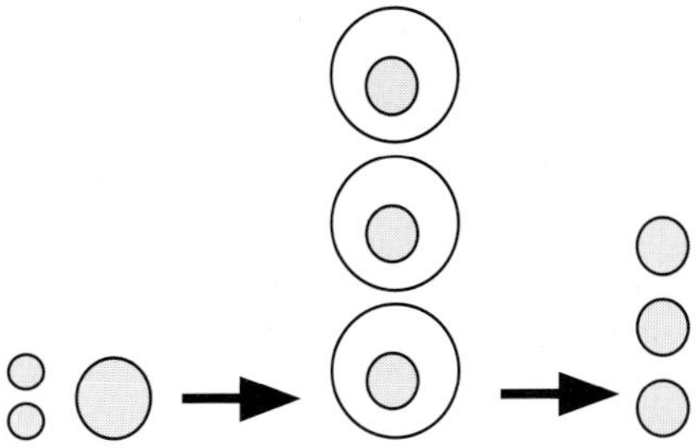

Figure 13.5 Demonstration of a problem with the full-stationary size structure. Suppose two particles exist in one size bin, one particle exists in a second size bin, and all three particles are composed of the same material. Suppose also that liquid water condenses on all particles, causing them to grow to the same total size. In a full-stationary size structure, the three particles are moved to one size bin with volume υ_i. Since the composition of all particles in a bin must be the same, the core volume of each particle must be averaged with those of all other particles in the bin. If the water now evaporates, the resulting distribution differs from the original distribution although it should not.

assumed to have the same average diameter, d_B, as all other particles in the bin. In reality, the new particles in bin B may be smaller than d_B. During the next time step, new particles at the average diameter in bin B grow to the next larger bin, C, where the average diameter is d_C. In reality, they should have grown from a diameter smaller than d_B to a diameter smaller than d_C and that may not even be in bin C. The artificial spreading of the distribution into bin C is numerical diffusion.

13.5.2 Full-moving structure

The extreme alternative to the full-stationary size structure is a **full-moving size structure** (e.g., Mordy 1959; Gelbard 1990). This structure may be initialized with a volume ratio or another size distribution. During growth, the number concentration (n_i) of particles in a size bin does not change. Instead, the single-particle volume (υ_i) changes. Once υ_i changes, the average volume of particles in one size bin no longer equals V_rat multiplied by the average volume of particles in a previous size bin and particle volumes no longer necessarily increase with increasing size-bin number. The full-moving structure is analogous to Lagrangian horizontal advection, just as the full-stationary size structure is analogous to Eulerian horizontal advection.

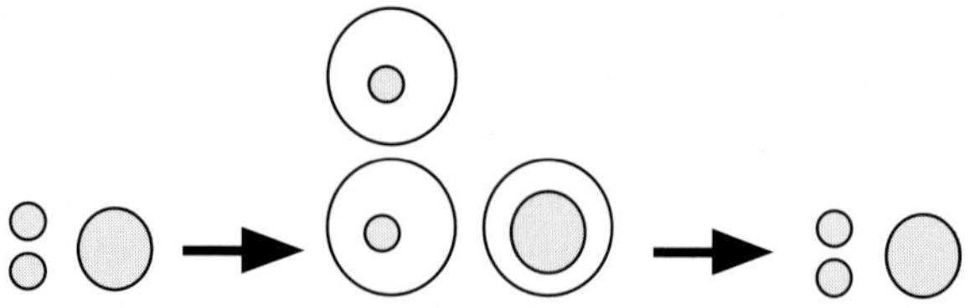

Figure 13.6 Preservation of an aerosol distribution upon growth and evaporation in a full-moving structure. When water condenses onto particles of different size, all particles may grow to the same total volume, but the core within each particle is not averaged with those of other particles of the same size. Thus, upon evaporation, the original core distribution is preserved.

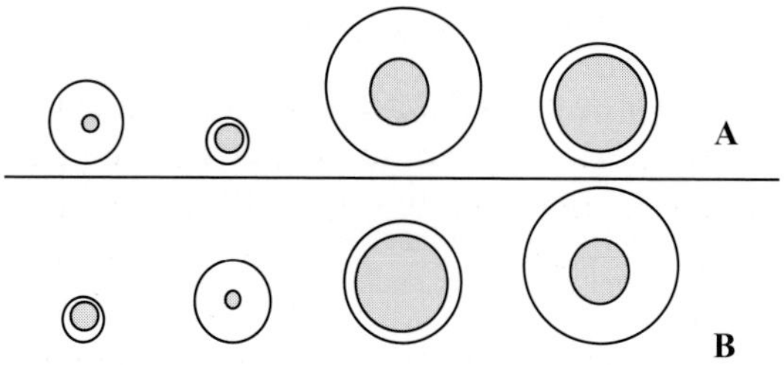

Figure 13.7 Reordering of particle size bins for coagulation in a full-moving structure. After growth, particles in each bin have new volumes, as illustrated by row A (the inner circle in each particle is core material, and the outer shell is volatile material). During coagulation, particle bins are reordered from smallest to largest total volume (row B), and coagulation proceeds in the same way as with a full-stationary structure.

One advantage of the full-moving size structure is that core particle material is preserved during growth. The problem shown in Fig. 13.5 for a full-stationary structure is not a problem for the full-moving structure, as shown in Fig. 13.6.

The second advantage of the full-moving structure is that it eliminates numerical diffusion during growth. When one particle grows from $d_i = 0.1$ to 0.15 μm in diameter in a full-moving structure, the particle physically attains this diameter and is not partitioned by volume between adjacent size bins. Particles in a full-moving structure grow to their exact size, eliminating numerical diffusion. When two particles coagulate, they form a larger particle whose total volume may be partitioned between the total volume of particles in two adjacent size bins (Jacobson 1994, 1997a). Thus, coagulation may be treated in the same way as in the full-stationary structure. During coagulation in the full-moving structure, size bins are reordered from smallest to largest each time step, since growth changes particle volume. Figure 13.7 illustrates bin reordering during coagulation.

The full-moving structure has disadvantages. Because it contains a finite number of size bins, each of which can grow to any volume, it has problems during nucleation, coagulation, and transport. Suppose all particles grow to fog-sized drops (around 10 μm diameter). In such a case, a size bin no longer exists for homogeneously nucleated particles. Similarly, if all particles grow to the same large size, no larger size bin may remain for particles to coagulate into. Transport also causes problems, since if particles in grid cell X and size bin i grow to volume $\upsilon_{X,i}$, and particles in cell Y and bin i grow to volume $\upsilon_{Y,i}$, then transport of particles from cell X to Y results in new particles in cell Y with volumes averaged between $\upsilon_{X,i}$ and $\upsilon_{Y,i}$. Averaging diminishes benefits of nondiffusive growth. Because of the problems described, the full-moving structure is not used in three-dimensional models.

13.5.3 Quasistationary structure

The **quasistationary size structure** (Jacobson 1997a) is similar to the full-stationary structure in most respects. With a quasistationary structure, particle volume grows to its exact size during one time step, but the adjusted volume is fitted back onto a stationary grid at each step. The simplest way to partition an adjusted volume is between two adjacent stationary size bins in a number- and volume-concentration-conserving manner. Suppose that particles in bin i of number concentration n_i and volume υ_i grow by condensation to actual volume υ_i' which lies between the volumes of two adjacent fixed size bins, j and k (i.e., $\upsilon_j \le \upsilon_i' < \upsilon_k$). The particles in bin i can be partitioned between those in bins j and k by solving (1) the number-conservation relationship

$$n_i = \Delta n_j + \Delta n_k \tag{13.32}$$

where Δn_j and Δn_k are the portions of number concentration n_i of grown particles added to size bins j and k, respectively, and (2) the volume-concentration-conservation relationship

$$n_i \upsilon_i' = \Delta n_j \upsilon_j + \Delta n_k \upsilon_k \tag{13.33}$$

The solution to this set of two equations and two unknowns is

$$\Delta n_j = n_i \frac{\upsilon_k - \upsilon_i'}{\upsilon_k - \upsilon_j} \qquad \Delta n_k = n_i \frac{\upsilon_i' - \upsilon_j}{\upsilon_k - \upsilon_j} \tag{13.34}$$

Like the full-stationary structure, the quasistationary structure allows practical treatment of nucleation, emission, and transport in three dimensions. Like the full-moving structure, it allows particles to grow to their exact sizes during a time step. However, because it partitions number and volume concentration after growth, the quasistationary structure is **numerically diffusive** and should be used cautiously.

13.5.4 Hybrid structure

The **hybrid structure** (Jacobson 1994; Jacobson and Turco 1995) is a method that divides particles into a core and a shell and treats the core with a full-stationary structure and the shell with a full-moving structure. The core is assumed to consist of relatively involatile material (e.g., black carbon, soil components, sodium) and the shell, of relatively volatile material (e.g., liquid water). The average core volume of particles in a size bin is held constant, but total particle (core plus shell) volume varies.

When two particles coagulate under this structure, they become one particle whose core volume (not total volume) is partitioned between two adjacent size bins in a number- and volume-concentration-conserving manner, such as with (13.34). During condensation of volatile material, particles grow to their exact sizes, eliminating numerical diffusion, but particles do not move out of the size bin that they start in since the size bin is controlled by the core volume, not the total volume. Because particles do not move to a different size bin during growth, their core volumes are preserved upon evaporation of the shell material.

The hybrid structure has a disadvantage with respect to transport over an Eulerian grid. Suppose particles in a small core bin in one grid cell grow by condensation to cloud-sized drops while those in the same bin in an adjacent cell do not. Transport from one cell to the next results in unrealistic averaging of the volatile material (but a realistic averaging of the core material). Thus, the hybrid structure, which is useful for box or Lagrangian modeling, is less useful for Eulerian modeling.

13.5.5 Moving-center structure

The **moving-center structure** (Jacobson 1997a) is a size structure in which size-bin edges are fixed but size-bin centers vary between the low and high edges of the bin. Thus, for example, $\upsilon_{i,\mathrm{hi}}$, $\upsilon_{i,\mathrm{lo}}$, and $\Delta\upsilon_i$ are fixed, and υ_i varies between $\upsilon_{i,\mathrm{hi}}$ and $\upsilon_{i,\mathrm{lo}}$. Because size-bin edges are fixed, homogeneous nucleation, emission, coagulation, and transport are treated in the same way as in the full-stationary structure. When growth occurs, though, particles in a size bin are allowed to grow to their exact size. If the average volume of a particle in a bin grows larger than the high-edge volume of the bin, all particles in the bin are moved to a single size bin bounding the average volume.

Because all particles are moved to the same size bin and not fractionated among two or more size bins, numerical diffusion is reduced with the moving-center structure. Some diffusion can occur, since the volume of particles moved to the larger (or smaller) bin is averaged with the volume of particles already in the bin. Figure 13.8 shows, though, that numerical diffusion from the moving-center structure was very small relative to that from the full-moving structure during a case of cloud-drop growth. The figure also shows that the quasistationary structure was very diffusive. Zhang *et al.* (1999) similarly found that the simulation of condensational growth with the moving-center structure resulted in the accurate predictions among several techniques compared.

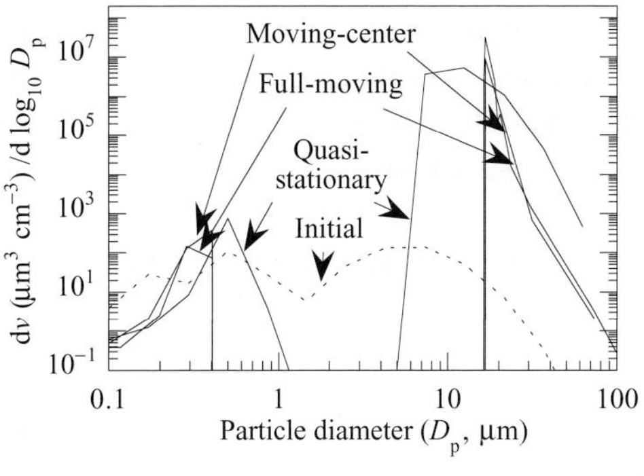

Figure 13.8 Comparison of moving-center, full-moving, and quasistationary size structure after growth of water onto aerosol particles to form cloud-sized drops. Growth occurred when the relative humidity was increased to 100.002 percent and replenished every 60-s time step for a 10-min period. Results from the moving-center and full-moving distributions were nondiffusive and nearly identical. The quasistationary structure was very diffusive.

When nucleation or emission occurs with the moving-center structure, new particles enter the size bin with a high- and low-edge volume surrounding that of the new particles, and the new and existing particles in the bin are averaged by volume. When transport occurs, particles in a bin move and replace particles in adjacent grid cells with the same high- and low-edge volumes. When two particles coagulate, they form one particle that is partitioned by volume between two adjacent size bins. A disadvantage of the moving-center structure is that evaporation following growth results in a distribution of core aerosol material different from the initial distribution prior to growth. This is the same problem that occurs with the full-stationary structure.

In sum, the moving-center structure maintains the main advantages of the full-stationary structure but also nearly eliminates numerical diffusion during growth. Because the moving-center structure treats nucleation, emission, coagulation, and transport realistically, it is more useful than the full-moving structure in three dimensions.

13.6 SUMMARY

In this chapter, particle composition and size distributions were introduced. Particles can contain hundreds of components, many of which form from gas-to-particle conversion and chemical reaction. Size distributions are used to describe number, surface area, volume, or mass concentrations of particles versus diameter. Size distributions may be continuous or discrete. Some types of continuous size distributions include the lognormal, Marshall–Palmer, and the modified gamma distributions. A model size structure simulates the change of a discrete size distribution over time. Size structures discussed include the full-stationary, full-moving, quasistationary, hybrid, and moving-center structures. The full-moving structure eliminates numerical diffusion during growth, but treats nucleation, emission, and transport unrealistically and is not used for Eulerian simulations. The full-stationary and quasistationary structures treat all these processes realistically, but both lead to numerical diffusion during growth. The hybrid structure eliminates numerical diffusion during growth but treats transport unrealistically. The moving-center structure treats nucleation, coagulation, and transport like the full-stationary structure, but minimizes numerical diffusion during growth. All structures have advantages and drawbacks.

13.7 PROBLEMS

13.1 Using the diameter information from Table 13.1, calculate how many (a) gas molecules, (b) medium aerosol particles, and (c) fog drops make up the volume of a single medium raindrop. Use average diameters where applicable.

13.2 Calculate the high-edge diameter of the largest size bin of a volume-ratio size distribution with 25 size bins and a volume ratio of adjacent size bins equal to 2 when the average diameter of the smallest bin is 1 nm.

13.3 If a population of 100-nm diameter particles has a number concentration of 10^5 particles cm^{-3}-air and a mass density of 1.5 g cm^{-3}-particle, calculate the (a) area concentration, (b) volume concentration, and (c) mass concentration of particles. If the same mass concentration were distributed over particles 500-nm in diameter, what would be the number concentration of the population?

13.4 Calculate the number concentration of raindrops from the Marshall–Palmer distribution in the diameter ranges (a) 200–300 μm and (b) 800–900 μm, when $R = 25$ mm h^{-1}. Why are the number concentrations different in the two size ranges?

13.5 Compare the number concentrations of drops between 18 and 22 μm in diameter at the base and at the top of a nimbostratus cloud using a modified gamma distribution. Why do you think the concentrations differ in the two cases?

13.6 Suppose 10^5 particles cm^{-3} grow by condensation to a diameter of 250 nm. If this size falls between two fixed size bins of respective diameters 100 nm and 200 nm, calculate the number concentration of the grown particles partitioned to each of the two fixed bins with the quasistationary size structure assuming all particles are partitioned and number concentration and volume concentration are conserved.

13.8 COMPUTER PROGRAMMING PRACTICE

13.7 Write a program to find υ_i, $\upsilon_{i,\mathrm{lo}}$, $\upsilon_{i,\mathrm{hi}}$, $\mathrm{d}\upsilon_i$, and $\mathrm{d}d_i$ for a volume-ratio size distribution when $d_1 = 0.005$ μm, $d_{N_B} = 500$ μm, and $V_{rat} = 1.5$. Print a table of results.

13.8 **(a)** Calculate $\bar{D}_M$ and σ_g for the size distribution resulting from the data in the accompanying table. The data give a typical distribution of emitted soot particles over a freeway.

Stage j	Stage diameter d_j (μm)	Mass between d_j and d_{j+1} (μg m^{-3})	Stage j	Stage diameter d_j (μm)	Mass between d_j and d_{j-1} (μg m^{-3})
8	4		4	0.26	0
7	2	0	3	0.12	2.61
6	1	0	2	0.075	47.1
5	0.5	0	1	0.05	10.5

(b) Using $\bar{D}_M$ and σ_g from (a), replot the size distribution on a curve of d_i versus $\mathrm{d}m/(M_L \mathrm{d}\log_{10} d_p)$, where $\mathrm{d}\log_{10} d_p = \mathrm{d}\log_{10} d_i = \log_{10} d_{i+1} - \log_{10} d_i$, M_L is the total mass concentration (μg m^{-3}), and $\mathrm{d}m$ is the mass concentration in each size bin (subscript i refers to the model distribution size bin and subscript j refers to the data distribution size bin). To obtain the curve, use a volume-ratio size distribution with $V_{rat} = 1.5$, and assume the particles are spherical. Choose appropriate values for d_1 and N_B.

(c) If soot is emitted primarily in small particles, why is it observed in larger particles away from emission sources?

14

Aerosol emission and nucleation

THE two processes that increase the number concentration of particles in the atmosphere are emission and homogeneous nucleation. Horizontal advection, vertical convection/eddy diffusion, and sedimentation also move particles into or out of a region. Particle emission originates from natural and anthropogenic sources. Natural sources include wind uplift of sea spray, soil dust, pollen, spores, and bacteria, volcanic outgassing, natural biomass fires, and lightning. Anthropogenic sources include fossil-fuel combustion, biofuel burning, biomass burning, and wind uplift of soil dust over eroded land. Globally, particle emission rates from natural sources exceed those from anthropogenic sources. In urban areas, the reverse is true. Nucleation may occur on the surface of an existing particle (heterogeneous nucleation) or in the absence of pre-existing surfaces (homogeneous nucleation). Nucleation may also involve the agglomeration of one, two, or more types of molecules. This chapter discusses aerosol emission and nucleation.

14.1 AEROSOL EMISSION

Emission is a source of primary particles in the atmosphere. In the subsections below, several types of natural and anthropogenic particle emission sources are discussed.

14.1.1 Sea-spray emission

The most abundant natural aerosol particles in the atmosphere, in terms of mass, are sea-spray drops. **Sea spray** forms when winds and waves force air bubbles to burst at the sea surface (Woodcock 1953). Sea spray is emitted primarily in the coarse mode of the particle size distribution. Winds also tear off wave crests to form larger drops, called **spume drops**, but these drops fall back to the ocean quickly. Sea spray initially contains all the components of seawater. About 96.8 percent of sea-spray weight is water and 3.2 percent is sea salt, most of which is sodium chloride. Table 14.1 shows the relative composition of the major constituents in seawater. The chlorine-to-sodium mass ratio in seawater, which can be obtained from the table, is about 1.8 to 1.

When seawater is emitted as sea spray or spume drops, the chlorine-to-sodium mass ratio, originally 1.8:1, sometimes decreases because the chlorine is removed by sea-spray acidification (e.g., Eriksson 1960; Duce 1969; Martens *et al.* 1973; Hitchcock *et al.* 1980; Kritz and Rancher 1980). **Sea-spray acidification** occurs

Table 14.1 Concentrations and mass percentages of major constituents in seawater

Constituent	Concentration ($mg\,L^{-1}$)	Mass percentage	Constituent	Concentration ($mg\,L^{-1}$)	Mass percentage
Water	1 000 000	96.78	Sulfur	905	0.0876
Sodium	10 800	1.05	Calcium	412	0.0398
Chlorine	19 400	1.88	Potassium	399	0.0386
Magnesium	1290	0.125	Carbon	28	0.0027

Source: Lide (2003). The total concentration of all constituents is 1 033 234 $mg\,L^{-1}$.

when sulfuric or nitric acid enters a sea-spray drop and forces chloride to evaporate as hydrochloric acid. Some sea-spray drops lose all of their chloride in the presence of sulfuric or nitric acid.

The size of a sea-spray drop is also affected by dehydration. **Dehydration** (loss of water) occurs when water from a drop evaporates due to a decrease in the relative humidity between the air just above the ocean surface and that a few meters higher. Dehydration increases the concentration of solute in a drop.

Sea-spray and spume drop emission parameterizations have been developed by Monahan *et al.* (1986), Andreas (1992), Wu (1993), Smith and Harrison (1998), de Leeuw *et al.* (2000), and Martensson *et al.* (2003), among others, from measurements of drop concentration as a function of drop size and wind speed. For example, the parameterization of Monahan *et al.* (1986) gives the flux of sea-spray drops, $\Delta F_i^{\rm spray}$ (particles $cm^{-2}\,s^{-1}$), per unit radius interval $\Delta r_i = \Delta d_i/2$ (cm), where Δd_i is defined in (13.9), as

$$\frac{\Delta F_i^{\rm spray}}{\Delta r_i} = 1.373|\mathbf{v}_{h,10}|^{3.41}(1 + 0.057 r_i^{1.05})10^{1.19e^{-B^2}} r_i^{-3} \quad (14.1)$$

where r_i is the mean radius of the particle (μm) at the ambient relative humidity in the radius interval, $|\mathbf{v}_{h,10}|$ is the mean horizontal wind speed at 10 m ($m\,s^{-1}$), and

$$B = (0.38 - \log_{10} r_i)/0.65 \quad (14.2)$$

The parameterization is valid for $0.8 \le r_i \le 10$ μm and $|\mathbf{v}_{h,10}| < 20\,m\,s^{-1}$. The corresponding parameterization for spume drops is

$$\frac{\Delta F_i^{\rm spume}}{\Delta r_i} = \begin{cases} 0 & r_i < 10\,\mu m \\ 8.6 \times 10^{-6} e^{2.08|\mathbf{v}_{h,10}|} r_i^{-2} & 10 \le r_i < 75\,\mu m \\ 4.83 \times 10^{-2} e^{2.08|\mathbf{v}_{h,10}|} r_i^{-4} & 75 \le r_i < 100\,\mu m \\ 4.83 \times 10^{6} e^{2.08|\mathbf{v}_{h,10}|} r_i^{-8} & 100\,\mu m \le r_i \end{cases} \quad (14.3)$$

Figure 14.1(a) shows (14.1) and (14.3) as a function of wind speed for 10-μm-radius drops. Below wind speeds of about 9 $m\,s^{-1}$, most of the flux of drops

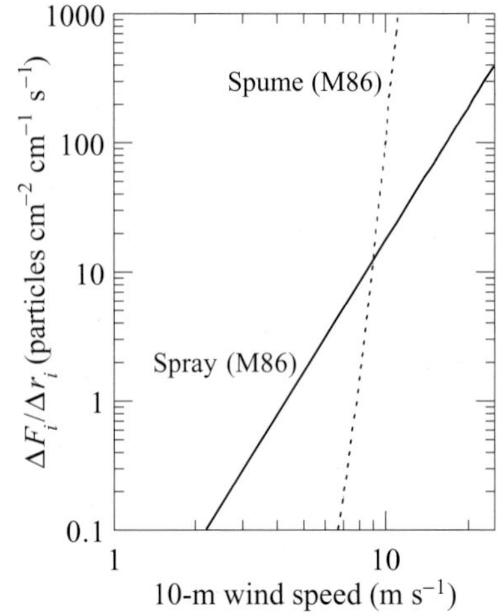

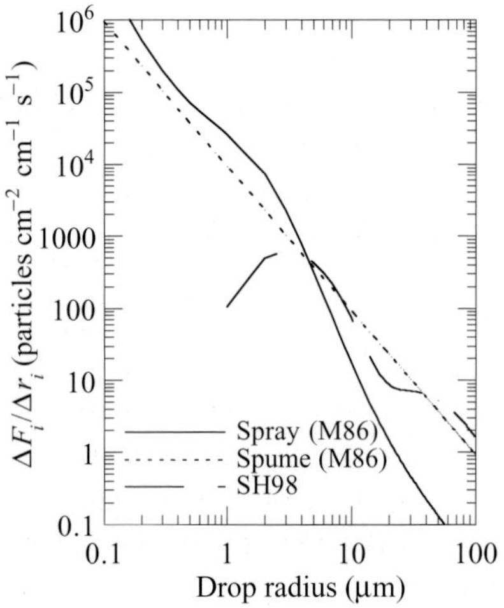

Figure 14.1 Fluxes of sea-spray and spume drops per radius interval as a function of (a) wind speed from (14.1) and (14.3) when the average radius is 10 μm and (b) drop radius from (14.1), (14.3), and (14.4) when the wind speed is 10 m s^{-1}. M86 is Monahan *et al.* (1986); SH98 is Smith and Harrison (1998).

is due to sea-spray generation (bursting of whitecap bubbles). Above 9 m s^{-1}, most generation is due to spume-drop generation (mechanical disruption of wave crests).

Guelle *et al.* (2001) suggested that the sea-spray parameterization of Monahan *et al.* (1986) is accurate for predicting observed concentrations of ocean-generated aerosol particles up to about 4-μm radius. They also suggested that, at higher radii, the parameterization of Smith and Harrison (1998), given as

$$\frac{\Delta F_i^{\text{spray}}}{\Delta r_i} = 0.2|\mathbf{v}_{\text{h},10}|^{3.5}\text{e}^{-1.5\ln(r_i/3)^2} + 0.0068|\mathbf{v}_{\text{h},10}|^{3}\text{e}^{-\ln(r_i/30)^2} \tag{14.4}$$

(where r_i is in μm and $|\mathbf{v}_{\text{h},10}|$ is in m s^{-1}) may be more accurate than Monahan *et al.*'s sea-spray parameterization. The valid ranges of this parameterization are 1–150 μm radius and 0–20 m s^{-1} wind speed. Figure 14.1(b) shows that the parameterization of Smith and Harrison produces results close to those of the spume-drop parameterization of Monahan *et al.* (1986) for particles larger than 4 μm diameter when the wind speed is 10 m s^{-1}, suggesting that spume drops may be important between 4 and 10 μm and the use of the Monahan *et al.* parameterization of spume drops in this size range may work just as well as the Smith and Harrison parameterization. At higher wind speeds (e.g., 20 m s^{-1}), though, (14.3) predicts much greater fluxes than does (14.4) at all sizes, so the use of (14.4), which was derived in part from high-wind-speed data, may prevent unrealistic buildup of ocean aerosol particles in such cases.

From sea-spray and spume-drop flux expressions, the number concentration of particles (particles cm^{-3}) emitted to the lowest model layer of thickness Δz (cm) during a time step h can be calculated with

$$\Delta n_i = \left(\frac{\Delta F_i^{\text{spray}}}{\Delta r_i} + \frac{\Delta F_i^{\text{spume}}}{\Delta r_i}\right)\frac{\Delta r_i}{\Delta z}h \tag{14.5}$$

The danger of this equation is that particles in a model are generally emitted into the middle of the lowest model layer, which may be 10–50 m above the surface. Most spume drops are emitted only a few meters before they fall back to the ocean surface. Thus, the use of (14.3) without considering the height of emission of spume drops can lead to a significant overestimate of spume drop concentrations in the mid boundary layer, which is why the use of (14.4), which predicts lower spume drop emission at high wind speeds, may be advantageous.

14.1.2 Soil- and fugitive-dust emission

A major emission source of aerosol particles in the atmosphere is the lifting of soil dust and road dust by the wind. **Soil** is the natural, unconsolidated mineral and organic matter lying above bedrock on the surface of the Earth. A **mineral** is a natural, homogeneous, inorganic, solid substance with a characteristic chemical composition, crystalline structure, color, and hardness. Minerals in soil originate from the breakdown of rocks. Soil organic matter originates from the decay of dead plants and animals.

Breakdown of rocks to soil material occurs primarily by **physical weathering**, which is the disintegration of rocks and minerals by processes that do not involve chemical reactions. Disintegration may occur when a stress is applied to a rock, causing it to break into blocks or sheets of different sizes and ultimately into fine soil minerals. Stresses arise when rocks are subjected to high pressure by soil or other rocks lying above. Stresses also arise when rocks freeze then thaw or when saline solutions enter cracks and cause rocks to disintegrate or fracture. Salts have higher thermal expansion coefficients than do rocks, so when temperatures increase, salts within rock fractures expand, forcing the rock to open and break apart. One source of salt for rock disintegration is sea spray transported from the oceans. Another source is desert salt (from deposits) transported by winds. Physical weathering of rocks on the Earth's surface can occur by their constant exposure to winds or running water. Rocks also disintegrate by **chemical weathering**, which is the breakdown of rocks by chemical reaction, often in the presence of water.

Soil-dust particles consist of the minerals and organic material making up soil. The Earth's crust contains several types of minerals. Pure **quartz** ($SiO_2(s)$) is a clear, colorless mineral that is resistant to chemical weathering. **Feldspars**, which make up at least 50 percent of the rocks on the Earth's surface, are by far the most abundant minerals on Earth. Two common types of feldspars are **potassium feldspar** ($KAlSi_3O_8(s)$) and **plagioclase feldspar** ($NaAlSi_3O_3$–$CaAl_2Si_2O_8(s)$). **Hematite** ($Fe_2O_3(s)$) is an oxide mineral because it includes a metallic element bonded with oxygen. The iron within it causes it to appear reddish-brown. **Calcite** ($CaCO_3(s)$) and **dolomite** ($CaMg(CO_3)_2(s)$) are carbonate minerals. **Gypsum** ($CaSO_4{\cdot}2H_2O(s)$) and **epsomite** ($MgSO_4{\cdot}7H_2O(s)$) are two of only a handful of sulfate-containing minerals. **Clays** are odorous minerals resulting from the weathering of rocks. They are usually soft, compact, and composed of aggregates of small crystals. Several types of clays include **kaolinite, illite, smectite, vermiculite**, and **chlorite**. In addition to containing minerals, soil dust contains organic matter, such as plant litter or animal tissue broken down by bacteria.

Soil dust is lifted into the air by winds. The extent of lifting depends on the wind speed and particle mass. Most mass of soil dust lifted into the air is in particles larger than 1 μm in diameter; thus, soil-dust particles, in terms of mass, are predominantly coarse mode particles. Those larger than 10 μm fall out quite rapidly, but those between 1 and 10 μm can stay in the air for days to weeks or longer, depending on the height to which they are originally lifted. Particles 1 μm in diameter fall 1 km in about 328 days, whereas those 10 μm in diameter fall 1 km in about 3.6 days. Although most soil-dust particles are not initially lofted more than 1 km in the air, many are, and these particles can travel long distances before falling to the ground.

Source regions of soil dust on a global scale include deserts (e.g., the Sahara in North Africa, Gobi in Mongolia and northern China, Mojave in southeastern California) and regions where foliage has been cleared by biomass burning and plowing. Deserts are continuously expanding due to land clearing and erosion on desert borders. Up to half of all wind-driven soil-dust emission worldwide may be anthropogenic (Tegen *et al.* 1996).

Wind-driven soil-dust emission parameterizations have been developed or described by Bagnold (1941); Gillette (1974); Greeley and Iversen (1985); Marticorena and Bergametti (1995); Shao *et al.* (1996); Alfaro and Gomes (2001); Shao (2001), among others. Below, the parameterization of Marticorena and Bergametti

(1995) is briefly discussed. This treatment assumes that soil grains resting on the surface experience three major forces: the downward force of gravity, interparticle cohesion forces, and wind-shear stress on their top surface. The effect of the first two forces depends on particle size; that of the third depends on the wind speed and the presence of roughness elements.

Gravity generally prevents particles larger than about 2000 μm in diameter from lifting off the ground. These particles, though, roll or **creep** along the ground in a strong wind. Particles 60–2000 μm in diameter may be lifted a few centimeters to a meter off the ground by the wind before falling back down. The wind may pick these particles up multiple times and transport them a little further each time. The transport of soil dust by intermittent, short leaps and bounces off the ground is called **saltation**, and the 1-m layer above the soil in which saltation occurs is the **saltation layer**. Particles smaller than 60 μm in diameter can be lifted above the saltation layer. These are the ones that enter a model atmosphere. Particles smaller than 10 μm in diameter, though, are the ones most likely to travel long distances.

The total **horizontal flux, integrated vertically** (g cm^{-1} s^{-1}), **of soil particles of all sizes in the saltation layer** can be calculated as

$$G = 2.61\frac{\rho_a}{g}u_*^3\sum_{i=1}^{N_s} f_{sa,i}\left(1+\frac{u_{*,i}^t}{u_*}\right)\left[1-\left(\frac{u_{*,i}^t}{u_*}\right)^2\right]\Delta d_{s,i} \qquad \text{for } u_* > u_{*,i}^t \tag{14.6}$$

where ρ_a is air density (g cm^{-3}), g is gravity (cm s^{-2}), u_* is the friction wind speed (cm s^{-1}), 2.61 is a constant of proportionality determined from wind-tunnel experiments, N_s is the number of soil grain size bins considered, $\Delta d_{s,i}$ is the diameter width of a soil grain size bin (cm) (from (14.9) for a volume-ratio size distribution), $f_{sa,i}$ is the fractional cross-sectional area concentration in soil size bin i among all soil particles, and $u_{*,i}^t$ is the threshold friction wind speed (cm s^{-1}).

The size distribution of soil grains may be described by 1–3 lognormal modes. Geometric mean diameters and standard deviations for sand in the first mode range from 80 to 200 μm and 1.2 to 2.7, respectively (e.g., Marticorena *et al.* 1997). Ranges in this mode differ slightly for silt, clay, and gravel. When multiple lognormal soil modes are considered, the **fractional cross-sectional area concentration of all soil grains in a size bin** is determined by

$$f_{sa,i} = \sum_{k=1}^{N_m}\frac{A_{L,s,k}}{A_{L,s,T}}\left(\frac{a_{s,i,k}}{A_{L,s,k}}\right) \tag{14.7}$$

where N_m is the number of lognormal modes, $A_{L,s,k}$ is the total surface-area concentration (cm^2 cm^{-3}) of soil particles in mode k, $A_{L,s,T}$ is the total surface-area concentration summed over all modes, $A_{L,s,k}/A_{L,s,T}$ is the **fractional area concentration of soil grains in mode *k* among all modes** (determined from soil data), $a_{s,i,k}$ is the surface-area concentration of particles in bin i of mode k, and

$$\frac{a_{s,i,k}}{A_{L,s,k}} = \frac{\Delta d_{s,i}}{d_{s,i}\sqrt{2\pi}\ln\sigma_{g,s,k}}\exp\left[-\frac{\ln^2(d_{s,i}/\bar{D}_{A,s,k})}{2\ln^2\sigma_{g,s,k}}\right] \tag{14.8}$$

is the fractional area concentration of particles in size bin i in mode k, obtained from (13.23). In this equation, $d_{s,i}$ is the diameter (cm) of soil particles in size bin i, $\bar{D}_{A,s,k}$ is the geometric mean cross-sectional area diameter in mode k (equal to the geometric-mean surface area diameter), and $\sigma_{g,s,k}$ is the geometric standard deviation of particles in the mode.

The **threshold friction wind speed** is the friction wind speed below which no horizontal soil flux occurs. In (14.6), only terms for which $u_* > u^{t}_{*,i}$ are treated. The threshold friction wind speed (cm s^{-1}) can be calculated with

$$u^{t}_{*,i} = \frac{u^{ts}_{*,i}}{f_{eff,i}} \tag{14.9}$$

where

$$u^{ts}_{*,i} = \begin{cases} \dfrac{0.129K_i}{\sqrt{1.928\mathrm{Re}_{s,i}^{0.092} - 1}} & 0.03 < \mathrm{Re}_{s,i} \le 10 \\ 0.129K_i[1 - 0.0858e^{-0.0617(\mathrm{Re}_{s,i}-10)}] & \mathrm{Re}_{s,i} > 10 \end{cases} \tag{14.10}$$

is the **threshold friction wind speed over a smooth surface** (cm s^{-1}) and

$$f_{eff,i} = 1 - \frac{\ln\left(\dfrac{z_0}{z_{0,s,i}}\right)}{\ln\left[0.35\left(\dfrac{10\,\mathrm{cm}}{z_{0,s,i}}\right)^{0.8}\right]} \tag{14.11}$$

is the ratio of the **friction wind speed over a smooth surface to that over the actual surface.** In these equations,

$$\mathrm{Re}_{s,i} = 1331d_{s,i}^{1.56} + 0.38 \tag{14.12}$$

is the dimensionless **friction Reynolds number,** which is the ratio of inertial forces to friction forces in the saltation layer. The diameter in the equation is in centimeters. In addition,

$$K_i = \sqrt{\frac{\rho_s g d_{s,i}}{\rho_a}\left(1 + \frac{0.006}{\rho_s g d_{s,i}^{2.5}}\right)} \tag{14.13}$$

is a prefactor (cm s^{-1}), where ρ_s is the density of soil (g cm^{-3}). Finally, z_0 is the surface roughness length for momentum (cm) over the actual terrain, and

$$z_{0,s,i} \approx \frac{d_{s,i}}{30} \tag{14.14}$$

is the surface roughness length over the terrain as if the terrain consisted of a smooth soil surface (cm) and where $d_{s,i}$ is in cm. Once the horizontal dust flux G is known, the **vertical soil dust flux** (g cm^{-2} s^{-1}) that escapes the saltation layer is estimated as

$$\Delta F_{\text{vert}}^{\text{soil}} = G\alpha_{\text{s}} \tag{14.15}$$

where

$$\alpha_{\text{s}} \approx 10^{13.4 f_{\text{clay}} - 6} \tag{14.16}$$

(cm^{-1}) is **the ratio of the vertical to horizontal dust flux**. This expression assumes that the ratio is a function of f_{clay}, the mass fraction of soil containing clay, which has a maximum range of 0–0.2 for use in this equation.

Finally, the change in the number concentration of soil dust particles (particles cm^{-3}) in aerosol size bin (not soil size bin) i in the bottom model layer of thickness Δz (m) over a time step h (s) due to emission is

$$\Delta n_i = f_{\text{em},i} \frac{\Delta F_{\text{vert}}^{\text{soil}}}{\rho_{\text{s}} \upsilon_i} \frac{h}{\Delta z} \tag{14.17}$$

where υ_i is single-particle volume (cm^3), and $f_{\text{em},i}$ is the mass fraction of the total vertical mass flux of emitted soil particles partitioned into aerosol bin i. The expression for $f_{\text{em},i}$ is similar to that for $f_{\text{sa},i}$ given in (14.7), except that, for $f_{\text{em},i}$, geometric mean mass diameters are used. The geometric mean mass diameter of the smallest mode is about 6–10 μm. Since only particles < 60 μm in diameter escape the saltation layer, the size distribution of vertically emitted particles is smaller than that of particles moving horizontally in the saltation layer.

Soil dust also enters the air from fugitive sources. **Fugitive emission** is the emission of particles during road paving, building construction and demolition, and operation of agricultural machines, trucks, and passenger vehicles. For example, tractors kick up soil dust with their treads. Passenger vehicles create turbulence that **resuspends** particles resting on paved and unpaved roads.

14.1.3 Volcanic eruptions

Over 1500 volcanos have been active in the last 10 000 years. Between 1975 and 1985, 158 volcanos erupted, some sporadically and others continuously (e.g., McClelland *et al.* 1989; Andres and Kasgnoc 1998). Volcanic eruptions result from the sudden release of gas dissolved in magma, which contains 1 to 4 percent gas by mass. Water vapor is the most abundant gas in magma, making up 50 to 80 percent of gas mass. Other important gases include carbon dioxide, sulfur dioxide, hydrogen sulfide (H_2S), carbonyl sulfide (OCS), and molecular nitrogen. Minor gases include carbon monoxide, molecular hydrogen, molecular sulfur, hydrochloric acid, molecular chlorine, and molecular fluorine.

Volcanos also emit particles that contain the elements of the Earth's mantle. The most abundantly emitted particle components are silicate minerals (minerals containing Si). Other components include aluminum, iron, sulfate, nitrate, chloride, sodium, calcium, magnesium, potassium, black carbon, and organic matter. The emission rate of sulfate particles is only about 3 percent that of sulfur dioxide gas (Andres and Kasgnoc 1998). The sizes of emitted volcanic particles range from

<0.1 to >100 μm in diameter. Particles 100 μm in diameter take about an hour to fall 1 km. The only volcanic particles that survive more than a few months before falling to the ground are those smaller than 4 μm. Such particles require no fewer than 23 days to fall 1 km. Volcanic particles of all sizes are also removed by rain.

Volcanic gases, such as carbonyl sulfide and sulfur dioxide are sources of new particles. When OCS is injected volcanically into the stratosphere, some of it photolyzes. The products react to form sulfur dioxide gas, which oxidizes to form sulfuric acid gas, which homogeneously nucleates to form new sulfuric acid–water aerosol particles. A layer of such particles, called the **Junge layer**, has formed in the stratosphere by this mechanism (Junge 1961). The average diameter of these particles is about 0.14 μm. Sulfuric acid, formed substantially in this manner in the stratosphere, is the dominant particle constituent, aside from liquid water, in the stratosphere and upper troposphere. More than 97 percent of particles in the lower stratosphere and 91 to 94 percent of particles in the upper troposphere contain oxygen and sulfur (Sheridan *et al.* 1994).

14.1.4 Biomass burning and forest fires

Biomass burning is the anthropogenic burning of evergreen forests, deciduous forests, woodlands, grassland, and agricultural land, either to clear land for other use, stimulate grass growth, manage forest growth, or satisfy a ritual. **Forest fires** are predominantly natural fires set by lightning but also unintentional or arson fires set by humans. About 90 percent of all fires worldwide are biomass burning fires; the rest are forest fires.

Biomass burning produces gases, such as carbon dioxide, carbon monoxide, methane, nitrous oxide, oxides of nitrogen, and reactive organic gases. It also produces particles, such as ash, plant fibers, soil dust, inorganic compounds, organic matter, and soot (e.g., Cooke and Wilson 1996; Liousse *et al.* 1996; Ferek *et al.* 1998; Reid *et al.* 1998; Andreae and Merlet 2001; Bond *et al.* 2004). **Ash** is the primarily inorganic solid or liquid residue left after biomass burning, but may also contain organic compounds oxidized to different degrees. **Organic matter** (OM) consists of carbon- and hydrogen-based compounds and often contains oxygen (O), nitrogen (N), etc., as well. It often appears white or sandy although it sometimes appears yellow, brown, or reddish due to absorption of ultraviolet radiation by some organic compounds.

Soot contains spherules of **black carbon** (BC) (also called elemental or graphitic carbon) bonded together in an amorphous shape and coated by organic matter. Pure black carbon contains only carbon atoms. The OM in soot contains aliphatic hydrocarbons, polycyclic aromatic hydrocarbons (PAHs), and small amounts of O and N (Chang *et al.* 1982; Kittelson 1998; Reid and Hobbs 1998; Fang *et al.* 1999). BC strongly absorbs visible light and gives soot its blackish color. The ratio of BC to OM produced in soot depends on temperature. High-temperature flames produce more BC than OM; low-temperature flames (such as in smoldering biomass) produce less. Because BC is black, the more BC produced, the blacker the smoke. Biomass burning produces about half of all soot worldwide. Most of the rest is emitted during fossil-fuel combustion.

Vegetation contains low concentrations of metals, including titanium (Ti), manganese (Mn), zinc (Zn), lead (Pb), cadmium (Cd), copper (Cu), cobalt (Co), antimony (Sb), arsenic (As), nickel (Ni), and chromium (Cr). These substances vaporize during burning then quickly recondense onto soot or ash particles. Young smoke also contains K^+, Ca^{2+}, Mg^{2+}, Na^+, NH_4^+, Cl^-, NO_3^-, and SO_4^{2-}. Young smoke particles are found in the upper-nucleation mode and lower-accumulation mode.

14.1.5 Biofuel burning

Biofuel burning is the anthropogenic burning of wood, grass, dung, and other biomass for electric power, home heating, and home cooking (e.g., Streets and Waldhoff 1998, 1999; Andreae and Merlet 2001; Bond *et al.* 2004). Biofuel burning releases most of the same gas and aerosol constituents as does biomass burning but often in different percentages because of different composition of fuel and temperature of burning. Home biofuel burning occurs primarily in developing countries where electric stoves and gas/electric heaters are uncommon. In developed countries, most biofuel burning is for power generation at centralized power plants.

14.1.6 Fossil-fuel combustion

Fossil-fuel combustion is another anthropogenic emission source. Major fossil fuels that produce aerosol particles include coal, oil, natural gas, gasoline, kerosene, diesel, and jet fuel.

Coal is a combustible brown-to-black carbonaceous sedimentary rock formed by compaction of partially decomposed plant material. The conversion of plant material to black coal goes through several stages: unconsolidated brown-black **peat** evolves to consolidated, brown-black **peat coal,** to brown-black **lignite coal,** to dark-brown-to-black **bituminous (soft) coal,** then to black **anthracite (hard) coal.**

Oil (or **petroleum**) is a natural greasy, viscous, combustible liquid (at room temperature) that is insoluble in water. It forms from the geological-scale decomposition of plants and animals and is made of hydrocarbon complexes. Oil is found in the Earth's continental and ocean crusts.

Natural gas is a colorless, flammable gas, made primarily of methane but also of other hydrocarbon gases, that is often found near petroleum deposits. As such, natural gas is often mined along with petroleum. **Gasoline** is a volatile mixture of liquid hydrocarbons derived by refining petroleum. **Kerosene** is a combustible, oily, water-white liquid with a strong odor that is distilled from petroleum and used as a fuel and solvent. **Diesel fuel** is a combustible liquid distilled from petroleum after kerosene. **Jet fuel** is a clear, colorless or yellowish combustible liquid distilled from kerosene and enhanced with additives.

Particle components emitted during the combustion of fossil fuels include soot (black carbon and organic matter), organic matter alone, sulfate, metals, and **fly**

Table 14.2 Metals present in fly ash of different industrial origin

Source	Metals present in fly ash
Smelters	Fe, Cd, Zn
Oil-fired power plants	V, Ni, Fe
Coal-fired power plants	Fe, Zn, Pb, V, Mn, Cr, Cu, Ni, As, Co, Cd, Sb, Hg
Municipal waste incineration	Zn, Fe, Hg, Pb, Sn, As, Cd, Co, Cu, Mn, Ni, Sb
Open-hearth furnaces at steel mills	Fe, Zn, Cr, Cu, Mn, Ni, Pb

Sources: Henry and Knapp 1980; Schroeder *et al.* 1987; Ghio and Samet 1999; Pooley and Mille 1999.

ash. Coal combustion results in the emission of submicrometer black carbon, organic matter, and sulfate and sub/supermicrometer fly ash and organic matter. The fly ash consists of oxygen, silicon, aluminum, iron, calcium, and magnesium in the form of quartz, hematite, gypsum, and clays. Combustion in gasoline engines usually results in the emission of submicrometer organic matter, some black carbon, sulfate, and elemental silicon, iron, zinc, and sulfur. Combustion of diesel fuel results in the emission of these components plus ammonium and greater amounts of black carbon and organic matter. Diesel-powered vehicles without particle emission-control devices emit 10 to 100 times more particulate mass than do gasoline-powered vehicles. Jet fuel combustion emits significant quantities of black carbon and sulfate.

Most soot from fossil-fuels originates from diesel-fuel, coal, and jet-fuel combustion (e.g., Cooke and Wilson 1996; Liousse *et al.* 1996; Cooke *et al.* 1999; Bond *et al.* 2004). Soot emitted during fossil-fuel combustion contains BC covered with a layer of polycyclic aromatic hydrocarbons and coated by a shell of organic and inorganic compounds (Steiner *et al.* 1992). In the case of diesel exhaust, most of the organic coating is lubricating oil (Kittelson 1998). Most fossil-fuel soot from vehicles is emitted in particles less than 0.2 μm in diameter (Venkataraman *et al.* 1994; ACEA 1999; Maricq *et al.* 1999). Thus, fossil-fuel combustion usually results in the emission of upper nucleation mode and lower accumulation mode particles.

14.1.7 Industrial sources

Many industrial processes involve burning of fossil fuels with metals. As such, industrial processes emit soot, sulfate, fly ash, and metals. Fly ash from industrial processes usually contains $Fe_2O_3(s)$, $Fe_3O_4(s)$, $Al_2O_3(s)$, $SiO_2(s)$, and various carbonaceous compounds that have been oxidized to different degrees (Greenberg *et al.* 1978). Fly ash is emitted primarily in the coarse mode (>2 μm in diameter). Metals are emitted during high-temperature industrial processes, such as waste incineration, smelting, cement kilning, and power-plant combustion. In such cases, heavy metals vaporize at high temperatures, then recondense onto soot and fly-ash particles that are emitted simultaneously. Table 14.2 lists some metals present in fly ash of different industrial origin. Of all the metals emitted industrially, iron is by far the most abundant. Lead is no longer used as an additive in gasoline in nearly every country, but it is still emitted industrially during lead-ore smelting, lead-acid battery manufacturing, lead-ore crushing, and solid waste disposal.

14.1.8 Miscellaneous sources

Several other particle types are emitted anthropogenically and naturally. **Tire-rubber particles** are emitted during the constant erosion of a tire at the tire-road interface. Such particles are mostly >2 μm in diameter, although some are smaller.

Pollen grains are moderately large (10–120 μm in diameter) granules containing male genetic material released from flowers and carried by the wind to other flowers for fertilization. They originate from trees, plants, grasses, and weeds. Pollen-grain mass density is about 0.9–1.1 g cm^{-3}. Pollen grains are generally released in the morning during a 2–14 day period in the spring. A single corn plant can emit 10–50 million pollen grains during the period. Of 25 million grains released by a plant and carried by the wind, about 0.5–1 percent may stay in the air 500 m downwind.

Spores are small (2–6 μm in diameter) one-celled or multicellular, asexual, reproductive or resting organisms that can develop into an adult on their own when the environment is favorable. They are usually released by fungi and algae and may be carried by the wind. Spore mass density is 0.5–1.5 g cm^{-3}. Spores are generally released in the afternoon, when the relative humidity is low. At relative humidities above 95 percent, spores are generally not released. Because they are smaller than are pollen grains, spores survive longer and have higher concentrations in the air than do pollen grains.

Bacteria are unicellular microorganisms found in nearly all environments. Their largest dimension is an average of 1 μm, with a range of 0.1–600 μm. **Viruses** are noncellular organisms that consist of a core of RNA or DNA enclosed by a coat of protein and/or an outer protective membrane. Their largest dimension ranges from 0.02 to 1 μm. **Plant debris** forms from the bacterial decay of plant matter. Bacteria, viruses, and plant debris can be lifted by the wind and carried long distances.

A stratosphere source of new particles is **meteoric debris.** Most meteorites disintegrate before they drop to an altitude of 80 km. Those that reach the stratosphere contain iron, titanium, and aluminum, among other elements. The net contribution of meteorites to particles in the stratosphere is small (Sheridan *et al.* 1994; Pruppacher and Klett 1997).

14.1.9 Summary of the emission sources of aerosol particles

Table 14.3 summarizes many of the components present in each of the nucleation, accumulation, and coarse particle modes and their major sources. Processes aside from emission will be discussed in subsequent sections and chapters.

Table 14.4 shows an estimate of particle component mass emission rates over a 400 × 150-km region of the Los Angeles basin for a day in August 1987. Emission rates are given for four size regimes: <1, 1–2.5, 2.5–10, and >10 μm in diameter.

Of the emission in the table, about 90 percent originated from area sources, 7 percent originated from mobile sources, and 3 percent originated from point

Table 14.3 Dominant sources of chemicals in nucleation, accumulation, and coarse mode particles

Nucleation mode	Accumulation mode	Coarse mode
Homogeneous nucleation $H_2O(aq)$, SO_4^{2-}, NH_4^+	Fossil-fuel emission BC, OM, SO_4^{2-}, Fe, Zn	Sea-spray emission H_2O, Na^+, Ca^{2+}, Mg^{2+}, K^+, Cl^-, SO_4^{2-}, Br^-, OM
Fossil-fuel emission BC, OM, SO_4^{2-}, Fe, Zn	Biomass-burning emission BC, OM, K^+, Na^+, Ca^{2+}, Mg^{2+}, SO_4^{2-}, NO_3^-, Cl^-, Fe, Mn, Zn, Pb, V, Cd, Cu, Co, Sb, As, Ni, Cr	Soil-dust emission Si, Al, Fe, Ti, P, Mn, Co, Ni, Cr, Na^+, Ca^{2+}, Mg^{2+}, K^+, SO_4^{2-}, Cl^-, CO_3^{2-}, OM
Biomass-burning emission BC, OM, K^+, Na^+, Ca^{2+}, Mg^{2+}, SO_4^{2-}, NO_3^-, Cl^-, Fe, Mn, Zn, Pb, V, Cd, Cu, Co, Sb, As, Ni, Cr	Industrial emission BC, OM, Fe, Al, S, P, Mn, Zn, Pb, Ba, Sr, V, Cd, Cu, Co, Hg, Sb, As, Sn, Ni, Cr, H_2O, NH_4^+, Na^+, Ca^{2+}, K^+, SO_4^{2-}, NO_3^-, Cl^-, CO_3^{2-}	Biomass-burning emission Ash, OM Industrial emission Fly-ash Tire-particle, pollen, spore, bacteria, plant debris emission
Condensation/dissolution $H_2O(aq)$, SO_4^{2-}, NH_4^+, OM	Condensation/dissolution $H_2O(aq)$, SO_4^{2-}, NH_4^+, OM	Condensation/dissolution $H_2O(aq)$, NO_3^-
	Coagulation from nucleation mode	Coagulation from smaller modes

sources. Some of the largest sources of PM_{10} (particles smaller than 10 μm in diameter) were fugitive dust (containing silicon, aluminum, and iron) from paved (32 percent) and unpaved (29.7 percent) roads, and construction/demolition emission (23.5 percent).

Table 14.4 shows that about 21 tons per day of particulate chloride were emitted in Los Angeles in August 1987. The important sources of land-based chloride are forest burning, gasoline combustion, agricultural burning, fireplace burning, chemical manufacturing, and soil dust (CARB 1988; Saxena *et al.* 1993). Almost all (98 percent) of anthropogenic hydrochloric acid gas emission is from coal combustion; much of the rest is from waste incineration (Saxena *et al.* 1993).

Table 14.4 shows that only 3 tons of particulate nitrate were emitted per day in 1987 in Los Angeles. The major atmospheric source of particulate nitrate is not emission but gas-to-particle conversion of nitric acid gas. The table also shows that about 28 tons per day of black carbon <1 μm in diameter were emitted. Much of this originated from automobile exhaust.

Globally, natural sources contribute more particle mass to the air than do anthropogenic sources. Natural sources emit 250–1610 Tg/yr of primary particles and enough gases to produce 345–2080 Tg/yr of secondary-particle material. Anthropogenic sources emit 6–224 Tg/yr of primary particles and enough gases to produce 140–396 Tg/yr of secondary-particle material (Jaenicke 1988; Pruppacher and Klett 1997; Wolf and Hidy 1997). The relative contribution of anthropogenic sources to global particle loading is increasing in many countries. In urban regions, anthropogenic particle emission exceeds natural emission.

Table 14.4 Emission rates of aerosol constituents in the Los Angeles basin as a function of particle diameter

	Emission rate (kg day^{-1})					
Substance	<1 μm	1–2.5 μm	2.5–10 μm	>10 μm	All sizes	Total %
Other (O, H, etc.)	147 884	86 431	380 221	712 049	1 326 585	53.167
Silicon	37 312	37 086	183 641	166 527	424 566	17.015
Organic carbon	28 462	9 363	69 967	58 111	165 903	6.649
Aluminum	14 550	14 644	67 991	58 216	155 401	6.228
Iron	7 090	7 189	37 210	38 947	90 436	3.624
Calcium	5 587	5 511	32 619	34 028	77 745	3.116
Sulfates	45 922	894	3 998	3 122	53 936	2.162
Potassium	7 364	3 586	16 266	18 989	46 205	1.852
Black carbon	28 467	1 095	7 247	7 429	44 238	1.773
Unknown	9 919	6 745	11 110	11 903	39 677	1.590
Chloride	11 318	814	4 535	4 796	21 463	0.860
Titanium	1 048	877	4 241	4 716	10 882	0.436
Sulfur	618	573	3 216	2 129	6 536	0.262
Carbonate ion	306	162	2 514	1 879	4 861	0.195
Sodium	569	233	2 080	1 916	4 798	0.192
Manganese	899	521	1 511	1 824	4 755	0.191
Phosphorus	130	286	1 660	1 148	3 224	0.129
Nitrates	1 237	147	935	782	3 101	0.124
Zinc	226	154	729	674	1 783	0.071
Lead	173	156	758	653	1 740	0.070
Barium	79	88	544	856	1 567	0.063
Ammonium	841	51	120	136	1 148	0.046
Strontium	25	42	308	364	739	0.030
Vanadium	94	66	274	280	714	0.029
Copper	132	60	203	208	603	0.024
Cobalt	127	52	158	212	549	0.022
Nickel	130	48	132	158	468	0.019
Chromium	87	26	158	176	447	0.018
Rubidium	11	12	91	100	214	0.009
Zirconium	6	10	80	110	206	0.008
Lanthanum	26	7	52	65	150	0.006
Bromine	68	5	30	24	127	0.005
Arsenic	26	3	10	19	58	0.002
Cadmium	9	2	23	16	50	0.002
Antimony	5	3	15	23	46	0.002
Yttrium	2	3	17	22	44	0.002
Tin	6	7	15	14	42	0.002
Indium	5	1	11	12	29	0.001
Mercury	2	1	11	13	27	0.001
Molybdenum	3	1	7	9	20	<0.001
Silver	5	2	6	6	19	<0.001
Palladium	3	1	5	10	19	<0.001
Selenium	3	0	1	2	6	<0.001
Gallium	0	0	5	0	5	<0.001
Totals	350 776	176 958	834 725	1 132 673	2 495 132	100.00
% of total	14.06	7.09	33.45	45.40	100.00	

Emission rates of several species may be significantly overestimated. The component "other" comprises oxygen, hydrogen, and other components not included in the aerosol analysis (e.g., the oxygen atoms in oxides of silicon, aluminum, and iron). Data were provided by Allen and Wagner (1992).

14.2 NUCLEATION

Nucleation is a process by which gas molecules aggregate to form a cluster that condenses to form a small liquid drop. If the radius of a cluster reaches a critical size, the cluster becomes stable and can grow further by condensation. Nucleation is either homogeneous or heterogeneous and either homomolecular, binary, or ternary. **Homogeneous nucleation** occurs when one or more gases nucleates without the aid of a pre-existing surface. **Heterogeneous nucleation** occurs when one or more gases nucleates on a pre-existing surface. **Homomolecular nucleation** occurs when a single gas, such as a high-molecular-weight condensable organic gas, nucleates. **Binary nucleation** occurs when two gases, such as sulfuric acid and water, nucleate, and **ternary nucleation** occurs when three gases, such as sulfuric acid, water, and ammonia, nucleate. In the sections below, nucleation parameterizations are discussed.

14.2.1 Homogeneous nucleation rates from classical theory

Equations based on **classical nucleation theory** are used to estimate homogeneous and heterogeneous nucleation rates. Although other theories exist and classical theory has been criticized for inaccurately simulating nucleation in the atmosphere, other theories are often too computationally intensive to be used in a large model. With respect to homogeneous nucleation, classical theory predicts a production rate of new particles. The rate is derived from

$$\Delta G = 4\pi r_p^2 \sigma_p - \frac{4}{3}\pi r_p^3 \rho_p \frac{R^* T}{m_q} \ln S_q \tag{14.18}$$

which states that the **change in Gibbs free energy** (ΔG) (J) during nucleation of a cluster of molecules is affected primarily by surface tension and the saturation ratio of the gas over the cluster surface. In the equation, r_p is the radius of the nucleating cluster (cm), ρ_p is the mass density of the condensed cluster (g cm^{-3}), R^* is the gas constant (8.314 51 J mol^{-1} K^{-1}), m_q is the molecular weight of the condensing gas (g mol^{-1}), σ_p is the surface tension of the condensed cluster (J cm^{-2}), and S_q is the saturation ratio of the gas. The **saturation ratio** of a gas is its partial pressure divided by its saturation vapor pressure over the surface ($S_q = p_q/p_{q,s}$).

Surface tension is a force per unit distance that tends to minimize the surface area of a body. The force arises because molecules in a three-dimensional body are equally attracted in all directions by other molecules, except at the surface, where a net inward attraction exists. Work needs to be done to increase the surface area against the surface-tension force arising from the inward attraction. The greater the tension, the greater the work, or **surface energy**, required to bring molecules to the surface from within the body.

Surface tension acts parallel to the boundary surface and is expressed in units of energy per unit area, which simplifies to force per unit distance. An expression

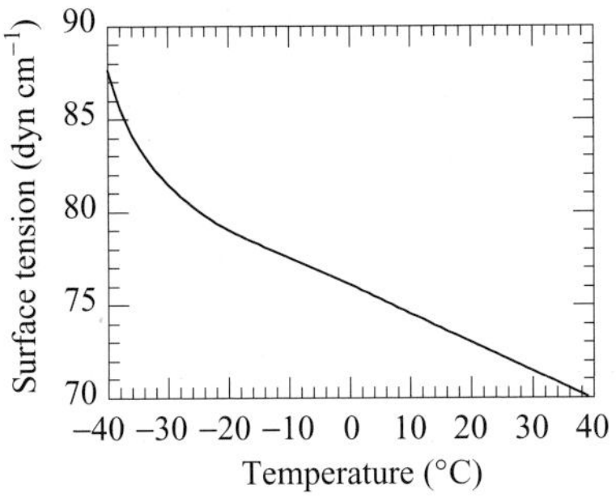

Figure 14.2 Surface tension of liquid water against air, obtained from (14.19).

for the surface tension (dyn cm^{-1}) of liquid water against air for the temperature range −40 to 40 °C is

$$\sigma_{w/a} = \sigma_p = \begin{cases} \sum_{n=0}^{6} a_n T_c^n & -40 \le T_c < 0 \\ 76.1 - 0.155 T_c & 0 \le T_c < 40 \end{cases} \tag{14.19}$$

(Dorsch and Hacker 1951; Gittens 1969; Pruppacher and Klett 1997), where T_c is the temperature in degrees Celsius, $a_0 = 75.93$, $a_1 = 0.115$, $a_2 = 0.068\,18$, $a_3 = 6.511 \times 10^{-3}$, $a_4 = 2.933 \times 10^{-4}$, $a_5 = 6.283 \times 10^{-6}$, and $a_6 = 5.285 \times 10^{-8}$. Figure 14.2 shows a plot of (14.19). The figure shows that surface tension decreases with increasing temperature. It should also theoretically decrease with decreasing radius of curvature, but researchers have found that the surface tension of a liquid water cluster with 13 molecules is anywhere from 0 to 40 percent less than that of a flat liquid water surface (Pruppacher and Klett 1997).

Figure 14.3 illustrates the saturation-ratio and surface-tension terms on the right side of (14.18). When molecules are added to a cluster, they condense, releasing latent heat. The decrease in the change in energy with increasing particle radius depicted by the saturation-ratio term shown in the figure represents the energy available from latent-heat release.

The addition of molecules to the cluster increases the energy required to expand the surface area of the cluster against the force of surface tension. The surface-tension term in Fig. 14.3 represents the energy required to expand the surface.

When the sum of the saturation-ratio and surface-tension terms is increasing, the addition of a molecule to the cluster consumes more energy than it adds, and the cluster becomes more unstable. When the sum is decreasing, the addition of a molecule adds more energy than it consumes, and the cluster becomes more stable. When the sum reaches a maximum, the addition of a molecule causes no net energy change. The radius at which the maximum occurs is the **critical radius**. Nucleation occurs if a cluster contains enough molecules for the cluster radius to exceed the critical radius. At larger radii, condensation occurs readily on the nucleated cluster. At smaller radii, the cluster is unstable and may evaporate.

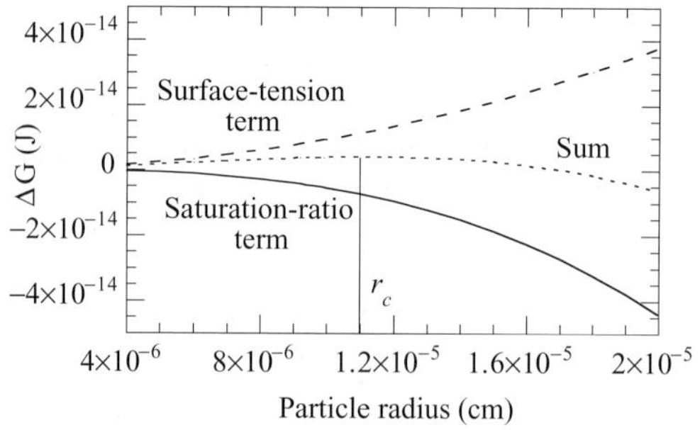

Figure 14.3 The net change in free energy as a function of radius is the sum of the surface-tension and the supersaturation terms, as described in the text. The critical radius is the radius at which the net change in free energy with radius is maximized. Conditions for this curve are $T = 288$ K, $\sigma_p = 7.5 \times 10^{-6}$ J cm^{-2}, $S = 1.01$, $\rho_p = 1.0$ g cm^{-3}, $m_q = 18$ g mol^{-1} (water).

The critical radius can be found by setting $\mathrm{d}\Delta G/\mathrm{d}r_p = 0$ in (14.18). The result is

$$\frac{\mathrm{d}\Delta G}{\mathrm{d}r_p} = 8\pi r_p \sigma_p - 4\pi r_p^2 \rho_p \frac{R^* T}{m_q} \ln S_q = 0 \tag{14.20}$$

When $\mathrm{d}\Delta G/\mathrm{d}r_p$ is positive, the cluster is unstable; when it is zero or negative, the cluster is stable. Solving (14.20) for the radius gives the **critical radius** as

$$r_c = \frac{2\sigma_p m_q}{\rho_p R^* T \ln S_q} \tag{14.21}$$

Dividing the volume of a critical cluster $(4\pi r_c^3/3)$ by the volume of one gas molecule $[m_q/(\rho_p A)]$, where A is Avogadro's number (molec. mol^{-1}), gives the number of molecules in a nucleated cluster as

$$n_c = \frac{32\pi \sigma_p^3 m_q^2 A}{3\rho_p^2 (R^* T \ln S_q)^3} \tag{14.22}$$

Table 14.5 shows critical radii and number of water molecules in a nucleated cluster for several saturation ratios.

An expression for the **homogeneous homomolecular nucleation rate** of particles (particles cm^{-3} s^{-1}) from classical theory is

$$J_{\mathrm{hom}} = 4\pi r_c^2 \beta_x Z_n N_x \exp\left(\frac{-\Delta G^*_{\mathrm{hom}}}{k_B T}\right) \tag{14.23}$$

where

$$\Delta G^*_{\mathrm{hom}} = 4\pi r_c^2 \sigma_p / 3 \tag{14.24}$$

Table 14.5 Critical radii and number of water molecules in a critical cluster

Saturation ratio S	Critical radius (μm)	Number of molecules
1	∞	∞
1.01	0.11	2.03×10^8
1.02	0.055	2.58×10^7
1.10	0.011	2.32×10^5
1.5	0.0028	3010
2	0.0016	603
5	0.0007	48
10	0.00048	16

$T = 288$ K, $\sigma_p = 7.5 \times 10^{-6}$ J cm^{-2}, $\rho_p = 1.0$ g cm^{-3}, and $m_q =$ 18 g mol^{-1}.

(J) is the critical change in free energy, which is found by combining (14.18) with (14.20), k_B is Boltzmann's constant (1.3807×10^{-23} J K^{-1}), N_x is the number concentration (molec. cm^{-3}) of nucleating gas x,

$$\beta_x = N_x \sqrt{\frac{k_B T}{2\pi \bar{M}_x}} \qquad (14.25)$$

is the number of gas molecules striking a unit surface area per second (molec. cm^{-2} s^{-1}) (where $\bar{M}_x = m_x/A$ is the mass, in grams, of one molecule of x and $k_B = 1.380\,658 \times 10^{-16}$ g cm^2 s^{-2} K^{-1} molec.$^{-1}$ in this expression), and $N_x \exp(-\Delta G^*_{\text{hom}}/k_B T)$ is an equilibrium number concentration (particles cm^{-3}) of nucleated clusters of critical radius r_c.

The equilibrium cluster concentration is an ideal cluster number per unit volume of air that exists when the number of molecules added to a cluster equals the number lost from the cluster. At theoretical equilibrium, the number of nucleated clusters of radius r_c does not change. In homogeneous nucleation theory, the system is not always in equilibrium, so some clusters form and others evaporate over time, changing the number of nucleated clusters with time. Equation (14.23) accounts for the overall rate of change of cluster concentration through the term β_x and through the **Zeldovich nonequilibrium factor** (Zeldovich 1942),

$$Z_n = \frac{\bar{M}_x}{2\pi r_c^2 \rho_x} \sqrt{\frac{\sigma_p}{k_B T}} \qquad (14.26)$$

which takes account of the remaining differences between equilibrium and nonequilibrium cluster concentration. Its value is less than unity; thus, time-dependent nucleation results in a cluster concentration below the equilibrium level (Pruppacher and Klett 1997).

As illustrated in Example 14.1, water vapor, alone, does not homogeneously nucleate in the atmosphere, even when its saturation ratio is near its highest observed level, 1.02. Under such conditions, the critical radius, and thus the critical change in free energy, is too large, causing the exponent in (14.23) to become zero. Water can only homogeneously nucleate if the critical radius decreases to <1 nm, which occurs when the saturation ratio is 5 (the relative humidity is 400 percent). Water vapor readily nucleates heterogeneously, since the free energy change required to nucleate heterogeneously is much less than that required to nucleate homogeneously. The free energy change required for heterogeneous nucleation is discussed shortly.

Example 14.1

Calculate the homogeneous nucleation rate of water vapor when its saturation ratio is 1.02 and $T = 288$ K.

SOLUTION

From Table 14.5, the critical radius for new-particle formation with this saturation ratio is 0.055 μm. Substituting the critical radius into (14.24) with a surface tension of 73.8 dyn cm^{-1} (7.38×10^{-6} J cm^{-2}) from (14.19) gives the critical change in free energy as 9.35×10^{-16} J. Substituting this into the exponential term of (14.23) with $T = 288$ K gives the argument of the exponential term as -2.35×10^5. The exponent of this number gives zero. Thus, water vapor cannot homogeneously nucleate in the atmosphere under even its highest observed saturation ratio.

In (14.23), small changes in free energy affect the nucleation rate exponentially. Changes in the prefactor affect the rate linearly. As such, changes in the free energy have a much greater effect on the nucleation rate than do changes in the prefactor term. Because the saturation ratio, which affects the free energy change, is difficult to calculate precisely in the atmosphere, model-predicted free energy changes are often slightly off. Since small errors in the free energy are amplified in the exponential term of (14.23), they result in nucleation rate predictions from classical theory that are often several orders of magnitude off when applied to the atmosphere.

Classical homogeneous nucleation theory can be extended to systems in which two gases nucleate. The resulting **homogeneous binary nucleation rate** is

$$J_{\text{hom}} = 4\pi r_c^2 \beta_y N_x \exp\left(\frac{-\Delta G^*_{\text{hom}}}{k_B T}\right) \qquad (14.27)$$

(particles cm^{-3} s^{-1}) where the two gases, x and y, are assumed to nucleate together. Gas x is assumed to have a much larger concentration than gas y ($N_x \gg N_y$). In the atmosphere, sulfuric acid commonly nucleates homogeneously with water vapor, which has a much higher number concentration than does sulfuric acid. For sulfuric acid–water binary nucleation, gas x is water vapor and gas y is sulfuric acid.

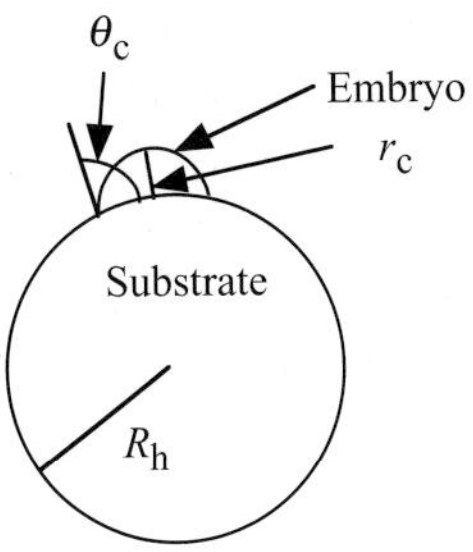

Figure 14.4 Formation of a critical embryo on the surface of an existing particle. θ_c is the contact angle, R_h is the radius of the particle, and r_c is the critical radius.

In equilibrium, the number concentration of nucleated clusters of critical radius, r_c, is about $(N_x + N_y)\exp(-\Delta G^*_{hom}/k_B T)$. Since $N_x \gg N_y$, this expression reduces to $N_x \exp(-\Delta G^*_{hom}/k_B T)$, which is used in (14.27). Because the flux of molecules to a nucleating cluster is limited by the flux of the gas with the lowest concentration, β is written in terms of gas y in (14.27). The equation for β_y (molec. cm^{-2} s^{-1}) is

$$\beta_y = N_y \sqrt{\frac{k_B T}{2\pi \bar{M}_y}} \tag{14.28}$$

where $\bar{M}_y$ is the mass of one molecule of gas y.

For basic homogeneous binary nucleation calculations, (14.27) does not include a Zeldovich nonequilibrium factor. The saturation ratio, surface tension, and particle density used to calculate $\Delta G/^*_{hom}$ must be estimated as weighted averages between the two nucleating gases (e.g., Zhao and Turco 1995; Kulmala *et al.* 1998; Vehkamaki *et al.* 2002).

For **homogeneous ternary nucleation** among gases x, y, and z, (e.g., water, sulfuric acid, and ammonia, respectively) where $N_x \gg N_y \approx N_z$, Equation (14.27) can be modified by replacing β_y with $\beta_y + \beta_z$, since the flux of the lower-concentrated gases limits the rate of nucleation. In addition, the surface tension, cluster density, and saturation ratio must be weighted by the relative composition of the cluster. Ternary-nucleation parameterizations have been developed for the sulfuric acid–water–ammonia system (e.g., Coffman and Hegg 1995; Napari *et al.* 2002), the hydrochloric acid–water–ammonia system (e.g., Arstila *et al.* 1999), and the sulfuric acid–water–methanesulfonic acid system (e.g., Van Dingenen and Raes 1993).

14.2.2 Heterogeneous nucleation rates from classical theory

During **heterogeneous nucleation**, a critical embryo forms on the surface of an existing particle, as shown in Fig. 14.4. The ability of the embryo to survive breakup

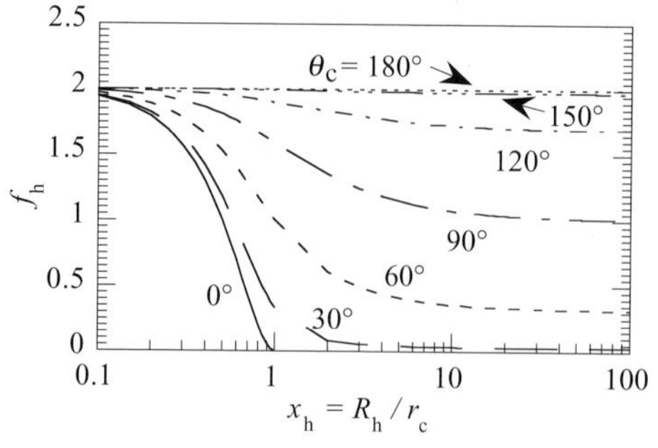

Figure 14.5 Values of Fletcher's correction factor as a function of x_h for seven different contact angles θ_c.

on the surface depends on the **contact angle** θ_c, which is the angle at which an embryo contacts a substrate at its surface. If the contact angle is 0°, the surface is completely covered, or **wetted**, with the embryo. If the contact angle is 180°, the surface is nonwettable, and no embryo forms. If water readily wets a surface, the surface is **hydrophilic**. Otherwise, it is **hydrophobic**. Contact angles of water vapor vary with surface composition. The contact angle of water over sand is 43–52°, over soil is 65.2–68.9°, and over silver iodide is 9–17° (Pruppacher and Klett 1997). The contact angle of water over a substrate is calculated from **Young's relation**,

$$\theta_c = \cos^{-1}\left(\frac{\sigma_{S,a} - \sigma_{S,w}}{\sigma_{w,a}}\right) \tag{14.29}$$

where $\sigma_{S,a}$, $\sigma_{S,w}$, and $\sigma_{w,a}$ are the surface tensions of the substrate against air, the substrate against water, and water against air, respectively.

The free energy change required to heterogeneously nucleate an embryo on a surface is approximated with

$$\Delta G^*_{het} = \Delta G^*_{hom} f_h(x_h, m_h) \tag{14.30}$$

where f_h is a correction factor that depends on $x_h = R_h/r_c$, the ratio of the radius of the host particle to the critical radius, and on $m_h = \cos\theta_c$, the cosine of the contact angle. Fletcher (1958) parameterized the correction factor as

$$f_h(x_h, m_h) = 1 + \left(\frac{1 - m_h x_h}{g_h}\right)^3 + x_h^3\left[2 - 3\left(\frac{x_h - m_h}{g_h}\right) + \left(\frac{x_h - m_h}{g_h}\right)^3\right] + 3m_h x_h^2\left(\frac{x_h - m_h}{g_h} - 1\right) \tag{14.31}$$

where $g_h = \sqrt{1 + x_h^2 - 2m_h x_h}$. Figure 14.5 shows a plot of f_h versus x_h and θ_c. When $\theta_c = 0°$, f_h and ΔG^*_{het} are minimized. When $\theta_c = 180°$, f_h and ΔG^*_{het} are maximized. The factors f_h and ΔG^*_{het} can have small values even when $x_h < 1$, indicating that clusters may form when the host particle is smaller than the nucleating cluster.

A classical-theory expression for the **heterogeneous nucleation rate** (number of embryos cm^{-2} s^{-1}) of new particles on a surface is

$$J_{het} = 4\pi r_c^2 \beta_y \beta_x \tau \exp\left(\frac{-\Delta G_{het}^*}{k_B T}\right) \tag{14.32}$$

where

$$\tau = \tau_0 \exp\left(\frac{E}{R^* T}\right) \tag{14.33}$$

is the characteristic time (s) that a gas molecule spends on the surface before bouncing off. In this expression, τ_0 is the characteristic time of adsorption (s), which is the inverse of the characteristic frequency of vibration of the molecules, and depends on the nature of the surface and of the condensing gas(es). The term E is the enthalpy of adsorption (J mol^{-1}), which depends on the surface type and properties of the condensed gas. Hamill *et al.* (1982) calculated $\tau_0 \approx 2.4 \times 10^{-16}$ s and $E \approx 45\,188$ J mol^{-1} for water nucleating on solid carbon particles. Korhonen *et al.* (2003) calculated $\tau_0 \approx 2.55 \times 10^{-13}$ s and $E \approx 44\,970$ J mol^{-1} for liquid water, $\tau_0 \approx 8.62 \times 10^{-13}$ s and $E \approx 84\,340$ J mol^{-1} for sulfuric acid, and $\tau_0 \approx 2.55 \times 10^{-13}$ s and $E \approx 21\,750$ J mol^{-1} for ammonia nucleating on soil particles.

In (14.32), β_x and β_y are the numbers of molecules of gases x and y, respectively, striking the substrate surface per second (molec. cm^{-2} s^{-1}) and can be found from (14.25) and (14.28). If gas x differs from gas y, the nucleation is heterogeneous binary. If the two are the same, it is heterogeneous homomolecular. Common binary-nucleating gases are sulfuric acid and water vapor, since sulfuric acid has a low saturation vapor pressure, and water readily equilibrates with sulfuric acid when sulfuric acid condenses.

J_{het} gives the rate per unit surface area at which embryos form when molecules of gas y transfer to a surface at rate β_y and combine with molecules of x adsorbed to the surface. The product $\beta_x \tau \leq n_m$ is the number of molecules of gas x that adsorb to a unit area of the particle surface (molec. cm^{-2}), where $n_m \approx 10^{15}$ sites cm^{-2} is the maximum number of adsorption sites available per unit area of surface (Section 11.3.6.3). Species x is chosen as the chemical that limits surface coverage because, in the case of binary nucleation, it is assumed to be the more highly concentrated species.

14.2.3 Parameterized homogeneous nucleation rate

The most important homogeneous nucleation process in the free atmosphere is binary nucleation of sulfuric acid and water. Homogeneously nucleated sulfuric acid–water particles are typically 3–20 nm in diameter. In the remote atmosphere (e.g., over the ocean), homogeneous nucleation events can produce $>10^4$ particles cm^{-3} in this size range.

Because classical theory is often unreliable for predicting nucleation rates in the atmosphere, nucleation parameterizations have been developed to replace classical theory. Zhang *et al.* (1999) have compared several such parameterizations. One empirical parameterization, valid for a temperature range 230–300 K, a relative humidity range 0.01–100 percent, and a sulfuric-acid concentration range of 10^4–10^{11} molec. cm^{-3}, is given in Vehkamaki *et al.* (2002).

A two-parameter parameterization, based on calculations of binary nucleation of sulfuric acid and water for the remote boundary layer by Jaecker-Voirol and Mirabel (1989), is

$$J_{\text{hom}} = 10^{7.0-(64.24+4.7 f_r)+(6.13+1.95 f_r)\log_{10} N_{\text{H}_2\text{SO}_4}} \tag{14.34}$$

(Pandis *et al.* 1994; Russell *et al.* 1994), where J_{hom} is the rate of production of new sulfuric acid–water particles (particles cm^{-3} s^{-1}), f_r is the relative humidity (expressed as a fraction), and $N_{\text{H}_2\text{SO}_4}$ is the number concentration of gas-phase sulfuric acid molecules (molec. cm^{-3}). This parameterization is independent of temperature, so it cannot be used for the upper troposphere or stratosphere.

Example 14.2

Estimate the homogeneous nucleation rate of sulfuric acid–water particles over the remote ocean when $f_r = 0.9$ and when H_2SO_4(g) concentrations are (a) 0.005 and (b) 0.05 μg m^{-3}.

SOLUTION

In the first case, the number concentration of gas-phase sulfuric acid molecules is

$$N_{\text{H}_2\text{SO}_4} = \left(\frac{0.005\ \mu\text{g}}{\text{m}^3}\right)\left(\frac{\text{m}^3}{10^6\ \text{cm}^3}\right)\left(\frac{\text{mol}}{98\times10^6\ \mu\text{g}}\right)\left(\frac{6.02\times10^{23}\ \text{molec.}}{\text{mol}}\right)$$
$$= 3.1\times10^7\ \frac{\text{molec.}}{\text{cm}^3}$$

Substituting this result into (14.34) gives $J_{\text{hom}} = 0.004$ particles cm^{-3} s^{-1}. In the second case, $J_{\text{hom}} = 3.1 \times 10^5$ particles cm^{-3} s^{-1}. A factor-of-10 increase in $N_{\text{H}_2\text{SO}_4}$ between the two cases increased the nucleation rate by eight orders of magnitude. In reality, H_2SO_4(g) condenses on particles as they nucleate, reducing the amount of gas available for further nucleation.

14.3 SUMMARY

In this chapter, aerosol emission and nucleation were discussed. Emissions originate from human and natural sources. In urban regions, human emission dominates. Globally, natural sources dominate. Some natural emission sources include volcanos and sea-spray emission. Homogeneous nucleation is a second means of new particle production. Such nucleation can be homomolecular, binary, or ternary.

Homogeneous and heterogeneous nucleation rates can be estimated with classical nucleation theory, but this theory may lead to large errors in nucleation rates when applied to the atmosphere. Empirical parameterizations of nucleation rates may be better in some cases. Once nucleated, small particles collide and coalesce during coagulation, discussed in Chapter 15.

14.4 PROBLEMS

14.1 Calculate the average flux of sea-spray drops (particles cm^{-2} s^{-1}) between 0.8 and 0.9 μm in radius when the wind speed at 10 m is 10 m s^{-1}.

14.2 Calculate the threshold friction wind speed for soil particles 30 μm in diameter when the average surface roughness over the overall terrain is 0.001 m, soil density is 2.5 g cm^{-3}, and air density is 0.001225 g cm^{-3}.

14.3 Given identical vapor concentrations of water at both locations, would you expect more particles to nucleate heterogeneously over the ocean or over land? Why? Once particles have nucleated, would you expect the particles to grow larger over the ocean or over land? Why?

14.4 If the gas concentrations of sulfuric acid and nitric acid are the same, why should sulfuric acid and water homogeneously nucleate more readily than nitric acid and water?

14.5 **(a)** Calculate the critical radius of a homogeneously nucleating water drop when the saturation ratio is (i) 4.0 and (ii) 1.1. Assume $\rho_p =$ 1.0 g cm^{-3} and $T = 298.15$ K in both cases.

(b) Using the information from (a), calculate the classical-theory homogeneous nucleation rate of liquid water drops in both cases.

(c) If the air temperature in part (a) (i) decreased from 298.15 to 283.15 K, what would the new nucleation rate be? (Be sure to calculate the change in saturation ratio.)

14.6 For the two cases in Problem 14.5(a), calculate heterogeneous nucleation rates when $\theta_c = 90°$, $R_h = 0.5$ μm, $\tau_0 = 2.4 \times 10^{-16}$ s, $E = 45\,188$ J mol^{-1}, and $T = 298.15$ K.

14.5 COMPUTER PROGRAMMING PRACTICE

14.7 Write a computer script to calculate r_c, J_{hom}, and J_{het} from classical nucleation theory for water. Perform calculations at $T = 298$ K for $f_r = 100$–400 percent and at $f_r = 400$ percent for $T = 233$–315 K. Plot the results. Use (2.62) to calculate $p_{v,s}$. Use information from Problems 14.5 and 14.6 where needed.

15

Coagulation

COAGULATION occurs when two particles collide and stick together (coalesce), reducing the number concentration but conserving the volume concentration of particles in the atmosphere. Coagulation is a mechanism by which externally mixed particles become internally mixed. More small particles than large particles are lost to larger sizes by coagulation, because the atmosphere contains many more small particles than large particles (Table 13.1). Simulating coagulation in a model is important, since if coagulation is neglected, erroneously large aerosol number concentrations will be predicted. Even if the total aerosol mass concentration in a model is correct, the mass concentration will be spread among too many particles. Coagulation is also important for determining the number concentration of particles that exceed the critical radius required for cloud-drop activation and determining the evolution of nucleated and newly emitted particles. In fogs and many clouds, coagulation (collision plus coalescence) is the main mechanism by which cloud drops evolve into raindrops. To simulate coagulation in an Eulerian model, a numerical scheme must be fast and accurate. One scheme is described in this chapter. A solution for homogeneous particles over a single size distribution is first derived. The solution is then expanded to include any number of size distributions and particles of any composition.

15.1 IMPLICIT COAGULATION

The derivation of numerical methods for solving coagulation begins with the **integrodifferential coagulation equation** (Muller 1928)

$$\frac{\partial n_\upsilon}{\partial t} = \frac{1}{2}\int_0^{\upsilon} \beta_{\upsilon-\bar{\upsilon},\bar{\upsilon}} n_{\upsilon-\bar{\upsilon}}\, n_{\bar{\upsilon}}\, \mathrm{d}\bar{\upsilon} - n_\upsilon \int_0^{\infty} \beta_{\upsilon,\bar{\upsilon}} n_{\bar{\upsilon}}\, \mathrm{d}\bar{\upsilon} \qquad (15.1)$$

where $\upsilon - \bar{\upsilon}$ and $\bar{\upsilon}$ are the volumes of two coagulating particles, υ is the volume of the new, coagulated particle, n is the time-dependent number concentration (particles cm^{-3}) of particles of volume υ, $\upsilon - \bar{\upsilon}$, or $\bar{\upsilon}$, and β is the coagulation kernel (rate coefficient) of the two colliding particles (cm^3 particle^{-1} s^{-1}). Equation (15.1) states that the change in number concentration of particles of volume υ equals the rate at which particles of volume $\upsilon - \bar{\upsilon}$ coagulate with particles of volume $\bar{\upsilon}$ minus

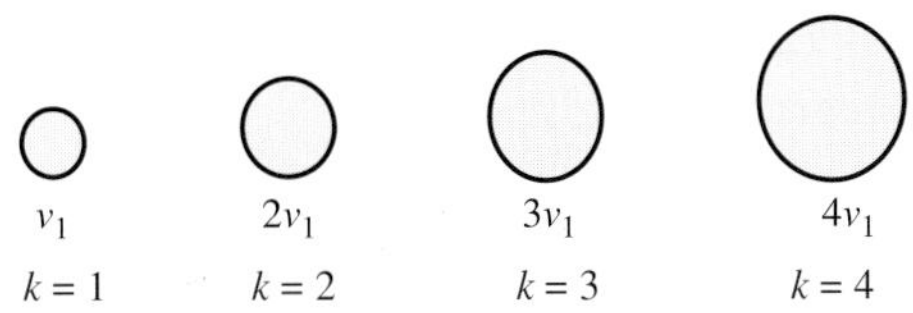

Figure 15.1 A monomer size distribution. The volume of particles in any size bin k equals the volume of particles in the smallest bin multiplied by k.

the rate at which particles of volume υ are lost due to coagulation with particles of all sizes. The first integral in (15.1) is multiplied by one-half to eliminate double counting of production terms.

A few analytical solutions to (15.1) exist. Some are described in Section 15.3. All analytical solutions require a particular form for the initial aerosol size distribution. Other solutions to the equation are numerical. Some numerical solutions are based on the assumption that the aerosol size distribution is represented by lognormal modes (e.g., Pratsinis 1988; Binkowski and Shankar 1995; Binkowski *et al.* 2003). Others are based on the assumption that the aerosol distribution is discrete. Solutions to the discrete form of the coagulation equation are discussed below.

The coagulation equation can be discretized in terms of a **monomer size distribution.** In a monomer distribution, the volume of each particle in size bin k equals the volume of a particle in the smallest size bin multiplied by k, as shown in Fig. 15.1. In the monomer distribution, every possible particle volume that is a multiple of the smallest volume is accounted for. Thus, coagulation always results in particles moving to an exact bin instead of being partitioned between two bins. For monomer size bins, the coagulation equation is discretized as

$$\frac{\Delta n_k}{\Delta t} = \frac{1}{2}\sum_{j=1}^{k-1} \beta_{k-j,j} n_{k-j} n_j - n_k \sum_{j=1}^{\infty} \beta_{k,j} n_j \tag{15.2}$$

where j and k are size-bin indices. Particles in bin k are produced when particles in bin $k - j$ coagulate with particles in bin j. Bin k corresponds to particles with volume $\upsilon_1 k$, and bin j corresponds to particles with volume $\upsilon_1 j$.

Equation (15.2) can be written in fully implicit finite-difference form (all variables on the right side are evaluated at time t) as

$$\frac{n_{k,t} - n_{k,t-h}}{h} = \frac{1}{2}\sum_{j=1}^{k-1} P_{k,j} - \sum_{j=1}^{\infty} L_{k,j} \tag{15.3}$$

where h is the time step, subscripts t and $t - h$ are the final and initial times, respectively, and

$$P_{k,j} = \beta_{k-j,j} n_{k-j,t} n_{j,t} \qquad L_{k,j} = \beta_{k,j} n_{k,t} n_{j,t} \tag{15.4}$$

are production and loss rates, respectively. Note that $P_{k,j} = L_{k-j,j}$. Substituting (15.4) into (15.3) and rearranging gives

$$n_{k,t} = n_{k,t-h} + \frac{1}{2}h\sum_{j=1}^{k-1}\beta_{k-j,j}n_{k-j,t}n_{j,t} - h\sum_{j=1}^{\infty}\beta_{k,j}n_{k,t}n_{j,t} \tag{15.5}$$

which represents a coagulation equation for one monomer size bin k.

15.2 SEMIIMPLICIT COAGULATION

Equation (15.5) can be solved with an iterative ordinary differential equation solver (e.g., Jacobson and Turco 1995) or with one of several other techniques (e.g., Suck and Brock 1979; Gelbard and Seinfeld 1980; Tsang and Brock 1982; Friedlander 1983; Strom *et al.* 1992; Fassi-Fihri *et al.* 1997; Trautmann and Wanner 1999; Fernandez-Diaz *et al.* 2000; Sandu 2002). The scheme discussed here (Jacobson *et al.* 1994; Toon *et al.* 1988; Turco *et al.* 1979) is called the **semiimplicit coagulation solution.** It is named such because it is based on the substitution of $n_{j,t-h}$ for $n_{j,t}$ in the production and loss terms of (15.4). The terms are now written in semiimplicit form as

$$P_{k,j} = \beta_{k-j,j}n_{k-j,t}n_{j,t-h} \qquad L_{k,j} = \beta_{k,j}n_{k,t}n_{j,t-h} \tag{15.6}$$

respectively. The use of these terms allows a noniterative solution to coagulation. Combining (15.3) and (15.6) gives the semiimplicit solution for the number concentration of monomer particles after a time step as

$$n_{k,t} = n_{k,t-h} + \frac{1}{2}h\sum_{j=1}^{k-1}\beta_{k-j,j}n_{k-j,t}n_{j,t-h} - h\sum_{j=1}^{\infty}\beta_{k,j}n_{k,t}n_{j,t-h} \tag{15.7}$$

which can be rewritten as

$$n_{k,t} = \frac{n_{k,t-h} + \frac{1}{2}h\sum\limits_{j=1}^{k-1}\beta_{k-j,j}n_{k-j,t}n_{j,t-h}}{1 + h\sum\limits_{j=1}^{\infty}\beta_{k,j}n_{j,t-h}} \tag{15.8}$$

In this equation, k varies from 1 to any desired number of monomer size bins. Although (15.8) correctly accounts for the reduction in particle number when two particles coagulate (reducing the number by one-half), it does not conserve volume. Equation (15.5) correctly accounts for number and volume, but it is fully implicit so requires iteration to solve. In order **to conserve volume and volume concentration** (which coagulation physically does) while giving up some accuracy in number concentration, (15.8) can be rederived in terms of volume concentrations of monomer particles and rewritten as

$$v_{k,t} = \frac{v_{k,t-h} + h \sum_{j=1}^{k-1} \beta_{k-j,j} v_{k-j,t} n_{j,t-h}}{1 + h \sum_{j=1}^{\infty} \beta_{k,j} n_{j,t-h}} \tag{15.9}$$

where $v_{k,t} = \upsilon_k n_{k,t}$ from (13.10). Equation (15.9) satisfies the volume-conservation requirement, $\upsilon_{k-j} P_{k,j} = \upsilon_{k-j} L_{k-j,j}$, for each k and j.

The solution just presented is applicable to a monomer size distribution. It is usually desirable to solve coagulation over an arbitrary size distribution, such as the volume-ratio distribution. As discussed in Chapter 13, a volume-ratio size distribution is defined so that the volume of a particle in one size bin equals the volume of a particle in the next smallest size bin multiplied by a constant, V_{rat}.

For the volume ratio or any other type of discretized size distribution, the semiimplicit solution can be modified to treat coagulation. The solution is found by first defining the volume of an intermediate particle that results when a particle in size bin i collides and sticks to a particle in size bin j as

$$V_{i,j} = \upsilon_i + \upsilon_j \tag{15.10}$$

The intermediate particle has volume between those of two arbitrary model size bins, k and $k+1$, and needs to be partitioned between the two bins. This is done by defining the volume fraction of $V_{i,j}$ that is partitioned to each model bin k as

$$f_{i,j,k} = \begin{cases} \left(\dfrac{\upsilon_{k+1} - V_{i,j}}{\upsilon_{k+1} - \upsilon_k}\right)\dfrac{\upsilon_k}{V_{i,j}} & \upsilon_k \le V_{i,j} < \upsilon_{k+1} \quad k < N_B \\ 1 - f_{i,j,k-1} & \upsilon_{k-1} < V_{i,j} < \upsilon_k \quad k > 1 \\ 1 & V_{i,j} \ge \upsilon_k \quad k = N_B \\ 0 & \text{all other cases} \end{cases} \tag{15.11}$$

(Jacobson *et al.* 1994). The volume fractions in (15.11) are independent of the size distribution. They work for a monomer distribution (where all values of f are 1 or 0), a volume-ratio distribution with any value of $V_{rat} > 1$, or an arbitrary distribution. Note that the volume fractions in (15.11) differ from the number fractions used for partitioning number concentration with the quasistationary size structure, given in (13.34).

When volume fractions are used, the semiimplicit, volume- and volume-concentration conserving, noniterative, positive-definite for any time step coagulation solution for total particle volume concentration becomes

$$v_{k,t} = \frac{v_{k,t-h} + h \sum_{j=1}^{k} \left(\sum_{i=1}^{k-1} f_{i,j,k} \beta_{i,j} v_{i,t} n_{j,t-h} \right)}{1 + h \sum_{j=1}^{N_B} (1 - f_{k,j,k}) \beta_{k,j} n_{j,t-h}} \tag{15.12}$$

The series of equations must be solved in the order $k = 1, \ldots, N_B$. No production occurs into the first bin, $k = 1$, since for it, $k - 1 = 0$ in the numerator of (15.12). Thus, all $v_{i,t}$ terms are known when $v_{k,t}$ is calculated. Values of $f_{i,j,k}$ in (15.12) are frequently zero. During computer simulations, all such multiplications by zero are eliminated.

The final number concentration of particles due to coagulation is now simply

$$n_{k,t} = \frac{v_{k,t}}{\upsilon_k} \tag{15.13}$$

If particles have multiple components, the change in volume concentration of each component q is calculated analogously to (15.12) with

$$v_{q,k,t} = \frac{v_{q,k,t-h} + h \sum_{j=1}^{k} \left(\sum_{i=1}^{k-1} f_{i,j,k}\beta_{i,j} v_{q,i,t} n_{j,t-h} \right)}{1 + h \sum_{j=1}^{N_B} [(1 - f_{k,j,k})\beta_{k,j} n_{j,t-h}]} \tag{15.14}$$

where the sum of $v_{q,k,t}$ over all components $q = 1, \ldots, N_V$ is $v_{k,t}$.

The advantage of a semiimplicit over a fully implicit equation is that the semi-implicit equation yields a noniterative volume-conserving, volume concentration-conserving stable solution to coagulation for any time step. A fully implicit equation requires the use of an iterative ordinary-differential-equation solver. The disadvantage of a semiimplicit equation is that particle number is not calculated exactly. By increasing the resolution of the size distribution, the error in number concentration approaches zero while the solution remains noniterative and volume- and volume concentration-conserving.

15.3 COMPARISON WITH ANALYTICAL SOLUTIONS

In this section, results from the semiimplicit coagulation solution are compared with analytical and numerical solutions. Smoluchowski's (1918) analytical solution assumes that all particles are initially **monodisperse** (have the same size), and the coagulation kernel is constant for all interactions. After initialization, coagulation occurs over a monomer size distribution (Section 15.1). Atmospheric particles are really **polydisperse** (particle number concentration varies with size). Nevertheless, Smoluchowski's analytical solution is useful for checking the accuracy of numerical coagulation schemes. **Smoluchowski's solution** is

$$n_{k,t} = \frac{n_{T,t-h}(0.5h\beta n_{T,t-h})^{k-1}}{(1 + 0.5h\beta n_{T,t-h})^{k+1}} \tag{15.15}$$

where the subscripts t and $t - h$ refer to the final and initial times, respectively, $n_{T,t-h}$ is the initial number concentration of monodisperse particles in bin $k = 1$,

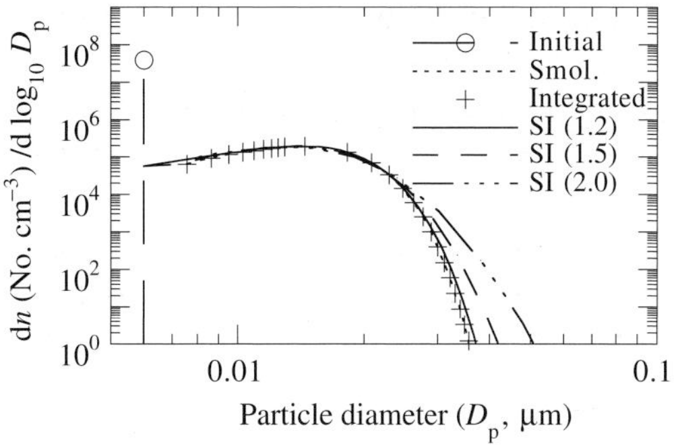

Figure 15.2 Comparison of Smoluchowski's analytical solution with an integrated numerical solution and with three semiimplicit solutions (SI), each with a different value of V_{rat} (given in parentheses). The initial number concentration of monodisperse particles was 10^6 particles cm^{-3} and $T = 298$ K. This simulation time was 12 h. Remaining conditions are described in the text.

$n_{k,t}$ is the final concentration of particles in any monomer size bin k after time h, the exponents $k - 1$ and $k + 1$ refer to the size-bin number, and

$$\beta = \frac{8k_B T}{3\eta_a} \tag{15.16}$$

(cm^3 particle^{-1} s^{-1}) is the constant coagulation kernel. Equation (15.15) is solved over one time step h of any size. Figure 15.2 compares Smoluchowski's solution with an integrated numerical solution and three semiimplicit solutions to the same problem over a full-stationary size structure with a volume-ratio size distribution. The semiimplicit solutions improved with higher resolution (smaller V_{rat}). The integrated solution matched the analytical solution exactly, but the computer times required for the semiimplicit solutions were much less than that required for the integrated solution.

Another analytical solution is that for a **self-preserving size distribution**. Such a distribution has initial number concentration in each size bin i of

$$n_{i,t-h} = \frac{n_{T,t-h}\Delta\upsilon_i}{\upsilon_p}\exp\left(-\frac{\upsilon_i}{\upsilon_p}\right) \tag{15.17}$$

where $n_{T,t-h}$ is the initial total number concentration of particles among all size bins, υ_i is the volume of a single particle in size bin i, υ_p is the volume at which the initial peak number concentration occurs in the distribution, and $\Delta\upsilon_i$ is the volume width of a size bin. For a volume-ratio size distribution, $\Delta\upsilon_i$ is calculated with (13.8).

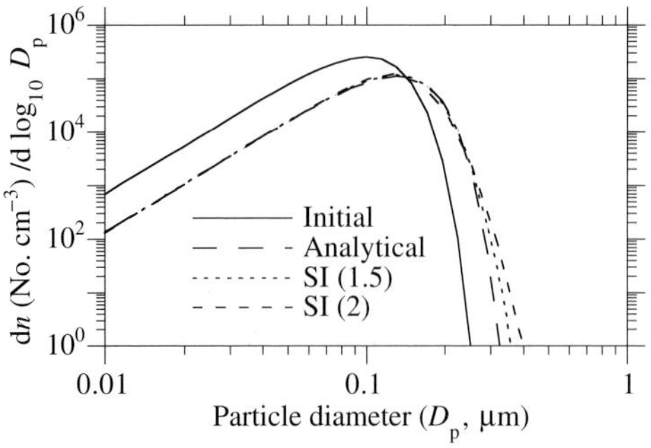

Figure 15.3 Comparison of self-preserving analytical solution with two semiimplicit solutions (SI), each with a different value of V_{rat} (given in parentheses). The conditions were $n_{\mathrm{T},t-h} = 10^5$ cm^{-3}, $T = 298$ K, and $\upsilon_{\mathrm{p}} = 0.0005236$ μm^3. The simulation time was 12 h.

After time h, the solution for coagulation over the self-preserving distribution is

$$n_{i,t} = \frac{n_{\mathrm{T},t-h}\Delta\upsilon_i/\upsilon_{\mathrm{p}}}{(1+0.5h\beta n_{\mathrm{T},t-h})^2}\exp\left(-\frac{\upsilon_i/\upsilon_{\mathrm{p}}}{1+0.5h\beta n_{\mathrm{T},t-h}}\right) \tag{15.18}$$

where the coagulation kernel is given in (15.16). Figure 15.3 compares results from the self-preserving analytical solution with those from two semiimplicit solutions.

15.4 COAGULATION AMONG MULTIPLE PARTICLE DISTRIBUTIONS

The semiimplicit solution can be used to solve coagulation, not only over a single size distribution with multiple components, but also among multiple size distributions, each with multiple components. Figure 15.4 illustrates coagulation interactions among three externally mixed size distributions (A, B, and C, each represented by one particle) that result in internally mixed particles of composition AB, AC,

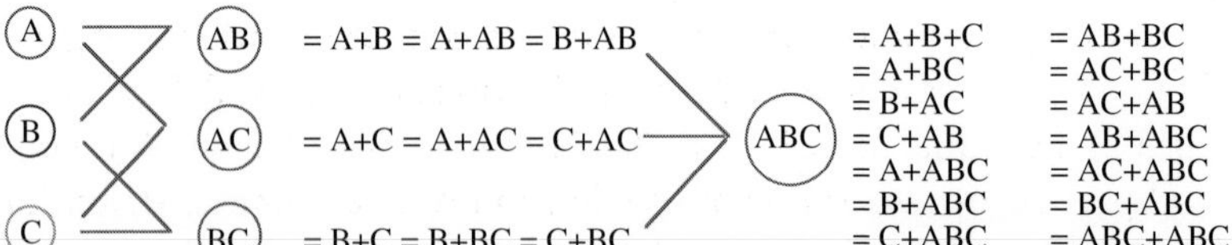

Figure 15.4 Example of internal mixing among three externally mixed size distributions (A, B, and C, each represented by one particle). The figure shows many (but not all) combinations resulting in internally mixed particles (AB, AC, BC, and ABC).

BC, or ABC. The figure and Example 15.1 show that many such combinations are possible.

Example 15.1

What other possible combinations, not shown in Fig. 15.4, of three separate particles can result in a particle of composition ABC?

SOLUTION

A+A+BC; B+B+AC; C+C+AB; A+AB+BC; B+AB+BC; C+AB+BC; A+B+BC; A+B+AC; A+C+AB; A+C+BC; B+C+AB; B+C+AC.

Here, the semiimplicit coagulation solution is expanded to treat coagulation among any number of size distributions (N_T), size bins per distribution (N_B), and components per size bin per distribution (N_V) (Jacobson 2002). The scheme remains noniterative, positive-definite for any time step, and volume- and volume-concentration-conserving.

With the scheme, the final volume concentration ($cm^3\ cm^{-3}$) of component $q = 1, \ldots, N_V$ in size bin $k = 1, \ldots, N_B$ of size distribution $N = 1, \ldots, N_T$ at time t is found by solving

$$v_{q,Nk,t} = \frac{v_{q,Nk,t-h} + h(T_{q,Nk,t,1} + T_{q,Nk,t,2})}{1 + hT_{q,Nk,t,3}} \tag{15.19}$$

$$T_{q,Nk,t,1} = \sum_{M=1}^{N_T} \left[P_{N,M} \sum_{j=1}^{k} \left(n_{Mj,t-h} \sum_{i=1}^{k-1} f_{Ni,Mj,Nk,t-h} \beta_{Ni,Mj,t-h} v_{q,Ni,t} \right) \right]$$

$$T_{q,Nk,t,2} = \sum_{M=1}^{N_T} \sum_{I=1}^{N_T} \left[Q_{I,M,N} \sum_{j=1}^{k} \left(n_{Mj,t-h} \sum_{i=1}^{k} f_{Ii,Mj,Nk,t-h} \beta_{Ii,Mj,t-h} v_{q,Ii,t} \right) \right]$$

$$T_{q,Nk,t,3} = \sum_{j=1}^{N_B} \left[\sum_{M=1}^{N_T} [(1 - L_{N,M})(1 - f_{Nk,Mj,Nk,t-h}) + L_{N,M}] \beta_{Nk,Mj,t-h} n_{Mj,t-h} \right]$$

where term T_1 accounts for production of larger particles in distribution N from self-coagulation and from heterocoagulation of distribution N with distribution $M \neq N$. Term T_2 accounts for production of particles in distribution N from heterocoagulation of two independent distributions ($I \neq N$ and $M \neq N$). The first part of term T_3 accounts for self-coagulation loss in distribution N to larger sizes. The second part accounts for loss of distribution N to all other distributions aside from N due to heterocoagulation of N with the other distributions.

Equation (15.19) must be solved in a special order for all implicit (at time t) values on the right side of the equation to be known when they are used. Distributions that have no coagulation production from other distributions are solved first, followed by distributions with production terms from previously solved distributions.

Take, for example, the set of 16 distributions shown in Table 15.1. This table lists five externally mixed distributions (A, B, D, E, and F), 10 binary distributions (AB, AD, AE, AF, BD, BE, BF, DE, DF, and EF), and one well-mixed distribution (MX = ABDEF). Due to the computational burden, ternary (e.g., ABD, ADE) and quartenary (e.g., ABDE, ABDF) combinations are included in the MX distribution. In other words, coagulation between distributions AB and D, for example, is assumed to produce mixed particles in distribution ABDEF.

In this example set of distributions, the externally mixed distributions are solved first (in any order), followed by the binary distributions (in any order), followed by the MX distribution. Within each distribution, equations are solved from bin $k = 1, \ldots, N_B$. The order of solving each volume component, $q = 1, \ldots, N_V$, does not matter. To conserve volume at the upper boundary, (15.19) assumes that particles cannot self-coagulate out of the largest size bin of any size distribution. For example, when two particles self-coagulate to a size larger than the largest bin in the distribution, they are placed in the largest bin but particle number is increased to conserve volume concentration. Similarly, when two particles heterocoagulate to form a new particle larger than the largest bin in the new distribution that the particle enters, the new particle is placed in the largest bin of the new distribution and number concentration is adjusted. These boundary conditions have virtually no effect on results so long as the largest bin is much larger than that of the median particle size in the distribution.

The parameters P, Q, and L in (15.19) are either 1 or 0, depending on the coagulation interactions accounted for. The values are illustrated by considering Table 15.1. The parameter $P_{N,M} = 1$ if particles in distribution N coagulating with particles in distribution M produce larger particles in distribution N. For example, sea spray–soil (AB) plus sea spray (A) produces more sea spray–soil (AB); thus, $P_{AB,A} = 1$. The parameter $Q_{I,M,N} = 1$ if particles in distribution I coagulating with particles in distribution M produce particles in distribution N, where $I \neq M$ and $I \neq N$. For example, soil (B) plus sulfate (D) produces soil–sulfate (BD); thus, $Q_{B,D,BD} = 1$. The parameter $L_{N,M} = 1$ if particles in distribution N coagulating with particles in distribution M do not produce particles in distribution N. For example, soil (B) plus sulfate (D) does not produce soil (B); thus, $L_{B,D} = 1$. On the other hand, soil–sulfate (BD) plus sulfate (D) produces soil–sulfate (BD); thus, $L_{BD,D} = 0$ in that case. Further, in (15.19)

$$f_{Ii,Mj,Nk} = \begin{cases} \left(\dfrac{\upsilon_{Nk+1} - V_{Ii,Mj,t-h}}{\upsilon_{Nk+1} - \upsilon_{Nk,t-h}}\right)\dfrac{\upsilon_{Nk}}{V_{Ii,Mj}} & \upsilon_{Nk} \le V_{Ii,Mj} < \upsilon_{Nk+1} \quad k < N_B \\ 1 - f_{Ii,Mj,Nk-1} & \upsilon_{Nk-1} < V_{Ii,Mj} < \upsilon_{Nk} \quad k > 1 \\ 1 & V_{Ii,Mj} \ge \upsilon_{Nk} \quad k = N_B \\ 0 & \text{all other cases} \end{cases} \tag{15.20}$$

is the volume fraction of a coagulated pair, $V_{Ii,Mj} = \upsilon_{Ii} + \upsilon_{Mj}$, that is partitioned into bin k of distribution N. To minimize computer time, all calculations involving a zero value of f, P, Q, or L are eliminated ahead of time.

Table 15.1 Coagulation interactions for the simulation in Fig. 15.5

Size distribution name/symbol		Symbol of second distribution															
		A	**B**	**D**	**E**	**F**	**AB**	**AD**	**AE**	**AF**	**BD**	**BE**	**BF**	**DE**	**DF**	**EF**	**MX**
Sea spray	A	A	AB	AD	AE	AF	AB	AD	AE	AF	MX	MX	MX	MX	MX	MX	MX
Soil	B	AB	B	BD	BE	BF	AB	MX	MX	MX	BD	BE	BF	MX	MX	MX	MX
Sulfate	D	AD	BD	D	DE	DF	MX	AD	MX	MX	BD	MX	MX	DE	DF	MX	MX
Emitted soot (ES)	E	AE	BE	DE	E	EF	MX	MX	AE	MX	MX	BE	MX	DE	MX	EF	MX
Background soot (BS)	F	AF	BF	DF	EF	F	MX	MX	MX	AF	MX	MX	BF	MX	DF	EF	MX
Sea spray–soil	AB	AB	AB	MX	MX	MX	AB	MX	MX	MX	MX	MX	MX	MX	MX	MX	MX
Sea spray–sulfate	AD	AD	MX	AD	MX	MX	MX	AD	MX	MX	MX	MX	MX	MX	MX	MX	MX
Sea spray–ES	AE	AE	MX	MX	AE	MX	MX	MX	AE	MX	MX	MX	MX	MX	MX	MX	MX
Sea spray–BS	AF	AF	MX	MX	MX	AF	MX	MX	MX	AF	MX	MX	MX	MX	MX	MX	MX
Soil–sulfate	BD	MX	BD	BD	MX	MX	MX	MX	MX	MX	BD	MX	MX	MX	MX	MX	MX
Soil–ES	BE	MX	BE	MX	BE	MX	MX	MX	MX	MX	MX	BE	MX	MX	MX	MX	MX
Soil–BS	BF	MX	BF	MX	MX	BF	MX	MX	MX	MX	MX	MX	BF	MX	MX	MX	MX
Sulfate–ES	DE	MX	MX	DE	DE	MX	MX	MX	MX	MX	MX	MX	MX	DE	MX	MX	MX
Sulfate–BS	DF	MX	MX	DF	MX	DF	MX	MX	MX	MX	MX	MX	MX	MX	DF	MX	MX
ES–BS	EF	MX	MX	MX	EF	EF	MX	MX	MX	MX	MX	MX	MX	MX	MX	EF	MX
Internal mixture	MX	MX	MX	MX	MX	MX	MX	MX	MX	MX	MX	MX	MX	MX	MX	MX	MX

The table gives the symbol of the size distribution into which two particles from either the same or different distributions coagulate. For example, when a sea spray particle (A) coagulates with a soil particle (B), the resulting particle is a sea spray–soil (AB) particle. When three or more externally mixed distributions (A, . . . , F) or an externally mixed and binary distribution (AB, . . . , EF) or two binary distributions combine, they enter the mixed distribution (MX). Thus, the combination of background soot (F) with sea spray–soil (AB) produces a mixed particle (MX).

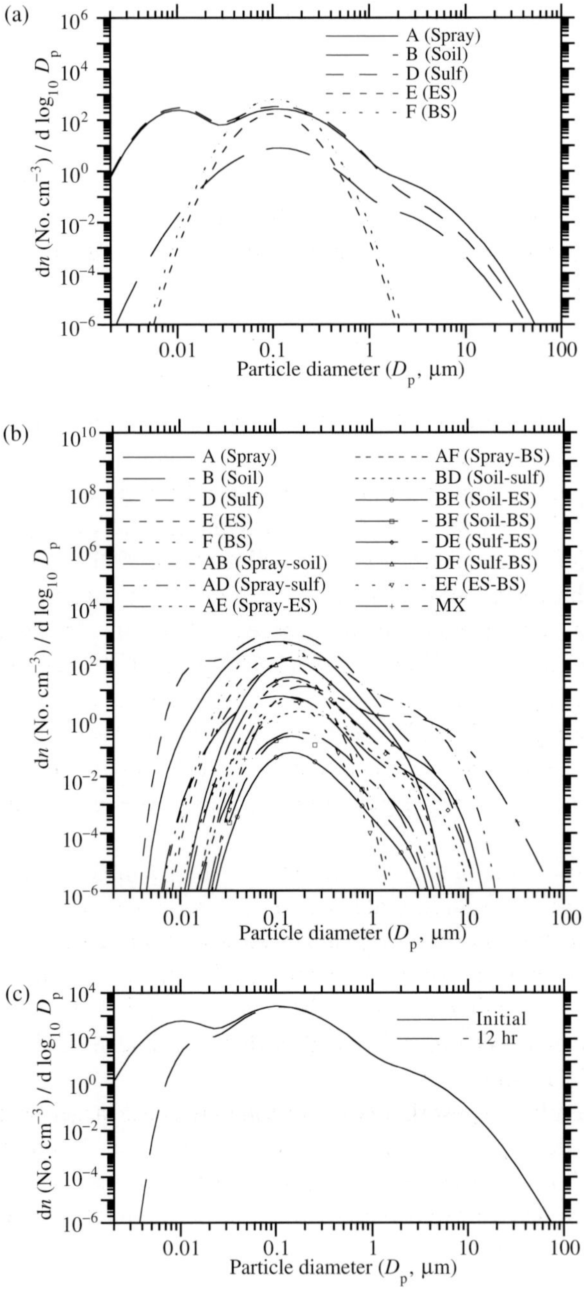

Figure 15.5 (a) Initial number concentration of five externally mixed size distributions. (b) Number concentrations of 16 size distributions listed in Table 15.1 following 12 hours of coagulation starting with the five distributions in (a). (c) The initial and final number concentrations, summed over all distributions, from (a) and (b), respectively. Following Jacobson (2002).

The final particle total volume concentration in each bin of each distribution is

$$v_{Nk,t} = \sum_{q=1}^{N_V} v_{q,Nk,t} \tag{15.21}$$

Alternatively, the total volume concentration can be solved by summing the component volume concentrations at time $t - h$ then solving (15.19) after replacing $v_{q,Nk,t}$ with $v_{Nk,t}$. Finally, the final number concentration of particles in each bin of each distribution is

$$n_{Nk,t} = \frac{v_{Nk,t}}{\upsilon_{Nk}} \tag{15.22}$$

Figure 15.5 illustrates the effect of coagulation among multiple size distributions on the internal mixing of aerosol particles. Coagulation alone was simulated among the 16 distributions in Table 15.1 for a 12-hour period. Five of the distributions (A, B, D, E, F) were initialized, as shown in Fig. 15.5(a). Figure 15.5(b) shows all 16 distributions after coagulation, and Fig. 15.5(c) shows the sum of all initial and final distributions. The last figure appears to indicate, as coagulation simulations of one size distribution have indicated, that coagulation among multiple size distributions has little effect on particles larger than 0.2 μm in diameter over a 12-hour period. Figure 15.5(b) clarifies this misconception. It shows that, although coagulation among multiple distributions has little effect on the total number concentration of large particles summed over all distributions, it internally mixes particles of different original compositions, affecting the composition of each size distribution.

The results also suggest that, in an air mass containing a moderate loading of particles, coagulation alone can internally mix most particles to some degree within 12 hours. In the simulation, coagulation internally mixed almost all the largest particles and a smaller percentage of the smaller particles, which may explain a result of Okada and Hitzenberger (2001), who found that the number fraction of mixed particles in Vienna increased with increasing particle radius and with increasing particle abundance.

Finally, the results suggest that coagulation may cause larger particles to become mixed with more components than smaller particles. Since the combination of even the smallest with the largest particle is considered a mixture, though, an internally mixed large particle can often contain a trivial amount of a second and/or third component.

15.5 PARTICLE FLOW REGIMES

For most of the rest of this chapter, physical processes causing particles to collide and coalesce (coagulate) are discussed. The rate of particle collision during coagulation depends on several parameters, including the Knudsen number for air, the mean free path of an air molecule, the thermal speed of an air molecule, and the particle Reynolds number. In this subsection, these parameters are described.

The **Knudsen number for air**, a dimensionless number, is defined as

$$\mathrm{Kn}_{\mathrm{a},i} = \frac{\lambda_\mathrm{a}}{r_i} \tag{15.23}$$

where λ_a is the mean free path of an air molecule, and r_i is the radius of a particle of size i. The **mean free path of an air molecule** is the average distance an air molecule can travel before it encounters another air molecule by random motion and exchanges momentum with it. It is analogous to the mixing length of an eddy, defined in Chapter 8, and may be calculated as

$$\lambda_\mathrm{a} = \frac{2\eta_\mathrm{a}}{\rho_\mathrm{a}\bar{v}_\mathrm{a}} = \frac{2\nu_\mathrm{a}}{\bar{v}_\mathrm{a}} \tag{15.24}$$

where η_a is the dynamic viscosity of air (g cm^{-1} s^{-1}), ρ_a is the density of air (g cm^{-3}), $\bar{v}_\mathrm{a}$ is thermal speed of an air molecule (cm s^{-1}), and $\nu_\mathrm{a} = \eta_\mathrm{a}/\rho_\mathrm{a}$ is the kinematic viscosity of air, defined in (4.55). The greater the air density, the closer molecules are to each other, and the shorter the mean free path of a molecule. The dynamic viscosity of air was given empirically as a function of temperature in (4.54). The average **thermal speed of an air molecule**, from (2.3), is

$$\bar{v}_\mathrm{a} = \sqrt{\frac{8k_\mathrm{B}T}{\pi \bar{M}}} \tag{15.25}$$

The higher the air temperature, the greater the kinetic energy and average speed of an air molecule.

Example 15.2

Find λ_a and $\mathrm{Kn}_{\mathrm{a},i}$ when $T = 288$ K, $p_\mathrm{a} = 1013$ hPa, the air is dry, and $r_i = 0.1$ μm.

SOLUTION

From (15.25), $\bar{v}_\mathrm{a} = 4.59 \times 10^4$ cm s^{-1}. From (4.54), $\eta_\mathrm{a} = 1.79 \times 10^{-4}$ g cm^{-1} s^{-1}. Since the air is dry, $R_\mathrm{m} = R'$ and $\rho_\mathrm{a} = 0.00123$ g cm^{-3} from (2.36). Substituting these values into (15.24) gives $\lambda_\mathrm{a} = 6.34 \times 10^{-6}$ cm. From (15.23), $\mathrm{Kn}_{\mathrm{a},i} = 0.63$.

When a particle is large relative to the mean free path of an air molecule ($\mathrm{Kn}_{\mathrm{a},i} \ll 1$), the particle is likely to be intercepted by many air molecules, and its resistance to motion is due primarily to air viscosity. This Knudsen-number regime is called the **continuum regime** because particles see air as a continuum.

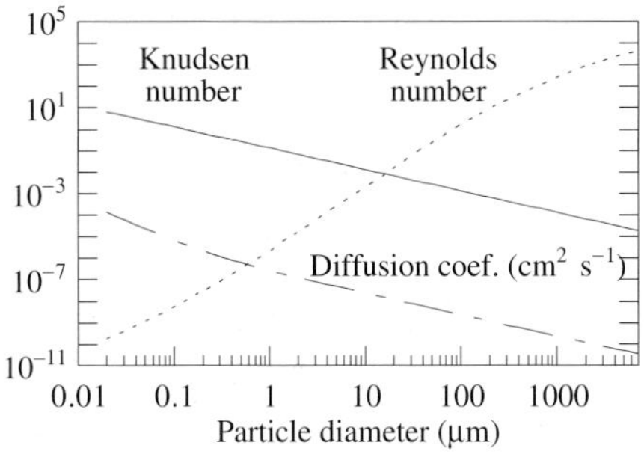

Figure 15.6 Knudsen number for air, Reynolds number, and diffusion coefficient of particles falling as a function of diameter when $T = 292$ K, $p_a = 999$ hPa, and $\rho_p = 1.0$ g cm^{-3}.

When the mean free path of an air molecule is large relative to particle size ($Kn_{a,i} \gg 10$), a particle is likely to be intercepted by relatively few air molecules, and its resistance to motion is due primarily to the inertia of air molecules hitting it. This Knudsen-number regime is called the **free-molecular regime** because particle motion is governed by free-molecular kinetics. Between the continuum and free-molecular regimes is the **transition regime.**

The **particle Reynolds number** gives the ratio of the inertial force exerted by a particle to the viscous force exerted by air. If the inertial force is due to the particle's fall speed, the Reynolds number (dimensionless) is

$$\mathrm{Re}_i = 2r_i V_{f,i}/\nu_a \tag{15.26}$$

where $V_{f,i}$ is the **terminal fall speed of the particle** (cm s^{-1}), calculated in Section 20.1. Figure 15.6 shows Reynolds and Knudsen numbers for particles falling under specified conditions.

When particle radius is less than the mean free path of a gas molecule, the particle motion through the air is called **slip flow**, because particles slip through the air with little viscous resistance. Slip flow occurs mostly in the free-molecular regime. Pure slip flow for falling particles occurs at low Reynolds numbers (e.g., $\mathrm{Re}_i < 10^{-6}$), since small, falling particles have a low terminal fall speed, as shown in Fig. 15.6.

When the radius of a particle is larger than that of an air molecule, but the inertial force imparted by the particle is negligible compared with the viscous force exerted by air, the flow of the particle through air is **Stokes flow**, named after George Stokes (1819–1903). Stokes flow occurs within the continuum regime, since in Stokes flow the particle radius is larger than the mean free path of an air molecule. Stokes flow for falling particles occurs at moderate Reynolds numbers (e.g., $0.01 < \mathrm{Re}_i < 1$). At higher Reynolds numbers, the inertial force imparted by a particle becomes larger than the viscous force exerted by air. At smaller Reynolds numbers, Stokes flow approaches slip flow.

15.6 COAGULATION KERNEL

The **coagulation rate coefficient** (kernel),

$$\beta_{i,j} = E_{\mathrm{coal},i,j} K_{i,j} \tag{15.27}$$

(cm^3 particle^{-1} s^{-1}) is the product of a **coalescence efficiency** ($E_{\mathrm{coal},i,j}$, dimensionless) and a **collision kernel** ($K_{i,j}$, cm^3 particle^{-1} s^{-1}), where the subscripts indicate particles in size bin i are coagulating with particles in size bin j. Equation (15.27) assumes coagulation is occurring in only one size distribution, but the equation is easily extended to multiple distributions.

For aerosol–aerosol coagulation, the collision kernel is the sum of the kernels due to the following physical processes: Brownian motion, convective Brownian motion enhancement, gravitational collection, turbulent inertial motion, turbulent shear, and van der Waals forces. The kernel is also affected by particle shape. For aerosol–hydrometeor coagulation and hydrometeor–hydrometeor coagulation, the kernels are also affected by thermophoresis, diffusiophoresis, and electric charge. Collision kernels and coalescence efficiencies for aerosol–aerosol, aerosol–hydrometeor, and hydrometeor–hydrometeor coagulation are discussed in this section. Numerical solutions for solving aerosol–hydrometeor and hydrometeor coagulation are given in Chapter 18.

15.6.1 Brownian diffusion

Brownian motion is the irregular motion of a particle suspended in a gas due to the random bombardment of the particle by gas molecules. Brownian motion leads to particle scatter and is represented by a particle diffusion coefficient. As such, it is often referred to as **Brownian diffusion**. However, **particle diffusion** is the net transport of particles from a region of high concentration to one of low concentration, whereas Brownian diffusion can occur in the absence of a concentration gradient.

Brownian coagulation is the process by which particles scatter, collide, and coalesce in the atmosphere due to Brownian motion. In the **continuum regime**, the **Brownian collision kernel** (cm^3 particle^{-1} s^{-1}) for particles of size i coagulating with particles of size j is

$$K^{\mathrm{B}}_{i,j} = 4\pi(r_i + r_j)(D_{\mathrm{p},i} + D_{\mathrm{p},j}) \tag{15.28}$$

where $D_{\mathrm{p},i}$ and $D_{\mathrm{p},j}$ are the diffusion coefficients of particles i and j, respectively.

An expression for the **particle diffusion coefficient** is

$$D_{\mathrm{p},i} = \frac{k_{\mathrm{B}} T}{6\pi r_i \eta_{\mathrm{a}}} G_i \tag{15.29}$$

where k_{B} is Boltzmann's constant, T is the Kelvin temperature, and $G_i = 1 + \mathrm{Kn}_{\mathrm{a},i}\alpha_{\mathrm{a},i}$ is the **Cunningham slip-flow correction** (Cunningham 1910). When

$G_i = 1$, (15.29) is called the **Stokes–Einstein equation**, which is valid for Stokes flow only ($0.01 < \mathrm{Re}_i < 1$). When $\mathrm{Re}_i < 0.01$ (reaching slip flow at $\mathrm{Re}_i < 10^{-6}$), G_i and the diffusion coefficient increase because particle resistance to motion decreases, allowing particles to move more freely through the air.

Knudsen and Weber (1911) approximated $\alpha_{a,i} = \mathrm{A}' + \mathrm{B}' \exp(-\mathrm{C}'/\mathrm{Kn}_{a,i})$, giving the Cunningham slip-flow correction as

$$G_i = 1 + \mathrm{Kn}_{a,i}[\mathrm{A}' + \mathrm{B}' \exp(-\mathrm{C}'/\mathrm{Kn}_{a,i})] \tag{15.30}$$

Knusden and Weber estimated A′, B′, and C′ based on experiments with glass beads falling through air. In a more relevant experiment, Millikan (1923) estimated A′, B′, and C′ as 0.864, 0.29, and 1.25, respectively, based on measurements of oil drops falling through air. Kasten (1968) gives a more recent set of values for A′, B′, and C′ as 1.249, 0.42, and 0.87, respectively. Equation (15.30) predicts that, at low Knudsen number (large particle radius and high Reynolds number), G_i approaches unity, and the particle diffusion coefficient approaches that from the Stokes–Einstein equation. At high Knudsen number (small particle radius and low Reynolds number), G_i increases, increasing the particle diffusion coefficient. Figure 15.6 shows the corrected particle diffusion coefficient as a function of particle diameter. The figure shows that, at very low particle diameter (and Reynolds number), diffusion coefficients are slightly enhanced by the reduced resistance to particle motion.

In the **free-molecular regime**, the **Brownian collision kernel** (cm^3 particle^{-1} s^{-1}) for particles of size i coagulating with particles of size j is based on the kinetic theory of gases and is given by

$$K^{\mathrm{B}}_{i,j} = \pi(r_i + r_j)^2 \sqrt{\bar{v}^2_{p,i} + \bar{v}^2_{p,j}} \tag{15.31}$$

where

$$\bar{v}_{p,i} = \sqrt{\frac{8k_{\mathrm{B}}T}{\pi \bar{M}_{p,i}}} \tag{15.32}$$

is the **thermal speed of a particle in air**, in which $\bar{M}_{p,i}$ is the mass (g) of one particle of size i. Equation (15.32) indicates that the thermal speed of a light (small) particle exceeds that of a heavy (large) particle at a given temperature.

For particles in the **transition regime**, the Brownian collision kernel can be calculated with the interpolation formula of Fuchs (1964),

$$K^{\mathrm{B}}_{i,j} = \frac{4\pi(r_i + r_j)(D_{p,i} + D_{p,j})}{\dfrac{r_i + r_j}{r_i + r_j + \sqrt{\delta_i^2 + \delta_j^2}} + \dfrac{4(D_{p,i} + D_{p,j})}{\sqrt{\bar{v}^2_{p,i} + \bar{v}^2_{p,j}}\,(r_i + r_j)}} \tag{15.33}$$

where δ_i is the mean distance from the center of a sphere reached by particles leaving the sphere's surface and traveling a distance of particle mean free path $\lambda_{p,i}$.

These two parameters are, respectively,

$$\delta_i = \frac{(2r_i + \lambda_{p,i})^3 - \left(4r_i^2 + \lambda_{p,i}^2\right)^{3/2}}{6r_i\lambda_{p,i}} - 2r_i \qquad \lambda_{p,i} = \frac{8D_{p,i}}{\pi \bar{v}_{p,i}} \tag{15.34}$$

Equation (15.33) simplifies to (15.28) for small Knudsen numbers and to (15.31) for large Knudsen numbers.

15.6.2 Convective Brownian diffusion enhancement

When a large particle falls through the air, eddies created in its wake enhance diffusion of other particles to its surface. This process, called **convective Brownian diffusion enhancement**, has the collision kernel

$$K_{i,j}^{DE} = \begin{cases} K_{i,j}^{B} 0.45 Re_j^{1/3} Sc_{p,i}^{1/3} & Re_j \le 1, r_j \ge r_i \\ K_{i,j}^{B} 0.45 Re_j^{1/2} Sc_{p,i}^{1/3} & Re_j > 1, r_j \ge r_i \end{cases} \tag{15.35}$$

(Pruppacher and Klett 1997), where $Sc_{p,i}$ is the **particle Schmidt number,**

$$Sc_{p,i} = \frac{\nu_a}{D_{p,i}} \tag{15.36}$$

which gives the ratio of viscous to diffusive forces. Equation (15.35) states that the larger the particle and the higher its fall speed, the greater the effect of diffusion enhancement.

15.6.3 Gravitational collection

A third collision kernel is the gravitational collection (differential fall speed) kernel. When two particles of different size fall, one may catch up with and collide with the other. The collision kernel for **gravitational collection** accounts for this process and is given as

$$K_{i,j}^{GC} = E_{coll,i,j}\pi(r_i + r_j)^2 |V_{f,i} - V_{f,j}| \tag{15.37}$$

where $E_{coll,i,j}$ is a **collision efficiency**. A parameterization of the collision efficiency of falling raindrops is

$$E_{coll,i,j} = \frac{60E_{V,i,j} + E_{A,i,j}Re_j}{60 + Re_j} \qquad r_j \ge r_i \tag{15.38}$$

$$E_{V,i,j} = \begin{cases} \left[1 + \frac{0.75\ln(2St_{i,j})}{St_{i,j} - 1.214}\right]^{-2} & St_{i,j} > 1.214 \\ 0 & St_{i,j} \le 1.214 \end{cases}$$

$$E_{A,i,j} = \frac{St_{i,j}^2}{(St_{i,j} + 0.5)^2} \tag{15.39}$$

(Ludlum 1980) where $\mathrm{St}_{i,j} = V_{\mathrm{f},i}|V_{\mathrm{f},j} - V_{\mathrm{f},i}|/r_j g$ (for $r_j \geq r_i$) is the dimensionless **Stokes number**. Equation (15.38) simplifies to $E_{\mathrm{V},i,j}$ when $\mathrm{Re}_j \ll 1$ and to $E_{\mathrm{A},i,j}$ when $\mathrm{Re}_j \gg 1$. Gravitational collection is important where at least one particle is large, such as in a cloud or in the presence of sea spray or soil dust particles. It is the most important mechanism by which cloud drops form raindrops in warm clouds (clouds with temperatures above 0 °C).

Beard and Grover (1974) provide a set of measurements from a cloud-chamber experiment of the collision kernel for liquid water drops resulting from Brownian motion, Brownian diffusion enhancement, and gravitational collection simultaneously. Such measurements can be used instead of the model kernels provided above, but are limited in the range of drop size interactions measured.

15.6.4 Turbulent inertial motion and turbulent shear

Two additional kernels include those for turbulent inertial motion and turbulent shear. **Turbulent inertial motion** causes particles of different size to accelerate differentially. The resulting collision kernel is

$$K_{i,j}^{\mathrm{TI}} = \frac{\pi \varepsilon_{\mathrm{d}}^{3/4}}{g \nu_{\mathrm{a}}^{1/4}} (r_i + r_j)^2 |V_{\mathrm{f},i} - V_{\mathrm{f},j}| \tag{15.40}$$

(Saffman and Turner 1956), where ε_{d} is the rate of dissipation of turbulent kinetic energy per gram of medium (cm^2 s^{-3}), defined in Section 8.4. Values for ε_{d} vary between 3 and 2000 cm^2 s^{-3}, with lower values corresponding to clear or partly cloudy air and larger values corresponding to strong cumulus convection. A typical value for clear air is 5 cm^2 s^{-3} (Pruppacher and Klett 1997).

When wind shear in turbulent air causes particles moving with the air to collide, the collision kernel is parameterized with the kernel for **turbulent shear,**

$$K_{i,j}^{\mathrm{TS}} = \left(\frac{8\pi \varepsilon_{\mathrm{d}}}{15 \nu_{\mathrm{a}}} \right)^{1/2} (r_i + r_j)^3 \tag{15.41}$$

(Saffman and Turner 1956). Two small particles are more likely to collide because of turbulent shear than because of turbulent inertial motion. A large particle is more likely to collide with a small particle because of turbulent inertial motion than because of turbulent shear. Brownian motions dominate both processes so long as at least one particle is small.

Figures 15.7 (a) and (b) show how each of the five coagulation kernels discussed so far varies when particles of 0.01 and 10 μm, respectively, coagulate with particles of all sizes. The curves indicate that, for small particles, Brownian coagulation is always dominant. For larger particles, the other kernels become important. When two particles have the same size, their fall speeds are identical, and the coagulation

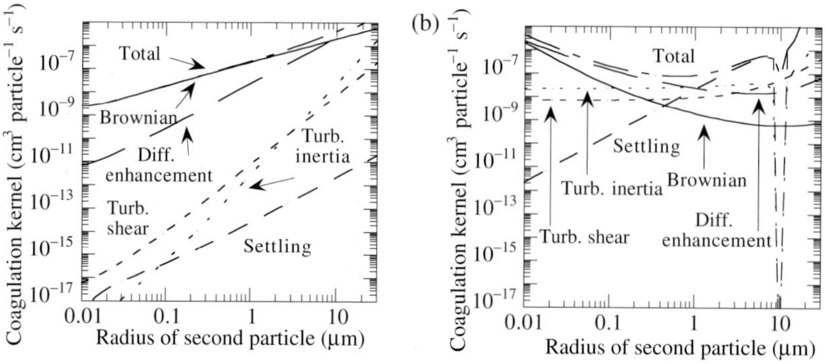

Figure 15.7 (a) Coagulation kernels for five processes when a particle 0.01 μm in radius coagulates with particles of different size at 298 K. (b) Coagulation kernels for five processes when a particle 10 μm in radius coagulates with particles of different size at 298 K. The dip at 10 μm results because the difference in fall speed is zero at that point. The real width of the dip is narrower, but the resolution of the size bins used for the graph was coarse.

kernels for turbulent inertial motions and gravitational collection are zero, as seen in Fig. 15.7(b).

15.6.5 Van der Waals and viscous forces

Two forces that affect primarily small aerosol particles are van der Waals and viscous forces. **Van der Waals forces** are weak dipole–dipole attractions caused by brief, local charge fluctuations in nonpolar molecules having no net charge. In other words, uncharged particles experience random charge fluctuations that cause one part of the particle to experience a brief positive charge and the other, a brief negative charge, so that the particle still exhibits no net charge. When a particle experiencing a brief charge fluctuation approaches another, the first induces a charge of the opposite sign on the closest end of the second particle. The opposite charge between the two particles causes an attraction, enhancing the rate of coagulation between the particles.

Viscous forces arise because two particles moving toward each other in a viscous medium have a net diffusion coefficient smaller than that of the sum of the two individual coefficients since velocity gradients generated by one particle affect those of the other. Whereas van der Waals forces enhance the rate of coagulation of small particles, particularly of **nanoparticles** (<50 nm diameter), in the free-molecular regime, viscous forces retard the rate of van der Waals force enhancement in the transition and continuum regimes (e.g., Marlow 1981; Schmidt-Ott and Burtscher 1982; Alam 1987; Seinfeld and Pandis 1998).

An interpolation formula for the **van der Waals/viscous collision kernel** between the free-molecular and continuum regimes (Alam 1987; Jacobson and Seinfeld

2004) is

$$K_{i,j}^{V} = K_{i,j}^{B}\left(V_{E,i,j} - 1\right) = K_{i,j}^{B}\left\{\frac{W_{c,i,j}\left[1 + \dfrac{4(D_{p,i} + D_{p,j})}{\sqrt{\bar{v}_{p,i}^2 + \bar{v}_{p,j}^2}(r_i + r_j)}\right]}{1 + \dfrac{W_{c,i,j}}{W_{k,i,j}}\dfrac{4(D_{p,i} + D_{p,j})}{\sqrt{\bar{v}_{p,i}^2 + \bar{v}_{p,j}^2}(r_i + r_j)}} - 1\right\} \quad (15.42)$$

where $V_{E,i,j}$ is the Van der Waals/viscous collision correction factor (which may be less than or greater than unity). Although $K_{i,j}^{V}$ may be less than zero, $K_{i,j}^{V} + K_{i,j}^{B} > 0$.

The individual correction factors in (15.42) for the free-molecular and continuum regimes are

$$W_{k,i,j} = \frac{-1}{2(r_i + r_j)^2 k_B T}\int_{r_i+r_j}^{\infty} \times\left(\frac{dE_{P,i,j}(r)}{dr} + r\frac{d^2 E_{P,i,j}(r)}{dr^2}\right)\exp\left[\frac{-1}{k_B T}\left(\frac{r}{2}\frac{dE_{P,i,j}(r)}{dr} + E_{P,i,j}(r)\right)\right]r^2\,dr \quad (15.43)$$

$$W_{c,i,j} = \frac{1}{(r_i + r_j)\displaystyle\int_{r_i+r_j}^{\infty}\frac{D_{i,j}^{\infty}}{D_{r,i,j}}(r)\exp\left[\frac{E_{P,i,j}(r)}{k_B T}\right]\frac{dr}{r^2}} \quad (15.44)$$

respectively. In these equations,

$$E_{P,i,j}(r) = -\frac{A_H}{6}\left[\frac{2r_i r_j}{r^2 - (r_i + r_j)^2} + \frac{2r_i r_j}{r^2 - (r_i - r_j)^2} + \ln\frac{r^2 - (r_i + r_j)^2}{r^2 - (r_i - r_j)^2}\right] \quad (15.45)$$

is the **van der Waals interaction potential,** in which A_H is the Hamaker constant, which depends on van der Waals properties of each substance (e.g., Seinfeld and Pandis 1998). Values of $A_H/k_B T$ generally range from <20 to 200. Also,

$$\frac{D_{i,j}^{\infty}}{D_{r,i,j}}(r) = 1 + \frac{2.6 r_i r_j}{(r_i + r_j)^2}\sqrt{\frac{r_i r_j}{(r_i + r_j)(r - r_i - r_j)}} + \frac{r_i r_j}{(r_i + r_j)(r - r_i - r_j)} \quad (15.46)$$

is a **viscous-force correction factor to the diffusion coefficient in the continuum regime.** In the equation, $D_{r,i,j}$ is a "relative" diffusion coefficient between particles i and j, and $D_{i,j}^{\infty} = D_{p,i} + D_{p,j}$ is the sum of the individual diffusion coefficients of the two particles.

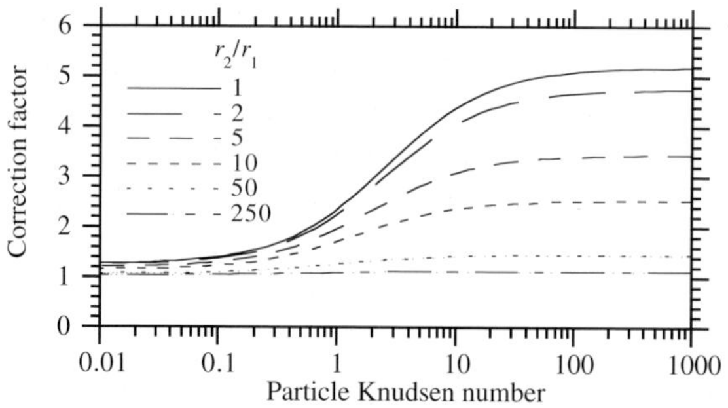

Figure 15.8 Van der Waals/viscous force correction factor ($V_{E,i,j}$) as a function of particle Knudsen number (Kn_p) and different ratios of the radii of two interacting particles when $T = 300$ K, $p = 1$ atm, particle density = 1 g/cm^3, and $A_H/k_BT = 200$. From Jacobson and Seinfeld (2004).

Figure 15.8 shows the van der Waals/viscous force correction factor $V_{E,i,j}$ versus particle pair Knudsen number,

$$Kn_p = \frac{\sqrt{\lambda_{p,i}^2 + \lambda_{p,j}^2}}{r_i + r_j} \tag{15.47}$$

assuming $A_H/k_BT = 200$. The figure shows that van der Waals forces enhanced the coagulation rate of small particles (high Kn_p) by up to a factor of five. The calculated enhancement factor for two 14-nm diameter particles ($Kn_p = 2.26$) from the figure is about 3.25, and that for two 28-nm particles ($Kn_p = 0.983$) is about 2.3. These numbers compare with measured enhancement factors for 14- and 28-nm graphitic carbon particles of 6–8 and 1.4–3, respectively (Burtscher and Schmidt-Ott 1982). Although a value of $A_H/k_BT = 200$ is higher than typical measured values for many chemicals (e.g., Visser 1972), the comparison above illustrates that $A_H/k_BT = 200$ results in conservative enhancement factors for graphite. Similarly, silver particles have measured enhancement factors much higher than those predicted using their Hamaker constant (Burtscher and Schmidt-Ott 1982; Okuyama *et al.* 1984).

15.6.6 Fractal geometry

Fractals are particles of irregular, fragmented shape. Many aerosol particles containing liquids, such as water, are roughly spherical. Soot aggregates, though, are fractal in nature. Treatment of aggregates as fractal particles increases the rate of coagulation (e.g., Mountain *et al.* 1986; Mulholland *et al.* 1988; Rogak and Flagan 1992; Harris and Maricq 2001; Artelt *et al.* 2003).

The effect of fractal geometry on collisions can be treated by considering the effect of shape on radius, the diffusion coefficient, and the Knudsen number in previous collision kernels. For example, when an amorphous-shaped soot particle consisting of an agglomerate of individual **spherules** is considered, the **volume-equivalent radius** (r_i, the radius of a sphere with the same volume and density as the agglomerate) is replaced in the Brownian collision kernel of (15.33) by different terms in different parts of the equation.

For example, some terms are replaced with the **fractal radius** (outer radius) of the agglomerate,

$$r_{f,i} = r_s N_i^{1/D_f} \tag{15.48}$$

In this equation, r_s is the radius (cm) of each individual spherule (each assumed to be the same size), D_f is the **fractal dimension**, which relates the outer volume of an aggregate to the summed volume of individual spherules, and

$$N_{s,i} = \frac{\upsilon_i}{\upsilon_s} \tag{15.49}$$

is the **number of individual spherules in the soot aggregate**, where $\upsilon_s = 4\pi r_s^3/3$ is the volume of an individual spherule and $\upsilon_i = 4\pi r_i^3/3$ is the volume of the aggregate as if it were a sphere of uniform density. For diesel soot, individual spherules have typical radii of about $r_s \approx 10$–15 nm (e.g., Weingartner *et al.* 1997; Naumann 2003; Schnaiter *et al.* 2003; Wentzel *et al.* 2003). For pure spheres, $D_f = 3$. For diesel soot aggregates measured in the laboratory and in ambient air, D_f ranges from 1.5 to 2.5, with a typical value near 1.7 (e.g., Klingen and Roth 1989; Katrlnak *et al.* 1993; Lee 2001; Xiong and Friedlander 2001; Wentzel *et al.* 2003). As material condenses upon a soot aggregate, the material tends to compress the aggregate closer to a spherical shape, increasing D_f (Schnaiter *et al.* 2003; Wentzel *et al.* 2003).

Another type of radius used to calculate the effect of fractal geometry is the **mobility radius**, which is the radius of a sphere moving at the same speed and experiencing the same drag force as the agglomerate. This differs from the **Stokes radius**, which is the radius of a sphere with the same density and moving at the same fall speed as the agglomerate. It also differs from the **aerodynamic radius**, which is the radius of a unit-density sphere that has the same fall speed as the agglomerate. An expression for the mobility radius is

$$r_{m,i} = \text{MAX}\left\{\frac{r_{f,i}}{\ln(r_{f,i}/r_s)+1}, r_{f,i}\left(\frac{D_f-1}{2}\right)^{0.7}, r_{A,i}\right\} \tag{15.50}$$

(Rogak and Flagan 1992) where

$$r_{A,i} = r_s\sqrt{\text{MAX}\left\{N_{s,i}^{2/3}, \text{MIN}\left[1+\frac{2}{3}(N_{s,i}-1), \frac{1}{3}D_f N_{s,i}^{2/D_f}\right]\right\}} \tag{15.51}$$

is the **area-equivalent radius** (cm) (the radius of a sphere with the same surface area as the agglomerate).

For determining collision rates among fractals, an addition radius, the **collision radius** ($r_{c,i}$), needs to be considered. Rogak and Flagan (1992) suggest that this radius should lie between the fractal radius and the mobility radius. They also found that their modeled coagulation kernel matched measurements the best when the fractal (outer) radius was used as the collision radius ($r_{c,i} = r_{f,i}$), which is the assumption used here. For spheres ($D_f = 3$), the fractal radius, mobility radius, area-equivalent radius, and collision radius are identical and equal the volume-equivalent radius.

With the definitions above, the **Brownian collision kernel** (cm^3 particle^{-1} s^{-1}) from (15.33) can be **modified for fractal geometry** with

$$K^{B}_{i,j} = \frac{4\pi(r_{c,i} + r_{c,j})(D_{m,i} + D_{m,j})}{\dfrac{r_{c,i} + r_{c,j}}{r_{c,i} + r_{c,j} + \sqrt{\delta^2_{m,i} + \delta^2_{m,j}}} + \dfrac{4(D_{m,i} + D_{m,j})}{\sqrt{\bar{v}^2_{p,i} + \bar{v}^2_{p,j}}(r_{c,i} + r_{c,j})}} \qquad (15.52)$$

(Jacobson and Seinfeld 2004) where D_m is a particle diffusion coefficient (cm^2 s^{-1}) from (15.29) except that it and the Knudsen number for air, used within it, are evaluated at the mobility radius, and δ_m is the mean distance (cm) from (15.34), except that it and the mean free path of a particle, used within it, are evaluated at the mobility radius (Rogak and Flagan 1992). The mean thermal speed of a particle ($\bar{v}_p$), used in (15.52) and (15.34), is unaffected by particle radius. Instead, it is affected by particle mass, which is independent of fractal dimension. Equation (15.52) simplifies to the Brownian kernel for spheres when $D_f = 3$.

Figure 15.9 shows the effect of van der Waals/viscous forces and fractal geometry on the Brownian collision kernel. Van der Waals forces enhanced the kernel less for fractals than for spherical particles because, for the same volume-equivalent diameter, a fractal is more likely than a spherical particle to be in the transition or continuum regime than the free-molecular regime, and van der Waals forces are weakest in the continuum regime. The figure also shows that, when one particle was small (10 nm), fractal geometry increased the kernel with increasing size of the second particle. When one particle was larger (100 nm), fractal geometry enhancement was minimized when the second particle was also 100 nm.

Figure 15.10(a) shows the evolution, due to coagulation alone, of a size distribution dominated by nanoparticles emitted near a freeway. The coagulation kernel accounted for Brownian motion, Brownian diffusion enhancement, gravitational collection, turbulent inertial motion, and turbulent shear but not van der Waals forces, viscous forces, or fractal geometry. Of the kernels treated, Brownian motion was by far the most important, causing over 97 percent of particle loss. The figure shows that the first peak disappeared in favor of the second peak of 40 nm after 30–45 minutes.

Figure 15.10(b) shows the effect of van der Waals/viscous forces and fractal geometry on the size distribution of aerosol particles simulated in Fig. 15.10(a). The figure shows that treating these additional processes decreased the time of removal of the first peak from 30–45 minutes to about 10 minutes. Thus, van der Waals/viscous forces and fractal geometry may account for a greater share of the evolution of nanoparticle size than Brownian motion alone (Jacobson and Seinfeld 2004).

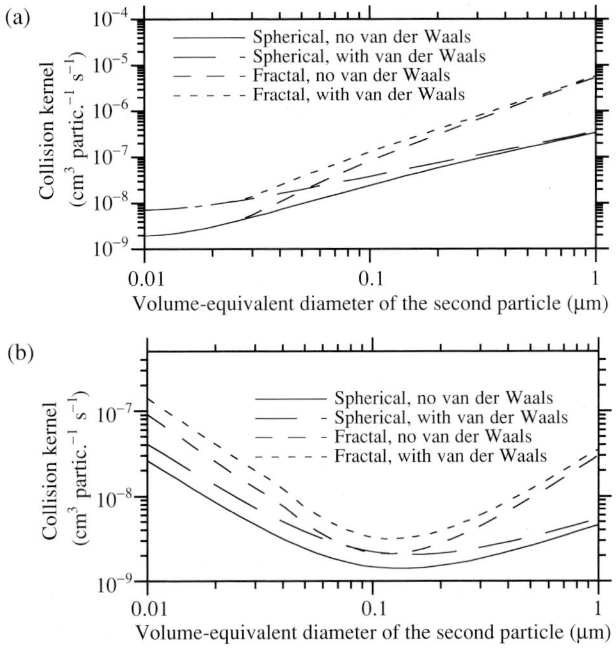

Figure 15.9 Brownian collision kernel when the volume-equivalent diameter of the first particle is (a) 10 nm and (b) 100 nm and the volume-equivalent diameter of the second particle varies from 10 to 1000 nm. The four curves shown in each figure account for when particles are spherical ($D_f = 3$) or fractal ($D_f = 1.7$ above 27 nm diameter and = 3.0 below) and when van der Waals and viscous forces are or are not included. The collision radius was set to the fractal radius. Also, $A_H/k_B T = 200$, $T = 300$ K, $p = 1$ atm, and $r_s = 13.5$ nm.

15.6.7 Diffusiophoresis, thermophoresis, and electric charge

Aerosol–hydrometeor coagulation and hydrometeor–hydrometeor coagulation are affected by three additional processes not so important during aerosol–aerosol coagulation: diffusiophoresis, thermophoresis, and electric charge.

Diffusiophoresis is the flow of aerosol particles down a concentration gradient of a gas due to the bombardment of the particles by the gas as the gas diffuses down the same gradient. For example, when water vapor's partial pressure exceeds its saturation vapor pressure over a liquid cloud drop surface, water vapor flows to the drop surface, enhancing the flow of aerosol particles in its path to the surface as well. The increased flux of particles to the drop increases the rate of collision by diffusiophoresis between the particles and the liquid drop. Evaporation of a drop causes water vapor and particles to flow away from the drop surface, decreasing the aerosol–hydrometeor collision rate.

Thermophoresis is the flow of aerosol particles from warm air to cool air due to bombardment of the particles by gases in the presence of a temperature gradient.

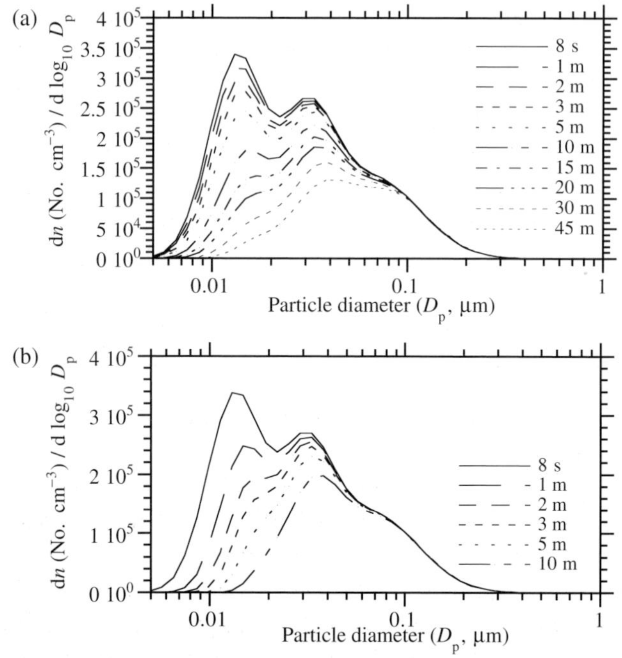

Figure 15.10 Box-model calculation of the evolution of a size distribution by coagulation when (a) particles are assumed spherical and no van der Waals or viscous forces are acting and (b) particles are assumed fractal ($D_f = 1.7$ above 27 nm diameter and $= 3.0$ below) and van der Waals and viscous forces are acting ($A_H/k_B T = 200$). Other conditions were $T = 300$ K and $p = 1$ atm. No dilution was accounted for. m is minutes and s is seconds. The results were obtained by treating coagulation among 10 size distributions and summing the results among all. From Jacobson and Seinfeld (2004).

When a temperature gradient exists, one side of an aerosol particle is warmer than the other side. On the warm side, gas molecules have a greater thermal speed, thus impart a greater force than they do on the cool side, pushing the particle from warm to cool air. Near a cloud-drop surface, release of latent heat due to condensation of water vapor causes the surface to be warmer than the air around it, enhancing flow of aerosol particles away from the surface, decreasing the collision rate of aerosol particles with the surface. Evaporation has the opposite effect, increasing the rate of aerosol–hydrometeor collisions due to thermophoresis.

The rate of coagulation of aerosol particles with cloud drops is also affected by **electric charge**. Ice crystals, cloud drops, and aerosol particles are often charged. Ice crystals, in particular, become charged when they collide with but bounce off other ice crystals. The bounceoff often causes a charge separation, especially if a temperature difference exists between colliding particles. Alternatively, a charged ice crystal may induce a charge on a nearby aerosol particle or a cloud drop due to electrostatic forces. Similarly, a charged aerosol particle may induce a charge upon a cloud drop. Electrostatic forces, though, are important only when charge differences are large, such as between an aerosol particle and a cloud drop or two cloud drops, but not, for example, between two submicrometer aerosol particles (e.g., Seinfeld and Pandis 1998).

A collision kernel for aerosol particles with cloud drops that accounts for diffusiophoresis, thermophoresis, electric charge, Brownian motion, Brownian diffusion enhancement, turbulent shear, and turbulent inertial motion simultaneously is

$$K_{i,j} = \frac{4\pi B_{\mathrm{P},i} C_{i,j}}{\exp\left(4\pi B_{\mathrm{P},i} C_{i,j} / \left[K^{\mathrm{B}}_{i,j} + K^{\mathrm{DE}}_{i,j} + K^{\mathrm{TI}}_{i,j} + K^{\mathrm{TS}}_{i,j}\right]\right) - 1} \tag{15.53}$$

(Wang *et al.* 1978; Martin *et al.* 1980 with addition of the turbulence terms by Jacobson 2003). In this equation, $B_{\mathrm{P},i}$ is the **mobility** (s g^{-1}) of an aerosol particle, which is the particle's drift speed per unit external force acting on it. The **drift speed** is the mean speed of a population of particles arising from the external force. For example, if the only external force acting on an aerosol particle of spherical radius r_i (cm) falling through the air is the force of gravity, the drift speed of the particle is its **terminal fall speed,** $V_{\mathrm{f},i}$ (Section 20.1) (cm s^{-1}), and the resulting mobility of the particle is its terminal fall speed divided by the force of gravity,

$$B_{\mathrm{P},i} = \frac{V_{\mathrm{f},i}}{F_{\mathrm{G}}} = \frac{V_{\mathrm{f},i}}{F_{\mathrm{D}}} = \frac{G_i}{6\pi \eta_{\mathrm{a}} r_i} = \frac{D_{\mathrm{p},i}}{k_{\mathrm{B}} T} \tag{15.54}$$

In this equation,

$$F_{\mathrm{G}} = \bar{M}_{\mathrm{p},i} g \tag{15.55}$$

is the **force of gravity** (g cm s^{-2}), where $\bar{M}_{\mathrm{p},i}$ is **single-particle mass** (g), and

$$F_{\mathrm{D}} = \frac{6\pi r_i \eta_{\mathrm{a}} V_{\mathrm{f},i}}{G_i} \tag{15.56}$$

(g cm s^{-2}) is the **force of drag**, which is equal in magnitude to but opposite in direction from the force of gravity (Section 20.1). In addition, G_i is the dimensionless Cunningham slip-flow correction from (15.30), η_{a} is the dynamic viscosity of air from (4.54) converted from units of kg m^{-1} s^{-1} to g cm^{-1} s^{-1}, and $D_{\mathrm{p},i}$ is the particle diffusion coefficient from (15.29) (cm^2 s^{-1}). From (15.54), the diffusion coefficient of the particle can be rewritten as

$$D_{\mathrm{p},i} = B_{\mathrm{P},i} k_{\mathrm{B}} T \tag{15.57}$$

which is the **Einstein relation.**

The second new term in (15.53) is

$$C_{i,j} = C^{\mathrm{Th}}_{i,j} + C^{\mathrm{Df}}_{i,j} + C^{\mathrm{e}}_{i,j} \tag{15.58}$$

(g cm^3 s^{-2}), which is a parameter that is a function of the thermophoretic, diffusiophoretic, and electric forces acting on an aerosol particle in the presence of water vapor and a cloud drop. The individual terms in the parameter are

$$C_{i,j}^{\mathrm{Th}} = -\frac{12\pi r_i \eta_a (\kappa_a + 2.5\kappa_p \mathrm{Kn}_{a,i})\kappa_a r_j (T_\infty - T_{s,j}) F_{h,L,j}}{5(1 + 3\mathrm{Kn}_{a,i})(\kappa_p + 2\kappa_a + 5\kappa_p \mathrm{Kn}_{a,i}) p_a} \tag{15.59}$$

$$C_{i,j}^{\mathrm{Df}} = -6\pi \eta_a r_i \frac{0.74 D_v m_d r_j (\rho_v - \rho_{v,s}) F_{v,L,j}}{G_i m_v \rho_a} \tag{15.60}$$

$$C_{i,j}^{\mathrm{e}} = Q_i Q_j \tag{15.61}$$

where r_i and r_j are the aerosol particle and cloud drop radii, respectively (cm), $\mathrm{Kn}_{a,i}$ is the dimensionless Knudsen number for air, κ_a is the thermal conductivity of moist air from (2.7) converted to units of J cm^{-1} s^{-1} K^{-1}, $\kappa_p \approx 0.00419$ J cm^{-1} s^{-1} K^{-1} is the thermal conductivity of the aerosol particle, T_∞ is the air temperature far from the cloud drop (K), $T_{s,j}$ is the cloud-drop surface temperature (K), p_a is the air pressure (J cm^{-3}), ρ_v is the mass density of water vapor away from the drop surface (g cm^{-3}), $\rho_{v,s}$ is the saturation mass density of water over the surface from Section 2.5.3 (g cm^{-3}), ρ_a is air density (g cm^{-3}), $F_{h,L,j}$ is the dimensionless thermal ventilation factor from (16.31), $F_{v,L,j}$ is the ventilation coefficient for water vapor from (16.24), D_v is the diffusion coefficient of water vapor through air from (16.17) (cm^2 s^{-1}), and m_d and m_v are the molecular weights of dry air and water vapor, respectively (g mol^{-1}).

In addition, $Q_i = q_i r_i^2$ and $Q_j = q_j r_j^2$ are the electric charge on the aerosol particle and water drop, respectively (esu, where 1 esu = 1 g$^{1/2}$ cm$^{3/2}$ s^{-1} = 3.33 × 10^{-10} C). In these terms, q_i and q_j are the charge per square centimeter of surface (esu cm^{-2}). The theoretical maximum charge a particle can carry is the **Rayleigh limit for disruption,**

$$Q_{j,\max} = \sqrt{16\pi\sigma_p r_j^3} \tag{15.62}$$

(esu) where σ_p is the surface tension of the particle (dyn cm^{-1} = g s^{-2}). An empirical fit to the measured **charge on cloud drops in highly electrified, precipitating clouds** is

$$Q_j = 2r_j^2 \tag{15.63}$$

(esu), where r_j is in cm. This equation gives q_j = 2 esu cm^{-2}. An empirical fit to measured charges in warm clouds, which have weak electrification, is

$$Q_j = 0.0005 r_j^{1.3} \tag{15.64}$$

(Pruppacher and Klett 1997).

$C_{i,j}$ is valid for $0.001 < r_i < 10$ μm and $42 < r_j < 310$ μm. If gravitational forces, charge, thermophoresis, and diffusiophoresis are ignored, (15.53) simplifies to $K_{i,j} = K_{i,j}^{\mathrm{B}} + K_{i,j}^{\mathrm{DE}} + K_{i,j}^{\mathrm{TI}} + K_{i,j}^{\mathrm{TS}}$.

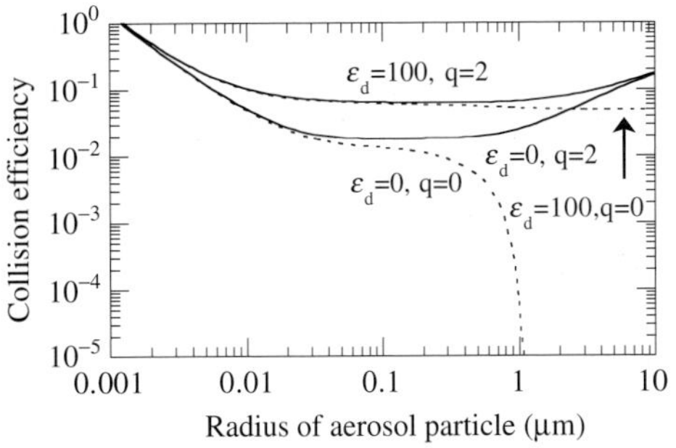

Figure 15.11 Collision efficiency from (15.53) of a 42-μm radius cloud drop (large particle) interacting with smaller aerosol particles for two dissipation rates (ε_d, cm^2 s^{-3}) and two charge conditions (q, esu cm^{-2}, set to the same value for both particles). Other conditions were $T = 283.15$ K, $p_a = 900$ hPa, and RH = 75%. The collision efficiency equaled the total coagulation kernel divided by $\pi(r_i + r_j)^2|V_{f,j} - V_{f,i}|$. Eight collision kernels were accounted for (see text). (Compare $\varepsilon_d = 0$ cases with Curves 2 and 6 of Fig. 4a of Wang *et al.* 1978.)

Figure 15.11 shows modeled collision efficiencies from (15.53), accounting for Brownian motion, Brownian diffusion enhancement, gravitational collection, turbulent shear, turbulent inertial motion, diffusiophoresis, thermophoresis, and electric charge for a 42-μm radius cloud drop interacting with smaller aerosol particles for two turbulent dissipation rates (ε_d, cm^2 s^{-3}) and two charge conditions (q, esu cm^{-2}). The addition of turbulence affected the collision efficiency the greatest when electric charge was zero. For nonzero or zero charge, turbulence increased collision efficiencies more for large particles than for small particles.

Figure 15.12 shows individual collision kernels when $\varepsilon_d = 100$ cm^2 s^{-3}, q = 2 esu cm^{-2} (electrified thunderstorm cloud) and other conditions were the same as those in Fig. 15.11. Brownian motion and diffusion enhancement dominated collision with a 42-μm cloud drop when the aerosol-particle radius was < 0.01 μm. Turbulent inertia dominated when the aerosol-particle radius was 0.01–4 μm. Gravitational collection dominated when the aerosol-particle radius was > 4 μm.

15.6.8 Coalescence efficiencies

When two aerosol particles collide, they may or may not stick together, depending on the efficiency of coalescence. The efficiency of coalescence depends on particle shape, composition, ambient relative humidity, and other factors. Because the kinetic energy of collision between one small particle (<2-μm-radius) and a small or large particle is relatively small in comparison with that between

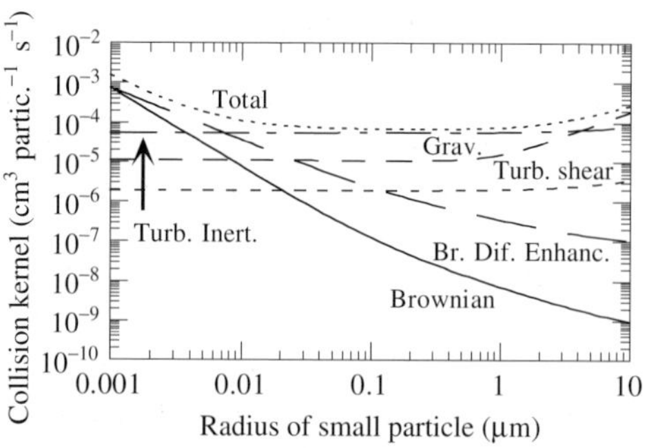

Figure 15.12 Individual and total collision kernels for the q = 2 esu cm^{-2} and ε_d = 100 cm^2 s^{-3} case in Fig. 15.11.

two large particles, the likelihood that bounceoff occurs when a small particle collides with another particle is low, and the coalescence efficiency is approximately unity (Beard and Ochs 1984; Pruppacher and Klett 1997). To the contrary, van der Waals forces enhance coalescence between two small particles. This enhancement is quantified by van der Waals/viscous force correction factor, $V_{E,i,j}$. Thus, van der Waals/viscous forces can be treated either in terms of a collision kernel (15.42) or in terms of a coalescence efficiency ($E_{coal,i,j} = V_{E,i,j}$).

When two colliding particles are large, the kinetic energy of collision is high, increasing the chance of a bounceoff following collision, and the coalescence efficiency is less than 1. Some parameterizations of coalescence efficiency between large particles, derived from cloud-chamber experiments, are given in Beard and Ochs (1984) and Pruppacher and Klett (1997).

15.7 SUMMARY

The process of coagulation is described by the integrodifferential coagulation equation. This equation can be solved exactly with analytical solutions for a few idealized cases and with iterative numerical techniques for all cases. It can also be solved approximately with a semiimplicit solution technique that conserves volume and volume concentration and is noniterative and positive-definite. The technique also allows simulation of coagulation among any number of particle size distributions and among particles with multiple components. The rate of coagulation depends on Brownian diffusion, enhancement of Brownian motion due to eddies, differences in fall speeds, turbulent shear, turbulent inertial motion, van der Waals forces, viscous forces, particle shape, diffusiophoresis, thermophoresis, and electric charge. Coagulation kernels were described for each of these processes. Effects of coagulation coupled to growth are discussed in Chapter 16. Coagulation in clouds is discussed in more detail in Chapter 18.

15.8 PROBLEMS

15.1 If an air mass contains six homogeneous, externally-mixed particle types, how many possible combinations of internal mixtures of 2, 3, 4, 5, and 6 particle types exist? What is the total number of externally-plus internally-mixed particle types possible?

15.2 Suppose two particles of diameter 100 nm and 200 nm coagulate with each other. (a) Calculate the volume of the resulting coagulated pair. (b) If a model has fixed size bins with mean diameters 200 nm and 300 nm, calculate the volume fraction of the coagulated pair going into each fixed bin. (c) Calculate the number fraction of particles going into each fixed size bin.

15.3 Use Smoluchowski's solution to calculate the number concentration of particles with volume 10 times that of the initial monodisperse distribution after 12 hours of coagulation when the temperature is 298 K assuming the initial concentration of monodisperse particles is 10^6 particles cm^{-3}.

15.4 Calculate the coagulation kernel due to Brownian motion when the radius of one particle is 0.1 μm and the radius of the other is 1.0 μm. Compare the result with the kernel obtained when both particles are 1.0 μm in radius. Assume $T = 298$ K, $p_d = 990$ hPa, the air is dry, and the particle density is $\rho_p = 1.2$ g cm^{-3}.

15.5 Calculate the turbulent inertial motion coagulation kernel between 0.5-μm and 10-μm radius liquid water drops assuming a dissipation rate for turbulent kinetic energy of (a) 5 cm^2 s^{-3} and (b) 2000 cm^2 s^{-3}. Assume the temperature is 298 K, the air density is 0.001 225 g cm^{-3}, and the fall speed can be approximated by $V_{f,i} \approx 2r_i^2\rho_p/9\eta_a$, where ρ_p = drop density (1 g cm^{-3}).

15.6 If the volume-equivalent radius of a soot agglomerate with spherule radius of 27 nm is 100 nm, and the agglomerate's fractal dimension is 1.7, calculate its (a) fractal radius, (b) area-equivalent radius, and (c) mobility radius.

15.7 Give two reasons why the coalescence efficiency of two nanoparticles is near unity.

15.9 COMPUTER PROGRAMMING PRACTICE

15.8 Using the constant coagulation kernel from (15.16), plot Smoluchowski's analytical solution on a $dn/d\log_{10} D_p$ graph when $T = 298$ K and $h = 12$ h ($dn = n_j$ and $d\log_{10} D_p = \log_{10} d_{i+1} - log_{10} d_{ij}$ for each i). Assume that the initial total particle number concentration is 10^6 particles cm^{-3} and all initial particles have diameter 0.006 μm.

15.9 Write a computer script to calculate the coagulation kernel due to Brownian motion as a function of temperature and particle size. Use a volume ratio size distribution, with $d_1 = 0.001$ μm, $d_{N_B} = 10$ μm, and $V_{rat} = 1.5$. Assume $p_d = 1013$ hPa, the air is dry, and the particle mass density is $\rho_p = 1.5$ g cm^{-3}. Calculate the coagulation kernels when (a) $T = 298$ K, the diameter of one particle (d_a) is 0.001 μm, and the diameter of the other particle (d_b) varies from 0.001 to 10 μm; (b) $T = 298$ K, $d_a = 10$ μm, and d_b varies from 0.001 to 10 μm; (c) the diameters of both particles are 0.001 μm and T varies from 190 to 320 K in increments of 5 K; and (d) $d_a = 0.001$ μm, $d_b = 10$ μm, and T varies from 190 to 320 K in increments of 5 K. Plot and interpret the results.

15.10 Write a computer program to simulate semiimplicit coagulation with (15.12). Use the program with the coagulation kernel and conditions described in Problem 15.8 to simulate Smoluchowski's solution. Compare the results with those from Problem 15.8.

15.11 Write a computer program to calculate the collision kernel for a 0.5-μm aerosol particle with a 10-μm radius liquid water drop using (15.53), neglecting turbulent shear and turbulent inertial motion. Assume the temperature away from cloud drop surfaces is 298 K and that at drop surfaces is 300 K. Also assume that the air density is 0.001 225 g cm^{-3}, the relative humidity away from cloud drop surfaces is 90 percent, and the fall speed is approximately $V_{f,i} \approx 2r_i^2 \rho_p / 9\eta_a$, where ρ_p = drop density (1 g cm^{-3} for both the aerosol particle and cloud drop). Ignore ventilation coefficients and electric charge. Once the code has been tested for one pair of particles, plot the kernel for coagulation between the 10-μm cloud drop and aerosol particles varying from 0.001 to 10 μm in radius.

16

Condensation, evaporation, deposition, and sublimation

A PARTICLE grows in size by the mass transfer of gas to its surface followed by the change in state or chemical conversion of the gas at the surface by condensation, solid deposition, chemical reaction, and/or dissolution. In this chapter, equations for vapor condensation and ice deposition on particle surfaces are discussed. Equations for water-vapor condensation onto a single homogeneous particle are first derived. The growth equations are extended to gases other than water vapor, to a population of particles, and to particles with multiple components. Numerical solutions to growth equations are given. A solution to growth coupled with nucleation is also described. Dissolution and surface reaction are discussed in Chapter 17. Formation of clouds by condensation is discussed further in Chapter 18.

16.1 FLUXES TO AND FROM A SINGLE DROP

Condensation occurs when a gas diffuses to a homogeneously or heterogeneously nucleated particle surface and changes state to a liquid on the surface. Common condensable gases include water vapor, sulfuric acid, and by-products of toluene, xylene, alkylbenzene, alkanes, alkenes, and biogenic hydrocarbon oxidation (e.g., Pandis *et al.* 1992). **Solid deposition** occurs when a gas diffuses to and changes state to a solid on a nucleated surface. Deposition of water vapor onto particles to form ice is an important mechanism of ice crystal formation in the atmosphere. **Evaporation** and **sublimation** are the reverse of condensation and ice deposition, respectively. In this section, growth equations for the condensation of water vapor onto a single homogeneous liquid water drop are derived.

From **Fick's first law of diffusion**, the rate of change of mass of a single, spherical, homogeneous liquid water drop can be described with

$$\frac{\mathrm{dm}}{\mathrm{d}t} = 4\pi R^2 D_\mathrm{v} \frac{\mathrm{d}\rho_\mathrm{v}}{\mathrm{d}R} \tag{16.1}$$

where m is the mass of the drop (g), R is the radial distance from the center of the drop (cm), D_v is the molecular diffusion coefficient of water vapor in air (cm^2 s^{-1}), ρ_v is the density of water vapor (g cm^{-3}), and $\mathrm{d}\rho_\mathrm{v}/\mathrm{d}R$ is the radial gradient of vapor density. Equation (16.1) applies to large particles in the continuum regime (Kn $\ll$ 1).

Integrating the equation from the drop surface, where $\rho_v = \rho_{v,r}$ and $R = r$, to infinity, where $\rho_v = \rho_v$ and $R = \infty$, gives

$$\frac{dm}{dt} = 4\pi r D_v(\rho_v - \rho_{v,r}) \tag{16.2}$$

If $\rho_v > \rho_{v,r}$, vapor diffuses to and condenses on the drop surface. If $\rho_v < \rho_{v,r}$, condensate evaporates from and diffuses away from the surface.

When water vapor condenses, latent heat is released, increasing the temperature at the drop surface, creating a gradient between the surface and the surrounding air. The gradient is reduced by the conduction of energy away from the drop surface by air molecules. The equation for the **cooling rate at the drop surface** due to conduction is

$$\frac{dQ_r^*}{dt} = -4\pi R^2 \kappa_a \frac{dT}{dR} \tag{16.3}$$

(J s^{-1}), where κ_a is the thermal conductivity of moist air (J cm^{-1} s^{-1} K^{-1}), and the temperature gradient is negative since temperature decreases with increasing distance from the drop during condensation. Integrating (16.3) from the drop surface, where $T = T_r$, to infinity where $T = T$, gives

$$\frac{dQ_r^*}{dt} = 4\pi r \kappa_a (T_r - T) \tag{16.4}$$

Equations (16.2) and (16.4) were first written in a similar form by James Clerk Maxwell in 1877 (Maxwell 1890). The two equations are related to temperature at the drop surface by

$$m c_W \frac{dT_r}{dt} = L_e \frac{dm}{dt} - \frac{dQ_r^*}{dt} \tag{16.5}$$

where c_W is the specific heat of liquid water (J g^{-1} K^{-1}), and L_e is the latent heat of evaporation of water (J g^{-1}). Combining (16.4) with (16.5) under steady-state conditions ($dT_r/dt = 0$) gives

$$L_e \frac{dm}{dt} = 4\pi r \kappa_a (T_r - T) \tag{16.6}$$

The remainder of this derivation originates from Mason (1971) (see also Rogers and Yau 1989). Combining the equation of state at saturation, $p_{v,s} = \rho_{v,s} R_v T$, with the Clausius–Clapeyron equation, $dp_{v,s}/dT = \rho_{v,s} L_e / T$ from (2.57), gives

$$\frac{d\rho_{v,s}}{\rho_{v,s}} = \frac{L_e}{R_v} \frac{dT}{T^2} - \frac{dT}{T} \tag{16.7}$$

Integrating this equation from infinity, where $T = T$ and $\rho_{v,s} = \rho_{v,s}(T)$, to the drop surface, where $T = T_s$ and $\rho_{v,s} = \rho_{v,s}(T_r)$, yields

$$\ln \frac{\rho_{v,s}(T_r)}{\rho_{v,s}(T)} = \frac{L_e}{R_v} \frac{(T_r - T)}{T T_r} - \ln \frac{T_r}{T} \tag{16.8}$$

Since $T \approx T_r$, (16.8) simplifies to

$$\frac{\rho_{v,s}(T_r) - \rho_{v,s}(T)}{\rho_{v,s}(T)} = \frac{L_e}{R_v}\frac{(T_r - T)}{T^2} - \frac{T_r - T}{T} \tag{16.9}$$

Substituting $T_r - T$ from (16.6) into (16.9) gives

$$\frac{\rho_{v,s}(T_r) - \rho_{v,s}(T)}{\rho_{v,s}(T)} = \frac{L_e}{4\pi r \kappa_a T}\left(\frac{L_e}{R_v T} - 1\right)\frac{dm}{dt} \tag{16.10}$$

Dividing (16.2) by $\rho_{v,s}(T)$, rearranging, adding the result to (16.10), and assuming $\rho_{v,r} \approx \rho_{v,s}(T_r)$ results in

$$\frac{\rho_v - \rho_{v,s}(T)}{\rho_{v,s}(T)} = \left[\frac{L_e}{4\pi r \kappa_a T}\left(\frac{L_e}{R_v T} - 1\right) + \frac{1}{4\pi r D_v \rho_{v,s}(T)}\right]\frac{dm}{dt} \tag{16.11}$$

Substituting $\rho_{v,s} = \rho_{v,s}(T)$, $\rho_{v,s} = p_{v,s}/R_v T$, and $\rho_v = p_v/R_v T$ into (16.11) and solving for dm/dt gives the **mass-flux form** of the growth equation for a single, homogeneous liquid water drop as

$$\frac{dm}{dt} = \frac{4\pi r D_v (p_v - p_{v,s})}{\dfrac{D_v L_e p_{v,s}}{\kappa_a T}\left(\dfrac{L_e}{R_v T} - 1\right) + R_v T} \tag{16.12}$$

This equation can be modified to treat any gas by replacing L_e, D_v, p_v, $p_{v,s}$, and R_v with $L_{e,q}$, D_q, p_q, $p_{q,s}$, and R^*/m_q, respectively, and to treat size dependence by adding the subscript i to m and r. The molecular diffusion coefficient (D_q) and thermal conductivity (κ_a) can be modified by considering the collision geometry between vapor molecules and a particle surface, the probability that molecules stick to a surface upon collision, and eddies created by the particle. The saturation vapor pressure ($p_{q,s}$) can be modified by considering surface curvature, the presence of solute on a surface, and radiative heating/cooling. The corrections will be discussed shortly, but for now, a subscript i and a prime are added to D_q, κ_a, $p_{q,s}$ to denote size dependence and a modified value, respectively. With the above substitutions in (16.12), the mass-flux form of the growth equation of substance q to one homogeneous drop of size i becomes

$$\frac{dm_i}{dt} = \frac{4\pi r_i D'_{q,i}(p_q - p'_{q,s,i})}{\dfrac{D'_{q,i} L_{e,q} p'_{q,s,i}}{\kappa'_{a,i} T}\left(\dfrac{L_{e,q} m_q}{R^* T} - 1\right) + \dfrac{R^* T}{m_q}} \tag{16.13}$$

Sometimes, the assumption $L_{e,q}m_q/R^*T - 1 \approx L_{e,q}m_q/R^*T$ is made. If $T = 273$ K, $L_{e,q}m_q/R^*T \approx 19.8$ for water vapor, and the assumption induces an error of 1/19.8. Because (16.13) is a better approximation, it is used hereafter.

The change in mass of a spherical drop is related to its change in radius by

$$\frac{dm_i}{dt} = 4\pi r_i^2 \rho_{p,i}\frac{dr_i}{dt} \tag{16.14}$$

where $\rho_{p,i}$ is the drop density (g cm^{-3}). Combining (16.14) with (16.13) gives the **radius flux form** of the growth equation for a single, homogeneous, spherical drop as

$$r_i \frac{dr_i}{dt} = \frac{D'_{q,i}(p_q - p'_{q,s,i})}{\dfrac{D'_{q,i} L_{e,q} \rho_{p,i} p'_{q,s,i}}{\kappa'_{a,i} T}\left(\dfrac{L_{e,q} m_q}{R^* T} - 1\right) + \dfrac{R^* T \rho_{p,i}}{m_q}} \tag{16.15}$$

The time rate of change of mass of a spherical drop is related to its change of volume by $dm_i/dt = \rho_{p,i} dv_i/dt$, and radius is related to volume by $r_i = (3v_i/4\pi)^{1/3}$. Combining these expressions with (16.13) gives the **volume flux form** of the growth equation for a single, homogeneous, spherical drop as

$$\frac{dv_i}{dt} = \frac{(48\pi^2 v_i)^{1/3} D'_{q,i}(p_q - p'_{q,s,i})}{\dfrac{D'_{q,i} L_{e,q} \rho_{p,i} p'_{q,s,i}}{\kappa'_{a,i} T}\left(\dfrac{L_{e,q} m_q}{R^* T} - 1\right) + \dfrac{R^* T \rho_{p,i}}{m_q}} \tag{16.16}$$

16.2 CORRECTIONS TO GROWTH PARAMETERS

In this section, corrections to the molecular diffusion coefficient, thermal conductivity of air, and the saturation vapor pressure used in the growth equation are described.

16.2.1 Corrections to the molecular diffusion coefficient

Molecular diffusion is the movement of molecules due to their kinetic energy and redirection of the molecules due to their collision with other molecules. Molecules in a body with a temperature greater than 0 K have an average kinetic energy and thermal speed given by (11.151). When a molecule moves, it collides with other molecules, which redirect it along an arbitrary path. The pace at which a molecule spreads by diffusion is proportional to its thermal speed and the distance between collisions. Thermal speed is proportional to the square root of temperature, and the distance between collisions is inversely proportional to the density of air. An expression for the **molecular diffusion coefficient** of a trace gas in air (cm^2 s^{-1}) with these characteristics is

$$D_q = \frac{5}{16 A d_q^2 \rho_a} \sqrt{\frac{R^* T m_a}{2\pi}\left(\frac{m_q + m_a}{m_q}\right)} \tag{16.17}$$

(e.g., Chapman and Cowling 1970; Davis 1983), where A is Avogadro's number (6.022 136 7 × 10^{23} molec. mol^{-1}), d_q is the **collision diameter** (cm) of gas molecule q, ρ_a is air density (g cm^{-3}), R^* is the universal gas constant (8.314 51 × 10^7 g cm^2 s^{-2} mol^{-1} K^{-1}), T is absolute temperature (K), m_a is the molecular weight of air (28.966 g mol^{-1}), and m_q is the molecular weight of gas q (g mol^{-1}).

Table 16.1 Collision diameters and diffusion coefficients of several gases at 288 K and 1013 hPa

Gas	Collision diameter (Å)	Molecular weight ($g\ mol^{-1}$)	Diffusion coefficient ($cm^2\ s^{-1}$)
Air	3.67	28.966	0.147
Ar	3.58	39.948	0.144
CO_2	4.53	44.011	0.088
H_2	2.71	2.014	0.751
He	2.15	4.002	0.871
Xe	4.78	131.29	0.067
Kr	4.08	83.80	0.098
N_2	3.70	28.013	0.146
NH_3	4.32	17.031	0.123
Ne	2.54	20.180	0.339
O_2	3.54	31.999	0.154
H_2O	3.11	18.015	0.234

Collision diameters are from Lide (2003), except that for water vapor, the collision diameter is derived by equating (16.17) with (8.14). Air density at the given temperature and pressure is 0.001 23 g cm^{-3}.

Equation (16.17) simplifies to the expression for the kinematic viscosity of air from (4.55)/(4.54) when the gas considered is air. Table 16.1 gives collision diameters and diffusion coefficients for several gases from (16.17). Example 16.1 shows that the diffusion coefficient increases with increasing altitude (lower pressure).

Example 16.1

Calculate the molecular diffusion coefficient of carbon dioxide in air when the temperature is $T = 220$ K and air pressure is $p_a = 25$ hPa. Assume the air is dry.

SOLUTION

Since the air is dry, $R_m = R'$. From the equation of state, (2.36), $\rho_a = 0.000\,039\,6$ g cm^{-3}. From Table 16.1, the molecular weight and molecular diameter of carbon dioxide are $m_q = 44.011$ g mol^{-1} and 4.53×10^{-8} cm, respectively. Substituting these values into (16.17) gives the molecular diffusion coefficient of CO_2 as $D_q = 2.39$ cm^2 s^{-1}. Comparing this result with that in Table 16.1 suggests that the diffusion coefficient increases with decreasing air pressure.

Near a particle surface, the diffusion coefficient of a gas is affected by collision geometry, sticking probability, and ventilation. A **corrected molecular diffusion coefficient** (cm^2 s^{-1}) that takes these effects into account is

$$D'_{q,i} = D_q \omega_{q,i} F_{q,L,i} \qquad (16.18)$$

where $\omega_{q,i}$ is the correction for collision geometry and sticking probability, and $F_{q,i}$ is the correction for ventilation.

The **correction for collision geometry** accounts for noncontinuum-regime growth and is required, since (16.1) applies only to the continuum regime. The correction is a function of the ratio of the distance between molecular collisions of a diffusing gas and the size of the particle it is diffusing to. The **correction for sticking probability** takes account of the ability of a gas to stick to a particle surface once the gas has diffused to the surface. An expression for the two corrections, together, is

$$\omega_{q,i} = \left\{1 + \left[\frac{1.33 + 0.71\mathrm{Kn}_{q,i}^{-1}}{1 + \mathrm{Kn}_{q,i}^{-1}} + \frac{4(1-\alpha_{q,i})}{3\alpha_{q,i}}\right]\mathrm{Kn}_{q,i}\right\}^{-1} \tag{16.19}$$

(Fuchs and Sutugin 1971; Pruppacher and Klett 1997), where

$$\mathrm{Kn}_{q,i} = \frac{\lambda_q}{r_i} \tag{16.20}$$

is the **Knudsen number of the condensing gas** with respect to particles of size i. This Knudsen number gives the ratio of the mean free path of the condensing gas to the size of the particle of interest. The mean free path of a gas molecule is the average distance it travels before colliding and exchanging momentum with any other gas molecule.

If the Knudsen number is large, the particle of interest is small relative to the distance between molecular collisions, and the particle is intercepted relatively infrequently by gas molecules. If the Knudsen number is small, the particle is large relative to the distance between molecular collisions, and the particle is frequently intercepted by gas molecules. In sum, (16.19) states that, in the absence of sticking probability,

$$\omega_{q,i} \to \begin{cases} 0 & \text{as} \quad \mathrm{Kn}_{q,i} \to \infty \quad \text{(small particles)} \\ 1 & \text{as} \quad \mathrm{Kn}_{q,i} \to 0 \quad \text{(large particles)} \end{cases} \tag{16.21}$$

In a dilute mixture, where the number concentration of a trace-gas is much less than that of air, the **mean free path of a gas molecule** is

$$\lambda_q = \frac{m_\mathrm{a}}{\pi A d_q^2 \rho_\mathrm{a}}\sqrt{\frac{m_\mathrm{a}}{m_\mathrm{a} + m_q}} \tag{16.22}$$

(Jeans 1954; Davis 1983). In (16.22), the mean free path is inversely proportional to the air density. The greater the air density, the shorter the distance between collisions of a gas molecule with an air molecule. Combining (16.17) and (16.22) with the **thermal speed of a gas molecule** ($\bar{v}_q$, cm s^{-1}) from (11.151) gives

$$\lambda_q = \frac{64 D_q}{5\pi \bar{v}_q}\left(\frac{m_\mathrm{a}}{m_\mathrm{a} + m_q}\right) \tag{16.23}$$

In (16.19), $\alpha_{q,i}$ is the **mass accommodation (sticking) coefficient** of the gas, which is the fractional number of collisions of gas q with particles of size i that results in the gas sticking to the surface of the particle. Mass accommodation coefficients differ from **reaction probabilities**, discussed in Chapter 11, in that reaction probabilities take account of molecular diffusion to and reaction with a material on a particle surface. Mass accommodation coefficients take account of only adsorption of a gas to the surface.

Accommodation coefficients for soluble species range from 0.01 to 1.0 (Mozurkewich *et al.* 1987; Jayne *et al.* 1990; Van Doren *et al.* 1990; Chameides and Stelson 1992). Van Doren *et al.* (1990) found that the mass accommodation coefficient for nitric acid (HNO_3) on liquid water varied from 0.07 at 268 K to 0.193 at 293 K, the coefficient for hydrochloric acid (HCl) on liquid water varied from 0.064 at 294 K to 0.177 at 274 K, and the coefficient for dinitrogen pentoxide (N_2O_5) on liquid water varied from 0.04 at 282 K to 0.061 at 271 K.

To correct for the increased rate of vapor and energy transfer to the upstream surface of a large, falling particle, a **ventilation factor for vapor** is included in the corrected molecular diffusion coefficient. The ventilation factor is the vapor mass flux to (or from) a particle when it is moving, divided by the flux when it is at rest. Increased vapor transfer occurs when a large particle falls, creating eddies that sweep additional vapor toward the particle. The ventilation factor is small for small, falling particles because such particles move too slowly to create eddies. An expression for the ventilation factor for a gas q interacting with a liquid water drop is

$$F_{q,L,i} = \begin{cases} 1 + 0.108x_{q,i}^2 & x_{q,i} \le 1.4 \\ 0.78 + 0.308x_{q,i} & x_{q,i} > 1.4 \end{cases} \qquad x_{q,i} = \mathrm{Re}_i^{1/2}\mathrm{Sc}_q^{1/3} \tag{16.24}$$

(Pruppacher and Klett 1997). The ventilation factor depends on the particle Reynolds number, defined in (15.26), and the **gas Schmidt number,**

$$\mathrm{Sc}_q = \frac{\nu_a}{D_q} \tag{16.25}$$

which is the ratio of the kinematic viscosity of air to the molecular diffusion coefficient of a gas. Since the particle Reynolds number is proportional to a particle's radius and its fall speed, the larger a particle and the faster it falls, the greater the ventilation factor.

16.2.2 Corrections to the thermal conductivity of air

Conductive energy transfer to and from a particle surface is affected by some of the same correction factors that affect vapor transfer. These factors are embodied in the **corrected thermal conductivity** term

$$\kappa'_{a,i} = \kappa_a \omega_{h,i} F_{h,L,i} \tag{16.26}$$

where κ_a is the uncorrected thermal conductivity of moist air (J cm^{-1} s^{-1} K^{-1}), from (2.7), $\omega_{h,i}$ is the correction factor for collision geometry and sticking probability, and $F_{h,L,i}$ is the thermal ventilation factor. The correction factor for collision geometry and sticking probability is

$$\omega_{h,i} = \left\{ 1 + \left[\frac{1.33 + 0.71\mathrm{Kn}_{h,i}^{-1}}{1 + \mathrm{Kn}_{h,i}^{-1}} + \frac{4(1-\alpha_h)}{3\alpha_h} \right] \mathrm{Kn}_{h,i} \right\}^{-1} \tag{16.27}$$

where

$$\mathrm{Kn}_{h,i} = \frac{\lambda_h}{r_i} \tag{16.28}$$

is the **Knudsen number for energy** with respect to particles of size i. In this equation, λ_h is the **thermal mean free path** (cm), which is the average distance that an air molecule travels before exchanging energy by conduction with another molecule. It is quantified as

$$\lambda_h = \frac{3D_h}{\bar{v}_\mathrm{a}} \tag{16.29}$$

where D_h is the **molecular thermal diffusivity**, defined in (8.14), and $\bar{v}_\mathrm{a}$ is the thermal speed of an air molecule, defined in (15.25). The molecular thermal diffusivity depends on the thermal conductivity of air, the specific heat of air, and the air density.

In (16.27), α_h is the **thermal accommodation coefficient**, interpreted as the fractional number of molecules bouncing off the surface of a drop that have acquired the temperature of the drop (Pruppacher and Klett 1997). An equation for α_h is

$$\alpha_h = \frac{T_\mathrm{m} - T}{T_\mathrm{s} - T} \tag{16.30}$$

where T_m is the temperature of vapor molecules leaving the surface of a drop (K), T_s is the temperature of the surface (K), and T is the temperature of the ambient vapor (K). For water-vapor growth, α_h has a typical value of 0.96.

The **thermal ventilation factor** has a meaning similar to that for vapor. When a large particle falls through the air, it creates eddies that sweep additional energy to the surface of the particle. An empirical equation for the thermal ventilation factor, obtained from experiments with liquid water drops, is

$$F_{h,L,i} = \begin{cases} 1 + 0.108x_{h,i}^2 & x_{h,i} \le 1.4 \\ 0.78 + 0.308x_{h,i} & x_{h,i} > 1.4 \end{cases} \qquad x_{h,i} = \mathrm{Re}_i^{1/2}\mathrm{Pr}^{1/3} \tag{16.31}$$

(Pruppacher and Klett 1997), where the dimensionless **Prandtl number**,

$$\mathrm{Pr} = \frac{\eta_\mathrm{a} c_{p,\mathrm{m}}}{\kappa_\mathrm{a}} \tag{16.32}$$

is proportional to the ratio of the dynamic viscosity of air to its thermal conductivity.

16.2.3 Corrections to the saturation vapor pressure

Expressions for the saturation vapor pressure of water over dilute, flat liquid water and ice surfaces were given in (2.62) and (2.65), respectively. Curvature, solutes, and radiative cooling modify these saturation vapor pressures.

16.2.3.1 *Curvature effect*

The saturation vapor pressure increases over a curved surface relative to over a flat surface due to the **curvature (Kelvin) effect.** The surface of a small spherical particle is more curved than that of a large spherical particle, causing the saturation vapor pressure over a small particle to be greater than that over a large particle. The saturation vapor pressure over a curved surface is greater than that over a flat surface because molecules desorb more readily from a curved surface than from a flat surface. Surface tension also plays a role in the curvature effect. The greater the surface tension of a particle against air, the more likely a molecule desorbs from the surface, and the greater the saturation vapor pressure.

The saturation vapor pressure over a curved, dilute surface relative to that over a flat, dilute surface is

$$\frac{p'_{q,\mathrm{s},i}}{p_{q,\mathrm{s}}} = \exp\left(\frac{2\sigma_\mathrm{p} m_\mathrm{p}}{r_i R^* T \rho_{\mathrm{p},i}}\right) \approx 1 + \frac{2\sigma_\mathrm{p} m_\mathrm{p}}{r_i R^* T \rho_{\mathrm{p},i}} \qquad (16.33)$$

where $p'_{q,\mathrm{s},i}$ is the saturation vapor pressure of gas q over a curved, dilute surface (hPa), $p_{q,\mathrm{s}}$ is the saturation vapor pressure over a flat, dilute surface (hPa), σ_p is the average particle surface tension (dyn cm^{-1} = g s^{-2}), m_p is the average particle molecular weight (g mol^{-1}), r_i is particle radius (cm), R^* is the universal gas constant (8.31451×10^7 g cm^2 s^{-2} mol^{-1} K^{-1}), T is absolute temperature (K), and $\rho_{\mathrm{p},i}$ is the average particle density (g cm^{-3}). The second expression was obtained by noting that the exponent in the first term is small, and $e^x \approx 1 + x$ for small x.

For liquid water and ice, $p_{q,\mathrm{s}} = p_{v,\mathrm{s}}$, and $p_{q,\mathrm{s}} = p_{v,\mathrm{I}}$, respectively, equations for which were given in Section 2.5.3. For sulfuric acid, parameterizations of the saturation vapor pressures are available from Bolsaitis and Elliott (1990) and Vehkamaki *et al.* (2002). Vapor pressure information for several organic gases can be obtained from McMurry and Grosjean (1985), Tao and McMurry (1989), Odum *et al.* (1996), Makar (2001), and Makar *et al.* (2003).

An expression for the surface tension of pure liquid water against air ($\sigma_\mathrm{p} = \sigma_{\mathrm{w/a}}$) was given in (14.19). The surface tension of water containing dissolved organic material is less than that of pure water (e.g., Li *et al.* 1998; Facchini *et al.* 1999). An empirical expression for the **surface tension of water containing dissolved**

organics is

$$\sigma_p = \sigma_{w/a} - 0.0187T\ln(1 + 628.14\mathbf{m}_C) \tag{16.34}$$

(Facchini *et al.* 1999), where σ_p is in units of dyn cm^{-1} (g s^{-2}), T is absolute temperature (K), and $\mathbf{m}_C$ is the molality of carbon dissolved in water (moles carbon per kilogram of water). Aerosol particles often also contain dissolved inorganic solutes, such as sodium chloride, ammonium sulfate, potassium nitrate, etc. These solutes frequently dissociate into ions, such as Na^+, Cl^-; NH_4^+, SO_4^{2-}; K^+, NO_3^-, respectively. Dissolved inorganic ions have the opposite effect to dissolved organics; they increase the surface tension relative to pure water. An empirical expression for the **surface tension of water containing dissolved inorganic ions** is

$$\sigma_p = \sigma_{w/a} + 1.7\mathbf{m}_I \tag{16.35}$$

(e.g., Pruppacher and Klett 1997), where σ_p is in units of dyn cm^{-1} (g s^{-2}) and $\mathbf{m}_I$ is the molality of total dissolved ions in water. Thus, if a solute, such as sodium chloride, dissociates, $\mathbf{m}_I$ is the molality of the sodium ion plus that of the chloride ion (or twice the molality of undissociated sodium chloride). Equations (16.34) and (16.35) suggest that the addition of dissolved organics decreases the saturation vapor pressure and the addition of dissolved ions increases the saturation vapor pressure of water over a curved surface when only the effect of surface tension on the curvature effect is considered.

Example 16.2

Compare the surface tensions of a (a) pure water drop, (b) water drop containing 0.03 mol-C kg^{-1} of an undissociated organic compound, and (c) water drop containing 0.03 mol kg^{-1} of dissociated ammonium sulfate $(NH_4)_2SO_4$ at $T = 298.15$ K.

SOLUTION

From (14.19), (16.34), and (16.35), the surface tensions in the three respective cases are

$$\sigma_{w/a} = 76.1 - 0.155 \times 25\,^\circ\mathrm{C} = 72.225\ \mathrm{dyn\,cm^{-1}}$$

$$\sigma_C = \sigma_{w/a} - 0.0187 \times 298.15\,\mathrm{K} \times \ln(1 + 628.14 \times 0.03\ \mathrm{mol\text{-}C\,kg^{-1}})$$

$$= 55.56\ \mathrm{dyn\,cm^{-1}}$$

$$\sigma_I = \sigma_{w/a} + 1.7 \times 3 \times 0.03\ \mathrm{mol\,kg^{-1}} = 72.378\ \mathrm{dyn\,cm^{-1}}$$

The factor of three in the last equation accounts for the fact that ammonium sulfate dissociates into three ions. The organic compound reduced the surface tension of water, whereas the dissolved ions increased it.

16.2.3.2 *Solute effect*

A second factor that affects the saturation vapor pressure is that of dissolved solute (dissociated or undissociated). When a solute dissolves in solution, the saturation vapor pressure of the solvent is reduced to the saturation vapor pressure of the pure solvent multiplied by its mole fraction in solution. Thus, if n_s is the number of moles of solute dissolving in n_w moles of water (the solvent) in a liquid drop of size i, the **saturation vapor pressure of water (species q) over the solution surface** is

$$\frac{p'_{q,s,i}}{p_{q,s}} = \frac{n_w}{n_w + n_s} \tag{16.36}$$

This is known as **Raoult's law** (Raoult 1887) (also called the **solute effect**), named after François-Marie Raoult who studied the vapor pressures and freezing points of solutions. The reason for the reduction in the solvent's vapor pressure is that solute molecules replace solvent molecules on the solution surface, preventing solvent molecules from evaporating. According to Raoult's law, the vapor pressure of the dissolved solute over the solution similarly equals the vapor pressure of the pure solute multiplied by its mole fraction in solution. Thus, although the solvent's vapor pressure is reduced, it is also replaced by solute vapor pressure. Thus, the total vapor pressure over the solution is the sum of the vapor pressures of the solvent and solute over the solution. As such, if the vapor pressure of a pure solute is zero (e.g., if the solute is **involatile** – it does not evaporate), the total vapor pressure over the solution is less than that of the pure solvent alone.

For dilute solutions, $n_w \gg n_s$, and (16.36) simplifies to

$$\frac{p'_{q,s,i}}{p_{q,s}} \approx 1 - \frac{n_s}{n_w} \tag{16.37}$$

The number of moles of solute in solution is approximately $n_s = i_v M_s / m_s$, where M_s is the mass of solute in the particle (g), m_s is the molecular weight of the solute (g mol^{-1}), and i_v is the degree of dissociation of the solute into ions, called the **van't Hoff factor**. The factor i_v gives the actual number of moles of ions that dissociate from one mole of solute. For sodium chloride (NaCl), which dissociates in solution to Na^+ and Cl^-, $i_v = 2$. For ammonium sulfate ($(NH_4)_2SO_4$), which dissociates to $2NH_4^+$ and SO_4^{2-}, $i_v = 3$. For a nondissociating solute, $i_v = 1$. The more ions present on the surface, the greater the reduction in saturation vapor pressure.

The number of moles of liquid water in a drop, used in (16.37), is approximately

$$n_w = \frac{M_w}{m_v} \approx \frac{4\pi r_i^3 \rho_w}{3m_v} \tag{16.38}$$

where M_w is the mass of liquid water in the drop (g), m_v is the molecular weight of water (g mol^{-1}), and ρ_w is the density of liquid water (g cm^{-3}). This equation assumes that $M_w \gg M_s$. Substituting $n_s = i_v M_s/m_s$ and (16.38) into (16.37) gives the solute effect over a dilute liquid water solution as

$$\frac{p'_{q,s,i}}{p_{q,s}} \approx 1 - \frac{3m_v i_v M_s}{4\pi r_i^3 \rho_w m_s} \tag{16.39}$$

The solute effect affects primarily small particles, since small particles have a higher concentration of solute than do large particles. Equation (16.39) can be modified to treat multiple dissolved gases simultaneously, as described in Section 18.8.1.

16.2.3.3 *Köhler equation*

Taking the product of the right sides of (16.33) and (16.39) and eliminating the last term of the result, which is small, gives the **Köhler equation** (Köhler 1936),

$$S'_{q,i} = \frac{p'_{q,s,i}}{p_{q,s}} \approx 1 + \frac{2\sigma_p m_p}{r_i R^* T \rho_{p,i}} - \frac{3m_v i_v M_s}{4\pi r_i^3 \rho_w m_s} \tag{16.40}$$

where $S'_{q,i}$ is the **saturation ratio at equilibrium**. The Köhler equation relates the saturation vapor pressure of water over a curved surface containing solute to that over a flat surface without solute.

The Köhler equation is useful for determining whether a cloud drop can form from an aerosol particle under ambient conditions. All cloud drops in the atmosphere form on aerosol particles. As discussed in Section 14.2.1, water vapor cannot homogeneously nucleate under atmospheric conditions but can readily heterogeneously nucleate on an existing surface. The most prevalent surfaces in the atmosphere are aerosol particle surfaces. Following heterogeneous nucleation on an aerosol particle, water vapor readily grows onto the particle if the air is supersaturated with water over the particle's surface.

Water does not easily heterogeneously nucleate on all aerosol particle surfaces. Those particles that water can heterogeneously nucleate on to form liquid drops are called **cloud condensation nuclei (CCN)**. Most aerosol particles can serve as a CCN although some, such as diesel soot particles (mostly black carbon plus lubricating oil), are relatively hydrophobic, so water does not readily condense upon them (e.g., Ishizaka and Adhikari 2003). As soot ages, though, other chemicals coat it through condensation and coagulation. If the coating chemicals are soluble (e.g., sulfuric acid, nitric acid, hydrochloric acid, hydroscopic organic matter), soot's ability to act as a CCN increases.

The heterogeneous nucleation and growth of a CCN to a cloud drop is called **CCN activation**. Whether a single CCN of a given size activates into a cloud drop depends substantially on three parameters: the ambient saturation ratio, the critical saturation ratio, and the critical radius for growth. The first parameter depends on ambient conditions, and the last two are derived from the Köhler equation.

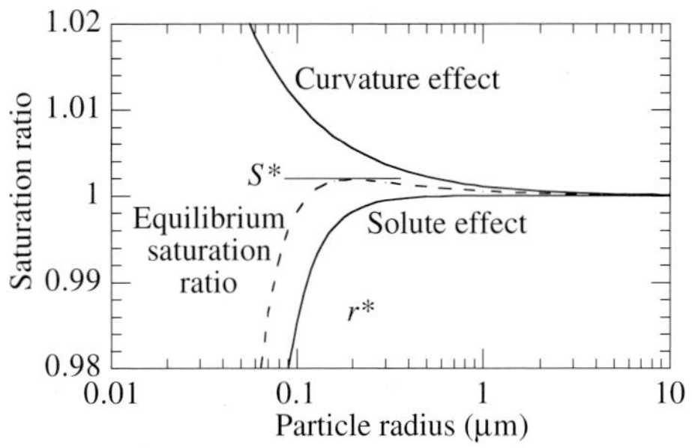

Figure 16.1 Example of how curvature and solute effects affect the equilibrium saturation ratio $S'_{q,i}$. The sum of the curvature and solute curves (dashed line) is $S'_{q,i}$. The curvature effect increases the saturation vapor pressure over small aerosol particles, increasing $S'_{q,i}$. The solute effect decreases the saturation vapor pressure over small aerosol particles, decreasing $S'_{q,i}$. The maximum $S'_{q,i}$ is the critical saturation ratio S^*, which occurs at the critical radius r^*.

The **ambient saturation ratio** of a gas is the ratio of the partial pressure of the gas (p_q) to the gas's saturation vapor pressure over a flat, dilute surface ($p_{q,s}$):

$$S_q = \frac{p_q}{p_{q,s}} \tag{16.41}$$

When $S_q > S'_{q,i}$, the air is supersaturated with respect to particles of size i, and vapor condenses onto these particles. Figure 16.1, though, shows that the equilibrium saturation ratio $S'_{q,i}$ (from 16.40) varies with particle radius. Thus, although $S_q > S'_{q,i}$ initially in many cases, the condition does not necessarily hold as the particle grows. Suppose the ambient saturation ratio $S_q = 1.001$ and a particle is small (e.g., < 0.1 μm in radius). At that radius, $S_q > S'_{q,i}$, and the particle starts to grow. As it grows, $S'_{q,i}$ increases along the "equilibrium saturation ratio" curve in Fig. 16.1. When the particle exceeds about 0.15 μm radius, $S_q \le S'_{q,i}$ and the particle cannot grow further. If, instead, S_q were about 1.003, the particle would grow until it reached cloud-drop size (about 5–10 μm). Alternatively, if the initial particle were large (e.g., > 0.6 μm in radius) and $S_q = 1.001$, the particle would also grow to cloud-drop size.

In Fig. 16.1, the radius at which the equilibrium saturation ratio is maximum is the **critical radius for growth** (r^*). The equilibrium saturation ratio at this radius is the **critical saturation ratio** (S^*). Aerosol particles that start smaller than the critical radius cannot grow (activate) to cloud-drop size unless the ambient saturation ratio exceeds the critical saturation ratio. Aerosol particles initially larger than the critical radius can activate if $S_q > S'_{q,i}$ over such drops. In sum, activation of a CCN can occur if $r_i < r^*$ and $S_q > S^*$ or $r_i > r^*$ and $S_q > S'_{q,i}$. When $r_i < r^*$ and $S^* > S_q > S'_{q,i}$, a CCN cannot activate, and it grows no larger than r^*. Also, when

Table 16.2 Critical radii for growth and critical supersaturations for aerosol particles containing dissolved sodium chloride or ammonium sulfate when solute mass varies at 275 K. The effect of solute on surface tension is ignored.

	Sodium chloride		Ammonium sulfate	
Solute mass (g)	r^* (μm)	$S^* - 1$ (%)	r^* (μm)	$S^* - 1$ (%)
0	0	∞	0	∞
10^{-18}	0.019	4.1	0.016	5.1
10^{-16}	0.19	0.41	0.16	0.51
10^{-14}	1.9	0.041	1.6	0.051
10^{-12}	19	0.0041	16	0.0051

$S_q < S'_{q,i}$, a CCN does not activate. When $r_i > r^*$, large particles are more likely to grow than small particles because the growth rate is proportional to $S_q - S'_{q,i}$, and this difference is larger for large particles than for small particles.

The critical radius and critical saturation ratio are found by rewriting (16.40) as

$$S'_{q,i} = 1 + \frac{a}{r_i} - \frac{b}{r_i^3} \qquad a = \frac{2\sigma_p m_p}{R^* T \rho_{p,i}} \qquad b = \frac{3 m_v i_v M_s}{4\pi \rho_w m_s} \tag{16.42}$$

where a has units of cm and b has units of cm^3. Taking the derivative of $S'_{q,i}$, setting the derivative to zero, solving for the radius, and substituting the result back into (16.42) gives the critical radius for growth and critical saturation ratio as

$$r^* = \sqrt{\frac{3b}{a}} \qquad S^* = 1 + \sqrt{\frac{4a^3}{27b}} \tag{16.43}$$

respectively.

Supersaturation is the saturation ratio minus one. Table 16.2 gives critical radii and critical supersaturations (expressed as percentages) when liquid particles contain either dissolved sodium chloride or dissolved ammonium sulfate. In the atmosphere, ambient supersaturations rarely exceed 2 percent (saturation ratio of 1.02) and are usually much smaller. Since the number concentration of cloud drops (100–1000 cm^{-3}) is less than that of CCN (10^3–10^6 cm^{-3}), only a fraction of CCN activate into cloud drops. In the absence of solute, $r^* = 0$, so all particles for which $S_q > S'_{q,i}$ have potential to activate to cloud drops.

16.2.3.4 *Radiative cooling effect*

When a large water-containing particle releases thermal-infrared radiation, it cools, reducing the temperature-dependent saturation vapor pressure of water over its surface. Large water-containing particles also warm by absorbing solar radiation, but their thermal-infrared cooling rates are generally faster than are their solar heating rates. Small liquid water particles do not absorb much solar radiation, because their solar absorption efficiencies are negligible, as shown in Fig. 9.21. The net change in saturation vapor pressure over a liquid drop surface due to radiative cooling or heating is called the **radiative cooling effect** since cooling is the predominant result of this effect.

The saturation vapor pressure over a flat, dilute surface that experiences net radiative cooling relative to that over a flat, dilute surface that does not is

$$\frac{p'_{q,\mathrm{s},i}}{p_{q,\mathrm{s}}} \approx 1 + \frac{L_{\mathrm{e},q} m_q H_{\mathrm{r},i}}{4\pi r_i R^* T^2 \kappa'_{\mathrm{d},i}} \tag{16.44}$$

where

$$H_{\mathrm{r},i} = (\pi r_i^2) 4\pi \int_0^\infty Q_\mathrm{a}(m_\lambda, \alpha_{i,\lambda})(I_\lambda - B_\lambda)\mathrm{d}\lambda \tag{16.45}$$

is the **radiative cooling rate** (W) of the particle (e.g., Toon *et al.* 1989b). In this equation, Q_a is the single-particle absorption efficiency, I_λ is the incoming spectral radiance, and B_λ is the Planck spectral radiance emitted by the particle surface. Absorption efficiencies, which depend on the complex index of refraction and size parameter (m_λ and $\alpha_{i,\lambda}$), were discussed in Section 9.4.3. A method of determining radiance was given in Section 9.8.3. The Planck function was given in (9.4). Radiative cooling rates can be positive or negative.

16.2.3.5 *Overall effects*

For condensation of water vapor onto a liquid-water surface, the overall equilibrium saturation ratio is obtained by taking the product of the right sides of (16.40) and (16.44) and eliminating small terms. The result is

$$S'_{q,i} = \frac{p'_{q,\mathrm{s},i}}{p_{q,\mathrm{s}}} \approx 1 + \frac{2\sigma_\mathrm{p} m_\mathrm{p}}{r_i R^* T \rho_{\mathrm{p},i}} - \frac{3 m_\mathrm{v} i_\mathrm{v} M_\mathrm{s}}{4\pi r_i^3 \rho_\mathrm{w} m_\mathrm{s}} + \frac{L_{\mathrm{e},q} m_q H_{\mathrm{r},i}}{4\pi r_i R^* T^2 \kappa'_{\mathrm{d},i}} \tag{16.46}$$

The solute-effect term in (16.46) is most applicable to water vapor growth onto liquid drops. For growth of vapor onto small particles, the radiative cooling term is ignored, since small particles have masses too small to affect the saturation vapor pressure noticeably through this term. Since most gases other than water vapor grow primarily onto small particles, the term is also ignored for such gases. Removing the solute and radiative terms from (16.46) gives an estimate of the overall equilibrium saturation ratio for the growth of gases other than water vapor as

$$S'_{q,i} = \frac{p'_{q,\mathrm{s},i}}{p_{q,\mathrm{s}}} \approx 1 + \frac{2\sigma_\mathrm{p} m_\mathrm{p}}{r_i R^* T \rho_{\mathrm{p},i}} \tag{16.47}$$

16.3 FLUXES TO A PARTICLE WITH MULTIPLE COMPONENTS

The equations in Section 16.1 give the change in mass, volume, and radius of a single, homogeneous particle. With one component present, the volume of a spherical drop changes proportionately to the one-third power of the volume of the condensed material already in the drop, as shown in (16.16). When multiple components are present within a drop, the volume changes proportionately to the one-third power of the summed volume of all components. The growth equation for individual components in a drop containing multiple components is derived below.

The total volume and mass of one particle in which one component grows can be defined as

$$\upsilon_{i,t} = \upsilon_{q,i,t} + \upsilon_{i,t-h} - \upsilon_{q,i,t-h} \tag{16.48}$$

$$m_{i,t} = \rho_{\mathrm{p},i,t}\upsilon_{i,t} = \rho_{\mathrm{p},q}\upsilon_{q,i,t} + \rho_{\mathrm{p},i,t-h}\upsilon_{i,t-h} - \rho_{\mathrm{p},q}\upsilon_{q,i,t-h} \tag{16.49}$$

respectively, where $\upsilon_{q,i,t}$ is the volume of growing component q in size bin i at current time t, $\upsilon_{i,t-h}$ is the initial total volume of the particle before growth, $\upsilon_{q,i,t-h}$ is the initial volume of component q in the particle before growth (time $t-h$), $\rho_{\mathrm{p},i,t-h}$ is the mass density (g cm^{-3}) of the total particle before growth, and $\rho_{\mathrm{p},q}$ is the mass density (g cm^{-3}) of component q, which is assumed to be constant. In reality, the mass density of a component in a particle may vary. The time derivative of (16.49) is

$$\frac{\mathrm{d}m_{i,t}}{\mathrm{d}t} = \rho_{\mathrm{p},i,t}\frac{\mathrm{d}\upsilon_{i,t}}{\mathrm{d}t} = \rho_{\mathrm{p},q}\frac{\mathrm{d}\upsilon_{q,i,t}}{\mathrm{d}t} \tag{16.50}$$

where $\mathrm{d}\upsilon_{i,t-h}/\mathrm{d}t = \mathrm{d}\upsilon_{q,i,t-h}/\mathrm{d}t = 0$. Combining (16.50) and (16.48) with (16.16) gives the volume rate of change of one component in a multicomponent particle as

$$\frac{\mathrm{d}\upsilon_{q,i,t}}{\mathrm{d}t} = \frac{[48\pi^2(\upsilon_{q,i,t} + \upsilon_{i,t-h} - \upsilon_{q,i,t-h})]^{1/3} D'_{q,i}(p_q - p'_{q,\mathrm{s},i})}{\dfrac{D'_{q,i}L_{\mathrm{e},q}\rho_{\mathrm{p},q}p'_{q,\mathrm{s},i}}{\kappa'_{\mathrm{a},i}T}\left(\dfrac{L_{\mathrm{e},q}m_q}{R^*T} - 1\right) + \dfrac{R^*T\rho_{\mathrm{p},q}}{m_q}} \tag{16.51}$$

16.4 FLUXES TO A POPULATION OF PARTICLES

Equation (16.51) gives the flux of one gas to **one** particle that contains many components. The fluxes of individual gases to **many** particles, each with different size and composition, are now considered. In a model, particles are often segregated into size bins, and each particle in a bin is assumed to have the same volume and composition as each other particle in the bin. The volume (cm^3) of a single particle is related to its volume concentration (cm^3 cm^{-3}) and number concentration (particles cm^{-3}) by $\upsilon_i = v_i/n_i$. The volume of an individual component in a particle is related to its volume concentration and the number concentration of particles by $\upsilon_{q,i,t} = v_{q,i,t}/n_{i,t-h}$. Substituting expressions of this type into (16.51) gives the

volume-concentration form of the growth equation for a component in particles of size i as

$$\frac{\mathrm{d}v_{q,i,t}}{\mathrm{d}t} = \frac{n_{i,t-h}^{2/3}[48\pi^2(v_{q,i,t} + v_{i,t-h} - v_{q,i,t-h})]^{1/3} D'_{q,i}(p_q - p'_{q,s,i})}{\frac{D'_{q,i} L_{e,q} \rho_{p,q} p'_{q,s,i}}{\kappa'_{a,i} T}\left(\frac{L_{e,q} m_q}{R^* T} - 1\right) + \frac{R^* T \rho_{p,q}}{m_q}} \qquad (16.52)$$

(Jacobson and Turco 1995). Equation (16.52) can be simplified by defining

$$p_q = C_q R^* T \qquad p'_{q,s,i} = C'_{q,s,i} R^* T \qquad (16.53)$$

where C_q is the gas-phase mole concentration (mol cm^{-3}) of component q, and $C'_{q,s,i}$ is an effective saturation vapor concentration (mol cm^{-3}). Combining (16.52) with (16.53) and adding more time subscripts give

$$\frac{\mathrm{d}v_{q,i,t}}{\mathrm{d}t} = n_{i,t-h}^{2/3}[48\pi^2(v_{q,i,t} + v_{i,t-h} - v_{q,i,t-h})]^{1/3} D_{q,i,t-h}^{\mathrm{eff}} \frac{m_q}{\rho_{p,q}}(C_{q,t} - C'_{q,s,i,t-h}) \qquad (16.54)$$

where

$$D_{q,i,t-h}^{\mathrm{eff}} = \frac{D'_{q,i}}{\frac{m_q D'_{q,i} L_{e,q} C'_{q,s,i,t-h}}{\kappa'_{a,i} T}\left(\frac{L_{e,q} m_q}{R^* T} - 1\right) + 1} \qquad (16.55)$$

is an **effective molecular diffusion coefficient** (cm^2 s^{-1}). For water vapor, $C'_{q,s,i,t-h}$ is large, and the left side of the denominator of (16.55) cannot be ignored. For other gases, $C'_{q,s,i,t-h}$ is usually orders of magnitude smaller than for water vapor, and the left side of the denominator is much less than 1. Thus, for gases other than water vapor, the left side of the denominator can be ignored, and (16.55) simplifies to

$$D_{q,i,t-h}^{\mathrm{eff}} \approx D'_{q,i} = D_q \omega_{q,i} F_{q,L,i} = \frac{D_q F_{q,L,i}}{1 + \left[\frac{1.33 + 0.71\mathrm{Kn}_{q,i}^{-1}}{1 + \mathrm{Kn}_{q,i}^{-1}} + \frac{4(1 - \alpha_{q,i})}{3\alpha_{q,i}}\right]\mathrm{Kn}_{q,i}} \qquad (16.56)$$

Equation (16.54) is the growth equation for a single gas to a population of spherical particles of size i. The equation yields the change in volume concentration of a particle with many components when one component in the particle grows or evaporates. The equation is applied to each particle size bin and each volatile component q. Since gases condense on particles in several size bins, a **gas conservation equation** must be written to accompany (16.54). Such an equation is

$$\frac{\mathrm{d}C_{q,t}}{\mathrm{d}t} = -\frac{\rho_{p,q}}{m_q}\sum_{i=1}^{N_B}\frac{\mathrm{d}v_{q,i,t}}{\mathrm{d}t} \qquad (16.57)$$

16.5 SOLUTIONS TO GROWTH EQUATIONS

Many methods have been developed to solve time-dependent condensation and evaporation equations. These include finite-element methods (Varoglu and Finn 1980; Tsang and Brock 1986; Tsang and Huang 1990), discrete-size-bin methods (e.g., Gelbard and Seinfeld 1980; Toon *et al.* 1988; Rao and McMurry 1989; Lister *et al.* 1995; Gelbard *et al.* 1998), the cubic spline method (e.g., Middleton and Brock 1976; Nguyen and Dabdub 2001), modified upwind difference methods (e.g., Smolarkiewicz 1983; Tsang and Korgaonkar 1987), trajectory-grid methods (e.g., Chock and Winkler 2000; Gaydos *et al.* 2003), semi-Lagrangian methods (e.g., Bott 1989; Nguyen and Dabdub 2002), and moment methods (e.g., Friedlander 1983; Lee 1985; Whitby 1985; Brock *et al.* 1986; Binkowski and Shankar 1995). Some of these methods conserve mass between the gas and particle phases, and others do not. Some methods reduce or eliminate numerical diffusion during growth, and others are more diffusive. Some methods are computationally more efficient than others. Zhang *et al.* (1999) compared several condensation schemes.

Below, two solutions to condensation/evaporation equations are given. Both solutions conserve mass between the gas and aerosol phases and reduce or eliminate numerical diffusion during growth when used in conjunction with a hybrid, full-moving, or moving-center size structure. Although the first method gives exact solutions, it is iterative and requires more computational time than the second. Both schemes have been used in three-dimensional models. The second method, which is noniterative but unconditionally stable, is useful when computer time is limited.

16.5.1 Integrated numerical solution

Together, (16.54) and (16.57) constitute $N_B + 1$ nonlinear, first-order ordinary differential equations (ODEs). A method of solving these equations is to integrate them with a sparse-matrix ODE solver, such as that discussed in Section 12.8. An advantage of such a solver is that the matrix of partial derivatives arising from (16.54) and (16.57) is sparse. When $N_B = 3$, a matrix of partial derivatives arising from this set of equations is

$$\begin{array}{c} \\ v_{q,1,t} \\ \\ v_{q,2,t} \\ \\ v_{q,3,t} \\ \\ C_{q,t} \end{array}
\begin{array}{c} \begin{array}{cccc} v_{q,1,t} & v_{q,2,t} & v_{q,3,t} & C_{q,t} \end{array} \\
\begin{bmatrix}
1 - h\beta_s \dfrac{\partial^2 v_{q,1,t}}{\partial v_{q,1,t}\partial t} & 0 & 0 & -h\beta_s \dfrac{\partial^2 v_{q,1,t}}{\partial C_{q,t}\partial t} \\
0 & 1 - h\beta_s \dfrac{\partial^2 v_{q,2,t}}{\partial v_{q,2,t}\partial t} & 0 & -h\beta_s \dfrac{\partial^2 v_{q,2,t}}{\partial C_{q,t}\partial t} \\
0 & 0 & 1 - h\beta_s \dfrac{\partial^2 v_{q,3,t}}{\partial v_{q,3,t}\partial t} & -h\beta_s \dfrac{\partial^2 v_{q,3,t}}{\partial C_{q,t}\partial t} \\
-h\beta_s \dfrac{\partial^2 C_{q,t}}{\partial v_{q,1,t}\partial t} & -h\beta_s \dfrac{\partial^2 C_{q,t}}{\partial v_{q,2,t}\partial t} & -h\beta_s \dfrac{\partial^2 C_{q,t}}{\partial v_{q,3,t}\partial t} & 1 - h\beta_s \dfrac{\partial^2 C_{q,t}}{\partial C_{q,t}\partial t}
\end{bmatrix} \end{array} \tag{16.58}$$

Table 16.3 Reduction in array space and number of matrix operations before and after sparse-matrix techniques are applied to growth/evaporation ODEs

	Value	
Quantity	Initial	After sparse reductions
Order of matrix ($N_B + 1$)	17	17
No. of initial matrix positions filled	289	49
Percent of initial positions filled	100	17
No. of final matrix positions filled	289	49
Percent of final positions filled	100	17
No. of operations in decomp. 1	1496	16
No. of operations in decomp. 2	136	16
No. of operations in backsub. 1	136	16
No. of operations in backsub. 2	136	16

The last four rows show the number of operations in each of four loops of decomposition and backsubstitution.

The partial derivatives of this matrix are

$$\frac{\partial^2 v_{q,i,t}}{\partial v_{q,i,t}\partial t} = \frac{1}{3}\left(\frac{n_{i,t-h}}{v_{q,i,t}}\right)^{2/3}\left(48\pi^2\right)^{1/3} D^{\text{eff}}_{q,i,t-h}\frac{m_q}{\rho_{\text{p},q}}\left(C_{q,t} - C'_{q,\text{s},i,t-h}\right) \qquad i = 1,\ldots,N_B \tag{16.59}$$

$$\frac{\partial^2 v_{q,i,t}}{\partial C_{q,t}\partial t} = n^{2/3}_{i,t-h}[48\pi^2(v_{q,i,t} + v_{i,t-h} - v_{q,i,t-h})]^{1/3} D^{\text{eff}}_{q,i,t-h}\frac{m_q}{\rho_{\text{p},q}} \qquad i = 1,\ldots,N_B \tag{16.60}$$

$$\frac{\partial^2 C_{q,t}}{\partial v_{q,i,t}\partial t} = -\frac{\rho_{\text{p},q}}{m_q}\frac{\partial^2 v_{q,i,t}}{\partial v_{q,i,t}\partial t} \qquad i = 1,\ldots,N_B \tag{16.61}$$

$$\frac{\partial^2 C_{q,t}}{\partial C_{q,t}\partial t} = -\frac{\rho_{\text{p},q}}{m_q}\sum_{i=1}^{N_B}\frac{\partial^2 v_{q,i,t}}{\partial C_{q,t}\partial t} \tag{16.62}$$

Matrix (16.58) does not require reordering, since it is already ordered from the fewest partial-derivative entries at the top of the matrix to the most at the bottom. Table 16.3 confirms that the current ordering of the matrix is optimal. It shows statistics from a case with 16 size bins ($N_B + 1 = 17$). The table shows that no matrix fill-in was required during matrix decomposition and backsubstitution.

An ODE integrator gives an accurate and relatively rapid solution to the growth equations if sparse-matrix techniques are used. An integrator still requires iteration; thus, in a model with 10^4–10^6 grid cells, faster solutions may be desired.

16.5.2 Analytical predictor of condensation

Another method of solving growth equations is the **analytical predictor of condensation** (APC) scheme (Jacobson 1997c, 2002). This scheme requires no iteration, conserves mass exactly, and is unconditionally stable. The solution is obtained by holding the volume on the right side of (16.54) constant for a time step. Thus,

$$\frac{dv_{q,i,t}}{dt} = n_{i,t-h}^{2/3}(48\pi^2 v_{i,t-h})^{1/3} D_{q,i,t-h}^{\text{eff}} \frac{m_q}{\rho_{\text{p},q}}(C_{q,t} - C'_{q,\text{s},i,t-h}) \tag{16.63}$$

This equation is modified by defining a mass-transfer coefficient (s^{-1}), an effective saturation vapor mole concentration, and particle volume concentration as

$$k_{q,i,t-h} = n_{i,t-h}^{2/3}(48\pi^2 v_{i,t-h})^{1/3} D_{q,i,t-h}^{\text{eff}} = n_{i,t-h} 4\pi r_{i,t-h} D_{q,i,t-h}^{\text{eff}} \tag{16.64}$$

$$C'_{q,\text{s},i,t-h} = S'_{q,i,t-h} C_{q,\text{s},i,t-h} \tag{16.65}$$

$$v_{q,i,t} = m_q c_{q,i,t}/\rho_{\text{p},q} \tag{16.66}$$

respectively. In (16.65), $C_{q,\text{s},i,t-h} = p_{q,\text{s},t-h}/R^*T$ is an uncorrected saturation vapor mole concentration (mol cm^{-3}), and $S'_{q,i,t-h}$ is the equilibrium saturation ratio of the condensing gas. In (16.66), $c_{q,i,t}$ is the particle-phase mole concentration of component q (moles per cubic centimeter of air). Equilibrium saturation ratios were defined for liquid water in (16.46) and other gases in (16.47). Substituting (16.64)–(16.66) into (16.63) and (16.57) gives

$$\frac{dc_{q,i,t}}{dt} = k_{q,i,t-h}(C_{q,t} - S'_{q,i,t-h} C_{q,\text{s},i,t-h}) \qquad i = 1, \ldots, N_B \tag{16.67}$$

$$\frac{dC_{q,t}}{dt} = -\sum_{i=1}^{N_B} [k_{q,i,t-h}(C_{q,t} - S'_{q,i,t-h} C_{q,\text{s},i,t})] \tag{16.68}$$

respectively. Equations (16.67) and (16.68) represent $N_B + 1$ ODEs.

The noniterative solution to the growth equations is obtained by integrating (16.67) for a final aerosol concentration. The result is

$$c_{q,i,t} = c_{q,i,t-h} + hk_{q,i,t-h}(C_{q,t} - S'_{q,i,t-h} C_{q,\text{s},i,t-h}) \tag{16.69}$$

where the final gas mole concentration, $C_{q,t}$, is currently unknown. Final aerosol and gas concentrations are constrained by the mass-balance equation,

$$C_{q,t} + \sum_{i=1}^{N_B}(c_{q,i,t}) = C_{q,t-h} + \sum_{i=1}^{N_B}(c_{q,i,t-h}) = C_{\text{tot}} \tag{16.70}$$

Substituting (16.69) into (16.70) and solving for the gas concentration give

$$C_{q,t} = \frac{C_{q,t-h} + h\sum_{i=1}^{N_B}(k_{q,i,t-h} S'_{q,i,t-h} C_{q,\text{s},i,t-h})}{1 + h\sum_{i=1}^{N_B} k_{q,i,t-h}} \tag{16.71}$$

Table 16.4 Demonstration of unconditional stability of the APC scheme

Time step size (s)	$C_{q,t}$ (μg m^{-3})	$c_{q,1,t}$ (μg m^{-3})	$c_{q,2,t}$ (μg m^{-3})	$c_{q,3,t}$ (μg m^{-3})
0.1	1.0	2.18	5.36	1.46
10	1.0	2.18	5.36	1.46
60	1.0	2.18	5.36	1.46
600	1.0	2.18	5.36	1.46
7200	1.0	2.18	5.36	1.46

Assume a gas transfers to and from three particle size bins, where $k_{q,i,t-h} = 0.000833$, 0.001667, and 0.00667 s^{-1} for the respective bins, $S'_{q,i,t-h} = 1.0$ for all three bins, $c_{q,i,t-h} = 2$, 5, and 0 μg m^{-3} for the respective bins, $C_{q,t-h} = 3$ μg m^{-3}, and $C_{q,s,i,t-h} = 1$ μg m^{-3}. The table shows concentrations after 4 h predicted by the APC scheme when each of five time step sizes was used. Regardless of the time step size, the APC solution was stable and mass-conserving.

The concentration from this equation cannot fall below zero, but can increase above the total mass of the species in the system. In such cases, gas concentration is limited by $C_{q,t} = \min(C_{q,t}, C_{\text{tot}})$. This value is substituted into (16.69) to give the final aerosol concentration. Since (16.69) can result in a negative concentration or a concentration above the maximum, two more limits are placed sequentially after (16.69) has been solved among all size bins. The first is $c_{q,i,t} = \max(c_{q,i,t}, 0)$ and the second is

$$c_{q,i,t} = \frac{\left[C_{q,t-h} - C_{q,t} + \sum_{i=1}^{N_B} \max(c_{q,i,t-h} - c_{q,i,t}, 0)\right]}{\sum_{i=1}^{N_B} [\max(c_{q,i,t} - c_{q,i,t-h}, 0)]} (c_{q,i,t} - c_{q,i,t-h}) \quad (16.72)$$

where the $c_{q,i,t}$ values on the right side of the equation are determined after the first limit has been solved over all size bins. The APC scheme is unconditionally stable, since all final concentrations are bounded between 0 and C_{tot}, regardless of the time step. Table 16.4 demonstrates the unconditional stability and mass conservation of the scheme.

Figure 16.2 compares an APC scheme prediction with an exact solution for a case of cloud drop growth from aerosol particles. The figure shows that the APC scheme matched the exact solution almost exactly for this set of conditions. The APC scheme can be extended to treat any number of aerosol size distributions simultaneously, as described in Jacobson (2002).

16.6 SOLVING HOMOGENEOUS NUCLEATION WITH CONDENSATION

Homogeneous nucleation competes with condensation for a limited amount of available vapor. If the numerical solution for condensation is operator-split from that for homogeneous nucleation, the process solved first may incorrectly consume more than its share of vapor unless the operator-splitting time interval is short

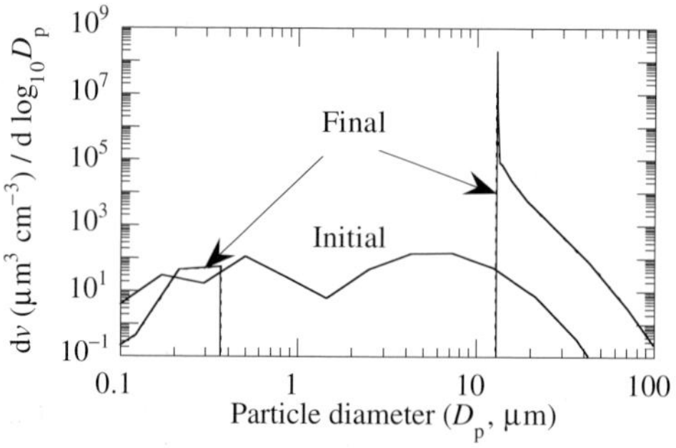

Figure 16.2 Comparison of APC solution, when $h = 10$ s, with an exact solution of condensational growth over 16 size bins when the moving-center size structure was used. The two solutions lie almost exactly on top of each other, thus are indistinguishable. At the start of the simulation, the relative humidity was increased from 90 to 100.002 percent and replenished every 10 s back to 100.002 percent for five minutes.

($\ll$1 s). Solving homogeneous nucleation simultaneously with condensation eliminates this problem.

Condensation can be solved simultaneously with homogeneous nucleation with the APC scheme by treating the mass transfer rate (s^{-1}) to the first size bin as the **sum of a condensation and a homogeneous nucleation mass transfer rate:**

$$k_{q,1,t-h} = k_{q,\mathrm{cond},1,t-h} + k_{q,\mathrm{hom},1,t-h} \tag{16.73}$$

where the condensation mass transfer rate is the rate from (16.64), and the **homogeneous nucleation mass transfer rate** is

$$k_{q,\mathrm{hom},1,t-h} = \frac{\rho_q \upsilon_1}{m_q}\left(\frac{J_{\mathrm{hom},q}}{C_{q,t-h} - S'_{q,1,t-h}C_{q,\mathrm{s},1,t-h}}\right) \tag{16.74}$$

In this equation, $J_{\mathrm{hom},q}$ is the homogeneous nucleation rate (new particles per cubic centimeter per second) from (14.23), ρ_q is the mass density (g cm^{-3}) and m_q is the molecular weight (g mol^{-1}) of a newly nucleated particle, υ_1 is the volume (cm^3) of a single nucleated particle, and the denominator on the right side is the maximum gas available for homogeneous nucleation at the beginning of a time step (mol cm^{-3}). Equation (16.74) is nonzero only if $J_{\mathrm{hom},q} > 0$ and $C_{q,t-h} > S'_{q,1,t-h}C_{q,\mathrm{s},1,t-h}$. After gas and size-resolved aerosol mole concentrations are solved with (16.67)–(16.72), the **new number concentration of particles in the smallest size bin** is

$$n_{1,t} = n_{1,t-h} + \mathrm{MAX}\left[(c_{q,1,t} - c_{q,1,t-h})\frac{m_q}{\rho_q \upsilon_1}\frac{k_{q,\mathrm{hom},1,t-h}}{k_{q,1,t-h}}, 0\right] \tag{16.75}$$

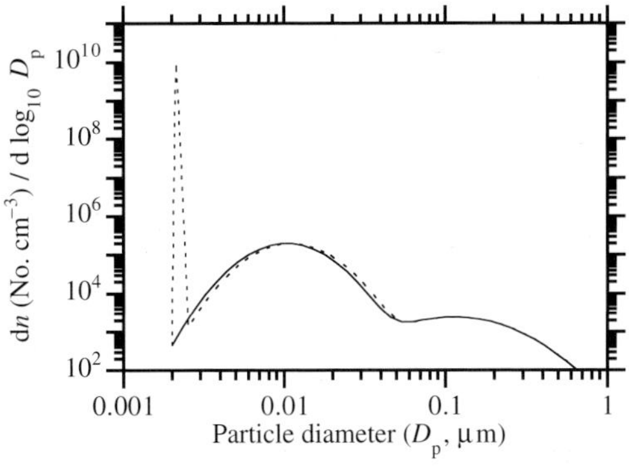

Figure 16.3 Number concentration distribution initially, —, and after 8 seconds, ----, of simultaneous nucleation and condensation of sulfuric acid–water with the APC scheme. The initial gas concentration of sulfuric acid was 15 μg m^{-3}, the relative humidity was 90 percent, and the temperature was 300 K.

(particles cm^{-3}). Figure 16.3 shows results from an 8-s simulation of simultaneous homogeneous nucleation and condensation when a limited amount of gas was available for both. The figure suggests that, not only did homogeneous nucleation produce many new particles in this case, but many of the nucleated particles and the pre-existing particles grew slightly by condensation. In a competition for vapor between homogeneous nucleation and condensation, the relative importance of condensation increases with an increasing number of background particles.

16.7 EFFECTS OF CONDENSATION ON COAGULATION

Condensation increases particle volume, and coagulation reduces particle number. Because both processes occur simultaneously, it is difficult to estimate the effect of each experimentally. A model can separate the relative effects. Results from one such simulation are shown in Fig. 16.4. Simulations of coagulation alone, coagulation coupled with growth, and growth alone were performed for an urban size distribution consisting of four lognormal modes. Two modes had geometric-mean volume diameters near 0.2 and 0.5–0.7 μm, respectively (Hering and Friedlander 1982; John *et al.* 1989). Simulations were performed over an 8-h period. The only growth process considered was sulfuric acid condensation.

The figures show that coagulation alone decreased the number concentration of particles smaller than 0.2 μm in diameter and slightly increased the number concentration of particles near 0.3 μm in diameter. Growth alone caused the initial size distribution to shift to the right. The shift of particles to larger diameters caused the volume concentration of those particles to increase as well, as shown in Fig. 16.4(b). When coagulation was combined with growth, the number concentration of particles smaller than 0.2 μm in diameter decreased, and the number concentration

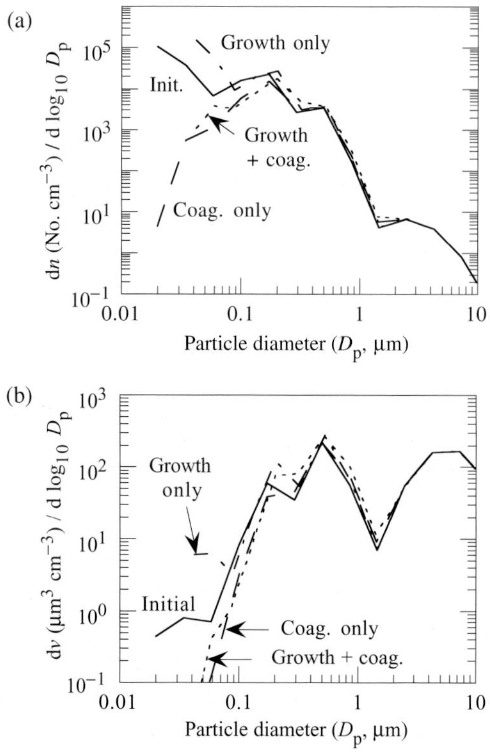

Figure 16.4 Change in aerosol (a) number concentration and (b) volume concentration after 8 h when coagulation alone, growth alone, and growth coupled to coagulation were considered. Only H_2SO_4(g) condensed. Initial H_2SO_4(g), 50 μg m^{-3}. The initial particle number concentration was 45 070 particles cm^{-3}. T = 298 K. From Jacobson (1997a).

of particles between 0.2 and 0.5 μm in diameter increased. In sum, growth plus coagulation pushed particles to slightly larger sizes than did growth alone or coagulation alone.

Figure 16.5 shows the growth-only and growth plus coagulation solutions from Fig. 16.4(a) when the moving-center and full-moving size structures (Chapter 13) were used. Since the full-moving structure is nondiffusive during growth, the figure suggests that the moving-center size structure was also relatively nondiffusive.

16.8 ICE CRYSTAL GROWTH

When the temperature drops below 0 °C, water vapor may deposit directly onto aerosol particles to form ice crystals. Whereas most aerosol particle types can serve as CCN, only a few can serve as **ice deposition nuclei (IDN)**, which are aerosol particles that water vapor can potentially heterogeneously nucleate on

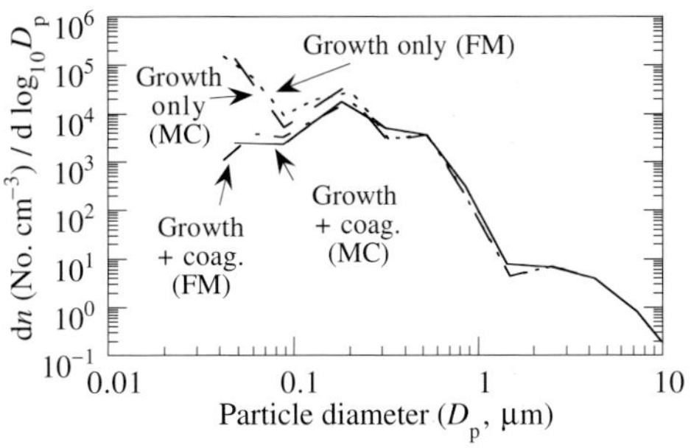

Figure 16.5 Comparison of growth-only and coagulation plus growth results from Fig. 16.4(a), obtained with the full-moving (FM) size structure, with those obtained with the moving-center (MC) size structure.

to form ice crystals. Whereas the concentration of activated CCN can approach 10^3 cm^{-3} (10^6 L^{-1}), the concentration of activated IDN is on the order of 0.5–25 L^{-1}. Common IDN include large hydrophobic particles, soil dust particles, a small fraction of organic matter, certain viruses and bacteria, and a small fraction of sea salt. Soot is not a good IDN, except below -15 °C and at high supersaturations. Some pollutants deactivate ice nuclei (Pruppacher and Klett 1997).

The rate of mass growth of a single ice crystal, obtained by modifying (16.13), is

$$\frac{\mathrm{d}m_i}{\mathrm{d}t} = \frac{4\pi \chi_i D'_{\mathrm{v},i}(p_q - p'_{\mathrm{v,I},i})}{\dfrac{D'_{\mathrm{v},i} L_{\mathrm{s}} p'_{\mathrm{v,I},i}}{\kappa'_{\mathrm{a},i} T}\left(\dfrac{L_{\mathrm{s}}}{R_{\mathrm{v}} T} - 1\right) + R_{\mathrm{v}} T} \tag{16.76}$$

where χ_i is the crystal electrical capacitance (cm), $p'_{\mathrm{v,I},i}$ is the effective saturation vapor pressure of water over ice, and L_{s} is the latent heat of sublimation (J g^{-1}).

The **electrical capacitance**, analogous to the radius in the liquid drop growth equation, is a function of ice crystal shape only. Ice crystal shape changes with temperature, relative humidity, and crystal size. The electrical capacitance is indirectly a function of these variables. Parameterizations of electrical capacitance as a function of crystal shape include

$$\chi_i = \begin{cases} a_{c,i}/2 & \text{sphere} \\ a_{c,i}e_{c,i}/\ln[(1+e_{c,i})a_{c,i}/b_{c,i}] & \text{prolate spheroid} \\ a_{c,i}e_{c,i}/\sin^{-1} e_{c,i} & \text{oblate spheroid} \\ a_{c,i}/\ln\left(4a_{c,i}^2/b_{c,i}^2\right) & \text{needle} \\ a_{c,i}e_{c,i}/\ln[(1+e_{c,i})/(1-e_{c,i})] & \text{column} \\ a_{c,i}e_{c,i}/(2\sin^{-1} e_{c,i}) & \text{hexagonal plate} \\ a_{c,i}/\pi & \text{thin plate} \end{cases} \tag{16.77}$$

(e.g., Harrington *et al.* 1995), where $a_{c,i}$ and $b_{c,i}$ are the lengths of the **major** and **minor** semiaxes (cm), respectively, and $e_{c,i} = \left(1 - b_{c,i}^2/a_{c,i}^2\right)^{1/2}$. The **major** and **minor semiaxes** are the maximum and minimum dimensions, respectively, across a crystal. A prolate spheroid is elongated at the poles and thin at the equator. An oblate spheroid is flattened at the poles and bulging at the equator. The electrical capacitance of a sphere is its radius.

The effective saturation vapor pressure in (16.76) is $p'_{v,I,i} = S'_{I,i}\,p_{v,I}$, where $p_{v,I}$ is the saturation vapor pressure over a flat ice surface, given in (2.65), and $S'_{I,i}$ is the equilibrium saturation ratio over ice, analogous to (16.46). Equation (16.76) requires an effective molecular diffusion coefficient and thermal conductivity, given by (16.18) and (16.26), respectively. The ventilation factor used in these equations must be modified for ice crystals. An empirical expression for the ventilation factors of a falling oblate spheroid ice crystal is

$$F_{q,I,i}, F_{h,I,i} = \begin{cases} 1 + 0.14x^2 & x < 1.0 \\ 0.86 + 0.28x & x \geq 1.0 \end{cases} \tag{16.78}$$

where $x = x_{q,i}$ for vapor and $x = x_{h,i}$ for energy (Pruppacher and Klett 1997), given in (16.24) and (16.31), respectively. Equation (16.76) can be solved by integration or with the APC scheme. A solution to growth of liquid drops and ice crystals simultaneously in a cloud is given in Section 18.8.1.

16.9 SUMMARY

In this chapter, the growth equation for a single, homogeneous, spherical liquid drop was derived. The equation assumed that, during condensation, a gas diffuses to and energy conducts away from a particle surface. The reverse occurs during evaporation. The equation was expanded to a population of particles and to particles with multiple components. Two numerical solutions to growth equations – an exact numerical solution and a semianalytical solution – were discussed. The growth equations were used to test the importance of condensation relative to coagulation. The conclusion was that growth plus coagulation moved particles to larger sizes than did growth alone or coagulation alone. A method of solving homogeneous nucleation simultaneously with condensation was also given. Finally, an ice crystal-depositional growth equation was described. In the next chapter, the discussion of growth is extended to particles in which trace gases dissolve.

16.10 PROBLEMS

16.1 Calculate the diffusion coefficient of ammonia gas in air assuming the air temperature is 288 K, the air pressure is 1013 hPa, and the air is dry.

16.2 Calculate the diffusion-coefficient correction for collision geometry and sticking probability for ammonia gas interacting with a particle of (a) 0.1 μm in diameter and (b) 10 μm in diameter when the air temperature is 288 K, the air pressure is 1013 hPa, and the accommodation coefficient is 0.1.

Assume the air is dry. Explain reasons for the difference in results between the two diameters.

16.3 Calculate the ventilation factor for water vapor interacting with liquid drops of the following radii: (a) 10 μm, (b) 100 μm, (c) 1000 μm. Assume the temperature is 298 K, the air pressure is 1010 hPa, and the fall speed of the liquid drop is approximately $V_{f,i} \approx 2r_i^2\rho_p/9\eta_a$, where ρ_p = drop density (1 g cm^{-3}). Ignore water vapor effects on air density.

16.4 Calculate the critical radius and critical supersaturation when a 0.1-μm radius particle contains (a) 0.05 mol kg^{-1} of an organic compound of molecular weight 120 g mol^{-1}, and (b) 0.05 mol kg^{-1} of sodium chloride of molecular weight 58.44 g mol^{-1}. Assume the organic compound does not dissociate, sodium chloride dissociates completely, and $T = 298.15$ K. Which of the two solutes enhances CCN activation the most? Explain.

16.5 **(a)** Solve (16.67) and (16.68) simultaneously (analytically) for $c_{q,i,t}$ and $C_{q,t}$. Assume one size bin is present and all variables are constant, except $c_{q,i,t}$ and $C_{q,t}$. (Hint: $C_{q,t} + c_{q,i,t} = C_{q,t-h} + c_{q,i,t-h}$, since $i = 1$.)

(b) Discretize the time derivative in the equations from part (a) in first-order, backward-difference form, and write out the right side of both equations implicitly. Solve the resulting equations for $c_{q,i,t}$ and $C_{q,t}$.

16.6 Can condensation of a species occur on particles of some sizes while evaporation occurs on particles of other sizes? If so, under what conditions would this occur?

16.7 In Fig. 16.1, particles above what radius can grow to cloud-sized drops? Under what condition can all particles grow larger than the critical radius? Is this condition likely to occur in the atmosphere? Why or why not?

16.8 For what temperature range is (16.76) valid? Can (16.13) be used to simulate the growth of supercooled liquid water?

16.11 COMPUTER PROGRAMMING PRACTICE

16.9 Write a computer script to calculate the equilibrium saturation ratio versus diameter and temperature for water vapor over a pure liquid water drop from (16.46). Ignore the solute and radiative heating effects. Calculate the Kelvin term for $d_i = 0.001$ to 10 μm, when $T = 298$ K, and plot the results. Calculate the term for $d_i = 0.01$ μm when T varies from 233 K to 315 K. Discuss the results.

16.10 **(a)** Write a computer script to calculate ρ_a, η_a, ν_a, D_q, $\omega_{q,i}$, $\bar{v}_q$, λ_q, $\mathrm{Kn}_{q,i}$, $N_{\mathrm{Re},i}$, $N_{\mathrm{Sc},q}$, $F_{q,i}$, and $D'_{q,i}$ assuming that water vapor condenses onto a particle of radius $r_i = 0.2$ μm. Assume also that $T = 298$ K and $p_a = 1013$ hPa. For the fall speed required in the Reynolds-number calculation, assume $V_{f,i} = 2r_i^2\rho_p g/(9\eta_a)$. Use additional data from the text, where necessary. Assume the air is dry.

(b) Using information from Problem 16.10(a), calculate κ_a, $\omega_{h,i}$, $\bar{v}_a$, λ_h, $\mathrm{Kn}_{h,i}$, N_{Pr}, $F_{h,i}$, and $\kappa'_{d,i}$ with the computer script.

(c) Using information from Problem 16.10 (a) and (b), calculate L_e, $p_{q,s}$, $S'_{q,i}$, $p'_{q,s,i}$, $C'_{q,s,i}$, and $D^{\mathrm{eff}}_{q,i}$ with the computer script. Assume that only curvature affects $S'_{q,i}$.

(d) Repeat Problem 16.10 (a), (b), and (c) for $T = 273$ K. Comment on the results.

(e) Repeat Problem 16.10 (a), (b), and (c) for $r_i = 1.0$ μm. Comment on the results.

16.11 Write a computer script to solve (16.67) and (16.68) with the APC scheme. Assume three size bins are present ($N_B = 3$). Also assume $k_{q,1,t-h} = 0.000\,833$ s^{-1}, $k_{q,2,t-h} = 0.001\,667$ s^{-1}, $k_{q,3,t-h} = 0.006\,67$ s^{-1}, $c_{q,i,t-h} = 0$ for all i, $S'_{q,i,t-h} = 1$ for all i, $C_{q,t-h} = 10$, $C_{q,s,i,t-h} = 2$, and concentration units are μg m^{-3}. Plot the mass concentration of particles in each size bin versus time for a 50-min simulation period, using $h = 1$ min. Compare the result with an explicit solution to (16.67) and (16.68), when $h = 0.01$ s. Is this simulation representative of growth under atmospheric conditions? Why or why not?

17

Chemical equilibrium and dissolution processes

THE change in size and composition of an aerosol depends on several processes, including nucleation, emission, coagulation, condensation, ice deposition, dissolution, heterogeneous reaction, molecular dissociation, solid precipitation, irreversible chemistry, interactions with clouds, sedimentation, dry deposition, and advection. Many of these processes are tightly coupled. Condensation and ice deposition at the gas–particle interface were discussed in Chapter 16, interactions of aerosol particles with clouds are discussed in Chapter 18, and irreversible reactions are examined in Chapter 19. In this chapter, dissolutional growth, molecular dissociation, solid precipitation, and heterogeneous chemical reaction are described. These four processes can be simulated with reversible (equilibrium) chemical reactions. Dissolution can also be simulated as a nonequilibrium process. In the sections that follow, terminology, equilibrium equations, activity-coefficient equations, the water equation, and methods of solving equilibrium and nonequilibrium equations are given.

17.1 DEFINITIONS

An important process that takes place at gas–particle interfaces is dissolution. **Dissolution** is the process by which a gas, suspended over an aerosol particle surface, diffuses to and dissolves in a liquid on the surface. The liquid in which a gas dissolves is a **solvent**. A solvent makes up the bulk of a solution, and in aerosol and hydrometeor particles, liquid water is most often the solvent. In some cases, such as when sulfuric acid–water particles form, the concentration of sulfuric acid exceeds that of liquid water, and sulfuric acid is the solvent. In this chapter, liquid water is assumed to be the solvent.

Any gas, liquid, or solid that dissolves in solution is a **solute**. One or more solutes plus the solvent make up a **solution**, which is a homogeneous mixture of substances that can be separated into individual components upon a change of state (e.g., freezing). A solution may contain many solutes. Suspended material (e.g., solids) may also be mixed throughout a solution. Such material is not considered part of the solution.

The ability of a gas to dissolve in water depends on the **solubility** of the gas, which is the maximum amount of a gas that can dissolve in a given amount of solvent at a given temperature. Solutions usually contain solute other than the dissolved gas. The solubility of a gas depends on the quantity of other solutes, since such solutes affect the thermodynamic activity of the dissolved gas in solution. (Thermodynamic activity is discussed shortly.) If water is saturated with a dissolved gas, and if the solubility of the gas changes due to a change in composition or temperature of the solution, the dissolved gas can **evaporate** from the solution to the gas phase.

In a solution, dissolved molecules (e.g., HNO_3, HCl, H_2SO_4) may **dissociate** (break into simpler components, namely ions). Positive ions, such as H^+, Na^+, K^+, Ca^{2+}, and Mg^{2+}, are **cations**. Negative ions, such as OH^-, Cl^-, NO_3^-, HSO_4^-, SO_4^{2-}, HCO_3^-, and CO_3^{2-}, are **anions**. Dissociation is reversible, meaning that ions can reform a dissolved molecule. Cations and anions may combine to form solids (crystallize), which precipitate out of solution (e.g., they are no longer part of the solution). **Precipitation** is the formation of an insoluble solid compound due to the buildup in concentration of dissolved ions in a solution. Solids may also dissociate back to ions, which reenter the solution.

A chemical process that occurs at the surface of a liquid or solid particle is **heterogeneous reaction** (Sections 11.1.4 and 11.3.6). A heterogeneous reaction may occur when a gas molecule collides with another molecule adsorbed to a solid particle surface or when a gas molecule adsorbs to a surface, diffuses on the surface, then reacts with another adsorbed molecule on the surface.

Gas dissolution, molecular dissociation, solid precipitation, and heterogeneous reaction can be described with **reversible chemical reactions**, also called thermodynamic equilibrium reactions. The rates of these reactions in the forward and reverse directions are usually fast. In this chapter, equilibrium reactions and the coupling of equilibrium reactions to nonequilibrium transfer of gases between gases and particles, are discussed.

17.2 EQUILIBRIUM REACTIONS

In this section, equilibrium equations are discussed, and temperature-dependent equilibrium coefficients are derived. An **equilibrium reaction** describes a reversible chemical process among solids, liquids, ions, and/or gases. The change in the number of moles dn_i of each reactant i = A, B, D, E, etc. during such a process is described by

$$\mathrm{d}n_\mathrm{D}\mathrm{D} + \mathrm{d}n_\mathrm{E}\mathrm{E} + \cdots \rightleftharpoons \mathrm{d}n_\mathrm{A}\mathrm{A} + \mathrm{d}n_\mathrm{B}\mathrm{B} + \cdots \tag{17.1}$$

Each equilibrium reaction must conserve mass. Thus,

$$\sum_i k_i(\mathrm{d}n_i)m_i = 0 \tag{17.2}$$

where m_i is the molecular weight of species i, $k_i = +1$ for products, and $k_i = -1$ for reactants. Dividing each dn_i by the smallest dn_i among all species in (17.1) gives a set of dimensionless stoichiometric coefficients ν_i that can be substituted into (17.1) to yield

$$\nu_D D + \nu_E E + \cdots \rightleftharpoons \nu_A A + \nu_B B + \cdots \quad (17.3)$$

which is the general form of an equilibrium reaction. In this subsection, several types of equilibrium reactions are discussed.

17.2.1 Dissolution reactions

Dissolution occurs when a gas dissolves in a liquid. The gas–liquid interface may be at the surface of the ocean, of an aerosol particle, or of a cloud drop. Dissolution is often (but not always) a reversible process that can be described by reactions of the form

$$AB(g) \rightleftharpoons AB(aq) \quad (17.4)$$

where (g) indicates a gas, (aq) indicates dissolved in solution, and the stoichiometric coefficients are unity. In (17.4), the gas and dissolved (solution) phases of species AB are assumed to be in equilibrium with each other at the gas–solution interface; thus, the number of molecules of AB transferring from the gas to the solution equals the number of molecules transferring in the reverse direction. Examples of **dissolution reactions** include

$$\begin{aligned} H_2SO_4(g) &\rightleftharpoons H_2SO_4(aq) \\ HNO_3(g) &\rightleftharpoons HNO_3(aq) \\ HCl(g) &\rightleftharpoons HCl(aq) \\ CO_2(g) &\rightleftharpoons CO_2(aq) \\ NH_3(g) &\rightleftharpoons NH_3(aq) \end{aligned} \quad (17.5)$$

17.2.2 Dissociation reactions

Once in solution, dissolved gases often dissociate reversibly into ions. Substances that undergo partial or complete dissociation in solution are **electrolytes**. The degree of dissociation of an electrolyte depends on the acidity of solution, the strength of the electrolyte, and the concentrations of other ions in solution.

The **acidity** of a solution is a measure of the concentration of **hydrogen ions** (protons or H^+ ions) in solution. Acidity is measured in terms of **pH**, where

$$pH = -\log_{10}[H^+] \quad (17.6)$$

Here, $[H^+]$ is the **molarity** of H^+ (moles of H^+ per liter of solution). The pH scale, shown in Fig. 17.1 for a limited range, varies from less than 0 (lots of H^+ and very acidic) to greater than 14 (very little H^+ and very basic or alkaline). Neutral pH,

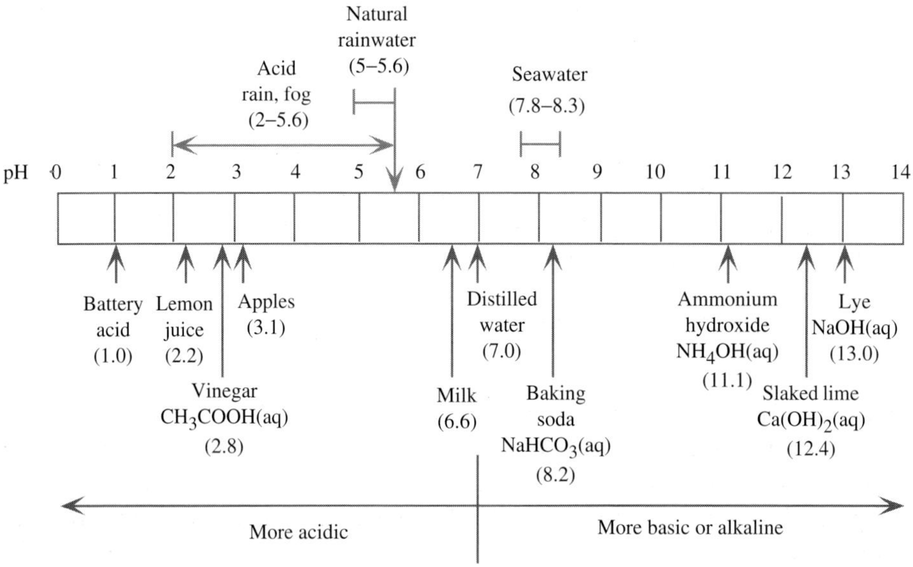

Figure 17.1 Diagram of the pH scale and the pH level of selected solutions.

the pH of distilled water, is 7.0. At this pH, the molarity of H^+ is 10^{-7} mol L^{-1}. A pH of 4 means that the molarity of H^+ is 10^{-4} mol L^{-1}. Thus, water at a pH of 4 is 1000 times more acidic (contains 1000 times more H^+ ions) than is water at a pH of 7.

The more acidic a solution, the higher the molarity of H^+, and the lower the pH. Hydrogen ions in solution are donated by acids that dissolve. In the atmosphere, such acids are primarily sulfuric acid, nitric acid, hydrochloric acid, and carbonic acid.

The ability of an acid to dissociate into protons and anions varies among acids. At low pH ($\approx$1), $H_2SO_4(aq)$, $HNO_3(aq)$, and $HCl(aq)$ dissociate readily, whereas $H_2CO_3(aq)$ does not. The former acids are **strong acids**, and the latter is a **weak acid.** Since all acids are electrolytes, a strong acid is a **strong electrolyte** (i.e., it dissociates significantly) and a weak acid is a **weak electrolyte**.

Sulfuric acid is a strong acid and electrolyte. At pH > -3, it dissociates to another strong acid, the **bisulfate ion**, which, at pH $> +2$, dissociates to the **sulfate ion**. These two reactions are, respectively,

$$\begin{aligned} H_2SO_4(aq) &\rightleftharpoons H^+ + HSO_4^- \\ HSO_4^- &\rightleftharpoons H^+ + SO_4^{2-} \end{aligned} \tag{17.7}$$

Nitric acid is also a strong acid and dissociates significantly to the **nitrate ion** above a pH of -1 by

$$HNO_3(aq) \rightleftharpoons H^+ + NO_3^- \tag{17.8}$$

Hydrochloric acid is an extremely strong acid because it almost always dissociates completely to the **chloride ion** above a pH of −6 by

$$HCl(aq) \rightleftharpoons H^+ + Cl^- \qquad (17.9)$$

Carbonic acid, which originates from **dissolved carbon dioxide,** is a weak acid. It dissociates to the **bicarbonate ion** above a pH of only +6. The bicarbonate ion further dissociates to the **carbonate ion** above a pH of +10. Carbonic acid and bicarbonate ion dissociation reactions are

$$CO_2(aq) + H_2O(aq) \rightleftharpoons H_2CO_3(aq) \rightleftharpoons H^+ + HCO_3^- \qquad (17.10)$$

$$HCO_3^- \rightleftharpoons H^+ + CO_3^{2-}$$

respectively. In the background air, the mixing ratio of $CO_2(g)$ is about 375 ppmv. A fraction of this $CO_2(g)$ always dissolves in rainwater. Thus, rainwater, even in the cleanest environment on Earth, is naturally acidic due to the presence of background carbonic acid in it. The **pH of rainwater affected by only carbonic acid is about 5.6,** indicating that its hydrogen-ion molarity is 25 times that of distilled water. Other natural acids may reduce the pH of natural rainwater slightly below 5.6. Anthropogenic acids may reduce the pH of rainwater to 2.5 (Fig. 17.1).

Whereas acids provide hydrogen ions, **bases** provide **hydroxide ions** (OH^-). Such ions react with hydrogen ions to form neutral water via

$$H_2O(aq) \rightleftharpoons H^+ + OH^- \qquad (17.11)$$

Thus, the addition of a base removes hydrogen ions, increasing the pH. An important basic substance in aerosol particles is ammonia. In solution, **dissolved ammonia** reacts with water to form the **ammonium ion** and the hydroxide ion by

$$NH_3(aq) + H_2O(aq) \rightleftharpoons NH_4^+ + OH^- \qquad (17.12)$$

Since some strong electrolytes, such as hydrochloric acid and nitric acid, dissociate nearly completely in atmospheric particles, the undissociated forms of these species are sometimes ignored. Instead, gas–ion equilibrium reactions, such as

$$HCl(g) \rightleftharpoons H^+ + Cl^- \qquad (17.13)$$

$$HNO_3(g) \rightleftharpoons H^+ + NO_3^- \qquad (17.14)$$

replace the combination of the dissolution and dissociation reactions.

17.2.3 Solid precipitation reactions

At high enough concentrations, ions may precipitate from solution to form **solid electrolytes** that are in equilibrium with the ions that formed them in solution. If ion concentrations decrease (e.g., if the liquid-water content of a solution increases, causing dilution), a solid electrolyte may also dissociate back to ions. These processes are represented by reversible solid–ion equilibrium reactions. Some such reactions for ammonium-containing electrolytes include

$$\begin{aligned} NH_4Cl(s) &\rightleftharpoons NH_4^+ + Cl^- \\ NH_4NO_3(s) &\rightleftharpoons NH_4^+ + NO_3^- \\ (NH_4)_2SO_4(s) &\rightleftharpoons 2NH_4^+ + SO_4^{2-} \end{aligned} \tag{17.15}$$

Others, for sodium-containing electrolytes, include

$$\begin{aligned} NaCl(s) &\rightleftharpoons Na^+ + Cl^- \\ NaNO_3(s) &\rightleftharpoons Na^+ + NO_3^- \\ (Na)_2SO_4(s) &\rightleftharpoons 2Na^+ + SO_4^{2-} \end{aligned} \tag{17.16}$$

17.2.4 Gas–solid heterogeneous reactions

Gas–solid heterogeneous reactions may also be represented by reversible reactions. For example, suppose nitric acid gas collides with and adsorbs to a particle surface, and an ammonia gas molecule collides with and reacts with the adsorbed nitric acid. This reaction may result in the formation of solid ammonium nitrate ($NH_4NO_3(s)$), which is adsorbed to the surface. The equilibrium reaction representing this process and a similar one for the formation of solid ammonium chloride ($NH_4Cl(s)$) is

$$\begin{aligned} NH_4NO_3(s) &\rightleftharpoons NH_3(g) + HNO_3(g) \\ NH_4Cl(s) &\rightleftharpoons NH_3(g) + HCl(g) \end{aligned} \tag{17.17}$$

In sum, equilibrium relationships can describe gas–liquid, gas–ion, gas–solid, liquid–liquid, liquid–ion, liquid–solid, ion–ion, and ion–solid reversible reactions. Appendix Table B.7 lists major equilibrium reactions of atmospheric interest.

17.3 EQUILIBRIUM RELATION AND COEFFICIENTS

The relative quantities of reactants and products in an equilibrium equation are determined from the equilibrium relation,

$$\prod_i \{a_i\}^{k_i \nu_i} = \frac{\{A\}^{\nu_A}\{B\}^{\nu_B}\ldots}{\{D\}^{\nu_D}\{E\}^{\nu_E}\ldots} = K_{eq}(T) \tag{17.18}$$

where $K_{eq}(T)$ is a temperature-dependent **equilibrium coefficient**, $\{a_i\}$ is the **thermodynamic activity** of species i, $\{A\}$, etc., are individual thermodynamic activities, $k_i = +1$ for products, and $k_i = -1$ for reactants. Below, thermodynamic activities and a derivation of the equilibrium relation are described.

17.3.1 Thermodynamic activities

An **activity** is a relative quantity of a substance and is determined differently for each phase state. The activity of a gas over a particle surface is its saturation vapor pressure (atm). Thus,

$$\{A(g)\} = p_{A,s} \tag{17.19}$$

The activity of an ion in solution or of an undissociated electrolyte is its **molality** $\mathbf{m}_A$ (m, moles of solute per kilogram of solvent) multiplied by its **activity coefficient** γ (dimensionless). Thus,

$$\{A^+\} = \mathbf{m}_{A^+}\gamma_{A^+} \qquad \{A(aq)\} = \mathbf{m}_A\gamma_A \tag{17.20}$$

respectively. A **solute activity coefficient** represents the deviation from ideal behavior of a solute in solution. It is a dimensionless parameter by which the molality of a species in solution is multiplied to give the species' thermodynamic activity. In an ideal, infinitely dilute solution, the solute activity coefficient is unity. In a nonideal, concentrated solution, solute activity coefficients may exceed or be less than unity. **Debye and Hückel** showed that, in sufficiently dilute solutions, where ions are far apart, the deviation of molality from thermodynamic activity is caused by Coulombic (electric) forces of attraction and repulsion. At high concentrations, ions are close together, and ion–ion interactions affect solute activity coefficients more than do Coulombic forces.

The activity of liquid water in a particle is the ambient relative humidity. Thus,

$$\{H_2O(aq)\} = a_w = \frac{p_v}{p_{v,s}} = f_r \tag{17.21}$$

where a_w denotes the activity of water, p_v is the partial pressure of water vapor, $p_{v,s}$ is the saturation vapor pressure of water over a bulk liquid surface, and f_r is the relative humidity, expressed as a fraction. For pure liquid water in equilibrium with water vapor, $p_v = p_{v,s}$, and $a_w = 1$, which is the water activity in a dilute solution. When liquid water contains solute, some of the solute replaces liquid water on the surface of the drop. Water molecules often bond to the solute (Section 17.8), requiring more water to condense from the vapor phase to maintain saturation over the solution, lowering the partial pressure of water vapor over the drop surface, decreasing the relative humidity over the surface, and decreasing the water activity.

Solids are not in solution, and their concentrations do not affect molalities or activity coefficients of solutes in solution. The activity of a pure solid is the mole fraction of the pure solid in itself. In other words,

$$\{A(s)\} = 1 \tag{17.22}$$

17.3.2 Derivation of the equilibrium relation

The equilibrium relation (17.18) is derived by minimizing the change in Gibbs free energy of a system. The Gibbs free energy (J) is defined as

$$G^* = H^* - TS^* = U^* + p_a V - TS^* \tag{17.23}$$

where $H^* = U^* + p_a V$ is the **enthalpy** (J), T is the absolute temperature (K), S^* is the **entropy** (J K^{-1}), U^* is the internal energy (J), p_a is the pressure, and V is the volume of a system. The change in Gibbs free energy is a measure of the maximum amount of useful work that may be obtained from a change in enthalpy or entropy in the system. It is calculated as

$$dG^* = d(H^* - TS^*) = dU^* + p_a\, dV + V\, dp_a - T\, dS^* - S^*\, dT \tag{17.24}$$

where $dS^* = dQ^*/T$. Thus, the change in entropy of a system equals the incremental energy added to or removed from the system divided by the absolute temperature. The internal energy was introduced in Chapter 2. For a system in which no changes in chemical composition occur, it is written as

$$dU^* = dQ^* - p_a\, dV = T\, dS^* - p_a\, dV \tag{17.25}$$

For a system in which reversible chemical reactions of the form shown in (17.1) occur, it is modified to

$$dU^* = T\, dS^* - p_a\, dV + \sum_i k_i (dn_i)\, \mu_i \tag{17.26}$$

where μ_i (J mol^{-1}) is the chemical potential of species i. The **chemical potential** is a measure of the intensity of a substance and is a function of temperature and pressure. Differences in chemical potential between two substances can lead to chemical reactions between them. Substituting (17.26) into (17.24) gives

$$dG^* = V\, dp_a - S^*\, dT + \sum_i k_i (dn_i)\, \mu_i \tag{17.27}$$

When temperature and pressure are held constant, (17.27) simplifies to

$$dG^* = \sum_i k_i\, (dn_i)\, \mu_i \tag{17.28}$$

From this equation, it can be seen that the chemical potential of a substance is its change in Gibbs free energy per unit change in its number of moles, or its partial molar free energy, when temperature and pressure are held constant. The chemical potential can also be defined in terms of a standard chemical potential and perturbation term. Thus,

$$\mu_i = \left(\frac{\partial G_i^*}{\partial n_i}\right)_{T,p_a} = \mu_i^o(T) + R^* T \ln\{a_i\} \tag{17.29}$$

where dG_i^* is the change in Gibbs free energy of an individual species at constant temperature and pressure, R^* is the universal gas constant (J mol^{-1} K^{-1}), and T is absolute temperature (K). In the second expression, μ_i^o (J mol^{-1}), which varies with temperature, is the standard chemical potential of a substance, and $R^* T \ln\{a_i\}$ accounts for the deviation in chemical potential from the standard state due to the activity of the substance.

Thermodynamic equilibrium occurs when $dG^* = 0$. Setting (17.28) equal to zero and dividing through by the smallest value of dn_i among all species in the reaction under consideration give

$$\sum_i k_i \nu_i \mu_i = 0 \tag{17.30}$$

where ν_i is the dimensionless stoichiometric coefficient. Substituting (17.29) into (17.30), assuming $T = T_0 = 298.15$ K for now, defining $\Delta_f G_i^o = \mu_i^o(T_0)$ as the **standard molal Gibbs free energy of formation** (J mol^{-1}) of substance i at $T = T_0$, and noting that $\sum_i k_i \nu_i \ln\{a_i\} = \ln \prod_i \{a_i\}^{k_i \nu_i}$ give

$$\sum_i k_i \nu_i \mu_i^o(T_0) + R^* T_0 \sum_i k_i \nu_i \ln\{a_i\} = \sum_i k_i \nu_i \Delta_f G_i^o + R^* T_0 \ln \prod_i \{a_i\}^{k_i \nu_i} = 0 \tag{17.31}$$

Rearranging (17.31) yields

$$\prod_i \{a_i\}^{k_i \nu_i} = \exp\left(-\frac{1}{R^* T_0} \sum_i k_i \nu_i \Delta_f G_i^o\right) \tag{17.32}$$

The right side of (17.32) is the **equilibrium coefficient** at $T = T_0$. Thus,

$$K_{eq}(T_0) = \exp\left(-\frac{1}{R^* T_0} \sum_i k_i \nu_i \Delta_f G_i^o\right) \tag{17.33}$$

which is independent of species concentration. Substituting (17.33) into (17.32) when $T = T_0$ gives the equilibrium-coefficient equation shown in (17.18). Values of $\Delta_f G_i^o$ are given in Appendix Table B.6 for some substances.

17.3.3 Temperature dependence of the equilibrium coefficient

The temperature dependence of the equilibrium coefficient can be estimated by solving the **van't Hoff equation,**

$$\frac{d \ln K_{eq}(T)}{dT} = \frac{1}{R^* T^2} \sum_i k_i \nu_i \Delta_f H_i \tag{17.34}$$

where $\Delta_f H_i$ is the **molal enthalpy of formation** (J mol^{-1}) of a substance. The van't Hoff equation is similar in form to the Arrhenius equation given in Chapter 10. $\Delta_f H_i$ can be approximated as a function of temperature with

$$\Delta_f H_i \approx \Delta_f H_i^o + c_{p,i}^o (T - T_0) \tag{17.35}$$

where $\Delta_f H_i^o$ is the **standard molal enthalpy of formation** (J mol^{-1}) (at $T_0 =$ 298.15 K) and $c_{p,i}^o$ is the **standard molal heat capacity** (standard specific heat) **at constant pressure** (J mol^{-1} K^{-1}). Although $c_{p,i}^o$ varies slightly with temperature, (17.35) assumes that it does not. Combining (17.34) and (17.35) and writing the result in integral form give

$$\int_{T_0}^{T} d \ln K_{eq}(T) = \int_{T_0}^{T} \frac{1}{R^* T^2} \sum_i k_i \nu_i \left[\Delta_f H_i^o + c_{p,i}^o (T - T_0)\right] dT \tag{17.36}$$

Integrating yields the temperature-dependent expression for the equilibrium coefficient,

$$K_{eq}(T) = K_{eq}(T_0)\exp\left\{-\sum_i k_i \nu_i\left[\frac{\Delta_f H_i^o}{R^* T_0}\left(\frac{T_0}{T}-1\right)+\frac{c_{p,i}^o}{R^*}\left(1-\frac{T_0}{T}+\ln\frac{T_0}{T}\right)\right]\right\} \quad (17.37)$$

where $K_{eq}(T_0)$ is the equilibrium coefficient at $T = T_0$ found in (17.33). Values of $\Delta_f H_i^o$ and $c_{p,i}^o$ are measured experimentally and given in Appendix Table B.6 for some substances. Appendix Table B.7 gives temperature-dependent equilibrium-coefficient expressions for some reactions of the form shown in (17.37) derived from the data in Table B.6.

Example 17.1

Determine the equilibrium-coefficient expression for $Na_2SO_4(s) \rightleftharpoons 2Na^+ + SO_4^{2-}$.

Applying data from Appendix Table B.6 gives

$$\sum_i k_i \nu_i \Delta_f G_i^o = 1820\,\frac{\text{J}}{\text{mol}} \qquad \sum_i k_i \nu_i \Delta_f H_i^o = -2430\,\frac{\text{J}}{\text{mol}}$$

$$\sum_i k_i \nu_i c_{p,i}^o = -328.4\,\frac{\text{J}}{\text{mol K}}$$

Substituting these expressions into (17.37) and (17.33) yields $K_{eq}(T) = 0.4799\exp\left\{0.9802\left(\frac{T_0}{T}-1\right)+39.497\left[1+\ln\left(\frac{T_0}{T}\right)-\frac{T_0}{T}\right]\right\}\frac{\text{mol}^2}{\text{kg}^2}$

17.4 FORMS OF EQUILIBRIUM-COEFFICIENT EQUATIONS

Equation (17.18) expresses an equilibrium reaction in terms of thermodynamic activities and an equilibrium coefficient. In this section, activity expressions are substituted for activities to give equilibrium-coefficient equations for several types of reactions.

17.4.1 Dissolution reactions

For a dissolution reaction, such as $HNO_3(g) \rightleftharpoons HNO_3(aq)$, the equilibrium coefficient relates the pressure exerted by a gas at the gas–liquid interface to the molality of the dissolved gas in solution. The equilibrium coefficient expression for HNO_3 dissolution is

$$\frac{\{HNO_3(aq)\}}{\{HNO_3(g)\}} = \frac{\mathbf{m}_{HNO_3(aq)}\gamma_{HNO_3(aq)}}{p_{HNO_3(g),s}} = K_{eq}(T)\frac{\text{mol}}{\text{kg atm}} \quad (17.38)$$

where $p_{HNO_3(g),s}$ is the saturation vapor pressure of nitric acid (atm), $\mathbf{m}_{HNO_3(aq)}$ is the molality of nitric acid in solution (mol kg^{-1}), and $\gamma_{HNO_3(aq)}$ is the solute activity coefficient of dissolved, undissociated nitric acid (dimensionless). In Chapter 2, a saturation vapor pressure was defined as the partial pressure of a gas when the gas is in equilibrium with a particle surface at a given temperature. Since $HNO_3(g)$ is in equilibrium with the particle surface in this case, its saturation vapor pressure, not its partial pressure, is used in (17.38).

Over a dilute solution, the partial pressure exerted by a gas is proportional to the molality of the dissolved gas in solution. This is **Henry's law.** For a dilute solution, $\gamma_{HNO_3(aq)} = 1$, and (17.38) obeys Henry's law. In a dilute or concentrated solution, $K_{eq}(T)$ in (17.38) is called a **Henry's constant.** Henry's constants, like other equilibrium coefficients, are temperature dependent. Henry's constants for gases dissolving in liquid water are given in Appendix Table B.7. Species with moderate or large Henry's constants include NH_3, HNO_3, HCl, H_2SO_4, H_2O_2, and SO_2. For these and many other gases $\sum_i k_i \nu_i \Delta_f H_i^o < 0$. In such cases, (17.37) predicts that solubilities and Henry's constants increase with decreasing temperature.

17.4.2 Dissociation and other reactions

In solution, many substances dissociate. The dissociation reaction for nitric acid has the form $HNO_3(aq) \rightleftharpoons H^+ + NO_3^-$. The equilibrium coefficient expression for this reaction is

$$\frac{\{H^+\}\{NO_3^-\}}{\{HNO_3(aq)\}} = \frac{\mathbf{m}_{H^+}\gamma_{H^+}\mathbf{m}_{NO_3^-}\gamma_{NO_3^-}}{\mathbf{m}_{HNO_3(aq)}\gamma_{HNO_3(aq)}} = \frac{\mathbf{m}_{H^+}\mathbf{m}_{NO_3^-}\gamma^2_{H^+,NO_3^-}}{\mathbf{m}_{HNO_3(aq)}\gamma_{HNO_3(aq)}} = K_{eq}(T)\frac{\text{mol}}{\text{kg}} \tag{17.39}$$

The solute activity coefficients in (17.39) are mixed activity coefficients because they are determined by considering a mixture of all dissociated and undissociated electrolytes in solution. The parameters γ_{H^+} and $\gamma_{NO_3^-}$ are single-ion mixed solute activity coefficients, and γ_{H^+,NO_3^-} is a **mean** (geometric-mean) **mixed solute activity coefficient.** When $HNO_3(aq)$, H^+, and NO_3^- are alone in solution, γ_{H^+} and $\gamma_{NO_3^-}$ are single-ion binary solute activity coefficients, and γ_{H^+,NO_3^-} is a **mean** (geometric-mean) **binary solute activity coefficient.** Activity coefficients for single ions are difficult to measure, because single ions cannot be isolated from a solution. Single-ion solute activity coefficients are generally estimated mathematically. Mean binary solute activity coefficients are measured in the laboratory. Mean mixed solute activity coefficients are estimated from mean binary solute activity coefficient data with a mixing rule.

A geometric mean solute activity coefficient, $\gamma_\pm$, is related to single-ion solute activity coefficients by

$$\gamma_\pm = \left(\gamma_+^{\nu_+}\gamma_-^{\nu_-}\right)^{1/(\nu_+ + \nu_-)} \tag{17.40}$$

where γ_+ and γ_- are the activity coefficients of a cation and anion, respectively, and ν_+ and ν_- are the stoichiometric coefficients of the cation and anion, respectively.

When $\nu_+ = 1$ and $\nu_- = 1$, the electrolyte is **univalent.** When $\nu_+ > 1$ or $\nu_- > 1$, the electrolyte is **multivalent.** When $\nu_+ = \nu_-$, the electrolyte is **symmetric**; otherwise, it is **nonsymmetric.** In all cases, a dissociation reaction must satisfy the charge balance requirement $z_+\nu_+ + z_-\nu_- = 0$, where z_+ is the positive charge on the cation and z_- is the negative charge on the anion. Rearranging (17.40) gives

$$\gamma_\pm^{\nu_+ + \nu_-} = \gamma_+^{\nu_+}\gamma_-^{\nu_-} \tag{17.41}$$

which is the form of the mean mixed solute activity coefficient used in (17.39).

Example 17.2

$HNO_3(aq)$ dissociates by the reaction $HNO_3(aq) \rightleftharpoons H^+ + NO_3^-$. Since $\nu_+ = 1$ and $\nu_- = 1$, $HNO_3(aq)$ is a univalent, symmetric electrolyte. Since $z_+ = +1$ and $z_- = -1$, nitric acid dissociation satisfies $z_+\nu_+ + z_-\nu_- = 0$. $Na_2SO_4(s)$ dissociates by the reaction, $Na_2SO_4(s) \rightleftharpoons 2Na^+ + SO_4^{2-}$. Since $\nu_+ = 2$ and $\nu_- = 1$, $Na_2SO_4(s)$ is a multivalent, nonsymmetric electrolyte. Since $z_+ = +1$ and $z_- = -2$, sodium sulfate dissociation satisfies $z_+\nu_+ + z_-\nu_- = 0$.

Below, equilibrium coefficient expressions for several additional reactions are given. In the case of the solid-forming reaction (17.42), the equilibrium coefficient is called a **solubility product.**

$$\begin{aligned} Na_2SO_4(s) &\rightleftharpoons 2Na^+ + SO_4^{2-} \\ \frac{\{Na^+\}^2\{SO_4^{2-}\}}{\{Na_2SO_4(s)\}} &= \frac{\mathbf{m}_{Na^+}^2\gamma_{Na^+}^2\mathbf{m}_{SO_4^{2-}}\gamma_{SO_4^{2-}}}{1.0} \\ &= \mathbf{m}_{Na^+}^2\mathbf{m}_{SO_4^{2-}}\gamma_{2Na^+,SO_4^{2-}}^3 = K_{eq}(T)\frac{\text{mol}^3}{\text{kg}^3} \end{aligned} \tag{17.42}$$

$$\begin{aligned} HSO_4^- &\rightleftharpoons H^+ + SO_4^{2-} \\ \frac{\{H^+\}^2\{SO_4^{2-}\}}{\{H^+\}\{HSO_4^-\}} &= \frac{\mathbf{m}_{H^+}^2\gamma_{H^+}^2\mathbf{m}_{SO_4^{2-}}\gamma_{SO_4^{2-}}}{\mathbf{m}_{H^+}\gamma_{H^+}\mathbf{m}_{HSO_4^-}\gamma_{HSO_4^-}} \\ &= \frac{\mathbf{m}_{H^+}\mathbf{m}_{SO_4^{2-}}\gamma_{2H^+,SO_4^{2-}}^3}{\mathbf{m}_{HSO_4^-}\gamma_{H^+,HSO_4^-}^2} = K_{eq}(T)\frac{\text{mol}}{\text{kg}} \end{aligned} \tag{17.43}$$

$$\begin{aligned} NH_3(g) + HNO_3(g) &\rightleftharpoons NH_4^+ + NO_3^- \\ \frac{\{NH_4^+\}\{NO_3^-\}}{\{NH_3(g)\}\{HNO_3(g)\}} &= \frac{\mathbf{m}_{NH_4^+}\gamma_{NH_4^+}\mathbf{m}_{NO_3^-}\gamma_{NO_3^-}}{p_{NH_3(g),s}\,p_{HNO_3(g),s}} \\ &= \frac{\mathbf{m}_{NH_4^+}\mathbf{m}_{NO_3^-}\gamma_{NH_4^+,NO_3^-}^2}{p_{NH_3(g),s}\,p_{HNO_3(g),s}} = K_{eq}(T)\frac{\text{mol}^2}{\text{kg}^2\,\text{atm}^2} \end{aligned} \tag{17.44}$$

$$\begin{aligned} NH_3(aq) + H_2O(aq) &\rightleftharpoons NH_4^+ + OH^- && (17.45) \\ \frac{\{NH_4^+\}\{OH^-\}}{\{NH_3\,(aq)\}\{H_2O\,(aq)\}} &= \frac{\mathbf{m}_{NH_4^+}\gamma_{NH_4^+}\mathbf{m}_{OH^-}\gamma_{OH^-}}{\mathbf{m}_{NH_3(aq)}\gamma_{NH_3(aq)}\,f_r} \\ &= \frac{\mathbf{m}_{NH_4^+}\mathbf{m}_{OH^-}\gamma^2_{NH_4^+,OH^-}}{\mathbf{m}_{NH_3(aq)}\gamma_{NH_3(aq)}\,f_r} = K_{eq}(T)\frac{\text{mol}}{\text{kg}} \end{aligned}$$

17.5 MEAN BINARY SOLUTE ACTIVITY COEFFICIENTS

Mean binary solute activity coefficients, which are used in mixing rules to determine mean mixed solute activity coefficients, can be determined from measurement or theory. Measurements of mean binary solute activity coefficients for several species have been performed at 298.15 K. Theoretical parameterizations of solute activity coefficients and their temperature dependences have also been developed.

One parameterization is **Pitzer's method** (Pitzer and Mayorga 1973; Pitzer 1991), which estimates the mean binary solute activity coefficient of an electrolyte at 298.15 K with

$$\ln \gamma^0_{12b} = Z_1 Z_2 f^\gamma + \mathbf{m}_{12}\frac{2\nu_1\nu_2}{\nu_1+\nu_2}B^\gamma_{12} + \mathbf{m}^2_{12}\frac{2(\nu_1\nu_2)^{3/2}}{\nu_1+\nu_2}C^\gamma_{12} \qquad (17.46)$$

where γ^0_{12b} is the mean binary solute activity coefficient of electrolyte 1–2 (cation 1 plus anion 2) at reference temperature 298.15 K, Z_1 and Z_2 are the absolute values of the charges of cation 1 and anion 2, respectively; $\mathbf{m}_{12}$ is the molality of the electrolyte dissolved in solution, and ν_1 and ν_2 are the stoichiometric coefficients of the dissociated ions. The remaining quantities in (17.46) are

$$f^\gamma = -0.392\left[\frac{\mathbf{I}^{1/2}}{1+1.2\mathbf{I}^{1/2}} + \frac{2}{1.2}\ln(1+1.2\mathbf{I}^{1/2})\right] \qquad (17.47)$$

$$B^\gamma_{12} = 2\beta^{(1)}_{12} + \frac{2\beta^{(2)}_{12}}{4\mathbf{I}}\left[1 - e^{-2\mathbf{I}^{1/2}}\left(1 + 2\mathbf{I}^{1/2} - 2\mathbf{I}\right)\right] \qquad (17.48)$$

where **I** is the solution's ionic strength (mol kg^{-1}). **Ionic strength** is a measure of interionic effects resulting from attraction and repulsion among ions. It is calculated from

$$\mathbf{I} = \frac{1}{2}\left(\sum_{i=1}^{N_C}\mathbf{m}_{2i-1}Z^2_{2i-1} + \sum_{i=1}^{N_A}\mathbf{m}_{2i}Z^2_{2i}\right) \qquad (17.49)$$

where N_C and N_A are the numbers of cations and anions, respectively, in solution. Odd-numbered subscripts refer to cations, and even-numbered subscripts refer to anions. In the case of one electrolyte, such as HCl(aq) alone in solution, $N_C = 1$ and $N_A = 1$. The quantities, $\beta^{(1)}_{12}$, $\beta^{(2)}_{12}$, and C^γ_{12} are empirical parameters derived from measurements. **Pitzer parameters** for three electrolytes are shown in Table 17.1.

Table 17.1 Pitzer parameters for three electrolytes

Electrolyte	$\beta_{12}^{(1)}$	$\beta_{12}^{(2)}$	C_{12}^{γ}
HCl	0.17750	0.2945	0.0012
HNO_3	0.1119	0.3206	0.0015
NH_4NO_3	−0.0154	0.112	−0.000045

Source: Pilinis and Seinfeld (1987).

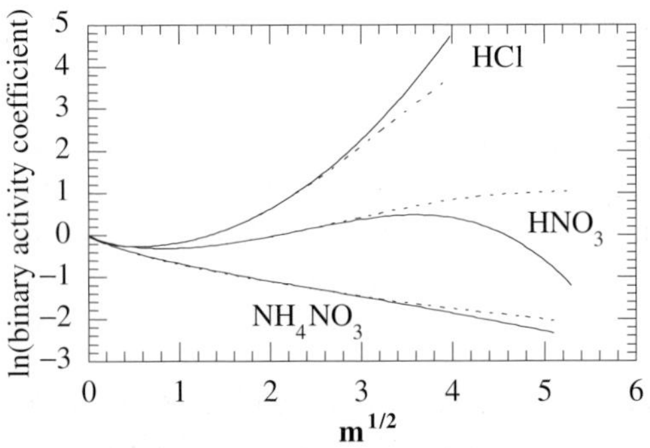

Figure 17.2 Comparison of mean binary solute activity coefficients measured by Hamer and Wu (1972), - - -, with those computed from Pitzer's method, —, as a function of the square root of molality.

Pitzer's method predicts mean binary solute activity coefficients at 298.15 K from physical principles. Its limitation is that the coefficients are typically accurate only up to about 6 m. Figure 17.2 shows a comparison of mean binary solute activity coefficients predicted by Pitzer's method to those measured by Hamer and Wu (1972). The measured data are accurate to higher molalities.

Figure 17.2 provides insight into why the chloride ion is forced out of aerosol-particle solutions in the presence of nitric acid. When sea-spray particles containing the chloride ion encounter nitric acid gas, the nitric acid enters the particles and dissociates to the nitrate ion, displacing the chloride to the gas phase. This process can remove nearly all chloride from sea-spray particles over the ocean, as discussed in Section 14.1.1. Upon initial observation of the effective Henry's constants of $HCl(g) \rightleftharpoons H^+ + Cl^-$ and $HNO_3(g) \rightleftharpoons H^+ + NO_3^-$, it is not obvious that the nitrate ion should force the chloride ion out of solution. The effective Henry's-constant expressions for the two reactions at 298.15 K, obtained from Appendix Table B.7, are

$$\frac{m_{H^+} m_{Cl^-} \gamma_{H^+,Cl^-}^2}{p_{HCl(g),s}} = 1.97 \times 10^6 \qquad \frac{m_{H^+} m_{NO_3^-} \gamma_{H^+,NO_3^-}^2}{p_{HNO_3(g),s}} = 2.51 \times 10^6 \quad (17.50)$$

The equilibrium coefficients are comparable for the two reactions, but the binary solute activity coefficient data in Fig. 17.2 together with (17.50) indicate that, at the same molality, $\mathbf{m}_{H^+}\mathbf{m}_{Cl^-}/p_{HCl(g)}$ is much smaller than $\mathbf{m}_{H^+}\mathbf{m}_{NO_3^-}/p_{HNO_3(g)}$. The large binary solute activity coefficient of HCl decreases HCl's solubility, causing it to evaporate. Thus, if NO_3^- and Cl^- are initially present in solution at the same time, the nitrate ion is more likely to stay in solution than is the chloride ion.

Whether mean binary solute activity coefficients at 298.15 K are determined from measurements or theory, they can be parameterized with a polynomial fit of the form

$$\ln \gamma_{12b}^0 = B_0 + B_1\mathbf{m}_{12}^{1/2} + B_2\mathbf{m}_{12} + B_3\mathbf{m}_{12}^{3/2} + \cdots \tag{17.51}$$

where B_0, B_1, . . . are fitting coefficients. Polynomial coefficients for several electrolytes are given in Appendix Table B.9. Polynomial fits are used to simplify and speed up the use of binary solute activity coefficient calculations in computer programs.

17.6 TEMPERATURE DEPENDENCE OF BINARY SOLUTE ACTIVITY COEFFICIENTS

The temperature dependence of mean binary solute activity coefficients can be derived from thermodynamic principles. An expression for the binary solute activity coefficient of electrolyte 1–2 at temperature T is (Harned and Owen 1958)

$$\ln \gamma_{12b}(T) = \ln \gamma_{12b}^0 + \frac{T_L}{(\nu_1 + \nu_2)R^* T_0}\left(\phi_L + \mathbf{m}\frac{\partial \phi_L}{\partial \mathbf{m}}\right) + \frac{T_C}{(\nu_1 + \nu_2)\,R^*}\left(\phi_{c_p} + \mathbf{m}\frac{\partial \phi_{c_p}}{\partial \mathbf{m}} - \phi_{c_p}^{\text{o}}\right) \tag{17.52}$$

where T_0 is the reference temperature (298.15 K), R^* is the universal gas constant (J mol^{-1} K^{-1}), ϕ_L is the relative apparent molal enthalpy (J mol^{-1}) of the electrolyte at molality $\mathbf{m}$ (with subscript 12 omitted), ϕ_{c_p} is the apparent molal heat capacity (J mol^{-1} K^{-1}) at molality $\mathbf{m}$, $\phi_{c_p}^{\text{o}}$ is the apparent molal heat capacity at infinite dilution, and

$$T_L = \frac{T_0}{T} - 1 \qquad T_C = 1 + \ln\left(\frac{T_0}{T}\right) - \frac{T_0}{T} \tag{17.53}$$

The relative apparent molal enthalpy equals the negative of the heat of dilution. Appendix Table B.9 lists several sources for heat of dilution and apparent molal heat capacity experimental data. With these data, polynomials of the form

$$\begin{aligned}\phi_L &= U_1\mathbf{m}^{1/2} + U_2\mathbf{m} + U_3\mathbf{m}^{3/2} + \cdots \\ \phi_{c_p} &= \phi_{c_p}^0 + V_1\mathbf{m}^{1/2} + V_2\mathbf{m} + V_3\mathbf{m}^{3/2} + \cdots\end{aligned} \tag{17.54}$$

can be constructed, where $U_{1,\ldots}$ and $V_{1,\ldots}$ are polynomial coefficients. The relative apparent molal enthalpy is defined as $\phi_L = \phi_H - \phi_H^0$, where ϕ_H is the apparent

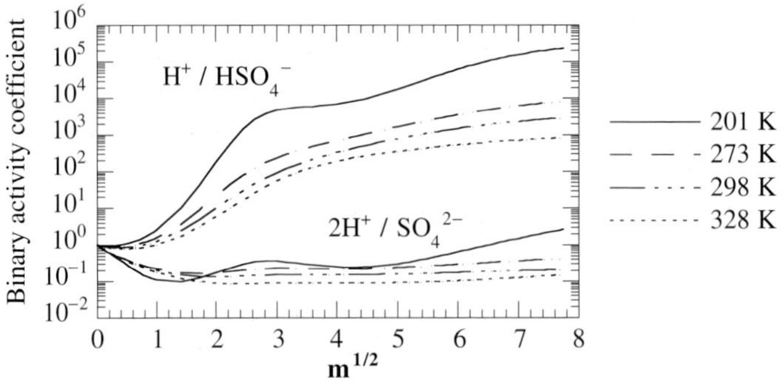

Figure 17.3 Binary solute activity coefficients of sulfate and bisulfate, alone in solution as a function of square root of molality (Jacobson *et al.* 1996b). The curves were obtained by combining equations from Clegg and Brimblecombe (1995) with Equations (72) and (73) of Stelson *et al.* (1984) in a Newton–Raphson iteration. The results are valid for 0–40 m total H_2SO_4.

molal enthalpy and ϕ_H^0 is the apparent molal enthalpy at infinite dilution, which occurs when $\mathbf{m} = 0$. At $\mathbf{m} = 0$, $\phi_H = \phi_H^0$ and $\phi_L = 0$.

Equations (17.51)–(17.54) can be combined to give temperature-dependent mean-binary-solute-activity-coefficient polynomials of the form

$$\ln \gamma_{12\mathrm{b}}(T) = F_0 + F_1\mathbf{m}^{1/2} + F_2\mathbf{m} + F_3\mathbf{m}^{3/2} + \cdots \tag{17.55}$$

where

$$F_0 = B_0 \qquad F_j = B_j + G_j T_L + H_j T_C \quad \text{for } j = 1, \ldots \tag{17.56}$$

$$G_j = \frac{0.5(j+2)U_j}{(\nu_1 + \nu_2)R^* T_0} \qquad H_j = \frac{0.5(j+2)V_j}{(\nu_1 + \nu_2)R^*} \tag{17.57}$$

Appendix Table B.9 lists *B*, *G*, and *H* for 10 electrolytes and maximum molalities associated with the fits for calculating temperature- and molality-dependent mean binary solute activity coefficients from (17.55)–(17.57). Since polynomial fits wander when the molality increases much beyond the limit of the given data, activity coefficients from beyond the last valid molality should be used cautiously when modeling.

Determining temperature-dependent binary solute activity coefficients of bisulfate and sulfate is more difficult. Figure 17.3 shows calculated solute activity coefficients for these electrolytes as a function of temperature and molality and discusses how they were determined. Such coefficients can be written in polynomial or tabular form.

17.7 MEAN MIXED SOLUTE ACTIVITY COEFFICIENTS

In aerosol particles, multiple electrolytes, such as dissolved sulfuric acid, nitric acid, hydrochloric acid, ammonia, and sodium chloride, often coexist in solution. In such cases, mean mixed solute activity coefficients are needed. They can be approximated with an empirical mixing rule that is a function of mean binary solute activity coefficients and accounts for interactions among ions. Two common mixing rules are **Bromley's method** (Bromley 1973) and the **Kusik–Meissner method** (Kusik and Meissner 1978). Bromley's method, discussed here, gives the mean activity coefficient of electrolyte 1–2 in a mixture as

$$\log_{10} \gamma_{12m}(T) = -A_\gamma \frac{Z_1 Z_2 \mathbf{I}_m^{1/2}}{1 + \mathbf{I}_m^{1/2}} + \frac{Z_1 Z_2}{Z_1 + Z_2}\left(\frac{W_1}{Z_1} + \frac{W_2}{Z_2}\right) \tag{17.58}$$

where A_γ is the **Debye–Hückel parameter** (0.392 at 298 K), Z_1 and Z_2 are the absolute-value charges of cation 1 and anion 2, respectively, $\mathbf{I}_m$ is the total ionic strength of the mixture, and

$$W_1 = Y_{21}\left(\log_{10} \gamma_{12b}(T) + A_\gamma \frac{Z_1 Z_2 \mathbf{I}_m^{1/2}}{1 + \mathbf{I}_m^{1/2}}\right) + Y_{41}\left(\log_{10} \gamma_{14b}(T) + A_\gamma \frac{Z_1 Z_4 \mathbf{I}_m^{1/2}}{1 + \mathbf{I}_m^{1/2}}\right) + \cdots \tag{17.59}$$

$$W_2 = X_{12}\left(\log_{10} \gamma_{12b}(T) + A_\gamma \frac{Z_1 Z_2 \mathbf{I}_m^{1/2}}{1 + \mathbf{I}_m^{1/2}}\right) + X_{32}\left(\log_{10} \gamma_{32b}(T) + A_\gamma \frac{Z_3 Z_2 \mathbf{I}_m^{1/2}}{1 + \mathbf{I}_m^{1/2}}\right) + \cdots \tag{17.60}$$

are functions of all electrolytes in solution. In the last equations,

$$Y_{21} = \left(\frac{Z_1 + Z_2}{2}\right)^2 \frac{\mathbf{m}_{2,m}}{\mathbf{I}_m} \qquad X_{12} = \left(\frac{Z_1 + Z_2}{2}\right)^2 \frac{\mathbf{m}_{1,m}}{\mathbf{I}_m} \tag{17.61}$$

Similar expressions can be written for X_{32}, X_{52}, . . . , Y_{41}, Y_{61}, . . . In these equations, $\gamma_{12b}(T)$, $\gamma_{14b}(T)$, $\gamma_{32b}(T)$, ... are temperature-dependent mean binary solute activity coefficients, odd-numbered subscripts refer to cations, and even-numbered subscripts refer to anions. Thus, $\mathbf{m}_{1,m}$ and $\mathbf{m}_{2,m}$ are the molalities of a cation and anion, respectively, in the mixture.

When a mean binary solute activity coefficient is used in (17.59) and (17.60), it is determined at a binary electrolyte molality $\mathbf{m}_{12,b}$ that results in the current ionic strength of the mixture, $\mathbf{I}_m$. The molality of binary electrolyte 1–2 is found from

$$\mathbf{I}_m = \frac{1}{2}\left(\mathbf{m}_{1,b} Z_1^2 + \mathbf{m}_{2,b} Z_2^2\right) = \frac{1}{2}\left(\nu_+ \mathbf{m}_{12,b} Z_1^2 + \nu_- \mathbf{m}_{12,b} Z_2^2\right) \tag{17.62}$$

where $\mathbf{m}_{1,b} = \nu_+ \mathbf{m}_{12,b}$ and $\mathbf{m}_{2,b} = \nu_- \mathbf{m}_{12,b}$ are the molalities of a cation and an anion, alone in solution, that results in the total ionic strength of the mixture. Solving (17.62) for $\mathbf{m}_{12,b}$ gives

$$\mathbf{m}_{12,b} = \frac{2\mathbf{I}_m}{\nu_+ Z_1^2 + \nu_- Z_2^2} \tag{17.63}$$

The molality of the binary electrolyte in (17.63) is used to determine $\gamma_{12b}(T)$ from the polynomial fit given in (17.55).

17.8 THE WATER EQUATION

Aerosol particles containing solute often absorb liquid water at relative humidities less than 100 percent, causing the particles to swell. Solutes attracting water include inorganic and organic ions, undissociated electrolytes, and nonelectrolytes. Some inorganic electrolytes that attract water include sulfuric acid, nitric acid, ammonium sulfate, and sodium chloride. Some organic compounds that attract water include succinic acid, glutaric acid, oxalic acid, and malonic acid.

The bonding of a solute to a solvent, such as liquid water, in solution is called **solvation**. When the solvent is liquid water, the bonding is called **hydration**. The solute may be a cation, anion, undissociated electrolyte, or a nonelectrolyte (e.g., sucrose). During hydration of an anion, the water molecule attaches to the anion end of the ion dipole via hydrogen bonding. Several water molecules can hydrate to each ion. During hydration of a cation, the lone pair of electrons on the oxygen atom of a water molecule bonds to the cation end of the ion dipole.

When the relative humidity is less than 100 percent in the atmosphere, all aerosol liquid water content is due to hydration. Thus, for example, if a sea-spray drop emitted from the ocean surface did not contain sodium, chloride, and other ions that hydrate, the drop would evaporate entirely when it entered sub-100-percent relative humidity air. The quantity of water taken up by hydration depends on the molality of the solute and the relative humidity. The higher the molality and the higher the relative humidity, the greater the water uptake. Conversely, when the relative humidity decreases, the liquid water content decreases as well. Thus, when a sea-spray drop travels from near the ocean surface, where the relative humidity is near 100 percent, to the boundary layer, where it may be 80 percent, the liquid water content of the drop decreases. For some solutes, such as sulfuric acid, hydration can occur down to relative humidities near 0 percent. For other solutes, such as sodium chloride, water uptake stops when the relative humidity drops below a certain relative humidity. In all cases, as the relative humidity approaches 100 percent, the rate of water uptake increases superlinearly with increasing relative humidity.

When a gas, such as sulfuric acid, condenses onto a dry particle surface, the saturation vapor pressure of water over the liquid sulfuric acid is so low that water vapor condenses immediately. Once condensed, the liquid water hydrates with the sulfuric acid, creating a sulfuric acid–water solution. In the solution, other gases may dissolve and absorb more water from the surrounding air, causing the particle to swell further. The reduction of the saturation vapor pressure of water over a particle surface due to the presence of a solute is Raoult's law (the **solute effect**), discussed in Section 16.2.3.2. The solute effect, through hydration, allows particles to absorb liquid water when the relative humidity is less than 100 percent.

Above 100 percent relative humidity, water vapor cascades onto particles by condensation. In such cases, the volume of water added to particles by hydration is

Table 17.2 Experimental test of ZSR prediction accuracy for a mixture of sucrose (species a) and mannitol (species b) at two water activities

Case	$\mathbf{m}_{x,\mathrm{a}}$	$\mathbf{m}_{y,\mathrm{a}}$	$\mathbf{m}_{x,\mathrm{m}}$	$\mathbf{m}_{y,\mathrm{m}}$	$\frac{\mathbf{m}_{x,\mathrm{m}}}{\mathbf{m}_{x,\mathrm{a}}}+\frac{\mathbf{m}_{y,\mathrm{m}}}{\mathbf{m}_{y,\mathrm{a}}}$
1	0.7751	0.8197	0.6227	0.1604	0.9990
2	0.9393	1.0046	0.1900	0.8014	1.0000

Source: Stokes and Robinson (1966).

small compared with that added by condensation. Thus, hydration does not affect water content much when the relative humidity exceeds 100 percent.

A method of estimating aerosol liquid-water content as a function of electrolyte molality and sub-100-percent relative humidity is the Zdanovskii–Stokes–Robinson (**ZSR**) equation (Stokes and Robinson 1966; Clegg *et al.* 2003; Clegg and Seinfeld 2004). The equation can be applied to electrolytes or nonelectrolytes. The simplest form of the equation is

$$\frac{\mathbf{m}_{x,\mathrm{m}}}{\mathbf{m}_{x,\mathrm{a}}}+\frac{\mathbf{m}_{y,\mathrm{m}}}{\mathbf{m}_{y,\mathrm{a}}}=1 \tag{17.64}$$

where $\mathbf{m}_{x,\mathrm{a}}$ and $\mathbf{m}_{y,\mathrm{a}}$ are the molalities of solutes x and y each alone in solution at a given water activity, and $\mathbf{m}_{x,\mathrm{m}}$ and $\mathbf{m}_{y,\mathrm{m}}$ are the molalities of x and y mixed together in solution at the same water activity. The **water activity** for atmospheric particles was defined in (17.21) as the relative humidity. Table 17.2 shows experimental data validating (17.64).

Equation (17.64) can be generalized for a mixture of any number of solutes with

$$\sum_k \frac{\mathbf{m}_{k,\mathrm{m}}}{\mathbf{m}_{k,\mathrm{a}}}=1 \tag{17.65}$$

where $\mathbf{m}_{k,\mathrm{m}}$ is the molality of solute k in a solution containing all solutes at the ambient relative humidity, and $\mathbf{m}_{k,\mathrm{a}}$ is the molality of solute k alone in solution at the ambient relative humidity.

Experimental data for water activity as a function of electrolyte molality are available (e.g., Robinson and Stokes 1955; Pitzer and Mayorga 1973; Cohen *et al.* 1987a,b; Tang and Munkelwitz 1994; Tang 1997; Peng *et al.* 2001). Such data can be fitted to polynomial expressions of the form

$$\mathbf{m}_{k,\mathrm{a}}=Y_{0,k}+Y_{1,k}a_{\mathrm{w}}+Y_{2,k}a_{\mathrm{w}}^2+Y_{3,k}a_{\mathrm{w}}^3+\cdots \tag{17.66}$$

where a_{w} is the water activity (relative humidity expressed as a fraction), and the Y's are polynomial coefficients, listed in Appendix Table B.10 for several electrolytes.

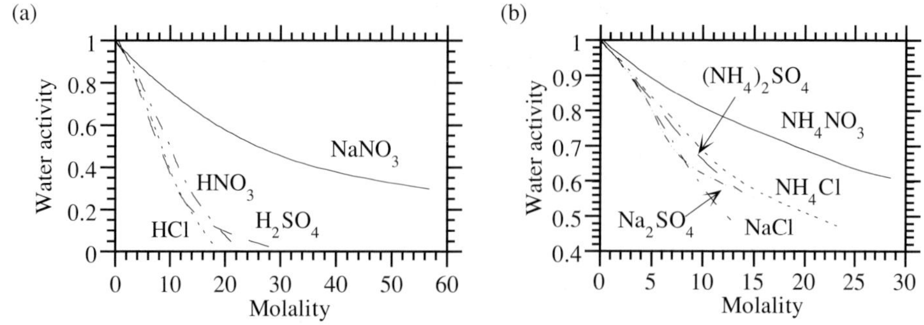

Figure 17.4 Water activity as a function of molality for several electrolytes at 298.15 K. The curves were obtained from (17.66) and the coefficients in Appendix Table B.10.

Figures 17.4 (a) and (b) show molality versus water activity for some of these electrolytes.

In comparison with the temperature dependence of binary solute activity coefficients, the temperature dependence of the water activity coefficients is small under lower tropospheric conditions. This temperature dependence of the water activity is

$$\ln a_{\rm w}(T) = \ln a_{\rm w}^0 - \frac{m_{\rm v}\mathbf{m}_{k,{\rm a}}^2}{R^*}\left(\frac{T_L}{T_0}\frac{\partial \phi_L}{\partial \mathbf{m}_{k,{\rm a}}} + T_{\rm C}\frac{\partial \phi_{c_{\rm P}}}{\partial \mathbf{m}_{k,{\rm a}}}\right) \tag{17.67}$$

(e.g., Harned and Owen 1958; Jacobson *et al.* 1996b) where $m_{\rm v}$ is the molecular weight of water (0.01802 kg mol^{-1} here) and the remaining parameters were defined in (17.52). If the water activity at the reference temperature is extracted from (17.66) and expressed as

$$\ln a_{\rm w}^0 = A_0 + A_1\mathbf{m}_{k,{\rm a}}^{1/2} + A_2\mathbf{m}_{k,{\rm a}} + A_3\mathbf{m}_{k,{\rm a}}^{3/2} + \cdots \tag{17.68}$$

then (17.67), (17.68), and (17.54) can be combined to form the polynomial

$$\ln a_{\rm w}(T) = A_0 + A_1\mathbf{m}_{k,{\rm a}}^{1/2} + A_2\mathbf{m}_{k,{\rm a}} + E_3\mathbf{m}_{k,{\rm a}}^{3/2} + E_4\mathbf{m}_{k,{\rm a}}^2 + \cdots \tag{17.69}$$

where

$$E_l = A_l - \frac{0.5(l-2)m_{\rm v}}{R^*}\left(\frac{T_L}{T_0}U_{l-2} + T_{\rm C}V_{l-2}\right) \qquad \text{for } l = 3, \ldots \tag{17.70}$$

and the U's and V's are polynomial coefficients from (17.54). Since the A-terms in (17.69) are not affected by temperature, temperature affects the water activity polynomial starting with the fourth term, E_3. In (17.55) temperature affected the solute activity polynomial starting with the second term. Thus, the temperature dependence of solute activity coefficients is greater than that of water activity coefficients (Jacobson *et al.* 1996b).

Table 17.3 Three sets of hypothetical electrolyte concentrations that may result when 6 μmol m^{-3} of H^+, 6 μmol m^{-3} Na^+, 7 μmol m^{-3} of Cl^-, and 5 μmol m^{-3} of NO_3^- are combined

Case	$c_{HCl,m}$	$c_{HNO_3,m}$	$c_{NaCl,m}$	$c_{NaNO_3,m}$
1	6	0	1	5
2	4	2	3	3
3	1	5	6	0

Case 1 was obtained from (17.72). Units are micromoles per cubic meter.

Example 17.3

At high molalities (>10) and ambient temperatures (273–310 K), temperature affects water activity only slightly. For example, at $\mathbf{m} = 16$ m, the water activity of HCl increases from 0.09 to only 0.11 when the temperature increases from 273 K to 310 K. At low molalities, temperature has even less of an effect.

The water equation for mixed aerosol particles can be rewritten from (17.65) to

$$c_w = \frac{1}{m_v}\sum_{i=1}^{N_C}\left(\sum_{j=1}^{N_A}\frac{c_{i,j,m}}{\mathbf{m}_{i,j,a}}\right) \tag{17.71}$$

where c_w is the **aerosol liquid-water content** in units of mole concentration (moles of $H_2O(aq)$ per cubic centimeter of air), m_v is the molecular weight of water (0.018 02 kg mol^{-1}), $c_{i,j,m} = \mathbf{m}_{i,j,a}c_w m_v$ is the number of moles of electrolyte pair *i*, *j* per cubic centimeter of air in a solution containing all solutes at the ambient relative humidity, and $\mathbf{m}_{i,j,a}$ is the molality (mol kg^{-1}) of the electrolyte pair alone in solution at the ambient relative humidity, found from (17.66). The subscript *i*, *j* in (17.71) replaces the subscript *k* in (17.65).

In a model, the mole concentrations of electrolyte pairs are not usually known, but those of individual ions are. Thus, ions must be combined into hypothetical electrolyte pairs for (17.71) to be solved. The easiest way to recombine ions into hypothetical electrolytes in a mixture is to execute the equations

$$c_{i,j,m} = \min\left(\frac{c_{i,m}}{\nu_i}, \frac{c_{j,m}}{\nu_j}\right) \qquad c_{i,m} = c_{i,m} - \nu_i c_{i,j,m} \qquad c_{j,m} = c_{j,m} - \nu_j c_{i,j,m} \tag{17.72}$$

in succession, for each undissociated electrolyte pair *i*, *j*, where $c_{i,j,m}$ is the mole concentration of undissociated electrolyte *i*, *j* (e.g., HNO_3, HCl), and $c_{i,m}$ and $c_{j,m}$ are cation and anion mole concentrations, respectively. Applying (17.72) to Example 17.4 gives case 1 of Table 17.3.

Example 17.4

In a solution containing a mixture of ions, the combination of ions to form hypothetical electrolytes can be arbitrary for use in the water equation. Suppose a solution contains 6 μmol m^{-3} of H^+, 6 μmol m^{-3} Na^+, 7 μmol m^{-3} of Cl^-, and 5 μmol m^{-3} of NO_3^-. This set of concentrations is charge conserving (i.e., the sum, over all ions, of mole concentration multiplied by ion charge equals zero), which is a requirement for any aqueous system. The four ions can be combined in many ways to form electrolytes, HNO_3, HCl, $NaNO_3$, and NaCl. Recombination is limited by the mole-balance constraints

$$c_{H^+,m} = c_{HNO_3,m} + c_{HCl,m} \qquad c_{Na^+,m} = c_{NaNO_3,m} + c_{NaCl,m}$$

$$c_{Cl^-,m} = c_{HCl,m} + c_{NaCl,m} \qquad c_{NO_3^-,m} = c_{HNO_3,m} + c_{NaNO_3,m}$$

Table 17.3 shows three sets of electrolyte concentrations that satisfy these constraints.

17.9 SOLID FORMATION AND DELIQUESCENCE RELATIVE HUMIDITY

Solid electrolytes form by precipitation of ions or liquids from a particle solution or by chemical reaction of gases on a particle surface. **Precipitation** is the formation of an insoluble compound from a solution, and it can be simulated as a reversible equilibrium process. The equilibrium coefficient of a precipitation reaction is a **solubility product.** A solid precipitates from solution when the product of its reactant ion concentrations and mean solute activity coefficient exceeds its solubility product. For the reaction $NH_4NO_3(s) \rightleftharpoons NH_4^+ + NO_3^-$, precipitation of ammonium nitrate from solution occurs when

$$\mathbf{m}_{NH_4^+}\mathbf{m}_{NO_3^-}\gamma^2_{NH_4^+,NO_3^-} > K_{eq}(T) \tag{17.73}$$

A solid may also form as a result of heterogeneous reaction of gases on a particle surface. The solid may form when one gas adsorbs to a surface and the other gas collides and reacts with the adsorbed gas, or when both gases adsorb to the surface, diffuse on the surface, and collide. Solid formation by gas interaction with a particle surface can be simulated with an equilibrium reaction, such as $NH_4NO_3(s) \rightleftharpoons NH_3(g) + HNO_3(g)$. In this case, a solid forms when

$$p_{NH_3(g),s}\,p_{HNO_3(g),s} > K_{eq}(T) \tag{17.74}$$

Whether a solid forms or dissolves depends on the relative humidity and whether the relative humidity is increasing or decreasing. If an aerosol particle consists of an initially solid electrolyte at a given relative humidity, and the relative humidity increases, the electrolyte does not take on liquid water by hydration until the **deliquescence relative humidity** (DRH) is reached. **Deliquescence** is the process by which an initially dry particle lowers its saturation vapor pressure and takes up

Table 17.4 DRHs and CRHs of several electrolytes at 298 K

Electrolyte	DRH (percent)	CRH (percent)	Electrolyte	DRH (percent)	CRH (percent)
NaCl	75.28[a]	47[c]	NH_4HSO_4	40.0[b]	<5–22[b]
Na_2SO_4	84.2[b]	57–59[b]	NH_4NO_3	61.83[a]	25–32[d]
$NaHSO_4$	52.0[d]	<5[d]	$(NH_4)_3H(SO_4)_2$	69[b]	35–44[b]
$NaNO_3$	74.5[d]	<5–30[b]	KCl	84.26[a]	62[c]
NH_4Cl	77.1[a]	47[e]	Oxalic acid	97.3[f]	51.8–56.7[f]
$(NH_4)_2SO_4$	79.97[a]	37–40[b]	Succinic acid	98.8[f]	55.2–59.3[f]

[a]Robinson and Stokes (1955), [b]Tang and Munkelwitz (1994), [c]Tang (1997), [d]Tang (1996), [e]Cohen *et al.* (1987a), [f]Peng *et al.* (2001)

liquid water. At the DRH, water rapidly hydrates with the electrolyte, dissolving the solid, and increasing the liquid-water content of the particle. Above the DRH, the solid no longer exists, and the particle takes up liquid water to maintain equilibrium.

If the particle initially consists of an electrolyte dissolved in water, and the relative humidity decreases below the DRH, liquid water evaporates, but dissolved ions in solution do not necessarily **precipitate** (crystallize) immediately. Instead, the solution becomes supersaturated with respect to water and remains so until nucleation of a solid crystal occurs within the solution. The relative humidity at which nucleation occurs and an initially aqueous electrolyte becomes crystalline is the **crystallization relative humidity** (CRH). The CRH is always less than or equal to the DRH. Table 17.4 shows the DRHs and CRHs of several electrolytes at 298 K.

Some electrolytes, such as NH_3, HNO_3, HCl, and H_2SO_4 do not have a solid phase at room temperature. These substances, therefore, do not have a DRH or a CRH.

For an initially dry mixture of two or more electrolytes, the relative humidity above which the mixture takes on liquid water is the **mutual deliquescence relative humidity** (MDRH). Above the MDRH, mixed particles may consist of one or more solids in equilibrium with the solution phase. The MDRH of a mixture is always lower than the DRH of any electrolyte within the mixture. (Wexler and Seinfeld 1991; Tang and Munkelwitz 1993).

17.10 EXAMPLE EQUILIBRIUM PROBLEM

Equilibrium equations, activity-coefficient equations, and the water equation are solved together with other equations in an atmospheric model to determine aerosol composition and liquid-water content. In this section, a basic equilibrium problem and the set of equations required to solve it are given.

In this problem, aerosol particles are assumed to contain a solution of water, H^+, Cl^-, HSO_4^- and SO_4^{2-}, all of which are in equilibrium with each other. HCl(g) is also assumed to be in equilibrium with the particles. The two equilibrium reactions describing this situation are

$$HCl(g) \rightleftharpoons H^+ + Cl^- \qquad HSO_4^- \rightleftharpoons H^+ + SO_4^{2-} \tag{17.75}$$

The **equilibrium-coefficient expressions** for these reactions are

$$\frac{\mathbf{m}_{H^+,eq}\mathbf{m}_{Cl^-,eq}\gamma^2_{H^+,Cl^-,eq}}{p_{HCl,s,eq}} = K_{eq}(T) \qquad \frac{\mathbf{m}_{H^+,eq}\mathbf{m}_{SO_4^{2-},eq}\gamma^3_{2H^+,SO_4^{2-},eq}}{\mathbf{m}_{HSO_4^-,eq}\gamma^2_{H^+,HSO_4^-,eq}} = K_{eq}(T) \tag{17.76}$$

respectively, where the subscript eq indicates that the value is at equilibrium, and the subscript s indicates a saturation vapor pressure. The activity coefficients are nonlinear functions of H^+, Cl^-, HSO_4^-, and SO_4^{2-} mole concentrations.

When the reactions in (17.75) are solved, moles of individual atoms (e.g., chlorine, sulfur) must be conserved. The **mole balance constraints** are

$$C_{HCl(g),eq} + c_{Cl^-,eq} = C_{HCl(g),t-h} + c_{Cl^-,t-h} \tag{17.77}$$

$$c_{HSO_4^-,eq} + c_{SO_4^{2-},eq} = c_{HSO_4^-,t-h} + c_{SO_4^{2-},t-h} \tag{17.78}$$

where C is the mole concentration of a gas (mol cm^{-3}), c is the mole concentration of a particle component, and $t-h$ indicates an initial value. Saturation vapor pressures and molalities are related to mole concentrations by

$$p_{HCl,s,eq} = C_{HCl(g),s,eq}R^*T \qquad \mathbf{m}_{Cl^-,eq} = \frac{c_{Cl^-,eq}}{c_{w,eq}m_v} \tag{17.79}$$

respectively. The **charge balance equation** for this problem requires

$$c_{Cl^-,eq} + c_{HSO_4^-,eq} + 2c_{SO_4^{2-},eq} = c_{H^+,eq} \tag{17.80}$$

and the ZSR equation giving the water content is

$$c_{w,eq} = \frac{1}{m_v}\left(\frac{c_{H^+,Cl^-,m}}{\mathbf{m}_{H^+,Cl^-,a}} + \frac{c_{H^+,HSO_4^-,m}}{\mathbf{m}_{H^+,HSO_4^-,a}} + \frac{c_{2H^+,SO_4^{2-},m}}{\mathbf{m}_{2H^+,SO_4^{2-},a}}\right) \tag{17.81}$$

The molality of a species alone in solution ($\mathbf{m}_a$) is determined from an empirical function of the relative humidity, such as that given in (17.66). The quantities $c_{H^+,Cl^-,m}$, etc., are hypothetical mole concentrations of electrolyte pairs, constrained by

$$c_{H^+,eq} = c_{H^+,Cl^-,m} + c_{H^+,HSO_4^-,m} + 2c_{2H^+,SO_4^{2-},m} \tag{17.82}$$

$$c_{Cl^-,eq} = c_{H^+,Cl^-,m} \qquad c_{HSO_4^-,eq} = c_{H^+,HSO_4^-,m} \qquad c_{SO_4^{2-},eq} = c_{2H^+,SO_4^{2-},m}$$

This problem requires the solution of two equilibrium equations, two mole balance equations, a charge balance equation, a water equation, and activity-coefficient equations. These equations or a variation of them can be solved with an iterative Newton–Raphson method (e.g., Press *et al.* 1992), an iterative bisectional Newton method (e.g., Pilinis and Seinfeld 1987; Kim *et al.* 1993a,b; Nenes *et al.* 1998, 1999; Makar *et al.* 2003), an iterative method that minimizes the Gibbs free energy (Bassett and Seinfeld 1983, 1984; Wexler and Seinfeld 1990, 1991; Wexler and Clegg 2002), approximation method (e.g., Saxena *et al.* 1986; Metzger *et al.* 2002), or a mole-fraction-based thermodynamic model that parameterizes ion pair and triplet interactions (Clegg *et al.* 1997). Most of these methods are mole- and charge-conserving. Zhang *et al.* (2000) compare several aerosol equilibrium models.

17.11 MASS-FLUX ITERATION METHOD

Another method used to solve equilibrium problems is the **mass-flux iteration** (MFI) method (Jacobson 1994; Jacobson *et al.* 1996b), which relies in part on the technique of Villars (1959). This method can converge thousands of equilibrium equations simultaneously, cannot produce negative concentrations, and is mole- and charge-conserving at all times. The only constraints are that the equilibrium equations must be mole- and charge-conserving, and the system must start in charge balance. For example, the equation $HNO_3(aq) = H^+ + NO_3^-$ conserves moles of H, O, and N and charge. The charge balance constraint allows initial charges to be distributed among all dissociated ions, but the initial sum, over all species, of charge multiplied by molality must equal zero. The simplest way to initialize charge is to set all ion molalities to zero. Initial mass in the system can be distributed arbitrarily, subject to the charge balance constraint. If the total nitrate in the system is known to be, say, 20 μg m^{-3}, the nitrate can be distributed initially in any proportion among $HNO_3(aq)$, NO_3^-, $NH_4NO_3(s)$, etc.

The MFI method requires the solution of one equilibrium equation at a time by iteration. A system of equations is solved by iterating all equations many times. Suppose a system consists of a single aerosol size bin and 15 equations representing the equilibrium chemistry within that bin. At the start, the first equation is iterated. When the first equation converges, the updated and other initial concentrations are used as inputs into the second equation. This continues until the last equation has converged. At that point, the first equation is no longer converged, since concentrations used in it have changed. The iteration sequence must be repeated over all equations several times until the concentrations no longer change upon more iteration.

Equilibrium among multiple particle size bins and the gas phase is solved in a similar manner. Suppose a system consists of several size bins, equations per bin, and gases that equilibrate with dissolved molecules in each size bin. This system can be solved with equilibrium equations if it is assumed that each gas's saturation vapor pressure over each particle surface equals the gas's partial pressure, which is a single value. In reality, a gas's saturation vapor pressure differs over each particle surface. In order to take variations in saturation vapor pressure over particle surfaces into account, nonequilibrium gas–aerosol transfer equations, discussed in Section 17.14, must be solved.

Gas–particle equilibrium over multiple size bins can be solved by iterating each equilibrium equation, including gas–particle equations, starting with the first size bin. Updated gas concentrations from the first bin affect the equilibrium distribution in subsequent bins. After the last size bin has been iterated, the sequence is repeated in reverse order (to speed convergence), from the last to first size bin. The marches back and forth among size bins continue until gas and aerosol concentrations do not change upon further iteration.

To demonstrate the MFI method, an example in which two gases equilibrate with two ions in a single equilibrium equation is discussed. The sample equation has the form of (17.3), with the gases on the left side of the equation. The first step in the solution method is to calculate Q_d and Q_n, the smallest ratio of mole concentration to the stoichiometric coefficient among species appearing in the denominator and numerator, respectively, of (17.18). Thus,

$$Q_d = \min\left(\frac{C_{D,0}}{\nu_D}, \frac{C_{E,0}}{\nu_E}\right) \qquad Q_n = \min\left(\frac{c_{A,0}}{\nu_A}, \frac{c_{B,0}}{\nu_B}\right) \tag{17.83}$$

where C and c are gas- and particle-phase mole concentrations (mol cm^{-3}), respectively, and the subscript 0 indicates an initial value. If an equilibrium equation contains a solid, the solid's concentration is included in (17.83).

After two parameters, $z_1 = 0.5(Q_d + Q_n)$ and $\Delta x_1 = Q_d - z_1$, where the subscript 1 refers to the iteration number, are initialized, the iteration sequence begins by adding the mass (mole) flux factor Δx, which may be positive or negative, to each mole concentration in the numerator and subtracting it from each mole concentration in the denominator of the equilibrium equation. Thus,

$$\begin{aligned} c_{A,l+1} &= c_{A,l} + \nu_A \Delta x_l \qquad & c_{B,l+1} &= c_{B,l} + \nu_B \Delta x_l \\ C_{D,l+1} &= C_{D,l} - \nu_D \Delta x_l \qquad & C_{E,l+1} &= C_{E,l} - \nu_E \Delta x_l \end{aligned} \tag{17.84}$$

Starting with (17.84), iteration numbers are referred to by subscripts l and $l+1$. If the equation contains a solid, the change in the solid's concentration is calculated with (17.84) (solid, aqueous, and ionic mole concentrations are identified with a c). Equation (17.84) shows that, during each iteration, moles and charge are transferred from either reactants to products or vice versa. This transfer continues until $\Delta x = 0$. Thus, the scheme conserves moles and charge each iteration.

Third, a ratio comparing activities to the equilibrium coefficient is calculated as

$$F = \frac{\mathbf{m}_{A,l+1}^{\nu_A} \mathbf{m}_{B,l+1}^{\nu_B} \gamma_{AB,l+1}^{\nu_A+\nu_B}}{p_{D,l+1}^{\nu_D} p_{E,l+1}^{\nu_E}} \frac{1}{K_{eq}(T)} \tag{17.85}$$

For this calculation, mole concentrations are converted to molalities (for solutes) or atmospheres (for gases). In the case of solids, the activities are unity and do not appear in (17.85). Mean mixed solute activity coefficients (e.g., $\gamma_{AB,l+1}$) are updated before each iteration of an equilibrium equation and converge over time. The liquid-water content c_w, which also converges, is updated with (17.71) before each iteration of an equilibrium equation or more sporadically.

The fourth step in the procedure is to recalculate z for the next iteration as $z_{l+1} = 0.5z_l$. The final step is to check convergence with the criterion

$$F = \begin{cases} > 1 & \rightarrow & \Delta x_{l+1} = -z_{l+1} \\ < 1 & \rightarrow & \Delta x_{l+1} = +z_{l+1} \\ = 1 & \rightarrow & \text{convergence} \end{cases} \tag{17.86}$$

At each nonconvergence, Δx is updated, the iteration number is advanced, and the iteration sequence restarts with (17.84). All molalities eventually converge to positive numbers.

When a solid forms during iteration of a gas–solid or ion–solid equilibrium reaction, F converges to 1, and the solid concentration is found from (17.84). When a solid does not form, F cannot converge, and (17.84) correctly predicts no net change in concentration of any species involved in the reaction.

17.12 ANALYTICAL EQUILIBRIUM ITERATION METHOD

An advantage of the MFI method is that it can be used to solve equilibrium equations of any form. A disadvantage is that it requires iteration of each equilibrium equation as well as iteration among all equations. A method that eliminates iteration of some individual equilibrium equations and reduces the number of iterations of others (but does not eliminate iteration among all equations) is the analytical equilibrium iteration (AEI) method (Jacobson 1999b). With this method, most individual equilibrium equations are solved analytically instead of iteratively. Solutions from this method are, to within several decimal places, identical to those from the MFI method but are obtained significantly faster. Below, the AEI method is described for three types of equilibrium equations. Solutions for other types of reactions are given in Jacobson (1999b).

17.12.1 Reactions of the form D $\rightleftharpoons$ A

For equations of this form, the analytical solution is found by solving the equilibrium relation,

$$\frac{c_{A,c}}{c_{D,c}} = \frac{c_{A,0} + \Delta x_{fin}}{c_{D,0} - \Delta x_{fin}} = K_r \tag{17.87}$$

where the subscript 0 indicates an initial value, the subscript c indicates a converged value, Δx_{fin} is the change in mole concentration required to converge the solution, K_r is the equilibrium coefficient (mol mol^{-1}), which contains activity coefficients, if treated, and concentrations are in units of mole concentration (mol cm^{-3}-air). The analytical solution to this equation is

$$\Delta x_{fin} = \frac{c_{D,0}K_r - c_{A,0}}{1 + K_r} \tag{17.88}$$

which is substituted into $c_{A,c} = c_{A,0} + \Delta x_{fin}$ and $c_{D,c} = c_{D,0} - \Delta x_{fin}$ to obtain converged concentrations.

Example 17.5

Calculate the equilibrium mole concentrations of $O_3(g)$ and $O_3(aq)$ in the presence of a dilute solution of 1 g m^{-3} of liquid water (representing the liquid water content of a cloud) if the only reaction considered is $O_3(g) \rightleftharpoons O_3(aq)$, the atmosphere initially contains 40 μg m^{-3} $O_3(g)$, and no $O_3(aq)$, and $T =$ 298.15 K.

SOLUTION

From Appendix Table B.7, the equilibrium coefficient for the reaction $O_3(g) \rightleftharpoons O_3(aq)$ is $K_{eq} = 1.13 \times 10^{-2}$ mol kg^{-1} atm^{-1}. This is converted to moles of dissolved ozone per mole of gas-phase ozone with

$$K_r = K_{eq} R^* T \mathbf{m}_L$$
$$= 0.0113 \frac{\text{mol}}{\text{kg atm}} \times 0.08206 \frac{\text{L atm}}{\text{mol K}} \times 298.15\,\text{K} \times 10^{-6} \frac{\text{kg}}{\text{L}}$$
$$= 2.76 \times 10^{-7} \frac{\text{mol}}{\text{mol}}$$

where $\mathbf{m}_L = 10^{-6}$ kg-H_2O L^{-1}-air = 1 g m^{-3} is the liquid water content. The initial mole concentration of gas-phase ozone is

$$c_{O_3(g),0} = 40 \frac{\mu\text{g}}{\text{m}^3} \times 10^{-6} \frac{\text{m}^3}{\text{cm}^3} \times 10^{-6} \frac{\text{g}}{\mu\text{g}} \times \frac{1}{48} \frac{\text{mol}}{\text{g}}$$
$$= 8.33333 \times 10^{-13} \frac{\text{mol}}{\text{cm}^3}$$

The initial aqueous mole concentration is $c_{O_3(aq),0} = 0$. Substituting these variables into (17.88) gives

$$\Delta x_{\text{fin}} = \frac{8.333333 \times 10^{-13} \dfrac{\text{mol}}{\text{cm}^3} \times 2.76 \times 10^{-7} \dfrac{\text{mol}}{\text{mol}} - 0}{1 + 2.76 \times 10^{-7} \dfrac{\text{mol}}{\text{mol}}}$$
$$= 2.299999356 \times 10^{-19} \frac{\text{mol}}{\text{cm}^3}$$

Final concentrations are now

$$c_{O_3(aq),c} = c_{O_3(aq),0} + \Delta x_{\text{fin}} = 2.299999356 \times 10^{-19} \frac{\text{mol}}{\text{cm}^3} = 1.104 \times 10^{-5} \frac{\mu\text{g}}{\text{m}^3}$$
$$c_{O_3(g),c} = c_{O_3(g),0} - \Delta x_{\text{fin}} = 8.33333103 \times 10^{-13} \frac{\text{mol}}{\text{cm}^3} = 39.999988 \frac{\mu\text{g}}{\text{m}^3}$$

The solution can be checked by substituting it into (17.87):

$$\frac{c_{O_3(aq),c}}{c_{O_3(g),c}} = \frac{2.299999365 \times 10^{-19}}{8.33333103 \times 10^{-13}} = 2.76 \times 10^{-7}$$

Since the ratio yields the equilibrium coefficient, the solution is correct. The solution implies that relatively little ozone partitions into cloud drops. If aqueous reactions destroy ozone, though, the removal rate of ozone by clouds can increase (Chapter 19).

17.12.2 Nonprecipitation reactions of the form $D + E \rightleftharpoons A + B$

For equations of this form, the analytical solution is found by solving the equilibrium relation,

$$\frac{c_{A,c}c_{B,c}}{c_{D,c}c_{E,c}} = \frac{(c_{A,0} + \Delta x_{fin})(c_{B,0} + \Delta x_{fin})}{(c_{D,0} - \Delta x_{fin})(c_{E,0} - \Delta x_{fin})} = K_r \tag{17.89}$$

where K_r is in units of (mol^2 mol^{-2}). The analytical solution to this equation is

$$\Delta x_{fin} = \frac{-c_{A,0} - c_{B,0} - c_{D,0}K_r - c_{E,0}K_r + \sqrt{\begin{array}{l}(c_{A,0} + c_{B,0} + c_{D,0}K_r + c_{E,0}K_r)^2 \\ -4(1 - K_r)(c_{A,0}c_{B,0} - c_{D,0}c_{E,0})\end{array}}}{2(1 - K_r)} \tag{17.90}$$

which is substituted into $c_{A,c} = c_{A,0} + \Delta x_{fin}$, $c_{B,c} = c_{B,0} + \Delta x_{fin}$, $c_{D,c} = c_{D,0} - \Delta x_{fin}$, and $c_{E,c} = c_{E,0} - \Delta x_{fin}$ to obtain converged concentrations.

17.12.3 Solid reactions of the form $D(s) \rightleftharpoons 2A + B$

For this type of equation, a solid can form only when

$$(c_{A,0} + 2c_{D,0})^2 (c_{B,0} + 2c_{D,0}) > K_r \tag{17.91}$$

where K_r is in units of (mol^3 cm^{-9}). If this criterion is met, the equilibrium equation

$$c_{A,c}^2 c_{B,c} = (c_{A,0} + 2\Delta x_{fin})^2 (c_{B,0} + \Delta x_{fin}) = K_r \tag{17.92}$$

is solved, where $c_{A,c} = c_{A,0} + 2\Delta x_{fin}$, $c_{B,c} = c_{B,0} + \Delta x_{fin}$, and $c_{D,c} = c_{D,0} - \Delta x_{fin}$, are the final, converged concentrations. The term Δx_{fin} must still be determined. If the criterion is not met, (17.92) is not solved and the solid concentration is set to zero.

Due to their complexity, reactions of this form must be solved iteratively. One solution is with the MFI method. A method that solves twice as fast for this form of reaction is the Newton–Raphson method. The solution is obtained by rewriting (17.92) as

$$f_n(x) = \Delta x_{fin,n}^3 + q\Delta x_{fin,n}^2 + r\Delta x_{fin,n} + s = 0 \tag{17.93}$$

where $q = c_{A,0} + c_{B,0}$, $r = c_{A,0}c_{B,0} + 0.25c_{A,0}^2$, $s = c_{A,0}^2 c_{B,0} - K_r$, and n is the iteration number. The iterative solution to (17.93) is

$$\Delta x_{fin,n+1} = \Delta x_{fin,n} - \frac{f_n(x)}{f_n'(x)} \tag{17.94}$$

where $f_n'(x) = 3\Delta x_{fin,n}^2 + 2qx_{fin,n} + r$.

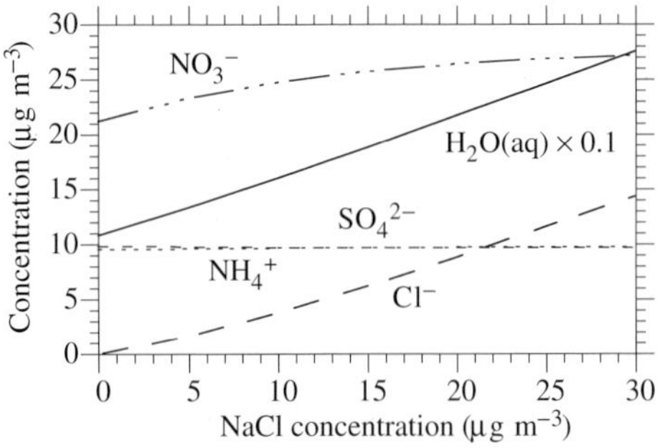

Figure 17.5 Modeled aerosol composition versus NaCl concentration when the relative humidity was 90 percent. Other initial conditions were $H_2SO_4(aq)$, 10 μg m^{-3}; HCl(g), 0 μg m^{-3}; $NH_3(g)$, 10 μg m^{-3}; $HNO_3(g)$, 30 μg m^{-3}; and T = 298 K. NaCl dissolves and dissociates completely.

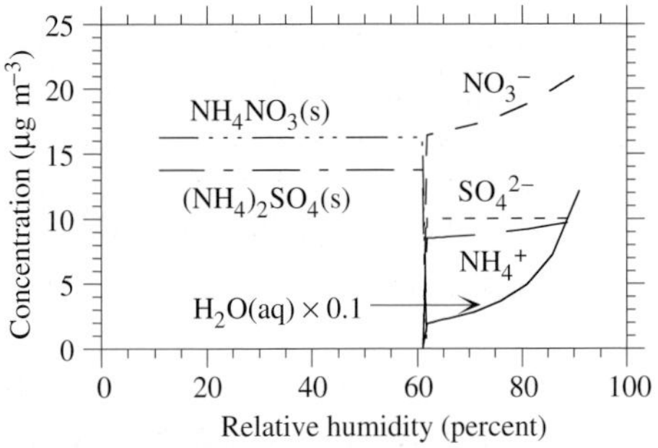

Figure 17.6 Modeled aerosol composition versus relative humidity. Initial conditions were $H_2SO_4(aq)$, 10 μg m^{-3}; HCl(g), 0 μg m^{-3}; $NH_3(g)$, 10 μg m^{-3}; $HNO_3(g)$, 30 μg m^{-3}; and T = 298 K.

17.13 EQUILIBRIUM SOLVER RESULTS

Figures 17.5 and 17.6 show equilibrium aerosol composition determined by a computer program that used the MFI method. Figure 17.5 illustrates equilibrium aerosol composition as a function of NaCl concentration. The figure shows that nearly all ammonia gas dissolved and formed ammonium at low NaCl concentration. As NaCl concentration increased, liquid-water and nitrate concentrations increased since an increase in NaCl increases hydration, enabling more dissolution and dissociation of nitric acid. Sulfate existed in the solution at all times.

Figure 17.6 shows the simulated change in aerosol composition as a function of relative humidity. At low relative humidity, aerosol particles contained solid ammonium nitrate and solid ammonium sulfate. When the relative humidity increased to the MDRH of just less than 62 percent (in this case), which is near the DRH of ammonium nitrate, liquid water condensed, dissolving both solids. Although its DRH is about 80 percent, ammonium sulfate dissolved at the MDRH because the MDRH of a mixture is less than the DRH of any electrolyte in the mixture. As the relative humidity increased past the MDRH, the liquid-water content of the solution increased, increasing the rate of dissolution of nitric acid and ammonia from the gas phase.

17.14 NONEQUILIBRIUM BETWEEN GASES AND PARTICLES

Transfer between a gas and a particle is time-dependent and not instantaneous. Thus, a real gas's saturation vapor pressure over a particle's surface does not equal the gas's partial pressure. As such, under many conditions, gas–particle equilibrium equations should not be used to solve for the partial pressure of a gas. Instead, transfer between gases and particles should be simulated with time-dependent (nonequilibrium) growth equations that account for gas dissolution into aerosol particles.

Generally, the equilibrium assumption between gases and particles holds better for nucleation- and accumulation-mode particles than for coarse-mode particles (e.g., Meng and Seinfeld 1996; Capaldo *et al.* 2000; Pilinis *et al.* 2000; Fridlind and Jacobson 2000; Moya *et al.* 2001). Although the rate of transfer of a gas to the surface of a large particle is greater than is that to the surface of a small particle, many more small particles than large particles are generally present. As such, gas–particle fluxes are faster to a population of small particles than to a population of large particles.

In Chapter 16, solutions to nonequilibrium condensational growth and ice depositional growth equations were given and solved. In the following subsections, nonequilibrium dissolutional growth equations and a solution to them are described.

17.14.1 Solution to dissolution growth for nondissociating species

When a gas transfers to a particle surface and dissolves in liquid water on the surface, the process is called **dissolutional growth**. Dissolutional growth is treated here as a nonequilibrium (time-dependent) process. Equations that describe dissolutional growth are similar to those that describe condensational growth, (16.67) and (16.68), except that, for dissolutional growth, the saturation vapor pressure depends on the molality of gas in solution, whereas for condensational growth, the saturation vapor pressure is held constant for a given temperature. Thus, condensation can occur on a solid or liquid surface, whereas dissolution must occur in liquid on a surface.

A saturation vapor pressure was defined earlier as the partial pressure of a gas over a particle surface when the gas is in equilibrium with the surface at a given temperature. During dissolution, the saturation vapor pressure of gas A can be determined from the equilibrium relationship $A(g) \rightleftharpoons A(aq)$. In terms of an arbitrary nondissociating and nonreacting gas q, the Henry's law constant equation for this reaction is

$$p_{q,\mathrm{s},i} = \frac{\mathrm{m}_{q,i}}{H_q} \tag{17.95}$$

where $p_{q,\mathrm{s},i}$ is the saturation vapor pressure of gas q over a particle in size bin i (atm), $\mathrm{m}_{q,i}$ is the molality of dissolved gas q in particles of size i (mol kg^{-1}), and $H_q = K_{\mathrm{eq},q}$ is the Henry's law constant (mol kg^{-1} atm^{-1}). Activity coefficients of undissociated, dissolved gases are often unity. Equation (17.95) states that, as a gas dissolves (as its molality increases in solution), the gas's saturation vapor pressure increases.

Saturation vapor pressure (atm) is related to **saturation vapor mole concentration,** $C_{q,\mathrm{s},i}$ (mol cm^{-3}), by

$$p_{q,\mathrm{s},i} = C_{q,\mathrm{s},i} R^* T \tag{17.96}$$

where R^* is the universal gas constant (82.06 cm^3 atm mol^{-1} K^{-1}), and T is absolute temperature (K). Molality (mol kg^{-1}) is related to mole concentration in solution by

$$\mathrm{m}_{q,i} = \frac{c_{q,i}}{m_\mathrm{v} c_{\mathrm{w},i}} \tag{17.97}$$

where $c_{q,i}$ is the mole concentration of dissolved gas q in size bin i (mol cm^{-3}), $c_{\mathrm{w},i}$ is the mole concentration of liquid water in bin i, and m_v is the molecular weight of water (0.01802 kg mol^{-1}). Substituting (17.95) and (17.97) into (17.96) gives the saturation mole concentration of a nondissociating soluble gas over a dilute solution as a function of its molality in solution as

$$C_{q,\mathrm{s},i} = \frac{p_{q,\mathrm{s},i}}{R^* T} = \frac{\mathrm{m}_{q,i}}{R^* T H_q} = \frac{c_{q,i}}{m_\mathrm{v} c_{\mathrm{w},i} R^* T H_q} = \frac{c_{q,i}}{H'_{q,i}} \tag{17.98}$$

where

$$H'_{q,i} = m_\mathrm{v} c_{\mathrm{w},i} R^* T H_q \tag{17.99}$$

is the **dimensionless Henry's constant** of gas q. If the gas does not dissociate or chemically react in solution, its **time-rate-of-change in concentration due to dissolutional growth/evaporation** can be found by substituting (17.98) into (16.67). The result is

$$\frac{\mathrm{d}c_{q,i,t}}{\mathrm{d}t} = k_{q,i,t-h}\left(C_{q,t} - S'_{q,i,t-h}\frac{c_{q,i,t}}{H'_{q,i,t-h}}\right) \tag{17.100}$$

where the time subscript for $c_{q,i,t}$ is implicit (t) and that for $H'_{q,i,t-h}$ is explicit ($t-h$), allowing for an implicit solution to concentration. The mass transfer rate $k_{q,i,t-h}$ was given in (16.64). The **equation for mole conservation between the gas and all aerosol-particle size bins,** corresponding to (17.100), is

$$\frac{\mathrm{d}C_{q,t}}{\mathrm{d}t} = -\sum_{i=1}^{N_\mathrm{B}}\left[k_{q,i,t-h}\left(C_{q,t} - S'_{q,i,t-h}\frac{c_{q,i,t}}{H'_{q,i,t-h}}\right)\right] \tag{17.101}$$

Equations (17.100) and (17.101) represent $N_B + 1$ growth/evaporation expressions. Like condensation equations, dissolution equations may be solved with an iterative ODE solver. Here, a noniterative, mole-conserving, and unconditionally stable method of solving these equations, called the **analytical predictor of dissolution** (APD) method (Jacobson 1997c), is described.

The noniterative solution to dissolutional growth with the APD scheme is obtained by assuming that the final concentration of component q in size bin i can be calculated by integrating (17.100). The result is

$$c_{q,i,t} = \frac{H'_{q,i,t-h} C_{q,t}}{S'_{q,i,t-h}} + \left(c_{q,i,t-h} - \frac{H'_{q,i,t-h} C_{q,t}}{S'_{q,i,t-h}} \right) \exp\left(-\frac{h S'_{q,i,t-h} k_{q,i,t-h}}{H'_{q,i,t-h}} \right) \tag{17.102}$$

where $C_{q,t}$ is currently unknown. As with the APC scheme (Chapter 16), the final aerosol and gas concentrations are constrained by

$$C_{q,t} + \sum_{i=1}^{N_B} c_{q,i,t} = C_{q,t-h} + \sum_{i=1}^{N_B} c_{q,i,t-h} \tag{17.103}$$

Substituting (17.102) into (17.103) and solving for $C_{q,t}$ give

$$C_{q,t} = \frac{C_{q,t-h} + \sum_{i=1}^{N_B} \left\{ c_{q,i,t-h} \left[1 - \exp\left(-\frac{h S'_{q,i,t-h} k_{q,i,t-h}}{H'_{q,i,t-h}} \right) \right] \right\}}{1 + \sum_{i=1}^{N_B} \left\{ \frac{H'_{q,i,t-h}}{S'_{i,q,t-h}} \left[1 - \exp\left(-\frac{h S'_{q,i,t-h} k_{q,i,t-h}}{H'_{q,i,t-h}} \right) \right] \right\}} \tag{17.104}$$

which is the final gas concentration. This concentration is substituted back into (17.102) to give the final particle component concentration for a time step. The APD scheme is unconditionally stable, since all final concentrations are bounded between 0 and C_{tot}, regardless of the time step and liquid water content, and it is exactly mole- and mass-conserving. Table 17.5 illustrates the unconditional stability and mass conservation of the APD scheme. An extension of the APD scheme to multiple aerosol size distributions is discussed in Jacobson (2002).

17.14.2 Solution to acid/base dissolution growth

When a strong acid, such as nitric acid, hydrochloric acid, or sulfuric acid, dissolves in solution, it usually dissociates, and the resulting anion may react with a cation, such as ammonium, sodium, calcium, potassium, or magnesium, in solution, to form a solid. The simultaneous transfer of acids and bases from the gas phase to particle surfaces results in instantaneous adjustment to pH, particle composition,

Table 17.5 Stability test of the APD scheme

Time step size (s)	$C_{q,t}$ (μg m^{-3})	$c_{q,1,t}$ (μg m^{-3})	$c_{q,2,t}$ (μg m^{-3})	$c_{q,3,t}$ (μg m^{-3})
0.1	0.769	3.08	3.08	3.08
10	0.769	3.08	3.08	3.08
60	0.769	3.08	3.08	3.08
600	0.769	3.08	3.08	3.08
7200	0.769	3.08	3.08	3.08

Assume a gas dissolves in three particle size bins, where $k_{q,i,t-h} = 0.00333$, 0.00833, and 0.0117 s^{-1} for the respective bins. Also, $H'_{q,i} = 4.0$, $S'_{q,i,t-h} = 1.0$, and $c_{q,i,t-h} = 0$ μg m^{-3} for the bins, and $C_{q,t-h} = 10$ μg m^{-3}. The table shows concentrations after 2 h predicted by the APD scheme when five time step sizes were used. In all cases, the solution was stable and mass-conserving.

and vapor pressure that renders the resulting set of equations stiff and difficult to solve at long time step without an oscillatory solution.

In this subsection, a method of solving nonequilibrium dissolutional growth of acids and bases at high and moderate liquid water contents and long time step is discussed. In Section 17.14.3, a method of solving such equations at low liquid water contents is described. In both cases, nonequilibrium growth calculations are coupled with intra-aerosol equilibrium chemistry calculations. The growth portion of the algorithm described here is referred to as the **predictor of nonequilibrium growth** (PNG) scheme. The equilibrium algorithm is EQUISOLV II. Thus, the coupled scheme is PNG-EQUISOLV II (Jacobson 2005b). Other techniques of solving nonequilibrium acid/base transfer to aerosol particle surfaces are given in Sun and Wexler (1998), Capaldo *et al.* (2000), and Pilinis *et al.* (2000). In Chapter 19, a method of solving nonequilibrium dissolutional growth with nonequilibrium aqueous chemistry is given.

The **growth equation for an acid gas**, such as hydrochloric acid (HCl), that dissolves then dissociates (in this case to the hydrogen ion, H^+, and the chloride ion, Cl^-) in solution, can be written as

$$\frac{dc_{Cl,i,t}}{dt} = k_{HCl,i,t-h}(C_{HCl,t} - S'_{HCl,i,t-h}C_{HCl,s,i,t}) \tag{17.105}$$

where the subscript t identifies values that are currently unknown, the subscript $t - h$ identifies values that are known from the beginning of the time step, and

$$c_{Cl,i,t} = c_{HCl(aq),i,t} + c_{Cl^-,i,t} \tag{17.106}$$

is the mole concentration (mol cm^{-3}-air) of dissolved, undissociated hydrochloric acid plus that of the chloride ion in particles of size i at time t. In addition, $C_{HCl,s,i,t}$ is the saturation mole concentration (SMC) over the surface of particles of size i and is currently unknown, and all other variables were defined previously. A similar equation can be written for nitric acid and sulfuric acid. Sulfuric acid, though, has such a low SMC that its SMC is held constant here at time $t - h$, and its growth is solved with the APC scheme, discussed in Chapter 16. Ammonia is a strong base, and its growth solution is coupled with the growth solutions of all dissolving acids (e.g., HCl, HNO_3) and condensing acids (e.g., H_2SO_4), as discussed shortly.

The saturation mole concentration of HCl in (17.105) can be approximated from the gas–aerosol equilibrium equations and their corresponding equilibrium coefficient relationships,

$$\mathrm{HCl} \rightleftharpoons \mathrm{HCl(aq)} \qquad \frac{\mathbf{m}_{\mathrm{HCl(aq)},i}}{p_{\mathrm{HCl,s},i}} = H_{\mathrm{HCl}} \frac{\mathrm{mol}}{\mathrm{kg\ atm}} \tag{17.107}$$

$$\mathrm{HCl(aq)} \rightleftharpoons \mathrm{H}^+ + \mathrm{Cl}^- \qquad \frac{\mathbf{m}_{\mathrm{H}^+,i}\mathbf{m}_{\mathrm{Cl}^-,i}\gamma^2_{i,\mathrm{H}^+,\mathrm{Cl}^-}}{\mathbf{m}_{\mathrm{HCl(aq)},i}} = K_{\mathrm{HCl}} \frac{\mathrm{mol}}{\mathrm{kg}} \tag{17.108}$$

where $r_{i,\mathrm{H}^+,\mathrm{Cl}^-}$ is a mean mixed solute activity coefficient (Section 17.7). Combining (17.96), (17.97), and (17.106)–(17.108) gives the saturation mole concentration of hydrochloric acid as

$$C_{\mathrm{HCl,s},i} = \frac{c_{\mathrm{Cl},i}}{K'_{\mathrm{HCl},i}} \tag{17.109}$$

where

$$K'_{\mathrm{HCl},i} = H_{\mathrm{HCl}}\left[1 + \frac{K_{\mathrm{HCl}}(m_v c_{\mathrm{w},i})^2 R^* T}{c_{\mathrm{H}^+,i}\gamma^2_{i,\mathrm{H}^+,\mathrm{Cl}^-}}\right] \tag{17.110}$$

(mol cm^{-3}) is the adjusted equilibrium coefficient determined at the beginning of the growth step from known parameters, including the hydrogen ion concentration. Substituting (17.109) into (17.105) and adding time subscripts gives the **growth equation for HCl** as

$$\frac{\mathrm{d}c_{\mathrm{Cl},i,t}}{\mathrm{d}t} = k_{\mathrm{HCl},i,t-h}\left(C_{\mathrm{HCl},t} - S'_{\mathrm{HCl},i,t-h}\frac{c_{\mathrm{Cl},i,t}}{K'_{\mathrm{HCl},i,t-h}}\right) \tag{17.111}$$

This equation has two unknowns for each size bin: the gas concentration and the total dissolved chlorine concentration, and it is identical in form to (17.100).

The hydrochloric acid gas concentration is linked to dissolved chlorine concentrations among N_B particle size bins with the **mole conservation relationship**,

$$C_{\mathrm{HCl},t} + \sum_{i=1}^{N_B} c_{\mathrm{Cl},i,t} = C_{\mathrm{HCl},t-h} + \sum_{i=1}^{N_B} c_{\mathrm{Cl},i,t-h} \tag{17.112}$$

The solution to (17.111) and (17.112) over one time step is similar to that for a nondissociating species, except that, each time step, the solution among all growing and dissociating acids is tied to the growth onto aerosol particles of the base, ammonia, as described shortly.

First, integrating (17.111) gives the **growth solution for the chloride ion to a size bin** as

$$c_{\mathrm{Cl},i,t} = \frac{K'_{\mathrm{HCl},i,t-h} C_{\mathrm{HCl},t}}{S'_{\mathrm{Cl}^-,i,t-h}} + \left(c_{\mathrm{Cl},i,t-h} - \frac{K'_{\mathrm{HCl},i,t-h} C_{\mathrm{HCl},t}}{S'_{\mathrm{HCl},i,t-h}} \right) \times \exp\left(- \frac{h k_{\mathrm{HCl},i,t-h} S'_{\mathrm{HCl},i,t-h}}{K'_{\mathrm{HCl},i,t-h}} \right) \qquad (17.113)$$

where the gas concentration in the equation is still unknown. Substituting (17.113) into (17.112) and solving give the **final gas concentration** as

$$C_{\mathrm{HCl},t} = \frac{C_{\mathrm{HCl},t-h} + \sum_{i=1}^{N_B} \left\{ c_{\mathrm{Cl},i,t-h} \left[1 - \exp\left(- \frac{h k_{\mathrm{HCl},i,t-h} S'_{\mathrm{HCl},i,t-h}}{K'_{\mathrm{HCl},i,t-h}} \right) \right] \right\}}{1 + \sum_{i=1}^{N_B} \left\{ \frac{K'_{\mathrm{HCl},i,t-h}}{S'_{\mathrm{HCl},i,t-h}} \left[1 - \exp\left(- \frac{h k_{\mathrm{HCl},i,t-h} S'_{\mathrm{HCl},i,t-h}}{K'_{\mathrm{HCl},i,t-h}} \right) \right] \right\}} \qquad (17.114)$$

Substituting (17.114) back into (17.113) gives the final total dissolved chlorine concentration in each size bin.

Growth equations for other semivolatile acid gases (e.g., HNO_3, CO_2) are solved in the same manner as those for HCl. In the case of carbon dioxide, the adjusted equilibrium coefficient (17.110) must include a second dissociation constant. Growth equations for involatile acid gases (e.g., H_2SO_4) are solved for with the APC scheme (Chapter 16). Following the growth calculation for all acid gases, the final gas and aerosol concentrations of ammonia/ammonium for the time step are solved assuming ammonia equilibrates with all acids simultaneously after the acids have grown onto each aerosol size bin. This method produces remarkable stable, nonoscillatory, and accurate solutions at a long time step (e.g., hundreds of seconds), as shown shortly.

The ammonia/ammonium solution is obtained by solving a charge balance equation within each size bin, a mole balance equation between the gas phase and each bin, and an equilibrium relation between the gas phase and each bin, simultaneously and after the growth step for acids. With this method, the quantity of ammonium added to each size bin each time step is the exact amount necessary to balance charge among all ions in solution following acid growth. Because the equilibration of ammonia is calculated after the diffusion-limited growth of all acids, ammonia growth is effectively a nonequilibrium growth process. The technique allows smooth solutions and convergence at long time step among all growing species. The alternative to equilibrating ammonia is to solve ammonia with (17.113) and

(17.114), but such a solution, when coupled with the growth solutions for acids, is oscillatory and inaccurate unless the time step is short (e.g., a few seconds or less). A solution with a short time step, though, requires substantial computer time in three dimensions.

With the present method, the **charge balance equation** for ammonium within each size bin is

$$c_{NH_4^+,i,t} + c_{H^+,i,t} + c_{\pm,i,t} = 0 \tag{17.115}$$

where $c_{NH_4^+,i,t}$ and $c_{H^+,i,t}$, are unknown, and

$$c_{\pm,i,t} = -c_{NO_3^-,i,t} - c_{Cl^-,i,t} - c_{HSO_4^-,i,t} - 2c_{SO_4^{2-},i,t} + \sum_q zc_{q,i,t-h} \tag{17.116}$$

consists of known values. The terms outside the summation in (17.116) are concentrations of the ions, NO_3^-, Cl^-, HSO_4^-, and SO_4^{2-}, from the end of the current time step (time *t*) of acid growth. The dissolutional and condensational growth solutions gave total nitrate [$HNO_3(aq) + NO_3^-$], total chloride [$HCl(aq) + Cl^-$], and total sulfate [$S(VI) = H_2SO_4(aq) + HSO_4^- + SO_4^{2-}$] in solution. The partitioning of total S(VI) into ions following S(VI) growth, for use in (17.116), is obtained by applying the ratios, $HSO_4^-/S(VI)$ and $SO_4^{2-}/S(VI)$, determined from the equilibrium calculation at the beginning of the operator-split time interval, to the total S(VI) following growth. The partitionings of total chloride and total nitrate to the chloride ion and the nitrate ion are calculated in a similar manner. The terms inside the summation in (17.116) are mole concentrations (*c*) multiplied by charge ($z = +/- 1, 2, 3$) of all ions, except those outside the summation, that are present in solution. The exact charge distribution within particles is recalculated with an internal-aerosol equilibrium calculation, discussed shortly.

The **mole balance equation** between gas-phase ammonia and the ammonium ion within each aerosol size bin is

$$C_{NH_3,t} + \sum_{i=1}^{N_B} (c_{NH_3(aq),i,t} + c_{NH_4^+,i,t}) = C_{NH_3,t-h} + \sum_{i=1}^{N_B} (c_{NH_3(aq),i,t-h} + c_{NH_4^+,i,t-h})$$
$$= C_{tot} \tag{17.117}$$

Finally, the **gas–aerosol equilibrium relations** for ammonia are

$$NH_3(g) \rightleftharpoons NH_3(aq) \qquad \frac{m_{NH_3(aq),i}}{p_{NH_3}} = H_{NH_3} \frac{\text{mol}}{\text{kg atm}} \tag{17.118}$$

$$NH_3(aq) + H^+ \rightleftharpoons NH_4^+ \qquad \frac{m_{NH_4^+,i}\gamma_{i,NH_4^+}}{m_{NH_3(aq),i}m_{H^+,i}\gamma_{i,H^+}} = K_{NH_3} \frac{\text{kg}}{\text{mol}} \tag{17.119}$$

respectively, where the pressure in the denominator of (17.118) is a partial pressure, rather than a saturation vapor pressure. The **single-ion activity-coefficient ratio** in (17.119) is calculated in terms of binary solute activity coefficients in a mixture with, for example,

$$\frac{\gamma_{i,NH_4^+}}{\gamma_{i,H^+}} = \frac{\gamma_{i,NH_4^+}\gamma_{i,NO_3^-}}{\gamma_{i,H^+}\gamma_{i,NO_3^-}} = \frac{\gamma^2_{i,NH_4^+,NO_3^-}}{\gamma^2_{i,H^+,NO_3^-}} = \frac{\gamma_{i,NH_4^+}\gamma_{i,Cl^-}}{\gamma_{i,H^+}\gamma_{i,Cl^-}} = \frac{\gamma^2_{i,NH_4^+,Cl^-}}{\gamma^2_{i,H^+,Cl^-}} \tag{17.120}$$

all of which are calculated during the operator-split internal aerosol equilibrium calculation. In mole concentration units, (17.118) and (17.119) simplify to

$$\frac{c_{NH_3(aq),i}}{C_{NH_3}} = H'_{NH_3,i}\ \frac{\text{mol}}{\text{mol}} \qquad H'_{NH_3,i} = H_{NH_3}R^*Tm_v c_{w,i} \tag{17.121}$$

$$\frac{c_{NH_4^+,i}}{c_{NH_3(aq),i}c_{H^+,i}} = K'_{NH_3,i}\ \frac{\text{cm}^3}{\text{mol}} \qquad K'_{NH_3,i} = K_{NH_3}\frac{1}{m_v c_{w,i}}\frac{\gamma_{i,H^+}}{\gamma_{i,NH_4^+}} \tag{17.122}$$

where all three parameters on the left side of the equation are unknown. Equations (17.115), (17.117), (17.121), and (17.122) represent a system of $3N_B+1$ equations and unknowns. These equations are solved exactly as follows.

Combining (17.121) and (17.122) gives an expression for the **hydrogen ion concentration in each size bin** as

$$c_{H^+,i} = \frac{c_{NH_4^+,i}}{C_{NH_3}H'_{NH_3,i}K'_{NH_3,i}} \tag{17.123}$$

Substituting this expression into (17.115) gives

$$c_{NH_4^+,i,t} = \frac{-c_{\pm,i,t}C_{NH_3,t}H'_{NH_3,i,t-h}K'_{NH_3,i,t-h}}{C_{NH_3,t}H'_{NH_3,i,t-h}K'_{NH_3,i,t-h}+1} \tag{17.124}$$

where the gas concentration on the right side of this equation is still unknown. This equation requires $c_{\pm,i,t} = \min(c_{\pm,i,t},0)$. If $c_{\pm,i,t} > 0$, then the size bin, following acid growth and before re-equilibration, has a net positive charge without H^+ or NH_4^+, which is not physically possible unless acid evaporates from the bin in a quantity greater than the H^+ + NH_4^+ present in the bin (since at the beginning of each growth step, the bin is in charge balance). If $c_{\pm,i,t} = 0$ due to such an occurrence, $c_{NH_4^+,i,t} = 0$ for the bin and time step.

Substituting (17.124) and $c_{NH_3(aq),i,t} = C_{NH_3,t}H'_{NH_3,i,t-h}$ into the mole balance equation (17.117), gives

$$C_{NH_3,t} + \sum_{i=1}^{N_B}\left(C_{NH_3,t}H'_{NH_3,i,t-h} - \frac{c_{\pm,i,t}C_{NH_3,t}H'_{NH_3,i,t-h}K'_{NH_3,i,t-h}}{C_{NH_3,t}H'_{NH_3,i,t-h}K'_{NH_3,i,t-h}+1}\right) - C_{tot} = 0 \tag{17.125}$$

This nonlinear equation has only one unknown, $C_{NH_3,t}$. It can be solved with a Newton–Raphson iteration,

$$C_{NH_3,t,n+1} = C_{NH_3,t,n} - \frac{f_n(C_{NH_3,t,n})}{f'_n(C_{NH_3,t,n})} \tag{17.126}$$

where n is the iteration number and

$$f_n(C_{NH_3,t,n}) = C_{NH_3,t,n} + \sum_{i=1}^{N_B} \left(C_{NH_3,t,n} H'_{NH_3,i,t-h} - \frac{c_{\pm,i,t} C_{NH_3,t,n} H'_{NH_3,i,t-h} K'_{NH_3,i,t-h}}{C_{NH_3,t,n} H'_{NH_3,i,t-h} K'_{NH_3,i,t-h} + 1} \right) - C_{tot} \quad (17.127)$$

$$f'_n(C_{NH_3,t,n}) = 1 + \sum_{i=1}^{N_B} \left[\begin{array}{l} H'_{NH_3,i,t-h} - \dfrac{c_{\pm,i,t} H'_{NH_3,i,t-h} K'_{NH_3,i,t-h}}{C_{NH_3,t,n} H'_{NH_3,i,t-h} K'_{NH_3,i,t-h} + 1} \\ + \dfrac{c_{\pm,i,t} C_{NH_3,t,n} \left(H'_{NH_3,i,t-h} K'_{NH_3,i,t-h} \right)^2}{\left(C_{NH_3,t,n} H'_{NH_3,i,t-h} K'_{NH_3,i,t-h} + 1 \right)^2} \end{array} \right] \quad (17.128)$$

This iteration always converges to a positive number and almost always within ≤ 12 iterations when the first guess of the ammonia gas concentration is zero ($C_{NH_3,t,0} = 0$). The computer time for iteration is relatively trivial compared with the computer time saved by taking a long time step for growth among all growing acids and between growth and equilibrium calculations. Once $C_{NH_3,t}$ is obtained, it is substituted back into (17.124) to obtain the updated ammonium concentration, into $c_{NH_3(aq),i,t} = C_{NH_3,t} H'_{NH_3,i,t-h}$ to update the liquid ammonia concentration, and into (17.123) with ammonium to obtain the updated pH. However, pH is updated more completely during the operator-split equilibrium calculation.

Following the ammonia calculation, a new operator-split internal aerosol equilibrium calculation is performed to recalculate aerosol ion, liquid, and solid composition, activity coefficients, and liquid water content, accounting for all species in solution. After the equilibrium calculation and after particle size and growth rates have been updated, another time interval of growth is taken.

Figure 17.7 shows results from a simulation in which nonequilibrium growth was coupled with equilibrium chemistry when the relative humidity was 90 percent. During the simulation period, H_2SO_4 condensation, HCl and HNO_3 dissolution, and NH_3 equilibration were solved. Water uptake was calculated during the operator-split equilibrium calculation. Figures 17.7 (b) and (c) show time-series of the aerosol concentrations, summed over all size bins, when the operator-split time interval between growth and equilibrium was 5 s and 300 s, respectively. The figures also show results from an equilibrium-only calculation. Results from both operator-split time steps matched each other and the equilibrium solution fairly accurately, although the longer time step resulted in a small error during the first 15 minutes. Equilibrium was reached within about six hours in both simulations.

17.14.3 Solution to growth in the presence of solid formation

As the liquid water content of an aerosol population decreases, solid formation increases due to either ion crystallization or gas heterogeneous reaction on particle surfaces. At high and moderate liquid water contents, solid formation, if it

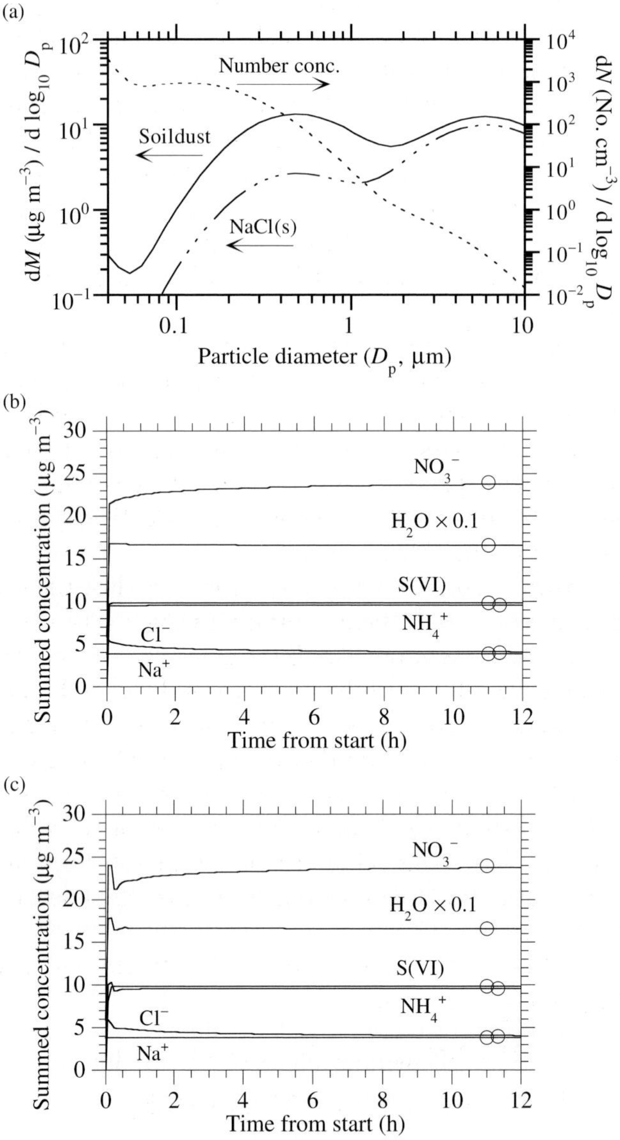

Figure 17.7 (a) Initial model size distribution, consisting of 10 μg m^{-3} NaCl(s) and 20.4 μg m^{-3} of nonreactive soildust, spread over 60 size bins. (b) Time series of aerosol component concentrations, summed over all size bins, from a simulation of dissolutional growth coupled with chemical equilibrium (solved with PNG-EQUISOLV II) when the operator-split time interval was $h = 5$ s. (c) Same as (b) but when $h = 300$ s. The circles are equilibrium solutions, found from EQUISOLV II alone. The initial gas concentrations were 30 μg m^{-3} HNO_3(g); 10 μg m^{-3} NH_3(g); 0 μg m^{-3} HCl(g); and 10 μg m^{-3} H_2SO_4(g). The relative humidity and temperature were 90 percent and 298 K, respectively. From Jacobson (2005b).

occurs, is represented well by the dissolutional growth/chemical equilibrium scheme described in Section 17.14.2. In that case, solid formation is calculated during the operator-split equilibrium calculation between growth time steps. When the total (over all sizes) aerosol liquid water content decreases to below about 0.01 $\mu g\ m^{-3}$, though, solid formation by the dissolutional growth/equilibrium procedure proceeds slowly at long time steps since solids form only during the operator-splitting equilibrium calculation. If the time interval between growth and equilibrium is long, the little gas that dissolves in the small amount of water each growth time step during the interval cannot form much solid when the equilibrium calculation is finally performed.

The problem of solid formation at low liquid water content is addressed here by treating HCl and HNO_3 growth as condensation processes when the liquid water content is low. At low water contents, growth of CO_2 and other acids are still treated as dissolution processes since (1) such acids are less important with respect to solid formation than is HCl or HNO_3, and (2) the dissolution scheme, which still allows solids to form through the equilibrium calculation, is unconditionally stable for any nonzero liquid water content.

When acid growth is solved as a condensation process, the saturation mole concentration (SMC) of a gas is held constant during a time step of growth. In addition, the condensed gases are treated as liquids, except that their saturation mole concentrations are determined based on gas–solid equilibrium to ensure the correct amount of acidic gas transfers to particle surfaces for solid formation. The liquids/ions are assumed to enter into the small (nonzero) amount of liquid water on the particle surface. Once the acids are grown, the amount of ammonia that condenses simultaneously is calculated in the same manner as in Section 17.14.2. Because acids and the base ammonium are now supersaturated in solution with respect to the solid phase, solids readily form during the subsequent internal-aerosol equilibrium calculation.

The SMCs of HCl and HNO_3 are solved here simultaneously by considering the two gas–solid equilibrium reactions

$$NH_4NO_3(s) \rightleftharpoons NH_4(g) + HNO_3(g) \tag{17.129}$$

$$NH_4Cl(s) \rightleftharpoons NH_4(g) + HCl(g) \tag{17.130}$$

Either solid may form within aerosol particles when either (a) the relative humidity is increasing and less than the DRH of the solid or (b) the relative humidity is decreasing and less than the CRH of the solid. DRHs and CRHs were given in Table 17.4. Ammonium nitrate and ammonium chloride also form only when

$$p_{NH_3} p_{HNO_3} > K_{NH_4NO_3} \tag{17.131}$$

$$p_{NH_3} p_{HCl} > K_{NH_4Cl} \tag{17.132}$$

respectively, where the p's are partial pressures of the gases (atm) and the K's are gas–solid equilibrium coefficients (atm^2). At equilibrium, the relation between the SMC and the equilibrium coefficients of the two reactions can be obtained by substituting (17.96) into (17.131) and (17.132) and rewriting the equation as an equality, giving

$$C_{NH_3,s,t}C_{HNO_3,s,t} = K_{NH_4NO_3}(R^*T)^{-2} \quad (17.133)$$

$$C_{NH_3,s,t}C_{HCl,s,t} = K_{NH_4Cl}(R^*T)^{-2} \quad (17.134)$$

respectively. If (17.131) and (17.132) are satisfied, SMCs are calculated by solving (17.133) and (17.134) together with the mole-balance equation

$$C_{NH_3,t-h} - C_{NH_3,s,t} = C_{HNO_3,t-h} - C_{HNO_3,s,t} + C_{HCl,t-h} - C_{HCl,s,t} \quad (17.135)$$

The analytical solution to these three equations and three unknowns for the ammonia gas saturation mole concentration is

$$C_{NH_3,s,t} = \frac{1}{2}C_0 + \frac{1}{2}\sqrt{C_0^2 + 4[K_{NH_4NO_3} + K_{NH_4Cl}](R^*T)^{-2}} \quad (17.136)$$

where

$$C_0 = C_{NH_3,t-h} - C_{HNO_3,t-h} - C_{HCl,t-h} \quad (17.137)$$

The saturation mole concentrations of HNO_3 and HCl are now trivially found from (17.133) and (17.134), respectively. The solution assumes that all three SMCs are independent of particle size.

If only ammonium nitrate can form, (17.136) is solved by removing K_{NH_4Cl} and setting $C_0 = C_{NH_3,t-h} - C_{HNO_3,t-h}$. If only ammonium chloride can form, the equation is solved by removing $K_{NH_4NO_3}$ and setting $C_0 = C_{NH_3,t-h} - C_{HCl,t-h}$.

Once the saturation mole concentrations for HNO_3 and HCl are calculated, they are substituted back into growth equations of the form (16.67)/(16.68) and solved with the APC scheme (16.69)–(16.72). Once the acids are grown, the amount of ammonia that condenses simultaneously is calculated with the method given in Section 17.14.2. Subsequently, an internal-aerosol equilibrium calculation is performed during which the ions may crystallize to form solids through the ion–solid equilibrium equations, $NH_4NO_3(s) \rightleftharpoons NH_4^+ + NO_3^-$ and $NH_4Cl(s) \rightleftharpoons NH_4^+ + Cl^-$. Alternatively, the ions may hydrate liquid water if conditions are right. Thus, the resulting particle can consist of both aqueous and solid components.

Figure 17.8 shows the time series of aerosol component concentrations, summed over all size bins, during a 12-hour nonequilibrium growth–equilibrium chemistry calculation when the relative humidity was 10 percent. The initial aerosol distribution was the same as that in Fig. 17.7(a). During the period, sulfuric acid, nitric acid, and hydrochloric acid gas–aerosol transfer were treated as nonequilibrium condensation processes. Ammonia gas equilibrated with acids after they grew. Solids formed during the operator-split equilibrium calculation. The figure shows that the result with a 300 s time step was nearly the same as that with a 5 s step.

17.15 SUMMARY

In this chapter, chemical equilibrium, activity-coefficient, and water content equations were discussed and applied. Two methods of solving equilibrium,

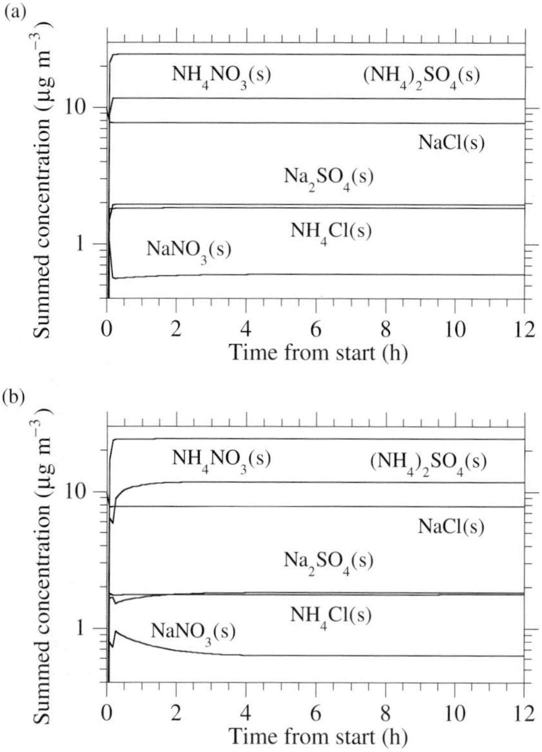

Figure 17.8 (a) Time series of aerosol component concentrations, summed over all size bins, from a simulation of dissolutional growth coupled with chemical equilibrium (solved with PNG-EQUISOLV II) when the operator-split time interval between growth and equilibrium was h = 5 s. Initial gas concentrations were 30 μg m^{-3} $HNO_3(g)$; 10 μg m^{-3} $NH_3(g)$; 0 μg m^{-3} HCl(g); and 10 μg m^{-3} $H_2SO_4(g)$. In addition, the relative humidity was 10 percent and $T = 298$ K. (b) Same as part (a), but with $h = 300$s. From Jacobson (2005b).

activity-coefficient, and water content equations simultaneously were also given, and temperature-dependent expressions for mean binary solute activity coefficients were shown. Mean mixed solute activity coefficients were calculated from mean binary solute activity coefficients with a mixing rule. Techniques for solving nonequilibrium dissolutional growth among multiple size bins and coupling growth to equilibrium were also discussed. The growth techniques included those for nondissociating species, dissociating species, and species that form solids. In cloud drops and aerosol particles containing substantial water, irreversible aqueous chemical reaction should also be accounted for. Cloud formation and development are described in Chapter 18. Irreversible chemistry within aerosol particles and clouds is discussed in Chapter 19.

17.16 PROBLEMS

17.1 Write the equilibrium coefficient expression, including a solute activity coefficient term, for the reaction $NaCl(s) \rightleftharpoons Na^+ + Cl^-$.

17.2 **(a)** Calculate the equilibrium coefficient for the reaction $NH_4NO_3(s) + NaCl(s) \rightleftharpoons NH_4Cl(s) + NaNO_3(s)$ using thermodynamic data from Appendix Table B.6.

(b) Calculate the equilibrium coefficient for the same reaction by combining rate coefficient terms from Appendix Table B.7 for four separate solid–ion equilibrium reactions.

17.3 What is the purpose of calculating mean mixed solute activity coefficients? Why can't binary solute activity coefficients be used for all calculations?

17.4 How are the thermodynamic activities of gases, liquids, ions, and solids related to their mole concentrations? Write out expressions for each case.

17.5 List the experimental data required to solve an equilibrium problem that includes gas, solution, and solid phases. Ignore coupling to growth equations.

17.6 What is the mathematical relationship between Q_d and Q_n in Section 17.11 when $\Delta x = 0$?

17.7 **(a)** Solve (17.91) and (17.92) together analytically, excluding the equilibrium term, for $c_{q,i,t}$ and $C_{q,t}$. Assume only one size bin is present and all variables, except $c_{q,i,t}$ and $C_{q,t}$, are constant. (Hint: $C_{q,t} + c_{q,i,t} = C_{q,t-h} + c_{q,i,t-h}$, since $i = 1$.)

(b) Discretize the time derivatives in the equations from part (a) in first-order, backward-difference form, and write out the right sides of both equations implicitly. Find the resulting solutions for $c_{q,i,t}$ and $C_{q,t}$.

17.8 Calculate the equilibrium mole concentrations of $HNO_3(g)$ and $HNO_3(aq)$ in the presence of a dilute solution of 2000 μg m^{-3} of liquid water if the only reaction considered is $HNO_3(g) \rightleftharpoons HNO_3(aq)$, the atmosphere initially contains 10 μg m^{-3} $HNO_3(g)$ and no $HNO_3(aq)$, and $T = 298.15$ K.

17.9 Calculate the equilibrium mole concentrations of $HNO_3(g)$ and NO_3^- in the presence of a dilute solution of 3000 μg m^{-3} of liquid water if the only reaction considered is $NH_3(g) + HNO_3(g) \rightleftharpoons NH^{+++}NO_3^-$, the atmosphere initially contains 10 μg m^{-3} $NH_3(g)$, 5 μg m^{-3} HNO_3 and no ions, and $T = 298.15$ K. Use at least seven digits of accuracy.

17.17 COMPUTER PROGRAMMING PRACTICE

17.10 Write a computer script to solve the following problem. Assume uniformly sized atmospheric particles contain dissolved nitric acid (H^+, NO_3^-), dissolved hydrochloric acid (H^+, Cl^-), and liquid water. Assume $f_r = 82$ percent, $T = 298$ K, and the mass concentrations of nitrate (NO_3^-) and chloride (Cl^-) at equilibrium in each particle are both 4 micrograms per cubic meter of air.

(a) Convert the units of chloride and nitrate to moles per cubic centimeter (mole concentration), and determine the mole concentration of H^+ by charge balance, assuming H^+ is the only cation and Cl^- and NO_3^- are the only anions in solution. Use the mole balance to determine the mole concentrations of HNO_3 and HCl dissolved in solution.

(b) Calculate the liquid-water content of air (micrograms of H_2O per cubic meter of air) from the water equation using the water activity data in Table B.10.

(c) Using the liquid-water content, convert all particle component concentrations to units of molality (moles per kilogram of water), and calculate the pH of the mixed solution, assuming molarity equals molality for convenience.

(d) Calculate the binary solute activity coefficients, γ_{H^+,NO_3^-} and γ_{H^+,Cl^-}, using Pitzer's method. For each calculation, assume H^+–NO_3^- and H^+–Cl^- pairs are alone in solution. Assume $\mathbf{m}_{NO_3^-}$ and $\mathbf{m}_{Cl^-}$ are the same as in part (c). Assume $\mathbf{m}_{H^+} = \mathbf{m}_{NO_3^-}$ and $\mathbf{m}_{H^+} = \mathbf{m}_{Cl^-}$ in each of the two cases, since the pairs are alone in solution.

(e) Assume H^+–NO_3^- and H^+–Cl^- pairs are alone in solution. Calculate the gas-phase partial pressures (in atmospheres) of HCl and HNO_3 from (17.50).

17.11 Assume for some anion B^- that HB(g) dissolves in particles of three sizes, and its effective Henry's constant for each size bin is $H'_{i,HB} = 4$. In solution, HB(aq) dissociates into H^+ and B^- with a dissociation constant $K_{eq} = 8$, written in generic units. Assume that the activity coefficients are unity, and ignore the Kelvin effect.

(a) Write the Henry's law and dissociation equilibrium-coefficient relationships for each size bin.

(b) Write the charge-balance constraint for each size bin and the mole-balance equation for B between the gas and aerosol phases. Assume $C_{tot} = 10\ \mu g\ m^{-3}$ is the total amount of B in the system.

(c) Solve for the equilibrium gas mole concentration of HB(g) and aerosol mole concentrations of H^+, B^-, and HB(aq) in each size bin.

(d) Describe how you would solve this problem if the gas phase were not in equilibrium with the particle phase (i.e., write out and describe the equations you would solve). You do not have to solve the equations.

17.12 Write a computer program to solve the three equilibrium equations

$$\begin{aligned} HNO_3(g) &\rightleftharpoons H^+ + NO_3^- \\ NH_3(g) + HNO_3(g) &\rightleftharpoons NH_4^+ + NO_3^- \\ HCl(g) &\rightleftharpoons H^+ + Cl^- \end{aligned}$$

plus the water equation with the AEI method. Assume the gases equilibrate with the ions in a single dilute size bin (all solute activity coefficients are unity), and $T = 298.15$ K. Assume initial gas concentrations of nitric acid, ammonia, and hydrochloric acid are 5, 15, and 3 $\mu g\ m^{-3}$, respectively. Plot equilibrium concentrations of all species as a function of relative humidity between 1 and 99 percent.

18

Cloud thermodynamics and dynamics

Clouds affect the atmosphere in several ways. They absorb and reflect radiation, modify local air temperatures, pressures, and winds, produce precipitation, mix and remove gases and particles, and alter photolysis coefficients. In this chapter, the thermodynamics and microphysics of clouds are discussed. The thermodynamic energy equation and the vertical momentum equation for a cloud are first derived. Cumulus parameterizations for simulating effects of subgrid scale clouds in a model are then described. Finally, numerical techniques for size-resolved liquid and ice cloud microphysics are given. The techniques include those for growth of water vapor onto aerosol particles to form liquid drops and ice crystals, hydrometeor–hydrometeor coagulation, breakup of large liquid drops, contact freezing, heterogeneous/homogeneous freezing, evaporation/sublimation, evaporative freezing, ice crystal melting, and aerosol–hydrometeor coagulation. Finally, numerical methods of treating removal of aerosol particles by rainout and washout and of treating lightning generation are given.

18.1 FOG AND CLOUD TYPES AND FORMATION MECHANISMS

Clouds are a type of hydrometeor. A **hydrometeor** is an ensemble of liquid or solid water particles suspended in or falling through the air. In this section, cloud types and their formation mechanisms are described. Additional discussions of clouds can be found in Cotton and Anthes (1989), Rogers and Yau (1989), and Houze (1993).

18.1.1 Cloud classification

Clouds form primarily in the troposphere. In the tropics, the highest clouds reach heights of 18 km. In midlatitudes and at the poles, tropospheric clouds reach maximum heights of 13 and 8 km, respectively. Water ice and nitric acid ice clouds can form in the polar stratosphere. These clouds are discussed in Chapter 11. For purposes of cloud classification, the troposphere is divided into three altitude ranges or **étages**, in which clouds of different types (genera) form most frequently. These étages encompass high, middle, and low altitude ranges, as shown in Table 18.1.

In 1802, Jean Baptiste Lamarck (1744–1829) proposed a cloud classification scheme, but the cloud types he suggested were not generally accepted. In 1803,

Table 18.1 Altitude range of cloud étages

	Range (km)		
Étage	Polar regions	Temperate regions	Tropical regions
High	3–8	5–13	6–9
Middle	2–4	2–7	2–8
Low	0–2	0–2	0–2

Source: WMO (1975).

Luke Howard proposed an alternative identification scheme that used Latin roots in the cloud names. Sheetlike clouds were called **stratus** (Latin for *layer*). Puffy clouds were called **cumulus** (*heap*), featherlike clouds were called **cirrus** (*curl of hair*), and rain clouds were called **nimbus** (*violent rain*).

Cloud types are now categorized by 10 **genera**. Table 18.2 describes the genera and identifies the étage(s) in which each genus is most commonly observed. A given cloud belongs to one genus only.

A **fog** is essentially a cloud touching the ground, but a fog is not classified as a cloud. Instead, it has its own classification and is defined as a suspension of liquid water drops that causes visibility to be reduced to less than 1 km (WMO 1975). If visibility is greater than 1 km, the fog is called a **mist**. If the fog contains ice crystals instead of liquid water, it is an **ice fog**.

Other hydrometeors suspended in the air include rain, supercooled rain, drizzle, supercooled drizzle, snow, snow grains, snow pellets, diamond dust, hail, small hail, ice pellets, drifting snow, blowing snow, and spray. **Rain** is precipitation of water drops falling from a cloud. **Supercooled rain** is liquid rain at a temperature below 0 °C. **Drizzle** is precipitation of water drops with diameter <0.5 mm. **Supercooled drizzle** is liquid drizzle at a temperature below 0 °C. **Snow** is precipitation from a cloud of single or agglomerated ice crystals. **Snow grains** are opaque ice crystals of diameter <1 mm. **Snow pellets** are rounded white and opaque ice particles with diameters up to 5 mm. **Diamond dust** is precipitation of small ice crystals from the clear sky. **Hail** is precipitation of clumpy, spheroidal ice particles 5–50 mm in diameter. **Small hail** is precipitation of translucent, spherical ice particles near 5 mm in diameter. **Ice pellets** are transparent ice particles, either spheroidal or irregular, with diameter <5 mm.

Hydrometeors affixed to a surface include fog deposits, dew, frozen dew, hoar frost, rime, glaze, and freezing rain. **Fog deposits** are fog drops fixed to a surface when the surface temperature exceeds 0 °C. **Dew** is liquid water on a surface produced from condensation of water vapor. **Frozen dew** forms from dew if the temperature drops below 0 °C. **Hoar frost** or **frost** is ice produced by sublimation of water vapor on a surface. **Rime** is ice produced by the freezing of supercooled fog droplets on a surface when the surface temperature is below 0 °C. **Glaze** is transparent ice that forms when supercooled drizzle or rain freezes on contact with a surface. **Freezing rain** occurs when nonsupercooled drizzle or rain freezes on contact with a surface.

Table 18.2 Cloud classification

Genera	Description	Étage
Cirrus (Ci)	Detached with white, delicate filaments, whitish patches, or narrow bands. Fibrous appearance and/or silky sheen.	High
Cirrocumulus (Cc)	Thin, white patch, sheet, or layer cloud, made of small elements with grains or ripples, merged or separate, but regularly arranged.	High
Cirrostratus (Cs)	Transparent, whitish cloud veil. Fibrous or smooth, totally or partly covering the sky. Produces a halo.	High
Altocumulus (Ac)	White and/or gray patch, sheet, or layer, generally with shading. Contains rounded masses, sometimes fibrous, which may or may not be merged.	Middle
Altostratus (As)	Grayish sheet or layer, fibrous or uniform, totally or partly covers the sky. Has parts thin enough to reveal the Sun vaguely. No halo.	Middle High
Nimbostratus (Ns)	Gray, often dark. Contains rain that almost always reaches the ground. Thick enough to block the Sun. Low, ragged clouds appear below the cloud.	Low Middle High
Stratocumulus (Sc)	Gray and/or whitish, patch, sheet, or layer. Almost always has dark parts and rolls, which may or may not be merged.	Low
Stratus (St)	Gray cloud layer with uniform base that may produce drizzle, ice, or snow. When the Sun is visible through the cloud, its outline is clear. Stratus does not produce a halo, except at low temperatures.	Low
Cumulus (Cu)	Detached clouds, usually dense, with sharp outlines. Vertically developed as a rising mound, dome, or tower, of which the upper part often looks like a cauliflower. The sunlit portion appears white. Its base is dark and horizontal.	Low Middle
Cumulonimbus (Cb)	Heavy, dense cloud with great vertical extent, in the form of a mountain or tower. Part of upper cloud is smooth, fibrous, or striated, and almost always flattened, spreading out like an anvil. Under its dark base, precipitation occurs.	Low Middle High

Source: WMO (1975).

18.1.2 Cloud formation

Clouds form by one of several mechanisms. Clouds that form by surface heating and free convection are **convective clouds**. When the ground is exposed to intense sunlight, air immediately above the ground heats by conduction. The resulting unstable temperature profile causes buoyancy and lifting (free convection). The altitude at which lifting by free convection starts is the **level of free convection** (LFC). As warm air near the surface rises, it expands and cools dry-adiabatically. If the buoyant parcel contains water vapor, and if the parcel temperature cools to the **isentropic condensation temperature** (ICT), the vapor condenses, forming a cloud base. The altitude of the cloud base is the **lifting condensation level** (LCL). Release

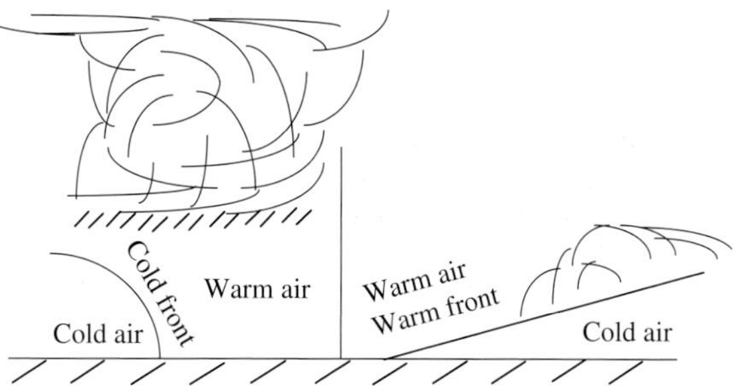

Figure 18.1 Formation of clouds along a cold and a warm front. The cold air behind a cold front forces warm air vertically. The warm air behind a warm front slides up over cold air ahead of the front. In both cases, the rising air cools and may result in cloud formation.

of latent heat during condensation at the cloud base provides buoyancy for further lifting, cooling, and cloud development.

Orographic clouds appear when a horizontal wind encounters a topographic barrier, such as a mountain, which forces the wind to shift slantwise or vertically (orographic uplifting). As the air rises, it expands and cools adiabatically. If the parcel rises to the LCL, a cloud forms.

Clouds can also form when winds converge horizontally, such as around a center of a low pressure. Convergence forces air to rise vertically (forced convection). Again, as the air rises, it expands and cools, eventually forming a cloud.

Clouds also form along weather fronts, as shown in Fig. 18.1. Along a **cold front**, cold, dense air pushes warm, moist air vertically, causing the warm air to expand and cool until condensation occurs. Along a **warm front**, warm, moist air slides over cold air and expands and cools until a cloud forms. The types of clouds that form along a cold front differ from those that form along a warm front. Ahead of a cold front, cumulonimbus, altocumulus, cirrostratus, and cirrus are the most prevalent cloud types. In front of a warm front, stratocumulus, stratus, nimbostratus, altostratus, cirrostratus, and cirrus are the most likely clouds to form.

Atmospheric stability above the LCL affects cloud type. If a cloud base is below 2 km, and the air is stable above the LCL, the cloud that forms is a cumulus humilis (cumulus cloud of slight vertical extent), stratus, or stratocumulus. If the atmosphere is unstable up to the tropopause, the cloud that forms is a cumulonimbus. Some cloud types form from interactions between other cloud types and the environment. When the top of a cirrostratus, altostratus, or stratus cloud cools and the bottom warms radiatively, the resulting buoyant motions within the cloud produce pockets of air that rise or sink to form cirrocumulus, altocumulus, or stratocumulus clouds, respectively. Cirrus clouds often form as the top of a cumulonimbus cloud dissipates.

18.1.3 Fog formation

Fogs form in one of several ways. A **radiation fog** forms when air near the ground cools radiatively during the night to the dew point. An **advection fog,** which generally appears near the coast, forms when the wind advects warm, moist air (usually ocean air) over a cold surface, cooling the air to the dew point. An **upslope fog** forms when warm, moist air flows up a topographical barrier and cools adiabatically to the dew point. The formation of this type of fog is similar to the formation of an orographic cloud. An **evaporation fog** forms when warm water evaporates and the vapor recondenses when it mixes with cool, dry air. Two types of evaporation fogs are steam fogs and frontal fogs. A **steam fog** occurs when warm surface water (e.g., from a lake or the ocean) evaporates and the vapor recondenses as it rises into cooler air, giving the appearance of rising steam. A **frontal fog** occurs when warm raindrops evaporate as they fall through a cold air mass and the vapor recondenses in the cold air. These conditions often exist ahead of an approaching front.

Fogs generally burn off from the top and bottom simultaneously. Solar energy evaporates drops at the fog top, and thermal-infrared energy emitted by the ground evaporates drops at the fog base. When the fog no longer touches the ground, it is a stratus cloud.

18.2 MOIST ADIABATIC AND PSEUDOADIABATIC PROCESSES

The extent to which a cloud grows vertically depends on atmospheric stability and moisture. In Chapter 2, stability in unsaturated air was discussed. Here, the concept of stability is extended to saturated air.

An unsaturated parcel of air rising dry-adiabatically cools at the rate of -9.8 K km^{-1}. If condensation occurs as the parcel rises, latent heat is released, offsetting the dry-adiabatic cooling, typically by $+4$ K km^{-1} but by up to $+8$ K km^{-1} in the tropics. When the released latent heat is absorbed by dry air, water vapor, and liquid water, the process is **moist adiabatic.** When, by assumption, it is absorbed by only dry air and water vapor, the process is **pseudoadiabatic.**

18.2.1 Pseudoadiabatic lapse rate

In this subsection, the pseudoadiabatic lapse-rate equation is derived. During a pseudoadiabatic ascent, the latent heat per unit mass of air (J kg^{-1}) released during condensation is

$$\mathrm{d}Q = -L_e\,\mathrm{d}\omega_{v,s} \tag{18.1}$$

where L_e is the latent heat of evaporation (J kg^{-1}) from (2.53), and $\omega_{v,s} = \varepsilon p_{v,s}/p_d$ is the mass mixing ratio (kilograms of vapor per kilogram of dry air) of water vapor at saturation over a liquid surface, defined in (2.67). Combining (18.1) with the first law of thermodynamics from (2.87) yields

$$-L_e\, d\omega_{v,s} = c_{p,m}\, dT - \alpha_a\, dp_a \tag{18.2}$$

Substituting $\alpha_a = 1/\rho_a = R_m T/p_a$ from (2.36) into (18.2) and rearranging give

$$dT = \frac{R_m T}{c_{p,m} p_a} dp_a - \frac{L_e}{c_{p,m}} d\omega_{v,s} \tag{18.3}$$

Differentiating (18.3) with respect to altitude and combining the result with $\partial p_a/\partial z = -\rho_a g$ and $p_a = \rho_a R_m T$ yield

$$\left(\frac{\partial T}{\partial z}\right)_w = \frac{R_m T}{c_{p,m} p_a}\frac{\partial p_a}{\partial z} - \frac{L_e}{c_{p,m}}\frac{\partial \omega_{v,s}}{\partial z} = -\frac{g}{c_{p,m}} - \frac{L_e}{c_{p,m}}\frac{\partial \omega_{v,s}}{\partial z} \tag{18.4}$$

where subscript w signifies pseudoadiabatic. Equation (18.4) simplifies to (2.90), the equation for the dry adiabatic lapse rate, when $\partial\omega_{v,s}/\partial z = 0$. Differentiating $\omega_{v,s} = \varepsilon p_{v,s}/p_d$ with respect to altitude, then substituting $dp_{v,s} = L_e p_{v,s}\, dT/R_v T^2$ (Clausius–Clapeyron equation), $\omega_{v,s} = \varepsilon p_{v,s}/p_d$, $R' = \varepsilon R_v$, and $\partial p_d/\partial z = -p_d g/R'T$ yield

$$\frac{\partial \omega_{v,s}}{\partial z} = \frac{\varepsilon}{p_d}\left(\frac{\partial p_{v,s}}{\partial z} - \frac{p_{v,s}}{p_d}\frac{\partial p_d}{\partial z}\right) = \frac{L_e \varepsilon \omega_{v,s}}{R' T^2}\frac{\partial T}{\partial z} + \frac{\omega_{v,s} g}{R'T} \tag{18.5}$$

Substituting (18.5) and $\Gamma_{d,m} = g/c_{p,m}$ from (2.90) into (18.4) and rearranging give the **pseudoadiabatic temperature change with altitude** as

$$\left(\frac{\partial T}{\partial z}\right)_w = -\Gamma_w = -\Gamma_{d,m}\left(1 + \frac{L_e \omega_{v,s}}{R'T}\right)\Bigg/\left(1 + \frac{L_e^2 \varepsilon \omega_{v,s}}{R' c_{p,m} T^2}\right) \tag{18.6}$$

where Γ_w is the **pseudoadiabatic lapse rate** (K km^{-1}).

Example 18.1

Find Γ_w if $p_d = 950$ hPa and (a) $T = 283$ K; (b) $T = 293$ K.

SOLUTION

(a) From (2.62), $p_{v,s} = 12.27$ hPa; from (2.67), $\omega_{v,s} = 0.00803$ kg kg^{-1}; from (2.80), $c_{p,m} = 1011.6$ J kg^{-1} K^{-1} (assuming the air is saturated with water vapor); from (2.54), $L_e = 2.4761 \times 10^6$ J kg^{-1}; and from (18.6) $\Gamma_w = 5.21$ K km^{-1}.

(b) When $T = 293$ K, $p_{v,s} = 23.37$ hPa, $\omega_{v,s} = 0.0153$ kg kg^{-1}, $c_{p,m} = 1017.9$ J kg^{-1} K^{-1}, $L_e = 2.4522 \times 10^6$ J kg^{-1}, and $\Gamma_w = 4.27$ K km^{-1}. Thus, the pseudoadiabatic lapse rate changes with temperature and pressure.

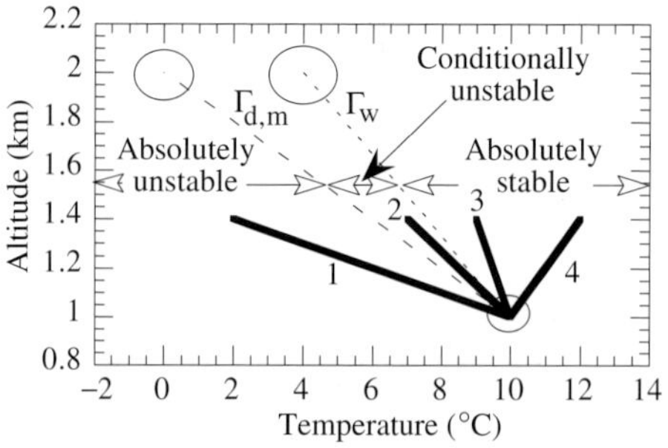

Figure 18.2 Stability criteria for unsaturated and saturated air. If air is saturated, the environmental temperature profile (thick solid lines) is compared with the pseudoadiabatic profile (right dashed line) to determine stability. Environmental temperature profiles 3 and 4 are stable and 1 and 2 are unstable with respect to saturated air. Profiles 2, 3, and 4 are stable and 1 is unstable with respect to unsaturated air. A rising or sinking air parcel follows the $\Gamma_{d,m}$-line when the air is unsaturated and the Γ_w-line when the air is saturated.

18.2.2 Stability criteria

In (2.100), stability criteria were given for unsaturated air. These criteria can be expanded to account for saturated and unsaturated air with

$$\begin{cases} \Gamma_e > \Gamma_{d,m} & \text{absolutely unstable} \\ \Gamma_e = \Gamma_{d,m} & \text{unsaturated neutral} \\ \Gamma_{d,m} > \Gamma_e > \Gamma_w & \text{conditionally unstable} \\ \Gamma_e = \Gamma_w & \text{saturated neutral} \\ \Gamma_e < \Gamma_w & \text{absolutely stable} \end{cases} \tag{18.7}$$

where $\Gamma_e = -\partial T/\partial z$ is the environmental lapse rate from Section 2.6.2.1. If, for example, $\Gamma_w = +6.0$ K km^{-1} and $\Gamma_e = +8.0$ K km^{-1} (and $\Gamma_{d,m} = 9.8$ K km^{-1}), the atmosphere is said to be **conditionally unstable**, meaning it is stable if the air is unsaturated but unstable if air is saturated (if a cloud can form). If the atmosphere is **absolutely stable**, it is stable in both unsaturated and saturated air. Figure 18.2 shows stability criteria for unsaturated and saturated air, and Fig. 18.3 shows how stability is determined in atmospheric layers when each layer has a different ambient environmental temperature profile.

Another method of estimating stability is with **equivalent potential temperature**, which is the potential temperature a parcel of air would have if all its water vapor were condensed and if the latent heat released were used to heat the parcel. It can be found by lifting a parcel from 1000 hPa to low pressure until all water vapor is condensed out, then lowering the parcel dry-adiabatically back to 1000 hPa, as shown in Fig. 18.4. In reality, not all water vapor in a parcel condenses because

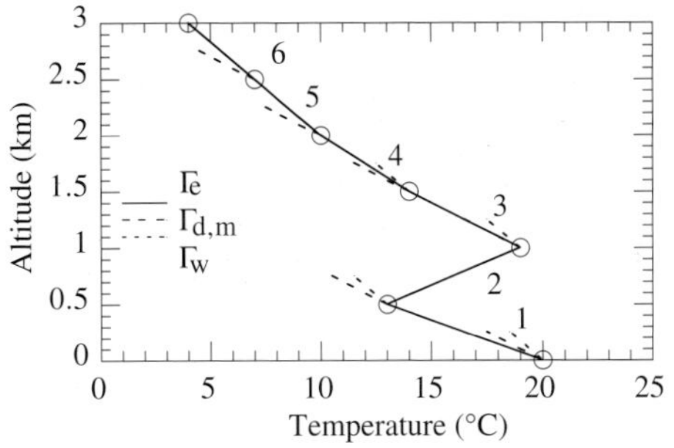

Figure 18.3 Determination of stability in multiple layers of air. Layer 1 is absolutely unstable, layer 2 is absolutely stable and an inversion, layer 3 is dry neutral, layer 4 is conditionally unstable, and layers 5 and 6 are moist neutral. Stability is determined by comparing the slope of the environmental temperature profile (solid line) with the slopes of the dry and pseudoadiabatic lapse-rate profiles.

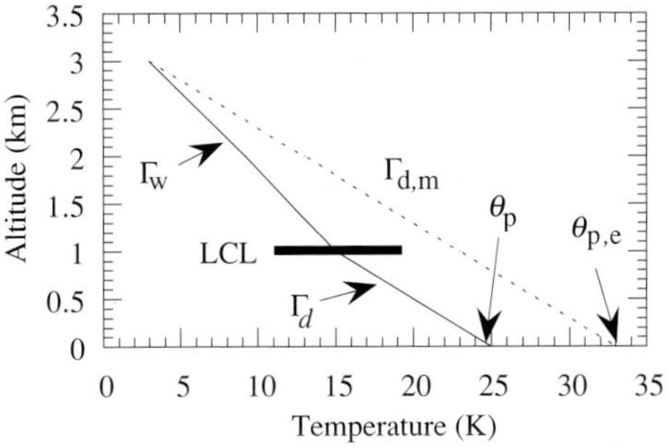

Figure 18.4 Schematic showing the relationship between potential temperature and equivalent potential temperature. Suppose a parcel with θ_p = 298.15 K (25 °C) rises to the LCL dry-adiabatically then to 3 km pseudoadiabatically, at which point it has lost its water vapor. If the parcel descends dry-adiabatically back to the surface, its final potential temperature is $\theta_{p,e}$ = 306.15 K (33 °C).

the saturation mass mixing ratio of water always exceeds zero. At high altitudes, though, the additional condensation from an incremental decrease in temperature is negligible, and the pseudoadiabatic lapse rate approaches the dry adiabatic lapse rate.

If the air is initially saturated, as it is above the LCL in Fig. 18.4, the equivalent potential temperature is

$$\theta_{p,e} \approx \theta_p \exp\left(\frac{L_e}{c_{p,d}T}\omega_{v,s}\right) \tag{18.8}$$

where T is the actual initial temperature of the saturated parcel (K), and $\omega_{v,s}$ is the saturation mass mixing ratio of water vapor at that temperature (kg kg^{-1}). If the air is initially unsaturated, as it is below the LCL, equivalent potential temperature is

$$\theta_{p,e} \approx \theta_p \exp\left(\frac{L_e}{c_{p,d} T_{LCL}} \omega_v\right) \tag{18.9}$$

where T_{LCL} is the temperature (K) of the parcel if it were lifted dry-adiabatically to the LCL, and ω_v is the initial, unsaturated mass mixing ratio of water vapor in the parcel.

Stability in unsaturated or saturated air can be found from equivalent potential temperature by defining a variable, $\hat{\theta}_{p,e}$ that equals $\theta_{p,e}$ from (18.8) for either saturated or unsaturated air. For unsaturated air, the temperature in (18.8) is the initial temperature of the environment, not T_{LCL}. In saturated air, the temperature is the saturated-parcel temperature. In both cases, $\omega_{v,s}$ is the saturation mixing ratio at the temperature used. Conceptually, $\hat{\theta}_{p,e}$ is the value of $\theta_{p,e}$ in a hypothetically saturated parcel at the temperature of the parcel. The **stability criteria in terms of $\hat{\theta}_{p,e}$** are

$$\frac{\partial \hat{\theta}_{p,e}}{\partial z} \begin{cases} < 0 & \text{saturated unstable} \\ = 0 & \text{saturated neutral} \\ > 0 & \text{saturated stable} \end{cases} \tag{18.10}$$

18.3 CLOUD DEVELOPMENT BY FREE CONVECTION

A fog or cloud forms when the air temperature drops below the dew point. Figure 18.5 shows an example of how the environmental, dry-adiabatic, and pseudoadiabatic lapse rates can be applied together with the dew point to estimate the extent of cloud development. If air at the surface is moist but unsaturated, the environment is unstable with respect to dry air. If an external forcing, such as forced convection, occurs, surface air rises. As the parcel rises, the dry-air pressure (p_d) in the parcel decreases because pressure decreases with increasing altitude. If the water-vapor mixing ratio (ω_v) is constant in the parcel, p_v must decrease with increasing height to satisfy $\omega_v = \varepsilon p_v / p_d$. Because p_v decreases, the dew point must also decrease, since it is an increasing function of p_v, as seen from (2.68). The decrease in dew point with increasing altitude is relatively modest (e.g., around 2 K km^{-1}).

The temperature in an unsaturated rising parcel decreases at the dry adiabatic lapse rate. When the dew point and the parcel temperature meet, the air is saturated, condensation occurs, and a cloud forms, as shown in Fig. 18.5. The altitude of cloud formation is the LCL. The temperature at saturation is the **isentropic condensation temperature** (ICT) (K), which usually occurs at the LCL. The ICT is more formally defined as the temperature at which saturation is reached when unsaturated air is cooled adiabatically at a constant mass mixing ratio of water vapor. The ICT can be approximated by substituting T_{IC} and $p_{d,IC}$ for T_D and p_d, respectively, into (2.68),

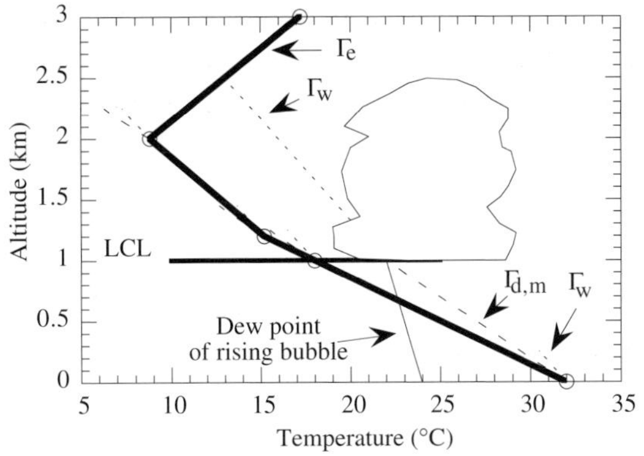

Figure 18.5 Simplified model of cumulus-cloud formation. An unsaturated surface parcel of air at 32 °C in an unstable environment (with environmental lapse rate (Γ_e) is perturbed and starts rising. As it rises, it cools at the dry adiabatic lapse rate ($\Gamma_{d,m}$). The dew point in the parcel decreases at a slower rate. At the lifting condensation level (LCL), the dew point equals the parcel temperature, and a cloud begins to form. Because the parcel temperature is still greater than that of the environment, the parcel continues to rise but cools at the pseudoadiabatic lapse rate (Γ_w). When the parcel temperature equals that of the environment (cloud top), the cloud parcel stops rising.

combining the result with $p_{d,IC} = p_{d,0}\,(T_{IC}/T_0)^{1/\kappa}$ from (2.96), where $p_{d,0}$ is the surface air pressure (not necessarily 1000 hPa) and T_0 is the surface temperature (K), then solving for T_{IC}. The result is

$$T_{IC} \approx \frac{4880.357 - 29.66 \ln\left[\dfrac{\omega_v p_{d,0}}{\varepsilon}\left(\dfrac{T_{IC}}{T_0}\right)^{1/\kappa}\right]}{19.48 - \ln\left[\dfrac{\omega_v p_{d,0}}{\varepsilon}\left(\dfrac{T_{IC}}{T_0}\right)^{1/\kappa}\right]} \tag{18.11}$$

where ω_v is assumed to be constant between the surface and LCL. The solution must be found iteratively.

Condensation releases latent heat, giving a parcel buoyancy. As a parcel rises above the LCL, as in Fig. 18.5, its temperature and dew point decrease pseudoadiabatically. So long as the temperature of the cloudy air exceeds that of the ambient air, the parcel remains buoyant and continues to rise and expand. When the cloud temperature and ambient temperature equalize, the cloud top decelerates. As shown in Fig. 18.5, the environmental temperature profile, which was unstable below the cloud base and within part of the cloud, must become stable for the cloud top to decelerate.

18.4 ENTRAINMENT

Entrainment is the mixing of relatively cool, dry air from outside a cloud with warm, moist air within the cloud, causing a cloud to evaporate and cool at its edges. Cool air sinks, creating downdrafts at the edges and reducing the height to which the cloud penetrates. Entrainment may also occur at the top or base of a cloud. **Detrainment** is the opposite of entrainment. During detrainment, air leaves the cloud and mixes with air around the cloud, slightly increasing the moisture and heat content of the outside air.

A simple model of the effects of entrainment on a cumulus cloud was developed by Stommel (1947). This model assumes that entrained air continuously enters a rising cloud from its sides and instantaneously mixes uniformly throughout the cloud. In reality, entrainment is not entirely lateral, instantaneous, or continuous (Houze 1993). Stommel's model is a useful tool for one-dimensional studies of clouds and can be used to derive a simplified entrainment term in the thermodynamic energy equation.

The model assumes that entrainment affects cloud temperatures in two ways. First, entrainment forces the cloud to expend energy to heat cool, entrained air to the cloud virtual temperature. The energy (J) used by the cloud for this purpose is

$$\mathrm{d}Q_1^* = -c_{p,\mathrm{d}}(T_\mathrm{v} - \hat{T}_\mathrm{v})\mathrm{d}M_\mathrm{c} \tag{18.12}$$

where T_v is the cloud virtual temperature, $\hat{T}_\mathrm{v}$ is the ambient virtual temperature, and $\mathrm{d}M_\mathrm{c}$ is the mass of ambient dry air plus water vapor entrained in the cloud.

Second, entrainment forces the cloud to expend energy evaporating liquid water to maintain saturation of dry, entrained air. This energy loss (J) is

$$\mathrm{d}Q_2^* = -L_\mathrm{e}(\omega_{\mathrm{v,s}} - \hat{\omega}_\mathrm{v})\mathrm{d}M_\mathrm{c} \tag{18.13}$$

where $\omega_{\mathrm{v,s}}$ is the saturation mass mixing ratio of water vapor in the cloud (kg kg^{-1}), and $\hat{\omega}_\mathrm{v}$ is the mass mixing ratio of water vapor outside the cloud.

Third, the entrainment region gains latent heat energy when rising water vapor condenses. The energy gained (J) is

$$\mathrm{d}Q_3^* = -M_\mathrm{c} L_\mathrm{e}\, \mathrm{d}\omega_{\mathrm{v,s}} \tag{18.14}$$

where M_c is the total mass of air within an entrainment region of the cloud, which consists of dry air, water vapor, and liquid water.

The sum of the three sources and sinks of energy is $\mathrm{d}Q^* = \mathrm{d}Q_1^* + \mathrm{d}Q_2^* + \mathrm{d}Q_3^*$. Substituting (18.12)–(18.14) into this equation gives the total change of energy in the entrainment region of a cloud as

$$\mathrm{d}Q^* = -c_{p,\mathrm{d}}(T_\mathrm{v} - \hat{T}_\mathrm{v})\mathrm{d}M_\mathrm{c} - L_\mathrm{e}(\omega_{\mathrm{v,s}} - \hat{\omega}_\mathrm{v})\mathrm{d}M_\mathrm{c} - M_\mathrm{c} L_\mathrm{e}\, \mathrm{d}\omega_{\mathrm{v,s}} \tag{18.15}$$

From (2.82) and (2.70), the first law of thermodynamics for this problem requires

$$\mathrm{d}Q^* = M_\mathrm{c}(c_{p,\mathrm{d}}\, \mathrm{d}T_\mathrm{v} - \alpha_\mathrm{a}\, \mathrm{d}p_\mathrm{a}) \tag{18.16}$$

Subtracting (18.16) from (18.15) and rearranging give

$$c_{p,\mathrm{d}}\,\mathrm{d}T_\mathrm{v} - \alpha_\mathrm{a}\,\mathrm{d}p_\mathrm{a} = -[c_{p,\mathrm{d}}(T_\mathrm{v} - \hat{T}_\mathrm{v}) + L_\mathrm{e}(\omega_\mathrm{v,s} - \hat{\omega}_\mathrm{v})]\frac{\mathrm{d}M_\mathrm{c}}{M_\mathrm{c}} - L_\mathrm{e}\,\mathrm{d}\omega_\mathrm{v,s} \qquad (18.17)$$

Dividing (18.17) by $c_{p,\mathrm{d}}T_\mathrm{v}$ and substituting $\alpha_\mathrm{a} = R'T_\mathrm{v}/p_\mathrm{a}$ result in

$$\frac{\mathrm{d}T_\mathrm{v}}{T_\mathrm{v}} - \frac{R'}{c_{p,\mathrm{d}}}\frac{\mathrm{d}p_\mathrm{a}}{p_\mathrm{a}} = -\left[\frac{T_\mathrm{v} - \hat{T}_\mathrm{v}}{T_\mathrm{v}} + \frac{L_\mathrm{e}(\omega_\mathrm{v,s} - \omega'_\mathrm{v})}{c_{p,\mathrm{d}}T_\mathrm{v}}\right]\frac{\mathrm{d}M_\mathrm{c}}{M_\mathrm{c}} - \frac{L_\mathrm{e}\,\mathrm{d}\omega_\mathrm{v,s}}{c_{p,\mathrm{d}}T_\mathrm{v}} \qquad (18.18)$$

Differentiating (18.18) with respect to height and then substituting $\partial p_\mathrm{a}/\partial z = -\rho_\mathrm{a} g$ and $p_\mathrm{a} = \rho_\mathrm{a} R'T_\mathrm{v}$ give the virtual temperature change with altitude in an entraining cloud as

$$\frac{\partial T_\mathrm{v}}{\partial z} = -\frac{g}{c_{p,\mathrm{d}}} - \left[(T_\mathrm{v} - \hat{T}_\mathrm{v}) + \frac{L_\mathrm{e}}{c_{p,\mathrm{d}}}(\omega_\mathrm{v,s} - \hat{\omega}_\mathrm{v})\right]\frac{1}{M_\mathrm{c}}\frac{\partial M_\mathrm{c}}{\partial z} - \frac{L_\mathrm{e}}{c_{p,\mathrm{d}}}\frac{\partial \omega_\mathrm{v,s}}{\partial z} \qquad (18.19)$$

When no entrainment occurs ($\mathrm{d}M_\mathrm{c} = 0$), (18.19) simplifies to (18.4).

Rearranging (2.103) as

$$\frac{\partial T_\mathrm{v}}{\partial z} = \frac{T_\mathrm{v}}{\theta_\mathrm{v}}\frac{\partial \theta_\mathrm{v}}{\partial z} + \frac{R'T_\mathrm{v}}{c_{p,\mathrm{d}}p_\mathrm{a}}\frac{\partial p_\mathrm{a}}{\partial z} = \frac{T_\mathrm{v}}{\theta_\mathrm{v}}\frac{\partial \theta_\mathrm{v}}{\partial z} - \frac{g}{c_{p,\mathrm{d}}} \qquad (18.20)$$

and substituting the result into (18.19) give the change in potential virtual temperature with altitude in an entrainment region of a cloud as

$$\frac{\partial \theta_\mathrm{v}}{\partial z} = -\frac{\theta_\mathrm{v}}{T_\mathrm{v}}\left[(T_\mathrm{v} - \hat{T}_\mathrm{v}) + \frac{L_\mathrm{e}}{c_{p,\mathrm{d}}}(\omega_\mathrm{v,s} - \hat{\omega}_\mathrm{v})\right]\frac{1}{M_\mathrm{c}}\frac{\partial M_\mathrm{c}}{\partial z} - \frac{\theta_\mathrm{v}}{T_\mathrm{v}}\frac{L_\mathrm{e}}{c_{p,\mathrm{d}}}\frac{\partial \omega_\mathrm{v,s}}{\mathrm{d}z} \qquad (18.21)$$

Multiplying through by $\mathrm{d}z$ and dividing through by $\mathrm{d}t$ give the time rate of change of potential virtual temperature in a cloud as

$$\frac{\mathrm{d}\theta_\mathrm{v}}{\mathrm{d}t} = -\frac{\theta_\mathrm{v}}{T_\mathrm{v}}\left[(T_\mathrm{v} - \hat{T}_\mathrm{v}) + \frac{L_\mathrm{e}}{c_{p,\mathrm{d}}}(\omega_\mathrm{v,s} - \hat{\omega}_\mathrm{v})\right]\mathrm{E} - \frac{\theta_\mathrm{v} L_\mathrm{e}}{c_{p,\mathrm{d}}T_\mathrm{v}}\frac{\mathrm{d}\omega_\mathrm{v,s}}{\mathrm{d}t} \qquad (18.22)$$

where $\mathrm{E} = (1/M_\mathrm{c})\,\mathrm{d}M_\mathrm{c}/\mathrm{d}t$ is the **entrainment rate** (s^{-1}) of outside air into the cloud. If a thermal is modeled as a spherical bubble with radius $r_\mathrm{t} \approx 0.2z_\mathrm{c}$ (m), where z_c is the center altitude (m) of the thermal above its starting point, then the rate of entrainment can be approximated with (Houze 1993)

$$\mathrm{E} = \frac{1}{M_\mathrm{c}}\frac{\mathrm{d}M_\mathrm{c}}{\mathrm{d}t} \approx \frac{3}{4\pi r_\mathrm{t}^3}\frac{\mathrm{d}}{\mathrm{d}t}\left(\frac{4\pi r_\mathrm{t}^3}{3}\right) \qquad (18.23)$$

Equation (3.76) gave the thermodynamic energy equation with diabatic source and sink terms. One such term is energy release due to condensation, $\mathrm{dQ}_{\mathrm{c/e}} = -L_\mathrm{e}\,\mathrm{d}\omega_\mathrm{v,s}$. Two other terms are energy release due to freezing of liquid water and deposition of vapor to ice. These terms can be quantified as $\mathrm{dQ}_{\mathrm{f/m}} = -L_\mathrm{m}\,\mathrm{d}\omega_L$ and $\mathrm{dQ}_{\mathrm{dp/s}} = -L_\mathrm{s}\,\mathrm{d}\omega_\mathrm{v,I}$, respectively, where L_m is the latent heat of melting from (2.55), $\mathrm{d}\omega_L$ is the change in mass mixing ratio (kg kg^{-1}) of liquid water upon freezing, L_s

is the latent heat of sublimation from (2.56), $\omega_{v,I} \approx \varepsilon p_{v,I}/p_d$ is the saturation mass mixing ratio of water vapor over an ice surface, which is analogous to (2.67), and $d\omega_{v,I}$ is the change in saturation mass mixing ratio upon sublimation.

Adding these and remaining terms from (3.76) to (18.22) gives the **thermodynamic energy equation in a cloud** as

$$\frac{d\theta_v}{dt} = -\frac{\theta_v}{T_v}\left[(T_v - \hat{T}_v) + \frac{L_e}{c_{p,d}}(\omega_{v,s} - \hat{\omega}_v)\right]E + \frac{1}{\rho_a}(\nabla \bullet \rho_a \mathbf{K}_h \nabla)\theta_v + \frac{\theta_v}{c_{p,d}T_v}\left(-L_e\frac{d\omega_{v,s}}{dt} - L_m\frac{d\omega_L}{dt} - L_s\frac{d\omega_{v,I}}{dt} + \frac{dQ_{solar}}{dt} + \frac{dQ_{ir}}{dt}\right) \quad (18.24)$$

Water categories in a cloud include water vapor, liquid water, drizzle, rainwater, cloud ice, snow, graupel, and hail. Conversion from one form of water to another results in a gain or loss of energy, accounted for in the latent-heat terms of (18.24). Bulk parameterizations of the conversion processes are given in Houze (1993), Fowler *et al.* (1996), and Pruppacher and Klett (1997). Size-resolved calculations of these conversion processes are discussed in Section 18.8.

18.5 VERTICAL MOMENTUM EQUATION IN A CLOUD

In a cloud, vertical scalar velocities are affected by local acceleration, gravity, pressure gradients, and turbulence. From (4.75), the vertical momentum equation in Cartesian-altitude coordinates was

$$\frac{dw}{dt} = -g - \frac{1}{\rho_a}\frac{\partial p_a}{\partial z} + \frac{1}{\rho_a}(\nabla \bullet \rho_a \mathbf{K}_m \nabla)w \quad (18.25)$$

If $\hat{p}_a$ and $\hat{\rho}_a$ are the pressure and density, respectively, of the air outside a cloud, and if the ambient air is in hydrostatic balance, $\partial \hat{p}_a/\partial z = -\hat{\rho}_a g$. Adding this equation to (18.25) gives

$$\frac{dw}{dt} = -g\frac{\rho_a - \hat{\rho}_a}{\rho_a} - \frac{1}{\rho_a}\frac{\partial(p_a - \hat{p}_a)}{\partial z} + \frac{1}{\rho_a}(\nabla \bullet \rho_a \mathbf{K}_m \nabla)w \quad (18.26)$$

where $p_a - \hat{p}_a$ and $\rho_a - \hat{\rho}_a$ are the deviations of cloud pressure and density from ambient pressure and density, respectively.

The **buoyancy factor** is defined as

$$B = -\frac{\rho_a - \hat{\rho}_a}{\rho_a} = -\frac{p_a\hat{T}_v - \hat{p}_a T_v}{p_a\hat{T}_v} = -\frac{\hat{T}_v - T_v}{\hat{T}_v} + \left(\frac{T_v}{\hat{T}_v}\right)\frac{\hat{p}_a - p_a}{p_a} \approx -\frac{\hat{\theta}_v - \theta_v}{\hat{\theta}_v} \quad (18.27)$$

(e.g., Rogers and Yau 1989) where T_v, θ_v, and $\rho_a = p_a/R'T_v$ are the virtual temperature, potential virtual temperature, and density, respectively, of cloudy air, and $\hat{T}_v$, $\hat{\theta}_v$, and $\hat{\rho}_a = \hat{p}_a/R'\hat{T}_v$ are the analogous variables for ambient air. The approximation on the right side of (18.27) was obtained by assuming $(\hat{p}_a - p_a)/p_a$ is small.

If a parcel contains liquid water, as it does above the lifting condensation level, condensate (condensed water) adds a downward force to the parcel. Condensate in the surrounding air also adds a downward force to the surrounding air. To allow for condensate, the buoyancy factor can be modified to

$$B = -\frac{\rho_a - \hat{\rho}_a}{\rho_a} = -\frac{\hat{\theta}_v(1+\omega_L) - \theta_v(1+\hat{\omega}_L)}{\hat{\theta}_v} \approx \frac{\theta_v - \hat{\theta}_v}{\hat{\theta}_v} - \omega_L \quad (18.28)$$

where ω_L and $\hat{\omega}_L$ are the mass mixing ratios of liquid water in the parcel and ambient air, respectively (kilograms of liquid water per kilogram of dry air). The second expression assumes the liquid-water content of the ambient air is small compared with that in a cloud.

Substituting (18.28) into (18.26) gives the vertical momentum equation as

$$\frac{dw}{dt} = g\left(\frac{\theta_v - \hat{\theta}_v}{\hat{\theta}_v} - \omega_L\right) - \frac{1}{\rho_a}\frac{\partial(p_a - \hat{p}_a)}{\partial z} + \frac{1}{\rho_a}(\nabla \cdot \rho_a \mathbf{K}_m \nabla)w \quad (18.29)$$

From (2.40), (4.48), and (5.38),

$$\frac{1}{\rho_a}\frac{\partial p_a}{\partial z} = -g = -\frac{\partial \Phi}{\partial z} = c_{p,d}\theta_v \frac{\partial P}{\partial z} \quad (18.30)$$

Substituting (18.30) into (18.29) for cloudy and ambient air gives the **vertical momentum equation in a cloud** as

$$\frac{dw}{dt} = g\left(\frac{\theta_v - \hat{\theta}_v}{\hat{\theta}_v} - \omega_L\right) - c_{p,d}\theta_v \frac{\partial(P - \hat{P})}{\partial z} + \frac{1}{\rho_a}(\nabla \cdot \rho_a \mathbf{K}_m \nabla)w \quad (18.31)$$

This equation is similar in several respects to (5.4), the nonhydrostatic vertical momentum equation. As with (5.4), (18.31) can be solved directly, used to derive a diagnostic equation for nonhydrostatic pressure, or solved implicitly.

If the pressure perturbation and the eddy diffusion term are ignored, (18.31) becomes

$$\frac{dw}{dt} = \frac{dw}{dz}\frac{dz}{dt} = \frac{dw}{dz}w = g\left(\frac{\theta_v - \hat{\theta}_v}{\hat{\theta}_v} - \omega_L\right) = gB \quad (18.32)$$

where $w = dz/dt$. Rearranging (18.32) gives $w\,dw = gB\,dz$. Integrating this equation from a reference height z_a, where the vertical scalar velocity is w_a, to height z yields a simplified expression for the vertical scalar velocity in a cloud,

$$w^2 = w_a^2 + 2g\int_{z_a}^{z}\left(\frac{\theta_v - \hat{\theta}_v}{\hat{\theta}_v} - \omega_L\right)dz = w_a^2 + 2g\int_{z_a}^{z} B\,dz \quad (18.33)$$

In this expression, w is determined from the integral of buoyancy between the base of the cloud and the altitude of interest.

18.6 CONVECTIVE AVAILABLE POTENTIAL ENERGY

Equation (18.33) can be modified to give an expression for the **convective available potential energy** (CAPE), which describes the growth potential of a cloud. The CAPE determines the buoyant stability of the atmosphere and correlates positively with growth of and rainfall production in cumulus clouds (Zawadski *et al.* 1981). CAPE (m^2 s^{-2}) is defined as

$$\mathrm{CAPE} = g\int_{z_{\mathrm{LFC}}}^{z_{\mathrm{LNB}}} B\,\mathrm{d}z \approx g\int_{z_{\mathrm{LFC}}}^{z_{\mathrm{LNB}}} \left(\frac{\theta_{\mathrm{v}} - \hat{\theta}_{\mathrm{v}}}{\hat{\theta}_{\mathrm{v}}}\right)\mathrm{d}z \tag{18.34}$$

where z_{LFC} is the **level of free convection** (LFC), z_{LNB} is the **level of neutral buoyancy** (LNB), θ_{v} is the potential virtual temperature of a rising parcel of air, and $\hat{\theta}_{\mathrm{v}}$ is the potential virtual temperature of the environment. The LFC is the altitude at which a parcel of rising air first becomes warmer than the environment. It may be below or above the LCL. The LNB is the altitude near the cloud top at which environmental and cloud temperatures equalize and the cloud is no longer buoyant.

Example 18.2

Estimate CAPE and w for a 10 km-thick cumulonimbus cloud over the ocean and over land if $\theta_{\mathrm{v}} - \hat{\theta}_{\mathrm{v}} \approx 1.5$ K over the ocean, $\theta_{\mathrm{v}} - \hat{\theta}_{\mathrm{v}} \approx 8$ K over land, and the average ambient virtual temperature between 0 and 10 km is $\theta_{\mathrm{v}} = 288$ K in both cases.

SOLUTION

From (18.34), CAPE ≈ 511 m^2 s^{-2} for the ocean case and 2725 m^2 s^{-2} for the land case. If ω_{L} is ignored, (18.33) gives $w = 32$ m s^{-1} for the ocean case and $w = 74$ m s^{-1} for the land case. These values are higher than observed maximums over the ocean and land, which are 10 and 50 m s^{-1}, respectively, in thunderstorm clouds.

18.7 CUMULUS PARAMETERIZATIONS

Clouds form over horizontal scales of tens to hundreds of meters. When a model's horizontal resolution is smaller than this, the vertical momentum equation can be used to reproduce cloud structure. The vertical momentum equation can also be used to reproduce much of the structure but not the details of a squall-line convective system for grid resolution of up to 4 km (Weisman *et al.* 1997). Many mesoscale and global models have horizontal resolutions of 4–50 km and 100–600 km, respectively. In both types of models, cloud development is a subgrid-scale phenomenon and must be parameterized.

Several techniques have been developed to estimate the effects of subgrid-scale cumulus clouds on the model-scale environment. These techniques are called **cumulus parameterizations** and require input variables from the model-scale environment. Important model-scale variables used to predict subgrid effects are horizontal and vertical wind speeds, potential temperatures, and total water mixing ratios. Cumulus parameterizations use these variables to adjust potential temperature, total water, and momentum fields and to predict precipitation rates. The effects of a cumulus parameterization on the model-scale environment are **feedbacks**.

Cumulus parameterizations include moist convective adjustment schemes (e.g., Manabe *et al.* 1965; Miyakoda *et al.* 1969; Krishnamurti and Moxim 1971; Kurihara 1973), Kuo schemes (Kuo 1965, 1974; Anthes 1977; Krishnamurti *et al.* 1980; Molinari 1982), the Arakawa–Schubert scheme (Arakawa and Schubert 1974; Lord and Arakawa 1980; Kao and Ogura 1987; Moorthi and Suarez 1992; Cheng and Arakawa 1997; Ding and Randall 1998), and other schemes (Ooyama 1971; Kreitzberg and Perkey 1976; Fritsch and Chappel 1980; Betts 1986; Betts and Miller 1986; Frank and Cohen 1987; Tiedtke 1989; Kain and Fritsch 1990; Emanuel 1991; Grell 1993; Hack 1994; Kain 2004). Cotton and Anthes (1989) discuss several schemes in detail. Three are briefly described below.

Moist convective adjustment schemes are the most basic cumulus parameterizations. In these schemes, the model-scale vertical temperature profile is adjusted to a critical, stable profile when the relative humidity exceeds a specified value and the temperature profile is unstable with respect to moist air. During adjustment, the temperature profile is adjusted to the pseudoadiabatic rate, the large-scale relative humidity is unchanged, condensed water vapor precipitates, and total moist enthalpy is conserved.

In **Kuo schemes**, rainfall from cumulus convection is assumed to occur following model-scale convergence of moisture. Part of the moisture condenses, releasing latent heat and increasing rainfall. The rest is used to increase the relative humidity of the environment. Cloud dynamics and microphysics are not computed in Kuo schemes, cloud types are not classified, and the altitudes of cloud bases and tops cannot be found.

In the **Arakawa–Schubert scheme** the model-scale environment is divided into a cloud layer, where clouds form, and a subcloud mixed layer. Within the cloud layer, multiple individual clouds are allowed to form. The sum of the individual clouds in a column makes up a cloud ensemble. The cloud ensemble occupies a horizontal area much smaller than the horizontal area of a grid cell. Each cloud in an ensemble has its own fractional entrainment rate, vertical mass flux across the cloud base, and cloud top height. The fractional entrainment rate is defined as the entrainment rate per unit height divided by the vertical mass flux. The ensemble is divided into subensembles, which consist of clouds with similar fractional entrainment rates. Equations are derived for each subensemble and summed over all subensembles to obtain the net effect of the ensemble on the model-scale environment.

Cloud ensembles affect the model-scale environment in two ways. First, when saturated air containing liquid water detrains (escapes) from cloud tops and evaporates, it cools the model-scale environment. Evaporation increases the water-vapor content in the environment. Rates of detrainment differ for different cloud types and cloud-top heights. Second, cumulus convection, which occurs when clouds grow vertically, induces subsidence between clouds. During subsidence, the model-scale temperature increases and relative humidity decreases.

An advantage of the Arakawa–Schubert scheme is its sophisticated treatment of subgrid-scale clouds. Disadvantages of the original scheme were the omission of convective downdrafts and the assumption that all cloud bases existed in the lowest model layer. Convective downdrafts were added in Cheng and Arakawa (1997). Some versions of the model now permit cloud bases to appear at any altitude (Ding and Randall 1998).

18.8 CLOUD MICROPHYSICS

Prior sections of this chapter described cloud thermodynamics. Here, cloud microphysics, including microphysical interactions of hydrometeor particles with aerosol particles, is discussed. Numerical techniques are given for condensation/deposition of water vapor onto aerosol particles to form liquid and ice cloud particles, coagulation among liquid and ice hydrometeors to form precipitation-sized drops, liquid drop breakup, contact freezing, homogeneous-heterogeneous freezing, evaporation/sublimation of falling drops, evaporative freezing, ice crystal melting, aerosol–hydrometeor coagulation, gas washout, and lightning. Most techniques given here are described in greater detail in Jacobson (2003).

18.8.1 Condensation and ice deposition onto aerosol particles

Liquid cloud drops first form when water vapor condenses onto pre-existing aerosol particles in supersaturated air. Liquid drop formation can occur at temperatures down to −40 °C. Liquid drops that exist below 0 °C are called **supercooled** drops. At subfreezing temperatures, supercooled drop formation competes with ice crystal formation (ice deposition) for the limited amount of water vapor available. In Chapter 16, the equations for condensation and ice deposition of water vapor onto a single aerosol particle and a population of particles were derived. Numerical solutions were given for growth onto one aerosol size distribution. Here, the technique is extended to growth of liquid drops and ice crystals simultaneously onto multiple aerosol size distributions, such as those described by Table 15.1.

The ordinary differential equations for water vapor condensation/evaporation and deposition/sublimation onto multiple aerosol size distributions are

$$\frac{\mathrm{d}c_{\mathrm{L},Ni,t}}{\mathrm{d}t} = k_{\mathrm{L},Ni,t-h}(C_{\mathrm{v},t} - S'_{\mathrm{L},Ni,t-h}C_{\mathrm{L,s},t-h}) \tag{18.35}$$

$$\frac{\mathrm{d}c_{\mathrm{I},Ni,t}}{\mathrm{d}t} = k_{\mathrm{I},Ni,t-h}(C_{\mathrm{v},t} - S'_{\mathrm{I},Ni,t-h}C_{\mathrm{I,s},t-h}) \tag{18.36}$$

respectively. The corresponding vapor–hydrometeor mole balance equation is

$$\frac{\mathrm{d}C_{\mathrm{v},t}}{\mathrm{d}t} = -\sum_{N=1}^{N_\mathrm{T}}\sum_{i=1}^{N_\mathrm{B}}\begin{bmatrix} k_{\mathrm{L},Ni,t-h}(C_{\mathrm{v},t} - S'_{\mathrm{L},Ni,t-h}C_{\mathrm{L,s},t-h}) \\ +\,k_{\mathrm{I},Ni,t-h}(C_{\mathrm{v},t} - S'_{\mathrm{I},Ni,t-h}C_{\mathrm{I},s,t-h}) \end{bmatrix} \tag{18.37}$$

where t and $t-h$ indicate the end and beginning, respectively, of a time step of h seconds, subscripts L and I indicate liquid and ice, respectively, subscripts N

and i indicate the aerosol distribution and size bin in the distribution, respectively, from which the hydrometeor originates, N_T and N_B are the number of aerosol size distributions and the number of size bins in each distribution, respectively, $c_{L,Ni}$ and $c_{I,Ni}$ are mole concentrations (moles per cubic centimeter of air) of liquid water and ice, respectively, in size bin i of aerosol distribution N, C_v is water vapor mole concentration, $C_{L,s}$ and $C_{I,s}$ are saturation vapor mole concentrations over flat, dilute liquid water and ice surfaces, respectively (so are independent of the aerosol distribution and size bin), $S'_{L,Ni}$ and $S'_{I,Ni}$ are the saturation ratios at equilibrium of water vapor over a liquid solution and over an ice surface, respectively, in size bin i of distribution N, and $k_{L,Ni}$ and $k_{I,Ni}$ are the growth rates (s^{-1}) of water vapor to activated **cloud condensation nuclei** (CCN) and **ice deposition nuclei** (IDN), respectively. CCN were defined in Section 16.2.3.3 and IDN were defined in Section 16.8. Briefly, CCN are aerosol particles that can potentially activate into cloud drops if the air is supersaturated with respect to liquid over the particle surface. IDN are aerosol particles that can potentially activate into ice crystals if the air is supersaturated with respect to ice over the particle surface.

Expressions for the **growth rates** are

$$k_{L,Ni} = \frac{n_{lq,Ni} 4\pi r_{Ni} D_v \omega_{v,L,Ni} F_{v,L,Ni}}{\dfrac{m_v D_v \omega_{v,L,Ni} F_{v,L,Ni} L_e S'_{L,Ni} C_{L,s}}{\kappa_a \omega_{h,Ni} F_{h,L,Ni} T}\left(\dfrac{L_e m_v}{R^* T} - 1\right) + 1} \tag{18.38}$$

$$k_{I,Ni} = \frac{n_{ic,Ni} 4\pi \chi_{Ni} D_v \omega_{v,I,Ni} F_{v,I,Ni}}{\dfrac{m_v D_v \omega_{v,I,Ni} F_{v,I,Ni} L_s S'_{I,Ni} C_{I,s}}{\kappa_a \omega_{h,Ni} F_{h,I,Ni} T}\left(\dfrac{L_s m_v}{R^* T} - 1\right) + 1} \tag{18.39}$$

which were derived from (16.64) and (16.79), respectively. All terms are evaluated at time $t - h$. In these equations, $n_{lq,Ni}$ and $n_{ic,Ni}$ are the number concentrations (particles cm^{-3}) of CCN and IDN, respectively, in each size bin of each aerosol distribution.

Other terms in (18.38) and (18.39) include the following: m_v is the molecular weight of water vapor (g mol^{-1}), D_v is the diffusion coefficient of water vapor in air (cm^2 s^{-1}) (from (16.17) with subscript v substituted for subscript q to indicate water vapor), r_{Ni} is particle radius (cm), χ_{Ni} is the ice crystal electrical capacitance (cm) (from (16.77)), ω_v and ω_h are dimensionless factors for water vapor and energy, respectively, that account for corrections for collision geometry and sticking probability during growth (from (16.19) and (16.27), respectively), F_v is the dimensionless ventilation coefficient for vapor (from (16.24) for liquid and (16.78) for ice), F_h is the dimensionless ventilation coefficient for energy (from (16.31) for liquid and (16.78) for ice), L_e and L_s are latent heats of evaporation and sublimation, respectively (J g^{-1}), κ_a is the thermal conductivity of moist air (J cm^{-1} s^{-1} K^{-1}), T is temperature (K), and R^* is the universal gas constant (8.31451 J mol^{-1} K^{-1}).

The size bins included in (18.36) and (18.37) depend not only on whether CCN or IDN are present but also on whether conditions for activation of particles in a given size bin are met. CCN/IDN activation is determined by solving a Köhler equation similar to (16.40) but assuming (a) the Kelvin and solute effects affect the saturation ratio at equilibrium over liquid water, (b) multiple solutes dissolved in solution affect the saturation ratio over liquid water, and (c) only the Kelvin effect affects the saturation ratio at equilibrium over ice. The **Köhler equations** under such conditions are

$$S'_{\mathrm{L},Ni,t-h} \approx 1 + \frac{2\sigma_{\mathrm{L},Ni,t-h} m_{\mathrm{v}}}{r_{Ni} R^* T \rho_{\mathrm{L}}} - \frac{3m_{\mathrm{v}}}{4\pi r_{Ni}^3 \rho_{\mathrm{L}} n_{Ni,t-h}} \sum_{q=1}^{\mathrm{N_s}} c_{q,Ni,t-h} \tag{18.40}$$

$$S'_{\mathrm{I},Ni,t-h} \approx 1 + \frac{2\sigma_{\mathrm{I},Ni,t-h} m_{\mathrm{v}}}{r_{Ni} R^* T \rho_{\mathrm{I}}} \tag{18.41}$$

where $\sigma_{\mathrm{L},Ni}$ and $\sigma_{\mathrm{I},Ni}$ are the surface tensions over liquid and ice, respectively (dyn cm^{-1} = g s^{-2}), m_{v} is the molecular weight of water (g mol^{-1}), ρ_{L} and ρ_{I} are the densities of liquid water and ice, respectively (g cm^{-3}), r_{Ni} is particle radius (cm), R^* is the universal gas constant ($8.314\,51 \times 10^7$ g cm^2 s^{-2} mol^{-1} K^{-1}), T is temperature (K), n_{Ni} is the total number concentration of aerosol particles of size i in distribution N (particles cm^{-3}-air), $\mathrm{N_s}$ is the number of soluble components in an aerosol particle, and $c_{q,Ni}$ is the mole concentration of soluble component q in aerosol particles of size i in distribution N. Soluble components include dissociated or undissociated electrolytes and undissociated soluble molecules. The surface tension itself depends on solute concentration, as discussed in Section 16.2.3.1.

Equation (18.40) can be rewritten as

$$S'_{\mathrm{L},Ni,t-h} \approx 1 + \frac{a_{\mathrm{L},Ni,t-h}}{r_{Ni}} - \frac{b_{\mathrm{L},Ni,t-h}}{r_{Ni}^3} \tag{18.42}$$

where

$$a_{\mathrm{L},Ni,t-h} = \frac{2\sigma_{\mathrm{L},Ni,t-h} m_{\mathrm{v}}}{R^* T \rho_{\mathrm{L}}} \qquad b_{\mathrm{L},Ni,t-h} = \frac{3m_{\mathrm{v}}}{4\pi \rho_{\mathrm{L}} n_{Ni,t-h}} \sum_{q=1}^{\mathrm{N_s}} c_{q,Ni,t-h} \tag{18.43}$$

Taking the partial derivative of (18.42) with respect to radius and setting it to zero, solving for the radius, and substituting the result back into (18.42) give the **critical radius for growth** (cm) and **critical saturation ratio** for liquid cloud drop activation as

$$r^*_{\mathrm{L},Ni,t-h} = \sqrt{\frac{3b_{\mathrm{L},Ni,t-h}}{a_{\mathrm{L},Ni,t-h}}} \qquad S^*_{\mathrm{L},Ni,t-h} = 1 + \sqrt{\frac{4a^3_{\mathrm{L},Ni,t-h}}{27b_{\mathrm{L},Ni,t-h}}} \tag{18.44}$$

respectively. A CCN activates into a liquid cloud drop under the following conditions:

$$\text{CCN activation} \begin{cases} r_{Ni} > r^*_{\mathrm{L},Ni} \quad \text{and} \quad C_{\mathrm{v},t-h} > S'_{\mathrm{L},Ni,t-h} C_{\mathrm{L,s},t-h} \\ \qquad\qquad\qquad \text{or} \\ r_{Ni} \le r^*_{\mathrm{L},Ni} \quad \text{and} \quad C_{\mathrm{v},t-h} > S^*_{\mathrm{L},Ni,t-h} C_{\mathrm{L,s},t-h} \end{cases} \tag{18.45}$$

An IDN activates when

$$\text{IDN activation} \quad C_{v,t-h} > S'_{I,Ni,t-h} C_{I,s,t-h} \tag{18.46}$$

Above 0 °C, no ice can form and only (18.45) applies. Between −40 and 0 °C, condensation competes with ice deposition, but the saturation vapor mole concentration over ice is less than that over liquid ($C_{I,s} < C_{L,s}$), so ice growth is favored. Below −40 °C, no liquid drops can form, and only (18.46) applies.

Example 18.3

Calculate the critical radius and critical supersaturation when a population of 100 aerosol particles cm^{-3}, all with initial radius of 0.3 μm, contains 2×10^{-16} mol cm^{-3}-air of an organic compound that does not dissociate. Assume $T = 298.15$ K.

SOLUTION

The initial molality of the organic compound in solution is

$$\mathbf{m} = \frac{c}{n\upsilon\rho_W} = \frac{2 \times 10^{-16}\,\dfrac{\text{mol}}{\text{cm}^3} \times 1000\,\dfrac{\text{g}}{\text{kg}}}{100\,\dfrac{\text{partic.}}{\text{cm}^3} \times \dfrac{4}{3}\pi\left(3 \times 10^{-5}\,\text{cm}\right)^3 \times 1\,\dfrac{\text{g}}{\text{cm}^3}} = 0.0177\,\frac{\text{mol}}{\text{kg}}$$

From (14.19), the surface tension of liquid water at 298.15 K is 72.225 dyn cm^{-1}. From (16.34), the surface tension of water containing 0.0177 mol kg^{-1}of an organic compound is 58.31 dyn cm^{-1}. From (18.43),

$$a_L = \frac{2\sigma_L m_v}{R^* T \rho_L} = \frac{2 \times 58.31\,\dfrac{\text{dyn}}{\text{cm}} \times 18.02\,\dfrac{\text{g}}{\text{mol}}}{8.31451 \times 10^7\,\dfrac{\text{g cm}^2}{\text{s}^2\,\text{mol K}} \times 298.15\,\text{K} \times 1\,\dfrac{\text{g}}{\text{cm}^3}} 1\frac{\text{g cm}}{\text{dyn s}^2}$$

$$= 8.478 \times 10^{-8}\,\text{cm}$$

$$b_L = \frac{3m_w c}{4\pi\rho_L n} = \frac{3 \times 18.02\,\dfrac{\text{g}}{\text{mol}} \times 2 \times 10^{-16}\,\dfrac{\text{mol}}{\text{cm}^3}}{4\pi \times 1\,\dfrac{\text{g}}{\text{cm}^3} \times 100\,\dfrac{\text{partic.}}{\text{cm}^3}} = 8.604 \times 10^{-18}\,\text{cm}^3$$

respectively. Finally, from (18.44), the critical radius and critical saturation ratio are

$$r_L^* = \sqrt{\frac{3b_L}{a_L}} = 1.744 \times 10^{-5}\,\text{cm} = 0.174\,\mu\text{m}$$

$$S_L^* = 1 + \sqrt{\frac{4a_L^3}{27b_L}} = 1.0032$$

respectively. Since the radius of the population of aerosol particles exceeds the critical radius, the population will activate for any saturation ratio greater than that over an individual particle surface.

A noniterative numerical solution to (18.35)–(18.37) is obtained by integrating (18.35) and (18.36) for one size bin over a time step h, yielding

$$c_{\mathrm{L},Ni,t} = c_{\mathrm{L},Ni,t-h} + hk_{\mathrm{L},Ni,t-h}(C_{\mathrm{v},t} - S'_{\mathrm{L},Ni,t-h}C_{\mathrm{L,s},t-h}) \tag{18.47}$$

$$c_{\mathrm{I},Ni,t} = c_{\mathrm{I},Ni,t-h} + hk_{\mathrm{I},Ni,t-h}(C_{\mathrm{v},t} - S'_{\mathrm{I},Ni,t-h}C_{\mathrm{I,s},t-h}) \tag{18.48}$$

respectively, where the final gas mole concentration in both cases, $C_{\mathrm{v},t}$, is currently unknown. Final hydrometeor and gas concentrations are constrained by the gas–hydrometeor mole-balance equation,

$$C_{\mathrm{v},t} + \sum_{N=1}^{N_\mathrm{T}}\sum_{i=1}^{N_\mathrm{B}}(c_{\mathrm{L},Ni,t} + c_{\mathrm{I},Ni,t}) = C_{\mathrm{v},t-h} + \sum_{N=1}^{N_\mathrm{T}}\sum_{i=1}^{N_\mathrm{B}}(c_{\mathrm{L},Ni,t-h} + c_{\mathrm{I},Ni,t-h}) = C_{\mathrm{tot}} \tag{18.49}$$

Substituting (18.47) and (18.48) into (18.49) and solving for $C_{\mathrm{v},t}$ give a generalized solution for simultaneous condensation/evaporation and deposition/sublimation,

$$C_{\mathrm{v},t} = \frac{C_{\mathrm{v},t-h} + h\sum_{N=1}^{N_\mathrm{T}}\sum_{i=1}^{N_\mathrm{B}}(k_{\mathrm{L},Ni,t-h}S'_{\mathrm{L},Ni,t-h}C_{\mathrm{L,s},t-h} + k_{\mathrm{I},Ni,t-h}S'_{\mathrm{I},Ni,t-h}C_{\mathrm{I,s},t-h})}{1 + h\sum_{N=1}^{N_\mathrm{T}}\sum_{i=1}^{N_\mathrm{B}}(k_{\mathrm{L},Ni,t-h} + k_{\mathrm{I},Ni,t-h})} \tag{18.50}$$

Since $C_{\mathrm{v},t}$ from (18.50) can exceed the maximum gas concentration in the system, C_{tot}, $C_{\mathrm{v},t}$ must be set to the smaller of itself and C_{tot}. Equation (18.50) does not allow $C_{\mathrm{v},t}$ to fall below zero in any situation. Once $C_{\mathrm{v},t}$ is solved, it is substituted back into (18.47) and (18.48) to give the final hydrometeor concentration in each size bin of each distribution. Since (18.47) and (18.48) can result in negative concentrations or concentrations above the maximum, two limits are placed sequentially after both equations are solved among all size bins. The first is

$$c_{\mathrm{L},Ni,t} = \mathrm{MAX}\,(c_{\mathrm{L},Ni,t}, 0) \qquad c_{\mathrm{I},Ni,t} = \mathrm{MAX}\,(c_{\mathrm{I},Ni,t}, 0) \tag{18.51}$$

which is solved for all size bins. The second, shown for liquid (the equation for ice uses $c_{\mathrm{I},Ni,t} - c_{\mathrm{I},Ni,t-h}$ instead of $c_{\mathrm{L},Ni,t} - c_{\mathrm{L},Ni,t-h}$ in the rightmost term) is

$$c_{\mathrm{L},Ni,t} = \frac{\left\{C_{\mathrm{v},t-h} - C_{\mathrm{v},t} + \sum_{N=1}^{N_\mathrm{T}}\sum_{i=1}^{N_\mathrm{B}}\left\{\begin{array}{l}\mathrm{MAX}\,[c_{\mathrm{L},Ni,t-h} - c_{\mathrm{L},Ni,t}, 0]\\ +\mathrm{MAX}\,[c_{\mathrm{I},Ni,t-h} - c_{\mathrm{I},Ni,t}, 0]\end{array}\right\}\right\}}{\sum_{N=1}^{N_\mathrm{T}}\sum_{i=1}^{N_\mathrm{B}}\left\{\begin{array}{l}\mathrm{MAX}\,[c_{\mathrm{L},Ni,t} - c_{\mathrm{L},Ni,t-h}, 0]\\ +\mathrm{MAX}\,[c_{\mathrm{I},Ni,t} - c_{\mathrm{I},Ni,t-h}, 0]\end{array}\right\}} \times (c_{\mathrm{L},Ni,t} - c_{\mathrm{L},Ni,t-h}) \tag{18.52}$$

where all $c_{\mathrm{L},N,i,t}$ and $c_{\mathrm{I},N,i,t}$ values on the right side of the equation are determined from (18.51) after it has been solved for all size bins. The solution in (18.47)–(18.52) is exactly mole conserving between water vapor and all liquid and ice

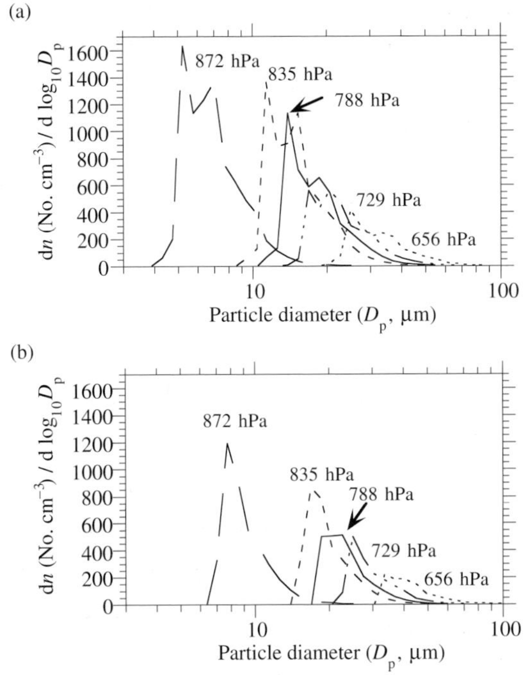

Figure 18.6 Condensed water onto 16 CCN distributions simultaneously in each of five layers in which the temperature was above 0 °C during a one-dimensional simulation when cloud activation was (a) determined from Equation (18.45) and (b) assumed to occur for all particles >0.2 μm in diameter regardless of their composition. From Jacobson (2003).

hydrometeors in all size bins of all distributions under all conditions and is noniterative.

Figure 18.6(a) shows a calculation of cloud drop growth at five altitudes. At each altitude, water vapor condensed onto the 16 aerosol distributions similar to those shown in Table 15.1, where CCN activation in each distribution was determined from (18.45). Liquid drop concentrations were summed over the 16 distributions in each layer to generate the figure. Figure 18.6(b) shows a similar calculation, but when all particles >0.2 μm in diameter were allowed to activate, regardless of their compositions. A comparison of the figures suggests that treatment of multiple distributions, each with different activation characteristics, produced dual peaks in the resulting cloud drop size distribution. Such dual peaks, which have also been observed (e.g., Pruppacher and Klett 1997, Figs. 2.25 and 2.12(a)), arose because different distributions activate at different diameters, causing discontinuities in the summed size distribution. When the activation diameter was held constant over all distributions (Fig. 18.6(b)), only one peak arose for each layer. Single peaks are also widely observed, suggesting that whether one or two peaks exist may depend on the activation properties of the underlying aerosol particles, which depend on composition, surface tension, and density.

18.8.2 Hydrometeor–hydrometeor coagulation

In Chapter 15, the integrodifferential coagulation equation and a numerical solution to it were given. Here, the solution is extended to liquid and ice hydrometeor distributions. Hydrometeor–hydrometeor coagulation (also called collision-coalescence) is the main process producing precipitation in warm clouds.

The scheme described here is semiimplicit, volume conserving, volume-concentration conserving, positive-definite, unconditionally stable, and noniterative, with no limitation on time step. Other schemes that have been used to solve coagulation in clouds include those of Tzivion *et al.* (1987), Hounslow *et al.* (1988), Lister *et al.* (1995), and Bott (2000). All such schemes conserve various properties, but are explicit; thus, their time step is limited by stability constraints.

The interactions considered here include liquid–liquid, ice–ice, and graupel–graupel self coagulation, liquid–ice, liquid–graupel, and ice–graupel heterocoagulation, and coagulation of aerosol components contained within the hydrometeor distributions. Whereas the initial sources of liquid and ice hydrometeors in the atmosphere are condensation and ice deposition, respectively, the initial source of graupel is ice–liquid coagulation. The solution method here accounts for all coagulation interactions simultaneously; thus ice–liquid heterocoagulation for example, is solved together with liquid–liquid self-coagulation.

The final volume concentration of component x, in particles of hydrometeor distribution Y in bin k at time t due to coagulation is determined with

$$v_{x,Yk,t} = \frac{v_{x,Yk,t-h} + h\left(T_{x,Yk,t,1} + T_{x,Yk,t,2}\right)}{1 + hT_{x,Yk,t,3}} \tag{18.53}$$

$$T_{x,Yk,t,1} = \sum_{M=1}^{N_H}\left[P_{Y,M}\sum_{j=1}^{k}\left(n_{Mj,t-h}\sum_{i=1}^{k-1} f_{Yi,Mj,Yk}\beta_{Yi,Mj,t-h}v_{x,Yi,t}\right)\right]$$

$$T_{x,Yk,t,2} = \sum_{M=1}^{N_H}\sum_{I=1}^{N_H}\left[Q_{I,M,Y}\sum_{j=1}^{k}\left(n_{Mj,t-h}\sum_{i=1}^{k} f_{Ii,Mj,Yk}\beta_{Ii,Mj,t-h}v_{x,Ii,t}\right)\right]$$

$$T_{x,Yk,t,3} = \sum_{j=1}^{N_C}\left[\sum_{M=1}^{N_H}\left[(1 - L_{Y,M})(1 - f_{Yk,Mj,Yk}) + L_{Y,M}\right]\beta_{Yk,Mj,t-h}n_{Mj,t-h}\right]$$

where N_H is the total number of hydrometeor distributions and N_C is the number of size bins in each hydrometeor distribution. In this case, $N_H = 3$, where distributions $Y = \mathrm{lq} = 1$ for liquid, $Y = \mathrm{ic} = 2$ for ice, and $Y = \mathrm{gr} = 3$ for graupel. The equation applies when v_x is either the total hydrometeor volume concentration ($v_{T,lq}$, $v_{T,ic}$, or $v_{T,gr}$), the volume concentration of liquid water or ice in a hydrometeor

Table 18.3 Values of $P_{Y,M}$ and $L_{Y,M}$ when coagulation is treated among liquid (lq), ice (ic), and graupel (gr) size distributions.

	M		
Y	lq	ic	gr
$P_{Y,M}$			
lq	1	0	0
ic	0	1	0
gr	1	1	1
$L_{Y,M}$			
lq	0	1	1
ic	1	0	1
gr	0	0	0

($v_{\mathrm{L,lq}}$, $v_{\mathrm{I,ic}}$, or $v_{\mathrm{I,gr}}$), the volume concentration of an aerosol solution incorporated within a hydrometeor ($v_{\mathrm{s,lq}}$, $v_{\mathrm{s,ic}}$, or $v_{\mathrm{s,gr}}$), or the volume concentration of any individual aerosol component incorporated within a hydrometeor ($v_{q,\mathrm{lq}}$, $v_{q,\mathrm{ic}}$, or $v_{q,\mathrm{gr}}$). If it is total volume concentration, then total hydrometeor number concentration (n_{lq}, n_{ic}, or n_{gr}) equals total hydrometeor volume concentration divided by the single-particle volume in the bin. For example,

$$n_{\mathrm{lq},k,t} = \frac{v_{\mathrm{T,lq},k,t}}{\upsilon_{\mathrm{lq},k}} \tag{18.54}$$

Alternatively, both sides of (18.53) can be divided by single-particle volume to solve for number concentration directly, and the result is identical.

In (18.53), $f_{Ii,Mj,Yk}$ is the volume fraction of the summed volume of two single particles, $V_{Ii,Mj} = \upsilon_{Ii} + \upsilon_{Mj}$, from distributions I and M, partitioned to a fixed size bin k of distribution Y. Each $V_{Ii,Mj}$ is fractionated between two fixed bins in a volume- and number-conserving manner. The volume fractions are

$$f_{Ii,Mj,Yk} = \begin{cases} \left(\dfrac{\upsilon_{Yk+1} - V_{Ii,Mj}}{\upsilon_{Yk+1} - \upsilon_{Yk}}\right)\dfrac{\upsilon_{Nk}}{V_{Ii,Mj}} & \upsilon_{Yk} \le V_{Ii,Mj} < \upsilon_{Yk+1} \quad k < N_{\mathrm{C}} \\ 1 - f_{Ii,Mj,Yk-1} & \upsilon_{Yk-1} < V_{Ii,Mj} < \upsilon_{Yk} \quad k > 1 \\ 1 & V_{Ii,Mj} \ge \upsilon_{Yk} \quad k = N_{\mathrm{C}} \\ 0 & \text{all other cases} \end{cases} \tag{18.55}$$

Finally, P, Q, and L are either 1 or 0, depending on the coagulation interactions accounted for. The parameter $P_{Y,M} = 1$ if particles in distribution Y coagulating with particles in distribution M produce larger particles in distribution Y. The upper part of Table 18.3 lists the values of $P_{Y,M}$ when liquid, ice, and graupel distributions are considered. The parameter $Q_{I,M,Y} = 1$ if particles in distribution I coagulating with particles in distribution M produce particles in distribution Y, and $I \ne M$ and $I \ne Y$. For example, $Q_{\mathrm{lq,ic,gr}}$, $Q_{\mathrm{lq,gr,gr}}$, $Q_{\mathrm{ic,lq,gr}}$, and $Q_{\mathrm{ic,gr,gr}} = 1$, but all other interactions are zero. The parameter $L_{Y,M} = 1$ if particles in distribution Y

coagulating with particles in distribution M do not produce particles in distribution Y. The lower part of Table 18.3 lists the values of $L_{Y,M}$.

In (18.53), term T_1 accounts for production of larger liquid, ice, and graupel particles from self-coagulation and production of larger graupel from liquid and ice heterocoagulation with graupel. Term T_2 accounts for production of graupel from liquid heterocoagulation with ice. The first part of term T_3 accounts for self-coagulation loss of liquid, ice, and graupel to form larger sizes. The second part of the same term accounts for loss of liquid and ice by heterocoagulation with graupel to form more graupel.

Equation (18.53) is solved in a special order. Distributions that have no coagulation production from other distributions (e.g., the liquid and ice distributions) are solved first, followed by distributions with production terms from previously solved distributions (e.g., the graupel distribution). Within each distribution, equations are solved from bin $k = 1 \ldots N_C$. The volume concentrations of individual components within a distribution can be solved in any order. To minimize computer time, all calculations involving a zero value of f, P, Q, or L are eliminated ahead of time.

The total hydrometeor–hydrometeor coagulation kernel (cm^3 particle^{-1} s^{-1}) is

$$\beta_{Ii,Jj} = E_{\mathrm{coal},Ii,Jj} K_{Ii,Jj} \tag{18.56}$$

where $E_{\mathrm{coal},Ii,Jj}$ is a coalescence efficiency (dimensionless) and $K_{Ii,Jj}$ is a collision kernel (cm^3 particle^{-1} s^{-1}) accounting for several physical processes causing collision in the atmosphere. Collision kernels are given in Section 15.6 for Brownian motion, convective Brownian motion enhancement, gravitational collection, turbulent inertial motion, turbulent shear, van der Waals/viscous forces, particle shape, thermophoresis, diffusiophoresis, and electric charge. Coalescence efficiencies are discussed in Section 15.6.8.

18.8.3 Drop breakup

When liquid raindrops grow sufficiently large by coagulation, they become unstable and break up, either spontaneously or upon collision with other hydrometeor particles. One method of solving for drop breakup is to add breakup terms to the coagulation solution (e.g., List and Gillespie 1976). This method requires a fragment probability distribution for the interaction of each hydrometeor pair. A second method is simply to assume drops break up once they reach a critical size (e.g., Danielsen *et al.* 1972). This method requires a breakup distribution, which is similar to a fragment probability distribution, except that a breakup distribution does not consider the physics behind breakup. In addition, it assumes all drops break up once they reach a certain size whereas, in reality, some do not breakup until they are larger. On the other hand, fragment probabilities contain uncertainties as well.

Table 18.4 Polynomial coefficients for drop breakup for use in (18.57)

	300 < D ≤ 1290 μm	1290 < D ≤ 5160 μm
A_0	0.53098621799986	6.5418838298481
A_1	−0.0036655403240035	−0.0043878127949574
A_2	0.0000077765141976619	0.0000010066406670884
A_3	$-2.9695029431377 \times 10^{-9}$	$-7.771123366063 \times 10^{-11}$

Source: Jacobson (2003).

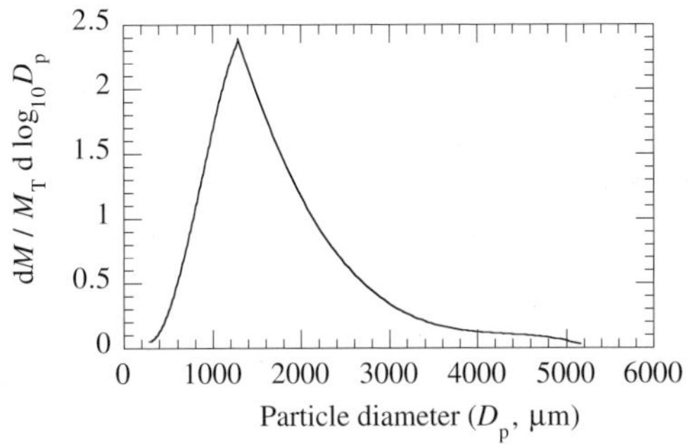

Figure 18.7 Large-drop breakup distribution from (18.57).

Figure 18.7 shows a raindrop breakup distribution obtained by curve-fitting a measured breakup distribution from Danielsen *et al.* (1972). The polynomial fit used to produce the figure is

$$\frac{\mathrm{d}M}{M_\mathrm{T}\,\mathrm{d}\log_{10} D} = A_0 + D[A_1 + D(A_2 + DA_3)] \tag{18.57}$$

where $\mathrm{d}M$ is incremental liquid water mass in a size increment, M_T is liquid water mass summed over all sizes, D is particle diameter (μm), and Table 18.4 gives the coefficients used in the equation.

In a model, drop breakup may be assumed to occur when drops exceed 5-mm in diameter. Upon breakup, mass fractions of the breakup drop are assigned to each of several hydrometeor size bins. In such a case, the mass fraction of each breakup drop going to model size bin k of diameter $D = d_k$ is

$$f_{\mathrm{M},k} = \frac{\dfrac{\mathrm{d}M}{M_\mathrm{T}\,\mathrm{d}\log_{10} d_k}\mathrm{d}\log_{10} d_k}{\sum\limits_{k=1}^{N_\mathrm{C}}\left(\dfrac{\mathrm{d}M}{M_T\,\mathrm{d}\log_{10} d_k}\mathrm{d}\log_{10} d_k\right)} = \frac{[A_0 + d_k(A_1 + d_k(A_2 + d_k A_3))]\mathrm{d}\log_{10} d_k}{\sum\limits_{k=1}^{N_\mathrm{C}}([A_0 + d_k(A_1 + d_k(A_2 + d_k A_3))]\mathrm{d}\log_{10} d_k)} \tag{18.58}$$

Finally, the number concentration of drops added to each bin of diameter-width $\mathrm{d}\log_{10} d_k$ due to the breakup of a single drop of diameter $d_\mathrm{orig} > 5\mathrm{mm}$ is $\Delta n_k = f_{\mathrm{M},k}(d_\mathrm{orig}/d_k)^3$.

18.8.4 Contact freezing

Contact freezing is a mechanism by which an aerosol ice contact nucleus (ICN) collides with the surface of a liquid drop at a subfreezing temperature and causes the drop to freeze spontaneously. This process can be treated numerically by coagulating size-resolved ICN with size-resolved liquid hydrometeors. As applied here, the calculation assumes that if an ICN collides with a liquid drop, the drop and its aerosol inclusions are transferred to the graupel distribution.

The loss of liquid hydrometeor total volume concentration ($x = T$) and aerosol component volume concentration ($x = q$) due to contact freezing is

$$v_{x,\mathrm{lq},k,t} = \frac{v_{x,\mathrm{lq},k,t-h}}{1 + hT_{x,k,t,3}} \tag{18.59}$$

and the corresponding gain of graupel is

$$v_{x,\mathrm{gr},k,t} = v_{x,\mathrm{gr},k,t-h} + v_{x,\mathrm{lq},k,t} h T_{x,k,t,3} \tag{18.60}$$

In these equations, the implicit loss coefficient (s^{-1}) of drops due to their heterocoagulation with ICN is

$$T_{x,k,t,3} = F_T \sum_{j=1}^{N_\mathrm{C}} \left[\sum_{N=1}^{N_\mathrm{T}} \beta_{Yk,Nj,t-h} F_{\mathrm{ICN},Nj} n_{Nj,t-h} \right] \tag{18.61}$$

where $F_{\mathrm{ICN},Nj}$ is the ratio of the number concentration of contact nuclei to that of total particles in an aerosol size bin, F_T is a temperature-dependent parameter that reduces the rate of contact nucleation at high temperatures, and $\beta_{Yk,Nj,t-h}$ is the kernel (cm^3 particle^{-1} s^{-1}) for coagulation between liquid drops and aerosol particles (Section 15.6). Equations (18.59) and (18.60) are solved in the order $k = 1 \ldots N_\mathrm{C}$, where N_C is the number of hydrometeor size bins per hydrometeor distribution. The **final number concentrations** of liquid and graupel hydrometeors in each bin of each distribution are

$$n_{\mathrm{lq},k,t} = \frac{v_{T,\mathrm{lq},k,t}}{\upsilon_{\mathrm{lq},k}} \tag{18.62}$$

$$n_{\mathrm{gr},k,t} = \frac{v_{T,\mathrm{gr},k,t}}{\upsilon_{\mathrm{gr},k}} \tag{18.63}$$

respectively. **The temperature-dependence parameter** in (18.61) is a fraction,

$$F_T = \begin{cases} 0 & T > -3\,^\circ\mathrm{C} \\ -(T+3)/15 & -18 < T < -3\,^\circ\mathrm{C} \\ 1 & T < -18\,^\circ\mathrm{C} \end{cases} \tag{18.64}$$

where T is in °C. This equation was obtained by noting that Fig. 2 of Pitter and Pruppacher (1973) shows that kaolinite and montmorillonite contact freeze 100 percent of drops at −18 °C and 0 percent of drops at −3 °C. The product, $F_T F_{\mathrm{ICN},Nj} n_{Nj,t-h}$, is the number concentration of ICN in a given size bin of a given aerosol size distribution at a given temperature. The fractional number of ICN ($F_{\mathrm{ICN},Nj}$) should be larger than the fractional number of IDN in each size bin of a distribution since only few aerosol types can serve as IDN but many can serve as ICN.

Contact freezing freezes a higher proportion of small drops than large drops. Although the coagulation rate coefficient of an aerosol particle with a liquid drop generally increases with increasing size, the number concentration of large drops decreases at a greater rate than the coagulation rate coefficient increases with increasing size, causing the aerosol–liquid drop coagulation rate (rate coefficient multiplied by number concentrations of colliding particles) to decrease with increasing size.

Even though contact freezing freezes primarily small drops, the number concentration of aerosol particles in the upper troposphere is sufficiently small that contact freezing has relatively little effect on the size distribution of liquid or graupel in the upper troposphere. In cold, lower-tropospheric regions in which aerosol particle concentrations are high, though (e.g., in populated northern latitude regions during winter), contact freezing is a more important freezing mechanism than in the upper troposphere.

Example 18.4

Calculate the fraction of the number concentration of 40-μm-radius liquid drops that contact freeze during one hour in the presence of 10^3 aerosol particles cm^{-3}. Assume the temperature is $-20\,^\circ$C, the coagulation rate coefficient is 10^{-4} cm^3 particle^{-1} s^{-1}, and a maximum of 5 percent of the aerosol particles present can serve as contact nuclei.

SOLUTION

The fractional number concentration of liquid drops that freeze is

$$1-\frac{n_{\mathrm{lq},t}}{n_{\mathrm{lq},t-h}}=1-\frac{v_{T,\mathrm{lq},t}}{v_{T,\mathrm{lq},t-h}}=1-\frac{1}{1+hF_T F_{\mathrm{ICN}}\beta}$$

$$=1-\frac{1}{1+3600\mathrm{s}\times 1\times 0.05\times 10^{-4}\,\dfrac{\mathrm{cm}^3}{\mathrm{partic.\,s}}\times 10^3\,\dfrac{\mathrm{partic.}}{\mathrm{cm}^3}}=0.947$$

Thus, nearly 95 percent of liquid drops contact freeze within an hour under the conditions specified.

18.8.5 Homogeneous and heterogeneous freezing

Two other methods of freezing liquid drops are homogeneous and heterogeneous freezing. **Heterogeneous freezing** is triggered when an aerosol particle, called an **ice immersion nucleus** (IIN) in this case, enters a liquid drop. Once immersed in the drop, the IIN's surface serves as a site for liquid water heterogeneous nucleation to ice. Once an ice cluster forms on the IIN surface, it grows rapidly, engulfing the IIN and eventually converting all the liquid water in the drop to ice. Often, multiple IIN act simultaneously, enhancing the freezing rate of the drop. **Homogeneous freezing** is caused by the homogeneous nucleation of ice crystals within or on the surface of a liquid drop in the absence of an IIN. Once a homogeneously nucleated cluster forms, it grows until the entire drop freezes.

One method of modeling homogeneous freezing together with heterogeneous freezing is to fit an equation of drop freezing to laboratory data. This treatment is uncertain, because it assumes that the composition of a drop in the air is similar to that in a laboratory experiment, which is often performed with tap water. Thus, it is uncertain whether the experiments simulated heterogeneous or homogeneous freezing or both.

Several studies have suggested that the homogeneous–heterogeneous freezing (HHF) temperature is related logarithmically to drop volume (e.g., Bigg 1953; Vali 1971; Pitter and Pruppacher 1973). For heterogeneous freezing, this assumption is physical since IIN can be randomly distributed within a drop. Under this theory, the **fractional number of drops that freeze** at a given temperature and particle size is

$$F_{\mathrm{Fr},k,t} = \min\{v_{\mathrm{lq},k}\exp[-B\,(T_\mathrm{c} - T_\mathrm{r})]\,,\,1\} \tag{18.65}$$

where $v_{\mathrm{lq},k}$ is the volume of a single liquid drop in bin k (cm^3), T_c is the temperature of a population of drops (°C), T_r is a reference temperature (°C), and B is a fitting coefficient (°C^{-1}). Since the equation is empirical, units do not equate. Equation (18.65) suggests that, the lower the temperature and the larger the single-particle volume, the greater the fractional number of drops that freeze. From the equation, the **equilibrium median freezing temperature** (°C), which is the temperature at which 50 percent of drops freeze ($F_\mathrm{Fr} = 0.5$), is

$$T_\mathrm{mf} = T_\mathrm{r} - \frac{1}{B}\ln\left(\frac{0.5}{v_{\mathrm{lq},k}}\right) \quad \begin{cases} B = 0.475\,^\circ\mathrm{C}^{-1};\ T_\mathrm{r} = 0\,^\circ\mathrm{C} & T_m < -15\,^\circ\mathrm{C} \\ B = 1.85\,^\circ\mathrm{C}^{-1};\ T_\mathrm{r} = -11.14\,^\circ\mathrm{C} & -15\,^\circ\mathrm{C} \le T_m < -10\,^\circ\mathrm{C} \end{cases} \tag{18.66}$$

where the parameter values were derived in Jacobson (2003) from a fit to data of Pitter and Pruppacher (1973). Figure 18.8 compares the fits with the original data. The data suggest that the freezing rate follows logarithmically with drop volume only for temperatures below –15 °C. Above –10 °C, no drops freeze. Other values of B and T_r derived from laboratory data can be found in Danielsen *et al.* (1972) and Orville and Kopp (1977).

Example 18.5

Calculate the fraction of the number concentration of supercooled 10-μm- and 100-μm-radius liquid drops that freeze due to homogeneous-heterogeneous freezing at a temperature of −25 °C.

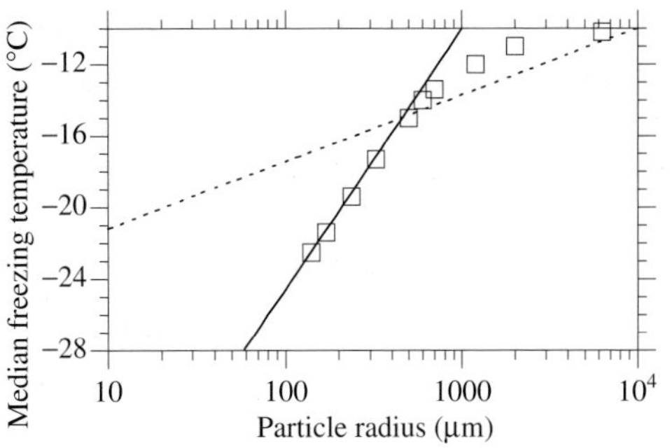

Figure 18.8 Comparison of fitted equation (18.66) for median freezing temperature with data of Pitter and Pruppacher (1973, Fig. 1), from which the fits were derived. From Jacobson (2003).

SOLUTION

From (18.65),

$$F_{\mathrm{Fr},10\mu\mathrm{m}} = 4.19 \times 10^{-9}\,\mathrm{cm}^3 \exp[-0.475\,^\circ\mathrm{C}^{-1}(-25\,^\circ\mathrm{C} - 0\,^\circ\mathrm{C})] = 0.000602$$

$$F_{\mathrm{Fr},100\mu\mathrm{m}} = 4.19 \times 10^{-6}\,\mathrm{cm}^3 \exp[-0.475\,^\circ\mathrm{C}^{-1}(-25\,^\circ\mathrm{C} - 0\,^\circ\mathrm{C})] = 0.602$$

Thus, a greater fraction of larger particles than smaller particles freeze at the same subfreezing temperature.

Equation (18.65) is an equilibrium equation. A **time-dependent freezing-rate** equation is

$$\frac{\mathrm{d}n_{\mathrm{gr},k,t}}{\mathrm{d}t} = n_{\mathrm{lq},k,t-h} \upsilon_{\mathrm{lq},k} A \exp[-B\,(T_\mathrm{c} - T_\mathrm{r})] \tag{18.67}$$

where $A = 10^{-4}\mathrm{cm}^{-3}\mathrm{s}^{-1}$ (e.g., Orville and Kopp 1977; Reisin *et al.* 1996). Integrating this equation from $t = 0$ to h gives the **number concentration of liquid drops and graupel particles**, respectively, after homogeneous–heterogeneous freezing, as

$$n_{\mathrm{lq},k,t} = n_{\mathrm{lq},k,t-h}\,(1 - F_{\mathrm{Fr},k,t}) \tag{18.68}$$

$$n_{\mathrm{gr},k,t} = n_{\mathrm{gr},k,t-h} + n_{\mathrm{lq},k,t-h} F_{\mathrm{Fr},k,t} \tag{18.69}$$

where

$$F_{\mathrm{Fr},k,t} = 1 - \exp\{-h A \upsilon_{\mathrm{lq},k} \exp[-B(T_\mathrm{c} - T_\mathrm{r})]\} \tag{18.70}$$

is the **fractional number of drops that freeze**. Component volume concentrations are similarly calculated for liquid and graupel. Setting $F_{\mathrm{Fr},k,t} = 0.5$ and solving for

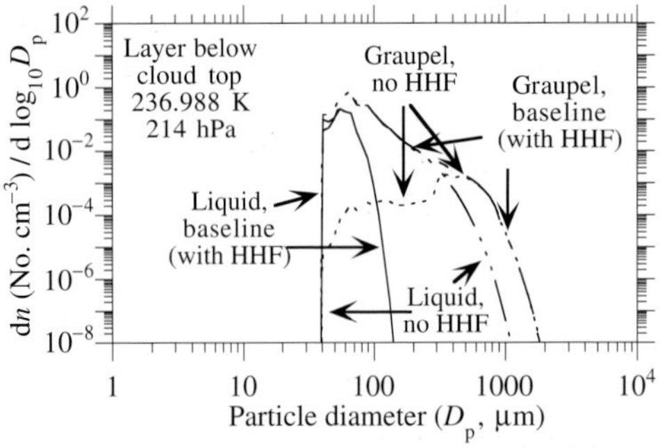

Figure 18.9 Comparison of modeled liquid and graupel distributions in the layer below cloud top from a simulation in which homogeneous-heterogeneous freezing (HHF) was calculated and one in which it was not. The time over which freezing occurred was one hour. The temperature is the initial in-cloud temperature. From Jacobson (2003).

T in (18.70) give the **time-dependent median freezing temperature** as

$$T_{\text{mf}} = T_{\text{r}} - \frac{1}{B}\ln\left(-\frac{\ln 0.5}{h A v_{\text{lq},k}}\right) \qquad (18.71)$$

Equating Equation (18.71) with (18.66) suggests that the time for equilibration is $h_{\text{eq}} = -\ln 0.5/(0.5A) = 13\,862$ s when $A = 10^{-4}$ cm^{-3} s^{-1}. This time is much greater than the time step for typical atmospheric model processes.

Figure 18.9 illustrates the modeled effect on liquid and graupel size distributions of HHF over one hour. The figure shows that HHF (baseline case) converted primarily large liquid drops to graupel. The major effect of HHF was to increase 50–300 μm graupel particles. Since graupel in this size range merely replaces liquid, and because much of the graupel sublimates as it falls, the production of graupel due to HHF had little effect on hydrometeor or aerosol removal below the layer shown in the figure.

18.8.6 Drop surface temperature and evaporation/sublimation

When a hydrometeor particle falls below a cloud, where the air is subsaturated with respect to water vapor, the particle begins to evaporate or sublimate. The shrinkage rate of a small hydrometeor is sufficiently fast that the entire hydrometeor may dissipate to its aerosol-particle core. Large hydrometeors, though, often survive, possibly reaching the ground as rain or snow. Evaporation/sublimation reduces the surface temperature of a hydrometeor particle.

Here, equations are given to describe the final hydrometeor surface temperature and evaporation/sublimation rate when a hydrometeor falls through subsaturated air. The procedure is performed by first calculating the equilibrium surface temperature iteratively then solving evaporation/sublimation using information from this information. This procedure is based on the method of Beard and Pruppacher (1971) and Pruppacher and Rasmussen (1979) with modification.

In the case of a liquid drop, the iteration involves solving the following equations:

$$
\begin{aligned}
p_{s,n} &= p_{v,s}(T_{s,n}) \\
\Delta p_{v,n} &= 0.3(p_{s,n} - p_{v,n}) \\
p_{f,n} &= 0.5(p_{s,n} + p_{v,n}) \\
T_{f,n} &= 0.5(T_{s,n} + T_a) \\
T_{s,n+1} &= T_{s,n} - \frac{D_v L_e}{\kappa_a(1 - p_{f,n}/p_a)} \frac{\Delta p_{v,n}}{R_v T_{f,n}} \\
p_{v,n+1} &= p_{v,n} + \Delta p_{v,n}
\end{aligned}
\qquad (18.72)
$$

where the subscript n is the iteration number, $T_{s,n}$ is the **drop surface temperature** (K) at iteration n, initialized at the ambient temperature T_a (K), which stays constant, $p_{s,n}$ is the saturation vapor pressure (hPa) over the drop surface, evaluated at the drop surface temperature, $\Delta p_{v,n}$ is an estimated change in water vapor partial pressure at iteration n, $p_{f,n}$ and $T_{f,n}$ are average values of water vapor partial pressure and of temperature between the drop surface and ambient air, D_v is the diffusion coefficient of water vapor in air (cm^2 s^{-1}), L_e is the latent heat of evaporation (J g^{-1}), κ_a is the thermal conductivity of moist air (J cm^{-1} s^{-1} K^{-1}), R_v is the gas constant for water vapor (4614 cm^3 hPa g^{-1} K^{-1}), and p_a is the ambient air pressure (hPa). The above set of equations is iterated a minimum of three times. The factor of 0.3 is included in $\Delta p_{v,n}$ to ensure convergence without overshooting.

The equations differ from Beard and Pruppacher (1971) only in that the present equations ignore ventilation coefficients for energy and vapor and radiative heating to eliminate the size-dependence of the iterative calculation, which would add to computer time without affecting the result significantly. For example, when size-dependent parameters are included, the difference in final drop surface temperature between a 0.001 and 1000 μm drop is only 2.5 percent (0.05 K/2 K) of the mean drop surface temperature depression when the relative humidity (RH) is 80 percent and the ambient temperature is 283.15 K. Even at RH = 1 percent, the difference is only 4 percent (0.4 K/10 K). Temperature equations for sublimation are analogous to those for evaporation.

Figure 18.10 shows the variation in equilibrium drop surface temperature and other parameters for initial relative humidities of 1 to 100 percent under (a) lower- (b) mid- and (c) upper-tropospheric conditions. The figure shows that, under lower-tropospheric conditions (Fig. 18.10(a)), drop surface temperatures can decrease by as much as 10 K when RH = 1 percent. At RH = 80 percent, which is more typical

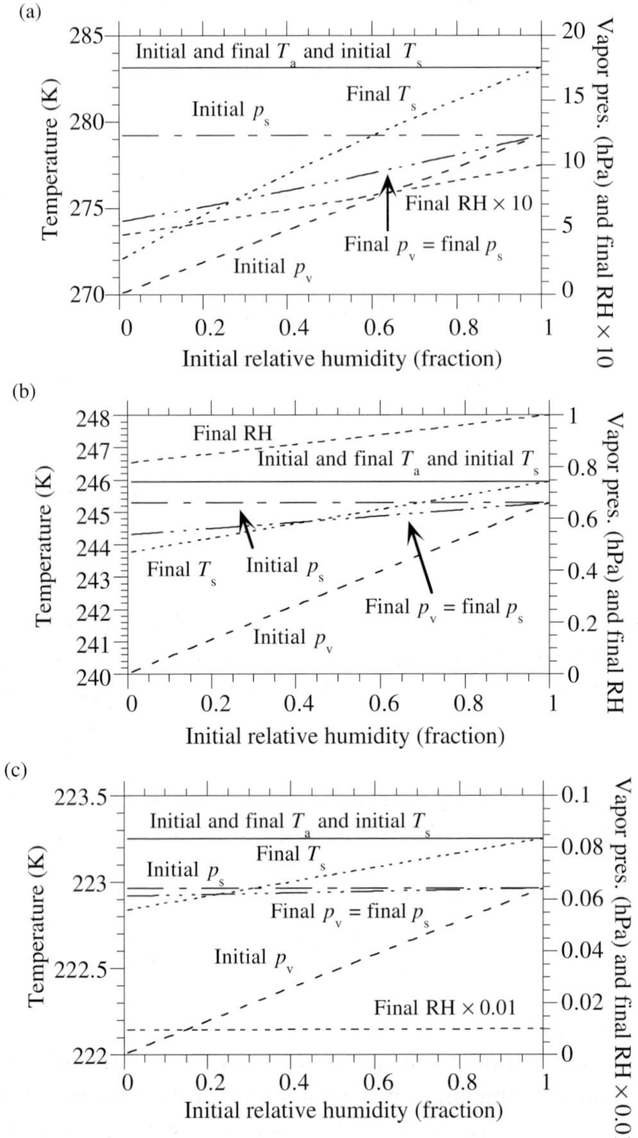

Figure 18.10 Variation in equilibrium liquid drop surface temperature (T_s, K), saturation vapor pressure over the drop surface (p_s, hPa), partial pressure of water away from the drop surface (p_v, hPa), and relative humidity (RH) near the drop surface for initial ambient relative humidities of 1 to 100 percent when (a) the ambient temperature (T_a) = 283.15 K and the air pressure (p_a) = 900 hPa (lower-tropospheric conditions), (b) T_a = 245.94 K and p_a = 440.7 hPa (middle-tropospheric conditions), and (c) T_a = 223.25 K and p_a = 265 hPa (upper-tropospheric conditions). The figure was obtained by solving Equation (18.72). From Jacobson (2003).

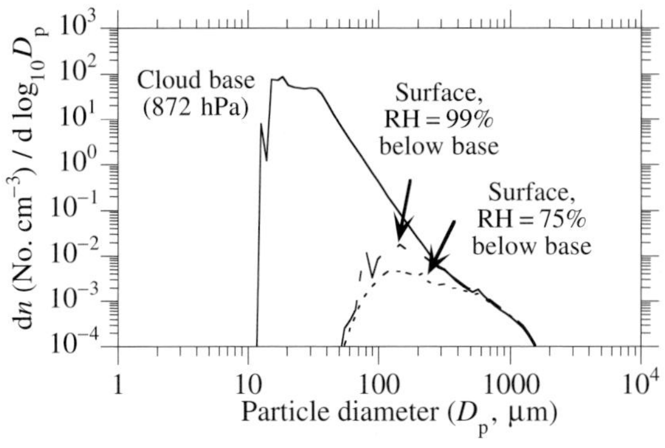

Figure 18.11 Effect of evaporation on the size distribution of precipitation drops reaching the surface after falling from a cloud base at about 1.25 km through subsaturated air in which the relative humidity (RH) was 99% and 75%, respectively. From Jacobson (2003).

below a cloud, the temperature depression is close to 2 K. Under mid-tropospheric conditions (Fig. 18.10(b)), supercooled drop temperatures can decrease by as much as 2.2 K when RH = 1 percent or 1 K when RH = 50 percent. Under upper-tropospheric conditions (Fig. 18.10(c)), the maximum temperature depression is about 0.5 K.

From the drop surface temperature and other parameters found iteratively, the change in drop liquid water volume concentration ($cm^3\ cm^{-3}$) due to evaporation is

$$v_{L,lq,k,t,m} = \text{MAX}\left[v_{L,lq,k,t-h} - \frac{n_{lq,k}4\pi r_k D_v}{(1 - p_{f,nf}/p_a)}\frac{(p_{v,s,0} - p_{v,nf})}{\rho_L R_v T_{f,nf}}\frac{\Delta z}{V_{f,lq,k}}, 0\right]_m \tag{18.73}$$

where the subscripts 0 and nf indicate initial and final iterated values, respectively, from (18.72), and the time step over which evaporation occurs is determined as the layer thickness (cm) divided by the fall speed (cm s^{-1}) ($\Delta z/V_{f,lq,k}$); thus it is the time that the hydrometeor can last in the layer before falling to the next layer. This time step is physical since, if the same time step were constant for all drop sizes, big drops with short lifetimes would evaporate for a longer period than they would last in the layer. The fall speed of a hydrometeor is discussed in Chapter 20. The sublimation rates of ice crystals falling through subsaturated air are calculated in a manner similar to those of liquid drops.

Figure 18.11 shows the modeled effect of evaporation on an idealized size distribution of liquid drops falling about 1.25 km from cloud base to the ground when the relative humidity is 99 percent and 75 percent. Evaporation eliminated almost all small drops in both cases and shrank drops more effectively at 75 percent than at 99 percent relative humidity.

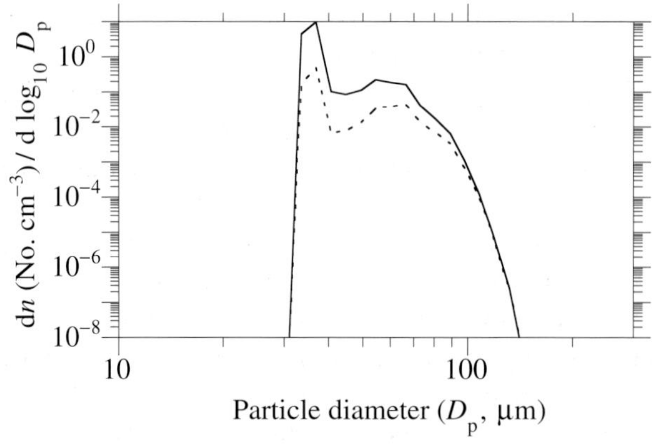

Figure 18.12 Incremental homogeneous-heterogeneous freezing due to evaporative cooling. The solid line (—) shows a liquid size distribution at an ambient pressure of 214 hPa, an ambient temperature of 236.988 K, and a relative humidity (RH) of 100%. When the RH decreases to 80%, the drop surface temperature decreases to 236.617 K, causing 5% more liquid drops to freeze by homogeneous-heterogeneous nucleation than would otherwise have frozen at 236.988 K. The incremental 5% is represented by the dashed line (----).

18.8.7 Evaporative freezing

As liquid drops falling through subsaturated air evaporate, their surfaces cool (Fig. 18.10). As temperatures decrease below −10 °C, the fractional number of liquid drops that freeze at a given size increases (Equation (18.65)). As such, the cooling of a liquid drop due to evaporation at its surface is a mechanism of drop freezing, termed **evaporative freezing** (Jacobson 2003). The initial temperature of freezing and the median freezing temperature of small drops are lower than are those of large drops, but even for small drops, evaporative cooling may enhance the probability of freezing under the right conditions.

Figure 18.12 illustrates evaporative freezing. The figure shows a liquid hydrometeor size distribution at a relative humidity of 100 percent and $T = 236.988$ K. Under these conditions, a portion of the distribution, determined from (18.65), freezes due to homogeneous-heterogeneous freezing. When the relative humidity drops to 80 percent, (18.72) predicts a decrease in the surface temperature of all drops by 0.35 K. At this lower temperature, (18.65) predicts an increase in the number concentration of drops that freeze by 5 percent. The dashed curve in Fig. 18.12 shows this incremental change in the number distribution of frozen drops due to evaporative freezing.

The theory that evaporative cooling at a drop's surface as the drop falls through subsaturated air may enhance its rate of freezing is analogous to the theory, shown experimentally by Rasmussen and Pruppacher (1982), that evaporative cooling delays the onset of melting.

The evaporative freezing theory also implies that homogeneous-heterogeneous drop freezing in subsaturated air may occur first on a drop surface rather than in the bulk of the drop. Homogeneous-heterogeneous freezing may also occur more readily at a drop surface than in the bulk of a drop for other reasons. Tabazadeh *et al.* (2002), for example, found that ice nucleation is thermodynamically favored on the surface in comparison with in the bulk of a liquid drop, even when no temperature gradient exists. Stuart (2002) found that, if a frozen nucleus forms in the middle of a drop, thin dendrites shoot out to the surface and warm the entire drop to the melting point within less than 0.1 s. Heat loss by the partially frozen drop occurs only by conduction and evaporation at the surface, causing the surface temperature to be cooler than the rest of the drop but warmer than the drop was originally. Due to the cooler surface relative to the interior, remaining freezing proceeds from the surface, inward. Thus, although freezing may initiate in the center, the rate of freezing quickly increases at the surface relative to the interior. The main difference between the mechanism of Stuart (2002) and pure evaporative freezing is that, with the former, surface freezing is triggered by freezing within the drop itself; with the latter, surface freezing is triggered by surface cooling as a drop falls through subsaturated air. Evaporative freezing can occur on its own or with one of these other mechanisms to enhance the rate of surface freezing.

18.8.8 Ice crystal melting

If the temperature of an ice crystal or graupel particle increases above the ice melting point (nominally $T_0 = 273.15$ K), the frozen hydrometeor begins to melt. But, when an ice crystal melts in subsaturated air (e.g., when the partial pressure of water away from the crystal surface is lower than the saturation vapor pressure over meltwater on the surface), simultaneous evaporation of the meltwater cools the particle surface, retarding the rate of melting. Thus, the melting temperature must exceed T_0 for the final drop surface temperature to equal T_0. The lower the relative humidity (partial pressure divided by saturation vapor pressure over liquid water), the greater the melting temperature.

The **melting point** of a population of ice particles can be found from

$$T_{melt} = T_0 + \mathrm{MAX}\left\{\frac{D_v L_e}{\kappa_a R_v}\left[\frac{p_{v,s}(T_0)}{T_0} - \frac{p_v}{T_a}\right], 0\right\} \tag{18.74}$$

(Rasmussen and Pruppacher 1982), where $p_{v,s}(T_0)$ is the saturation vapor pressure over liquid water (hPa) at T_0, T_a is the ambient air temperature (K), p_v is the ambient partial pressure of water vapor (hPa), D_v is the diffusion coefficient of water vapor in air (cm^2 s^{-1}), L_e is the latent heat of evaporation (J g^{-1}), κ_a is the thermal conductivity of moist air (J cm^{-1} s^{-1} K^{-1}), and R_v is the gas constant for water vapor (4614 cm^3 hPa g^{-1} K^{-1}). When melting occurs, the change in mass (g) of ice in a particle of a given size is

$$m_{\text{ic},Ni,t} = m_{\text{ic},Ni,t-h} - \text{MAX}\left\{h\frac{4\pi r_{Ni}}{L_\text{m}}\left[\kappa_\text{a}(T_\text{a} - T_0)F_{h,\text{I},Ni} - \frac{D_\text{v}L_\text{e}}{R_\text{v}}\left(\frac{p_{\text{v,s}}(T_0)}{T_0} - \frac{p_\text{v}}{\text{T}_\text{a}}\right)F_{\text{v,I},Ni}\right], 0\right\} \quad (18.75)$$

(e.g., Rasmussen *et al.* 1984), where L_m is the latent heat of melting (J g^{-1}) and h is the time step (s). Although the rate of melting increases linearly with increasing particle radius, mass increases with radius cubed, so small crystals melt completely much faster than do large crystals.

Example 18.6

Calculate the melting temperature of a frozen particle when the ambient temperature is 273.15 K, total air pressure is 970 hPa, and the relative humidity is 80 percent.

SOLUTION

From (8.14), the diffusion coefficient of water vapor in air is $D_\text{v} = 2.204 \times 10^{-5}$ m^2 s^{-1}. From (2.62), the saturation vapor pressure at 273.15 K is $p_{\text{s,v}} = 6.112$ hPa. From (2.66), the partial pressure of water vapor at a relative humidity of 80 percent and at 273.15 K is $p_\text{v} = 4.89$ hPa. From (2.54), the latent heat of evaporation at 273.15 K is $L_\text{e} = 2.501 \times 10^6$ J kg^{-1}. From (2.5) and (2.6), the thermal conductivities of dry air and water vapor are $\kappa_\text{d} = 0.023807$ and $\kappa_\text{v} = 0.015606$ J m^{-1} s^{-1} K^{-1}, respectively, at 273.15 K. Since $n_\text{v}/(n_\text{v} + n_\text{d}) = p_\text{v}/p_\text{a}$, the thermal conductivity of moist air from (2.7) is then $\kappa_\text{a} = 0.023746$ J m^{-1} s^{-1} K^{-1}. Substituting these values into (18.74) gives the melting point of an ice crystal as

$$T_\text{melt} = 273.15\,\text{K} + \frac{2.204 \times 10^{-5}\,\dfrac{\text{m}^2}{\text{s}} \times 2.501 \times 10^6\,\dfrac{\text{J}}{\text{kg}}}{0.023746\,\dfrac{\text{J}}{\text{m s K}} \times 461.4\,\dfrac{\text{J}}{\text{kg K}}} \times \left[\frac{6.112\,\text{hPa}}{273.15\,\text{K}} - \frac{4.89\,\text{hPa}}{273.15\,\text{K}}\right]\frac{100\,\text{J}}{\text{hPa m}^3} = 275.40\,\text{K}$$

Thus, the melting point of an ice crystal at 80 percent relative humidity is greater than that at 100 percent relative humidity.

18.8.9 Aerosol–hydrometeor coagulation: aerosol washout

The two major mechanisms of aerosol removal by precipitation are **nucleation scavenging** (rainout) and **aerosol–hydrometeor coagulation** (washout). **Rainout** occurs when a CCN activates to form a liquid drop, and the drop coagulates with other drops to become rain or graupel, which falls to the surface, removing the

CCN inclusions. **Washout** occurs when growing or falling precipitation particles coagulate with interstitial aerosol particles and fall to the surface, bringing the aerosol particles with them.

Together, rainout and washout are the most important mechanisms removing aerosol particles globally. The other mechanisms of aerosol removal, gravitational settling and dry deposition, are important for very large particles and over long periods, but not so important relative to rainout and washout for small particles over short periods.

Aerosol–hydrometeor coagulation results in the coalescence of aerosol particles with liquid or frozen hydrometeors, reducing aerosol particle number. The final aerosol-particle total volume concentration ($x = T$), solution volume concentration ($x = s$), or individual component concentration ($x = q$) in bin k of aerosol distribution N after one time step of aerosol–hydrometeor coagulation can be calculated as

$$v_{x,Nk,t} = \frac{v_{x,Nk,t-h}}{1 + hT_{x,Nk,t,3}}$$

$$T_{x,Nk,t,3} = \sum_{j=1}^{N_C}\left[\sum_{M=1}^{N_H} \beta_{Nk,Mj,t-h} n_{Mj,t-h}\right] \tag{18.76}$$

where N_H is the total number of hydrometeor distributions and N_C is the number of bins in each hydrometeor distribution. Equation (18.76) is solved for all aerosol distributions $N = 1 \ldots N_T$ and size bins $k = 1 \ldots N_B$ per distribution. The corresponding final volume concentrations of hydrometeor particles and their aerosol inclusions within size bin k of hydrometeor distribution Y after one time step is

$$v_{x,Yk,t} = \frac{v_{x,Yk,t-h} + h\left(T_{x,Yk,t,1} + T_{x,Yk,t,2}\right)}{1 + hT_{x,Yk,t,3}}$$

$$T_{x,Yk,t,1} = \sum_{N=1}^{N_T}\left[\sum_{j=1}^{k}\left(n_{Nj,t-h}\sum_{i=1}^{k-1} f_{Yi,Nj,Yk}\beta_{Yi,Nj,t-h}v_{x,Yi,t}\right)\right]$$

$$T_{x,Yk,t,2} = \sum_{N=1}^{N_T}\left[\sum_{j=1}^{k}\left(n_{Yj,t-h}\sum_{i=1}^{k} f_{Ni,Yj,Yk}\beta_{Ni,Yj,t-h}v_{x,Ni,t}\right)\right]$$

$$T_{x,Yk,t,3} = \sum_{j=1}^{N_B}\left[\sum_{N=1}^{N_T}\left(1 - f_{Yk,Nj,Yk}\right)\beta_{Yk,Nj,t-h} n_{Nj,t-h}\right] \tag{18.77}$$

which is solved for all hydrometeor distributions $Y = 1 \ldots N_H$ and size bins $k = 1 \ldots N_C$. The volume fractions (e.g., $f_{Yi,Nj,Yk}$) in (18.77) are calculated in the same manner as in (18.55) but, here, represent the fraction of a coagulated hydrometeor–aerosol particle partitioned into a hydrometeor bin. The final number concentrations of aerosol particles and hydrometeor particles following aerosol–hydrometeor coagulation are

$$n_{Nk,t} = \frac{v_{T,Nk,t}}{\upsilon_{Nk}} \qquad N = 1 \ldots N_T;\ k = 1 \ldots N_B \tag{18.78}$$

$$n_{Yk,t} = \frac{v_{T,Yk,t}}{\upsilon_{Yk}} \qquad Y = 1 \ldots N_H;\ k = 1 \ldots N_C \tag{18.79}$$

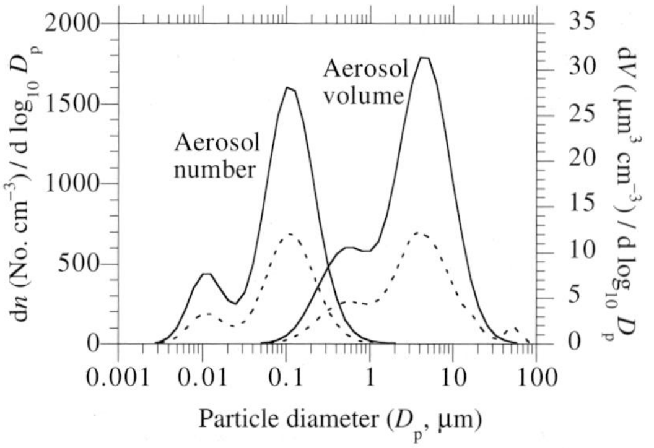

Figure 18.13 Below-cloud-base aerosol number and volume concentration (at 902 hpa), summed over 16 size distributions, before (solid lines) and after (short-dashed lines) aerosol–hydrometeor coagulation. The simulation period was one hour.

respectively. The scheme described above is volume and volume-concentration conserving, positive definite, and noniterative. Aerosol–hydrometeor coagulation kernels are given in Section 15.6.

Within a cloud, rainout removes >50 percent, whereas washout removes <0.1 percent of aerosol mass (e.g., Kreidenweis *et al.* 1997; Jacobson 2003). Rainout scavenges all large and most midsize particles, which have large mass, before washout has a chance to remove particles within a cloud. Within a cloud, rainout may remove 30 to >50% of aerosol number (e.g., Flossmann *et al.* 1985; Jacobson 2003; Kreidenweis *et al.* 2003). However, below a cloud, rainout does not remove aerosol particles since no activation of new drops occurs below a cloud. Thus, all below-cloud removal of aerosol particles is by washout. Figure 18.13 shows the modeled effect of below-cloud washout on the size distribution of aerosol particles. Washout removed aerosol particles across the entire size distribution. Because washout removes small particles within a cloud (large particles in a cloud are removed by rainout) and washout removes small and large particles below a cloud, washout generally removes more particles within plus below cloud by number than does rainout (Jacobson 2003).

18.8.10 Gas washout

Precipitation removes soluble gases by washout. As a raindrop falls through air containing a soluble gas, the gas may dissolve in the drop. As the drop falls further, more gas will dissolve if the air is supersaturated with the gas (i.e., if the partial pressure of the gas exceeds its saturation vapor pressure as determined by the molality of the dissolved gas in rainwater divided by its Henry's law constant). If the air is undersaturated some of the gas in the drop will evaporate to maintain saturation at the drop surface. To complicate matters, a gas may dissolve in drops of one size and evaporate from drops of a different size in a layer.

Below, a parameterization for gas washout is given. It accounts for the changing amount of solute, summed over drops of all sizes, as the drops fall through the air. The numerical solution is derived by considering the **gas–hydrometeor equilibrium relation,**

$$\frac{c_{q,\mathrm{lq},t,m}}{C_{q,t,m}} = H'_q R^* T \sum_{k=1}^{N_\mathrm{C}} p_{L,\mathrm{lq},t,k,m} \tag{18.80}$$

and the **gas–hydrometeor mole-balance equation**

$$C_{q,t,m} + c_{q,\mathrm{lq},t,m} = C_{q,t-h,m} + c_{q,\mathrm{lq},t,m-1}\frac{\Delta z_{m-1}}{\Delta z_m} \tag{18.81}$$

where m is the layer of the model atmosphere (increasing from 1 at the top to the number of model layers), Δz_m is the thickness (cm) of a layer, $C_{q,t,m}$ is the mole concentration of gas q (mol cm^{-3}), $c_{q,\mathrm{lq},t,m}$ is the mole concentration of the dissolved gas, summed over all size bins $k = 1 \ldots N_\mathrm{C}$ in the liquid hydrometeor size distribution, $p_{L,\mathrm{lq},t,k,m}$ is liquid precipitation passing through layer m in bin k (cm^3 cm^{-3}), R^* is the ideal gas constant (0.08206 L atm mol^{-1} K^{-1}), H'_q is an effective Henry's constant for species q (mol L^{-1} atm^{-1}), and T is temperature (K). Equation (18.81) states that the final gas plus aqueous species concentration after dissolution/evaporation equals the initial gas concentration in the layer plus the aqueous concentration in precipitation from the layer above. For the top cloud layer, $c_{q,\mathrm{lq},t,m-1} = 0$. Combining (18.80) and (18.81) gives the **final gas concentration** in any layer m as

$$C_{q,t,m} = \frac{C_{q,t-h,m} + c_{q,\mathrm{lq},t,m-1}\dfrac{\Delta z_{m-1}}{\Delta z_m}}{1 + H'_q R^* T \sum\limits_{k=1}^{N_\mathrm{C}} p_{L,\mathrm{lq},t,m}} \tag{18.82}$$

The **final aqueous mole concentration,** used for the calculation in the next layer, is

$$c_{q,\mathrm{lq},t,m} = C_{q,t-h,m} + c_{q,\mathrm{lq},t,m-1}\frac{\Delta z_{m-1}}{\Delta z_m} - C_{q,t,m} \tag{18.83}$$

This solution is exactly mole conserving, noniterative, unconditionally stable, and positive definite. It states that if rainwater passing through a model layer is already saturated with dissolved gas (the air is saturated or undersaturated with the gas), no additional gas can enter the rainwater, but some may evaporate in the current layer. This mechanism not only removes gases but also transfers gases from a supersaturated layer to a subsaturated layer. The formulation given here applies ideally to soluble gases that are not chemically reactive in solution. For reactive gases, particularly SO_2, aqueous chemical reactions often occur on time scales shorter than the time a drop takes to fall from one layer to the next. Irreversible aqueous chemical reactions occurring within drops are discussed in Chapter 19.

18.8.11 Lightning

Lightning is a bolt of electricity that travels between two regions of a cloud, between two clouds, or between a cloud and the ground. A lightning stroke heats the air to greater than 30 000 °C. The heating causes the air to expand violently, creating a shock wave that propagates at the speed of sound (about 330 m s^{-1}), producing **thunder.** The energy of the lightning bolt is also intense enough to split molecular nitrogen (N_2) and molecular oxygen (O_2) to N and O, respectively, which together react to form nitric oxide (NO). Thus, lightning is a natural source of NO in the atmosphere.

Lightning occurs following the buildup of opposite charges between one part of a cloud and either another part of the same cloud, a part of another cloud, or the ground. The opposing charges create an attractive electrostatic force that becomes strong enough to cause a discharge in electricity.

The force (N) between two point charges Q_0 (C) and Q_1 (C) separated by distance r_{01} (m) is called the **electrostatic force** and is determined by **Coulomb's law,**

$$F_e = \frac{k_C Q_0 Q_1}{r_{01}^2} \tag{18.84}$$

which states that the force between two point charges varies inversely as the square of the distance separating the charges and is proportional to the magnitude of each charge. The force is attractive if the point charges have opposite sign. It is repulsive if they have the same sign. The constant in the equation is **Coulomb's constant,**

$$k_C = \frac{1}{4\pi\varepsilon_0} = 8.98755 \times 10^9\ \mathrm{N\,m^2\,C^{-2}} = 8.98755 \times 10^{11}\ \mathrm{V\,cm\,C^{-1}} \tag{18.85}$$

which is measured experimentally and is usually written in terms of the **permittivity of free space,** $\varepsilon_0 = 8.85419 \times 10^{-12}\ \mathrm{C^2\,N^{-1}\,m^{-2}} = 8.85419 \times 10^{-14}\ \mathrm{C\,V^{-1}\,cm^{-1}}$, where 1 V = 1 J C^{-1} and 1 J = 1 N m. The smallest possible charge is that on an electron, $Q = 1.6 \times 10^{-19}$ C.

The **electric field strength** is defined for a specific location as the sum of electrostatic forces between a point charge at that location and point charges at all surrounding locations, divided by the charge at the location of interest. For example, the electric field strength at the location of charge Q_0 due to forces between Q_0 and all charges Q_i separated by distance r_{0i} is

$$E_f = \sum_i \frac{F_{e,0i}}{Q_0} = \sum_i \frac{k_C Q_i}{r_{0i}^2} \tag{18.86}$$

which has units of N C^{-1} (V m^{-1}).

Lightning occurs only when the electric field strength exceeds the **threshold electric field strength,** E_{th}, which ranges from 100 to 400 kV m^{-1}, with an average of 300 kV m^{-1}. In the absence of clouds, the upper atmosphere is charged slightly positive and the surface of the Earth is charged slightly negative, resulting in an electric field strength near the surface of only about 130 V m^{-1}. The average charge density at the ground is about 21 electrons cm^{-3}. The charge density averaged in the first kilometer above the ground is about 5 electrons cm^{-3} (Pruppacher and Klett 1997).

One theory of how clouds become electrified enough for lightning to form is the theory of **particle rebound charging**. According to this theory, large liquid drops, ice crystals, and graupel particles that are polarized with a net negative charge on their top and positive charge on their bottom may become negatively charged overall when they collide with and bounce off smaller liquid or ice hydrometeors that are also polarized. Since large particles fall faster than small particles, a falling large particle, which has a positive charge, is hit on its bottom by a small particle in its path. During the brief collision, the positive charge from the bottom of the large particle transfers to the small particle, giving the large particle a net negative charge and the small one a net positive charge. Since the large particle is heavy, it continues falling toward the bottom of the cloud whereas the small particle, which is light, stays suspended near the middle or top of the cloud.

Rebound charging can be enhanced by the **thermo-electric effect**. As graupel and hail fall through the air, they grow by collision and coalescence with supercooled liquid water drops, which freeze upon contact. The release of latent heat due to the freezing warms the surface of the graupel or hail. When these warm hydrometeor particles subsequently collide with and bounce off smaller ice crystals, a net transfer of positively charged hydrogen ions (H^+) occurs from the warmer graupel and hail to the colder crystals. Negatively charged OH^- ions try to diffuse in the opposite direction, but their mobility is lower than that of H^+ ions.

As a thunderstorm cloud gains strength, the charge differential between its top and base grows, due in part to the thermo-electric rebound charging mechanism, and intracloud lightning may occur. Since the cloud base is now predominantly negatively charged, it induces a positive charge on the Earth's surface, which is normally slightly negatively charged. If the charge difference between the cloud base and ground increases above the threshold electric field strength, cloud-to-ground lightning occurs. Several other possible cloud charging mechanisms are described in Pruppacher and Klett (1997).

Lightning generation by the **rebound charging mechanism** can be modeled by considering the collision and bounceoff of size-resolved liquid and solid hydrometeors. The **rate coefficient for bounceoff** (bounceoff kernel) (cm^3 particle^{-1} s^{-1}) of a particle in size bin i of hydrometeor distribution I interacting with a particle in size bin j of hydrometeor distribution J in any model layer m of a cloud is

$$B_{Ii,Jj,m} = (1 - E_{\text{coal},Ii,Jj,m})K_{Ii,Jj,m} \tag{18.87}$$

where $E_{\text{coal},Ii,Jj,m}$ is the dimensionless coalescence efficiency, discussed in Section 15.6.8, and $K_{Ii,Jj,m}$ is the collision kernel (cm^3 particle^{-1} s^{-1}) discussed in Section 15.6. Combining the bounceoff kernel with other terms gives the **charge separation**

rate per unit volume of air (C cm^{-3} s^{-1}) in layer m,

$$\frac{\mathrm{d}Q_{b,m}}{\mathrm{d}t} = \left[\sum_{J=2}^{N_H}\sum_{j=1}^{N_C}\sum_{I=J}^{N_H}\sum_{i=j}^{N_C} B_{Ii,Jj}\frac{(\upsilon_{Ii}n_{Ii,t}n_{Jj,t-h} + \upsilon_{Jj}n_{Ii,t-h}n_{Jj,t})}{\upsilon_{Ii}+\upsilon_{Jj}}\Delta Q_{Ii,Jj}\right]_m \tag{18.88}$$

where N_{H} is the number of hydrometeor distributions, n is the number concentration of hydrometeor particles (particles cm^{-3}), υ is the volume of a single particle (cm^3 particle^{-1}), and $\Delta Q_{Ii,Jj}$ is the **charge separation per collision** (coulombs per collision). The charge separation per bounceoff due to the thermo-electric rebound charging mechanism is in the range $1–5 \times 10^{-14}$ C per collision, with an average of 3.33×10^{-14} C per collision (Pruppacher and Klett 1997). The charge separated during a collision is a complex function of the available charge on each colliding particle, the angle and surface area of collision, the time during which contact occurs, and the temperature. The charge separation (C per collision), though, must be limited by

$$\Delta Q_{Ii,Jj} = \min[3.33 \times 10^{-16}\,\mathrm{C}, 0.5(Q_{Ii} + Q_{Jj})] \tag{18.89}$$

where

$$Q_{Ii} = 3.333 \times 10^{-10} \times 2r_{Ii}^2 \tag{18.90}$$

(and a similar expression for Q_{Jj}) is the total charge (C) on a single particle in a highly electrified, precipitating cloud from (15.63), where r_{Ii} is the radius in cm.

In the equations discussed in this chapter, $N_{\mathrm{H}} = 3$, where distributions 1–3 are liquid, ice, and graupel distributions, respectively. Since the thermo-electric rebound charging mechanism occurs among ice crystals and graupel, the first summation in (18.88) considers interactions among only distributions 2 and 3. Thus, (18.88) treats bounceoffs during size-resolved ice–ice, ice–graupel, and graupel–graupel interactions.

Summing the charge separation rate from (18.88) over all cloud layers and multiplying the result by the cloudy-sky area of the model column give the **overall charge separation rate** in the cloudy region of a model column as

$$\frac{\mathrm{d}Q_{\mathrm{b,c}}}{\mathrm{d}t} = F_{\mathrm{c}}A_{\mathrm{cell}}\sum_{m=K_{\mathrm{top}}}^{K_{\mathrm{bot}}}\frac{\mathrm{d}Q_{\mathrm{b},m}}{\mathrm{d}t}\Delta z_m \tag{18.91}$$

(C s^{-1}) where F_{c} is the cloud fraction in a model column, A_{cell} is the total horizontal area of the grid cell (cm^2), and Δz_m is the vertical thickness of each model layer (cm).

The overall charge separation rate is used to calculate the **time-rate-of-change of the in-cloud electric-field strength** (V cm^{-1} s^{-1}),

$$\frac{\mathrm{d}E_{\mathrm{f}}}{\mathrm{d}t} = \frac{2k_{\mathrm{C}}}{Z_{\mathrm{c}}\sqrt{Z_{\mathrm{c}}^2 + R_{\mathrm{c}}^2}}\frac{\mathrm{d}Q_{\mathrm{b,c}}}{\mathrm{d}t} \tag{18.92}$$

(e.g., Wang and Prinn 2000), where E_f is in units of V cm^{-1}, k_C is Coulomb's constant from (18.85),

$$Z_c = \sum_{m=K_{top}}^{K_{bot}} \Delta z_m \tag{18.93}$$

is the **summed vertical thickness of cloud layers** between the bottom layer (K_{bot}) and top layer (K_{top}) of the cloud (cm), and

$$R_c = \sqrt{F_c A_{cell}/\pi} \tag{18.94}$$

is the **horizontal radius of the cloudy region** (cm).

The number of intracloud flashes per centimeter per second can now be estimated with

$$\frac{dF_r}{dt} = \frac{1}{Z_c E_{th}} \frac{dE_f}{dt} \tag{18.95}$$

where F_r is the number of flashes per centimeter and E_{th} is the **threshold electric field strength** (V cm^{-1}) defined previously. The cloud-to-ground flash rate is approximately 30–45 percent of the intracloud flashrate (Price *et al.* 1997; Boccippio *et al.* 2001).

Example 18.7

Calculate the number of flashes per hour in a cylindrical cloud of radius 0.5 km and thickness 5 km when two populations of particles, with number concentrations 1000 and 0.05 particles cm^{-3}, respectively, are present. Assume the collision kernel between the populations is 10^{-4} cm^3 $particle^{-1}$ s^{-1} and that the coalescence efficiency is 40 percent.

SOLUTION

The charge separation rate for this problem is

$$\frac{dQ_{b,c}}{dt} = B_{1,2} n_1 n_2 \Delta Q_{1,2} V_c$$

$$= (1 - 0.4) \times 10^{-4} \frac{cm^3}{partic.\ s} \times \frac{1000\ partic.}{cm^3} \times \frac{0.05\ partic.}{cm^3}$$

$$\times \frac{3.33 \times 10^{-14}\ C}{collision} \times 1.25 \times 10^{15}\ cm^3 = 0.125 \frac{C}{s}$$

where $V_c = 1.25 \times 10^{15}$ cm^3 is the volume of the cylindrical cloud. From (18.92), the rate of change of the in-cloud electric field strength is

$$\frac{dE_f}{dt} = \frac{2 \times 8.98755 \times 10^{11} \dfrac{V\ cm}{C}}{5 \times 10^5\ cm \sqrt{(5 \times 10^5\ cm)^2 + (0.5 \times 10^5\ cm)^2}} \times 0.125 \frac{C}{s} = 0.894 \frac{V}{cm\ s}$$

The number of intracloud flashes per hour in the cloud is then

$$\frac{1}{E_{\mathrm{th}}}\frac{\mathrm{d}E_{\mathrm{f}}}{\mathrm{d}t} = \frac{1}{3000\dfrac{\mathrm{V}}{\mathrm{cm\,flash}}}0.894\frac{\mathrm{V}}{\mathrm{cms}} = 2.98 \times 10^{-4}\frac{\mathrm{flashes}}{\mathrm{s}} = 1.07\frac{\mathrm{flashes}}{\mathrm{hr}}$$

Finally, the **number of nitric oxide (NO) molecules produced per cubic centimeter of air per second** within a lightning region of a cloud or between the cloud and the ground is

$$E_{\mathrm{NO}} = \frac{E_{\mathrm{l}}F_{\mathrm{NO}}}{A_{\mathrm{cell}}}\frac{\mathrm{d}F_{\mathrm{r}}}{\mathrm{d}t} \tag{18.96}$$

where E_{l} is the **number of joules per lightning flash** and F_{NO} is the **number of NO molecules produced per joule** of energy released. Values of E_{l} for cloud-to-ground lightning range from 1.8 to 11 GJ/flash with a mean of 6.7 GJ/flash (Price *et al.* 1997). The energy released by intracloud lightning is about 10 percent that of cloud-to-ground lightning. Values of F_{NO} are in the range 5–15 $\times$ 10^{16} molecules NO/J with an average of 10^{17} molecules NO/J (Price *et al.* 1997). These values give the **number of NO molecules per flash** as 9×10^{25}–1.7×10^{27} for cloud-to-ground lightning and 9×10^{24}–1.7×10^{26} for intracloud lightning. An observed NO production rate for intracloud lightning is 2.6×10^{25} molecules NO/flash (Skamarock *et al.* 2003). Estimates of the total NO produced by lightning range from 5 to 20 Tg-N per year (Price *et al.* 1997; Bond *et al.* 2002).

18.9 SUMMARY

In this chapter, cloud, fog, and precipitation classification, formation, development, microphysics, and interactions with aerosol particles were discussed. Clouds form by free convection, orographic uplifting, forced convection, and lifting along frontal boundaries. Fogs form by radiational cooling, advection of warm moist air over a cool surface, orographic uplifting, and evaporation of warm water into cold air. When air rises in a cloud, it expands and cools pseudoadiabatically. Condensation adds energy and buoyancy to a cloud, and entrainment of outside air causes cooling and downdrafts along its edges. The thermodynamic energy equation for a cloud takes account of convection, latent-heat release, entrainment, and radiative effects. Due to strong inertial accelerations in a cloud, the hydrostatic approximation is not valid; thus, a vertical momentum equation with an inertial acceleration term is needed. Microphysical processes affecting cloud evolution include condensation/deposition, coagulation, drop breakup, contact freezing, homogeneous-heterogeneous freezing, evaporative freezing, evaporation/sublimation, and melting. Aerosol particles are removed by clouds and precipitation through rainout (nucleation scavenging) and washout (aerosol–hydrometeor coagulation). Gases are removed by gas washout. Collision followed by bounceoff among frozen hydrometeors results in charge separation that can produce lightning, a major natural source of nitric oxide in the atmosphere.

18.10 PROBLEMS

18.1 Calculate the pseudoadiabatic lapse rate when p_d = 900 hPa and T = 273.15 K.

18.2 Calculate the dew-point lapse rate in an unsaturated parcel of air that rises from p_d = 1000 hPa to p_d = 900 hPa. Assume ω_v = 0.003 kg kg^{-1} and does not change in the parcel, and the average T_v = 279 K.

18.3 Calculate the critical radius and critical supersaturation when a population of 1500 particles cm^{-3}, all with radius of 0.1 μm, contains 1 × 10^{-16} mol cm^{-3}-air of an organic compound that does not dissociate. Assume T = 298.15 K.

18.4 Calculate the median freezing temperature of a liquid water drop of radius (a) 10 μm and (b) 1000 μm. At which size does a higher fraction of drops freeze at a constant temperature?

18.5 Explain the difference between contact freezing and heterogeneous freezing.

18.6 Calculate the time required for 40 percent of liquid drops to contact freeze in the presence of 1000 aerosol particles cm^{-3} if the temperature is −15 °C, the coagulation rate coefficient is 10^{-4} cm^3 particle^{-1} s^{-1}, and a maximum of 5 percent of the aerosol particles present can serve as contact nuclei.

18.7 Calculate the melting temperature of a frozen particle when the ambient temperature is 298.15 K, total air pressure is 1013.25 hPa, and the relative humidity is 50 percent.

18.8 **(a)** If the number concentrations of particles 10 and 100 μm in radius are 1000 and 0.1 particles cm^{-3}, respectively, the collision kernel between particles of these two sizes is 10^{-5} cm^3 particle^{-1} s^{-1}, and the coalescence efficiency is 30 percent, calculate the number of lightning flashes after one hour in a cloud column 1 km thick and 1 km in radius. Assume the charge separation per bounceoff is 3.33 × 10^{-16} C and the threshold electric field strength is 300 kV m^{-1}.

(b) What is the average number concentration of NO molecules resulting from the lightning after one hour assuming no molecules existed initially?

18.11 COMPUTER PROGRAMMING PRACTICE

18.9 Write a computer script to calculate the pseudoadiabatic lapse rate between $z = 0$ and 10 km. Assume $T = 288$ K at the surface and decreases 6.5 K km^{-1} and the air is saturated with water vapor. Assuming p_a = 1013.25 hPa at the surface, use (2.44) to estimate p_a at each subsequent altitude. Use the program to estimate the pseudoadiabatic lapse rate at 100-m increments.

18.10 Assume that, at the surface, $T = 285$ K, $T_D = 278$ K, $p_a = 998$ hPa, and $\partial T/\partial z =$ −11 K km^{-1}. Write a computer script to estimate p_a and T at the lifting condensation level. [Hint: First estimate p_v at the surface from (2.68), which gives $p_d = p_a - p_v$, then use the result to obtain ω_v which stays constant in the rising parcel of air. Estimate the decrease in p_a with altitude from (2.42). Use ω_v and p_d to estimate T_D at each altitude.]

18.11 Write a computer code to calculate the total number of drops, integrated over all sizes, that result from the breakup of one 5-mm diameter liquid water drop assuming the breakup distribution from (18.57). Use a volume ratio size distribution with 100 size bins between 10 and 5500 μm.

18.12 Write a computer program to replicate Fig. 18.10(a).

19

Irreversible aqueous chemistry

IRREVERSIBLE aqueous chemistry, or irreversible chemistry in liquid water, is an important step in the production of acid rain, acid fog, and acid haze. For example, it is responsible for converting sulfur dioxide gas to aqueous sulfuric acid in aerosol particles, cloud drops, and precipitation. Irreversible aqueous reactions are tightly coupled with reversible (equilibrium) reactions, such as those discussed in Chapter 17. Since many aqueous reactants originate as gases, aqueous chemistry is tightly linked to nonequilibrium gas-to-particle conversion as well. Since aqueous reactions are described by first-order, ordinary differential equations, they are solved with methods similar to those used to solve gas-phase reactions. In this chapter, the significance of aqueous chemistry, identification of important aqueous reactions, and a method of solving aqueous chemistry together with gas-to-particle conversion are discussed.

19.1 SIGNIFICANCE OF AQUEOUS CHEMICAL REACTIONS

Aqueous chemistry is irreversible chemistry that takes place in liquid water. It occurs following dissolution of gases into liquid aerosol particles, cloud drops, and precipitation as follows: First, soluble gases diffuse to and dissolve in liquid water on a particle surface. Some of the dissolved molecules dissociate reversibly into ions. The chemicals then diffuse through the water. Irreversible chemical reactions take place during diffusion. Some of the dissolved molecules and reaction products may diffuse back to the particle surface and evaporate, whereas others may stay within the particle.

Gas dissolution and irreversible chemistry affect particle composition at a rate proportional to the quantity of liquid water present. As such, aqueous chemistry proceeds faster in cloud drops, which generally have liquid-water contents $>10^5$ $\mu g\ m^{-3}$, than in aerosol particles, which have liquid-water contents $<500\ \mu g\ m^{-3}$.

When a gas dissolves in a dilute liquid-water drop and does not react chemically, the quantities of the gas remaining and dissolved in water are governed by Henry's law, which states that, over a dilute solution, the pressure exerted by a gas is proportional to the molality of the dissolved gas in solution (Section 17.4). When Henry's law holds, the quantity of gas over a drop surface is the saturation vapor pressure of the gas, and the gas is in equilibrium with its dissolved phase at the surface. As such, in the absence of aqueous chemistry, only a finite amount of a gas dissolves in a liquid drop.

Table 19.1 Names, formulae, and lewis structures of S(IV) and S(VI) species

S(IV) family		S(VI) family	
Chemical name and formula	Chemical structure	Chemical name and formula	Chemical structure
Sulfur dioxide (g,aq) SO_2(g,aq)	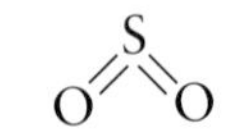		
Sulfurous acid (aq) H_2SO_3(aq)	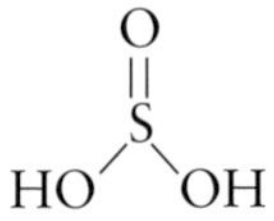	Sulfuric acid (aq) H_2SO_4(aq)	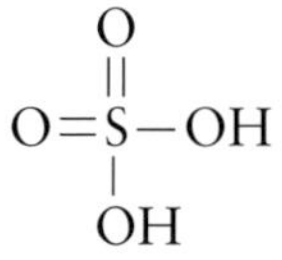
Bisulfite ion HSO_3^-	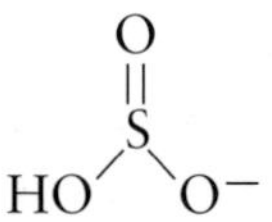	Bisulfate ion HSO_4^-	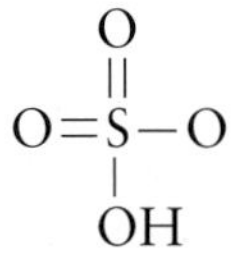
Sulfite ion SO_3^{2-}	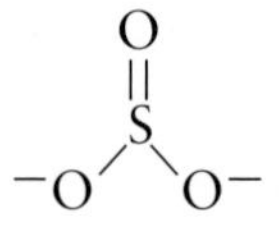	Sulfate ion SO_4^{2-}	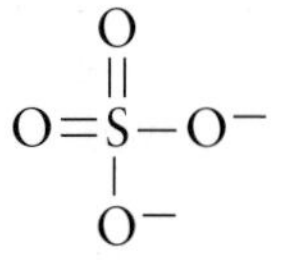

When irreversible chemistry occurs, the dissolved gas chemically reacts to produce a product, instantaneously decreasing the molality of the dissolved gas. As such, more gas must dissolve into the drop to maintain saturation over the drop surface (i.e., to maintain the Henry's law relation). Once additional gas dissolves, it reacts away irreversibly as well. The continuous dissolution and irreversible reaction that occur in a drop can deplete the gas phase of dissolving species, sometimes within minutes. Below, aqueous chemistry processes of importance are discussed.

19.2 MECHANISMS OF CONVERTING S(IV) TO S(VI)

The most important processes affected by irreversible chemistry are those that convert sulfur dioxide gas to dissolved sulfuric acid and its dissociation products, bisulfate and sulfate. These processes are important because sulfuric acid is the most abundant acid in the air and the main culprit in acid deposition. Much of its atmospheric production is due to irreversible chemistry within clouds, precipitation, and aerosol particles.

Sulfur dioxide is a member of the **S(IV) family**. Dissolved sulfuric acid and its products are members of the **S(VI) family** (Table 19.1). In these families, the IV and the VI represent the oxidation states (+4 and +6, respectively) of the members of the respective families. Thus, S(VI) members are more oxidized than are S(IV) members. Because sulfur dioxide is in the S(IV) family and sulfuric acid, the main source of acidity in rainwater, is in the S(VI) family, the oxidation of sulfur dioxide gas to aqueous sulfuric acid represents a conversion from the S(IV) family to the S(VI) family. This conversion occurs through two pathways, described next.

19.2.1 Gas-phase oxidation of S(IV)

The first conversion mechanism of S(IV) to S(VI) involves three steps. The initial step is gas-phase oxidation of sulfur dioxide to sulfuric acid by

$$\underset{\substack{\text{Sulfur}\\\text{dioxide gas}}}{SO_2(g)} \xrightarrow{+\dot{O}H(g),\,M} \underset{\text{Bisulfite}}{H\dot{S}O_3(g)} \xrightarrow[\ H\dot{O}_2(g)\ ]{+O_2(g)} \underset{\substack{\text{Sulfur}\\\text{trioxide}}}{SO_3(g)} \xrightarrow{+H_2O(g)} \underset{\substack{\text{Sulfuric}\\\text{acid gas}}}{H_2SO_4(g)} \tag{19.1}$$

The second step is condensation of sulfuric acid gas onto aerosol particles and simultaneous hydration (Section 17.8) of liquid water to the sulfuric acid to produce a sulfuric acid–water solution. Because sulfuric acid has a low saturation vapor pressure, nearly all of it produced by (19.1) condenses, and this process is relatively irreversible. The third step is dissociation of aqueous sulfuric acid. The second and third steps are represented by

$$\underset{\substack{\text{Sulfuric}\\\text{acid gas}}}{H_2SO_4(g)} \longrightarrow \underset{\substack{\text{Aqueous}\\\text{sulfuric acid}}}{H_2SO_4(aq)} \rightleftharpoons H^+ + \underset{\text{Bisulfate ion}}{HSO_4^-} \rightleftharpoons 2H^+ + \underset{\text{Sulfate ion}}{SO_4^{2-}} \tag{19.2}$$

At typical pHs of aerosol particles and cloud drops, nearly all aqueous sulfuric acid dissociates to the sulfate ion. The dissociation releases two protons, decreasing pH and increasing acidity of the solution.

Condensation accounts for much of the increase of S(VI) in particles when the relative humidity is below 100 percent. Below 70-percent relative humidity, nearly all S(VI) production in particles is due to condensation. Because the rate of change of mass of a particle due to condensation is proportional to the radius of the particle, as shown in (16.13), condensation is a **radius-limited** process. Whereas gas-phase oxidation followed by condensation is the dominant mechanism by which S(IV) produces S(VI) in aerosol particles, particularly at low relative humidity, a second mechanism produces S(VI) from S(IV) more rapidly in cloud drops and raindrops.

19.2.2 Aqueous-phase oxidation of S(IV)

The second conversion process of S(IV) to S(VI) involves several steps as well. The initial step is reversible dissolution of sulfur dioxide gas into liquid drops by

$$\underset{\substack{\text{Sulfur}\\\text{dioxide gas}}}{SO_2(g)} \rightleftharpoons \underset{\substack{\text{Dissolved}\\\text{sulfur dioxide}}}{SO_2(aq)} \tag{19.3}$$

Sulfur dioxide is less soluble in water than is hydrochloric acid or nitric acid but more soluble than are many other gases. Because the dissolution rate of sulfur dioxide or any other gas depends on the volume of liquid water present, dissolution is a **volume-limited** process.

The second step is reversible reaction of aqueous sulfur dioxide to sulfurous acid, followed by dissociation of sulfurous acid:

$SO_2(aq)$	$+ H_2O(aq) \rightleftharpoons$	$H_2SO_3(aq) \rightleftharpoons$	H^+	$+ HSO_3^- \rightleftharpoons$	$2H^+$	$+ SO_3^{2-}$
Dissolved sulfur dioxide	Liquid water	Sulfurous acid	Hydrogen ion	Bisulfite ion	Hydrogen ion	Sulfite ion

(19.4)

At pH of 2–7, the major dissociation product of sulfurous acid is the bisulfite ion (HSO_3^-). All sulfur species in (19.4) are in the S(IV) family, so the bisulfite ion is the major species present in the S(IV) family upon sulfur dioxide dissolution and dissociation.

The third step is irreversible aqueous oxidation of S(IV) to S(VI). Over a range of liquid-water contents (50 μg m^{-3} to 2 g m^{-3}), hydrogen peroxide and ozone are the major aqueous-phase oxidants of S(IV) at pH $\leq$ 6 and pH $>$ 6, respectively. Since oxidation of SO_2 by O_3 reduces pH, H_2O_2 often becomes the most important oxidant of SO_2 even when the initial pH $>$ 6 (e.g., Liang and Jacobson 1999).

When the pH $\leq$ 6, hydrogen peroxide oxidizes S(IV) by

HSO_3^-	$+ H_2O_2(aq)$	$+ H^+ \longrightarrow$	SO_4^{2-}	$+ H_2O(aq) + 2H^+$
Bisulfite ion	Hydrogen peroxide(aq)		Sulfate ion	

(19.5)

This reaction is written in terms of the bisulfite ion and the sulfate ion because, at pH of 2–6, these are the two most prevalent components of the S(IV) and S(VI) families, respectively. This reaction consumes S(IV) within tens of minutes if the $H_2O_2(aq)$ concentration exceeds that of S(IV). If the S(IV) concentration exceeds that of $H_2O_2(aq)$, then $H_2O_2(aq)$ is depleted within minutes (Jacob 1986). The rate coefficient of (19.5) multiplied by $[H^+]$ is 9 times greater at pH = 0 than at pH = 4 (Appendix Table B.8), suggesting that the coefficient increases with decreasing pH. A cloud drop typically has a pH of 2.5 to 5.6. An aerosol particle often has a pH of −1 to +2. The lower pH in aerosol particles speeds reaction rates in such particles relative to those in cloud drops. But, because aerosol particles contain less $H_2O_2(aq)$ and liquid water than do cloud drops, aerosol particles are still less efficient at converting S(IV) than are cloud drops.

Hydrogen peroxide is also photolyzed within an aqueous solution by

$$H_2O_2(aq) + h\nu \longrightarrow 2\dot{O}H(aq) \tag{19.6}$$

and is lost by reaction with the aqueous hydroxyl radical,

$$H_2O_2(aq) + \dot{O}H(aq) \longrightarrow H_2O(aq) + H\dot{O}_2(aq) \tag{19.7}$$

Sources of hydrogen peroxide in liquid water are its dissolution by

$$H_2O_2(g) \rightleftharpoons H_2O_2(aq) \tag{19.8}$$

and aqueous production from the hydroperoxy radical,

$$\underset{\substack{\text{Hydroperoxy}\\ \text{radical}}}{H\dot{O}_2(aq)} + \underset{\substack{\text{Peroxy}\\ \text{ion}}}{O_2^-} + H_2O(aq) \longrightarrow \underset{\substack{\text{Hydrogen}\\ \text{peroxide}}}{H_2O_2(aq)} + O_2(aq) + \underset{\substack{\text{Hydroxide}\\ \text{ion}}}{OH^-} \tag{19.9}$$

The aqueous hydroperoxy radical dissolves from the gas phase. At a pH above 5, much of it dissociates. Dissolution and dissociation are represented by

$$H\dot{O}_2(g) \rightleftharpoons H\dot{O}_2(aq) \rightleftharpoons H^+ + O_2^- \tag{19.10}$$

When the pH > 6, which occurs only in cloud drops that contain basic substances, such as ammonium or sodium, the most important reaction converting S(IV) to S(VI) is

$$\underset{\substack{\text{Sulfite}\\ \text{ion}}}{SO_3^{2-}} + \underset{\substack{\text{Dissolved}\\ \text{ozone}}}{O_3(aq)} \longrightarrow \underset{\substack{\text{Sulfate}\\ \text{ion}}}{SO_4^{2-}} + \underset{\substack{\text{Dissolved}\\ \text{oxygen}}}{O_2(aq)} \tag{19.11}$$

This reaction is written in terms of the sulfite ion and the sulfate ion because the HSO_3^-–O_3 reaction is relatively slow, and at pH levels greater than 6, most S(VI) exists as SO_4^{2-}. This reaction not only oxidizes S(IV), but also helps to deplete gas-phase ozone in a cloud (e.g., Liang and Jacob 1997; Walcek *et al.* 1997; Zhang *et al.* 1998). As the reaction proceeds, more ozone gas must dissolve to maintain saturation over a drop surface.

When the liquid-water content exceeds 0.2 g m^{-3} and the pH $\approx$ 5, or when H_2O_2 is depleted, OH can also oxidize S(IV) by

$$\underset{\substack{\text{Bisulfite}\\ \text{ion}}}{HSO_3^-} + \dot{O}H(aq) + O_2(aq) \longrightarrow \underset{\substack{\text{Peroxysulfate}\\ \text{ion}}}{S\dot{O}_5^-} + H_2O(aq) \tag{19.12}$$

The hydroxyl radical is not very soluble in water, so its production from aqueous chemistry generally exceeds its production from gas dissolution. The peroxysulfate ion catalyzes oxidation of S(IV) by

$$\underset{\substack{\text{Bisulfite}\\ \text{ion}}}{HSO_3^-} + \underset{\substack{\text{Dissolved}\\ \text{oxygen}}}{O_2(aq)} \xrightarrow{\dot{S}O_5^-} \underset{\substack{\text{Peroxymonosulfate}\\ \text{ion}}}{HSO_5^-} \quad (19.13)$$

The peroxymonosulfate ion produced from this reaction also oxidizes S(IV) by

$$\underset{\substack{\text{Bisulfite}\\ \text{ion}}}{HSO_3^-} + HSO_5^- + H^+ \longrightarrow \underset{\substack{\text{Sulfate}\\ \text{ion}}}{2SO_4^{2-}} + 3H^+ \quad (19.14)$$

Dissolved oxygen oxidizes S(IV) when a **transition metal**, such as Fe(III) or Mn(II), catalyzes the reaction (Jacob *et al.* 1989a). About 85–90 percent of soluble iron appears as Fe(II) in the form of Fe^{2+}, and about 10–15 percent appears as Fe(III), although these values are location-dependent. At pH < 12, most Fe(III) is present as Fe^{3+} and catalyzes S(IV) oxidation by

$$\underset{\text{Sulfite ion}}{SO_3^{2-}} + H_2O(aq) + O_2(aq) \xrightarrow{\text{Fe(III)}} \underset{\text{Sulfate ion}}{SO_4^{2-}} + H_2O_2(aq) \quad (19.15)$$

Fe(II) does not catalyze S(IV) oxidation. Mn(II) is usually in the form of Mn^{2+} and catalyzes S(IV) by

$$\underset{\text{Bisulfite ion}}{HSO_3^-} + H_2O(aq) + O_2(aq) \xrightarrow{\text{Mn(II)}} \underset{\text{Sulfate ion}}{SO_4^{2-}} + H_2O_2(aq) + H^+ \quad (19.16)$$

Mn(III) does not catalyze S(IV) oxidation significantly.

Slightly soluble carbonyl compounds, such as formaldehyde and acetaldehyde, also affect S(IV) oxidation in clouds and rain (e.g., Munger *et al.* 1989). These species originate from the gas phase, where they are produced by combustion and photochemistry. Dissolved formaldehyde reacts reversibly in solution to form **methylene glycol** by

$$\underset{\text{Formaldehyde}}{HCHO(aq)} + H_2O(aq) \rightleftharpoons \underset{\text{Methylene glycol}}{H_2C(OH)_2(aq)} \quad (19.17)$$

Methylene glycol is lost during the reaction,

$$\underset{\text{Methylene glycol}}{H_2C(OH)_2(aq)} + \dot{O}H(aq) + O_2(aq) \longrightarrow \underset{\text{Formic acid}}{HCOOH(aq)} + H\dot{O}_2(aq) + H_2O(aq) \quad (19.18)$$

At high pH, formaldehyde oxidizes the sulfite ion to form **hydroxymethanesulfonate** (HMSA) by

$$\underset{\text{Sulfite ion}}{SO_3^{2-}} + \underset{\text{Formaldehyde}}{HCHO(aq)} + H_2O(aq) \longrightarrow \underset{\text{HMSA}}{HOCH_2SO_3^-} + OH^- \quad (19.19)$$

which is subsequently lost during the reaction

$$\underset{\text{HMSA}}{HOCH_2SO_3^-} + \dot{O}H(aq) + O_2(aq) \longrightarrow HCHO(aq) + \underset{\text{Peroxysulfate ion}}{S\dot{O}_5^-} + H_2O(aq) \quad (19.20)$$

Other slightly soluble organic gases include the methylperoxy radical (CH_3O_2), methyl hydroperoxide (CH_3OOH), formic acid (HCOOH), acetic acid (CH_3COOH), and peroxyacetic acid ($CH_3C(O)OOH$). Relevant aqueous reactions for these species appear in Appendix Table B.8, and equilibrium reactions appear in Appendix Table B.7.

Other soluble inorganic gases include nitric acid, ammonia, and hydrochloric acid. Under most conditions, the dissolved products of nitric acid and ammonia, $HNO_3(aq)$, NO_3^-, $NH_3(aq)$, and NH_4^+ are unaffected significantly by irreversible reactions (Pandis and Seinfeld 1989). The chemistry of these species is governed by reversible reactions, discussed in Chapter 18. The chloride ion (Cl^-), the dissociation product of HCl(aq), reacts irreversibly in solution. It maintains equilibrium with several aqueous chlorine species through

$$\underset{\text{Dichloride ion}}{Cl_2^-} \rightleftharpoons \underset{\text{Chlorine atom}}{\dot{C}l(aq)} + \underset{\text{Chloride ion}}{Cl^-} \quad (19.21)$$

$$\underset{\text{Chlorine hydroxide radical}}{ClOH^-} \rightleftharpoons \underset{\text{Chloride ion}}{Cl^-} + \dot{O}H(aq) \quad (19.22)$$

(Jayson *et al.* 1973). The chlorine atom reacts irreversibly with hydrogen peroxide by

$$H_2O_2(aq) + \dot{C}l(aq) \longrightarrow Cl^- + H\dot{O}_2(aq) + H^+ \quad (19.23)$$

and the dichloride ion oxidizes S(IV) by

$$\underset{\text{Bisulfite ion}}{HSO_3^-} + \underset{\text{Dichloride ion}}{Cl_2^-} + O_2(aq) \longrightarrow \underset{\text{Peroxysulfate ion}}{S\dot{O}_5^-} + 2Cl^- + H^+ \quad (19.24)$$

but this reaction is less important than other reactions that oxidize S(IV).

In sum, the second conversion process of S(IV) to S(VI) involves dissolution of sulfur dioxide gas to sulfurous acid, reversible dissociation of sulfurous acid to the bisulfite ion and the sulfite ion, and irreversible oxidation of these ions to the sulfate ion. Oxidants of bisulfite and/or sulfite in solution include hydrogen peroxide, ozone, the hydroxyl radical, the peroxymonosulfate ion, formaldehyde, the dichloride ion, and molecular oxygen catalyzed by iron, manganese, and the peroxysulfate ion (e.g., Hoffmann and Calvert 1985; Jacob *et al.* 1989a; Pandis and Seinfeld 1989).

The conversion rate of S(IV) to S(VI) depends on liquid-water content, oxidant concentration, and pH. Aerosol liquid-water contents are generally less than 500 μg m^{-3}. Fog liquid-water contents are generally less than 0.2 g m^{-3}. Cloud liquid-water contents, which range from 0.2–2 g m^{-3}, exceed those of fogs. Thus, a cloud has 500–10 000 times more liquid water than does an aerosol plume.

The Henry's law constants of H_2O_2, HO_2, OH, and O_3 are moderate to low, as shown in Appendix Table B.7. Because of its high liquid-water content, a cloud can absorb all four gases. Because of its low liquid-water content, an aerosol plume absorbs proportionately less of all four gases and absorbs only H_2O_2 in significant quantities. Thus, S(IV) conversion in an aerosol plume is often limited by the concentration of H_2O_2. In a cloud, the greater abundance of liquid water, H_2O_2, HO_2, OH, and O_3 in solution permits S(IV) to convert to S(VI) faster than in an aerosol plume. Although S(IV) oxidation is slow in aerosol particles, it can be enhanced at low temperatures (268 K) if the relative humidity is high.

The faster that S(IV) oxidation occurs, the faster that sulfur dioxide is depleted from the gas phase, and the faster that S(VI) forms. In cloud drops, dissolution and aqueous reaction can convert 60 percent of gas-phase sulfur dioxide to the sulfate ion within 20 minutes (e.g., Jacobson 1997a; Liang and Jacobson 1999). Figure 19.1 illustrates this point. The figure compares modeled in-cloud concentrations of liquid water, the bisulfate ion, and the sulfate ion over time when sulfur dioxide dissolution and aqueous reaction were and were not included in the calculation. Within 20 min, cloud drops dissolved and converted more than 30 μg m^{-3} (58 percent) out of an initial 52 μg m^{-3} of $SO_2(g)$. Aerosol particles, on the other hand, require hours to days to convert the same quantity of sulfur dioxide.

Conversion of S(IV) to S(VI) in aerosol particles, cloud drops, and precipitation drops contributes to **acid deposition.** Acid deposition occurs to some extent in and downwind of all countries worldwide. It is particularly prevalent in provinces of eastern Canada, the northeastern United States (particularly the Adirondack Mountain region), southern Scandinavia, middle and eastern Europe, India, Korea, Russia, China, Japan, Thailand, and South Africa. The natural pH of cloud water and rainwater is 5.6 (Fig. 17.1). Aqueous oxidation of S(IV) to S(VI) generally reduces the pH in fog and rain to 2–5.6, creating acid deposition.

19.3 DIFFUSION WITHIN A DROP

When a gas dissolves in a liquid drop, it immediately begins to diffuse through the drop and reacts chemically at the same time. The characteristic **time scale (s) for**

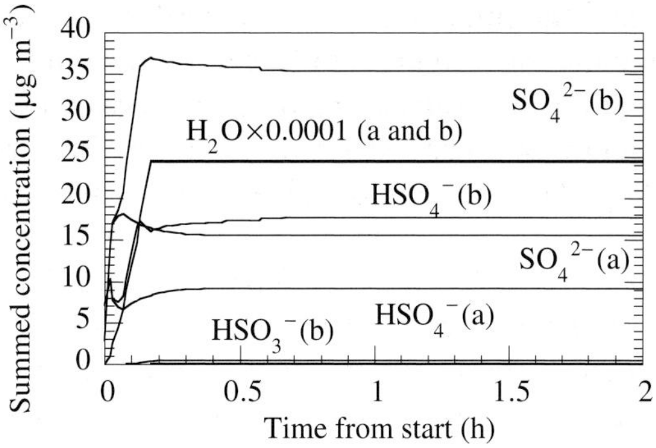

Figure 19.1 Time-series comparison of dissolved ion and water concentrations, summed over all cloud drop sizes, when cloud drops grew and sulfur dioxide (a) did not and (b) did dissolve in and react within the drops. H_2SO_4 condensed in both cases, but SO_2 dissolved and reacted in the second case only. Initial conditions were SO_2(g), 52 μg m^{-3}; H_2O_2(g), 10 μg m^{-3}; HNO_3(g), 30 μg m^{-3}; NH_3(g), 10 μg m^{-3}; HCl(g), 0 μg m^{-3}; H_2SO_4(g), 15 μg m^{-3}; H_2SO_4(aq), 10 μg m^{-3}; NaCl(aq), 15 μg m^{-3}; and $T = 298$ K. The cloud was formed by increasing the relative humidity to 100.01 percent at the start, solving the growth equations with a 1-s time step, and replenishing the relative humidity to 100.01 percent every time step for the first 10 min of simulation. After 10 min, the relative humidity was allowed to relax to its calculated value. From Jacobson (1997a).

diffusion of solute q through a liquid drop is

$$\tau_{\mathrm{ad},q} = \frac{r_i^2}{\pi^2 D_{\mathrm{aq},q}} \tag{19.25}$$

(Crank 1975) where $D_{\mathrm{aq},q}$ is the diffusion coefficient of the solute in water at 25 °C (cm^2 s^{-1}) and r_i is the radius of the drop (cm). A typical diffusion coefficient is $D_{\mathrm{aq},q} = 2 \times 10^{-5}$ cm^2 s^{-1} (e.g., Wilke and Chang 1955; Himmelblau 1964; Jacob 1986; Schwartz 1986).

Example 19.1

From (19.25), the characteristic time for diffusion within a 30-μm-diameter cloud drop is about 0.011 s. The characteristic time for diffusion within a 10-μm-diameter drop is about 0.0013 s. Thus, for many purposes, diffusion can be assumed instantaneous.

Although the time scale for diffusion is small, chemical reactions often occur on an even smaller time scale. Jacob (1986) found that $O_3(aq)$, $NO_3(aq)$, $OH(aq)$, $Cl(aq)$, SO_4^-, CO_3^-, and Cl_2^- had chemical lifetimes shorter than diffusion transport times. Such species react near the surface of a drop before they can diffuse to the center. When diffusion and chemical reaction are considered, the time rate of change of concentration of species q in size bin i as a function of radius r is

$$\left(\frac{\mathrm{d}c_{q,i,r}}{\mathrm{d}t}\right)_{\mathrm{ad,aq}} = D_{\mathrm{aq},q}\frac{1}{r^2}\frac{\partial}{\partial r}\left(r^2\frac{\partial c_{q,i,r}}{\partial r}\right) + P_{\mathrm{c},q,i,r} - L_{\mathrm{c},q,i,r} \tag{19.26}$$

where $P_{\mathrm{c},q,i,r}$ and $L_{\mathrm{c},q,i,r}$ are the net aqueous chemical production and loss rates (mol cm^{-3} s^{-1}), respectively, of q in size bin i at a distance r from the center of the drop, and the subscript ad, aq indicates that aqueous diffusion is coupled with aqueous chemistry. At the drop center, $\partial c_{q,i,r}/\partial r = 0$.

19.4 SOLVING GROWTH AND AQUEOUS CHEMICAL ODES

Aqueous chemical reactions are similar to gas-phase reactions in that both are described by first-order ODEs that can be solved with the methods discussed in Chapter 12. Aqueous chemistry ODEs are more difficult to solve than gas chemistry ODEs for two reasons. First, the e-folding lifetimes of aqueous species against chemical loss are usually shorter than are those of gases. Thus, aqueous reactions are stiffer than gas reactions. Second, when dissolved gases are destroyed by aqueous reactions, the dissolved gases are rapidly replenished by additional dissolution to maintain saturation over the particle surface. When $H_2O_2(aq)$ reacts in solution, for example, $H_2O_2(g)$ dissolves to replace lost $H_2O_2(aq)$ to satisfy (19.8). Thus, in a model that resolves aerosol particles and hydrometeor particles among multiple size bins, aqueous ODEs must be solved within each size bin, and growth ODEs must be solved between the gas phase and all bins simultaneously. Gas chemistry is effectively solved in one size bin.

To make matters more complex, many dissolving gases dissociate reversibly in solution. Thus, dissolution, equilibrium, and aqueous reaction processes must be solved nearly simultaneously. The rate of change of particle concentration due to these processes is

$$\left(\frac{\mathrm{d}c_{q,i,t}}{\mathrm{d}t}\right)_{\mathrm{ge,eq,aq}} = \left(\frac{\mathrm{d}c_{q,i,t}}{\mathrm{d}t}\right)_{\mathrm{ge}} + \left(\frac{\mathrm{d}c_{q,i,t}}{\mathrm{d}t}\right)_{\mathrm{eq}} + \left(\frac{\mathrm{d}c_{q,i,t}}{\mathrm{d}t}\right)_{\mathrm{aq}} \tag{19.27}$$

where $c_{q,i,t}$ is the mole concentration of species q in bin i at time t, and subscripts ge, eq, and aq identify growth/evaporation, equilibrium, and aqueous chemistry, respectively. The aqueous chemistry term in (19.27) is the same as that in (19.26), except that diffusion is assumed to be instantaneous and particles are assumed to be well mixed in (19.27). The gas-aqueous mole conservation equation corresponding to (19.27) is

$$\frac{\mathrm{d}C_{q,t}}{\mathrm{d}t} = -\sum_{i=1}^{N_B}\left(\frac{\mathrm{d}c_{q,i,t}}{\mathrm{d}t}\right)_{\mathrm{ge}} \tag{19.28}$$

where $C_{q,t}$ is the gas mole concentration of species q, and N_B is the number of size bins. Together, (19.27) and (19.28) make up a large set of stiff ODEs.

An exact solution to (19.27) and (19.28) is difficult to obtain. The equilibrium term is not an ordinary differential equation. It is a reversible equation solved with an iterative method, such as the MFI or AEI schemes discussed in Chapter 17. A reversible equation, of the form $D + E \rightleftharpoons A + B$, can be written as a combination of two ODEs, $D + E \rightarrow A + B$ and $A + B \rightarrow D + E$. These equations are very stiff and time-consuming to solve. Instead, the reaction $D + E \rightleftharpoons A + B$ is usually simulated as a reversible equation that may be operator-split from the other terms in (19.27).

The growth and aqueous-chemistry terms in (19.27) are often solved together in zero- and one-dimensional modeling studies (e.g., Chameides 1984; Schwartz 1984; Jacob 1986; Pandis and Seinfeld 1989; Bott and Carmichael 1993; Sander *et al.* 1995; Zhang *et al.* 1998). In some cases, the equations are solved with an integrator of stiff ODEs. In others, they are solved with a forward Euler method taking a small time step. Here, a method of solving the equations in a three-dimensional model is discussed.

With this method, species involved in dissociation/association reactions among themselves are grouped in **families**. Family species cycle rapidly among each other due to rapid and reversible dissociation/association. Changes in family concentrations are slower than changes in species concentrations. Here, the S(IV), S(VI), $HO_{2,T}$, $CO_{2,T}$, $HCHO_T$, $HCOOH_T$, and CH_3COOH_T families are considered. Family mole concentrations (moles per cubic centimeter of air) in size bin i are

$$c_{S(IV),i} = c_{SO_2(aq),i} + c_{HSO_3^-,i} + c_{SO_3^{2-},i} \tag{19.29}$$

$$c_{S(VI),i} = c_{H_2SO_4(aq),i} + c_{HSO_4^-,i} + c_{SO_4^{2-},i} \tag{19.30}$$

$$c_{HO_{2,T},i} = c_{HO_2(aq),i} + c_{O_2^-,i} \tag{19.31}$$

$$c_{CO_{2,T},i} = c_{CO_2(aq),i} + c_{HCO_3^-,i} + c_{CO_3^{2-},i} \tag{19.32}$$

$$c_{HCHO_T,i} = c_{HCHO(aq),i} + c_{H_2C(OH)_2,i} \tag{19.33}$$

$$c_{HCOOH_T,i} = c_{HCOOH(aq),i} + c_{HCOO^-,i} \tag{19.34}$$

$$c_{CH_3COOH_T,i} = c_{CH_3COOH(aq),i} + c_{CH_3COO^-,i} \tag{19.35}$$

respectively. Some species, such as $H_2O_2(aq)$ and $O_3(aq)$, are each self-contained in their own family since $H_2O_2(aq)$ dissociates significantly only at a pH higher than observed in cloud drops, and $O_3(aq)$ does not dissociate in solution. Growth–aqueous-chemical ODEs can be obtained for families by combining (17.99) with chemical production and loss terms. For the S(IV) family, the resulting ODE for a size bin is

$$\frac{dc_{S(IV),i}}{dt} = k_{S(IV),i}\left(C_{SO_2(g)} - S'_{S(IV),i}\frac{c_{SO_2(aq),i}}{H'_{SO_2(aq),i}}\right) + P_{c,S(IV),i} - L_{c,S(IV),i} \tag{19.36}$$

where $k_{S(IV),i}$ is the **coefficient for mass transfer** of SO_2 to particles of size i (s^{-1}) from (16.64), $C_{SO_2(g)}$ is the concentration of $SO_2(g)$ (mol cm^{-3}), $S'_{S(IV),i}$ is the equilibrium saturation ratio from (16.47), $P_{c,S(IV),i}$ is the **aqueous chemical production rate** of

S(IV) in size bin i (mol cm^{-3} s^{-1}), $L_{c,S(IV),i}$ is the **aqueous chemical loss rate** of S(IV) in bin i (mol cm^{-3} s^{-1}), and

$$H'_{SO_2(aq),i} = m_v c_w R^* T H_{SO_2} \tag{19.37}$$

is the **dimensionless Henry's constant** of SO_2. Parameters in (19.37) were defined in Section 17.14.1. The concentrations $c_{SO_2(aq),i}$ and $c_{S(IV),i}$ are related by

$$c_{SO_2(aq),i} = c_{S(IV),i} \frac{\mathrm{m}^2_{H^+,i}}{\mathrm{m}^2_{H^+,i} + \mathrm{m}_{H^+,i} K_{1,S(IV)} + K_{1,S(IV)} K_{2,S(IV)}} \tag{19.38}$$

derived from (19.29), (17.97), and

$$K_{1,S(IV)} = \frac{\mathrm{m}_{H^+,i} \mathrm{m}_{HSO_3^-,i} \gamma^2_{i,H^+/HSO_3^-}}{\mathrm{m}_{SO_2(aq),i}} \tag{19.39}$$

$$K_{2,S(IV)} = \frac{\mathrm{m}_{H^+,i} \mathrm{m}_{SO_3^{2-},i} \gamma^2_{i,H^+/SO_3^{2-}}}{\mathrm{m}_{HSO_3^-,i} \gamma^2_{i,H^+/HSO_3^-}} \tag{19.40}$$

which are the **first and second equilibrium dissociation coefficient expressions** for the reversible reactions,

$$SO_2(aq) + H_2O \rightleftharpoons H^+ + HSO_3^- \tag{19.41}$$

$$HSO_3^- \rightleftharpoons H^+ + SO_3^{2-} \tag{19.42}$$

respectively. For simplicity, the activity coefficients in (19.39) and (19.40) may be set to zero when dilute solutions (e.g., cloud drops) are considered.

Substituting (19.38) into (19.36) and adding time subscripts give

$$\frac{dc_{S(IV),i,t}}{dt} = k_{S(IV),i,t-h}\left(C_{SO_2(g),t} - S'_{S(IV),i,t-h}\frac{c_{S(IV),i,t}}{H'_{S(IV),i,t-h}}\right) + P_{c,S(IV),i,t} - L_{c,S(IV),i,t} \tag{19.43}$$

where

$$H'_{S(IV),i,t-h} = m_v c_{w,i,t-h} R^* T H_{SO_2}\left(1 + \frac{K_{1,S(IV)}}{\mathrm{m}_{H^+,i,t-h}} + \frac{K_{1,S(IV)} K_{2,S(IV)}}{\mathrm{m}^2_{H^+,i,t-h}}\right) \tag{19.44}$$

is the **dimensionless effective Henry's constant** of S(IV).

The chemical production and loss rates of a species q are determined from

$$P_{c,q,i,t} = \sum_{l=1}^{N_{prod,q}} R_{c,n_P(l,q),t} \qquad L_{c,q,i,t} = \sum_{l=1}^{N_{loss,q}} R_{c,n_L(l,q),t} \tag{19.45}$$

where $N_{prod,q}$ ($N_{loss,q}$) is the number of reactions in which species q is produced (lost), $R_{c,n_P(l,q)}$ [$R_{c,n_L(l,q)}$] is the lth production (loss) rate of q, and $n_P(l,q)$ [$n_L(l,q)$]

gives the reaction number corresponding to the lth production (loss) term of species q. The terms in (19.44) are similar to those in (12.53) and (12.55), respectively. S(IV) has no important aqueous chemical sources but several sinks.

Appendix Table B.8 lists S(IV) oxidation reactions. Two of the reactions

$$\mathrm{S(IV)} + \mathrm{H_2O_2(aq)} + \mathrm{H^+} \longrightarrow \mathrm{S(VI)} + 2\mathrm{H^+} + \mathrm{H_2O(aq)} \tag{19.46}$$

$$\mathrm{S(IV)} + \mathrm{HO}_{2,T} \longrightarrow \mathrm{S(VI)} + \mathrm{OH(aq)} + 2\mathrm{H^+} \tag{19.47}$$

are considered here to illustrate the calculation of the chemical loss rate of the S(IV) family for use in a chemical solver. Appendix Table B.8 shows that the only important S(IV) reactant in (19.46) is HSO_3^-:

$$\mathrm{HSO_3^-} + \mathrm{H_2O_2(aq)} + \mathrm{H^+} \tag{19.48}$$

In (19.47), HSO_3^- and SO_3^{2-} are important S(IV) reactants and $HO_2(aq)$ and O_2^- are important $HO_{2,T}$ reactants. As such, (19.47) is really a combination of the following four reactions:

$$\mathrm{HSO_3^-} + \mathrm{HO_2(aq)} \tag{19.49}$$

$$\mathrm{SO_3^{2-}} + \mathrm{HO_2(aq)} \tag{19.50}$$

$$\mathrm{HSO_3^-} + \mathrm{O_2^-} \tag{19.51}$$

$$\mathrm{SO_3^{2-}} + \mathrm{O_2^-} \tag{19.52}$$

One way to consider the loss of S(IV) in a chemical solver is to treat reactions (19.48)–(19.52) as separate reactions. Doing so requires the simultaneous solution of the equilibrium equations

$$\mathrm{SO_2(aq)} + \mathrm{H_2O(aq)} \rightleftharpoons \mathrm{H^+} + \mathrm{HSO_3^-} \rightleftharpoons 2\mathrm{H^+} + \mathrm{SO_3^{2-}} \tag{19.53}$$

$$\mathrm{HO_2(aq)} \rightleftharpoons \mathrm{H^+} + \mathrm{O_2^-} \tag{19.54}$$

Alternatively, the chemical loss of S(IV) or any other species can be written in terms of the family reactions, (19.46) and (19.47), where the families are defined in (19.29)–(19.35). In this case, the instantaneous equilibrium partitioning that occurs with (19.53) and (19.54) is accounted for in the loss rate coefficient. For example, in the case of S(IV) reacting in (19.46) and (19.47) only, the **chemical loss rate of S(IV)** is

$$L_{c,\mathrm{S(IV)},i,t} = k_a c_{\mathrm{S(IV)},i,t} c_{\mathrm{H_2O_2},i,t} c_{\mathrm{H^+},i,t} + k_b c_{\mathrm{S(IV)},i,t} c_{\mathrm{HO}_{2,T},i,t} \tag{19.55}$$

(mol cm^{-3} s^{-1}). In this equation,

$$k_a = k_{a,1}\alpha_{1,\mathrm{S(IV)}} \tag{19.56}$$

is a **third-order rate coefficient** ($cm^6\ mol^{-2}\ s^{-1}$), where $k_{a,1}$ is the rate coefficient for Reaction (19.48), given in Appendix Table B.8 (converted from ($M^{-2}\ s^{-1}$) to ($cm^6\ mol^{-2}\ s^{-1}$)), and

$$\alpha_{1,S(IV)} = \frac{\mathbf{m}_{H^+,i,t-h}K_{1,S(IV)}}{\mathbf{m}^2_{H^+,i,t-h} + \mathbf{m}_{H^+,i,t-h}K_{1,S(IV)} + K_{1,S(IV)}K_{2,S(IV)}} \tag{19.57}$$

is the **mole fraction of total S(IV) partitioned to HSO_3^-**. This fraction depends on pH, so $\mathbf{m}_{H^+}$ must be solved for simultaneously. The equilibrium coefficients in (19.57) are given in (19.39) and (19.40).

The second rate coefficient in (19.55) is

$$k_b = \left[k_{b,1}\alpha_{1,S(IV)} + k_{b,2}\alpha_{2,S(IV)}\right]\alpha_{0,HO_{2,T}} + \left[k_{b,3}\alpha_{1,S(IV)} + k_{b,4}\alpha_{2,S(IV)}\right]\alpha_{1,HO_{2,T}} \tag{19.58}$$

($cm^3\ mol^{-1}\ s^{-1}$) where $k_{b,1}$, $k_{b,2}$, $k_{b,3}$, and $k_{b,4}$ are the rate coefficients from Appendix Table B.8 for Reactions (19.49)–(19.52), respectively,

$$\alpha_{2,S(IV)} = \frac{K_{1,S(IV)}K_{2,S(IV)}}{\mathbf{m}^2_{H^+,i,t-h} + \mathbf{m}_{H^+,i,t-h}K_{1,S(IV)} + K_{1,S(IV)}K_{2,S(IV)}} \tag{19.59}$$

is the **mole fraction of total S(IV) partitioned to SO_3^{2-}**, and

$$\alpha_{0,HO_{2,T}} = \frac{\mathbf{m}_{H^+,i,t-h}}{\mathbf{m}_{H^+,i,t-h} + K_{1,HO_{2,T}}} \tag{19.60}$$

$$\alpha_{1,HO_{2,T}} = \frac{K_{1,HO_{2,T}}}{\mathbf{m}_{H^+,i,t-h} + K_{1,HO_{2,T}}} \tag{19.61}$$

are the **mole fractions of $HO_{2,T}$ partitioned to $HO_2(aq)$ and O_2^-**, respectively. In these expressions, $K_{1,HO_{2,T}}$ is the equilibrium coefficient for Reaction (19.54). The family method described above significantly reduces the number of reactions required for a solution to aqueous chemical equations.

The gas conservation equation corresponding to (19.43) is

$$\frac{dC_{SO_2(g),t}}{dt} = -\sum_{i=1}^{N_B} k_{S(IV),i,t-h}\left[C_{SO_2(g),t} - S'_{S(IV),i,t-h}\frac{c_{S(IV),i,t}}{H'_{S(IV),i,t-h}}\right] \tag{19.62}$$

Together, (19.43) and (19.62) make up a set of ODEs whose order depends on the number of dissolving gases, particle size bins, and species taking part in chemical reactions. This set of equations can be solved in a three-dimensional model with a sparse-matrix solver. Table 19.2 shows that such a solver reduced the number of multiplications during one call to a matrix decomposition subroutine by over two million (99.7 percent) for a given set of species, reactions, and size bins.

The last four rows show the number of operations in each of four loops of matrix decomposition and backsubstitution. The percentages are values in the second

Table 19.2 Reduction in array space and number of matrix operations before and after the use of sparse-matrix techniques for ODEs arising from 10 gases transferring to 16 size bins, and 11 aqueous chemistry reactions occurring among 11 species within each bin

	Initial	After sparse reductions
Order of matrix	186	186
No. init. matrix positions filled	34 596	1226 (3.5%)
No. final matrix positions filled	34 596	2164 (6.3%)
No. operations decomp. 1	2 127 685	6333 (0.3%)
No. operations decomp. 2	17 205	1005 (5.8%)
No. operations backsub. 1	17 205	1005 (5.8%)
No. operations backsub. 2	17 205	973 (5.6%)

column relative to those in the first. The solver used was SMVGEAR II (Jacobson 1998a).

19.5 SUMMARY

In this chapter, aqueous chemical reactions and their effects on aerosol and hydrometeor composition were discussed. Aqueous reactions are important because they affect the rate of chemical conversion of sulfur dioxide gas, a member of the S(IV) family, to the sulfate ion, a member of the S(VI) family. Production of the sulfate ion in a liquid drop increases the acidity of the drop. Thus, conversion of S(IV) to S(VI) in atmospheric particles enhances acid deposition. The rate of S(IV) conversion to S(VI) depends on liquid-water content, S(IV) oxidant concentrations, and pH. Conversion is faster in cloud drops than in aerosol particles because clouds have higher liquid-water contents than do aerosol particles. Numerical methods of solving aqueous chemistry ODEs are the same as those for solving gas-chemistry ODEs. Aqueous ODEs are more difficult to solve because they are stiffer and more tightly coupled to other processes than are gas-phase ODEs.

19.6 PROBLEMS

19.1 What are the two primary formation mechanisms of S(VI) in particles? How might changes in relative humidity affect the rate of growth of particles due to each mechanism?

19.2 Why do aerosol particles convert S(IV) species to S(VI) species at a lower rate than do cloud drops?

19.3 Draw approximate curves of relative concentration versus radius when the time scale for diffusion in a drop is (a) less than, (b) equal to, (c) greater than that for chemical reaction loss of an aqueous species. Assume the relative concentration at the drop center is unity.

19.4 Why are aqueous-phase reactions generally stiffer than gas-phase reactions?

19.5 Calculate the e-folding lifetimes of HSO_3^- against loss by reaction with $O_3(aq)$, $H_2O_2(aq)$, and $OH(aq)$, respectively. Assume $\chi_{O_3(g)}$, $\chi_{H_2O_2(g)}$, and $\chi_{OH(g)}$ are 0.1, 8×10^{-4}, and 4.0×10^{-8} ppmv, respectively. Estimate aqueous molalities of these species using Henry's constant from Appendix Table B.7. Assume that $T = 280$ K, $p_d = 950$ hPa, the solution is dilute, and pH = 4.5.

19.7 COMPUTER PROGRAMMING PRACTICE

19.6 Write a computer script to read reactions and rate coefficients. Use the script to calculate rate coefficients for all reactions in Appendix Table B.8 when $T = 298$ K and when 273 K.

20

Sedimentation, dry deposition, and air–sea exchange

Aerosol particles fall in the atmosphere due to the force of gravity. Their fall speeds are determined by a balance between the force of gravity and the opposing force of drag that arises due to air viscosity. The sinking of particles at their fall speed is called **sedimentation**. Gases also sediment, but their weights are so small that their fall speeds are negligible. A typical gas molecule has a diameter on the order of 0.5 nm (nanometer). Such diameters result in sedimentation (fall) speeds on the order of 10 cm per year. Because this speed is slow, and because the slightest turbulence moves gas molecules much further in seconds than sedimentation moves them in years, gas fall speeds are important only over time scales of thousands of years. Small-particle sedimentation is also of limited importance in the atmosphere over short periods, but sedimentation of large particles, particularly soil dust, sea spray, cloud drops, and raindrops, is important. **Dry deposition** removes gases and particles at air–surface interfaces. Dry deposition occurs when gases or particles contact a surface and stick to or react with the surface. Sedimentation is one mechanism by which particles contact surfaces. Others, which affect gases and particles, are molecular diffusion, turbulent diffusion, and advection. In this chapter, equations for sedimentation and dry deposition are described and applied. In addition, fluxes of gases between the ocean and atmosphere are discussed. Removal of gases and particles by precipitation (wet deposition) was outlined in Chapter 18.

20.1 SEDIMENTATION

The fall speed of a particle through air is determined by equating the downward force of gravity with the opposing drag force that arises from air viscosity, as illustrated in Fig. 20.1.

When a particle falling through the air is larger than the mean free path of an air molecule (about 63 nm at 288 K and 1013 hPa from (15.24)) but small enough so that the inertial force the particle imparts is less than the viscous force exerted by air, the flow of the particle through the air is called **Stokes flow**. During Stokes flow, the drag force of air on a sphere is governed by **Stokes' law**,

$$F_{\mathrm{D}} = 6\pi r_i \eta_{\mathrm{a}} V_{\mathrm{f},i} \tag{20.1}$$

where r_i is the particle radius (cm), η_{a} is the dynamic viscosity of air (g cm^{-1} s^{-1}), and $V_{\mathrm{f},i}$ is the **sedimentation (fall) speed** (cm s^{-1}). An expression for the dynamic

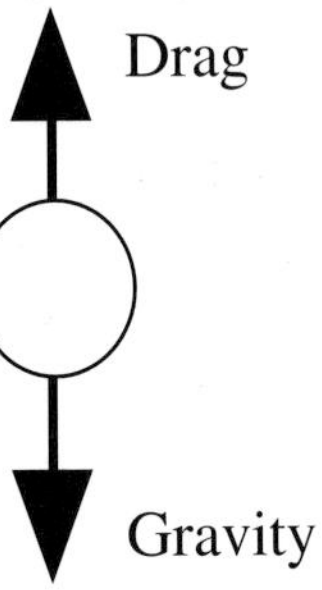

Figure 20.1 Vertical forces acting on a particle.

viscosity of air in units of kg m^{-1} s^{-1} was given in (4.54). Equation (20.1) illustrates that the drag force is proportional to the fall speed of the particle.

When a particle is smaller than the mean free path of an air molecule, motion through air is called **slip flow**. For such flow, Stokes' law can be modified with the Cunningham slip-flow correction and Knudsen–Weber term, defined in (15.30) as $G_i = 1 + \mathrm{Kn}_{a,i}[A' + B' \exp(-C'\mathrm{Kn}_{a,i}^{-1})]$. With this correction, the drag force becomes

$$F_D = \frac{6\pi r_i \eta_a V_{f,i}}{G_i} = \frac{6\pi r_i \eta_a V_{f,i}}{1 + \mathrm{Kn}_{a,i}\left[A' + B' \exp\left(-C'\mathrm{Kn}_{a,i}^{-1}\right)\right]} \tag{20.2}$$

The smaller the particle, the larger the Knudsen number for air, the larger G_i, and the smaller the force of drag on a falling sphere. For Stokes flow, which occurs for relatively larger particles, G_i approaches unity.

The downward force of gravity, net of buoyancy, acting on a spherical particle is

$$F_G = \frac{4}{3}\pi r_i^3(\rho_p - \rho_a)g \tag{20.3}$$

where ρ_p is the density of the particle (g cm^{-3}), ρ_a is the density of air (g cm^{-3}), and g is the gravitational acceleration (cm s^{-2}).

Equating drag and gravitational forces for a falling spherical particle and solving for the fall speed give

$$V_{f,i}^{est} = \frac{2r_i^2(\rho_p - \rho_a)g}{9\eta_a} G_i \tag{20.4}$$

In this equation, the superscript est indicates that the fall speed is an estimated value, since this equation will be used shortly in a corrective procedure. When $G_i > 1$, which occurs for relatively small particles in the slip flow regime, the fall speed is enhanced due to the low viscosity of the air. When $G_i = 1$, which occurs for larger particles, (20.4) simplifies to the **Stokes terminal fall speed**, which is accurate for Stokes flow only.

Example 20.1

Estimate the fall speed of a 1-μm-diameter particle of density 1.5 g cm^{-3} in the stratosphere, where $T = 220$ K and $p_a = 25$ hPa. Assume the particle is spherical and the air is dry. How much does the fall speed change at a pressure of 1013 hPa if all other conditions are the same?

SOLUTION

From (4.54), the dynamic viscosity of air is 1.436×10^{-4} g cm^{-1} s^{-1}. From (2.3), the thermal speed of an air molecule is 4.01×10^4 cm s^{-1}. From (2.36), the density of air is 3.96×10^{-5} g cm^{-3}. From (15.24), the mean free path of an air molecule is 1.769×10^{-4} cm. From (15.24), the Knudsen number for air is 3.54. From (15.30), the Cunningham slip-flow correction is 6.58. Finally, from (20.4), the estimated fall speed is

$$V_{f,0.05\mu m} \approx \frac{2(0.00005\ \text{cm})^2\left(1.5 - 0.0000396\ \dfrac{\text{g}}{\text{cm}^3}\right) \times 980.7\ \dfrac{\text{cm}}{\text{s}^2}}{9 \times 1.436 \times 10^{-4}\ \dfrac{\text{g}}{\text{cm s}}} 6.58$$

$$= 0.037\ \frac{\text{cm}}{\text{s}}$$

At an air pressure of 1013 hPa, the fall speed for this problem is 0.0063 cm s^{-1}. Thus, fall speed decreases with increasing pressure (decreasing altitude).

The fall speed is related to the particle Reynolds number, rewritten from (15.26) as

$$\text{Re}_i^{\text{est}} = \frac{2r_i V_{f,i}^{\text{est}}}{\nu_a} \tag{20.5}$$

Both slip flow and Stokes flow occur at Reynolds numbers <0.01, which correspond to particles <20 μm in diameter under near-surface conditions. At Reynolds numbers >0.01, particles are larger than those in the Stokes flow regime, and another correction to the drag force is necessary.

Beard (1976) studied the fall speeds of cloud drops and raindrops in air and found that the correction factor for such particles can be parameterized as a function of the physical properties of the drops and their surroundings. Such properties include temperature, pressure, air density, air viscosity, particle density, surface tension, and gravity. Beard's parameterization requires two steps. The first step is to estimate the fall speed and Reynolds number with (20.4) and (20.5), respectively. The second step is to recalculate the Reynolds number for one of three flow regimes: **slip flow** around a rigid sphere at low Reynolds number (<0.01), **continuum flow** past a rigid sphere at moderate Reynolds number (0.01–300), and **continuum flow** around a nonspherical, equilibrium-shaped drop at large Reynolds number (>300). These three regimes correspond to particle diameters of <20 μm (aerosol particles and cloud drops), 20 μm–1 mm (cloud drops to medium raindrops), and 1–5 mm

Table 20.1 Coefficients for polynomial fits used in (20.6)

$B_0 = -3.186\,57$	$E_0 = -5.000\,15$
$B_1 = 0.992\,696$	$E_1 = 5.237\,78$
$B_2 = -0.001\,531\,93$	$E_2 = -2.049\,14$
$B_3 = -0.000\,987\,059$	$E_3 = 0.475\,294$
$B_4 = -0.000\,578\,878$	$E_4 = -0.054\,281\,9$
$B_5 = 0.000\,085\,517\,6$	$E_5 = 0.002\,384\,49$
$B_6 = -0.000\,003\,278\,15$	

Source: Beard (1976).

(large raindrops), respectively. In sum,

$$\mathrm{Re}_i^{\mathrm{final}} = \begin{cases} 2r_i V_{\mathrm{f},i}^{\mathrm{est}}/\nu_{\mathrm{a}} & \mathrm{Re}_i^{\mathrm{est}} < 0.01 \\ G_i \exp(B_0 + B_1 X + B_2 X^2 + \cdots) & 0.01 \le \mathrm{Re}_i^{\mathrm{est}} < 300 \\ \mathrm{N}_{\mathrm{P}}^{1/6} G_i \exp(E_0 + E_1 Y + E_2 Y^2 + \cdots) & \mathrm{Re}_i^{\mathrm{est}} \ge 300 \end{cases} \tag{20.6}$$

where

$$X = \ln\left[\frac{32 r_i^3(\rho_{\mathrm{p}} - \rho_{\mathrm{a}})\rho_{\mathrm{a}} g}{3\eta_{\mathrm{a}}^2}\right] \qquad Y = \ln\left[\frac{4}{3}\mathrm{N}_{\mathrm{Bo}}\mathrm{N}_{\mathrm{P}}^{1/6}\right] \tag{20.7}$$

are functions of the physical properties of particles and the atmosphere. These equations contain the dimensionless **physical-property number** and **Bond number,**

$$\mathrm{N}_{\mathrm{P}} = \frac{\sigma_{\mathrm{w/a}}^3 \rho_{\mathrm{a}}^2}{\eta_{\mathrm{a}}^4(\rho_{\mathrm{p}} - \rho_{\mathrm{a}})g} \qquad \mathrm{N}_{\mathrm{Bo}} = \frac{4r_i^2(\rho_{\mathrm{p}} - \rho_{\mathrm{a}})g}{\sigma_{\mathrm{w/a}}} \tag{20.8}$$

respectively, where $\sigma_{\mathrm{w/a}}$ is the surface tension of water against air, defined in (14.19). The Bond number measures the relative strength of the drag and surface tension forces for a drop at terminal fall speed (Pruppacher and Klett 1997). The B and E coefficients in (20.6) are given in Table 20.1. After the final Reynolds number is found, a **final fall speed** is calculated with

$$V_{\mathrm{f},i} = V_{\mathrm{f},i}^{\mathrm{final}} = \frac{\mathrm{Re}_i^{\mathrm{final}}\nu_{\mathrm{a}}}{2r_i} \tag{20.9}$$

In the low-Reynolds-number regime (diameters < 20 μm), (20.9) simplifies to the Stokes terminal fall speed with the Cunningham correction factor. In the moderate-Reynolds-number regime (20 μm – 1 mm diameter), drops are essentially spherical, and neither surface tension nor internal viscosity affects their shape or sedimentation speed significantly. Large raindrops (>1 mm) are affected by turbulence and vary in shape from oblate to prolate spheroids.

Table 20.2 Time for a particle (or a gas for the smallest size) to fall 1 km in the atmosphere by sedimentation

Diameter (μm)	Time to fall 1 km	Diameter (μm)	Time to fall 1 km
0.0005	9630 y	4	23 d
0.02	230 y	5	14.5 d
0.1	36 y	10	3.6 d
0.5	3.2 y	20	23 h
1	328 d	100	1.1 h
2	89 d	1000	4 m
3	41 d	5000	1.8 m

Results were obtained from the fall speeds in Fig. 20.3

Beard's parameterization is valid for liquid water drops. Fall-speed parameterizations for ice crystals and aerosol particles of various shapes have also been derived. Some shapes that have been considered include prolate ellipsoids, oblate ellipsoids, circular cylinders, and rectangular parallelepipeds (Fuchs 1964); cubes, hexagons, and cylinders (Turco *et al.* 1982); and columnar crystals, platelike crystals, oblate spheroids, hexagonal crystals, hailstones, and graupel (Pruppacher and Klett 1997).

The time for a particle to fall a distance Δz in the air due to sedimentation is $\tau_{f,i} = \Delta z / V_{f,i}$. Table 20.2 shows the time required for particles to fall 1 km by sedimentation under near-surface conditions. Emitted volcanic particles range in diameter from <0.1 to >100 μm. Particles 100 μm in diameter take 1.1 h to fall 1 km. The only volcanic particles that survive more than a few months before falling out of the atmosphere are those smaller than 4 μm. Such particles require no less than 23 days to fall 1 km. Most aerosol particles of atmospheric importance are 0.1–1 μm in diameter. Table 20.2 indicates that these particles require 36 years–328 days, respectively, to drop 1 km. Gases, whose diameters are on the order of 0.5 nm (0.0005 μm), require about 9600 years to fall 1 km.

Fall speeds are useful for calculating not only downward fluxes of particles but also coagulation kernels (Section 15.5) and dry-deposition speeds (discussed next).

20.2 DRY DEPOSITION

Dry deposition occurs when a gas or particle is removed at an air–surface interface, such as on the surface of a tree, a building, a window, grass, soil, snow, or water (e.g., Sehmel 1980). A pollutant dry deposits when it impacts and sticks to the surface. A deposited pollutant may be **resuspended** into the air when wind or an eddy blows over a surface and lifts the pollutant.

Dry-deposition speeds are generally parameterized as the inverse sum of a series of resistances (Wesely and Hicks 1977; Slinn *et al.* 1978; McRae *et al.* 1982; Walcek *et al.* 1986; Russell *et al.* 1993). With the resistance model, the **dry-deposition speed**

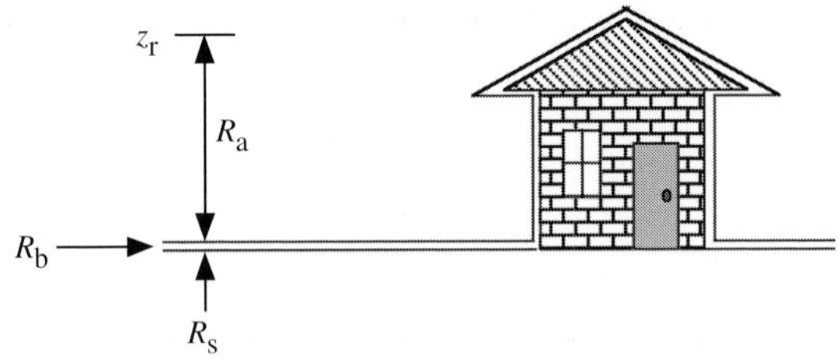

Figure 20.2 Locations where gas dry-deposition resistances apply. The figure is not to scale.

of a gas (m s^{-1}) is

$$V_{d,gas} = \frac{1}{R_a + R_b + R_s} \tag{20.10}$$

where R_a is the aerodynamic resistance between a reference height (about 10 m above the surface) and the laminar sublayer adjacent to the surface, R_b is the resistance to molecular diffusion through the 0.1 to 0.01-cm-thick laminar sublayer, and R_s is the resistance to chemical, biological, and physical interaction and sticking between the surface and the gas once the gas has collided with the surface. Because gases are so light and take so long to fall (Table 20.2), their dry-deposition speeds do not need to include a sedimentation term. Resistances have units of time per unit distance (e.g., s m^{-1}). Figure 20.2 depicts the region near the surface where the three resistances listed apply.

The aerodynamic resistance is a property of the turbulent transfer between a reference height and the laminar sublayer and depends on atmospheric stability. The resistance to molecular diffusion depends on the gas diffusion coefficient, air viscosity, air conductivity, air temperature, and air density. The resistance to surface interactions depends on physical and chemical properties of the surface and the depositing gas.

Dry-deposition speeds of particles differ from those of gases in two ways. First, particles are heavier than gases; so, particle sedimentation speeds must be included in the particle dry-deposition equation. Second, because of their weight, particles tend to stay on a surface once they deposit (in the absence of strong winds or turbulence creating resuspension); thus, the surface resistance term may be ignored. The **dry-deposition speed of a particle** (m s^{-1}) of size *i* is

$$V_{d,part,i} = \frac{1}{R_a + R_b + R_a R_b V_{f,i}} + V_{f,i} \tag{20.11}$$

where the fall speed is in meters per second.

For gas and particle dry deposition, the **aerodynamic resistance** (s m^{-1}) can be approximated from similarity theory with

$$R_a = \frac{\displaystyle\int_{z_{0,q}}^{z_r} \phi_h \frac{dz}{z}}{ku_*} \tag{20.12}$$

where k is the dimensionless von Kármán constant (0.40), u_* is the friction wind speed (m s^{-1}), z is the height above the surface (m), z_r is a reference height (10 m), $z_{0,q}$ is the surface roughness length of gas or particle q (m), and ϕ_h is a dimensionless potential temperature gradient. A dimensionless potential temperature gradient is used instead of a dimensionless wind shear because the transport of trace gases resembles the transport of energy more than it does that of momentum. The friction wind speed, defined in Chapter 8, is a measure of the vertical turbulent flux of horizontal momentum. A large value of u_* corresponds to high turbulence, low resistance to vertical turbulent motion, and a low dry deposition speed. The surface roughness length of species q, ($z_{0,q}$), can be found from (8.13) by substituting the diffusion coefficient of the gas or particle of interest for that of water vapor.

Integrals of ϕ_h were given in (8.38) as a function of the Monin–Obukhov length L, which relies on u_* and the turbulent temperature scale θ_*. A noniterative method of determining u_*, L, and θ_* was discussed in Section 8.4.2.4. During the day, winds are stronger, the atmosphere is more unstable, u_* is larger, the heat integral is smaller, and R_a is smaller than during the night. Because resistance to turbulent transfer is smaller during the day than night, dry-deposition speeds are generally larger during the day than night.

The reciprocal of an aerodynamic resistance is a **transfer speed**, which gives the rate at which a gradient in concentration of a species is transferred between $z_{0,q}$ and z_r. Equation (8.53) gave the vertical turbulent flux of moisture (kg m^{-2} s^{-1}) from Monin–Obukhov similarity theory as $E_f = -\rho_a q_* u_*$. Substituting (8.52) and (20.12) into (8.53) and assuming species q is water vapor, give another expression for the **turbulent moisture flux**, in terms of aerodynamic resistance, as

$$E_f = -\frac{\rho_a k u_* [q_v(z_r) - q_v(z_{0,v})]}{\displaystyle\int_{z_{0,v}}^{z_r} \phi_h\left(\frac{z}{L}\right)\frac{dz}{z}} = -\frac{\rho_a}{R_a}[q_v(z_r) - q_v(z_{0,v})] \tag{20.13}$$

where $1/R_{a,v}$ (m s^{-1}) is the transfer speed for water vapor. The transfer speed is larger under unstable than stable conditions.

The **resistance to molecular diffusion** in the laminar sublayer (s m^{-1}) is

$$R_b = \ln\left(\frac{z_{0,m}}{z_{0,q}}\right)\frac{(\mathrm{Sc}/\mathrm{Pr})^{2/3}}{ku_*} \tag{20.14}$$

where $z_{0,m}$ is the surface roughness length for momentum, defined in Section 8.3. For vegetated surfaces, $z_{0,m}/z_{0,q}$ is approximately 100. For snow, ice, water, and bare soil, $z_{0,m}/z_{0,q}$ ranges from < 1 to 3. For rough surfaces, $z_{0,m}/z_{0,q}$ can be as large as 10^5 (Garratt and Hicks 1973; Brutsaert 1991; Ganzeveld and Lelieveld 1995). The Schmidt number for particles is $\mathrm{Sc}_{pi} = \nu_a/D_{pi}$ and for gases is $\mathrm{Sc}_q = \nu_a/D_q$.

The Prandtl number is $\mathrm{Pr} = \eta_a c_{p,m}/\kappa_a$. These three terms were defined in (15.36), (16.25), and (16.32), respectively.

The resistance of gases to surface interactions is affected by chemical and biological processes, which vary with surface composition. A parameterization for gas deposition to vegetation is described here (Wesely 1989; Walmsley and Wesely 1996). In this treatment, the total **surface resistance** (s m^{-1}) is approximated with

$$R_s = \left(\frac{1}{R_{stom} + R_{meso}} + \frac{1}{R_{cut}} + \frac{1}{R_{conv} + R_{surf}} + \frac{1}{R_{canp} + R_{soil}} \right)^{-1} \tag{20.15}$$

where R_{stom} is the leaf stomata resistance, R_{meso} is the leaf mesophyll resistance, R_{cut} is the leaf cuticle resistance, R_{conv} is the resistance due to buoyant convection in canopies, R_{surf} is the resistance due to leaves, twigs, bark, or other exposed surfaces in the lower canopy, R_{canp} is the resistance due to canopy height and density, and R_{soil} is the resistance due to soil and leaf litter at the ground surface.

The stomata, mesophyll, and cuticle resistances account for deposition of gases into openings on leaf surfaces, into liquid water within leaves, and onto outer leaf surfaces, respectively. A plant **cuticle** is a waxy outer covering of a plant surface, formed from **cutin**. The cuticle is relatively impermeable to water vapor, preventing significant water loss from the leaf. Water vapor, carbon dioxide, oxygen, and other gases can pass through the cuticle into or out of plant leaves through **stomata**. Stomata are openings, scattered throughout the cuticle, that extend into the leaf's **epidermis**, or underlying skin. Stomata are important for allowing carbon dioxide to enter the middle leaf, called the **mesophyll**. The mesophyll contains empty spaces and chloroplasts. **Chloroplasts** are cells in the leaf in which photosynthesis takes place. During photosynthesis, carbon dioxide and water are converted to molecular oxygen and organic material. The oxygen escapes the mesophyll to the atmosphere through the stomata. Liquid water, brought through **xylem** from the plant's roots to the mesophyll, evaporates from the surfaces of chloroplasts into the open space of the mesophyll. This vapor diffuses, often through the stomata, to the open atmosphere. Gases from the atmosphere can also diffuse through the stomata to the mesophyll, where they may deposit onto chloroplast surfaces.

The **leaf stomata resistance** (s m^{-1}) accounts for the loss of a gas due to deposition within stomata. The actual stomata resistance depends on the plant species, but a parameterization that represents an average resistance for many species is

$$R_{stom,q} = R_{min} \left[1 + \left(\frac{200}{S_f + 0.1} \right)^2 \right] \frac{400}{T_{a,c}(40 - T_{a,c})} \frac{D_v}{D_q} \tag{20.16}$$

(Baldocchi *et al.* 1987; Wesely 1989), where R_{min} is the minimum bulk canopy stomata resistance for water vapor, S_f is the surface solar irradiance (W m^{-2}), $T_{a,c}$ is the surface air temperature (°C), and D_v/D_q is the ratio of the diffusion coefficient of water vapor to that of gas q. Minimal resistances for water vapor are given in Appendix Table B.11 for different land-use types and seasons. Diffusion-coefficient ratios for different gases are given in Appendix Table B.12. In (20.16), $T_{a,c}$ ranges from 0 to 40 °C. Outside this range, $R_{stom,q}$ is assumed to be infinite. During the

Table 20.3 Cuticle resistances R_{cut} (s m^{-1}) for SO_2, O_3, and any gas q

	R_{cut,SO_2}			
	Nonurban	Urban	R_{cut,O_3}	$R_{cut,q}$
Dry	R^{dry}_{cut,SO_2}	R^{dry}_{cut,SO_2}	R^{dry}_{cut,O_3}	$R^{dry}_{cut,q}$
Dew	100	50	$\left(\frac{1}{3000}+\frac{1}{3R_{cut,0}}\right)^{-1}$	$\left(\frac{1}{3R^{dry}_{cut,q}}+\frac{H^*_q}{10^7}+\frac{f_{0,q}}{R^{dew}_{cut,O_3}}\right)^{-1}$
Rain	$\left(\frac{1}{5000}+\frac{1}{3R_{cut,0}}\right)^{-1}$	50	$\left(\frac{1}{1000}+\frac{1}{3R_{cut,0}}\right)^{-1}$	$\left(\frac{1}{3R^{dry}_{cut,q}}+\frac{H^*_q}{10^7}+\frac{f_{0,q}}{R^{dew}_{cut,O_3}}\right)^{-1}$

"Urban" refers to areas with little vegetation, such as urban areas. The cuticle resistance for a dry surface is given in (20.18).
Sources: Wesely (1989), Walmsley and Wesely (1996).

night, stomata close, and resistance to deposition increases to infinity. The equation simulates this effect, predicting that $R_{stom,q}$ approaches 10^6 when $S_f = 0$.

Leaf mesophyll resistance depends on the ability of a gas to dissolve and react in liquid water on the surface of plant cells within the mesophyll of a leaf. The ability of a gas to dissolve depends on its effective Henry's constant H^*_q and on a reactivity factor for gases, $f_{0,q}$. The **effective Henry's law constant** is the ratio of the molarity of a dissolved substance plus its dissociated ions to the gas-phase partial pressure of the substance. It differs from a regular Henry's law constant (Section 17.4.1), which is the ratio of the molarity of the dissolved substance alone to its partial pressure. The reactivity factor accounts for the ability of the dissolved gas to oxidize biological substances once in solution. A value of $f_{0,q} = 1$ indicates the dissolved gas is highly reactive, and $f_{0,q} = 0$ indicates it is nonreactive. The **leaf mesophyll resistance** is estimated as

$$R_{meso,q} = \left(\frac{H^*_q}{3000} + 100 f_{0,q}\right)^{-1} \tag{20.17}$$

where values of H^*_q and $f_{0,q}$ are given in Appendix Table B.12 for several gases.

The **leaf cuticle resistance** accounts for the deposition of gases to leaf cuticles when vegetation is healthy, and to other vegetated surfaces in the canopy when it is not. Table 20.3 gives expressions of this resistance for sulfur dioxide, ozone, and any other gas when the leaf cuticle is dry, moist from dew, and wet from rain.

The leaf cuticle resistance over a sufficiently dry surface (s m^{-1}), required in Table 20.3, is approximated as

$$R^{dry}_{cut,q} = R_{cut,0}\left(\frac{H^*_q}{10^5} + f_{0,q}\right)^{-1} \tag{20.18}$$

where $R_{cut,0}$ is given in Appendix Table B.11 for several seasons and land-use categories.

The **resistance of gases to buoyant convection** (s m^{-1}) is estimated for all gases as

$$R_{\mathrm{conv}} = 100\left(1 + \frac{1000}{S_{\mathrm{f}} + 10}\right)\frac{1}{1 + 1000 s_{\mathrm{t}}} \tag{20.19}$$

where s_{t} is the slope of the local terrain, in radians. The **resistance due to deposition** on leaves, twigs, bark, and other exposed surfaces (s m^{-1}) is estimated for any gas q as

$$R_{\mathrm{surf},q} = \left(\frac{10^{-5} H_q^*}{R_{\mathrm{surf,SO_2}}} + \frac{f_{0,q}}{R_{\mathrm{surf,O_3}}}\right)^{-1} \tag{20.20}$$

where $R_{\mathrm{surf,SO_2}}$ and $R_{\mathrm{surf,O_3}}$ are given in Appendix Table B.11.

The **in-canopy resistance** (s m^{-1}) is a function of canopy height (h_{c}) and **one-sided leaf area index** (L_{T}), which was defined in Section 8.3 as the area of leaf surface between the ground and canopy per unit area of underlying ground. Appendix Table B.11 estimates values of in-canopy resistance. Another parameterization of this resistance is

$$R_{\mathrm{canp}} = \frac{b_{\mathrm{c}} h_{\mathrm{c}} L_{\mathrm{T}}}{u_*} \tag{20.21}$$

where $b_{\mathrm{c}} = 14$ m^{-1}, obtained from data (Erisman *et al.* 1994).

Finally, the **resistance due to soil and leaf litter at the ground surface** (s m^{-1}) is estimated with

$$R_{\mathrm{soil},q} = \left(\frac{10^{-5} H_q^*}{R_{\mathrm{soil,SO_2}}} + \frac{f_{0,q}}{R_{\mathrm{soil,O_3}}}\right)^{-1} \tag{20.22}$$

where $R_{\mathrm{soil,SO_2}}$ and $R_{\mathrm{soil,O_3}}$ are given in Appendix Table B.11.

20.3 DRY DEPOSITION AND SEDIMENTATION CALCULATIONS

Figure 20.3 shows particle sedimentation (fall) speed, particle dry-deposition speed in the absence of sedimentation, and particle total dry-deposition speed as a function of diameter. For small particles, the nonsedimentation component of the dry-deposition speed exceeds the sedimentation component, because particle diffusion affects small particles more than does gravity. For large particles, sedimentation dominates, because the effect of gravity exceeds that of particle diffusion. The total dry-deposition speed is its smallest for particles of about 1 μm in diameter; neither diffusion nor gravity dominates for particles near that size. The sedimentation speed levels off at large diameters due to the physical-property correction term in the sedimentation equation. At small particle diameters, the slope of the sedimentation speed also decreases due to the correction for particle resistance to motion, which accounts for the fact that particles can move more freely through air as their size decreases.

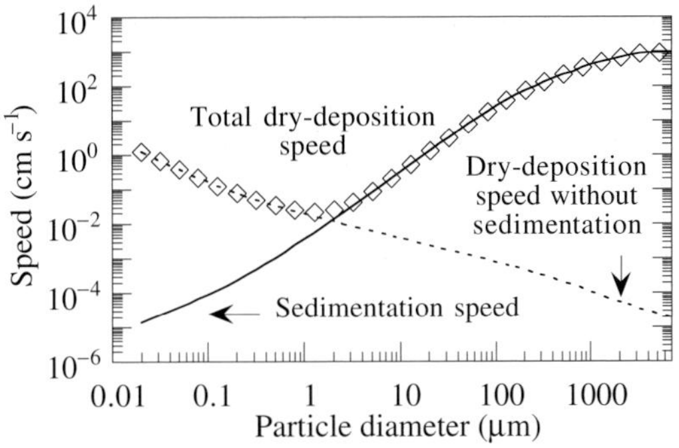

Figure 20.3 Particle total dry-deposition speed and its components versus particle diameter. Conditions were $z_r = 10$ m, $z_{0,m} = 0.0001$ m, $T(z_r) = 290.1$ K, $T(z_{0,m}) = 290$ K, $p_a(z_r) = 999$ hPa, $|\mathbf{v}_h(z_r)| = 10$ m s^{-1}, and $\rho_p = 1.0$ g cm^{-3}. The near-ground temperature was taken at $z_{0,m}$ instead of $z_{0,h}$. Surface resistance was ignored.

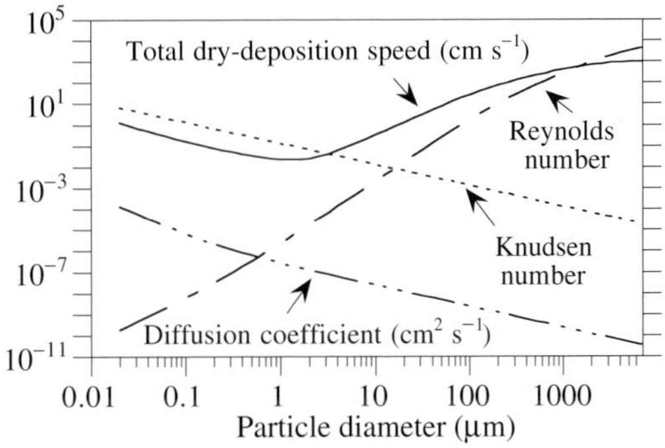

Figure 20.4 Particle total dry-deposition speed from (20.11), Reynolds number, Knudsen number for air, and particle diffusion coefficient versus particle diameter. Conditions were $z_r = 10$ m, $z_{0,m} = 0.0001$ m, $T(z_r) = 292$ K, $T(z_{0,m}) = 290$ K, $p_a(z_r) = 999$ hPa, $|\mathbf{v}_h(z_r)| = 10$ m s^{-1}, and $\rho_p = 1.0$ g cm^{-3}. The near-ground temperature was taken at $z_{0,m}$ instead of $z_{0,h}$.

Figure 20.4 shows the calculated particle diffusion coefficient, Knudsen number for air, Reynolds number, and particle total dry-deposition speed as a function of size, for the conditions described in the figure. The diffusion coefficient decreased with increasing particle size, as expected, since the diffusion coefficient is inversely proportional to the radius of the particle, as shown in (15.29). The Knudsen number also decreased with increasing size, since it is inversely proportional to the radius, as shown in (15.23). The Reynolds number increased with increasing size because it is proportional to the radius and fall speed, both of which increased with increasing

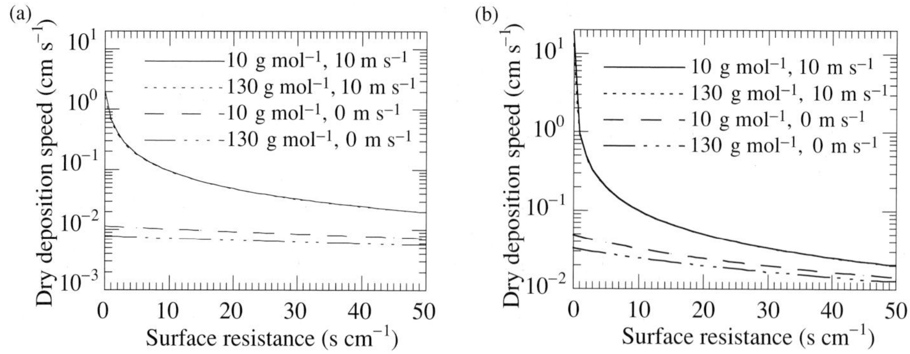

Figure 20.5 Gas dry-deposition speed versus surface resistance when (a) $z_{0,m} = 3$ m and (b) $z_{0,m} = 0.01$ m, and when the molecular weight and wind speed varied, as indicated in the figure legend. Also, $z_r = 10$ m, $T(z_r) = 288$ K, $T(z_{0,m}) = 290$ K, and $p_a(z_r) = 999$ hPa.

size. The dry-deposition speed decreased then increased with increasing size, as in Fig. 20.3.

Figures 20.5 (a) and (b) show gas deposition speeds versus surface resistance and as a function of molecular weight, wind speed, and surface roughness length for momentum in an unstable atmosphere. The curves show that surface resistance affected dry-deposition speed most when the surface roughness length was large and the wind speed was high. Molecular weight had relatively little effect on dry-deposition speed, except when the wind speed was zero. Surface roughness length affected dry-deposition speed primarily when the surface resistance was low. Wind speed affected dry-deposition speed at both roughness lengths.

20.4 AIR–SEA FLUX OF CARBON DIOXIDE AND OTHER GASES

The flux of carbon dioxide between the atmosphere and ocean is of particular importance on a global scale because atmospheric CO_2 has been increasing due to anthropogenic emission, and dissolution into the ocean is a major mechanism of its removal. As discussed in Chapter 17, Henry's law states that, over a dilute solution, the partial pressure exerted by a gas is proportional to the molality of the dissolved gas in solution. Thus, the more CO_2 in the atmosphere, the more that it dissolves in the ocean. The quantity of dissolved gas, though, is also a function of pH, temperature, and the concentration of other ions, such as calcium, in ocean water. Further, since the atmosphere and ocean are not in perfect equilibrium, the transfer of CO_2 to (from) the ocean following its increase (decrease) in the atmosphere is a time-dependent process that depends on the dry-deposition speed.

The fluxes of gases aside from CO_2 between the atmosphere and ocean are also important since the ocean serves as a sink for all gases and a source for some. An important gas emitted from the ocean, for example, is dimethylsulfide (DMS) (e.g., Kettle and Andreae 2000), and deposition to the ocean surface is one of the main removal mechanisms of, for example, ozone (e.g., Galbally and Roy 1980; Chang *et al.* 2004).

20.4.1 Numerical solution to air–sea exchange

In this section, a method of solving equations for the exchange of gases between the ocean and atmosphere is derived. The change in concentration of any gas q between the atmosphere and ocean due to surface dissolution and evaporation can be described by

$$C_{q,t} = C_{q,t-h} + \frac{hV_{\mathrm{d,gas},q}}{\Delta z_{\mathrm{a}}}\left(\frac{c_{q,T,t}}{H'_q} - C_{q,t}\right) \tag{20.23}$$

where the subscripts t and $t-h$ indicate the current time and one time step backward, respectively, h is the time step (s), C_q is the atmospheric mole concentration of the gas (mol cm^{-3}-air), $c_{q,T}$ is the mole concentration of the dissolved gas plus its dissociation products in seawater (mol cm^{-3}-air = mol cm^{-3}-seawater since seawater occupies the same volume as the air it displaces), $V_{\mathrm{d,gas},q}$ is the dry deposition speed of the gas (cm s^{-1}), Δz_{a} is the thickness of the atmospheric layer through which C_q is averaged (cm), and H'_q is a dimensionless effective Henry's law constant (mol mol^{-1}). Equation (20.23) is written implicitly; thus, $c_{q,T,t}$ and $C_{q,t}$ are unknown. If the terms were written explicitly (with time subscript $t-h$) on the right-most side of the equation, the final concentration $C_{q,t}$ could become negative at a long time step or high deposition speed.

The **mole concentration of a gas** is related to its partial pressure by

$$C_q = \frac{p_q}{R^*T} \tag{20.24}$$

where p_q is the partial pressure of the gas (atm), R^* is the universal gas constant (82.058 cm^3 atm mol^{-1} K^{-1}), and T is absolute air temperature (K). The **mole concentration of a gas dissolved in seawater** (mol cm^{-3}) is related to its molality in seawater by

$$c_{q,T} = \rho_{\mathrm{dw}}\mathbf{m}_{q,T} \tag{20.25}$$

where $\mathbf{m}_{q,T}$ is the molality (mol-solute per kilogram of dilute water) of the dissolved gas plus its dissociation products, if any, and ρ_{dw} is the density of dilute water, which is close to 0.001 kg cm^{-3}, but varies slightly with temperature. Dilute water instead of seawater is used in this equation since molality is defined as the moles of solute per kilogram of solvent (water), not of solution (seawater). If the gas dissolves and dissociates twice in solution, as carbon dioxide does by the following reactions,

$$CO_2(g) + H_2O(aq) \overset{H}{\rightleftharpoons} H_2CO_3(aq) \overset{K_1}{\rightleftharpoons} H^+ + HCO_3^- \overset{K_2}{\rightleftharpoons} 2H^+ + CO_3^{2-} \tag{20.26}$$

then the **dimensionless effective Henry's coefficient** of the gas in seawater is

$$H'_q = \rho_{\mathrm{dw}}R^*TH^{\mathrm{s}}_q\left(1 + \frac{K^{\mathrm{s}}_{1,q}}{\mathbf{m}_{\mathrm{H}^+}}\left[1 + \frac{K^{\mathrm{s}}_{2,q}}{\mathbf{m}_{\mathrm{H}^+}}\right]\right) \tag{20.27}$$

where H_q^s is the Henry's law coefficient of the gas measured in seawater but written in terms of dilute water (mol kg^{-1}-dw atm^{-1}), $K_{1,q}^s$ and $K_{2,q}^s$ are the first and second dissociation coefficients, respectively, of the dissolved gas (mol kg^{-1}-dw), and $\mathbf{m}_{H^+}$ is the molality of free (dissociated) hydrogen ion in dilute water (mol kg^{-1}-dw). The unit "mol kg^{-1}-dw" for a rate coefficient indicates that its units have been converted, if necessary, from moles per kilogram of seawater to moles per kilogram of dilute water. In the case of carbon dioxide dissolving in seawater, expressions for the Henry's law and first and second dissociation coefficients are

$$\ln H_{CO_2}^s\left(\frac{\text{mol}}{\text{kg-dw atm}}\right) = \left\{\begin{array}{l} -60.2409 + 93.4517\left(\frac{100}{T_w}\right) + 23.3585\ln\left(\frac{T_w}{100}\right) \\ +S\left[0.023517 - 0.023656\left(\frac{T_w}{100}\right) + 0.0047036\left(\frac{T_w}{100}\right)^2\right] \end{array}\right\} \tag{20.28}$$

$$\ln K_{1,CO_2}^s\left(\frac{\text{mol}}{\text{kg-dw}}\right) = \frac{\rho_{sw}}{\rho_{dw}}\frac{\mathbf{m}_{H^+}}{\mathbf{m}_{H^+,SWS}}\left(\begin{array}{l} 2.18867 - \frac{2275.0360}{T_w} - 1.468591\ln T_w \\ +\left(-0.138681 - \frac{9.33291}{T_w}\right)S^{0.5} + 0.0726483S \\ -0.00574938S^{1.5} \end{array}\right) \tag{20.29}$$

$$\ln K_{2,CO_2}^s\left(\frac{\text{mol}}{\text{kg-dw}}\right) = \frac{\rho_{sw}}{\rho_{dw}}\frac{\mathbf{m}_{H^+}}{\mathbf{m}_{H^+,SWS}}\left(\begin{array}{l} -0.84226 - \frac{3741.1288}{T_w} - 1.437139\ln T_w \\ +\left(-0.128417 - \frac{24.41239}{T_w}\right)S^{0.5} + 0.1195308S \\ -0.00912840S^{1.5} \end{array}\right) \tag{20.30}$$

respectively (Millero 1995 and references therein). In these equations, T_w is the absolute temperature (K) of water, S is seawater salinity (parts per thousand by mass), ρ_{sw} is the density of seawater (kg cm^{-3}), and $\mathbf{m}_{H^+,SWS}$ is the molality of total free plus associated hydrogen ion in solution based on the seawater scale (mol kg^{-1}-dw).

The salinity and density of seawater are

$$S = 1000\left[\left(\sum c_q m_q\right) - c_{H_2O}m_{H_2O}\right] / \left(\sum c_q m_q\right) \tag{20.31}$$

$$\rho_{sw} = \frac{\rho_{dw}}{1 - 0.001S} \tag{20.32}$$

respectively, where the summation in the salinity equation is over all components q in seawater, including dilute water, c is mole concentration in seawater (mol cm^{-3}-water), m is molecular weight (g mol^{-1}), and ρ_{dw} is the density of dilute water (kg cm^{-3}). A typical salinity of seawater is 35 ppth, giving a typical seawater density of 0.001036 kg cm^{-3}. The ratio $\rho_{sw}/\rho_{dw} = 1/(1 - 0.001S)$ is necessary to convert measured reaction rate coefficients from mol kg^{-1}-sw to mol kg^{-1}-dw.

The molality (mol kg^{-1}-dw) of free hydrogen in seawater is related to the pH of seawater by

$$\mathbf{m}_{H^+} = \frac{10^{-pH}}{\rho_{dw}1000} \tag{20.33}$$

where 1000 is the number of cubic centimeters per liter. The pH of seawater generally ranges from 7.8 to 8.3. The parameter $\mathbf{m}_{H^+,SWS}$ is approximately the sum of the molalities of the free hydrogen ion, the bisulfate ion, and dissolved hydrofluoric acid in seawater (Millero 1995). At $T_w = 298.15$ K and $S = 35$ ppth, the ratio $\mathbf{m}_{H^+}/\mathbf{m}_{H^+,SWS}$ is approximately 0.763. This ratio is needed in the rate equations since the equations were derived for seawater solutions.

The dry deposition speed in (20.23) is the same as that in (20.10), except that the **surface resistance over the ocean** is

$$R_{s,q} = \frac{1}{\alpha_{r,q} H'_q k_{w,q}} \tag{20.34}$$

where $\alpha_{r,q}$ is the dimensionless enhancement of gas transfer to seawater due to chemical reaction on the ocean surface, and $k_{w,q}$ is the transfer speed of a chemically unreactive gas through a thin film of water at the ocean surface to the mixed layer of the ocean (cm s^{-1}). For extremely soluble gases, such as HCl, H_2SO_4, HNO_3, and NH_3, the dimensionless effective Henry's constant H'_q is large and $\alpha_{r,q}$ may be large, so the surface resistance ($R_{s,q}$) is small, and the dry deposition speed is limited only by the aerodynamic resistance ($R_{a,q}$) and resistance to molecular diffusion ($R_{b,q}$). On the other hand, for slightly soluble gases, such as CO_2, CH_4, O_2, N_2, and N_2O, H'_q is relatively small, $\alpha_{r,q} \approx 1$, so $R_{s,q}$ is large and the dry deposition speed is controlled primarily by $k_{w,q}$.

The **transfer speed through a thin film of water on the ocean surface** is affected by the gas's dissolution in and molecular diffusion through the film and surfactants and bursting of bubbles on the surface of the film. Although the transfer speed depends on several physical processes, parameterizations of $k_{w,q}$ to date have been derived only in terms of wind speed. One such parameterization is

$$k_{w,q} = \frac{1}{3600}\begin{cases} 0.17|\mathbf{v}_{h,10}|S_{rw,q}^{2/3} & |\mathbf{v}_{h,10}| \le 3.6\,\mathrm{m\,s^{-1}} \\ 0.612S_{rw,q}^{2/3} + (2.85|\mathbf{v}_{h,10}| - 10.262)S_{rw,q}^{1/2} & 3.6\,\mathrm{m\,s^{-1}} < |\mathbf{v}_{h,10}| \le 13\,\mathrm{m\,s^{-1}} \\ 0.612S_{rw,q}^{2/3} + (5.9|\mathbf{v}_{h,10}| - 49.912)S_{rw,q}^{1/2} & 13\,\mathrm{m\,s^{-1}} < |\mathbf{v}_{h,10}| \end{cases} \tag{20.35}$$

(Liss and Merlivat 1986), where $k_{w,q}$ is in units of cm s^{-1}, the constant 3600 converts cm hr^{-1} to cm s^{-1}, $|\mathbf{v}_{h,10}|$ is the wind speed at 10 m above sea level (m s^{-1}), and

$$S_{rw,q} = \frac{Sc_{w,CO_2,20\,^\circ C}}{Sc_{w,q,T_{s,c}}} \tag{20.36}$$

is the ratio of the Schmidt number in water of carbon dioxide at 20 °C to that of gas q at the current temperature. An equation for the Schmidt number in air was given in (16.25). A second parameterization for the transfer speed of a gas through water is

$$k_{w,q} = \frac{0.31|\mathbf{v}_{h,10}|^2 S_{rw,q}^{1/2}}{3600} \tag{20.37}$$

(Wanninkhof 1992), where $k_{w,q}$ is in units of cm s^{-1}. An expression for the **dimensionless Schmidt number of CO_2 in seawater**, valid from 0 to 30 °C, is

$$\mathrm{Sc}_{w,CO_2} = 2073.1 - 147.12T_{s,c} + 3.6276T_{s,c}^2 - 0.043219T_{s,c}^3 \tag{20.38}$$

(Wanninkhof 1992), where $T_{s,c}$ is the temperature of seawater in degrees Celsius.

Equation (20.23) gives the time-dependent change in gas mole concentration due to transfer to and from ocean water. Here, a method of solving (20.23) is shown. Rearranging (20.23) to solve for gas concentration give

$$C_{q,t} = \frac{C_{q,t-h} + \dfrac{hV_{d,gas,q}}{\Delta z_a}\dfrac{c_{q,T,t}}{H'_q}}{1 + \dfrac{hV_{d,gas,q}}{\Delta z_a}} \tag{20.39}$$

where the mole concentration in seawater, $c_{q,T,t}$, is still unknown. Mole balance between the ocean and atmosphere requires

$$c_{q,T,t}D_l + C_{q,t}\Delta z_a = c_{q,T,t-h}D_l + C_{q,t-h}\Delta z_a \tag{20.40}$$

where D_l is the depth (cm) of the ocean mixed layer. Substituting (20.39) into (20.40) and solving give

$$c_{q,T,t} = \frac{c_{q,T,t-h} + \left[\dfrac{hV_{d,gas,q}C_{q,t-h}}{D_l}\bigg/\left(1 + \dfrac{hV_{d,gas,q}}{\Delta z_a}\right)\right]}{1 + \left[\dfrac{hV_{d,gas,q}}{D_l H'_q}\bigg/\left(1 + \dfrac{hV_{d,gas,q}}{\Delta z_a}\right)\right]} \tag{20.41}$$

(Jacobson 2005c). Substituting (20.41) into (20.39) gives the final gas concentration for the time step. Equations (20.39) and (20.41) are referred to as the **Ocean Predictor of Dissolution** (OPD) scheme. These equations conserve moles exactly, are noniterative, and cannot produce a negative gas or water concentration, regardless of the time step.

Once dissolved in ocean water, a chemical diffuses and chemically reacts. Many seawater reactions have been identified (e.g., Stumm and Morgan 1981; Butler 1982; Millero 1995). Here, just one is described. When carbon dioxide dissociates in water by (20.26), its dissociation products may combine with cations and crystallize, removing the ions from solution. The most common carbonate minerals in the ocean are calcite and aragonite, both of which have the chemical formula $CaCO_3(s)$. The reversible reaction forming calcite or aragonite is

$$Ca^{2+} + CO_3^{\ 2-} \rightleftharpoons CaCO_3(s) \tag{20.42}$$

Calcite or aragonite forms only when

$$\mathbf{m}_{Ca^{2+}}\mathbf{m}_{CO_3^{\ 2-}} > K^{s}_{CaCO_3}(T) \tag{20.43}$$

where

$$\mathbf{m}_{CO_3^{\ 2-}} = \frac{c_{CO_2,T}}{\rho_{dw}}\left(\frac{K^{s}_{1,CO_2}K^{s}_{2,CO_2}}{\mathbf{m}_{H^+}\mathbf{m}_{H^+} + K^{s}_{1,CO_2}\mathbf{m}_{H^+} + K^{s}_{1,CO_2}K^{s}_{2,CO_2}}\right) \tag{20.44}$$

is the molality (mol kg^{-1}-dw) of the carbonate ion in seawater, $\mathbf{m}_{Ca^{2+}} \approx 0.01028$ mol kg^{-1}-dw is the molality of the calcium ion in seawater (derived from Table 14.1), and $K^{s}_{CaCO_3}(T)$ is the solubility product (equilibrium coefficient) of the reaction in seawater. The **solubility product** of calcite at 25 °C and 35 parts per thousand salinity in seawater (with the solubility product converted to dilute water) is $K^{s}_{CaCO_3}(T) = 4.427 \times 10^{-7}$ mol^2 kg^{-2}-dw. That for aragonite is 6.717×10^{-7} mol^2 kg^{-2}-dw (derived from Millero 1995). Thus, the formation of calcite is favored over the formation of aragonite when only solubility products are considered.

Example 20.1

If the molality of total dissolved carbon is 0.0023 mol kg^{-1}-dw, pH = 8.0, $T = 298.15$ K, and $\rho_{dw} = 1$ g cm^{-3}, estimate the molality of aragonite formed in seawater assuming equilibrium.

SOLUTION

From (20.29), $K_{1,CO_2} = 1.125 \times 10^{-6}$ mol kg^{-1}-dw; from (20.30), $K_{2,CO_2} = 9.603 \times 10^{-10}$ mol kg^{-1}-dw; from (20.33), $\mathbf{m}_{H^+} = 1 \times 10^{-8}$ mol kg^{-1}; and from (20.44), $\mathbf{m}_{CO_3^{\ 2-}} = 1.999 \times 10^{-4}$ mol kg^{-1}-dw. Taking the difference $\mathbf{m}_{Ca^{2+}}\mathbf{m}_{CO_3^{\ 2-}} - K_{eq}(T)$ with $\mathbf{m}_{Ca^{2+}} = 0.01028$ mol kg^{-1}-dw gives 1.38×10^{-6} mol kg^{-1}-dw of aragonite in seawater.

The product of the measured molalities of the calcium ion and the carbonate ion in seawater exceed the solubility product of calcite near the ocean surface (e.g., Stumm and Morgan 1981; Butler 1982). Part of the reason is that magnesium poisons the surface of a growing calcite crystal, creating a magnesium-calcite crystal that is more soluble than calcite alone, increasing the dissolution rate of calcite. For this reason, aragonite is often found in higher concentration in the mixed layer than is calcite.

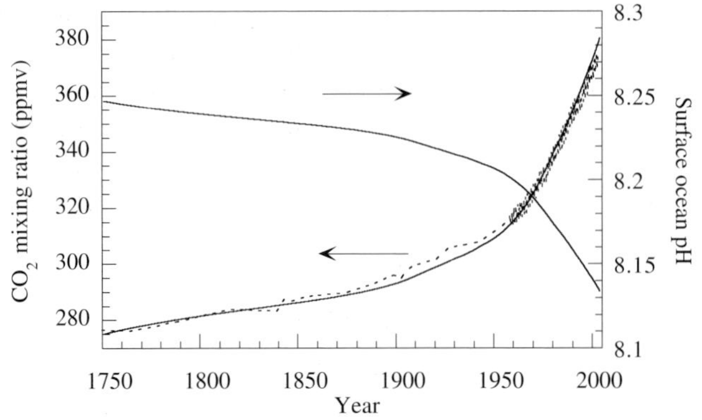

Figure 20.6 Modeled (—) versus measured (----) CO_2 mixing ratio and modeled surface ocean pH between 1750 and 2004 in a one-dimensional atmosphere–ocean calculation. The model treated 38 ocean layers of 100-m thickness and one atmospheric layer divided into two compartments: land–air and ocean–air. CO_2 was emitted in the land compartments, partitioned each time step between the land–air and ocean–air compartments, and transferred between the ocean–air compartment and surface ocean. Dissolved CO_2 diffused vertically in the ocean. Ocean chemistry was calculated in all layers. The CO_2 mixing ratio data were from Friedli *et al.* (1986) up to 1953 and from Keeling and Whorf (2003) for 1958–2003. From Jacobson (2005c).

20.4.2 Simulation of ocean acidification

Figure 20.6 shows results from a one-dimensional model simulation of historic and future ocean composition and pH in which the air–sea exchange algorithm just described was used. For the simulation, the ocean consisted of 38 100-m-thick layers (extending to 3800 m depth). The atmosphere, which consisted of one layer, was divided into a land–air and an ocean–air compartment. CO_2 was emitted into the land–air compartment. The added CO_2 was then instantaneously mixed between the ocean–air and land–air compartments in a mole-conserving manner assuming the ocean comprised 71.3 percent of the Earth's surface. Air–sea exchange equations were then solved with a time step of one day, changing ocean and air carbon concentrations and ocean pH. Each time step, vertical diffusion and equilibrium chemistry were solved in the ocean for dozens of chemicals.

Historic CO_2 emission from 1751 to near present was obtained from Marland *et al.* (2003). CO_2 was solved for prognostically; it was not nudged or assimilated. If the historic CO_2 results are correct, Fig. 20.6 suggests that ocean pH may have decreased from about 8.247 in 1751 to 8.136 in 2004, for a 26 percent increase in the hydrogen ion content of the ocean. A future simulation was also run using *Special Report on Emission Scenarios* (*SRES*), A1B CO_2 future emission factors (Nakicenovic *et al.* 2000). Based on the result, the hydrogen ion concentration in 2100 (pH = 7.846) may be a factor of 2.5 greater than in 1751 (pH = 8.247).

These results indicate that carbon dioxide buildup in the atmosphere leads to **ocean acidification**, a general finding consistent with that from other studies (e.g., Brewer 1997; Caldeira and Wickett 2003). The simulations also uncovered the possibility that ocean acidification due to CO_2 may slightly increase the atmospheric concentrations of non-CO_2 atmospheric acids, such as hydrochloric acid, nitric acid, and sulfurous acids, but may decrease the concentration of the base ammonia.

20.5 SUMMARY

In this chapter, particle sedimentation and gas and particle dry deposition were discussed. Sedimentation is affected by gravity and air viscosity. At very small and at very large particle diameters, sedimentation speeds are modified by particle resistance to motion and physical-property effects, respectively. Particle sedimentation speeds decrease with decreasing particle size, but dry-deposition speeds increase with decreasing size due to enhanced molecular diffusion at small size. Particle dry-deposition speeds are influenced by meteorological variables and diffusion at small diameters and by sedimentation at large diameters. Gas dry-deposition speeds are influenced by meteorological variables and surface interactions. Such interactions can be parameterized as a function of leaf-stomata, leaf-mesophyll, leaf-cuticle, buoyant-convective, exposed-surface, canopy, and soil resistances. Finally, a method of calculating the flux of gas between the atmosphere and ocean was discussed. Carbon dioxide fluxes change ocean pH and chemical composition and, in turn, are affected by them.

20.6 PROBLEMS

20.1 Calculate the fall speed of a 10-μm-radius soil particle of density 2.5 g cm^{-3} when $T = 230$ K and (a) $p_a = 300$ hPa and (b) $p_a = 10$ hPa. Assume particles are spherical and the air is dry.

20.2 Briefly describe the three resistances affecting gas dry-deposition speeds. What condition(s) give(s) maximum values for each resistance?

20.3 How do increases in surface heating and instability, respectively, affect individual components of surface resistance?

20.4 Compare the CO_2 transfer speed at 20 °C from (20.33) and (20.35) when the wind speed is (a) 3 m s^{-1} and (b) 10 m s^{-1}. Which parameterization should result in a perturbation to atmospheric CO_2 reaching equilibrium with the ocean faster?

20.7 COMPUTER PROGRAMMING PRACTICE

20.5 Terrain significantly affects particle dry-deposition speed. Write a computer script to calculate dry-deposition speeds of spherical particles 0.01 μm in diameter when $z_{0,m} = 0.0001$ m (ocean) and when $z_{0,m} = 5$ m (urban area). At $z_{0,m} = 0.0001$ m, assume $u_* = 0.48$ m s^{-1} and $L = +180$ m. At $z_{0,m} = 5$ m, assume $u_* = 0.8$ m s^{-1} and $L = +3000$ m. Also, assume $p_d = 1000$ hPa and $T = 292$ K at $z_r = 10$ m, and $\rho_v = 1$ g cm^{-3}.

20.6 Submicrometer atmospheric particles require significant time to fall out of the atmosphere. Write a computer script to calculate how long a 0.1-μm- and a 10-μm-diameter liquid water particle take to sediment from an altitude of 2 km to 1 km. Assume $T = 280$ K and $p_d = 850$ hPa are constant. Assume drops are spherical and $\rho_v = 1$ g cm^{-3}.

20.7 Write a computer script to calculate the time required for molecules of ozone and of carbon monoxide to dry-deposit to the surface from a reference height of $z_r = 10$ m, assuming $R_s = 500$ s m^{-1} for O_3 and $R_s = 3000$ s m^{-1} for CO. Assume $p_d = 1000$ hPa and $T = 290$ K at the ground surface. Also assume that, at $z_r = 10$ m, $p_d = 999$ hPa, $T = 288$ K, $u = 7$ m s^{-1}, and $v = 7$ m s^{-1}. Let $z_{0,m} = 1$ m. Use the temperature, pressure, etc. at 10 m to calculate other atmospheric variables.

20.8 Write a computer script to compare the dry-deposition speed of a 10-μm-diameter particle with that of a 0.1-μm-diameter particle to determine which is influenced more by surface resistance. Assume $V_f = 0.0031$ m s^{-1} for the 10-μm particle and $V_f = 8.6 \times 10^{-7}$ m s^{-1} for the 0.1-μm particle. Assume the other conditions are the same as in Problem 20.7.

20.9 Meteorological conditions significantly affect gas dry-deposition speeds when the resistance to surface interactions (R_s) is small. Write a computer script to calculate the difference in the dry-deposition speed of water vapor when $R_s = 100.0$ s m^{-1} in the following two cases: (a) when $u = 15$ m s^{-1}, $v = 15$ m s^{-1}, $p_d = 999$ hPa, and $T = 292$ K at $z_r = 10$ m, and (b) when $u = 2$ m s^{-1}, $v = 2$ m s^{-1}, $p_d = 999$ hPa, and $T = 288$ K at $z_r = 10$ m. In both cases, assume $p_d = 1000$ hPa and $T = 290$ K at the surface and $z_{0,m} = 0.1$ m.

20.10 Write a computer script to calculate surface resistance with equations in Section 20.2. (a) Calculate the total surface resistance for SO_2 during summer over a coniferous forest when $S_f = 600$ W m^{-2}, $T_{a,c} = 15$ °C, and $s_t = 0$. (b) Assume S_f varies as a sine function between 6 a.m. and 6 p.m., peaking with $S_f = 800$ W m^{-2} at noon and equaling zero at night. Plot surface resistance versus time for a 24-h period, assuming the same remaining conditions in part (a). (c) Repeat part (b) for O_3. Comment on the diurnal variations of surface resistances for both gases.

20.11 Write a computer script to solve (20.39) and (20.41). Use the program to calculate the change in concentration over time of ozone in water and air assuming $T = 288$ K, a wind speed of 10 m s^{-1}, a transfer speed from (20.37), a dry-deposition speed affected only by surface resistance, a Henry's constant from Appendix Table B.7, and a time step of one hour. Assume the atmosphere depth is 8 km and the ocean depth is 100 m. Assume the initial mixing ratio in air is 40 ppbv and zero in water. Assume air density is uniform through the atmospheric layer at 1.2 kg m^{-3}. Show results over 5 years.

21

Model design, application, and testing

In previous chapters, physical, chemical, radiative, and dynamical processes and numerical methods to simulate those processes were described. Here, steps in model design, application, and testing are discussed. The most important of these steps are to define the goals of a modeling study, select appropriate algorithms, obtain input and emission data, compare model predictions with data, and analyze results. To illustrate these steps, the design of a model is briefly discussed, and statistical and graphical comparisons of predictions of the model with data are shown. This chapter integrates numerical methods discussed in previous chapters with model development and analysis procedures.

21.1 STEPS IN MODEL FORMULATION

Model design, application, and testing require several steps. These include (1) defining and understanding the problem of interest, (2) determining the spatial and temporal scale of the problem, (3) determining the dimension of the model, (4) selecting the physical, chemical, and/or dynamical processes to simulate, (5) selecting variables, (6) selecting a computer architecture, (7) codifying and implementing algorithms, (8) optimizing the model on a computer architecture, (9) selecting time steps and intervals (10) setting initial conditions, (11) setting boundary conditions, (12) obtaining input data, (13) obtaining ambient data for comparison, (14) interpolating input data and model predictions, (15) developing statistical and graphical techniques, (16) comparing results with data, (17) running sensitivity tests and analyzing the results, and (18) improving algorithms. Each of these steps is discussed below.

21.1.1 Defining the purpose of the model

The first step in model development is to define the scientific, regulatory, or computational problem of interest. Example scientific topics include determining the effects of (1) gas and particle emission on ambient pollutant concentrations, (2) gas and particle emission on global and regional climate, (3) gas and particle emission on stratospheric ozone, (4) gas and particle emission on acid deposition, (5) global climate change on regional air quality, (6) aerosol particles on local weather patterns, (7) aerosol particles on photolysis rates of gases, (8) aerosol particles on cloud cover, (9) aerosol particles on atmospheric stability, (10) soil moisture on atmospheric stability, boundary layer height and pollutant concentrations, (11) global climate change on regional climate weather, (12) climate change on severe weather, (13) ocean circulation on global climate, and (15) ocean circulation on atmospheric carbon dioxide.

Some topics of interest to regulators and scientists include estimating the effects of nitrogen oxide and hydrocarbon emission controls on human exposure, chlorofluorocarbon emission regulations on stratospheric ozone, and carbon dioxide and aerosol particle emission controls on global warming. Computational topics of interest include determining the most efficient computer architecture, in terms of speed and memory, to run a model on, and developing numerical techniques to speed algorithms on existing computer architectures.

21.1.2 Determining scales of interest

The second step in model development is to determine the spatial and temporal scale of interest. Spatial scales include the molecular, micro-, meso-, synoptic, and global scales (Table 1.1). Urban air pollution, thunderstorms, and tornados are simulated over micro- to mesoscale domains. Acid deposition, regional climate change, and hurricanes are simulated over meso- to synoptic-scale domains. Global climate change is simulated over a global domain. With respect to time, urban air pollution and weather events are simulated over hours to weeks, acid deposition events are simulated over days to months, and regional and global climate change events are simulated over years to centuries.

21.1.3 Determining the dimension of the model

The third step in model development is to determine whether a zero-, one-, two-, or three-dimensional model is required and whether to nest the model. Three-dimensional models are ideal, but because such models require enormous computer time and memory, zero-, one-, and two-dimensional models are often used instead.

A **zero-dimensional** (0-D) **model** is a **box model** in which chemical and/or physical transformations occur. Gases and particles may enter or leave the box from any side. Since all material in the box is assumed to mix instantaneously, the concentration of each gas and particle is uniform throughout the box. A standard box model is fixed in space. A **parcel model** is a box model that moves through space along the direction of the wind. Emission enters the box at different points along the trajectory. Because a parcel model moves in a Lagrangian sense, it is also called a **Lagrangian trajectory model**. Box models have been used to simulate photochemical reactions that occur in smog chambers, fog production in a controlled environment, and chemical and physical interactions between aerosol particles and gases. Parcel models have been used to trace changes in an air parcel as it travels from ocean to land and through the polar vortex.

A **one-dimensional** (1-D) **model** is a set of adjacent box models, stacked vertically or horizontally. Vertical 1-D models may be used to study radiative transfer with photochemistry, gas and aerosol vertical transport, aerosol optical properties, aerosol sedimentation, or cloud convection. The main use of a horizontal 1-D model is to test the 1-D advection diffusion equation. The disadvantage of a 1-D model compared with a 2-D or 3-D model is that velocities in a 1-D model cannot be predicted but must be estimated crudely. In addition, a 1-D model either ignores gas, particle, and potential temperature fluxes through **lateral (horizontal) boundaries** or roughly parameterizes them. Figure 11.11 shows results from a 1-D photochemical-radiative model.

A **two-dimensional** (2-D) **model** is a set of 1-D models connected side by side. 2-D models can lie in the x–y, x–z, or y–z planes. Advantages of a 2-D over a 1-D model are that transport can be treated more realistically, and a larger spatial region can be simulated in a 2-D model. 2-D models have been used, for example, to simulate dynamics, transport, gas chemistry, and aerosol evolution (e.g., Garcia *et al.* 1992; Rahmes *et al.* 1998). A global 2-D model may stretch from the South to the North Pole and vertically. South–north and vertical winds in such a model are predicted or estimated from observations at each latitude and vertical layer. Zonally averaged winds (west–east winds, averaged over all longitudes for a given latitude and vertical layer) are needed in a 2-D global model. Such winds are found prognostically or diagnostically. Prognostic winds are obtained, for example, by writing an equation of motion for the average west–east scalar velocity at each latitude and altitude. Diagnostic winds are obtained, for example, by writing prognostic equations for south–north and vertical scalar velocities, then extracting the average west–east scalar velocity from the continuity equation for air.

A **three-dimensional** (3-D) **model** is a set of horizontal 2-D models layered on top of one another. The advantage of a 3-D over a 2-D model is that dynamics and transport can be treated more realistically in a 3-D model. The disadvantage of a 3-D model is that it requires significantly more computer time and memory than does a 2-D model. Nonetheless, studies of urban air pollution are readily carried out in 3-D, since simulation periods are generally only a few days. Computer-time limitations are most apparent for global simulations that last months to years. Because 3-D models represent dynamical and transport processes better than do 0-, 1-, and 2-D models, 3-D models should be used when computer time requirements are not a hindrance.

Four main types of 3-D models include global, hemispheric, limited-area, and nested models. A **global model** is a model that extends globally in the horizontal but is capped at its top and bottom by boundaries. A **hemispheric model** extends globally in one hemisphere in the west–east direction but has southern and/or northern boundaries and vertical boundaries. A **limited-area model** is a nonglobal model with west–east, south–north, and vertical boundaries. A **nested model** is one that contains two or more 3-D **grid domains** within it (where a grid domain is a 3-D model, itself) within it, where all except one of the grid domains (the coarsest-resolved) receives outside boundary information from a domain with coarser resolution.

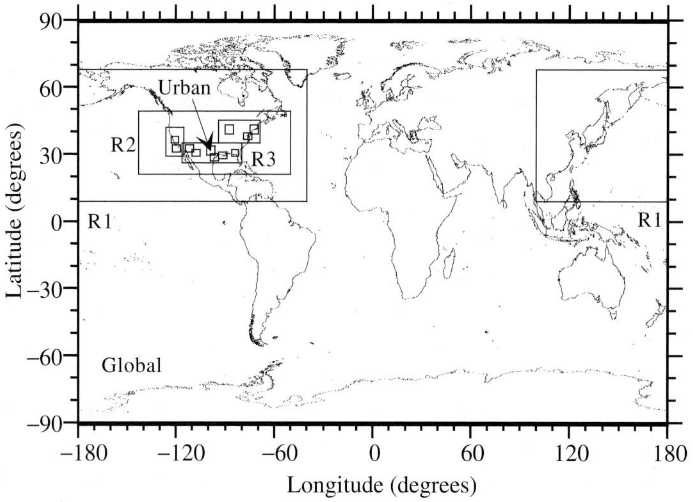

Figure 21.1 An example of nested domains. The largest domain, Global, covers the Earth. Domain R1 is nested inside the global domain and receives boundary data from it. Domain R2 is nested inside domain R1 and receives boundary data from domain R1. Three R3 domains are nested inside domain R2, and each receives boundary data from domain R2. Within each R3 domain are multiple urban domains that receive boundary information from the respective parent R3 domain.

Grid domains that provide boundary conditions to finer-resolved domains in a nested model are called **parent domains.** Domains that receive boundary conditions from coarser (parent) domains are **progeny domains.** Since several domains can be nested within each other (e.g., the Urban, R3, R2, R1, and global domains, respectively, in Fig. 21.1), a domain can serve as both a parent and a progeny domain. Each parent domain can also contain multiple progeny domains that are independent of one another (e.g., the multiple R3 domains within the R2 domain in Fig. 21.1).

Ideally, the coarsest-resolved domain in a nested model is the global domain, which does not require boundary information, except at its base and top. In some nested models, the coarsest domain is a limited-area domain, which needs lateral boundary information. Such information can be obtained from interpolated meteorological data or from stored global-model results. Variables passed at the boundaries in a nested model are discussed in Section 21.1.11.

Nesting can be one-way or two-way. A **one-way nested model** is a model in which information is passed only from a parent to a progeny domain. A **two-way nested model** is one in which information is passed in both directions. Most nested models nest one way. One problem with two-way nesting is that it results in perturbations to velocity, potential temperature, and other variables in the middle of the parent domain that degrade conservation properties (e.g., conservation of mass, energy, enstrophy, potential enstrophy, etc.) in the parent domains.

21.1.4 Selecting processes

The next step in model development is to select the physical, chemical, and dynamical equations for the model and the best available tools to solve the equations. An ideal model includes every conceivable process, each simulated with the most accurate solver. Because computer speed and memory are limited, either the number of processes simulated or the accuracy of individual solutions must be limited.

Seven major groups of processes simulated in atmospheric models are meteorological, transport, cloud, radiative, gas, aerosol, and surface processes. When a model is developed, it is necessary to decide whether one or more of these groups can be excluded from the simulation or replaced by measurements.

Many air pollution models interpolate meteorological fields from observations, ignore the effects of clouds, and/or ignore the effects of aerosol particles. The advantage of using interpolated meteorological fields is that, if sufficient data are available, an interpolated field is more accurate than is a prognostic field. The disadvantage is that measured meteorological data are usually available at only a few locations, most of which are near the surface. Data aloft are scarce or nonexistent. Also, a model that uses a pre-existing database of meteorology cannot be used to predict (**forecast**) weather or air pollution in the future. It can examine only events in the past (**hindcast**).

Finally, many models ignore the effects of hydrometeor particles and/or aerosol particles. In some urban locations, clouds do not form frequently during the summer; so, cloud formation is neglected. Aerosol particles often, although not always, have a minor impact on ozone formation, so studies of ozone often neglect aerosol particles. Modeling studies of clouds often neglect the effect of aerosol particles on cloud formation. This assumption may lead to errors since clouds form physically on aerosol particles.

21.1.5 Selecting variables

Once processes are selected, variables must be chosen. If a model treats meteorology, some of the variables solved for include air temperature, pressure, density, velocity, and geopotential, and the concentrations of water-vapor, liquid water, and ice. Meteorological models also calculate energy fluxes, moisture fluxes, and temperature at soil, vegetation, and ocean surfaces. If a model includes trace gases, it calculates concentrations of each gas. If the model includes aerosol particles, it may calculate particle number concentration and the concentration of each individual component in each particle size bin. If a model includes radiative processes, it often solves for heating rates. Many intermediate variables are also stored, but not permanently. Photolysis rate coefficients, extinction coefficients, particle growth rates, gas dry-deposition speeds, entrainment rates, and pressure-gradient forces are variables that are stored temporarily.

When a 3-D model is run at high spatial resolution and/or includes many variables, computer memory limits must be considered. For example, an atmospheric model that treats meteorology, gases, size-resolved aerosol particles, size-resolved hydrometeor particles, radiation, and surface processes, contains arrays with a minimum of

$$N_{min} = (N_M + N_G + N_D N_B N_V + N_R) N_{3D} + N_S N_{2D} \quad (21.1)$$

floating-point values, where N_M is the number of meteorological variables, N_G is the number of gases, N_D is the number of aerosol/hydrometeor size distributions (assuming aerosol particles can grow into hydrometeor particles), N_B is the number of size bins in each size distribution, N_V is the number of possible chemical components in particles in each size bin of each distribution, N_R is the number of radiative variables, N_{3D} is the total number of three-dimensional grid cells in the model, N_S is the number of surface variables (e.g., ground temperature, soil moisture), and N_{2D} is the total number of two-dimensional horizontal grid squares in the model. Whereas, N_{min} is the minimum summed array dimension in a model, additional memory space is needed for work arrays.

Example 21.1

If $N_M = 10$, $N_G = 100$, $N_D = 5$, $N_B = 20$, $N_V = 30$, $N_R = 2$, $N_{3D} = 50000$, $N_S = 6$, and $N_{2D} = 2500$, (21.1) implies that $N_{min} = 156$ million array points. Thus, the model requires over 156 million words of central memory (almost 1.25 gigabytes (GB), since 1 word = 8 bytes) just to store values of variables. If only meteorology is treated, $N_{min} = 515\,000$.

21.1.6 Selecting a computer architecture

The next step in model development is to select a computer architecture for building the model on. Example architectures are single-processor scalar, single-processor vector, multiprocessor shared-memory scalar, multiprocessor shared-memory vector, and multiprocessor distributed-memory scalar architectures.

A **scalar processor** is a processor that operates on variables in a loop, one at a time. For example, in the Fortran loop

```
      DO 100 I = 1, 150                          (21.2)
        DVAR(I) = AVAR(I) * BVAR(I)
100   CONTINUE
```

150 multiplications are performed in sequence on a scalar processor. A **single-processor scalar machine** has one processor performing scalar operations.

A **vector processor** is a processor that operates on several variables within a loop at the same time. In loop (21.2), a vector processor may break down variables AVAR and BVAR each into two registers of 64 and one register of 22 values. Elements from the first register of each AVAR and BVAR are loaded into a functional unit that multiplies values of AVAR and BVAR. At least seven values of AVAR and BVAR can be loaded into the functional unit at the same time. Thus, several operations are carried out simultaneously on a vector processor. Vector processors operate only on the inner loop of a nested loop. A **single-processor vector machine** has one processor performing vector operations. For a code to run fast on a vector machine, every inner loop must be optimized for the vector processor. Such an optimized code is often efficient on a scalar processor as well.

A **multiprocessor machine**, also called a parallel-processor machine, is one in which several processors operate on different parts of a program simultaneously. Separate single-processor machines can be linked together as a network to perform the same function as a single multiprocessor machine. On a multiprocessor machine, each of the 150 multiplications in loop (21.2) can be distributed to a separate processor. A **shared-memory multiprocessor machine** is a multiprocessor machine in which memory is common to all processors. A **distributed-memory multiprocessor machine** is a multiprocessor machine in which memory is allocated to each processor and not shared by all processors.

21.1.7 Coding the model

The next step is to codify the model on the chosen architecture in a manner that is easy to understand, follow, use, and modify. Although some coding may pre-exist, new coding is often required. A 3-D gas-phase chemical-transport model may require an emission algorithm, a chemical solver, a transport module, and a dry-deposition algorithm. If all but the emission algorithm pre-exist, the modeler must develop the emission algorithm and link it with the other modules.

Two methods of making the code easy to understand are to include many comments and references. Comments are important, not only for the model developer, who may need to edit the model after several years, but also for others who may use and/or modify the model. Comments include definitions, units of variables, and descriptions of equations. References are important, so that users of the model can check the origin of equations or numerical algorithms, either to understand them or to determine whether they should be replaced.

Two methods of making the code easy to follow are to use indentation frequently and to select appropriate variable names. Indenting nested loops (e.g., DO . . . CONTINUE loops or IF . . . THEN statements in Fortran) makes the code more readable than left-justifying all coding. Selecting descriptive variable names enables a user to follow the code easily. The variable names, TEMPK, PRESS, UWIND, and VWIND are easier to follow than are T, P, U, and V, respectively. Variables should consist of no fewer than three characters because one- or two-character names are difficult to search for. A search for all occurrences of T in a subroutine, for example, may generate a list of thousands of T's.

A method of making the code easy to use is to control portions of the code by **on–off switches**. For example, it may be desirable not to solve gas chemistry for a sensitivity test. This can be accomplished most readily by putting an IF statement, such as

$$\text{IF (IFCHEM.EQ.1) CALL SUBROUTA} \tag{21.3}$$

in the model, where IFCHEM = 1 for solving gas chemistry and 0 for ignoring it.

Two methods of making the code easy to modify are to create external datafiles for reading variables that are frequently changed and to allow the code to generate arrays from input data rather than **hardwiring** the arrays into the code. For example, if an on–off switch is frequently changed, it is less time consuming to change the switch within an external data file, which is read by a compiled code during runtime, than to change the switch within the code, then to recompile and run the code. In the second example, a chemical solver is easier to modify if its reactions are read from an external dataset and its first derivatives and partial derivatives are calculated automatically in the code rather than if the reactions, first derivatives, and partial derivatives are each hardwired in the code. In the former case, it is not necessary to change anything except the reaction information when a reaction is changed; in the latter, it is necessary to change up to hundreds of derivatives and partial derivatives when a reaction is changed, increasing the risk of error.

21.1.8 Optimizing the model

Another important step in model development is to increase the speed and reduce the memory requirements of the model. One method of increasing the speed of individual loops on a given computer is to vectorize the loop (Section 12.8). Another method is to use sparse-matrix techniques (Section 12.8).

Reducing memory requirements is important when computer memory is limited. Two methods of reducing memory requirements are to minimize the number of global arrays and to prevent work arrays from having a dimension equal to the size of the grid domain. **Global arrays** are arrays used to store information continuously over the entire domain. Example global arrays are arrays for temperature, pressure, wind speed, wind direction, relative humidity, gas concentration, and aerosol component concentration. **Work arrays** are arrays used temporarily. Global arrays must have one of their dimensions equal to the size of the grid domain. Thus, minimizing the number of global arrays is one method of reducing memory requirements. Although work arrays are often used to store information from global arrays temporarily, there is no reason that a work array ever has to have any dimension equal to the size of the grid domain. The reason is that the model domain can be divided into **blocks (groups) of grid cells**, where operations are carried out one block at a time rather than over the whole domain at once (Section 12.8). In such cases, one dimension of each work array is reduced from the number of grid cells in the model domain to the number of grid cells in a block.

A third method of reducing memory requirements is to write global-array information to a data file rather than to store the information in a global array. The tradeoff is that additional computer time is required to read and write the data, and additional disk space is required to store the data.

21.1.9 Time steps and intervals

The next step in model development is to select model time steps and time intervals. These parameters depend on the desired accuracy, the computer time available, and stability limitations of the code. For a 5-km × 5-km horizontal grid, a typical time step for hydrostatic dynamical calculations is 5 s. For a 5° × 5° global grid, it is around 300 s. For gas and aqueous chemistry, the time step is variable with some solution methods and fixed with others. The Gear chemical integration method (Chapter 12) uses a predicted time step that varies from $<10^{-4}$ to >900 s, depending on the stiffness of the system of equations. The MIE chemical solver, described in the same chapter, uses a fixed time step of 1, 10, 30, or 60 s. Generally, the longer the fixed time step, the less accurate the solution and the less computer time required. Cloud, aerosol, radiative, transport, and surface processes use fixed or variable time steps. Some processes, such as coagulation, are slow enough away from sources to allow time steps of minutes to days, depending on the application.

In Section 6.2, a time interval was defined as the period during which several time steps of a process are solved without interference by another process. For time-consuming processes in which the time step is constant (e.g., some aerosol processes), the time interval often equals the time step. For processes that require short time steps (e.g., dynamics) or variable time steps (e.g., chemistry), the time interval often exceeds the time step. In 3-D models, time intervals often range from 1 minute to 1 hour.

21.1.10 Initial conditions

A model calculates output parameters from a set of initial conditions, boundary conditions, input data, and model equations. This subsection discusses initial conditions. Boundary conditions and input data are discussed in the next two subsections.

Initial conditions are needed for meteorological, surface, gas, and aerosol variables in a model. Meteorological variables that are usually initialized in three dimensions include air temperature, air pressure, specific humidity, wind velocity, and geopotential height. Surface variables include initial soil moisture and ground temperature. Gas variables include gas mixing ratio. Aerosol variables include aerosol number concentration and component mole or volume concentration as a function of particle size.

For short-term simulations of air pollution and weather, accurate initialization is critical for producing accurate comparisons of model predictions with data. For long-term simulations of climate, initialization is less important than for short-term simulations for two reasons. First, the lifetimes of most (but not all) gases and aerosol particles are much shorter than are simulation times for climate studies, which are on the order of years to centuries. Second, time-dependent solar and thermal-infrared radiation drive the general circulation of the atmosphere, gradually diminishing the effect of initial conditions.

Two ways to initialize a model are to interpolate observations and to use output from another model. For simulations of regional air pollution episodes, for which many pollutant concentrations and meteorological variables are measured during a field campaign, initial conditions may be interpolated from measurements taken at the simulation start time. For global simulations, where data are sparse, meteorological variables are generally initialized with results from a previous global simulation. In some cases, the prior simulation may be run for many years to obtain a long-term-average result and to reduce the effect of initial conditions. In other cases, the prior simulation may be **nudged** toward observed values from a sparse network of soundings and surface stations. This practice is called **data assimilation.** A **sounding** is a vertical profile of variables such as temperature, pressure, specific humidity, and velocity as a function of height, measured from a balloon released from the surface. A **surface station** measures similar variables, but only near the surface. Data assimilation is performed by relaxing model values toward observed values in a manner similar to the relaxation technique described for treating boundary conditions in the next subsection.

21.1.11 Boundary conditions

Boundary conditions are needed in all models, including global, hemispheric, limited-area, and nested models.

In global and hemispheric models, west–east boundary conditions are **periodic** (the western edge of the westernmost grid cell is adjacent to the eastern edge of the easternmost grid cell); thus, complex lateral boundary conditions are not needed. At the poles in a spherical-coordinate global model, south–north velocities may be set to zero so that winds travel west–east around polar singularities.

In limited-area models and at the southern and/or northern boundary of a hemispheric model, lateral horizontal boundary conditions are needed. Some types of lateral boundary conditions for meteorological variables are described in Chapter 7. For gases and aerosol particles, many models assume that constant concentrations are advected into the model domain from outside a lateral boundary. In reality, gas concentrations outside a boundary vary during the day and night due to chemistry. One way to address this feature is to define a **virtual row or column** outside a lateral boundary wherein time-dependent chemical equations are solved for gases. The resulting virtual-boundary concentrations are then advected into the domain. Virtual boundaries, in which time-dependent physical and chemical processes are solved, can also be created for aerosol particles.

In a nested model, lateral boundary conditions in the global domain are the same as those in any other global model. In progeny domains, meteorological variables are relaxed toward values interpolated from the parent domain (e.g., Davies 1976; Kurihara and Bender 1983; Anthes *et al.* 1989; Giorgi *et al.* 1993; Lu and Turco 1994; Marbaix *et al.* 2003). **Relaxing** a progeny variable in this case means allowing it gradually to approach the interpolated parent variable.

Relaxing is necessary only if the time interval for passing variables between parent and progeny domains (the **nesting time interval,** h_{nest}) exceeds the time step of the process that affects the variable in the progeny domain (h). For example, if variables are passed between the parent and progeny domain every hour, and the time step for dynamical calculations in the progeny domain is five seconds, meteorological variables in the progeny domain may be relaxed toward parent variables. If the nesting time interval also equals five seconds, progeny variables can instead be set directly to interpolated parent variables each time step without relaxation.

Meteorological variables are generally relaxed in a **buffer zone** adjacent to each boundary in the progeny domain. The buffer zone generally consists of the $k = 1 \ldots N$ westernmost and easternmost columns and southernmost and northernmost rows of the progeny domain (where N is often set to 5 but may range from 1 to 10, and $k = 1$ represents outer-boundary columns and rows). One method of relaxing is first to calculate variable values at the end of a nesting time interval h_{nest} (s) in the parent domain. Then, during the mth time step, h (s), of the same nesting interval, **variable values in the buffer zone of the progeny domain** are updated with

$$\alpha^{\text{prog}}_{t_0+mh,k} = \frac{\alpha^{\text{par}}_{t_0+h_{\text{nest}},k} + F_k \alpha^{\text{prog}}_{t_0+(m-1)h,k}}{1 + F_k} \qquad (21.4)$$

where t_0 is the time (s) at the beginning of the current nesting interval, $t_0 + h_{\text{nest}}$ is the time at the end of the nesting interval, $t_0 + (m-1)h$ is the time at the beginning of a progeny-domain time step during the nesting interval (where m is an integer between 0 and h_{nest}/h), $t_0 + mh$ is the time at the end of the progeny-domain time step, α^{prog} is a parameter value in the progeny domain, α^{par} is a value of the same parameter interpolated from the parent domain to the location of α^{prog} in the progeny domain, and

$$F_k = \frac{(h_{\text{nest}} - h)}{h}\left[1 - e^{(1-k)/M}\right] \qquad (21.5)$$

is an expression for the **relaxation coefficient,** where M is a coefficient selected arbitrarily between 0.5 and 5. Equation (21.5) predicts that, when $h_{\text{nest}} = h$, $F_k = 0$, and $\alpha^{\text{prog}} = \alpha^{\text{par}}$ for all $k = 1 \ldots N$. When $h_{\text{nest}} > h$, $F_k = 0$ at the boundary, forcing $\alpha^{\text{prog}} = \alpha^{\text{par}}$ at the boundary. Because F_k increases with increasing distance from the boundary, the relative weight of the parent-domain value decreases with increasing distance from the boundary in (21.4). The product $F_k h$ is the e-folding lifetime for damping at the boundary.

Gas and aerosol concentrations do not need to be relaxed. Instead, values from the parent domain may be interpolated to a virtual boundary outside the progeny-domain boundary, and these concentrations may be advected into the progeny domain.

For nested and nonnested models on all scales, vertical boundary conditions are needed. At the model top and surface, vertical scalar velocities are usually set to zero, and variables are not transported through the model top or bottom boundaries by the wind. At the surface, energy and moisture fluxes are exchanged between the atmosphere and soil, vegetation, water, and other surfaces. In addition, gases and aerosol particles are emitted into the atmosphere. Dry deposition, sedimentation, and wet deposition fluxes remove gases and aerosol particles from the bottom atmospheric layer. Many global atmospheric models are coupled with a 2-D or 3-D **ocean model**. In such cases, the bottom boundary of the total model over the ocean is the bottom of the ocean mixed layer (for a 2-D ocean model) and the topographical bottom of the ocean (for a 3-D ocean model). Almost all 3-D atmospheric models that treat meteorology are coupled with a single- or multi-layer soil model. In such cases, the bottom boundary of the total model over soil is the bottom soil layer.

21.1.12 Input data

Models include some input data that are fixed and other data that vary in space and time. **Topographic data** are generally fixed in space and time. They are used to specify geopotential at the surface, which affects the hydrostatic equation in a sigma-pressure coordinate model (Section 7.6) and the definition of sigma in a sigma-altitude coordinate model (Section 5.4.1).

Solar radiation drives much of the general circulation of the atmosphere. In a model, yearly-averaged, top-of-the-atmosphere, spectral **solar irradiance data** (Appendix Table B.2) are used as a model input. The yearly-averaged data are scaled with day of the year (e.g., with the equations in Section 9.8.2), because the Earth–Sun distance varies with day of year.

Land-use data (which may vary with month or season) are used to estimate average grid-cell values of the surface roughness length for momentum, soil specific heat, soil density, soil porosity, and other parameters. For example, if each grid cell in a model is subdivided by fractional land-use type, an average surface roughness length for momentum in the grid cell can be calculated as

$$z_{0,m,\mathrm{c}} = f_{l,1} z_{0,m,1} + f_{l,2} z_{0,m,2} + \cdots + f_{l,i} z_{0,m,i} \cdots \tag{21.6}$$

where $z_{0,m,i}$ is the surface roughness length for momentum of land-use type i, and $f_{l,i}$ is the fractional area of the grid cell consisting of land-use type i. Similar equations can be written for other parameters. The fractions satisfy $f_{l,1} + f_{l,2} + \cdots + f_{l,i} + \cdots = 1$. Surface roughness lengths are used to estimate, among other parameters, eddy diffusion coefficients and dry-deposition speeds.

Chemical-rate coefficient data are essential for simulating gas, aqueous, or reversible chemistry. Such data include temperature- and/or pressure-dependent uni-, bi-, and termolecular rate coefficients for gas- or aqueous-phase chemical reactions (e.g., Appendix Tables B.4 and B.8) and equilibrium-coefficient data for reversible reactions (e.g., Appendix Table B.7).

Absorption cross-section data are used to calculate absorption coefficients of gases and liquids. Absorption coefficients are used for photolysis and optical depth calculations. Some sources of absorption cross-section data are given in Appendix Table B.4.

Activity-coefficient data are used for simulating reversible chemistry in concentrated aerosol particles. Some temperature-dependent activity-coefficient data are given in Appendix Table B.9.

Emission data are needed to simulate gas and particle pollution buildup. Some emission rates, such as of soildust particles, sea-spray drops, biogenic gases, and nitric oxide from lightning, are often calculated from physical principles in a model (e.g., Chapter 14; Chapter 18), so are not considered input data. Other emission rates, such as of volcanic, biomass burning, and fossil-fuel combustion sources are usually read in as a datafile. Such datafiles can be global or regional in space and may be hourly, daily, monthly, seasonally, or yearly in time.

21.1.13 Ambient data

A model's performance is judged by how its predictions compare with data. Data used for comparing results with are ideally an extension of the data used for initializing the model. In other words, if time-dependent measurements are available, data for the time corresponding to the beginning of the model simulation should be used to initialize the model, and data for all subsequent times should be used to compare with model predictions.

Model predictions are generally compared with gas, aerosol, cloud, radiative, meteorological, and surface data. Some gas parameters frequently compared include ozone, carbon monoxide, nitrogen dioxide, water vapor, sulfur dioxide, total reactive organic gases, and peroxyacetyl nitrate. Aerosol parameters compared include sulfate, nitrate, chloride, ammonium, sodium, black carbon, organic matter, and soil dust. Cloud parameters compared include cloud optical depth, cloud fraction, cloud liquid, cloud ice, and precipitation. Radiative parameters compared include aerosol optical depth, single-scattering albedo, scattering extinction coefficient, absorption extinction coefficient, surface albedo, solar irradiance, ultraviolet irradiance, thermal-infrared irradiance, and actinic flux. Meteorological parameters compared include air temperature (minimum, maximum, mean), air velocity, relative humidity, air pressure, and geopotential. Surface parameters compared include ground temperature, soil moisture, latent heat flux, and sensible heat flux.

Some models **assimilate data** (**nudge** results toward measurements) to improve their comparisons with data. Nudging is sometimes used when running a model in the past (**hindcasting**) in order to create an initial meteorological field for running other simulations in the future (Section 21.1.10). Nudging is also used to predict variables in the future (**forecasting**) by running a model prognostically for a period, nudging the prediction toward observations, then running the model prognostically for another period, and so on. A true forecast model, though, does not use nudging since, by definition, forecasting is the prediction of future events without knowledge of the future, and nudging assumes knowledge of the future. Another problem with

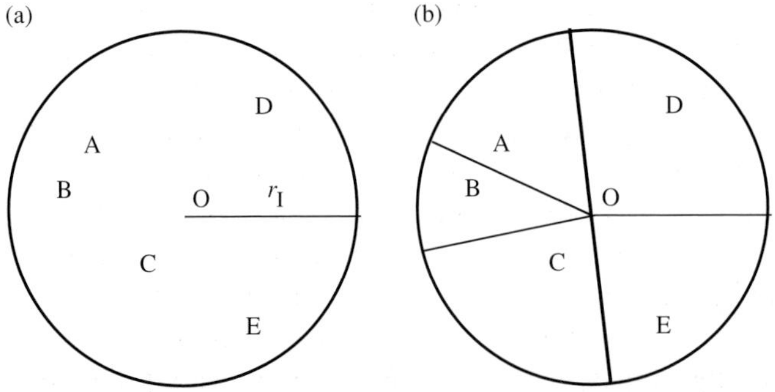

Figure 21.2 (a) Domain of influence around point O. The letters A, B, C, D, and E represent locations where data are available for interpolation to model point O. (b) Division of the domain of influence into sectors.

the use of nudging is that it impairs the evaluation of the accuracy of a model. If a model is nudged, errors are hidden.

21.1.14 Interpolating data and model results

Ambient, input, and output data are often interpolated in a model. Ambient data are interpolated to model grid cells to initialize the model. Emission data are sometimes interpolated between coordinate systems. Model predictions are interpolated to locations where ambient measurements are taken so that model results can be compared with data. A basic interpolation method is discussed for each of these cases.

21.1.14.1 *Interpolating from scattered points to a fixed point*

Scattered ambient data are often interpolated horizontally and vertically to each model grid cell for initialization. Goodin *et al.* (1979) provide a review of several interpolation methods. One method for horizontally interpolating scattered data to a fixed point is with **inverse-square interpolation**, whereby each datum value is weighed by the inverse square of the distance of the datum location to the point of interest. Only data within a predefined **domain of influence** are considered. The domain of influence is a circular area of a given **radius of influence**. Figure 21.2(a) shows an example domain of influence with radius of influence r_I, around a point O. The domain of influence contains five locations – A, B, C, D, and E – where data values are known. The value at point O is interpolated from the values at the data locations with

$$V_O \approx \frac{V_A d_{AO}^{-2} + V_B d_{BO}^{-2} + V_C d_{CO}^{-2} + V_D d_{DO}^{-2} + V_E d_{EO}^{-2}}{d_{AO}^{-2} + d_{BO}^{-2} + d_{CO}^{-2} + d_{DO}^{-2} + d_{EO}^{-2}} \tag{21.7}$$

where V_A, V_B, . . . , etc. are the known data values at points A, B, . . . , etc., and d_{AO}, d_{BO}, . . . , etc. are the distances from points A to O, B to O, . . . , etc.

Advantages of inverse-square interpolation are that it weighs nearby data points more than distant data points, and it is simple to implement. A disadvantage is that when two points, such as A and B in Fig. 21.2(a), are close together, they are each weighed the same as any other point of the same distance from O. A method to reduce this problem is to multiply each datum value by an angular distance (in radians) found by drawing a line from point O through the midpoint between each pair of data locations, as shown in Fig. 21.2(b). The angles (θ_A, θ_B, . . . , etc.) are constrained by $\theta_A + \theta_B + \theta_C + \theta_D + \theta_E = 2\pi$. Including the angle-dependence in (21.7) gives

$$V_O \approx \frac{\theta_A V_A d_{AO}^{-2} + \theta_B V_B d_{BO}^{-2} + \theta_C V_C d_{CO}^{-2} + \theta_D V_D d_{DO}^{-2} + \theta_E V_E d_{EO}^{-2}}{\theta_A d_{AO}^{-2} + \theta_B d_{BO}^{-2} + \theta_C d_{CO}^{-2} + \theta_D d_{DO}^{-2} + \theta_E d_{EO}^{-2}} \qquad (21.8)$$

With this equation, a datum value at point D now carries approximately the same weight as the data values at points A and B combined.

21.1.14.2 *Interpolating from one coordinate system to another*

When data are provided in one coordinate system but a model uses a different coordinate system, the data must be interpolated between coordinate systems. For example, emission data for urban modeling are often developed on a Universal Transverse Mercator (UTM) grid projection, which has rectangular grid cells. A model may be run on a spherical (geographic) grid projection. When a UTM grid lies on top of a geographic grid, UTM cell boundaries cross geographic cell boundaries at random locations; thus, interpolation is needed between the two grids.

The first interpolation step is to determine the area of each UTM grid cell that lies within a geographic grid cell. Since no formulae are available to calculate such areas, the areas must be obtained by physical integration. Each UTM cell, which may be 5 km × 5 km in area, can be divided into 10 000 or more smaller cells (subcells), each 50 m × 50 m in area. Each subcell is assumed to contain the same datum value as the larger UTM cell that the subcell lies in. The latitude and longitude of each subcell corner are found from UTM-to-geographic conversion formulae (US Department of the Army 1958). Once geographic boundaries of each subcell are known, the subcells from each large UTM cell U that fall within a geographic cell G are counted, and the sum is denoted by $N_{U,G}$. From the sum, two types of interpolated values can be obtained: cumulative and average. **Cumulative values** are those in which all data from the original UTM cells that fall within the geographic cell of interest are summed. Cumulative values are necessary for interpolating emission data. **Average values** are those in which all data from the original UTM cells that fall within the geographic cell of interest are averaged. Average values are necessary for interpolating most other data, such as land use and albedo data.

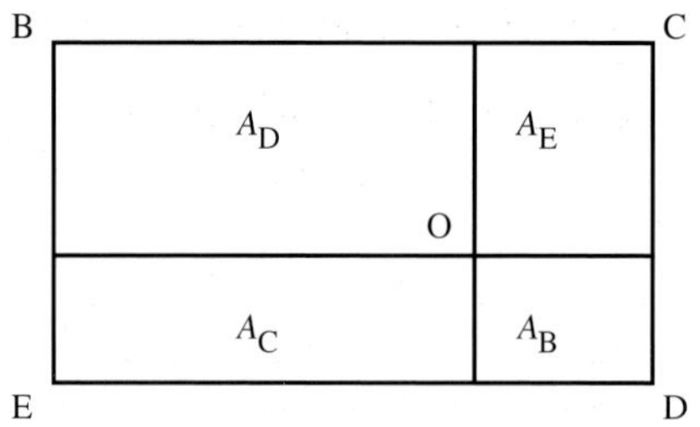

Figure 21.3 Location of point O in a rectangle with points B, C, D, and E at the corners.

The cumulative and average interpolated values of a variable in each geographical cell are determined with

$$V_{G,c} = \sum_{U=1}^{M_G} \left(\frac{N_{U,G} A_M V_U}{A_U} \right) \qquad V_{G,a} = \frac{1}{A_G} \sum_{U=1}^{M_G} (N_{U,G} A_M V_U) \tag{21.9}$$

respectively, where M_G is the number of large UTM cells overlapping part of geographic cell G, A_M is the area of a UTM subcell, A_U is the area of a large UTM cell, A_G is the area of a geographic cell, and V_U is the datum value in the large UTM cell. This interpolation method can be applied to any two-coordinate systems. The precision of the method improves with improved resolution of subcells.

21.1.14.3 *Interpolating from fixed points to a random point*

A model generally produces output at regularly spaced locations, such as at the horizontal center or edge of a grid cell. Observational data, used for comparison with model results, are located at scattered locations throughout a model grid. Thus, it is necessary to interpolate model predictions from fixed points to locations where data are measured. If grid cells are roughly rectangular, model values at data locations can be estimated using **bilinear interpolation** (e.g., Press *et al.* 1992).

Figure 21.3 shows the locations of known and unknown variables when bilinear interpolation is used. In this figure, values at points B, C, D, and E are known and produced by the model. Model values must be interpolated from these points to point O, and the resulting value is compared with a datum value at the same point.

With bilinear interpolation, the value interpolated to point O is

$$V_O \approx \frac{A_B V_B + A_C V_C + A_D V_D + A_E V_E}{A_B + A_C + A_D + A_E} \tag{21.10}$$

where V_B, V_C, V_D, and V_E are model values at points B, C, D, and E, respectively, and A_B, A_C, A_D, and A_E are the rectangular areas shown in Fig. 21.3. Equation (21.10) states that, as location O approaches location B, area A_B increases in size, increasing the weight of B on the value at O.

21.1.15 Statistics and graphics

Numerical models produce output that is difficult to analyze number by number. Statistics and graphics are useful for evaluating model performance, finding bugs, and studying model results.

Common statistical measures are the normalized gross error, normalized bias, paired peak estimation accuracy, and temporally paired peak estimation accuracy (e.g., Tesche 1988). The **overall normalized gross error** (NGE) compares the absolute value of a prediction with an observation, summed and averaged over all observations and times. Thus,

$$\mathrm{NGE} = \frac{1}{N_{\mathrm{tim}} N_{\mathrm{obs}}} \sum_{j=1}^{N_{\mathrm{tim}}} \left(\sum_{i=1}^{N_{\mathrm{obs}}} \frac{|P_{x_i,t_j} - O_{x_i,t_j}|}{O_{x_i,t_j}} \right) \tag{21.11}$$

where N_{tim} is the number of times that observations were taken, N_{obs} is the number of observations taken at each time, P is a predicted value, O is an observed value, x_i is the location of site i, and t_j is the jth time of observation. In this equation, all observations and predictions are paired in space and time; thus, they are compared at the same time and location. Because low measured values are often uncertain, (21.11) is applied only when the observation is larger than a threshold, called a **cutoff value.** With respect to ozone mixing ratios in urban air, cutoff values are typically between 0.02 and 0.05 ppmv.

Normalized gross errors can be measured at a single location for all times or at a single time for all locations. The **location-specific NGE** is

$$\mathrm{NGE}_x = \frac{1}{N_{\mathrm{tim}}} \sum_{j=1}^{N_{\mathrm{tim}}} \frac{|P_{x,t_j} - O_{x,t_j}|}{O_{x,t_j}} \tag{21.12}$$

where x is the location of interest. The **time-specific NGE** is

$$\mathrm{NGE}_t = \frac{1}{N_{\mathrm{obs}}} \sum_{i=1}^{N_{\mathrm{obs}}} \frac{|P_{x_i,t} - O_{x_i,t}|}{O_{x_i,t}} \tag{21.13}$$

where t is the time of interest.

The last two parameters are useful for isolating locations or times of poor prediction. If the location-specific NGEs at 29 out of 30 observations sites are 25 percent but the NGE is 1000 percent at the 30th site, the overall NGE is 57.5 percent. Thus, the overall NGE does not give complete information about the model performance. The location-specific NGE, in this case, shows that the model performance at 29 sites was much better than at the 30th site. With this information, the modeler can focus efforts on determining why the model performance at the 30th site was poor.

Many modeling studies compare monthly, seasonally, or yearly averaged data at a given location or at many locations with model values averaged in the same way. The resulting statistic, called an **unpaired-in-time, paired-in-space error,** has the form

$$\text{UTPSE} = \frac{1}{N_{\text{obs}}}\sum_{i=1}^{N_{\text{obs}}}\left(\left|\sum_{j=1}^{N_{\text{tim}}} P_{x_i,t_j} - \sum_{j=1}^{N_{\text{tim}}} O_{x_i,t_j}\right| \Big/ \sum_{j=1}^{N_{\text{tim}}} O_{x_i,t_j}\right) \tag{21.14}$$

Even though an absolute value is used in this equation, the UTPSE is not a true measure of error because the statistic does not pair prediction with measurements in both time and space, a necessary condition for the statistic to be a true measure of error. For example, suppose two values measured at the same location but at different times are 4 and 2 and the two corresponding model values are 1 and 5. The UTPSE for this set of statistics is 0, suggesting that the model has no error. However, the NGE for the same set of statistics is 112.5 percent, giving a true indication of model error.

Other modeling studies compare spatially and temporally averaged data with spatially-and-temporally averaged model values. Zonally averaged data for December–January–February and globally averaged data for the month of June are data of this type. The statistic that results from this type of comparison is the **unpaired-in-time, unpaired-in-space error**, and has the form,

$$\text{UTUSE} = \left|\sum_{j=1}^{N_{\text{tim}}}\sum_{i=1}^{N_{\text{obs}}} P_{x_i,t_j} - \sum_{j=1}^{N_{\text{tim}}}\sum_{i=1}^{N_{\text{obs}}} O_{x_i,t_j}\right| \Big/ \sum_{j=1}^{N_{\text{tim}}}\sum_{i=1}^{N_{\text{obs}}} O_{x_i,t_j} \tag{21.15}$$

Like with the UTPSE, the UTUSE is not a real measure of model error.

Another statistical measure is **normalized bias** (NB),

$$\text{NB} = \frac{1}{N_{\text{tim}} N_{\text{obs}}}\sum_{j=1}^{N_{\text{tim}}}\left(\sum_{i=1}^{N_{\text{obs}}} \frac{P_{x_i,t_j} - O_{x_i,t_j}}{O_{x_i,t_j}}\right) \tag{21.16}$$

The NB indicates whether a parameter is over- or underpredicted, on average, in comparison with the data. The NB does not indicate whether the model performance is accurate.

A measure of the magnitude of the spread around the mean value of a distribution is the variance. The **biased variance**, or square of the standard deviation of a distribution, is defined as

$$\sigma_u^2 = \frac{1}{N}\sum_{i=1}^{N}(V_i - \bar{V})^2 \tag{21.17}$$

where N is the total number of data values, $\bar{V}$ is the mean of all data values, and V_i is the ith data value. **The biased variance of the time-specific normalized gross error** is

$$\sigma_{u,\text{NGE}_t}^2 = \frac{1}{N_{\text{obs}}}\sum_{i=1}^{N_{\text{obs}}}\left(\frac{|P_{x_i,t} - O_{x_i,t}|}{O_{x_i,t}} - \text{NGE}_t\right)^2 \tag{21.18}$$

This value gives the spread of gross errors around the mean gross error taken at a specific time. The **unbiased variance** is the same as the biased variance, except that the summation in the unbiased variance is divided by $N_{\text{obs}} - 1$ instead of by N_{obs}.

The **paired peak accuracy** (PPA) identifies how well a model predicts a peak observed parameter value at the time and location of the peak. It is given by

$$\mathrm{PPA} = \frac{P_{\hat{x},\hat{t}} - O_{\hat{x},\hat{t}}}{O_{\hat{x},\hat{t}}} \tag{21.19}$$

where the circumflex (ˆ) indicates that a value is taken at the time and location of the peak observed value. The **temporally paired peak accuracy** (TPPA) identifies how well the model predicts the peak observed value at the same time of the peak, but at any other location. It is

$$\mathrm{TPPA} = \frac{P_{x,\hat{t}} - O_{x,\hat{t}}}{O_{x,\hat{t}}} \tag{21.20}$$

The TPPA is less useful than the PPA in that, even if the TPPA is zero, the location of the predicted peak may be far from that of the observed peak.

Another method of judging the accuracy of a model is with **time-series plots**, which are graphical comparisons of model predictions with data for one parameter at one location over the period of the simulation. Parameters frequently compared are gas concentration, particle concentration, temperature, relative humidity, wind speed, wind direction, and solar radiation, among others. Example time-series plots are shown in Section 21.2. Like location-specific NGEs, time-series plots are often better indicators of model performance than are overall NGEs.

A third method of judging the accuracy of a model is with spatial comparisons of model predictions with data at a given time. Two-dimensional contour plots of predictions can be laid on top of two-dimensional maps of measurements. Such comparisons allow a modeler to judge whether predictions at a given time are similar to or different from measurements. Although the interpretation of these plots is subjective, they are useful for estimating accuracy and whether serious flaws exist in the model.

The statistical and graphical techniques discussed above are used to judge model performance when data are available for comparison. Graphical displays are also useful for discovering bugs. A large and unrealistic perturbation in temperature in a three-dimensional plot may suggest that a programming bug has infected the model. Brilliant graphics, though, should not be used to argue the validity or performance of a model. Without evaluation against data, model results are often open to criticism.

21.1.16 Simulations

After a model has been developed and input and ambient data have been gathered, simulations can be run. When a simulation is first started, it usually does not run to completion because programming bugs still exist in the program. Debugging can take hours to weeks, depending on the number and severity of bugs and on the debugging experience of the programmer. Nevertheless, bugs are usually ironed out, and a baseline simulation can be performed.

Once a program has been debugged, it is ready for a **baseline simulation.** This type of simulation includes all model processes and input data, and results from it are often compared with ambient data. The primary purpose of model development is to study a scientific or regulatory issue, and the baseline simulation should be designed with such a study in mind. During a baseline simulation, model predictions and statistical comparisons with data are often stored and/or printed out.

21.1.17 Sensitivity tests

After the baseline simulation, sensitivity tests are often run to gauge the effect of different assumptions on model performance. The results of such tests should be compared with data and results from the baseline simulation.

For regional modeling, common sensitivity tests include testing changes in boundary conditions, initial conditions, and emissions. One test is to set all inflow gas and aerosol concentrations at horizontal boundaries equal to zero and compare the results with those from the baseline case and with data. Another test is to set all initial gas and aerosol concentrations to zero. A third test is to adjust the emission inventory to estimate the effect of possible under- or overprediction of emission on model results. On a global scale, sensitivity tests for emissions and initial conditions can also be run.

21.1.18 Improving the model

A modeler may find that simulation results deteriorate over time because of poor numerical treatment or physical representations in the model. In such cases, better numerical algorithms or sets of equations may be needed. Modelers are continually improving and updating their algorithms.

21.2 EXAMPLE MODEL SIMULATIONS

To demonstrate the steps involved in model design, application, and testing, a set of urban air pollution simulations is discussed. The purposes of the simulations were to test the effects of aerosol particles on surface air temperatures and to test the accuracy of a model against data. The model used included many of the processes shown in Fig. 1.1. Here, results from an application of the model to an air pollution episode in the Los Angeles basin for August 27–28, 1987 are shown (Jacobson 1997a,b). In the following subsections, model grids, model variables, ambient data, emission data, initial conditions, and boundary conditions are briefly discussed before results are analyzed.

21.2.1 Model grid

For the simulations, a nonnested limited-area grid with 55 west–east by 38 south–north grid cells was used. The southwest corner was at 33.06° N latitude and 119.1° W longitude. Grid spacing was 0.05 degrees west–east (about 4.6 km) by

0.045 degrees south–north (about 5.0 km). Horizontal spherical coordinates were used. In the vertical, 20 sigma-pressure coordinate layers were used for meteorological calculations, and 14 were used for all other calculations. The bottom eight layers (below 850 hPa – about 1.5 km) were the same for all processes. The model top was set to 250 hPa (about 10.3 km).

21.2.2 Model variables and time steps

Variables solved for in the model included horizontal and vertical scalar velocities, air pressure, air temperature, relative humidity, 106 gas concentrations, 16 aerosol size bin number concentrations, and 73 aerosol component volume concentrations per size bin. The time step for meteorology was 6 s, the time step for transport was 300 s, the time step for chemistry varied from $<10^{-4}$ to 900 s, the time step for radiation was 900 s, and the time steps for aerosol processes varied from $<10^{-4}$ to 900 s. The aerosol species included 18 solids, 24 liquids, and 30 ions, and one category of *residual* material. The 16 size bins used ranged from 0.014 to 74 μm in diameter.

21.2.3 Ambient and emissions data

August 27–28, 1987 was simulated because the number of available ambient measurements was large and a detailed emission inventory was prepared for this period. Ambient measurements were available for near-surface mixing ratios of ozone, carbon monoxide, nitrogen dioxide, nitric oxide, sulfur dioxide, reactive organic gases, methane, ammonia, and nitric acid. Surface data were available for temperature, dew point, relative humidity, sea-level pressure, wind speed, wind direction, and solar radiation. Aerosol measurements of black carbon, organic carbon, sodium, chloride, ammonium, nitrate, sulfate, and total aerosol mass in the sub-2.5-μm and sub-10-μm size regimes were also available. The aerosol and gas emission inventories extended over a region 325 km east–west by 180 km north–south, with a resolution of 5 km in each direction.

21.2.4 Initial conditions

Initial vertical and horizontal profiles of temperature, dew point, and pressure were interpolated from 12 sounding sites in and outside the basin for the early morning of August 27, 1987. Sea surface temperatures were interpolated each hour from buoy data. Winds were initialized with zero velocities to ensure mass conservation and to avoid startup waves near mountain regions. Differential heating and cooling over spatially varying topography created pressure gradients that forced winds to generate. The Coriolis force, pressure-gradient force, and turbulent fluxes affected the equations of motion over time. Diabatic heating and heat advection influenced changes in potential temperature, which affected pressure gradients. Initial gas and aerosol concentration were interpolated from data available at 04:30 on August 27.

Table 21.1 Normalized gross errors (NGEs) and normalized biases (NB) after 44 h for the baseline simulation

Parameter	No. of comparisons	NGE (%)	NB (%)	Parameter	No. of comparisons	NGE (%)	NB (%)
$Mass_{10}$	360	50.1	9.3	Na_{10}	360	36.0	−30.2
$Mass_{2.5}$	356	43.9	−8.1	Cl_{10}	190	46.8	16.0
BC_{10}	356	50.6	16.2	$O_3(g)$	571	27.8	−6.6
$BC_{2.5}$	356	57.5	29.9	$SO_2(g)$	339	35.4	−24.2
OC_{10}	352	45.4	0.33	$NH_3(g)$	269	69.3	−25.6
$OC_{2.5}$	352	49.0	−44.1	$HNO_3(g)$	109	54.6	22.3
$NH_{4\ 10}$	325	45.7	−30.2	HCHO(g)	61	45.8	32.9
$NH_{4\ 2.5}$	321	55.2	−0.15	Temperature	628	0.63	0.09
$SO_{4\ 10}$	304	26.3	−52.3	Rel. hum.	358	21.6	4.2
$SO_{4\ 2.5}$	360	28.4	−8.3	Solar rad.	50	7.9	−3.0
$NO_{3\ 10}$	360	69.8	3.7	σ_{sp}	125	43.0	−13.8
$NO_{3\ 2.5}$	360	67.8	18.4	σ_{ap}	255	39.8	16.3

The subscripts 10 and 2.5 indicate the mass of the species that resides in particles <10 μm and <2.5 μm in diameter, respectively. "Mass" is total particle mass, "BC" is black carbon, "OC" is organic carbon, "Solar rad." is the surface solar irradiance (W m^{-2}), σ_{sp} is the extinction coefficient due to particle scattering, and σ_{ap} is the extinction coefficient due to particle absorption. Cutoff mixing ratios were 50 ppbv for $O_3(g)$, 5 ppbv for $SO_2(g)$, 5 ppbv for HCHO(g), 1 ppbv for $NH_3(g)$, and 3 ppbv for $HNO_3(g)$. Other cutoff levels were 0.5 μg m^{-3} for sub-10-μm chloride, 2.0 μg m^{-3} for sub-2.5-μm ammonium, 0.02 km^{-1} for σ_{ap}, 10 W m^{-2} for solar irradiance, and 0 for all other parameters.

21.2.5 Boundary conditions

Outside the horizontal boundaries, initial gas and aerosol concentrations were interpolated from data. Photochemical calculations were performed on gas concentrations in virtual grid cells outside the boundary to simulate their time variation during the model run. Aerosol concentrations outside the boundary were fixed at low initial values.

21.2.6 Results from baseline simulation

A baseline simulation was run from 04:30 PST, August 27 to 0:30 PST, August 29, 1987. Table 21.1 shows statistical results from the simulation. The statistics indicate that the NGEs for sulfate, sodium, light absorption, surface solar radiation, temperature, relative humidity, sulfur dioxide gas, formaldehyde, and ozone were the lowest among the parameters compared. The NGEs for nitrate and ammonia gas were largest.

Figure 21.4 shows time-series comparisons of predictions with data for several parameters and locations. In some plots, three curves are shown. The third curve is a prediction from a sensitivity simulation in which aerosol processes were turned off in the model. The other two curves are model predictions from the

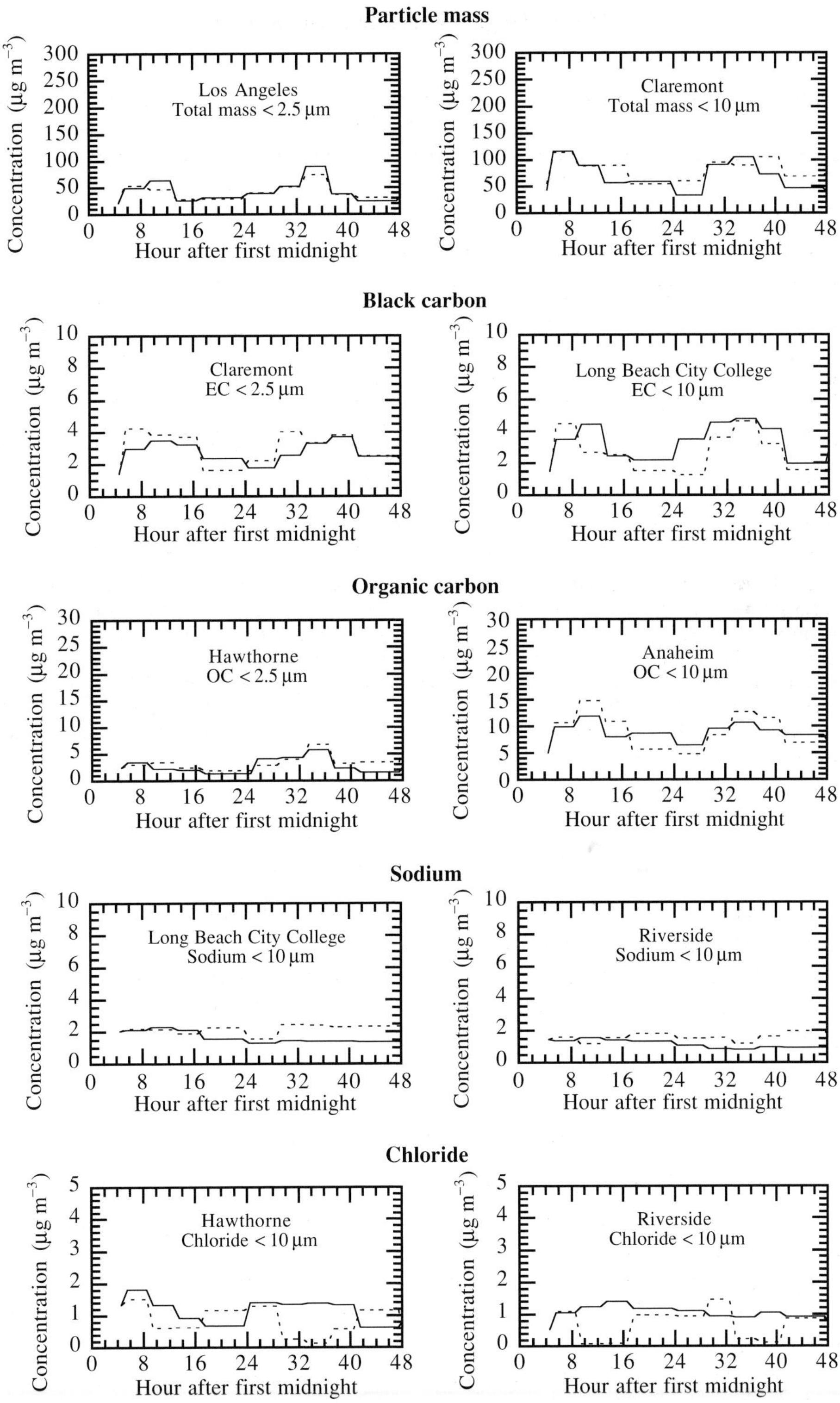

Figure 21.4 Time series comparisons of baseline (gas and aerosol processes included) model predictions (solid lines) with data (short-dashed lines) from 04:30 PST August 27 to 0:30 PST August 29, 1987. In the case of formaldehyde, circles are observed values. Most observational data were given as an average over a 4-h interval. In such cases, model predictions were averaged over the same interval. In plots with three curves, the third curves (long-dashed lines) are the predicted-value curves with gas, but not aerosol processes, turned on.

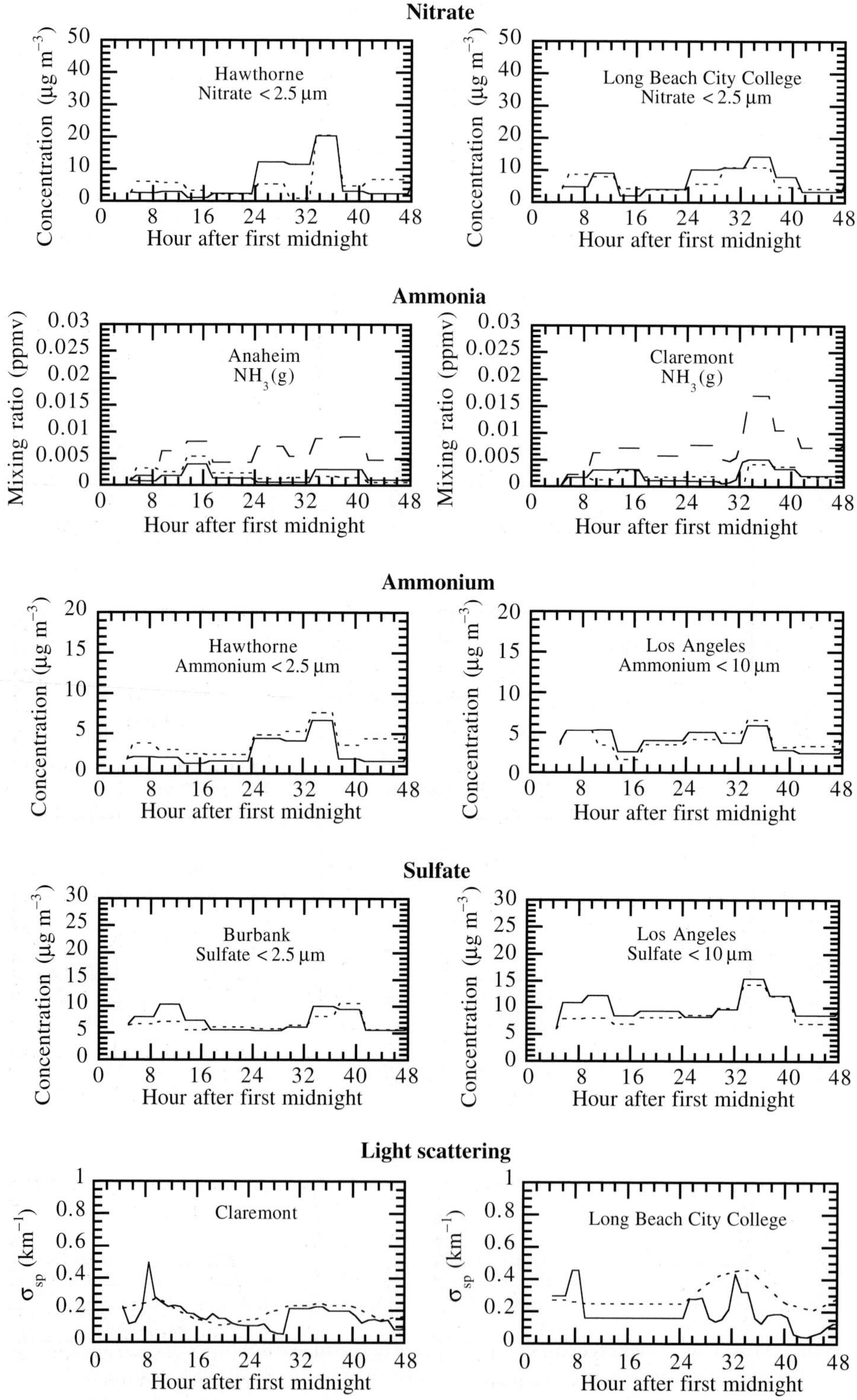

Figure 21.4 (*cont.*)

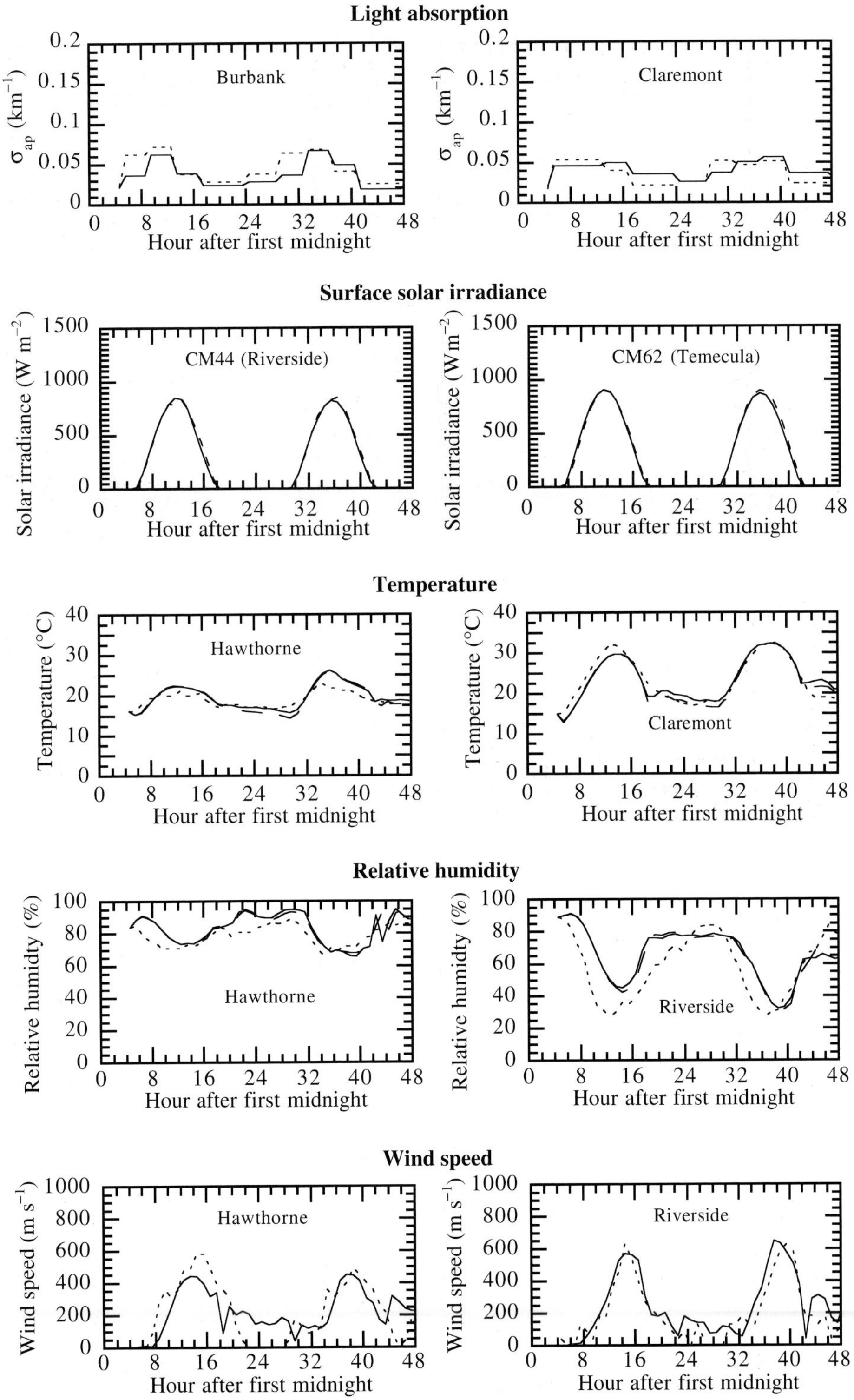

Figure 21.4 (*cont.*)

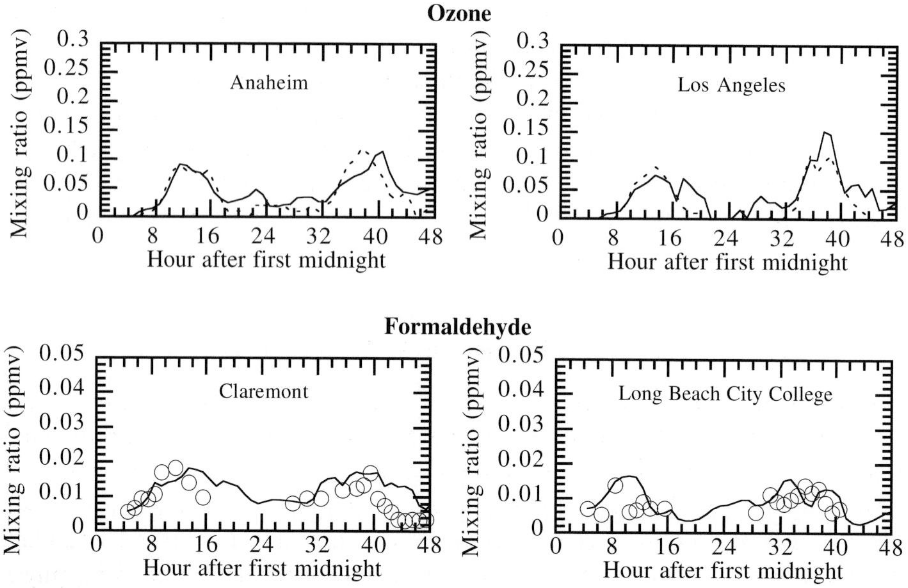

Figure 21.4 (*cont.*)

baseline simulation and data, respectively. The ammonia figures, for example, show that the inclusion of aerosol particles in the model was necessary for properly predicting ammonia gas mixing ratios.

Other statistics from the 44-h simulation indicate that aerosol particles increased nighttime surface air temperatures by about 0.77 K, decreased daytime temperatures by about 0.08 K, and increased overall temperatures (day plus night) by 0.43 K. Nighttime temperatures increased due to aerosol absorption of infrared radiation from the Earth's surface. Daytime temperatures decreased due to reduced solar radiation from aerosol scattering that exceeded warming due to aerosol absorption.

21.2.7 Results from sensitivity tests

Two additional sensitivity tests are discussed briefly. In the first, aerosol concentrations entering the model domain from all lateral boundaries were set to zero. In the second, lateral and initial concentrations of both gases and aerosol particles were set to zero.

The first sensitivity test was run to test the reliability of the boundary conditions. The test demonstrated that setting particle concentrations outside the lateral boundaries to zero did not change results significantly for most species, in part because most comparisons with data occurred far from the boundaries. The average change in model error was less than ±1 percent.

The second sensitivity test was run to test the effect of initial conditions on results. When initial concentrations of gases and aerosol particles were set to zero, model accuracy degraded for most pollutants. Ozone gross errors increased by over 50 percent, fine sulfate prediction errors doubled, and errors for several other particle components increased. The removal of initial values turned overpredictions into underpredictions for some species, but degraded prediction results for most. Thus, initializing concentrations with realistic values was important for maintaining accuracy during the two-day simulation.

21.3 SUMMARY

In this chapter, model design, application, and testing were discussed. Important steps in developing and applying a model are to define the purpose of the model, select appropriate algorithms for simulation, obtain sufficient input data, and compare model results with data. Other steps include determining an appropriate computer architecture, optimizing the computer code, setting initial and boundary conditions, running sensitivity tests, improving numerics, and improving physical parameterizations. Some of the most useful methods of comparing results with data include calculating normalized gross errors and developing time-series plot comparisons of predictions with observations. Simulations of a pollution episode indicated that a model can reasonably simulate atmospheric processes so long as data, such as initial and emission data, are available.

21.4 PROBLEMS

21.1 Interpolate data values 3, 7, 11, 4, and 6, located at geographic points (−119.2° W, 32.5° N), (−119.4° W, 32.6° N), (−118.9° W, 32.3° N), (−119.0° W, 32.4° N), and (−119.1° W, 32.3° N), respectively, to the location (−119.3° W, 32.4° N) with the inverse-square method. What are the main advantage and the main disadvantage of this method?

21.2 Interpolate data values 4, 8, 12, and 3, located at geographic points (−117.3° W, 34.5° N), (−117.3° W, 34.6° N), (−116.9° W, 34.6° N), and (−116.9° W, 34.5° N), respectively, to the location (−117.2° W, 34.53° N) with bilinear interpolation. For purposes of calculating areas, assume rectangular shapes. Interpolate the data with the inverse-square method. Discuss differences in results, if any.

21.5 COMPUTER PROGRAMMING PRACTICE

21.3 Write a computer script to calculate the time-specific normalized gross error, time-specific normalized bias, variance, standard deviation, peak prediction accuracy, and temporally paired peak prediction accuracy of the data in Table 21.2.

Table 21.2 Predicted and observed O_3 mixing ratios at 17 stations at 10:30 a.m.

Station	A	B	C	D	E	F	G	H	I	J	K	L	M	N	O	P	Q
Prediction	7.2	8.4	9.6	8.2	6.5	4.3	3.2	3.8	4.8	5.9	6.1	7.0	7.7	8.2	9.2	8.7	7.3
Observation	6.1	7.2	8.3	8.5	7.4	5.8	4.9	4.8	3.5	5.2	7.2	8.1	5.5	5.3	6.3	7.1	5.1

Mixing ratios are in units of parts per hundred million by volume.

APPENDIX A

Conversions and constants

Système Internationale (SI) and centimeter-gram-second (CGS) units are used. Table A.1 lists SI units and their conversion to CGS units. The conversions after the table include conversions to English units as well.

Table A.1 *Système Internationale (SI) and Centimeter-Gram-Second (CGS) Units*

Quantity	SI base or derived unit	CGS unit equivalent
Length	meter (m)	10^2 cm (centimeters)
Mass	kilogram (kg)	10^3 g (grams)
Temperature	kelvin (K)	K
Time	second (s)	s
Force	newton (N)	10^5 dyn (dynes)
Pressure	pascal (Pa)	10 dyn cm^{-2}
Energy	joule (J)	10^7 erg (erg)
Power	watt (W)	10^7 erg s^{-1}

A.1 DISTANCE CONVERSIONS

$$\begin{aligned} 1\ \text{m} &= 100\ \text{cm} = 1000\ \text{mm} = 10^6\ \mu\text{m} \\ &= 10^9\ \text{nm} = 10^{10}\ \text{Å} = 0.001\ \text{km} \\ &= 39.370\ \text{in} = 3.2808\ \text{ft} = 1.0936\ \text{yd} = 6.2138 \times 10^{-4}\ \text{mile} \end{aligned}$$

A.2 VOLUME CONVERSIONS

$$\begin{aligned} 1\ \text{m}^3 &= 1000\ \text{L} = 10^6\ \text{cm}^3 = 10^{18}\ \mu\text{m}^3 \\ &= 264.172\ \text{US gallon} = 35.313\ \text{ft}^3 \end{aligned}$$

A.3 MASS CONVERSIONS

$$\begin{aligned} 1\ \text{kg} &= 1000\ \text{g} = 10^6\ \text{mg} = 10^9\ \mu\text{g} \\ &= 10^{12}\ \text{ng} = 0.001\ \text{tonne (metric)} = 0.001\,102\,3\ \text{short ton} \\ &= 2.204\,62\ \text{lb} = 35.2739\ \text{oz} = 6.022 \times 10^{26}\ \text{amu} \end{aligned}$$

1 tonne	= 1000 kg	= 10^6 g	= 1.1023 short ton
	= 2204.623 lb		
1 Tg	= 10^{12} g	= 10^9 kg	= 10^6 Mg (megagram)
	= 1000 Gg (gigagram)	= 0.001 Pg (petagram)	= 10^6 tonne (metric)
	= 1 Mt (megatonne)	= 0.001 Gt (gigaton)	

A.4 TEMPERATURE CONVERSIONS

°C = K −273.15 = (°F −32)/1.8

A.5 FORCE CONVERSIONS

1 N	= 1 kg m s^{-2}	= 10^5 g cm s^{-2}	= 10^5 dyn
	= 0.2248 lbf (pound-force)		

A.6 PRESSURE CONVERSIONS

1 bar	= 10^3 mb	= 0.986 923 atm	= 10^5 N m^{-2}
	= 10^5 J m^{-3}	= 10^5 Pa	= 10^3 hPa
	= 10^5 kg m^{-1} s^{-2}	= 10^6 dyn cm^{-2}	= 10^6 g cm^{-1} s^{-2}
	= 750.06 torr	= 750.06 mm Hg	
1 atm	= 1.013 25 bar	= 760 torr	= 760 mm Hg
	= 29.92 in Hg	= 1013.25 hPa	= 14.696 psi (lbf in^{-2})

A.7 ENERGY CONVERSIONS

1 J	= 1 N m	= 10^7 erg	= 1 W s
	= 10^4 mb cm^3	= 10^7 dyn cm	= 0.238 902 cal
	= 1 kg m^2 s^{-2}	= 10^7 g cm^2 s^{-2}	= 10^{-5} bar m^3
	= 10^4 hPa cm^3	= 6.2415×10^{18} eV	= 1 C V
	= $2.777\,78 \times 10^{-7}$ kWh	= 0.009 869 L atm	= 0.737 3 lbf ft
	= $9.478\,2 \times 10^{-4}$ Btu		

A.8 POWER CONVERSIONS

1 W = 1 J s^{-1} = 3.412 52 Btu hr^{-1} = 0.013 407 horsepower

A.9 SPEED CONVERSIONS

1 m s^{-1}	= 100 cm s^{-1}	= 3.6 km h^{-1}	= 1.943 84 knots
	= 2.236 94 mi hr^{-1}		

A.10 CONSTANTS

Symbol	Quantity	Value
A	Avogadro's number	$6.022\,136\,7 \times 10^{23}$ molec. mol^{-1}
c	speed of light	$2.997\,92 \times 10^{8}$ m s^{-1}
$c_{p,d}$	specific heat of dry air at constant pressure	1004.67 J kg^{-1} K^{-1}
		1.004 67 J g^{-1} K^{-1}
		1004.67 m^2 s^{-2} K^{-1}
		0.240 cal g^{-1} K^{-1}
$c_{v,d}$	specific heat of dry air at constant volume	717.63 J kg^{-1} K^{-1}
		0.718 J g^{-1} K^{-1}
		717.63 m^2 s^{-2} K^{-1}
		0.171 cal g^{-1} K^{-1}
$c_{p,V}$	specific heat of water vapor at constant pressure (298.15 K)	1865.1 J kg^{-1} K^{-1}
$c_{v,V}$	specific heat of water vapor at constant volume (298.15 K)	1403.2 J kg^{-1} K^{-1}
c_W	specific heat of liquid water	4185.5 J kg^{-1} K^{-1}
		4.1855 J g^{-1} K^{-1}
		4185.5 m^2 s^{-2} K^{-1}
		1.00 cal g^{-1} K^{-1}
$\bar{F}_s$	mean solar constant	1365 W m^{-2}
g	effective gravity at surface of Earth	9.806 65 m s^{-2}
		980.665 cm s^{-2}
		32.174 03 ft s^{-2}
G	universal gravitational constant	$6.672\,0 \times 10^{-11}$ m^3 kg^{-1} s^{-2}
		$6.672\,0 \times 10^{-8}$ cm^3 g^{-1} s^{-2}
		$6.672\,0 \times 10^{-11}$ N m^2 kg^{-2}
h	Planck's constant	$6.626\,075\,5 \times 10^{-34}$ J s
k	von Kármán constant	0.40 (0.35–0.41)
k_B	Boltzmann's constant (R^*/A)	$1.380\,658 \times 10^{-23}$ J K^{-1}
		$1.380\,658 \times 10^{-23}$ kg m^2 s^{-2} K^{-1} molec.$^{-1}$
		$1.380\,658 \times 10^{-16}$ g cm^2 s^{-2} K^{-1} molec.$^{-1}$
		$1.380\,658 \times 10^{-19}$ cm^3 hPa K^{-1} molec.$^{-1}$
		$1.362\,603 \times 10^{-22}$ cm^3 atm K^{-1} molec.$^{-1}$
		$1.362\,603 \times 10^{-22}$ cm^3 atm K^{-1}
		$3.298\,419 \times 10^{-24}$ cal K^{-1}

(*cont.*)

Symbol	Quantity	Value
		$1.362\,603 \times 10^{-25}$ L atm K^{-1} molec.$^{-1}$
		$1.380\,658 \times 10^{-25}$ m^3 hPa K^{-1} molec.$^{-1}$
L_p	luminosity of the Sun's photosphere	3.9×10^{26} W
m_d	molecular weight of dry air	28.966 g mol^{-1}
m_v	molecular weight of water	18.02 g mol^{-1}
$\bar{M}$	mass of an air molecule (m_d/A)	4.8096×10^{-26} kg
		4.8096×10^{-23} g
M_e	mass of the Earth	5.98×10^{24} kg
		5.98×10^{27} g
R^*	universal gas constant	8.314 51 J mol^{-1} K^{-1}
		8.314 51 kg m^2 s^{-2} mol^{-1} K^{-1}
		$8.314\,51 \times 10^7$ g cm^2 s^{-2} mol^{-1} K^{-1}
		$8.314\,51 \times 10^4$ cm^3 hPa mol^{-1} K^{-1}
		82.06 cm^3 atm mol^{-1} K^{-1}
		0.083 145 1 m^3 hPa mol^{-1} K^{-1}
		0.082 06 L atm mol^{-1} K^{-1}
		$8.314\,51 \times 10^7$ erg mol^{-1} K^{-1}
		1.986 35 cal mol^{-1} K^{-1}
R'	gas constant for dry air (R^*/m_d)	287.04 J kg^{-1} K^{-1}
		0.287 04 J g^{-1} K^{-1}
		2.8704 m^3 hPa kg^{-1} K^{-1}
		2870.4 cm^3 hPa g^{-1} K^{-1}
		287.04 m^2 s^{-2} K^{-1}
		2.8704×10^6 cm^2 s^{-2} K^{-1}
R_e	radius of the Earth	6.371×10^6 m
R_p	radius of the Sun	6.96×10^8 m
$\bar{R}_{es}$	mean Earth–Sun distance	1.5×10^{11} m
R_v	gas constant for water vapor (R^*/m_v)	461.40 J kg^{-1} K^{-1}
		0.461 40 J g^{-1} K^{-1}
		4.6140 m^3 hPa kg^{-1} K^{-1}
		4614.0 cm^3 hPa g^{-1} K^{-1}
		461.40 m^2 s^{-2} K^{-1}
		4.6140×10^6 cm^2 s^{-2} K^{-1}
$T_{i,f}$	freezing point of seawater at 35 ppth salinity	271.23 K
$T_{i,m}$	melting point of ice	273.05 K

(*cont.*)

Symbol	Quantity	Value
T_p	temperature of the Sun's photosphere	5796 K
$T_{s,m}$	melting point of snow	273.15 K
γ	$c_{p,d}/c_{v,d}$	1.4
ε_0	permittivity of free space	$8.854\,19 \times 10^{-12}$ C^2 N^{-1} m^{-2}
κ	$R'/c_{p,d}$	0.286
κ_i	thermal conductivity of ice	2.20 W m^{-1} K^{-1}
κ_{sn}	thermal conductivity of snow	0.08 W m^{-1} K^{-1}
σ_B	Stefan–Boltzmann constant	$5.670\,51 \times 10^{-8}$ W m^{-2} K^{-4}
Ω	angular speed of the Earth	7.292×10^{-5} s^{-1}

APPENDIX B

Tables

B.1 STANDARD ATMOSPHERIC VARIABLES VERSUS ALTITUDE

Table B.1 Variation of gravitational acceleration, air pressure, air temperature, and air density with altitude in a standard atmosphere[a]

Alt. (km)	Gravity ($m\,s^{-2}$)	Press. (hPa)	Temp. (K)	Density ($kg\,m^{-3}$)	Alt. (km)	Gravity ($m\,s^{-2}$)	Press. (hPa)	Temp. (K)	Density ($kg\,m^{-3}$)
0	9.8072	1013.25	288.15	1.225	22	9.7396	40.5	218.57	0.0645
0.1	9.8069	1001.20	287.50	1.213	23	9.7365	34.7	219.57	0.0550
0.2	9.8066	989.45	286.85	1.202	24	9.7334	29.7	220.56	0.0469
0.3	9.8062	977.72	286.20	1.190	25	9.7304	25.5	221.55	0.0401
0.4	9.8059	966.11	285.55	1.179	26	9.7273	21.9	222.54	0.0343
0.5	9.8056	954.61	284.90	1.167	27	9.7243	18.8	223.54	0.0293
0.6	9.8053	943.22	284.25	1.156	28	9.7212	16.2	224.53	0.0251
0.7	9.8050	931.94	283.60	1.145	29	9.7182	13.9	225.52	0.0215
0.8	9.8047	920.77	282.95	1.134	30	9.7151	12.0	226.51	0.0184
0.9	9.8044	909.71	282.30	1.123	31	9.7121	10.3	227.50	0.0158
1	9.8041	898.80	281.65	1.112	32	9.7091	8.89	228.49	0.0136
1.5	9.8025	845.59	278.40	1.058	33	9.7060	7.67	230.97	0.0116
2	9.8010	795.0	275.15	1.007	34	9.7030	6.63	233.74	0.00989
2.5	9.7995	746.9	271.91	0.957	35	9.6999	5.75	236.51	0.00846
3	9.7979	701.2	268.66	0.909	36	9.6969	4.99	239.28	0.00726
3.5	9.7964	657.8	265.41	0.863	37	9.6939	4.33	242.05	0.00624
4	9.7948	616.6	262.17	0.819	38	9.6908	3.77	244.82	0.00537
4.5	9.7933	577.5	258.92	0.777	39	9.6878	3.29	247.58	0.00463
5	9.7917	540.5	255.68	0.736	40	9.6848	2.87	250.35	0.00400
5.5	9.7902	505.4	252.43	0.697	41	9.6817	2.51	253.11	0.00346
6	9.7887	472.2	249.19	0.660	42	9.6787	2.20	255.88	0.00299
6.5	9.7871	440.7	245.94	0.624	43	9.6757	1.93	258.64	0.00260
7	9.7856	411.1	242.70	0.590	44	9.6726	1.69	261.40	0.00226
7.5	9.7840	383.0	239.46	0.557	45	9.6696	1.49	264.16	0.00197
8	9.7825	356.5	236.22	0.526	46	9.6666	1.31	266.93	0.00171
8.5	9.7810	331.5	232.97	0.496	47	9.6636	1.16	269.68	0.0015
9	9.7794	308.0	229.73	0.467	48	9.6605	1.02	270.65	0.00132
9.5	9.7779	285.8	226.49	0.440	49	9.6575	0.903	270.65	0.00116
10	9.7764	265.0	223.25	0.414	50	9.6545	0.798	270.65	0.00103
11	9.7733	227.0	216.78	0.365	55	9.6394	0.425	260.77	5.7×10^{-4}
12	9.7702	194.0	216.65	0.312	60	9.6244	0.220	247.02	3.1×10^{-4}
13	9.7671	165.8	216.65	0.267	65	9.6094	0.109	233.29	1.6×10^{-4}
14	9.7641	141.7	216.65	0.228	70	9.5944	0.0522	219.59	8.3×10^{-5}

Table B.1 (*cont.*)

Alt. (km)	Gravity ($m\,s^{-2}$)	Press. (hPa)	Temp. (K)	Density ($kg\,m^{-3}$)	Alt. (km)	Gravity ($m\,s^{-2}$)	Press. (hPa)	Temp. (K)	Density ($kg\,m^{-3}$)
15	9.7610	121.1	216.65	0.195	75	9.5795	0.0239	208.40	4.0×10^{-5}
16	9.7579	103.5	216.65	0.166	80	9.5646	0.0105	198.64	1.8×10^{-5}
17	9.7549	88.5	216.65	0.142	85	9.5497	0.0045	188.89	8.2×10^{-6}
18	9.7518	75.7	216.65	0.122	90	9.5349	0.0018	186.87	3.4×10^{-6}
19	9.7487	64.7	216.65	0.104	95	9.5201	0.00076	188.42	7.5×10^{-7}
20	9.7457	55.3	216.65	0.0889	100	9.5054	0.00032	195.08	5.6×10^{-7}
21	9.7426	47.3	217.58	0.0757					

[a]*Source:* NOAA (1976), except gravity was calculated from (4.40) and (4.45), integrated globally. The globally averaged effective gravity at the surface of the Earth is 9.8060 $m\,s^{-2}$ and occurs at 0.231 km, the globally averaged topographical altitude above sea level.

B.2 SOLAR IRRADIANCE AT THE TOP OF THE ATMOSPHERE

Table B.2 Extraterrestrial solar irradiance, $F_{s,\lambda}$($Wm^{-2}\mu m^{-1}$) at the top of the Earth's atmosphere versus wavelength, λ(μm)

λ	$F_{s,\lambda}$	λ	$F_{s,\lambda}$	λ	$F_{s,\lambda}$	λ	$F_{s,\lambda}$	λ	$F_{s,\lambda}$
0.105	0.055	0.355	1125	0.605	1773	0.855	909	3.1	26
0.110	0.050	0.360	1077	0.610	1722	0.860	953	3.2	22.6
0.115	0.039	0.365	1274	0.615	1671	0.865	896	3.3	19.2
0.120	1.168	0.370	1359	0.620	1721	0.870	933	3.4	16.6
0.125	0.371	0.375	1219	0.625	1665	0.875	928	3.5	14.6
0.130	0.060	0.380	1340	0.630	1658	0.880	907	3.6	13.5
0.135	0.080	0.385	1113	0.635	1639	0.885	904	3.7	12.3
0.140	0.061	0.390	1345	0.640	1632	0.890	894	3.8	11.1
0.145	0.063	0.395	1096	0.645	1601	0.895	892	3.9	10.3
0.150	0.096	0.400	1796	0.650	1557	0.9	891	4	9.5
0.155	0.194	0.405	1643	0.655	1502	0.91	880	4.1	8.7
0.160	0.206	0.410	1768	0.660	1562	0.92	869	4.2	7.8
0.165	0.372	0.415	1810	0.665	1570	0.93	858	4.3	7.1
0.170	0.607	0.420	1760	0.670	1539	0.94	847	4.4	6.5
0.175	0.885	0.425	1719	0.675	1556	0.95	837	4.5	5.92
0.180	1.90	0.430	1615	0.680	1526	0.96	820	4.6	5.35
0.185	2.53	0.435	1798	0.685	1481	0.97	803	4.7	4.86
0.190	3.88	0.440	1829	0.690	1460	0.98	785	4.8	4.47
0.195	5.35	0.445	1951	0.695	1491	0.99	767	4.9	4.11
0.200	7.45	0.450	2048	0.700	1453	1.0	748	5	3.79
0.205	10.7	0.455	2043	0.705	1420	1.05	668	6	1.82
0.210	23.4	0.460	2054	0.710	1407	1.1	593	7	0.99
0.215	36.3	0.465	2012	0.715	1376	1.15	535	8	0.585
0.220	44.7	0.470	2007	0.720	1351	1.2	485	9	0.367
0.225	55.0	0.475	2042	0.725	1358	1.25	438	10	0.241
0.230	50.5	0.480	2061	0.730	1331	1.3	397	11	0.165
0.235	49.5	0.485	1867	0.735	1322	1.35	358	12	0.117
0.240	47.0	0.490	1943	0.740	1282	1.4	337	13	0.0851

(*cont.*)

Table B.2 (*cont.*)

λ	$F_{s,\lambda}$	λ	$F_{s,\lambda}$	λ	$F_{s,\lambda}$	λ	$F_{s,\lambda}$	λ	$F_{s,\lambda}$
0.245	62.2	0.495	1993	0.745	1276	1.45	312	14	0.0634
0.250	55.2	0.500	1892	0.750	1272	1.5	288	15	0.0481
0.255	69.5	0.505	1941	0.755	1262	1.55	267	16	0.0371
0.260	111	0.510	1937	0.760	1241	1.6	245	17	0.0291
0.265	212	0.515	1805	0.765	1220	1.65	223	18	0.0231
0.270	255	0.520	1811	0.770	1195	1.7	202	19	0.0186
0.275	197	0.525	1850	0.775	1179	1.75	180	20	0.0152
0.280	186	0.530	1907	0.780	1189	1.8	159	25	0.00617
0.285	317	0.535	1894	0.785	1183	1.85	142	30	0.00297
0.290	546	0.540	1840	0.790	1151	1.9	126	35	0.0016
0.295	573	0.545	1866	0.795	1142	1.95	114	40	0.000942
0.300	493	0.550	1845	0.800	1126	2	103	50	0.000391
0.305	669	0.555	1854	0.805	1112	2.1	90	60	0.00019
0.310	711	0.560	1801	0.810	1080	2.2	79	80	0.0000416
0.315	765	0.565	1828	0.815	1073	2.3	69	100	0.0000257
0.320	777	0.570	1824	0.820	1049	2.4	62	120	0.0000126
0.325	935	0.575	1851	0.825	1050	2.5	55	150	0.00000523
0.330	1041	0.580	1833	0.830	1027	2.6	48	200	0.00000169
0.335	950	0.585	1838	0.835	1012	2.7	43	250	0.0000007
0.340	1035	0.590	1760	0.840	1006	2.8	39	300	0.00000023
0.345	980	0.595	1791	0.845	983	2.9	35	400	0.00000011
0.350	1019	0.600	1752	0.850	951	3	31	1000	0

λ is the midpoint of a wavelength interval, and the irradiance at each λ is integrated from the lower to the upper edge of the interval. For example, the irradiance in the interval from $\lambda = 0.1475$ to 0.1525 μm, centered at $\lambda = 0.15$ μm is $\bar{F}_{s,\lambda}\Delta\lambda = 0.096$ W m^{-2} μm^{-1} × 0.005 μm = 0.000 48 W m^{-2}. The sum of irradiance over all bins is the solar constant.

The data were derived from Woods *et al.* (1996) for $\lambda < 0.275$ μm, Nicolet (1989) for $0.275 < \lambda < 0.9$ μm, and Thekaekara (1974) for $\lambda > 0.9$ μm.

B.3 CHEMICAL SYMBOLS AND STRUCTURES OF GASES

Table B.3 Inorganic and organic gases and their possible chemical structures

Chemical name
Molecular formula
Chemical structure

Inorganic

Hydrogen and oxygen species

Atomic hydrogen	Atomic oxygen (triplet)	Molecular oxygen	Ozone
H	O	O_2	O_3
H•		O=O	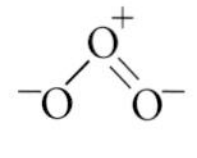

Table B.3 (*cont.*)

Chemical name Molecular formula Chemical structure			
	Inorganic		
Hydroxyl radical OH $\dot{O}$–H	Hydroperoxy radical HO_2 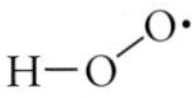	Hydrogen peroxide H_2O_2 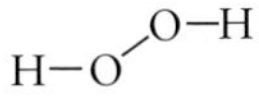	Water vapor H_2O 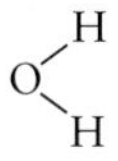
	Nitrogen species		
Molecular nitrogen N_2 N≡N	Nitric oxide NO $\dot{N}$=O	Nitrogen dioxide NO_2 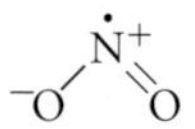	Nitrate radical NO_3 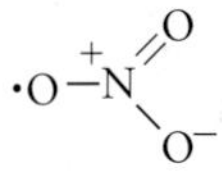
Nitrous acid HONO 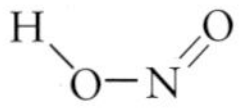	Nitric acid HNO_3 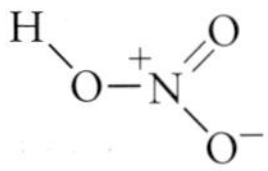	Peroxynitric acid HO_2NO_2 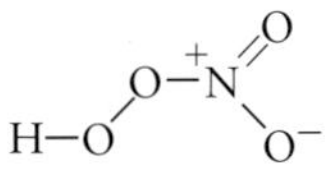	Nitrous oxide N_2O O=$\overset{+}{N}$=$\overset{-}{N}$
Dinitrogen pentoxide N_2O_5 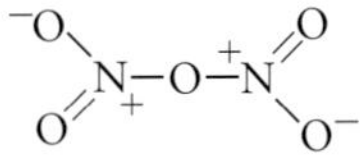			
	Sulfur species		
Atomic sulfur S 	Sulfur monoxide (sulfonyl or thionyl radical) SO S=O	Sulfur dioxide SO_2 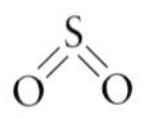	Sulfur trioxide SO_3 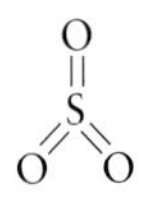
Bisulfite HSO_3 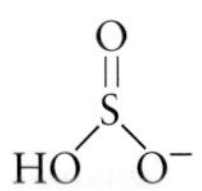	Sulfuric acid H_2SO_4 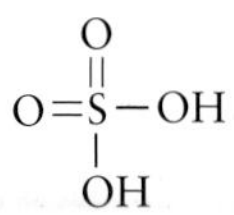	Carbonyl sulfide OCS O=C=S	Carbon monosulfide CS $^{-}C{\equiv}S^{+}$

(*cont.*)

Table B.3 (*cont.*)

Chemical name
Molecular formula
Chemical structure

Inorganic

Sulfur species

Carbon disulfide	Hydrogen sulfide radical	Hydrogen sulfide	Methanethiol
CS_2	HS	H_2S	(Methyl sulfide)
			MeSH
			CH_3SH
$S=C=S$			

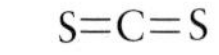

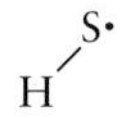

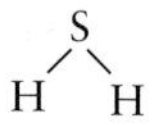

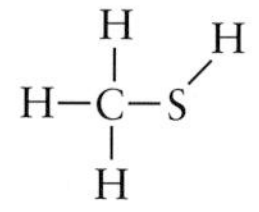

Methanethiolate radical	Methanethiolate oxy	Methanethiolate peroxy	Methanesulfenic acid
CH_3S	radical	radical CH_3SO_2	CH_3SOH
	CH_3SO		

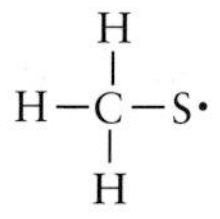

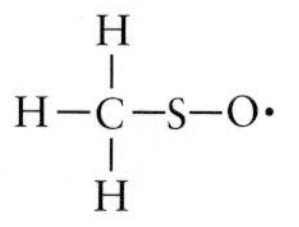

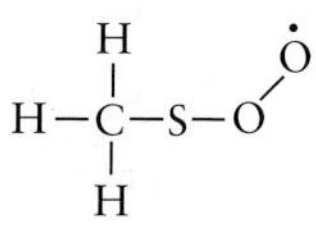

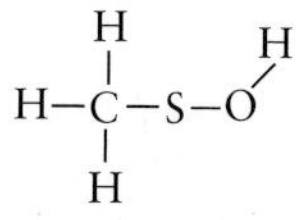

Dimethyl sulfide	Dimethyl sulfide radical	Dimethyl sulfide–OH	Dimethyl sulfone
DMS	CH_3SCH_2	adduct	$DMSO_2$
CH_3SCH_3		$CH_3S(OH)CH_3$	$CH_3S(O)_2CH_3$

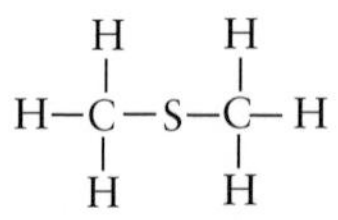

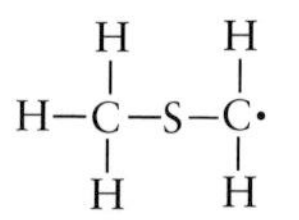

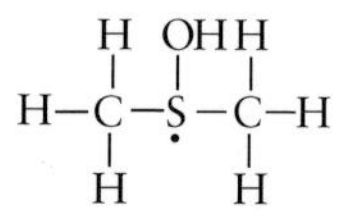

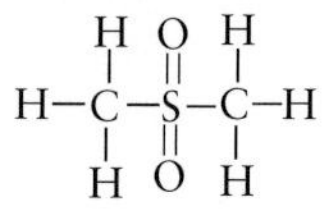

Dimethyl disulfide	Methane sulfonic acid	Hydroxymethane
DMDS	MSA	sulfonic acid
CH_3SSCH_3	$CH_3S(O)_2OH$	HMSA
		$HOCH_2S(O)_2OH$

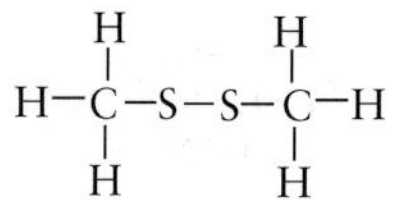

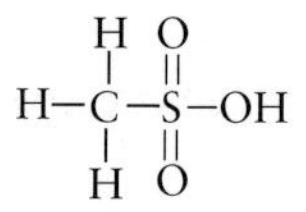

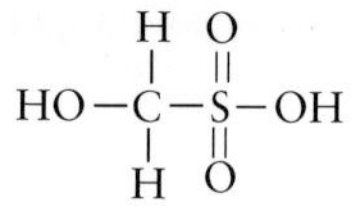

Chlorine species

Atomic chlorine	Molecular chlorine	Hydrochloric acid	Chlorine monoxide
Cl	Cl_2	HCl	ClO
	$Cl-Cl$	$H-Cl$	$Cl-\dot{O}$

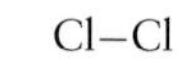

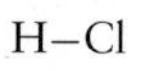

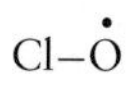

Hypochlorous acid	Chlorine peroxy radical	Chlorine peroxy radical	Dichlorine dioxide
HOCl	OClO	ClOO	Cl_2O_2

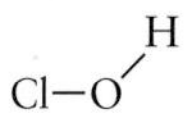

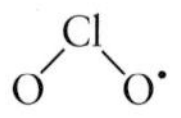

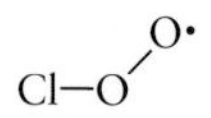

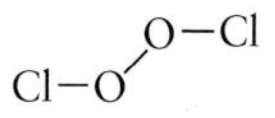

Table B.3 (*cont.*)

Chemical name Molecular formula Chemical structure

Inorganic

Chlorine species

Chlorine nitrate $ClONO_2$	Chlorine nitrite $ClNO_2$	Methyl chloride CH_3Cl	Methyl chloroform CH_3CCl_3

$$Cl-O\overset{+}{\underset{}{N}}\underset{O^-}{\overset{O}{\diagdown}}$$

$$Cl\overset{+}{\underset{}{N}}\underset{O^-}{\overset{O}{\diagdown}}$$

$$H-\underset{\underset{H}{|}}{\overset{\overset{H}{|}}{C}}-Cl$$

$$H-\underset{\underset{H}{|}}{\overset{\overset{H}{|}}{C}}-\underset{\underset{Cl}{|}}{\overset{\overset{Cl}{|}}{C}}-Cl$$

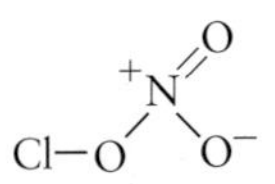

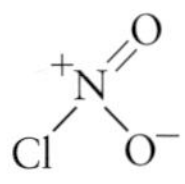

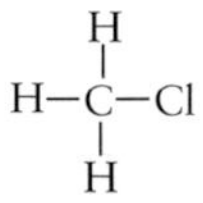

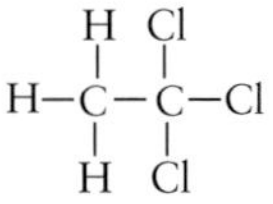

Trichlorofluoromethane (CFC-11) $CFCl_3$	Dichlorodifluoromethane (CFC-12) CF_2Cl_2	1-Fluorodichloro, 2-difluorochloroethane (CFC-113) $CFCl_2CF_2Cl$	Chlorodifluoromethane (HCFC-22) CF_2ClH

$$F-\underset{\underset{Cl}{|}}{\overset{\overset{Cl}{|}}{C}}-Cl$$

$$F-\underset{\underset{F}{|}}{\overset{\overset{Cl}{|}}{C}}-Cl$$

$$Cl-\underset{\underset{F}{|}}{\overset{\overset{F}{|}}{C}}-\underset{\underset{Cl}{|}}{\overset{\overset{Cl}{|}}{C}}-F$$

$$F-\underset{\underset{F}{|}}{\overset{\overset{Cl}{|}}{C}}-H$$

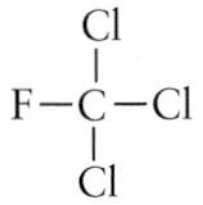

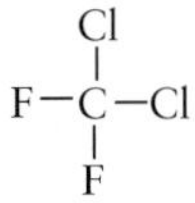

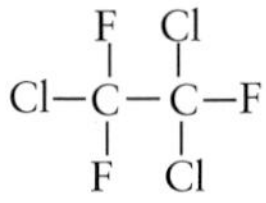

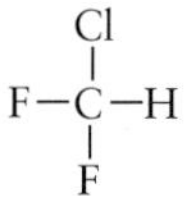

Carbon tetrachloride

CCl_4

$$Cl-\underset{\underset{Cl}{|}}{\overset{\overset{Cl}{|}}{C}}-Cl$$

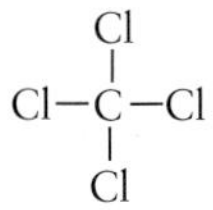

Bromine species

Atomic bromine Br	Molecular bromine Br_2	Hydrobromic acid HBr	Bromine chloride BrCl

$:\overset{\cdot\cdot}{\underset{\cdot\cdot}{Br}}\cdot$ | $Br-Br$ | $H-Br$ | $Br-Cl$

Hypobromous acid HOBr	Bromine monoxide BrO	Methyl bromide CH_3Br	Bromine nitrate $BrONO_2$

$$Br-O\overset{H}{\diagup}$$

$$Br-\overset{\cdot}{O}$$

$$H-\underset{\underset{H}{|}}{\overset{\overset{H}{|}}{C}}-Br$$

$$Br-O\overset{+}{\underset{O^-}{N}}\overset{O}{\diagup}$$

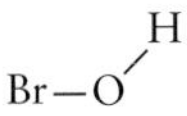

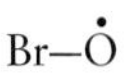

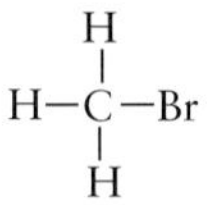

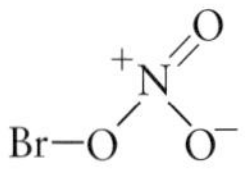

Inorganic carbon species

Carbon monoxide CO	Carbon dioxide CO_2

$^-C\equiv O^+$ | $O=C=O$

(*cont.*)

Table B.3 (*cont.*)

Chemical name
Molecular formula
Chemical structure

Organic

Alkanes

Methane
CH_4

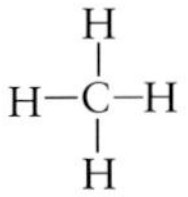

Ethane
C_2H_6

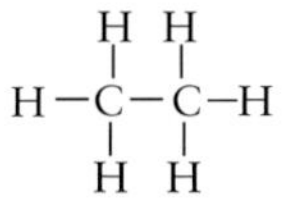

Propane
C_3H_8

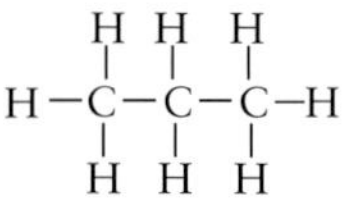

Butane
C_4H_{10}

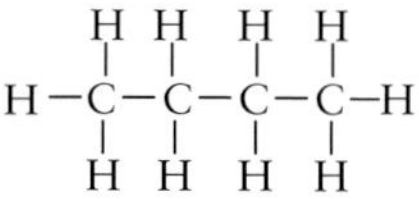

2,2-Dimethylpropane
(neopentane)
$C(CH_3)_4$

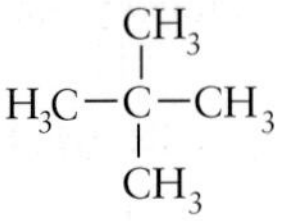

2,2,4-Trimethylpentane
(isooctane)
$(CH_3)_2CHCH_2C(CH_3)_3$

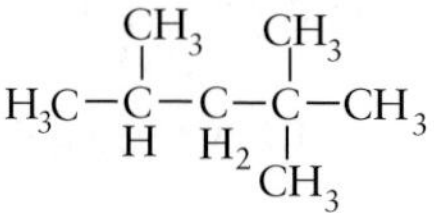

Cycloalkanes

Cyclopropane
$(CH_2)_3$

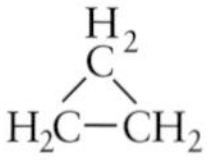

Cyclobutane
$(CH_2)_4$

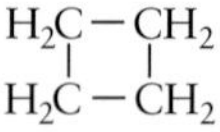

Cyclopentane
$(CH_2)_5$

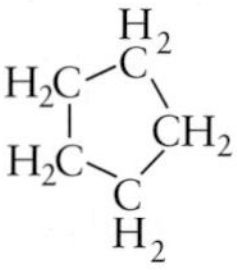

Alkenes

Ethene
(ethylene)
C_2H_4

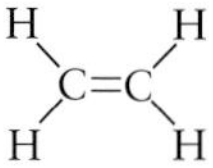

Propene
(propylene)
C_3H_6

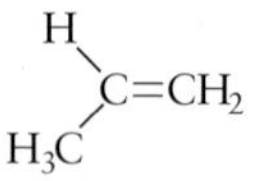

cis-2-Butene
C_4H_8

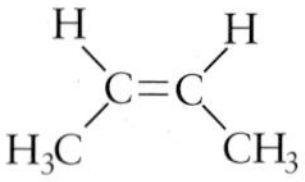

trans-2-Butene
C_4H_8

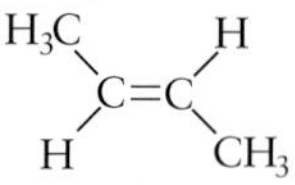

Cycloalkenes

Cyclopentene
C_5H_8

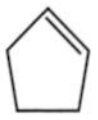

Cyclohexene
C_6H_{10}

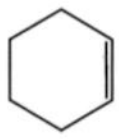

1-Methylcyclohexene
C_7H_{12}

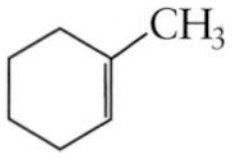

Table B.3 (*cont.*)

| Chemical name |
| Molecular formula |
| Chemical structure |

Organic

Alkynes

Ethyne (acetylene)
C_2H_2

$H-C\equiv C-H$

Propyne
C_3H_4

$$H-\overset{\overset{\displaystyle H}{|}}{\underset{\underset{\displaystyle H}{|}}{C}}-C\equiv C-H$$

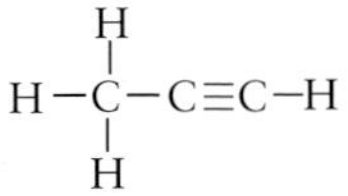

Aromatics

Benzene	Toluene	o-Xylene	m-Xylene
C_6H_6	(methylbenzene)	(1,2-dimethylbenzene)	(1,3-dimethylbenzene)
	TOL	**XYL**	**XYL**
	$C_6H_5CH_3$	$1,2\text{-}(CH_3)_2C_6H_4$	$1,3\text{-}(CH_3)_2C_6H_4$

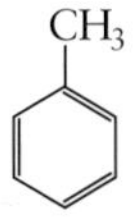

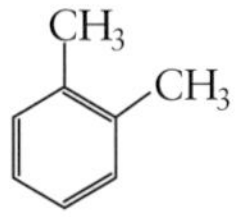

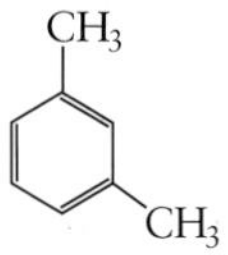

Ethylbenzene
$C_6H_5C_2H_5$

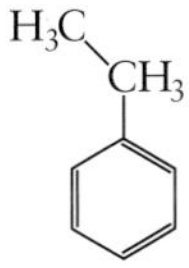

1,2,3-Trimethylbenzene
$1,2,3\text{-}(CH_3)_3C_6H_3$

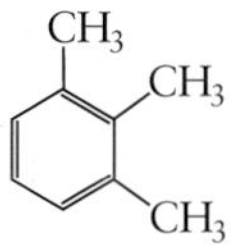

Terpenes

Isoprene	*d*-Limonene	α-Pinene
(2-methyl-1,3-butadiene)	$C_{10}H_{16}$	$C_{10}H_{16}$
ISOP		
C_5H_8		

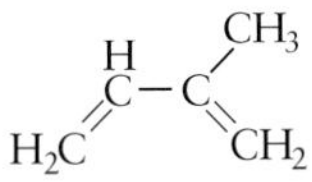

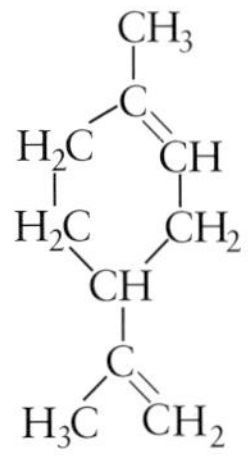

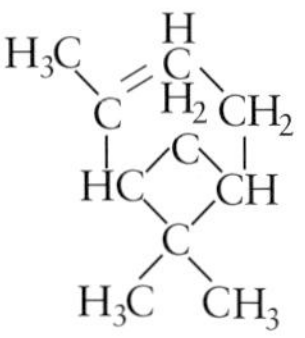

(*cont.*)

Table B.3 (*cont.*)

Chemical name
Molecular formula
Chemical structure

Organic

Alcohols

Methanol
(methyl alcohol)
CH_3OH

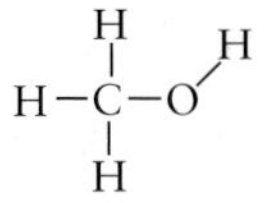

Ethanol
(ethyl alcohol)
C_2H_5OH

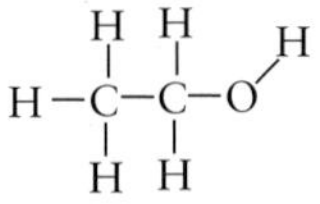

o-Cresol
CRES
2-$CH_3C_6H_4OH$

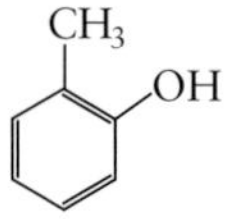

Phenol
(hydroxybenzene)
C_6H_5OH

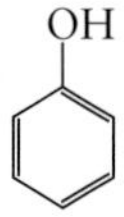

Aldehydes

Formaldehyde
(methanal)
HCHO

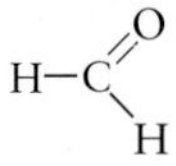

Acetaldehyde
(ethanal)
CH_3CHO

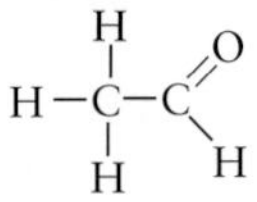

Propionaldehyde
(propanal)
CH_3CH_2CHO

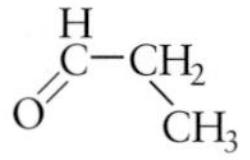

Benzaldehyde
BZA
C_6H_5CHO

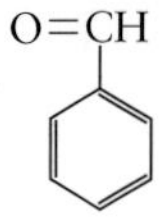

Glycol aldehyde
$HOCH_2CHO$

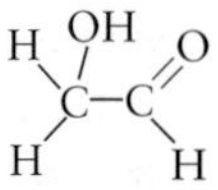

Acrolein
(propenal)
CH_2CHCHO

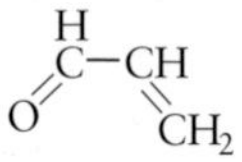

Methacrolein
(2-methyl-2propenal)
MACR
$CH_2{=}C(CH_3)CHO$

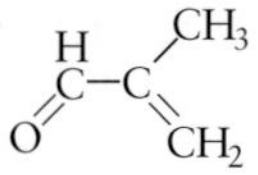

Ketones

Acetone
(2-propanone)
CH_3COCH_3

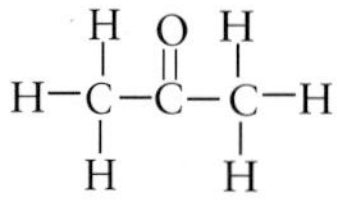

Methylethylketone
(2-butanone)
$CH_3CH_2COCH_3$

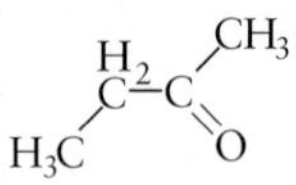

Methylvinylketone
(3-buten-2-one)
MVK
$CH_2 = CHCOCH_3$

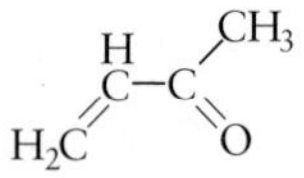

Carboxylic acids

Formic acid
(methanoic acid)
HCOOH

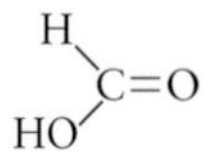

Acetic acid
(ethanoic acid)
CH_3COOH

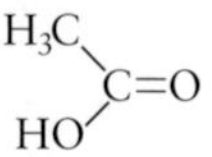

Propionic acid
(propanoic acid)
CH_3CH_2COOH

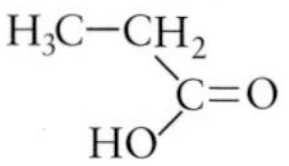

Table B.3 (*cont.*)

Chemical name
Molecular formula
Chemical structure

Organic

Alkyl radicals

Methyl radical
CH_3

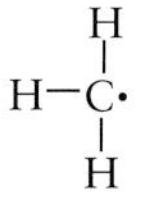

Ethyl radical
C_2H_5

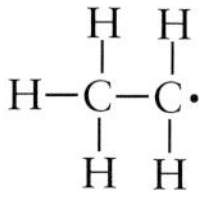

n-Propyl radical
C_3H_7

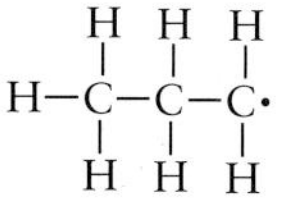

Isopropyl radical
C_3H_7

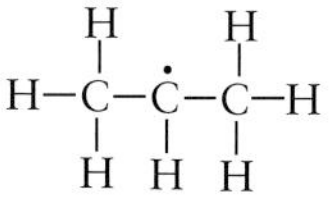

Hydroxyalkyl radicals

Hydroxyethyl radical
(ethanyl radical)
$HOCH_2CH_2$

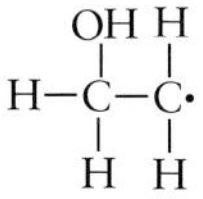

Acyl (alkanoyl) radicals

Formyl radical
(methanoyl radical)
HCO

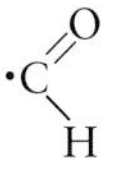

Acetyl radical
(ethanoyl radical)
CH_3CO

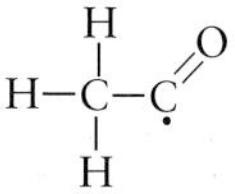

Aryl radicals

Phenyl radical
C_6H_5

Methylphenyl radical
$C_6H_4(CH_3)$

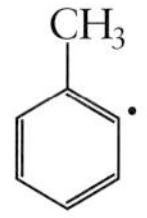

Benzyl radical
$C_6H_5CH_2$

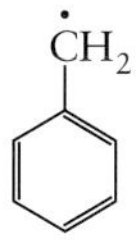

Alkoxy radicals

Methoxy radical
CH_3O

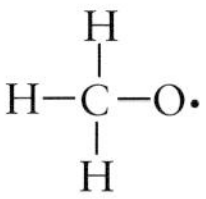

Ethoxy radical
C_2H_5O

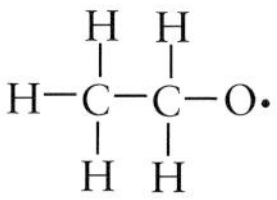

n-Propoxy radical
C_3H_7O

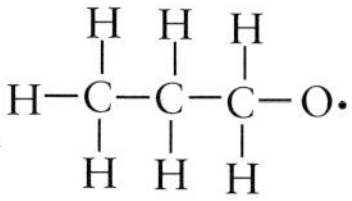

Hydroxyalkoxy radicals

Hydroxyethyloxy radical
(ethanoloxy radical)
$HOCH_2CH_2O$

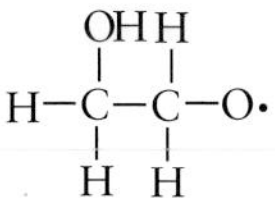

Acyloxy (alkanoyloxy) radicals

Acetyloxy radical
$CH_3C(O)O$

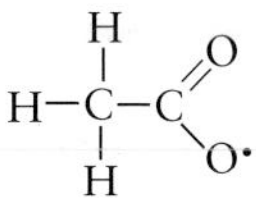

(*cont.*)

Table B.3 (*cont.*)

Chemical name
Molecular formula
Chemical structure

Organic

Aryloxy radicals

Phenoxy radical
PHO
C_6H_5O

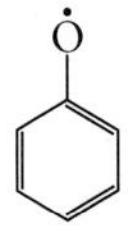

Methylphenoxy radical
CRO
$2\text{-}OC_6H_4CH_3$

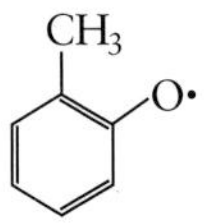

Benzoxy radical
$C_6H_5CH_2O$

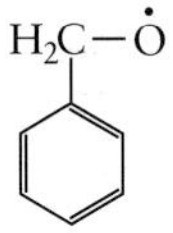

Alkylperoxy radicals

Methylperoxy radical
CH_3O_2

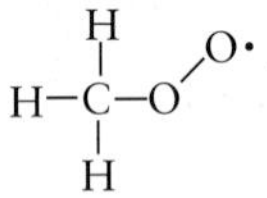

Ethylperoxy radical
$C_2H_5O_2$

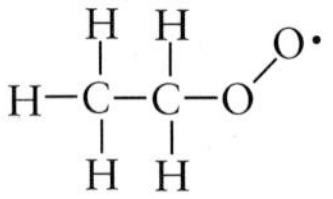

n-Propylperoxy radical
$C_3H_7O_2$

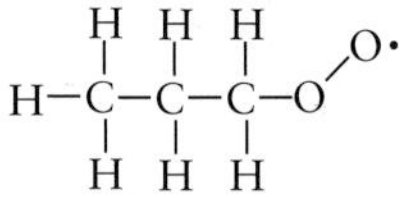

Alkylperoxy radical from OH addition to isoprene
ISOH

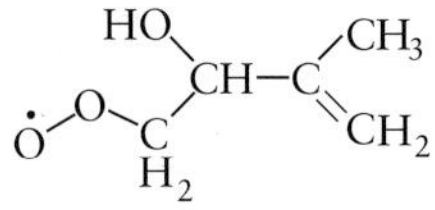

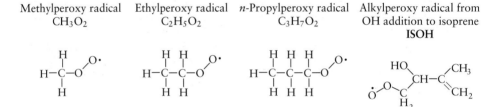

Hydroxyalkylperoxy radicals

Hydroxymethylperoxy radical (methanolperoxy radical)
$HOCH_2O_2$

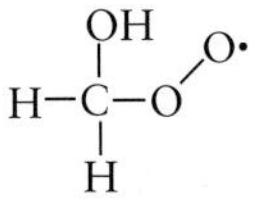

Hydroxyethylperoxy radical (ethanolperoxy radical)
$HOC_2H_4O_2$

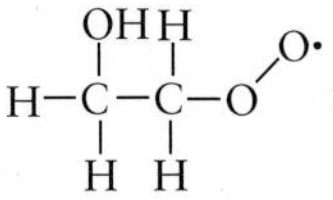

Hydroxypropylperoxy radical (propanolperoxy radical)
$HOC_3H_6O_2$

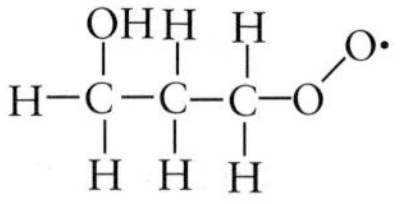

Acylperoxy (alkanoylperoxy) radicals

Peroxyacetyl radical
$CH_3C(O)OO$

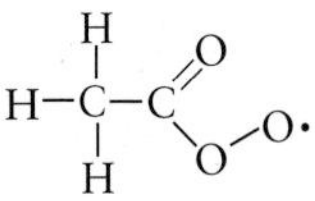

Acetylmethylperoxy radical
CH_3COCH_2OO

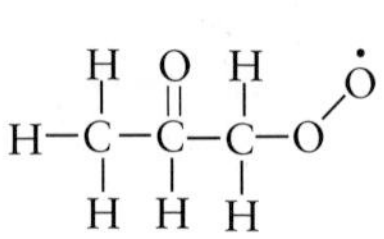

Table B.3 (*cont.*)

Chemical name Molecular formula Chemical structure			
Organic			
Arylperoxy radicals			
Phenylperoxy radical **PHO_2** C_6H_5OO 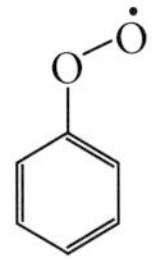	Methylphenylperoxy radical **CRO_2** 2-$O_2C_6H_4CH_3$ 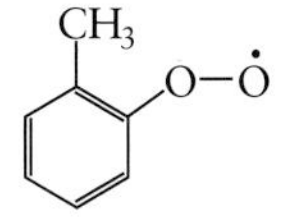	Benzylperoxy radical **BO_2** $C_6H_5CH_2OO$ 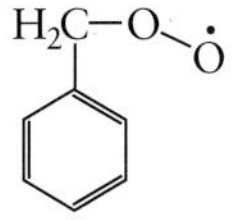	Methylbenzylperoxy radical **XLO_2** 2-$CH_2OOC_6H_5CH_3$ 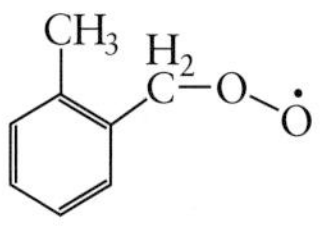
Other peroxy radicals			
Peroxybenzoyl radical **BZO_2** C_6H_5OCOO 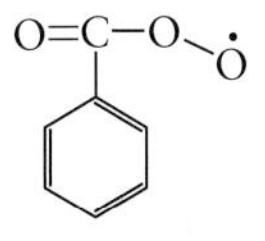	MVK+O_3 product $OOCCH_3CHCH_2$ 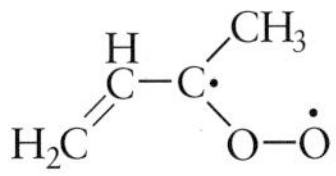	2-Methylperoxy propenoyl radical $CH_2CCH_3C(O)OO$ 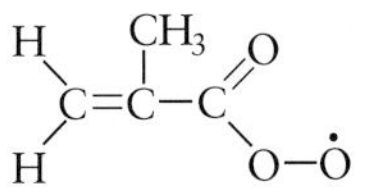	MACR+O_3 product CH_2CCH_3CHOO 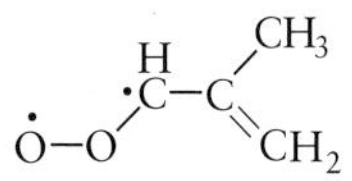
Peroxy radical of methylglyoxal **MGPX** $CH_3COC(O)OO$ 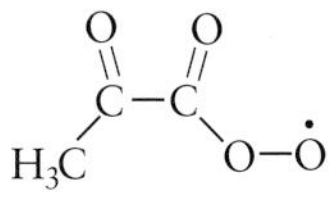	MACR+OH product $CH_2CCH_3C(O)OO$ 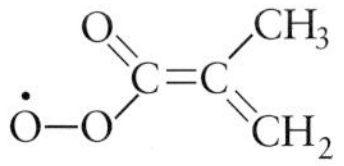	Nitrated organic peroxy radical **PNO_2** 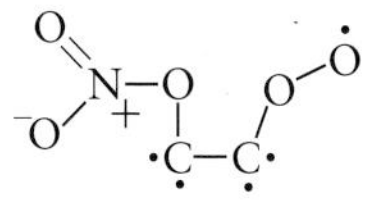	
OH adducts			
Methylvinylketone-OH adduct **MV1** $OOCH_2CH(OH)$ $C(CH_3)O$ 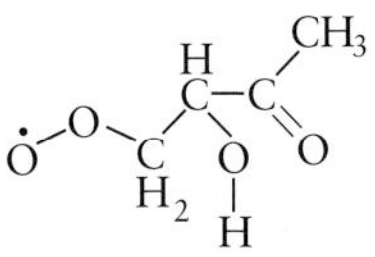	Methylvinylketone-OH adduct **MV2** $HOCH_2CH(OO)$ $C(CH_3)O$ 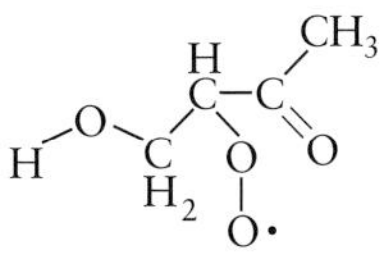	Methacrolein-OH adduct **MAC1** $OCHC(O_2)(CH_3)$ CH_2OH 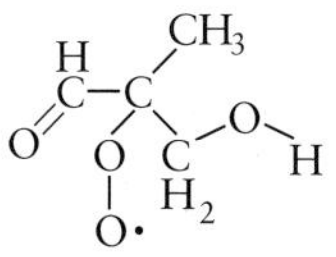	Methacrolein-OH adduct **MAC2** $OCHC(OH)(CH_3)$ CH_2OO 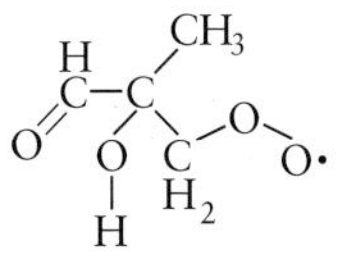

(*cont.*)

Table B.3 (*cont.*)

Chemical name
Molecular formula
Chemical structure

Organic

Toluene-OH adduct
TO_2
$2\text{-}OHC_6H_5^-(O_2)CH_3$

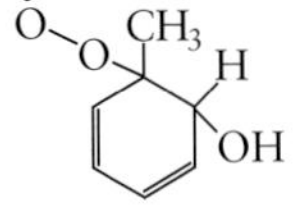

o-Xylene-OH adduct
XINT
$4\text{-}OH\text{-}3\text{-}O_2\text{-}2\text{-}CH_3C_6H_4CH_3$

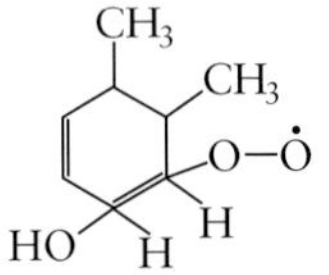

Molozonides

Ethene molozonide
OZD
$CH_2O_3CH_2$

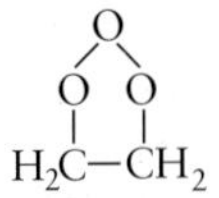

Propene molozonide
OZD
$CH_3CHO_3CH_2$

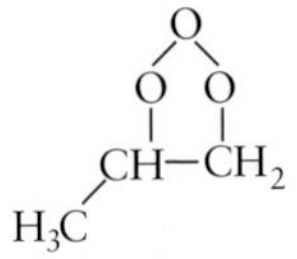

Biradicals

Criegee biradical
H_2COO

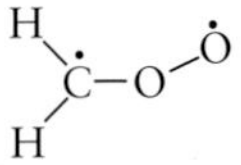

Methyl criegee biradical
CH_3HCOO

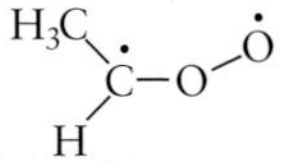

Dicarbonyls

Glyoxal (ethanedial)
$(CHO)_2$

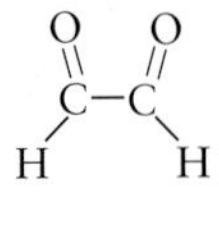

Methylglyoxal
MGLY
CH_3COCHO

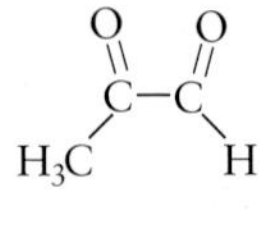

Organic peroxides

Methyl hydroperoxide
CH_3OOH

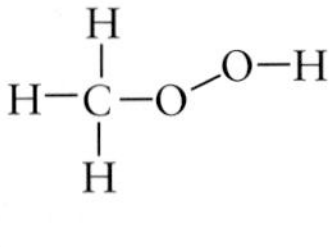

Nitrates

Methyl nitrate
CH_3ONO_2

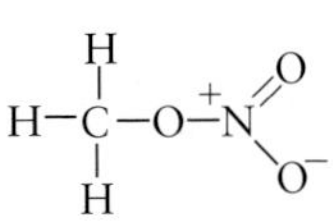

Ethyl nitrate
$C_2H_5ONO_2$

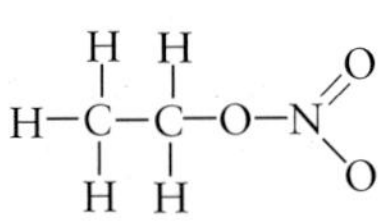

Propyl nitrate
$C_3H_7ONO_2$

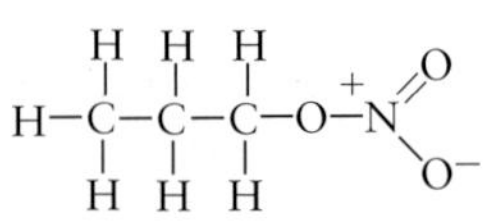

Benzyl nitrate
$C_6H_5CH_2ONO_2$

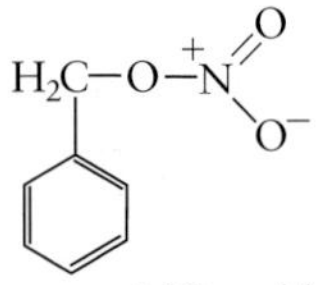

Peroxyacetyl nitrate
PAN
$CH_3C(O)OONO_2$

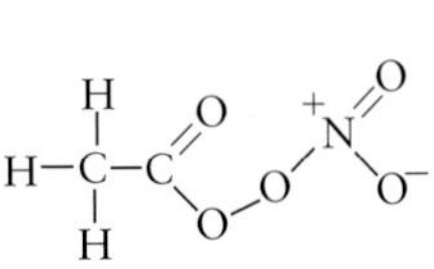

2-Methylperoxy propenoylnitrate
MPAN
$CH_2CCH_3C(O)OONO_2$

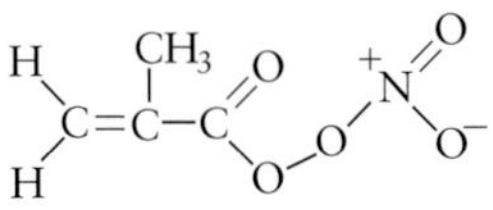

Peroxybenzoyl nitrate
PBZN
$C_6H_5C(O)OONO_2$

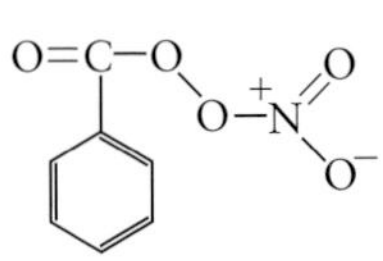

Isoprene-NO_3 adduct
ISNT
$C_5H_8ONO_2$

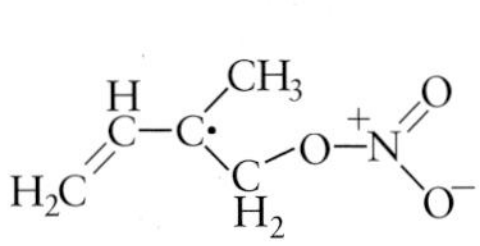

Table B.3 (*cont.*)

Chemical name
Molecular formula
Chemical structure

Organic

Nitric acids

Methylperoxy nitric acid
$CH_3O_2NO_2$

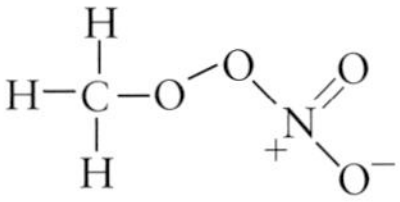

Ethylperoxy nitric acid
$C_2H_5O_2NO_2$

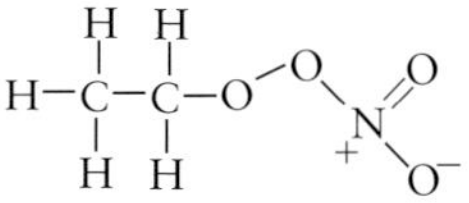

Nitrites

Methyl nitrite
CH_3ONO

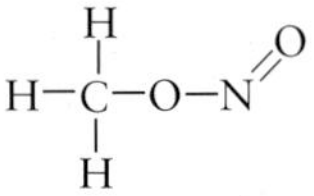

Ethyl nitrite
C_2H_5ONO

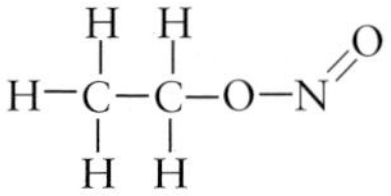

Propyl nitrite
C_3H_7ONO

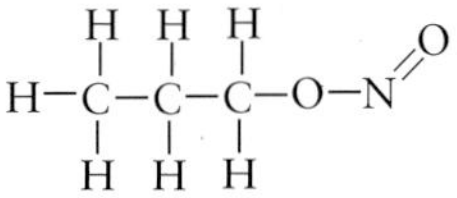

Benzyl nitrite
$C_6H_5CH_2ONO$

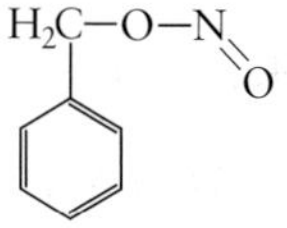

Nitro group

Nitrobenzene
$C_6H_5NO_2$

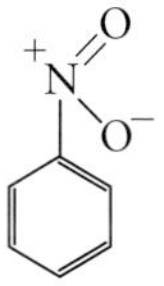

m-Nitrotoluene
3-$NO_2C_6H_4CH_3$

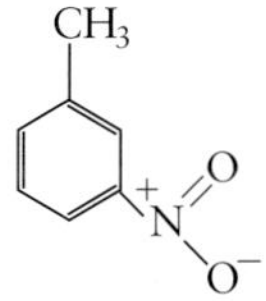

o-Nitrophenol
NPHN
2-$NO_2C_6H_4OH$

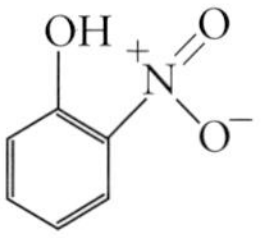

m-Nitrocresol
NCRE
3-NO_2-2-$OHC_6H_3CH_3$

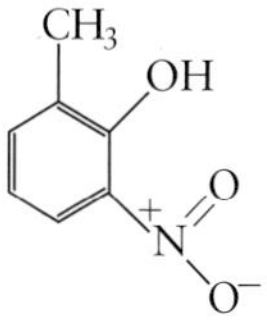

Carbon-bond groups

Paraffin carbon bond
PAR

Olefin carbon bond
OLE

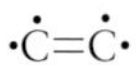

Ketone carbonyl group
KET

Organic peroxide
ROOH

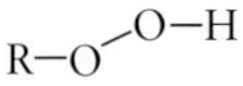

Primary organic peroxy radical
RO_2

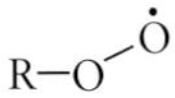

Secondary organic oxy radical
ROR

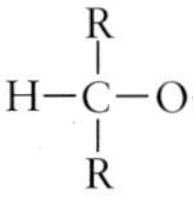

Secondary organic peroxy radical
RO_2R

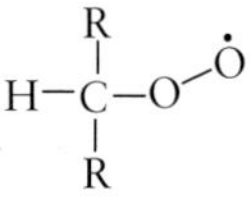

C_2 dinitrate group
DNIT

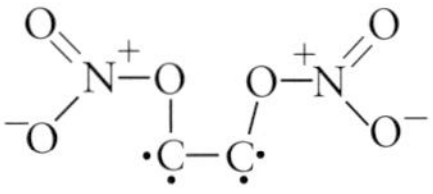

Organic nitrate
NTR

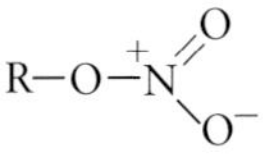

(*cont.*)

Table B.3 (*cont.*)

Chemical name Molecular formula Chemical structure			
Organic			
Miscellaneous species and groups			
Dimethyl 2nd organic peroxide radical AO_2	Peroxide radical of OPEN **OPPX**	High-molecular-weight aromatic oxidation ring fragments **OPEN**	Dinitrate of isoprene **DISN**
Group of hydroxy carbonyl alkenes from isoprene-OH reaction **IALD1**	Organic nitrates from OH addition to unsubstituted double bond **ISNI1**	Organic nitrates from OH addition to substituted double bond and OH reactions of primary products **ISNI2**	Alkyl peroxy radical from OH addition across the double bond of ISNI1 and ISNI2 **ISNIR**
Aromatic ring fragment acid **ACID**	Organic peroxide from isoprene **IPRX**	Paraffin to peroxy radical operator **DOP**	Paraffin loss operator **XOP**

Molecular formulae and abbreviated names (boldface) used in Appendix Table B.4 are defined here. Organic sulfur-, chlorine-, and bromine-containing species are included in the inorganic categories for convenience.

B.4 GAS-PHASE REACTIONS

Table B.4 Gas-phase chemical kinetic reactions, reaction rates, and photoprocesses

No.	Region	Kinetic reaction	F_c [a]	Rate coefficient (s^{-1}, $cm^3\ s^{-1}$, or $cm^6\ s^{-1}$)	Ref.[b]
		Inorganic chemistry			
1	All	$O + O_2 + M \longrightarrow O_3 + M$		$6.00 \times 10^{-34}(300/T)^{2.3}$	B
2	S	$O + O_3 \longrightarrow 2O_2$		$8.00 \times 10^{-12}e^{-2060/T}$	A
3	S	$O(^1D) + O_3 \longrightarrow 2O_2$		1.20×10^{-10}	B
4	S	$O(^1D) + O_3 \longrightarrow O_2 + 2O$		1.20×10^{-10}	B
5	All	$O(^1D) + O_2 \longrightarrow O + O_2$		$3.20 \times 10^{-11}e^{67/T}$	A
6	All	$O(^1D) + N_2 \longrightarrow O + N_2$		$1.80 \times 10^{-11}e^{107/T}$	A
7	S	$O(^1D) + N_2 + M \longrightarrow N_2O + M$		$3.50 \times 10^{-37}(300/T)^{0.6}$	B
8	S	$O(^1D) + N_2O \longrightarrow N_2 + O_2$		4.40×10^{-11}	A
9	S	$O(^1D) + N_2O \longrightarrow NO + NO$		7.20×10^{-11}	A
10	All	$O(^1D) + H_2 \longrightarrow OH + H$		1.10×10^{-10}	A
11	All	$O(^1D) + H_2O \longrightarrow OH + OH$		2.20×10^{-10}	A
12	All	$H + O_2 \xrightarrow{M} HO_2$	(P) 0.55	$5.40 \times 10^{-32}(300/T)^{1.8}$ 7.50×10^{-11}	A

Table B.4 (*cont.*)

No.	Region	Kinetic reaction	F_c [a]	Rate coefficient (s^{-1}, $cm^3\ s^{-1}$, or $cm^6\ s^{-1}$)	Ref.[b]
		Inorganic chemistry			
13	All	$H + O_3 \longrightarrow O_2 + OH$		$1.40 \times 10^{-10} e^{-470/T}$	B
14	All	$H + HO_2 \longrightarrow H_2 + O_2$		5.60×10^{-12}	A
15	All	$H + HO_2 \longrightarrow OH + OH$		7.20×10^{-11}	A
16	All	$H + HO_2 \longrightarrow H_2O + O$		2.40×10^{-12}	A
17	All	$OH + O \longrightarrow H + O_2$		$2.30 \times 10^{-11} e^{110/T}$	A
18	All	$OH + O_3 \longrightarrow HO_2 + O_2$		$1.90 \times 10^{-12} e^{-1000/T}$	A
19	All	$OH + H_2 \longrightarrow H_2O + H$		$7.70 \times 10^{-12} e^{-2100/T}$	A
20	All	$OH + OH \longrightarrow H_2O + O$		$4.20 \times 10^{-12} e^{-240/T}$	B
21	All	$OH + OH \xrightarrow{M} H_2O_2$	(P) 0.5	$6.90 \times 10^{-31}(300/T)^{0.8}$ 2.6×10^{-11}	A
22	All	$OH + HO_2 \longrightarrow H_2O + O_2$		$4.80 \times 10^{-11} e^{250/T}$	A
23	All	$OH + H_2O_2 \longrightarrow HO_2 + H_2O$		$2.90 \times 10^{-12} e^{-160/T}$	A
24	All	$OH + NO \xrightarrow{M} HONO$	(P) 0.90	$7.40 \times 10^{-31}(300/T)^{2.4}$ 4.50×10^{-11}	A
25	All	$OH + NO_2 \xrightarrow{M} HNO_3$	(P) 0.43	$2.60 \times 10^{-30}(300/T)^{2.9}$ $6.70 \times 10^{-11}(300/T)^{0.6}$	A
26	All	$OH + NO_3 \longrightarrow HO_2 + NO_2$		2.00×10^{-11}	A
27	All	$OH + HONO \longrightarrow H_2O + NO_2$		$1.80 \times 10^{-11} e^{-390/T}$	A
28	All	$OH + HNO_3 \longrightarrow H_2O + NO_3$		[c]	A
29	All	$OH + HO_2NO_2 \longrightarrow H_2O + NO_2 + O_2$		$1.50 \times 10^{-12} e^{360/T}$	A
30	All	$OH + CO \longrightarrow HO_2 + CO_2$		[d]	A
31	All	$HO_2 + O \longrightarrow OH + O_2$		$2.70 \times 10^{-11} e^{224/T}$	A
32	All	$HO_2 + O_3 \longrightarrow OH + 2O_2$		$1.40 \times 10^{-14} e^{-600/T}$	A
33	All	$HO_2 + HO_2 \longrightarrow H_2O_2 + O_2$		[e]	F
34	All	$HO_2 + NO \longrightarrow OH + NO_2$		$3.70 \times 10^{-12} e^{240/T}$	A
35	All	$HO_2 + NO_2 \xrightarrow{M} HO_2NO_2$	(P) 0.60	$1.80 \times 10^{-31}(300/T)^{3.2}$ $4.70 \times 10^{-12}(300/T)^{1.4}$	A
36	All	$HO_2 + NO_3 \longrightarrow HNO_3 + O_2$		4.00×10^{-12}	A
37	All	$H_2O_2 + O \longrightarrow OH + HO_2$		$1.40 \times 10^{-12} e^{-2000/T}$	A
38	All	$NO + O \xrightarrow{M} NO_2$	(P) 0.85	$1.00 \times 10^{-31}(300/T)^{1.6}$ $3.00 \times 10^{-11}(300/T)^{-0.3}$	A
39	All	$NO + O_3 \longrightarrow NO_2 + O_2$		$1.80 \times 10^{-12} e^{-1370/T}$	A
40	All	$NO_2 + O \longrightarrow NO + O_2$		$6.50 \times 10^{-12} e^{120/T}$	A
41	All	$NO_2 + O \xrightarrow{M} NO_3$	(P) 0.80	$9.00 \times 10^{-32}(300/T)^{2.0}$ 2.20×10^{-11}	A
42	All	$NO_2 + O_3 \longrightarrow NO_3 + O_2$		$1.20 \times 10^{-13} e^{-2450/T}$	A
43	All	$NO_3 + O \longrightarrow NO_2 + O_2$		1.70×10^{-11}	A
44	All	$NO_3 + NO \longrightarrow 2NO_2$		$1.80 \times 10^{-11} e^{110/T}$	A
45	All	$NO_3 + NO_2 \xrightarrow{M} N_2O_5$	(P) 0.33	$2.70 \times 10^{-30}(300/T)^{3.4}$ $2.00 \times 10^{-12}(300/T)^{-0.2}$	A
46	All	$N_2O_5 \xrightarrow{M} NO_3 + NO_2$	(P) 0.33	$1.00 \times 10^{-3}(300/T)^{3.5}$ $\times e^{-11000/T}$ $9.70 \times 10^{14}(300/T)^{-0.1}$ $\times e^{-11080/T}$	A
47	All	$N_2O_5 + H_2O \longrightarrow 2HNO_3$		2.00×10^{-21}	A
48	All	$HO_2NO_2 \xrightarrow{M} HO_2 + NO_2$	(P) 0.60	$5.00 \times 10^{-6} e^{-10000/T}$ $2.60 \times 10^{15} e^{-10900/T}$	A

(*cont.*)

Table B.4 (*cont.*)

No.	Region	Kinetic Reaction	F_c[a]	Rate coefficient (s^{-1}, $cm^3\ s^{-1}$, or $cm^6\ s^{-1}$)	Ref.[b]
		Organic Chemistry			
		Alkane, alkene, and aldehyde chemistry			
49	S	$CH_4 + O(^1D) \longrightarrow CH_3O_2 + OH$		1.40×10^{-10}	A
50	S	$CH_4 + O(^1D) \longrightarrow HCHO + H_2$		1.50×10^{-11}	A
51	All	$CH_4 + OH \longrightarrow CH_3O_2 + H_2O$		$2.30 \times 10^{-12}e^{-1765/T}$	A
52	All	$CH_3O + O_2 \longrightarrow HCHO + HO_2$		$7.20 \times 10^{-14}e^{-1080/T}$	A
53	U/T	$CH_3O + NO \longrightarrow HCHO + HO_2 + NO$		$4.00 \times 10^{-12}(300/T)^{0.7}$	A
54	U/T	$CH_3O + NO \xrightarrow{M} CH_3ONO$	(P) 0.60	$1.60 \times 10^{-29}(300/T)^{3.5}$ $3.60 \times 10^{-11}(300/T)^{0.6}$	A
55	U/T	$CH_3O + NO_2 \xrightarrow{M} CH_3ONO_2$	(P) 0.44	$2.80 \times 10^{-29}(300/T)^{4.5}$ 2.00×10^{-11}	A
56	U/T	$CH_3ONO_2 + OH \longrightarrow HCHO + NO_2 + H_2O$		$1.00 \times 10^{-14}e^{1060/T}$	A
57	All	$CH_3O_2 + HO_2 \longrightarrow CH_3OOH + O_2$		$3.80 \times 10^{-13}e^{780/T}$	A
58	All	$CH_3O_2 + NO \longrightarrow CH_3O + NO_2$		$4.20 \times 10^{-12}e^{180/T}$	A
59	U/T	$CH_3O_2 + NO_2 \xrightarrow{M} CH_3O_2NO_2$	(P) 0.36	$2.50 \times 10^{-30}(300/T)^{5.5}$ 7.50×10^{-12}	A
60	U/T	$CH_3O_2NO_2 \xrightarrow{M} CH_3O_2 + NO_2$	(P) 0.36	$9.00 \times 10^{-5}e^{-9690/T}$ $1.10 \times 10^{16}e^{-10560/T}$	A
61	U/T	$CH_3O_2 + CH_3O_2 \longrightarrow 2CH_3O + O_2$		$5.90 \times 10^{-13}e^{-509/T}$	A
62	U/T	$CH_3O_2 + CH_3O_2 \longrightarrow HCHO + CH_3OH$		$7.04 \times 10^{-14}e^{365/T}$	A
63	U/T	$CH_3O_2 + CH_3C(O)OO \longrightarrow CH_3O_2 + CH_3O + CO_2$		$5.10 \times 10^{-12}e^{272/T}$	A
64	All	$CH_3OOH + OH \longrightarrow CH_3O_2 + H_2O$		$1.90 \times 10^{-12}e^{190/T}$	A
65	U/T	$C_2H_6 + OH \longrightarrow C_2H_5O_2 + H_2O$		$7.90 \times 10^{-12}e^{-1030/T}$	A
66	U/T	$C_2H_5O_2 + NO \longrightarrow C_2H_5O + NO_2$		8.70×10^{-12}	A
67	U/T	$C_2H_5O_2 + NO_2 \xrightarrow{M} C_2H_5O_2NO$	(P) 0.31	$1.30 \times 10^{-29}(300/T)^{6.2}$ 8.80×10^{-12}	A
68	U/T	$C_2H_5O_2NO_2 \xrightarrow{M} C_2H_5O_2 + NO_2$	(P) 0.31	$4.80 \times 10^{-4}e^{-9285/T}$ $8.80 \times 10^{15}e^{-10440/T}$	A
69	U/T	$C_2H_5O_2 + HO_2 \longrightarrow ROOH + O_2$		$2.70 \times 10^{-13}e^{1000/T}$	A
70	U/T	$C_2H_5O + O_2 \longrightarrow CH_3CHO + HO_2$		$6.00 \times 10^{-14}e^{-550/T}$	A
71	U/T	$C_2H_5O + NO \longrightarrow C_2H_5ONO$		4.40×10^{-11}	A
72	U/T	$C_2H_5O + NO \longrightarrow CH_3CHO + HO_2 + NO$		1.30×10^{-11}	A
73	U/T	$C_2H_5O + NO_2 \longrightarrow C_2H_5ONO_2$		2.80×10^{-11}	A
74	U/T	$C_3H_8 + OH \longrightarrow C_3H_7O_2 + H_2O$		$8.00 \times 10^{-12}e^{-590/T}$	A
75	U/T	$C_3H_7O_2 + NO \longrightarrow C_3H_7O + NO_2$		4.80×10^{-12}	A
76	U/T	$C_3H_7O + O_2 \longrightarrow CH_3COCH_3 + HO_2$		$1.50 \times 10^{-14}e^{-200/T}$	A
77	U/T	$C_3H_7O + NO \longrightarrow C_3H_7ONO$		3.40×10^{-11}	A
78	U/T	$C_3H_7O + NO \longrightarrow CH_3COCH_3 + HO_2 + NO$		6.50×10^{-12}	A
79	U/T	$C_3H_7O + NO_2 \longrightarrow C_3H_7ONO_2$		3.50×10^{-11}	A
80	U/T	$C_2H_4 + OH \xrightarrow{M} HOC_2H_4O_2$	(P) 0.70	$7.00 \times 10^{-29}\ (300/T)^{3.1}$ 9.00×10^{-12}	A
81	U/T	$HOC_2H_4O_2 + NO \longrightarrow NO_2 + 2HCHO + H$		6.93×10^{-12}	C
82	U/T	$HOC_2H_4O_2 + NO \longrightarrow NO_2 + CH_3CHO + OH$		2.07×10^{-12}	C
83	U/T	$C_2H_4 + O_3 \longrightarrow HCHO + H_2COO$		$3.40 \times 10^{-15}e^{-2580/T}$	A
84	U/T	$C_2H_4 + O_3 \longrightarrow HCHO + HCOOH^*$		$5.70 \times 10^{-15}e^{-2580/T}$	A
85	U/T	$H_2COO + NO \longrightarrow NO_2 + HCHO$		7.00×10^{-12}	C

Table B.4 (*cont.*)

No.	Region	Kinetic reaction	F_c[a]	Rate coefficient (s^{-1}, $cm^3\ s^{-1}$, or $cm^6\ s^{-1}$)	Ref.[b]
		Organic chemistry			
		Alkane, alkene, and aldehyde chemistry			
86	U/T	$H_2COO + H_2O \longrightarrow HCOOH + H_2O$		4.00×10^{-16}	C
87	U/T	$H_2COO + HCHO \longrightarrow OZD$		2.00×10^{-12}	C
88	U/T	$H_2COO + CH_3CHO \longrightarrow OZD$		2.00×10^{-12}	C
89	U/T	$HCOOH^* \longrightarrow CO_2 + H_2$		0.21	A
90	U/T	$HCOOH^* \longrightarrow CO + H_2O$		0.60	A
91	U/T	$HCOOH^* \longrightarrow OH + HO_2 + CO$		0.19	A
92	U/T	$C_3H_6 + OH \xrightarrow{M} HOC_3H_6O_2$	(P) 0.50	$8.00 \times 10^{-27}(300/T)^{3.5}$ 3.00×10^{-11}	A
93	U/T	$HOC_3H_6O_2 + NO \longrightarrow NO_2 + CH_3CHO + HCHO + HO_2$		6.00×10^{-12}	C
94	U/T	$C_3H_6 + O_3 \longrightarrow HCHO + CH_3HCOO$		$4.10 \times 10^{-16}e^{-1880/T}$	A
95	U/T	$C_3H_6 + O_3 \longrightarrow HCHO + CH_3HCOO^*$		$2.34 \times 10^{-15}e^{-1880/T}$	A
96	U/T	$C_3H_6 + O_3 \longrightarrow CH_3CHO + H_2COO$		$1.03 \times 10^{-15}e^{-1880/T}$	A
97	U/T	$C_3H_6 + O_3 \longrightarrow CH_3CHO + H_2COO^*$		$1.72 \times 10^{-15}e^{-1880/T}$	A
98	U/T	$CH_3HCOO + NO \longrightarrow NO_2 + CH_3CHO$		7.00×10^{-12}	C
99	U/T	$CH_3HCOO + H_2O \longrightarrow CH_3COOH + H_2O$		4.00×10^{-16}	C
100	U/T	$CH_3HCOO + HCHO \longrightarrow OZD$		2.00×10^{-12}	C
101	U/T	$CH_3HCOO + CH_3CHO \longrightarrow OZD$		2.00×10^{-12}	C
102	U/T	$CH_3COOH^* \longrightarrow CH_4 + CO_2$		0.16	A
103	U/T	$CH_3COOH^* \longrightarrow CH_3O_2 + CO + OH$		0.64	A
104	U/T	$CH_3COOH^* \longrightarrow CH_3O + CO + HO_2$		0.20	A
105	All	$HCHO + OH \longrightarrow HO_2 + CO + H_2O$		$8.60 \times 10^{-12}e^{20/T}$	A
106	All	$HCHO + O \longrightarrow OH + HO_2 + CO$		$3.40 \times 10^{-11}e^{-1600/T}$	B
107	U/T	$HCHO + NO_3 \longrightarrow HNO_3 + HO_2 + CO$		5.80×10^{-16}	A
108	U/T	$HCHO + HO_2 \longrightarrow HOCH_2O_2$		$9.70 \times 10^{-15}e^{625/T}$	A
109	U/T	$HOCH_2O_2 \longrightarrow HO_2 + HCHO$		$2.40 \times 10^{12}e^{-7000/T}$	A
110	U/T	$HOCH_2O_2 + HO_2 \longrightarrow ROOH$		$5.60 \times 10^{-15}e^{2300/T}$	A
111	U/T	$HOCH_2O_2 + NO \longrightarrow NO_2 + HO_2 + HCOOH$		7.00×10^{-12}	C
112	U/T	$CH_3CHO + O \longrightarrow CH_3C(O)OO + OH$		$1.80 \times 10^{-11}e^{-1100/T}$	B
113	U/T	$CH_3CHO + OH \longrightarrow CH_3C(O)OO + H_2O$		$5.60 \times 10^{-12}e^{310/T}$	A
114	U/T	$CH_3CHO + NO_3 \longrightarrow CH_3C(O)OO + HNO_3$		$1.40 \times 10^{-12}e^{-1860/T}$	A
115	U/T	$CH_3C(O)OO + HO_2 \longrightarrow ROOH + O_2$		$1.14 \times 10^{-13}e^{1040/T}$	A
116	U/T	$CH_3C(O)OO + HO_2 \longrightarrow CH_3O_2 + OH + CO_2$		$3.16 \times 10^{-13}e^{1040/T}$	A
117	U/T	$CH_3C(O)OO + NO \longrightarrow NO_2 + CH_3O_2 + CO_2$		2.00×10^{-11}	A
118	U/T	$CH_3C(O)OO + NO_2 \xrightarrow{M} CH_3C(O)OONO_2$	(P) 0.30	$2.70 \times 10^{-28}(300/T)^{7.1}$ $1.20 \times 10^{-11}(300/T)^{0.9}$	A
119	U/T	$CH_3C(O)OO + CH_3C(O)OO \longrightarrow 2CH_3O_2 + O_2$		$2.80 \times 10^{-12}e^{530/T}$	A
120	U/T	$CH_3C(O)OONO_2 \xrightarrow{M} CH_3C(O)OO + NO_2$	(P) 0.30	$4.90 \times 10^{-3}e^{-12100/T}$ $5.40 \times 10^{16}e^{-13830/T}$	A
121	U/T	$CH_3COCH_3 + OH \longrightarrow CH_3COCH_2OO + H_2O$		$2.80 \times 10^{-12}e^{-760/T}$	A
122	U/T	$CH_3COCH_2OO + NO \longrightarrow CH_3C(O)OO + HCHO + NO_2$		8.10×10^{-12}	C

(*cont.*)

Table B.4 (*cont.*)

No.	Region	Kinetic reaction	F_c[a]	Rate coefficient (s^{-1}, $cm^3\ s^{-1}$, or $cm^6\ s^{-1}$)	Ref.[b]
		Organic chemistry			
		Alkane, alkene, and aldehyde chemistry			
123	U/T	$CH_3OH + OH \longrightarrow HCHO + HO_2 + H_2O$		$2.63 \times 10^{-12}e^{-360/T}$	A
124	U/T	$CH_3OH + OH \longrightarrow CH_3O + H_2O$		$4.66 \times 10^{-13}e^{-360/T}$	A
125	U/T	$C_2H_5OH + OH \longrightarrow CH_3CHO + HO_2 + H_2O$		$3.69 \times 10^{-12}e^{-70/T}$	A
126	U/T	$PAR + OH \longrightarrow RO_2 + H_2O$		9.20×10^{-14}	C
127	U/T	$PAR + OH \longrightarrow RO_2R + H_2O$		7.20×10^{-13}	C
128	U/T	$RO_2 + NO \longrightarrow NO_2 + HO_2 + CH_3CHO + XOP$		7.70×10^{-12}	C
129	U/T	$RO_2 + NO \longrightarrow NTR$		$4.40 \times 10^{-11}e^{-1400/T}$	C
130	U/T	$RO_2R + NO \longrightarrow NO_2 + ROR$		7.00×10^{-12}	C
131	U/T	$RO_2R + NO \longrightarrow NTR$		$1.20 \times 10^{-10}e^{-1400/T}$	C
132	U/T	$ROR + NO_2 \longrightarrow NTR$		1.50×10^{-11}	C
133	U/T	$ROR \longrightarrow KET + HO_2$		1.60×10^{3}	C
134	U/T	$ROR \longrightarrow KET + DOP$		$2.10 \times 10^{14}e^{-8000/T}$	C
135	U/T	$ROR \longrightarrow CH_3CHO + DOP + XOP$		$4.00 \times 10^{14}e^{-8000/T}$	C
136	U/T	$ROR \longrightarrow CH_3COCH_3 + DOP + 2XOP$		$4.40 \times 10^{14}e^{-8000/T}$	C
137	U/T	$XOP + PAR \longrightarrow$		6.80×10^{-12}	C
138	U/T	$DOP + PAR \longrightarrow RO_2$		5.10×10^{-12}	C
139	U/T	$DOP + PAR \longrightarrow AO_2 + 2XOP$		1.50×10^{-12}	C
140	U/T	$DOP + PAR \longrightarrow RO_2R$		1.70×10^{-13}	C
141	U/T	$DOP + KET \longrightarrow CH_3C(O)OO + XOP$		6.80×10^{-12}	C
142	U/T	$AO_2 + NO \longrightarrow NO_2 + CH_3COCH_3 + HO_2$		8.10×10^{-12}	C
143	U/T	$OLE + O \longrightarrow 2PAR$		$4.10 \times 10^{-12}e^{-324/T}$	C
144	U/T	$OLE + O \longrightarrow CH_3CHO$		$4.10 \times 10^{-12}e^{-324/T}$	C
145	U/T	$OLE + O \longrightarrow HO_2 + CO + RO_2$		$1.20 \times 10^{-12}e^{-324/T}$	C
146	U/T	$OLE + O \longrightarrow RO_2 + XOP + CO + HCHO + OH$		$2.40 \times 10^{-12}e^{-324/T}$	C
147	U/T	$OLE + OH \longrightarrow CH_3O_2 + CH_3CHO + XOP$		$5.20 \times 10^{-12}e^{504/T}$	C
148	U/T	$OLE + O_3 \longrightarrow CH_3CHO + H_2COO + XOP$		$2.80 \times 10^{-15}e^{-2105/T}$	C
149	U/T	$OLE + O_3 \longrightarrow HCHO + CH_3HCOO + XOP$		$2.80 \times 10^{-15}e^{-2105/T}$	C
150	U/T	$OLE + O_3 \longrightarrow CH_3CHO + HCOOH^* + XOP$		$4.30 \times 10^{-15}e^{-2105/T}$	C
151	U/T	$OLE + O_3 \longrightarrow HCHO + CH_3COOH^* + XOP$		$4.30 \times 10^{-15}e^{-2105/T}$	C
152	U/T	$OLE + NO_3 \longrightarrow PNO_2$		7.70×10^{-15}	C
153	U/T	$PNO_2 + NO \longrightarrow DNIT$		6.80×10^{-13}	C
154	U/T	$PNO_2 + NO \longrightarrow HCHO + CH_3CHO + XOP + 2NO_2$		6.80×10^{-12}	C
		Aromatic chemistry			
155	U	$TOL + OH \longrightarrow BO_2 + H_2O$		$1.70 \times 10^{-13}e^{322/T}$	C
156	U	$TOL + OH \longrightarrow CRES + HO_2$		$7.60 \times 10^{-13}e^{322/T}$	C
157	U	$TOL + OH \longrightarrow TO_2$		$1.20 \times 10^{-12}e^{322/T}$	C
158	U	$BO_2 + NO \longrightarrow NO_2 + BZA + HO_2$		8.10×10^{-12}	C
159	U	$BZA + OH \longrightarrow BZO_2 + H_2O$		1.30×10^{-11}	C
160	U	$BZO_2 + NO \longrightarrow NO_2 + PHO_2 + CO_2$		2.50×10^{-12}	C
161	U	$BZO_2 + NO_2 \longrightarrow PBZN$		8.40×10^{-12}	E
162	U	$PBZN \longrightarrow BZO_2 + NO_2$		$1.60 \times 10^{15}e^{-13033/T}$	E
163	U	$PHO_2 + NO \longrightarrow NO_2 + PHO$		8.10×10^{-12}	C
164	U	$PHO + NO_2 \longrightarrow NPHN$		$1.30 \times 10^{-11}e^{300/T}$	E
165	U	$CRES + OH \longrightarrow CRO + H_2O$		1.60×10^{-11}	C
166	U	$CRES + OH \longrightarrow CRO_2 + H_2O$		2.50×10^{-11}	C

Table B.4 (*cont.*)

No.	Region	Kinetic reaction	Rate coefficient $F_c{}^a$ (s^{-1}, $cm^3\ s^{-1}$, or $cm^6\ s^{-1}$)	Ref.[b]
		Organic chemistry		
		Aromatic chemistry		
167	U	$CRES + NO_3 \longrightarrow CRO + HNO_3$	2.20×10^{-11}	C
168	U	$CRO + NO_2 \longrightarrow NCRE$	1.40×10^{-11}	C
169	U	$CRO_2 + NO \longrightarrow NO_2 + OPEN + HO_2$	4.00×10^{-12}	C
170	U	$CRO_2 + NO \longrightarrow NO_2 + ACID + HO_2$	4.00×10^{-12}	C
171	U	$TO_2 + NO \longrightarrow NO_2 + OPEN + HO_2$	7.30×10^{-12}	C
172	U	$TO_2 + NO \longrightarrow NTR$	8.10×10^{-13}	C
173	U	$TO_2 \longrightarrow HO_2 + CRES$	4.20	C
174	U	$XYL + OH \longrightarrow CRES + PAR + HO_2$	$3.32 \times 10^{-12} e^{116/T}$	C
175	U	$XYL + OH \longrightarrow XLO_2 + H_2O$	$1.70 \times 10^{-12} e^{116/T}$	C
176	U	$XYL + OH \longrightarrow TO_2$	$5.00 \times 10^{-12} e^{116/T}$	C
177	U	$XYL + OH \longrightarrow XINT$	$6.60 \times 10^{-12} e^{116/T}$	C
178	U	$XLO_2 + NO \longrightarrow NO_2 + HO_2 + BZA + PAR$	8.10×10^{-12}	C
179	U	$XINT + NO \longrightarrow NO_2 + HO_2 + 2CH_3COCHO + PAR$	8.10×10^{-12}	C
180	U	$CH_3COCHO + OH \longrightarrow MGPX + H_2O$	1.50×10^{-11}	A
181	U	$MGPX + NO \longrightarrow NO_2 + CH_3C(O)OO + CO_2$	8.10×10^{-12}	C
182	U	$OPEN + OH \longrightarrow OPPX + CH_3C(O)OO + HO_2 + CO$	3.00×10^{-11}	C
183	U	$OPEN + O_3 \longrightarrow CH_3CHO + MGPX + HCHO + CO$	$1.60 \times 10^{-18} e^{-500/T}$	C
184	U	$OPEN + O_3 \longrightarrow HCHO + CO + OH + 2HO_2$	$4.30 \times 10^{-18} e^{-500/T}$	C
185	U	$OPEN + O_3 \longrightarrow CH_3COCHO$	$1.10 \times 10^{-17} e^{-500/T}$	C
186	U	$OPEN + O_3 \longrightarrow CH_3C(O)OO + HCHO + HO_2 + CO$	$3.20 \times 10^{-17} e^{-500/T}$	C
187	U	$OPEN + O_3 \longrightarrow$	$5.40 \times 10^{-18} e^{-500/T}$	C
188	U	$OPPX + NO \longrightarrow NO_2 + HCHO + HO_2 + CO$	8.10×10^{-12}	C
		Terpene chemistry		
189	U/T	$ISOP + OH \longrightarrow ISOH$	$2.55 \times 10^{-11} e^{410/T}$	D
190	U/T	$ISOP + O_3 \longrightarrow 0.67\ MACR + 0.26\ MVK + 0.3\ OH + 0.07\ PAR + 0.07\ OLE + 0.07\ H_2COO + 0.8\ HCHO + 0.06\ HO_2 + 0.15\ CO_2 + 0.05\ CO$	$1.23 \times 10^{-14} e^{-2013/T}$	D
191	U/T	$ISOP + O \longrightarrow 0.22\ MACR + 0.63\ MVK + 0.08\ ISOH$	3.30×10^{-11}	D
192	U/T	$ISOP + NO_3 \longrightarrow ISNT$	7.80×10^{-13}	D
193	U/T	$ISOH + NO \longrightarrow 0.364\ MACR + 0.477\ MVK + 0.840\ HCHO + 0.08\ ISNI1 + 0.08\ ISNI2 + 0.886\ HO_2 + 0.840\ NO_2$	$1.22 \times 10^{-11} e^{-180/T}$	D
194	U/T	$ISNT + NO \longrightarrow 1.1\ NO_2 + 0.8\ HO_2 + 0.80\ ISNI1 + 0.1\ MACR + 0.15\ HCHO + 0.05\ MVK + 0.05\ DISN$	$1.39 \times 10^{-11} e^{-180/T}$	D
195	U/T	$ISNI1 + OH \longrightarrow ISNIR$	3.35×10^{-11}	D
196	U/T	$ISNI2 + OH \longrightarrow ISNIR$	1.88×10^{-11}	D
197	U/T	$ISNIR + NO \longrightarrow 0.05\ DISN + 0.05\ HO_2 + 1.9\ NO_2 + 0.95\ CH_3CHO + 0.95\ CH_3COCH_3$	$1.39 \times 10^{-11} e^{-180/T}$	D
198	U/T	$ISNI1 + O_3 \longrightarrow 0.2\ O + 0.08\ OH + 0.5\ HCHO + 0.5\ IALD1 + 0.5\ ISNI2 + 0.5\ NO_2$	5.00×10^{-18}	D
199	U/T	$ISOH + ISOH \longrightarrow 0.6\ MACR + 0.6\ MVK + 1.2\ HCHO + 1.2\ HO_2$	2.00×10^{-13}	D
200	U/T	$ISOH + HO_2 \longrightarrow IPRX$	$6.15 \times 10^{-11} e^{-900/T}$	D

(*cont.*)

Table B.4 (*cont.*)

No.	Region	Kinetic reaction	F_c [a]	Rate coefficient (s^{-1}, $cm^3\ s^{-1}$, or $cm^6\ s^{-1}$)	Ref.[b]
		Organic chemistry			
		Terpene chemistry			
201	U/T	$IPRX + OH \longrightarrow ISOH$		2.00×10^{-11}	D
202	U/T	$IPRX + O_3 \longrightarrow 0.7\ HCHO$		8.00×10^{-18}	D
203	U/T	$MACR + O_3 \longrightarrow 0.8\ CH_3COCHO + 0.7\ HCHO + 0.2\ O + 0.09\ H_2COO + 0.2\ CO + 0.275\ HO_2 + 0.215\ OH + 0.16\ CO_2 + 0.15\ CH_2CCH_3CHOO$		$5.32 \times 10^{-15}e^{-2520/T}$	D
204	U/T	$MVK + O_3 \longrightarrow 0.82\ CH_3COCHO + 0.8\ HCHO + 0.2\ O + 0.11\ H_2COO + 0.05\ CO + 0.06\ HO_2 + 0.08\ OH + 0.04\ CH_3CHO + 0.07\ OOCCH_3CHCH_2$		$4.32 \times 10^{-15}e^{-2016/T}$	D
205	U/T	$MACR + OH \longrightarrow 0.42\ MAC1 + 0.08\ MAC2 + 0.5\ CH_2CCH_3C(O)OO$		$1.86 \times 10^{-11}e^{175/T}$	D
206	U/T	$MVK + OH \longrightarrow 0.28\ MV1 + 0.72\ MV2$		$4.11 \times 10^{-12}e^{453/T}$	D
207	U/T	$MAC1 + NO \longrightarrow 0.95\ HO_2 + 0.95\ CO + 0.95\ CH_3COCH_3 + 0.95\ NO_2 + 0.05\ ISNI2$		$1.39 \times 10^{-11}e^{-180/T}$	D
208	U/T	$MAC2 + NO \longrightarrow 0.95\ HO_2 + 0.95\ HCHO + 0.95\ CH_3COCHO + 0.95\ NO_2 + 0.05\ ISNI2$		$1.39 \times 10^{-11}e^{-180/T}$	D
209	U/T	$MV1 + NO \longrightarrow 0.95\ CH_3COCHO + 0.95\ HCHO + 0.05\ ISNI2 + 0.95\ NO_2 + 0.95\ HO_2$		$1.39 \times 10^{-11}e^{-180/T}$	D
210	U/T	$MV2 + NO \longrightarrow 0.95\ CH_3CHO + 0.95\ CH_3C(O)OO + 0.05\ ISNI2 + 0.95\ NO_2$		$1.39 \times 10^{-11}e^{-180/T}$	D
211	U/T	$MV1 + HO_2 \longrightarrow ROOH$		$6.15 \times 10^{-11}e^{-900/T}$	D
212	U/T	$MV2 + HO_2 \longrightarrow ROOH$		$6.15 \times 10^{-11}e^{-900/T}$	D
213	U/T	$MAC1 + HO_2 \longrightarrow ROOH$		$6.15 \times 10^{-11}e^{-900/T}$	D
214	U/T	$MAC2 + HO_2 \longrightarrow ROOH$		$6.15 \times 10^{-11}e^{-900/T}$	D
215	U/T	$CH_2CCH_3C(O)OO + NO_2 \longrightarrow MPAN$		8.40×10^{-12}	D
216	U/T	$MPAN \longrightarrow CH_2CCH_3C(O)OO + NO_2$		$1.58 \times 10^{16}e^{-13507/T}$	D
217	U/T	$CH_2CCH_3C(O)OO + NO \longrightarrow C_2H_4 + CH_3O_2 + NO_2 + CO_2$		1.40×10^{-11}	D
		Sulfur chemistry			
218	All	$SO_2 + OH \xrightarrow{M} HSO_3$	(P) 0.45	$4.00 \times 10^{-31}(300/T)^{3.3}$ 2.00×10^{-12}	A
219	All	$HSO_3 + O_2 \longrightarrow SO_3 + HO_2$		$1.30 \times 10^{-12}e^{-330/T}$	A
220	All	$SO_3 + H_2O \longrightarrow H_2SO_4$		6.00×10^{-15}	A
221	T	$CH_3SCH_3 + OH \longrightarrow CH_3SCH_2 + H_2O$		$1.13 \times 10^{-11}e^{-254/T}$	A
222	T	$CH_3SCH_3 + OH \longrightarrow CH_3S(OH)CH_3$		*f*	A
223	T	$CH_3SCH_2 + O_2 \longrightarrow CH_3SCH_2O_2$		7.30×10^{-13}	N
224	T	$CH_3SCH_2O_2 + NO \longrightarrow CH_3SCH_2O + NO_2$		8.00×10^{-12}	N
225	T	$CH_3SCH_2O \longrightarrow CH_3S + HCHO$		1.00×10^{1}	N

Table B.4 (*cont.*)

No.	Region	Kinetic reaction	F_c [a]	Rate coefficient (s^{-1}, $cm^3\ s^{-1}$, or $cm^6\ s^{-1}$)	Ref.[b]
		Sulfur chemistry			
226	T	$CH_3S + O_2 \longrightarrow CH_3SOO^*$		2.5×10^{-18}	A
227	T	$CH_3SOO^* + NO \longrightarrow CH_3SO + NO_2$		1.4×10^{-11}	N
228	T	$CH_3SOO^* \longrightarrow CH_3S + O_2$		6.0×10^{2}	N
229	T	$CH_3SO \longrightarrow CH_3O_2 + SO$		5.0×10^{-5}	N
230	T	$CH_3SO + O_3 \longrightarrow CH_3SO_2 + O_2$		2.0×10^{-12}	N
231	T	$CH_3SO_2 \longrightarrow CH_3O_2 + SO_2$		1.1×10^{1}	N
232	T	$CH_3S(OH)CH_3 \longrightarrow CH_3SOH + CH_3O_2$		5.0×10^{5}	N
233	T	$CH_3S(OH)CH_3 + OH \longrightarrow CH_3S(O)_2CH_3 + 2HO_2$		5.8×10^{-11}	N
234	T	$CH_3SOH + OH \longrightarrow CH_3SO + H_2O$		1.1×10^{-10}	N
235	T	$CH_3SSCH_3 + OH \longrightarrow CH_3SOH + CH_3S$		$7.00 \times 10^{-11} e^{350/T}$	A
236	T	$CH_3SH + OH \longrightarrow CH_3S + H_2O$		$9.90 \times 10^{-12} e^{356/T}$	A
237	T	$H_2S + OH \longrightarrow HS + H_2O$		$6.30 \times 10^{-12} e^{-80/T}$	A
238	T	$HS + O_2 \longrightarrow SO + OH$		4.00×10^{-19}	A
239	T/S	$SO + O_2 \longrightarrow SO_2 + O$		$1.60 \times 10^{-13} e^{-2280/T}$	A
240	T/S	$OCS + OH \longrightarrow HS + CO_2$		$1.10 \times 10^{-13} e^{-1200/T}$	A
241	T/S	$CS_2 + OH \longrightarrow HS + OCS$		2.00×10^{-15}	A
242	T/S	$CS + O_2 \longrightarrow OCS + O$		2.90×10^{-19}	A
243	T/S	$S + O_2 \longrightarrow SO + O$		2.10×10^{-12}	A
		Chlorine chemistry			
244	S	$Cl + O_2 + N_2 \longrightarrow ClOO + N_2$		$1.40 \times 10^{-33}(300/T)^{3.9}$	A
245	S	$Cl + O_3 \longrightarrow ClO + O_2$		$2.90 \times 10^{-11} e^{-260/T}$	A
246	S	$Cl + H_2 \longrightarrow HCl + H$		$3.70 \times 10^{-11} e^{-2300/T}$	A
247	S	$Cl + HO_2 \longrightarrow HCl + O_2$		$1.80 \times 10^{-11} e^{170/T}$	A
248	S	$Cl + H_2O_2 \longrightarrow HCl + HO_2$		$1.10 \times 10^{-11} e^{-980/T}$	A
249	S	$Cl + NO_2 \xrightarrow{M} ClNO_2$	(P) 0.60	$1.80 \times 10^{-31}(300/T)^{2.0}$ $1.00 \times 10^{-10}(300/T)^{1.0}$	B
250	S	$Cl + HNO_3 \longrightarrow HCl + NO_3$		2.00×10^{-16}	A
251	S	$Cl + NO_3 \longrightarrow ClO + NO_2$		2.40×10^{-11}	A
252	S	$Cl + CH_4 \longrightarrow HCl + CH_3O_2$		$9.60 \times 10^{-12} e^{-1350/T}$	A
253	S	$Cl + HCHO \longrightarrow HCl + CO + HO_2$		$8.20 \times 10^{-11} e^{-34/T}$	A
254	S	$Cl + HOCl \longrightarrow Cl_2 + OH$		$2.50 \times 10^{-12} e^{-130/T}$	B
255	S	$Cl + OClO \longrightarrow ClO + ClO$		$3.40 \times 10^{-11} e^{160/T}$	A
256	S	$Cl + ClOO \longrightarrow Cl_2 + O_2$		2.30×10^{-10}	B
257	S	$Cl + ClOO \longrightarrow ClO + ClO$		1.20×10^{-11}	B
258	S	$ClO + O \longrightarrow Cl + O_2$		$3.80 \times 10^{-11} e^{70/T}$	A
259	S	$ClO + O_3 \longrightarrow ClOO + O_2$		1.50×10^{-17}	A
260	S	$ClO + O_3 \longrightarrow OClO + O_2$		1.00×10^{-18}	A
261	S	$ClO + OH \longrightarrow Cl + HO_2$		$9.90 \times 10^{-12} e^{120/T}$	B
262	S	$ClO + OH \longrightarrow HCl + O_2$		$1.10 \times 10^{-12} e^{120/T}$	B
263	S	$ClO + HO_2 \longrightarrow HOCl + O_2$		$4.66 \times 10^{-13} e^{710/T}$	B
264	S	$ClO + HO_2 \longrightarrow HCl + O_3$		$1.44 \times 10^{-14} e^{710/T}$	B
265	S	$ClO + NO \longrightarrow Cl + NO_2$		$6.20 \times 10^{-12} e^{294/T}$	A
266	S	$ClO + NO_2 \xrightarrow{M} ClONO_2$	(P) 0.50	$1.60 \times 10^{-31}(300/T)^{3.4}$ $2.00 \times 10^{-11}(300/T)^{1.9}$	A
267	S	$ClO + ClO \longrightarrow Cl + ClOO$		$3.00 \times 10^{-11} e^{-2450/T}$	A

(*cont.*)

Table B.4 (*cont.*)

No.	Region	Kinetic reaction	F_c [a]	Rate coefficient (s^{-1}, $cm^3\ s^{-1}$, or $cm^6\ s^{-1}$)	Ref.[b]
		Chlorine chemistry			
268	S	$ClO + ClO \xrightarrow{M} Cl_2O_2$	(P) 0.60	$1.70 \times 10^{-32}(300/T)^{4.0}$ 5.40×10^{-12}	A
269	S	$HCl + OH \longrightarrow Cl + H_2O$		$2.40 \times 10^{-12}e^{-330/T}$	A
270	S	$ClONO_2 + O \longrightarrow Cl + NO_2$		$3.00 \times 10^{-12}e^{-800/T}$	A
271	S	$OClO + O \longrightarrow ClO + O_2$		$2.40 \times 10^{-12}e^{-960/T}$	A
272	S	$OClO + OH \longrightarrow HOCl + O_2$		$4.50 \times 10^{-13}e^{800/T}$	A
273	S	$OClO + NO \longrightarrow ClO + NO_2$		$2.50 \times 10^{-12}e^{-600/T}$	B
274	S	$ClOO + M \longrightarrow Cl + O_2 + M$		$2.80 \times 10^{-10}e^{-1820/T}$	A
275	S	$HOCl + O \longrightarrow ClO + OH$		$1.00 \times 10^{-11}e^{-1300/T}$	A
276	S	$HOCl + OH \longrightarrow ClO + H_2O$		$3.00 \times 10^{-12}e^{-500/T}$	A
277	S	$Cl_2O_2 \xrightarrow{M} ClO + ClO$	(P) 0.60	$1.00 \times 10^{-6}e^{-8000/T}$ $4.80 \times 10^{15}e^{-8820/T}$	A
		Bromine chemistry			
278	S	$Br + O_3 \longrightarrow BrO + O_2$		$1.70 \times 10^{-11}e^{-800/T}$	A
279	S	$Br + HO_2 \longrightarrow HBr + O_2$		$1.40 \times 10^{-11}e^{-590/T}$	A
280	S	$Br + H_2O_2 \longrightarrow HBr + HO_2$		$1.00 \times 10^{-11}e^{-3000/T}$	B
281	S	$Br + HCHO \longrightarrow HBr + CO + HO_2$		$1.70 \times 10^{-11}e^{-800/T}$	A
282	S	$BrO + O \longrightarrow Br + O_2$		$1.90 \times 10^{-11}e^{230/T}$	A
283	S	$BrO + OH \longrightarrow Br + HO_2$		7.5×10^{-11}	B
284	S	$BrO + HO_2 \longrightarrow HOBr + O_2$		$3.40 \times 10^{-12}e^{540/T}$	B
285	S	$BrO + NO \longrightarrow Br + NO_2$		$8.70 \times 10^{-12}e^{260/T}$	A
286	S	$BrO + NO_2 \xrightarrow{M} BrONO_2$	(P) 0.4	$4.70 \times 10^{-31}(300/T)^{3.1}$ $1.70 \times 10^{-11}(300/T)^{0.6}$	A
287	S	$BrO + ClO \longrightarrow Br + OClO$		$1.60 \times 10^{-12}e^{430/T}$	A
288	S	$BrO + ClO \longrightarrow Br + ClOO$		$2.90 \times 10^{-12}e^{220/T}$	A
289	S	$BrO + ClO \longrightarrow BrCl + O_2$		$5.80 \times 10^{-13}e^{170/T}$	A
290	S	$BrO + BrO \longrightarrow 2Br + O_2$		$2.40 \times 10^{-12}e^{40/T}$	B
291	S	$BrO + BrO \longrightarrow Br_2 + O_2$		$2.80 \times 10^{-14}e^{860/T}$	B
292	S	$BrO + O_3 \longrightarrow Br + 2O_2$		$1.00 \times 10^{-12}e^{-3200/T}$	B
293	S	$HBr + OH \longrightarrow Br + H_2O$		$1.10 \times 10^{-11}(298/T)^{0.8}$	A
294	S	$HOBr + O \longrightarrow BrO + OH$		$1.40 \times 10^{-10}e^{-430/T}$	H
295	S	$BrCl + O \longrightarrow BrO + Cl$		2.20×10^{-10}	I

Photoprocesses

	Region	Photolysis reaction	λ (μm)	J (25 km) (s^{-1})	J (0 km) (s^{-1})	Ref.[b]
296	S	$O_2 + h\nu \longrightarrow O + O$	<0.245	1.2×10^{-11}	4.36×10^{-26}	B
297	All	$O_3 + h\nu \longrightarrow O(^1D) + O_2$	<0.31	1.13×10^{-4}	5.08×10^{-5}	B
298	All	$O_3 + h\nu \longrightarrow O + O_2$	>0.31	4.95×10^{-4}	4.17×10^{-4}	B
299	S	$H_2O + h\nu \longrightarrow H + OH$	<0.21	6.61×10^{-11}	0	A
300	All	$H_2O_2 + h\nu \longrightarrow 2OH$	<0.355	1.18×10^{-5}	7.72×10^{-6}	A
301	All	$NO_2 + h\nu \longrightarrow NO + O$	<0.42	1.24×10^{-2}	8.82×10^{-3}	B
302	All	$NO_3 + h\nu \longrightarrow NO_2 + O$	0.41–0.67	3.04×10^{-1}	2.84×10^{-1}	A
303	All	$NO_3 + h\nu \longrightarrow NO + O_2$	0.59–0.63	2.65×10^{-2}	2.49×10^{-2}	A
304	S	$N_2O + h\nu \longrightarrow N_2 + O(^1D)$	<0.24	2.49×10^{-8}	1.81×10^{-23}	B
305	All	$N_2O_5 + h\nu \longrightarrow NO_2 + NO_3$	<0.385	4.73×10^{-5}	5.04×10^{-5}	B

Table B.4 (*cont.*)

Photoprocesses						
No.	Region	Photolysis reaction	λ (μm)	J (25 km) (s^{-1})	J (0 km) (s^{-1})	Ref.[b]
306	All	$HONO + h\nu \longrightarrow OH + NO$	<0.40	2.75×10^{-3}	1.94×10^{-3}	A
307	All	$HNO_3 + h\nu \longrightarrow OH + NO_2$	<0.335	5.61×10^{-6}	8.23×10^{-7}	A
308	All	$HO_2NO_2 + h\nu \longrightarrow HO_2 + NO_2$	<0.33	1.14×10^{-5}	3.68×10^{-6}	A
309	All	$HO_2NO_2 + h\nu \longrightarrow OH + NO_3$	<0.33	5.67×10^{-6}	1.84×10^{-6}	A
310	All	$HCHO + h\nu \longrightarrow 2HO_2 + CO$	<0.334	6.17×10^{-5}	3.29×10^{-5}	A
311	All	$HCHO + h\nu \longrightarrow CO + H_2$	<0.37	7.64×10^{-5}	4.40×10^{-5}	A
312	All	$CH_3OOH + h\nu \longrightarrow CH_3O + OH$	<0.36	9.92×10^{-6}	5.73×10^{-6}	A
313	U/T	$CH_3CHO + h\nu \longrightarrow CH_3O_2 + HO_2 + CO$	<0.325	1.30×10^{-5}	6.35×10^{-6}	A
314	U/T	$CH_3ONO_2 + h\nu \longrightarrow CH_3O + NO_2$	<0.33	2.54×10^{-6}	1.25×10^{-6}	A
315	U/T	$CH_3O_2NO_2 + h\nu \longrightarrow CH_3O_2 + NO_2$	<0.325	1.76×10^{-5}	6.43×10^{-6}	A
316	U/T	$C_2H_5ONO_2 + h\nu \longrightarrow C_2H_5O + NO_2$	<0.330	1.49×10^{-5}	1.79×10^{-6}	A
317	U/T	$C_3H_7ONO_2 + h\nu \longrightarrow C_3H_7O + NO_2$	<0.330	1.94×10^{-5}	3.44×10^{-6}	A
318	U/T	$CH_3CO_3NO_2 + h\nu \longrightarrow CH_3CO_3 + NO_2$	<0.300	3.54×10^{-6}	7.23×10^{-8}	A
319	U/T	$CH_3COCH_3 + h\nu \longrightarrow CH_3O_2 + CH_3C(O)OO$	<0.335	1.92×10^{-6}	8.39×10^{-7}	A
320	U/T	$KET + h\nu \longrightarrow CH_3C(O)OO + RO_2 + 2XOP$	<0.33	1.92×10^{-6}	8.39×10^{-7}	J
321	U/T	$MVK + h\nu \longrightarrow CH_3C(O)OO + C_2H_4 + HO_2$	<0.325	8.11×10^{-7}	4.12×10^{-7}	K
322	U/T	$MACR + h\nu \longrightarrow C_2H_4 + HO_2 + CO + CH_3O_2$	<0.375	2.59×10^{-6}	1.64×10^{-6}	A
323	U/T	$CH_3COCHO + h\nu \longrightarrow CH_3C(O)OO + CO + HO_2$	<0.465	3.82×10^{-4}	2.94×10^{-4}	A
324	U	$BZA + h\nu \longrightarrow PHO_2 + CO + HO_2$	<0.385	6.45×10^{-3}	4.31×10^{-5}	E
325	U	$OPEN + h\nu \longrightarrow CH_3C(O)OO + CO + HO_2$	<0.33	5.13×10^{-4}	2.94×10^{-4}	G
326	T	$CH_3SSCH_3 + h\nu \longrightarrow 2CH_3S$	<0.40	1.30×10^{-4}	6.28×10^{-5}	A
327	S	$OCS + h\nu \longrightarrow CO + S$	<0.26	2.83×10^{-8}	1.72×10^{-21}	A
328	T/S	$CS_2 + h\nu \longrightarrow CS + S$	<0.34	2.46×10^{-5}	1.46×10^{-5}	A
329	S	$HCl + h\nu \longrightarrow H + Cl$	<0.22	2.13×10^{-8}	9.25×10^{-24}	A
330	S	$ClO + h\nu \longrightarrow Cl + O$	<0.305	1.30×10^{-4}	4.06×10^{-5}	L
331	S	$ClOO + h\nu \longrightarrow ClO + O$	<0.280	6.22×10^{-8}	2.81×10^{-16}	B
332	S	$OClO + h\nu \longrightarrow ClO + O$	<0.45	1.36×10^{-1}	9.86×10^{-2}	A
333	S	$HOCl + h\nu \longrightarrow OH + Cl$	<0.375	4.10×10^{-4}	2.59×10^{-4}	A
334	S	$ClONO_2 + h\nu \longrightarrow Cl + NO_3$	<0.40	6.22×10^{-5}	4.75×10^{-5}	A
335	S	$Cl_2 + h\nu \longrightarrow Cl + Cl$	<0.45	3.51×10^{-3}	2.37×10^{-3}	A
336	S	$Cl_2O_2 + h\nu \longrightarrow Cl + ClOO$	<0.36	2.02×10^{-3}	1.21×10^{-3}	A
337	S	$ClNO_2 + h\nu \longrightarrow Cl + NO_2$	<0.372	6.50×10^{-4}	3.94×10^{-4}	A
338	S	$CH_3Cl + h\nu \longrightarrow CH_3O_2 + Cl$	<0.22	1.31×10^{-8}	2.26×10^{-24}	A
339	S	$CH_3CCl_3 + h\nu \longrightarrow 3Cl$	<0.24	7.20×10^{-7}	5.12×10^{-22}	A
340	S	$CCl_4 + h\nu \longrightarrow Cl + CCl_3$	<0.25	8.51×10^{-7}	2.12×10^{-21}	A
341	S	$CFCl_3 + h\nu \longrightarrow Cl + CFCl_2$	<0.25	5.10×10^{-7}	4.34×10^{-22}	A
342	S	$CF_2Cl_2 + h\nu \longrightarrow Cl + CF_2Cl$	<0.226	4.60×10^{-8}	1.65×10^{-23}	A
343	S	$CFCl_2CF_2Cl + h\nu \longrightarrow Cl + CFCl_2CF_2$	<0.230	8.78×10^{-8}	2.63×10^{-23}	B
344	S	$CF_2ClH + h\nu \longrightarrow Cl$	<0.205	5.68×10^{-10}	3.04×10^{-27}	B
345	S	$BrO + h\nu \longrightarrow Br + O$	<0.375	5.88×10^{-3}	3.82×10^{-3}	A
346	S	$HOBr + h\nu \longrightarrow Br + OH$	<0.48	9.05×10^{-4}	6.12×10^{-4}	A
347	S	$BrONO_2 + h\nu \longrightarrow Br + NO_3$	<0.39	1.61×10^{-3}	1.06×10^{-3}	A

(*cont.*)

Table B.4 *(cont.)*

Photoprocesses

No.	Region	Photolysis reaction	λ (μm)	*J* (25 km) (s^{-1})	*J* (0 km) (s^{-1})	Ref.[b]
348	S	$Br_2 + h\nu \longrightarrow 2Br$	<0.60	3.56×10^{-2}	3.02×10^{-2}	B
349	S	$CH_3Br + h\nu \longrightarrow CH_3O_2 + Br$	<0.26	1.13×10^{-6}	3.97×10^{-21}	A
350	S	$BrCl + h\nu \longrightarrow Br + Cl$	<0.57	5.88×10^{-3}	4.61×10^{-3}	M

Species in the reactions are listed in Table B.3. Species above reaction arrows are second or third bodies included in pressure- dependent reactions (footnote *a*). $M = N_2 + O_2$ is total air. The "region" column indicates whether the reaction is important in urban areas (U), the free-troposphere (T), the stratosphere (S), urban areas and the free-troposphere (U/T), or all three regions (All). The "Ref." column refers to sources of data for reaction rate coefficients, absorption cross sections, and quantum yields. Photolysis coefficients at 0 and 25 km were calculated for a nonpolluted sky, a zenith angle of 0°, and UV surface albedo of 0.03. The wavelengths given are valid for the listed photoprocesses.

[a](P) indicates a pressure-dependent reaction, for which the reaction rate coefficient is

$$k_r = \frac{k_{\infty,T} k_{0,T}[M]}{k_{\infty,T} + k_{0,T}[M]} F_c^{\left[1+\left(\log_{10}\frac{k_{0,T}[M]}{k_{\infty,T}}\right)^2\right]^{-1}}$$

where $k_{0,T}$ is the temperature-dependent three-body, low-pressure limit rate coefficient (the first rate listed), $k_{\infty,T}$ is the two-body, high-pressure limit rate coefficient (the second rate listed), $[M] = [N_2] + [O_2]$ is the concentration (molecules cm^{-3}) of the third body, and F_c is the broadening factor.

[b]A, Atkinson *et al.* (1997); B, DeMore *et al.* (1997); C, Gery *et al.* (1989); D, Paulson and Seinfeld (1992); E, Carter (1991); F, Stockwell (1995); G, Gery *et al.* (1988); H, Nesbitt *et al.* (1995); I, Clyne *et al.* (1976); J, assumed the same as for acetone; K, assumed the same as for methyl ethyl ketone; L, Watson (1977); M, Seery and Britton (1964); N, Yin *et al.* (1990).

[c]$k_r = k_1 + k_3[M]/\ (1 + k_3[M]/\ k_2)$, where $k_1 = 7.20 \times 10^{-15} e^{785/T}$, $k_2 = 4.10 \times 10^{-16} e^{1440/T}$, $k_3 = 1.90 \times 10^{-33} e^{725/T}$, and $[M] = [N_2] + [O_2]$ (molecules cm^{-3}).

[d]$k_r = 1.30 \times 10^{-13}(1 + 0.6 p_{a.bar})(300/T)^{1.0}$, where $p_{a.bar}$ is the ambient air pressure in bars.

[e]$k_r = (2.30 \times 10^{-13} e^{600/T} + 1.70 \times 10^{-33}[M]e^{1000/T})(1 + 1.40 \times 10^{-21}[H_2O]e^{2200/T})$, where $[M] = [N_2] + [O_2]$ and $[H_2O]$ are in units of molecules cm^{-3}.

[f]$k_r = 1.7 \times 10^{-42}[M]e^{7810/T}/(1.0 + 5.5 \times 10^{-31}[M]\ e^{7460/T})$, where $[M] = [N_2] + [O_2]$ (molecules cm^{-3}).

B.5 CHEMICALS INVOLVED IN EQUILIBRIUM AND AQUEOUS REACTIONS

Table B.5 List of species involved in equilibrium or aqueous-chemistry reactions

Chemical name	Chemical formula	Chemical name	Chemical formula
	Gas-phase precursors		
Nitrogen dioxide	$NO_2(g)$	Hydrochloric acid	$HCl(g)$
Nitrate radical	$NO_3(g)$	Carbon dioxide	$CO_2(g)$
Nitrous acid	$HONO(g)$	Formaldehyde	$HCHO(g)$
Nitric acid	$HNO_3(g)$	Formic acid	$HCOOH(g)$
Peroxynitric acid	$HO_2NO_2(g)$	Methanol	$CH_3OH(g)$
Hydroxyl radical	$OH(g)$	Methylperoxy radical	$CH_3O_2(g)$

Table B.5 (*cont.*)

Chemical name	Chemical formula	Chemical name	Chemical formula
Hydrogen peroxide	$H_2O_2(g)$	Methyl hydroperoxide	$CH_3OOH(g)$
Water vapor	$H_2O(g)$	Acetic acid	$CH_3COOH(g)$
Ozone	$O_3(g)$	Peroxyacetyl nitrate	$CH_3C(O)OONO_2(g)$
Sulfur dioxide	$SO_2(g)$	Peroxyacetic acid	$CH_3C(O)OOH(g)$
Sulfuric acid	$H_2SO_4(g)$	Methylglyoxal	$CH_3COCHO(g)$
Ammonia	$NH_3(g)$	Nitrocresol	$C_6H_3(CH_3)(OH)NO_2(g)$
		Dissociated ionic molecules	
Hydrogen ion	H^+	Peroxysulfate radical ion	SO_5^-
Ammonium ion	NH_4^+	Peroxymonosulfate ion	HSO_5^-
Sodium ion	Na^+	Hydromethanesulfonate ion	$HOCH_2SO_3^-$
Potassium ion	K^+	Oxymethanesulfonate ion	$^-OCH_2SO_3^-$
Magnesium ion	Mg^{2+}	Nitrate ion	NO_3^-
Calcium ion	Ca^{2+}	Nitrite ion	NO_2^-
Hydroxy ion	OH^-	Chloride ion	Cl^-
Hydroperoxy ion	HO_2^-	Dichloride ion	Cl_2^-
Peroxy ion	O_2^-	Chlorine hydroxide radical	$ClOH^-$
Bisulfate ion	HSO_4^-	Bicarbonate ion	HCO_3^-
Sulfate ion	SO_4^{2-}	Carbonate ion	CO_3^{2-}
Bisulfite ion	HSO_3^-	Carbonate radical ion	CO_3^-
Sulfite ion	SO_3^{2-}	Formate	$HCOO^-$
Sulfate radical ion	SO_4^-	Acetate	CH_3COO^-
		Undissociated molecules	
Nitrogen dioxide	$NO_2(aq)$	Carbon monoxide	$CO(aq)$
Nitrate radical	$NO_3(aq)$	Carbon dioxide	$CO_2(aq)$
Nitrous acid	$HONO(aq)$	Formaldehyde	$HCHO(aq)$
Nitric acid	$HNO_3(aq)$	Methylene glycol	$H_2C(OH)_2(aq)$
Peroxynitric acid	$HO_2NO_2(aq)$	Formic acid	$HCOOH(aq)$
Hydroxyl radical	$OH(aq)$	Methanol	$CH_3OH(aq)$
Hydroperoxy radical	$HO_2(aq)$	Methylperoxy radical	$CH_3O_2(aq)$
Hydrogen peroxide	$H_2O_2(aq)$	Methyl hydroperoxide	$CH_3OOH(aq)$
Water	$H_2O(aq)$	Acetic acid	$CH_3COOH(aq)$
Molecular oxygen	$O_2(aq)$	Peroxyacetyl nitrate	$CH_3C(O)OONO_2(aq)$
Ozone	$O_3(aq)$	Peroxyacetic acid	$CH_3C(O)OOH(aq)$
Sulfur dioxide	$SO_2(aq)$	Methylglyoxal	$CH_3COCHO(aq)$
Sulfuric acid	$H_2SO_4(aq)$	Hydrated methylglyoxal	$CH_3COCH_3O_2(aq)$
Ammonia	$NH_3(aq)$	Nitrocresol	$C_6H_3(CH_3)(OH)NO_2(aq)$
Hydrochloric acid	$HCl(aq)$	Soluble manganese(II) ion and complexes	Mn(II)
Chlorine atom	$Cl(aq)$	Soluble iron(III) ion and complexes	Fe(III)
		Solids	
Ammonium nitrate	$NH_4NO_3(s)$	Potassium carbonate	$K_2CO_3(s)$
Ammonium chloride	$NH_4Cl(s)$	Magnesium nitrate	$Mg(NO_3)_2(s)$
Ammonium bisulfate	$NH_4HSO_4(s)$	Magnesium chloride	$MgCl_2(s)$
Ammonium sulfate	$(NH_4)_2SO_4(s)$	Magnesium sulfate	$MgSO_4(s)$
Triammonium bisulfate	$(NH_4)_3H(SO_4)_2(s)$	Magnesium carbonate	$MgCO_3(s)$
Ammonium bicarbonate	$NH_4HCO_3(s)$	Calcium nitrate	$Ca(NO_3)_2(s)$

(*cont.*)

Table B.5 (*cont.*)

Chemical name	Chemical formula	Chemical name	Chemical formula
Sodium nitrate	$NaNO_3(s)$	Calcium chloride	$CaCl_2(s)$
Sodium chloride	$NaCl(s)$	Calcium sulfate	$CaSO_4{\cdot}2H_2O(s)$
Sodium bisulfate	$NaHSO_4(s)$	Calcium carbonate	$CaCO_3(s)$
Sodium sulfate	$Na_2SO_4(s)$	Manganese heptoxide	$Mn_2O_7(s)$
Sodium bicarbonate	$NaHCO_3(s)$	Silicon dioxide	$SiO_2(s)$
Sodium carbonate	$Na_2CO_3(s)$	Aluminum oxide	$Al_2O_3(s)$
Potassium nitrate	$KNO_3(s)$	Iron(III) oxide	$Fe_2O_3(s)$
Potassium chloride	$KCl(s)$	Lead suboxide	$Pb_2O(s)$
Potassium bisulfate	$KHSO_4(s)$	Elemental carbon	EC(s)
Potassium sulfate	$K_2SO_4(s)$	Organic carbon	OC(s)
Potassium bicarbonate	$KHCO_3(s)$		

B.6 THERMODYNAMIC DATA

Table B.6 Values of $\Delta_f G_i^o$, $\Delta_f H_i^o$, and $c_{p,i}^o$ for some substances

Substance	$\Delta_f G_i^o$ (kJ mol^{-1})	$\Delta_f H_i^o$ (kJ mol^{-1})	$c_{p,i}^o$ (J mol^{-1} K^{-1})
$NH_3(g)$	-16.45^A	-46.11^A	35.06^A
$HNO_3(g)$	-74.72^A	-135.06^A	53.35^A
$HCl(g)$	-95.299^A	-92.307^A	29.126^A
$SO_2(g)$	-300.194^A	-296.83^A	39.87^A
$H_2SO_4(g)$	-690.289^B	-814.21^B	–
$CO_2(g)$	-394.359^A	-393.509^A	37.11^A
$HCOOH(g)$	-351.0^A	-378.57^A	–
$CH_3COOH(g)$	-374.0^A	-432.25^A	66.5^A
$H_2O(aq)$	-237.129^A	-285.83^A	75.291^A
$NH_3(aq)$	-26.5^A	-80.29^A	35.06^E
$HNO_3(aq)$	-111.25^A	-207.36^A	-86.6^A
$H_2SO_3(aq)$	-537.81^A	-608.81^A	–
$H_2SO_4(aq)$	-690.003^A	-813.989^A	138.91^A
$CO_2(aq)$	-385.98^A	-413.8^A	277.64^H
$HCOOH(aq)$	-372.3^A	-425.43^A	–
$CH_3COOH(aq)$	-396.46^A	-485.76^A	–
H^+	0^A	0^A	0^A
NH_4^+	-79.31^A	-132.51^A	79.9^A
Na^+	-261.905^A	-240.12^A	46.4^A
K^+	-283.27^A	-252.38^A	21.8^A
Mg^{2+}	-454.8^A	-466.85^A	–
Ca^{2+}	-553.58^A	-542.83^A	–
OH^-	-157.244^A	-229.994^A	-148.5^A
NO_3^-	-111.25^F	-207.36^F	-86.6^F
Cl^-	-131.228^A	-167.159^A	-136.4^A
HSO_4^-	-755.91^A	-887.34^A	-84^A
SO_4^{2-}	-744.53^A	-909.27^A	-293.0^A
HSO_3^-	-527.73^A	-626.22^A	–

Table B.6 (*cont.*)

Substance	$\Delta_f G_i^o$ (kJ mol^{-1})	$\Delta_f H_i^o$ (kJ mol^{-1})	$c_{p,i}^o$ (J mol^{-1} K^{-1})
SO_3^{2-}	−486.5[A]	−635.5[A]	–
HCO_3^-	−586.77[A]	−691.99[A]	88.43[H]
CO_3^{2-}	−527.81[A]	−677.14[A]	−234.52[H]
$HCOO^-$	−351.0[A]	−425.55[A]	−87.9[A]
CH_3COO^-	−369.31[A]	−486.01[A]	−6.3[A]
NH_4NO_3(s)	−183.87[A]	−365.56[A]	139.3[A]
NH_4Cl(s)	−203.167[B]	−314.86[B]	84.1[A]
NH_4HSO_4(s)	−823.0[C]	−1026.96[A]	127.5[D]
$(NH_4)_2SO_4$(s)	−901.67[A]	−1180.85[A]	187.49[A]
$(NH_4)_3H(SO_4)_2$(s)	−1730.0[C]	−2207.0[C]	315.0[D]
NH_4HCO_3(s)	−665.9[A]	−849.4[A]	–
$NaNO_3$(s)	−367.0[A]	−467.85[A]	92.88[A]
NaCl(s)	−384.24[B]	−411.26[B]	50.5[A]
$NaHSO_4$(s)	−1003.81[B]	−1132.19[B]	85.0[D]
Na_2SO_4(s)	−1270.16[A]	−1387.08[A]	128.2[A]
$NaHCO_3$(s)	−851.0[A]	−950.81[A]	87.61[A]
Na_2CO_3(s)	−1044.44[A]	−1130.68[A]	112.3[A]
KNO_3(s)	−394.86[A]	−494.63[A]	96.4[A]
KCl(s)	−409.14[A]	−436.747[A]	51.3[A]
$KHSO_4$(s)	−1031.3[A]	−1160.6[A]	87.16[G]
K_2SO_4(s)	−1321.37[A]	−1437.79[A]	131.46[A]
$KHCO_3$(s)	−863.5[A]	−963.2[A]	89.27[H]
K_2CO_3(s)	−1063.5[A]	−1151.02[A]	114.43[A]
$Mg(NO_3)_2$(s)	−589.4[A]	−790.65[A]	141.92[A]
$MgCl_2$(s)	−591.79[A]	−641.32[A]	71.38[A]
$MgSO_4$(s)	−1170.6[A]	−1284.9[A]	96.48[A]
$MgCO_3$(s)	−1012.1[A]	−1095.8[A]	75.52[A]
$Ca(NO_3)_2$(s)	−743.07[A]	−938.39[A]	149.37[A]
$CaCl_2$(s)	−748.1[A]	−795.8[A]	72.59[A]
$CaSO_4 \cdot 2H_2O$(s)	−1797.28[A]	−2022.63[A]	186.02[A]
$CaCO_3$(s)	−1128.79[A]	−1206.92[A]	81.88[A]

A, Wagman *et al.* (1982); B, Zaytsev and Aseyev (1992); C, Bassett and Seinfeld (1983); D, Wexler and Seinfeld (1991); E, from A but for NH_2^+; F, from A but for HNO_3 since NO_3^- data may be incorrect; G, Kim and Seinfeld (1995); H, Meng *et al.* (1995).

B.7 EQUILIBRIUM REACTIONS AND RATE COEFFICIENTS

Table B.7 Equilibrium reaction, rate coefficients,and coefficient units

No.	Reaction	*A*	*B*	*C*	Units	Ref.
1	$SO_2(g) \rightleftharpoons SO_2(aq)$	1.22	10.55		mol kg^{-1} atm^{-1}	A
2	$H_2O_2(g) \rightleftharpoons H_2O_2(aq)$	7.45×10^4	22.21		mol kg^{-1} atm^{-1}	B
3	$O_3(g) \rightleftharpoons O_3(aq)$	1.13×10^{-2}	7.72		mol kg^{-1} atm^{-1}	C
4	$NO_2(g) \rightleftharpoons NO_2(aq)$	1.00×10^{-2}	8.38		mol kg^{-1} atm^{-1}	D

(*cont.*)

Table B.7 (*cont.*)

No.	Reaction		*A*	*B*	*C*	Units	Ref.
5	$NO_3(g)$	$\rightleftharpoons NO_3(aq)$	2.10×10^5	29.19		mol kg^{-1} atm^{-1}	E
6	$OH(g)$	$\rightleftharpoons OH(aq)$	2.50×10^1	17.72		mol kg^{-1} atm^{-1}	E
7	$HO_2(g)$	$\rightleftharpoons HO_2(aq)$	2.00×10^3	22.28		mol kg^{-1} atm^{-1}	E
8	$HONO(g)$	$\rightleftharpoons HONO(aq)$	4.90×10^1	16.04		mol kg^{-1} atm^{-1}	F
9	$HO_2NO_2(g)$	$\rightleftharpoons HO_2NO_2(aq)$	2.00×10^4			mol kg^{-1} atm^{-1}	G
10	$HNO_3(g)$	$\rightleftharpoons HNO_3(aq)$	2.10×10^5			mol kg^{-1} atm^{-1}	D
11	$NH_3(g)$	$\rightleftharpoons NH_3(aq)$	5.76×10^1	13.79	−5.39	mol kg^{-1} atm^{-1}	A
12	$HCHO(g)$	$\rightleftharpoons HCHO(aq)$	3.46	8.19		mol kg^{-1} atm^{-1}	H
13	$HCOOH(g)$	$\rightleftharpoons HCOOH(aq)$	5.39×10^3	18.9		mol kg^{-1} atm^{-1}	A
14	$CO_2(g)$	$\rightleftharpoons CO_2(aq)$	3.41×10^{-2}	8.19	−28.93	mol kg^{-1} atm^{-1}	A
15	$CH_3OH(g)$	$\rightleftharpoons CH_3OH(aq)$	2.20×10^2	16.44		mol kg^{-1} atm^{-1}	I
16	$CH_3O_2(g)$	$\rightleftharpoons CH_3O_2(aq)$	6.00	18.79		mol kg^{-1} atm^{-1}	E
17	$CH_3OOH(g)$	$\rightleftharpoons CH_3OOH(aq)$	2.27×10^2	18.82		mol kg^{-1} atm^{-1}	B
18	$CH_3COOH(g)$	$\rightleftharpoons CH_3COOH(aq)$	8.60×10^3	21.58		mol kg^{-1} atm^{-1}	A
19	$CH_3C(O)OOH(g)$	$\rightleftharpoons CH_3C(O)OOH(aq)$	4.73×10^2	20.70		mol kg^{-1} atm^{-1}	B
20	$CH_3C(O)OONO_2(g)$	$\rightleftharpoons CH_3C(O)OONO_2(aq)$	2.90	19.83		mol kg^{-1} atm^{-1}	J
21	$CH_3COCHO(g)$	$\rightleftharpoons CH_3COCHO(aq)$	3.70×10^3	25.33		mol kg^{-1} atm^{-1}	K
22	$HCHO(aq) + H_2O(aq)$	$\rightleftharpoons H_2C(OH)_2(aq)$	1.82×10^3	13.49		atm^{-1}	L
23	$SO_2(aq) + H_2O(aq)$	$\rightleftharpoons H^+ + HSO_3^-$	1.71×10^{-2}	7.04		mol kg^{-1}	A
24	$CO_2(aq) + H_2O(aq)$	$\rightleftharpoons H^+ + HCO_3^-$	4.30×10^{-7}	−3.08	31.81	mol kg^{-1}	A
25	$Cl(aq) + H_2O(aq)$	$\rightleftharpoons H^+ + ClOH^-$	6.20×10^{-8}			mol kg^{-1}	M
26	$NH_3(aq) + H_2O(aq)$	$\rightleftharpoons NH_4^+ + OH^-$	1.81×10^{-5}	−1.50	26.92	mol kg^{-1}	A
27	$HNO_3(aq)$	$\rightleftharpoons H^+ + NO_3^-$	1.20×10^1	29.17	16.83	mol kg^{-1}	N
28	$HCl(aq)$	$\rightleftharpoons H^+ + Cl^-$	1.72×10^6	23.15		mol kg^{-1}	O
29	$H_2O(aq)$	$\rightleftharpoons H^+ + OH^-$	1.01×10^{-14}	−22.52	26.92	mol kg^{-1}	A
30	$H_2SO_4(aq)$	$\rightleftharpoons H^+ + HSO_4^-$	1.00×10^3			mol kg^{-1}	R
31	$H_2O_2(aq)$	$\rightleftharpoons H^+ + HO_2^-$	2.20×10^{-12}	−12.52		mol kg^{-1}	S
32	$HO_2(aq)$	$\rightleftharpoons H^+ + O_2^-$	3.50×10^{-5}			mol kg^{-1}	R
33	$HONO(aq)$	$\rightleftharpoons H^+ + NO_2^-$	5.10×10^{-4}	−4.23		mol kg^{-1}	F
34	$HCOOH(aq)$	$\rightleftharpoons H^+ + HCOO^-$	1.86×10^{-4}	−0.05		mol kg^{-1}	A
35	$CH_3COOH(aq)$	$\rightleftharpoons H^+ + CH_3COO^-$	1.75×10^{-5}	0.10		mol kg^{-1}	A
36	$ClOH^-$	$\rightleftharpoons Cl^- + OH(aq)$	1.43			mol kg^{-1}	M
37	Cl_2^-	$\rightleftharpoons Cl(aq) + Cl^-$	5.26×10^{-6}			mol kg^{-1}	M
38	HSO_3^-	$\rightleftharpoons H^+ + SO_3^{2-}$	5.99×10^{-8}	3.74		mol kg^{-1}	A
39	HSO_4^-	$\rightleftharpoons H^+ + SO_4^{2-}$	1.02×10^{-2}	8.85	25.14	mol kg^{-1}	A
40	HCO_3^-	$\rightleftharpoons H^+ + CO_3^{2-}$	4.68×10^{-11}	−5.99	38.84	mol kg^{-1}	A
41	$HNO_3(g)$	$\rightleftharpoons H^+ + NO_3^-$	2.51×10^6	29.17	16.83	mol^2 kg^{-2} atm^{-1}	A
42	$HCl(g)$	$\rightleftharpoons H^+ + Cl^-$	1.97×10^6	30.19	19.91	mol^2 kg^{-2} atm^{-1}	A
43	$NH_3(g) + H^+$	$\rightleftharpoons NH_4^+$	1.03×10^{11}	34.81	−5.39	atm^{-1}	A
44	$NH_3(g) + HNO_3(g)$	$\rightleftharpoons NH_4^+ + NO_3^-$	2.58×10^{17}	64.02	11.44	mol^2 kg^{-2} atm^{-2}	A
45	$NH_3(g) + HCl(g)$	$\rightleftharpoons NH_4^+ + Cl^-$	2.03×10^{17}	65.05	14.51	mol^2 kg^{-2} atm^{-2}	A
46	$NH_4NO_3(s)$	$\rightleftharpoons NH_4^+ + NO_3^-$	1.49×10^1	−10.40	17.56	mol^2 kg^{-2}	A
47	$NH_4Cl(s)$	$\rightleftharpoons NH_4^+ + Cl^-$	1.96×10^1	−6.13	16.92	mol^2 kg^{-2}	A
48	$NH_4HSO_4(s)$	$\rightleftharpoons NH_4^+ + HSO_4^-$	1.38×10^2	−2.87	15.83	mol^2 kg^{-2}	A
49	$(NH_4)_2SO_4(s)$	$\rightleftharpoons 2NH_4^+ + SO_4^{2-}$	1.82	−2.65	38.57	mol^3 kg^{-3}	A
50	$(NH_4)_3H(SO_4)_2(s)$	$\rightleftharpoons 3NH_4^+ + HSO_4 + SO_4^{2-}$	2.93×10^1	−5.19	54.40	mol^5 kg^{-5}	A
51	$NH_4HCO_3(8)$	$\rightleftharpoons NH_4^+ + HCO_3^-$	1.08	−10.04		mol^2 kg^{-2}	A
52	$NaNO_3(s)$	$\rightleftharpoons Na^+ + NO_3^-$	1.20×10^1	−8.22	16.01	mol^2 kg^{-2}	A
53	$NaCl(s)$	$\rightleftharpoons Na^+ + Cl^-$	3.61×10^1	−1.61	16.90	mol^2 kg^{-2}	A

Table B.7 (*cont.*)

No.	Reaction		*A*	*B*	*C*	Units	Ref.
54	$NaHSO_4(s)$	$\rightleftharpoons Na^+ + HSO_4^-$	2.84×10^2	−1.91	14.75	$mol^2\ kg^{-2}$	A
55	$Na_2SO_4(s)$	$\rightleftharpoons 2Na^+ + SO_4^{2-}$	4.80×10^{-1}	0.98	39.50	$mol^3\ kg^{-3}$	A
56	$NaHCO_3(s)$	$\rightleftharpoons Na^+ + HCO_3^-$	3.91×10^{-1}	−7.54	−5.68	$mol^2\ kg^{-2}$	A
57	$Na_2CO_3(s)$	$\rightleftharpoons 2Na^+ + CO_3^{2-}$	1.81×10^1	10.77	30.55	$mol^3\ kg^{-3}$	A
58	$KNO_3(s)$	$\rightleftharpoons K^+ + NO_3^-$	8.72×10^{-1}	−14.07	19.39	$mol^2\ kg^{-2}$	A
59	$KCl(s)$	$\rightleftharpoons K^+ + Cl^-$	8.68	−6.94	19.95	$mol^2\ kg^{-2}$	A
60	$KHSO_4(s)$	$\rightleftharpoons K^+ + HSO_4^-$	2.40×10^1	−8.42	17.96	$mol^2\ kg^{-2}$	A
61	$K_2SO_4(s)$	$\rightleftharpoons 2K^+ + SO_4^{2-}$	1.57×10^{-2}	−9.59	45.81	$mol^3\ kg^{-3}$	A
62	$KHCO_3(s)$	$\rightleftharpoons K^+ + HCO_3^-$	1.40×10^1	−7.60	−2.52	$mol^2\ kg^{-2}$	A
63	$K_2CO_3(s)$	$\rightleftharpoons 2K^+ + CO_3^{2-}$	2.54×10^5	12.46	36.73	$mol^3\ kg^{-3}$	A
64	$Mg(NO_3)_2(s)$	$\rightleftharpoons Mg^{2+} + 2NO_3^-$	2.51×10^{15}			$mol^3\ kg^{-3}$	A
65	$MgCl_2$	$\rightleftharpoons Mg^{2+} + 2Cl^-$	9.55×10^{21}			$mol^3\ kg^{-3}$	A
66	$MgSO_4(s)$	$\rightleftharpoons Mg^{2+} + SO_4^{2-}$	1.08×10^5			$mol^2\ kg^{-2}$	A
67	$MgCO_3(s)$	$\rightleftharpoons Mg^{2+} + CO_3^{2-}$	6.82×10^{-6}			$mol^2\ kg^{-2}$	A
68	$Ca(NO_3)_2(s)$	$\rightleftharpoons Ca^{2+} + 2NO_3^-$	6.07×10^5			$mol^3\ kg^{-3}$	A
69	$CaCl_2$	$\rightleftharpoons Ca^{2+} + 2Cl^-$	7.97×10^{11}			$mol^3\ kg^{-3}$	A
70	$CaSO_4{\cdot}2H_2O(s)$	$\rightleftharpoons Ca^{2+} + SO_4^{2-} + 2H_2O(aq)$	4.32×10^{-5}			$mol^2\ kg^{-2}$	A
71	$CaCO_3(s)$	$\rightleftharpoons Ca^{2+} + CO_3^{2-}$	4.97×10^{-9}			$mol^2\ kg^{-2}$	A

The equilibrium coefficient reads,

$$K_{eq}(T) = A\exp\left\{B\left(\frac{T_0}{T} - 1\right) + C\left(1 - \frac{T_0}{T} + \ln\frac{T_0}{T}\right)\right\}$$

where

$$A = K_{eq}(T_0), \quad B = -\frac{1}{R^* T_0}\sum_i k_i \nu_i \Delta_f H_i^o, \quad C = -\frac{1}{R^*}\sum_i k_i \nu_i c_{p,i}^o, \quad \text{and } T_0 = 298.15\,\text{K}$$

The terms in *A*, *B*, and *C* are defined in Chapter 17.

A, Derived from Appendix Table B.6 using (17.37); B, Lind and Kok (1986); C, Kozac-Channing and Heltz (1983); D, Schwartz (1984); E, Jacob (1986); F, Schwartz and White (1981); G, Park and Lee (1987); H, Ledbury and Blair (1925); I, Snider and Dawson (1985); J, Lee (1984); K, Betterton and Hoffmann (1988); L, Le Henaff (1968); M, Jayson *et al.* (1973); N, derived from a combination of other rate coefficients in the table; O, Marsh and McElroy (1985); R, Perrin (1982); S, Smith and Martell (1976).

B.8 IRREVERSIBLE AQUEOUS REACTIONS

Table B.8 Aqueous-phase chemical kinetic reactions, rate coefficients, and photoprocesses

No.	Reaction	*A*	*B*	Ref.
1	$S(IV) + H_2O_2(aq) + H^+ \longrightarrow S(VI) + 2H^+ + H_2O(aq)$			
	$HSO_3^- + H_2O_2(aq) + H^+$	7.45×10^7 (1)	−15.96	A

(*cont.*)

Table B.8 *(cont.)*

No.	Reaction	*A*	*B*	Ref.
2	$S(IV) + HO_{2,T} \longrightarrow S(VI) + OH(aq) + H^+$			
	$HSO_3^- + HO_2(aq)$	4.35×10^5		A
	$SO_3^{2-} + HO_2(aq)$	5.65×10^5		A
	$HSO_3^- + O_2^-$	4.35×10^4		A
	$SO_3^{2-} + O_2^-$	5.65×10^4		A
3	$S(IV) + OH(aq) \longrightarrow SO_5^- + H_2O(aq)$			
	$HSO_3^- + OH(aq)$	4.20×10^9	−5.03	E
	$SO_3^{2-} + OH(aq)$	4.60×10^9	−5.03	E
4	$S(IV) + O_3(aq) \longrightarrow S(VI) + O_2(aq) + H^+$			
	$SO_2(aq) + O_3(aq)$	2.40×10^4		A
	$HSO_3^- + O_3(aq)$	3.70×10^5	−18.56	A
	$SO_3^{2-} + O_3(aq)$	1.50×10^9	−17.72	A
5	$S(IV) + HCHO_T \longrightarrow HOCH_2SO_3^-$			
	$HSO_3^- + HCHO(aq)$	7.90×10^2	−16.44	G
	$SO_3^{2-} + HCHO(aq)$	2.48×10^7	−6.04	G
6	$S(IV) + Cl_2^- \longrightarrow SO_5^- + 2Cl^- + H^+$			
	$HSO_3^- + Cl_2^-$	3.40×10^8	−5.03	D
	$SO_3^{2-} + Cl_2^-$	1.60×10^8	−5.03	D
7	$S(IV) + SO_5^- \longrightarrow HSO_5^- + SO_5^-$			
	$HSO_3^- + SO_5^-$	3.00×10^5	−10.40	D
	$SO_3^{2-} + SO_5^-$	1.30×10^7	−6.71	D
8	$S(IV) + SO_4^- \longrightarrow S(VI) + SO_5^- + H^+$			
	$HSO_3^- + SO_4^-$	1.30×10^9	−5.03	H
	$SO_3^{2-} + SO_4^-$	5.30×10^8	−5.03	H
9	$S(IV) + HSO_5^- \longrightarrow 2S(VI) + 2H^+$			
	$HSO_3^- + HSO_5^-$	7.10×10^6	−10.47	D
10	$S(IV) + CH_3OOH(aq) + H^+ \longrightarrow S(VI) + 2H^+ + CH_3OH(aq)$			
	$HSO_3^- + CH_3OOH(aq) + H^+$	1.90×10^7	−12.75	A
11	$S(IV) + CH_3C(O)OOH(aq) \longrightarrow S(VI) + H^+ + CH_3COOH_T$			
	$HSO_3^- + CH_3C(O)OOH(aq)$	3.60×10^7 (2)	−13.42	E
12	$S(IV) + Fe(III) \longrightarrow S(VI) + H_2O_2(aq) + Fe(III)$			
	$SO_3^{2-} + Fe(III)$	9.50×10^7	−20.27	B, C
13	$S(IV) + Mn(II) \longrightarrow S(VI) + H_2O_2(aq) + Mn(II) + H^+$			
	$HSO_3^- + Mn(II)$	1.00×10^3	−30.06	Cc
14	$H_2O_2(aq) + OH(aq) \longrightarrow H_2O(aq) + HO_{2,T}$	2.70×10^7	−5.70	J
15	$H_2O_2(aq) + Cl_2^- \longrightarrow 2Cl^- + HO_{2,T} + H^+$	1.40×10^5	−11.31	Q
16	$H_2O_2(aq) + Cl(aq) \longrightarrow Cl^- + HO_{2,T} + H^+$	4.50×10^7		R
17	$H_2O_2(aq) + SO_4^- \longrightarrow HO_{2,T} + S(VI) + H^+$	1.20×10^3	−6.71	P
18	$H_2O_2(aq) + CO_3^- \longrightarrow HO_{2,T} + CO_{2,T}$	8.00×10^5	−9.46	O
19	$OH(aq) + HO_{2,T} \longrightarrow H_2O(aq) + O_2(aq)$			
	$OH(aq) + HO_2(aq)$	7.00×10^9	−5.03	I
	$OH(aq) + O_2^-$	1.00×10^{10}	−5.03	I
20	$OH(aq) + HSO_5^- \longrightarrow SO_5^- + H_2O(aq)$	1.70×10^7	−6.38	H
21	$OH(aq) + CO_{2,T} \longrightarrow H_2O(aq) + CO_3^-$			
	$OH(aq) + HCO_3^-$	1.50×10^7	−6.41	M

Table B.8 (*cont.*)

No.	Reaction	*A*	*B*	Ref.
22	$OH(aq) + HCHO_T \rightarrow HO_{2,T} + HCOOH_T + H_2O(aq)$			
	$OH(aq) + H_2C(OH)_2(aq)$	2.00×10^9	-5.03	U
23	$OH(aq) + HOCH_2SO_3^- \longrightarrow SO_5^- + HCHO_T + H_2O(aq)$	1.40×10^9	-5.03	H
24	$OH(aq) + HCOOH_T \longrightarrow HO_{2,T} + CO_{2,T} + H_2O(aq)$			
	$OH(aq) + HCOOH(aq)$	1.60×10^8	-5.03	V
	$OH(aq) + HCOO^-$	2.50×10^9	-5.03	W
25	$OH(aq) + CH_3COOH_T \longrightarrow HO_{2,T} + CO(aq) + HCOOH_T + H_2O(aq)$			
	$OH(aq) + CH_3COOH(aq)$	2.00×10^7	-6.25	Y
	$OH(aq) + CH_3COO^-$	8.00×10^7	-5.07	Y
26	$OH(aq) + CH_3OH(aq) \longrightarrow HO_{2,T} + HCHO_T + H_2O(aq)$	4.50×10^8	-5.03	W
27	$OH(aq) + CH_3OOH(aq) \longrightarrow CH_3O_2(aq) + H_2O(aq)$	2.70×10^7	-5.70	H
28	$OH(aq) + CH_3OOH(aq) \longrightarrow HO_{2,T} + HCOOH_T + H_2O(aq)$	1.90×10^7	-6.04	H
29	$HO_{2,T} + O_3(aq) \longrightarrow OH(aq) + 2O_2(aq)$			
	$O_2^- + O_3(aq)$	1.50×10^9	-5.03	I
30	$HO_{2,T} + HO_{2,T} \longrightarrow H_2O_2(aq) + O_2(aq)$			
	$HO_2(aq) + HO_2(aq)$	8.60×10^5	-7.94	K
	$HO_2(aq) + O_2^-$	1.00×10^8	-5.03	K
31	$HO_{2,T} + CH_3O_2(aq) \longrightarrow CH_3OOH(aq) + O_2(aq)$			
	$HO_2(aq) + CH_3O_2(aq)$	4.30×10^5	-10.07	H
	$O_2^- + CH_3O_2(aq)$	5.00×10^7	-5.37	H
32	$HO_{2,T} + SO_4^- \longrightarrow S(VI) + H^+ + O_2(aq)$			
	$HO_2(aq) + SO_4^-$	5.00×10^9	-5.03	H
	$O_2^- + SO_4^-$	5.00×10^9	-5.03	H
33	$HO_{2,T} + SO_5^- \longrightarrow HSO_5^- + O_2(aq)$			
	$O_2^- + SO_5^-$	1.00×10^8	-5.03	H
34	$HO_{2,T} + Cl_2^- \longrightarrow 2Cl^- + O_2(aq)$			
	$O_2^- + Cl_2^-$	1.00×10^9	-5.03	P
35	$HO_{2,T} + CO_{2,T} \longrightarrow HO_2^- + CO_3^-$			
	$O_2^- + HCO_3^-$	1.50×10^6		N
36	$HO_{2,T} + CO_3^- \longrightarrow CO_{2,T} + O_2(aq) + OH^-$			
	$O_2^- + CO_3^-$	4.00×10^8	-5.03	O
37	$SO_4^- + HCOOH_T \longrightarrow S(VI) + CO_{2,T} + HO_{2,T}$			
	$SO_4^- + HCOO^-$	1.10×10^8	-5.03	AA
38	$SO_4^- + Cl^- \longrightarrow S(VI) + Cl(aq)$	2.60×10^8	-5.03	AA
39	$SO_5^- + HCOOH_T \longrightarrow HSO_5^- + CO_{2,T} + HO_{2,T}$			
	$SO_5^- + HCOO^-$	1.40×10^4	-13.42	H
40	$SO_5^- + SO_5^- \longrightarrow 2SO_4^- + O_2(aq)$	6.00×10^8	-5.03	D
41	$HCOOH_T + CO_3^- \longrightarrow 2CO_{2,T} + HO_{2,T} + OH^-$			
	$HCOO^- + CO_3^-$	1.10×10^5	-11.41	X

(*cont.*)

Table B.8 (*cont.*)

No.	Reaction	A	B	Ref.
42	$HCOOH_T + Cl_2^- \longrightarrow CO_{2,T} + HO_{2,T} + 2\ Cl^-$			
	$HCOO^- + Cl_2^-$	1.90×10^6	-8.72	Q
43	$H_2O_2(aq) + h\nu \longrightarrow 2OH(aq)$	Radiation dependent		BB
44	$NO_3^- + h\nu \longrightarrow NO_2(aq) + OH(aq) + OH^-$	Radiation dependent		BB

Rate coefficients for individual reactions in each family reaction have the form

$$K_{aq}(T) = A\exp\left[B\left(\frac{T_0}{T} - 1\right)\right] \qquad A = K_{aq}(T_0) \qquad B = -\frac{1}{R^* T_0}\sum_i k_i \nu_i \Delta_f H_i^o$$

where A has units of s^{-1}, $M^{-1}\ s^{-1}$, or $M^{-2}\ s^{-1}$, the terms in B are defined in Chapter 17, T_0 = 298.15 K, and T is in kelvin. Specialized rate coefficients have the form

$$K_{aq}(T) = \frac{A\exp\{B[(T_0/T) - 1]\}}{1 + 13[H^+]} \tag{1}$$

$$K_{aq}(T) = A\exp\left[B\left(\frac{T_0}{T} - 1\right)\right][H^+] + 7.0 \times 10^2 \tag{2}$$

where $K_{a1} = 0.0123$ mol L^{-1}, $K_{a2} = 6.61 \times 10^{-8}$ mol L^{-1}, and $[H^+]$ is in moles per liter. Rate coefficients for family reactions are described in (19.56)–(19.61).

A, Hoffmann and Calvert (1985); B, Conklin and Hoffmann (1988); C, Martin and Hill (1987a); Cc, Martin and Hill (1987b); D, Huie and Neta (1987); E, Lind *et al.* (1987); G, Boyce and Hoffmann (1984); H, Jacob (1986); I, Sehested *et al.* (1968); J, Christensen *et al.* (1982); K, Bielski (1978); M, Weeks and Rabani (1966); N, Schmidt (1972); O, Behar *et al.* (1970); P, Ross and Neta (1979); Q, Hagesawa and Neta (1978); R, Graedel and Goldberg (1983); U, Bothe and Schulte-Frohlinde (1980); V, Scholes and Willson (1967); W, Anbar and Neta (1967); X, Chen *et al.* (1973); Y, Farhataziz and Ross (1977); AA, Wine *et al.* (1989); BB, Graedel and Weschler (1981).

B.9 SOLUTE ACTIVITY COEFFICIENT DATA

Table B.9 Parameters for calculating electrolyte mean binary solute activity coefficients

Electrolyte	j	B_j	G_j	H_j
HCl		Hamer and Wu (1972)	Parker (1965)	Parker (1965)
		16 m	55.5 m	15.9 m
	0	-1.998104×10^{-2}	0	0
	1	$-7.959068 \times 10–1$	5.532198×10^{-1}	2.108728×10^{0}
	2	6.580198×10^{-1}	-2.571126×10^{-1}	8.542292×10^{-1}
	3	-7.409941×10^{-2}	2.790048×10^{-1}	-6.237459×10^{-1}
	4	1.345075×10^{-2}	-4.691631×10^{-2}	1.935911×10^{-1}
	5	-2.248651×10^{-3}	2.382485×10^{-3}	-2.037543×10^{-2}
HNO_3		Hamer and Wu (1972)	Parker (1965)	Parker (1965)
		28 m	55.5 m	55.5 m
	0	-2.388378×10^{-2}	0	0
	1	-7.777787×10^{-1}	5.785894×10^{-1}	-4.785171×10^{-1}

Table B.9 (*cont.*)

Electrolyte	j	B_j	G_j	H_j
	2	5.950086×10^{-1}	-9.860271×10^{-1}	6.521896×10^{0}
	3	-1.284278×10^{-1}	6.043012×10^{-1}	-2.605544×10^{0}
	4	1.291734×10^{-2}	-1.123169×10^{-1}	3.739984×10^{-1}
	5	-6.257155×10^{-4}	6.688134×10^{-3}	-1.832646×10^{-2}
NaCl		Hamer and Wu (1972) 6.1 m	Parker (1965) 6.1 m	Perron *et al.* (1981) 6.0 m
	0	-6.089937×10^{-3}	0	0
	1	-1.015184×10^{0}	5.808744×10^{-1}	2.261834×10^{0}
	2	9.345503×10^{-1}	-1.163239×10^{0}	3.622494×10^{0}
	3	-4.615793×10^{-1}	5.136893×10^{-1}	-1.608598×10^{0}
	4	1.431557×10^{-1}	-1.029523×10^{-1}	2.092972×10^{-1}
	5	-1.700298×10^{-2}	1.401488×10^{-2}	0
$NaNO_3$		Wu and Hamer (1980) 10.8 m	Parker (1965) 9.2 m	Parker (1965) 2.2 m
	0	-6.638145×10^{-3}	0	0
	1	-1.024329×10^{0}	5.678220×10^{-1}	7.232987×10^{-1}
	2	6.877457×10^{-1}	-2.136826×10^{0}	1.918907×10^{1}
	3	-3.336161×10^{-1}	1.145031×10^{0}	-2.382164×10^{1}
	4	8.387414×10^{-2}	-2.585350×10^{-1}	1.367081×10^{1}
	5	-8.154844×10^{-3}	2.390815×10^{-2}	-3.064556×10^{0}
$NaHSO_4$		Harvie *et al.* (1984) 6.0 m	Assumed same as for NaCl 6.1 m	Same as for NaCl 6.0 m
	0	-8.890979×10^{-3}	0	0
	1	-9.559487×10^{-1}	5.808744×10^{-1}	2.261834×10^{0}
	2	8.758970×10^{-1}	-1.163239×10^{0}	3.622494×10^{0}
	3	-4.607380×10^{-1}	5.136893×10^{-1}	-1.608598×10^{0}
	4	1.309144×10^{-1}	-1.029523×10^{-1}	2.092972×10^{-1}
	5	-1.398546×10^{-2}	1.401488×10^{-2}	0
Na_2SO_4		Goldberg (1981) 4.4 m	Wagman *et al.* (1982) 3.1 m	Holmes and Mesmer (1986) 2.0 m
	0	-2.323071×10^{-2}	0	0
	1	-3.321509×10^{0}	1.698182×10^{0}	9.410224×10^{0}
	2	3.388793×10^{0}	-5.160108×10^{0}	2.213823×10^{1}
	3	-2.402946×10^{0}	2.132810×10^{0}	-3.481895×10^{1}
	4	8.926764×10^{-1}	8.840108×10^{-1}	2.348397×10^{1}
	5	-1.225933×10^{-1}	-5.143058×10^{-1}	-6.471345×10^{0}
NH_4Cl		Hamer and Wu (1972) 7.4 m	Wagman *et al.* (1982) 7.0 m	Parker (1965) 7.4 m
	0	-5.022484×10^{-3}	0	0
	1	-1.037873×10^{0}	4.890513×10^{-1}	1.959107×10^{0}
	2	8.517483×10^{-1}	-7.013315×10^{-1}	9.894682×10^{-1}
	3	-4.225323×10^{-1}	4.682151×10^{-1}	-1.024499×10^{-1}
	4	1.214996×10^{-1}	-1.702461×10^{-1}	-2.354376×10^{-1}
	5	-1.471525×10^{-2}	2.502643×10^{-2}	6.600384×10^{-2}

(*cont.*)

Table B.9 (*cont.*)

Electrolyte	j	B_j	G_j	H_j
NH_4NO_3		Hamer and Wu (1972) 25.9 m	Vanderzee *et al.* (1980) 25.0 m	Roux *et al.* (1978) 22.4 m
	0	-1.044572×10^{-2}	0	0
	1	-1.004940×10^{0}	4.362921×10^{-1}	2.611682×10^{0}
	2	4.674064×10^{-1}	-1.455444×10^{0}	3.158677×10^{0}
	3	-1.750495×10^{-1}	6.282104×10^{-1}	-2.005748×10^{0}
	4	3.253844×10^{-2}	-1.123507×10^{-1}	4.113737×10^{-1}
	5	-2.276789×10^{-3}	7.438990×10^{-3}	-2.820677×10^{-2}
NH_4HSO_4		Bassett and Seinfeld (1983) 6.0 m	Same as for NH_4Cl 7.0 m	Same as for NH_4Cl 7.4 m
	0	-2.708121×10^{-3}	0	0
	1	-1.095646×10^{0}	4.890513×10^{-1}	1.959107×10^{0}
	2	1.042878×10^{0}	-7.013315×10^{-1}	9.894682×10^{-1}
	3	-6.289405×10^{-1}	4.682151×10^{-1}	-1.024499×10^{-1}
	4	2.079631×10^{-1}	-1.702461×10^{-1}	-2.354376×10^{-1}
	5	-2.776957×10^{-2}	2.502643×10^{-2}	6.600384×10^{-2}
$(NH_4)_2SO_4$		Filippov *et al.* (1985) 5.8 m	Wagman *et al.* (1982) 5.5 m	Sukhatme and Saikhedkar (1969) 5.5 m
	0	-2.163694×10^{-2}	2.297972×10^{-1}	0
	1	-3.377941×10^{0}	4.255129×10^{-1}	1.609902×10^{-3}
	2	3.118007×10^{0}	-2.220594×10^{0}	4.437758×10^{0}
	3	-1.920544×10^{0}	2.607601×10^{0}	6.101756×10^{-3}
	4	6.372975×10^{-1}	-1.243384×10^{0}	4.021805×10^{-1}
	5	-8.277292×10^{-2}	2.102563×10^{-1}	4.375833×10^{-4}

Parameters fit into (17.56), which fits into (17.55). B, which also fits into (17.51), is used to calculate binary solute activity coefficients at 298.15 K, G is a heat-of-dilution parameter, and H is a heat-capacity parameter. Molalities are the maximum molalities of the data used to derive the fits. From Jacobson *et al.* (1996b).

B.10 WATER ACTIVITY DATA

Table B.10 Parameters for calculating molalities of binary electrolytes as a function of relative humidity at 298.15 K

Parameter	^{A}HCl 0% r.h.; 18.5 m	$^{A}HNO_3$ 0% r.h.; 22.6 m	$^{B}H^+/HSO_4^-$ 0% r.h.; 30.4 m	$^{B}2H^+/SO_4^{2-}$ 0% r.h.; 30.4 m
Y_0	$1.874637647 \times 10^{1}$	$2.306844303 \times 10^{1}$	$3.0391387536 \times 10^{1}$	$3.0391387536 \times 10^{1}$
Y_1	$-2.052465972 \times 10^{1}$	$-3.563608869 \times 10^{1}$	$-1.8995058929 \times 10^{2}$	$-1.8995058929 \times 10^{2}$
Y_2	$-9.485082073 \times 10^{1}$	$-6.210577919 \times 10^{1}$	$9.7428231047 \times 10^{2}$	$9.7428231047 \times 10^{2}$
Y_3	$5.362930715 \times 10^{2}$	$5.510176187 \times 10^{2}$	$-3.1680155761 \times 10^{3}$	$-3.1680155761 \times 10^{3}$
Y_4	$-1.223331346 \times 10^{3}$	$-1.460055286 \times 10^{3}$	$6.1400925314 \times 10^{3}$	$6.1400925314 \times 10^{3}$
Y_5	$1.427089861 \times 10^{3}$	$1.894467542 \times 10^{3}$	$-6.9116348199 \times 10^{3}$	$-6.9116348199 \times 10^{3}$
Y_6	$-8.344219112 \times 10^{2}$	$-1.220611402 \times 10^{3}$	$4.1631475226 \times 10^{3}$	$4.1631475226 \times 10^{3}$
Y_7	1.90992437×10^{2}	$3.098597737 \times 10^{2}$	$-1.0383424491 \times 10^{3}$	$-1.0383424491 \times 10^{3}$

Table B.10 (*cont.*)

Parameter	[C]NaCl 47% r.h.; 13.5 m	[C]$NaNO_3$ 30% r.h.; 56.8 m	[D]$NaHSO_4$ 1.9% r.h.; 158 m	[C]Na_2SO_4 58% r.h.; 13.1 m
Y_0	5.875248×10^1	3.1021762×10^2	1.8457001681×10^2	5.5983158×10^2
Y_1	-1.8781997×10^2	-1.82975944×10^3	$-1.6147765817 \times 10^3$	-2.56942664×10^3
Y_2	2.7211377×10^2	5.13445395×10^3	8.444076586×10^3	4.47450201×10^3
Y_3	-1.8458287×10^2	-8.01200018×10^3	$-2.6813441936 \times 10^4$	-3.45021842×10^3
Y_4	4.153689×10^1	7.07630664×10^3	5.0821277356×10^4	9.8527913×10^2
Y_5		-3.33365806×10^3	$-5.5964847603 \times 10^4$	
Y_6		6.5442029×10^2	3.2945298603×10^4	
Y_7			-8.002609678×10^3	
Parameter	**[A]NH_4Cl 47% r.h.; 23.2 m**	**[A]NH_4NO_3 62% r.h.; 28 m**	**[D]NH_4HSO_4 6.5% r.h.; 165 m**	**[C]$(NH_4)_2SO_4$ 37% r.h.; 29.0 m**
Y_0	-7.110541604×10^3	3.983916445×10^3	2.9997156464×10^2	1.1065495×10^2
Y_1	7.217772665×10^4	1.153123266×10^4	$-2.8936374637 \times 10^3$	-3.6759197×10^2
Y_2	-3.071054075×10^5	-2.13956707×10^5	1.4959985537×10^4	5.0462934×10^2
Y_3	7.144764216×10^5	7.926990533×10^5	$-4.5185935292 \times 10^4$	-3.1543839×10^2
Y_4	-9.840230371×10^5	-1.407853405×10^6	8.110895603×10^4	6.770824×10^1
Y_5	8.03407288×10^5	1.351250086×10^6	$-8.4994863218 \times 10^4$	
Y_6	-3.603924022×10^5	-6.770046794×10^5	4.7928255412×10^4	
Y_7	6.856992393×10^4	1.393507324×10^5	$-1.1223105556 \times 10^4$	

Y coefficients fit into (17.66). The table also lists the lowest relative humidity (r.h.) and corresponding highest molality (**m**) for which each fit is valid.
A, derived from Jacobson (1996b) and references therein; B, derived from Robinson and Stokes (1955); C, obtained from Tang (1997); D, derived from Tang and Munkelwitz (1994).

B.11 SURFACE RESISTANCE DATA

Table B.11 Parameters used in surface resistance equations of Chapter 20

Parameter ($s\,m^{-1}$)	1	2	3	4	5	6	7	8	9	10	11
Seasonal Category 1: Midsummer with lush vegetation											
R_{min}	∞	60	120	70	130	100	∞	∞	80	100	150
$R_{cut,0,q}$	∞	2000	2000	2000	2000	2000	∞	∞	2500	2000	4000
R_{canp}	100	200	100	2000	2000	2000	0	0	300	150	200
R_{soil,SO_2}	400	150	350	500	500	100	0	1000	0	220	400
R_{soil,O_3}	300	150	200	200	200	300	2000	400	1000	180	200
R_{surf,SO_2}	∞	2000	2000	2000	2000	2000	∞	∞	2500	2000	4000
R_{surf,O_3}	∞	1000	1000	1000	1000	1000	∞	∞	1000	1000	1000
Seasonal Category 2: Autumn with unharvested cropland											
R_{min}	∞	∞	∞	∞	250	500	∞	∞	∞	∞	∞
$R_{cut,0,q}$	∞	9000	9000	9000	4000	8000	∞	∞	9000	9000	9000
R_{canp}	100	150	100	1500	2000	1700	0	0	200	120	140
R_{soil,SO_2}	400	200	350	500	500	100	0	1000	0	300	400
R_{soil,O_3}	300	150	200	200	200	300	2000	400	800	180	200
R_{surf,SO_2}	∞	9000	9000	9000	2000	4000	∞	∞	9000	9000	9000
R_{surf,O_3}	∞	400	400	400	1000	600	∞	∞	400	400	400

(*cont.*)

Table B.11 (*cont.*)

Parameter ($s\ m^{-1}$)	1	2	3	4	5	6	7	8	9	10	11
Seasonal Category 3: Late autumn after frost, no snow											
R_{min}	∞	∞	∞	∞	250	500	∞	∞	∞	∞	∞
$R_{cut,0,q}$	∞	∞	9000	9000	4000	8000	∞	∞	9000	9000	9000
R_{canp}	100	10	100	1000	2000	1500	0	0	100	50	120
R_{soil,SO_2}	400	150	350	500	500	200	0	1000	0	200	400
R_{soil,O_3}	300	150	200	200	200	300	2000	400	1000	180	200
R_{surf,SO_2}	∞	∞	9000	9000	3000	6000	∞	∞	9000	9000	9000
R_{surf,O_3}	∞	1000	400	400	1000	600	∞	∞	800	600	600
Seasonal Category 4: Winter, snow on ground and subfreezing											
R_{min}	∞	∞	∞	∞	400	800	∞	∞	∞	∞	∞
$R_{cut,0,q}$	∞	∞	∞	∞	6000	9000	∞	∞	9000	9000	9000
R_{canp}	100	10	10	1000	2000	1500	0	0	50	10	50
R_{soil,SO_2}	100	100	100	100	100	100	0	1000	100	100	50
R_{soil,O_3}	600	3500	3500	3500	3500	3500	2000	400	3500	3500	3500
R_{surf,SO_2}	∞	∞	∞	9000	200	400	∞	∞	9000	∞	9000
R_{surf,O_3}	∞	1000	1000	400	1500	600	∞	∞	800	1000	800
Seasonal Category 5: Transitional spring with partially green short annuals											
R_{min}	∞	120	240	140	250	190	∞	∞	160	200	300
$R_{cut,0,q}$	∞	4000	4000	4000	2000	3000	∞	∞	4000	4000	8000
R_{canp}	100	50	80	1200	2000	1500	0	0	200	60	120
R_{soil,SO_2}	500	150	350	500	500	200	0	1000	0	250	400
R_{soil,O_3}	300	150	200	200	200	300	2000	400	1000	180	200
R_{surf,SO_2}	∞	4000	4000	4000	2000	3000	∞	∞	4000	4000	8000
R_{surf,O_3}	∞	1000	500	500	1500	700	∞	∞	600	800	800

Column headings are the following land-use types: 1, urban land; 2, agricultural land; 3, range land; 4, deciduous forest; 5, coniferous forest; 6, mixed forest, including wetland; 7, water, both salt and fresh; 8, barren land, mostly desert; 9, nonforested wetland; 10, mixed agricultural and range land; 11, rocky open areas with low-growing shrubs.
Source: Wesely (1989).

B.12 MORE SURFACE RESISTANCE DATA

Table B.12 Parameters used in surface resistance calculations

Chemical formula	Chemical name	D_v/D_q	H_q^* (M atm^{-1})	$f_{0,q}$
SO_2	Sulfur dioxide	1.9	1×10^5	0
O_3	Ozone	1.6	1×10^{-2}	1
NO_2	Nitrogen dioxide	1.6	1×10^{-2}	0.1
NO	Nitric oxide	1.3	3×10^{-3}	0
HNO_3	Nitric acid	1.9	1×10^{14}	0
H_2O_2	Hydrogen peroxide	1.4	1×10^5	1
CH_3CHO	Acetaldehyde	1.6	1.5×10^1	0
HCHO	Formaldehyde	1.3	6×10^3	0
CH_3OOH	Methyl hydroperoxide	1.6	2.4×10^2	0.1
$CH_3C(O)OOH$	Peroxyacetic acid	2.0	5.4×10^2	0.1
HCOOH	Formic acid	1.6	4×10^6	0
NH_3	Ammonia	0.97	2×10^4	0
$CH_3C(O)OONO_2$	Peroxyacetyl nitrate	2.6	3.6×10^0	0.1
HONO	Nitrous acid	1.6	1×10^5	0.1

Source: Wesely (1989).

References

Abdella K. and McFarlane N. (1997) A new second-order turbulence closure scheme for the planetary boundary layer. *J. Atmos. Sci.* **54**, 1850–67.

ACEA (1999) *ACEA Programme on Emissions of Fine Particles from Passenger Cars*. Brussels, ACEA.

Adamson A. W. (1990) *Physical Chemistry of Surfaces*, 5th edn. New York, John Wiley and Sons, Inc.

Alam M. K. (1987) The effect of van der Waals and viscous forces on aerosol coagulation. *Aerosol Sci. Technol.* **6**, 41–52.

Alfaro S. C. and Gomes L. (2001) Modeling mineral aerosol production by wind erosion: Emission intensities and aerosol size distributions in source areas. *J. Geophys. Res.* **106**, 18075–84.

Allen P. and Wagner K. (1992) 1987 California Air Resources Board emissions inventory, magnetic tapes ARA806, ARA807.

Al Nakshabandi G. and Konhke H. (1965) Thermal conductivity and diffusivity of soils as related to moisture tension and other physical properties. *Agric. Meteor.* **2**, 271–9.

Ambartzumiam V. (1936) The effect of the absorption lines on the radiative equilibrium of the outer layers of the stars. *Publ. Obs. Astron. Univ. Leningrad* **6**, 7–18.

Anandakumar K. (1999) A study of the partition of net radiation into heat fluxes on a dry asphalt surface. *Atmos. Environ.* **33**, 3911–18.

Anbar M. and Neta P. (1967) A compilation of specific bimolecular rate constants for the reactions of hydrated electrons, hydrogen atoms, and hydroxyl radicals with inorganic and organic compounds in aqueous solution. *Int. J. Appl. Radiat. Isot.* **18**, 493–523.

Andre J. C., De Moor G., Lacarrere P., and Du Vachat R. (1978) Modeling the 24-hour evolution of the mean and turbulent structures of the planetary boundary layer. *J. Atmos. Sci.* **35**, 1861–83.

Andreae M. O. and Merlet P. (2001) Emission of trace gases and aerosols from biomass burning. *Global Biogeochemical Cycles* **15**, 955–66.

Andreas E. L. (1992) Sea spray and the turbulent air-sea heat fluxes. *J. Geophys. Res.* **97**, 11,429–41.

Andrén A. (1990) Evaluation of a turbulence closure scheme suitable for air pollution applications. *J. Appl. Math. Phys.* **29**, 224–39.

Andres R. J. and Kasgnoc A. D. (1998) A time-averaged inventory of subaerial volcanic sulfur emissions. *J. Geophys. Res.* **103**, 25251–61.

Angell C. A., Guni M. O., and Sichina W. J. (1982) Heat capacity of water at extremes of supercooling and superheating. *J. Phys. Chem.* **86**, 998–1002.

Anthes R. A. (1977) A cumulus parameterization scheme utilizing a one-dimensional cloud model. *Mon. Wea. Rev.* **105**, 270–86.

Anthes A., Kuo Y.-H., Hsie E.-Y., Low-Nam S., and Bettge T. W. (1989) Estimation of skill and uncertainty in regional numerical models. *Q. J. Roy. Meteor. Soc.* **115**, 763–806.

Apsley D. D. and Castro I. P. (1997) A limited-length-scale k-ε model for the neutral and stably-stratified atmospheric boundary layer. *Boundary-Layer Meteor.* **83**, 75–98.

Arakawa A. (1984) Boundary conditions in limited-area models. Course notes, Department of Atmospheric Sciences, University of California, Los Angeles.

(1997) Adjustment mechanisms in atmospheric models. *J. Meteor. Soc. Japan* **75**, 155–79.

Arakawa A. and Konor C. S. (1995) Vertical differencing of the primitive equations based on the Charney–Phillips grid in hybrid σ–p vertical coordinates. *Mon. Wea. Rev.* **124**, 511–28.

Arakawa A. and Lamb V. R. (1977) Computational design of the basic dynamical processes of the UCLA general circulation model. *Methods Comput. Phys.* **17**, 174–265.

Arakawa A. and Schubert W. H. (1974) Interaction of a cumulus cloud ensemble with large scale environment, Part I. *J. Atmos. Sci.* **31**, 674–701.

Arakawa A. and Suarez. M. J. (1983) Vertical differencing of the primitive equations in sigma coordinates. *Mon. Wea. Rev.* **111**, 34–45.

Archer C. L. and Jacobson M. Z. (2003) Spatial and temporal distributions of U.S. winds and wind power at 80 m derived from measurements. *J. Geophys. Res.* **108** (D9), 4289, doi:10.1029/2002JD002076.

Arking A. A. and Grossman K. (1972) The influence of line shape and band structure on temperatures in planetary atmospheres. *J. Atmos. Sci.* **29**, 937–49.

Arstila H., Korhonen P., and Kulmala M. (1999) Ternary nucleation: Kinetics and application to water–ammonia–hydrochloric acid system. *J. Aerosol Sci.* **30**, 131–8.

Artelt C., Schmid H.-J., and Peukert W. (2003) On the relevance of accounting for the evolution of the fractal dimension in aerosol process simulations. *J. Aerosol Sci.* **34**, 511–34.

Arya S. P. (1988) *Introduction to Micrometeorology*. San Diego, Academic Press, 307pp.

Asphalt Roofing Manufacturers Association (ARMA) (1999) *Roofing Basics*, http://www.asphaltroofing.org/basics.html.

Atkinson R., Lloyd A. C., and Winges L. (1982) An updated chemical mechanism for hydrocarbon/NO_x/SO_2 photooxidations suitable for inclusion in atmospheric simulation models. *Atmos. Environ.* **16**, 1341–55.

Atkinson R., Baulch D. L., Cox R. A., *et al.* (1997) Evaluated kinetic, photochemical, and heterogeneous data for atmospheric chemistry. Supplement V. *J. Phys. Chem. Ref. Data* **26**, 521–1011.

Austin J. (1991) On the explicit versus family solution of the fully diurnal photochemical equations of the stratosphere. *J. Geophys. Res.* **96**, 12,941–74.

Avissar R. and Mahrer Y. (1988) Mapping frost-sensitive areas with a three-dimensional local-scale numerical model. Part I: Physical and numerical aspects. *J. Appl. Meteor.* **27**, 400–13.

Bader G. and Deuflhard P. (1983) A semi-implicit mid-point rule for stiff systems of ordinary differential equations. *Numer. Math.* **41**, 373–98.

Bagnold R. A. (1941) *The Physics of Blown Sand and Desert Dunes*. New York, Methuen, 265pp.

Baldocchi D. D., Hicks B. B., and Camara P. (1987) A canopy stomatal resistance model for gaseous deposition to vegetated surfaces. *Atmos. Environ.* **21**, 91–101.

Bannon P. R. (1966) On the anelastic approximation for a compressible atmosphere. *J. Atmos. Sci.* **53**, 3618–28.

Bassett M. E. and Seinfeld J. H. (1983) Atmospheric equilibrium model of sulfate and nitrate aerosol. *Atmos. Environ.* **17**, 2237–52.

(1984) Atmospheric equilibrium model of sulfate and nitrate aerosol-II. Particle size analysis. *Atmos. Environ.* **18**, 1163–70.

Bates T. S., Kiene R. P., Wolfe G. V., *et al.* (1994) The cycling of sulfur in surface seawater of the Northeast Pacific. *J. Geophys. Res.* **99**, 7835–43.

Beard K. V. (1976) Terminal velocity and shape of cloud and precipitation drops aloft. *J. Atmos. Sci.* **33**, 851–64.

Beard K. V. and Grover S. N. (1974) Numerical collision efficiencies for small raindrops colliding with micron size particles. *J. Atmos. Sci.* **31**, 543–50.

Beard K. V. and Ochs III H. T. (1984) Collection and coalescence efficiencies for accretion. *J. Geophys. Res.* **89**, 7165–9.

Beard K. V. and Pruppacher H. R. (1971) A wind tunnel investigation of the rate of evaporation of small water drops falling at terminal velocity in air. *J. Atmos. Sci.* **28**, 1455–64.

Behar D., Czapski G., and Duchovny I. (1970) Carbonate radical in flash photolysis and pulse radiolysis of aqueous carbonate solutions. *J. Phys. Chem.* **74**, 2206–10.

Bermejo R. and Conde J. (2002) A conservative quasi-monotone semi-Lagrangian scheme. *Mon. Wea. Rev.* **130**, 423–30.

Berresheim H., Wine P. H., and Davis D. D. (1995) Sulfur in the atmosphere. In *Composition, Chemistry, and Climate of the Atmosphere*. H. B. Singh, ed., New York, Van Nostrand Reinhold, 251–307.

Betterton E. A. and Hoffmann M. R. (1988) Henry's law constants of some environmentally important aldehydes. *Environ. Sci. Technol.* **22**, 1415–18.

Betts A. K. (1986) A new convective adjustment scheme. Part I: Observational and theoretical basis. *Q. J. Roy. Meteor. Soc.* **112**, 677–91.

Betts A. K. and Miller M. J. (1986) A new convective adjustment scheme. Part II: Single column tests using GATE wave, BOMEX, ATEX, and arctic air-mass data sets. *Q. J. Roy. Meteor. Soc.* **112**, 693–709.

Bhumralkar C. M. (1975) Numerical experiments on the computation of ground surface temperature in an atmospheric general circulation model. *J. Appl. Meteor.* **14**, 67–100.

Bielski B. H. J. (1978) Reevaluation of the spectral and kinetic properties of HO_2 and O_2^- free radicals. *Photochem. Photobiol.* **28**, 645–9.

Bigg E. K. (1953) The formation of atmospheric ice crystals by the freezing of droplets. *Q. J. Roy. Meteor. Soc.* **79**, 510–19.

Binkowski F. S. and Roselle S. J. (2003) Models-3 Community Multiscale Air Quality (CMAQ) model aerosol component 1. Model description. *J. Geophys. Res.* **108** (D6), 4183, doi:10.1029/2001JD001409.

Binkowski F. S. and Shankar U. (1995) The regional particulate matter model 1. Model description and preliminary results. *J. Geophys. Res.* **100**, 26191–209.

Blackadar A. K. (1976) Modeling the nocturnal boundary layer. *Proceedings of the Third Symposium on Atmospheric Turbulence, Diffusion and Air Quality*, Boston, American Meteorological Society, 46–9.

(1978) Modeling pollutant transfer during daytime convection. *Proceedings of the Fourth Symposium on Atmospheric Turbulence, Diffusion, and Air Quality*, Reno, American Meteorological Society, 443–7.

Blumthaler M. and Ambach W. (1988) Solar UVB-albedo of various surfaces. *Photochem. Photobiol.* **48**, 85–8.

Boccippio D. J., Cummins K. L., Christian H. J., and Goodman S. J. (2001) Combined satellite- and surface-based estimation of the intracloud–cloud-to-ground lightning ratio over the continental United States. *Mon. Wea. Rev.* **129**, 108–22.

Bohren C. F. (1986) Applicability of effective-medium theories to problems of scattering and absorption by nonhomogeneous atmospheric particles. *J. Atmos. Sci.* **43**, 468–75.

Bohren C. F. and Huffman D. R. (1983) *Absorption and Scattering of Light by Small Particles.* New York, John Wiley and Sons, 530pp.

Bojkov R. D. and Fioletov V. E. (1995) Estimating the global ozone characteristics during the last 30 years. *J. Geophys. Res.* **100**, 16,537–51.

Bolsaitis P. and Elliott J. F. (1990) Thermodynamic activities and equilibrium partial pressures for aqueous sulfuric acid solutions. *J. Chem. Eng. Data* **35**, 69–85.

Bolton D. (1980) The computation of equivalent potential temperature. *Mon. Wea. Rev.* **108**, 1046–53.

Bond D. W., Steiger S., Zhang R., Tie X., and Orville R. E. (2002) The importance of NOx production by lightning in the tropics. *Atmos. Environ.* **36**, 1509–19.

Bond T. C., Streets D. G., Yarber K. F., Nelson S. M., Woo, J.-H., and Klimont, Z. (2004) A technology-based global inventory of black and organic carbon emissions from combustion. *J. Geophys. Res.*, **109**, (D1) 4203, doi: 10.1029/2003JD003697.

Bonsang B., Martin D., Lambert G., Kanakidou M., Le Roulley J. C., and Sennequier G. (1991) Vertical distribution of nonmethane hydrocarbons in the remote marine boundary layer. *J. Geophys. Res.* **96**, 7313–24.

Bothe E. and Schulte-Frohlinde D. (1980) Reaction of dihydroxymethyl radical with molecular oxygen in aqueous solution. *Z. Naturforsch. B, Anorg. Chem. Org. Chem.* **35**, 1035–9.

Bott A. (1989) A positive definite advection scheme obtained by nonlinear renormalization of the advective fluxes. *Mon. Wea. Rev.* **117**, 1006–15.

(2000) A flux method for the numerical solution of the stochastic collection equation: Extension to two-dimensional particle distributions. *J. Atmos. Sci.* **57**, 284–94.

Bott A. and Carmichael G. R. (1993) Multiphase chemistry in a microphysical radiation fog model – a numerical study. *Atmos. Environ.* **27A**, 503–22.

Boubel R. W., Fox D. L., Turner D. B., and Stern A. C. S. (1994) *Fundamentals of Air Pollution.* San Diego, Academic Press, Inc.

Boyce S. D. and Hoffmann M. R. (1984) Kinetics and mechanism of the formation of hydroxymethanesulfonic acid at low pH. *J. Phys. Chem.* **88**, 4740–6.

Brewer P. G. (1997) Ocean chemistry of the fossil fuel CO_2 signal: The haline signal of "business as usual." *Geophys. Res. Lett.* **24**, 1367–9.

Briere S. (1987) Energetics of daytime sea breeze circulation as determined from a two-dimensional and third-order closure mode. *J. Atmos. Sci.* **44**, 1455–74.

Brock J. R., Zehavi D., and Kuhn P. (1986) Condensation aerosol formations and growth in a laminar coaxial jet: Experimental. *J. Aerosol Sci.* **17**, 11–22.

Bromley L. A. (1973) Thermodynamic properties of strong electrolytes in aqueous solutions. *AIChE J.* **19**, 313–20.

Brownawell M. (2004) http://isa.dknet.dk/~innova/gemarkus.htm.

Bruggeman D. A. G. (1935) Berechnung verschiedener physikalischer Konstanten von heterogenen Substanzen. I. Dielektrizitätskonstanten und Leitfähigkeiten der Mischkörper aus isotropen Substanzen. *Ann. Phys. (Leipzig)* **24**, 639–79.

Brutsaert W. (1991) *Evaporation in the Atmosphere.* Dordrecht, Kluwer Academic Publishers, 299pp.

Burtscher H. and Schmidt-Ott A. (1982) Enormous enhancement of van der Waals forces between small silver particles. *Phys. Rev. Lett.* **48**, 1734–7.

Businger J. A., Wyngaard J. C., Izumi Y., and Bradley E. F. (1971) Flux-profile relationships in the atmospheric surface layer. *J. Atmos. Sci.* **28**, 181–9.

Butler J. N. (1982) *Carbon Dioxide Equilibria and Their Applications.* Reading MA, Addison-Wesley Publishing Co., 259pp.

Calder K. L. (1949) Eddy diffusion and evaporation in flow over aerodynamically smooth and rough surfaces: A treatment based on laboratory laws of turbulent flow with special reference to conditions in the lower atmosphere. *Q. J. Mech. Appl. Math.* **2**, 153–76.

Caldeira K. and Wickett M. E. (2003) Anthropogenic carbon and ocean pH. *Nature* **425**, 265.

California Air Resources Board (CARB) (1988) *Method Used to Develop a Size-Segregated Particulate Matter Inventory.* Technical Support Division, Emission Inventory Branch, California Air Resources Board, Sacramento, CA.

Campbell F. W. and Maffel L. (1974) Contrast and spatial frequency. *Sci. Am.* **231**, 106–14.

Capaldo K. P., Pilinis C., and Pandis S. N. (2000) A computationally efficient hybrid approach for dynamic gas/aerosol transfer in air quality models. *Atmos. Environ.* **34**, 3617–27.

Carmichael G. R., Peters L. K., and Kitada T. (1986) A second generation model for regional-scale transport/chemistry/deposition. *Atmos. Environ.* **20**, 173–88.

Carpenter R. L., Droegemeier K. K., Woodward P. R., and Hane C. E. (1990) Application of the piecewise parabolic method (PPM) to meteorological modeling. *Mon. Wea. Rev.* **118**, 586–612.

Carter W. P. L. (1990) A detailed mechanism for the gas-phase atmospheric reactions of organic compounds. *Atmos. Environ.* **24A**, 481–518.
(1991) *Development of Ozone Reactivity Scales for Volatile Organic Compounds.* EPA-600/3-91-050. U.S. Environmental Protection Agency, Research Triangle Park, NC.
(2000) *Documentation of the SAPRC-99 Chemical Mechanism for VOC Reactivity Assessment.* Final Report to the California Air Resources Board Under Contracts 92–329 and 95–308, California Air Resources Board, Sacramento, CA.
Cass G. R. (1979) On the relationship between sulfate air quality and visibility with examples in Los Angeles. *Atmos. Environ.* **13**, 1069–84.
Castro T., Madronich S., Rivale S., Muhlia A., and Mar B. (2001) The influence of aerosols on photochemical smog in Mexico City. *Atmos. Environ.* **35**, 1765–72.
Celia M. A. and Gray W. G. (1992) *Numerical Methods for Differential Equations.* Englewood Cliffs, Prentice-Hall.
Chameides W. L. (1984) The photochemistry of a remote marine stratiform cloud. *J. Geophys. Res.* **89**, 4739–55.
Chameides W. L. and Stelson A. W. (1992) Aqueous-phase chemical processes in deliquescent sea-salt aerosols: A mechanism that couples the atmospheric cycles of S and sea salt. *J. Geophys. Res.* **97**, 20, 565–80.
Chang E., Nolan K., Said M., Chico T., Chan S., and Pang E. (1991) *1987 Emissions Inventory for the South Coast Air Basin: Average Annual Day.* South Coast Air Quality Management District (SCAQMD), Los Angeles.
Chang, S., Brodzinsky G. R., Gundel L. A., and Novakov T. (1982) Chemical and catalytic properties of elemental carbon. In *Particulate Carbon: Atmospheric Life Cycle*, G. T. Wolff and R. L. Klimsch, eds., New York, Plenum Press, 158–81.
Chang W., Heikes B. G., and Lee M. (2004) Ozone deposition to the sea surface: chemical enhancement and wind speed dependence. *Atmos. Environ.* **38**, 1053–9.
Chapman S. (1930) A theory of upper-atmospheric ozone. *Mem. Roy. Meteor. Soc.* **3**, 104–25.
Chapman S. and Cowling T. G. (1970) *The Mathematical Theory of Nonuniform Gases.* Cambridge, Cambridge University Press.
Charney J. G. (1949) On a physical basis for numerical prediction of large-scale motions in the atmosphere. *J. Meteor.* **6**, 371–85.
(1951) Dynamical forecasting by numerical process. In *Compendium of Meteorology*, T. F. Malone, ed., Boston, American Meteorological Society, 470–82.
Charney J. G. and Phillips N. A. (1953) Numerical integration of the quasigeostrophic equations for barotropic and simple baroclinic flows. *J. Meteor.* **10**, 71–99.
Charnock H. (1955) Wind stress on a water surface. *Q. J. Roy. Meteor. Soc.* **81**, 639–40.
Chatfield R. B., Gardner E. P., and Calvert J. G. (1987) Sources and sinks of acetone in the troposphere: Behavior of reactive hydrocarbons and a stable product. *J. Geophys. Res.* **92**, 4208–16.
Chen C. (1991) A nested grid, nonhydrostatic, elastic model using a terrain-following coordinate transformation: The radiative-nesting boundary conditions. *Mon. Wea. Rev.* **119**, 2852–69.
Chen S., Cope V. W., and Hoffman M. Z. (1973) Behavior of CO_3^- radicals generated in the flash photolysis of carbonatoamines complexes of cobalt(III) in aqueous solution. *J. Phys. Chem.* 77, 1111–6.
Cheng M.-D. and Arakawa A. (1997) Inclusion of rainwater budget and convective downdrafts in the Arakawa-Schubert cumulus parameterization. *J. Atmos. Sci.* **54**, 1359–78.
Cheng Y., Canuto V. M., and Howard A. M. (2002) An improved model for the turbulent PBL. *J. Atmos. Sci.* **59**, 1550–65.
Chock D. P. (1991) A comparison of numerical methods for solving the advection equation – III. *Atmos. Environ.* **25A**, 853–71.

Chock D. P. and Winkler S. L. (1994) A comparison of advection algorithms coupled with chemistry. *Atmos. Environ.* **28**, 2659–75.

(2000) A trajectory-grid approach for solving the condensation and evaporation equations of aerosols. *Atmos. Environ.* **34**, 2957–73.

Chock D. P., Sun P., and Winkler S. L. (1996) Trajectory-grid: An accurate sign-preserving advection–diffusion approach for air quality modeling. *Atmos. Environ.* **30**, 857–68.

Christensen H., Sehested K., and Corfitzen H. (1982) Reactions of hydroxyl radicals with hydrogen peroxide at ambient and elevated temperatures. *J. Phys. Chem.* **86**, 1588–90.

Chylek P. (1977) A note on extinction and scattering efficiencies. *J. Appl. Meteor.* **16**, 321–2.

Chylek P., Srivastava V., Pinnick R. G., and Wang R. T. (1988) Scattering of electromagnetic waves by composite spherical particles: experiment and effective medium approximations. *Appl. Opt.* **27**, 2396–404.

Chylek P., Videen G., Ngo D., Pinnick R. G., and Klett J. D. (1995) Effect of black carbon on the optical properties and climate forcing of sulfate aerosols. *J. Geophys. Res.* **100**, 16,325–32.

Clapp R. B. and Hornberger G. M. (1978) Empirical equations for some soil hydraulic properties. *Water Resour. Res.* **14**, 601–4.

Cleaver B., Rhodes E., and Ubbelohde A. R. (1963) Studies of phase transformations in nitrates and nitrites I. Changes in ultra-violet absorption spectra on melting. *Proc. Roy. Soc. London* **276**, 437–53.

Clegg S. L. and Brimblecombe P. (1995) Application of a multicomponent thermodynamic model to activities and thermal properties of 0–40 mol kg^{-1} aqueous sulphuric acid from < 200 K to 328 K. *J. Chem. Eng. Data* **40**, 43–64.

Clegg S. L. and Seinfeld J. H. (2004) Improvement of the Zdanovskii–Stokes–Robinson model for mixtures containing solutes of different charge types. *J. Phys. Chem.* **108**, 1008–17.

Clegg S. L., Brimblecombe P., Liang Z., and Chan C. K. (1997) Thermodynamic properties of aqueous aerosols to high supersaturation: II – A model of the system Na^+–Cl^-–NO_3^-–SO_4^{2-}–H_2O at 298.15 K. *Aerosol. Sci. Technol.* **27**, 345–66.

Clegg S. L., Seinfeld J. H., and Edney E. O. (2003) Thermodynamic modeling of aqueous aerosols containing electrolytes and dissolved organic compounds. II. An extended Zdanovskii–Stokes–Robinson approach. *J. Aerosol Sci.* **34**, 667–90.

Clyne M. A. A., Monkhouse P. B., and Townsend L. W. (1976) Reactions of $O(^3P)$ atoms with halogens: The rate constants for the elementary reactions $O(^3P)$ + BrCl, $O(^3P) + Br_2$ and $O(^3P) + Cl_2$. *Int. J. Chem. Kinet.* **8**, 425–49.

Coffman D. J. and Hegg D. A. (1995) A preliminary study of the effect of ammonia on particle nucleation in the marine boundary layer. *J. Geophys. Res.* **100**, 7147–60.

Cohen M. D., Flagan R. C., and Seinfeld J. H. (1987a) Studies of concentrated electrolyte solutions using the electrodynamic balance. 1. Water activities for single-electrolyte solutions. *J. Phys. Chem.* **91**, 4563–74.

(1987b) Studies of concentrated electrolyte solutions using the electrodynamic balance. 2. Water activities for mixed-electrolyte solutions. *J. Phys. Chem.* **91**, 4575–82.

Coleman G. N. (1999) Similarity statistics from a direct numerical simulation of the neutrally stratified planetary boundary layer. *J. Atmos. Sci.* **56**, 891–9.

Collela P. and Woodward P. R. (1984) The piecewise parabolic method (PPM) for gas-dynamical simulations. *J. Comp. Phys.* **54**, 174–201.

Comes F. J., Forberich O., and Walter J. (1997) OH field measurements: A critical input into model calculations on atmospheric chemistry. *J. Atmos. Sci.* **54**, 1886–94.

Conklin M. H. and Hoffmann M. R. (1988) Metal ion-S(IV) chemistry III. Thermodynamics and kinetics of transient iron(III)-sulfur(IV) complexes. *Environ. Sci. Technol.* **22**, 891–8.

Cooke W. F. and Wilson J. J. N. (1996) A global black carbon aerosol model. *J. Geophys. Res.* **101**, 19,395–409.

Cooke, W. F., Liousse C., Cachier H., and Feichter J. (1999) Construction of a 1° × 1° fossil fuel emission data set for carbonaceous aerosol and implementation and radiative impact in the ECHAM4 model. *J. Geophys. Res.* **104**, 22,137–62.

Cotton W. R. and Anthes R. A. (1989) *Storm and Cloud Dynamics*. San Diego, Academic Press, Inc.

Courant R., Friedrichs K., and Lewy H. (1928) Über die partiellen Differenzengleichungen der mathematischen Physik. *Math. Ann.* **100**, 32–74.

Crank J. (1975) *The Mathematics of Diffusion*, 2nd edn. Oxford, Clarendon Press.

Crank J. and Nicolson P. (1947) A practical method for numerical evaluation of solutions of partial differential equations of the heat-conduction type. *Proc. Camb. Philos. Soc.* **43**, 50–67.

Crutzen P. J. (1971) Ozone production rates in an oxygen-hydrogen-nitrogen oxide atmosphere. *J. Geophys. Res.* **76**, 7311–27.

Cuenca R. H., Ek M., and Mahrt L. (1996) Impact of soil water property parameterization on atmospheric boundary layer simulation. *J. Geophys. Res.* **101**, 7269–77.

Cunningham E. (1910) On the velocity of steady fall of spherical particles through fluid medium. *Proc. Roy. Soc. London* **A83**, 357–65.

Curtiss C. F. and Hirschfelder J. O. (1952) Integration of stiff equations. *Proc. Nat. Acad. Sci. USA* **38**, 235–43.

Cuzzi J. N., Ackerman T. P., and Helmle L. C. (1982) The delta-four-stream approximation for radiative transfer. *J. Atmos. Sci.* **39**, 917–25.

Dabdub D. and Seinfeld J. H. (1994) Numerical advective schemes used in air quality models – sequential and parallel implementation. *Atmos. Environ.* **28**, 3369–85.

(1995) Extrapolation techniques used in the solution of stiff ODEs associated with chemical kinetics of air quality models. *Atmos. Environ.* **29**, 403–10.

Danielsen, E. F., Bleck R., and Morris D. A. (1972) Hail growth by stochastic collection in a cumulus model. *J. Atmos. Sci.* **29**, 135–55.

Davies H. C. (1976) A lateral boundary formulation for multi-level prediction models. *Q. J. Roy. Meteor. Soc.* **102**, 405–18.

Davis E. J. (1983) Transport phenomena with single aerosol particles. *Aerosol Sci. Technol.* **2**, 121–44.

Dean, J. A. (1992) *Lange's Handbook of Chemistry*. New York, McGraw-Hill, Inc.

Deardorff J. W. (1972) Numerical investigation of neutral and unstable planetary boundary layers. *J. Atmos. Sci.* **29**, 91–115.

(1977) A parameterization of ground surface moisture content for use in atmospheric prediction models. *J. Appl. Meteor.* **16**, 1182–5.

(1978) Efficient prediction of ground surface temperature and moisture with inclusion of a layer of vegetation. *J. Geophys. Res.* **83**, 1889–903.

De Arellano J. V., Duynkerke P., and van Weele M. (1994) Tethered-balloon measurements of actinic flux in a cloud-capped marine boundary layer. *J. Geophys. Res.* **99**, 3699–705.

De Leeuw G., Neele F. P., Hill M., Smith M. H., and Vignati E. (2000) Production of sea spray aerosol in the surf zone. *J. Geophys. Res.* **105**, 29397–409.

Deirmendjian D. (1969) *Electromagnetic Scattering on Spherical Polydispersions*. New York, Elsevier.

DeMore W. B., Sanders S. P., Golden D. M., *et al.* (1997) *Chemical Kinetics and Photochemical Data for Use in Stratospheric Modeling*. Evaluation number 12, JPL Publ. 97-4, Jet Propulsion Laboratory, Pasadena, CA.

Dickerson R. R., Kondragunta S., Stenchikov G., Civerolo K. L., Doddridge B. G., and Holben B. N. (1997) The impact of aerosols on solar UV radiation and photochemical smog. *Science* **278**, 827–30.

Dickinson R. E. (1984) Modeling evapotranspiration for three-dimensional global climate models. In *Climate Processes and Climate Sensitivity*, Geophys. Monogr. Ser., Vol. 29,

J. E. Hanson and T. Takahashi, eds., Washington, DC, American Geophysical Union, 58–72.

Ding P. and Randall D. A. (1998) A cumulus parameterization with multiple cloud-base levels. *J. Geophys. Res.* **103**, 11,341–53.

Donea J. (1984) A Taylor–Galerkin method for convective transport problems. *Int. J. Numer. Methods Engng.* **20**, 101–19.

Dorsch R. G. and Hacker P. (1951) *Experimental Values of Surface Tension of Supercooled Water.* National Advisory Committee for Aeronautics (NACA), Tech. Note 2510.

Duce R. A. (1969) On the source of gaseous chlorine in the marine atmosphere. *J. Geophys. Res.* **70**, 1775–9.

Dudhia J. (1993) A nonhydrostatic version of the Penn State-NCAR mesoscale model: Validation tests and simulation of an Atlantic cyclone and cold front. *Mon. Wea. Rev.* **121**, 1493–513.

Durran D. R. (1999) *Numerical Methods for Wave Equations in Geophysical Fluid Dynamics.* New York, Springer-Verlag.

Dyer A. J. (1974) A review of flux-profile relationships. *Boundary-Layer Meteor.* **7**, 363–72.

Dyer A. J. and Bradley E. F. (1982) An alternative analysis of flux-gradient relationships at the 1976 ITCE. *Boundary-Layer Meteor.* **22**, 3–19.

Easter R. C. (1993) Two modified versions of Bott's positive-definite numerical advection scheme. *Mon. Wea. Rev.* **121**, 297–304.

Eddington S. A. (1916) On the radiative equilibrium of the stars. *Mon. Not. Roy. Astron. Soc.* **77**, 16–35.

Edlen B. (1966) The refractive index of air. *Meteorology* **2**, 71–80.

Eliasen E., Machenhauer B., and Rasmussen E. (1970) *On a Numerical Method for Integration of the Hydrodynamical Equations with a Spectral Representation of the Horizontal Fields.* Report No. 2, Institut for Teoretisk Meteorologi, University of Copenhagen, 35pp.

Elliott D. L., Holladay C. G., Barchet W. R., Foote H. P., and Sandusky W. F. (1986) *Wind Energy Resource Atlas of the United States.* DOE/CH 10093-4 Natl. Renew. Energy Lab., Golden, CO.

Elliott S., Turco R. P., and Jacobson M. Z. (1993) Tests on combined projection/forward differencing integration for stiff photochemical family systems at long time step. *Computers Chem.* **17**, 91–102.

Emanuel K. A. (1991) A scheme for representing cumulus convection in large-scale models. *J. Atmos. Sci.* **38**, 1541–57.

Enger L. (1986) A higher order closure model applied to dispersion in a convective PGL. *Atmos. Environ.* **20**, 879–94.

Eriksson E. (1960) The yearly circulation of chloride and sulfur in nature; meteorological, geochemical and pedological implications. Part II. *Tellus* **12**, 63–109.

Erisman J. W., van Pul W. A. J., and Wyers P. (1994) Parameterization of surface resistance for the quantification of atmospheric deposition of acidifying pollutants and ozone. *Atmos. Environ.* **28**, 2595–607.

Facchini M. C., Mircea M., Fuzzi S., and Charlson R. J. (1999) Cloud albedo enhancement by surface-active organic solutes in growing droplets. *Nature* **401**, 257–9.

Fang, M., Zheng M., Wang F., To K. L., Jaafar A. B., and Tong S. L. (1999) The solvent-extractable organic compounds in the Indonesia biomass burning aerosols – characterization studies. *Atmos. Environ.* **33**, 783–95.

Farhataziz and Ross A. B. (1977) *Selected Specific Rates of Transients From Water in Aqueous Solutions, III. Hydroxyl Radical and Perhydroxyl Radical and Their Radical Ions*, Rep. NSRDBS-NBS 59, U.S. Department of Commerce, Washington, DC.

Farman J. C., Gardiner B. G., and Shanklin J. D. (1985). Large losses of total ozone in Antarctica reveal seasonal ClO_x/NO_x interaction. *Nature* **315**, 207–10.

Fassi-Fihri A., Suhre K., and Rosset R. (1997) Internal and external mixing in atmospheric aerosols by coagulation: Impact on the optical and hygroscopic properties of the sulphate–soot system. *Atmos. Environ.* **31**, 1392–402.

Ferek, R. J., Reid J. S., Hobbs P. V., Blake D. R., and Liousse C. (1998) Emission factors of hydrocarbons, halocarbons, trace gases, and particles from biomass burning in Brazil. *J. Geophys. Res.* **103**, 32,107–18.

Fernandez-Diaz J. M., Gonzalez-Pola Muniz C., Rodriguez Brana M. A., Arganza Garcia B., and Garcia Nieto P. J. (2000) A modified semi-implicit method to obtain the evolution of an aerosol by coagulation. *Atmos. Environ.* **34**, 4301–14.

Filippov V. K., Charykova M. V., and Trofimov Y. M. (1985) Thermodynamics of the system $NH_4H_2PO_4$-$(NH_4)_2SO_4$-H_2O at 25 °C. *J. Appl. Chem. USSR* **58**, 1807–11.

Finlayson-Pitts B. and Pitts Jr. J. N. (2000) *Chemistry of the Upper and Lower Atmosphere*, San Diego, Academic Press.

Fleming E. L., Chandra S., Schoeberl M. R., and Barnett J. J. (1988) *Monthly Mean Global Climatology of Temperature, Wind, Geopotential Height, and Pressure for 1–120 km.* Tech. Memo. 100697, NASA, 85pp.

Fletcher N. H. (1958) Size effect in heterogeneous nucleation. *J. Chem. Phys.* **29**, 572–76.

Flossmann A. I., Hall W. D., and Pruppacher H. R. (1985) A theoretical study of the wet removal of atmospheric pollutants. Part I: The redistribution of aerosol particles captured through nucleation and impaction scavenging by growing cloud drops. *J. Atmos. Sci.* **42**, 582–606.

Flubacher P., Leadbetter A. J., and Morrison J. A. (1960) Heat capacity of ice at low temperatures. *J. Chem. Phys.* **33**, 1751–5.

Foster V. G. (1992) Determination of the refractive index dispersion of liquid nitrobenzene in the visible and ultraviolet. *J. Phys. D* **25**, 525–9.

Fowler L. D., Randall D. A., and Rutledge S. (1996) Liquid and ice cloud microphysics in the CSU general circulation model. Part I: Model description and simulated microphysical processes. *J. Climate* **9**, 489–529.

Frank W. M. and Cohen C. (1987) Simulation of tropical convective systems. Part I: A cumulus parameterization. *J. Atmos. Sci.* **44**, 3787–99.

Freedman F. R. and Jacobson M. Z. (2002) Transport-dissipation analytical solutions to the E-ε turbulence model and their role in predictions of the neutral ABL. *Boundary-Layer Meteor.* **102**, 117–38.

(2003) Modification of the standard ε-equation for the stable ABL through enforced consistency with Monin–Obukhov similarity theory. *Boundary-Layer Meteor.* **106**, 383–410.

Fridlind A. M. and Jacobson M. Z. (2000) A study of gas–aerosol equilibrium and aerosol pH in the remote marine boundary layer during the First Aerosol Characterization Experiment (ACE 1). *J. Geophys. Res.* **105**, 17325–40.

(2003) Point and column aerosol radiative closure during ACE 1: Effects of particle shape and size. *J. Geophys. Res.* **108** (D3) doi:10.1029/2001JD001553.

Friedlander S. K. (1977) *Smoke, Dust, and Haze. Fundamentals of Aerosol Behavior.* New York, John Wiley & Sons, Inc.

(1983) Dynamics of aerosol formation by chemical reaction. *Ann. N. Y. Acad. Sci.* **404**, 354–64.

Friedli H., Lötscher H., Oeschger H., Siegenthaler U., and Stauffer B. (1996) Ice core record of 13C/12C ratio of atmospheric CO_2 in the past two centuries. *Nature* **324**, 237–8.

Fritsch J. M. and Chappel C. F. (1980) Numerical prediction of convectively driven mesoscale pressure systems. Part I: Convective parameterization. *J. Atmos. Sci.* **37**, 1722–33.

Fu Q. and Liou K. N. (1992) On the correlated k-distribution method for radiative transfer in nonhomogeneous atmospheres. *J. Atmos. Sci.* **49**, 2139–56.

(1993) Parameterization of the radiative properties of cirrus clouds. *J. Atmos. Sci.* **50**, 2008–25.

Fu Q., Liou K. N., Cribb M. C., Charlock T. P., and Grossman A. (1997) Multiple scattering parameterization in thermal infrared radiative transfer. *J. Atmos. Sci* **54**, 2799–812.

Fuchs N. A. (1964) *The Mechanics of Aerosols* (translated by R. E. Daisley and M. Fuchs). New York, Pergamon Press.

Fuchs N. A. and Sutugin A. G. (1971). Highly dispersed aerosols. In *Topics in Current Aerosol Research*, Vol. 2, G. M. Hidy and J. R. Brock, eds., New York, Pergamon Press, 1–60.

Fuller K. A. (1995) Scattering and absorption cross sections of compounded spheres. III. Spheres containing arbitrarily located spherical inhomogeneities. *J. Opt. Soc. Am. A* **12**, 893–904.

Fuller K. A., Malm W. C., and Kreidenweis S. M. (1999) Effects of mixing on extinction by carbonaceous particles. *J. Geophys. Res.* **104**, 15,941–54.

Galbally I. E. and Roy C. R. (1980) Destruction of ozone at the earth's surface. *Q. J. Roy. Meteor. Soc.* **106**, 599–620.

Ganzeveld L. and Lelieveld J. (1995) Dry deposition parameterization in a chemistry general circulation model and its influence on the distribution of reactive trace gases. *J. Geophys. Res.* **100**, 20,999–1,012.

Garcia R. R., Stordal F., Solomon S., and Kiehl J. T. (1992) A new numerical model of the middle atmosphere 1. Dynamics and transport of tropospheric source gases. *J. Geophys. Res.* **97**, 12,967–91.

Garratt J. R. (1992) *The Atmospheric Boundary Layer*. Cambridge, Cambridge University Press.

Garratt J. R. and Hicks B. B. (1973) Momentum, heat and water vapour transfer to and from natural and artificial surfaces. *Q. J. Roy. Meteor. Soc.* **99**, 680–7.

Gaydos, T. M., Koo B., Pandis S. N., and Chock D. P. (2003) *Atmos. Environ.* **37**, 3303–16.

Gazdag J. (1973) Numerical convective schemes based on accurate computation of space derivatives. *J. Comp. Phys.* **13**, 100–13.

Gear C. W. (1971) *Numerical Initial Value Problems in Ordinary Differential Equations*. Englewood Cliffs, NJ, Prentice-Hall.

Gelbard F. (1990) Modeling multicomponent aerosol particle growth by vapor condensation. *Aerosol Sci. Technol.* **12**, 399–412.

Gelbard F. and Seinfeld J. H. (1980) Simulation of multicomponent aerosol dynamics. *J. Colloid Interface Sci.* **78**, 485–501.

Gelbard F., Fitzgerald J. W., and Hoppel W. A. (1998) A one-dimensional sectional model to simulate multicomponent aerosol dynamics in the marine boundary layer. 3. Numerical methods and comparisons with exact solutions. *J. Geophys. Res.* **103**, 16,119–132.

Gerber H., Takano Y., Garrett T. J., and Hobbs P. V. (2000) Nephelometer measurements of the asymmetry parameter, volume extinction coefficient, and backscatter ratio in Arctic clouds. *J. Atmos. Sci.* **57**, 3021–33.

Gery M. W., Whitten G. Z., and Killus J. P. (1988) *Development and Testing of the CBM-IV for Urban and Regional Modeling*. Report EPA-600/3-88-012. U.S. Environmental Protection Agency, Research Triangle Park, NC.

Gery M. W., Whitten G. Z., Killus J. P., and Dodge M. C. (1989) A photochemical kinetics mechanism for urban and regional scale computer modeling. *J. Geophys. Res.* **94**, 12,925–56.

Ghio A. J. and Samet J. M. (1999) Metals and air pollution particles. In *Air Pollution and Health*, S. T. Holgate, J. M. Samet, H. S. Koren, and R. L. Maynard, eds., San Diego, Academic Press, 635–51.

Giauque W. F. and Stout J. W. (1936) The entropy of water and the third law of thermodynamics. The heat capacity of ice from 15 to 273 K. *J. Am. Chem. Soc.* **58**, 1144–50.

Gillette D. A. (1974) On the production of soil wind erosion aerosols having the potential for long range transport. *Atmos. Res.* **8**, 735–44.

Gillette D. A., Patterson Jr. E. M., Prospero J. M., and Jackson M. L. (1993) Soil aerosols. In *Aerosol Effects on Climate*, S. G. Jennings, ed., Tucson, University of Arizona Press, 73–109.

Giorgi F., Marinucci M. R., Bates G. T., and De Canio G. (1993) Development of a second-generation climate model (RegCM2) Part II: Convective processes and assimilation of lateral boundary conditions. *Mon. Wea. Rev.* **121**, 2814–32.

Gittens G. J. (1969) Variation of surface tension of water with temperature. *J. Colloid Interface Sci.* **30**, 406–12.

Goldberg R. N. (1981) Evaluated activity and osmotic coefficients for aqueous solutions: Thirty-six uni-bivalent electrolytes. *J. Phys. Chem. Ref. Data* **10**, 671–764.

Golding B. W. (1992) An efficient nonhydrostatic forecast model. *Meteor. Atmos. Phys.* **50**, 89–103.

Gong W. and Cho H.-R. (1993) A numerical scheme for the integration of the gas-phase chemical rate equations in three-dimensional atmospheric models. *Atmos. Environ.* **27A**, 2147–60.

Goodin W. R., McRae G. J., and Seinfeld J. H. (1979) A comparison of interpolations methods for sparse data: Application to wind and concentration fields. *J. Appl. Meteor.* **18**, 761–71.

Goody, R. M., West R., Chen L., and Crisp D. (1989) The correlated-k method for radiation calculations in nonhomogeneous atmospheres. *J. Quant. Spectrosc. Radiat. Transfer* **42**, 539–50.

Graedel T. E. and Goldberg K. I. (1983) Kinetic studies of raindrop chemistry, 1. Inorganic and organic processes. *J. Geophys. Res.* **88**, 10,865–82.

Graedel T. E. and Weschler C. J. (1981) Chemistry within aqueous atmospheric aerosols and raindrops. *Rev. Geophys.* **19**, 505–39.

Greeley R. and Iversen J. D. (1985) *Wind as a Geological Process on Earth, Mars, Venus, and Titan*. New York, Cambridge University Press, 333pp.

Greenberg R. R., Zoller W. H., and Gordon G. E. (1978) Composition and size distributions of articles released in refuse incineration. *Environ. Sci. Technol.* **12**, 566–73.

Grell G. A. (1993) Prognostic evaluation of assumptions used by cumulus parameterizations. *Mon. Wea. Rev.* **121**, 764–87.

Griffin R. J., Dabdub D., and Seinfeld J. H. (2002) Secondary organic aerosol 1. Atmospheric chemical mechanism for production of molecular constituents. *J. Geophys. Res.* **107** (D17), 4332, doi:10.1029/2001JD000541.

Groblicki P. J., Wolff G. T., and Countess R. J. (1981) Visibility-reducing species in the Denver "brown cloud" – I. Relationships between extinction and chemical composition. *Atmos. Environ.* **15**, 2473–84.

Guelle W., Schulz M., Balkanski Y., and Dentener F. (2001) Influence of source formulation on modeling the atmospheric global distribution of sea salt aerosol. *J. Geophys. Res.* **106**, 27509–24.

Hack J. J. (1992) Climate system simulation: Basic numerical and computational concepts. In *Climate System Modeling*, K. E. Trenberth, ed., Cambridge, Cambridge University Press, 283–318.

(1994) Parameterization of moist convection in the National Center for Atmospheric Research community climate model (CCM2). *J. Geophys. Res.* **99**, 5551–68.

Hagesawa K. and Neta P. (1978) Rate constants and mechanisms of reaction for Cl_2^- radicals. *J. Phys. Chem.* **82**, 854–7.

Hairer E. and Wanner G. (1991) *Solving Ordinary Differential Equations II. Stiff and Differential-Algebraic Problems*. Berlin, Springer-Verlag.

Hale G. M. and Querry M. R. (1973) Optical constants of water in the 200-nm to 200-μm wavelength region. *Appl. Opt.* **12**, 555–63.

Hamer W. J. and Wu Y.-C. (1972) Osmotic coefficients and mean activity coefficients of uni-univalent electrolytes in water at 25 °C. *J. Phys. Chem. Ref. Data* **1**, 1047–99.

Hamill P., Turco R. P., Kiang C. S., Toon O. B., and Whitten R. C. (1982) An analysis of various nucleation mechanisms for sulfate particles in the stratosphere. *J. Aerosol Sci.* **13**, 561–85.

Hansen J. E. (1969) Radiative transfer by doubling very thin layers. *Astrophys. J.* **155**, 565–73.

Harned H. S. and Owen B. B. (1958) *The Physical Chemistry of Electrolyte Solutions.* New York, Reinhold, Chapter 8.

Harrington J. Y., Meyers M. P., Walko R. L., and Cotton W. R. (1995) Parameterization of ice crystal conversion process due to vapor deposition for mesoscale models using double-moment basis functions. Part I: Basic formulation and parcel model results. *J. Atmos. Sci.* **52**, 4344–66.

Harris S. J. and Maricq, M. M. (2001) Signature size distributions for diesel and gasoline engine exhaust particulate matter. *J. Aerosol Sci.* **32**, 749–64.

Hartmann D. L. (1994) *Global Physical Climatology.* San Diego, Academic Press, Inc.

Harvey R. B., Stedman D. H., and Chameides W. (1977) Determination of the absolute rate of solar photolysis of NO_2. *J. Air Pol. Control Assn.* **27**, 663–6.

Harvie C. E., Moller N., and Weare J. H. (1984) The prediction of mineral solubilities in natural waters: The Na-K-Mg-Ca-H-Cl-SO_4-OH-HCO_3-CO_3-CO_2-H_2O system to high ionic strengths at 25 °C. *Geochim. Cosmochim. Acta* **48**, 723–51.

Henry W. M. and Knapp K. T. (1980) Compound forms of fossil fuel fly ash emissions. *Environ. Sci. Technol.* **14**, 450–6.

Henyey L. C. and Greenstein J. L. (1941) Diffuse radiation in the galaxy. *Astrophys. J.* **93**, 70–83.

Hering S. V. and Friedlander S. K. (1982) Origins of aerosol sulfur size distributions in the Los Angeles Basin. *Atmos. Environ.* **16**, 2647–56.

Hering S. V., Friedlander S. K., Collins J. C., and Richards L. W. (1979) Design and evaluation of a new low pressure impactor 2. *Environ. Sci. Technol.* **13**, 184–8.

Hertel O., Berkowicz R., and Christensen J. (1993) Test of two numerical schemes for use in atmospheric transport-chemistry models. *Atmos. Environ.* **27A**, 2591–611.

Hesstvedt E., Hov O., and Isaksen I. S. A. (1978) Quasi-steady-state approximations in air pollution modeling: Comparison of two numerical schemes for oxidant prediction. *Int. J. Chem. Kin.* **10**, 971–94.

Himmelblau P. M. (1964) Diffusion of dissolved gases in liquids. *Chem. Rev.* **64**, 527–50.

Hindmarsh A. C. (1983) ODEPACK, a systematized collection of ODE solvers. In *Scientific Computing*, R. S. Stepleman *et al.*, eds., Amsterdam, North-Holland, 55–74.

Hinze J. O. (1975) *Turbulence: An Introduction to its Mechanism and Theory*, 2nd edn. New York, McGraw-Hill, 790pp.

Hitchcock D. R., Spiller L. L., and Wilson W. E. (1980) Sulfuric acid aerosols and HCl release in coastal atmospheres: Evidence of rapid formation of sulfuric acid particulates. *Atmos. Environ.* **14**, 165–82.

Hoffmann M. R. and Calvert J. G. (1985) *Chemical Transformation Modules for Eulerian Acid Deposition Models, Vol. 2. The Aqueous-phase Chemistry.* EPA/600/3-85/017. U.S. Environmental Protection Agency, Research Triangle Park, NC.

Hogstrom U. (1988) Non-dimensional wind and temperature profiles in the atmospheric surface layer: A reevaluation. *Boundary-Layer Meteor.* **42**, 55–78.

Holmes H. F. and Mesmer R. E. (1986) Thermodynamics of aqueous solutions of the alkali metal sulfates. *J. Solution Chem.* **15**, 495–518.

Holton J. R. (1992) *An Introduction to Dynamical Meteorology.* San Diego, Academic Press, Inc.

Hounslow M. J., Ryall R. L., and Marshall V. R. (1988) A discretized population balance for nucleation, growth, and aggregation. *AIChE J.* **34**, 1821–32.

Houze R. A., Jr. (1993) *Cloud Dynamics.* San Diego, Academic Press, Inc.

Huang H.–C. and Chang J. S. (2001) On the performance of numerical solvers for a chemistry submodel in three-dimensional air quality models. 1. Box model simulations. *J. Geophys. Res.* **106**, 20,175–88.

Hughes T. J. R. and Brooks A. N. (1979) A multidimensional upwind scheme with no crosswind diffusion. In *Finite Element Methods for Convection Dominated Flows*, T. J. R. Hughes, ed., AMD Vol. 34, ASME, New York, 19–35.

Huie R. E. and Neta P. (1987) Rate constants for some oxidations of S(IV) by radicals in aqueous solutions. *Atmos. Environ.* **21**, 1743–7.

Hulburt H. M. and Katz S. (1964) Some problems in particle technology: A statistical mechanical formulation. *Chem. Eng. Sci.* **19**, 555–74.

Hynes A. J., Wine P. H., and Semmes D. H. (1986) Kinetic mechanism of OH reactions with organic sulfides. *J. Phys. Chem.* **90**, 4148–56.

Irvine W. M. (1968) Multiple scattering by large particles. II. Optically thick layers. *Astrophys. J.* **152**, 823–34.

(1975) Multiple scattering in planetary atmospheres. *Icarus* **25**, 175–204.

Ishizaka Y. and Adhikari M. (2003) Composition of cloud condensation nuclei. *J. Geophys. Res.* **108** (D4), 4138, doi:10.1029/2002JD002085.

Jackman C. H., Fleming E. L., Chandra S., Considine D. B., and Rosenfield J. E. (1996) Past, present, and future modeled ozone trends with comparisons to observed trends. *J. Geophys. Res.* **101**, 28,753–67.

Jacob D. J. (1986) Chemistry of OH in remote clouds and its role in the production of formic acid and peroxymonosulfonate. *J. Geophys. Res.* **91**, 9807–26.

Jacob D. J., Gottlieb E. W., and Prather M. J. (1989a) Chemistry of a polluted cloudy boundary layer. *J. Geophys. Res.* **94**, 12,975–13,002.

Jacob D. J., Sillman S., Logan J. A., and Wofsy S. C. (1989b) Least independent variables method for simulation of tropospheric ozone. *J. Geophys. Res.* **94**, 8497–509.

Jacobson M. Z. (1994) Developing, coupling, and applying a gas, aerosol, transport, and radiation model to study urban and regional air pollution. Ph.D. Thesis, Dept. of Atmospheric Sciences, University of California, Los Angeles.

(1995) Computation of global photochemistry with SMVGEAR II. *Atmos. Environ.* **29A**, 2541–6.

(1997a) Development and application of a new air pollution modeling system. Part II: Aerosol module structure and design. *Atmos. Environ.* **31A**, 131–44.

(1997b) Development and application of a new air pollution modeling system. Part III: Aerosol-phase simulations. *Atmos. Environ.* **31A**, 587–608.

(1997c) Numerical techniques to solve condensational and dissolutional growth equations when growth is coupled to reversible aqueous reactions. *Aerosol Sci. Technol.* **27**, 491–8.

(1998a) Vector and scalar improvement of SMVGEAR II through absolute error tolerance control. *Atmos. Environ.* **32**, 791–6.

(1998b) Studying the effects of aerosols on vertical photolysis rate coefficient and temperature profiles over an urban airshed. *J. Geophys. Res.* **103**, 10,593–604.

(1999a) Effects of soil moisture on temperatures, winds, and pollutant concentrations in Los Angeles. *J. Appl. Meteorol.* **38**, 607–16.

(1999b) Studying the effects of calcium and magnesium on size-distributed nitrate and ammonium with EQUISOLV II. *Atmos. Environ.* **33**, 3634–49.

(1999c) Isolating nitrated and aromatic aerosols and nitrated aromatic gases as sources of ultraviolet light absorption. *J. Geophys. Res.* **104**, 3527–42.

(2000) A physically-based treatment of elemental carbon optics: Implications for global direct forcing of aerosols. *Geophys. Res. Lett.* **27**, 217–20.

(2001a) GATOR-GCMM: A global- through urban-scale air pollution and weather forecast model 1. Model design and treatment of subgrid soil, vegetation, roads, rooftops, water, sea ice, and snow. *J. Geophys. Res.* **106**, 5385–401.

(2001b) Strong radiative heating due to the mixing state of black carbon in atmospheric aerosols. *Nature* **409**, 695–7.

(2002) Analysis of aerosol interactions with numerical techniques for solving coagulation, nucleation, condensation, dissolution, and reversible chemistry among multiple size distributions. *J. Geophys. Res.* **107** (D19), 4366, doi:10.1029/2001JD002044.

(2003) Development of mixed-phase clouds from multiple aerosol size distributions and the effect of the clouds on aerosol removal. *J. Geophys. Res.* **108** (D8), 4245, doi:10.1029/2002JD002691.

(2005a) A refined method of parameterizing absorption coefficients among multiple gases simultaneously from line-by-line data. *J. Atmos. Sci.* **62**, 506–17.

(2005b) Studying ocean acidification with conservative, stable numerical schemes for nonequilibrium air–ocean exchange and ocean equilibrium chemistry. *J. Geophys. Res.*, in review.

(2005c) A solution to the problem of nonequilibrium acid/base gas-particle transfer at long time step. *Aerosol Sci. Technol.* **39**, 92–103.

Jacobson M. Z. and Seinfeld J. H. (2004) Evolution of nanoparticle size and mixing state near the point of emission. *Atmos. Environ.* **38**, 1839–50.

Jacobson M. Z. and Turco R. P. (1994) SMVGEAR: A sparse-matrix, vectorized Gear code for atmospheric models. *Atmos. Environ.* **28A**, 273–84.

(1995) Simulating condensational growth, evaporation, and coagulation of aerosols using a combined moving and stationary size grid. *Aerosol Sci. Technol.* **22**, 73–92.

Jacobson M. Z., Turco R. P., Jensen E. J., and Toon O. B. (1994) Modeling coagulation among particles of different composition and size. *Atmos. Environ.* **28A**, 1327–38.

Jacobson M. Z., Tabazadeh A., and Turco R. P. (1996b) Simulating equilibrium within aerosols and non-equilibrium between gases and aerosols. *J. Geophys. Res.* **101**, 9079–91.

Jaecker-Voirol A. and Mirabel P. (1989) Heteromolecular nucleation in the sulfuric acid-water system. *Atmos. Environ.* **23**, 2033–57.

Jaenicke R. (1988) Aerosol physics and chemistry. In *Numerical Data and Functional Relationships in Science and Technology*. New Series Vol. 4, Meteorology Subvol. b, Physical and Chemical Properties of Air, G. Fischer, ed., Berlin, Springer-Verlag.

Jarvis P. G., James G. B., and Landsberg J. J. (1976) Coniferous forest. In *Vegetation and the Atmosphere*, Vol. 2. J. L. Monteight, ed., New York, Academic Press, 171–240.

Jayne J. T., Davidovits P., Worsnop D. R., Zahniser M. S., and Kolb C. E. (1990) Uptake of SO_2 by aqueous surfaces as a function of pH: The effect of chemical reaction at the interface. *J. Phys. Chem.* **94**, 6041–8.

Jayson G. G., Parsons B. J., and Swallow A. J. (1973) Some simple, highly reactive, inorganic chlorine derivatives in aqueous solution. *Trans. Faraday Soc.* **69**, 1597–607.

Jeans J. (1954) *The Dynamical Theory of Gases*. New York, Dover.

Jenkin M. E., Saunders S. M., Wagner V., and Pilling J. (2003) Protocol for the development of the Master Chemical Mechanism, MCM v3 (Part B): tropospheric degradation of aromatic volatile organic compounds. *Atmos. Chem. Phys.* **3**, 181–93.

John W., Wall S. M., Ondo J. L., and Winklmayr W. (1989) *Acidic Aerosol Size Distributions During SCAQS*. Final Report for the California Air Resources Board under Contract No. A6-112-32.

Joseph J. H., Wiscombe W. J., and Weinman J. A. (1976) The delta-Eddington approximation for radiative flux transfer. *J. Atmos. Sci.* **33**, 2452–9.

Joslin R. D., Streett C. L., and Chang C.-L. (1993) Spatial direct numerical simulation of boundary-layer transition mechanisms–validation of PSE theory. *Theor. Comput. Fluid Dyn.* **4**, 271–88.

Junge C. E. (1961) Vertical profiles of condensation nuclei in the stratosphere. *J. Meteor.* **18**, 501–9.

Kaimal J. C. and Finnigan J. J. (1994) *Atmospheric Boundary Layer Flows: Their Structure and Measurement*. New York, Oxford University Press.

Kain J. S. (2004) The Kain–Fritsch convective parameterization: An update. *J. Appl. Meteor.* **43**, 170–81.

Kain J. S. and Fritsch J. M. (1990) A one-dimensional entraining/detraining plume model and its application in convective parameterization. *J. Atmos. Sci.* **47**, 2784–802.

Kao C.-Y. J. and Ogura Y. (1987) Response of cumulus clouds to large-scale forcing using the Arakawa–Schubert cumulus parameterization. *J. Atmos. Sci.* **44**, 2437–548.

Kaps P. and Rentrop P. (1979) Generalized Runge–Kutta methods of order four with stepsize control for stiff ordinary differential equations. *Numer. Math.* **33**, 55–88.

Kasahara A. (1974) Various vertical coordinate systems used for numerical weather prediction. *Mon. Wea. Rev.* **102**, 509–22.

Kasten F. (1968) Falling speed of aerosol particles. *J. Appl. Meteor.* **7**, 944–7.

Katrinak K. A., Rez P., Perkes P. R., and Buseck P. R. (1993) Fractal geometry of carbonaceous aggregates from an urban aerosol. *Environ. Sci. Technol.* **27**, 539–47.

Kawata Y. and Irvine W. M. (1970) The Eddington approximation for planetary atmospheres. *Astrophys. J.* **160**, 787–90.

Keeling C. D. and Whorf T. P. (2003) Atmospheric CO_2 concentrations (ppmv) derived from in situ air samples collected at Mauna Loa Observatory, Hawaii. cdiac.esd.ornl.gov./ftp/maunaloa-co2/maunaloa.co2.

Kerker M. (1969) *The Scattering of Light and Other Electromagnetic Radiation.* New York, Academic Press.

Ketefian G. (2005) Development, testing, and application of a 3-D nonhydrostatic potential-energy-conserving, compressible atmospheric model. Ph.D. Dissertation, Stanford University.

Ketefian G. and Jacobson M. Z. (2005a) Development and application of a 2-D potential-enstrophy-, energy-, and mass-conserving mixed-layer ocean model with arbitrary boundaries, *Mon. Wea. Rev.*, in submission.

(2005b) Development and application of an orthogonal-curvilinear-grid nonhydrostatic potential-enstrophy-, energy-, and mass-conserving atmospheric model with arbitrary boundaries, *Mon. Wea. Rev.*, in submission.

Kettle A. J. and Andreae M. O. (2000) Flux of dimethylsulfide from the oceans: A comparison of updated data sets and flux models. *J. Geophys. Res.* **105**, 26,793–808.

Kim Y. P. and Seinfeld J. H. (1995) Atmospheric gas-aerosol equilibrium: III. Thermodynamics of crustal elements Ca^{2+}, K^+, and Mg^{2+}. *Aerosol Sci. Technol.* **22**, 93–110.

Kim J., Moin P., and Moser R. (1987) Turbulence statistics in fully developed channel flow at low Reynolds number. *J. Fluid Mech.* **177**, 133–66.

Kim Y. P., Seinfeld J. H., and Saxena P. (1993a) Atmospheric gas–aerosol equilibrium I. Thermodynamic model. *Aerosol Sci. Technol.* **19**, 157–81.

(1993b) Atmospheric gas–aerosol equilibrium II. Analysis of common approximations and activity coefficient calculation methods. *Aerosol Sci. Technol.* **19**, 182–98.

Kittelson D. B. (1998) Engine and nanoparticles: a review. *J. Aerosol Sci.* **6**, 443–51.

Klemp J. B. and Wilhelmson R. B. (1978) The simulation of three-dimensional convective storm dynamics. *J. Atmos. Sci.* **3**, 1070–96.

Klingen H.-J. and Roth P. (1989) Size analysis and fractal dimension of diesel particles based on REM measurements with an automatic imaging system. *J. Aerosol Sci.* **20**, 861–4.

Knudsen M. and Weber S. (1911) Luftwiderstand gegen die langsame Bewegung kleiner Kugeln. *Ann. Phys.* **36**, 981–94.

Köhler H. (1936) The nucleus in the growth of hygroscopic droplets. *Trans. Faraday Soc.* **32**, 1152–61.

Kondo J., Saigusa N., and Sato T. (1990) A parameterization of evaporation from bare soil surfaces. *J. Appl. Meteor.* **29**, 385–9.

Kondratyev K. Ya. (1969) *Radiation in the Atmosphere.* San Diego, Academic Press, 912pp.

Korhonen H., Napari I., Timmreck C., *et al.* (2003) Heterogeneous nucleation as a potential sulphate-coating mechanism of atmospheric mineral dust particles and implications of coated dust on new particle formation. *J. Geophys. Res.* **108** (D17), 4546, doi:10.1029/2003JD003553.

Koschmieder H. (1924) Theorie der horizontalen Sichtweite. *Beitr. Phys. Freien Atm.* **12**, 33–53, 171–81.

Kozac-Channing L. F. and Heltz G. R. (1983) Solubility of ozone in aqueous solutions of 0–0.6 M ionic strength at 5–30 °C. *Environ. Sci. Technol.* **17**, 145–9.

Kreidenweis S. M., Zhang Y., and Taylor G. R. (1997) The effects of clouds on aerosol and chemical species production and distribution 2. Chemistry model description and sensitivity analysis. *J. Geophys. Res.* **102**, 23,867–82.

Kreidenweis S. M., Walcek C., Kim C.-H., *et al.* (2003) Modification of aerosol mass and size distribution due to aqueous-phase SO_2 oxidation in clouds: comparison of several models. *J. Geophys. Res.* **108** (D7) doi:10.1029/2002JD002697.

Kreitzberg C. W. and Perkey D. (1976) Release of potential instability. Part I: A sequential plume model within a hydrostatic primitive equation model. *J. Atmos. Sci.* **33**, 456–75.

Krekov G. M. (1993) Models of atmospheric aerosols. In *Aerosol Effects on Climate*. S. G. Jennings, ed., Tucson, University of Arizona Press, 9–72.

Krishnamurti T. N. and Moxim W. J. (1971) On parameterization of convective and non-convective latent heat release. *J. Appl. Meteor.* **10**, 3–13.

Krishnamurti T. N., Bedi H. S., and Hardiker V. M. (1998) *An Introduction to Global Spectral Modeling*. New York, Oxford University Press.

Krishnamurti T. N., Pan H.-L., Pasch R. J., and Molinari J. (1980) Cumulus parameterization and rainfall rates I. *Mon. Wea. Rev.* **108**, 465–72.

Kritz M. A. and Rancher J. (1980) Circulation of Na, Cl, and Br in the tropical marine atmosphere. *J. Geophys. Res.* **85**, 1633–9.

Kulmala M., Laaksonen A., and Pirjola L. (1998) Parameterizations for sulfuric acid/water nucleation rates. *J. Geophys. Res.* **103**, 8301–7.

Kuo H. L. (1965) On formation and intensification of tropical cyclones through latent heat release by cumulus convection. *J. Atmos. Sci.* **22**, 40–63.

(1974) Further studies of the parameterization of the influence of cumulus convection on large-scale flow. *J. Atmos. Sci.* **31**, 1232–40.

Kurihara Y. (1973) A scheme of moist convective adjustment. *Mon. Wea. Rev.* **101**, 547–53.

Kurihara Y. and Bender M. A. (1983) A numerical scheme to treat the open lateral boundary of limited area model. *Mon. Wea. Rev.* **111**, 445–54.

Kusik C. L. and Meissner H. P. (1978) Electrolyte activity coefficients in inorganic processing. *AIChE J. Symp. Ser.* **173**, 14–20.

Lacis A. A. and Hansen J. E. (1974) A parameterization for the absorption of solar radiation in the Earth's atmosphere. *J. Atmos. Sci.* **31**, 118–33.

Lacis A., Wang W. C., and Hansen J. (1979) Correlated *k*-distribution method for radiative transfer in climate models: Application to effect of cirrus clouds on climate. *NASA Conf. Publ.* 2076, 309–314.

Lacis A. A. and Oinas V. (1991) A description of the correlated *k*-distribution method for modeling nongray gaseous absorption, thermal emission, and multiple scattering in vertically inhomogeneous atmospheres. *J. Geophys. Res.* **96**, 9027–63.

Lamb H. (1910) On atmospheric oscillations. *Proc. Roy. Soc. London* **84**, 551–72.

Langford A. O., Proffitt M. H., VanZandt T. E., and Lamarque J.-F. (1996) Modulation of tropospheric ozone by a propagating gravity wave. *J. Geophys. Res.* **101**, 26, 605–13.

Larson S., Cass G., Hussey K., and Luce F. (1984) *Visibility Model Verification by Image Processing Techniques*. Final report to the California Air Resources Board under Agreement A2-077-32.

Lary D. J. (1997) Catalytic destruction of stratospheric ozone. *J. Geophys. Res.* **102**, 21,515–26.

Lazrus A. L., Cadle R. D., Gandrud B. W., Greenberg J. P., Huebert B. J., and Rose W. I. (1979) Sulfur and halogen chemistry of the stratosphere and of volcanic eruption plumes. *J. Geophys. Res.* **84**, 7869.

Ledbury W. and Blair E. W. (1925) The partial formaldehyde vapour pressure of aqueous solutions of formaldehyde, II. *J. Chem. Soc.* **127**, 2832–39.

Lee H. D. P., translator (1951) *Meteorologica* by Aristotle, T. E. Page, ed., Cambridge, MA, Harvard University Press.

Lee K. O., Cole R., Sekar R., *et al.* (2001) Detailed characterization of morphology and dimensions of diesel particulates via thermophoretic sampling, SAF 2001-01-3572.

Lee K. W. (1985) Conservation of particle size distribution parameters during Brownian coagulation. *J. Colloid Interface Sci.* **108**, 199–206.

Lee T. J. and Pielke R. A. (1992) Estimating the soil surface specific humidity. *J. Appl. Meteor.* **31**, 480–4.

Lee Y.-N. (1984) Kinetics of some aqueous-phase reactions of peroxyacetyl nitrate. In *Gas-Liquid Chemistry of Natural Waters*, Vol.1, BNL 51757, pp. 21/1–21/7, Brookhaven National Laboratory, Brookhaven, NY.

Le Henaff P. (1968) Méthodes d'étude et propriétés des hydrates, hemiacétals et hemiacétals derivés des aldehydes et des cétones. *Bull. Soc. Chim. France*, **11**, 4687–700.

Lenschow D. H., Li X. S., Zhu C. J., and Stankov B. B. (1988) The stably stratified boundary layer over the great planes. *Boundary-Layer Meteor.* **42**, 95–121.

Lesins G., Chylek P., and Lohmann U. (2002) A study of internal and external mixing scenarios and its effect on aerosol optical properties and direct radiative forcing. *J. Geophys. Res.* **107** (D10), doi:10.1029/2001JD000973.

Lettau H. H. (1969) Note on aerodynamic roughness-parameter estimation on the basis of roughness element description. *J. Appl. Meteor.* **8**, 828–32.

Li Z., Williams A. L., and Rood M. J. (1998) Influence of soluble surfactant properties on the activation of aerosol particles containing inorganic solute. *J. Atmos. Sci.* **55**, 1859–66.

Liang J. and Jacob D. J. (1997) Effect of aqueous-phase cloud chemistry on tropospheric ozone. *J. Geophys. Res.* **102**, 5993–6002.

Liang J. and Jacobson M. Z. (1999) A study of sulfur dioxide oxidation pathways for a range of liquid water contents, pHs, and temperatures. *J. Geophys. Res.* **104**, 13, 749–69.

(2000) Comparison of a 4000-reaction chemical mechanism with the carbon bond IV and an adjusted carbon bond IV-EX mechanism using SMVGEAR II. *Atmos. Environ.* **34**, 3015–26.

Lide D. R., ed.-in-chief (2003) *CRC Handbook of Chemistry and Physics*. Boca Raton, FL, CRC Press, Inc.

Lilly D. K. (1996) A comparison of incompressible, anelastic and Boussinesq dynamics. *Atmos. Res.* **40**, 143–51.

Lind J. A. and Kok G. L. (1986) Henry's law determinations for aqueous solutions of hydrogen peroxide, methylhydroperoxide, and peroxyacetic acid. *J. Geophys. Res.* **91**, 7889–95.

Lind J. A., Kok G. L., and Lazrus A. L. (1987) Aqueous phase oxidation of sulfur(IV) by hydrogen peroxide, methylhydroperoxide, and peroxyacetic acid. *J. Geophys. Res.* **92**, 4171–7.

Lindzen R. S. (1981) Turbulence and stress due to gravity wave and tidal breakdown. *J. Geophys. Res.* **86**, 9707–14.

Liou K. N. (1974) Analytic two-stream and four-stream solutions for radiative transfer. *J. Atmos. Sci.* **31**, 1473–5.

(2002) *An Introduction to Atmospheric Radiation*. Amsterdam, Academic Press.

Liousse C., Penner J. E., Chuang C., Walton J. J., Eddleman H., and Cachier H. (1996) A global three-dimensional model study of carbonaceous aerosols. *J. Geophys. Res.* **101**, 19,411–32.

Liss P. S. and Merlivat L. (1986) Air–sea gas exchange rates: Introduction and synthesis. In *The Role of Air–Sea Exchange in Geochemical Cycling*. P. Buat-Menard, ed., Hingham, MA, D. Reidel Publishing Co., 113–127.

List R. J., ed. (1984) *Smithsonian Meteorological Tables*, 6th edn. Washington, DC, Smithsonian Institution Press.

List R. and Gillespie J. R. (1976) Evolution of raindrop spectra with collision-induced breakup. *J. Atmos. Sci.* **33**, 2007–13.

Lister J. D., Smit D. J., and Hounslow M. J. (1995) Adjustable discretized population balance for growth and aggregation. *AIChE Journal* **41**, 591–603.

Liu C. H. and Leung D. Y. C. (2001) Turbulence and dispersion studies using a three-dimensional second-order closure Eulerian model. *J. Appl. Meteorol.* **40**, 92–113.

Lord S. J. and Arakawa A. (1980) Interaction of a cumulus cloud ensemble with the large-scale environment. Part II. *J. Atmos. Sci.* **37**, 2677–92.

Lorentz H. A. (1906) The absorption and emission of lines of gaseous bodies. In *H. A. Lorentz Collected Papers* (The Hague, 1934–1939) **3**, 215–38.

Lorenz E. N. (1960) Energy and numerical weather prediction. *Tellus* **12**, 364–73.

Louis J.-F. (1979) A parametric model of vertical eddy fluxes in the atmosphere. *Boundary-Layer Meteor.* **17**, 187–202.

Lu R. (1994) Development of an integrated air pollution modeling system and simulations of ozone distributions over the Los Angeles Basin. Ph.D. Thesis, University of California, Los Angeles.

Lu R. and Turco R. P. (1994) Air pollution transport in a coastal environment. Part I: Two-dimensional simulations of sea-breeze and mountain effects. *J. Atmos. Sci.* **51**, 2285–308.

Ludlum F. H. (1980) *Clouds and Storms*. University Park, PA, The Pennsylvania State University Press.

Lurmann F. W., Carter W. P. L., and Coyner L. A. (1987) *A Surrogate Species Chemical Reaction Mechanism for Urban Scale Air Quality Simulation Models. Volume I: Adaption of the Mechanism*. EPA-600/3-87/014a, U.S. Environmental Protection Agency, Research Triangle Park, NC.

Lurmann F. W., Main H. H., Knapp K. T., Stockburger L., Rasmussen R. A., and Fung K. (1992) *Analysis of the Ambient VOC Data Collected in the Southern California Air Quality Study*, Final Report to the California Air Resources Board under Contract A832-130.

Madronich S. (1987) Photodissociation in the atmosphere 1. Actinic flux and the effects of ground reflections and clouds. *J. Geophys. Res.* **92**, 9740–52.

Madronich, S. and Calvert J. G. (1989) *The NCAR Master Mechanism of the Gas-phase Chemistry-Version 2.0*. Rep. NCAR/TN-333+STR, National Center for Atmospheric Research.

Mahfouf J.-F. and Noilhan J. (1991) Comparative study of various formulations of evaporation from bare soil using in situ data. *J. Appl. Meteor.* **30**, 1354–65.

(1996) Inclusion of gravitational drainage in a land surface scheme based on the force-restore method. *J. Appl. Meteor.* **35**, 987–92.

Mahrt L., Heald R. C., Lenschow D. H., Stankov B. B., and Troen I. (1979) An observational study of the structure of the nocturnal boundary layer. *Boundary-Layer Meteor.* **17**, 247–64.

Makar P. A. (2001) The estimation of organic gas vapour pressure. *Atmos. Environ.* **35**, 961–74.

Makar P. A. and Karpik S. R. (1996) Basis-spline interpolation on the sphere: Applications to semi-lagrangian advection. *Mon. Wea. Rev.* **124**, 182–99.

Makar P. A., Vouchet V. S., and Nenes A. (2003) Inorganic chemistry calculations using HETV – a vectorized solver for the SO_4^{2-}-NO_3^--NH_4^+ system based on the ISORROPIA algorithms. *Atmos. Environ.* **37**, 2279–94.

Makar P. A., Moran M. D., Scholtz M. T., and Taylor A. (2003) Speciation of volatile organic compound emissions for regional air quality modeling of particulate matter and ozone. *J. Geophys. Res.* **108** (D2), 4041, doi:10.1029/2001JD000797.

Manabe S. J., Smagorinsky J., and Strickler R. F. (1965) Simulated climatology of a general circulation model with a hydrological cycle. *Mon. Wea. Rev.* **93**, 769–98.

Marbaix P., Gallee H., Brasseur O., and Van Ypersele J.-P. (2003) Lateral boundary conditions in regional climate models: A detailed study of the relaxation procedure. *Mon. Wea. Rev.* **131**, 461–79.

Maricq, M. M., Chase R. E., Podsiadlik D. H., and Vogt R. (1999) *Vehicle Exhaust Particle Size Distributions: A Comparison of Tailpipe and Dilution Tunnel Measurements.* SAE Technical Paper 1999-01-1461, Warrendale, PA, USA.

Marland G., Boden T. A., and Andres R. J. (2003) Global CO_2 emissions from fossil-fuel burning, cement manufacture, and gas flaring: 1751–2000. In *Trends Online: A Compendium of Data on Global Change.* Carbon Dioxide Information Analysis Center, Oak Ridge National Laboratory, U.S. Department of Energy, Oak Ridge, TN, USA.

Marlow W. H. (1981) Size effects in aerosol particle interactions: The van der Waals potential and collision rates. *Surf. Sci.* **106**, 529–37.

Marsh A. R. W. and McElroy W. J. (1985) The dissociation constant and Henry's law constant of HCl in aqueous solution. *Atmos. Environ.* **19**, 1075–80.

Marshall J. S. and Palmer W. (1948) The distribution of raindrops with size. *J. Meteor.* **3**, 165–8.

Marshall S. F., Covert D.S., and Charlson R. J. (1995) Relationship between asymmetry parameter and hemispheric backscatter ratio: implications for climate forcing by aerosols. *Appl. Opt.* **34**, 6306–11.

Martens C. S., Wesolowski J. J., Hariss R. C., and Kaifer R. (1973) Chlorine loss from Puerto Rican and San Francisco Bay Area marine aerosols. *J. Geophys. Res.* **78**, 8778–92.

Martensson E. M., Nilsson E. D., de Leeuw G., Cohen L. H., and Hansson H.-C. (2003) Laboratory simulation and parameterization of the primary marine aerosol production. *J. Geophys. Res.* **108** (D9), 4297, doi:10.1029/2002JD002263.

Marticorena B. and Bergametti G. (1995) Modeling the atmospheric dust cycle: 1. Design of a soil-derived dust emission scheme. *J. Geophys. Res.* **100**, 16415–30.

Marticorena B., Bergametti G., Gillette D., and Belnap J. (1997) Factors controlling threshold friction velocity in semiarid and arid areas of the United States. *J. Geophys. Res.* **102**, 23277–87.

Martin J. J., Wang P. K., and Pruppacher H. R. (1980) A theoretical study of the effect of electric charges on the efficiency with which aerosol particles are collected by ice crystal plates. *J. Colloid Interface Sci.* **78**, 44–56.

Martin L. R. and Hill M. W. (1987a) The iron-catalyzed oxidation of sulfur: Reconciliation of the literature rates. *Atmos. Environ.* **21**, 1487–90.

(1987b) The effect of ionic strength on the manganese catalyzed oxidation of sulfur(IV). *Atmos. Environ.* **21**, 2267–70.

Mason B. J. (1971) *The Physics of Clouds.* Oxford, Clarendon Press.

Matsuno T. (1966) Numerical integrations of the primitive equations by simulated backward difference scheme. *J. Meteor. Soc. Japan* **44**, 76–84.

Mauna Loa Data Center (2001) Data for atmospheric trace gases. http://mloserv.mlo.hawaii.gov/.

Maxwell J. C. (1890) *The Scientific Papers of James Clerk Maxwell*, Vol. II. W. D. Niven, ed., Cambridge, Cambridge University Press, 636–40.

Maxwell Garnett J. C. (1904) Colours in metal glasses and in metallic films. *Philos. Trans. Roy. Soc.* **A203**, 385–420.

McClelland L., Simkin T., Summers M., Nielsen E., and Stein T. C. (eds.) (1989) *Global Volcanism 1975–1985.* Englewood Cliffs, NJ, Prentice-Hall, 655pp.

McCumber M. C. (1980) A numerical simulation of the influence of heat and moisture fluxes upon mesoscale circulations, Ph.D. Thesis, University of Virginia, Charlottesville.
McCumber M. C. and Pielke R. A. (1981) Simulation of the effects of surface fluxes of heat and moisture in a mesoscale numerical model. Part I: Soil layer. *J. Geophys. Res.* **86**, 9929–38.
McElroy M. B., Salawitch R. J., Wofsy S. C., and Logan J. A. (1986) Reduction of Antarctic ozone due to synergistic interactions of chlorine and bromine. *Nature* **321**, 759–62.
McGraw R. and Saunders J. H. (1984) A condensation feedback mechanism for oscillatory nucleation and growth. *Aerosol Sci. Technol.* **3**, 367–80.
McMurry P. H. and Grosjean D. (1985) Photochemical formation of organic aerosols: growth laws and mechanisms. *Atmos. Environ.* **19**, 1445–51.
McRae G. J., Goodin W. R., and Seinfeld J. H. (1982) Development of a second-generation mathematical model for urban air pollution – I. Model formulation. *Atmos. Environ.* **16**, 679–96.
Meador W. E. and Weaver W. R. (1980) Two-stream approximations to radiative transfer in planetary atmospheres: A unified description of existing methods and a new improvement. *J. Atmos. Sci.* **37**, 630–43.
Mellor G. L. and Yamada T. (1974) A hierarchy of turbulence closure models for planetary boundary layers. *J. Atmos. Sci.* **31**, 1791–806.
(1982) Development of a turbulence closure model for geophysical fluid problems. *Rev. Geophys. Space Phys.* **20**, 851–75.
Meng Z. and Seinfeld J. H. (1996) Time scales to achieve atmospheric gas–aerosol equilibrium for volatile species. *Atmos. Environ.* **30**, 2889–900.
Meng Z., Seinfeld J. H., Saxena P., and Kim Y. P. (1995) Atmospheric gas–aerosol equilibrium: IV. Thermodynamics of carbonates. *Aerosol. Sci. Technol.* **23**, 131–54.
Meng Z., Dabdub D., and Seinfeld J. H. (1998) Size-resolved and chemically resolved model of atmospheric aerosol dynamics. *J. Geophys. Res.* **103**, 3419–35.
Mesinger F. and Arakawa A. (1976) *Numerical Methods Used in Atmospheric Models.* GARP Publication Series. No. 17, 1. World Meteorological Organization, 64pp.
Metzger S., Dentener F., Pandis S., and Lelieveld J. (2002) Gas/aerosol partitioning: 1. A computationally efficient model. *J. Geophys. Res.* **107** (D16) 10.1029/2001JD001102.
Middleton P. and Brock J. R. (1976) Simulation of aerosol kinetics. *J. Colloid Interface Sci.* **54**, 249–64.
Middleton W. E. K. (1952) *Vision Through the Atmosphere.* Toronto, Canada, University of Toronto Press.
Mihailovic D. T., Rajkovic B., Lalic B., and Dekic L. (1995) Schemes for parameterizing evaporation from a non-plant covered surface and their impact on partitioning the surface energy in land–air exchange parameterization. *J. Appl. Meteor.* **34**, 2462–75.
Millero F. J. (1995) Thermodynamics of the carbon dioxide system in the oceans. *Geochim. Cosmochim. Acta* **59**, 661–7.
Millikan R. A. (1923) The general law of fall of a small spherical body through a gas, and its bearing upon the nature of molecular reflection from surfaces. *Phys. Rev.* **22**, 1–23.
Mitchell A. R. (1969) *Computational Methods in Partial Differential Equations.* New York, John Wiley.
Miyakoda K., Smagorinsky J., Strickler R. F., and Hembree G. D. (1969) Experimental extended predictions with a nine-level hemispheric model. *Mon. Wea. Rev.* **97**, 1–76.
Moeng C. H. (1984) A large-eddy simulation model for the study of planetary boundary-layer turbulence. *J. Atmos. Sci.* **41**, 2202–16.
Molina L. T. and Molina M. J. (1986) Production of Cl_2O_2 by the self reaction of the ClO radical. *J. Phys. Chem.* **91**, 433–6.
Molina M. J. and Rowland F. S. (1974) Stratospheric sink for chlorofluoromethanes: Chlorine atom catalysed destruction of ozone. *Nature* **249**, 810–2.

Molinari J. (1982) A method for calculating the effects of deep cumulus convection in numerical models. *Mon. Wea. Rev.* **11**, 1527–34.

Monahan E. C., Spiel D. E., and Davidson K. L. (1986) A model of marine aerosol generation via whitecaps and wave disruption. In *Oceanic Whitecaps and Their Role in Air–Sea Exchange Processes*. E. C. Monahan and G. MacNiocaill, eds., Norwell, MA, D. Reidel, 167–74.

Monin A. S. and Obukhov A. M. (1954) Basic laws of turbulent mixing in the ground layer of the atmosphere. *Trans. Geophys. Inst. Akad. Nauk USSR* **151**, 1963–87.

Monin A. S. and Yaglom (1971) *Statistical Fluid Mechanics*. Cambridge, MA, MIT Press.

Monteith J. L. and Szeicz G. (1962) Radiative temperature in the heat balance of natural surfaces. *Q. J. Roy. Meteor. Soc.* **88**, 496–507.

Moorthi S. and Suarez M. J. (1992) Relaxed Arakawa–Schubert: A parameterization of moist convection for general circulation models. *Mon. Wea. Rev.* **120**, 978–86.

Mordy W. (1959) Computations of the growth by condensation of a population of cloud droplets. *Tellus* 11, 16–44.

Mountain R. D., Mulholland G. W., and Baum H. (1986) Simulation of aerosol agglomeration in the free molecular and continuum flow regimes. *J. Colloid. Interface Sci.* **114**, 67–81.

Moya M., Pandis S. N., and Jacobson M. Z. (2001) Is the size distribution of urban aerosols determined by thermodynamic equilibrium? An application to Southern California. *Atmos. Environ.* **36**, 2349–65.

Mozurkewich M., McMurry P. H., Gupta A., and Calvert J. G. (1987) Mass accommodation coefficients for HO_2 radicals on aqueous particles. *J. Geophys. Res.* **92**, 4163–70.

Mulholland G. W., Samson R. J., Mountain R. D., and Ernst M. H. (1988) Cluster size distribution for free molecular agglomeration. *Energy and Fuels* **2**, 481–6.

Muller H. (1928) Zur allgemeinen Theorie der raschen Koagulation. Die koagulation von Stabchen- und Blattchen-kolloiden; die Theorie beliebig polydisperser Systeme und der Stromungskoagulation. *Kolloidbeihefte* **27**, 223–50.

Munger W. J., Collett J. Jr., Daube B. C., and Hoffmann M. R. (1989) Carboxylic acids and carbonyl compounds in southern California clouds and fogs. *Tellus* **41b**, 230–42.

Mylonas D. T., Allen D. T., Ehrman S. H., and Pratsinis S. E. (1991) The sources and size distributions of organonitrates in Los Angeles aerosols. *Atmos. Environ.* **25A**, 2855–61.

Nair R. D., Scroggs J. S., and Semazzi F. H. M. (2002) Efficient conservative global transport schemes for climate and atmospheric chemistry models. *Mon. Wea. Rev.* **130**, 2059–73.

Nakicenovic N. and Swart, R., eds. (2000) *Emissions Scenarios. A Special Report of the Intergovernmental Panel on Climate Change*. Cambridge, Cambridge University Press.

Napari I., Noppel M., Vehkamaki H., and Kulmala M. (2002) Parameterization of ternary nucleation rates for H_2SO_4-NH_3-H_2O vapors. *J. Geophys. Res.* **107** (D19), 4381, doi:10.1029/2002JD002132.

National Oceanic and Atmospheric Administration (NOAA) (1976) *U.S. Standard Atmosphere*. Washington, DC.

Naumann K.-H. (2003) COSIMA-A computer program simulating the dynamics of fractal aerosols. *J. Aerosol Sci.* **34**, 1371–97.

Nautical Almanac Office (NAO) and Her Majesty's Nautical Almanac Office (1993) *Astronomical Almanac*. Washington, DC, U.S. Government Printing Office.

Nebeker F. (1995) *Calculating the Weather*. San Diego, Academic Press, Inc.

Nenes A., Pandis S. N., and Pilinis C. (1998) ISORROPIA: A new thermodynamic equilibrium model for multiphase multicomponent inorganic aerosols. *Aquat. Geochem.* **4**, 123–52.

(1999) Continued development and testing of a new thermodynamic aerosol module for urban and regional air quality. *Atmos. Environ.* **33**, 1553–1560.

Nesbitt F. L., Monks P. S., Wayne W. A., Stief L. J., and Touni R. (1995) The reaction of $O(^3P)$ + HOBr: Temperature dependence of the rate constant and importance of the reaction as an HOBr loss process. *Geophys. Res. Lett.* **22**, 827–30.

Nguyen K. and Dabdub D. (2001) Two-level time-marching scheme using splines for solving the advection equation. *Atmos. Environ.* **35**, 1627–37.
(2002) Semi-Lagrangian flux scheme for the solution of the aerosol condensation/evaporation equation. *Aerosol Sci. Technol.* **36**, 407–418.
Nicolet M. (1989) Solar spectral irradiances with their diversity between 120 and 900 nm. *Planet. Space Sci.* **37**, 1249–89.
Noilhan J. and Planton S. (1989) A simple parameterization of land surface processes for meteorological models. *Mon. Wea. Rev.* **117**, 536–49.
Noll K. E., Fang K. Y. P., and Khalili E. (1990) Characterization of atmospheric coarse particles in the Los Angeles Basin. *Aerosol Sci. Technol.* **12**, 28–38.
Odum J. R., Hoffmann T., Bowman F., Collins T., Flagan R. C., and Seinfeld J. H. (1996) Gas-particle partitioning and secondary organic aerosol yields. *Environ. Sci. Technol.* **30**, 2580–5.
Ogura Y. and Phillips N. A. (1962) Scale analysis of deep and shallow convection in the atmosphere. *J. Atmos. Sci.* **19**, 173–9.
Okada K. and Hitzenberger R. (2001) Mixing properties of individual submicrometer particles in Vienna. *Atmos. Environ.* **35**, 5617–28.
Oke T. R. (1978) *Boundary Layer Climates*. London, Methuen.
Oke T. R., Spronken-Smith R. A., Jauregui E., and Grimmond C. S. B. (1999) The energy balance of central Mexico City during the dry season. *Atmos. Environ.* **33**, 3919–30.
Okuyama K., Kousaka Y., and Hayashi K. (1984) Change in size distribution of ultrafine aerosol particles undergoing Brownian coagulation. *J. Colloid Interface Sci.* **101**, 98–109.
Olscamp P. J., translator (1965) *Discourse on Method, Optics, Geometry, and Meteorology* by René Descartes. Indianapolis, Bobbs-Merrill Company, Inc.
Ooyama V. K. (1971) A theory on parameterization of cumulus convection. *J. Meteor. Soc. Japan* **49**, 744–56.
Orszag S. A. (1970) Transform method for calculation of vector coupled sums: Application to the spectral form of the vorticity equation. *J. Atmos. Sci.* **27**, 890–5.
(1971) Numerical simulation of incompressible flows within simple boundaries. I. Galerkin (spectral) representations. *Stud. Appl. Math.* **50**, 293–326.
Orville H. D. and Kopp F. J. (1977). Numerical simulations of the history of a hailstorm. *J. Atmos. Sci.* **34**, 1596–618.
Osborne N. S., Stimson H. F., and Ginnings D. C. (1939) Measurements of heat capacity and heat of vaporization of water in the range of 0 degrees to 100 degrees celsius. *J. Res. Nat. Bur. Stand.* **23**, 197–260.
Pandis S. N. and Seinfeld J. H. (1989) Sensitivity analysis of a chemical mechanism for aqueous-phase atmospheric chemistry. *J. Geophys. Res.* **94**, 1105–26.
Pandis S. N., Harley R. A., Cass G. R., and Seinfeld J. H. (1992) Secondary organic aerosol formation and transport. *Atmos. Environ.* **26A**, 2269–82.
Pandis S. N., Russell L. M., and Seinfeld J. H. (1994) The relationship between DMS flux and CCN concentration in remote marine regions. *J. Geophys. Res.* **99**, 16945–57.
Park J.-Y. and Lee Y.-N. (1987) Aqueous solubility and hydrolysis kinetics of peroxynitric acid. Paper presented at 193rd Meeting, American Chemical Society, Denver, CO, April 5–10.
Parker V. B. (1965) *Thermal Properties of Aqueous Uni-univalent Electrolytes*. National Standard Reference Data Series – NBS 2. U.S. Government Printing Office, Washington, DC.
Parkinson C. L. and Washington W. M. (1979) A large-scale numerical model for sea ice. *J. Geophys. Res.* **84**, 311–37.
Pasquill F. (1962) *Atmospheric Diffusion*. London, Van Nostrand.
Paulson S. E. and Seinfeld J. H. (1992) Development and evaluation of a photooxidation mechanism for isoprene. *J. Geophys. Res.* **97**, 20,703–15.

Peng C., Chan M. N., and Chan C. K. (2001) The hygroscopic properties of dicarboxylic and multifunctional acids: Measurements and UNIFAC predictions. *Environ. Sci. Technol.* **35**, 4495–501.

Pepper D. W., Kern C. D., and Long P. E. Jr. (1979) Modeling the dispersion of atmospheric pollution using cubic splines and chapeau functions. *Atmos. Environ.* **13**, 223–37.

Perrin D. D. (1982) *Ionization Constants of Inorganic Acids and Bases in Aqueous Solution*, 2nd edn. New York, Pergamon.

Perron G., Roux A., and Desnoyers J. E. (1981) Heat capacities and volumes of NaCl, $MgCl_2$, $CaCl_2$, and $NiCl_2$ up to 6 molal in water. *Can. J. Chem.* **59**, 3049–54.

Petersen R. L. (1997) A wind tunnel evaluation of methods for estimating surface roughness length at industrial facilities. *Atmos. Environ.* **31**, 45–57.

Philip J. R. (1957) Evaporation, and moisture and heat fields in the soil. *J. Meteor.* **14**, 354–66.

Phillips N. A. (1957) A coordinate system having some special advantages for numerical forecasting. *J. Meteor.* **14**, 184–5.

Pielke R. A. (1984) *Mesoscale Meteorological Modeling*. San Diego, Academic Press, Inc.

Pilinis C. and Seinfeld J. H. (1987) Continued development of a general equilibrium model for inorganic multicomponent atmospheric aerosols. *Atmos. Environ.* **21**, 2453–66.

(1988) Development and evaluation of an eulerian photochemical gas-aerosol model. *Atmos. Environ.* **22**, 1985–2001.

Pilinis C., Capaldo K. P., Nenes A., and Pandis S. N. (2000) MADM-A new multicomponent aerosol dynamics model. *Aerosol Sci. Technol.* **32**, 482–502.

Pinto J. P., Turco R. P., and Toon O. B. (1989) Self-limiting physical and chemical effects in volcanic eruption clouds. *J. Geophys. Res.* **94**, 11,165.

Pitter R. L. and Pruppacher H. R. (1973) A wind tunnel investigation of freezing of small water drops falling at terminal velocity in air. *Q. J. Roy. Meteor. Soc.* **99**, 540–50.

Pitzer K. S. (1991) Ion interaction approach: Theory and data correlation. In *Activity Coefficients in Electrolyte Solutions,* 2nd edn. K. S. Pitzer, ed., Boca Raton, FL, CRC Press, 75–153.

Pitzer K. S. and Mayorga G. (1973) Thermodynamics of electrolytes II. Activity and osmotic coefficients for strong electrolytes with one or both ions univalent. *J. Phys. Chem.* **77**, 2300–8.

Pollack J. B. and Cuzzi J. N. (1980) Scattering by nonspherical particles of size comparable to a wavelength: A new semi-empirical theory and its application to tropospheric aerosols. *J. Atmos. Sci.* **37**, 868–81.

Pooley F. D. and Mille M. (1999) Composition of air pollution particles. In *Air Pollution and Health*. S. T. Holgate, J. M. Samet, H. S. Koren, and R. L. Maynard, eds., San Diego, Academic Press, 619–34.

Potter J. F. (1970) The delta-function approximation in radiative transfer theory. *J. Atmos. Sci.* **27**, 943–9.

Prather M. J. (1986) Numerical advection by conservation of second-order moments. *J. Geophys. Res.* **91**, 6671–81.

Pratsinis S. E. (1988) Simultaneous nucleation, condensation, and coagulation in aerosol reactors. *J. Colloid Interface Sci.* **124**, 416–27.

Press W. H., Flannery B. P., Teukolsky, S. A., and Vetterling W. T. (1992). *Numerical Recipes: The Art of Scientific Computing*. Cambridge, Cambridge University Press.

Price C., Penner J., and Prather M. (1997) NOx from lightning 1. Global distribution based on lightning physics. *J. Geophys. Res.* **102**, 5929–41.

Price G. V. and MacPherson A. K. (1973) A numerical weather forecasting method using cubic splines on a variable mesh. *J. Appl. Meteor.* **12**, 1102–13.

Pruppacher H. R. and Klett J. D. (1997) *Microphysics of Clouds and Precipitation*, 2nd rev. and enl. edn., Dordrecht, Kluwer Academic Publishers.

Pruppacher H. R. and Rasmussen R. (1979) A wind tunnel investigation of the rate of evaporation of large water drops falling at terminal velocity in air. *J. Atmos. Sci.* **36**, 1255–60.

Purnell D. K. (1976) Solution of the advection equation by upstream interpolation with a cubic spline. *Mon. Wea. Rev.* **104**, 42–8.

Rahmes T. F., Omar A. H., and Wuebbles D. J. (1998) Atmospheric distributions of soot particles by current and future aircraft fleets and resulting radiative forcing on climate. *J. Geophys. Res.* **103**, 31,657–67.

Rao N. P. and McMurry P. H. (1989) Nucleation and growth of aerosol in chemically reacting systems. *Aerosol Sci. Technol.* **11**, 120–33.

Raoult F.-M. (1887) General law of the vapor pressure of solvents. *Comptes Rendus* **104**, 1430–3.

Rasch P. J. (1994) Conservative shape-preserving two-dimensional transport on a spherical grid. *Mon. Wea. Rev.* **122**, 1337–50.

Rasmussen R. and Pruppacher H. R. (1982) A wind tunnel and theoretical study of the melting behavior of atmospheric ice particles. I: A wind tunnel study of frozen drops of radius $<$500 mm. *J. Atmos. Sci.* **39**, 152–8.

Rasmussen R., Levizzani M. V., and Pruppacher H. R. (1984) A wind tunnel and theoretical study of the melting behavior of atmospheric ice particles. II: A theoretical study for frozen drops of radius $<$500 mm. *J. Atmos. Sci.* **41**, 374–80.

Reid J. S. and Hobbs P. V. (1998) Physical and optical properties of young smoke from individual biomass fires in Brazil. *J. Geophys. Res.* **103**, 32,013–30.

Reid, J. S., Hobbs P. V., Ferek R. J., *et al.* (1998) Physical, chemical, and optical properties of regional hazes dominated by smoke in Brazil. *J. Geophys. Res.* **103**, 32,059–80.

Reisin T., Levin Z., and Tzivion S. (1996) Rain production in convective clouds as simulated in an axisymmetric model with detailed microphysics. Part I: Description of the model. *J. Atmos. Sci.* **53**, 497–519.

Reynolds S. D., Roth P. M., and Seinfeld J. H. (1973) Mathematical modeling of photochemical air pollution – I: Formulation of the model. *Atmos. Environ.* **7**, 1033–61.

Richardson L. F. (1922) *Weather Prediction by Numerical Process.* Cambridge, Cambridge University Press, reprinted 1965, 236pp.

Robert A. (1982) A semi-Lagrangian and semi-implicit numerical integration scheme for the primitive meteorological equations. *Japan Meteor. Soc.* **60**, 319–25.

Robinson R. A. and Stokes R. H. (1955) *Electrolyte Solutions.* New York, Academic Press.

Rogak S. N. and Flagan R. C. (1992) Coagulation of aerosol agglomerates in the transition regime. *J. Colloid Interface Sci.* **151**, 203–24.

Rogers R. R. and Yau M. K. (1989) *A Short Course in Cloud Physics.* Oxford, Pergamon Press.

Rosenbaum J. S. (1976) Conservation properties of numerical integration methods for systems of ordinary differential equations. *J. Comp. Phys.* **20**, 259–67.

Ross A. B. and Neta P. (1979) *Rate Constants for Reactions of Inorganic Radicals in Aqueous Solutions.* NSRDS-NBS 65. National Bureau of Standards, U.S. Department of Commerce, Washington, DC.

Rossby C. and collaborators (1939) Relation between variations in the intensity of the zonal circulation of the atmosphere and the displacements of the semi-permanent centers of action. *J. Marine Res.* **2**, 38–55.

Rothman L. S., *et al.* (2003) The HITRAN molecular spectroscopic database: Edition of 2000 including updates of 2001. *J. Quant. Spectrosc. Radiat. Transfer* **82**, 5–44.

Roux A., Musbally G. M., Perron G., *et al.* (1978) Apparent molal heat capacities and volumes of aqueous electrolytes at 25 °C: $NaClO_3$, $NaClO_4$, $NaNO_3$, $NaBrO_3$, $NaIO_3$, $KClO_3$, $KBrO_3$, KIO_3, NH_4NO_3, NH_4Cl, and NH_4ClO_4. *Can. J. Chem.* **56**, 24–8.

Russell A. G., Winner D. A., Harley R. A., McCue K. F., and Cass G. R. (1993) Mathematical modeling and control of the dry deposition flux of nitrogen-containing air pollutants. *Environ. Sci. Technol.* **27**, 2772–82.

Russell L. M., Pandis S. N., and Seinfeld J. H. (1994) Aerosol production and growth in the marine boundary layer. *J. Geophys. Res.* **99**, 20,989–21,003.

Saffman P. G. and Turner J. S. (1956) On the collision of drops in turbulent clouds. *J. Fluid Mech.* **1**, 16–30.

Sander R., Lelieveld J., and Crutzen P. J. (1995) Modelling of nighttime nitrogen and sulfur chemistry in size resolved droplets of an orographic cloud. *J. Atmos. Chem.* **20**, 89–116.

Sandu A. (2001) Positive numerical integration methods for chemical kinetic systems. *J. Comp. Phys.* **170**, 589–602.

(2002) A Newton–Cotes quadrature approach for solving the aerosol coagulation equation. *Atmos. Environ.* **36**, 583–9.

Sandu A., Verwer J. G., van Loon M., *et al.* (1997) Benchmarking stiff ODE solvers for atmospheric chemistry problems I: Implicit versus explicit. *Atmos. Environ.* **31**, 3151–66.

San Jose R., Casanova J. L., Viloria R. E., and Casanova J. (1985) Evaluation of the turbulent parameters of the unstable surface boundary layer outside Businger's range. *Atmos. Environ.* **19**, 1555–61.

Saunders S. M., Jenkin M. E., Derwent R. G., and Pilling M. J. (2003) Protocol for the development of the Mater Chemical Mechanism, MCM v3 (Part A): tropospheric degradation of non-aromatic volatile organic compounds. *Atmos. Chem. Phys.* **3**, 161–80.

Saxena P., Hudischewskyj A. B., Seigneur C., and Seinfeld J. H. (1986) A comparative study of equilibrium approaches to the chemical characterization of secondary aerosols. *Atmos. Environ.* **20**, 1471–83.

Saxena P., Mueller P. K., and Hildemann L. M. (1993) Sources and chemistry of chloride in the troposphere: A review. In *Managing Hazardous Air Pollutants: State of the Art.* W. Chow and K. K. Connor, eds., Boca Raton, FL, Lewis Publishers, 173–90.

Schmidt K. H. (1972) Electrical conductivity techniques for studying the kinetics of radiation-induced chemical reactions in aqueous solutions. *Int. J. Radiat. Phys. Chem.* **4**, 439–68.

Schmidt-Ott A. and Burtscher H. (1982) The effect of van der Waals forces on aerosol coagulation. *J. Colloid Interface Sci.* **89**, 353–7.

Schnaiter M., Horvath H., Mohler O., Naumann K.-H., Saathoff H., and Schock O. W. (2003) UV-VIS-NIR spectral optical properties of soot and soot-containing aerosols. *J. Aerosol Sci.* **34**, 1421–44.

Schneider W., Moortgat G. K., Tyndall G. S., and Burrows J. P. (1987) Absorption cross-sections of NO_2 in the UV and visible region (200–700 nm) at 298 K. *J. Photochem. Photobiol, A: Chem.* **40**, 195–217.

Scholes G. and Willson R. L. (1967) γ-radiolysis of aqueous thymine solutions. Determination of relative reaction rates of OH radicals. *Trans. Faraday Soc.* **63**, 2982–93.

Schroeder W. H., Dobson M., Kane D. M., and Johnson N. D. (1987) Toxic trace elements associated with airborne particulate matter: a review. *J. Air Pollut. Control Assoc.* **37**, 1267–85.

Schwartz S. E. (1984) Gas- and aqueous-phase chemistry of HO_2 in liquid water clouds. *J. Geophys. Res.* **89**, 11,589–98.

(1986) Mass-transport considerations pertinent to aqueous phase reactions of gases in liquid-water clouds. In *Chemistry of Multiphase Atmospheric Systems*, NATO ASI Series, Vol. G6. W. Jaeschke, ed., Berlin, Springer-Verlag, 415–71.

Schwartz S. E. and White W. H. (1981) Solubility equilibria of the nitrogen oxides and oxyacids in aqueous solution. *Adv. Environ. Sci. Eng.* **4**, 1–45.

Seaman N. L., Ludwig F. L., Donall E. G., Warner T. T., and Bhumralkar C. M. (1989) Numerical studies of urban planetary boundary-layer structure under realistic synoptic conditions. *J. Appl. Meteor.* **28**, 760–81.

Seery D. J. and Britton D. (1964) The continuous absorption spectra of chlorine, bromine, bromine chloride, iodine chloride, and iodine bromide. *J. Phys. Chem.* **68**, 2263–6.

Sehested K., Rasmussen O. L., and Fricke H. (1968) Rate constants of OH with HO_2, O_2^-, and $H_2O_2^+$ from hydrogen peroxide formation in pulse-irradiated oxygenated water. *J. Phys. Chem.* **72**, 626–31.

Sehmel G. A. (1980) Particle and gas dry deposition: A review. *Atmos. Environ.* **14**, 983–1011.

Seinfeld J. H. and Pandis S. N. (1998) *Atmospheric Chemistry and Physics*. New York, Wiley-Interscience.

Sellers W. D. (1965) *Physical Climatology*. Chicago, University of Chicago Press, 272pp.

Sellers, P. J., Los S. O., Tucker C. J., *et al.* (1996) A revised land surface parameterization (SiB2) for atmospheric GCMs. Part II: The generation of global fields of terrestrial biophysical parameters from satellite data. *J. Clim.* **9**, 706–37.

Shao Y. (2001) A model for mineral dust emission. *J. Geophys. Res.* **106**, 20239–54.

Shao Y., Raupach M. R., and Leys J. F. (1996) A model for predicting Aeolian sand drift and dust entrainment on scales from paddock to region. *Aust. J. Soil Res.* **34**, 309–42.

Shen T.-L., Wooldridge P. J., and Molina M. J. (1995) Stratospheric pollution and ozone depletion. In *Composition, Chemistry, and Climate of the Atmosphere*. H. B. Singh, ed., New York, Van Nostrand Reinhold.

Sheridan P. J., Brock C. A., and Wilson J. C. (1994) Aerosol particles in the upper troposphere and lower stratosphere: Elemental composition and morphology of individual particles in northern midlatitudes. *Geophys. Res. Lett.* **21**, 2587–90.

Sherman A. H. and Hindmarsh A. C. (1980) *GEARS: A Package for the Solution of Sparse, Stiff Ordinary Differential Equations*. Report UCRL-84102, Lawrence Livermore Laboratory.

Shimazaki T. and Laird A. R. (1970) A model calculation of the diurnal variation in minor neutral constituents in the mesosphere and lower thermosphere including transport effects. *J. Geophys. Res.* **75**, 3221–35.

Shir C. C. and Bornstein R. D. (1976) Eddy exchange coefficients in numerical models of the planetary boundary layer. *Boundary-Layer Meteor.* **11**, 171–85.

Shuttleworth W. J. (1989) Micrometeorology of temperate and tropical forest. *Phil. Trans. Roy. Soc. London* **B324**, 299–334.

Siegel R. and Howell J. R. (1992) *Thermal Radiation Heat Transfer*. Washington, DC, Taylor and Francis.

Singh H. B. (1995) Halogens in the atmospheric environment. In *Composition, Chemistry, and Climate of the Atmosphere*. H. B. Singh, ed., New York, Van Nostrand Reinhold.

Singh H. B., Viezee W., and Salas L. J. (1988) Measurements of selected C_2-C_5 hydrocarbons in the troposphere: Latitudinal, vertical, and temporal variations. *J. Geophys. Res.* **93**, 15,861–78.

Singh H. B., Kanakidou M., Crutzen P. J., and Jacob D. J. (1995) High concentrations and photochemical fate of oxygenated hydrocarbons in the global troposphere. *Nature* **378**, 50–4.

Singh H. B., Herlth D., Kolyer R., *et al.* (1996) Reactive nitrogen and ozone over the western Pacific: Distributions, partitioning, and sources. *J. Geophys. Res.* **101**, 1793–808.

Skamarock W. C. and Klemp J. B. (1992) The stability of time-split numerical methods for the hydrostatic and the nonhydrostatic elastic equations. *Mon. Wea. Rev.* **120**, 2109–27.

Skamarock W. C., Dye J. E., Defer E., Barth M. C., Stith J. L., and Ridley B. A. (2003) Observational- and modeling-based budget of lightning-produced NOx in a continental thunderstorm. *J. Geophys. Res.* **108** (D10), 4305, doi:10.1029/2002JD002163.

Slinn W. G. N., Hasse L., Hicks B. B., *et al.* (1978) Some aspects of the transfer of atmospheric trace constituents past the air–sea interface. *Atmos. Environ.* **12**, 2055–87.

Smith M. H. and Harrison N. M. (1998) The sea spray generation function. *J. Aerosol Sci.* **29**, Suppl. 1, S189–S190.

Smith R. M. and Martell A. E. (1976) *Critical Stability Constants, Vol. 4: Inorganic Complexes*. New York, Plenum.

Smolarkiewicz P. K. (1983) A simple positive definite advection scheme with small implicit diffusion. *Mon. Wea. Rev.* **111**, 479–86.

Smoluchowski M. V. (1918) Versuch einer mathematischen Theorie der Koagulationskinetik kolloider Lösungen. *Z. Phys. Chem.* **92**, 129–68.

Snider J. R. and Dawson G. A. (1985) Tropospheric light alcohols, carbonyls, and acetonitrile: Concentrations in the southwestern United States and Henry's law data. *J. Geophys. Res.* **90**, 3797–805.
Snyder J. P. (1987) *Map Projections – A Working Manual.* U.S. Geological Survey professional paper 1395, U.S. Government Printing Office, Washington.
Sokolik I., Andronova A., and Johnson C. (1993) Complex refractive index of atmospheric dust aerosols. *Atmos. Environ.* **27A**, 2495–502.
Solomon, S., Garcia R. R., Rowland F. S., and Wuebbles D. J. (1986) On the depletion of Antarctic ozone. *Nature* **321**, 755–7.
Sommer L. (1989) *Analytical Absorption Spectrophotometry in the Visible and Ultraviolet.* Amsterdam, Elsevier.
Spencer J. W. (1971) Fourier series representation of the position of the Sun. *Search* **2**, 172.
Stam D. M., de Haan J. F., Hovenier J. W., and Stammes P. (2000) A fast method for simulating observations of polarized light emerging from the atmosphere applied to the oxygen-A band. *J. Quant. Spectrosc. Radiat. Transfer* **64**, 131–49.
Staniforth A. and Cote J. (1991) Semi-Lagrangian integration schemes for atmospheric models – a review. *Mon. Wea. Rev.* **119**, 2206–23.
Steiner D, Burtchnew H., and Grass H. (1992) Structure and disposition of particles from a spark ignition engine. *Atmos. Environ.* **26**, 997–1003.
Stelson A. W., Bassett M. E., and Seinfeld J. H. (1984) Thermodynamic equilibrium properties of aqueous solutions of nitrate, sulfate and ammonium. In *Chemistry of Particles, Fogs and Rain.* J. L. Durham, ed., Ann Arbor, MI, Ann Arbor Publication, 1–52.
Stephens E. R., Scott W. E., Hanst P. L., and Doerr R. C. (1956) Recent developments in the study of the organic chemistry of the atmosphere. *J. Air Pollut. Contr. Assoc.* **6**, 159–65.
Stockwell W. R. (1986) A homogeneous gas-phase mechanism for use in a regional acid deposition model. *Atmos. Environ.* **20**, 1615–32.
Stockwell W. R. (1995) On the $HO_2 + HO_2$ reaction: Its misapplication in atmospheric chemistry models. *J. Geophys. Res.* **100**, 11,695–8.
Stoer J. and Bulirsch R. (1980) *Introduction to Numerical Analysis.* New York, Springer-Verlag.
Stokes R. H. and Robinson R. A. (1966) Interactions in aqueous nonelectrolyte solutions. I. Solute–solvent equilibria. *J. Phys. Chem.* **70**, 2126–30.
Stommel H. (1947) Entrainment of air into a cumulus cloud. Part I. *J. Appl. Meteor.* **4**, 91–4.
Streets D. G. and Waldhoff S. T. (1998) Biofuel use in Asia and acidifying emissions, *Energy* **23**, 1029–42.
(1999) Greenhouse-gas emissions from biofuel combustion in Asia. *Energy* **24**, 841–55.
Strom J., Okada K., and Heintzenber J. (1992) On the state of mixing of particles due to Brownian coagulation. *J. Aerosol Sci.* **23** 467–80.
Stuart A. L. (2002) Volatile chemical partitioning during cloud hydrometeor freezing and its effects on tropospheric chemical distributions. Ph.D. Thesis, Stanford University.
Stull R. B. (1988) *An Introduction to Boundary Layer Meteorology.* Dordrecht, Kluwer Academic Publishers.
Stumm W. and Morgan J. J. (1981) *Aquatic Chemistry.* New York, Wiley Interscience, 780pp.
Suck S. H. and Brock J. R. (1979) Evolution of atmospheric aerosol particle size distributions via Brownian coagulation: Numerical simulation. *J. Aerosol Sci.* **10**, 581–90.
Sukhatme S. P. and Saikhedkar N. (1969) Heat capacities of glycerol-water mixtures and aqueous solutions of ammonium sulfate, ammonium nitrate and strontium nitrate. *Ind. J. Technol.* **7**, 1–4.
Sun Q. and Wexler A. S. (1998) Modeling urban and regional aerosols – condensation and evaporation near acid neutrality. *Atmos. Environ.* **32**, 3527–31.
Tabazadeh A. and Turco R. P. (1993a) Stratospheric chlorine injection by volcanic eruptions: HCl scavenging and implications for ozone. *Science* **260**, 1082–6.

(1993b) A model for heterogeneous chemical processes on the surfaces of ice and nitric acid trihydrate particles. *J. Geophys. Res.* **98**, 12,727–40.

Tabazadeh A., Turco R. P., Drdla K., and Jacobson M. Z. (1994) A study of Type I polar stratospheric cloud formation. *Geophys. Res. Lett.* **21**, 1619–22.

Tabazadeh A., Djikaev Y. S., and Reiss H. (2002) Surface crystallization of supercooled water in clouds. *Proc. Nat. Acad. Sci.* **99** 15,873–8.

Tang I. N. (1996) Chemical and size effects of hygroscopic aerosols on light scattering coefficients. *J. Geophys. Res.* **101**, 19,245–50.

(1997) Thermodynamic and optical properties of mixed-salt aerosols of atmospheric importance. *J. Geophys. Res.* **102**, 1883–93.

Tang I. N. and Munkelwitz H. R. (1993) Composition and temperature dependence of the deliquescence properties of hygroscopic aerosols. *Atmos. Environ.* **27A**, 467–73.

(1994) Water activities, densities, and refractive indices of aqueous sulfates and sodium nitrate droplets of atmospheric importance. *J. Geophys. Res.* **99**, 18,801–8.

Tang I. N., Wong W. T., and Munkelwitz H. R. (1981) The relative importance of atmospheric sulfates and nitrates in visibility reduction. *Atmos. Environ.* **15**, 2463–71.

Tanguay M., Robert A., and Laprise R. (1990) A semiimplicit semiLagrangian fully-compressible regional forecast model. *Mon. Wea. Rev.* **118**, 1970–80.

Tao Y. and McMurry P. H. (1989) Vapor pressures and surface free energies of C_{14}-C_{19} monocarboxylic acids and C_5-dicarboxylic and C_6-dicarboxylic acids. *Environ. Sci. Technol.* **25**, 1788–93.

Tapp M. C. and White P. W. (1976) A nonhydrostatic mesoscale model. *Quart. J. Roy. Meteor. Soc.* **102**, 277–96.

Tegen I., Lacis A. A., and Fung I. (1996) The influence on climate forcing of mineral aerosols from disturbed soils. *Nature* **380**, 419–22.

Terry D. A., McGraw R., and Rangel R. H. (2001) Method of moments solutions for a laminar flow aerosol reactor model. *Aerosol Sci. Technol.* **34**, 353–62.

Tesche T. W. (1988) Accuracy of ozone air quality models. *J. Environ. Eng.* **114**, 739–52.

Thekaekara M. P. (1974) Extraterrestrial solar spectrum, 3000–6100 Å at 1- Å intervals. *Appl. Opt.* **13**, 518–22.

Thompson N., Barrie N., and Ayles M. (1981) The meteorological office rainfall and evaporation calculation system: MORECS. *Hydrol. Memo.* **45**, 1–69.

Thuburn J. (1996) Multidimensional flux-limited advection schemes. *J. Comp. Phys.* **123**, 74–83.

(1997) TVD schemes, positive schemes, and the universal limiter. *Mon. Wea. Rev.* **125**, 1990–3.

Tiedtke M. (1989) A comprehensive mass flux scheme for cumulus parameterization in large-scale models. *Mon. Wea. Rev.* **117**, 1779–800.

Tjernstrom M. (1993) Turbulence length scales in stably stratified free shear flow analyzed from slant aircraft profiles. *J. Appl. Meteor.* **32**, 948–63.

Toon O. B. and Ackerman T. P. (1981) Algorithms for the calculation of scattering by stratified spheres. *Appl. Opt.* **20**, 3657–60.

Toon O. B., Hamill P., Turco R. P., and Pinto J. (1986) Condensation of HNO_3 and HCl in the winter polar stratospheres. *Geophys. Res. Lett. Nov. Supp.* **13**, 1284–7.

Toon O. B., Turco R. P., Westphal D., Malone R., and Liu M. S. (1988) A multidimensional model for aerosols: Description of computational analogs. *J. Atmos. Sci.* **45**, 2123–43.

Toon O. B., McKay C. P., and Ackerman T. P. (1989a) Rapid calculation of radiative heating rates and photodissociation rates in inhomogeneous multiple scattering atmospheres. *J. Geophys. Res.* **94**, 16,287–301.

Toon O. B., Turco R. P., Jordan J., Goodman J., and Ferry G. (1989b) Physical processes in polar stratospheric ice clouds. *J. Geophys. Res.* **94**, 11,359–80.

Trautmann T. and Wanner C. (1999) A fast and efficient modified sectional method for simulating multicomponent collisional kinetics. *Atmos. Environ.* **33**, 1631–40.

Tremback C. J., Powell J., Cotton W. R., and Pielke R. A. (1987) The forward-in-time upstream advection scheme: Extension to higher orders. *Mon. Wea. Rev.* **115**, 540–55.

Troe J. (1979) Predictive possibilities of unimolecular rate theory. *J. Phys. Chem.* **83**, 114–26.

Tsang T. H. and Brock J. R. (1982) Aerosol coagulation in the plume from a cross-wind line source. *Atmos. Environ.* **16**, 2229–35.

(1986) Simulation of condensation aerosol growth by condensation and evaporation. *Aerosol Sci. Technol.* **5**, 385–8.

Tsang, T. H. and Huang L. K. (1990) On a Petrov–Galerkin finite element method for evaporation of polydisperse aerosols. *Aerosol Sci. Technol.* **12**, 578–97.

Tsang T. H. and Korgaonkar N. (1987) Effect of evaporation on the extinction coefficient of an aerosol cloud. *Aerosol Sci. Technol.* **7**, 317–28.

Turco R. P. and Whitten R. C. (1974) A comparison of several computational techniques for solving some common aeronomic problems. *J. Geophys. Res.* **79**, 3179–85.

Turco R. P., Hamill P., Toon O. B., Whitten R. C., and Kiang C. S. (1979) *The NASA-Ames Research Center Stratospheric Aerosol Model: I. Physical Processes and Computational Analogs.* NASA Technical Publication (TP) 1362, iii–94.

Turco R. P., Toon O. B., Whitten R. C., Keesee R. G., and Hollenbach D. (1982) Noctilucent clouds: Simulation studies of their genesis, properties and global influence. *Planet. Space Sci.* **30**, 1147–81.

Turco R. P., Toon O. B., and Hamill P. (1989) Heterogeneous physiochemistry of the polar ozone hole. *J. Geophys. Res.* **94**, 16,493–510.

Twohy C. H., Clarke A. D., Warren S. G., Radke L. F., and Charlson R. J. (1989) Light-absorbing material extracted from cloud droplets and its effect on cloud albedo. *J. Geophys. Res.* **94**, 8623–31.

Tyndall G. S. and Ravishankara A. R. (1991) Atmospheric oxidation of reduced sulfur species. *Int. J. Chem. Kinet.* **23**, 483–527.

Tzivion S., Feingold G., and Levin Z. (1987) An efficient numerical solution to the stochastic collection equation. *J. Atmos. Sci.* **44**, 3139–49.

U.S. Department of the Army (1958) *Universal Transverse Mercator Grid. Tables for Transformation of Coordinates from Grid to Geographic; Clarke 1866 Spheroid.* U.S. Government Printing Office, Washington, DC.

U.S. Environmental Protection Agency (USEPA) (1978) *Air Quality Criteria for Ozone and Other Photochemical Oxidants.* Report No. EPA-600/8-78-004.

Vali G. (1971) Quantitative evaluation of experimental results on the heterogeneous freezing nucleation of supercooled liquids. *J. Atmos. Sci.* **28**, 402–9.

van de Hulst H. C. (1957) *Light Scattering by Small Particles.* New York, John Wiley and Sons, Inc.

Vanderzee C. E., Waugh D. H., and Haas N. C. (1980) Enthalpies of dilution and relative apparent molar enthalpies of aqueous ammonium nitrate. The case of a weakly hydrolysed (dissociated) salt. *J Chem. Thermodynam.* **12**, 21–5.

van Dingenen R. and Raes F. (1993) Ternary nucleation of methane sulphonic acid, sulphuric acid and water vapour. *J. Aerosol Sci.* **24**, 1–17.

van Doren J. M, Watson L. R., Davidovits P., Worsnop D. R., Zahniser S., and Kolb C. E. (1990) Temperature dependence of the uptake coefficients of HNO_3, HCl, and N_2O_5 by water droplets. *J. Phys. Chem.* **94**, 3256–69.

van Genuchten M. T. (1980) A closed-form equation for predicting the hydraulic conductivity of unsaturated soils. *Ann. Geophys.* **3**, 615–28.

van Weele M. and Duynkerke P. G. (1993) Effects of clouds on the photodissociation of NO_2: Observation and modelling. *J. Atmos. Chem.* **16**, 231–55.

van Zandt T. E. and Fritts D. C. (1989) A theory of enhanced saturation of the gravity wave spectrum due to increases in atmospheric stability. *Pure Appl. Geophys. Pageoph.* **130**, 399–420.

Varoglu E. and Finn W. D. L. (1980) Finite elements incorporating characteristics for one-dimensional diffusion-convection equation. *J. Comp. Phys.* **34**, 371–89.

Vehkamaki H., Kulmala M., Napari I., *et al.* (2002) An improved parameterization for sulfuric acid-water nucleation rates for tropospheric and stratospheric conditions. *J. Geophys. Res.* **107** (D22), 4622, doi:10.1029/2002JD002184.

Venkataraman C. and Friedlander S. K. (1994) Size distributions of polycyclic aromatic hydrocarbons and elemental carbon. 2. Ambient measurements and effects of atmospheric processes. *Environ. Sci. Technol.* **28**, 563–72.

Venkataraman C., Lyons J. M., and Friedlander S. K. (1994) Size distributions of polycyclic aromatic hydrocarbons and elemental carbon. 1. Sampling, measurement methods, and source characterization. *Environ. Sci. Technol.* **28**, 555–62.

Verwer J. G. (1994) Gauss–Seidel iteration for stiff ODEs from chemical kinetics. *SIAM J. Sci. Comput.* **15**, 1243–50.

Villars D. S. (1959) A method of successive approximations for computing combustion equilibria on a high speed digital computer. *J. Phys. Chem.* **63**, 521–5.

Visser J. (1972) On Hamaker constants: A comparison between Hamaker constants and Lifshitz–van der Waals constants. *Adv. Colloid Interface Sci.* **3**, 331–63.

Waggoner A. P., Weiss R. E., Ahlquist N. C., Covert D. S., Will S., and Charlson R. J. (1981) Optical characteristics of atmospheric aerosols. *Atmos. Environ.* **15**, 1891–909.

Wagman D. D., Evans W. H., Parker V. B., *et al.* (1982) The NBS tables of chemical thermodynamic properties: Selected values for inorganic and C_1 and C_2 organic substances in SI units. *J. Phys. Chem. Ref. Data* **11**, Suppl. 2.

Walcek C. (2000) Minor flux adjustment near mixing ratio extremes for simplified yet highly accurate monotonic calculation of tracer advection. *J. Geophys. Res.* **105**, 9335–48.

Walcek C. and Aleksic N. M. (1998) A simple but accurate mass conservative, peak-preserving, mixing ratio bounded advection algorithm with Fortran code. *Atmos. Environ.* **32**, 3863–80.

Walcek C. J., Brost R. A., and Chang J. S. (1986) SO_2, sulfate and HNO_3 deposition velocities computed using regional landuse and meteorological data. *Atmos. Environ.* **20**, 949–64.

Walcek C. J., Yuan H.-H., and Stockwell W. R. (1997) The influence of aqueous-phase chemical reactions on ozone formation in polluted and nonpolluted clouds. *Atmos. Environ.* **31**, 1221–37.

Walmsley J. L. and Wesely M. L. (1996) Modification of coded parameterizations of surface resistances to gaseous dry deposition. *Atmos. Environ.* **30A**, 1181–8.

Wang C. and Prinn R. G. (2000) On the roles of deep convective clouds in tropospheric chemistry. *J. Geophys. Res.* **105**, 22,269–97.

Wang P. K., Grover S. N., and Pruppacher H. R. (1978) On the effect of electric charges on the scavenging of aerosol particles by clouds and small raindrops. *J. Atmos. Sci.* **35**, 1735–43.

Wanninkhof R. (1992) Relationship between wind speed and gas exchange over the ocean. *J. Geophys. Res.* **97**, 7373–82.

Washington W. M. and Parkinson C. L. (1986) *An Introduction to Three-Dimensional Climate Modeling*. Mill Valley, CA, University Science Books.

Watson R. T. (1977) Rate constants for reactions of ClO_x of atmospheric interest. *J. Phys. Chem. Ref. Data* **6**, 871–917.

Weeks J. L. and Rabani J. (1966) The pulse radiolysis of deaerated aqueous carbonate solutions. *J. Phys. Chem.* **70**, 2100–6.

Weingartner E., Burtscher H., and Baltensperger U. (1997) Hygroscopic properties of carbon and diesel soot particles. *Atmos. Environ.* **31**, 2311–27.

Weisman M. L., Skamarock W. C., and Klemp J. B. (1997) The resolution dependence of explicitly modeled convective systems. *Mon. Wea. Rev.* **125**, 527–48.

Welch R. M., Cox S. K., and Davis J. M. (1980) *Solar Radiation and Clouds*, Meteorological Monograph 17. American Meteorological Society.

Wengle H. and Seinfeld J. H. (1978) Pseudospectral solution of atmospheric diffusion problems. *J. Comp. Phys.* **26**, 87–106.

Wentzel, M., Gorzawski, H., Naumann, K.-H., Saathoff, H., and Weinbruch, S. (2003) Transmission electron microscopical and aerosol dynamical characterization of soot aerosols. *J. Aerosol Sci.* **34**, 1347–70.

Wesely M. L. (1989) Parameterization of surface resistances to gaseous dry deposition in regional-scale numerical models. *Atmos. Environ.* **23**, 1293–304.

Wesely M. L. and Hicks B. B. (1977) Some factors that affect the deposition rates of sulfur dioxide and similar gases on vegetation. *J. Air Pollut. Control Ass.* **27**, 1110–6.

West R., Crisp D., and Chen L. (1990). Mapping transformation for broadband atmospheric radiation calculations. *J. Quant. Spectrosc. Radiat. Transfer* **43**, 191–9.

Wetzel P. J. and Chang J. (1987) Concerning the relationship between evapotranspiration and soil moisture. *J. Climate Appl. Meteor.* **26**, 18–27.

Wexler A. S. and Clegg S. L. (2002) Atmospheric aerosol models for systems including the ions H^+, $NH_4{}^+$, Na^+, $SO_4{}^{2-}$, NO^{3-}, Cl^-, Br^-, and H_2O. *J. Geophys. Res.* **107** (D14) 10.1029/2001JD000451.

Wexler A. S. and Seinfeld J. H. (1990) The distribution of ammonium salts among a size and composition dispersed aerosol. *Atmos. Environ.* **24A**, 1231–46.

(1991) Second-generation inorganic aerosol model. *Atmos. Environ.* **25A**, 2731–48.

Whitby E. R. (1985) *The Model Aerosol Dynamics Model. Part I.* Report to the U.S. Environmental Protection Agency. Department of Mechanical Engineering, University of Minnesota, Minneapolis.

Whitby K. T. (1978) The physical characteristics of sulfur aerosols. *Atmos. Environ.* **12**, 135–59.

White M. (2000) *Leonardo: The First Scientist.* London, Abacus, 370pp.

Whitten G. Z., Hogo H., and Killus J. P. (1980) The carbon bond mechanism: A condensed kinetic mechanism for photochemical smog. *Environ. Sci. Technol.* **14**, 690–700.

Wicker L. J. and Skamarock W. C. (1998) A time-splitting scheme for the elastic equations incorporating second-order Runge–Kutta time differencing. *Mon. Wea. Rev.* **126**, 1992–9.

(2002) Time-splitting methods for elastic models using forward time schemes. *Mon. Wea. Rev.* **130**, 2088–97.

Wilke C. R. and Chang P. (1955) Correlation of diffusion coefficients in dilute solutions. *Am. Inst. Chem. Eng. J.* **1**, 264–70.

Wine P. H., Tang Y., Thorn, R. P., Wells J. R., and Davis D. D. (1989) Kinetics of aqueous-phase reactions of the $SO_4{}^-$ radical with potential importance in cloud chemistry. *J. Geophys. Res.* **94**, 1085–94.

Wiscombe W. (1977) The delta-M method: Rapid yet accurate radiative flux calculations for strongly asymmetric phase functions. *J. Atmos. Sci.* **34**, 1408–22.

Wolf M. E. and Hidy G. M. (1997) Aerosols and climate: Anthropogenic emissions and trends for 50 years. *J. Geophys. Res.* **102**, 11, 113–21.

Woodcock A. H. (1953) Salt nuclei in marine air as a function of altitude and wind force. *J. Meteorol.* **10**, 362–71.

Woods T. N., Prinz D. K., Rottman, G. J., *et al.* (1996) Validation of the UARS solar ultraviolet irradiances: Comparison with the ATLAS 1 and 2 measurements. *J. Geophys. Res.* **101**, 9541–69.

World Meteorological Organization (WMO) (1975) *Manual on the Observation of Clouds and Other Meteors.* World Meteorological Organization, Geneva.

(1995) *Scientific Assessment of Ozone Depletion: 1994.* Report 25, Global Ozone Research and Monitoring Project, World Meteorological Organization, Geneva.

(1998) *Scientific Assessment of Ozone Depletion: 1998.* Report 44, WMO Global Ozone Research and Monitoring Project, World Meteorological Organization, Geneva.

Worsnop D. R., Fox L. E., Zahniser M. S., and Wofsy S. C. (1993). Vapor pressures of solid hydrates of nitric acid: Implications for polar stratospheric clouds. *Science* **259**, 71–4.

Wu J. (1993) Production of spume drops by the wind tearing of wave crests: The search for quantification. *J. Geophys. Res.* **98**, 18,221–7.

Wu Y.-C. and Hamer W. J. (1980) Revised values of the osmotic coefficients and mean activity coefficients of sodium nitrate in water at 25 °C. *J. Phys. Chem. Ref. Data* **9**, 513–8.

Xiong C. and Friedlander S. K. (2001) Morphological properties of atmospheric aerosol aggregates. *Proc. Natl. Acad. Sci.* **9**, 11,851–6.

Yabe T., Tanaka R., Nakamura T., and Xiao F. (2001) An exactly conservative semi-Lagrangian scheme (CIP-CSL) in one dimension. *Mon. Wea. Rev.* **129**, 332–44.

Yamamoto G., Tanaka M., and Asano S. (1970) Radiative transfer in water clouds in the infrared region. *J. Atmos. Sci.* **27**, 282–92.

Yamartino R. J. (1993) Nonnegative conserved scalar transport using grid-cell-centered spectrally constrained Blackman cubics for applications on a variable-thickness mesh. *Mon. Wea. Rev.* **121**, 753–63.

Yanenko N. A. (1971) *The Method of Fractional Steps*. Berlin, Springer-Verlag, 160pp.

Yin F., Grosjean D., and Seinfeld J. H. (1990) Photooxidation of dimethyl sulfide and dimethyl disulfide. I: Mechanism development. *J. Atmos. Chem.* **11**, 309–64.

Young A. T. (1980) Revised depolarization corrections for atmospheric extinction. *Appl. Opt.* **19**, 3427–8.

Young T. R. and Boris J. P. (1977) A numerical technique for solving stiff ordinary differential equations associated with the chemical kinetics of reactive-flow problems. *J. Phys. Chem.* **81**, 2424–7.

Zawadski I., Torlaschi E., and Sauvageau, R. (1981) The relationship between mesoscale thermodynamic variables and convective precipitation. *J. Atmos. Sci.* **38**, 1535–40.

Zaytsev I. D. and Aseyev G. G., eds. (1992) *Properties of Aqueous Solutions of Electrolytes* (translated by M. A. Lazarev and V. R. Sorochenko). Boca Raton, FL, CRC Press.

Zeldovich Y. B. (1942) Theory of new-phase formation: cavitation. *J. Exp. Theor. Phys. (USSR)* **12**, 525–38.

Zhang D. and Anthes R. A. (1982) A high-resolution model of the planetary boundary layer – sensitivity tests and comparisons with SESAME-79 data. *J. Appl. Meteor.* **21**, 1594–609.

Zhang Y., Bischof C. H., Easter R. C., and Wu P.-T. (1998) Sensitivity analysis of multi-phase chemical mechanism using automatic differentiation. *J. Geophys. Res.* **103**, 18, 953–79.

Zhang Y., Seigneur C., Seinfeld J. H., Jacobson M. Z., and Binkowski F. (1999) Simulation of aerosol dynamics: A comparative review of algorithms used in air quality models. *Aerosol Sci. Technol.* **31**, 487–514.

Zhang Y., Seigneur C., Seinfeld J. H., Jacobson M., Clegg S.L, and Binkowski F. (2000) A comparative review of inorganic aerosol thermodynamic equilibrium modules: Similarities, differences, and their likely causes. *Atmos. Environ.* **34**, 117–37.

Zhang Y., Pun. B., Wu S.-Y., *et al.* (2004) Development and application of the model for aerosol dynamics, reaction, ionization and dissolution (MADRID). *J. Geophys. Res.* **109**, D01202, doi: 10.1029/2003JD 003501.

Zhao J. and Turco R. P. (1995) Nucleation simulations in the wake of a jet aircraft in stratospheric flight. *J. Aerosol Sci.* **26**, 779–95.

Index

Index

G

H

I

O

T